MOLECULES TO MEDICINE WITH mTOR

MOLECULES TO MEDICINE WITH mTOR

Translating Critical Pathways into Novel Therapeutic Strategies

Edited by

KENNETH MAIESE
Cellular and Molecular Signaling, Newark, NJ, USA

AMSTERDAM • BOSTON • HEIDELBERG • LONDON
NEW YORK • OXFORD • PARIS • SAN DIEGO
SAN FRANCISCO • SINGAPORE • SYDNEY • TOKYO
Academic Press is an imprint of Elsevier

Academic Press is an imprint of Elsevier
125 London Wall, London EC2Y 5AS, UK
525 B Street, Suite 1800, San Diego, CA 92101-4495, USA
50 Hampshire Street, 5th Floor, Cambridge, MA 02139, USA
The Boulevard, Langford Lane, Kidlington, Oxford OX5 1GB, UK

Notices

Knowledge and best practice in this field are constantly changing. As new research and experience broaden our understanding, changes in research methods, professional practices, or medical treatment may become necessary.

Practitioners and researchers must always rely on their own experience and knowledge in evaluating and using any information, methods, compounds, or experiments described herein. In using such information or methods they should be mindful of their own safety and the safety of others, including parties for whom they have a professional responsibility.

To the fullest extent of the law, neither the Publisher nor the authors, contributors, or editors, assume any liability for any injury and/or damage to persons or property as a matter of products liability, negligence or otherwise, or from any use or operation of any methods, products, instructions, or ideas contained in the material herein.

ISBN: 978-0-12-802733-2

British Library Cataloguing-in-Publication Data
A catalogue record for this book is available from the British Library.

Library of Congress Cataloging-in-Publication Data
A catalog record for this book is available from the Library of Congress.

For Information on all Academic Press publications
visit our website at http://store.elsevier.com/

Publisher: Mica Haley
Acquisition Editor: Stacy Masucci
Editorial Project Manager: Sam Young
Project Manager: Edward Taylor
Designer: Matthew Limbert

Typeset by MPS Limited, Chennai, India
www.adi-mps.com

Printed and bound in the United States of America

Working together
to grow libraries in
developing countries

www.elsevier.com • www.bookaid.org

Contents

SECTION I

mTOR IN CELLULAR DEVELOPMENT, PROLIFERATION, AND SURVIVAL

1. Novel Stem Cell Strategies with mTOR

KENNETH MAIESE

2. mTOR: The Master Regulator of Conceptus Development in Response to Uterine Histotroph During Pregnancy in Ungulates

XIAOQIU WANG, GUOYAO WU AND FULLER W. BAZER

3. mTORC1 in the Control of Myogenesis and Adult Skeletal Muscle Mass

MARITA A. WALLACE, DAVID C. HUGHES AND KEITH BAAR

4. mTOR: A Critical Mediator of Articular Cartilage Homeostasis

AKIHIRO NAKAMURA AND MOHIT KAPOOR

5. The Role of mTOR, Autophagy, Apoptosis, and Oxidative Stress During Toxic Metal Injury

SARMISHTHA CHATTERJEE, CHAYAN MUNSHI AND SHELLEY BHATTACHARYA

SECTION II

mTOR IN GENETIC DISORDERS AND NEURODEGENERATIVE DISEASE

6. The mTOR Signaling Pathway in Neurodegenerative Diseases

ARNAUD FRANCOIS, JULIE VERITE, AGNÈS RIOUX BILAN,
THIERRY JANET, FRÉDÉRIC CALON, BERNARD FAUCONNEAU,
MARC PACCALIN AND GUYLÈNE PAGE

7. mTOR: Exploring a New Potential Therapeutic Target for Stroke

MAR CASTELLANOS, CARME GUBERN AND ELISABET KADAR

8. mTOR Signaling in Epilepsy and Epileptogenesis: Preclinical and Clinical Studies

ANTONIO LEO, ANDREW CONSTANTI, ANTONIETTA COPPOLA,
RITA CITRARO, GIOVAMBATTISTA DE SARRO AND EMILIO RUSSO

9. mTOR, Autophagy, Aminoacidopathies, and Human Genetic Disorders

GARRETT R. AINSLIE, K. MICHAEL GIBSON AND KARA R. VOGEL

SECTION III
mTOR IN MEMORY, BEHAVIOR, AND AGING

10. mTOR Involvement in the Mechanisms of Memory: An Overview of Animal Studies

MARIA GRAZIA GIOVANNINI AND DANIELE LANA

11. Mammalian Target of Rapamycin (mTOR), Aging, Neuroscience, and Their Association with Aging-Related Diseases

ERGUL DILAN CELEBI-BIRAND, ELIF TUGCE KAROGLU, FUSUN DOLDUR-BALLI AND MICHELLE M. ADAMS

12. The Role of mTOR in Mood Disorders Pathophysiology and Treatment

GISLAINE Z. RÉUS, MEAGAN R. PITCHER, CAMILA O. ARENT AND JOÃO QUEVEDO

13. mTOR and Drugs of Abuse

JACOB T. BECKLEY AND DORIT RON

19. mTOR and Neuroinflammation

FILIPE PALAVRA, ANTÓNIO FRANCISCO AMBRÓSIO AND FLÁVIO REIS

20. mTOR in Multiple Sclerosis: The Emerging Role in the Regulation of Glial Biology

CINZIA DELLO RUSSO, PIERLUIGI NAVARRA AND LUCIA LISI

SECTION VI

mTOR IN THE ENDOCRINE SYSTEM AND DISORDERS OF METABOLISM

21. mTOR in Metabolic and Endocrine Disorders

MARTA M. SWIERCZYNSKA AND MICHAEL N. HALL

22. Chronic mTOR Inhibition by Rapamycin and Diabetes: What is the Role of Mitochondria?

LIANG-JUN YAN AND ZHIYOU CAI

SECTION VII

mTOR AND CANCER

List of Contributors

Michelle M. Adams Interdisciplinary Graduate Program in Neuroscience, Bilkent University, Ankara, Turkey; Department of Psychology, Bilkent University, Ankara, Turkey

Garrett R. Ainslie Section of Experimental and Systems Pharmacology, College of Pharmacy, Washington State University, Spokane, WA, USA

António Francisco Ambrósio Laboratory of Pharmacology and Experimental Therapeutics, Institute for Biomedical Imaging and Life Sciences (IBILI), Faculty of Medicine, University of Coimbra, Coimbra, Portugal; Center for Neuroscience and Cell Biology, Institute for Biomedical Imaging and Life Sciences (CNC-IBILI) Consortium, University of Coimbra, Coimbra, Portugal

Camila O. Arent Laboratório de Neurociências, Programa de Pós-Graduação em Ciências da Saúde, Unidade Acadêmica de Ciências da Saúde, Universidade do Extremo Sul Catarinense, Criciúma, SC, Brazil

Athira A.P. Department of Biochemistry, University of Kerala, Thiruvananthapuram, Kerala, India

Keith Baar Department of Neurobiology, Physiology and Behavior, Functional Molecular Biology Lab, University of California Davis, Davis, CA, USA

Fusun Doldur-Balli Department of Molecular Biology and Genetics, Bilkent University, Ankara, Turkey

Fuller W. Bazer Center for Animal Biotechnology and Genomics, Texas A&M University, College Station, TX, USA; Department of Animal Science, Texas A&M University, College Station, TX, USA

Jacob T. Beckley Department of Neurology, University of California, San Francisco, CA, USA

Shelley Bhattacharya Environmental Toxicology Laboratory, Department of Zoology, Centre for Advanced Studies, Visva-Bharati (A Central University), Santiniketan, India

Agnès Rioux Bilan EA3808 Molecular Targets and Therapeutics of Alzheimer's Disease, Poitiers University Hospital, University of Poitiers, Poitiers, France

Binu S. Department of Biochemistry, University of Kerala, Thiruvananthapuram, Kerala, India

Ergul Dilan Celebi-Birand Interdisciplinary Graduate Program in Neuroscience, Bilkent University, Ankara, Turkey

Zhiyou Cai Department of Neurology, Renmin Hospital, Hubei University of Medicine, Shiyan, Hubei Province, People's Republic of China

Frédéric Calon Neurosciences Axis, Centre de recherche du CHU de Québec, Québec, QC, Canada; Faculty of Pharmacy, Université Laval, Québec, QC, Canada

Mar Castellanos Department of Neurology, University Hospital A Coruña, A Coruña, Spain

Sarmishtha Chatterjee Environmental Toxicology Laboratory, Department of Zoology, Centre for Advanced Studies, Visva-Bharati (A Central University), Santiniketan, India

Rita Citraro Science of Health Department, School of Medicine, University of Catanzaro, Italy

Andrew Constanti Department of Pharmacology, UCL School of Pharmacy, London, United Kingdom

Antonietta Coppola Epilepsy Centre; Department of Neuroscience, Reproductive and Odontostomatological Sciences, Federico II University, Naples, Italy

Anindita Das Pauley Heart Center, Division of Cardiology, Department of Internal Medicine, Virginia Commonwealth University Medical Center, Richmond, VA, USA

Giovambattista De Sarro Science of Health Department, School of Medicine, University of Catanzaro, Italy

Bernard Fauconneau EA3808 Molecular Targets and Therapeutics of Alzheimer's Disease, Poitiers University Hospital, University of Poitiers, Poitiers, France

Rosa Fernandes Laboratory of Pharmacology and Experimental Therapeutics, Institute for Biomedical Imaging and Life Sciences (IBILI), Faculty of Medicine, University of Coimbra, Coimbra, Portugal; Center for Neuroscience and Cell Biology, Institute for Biomedical Imaging and Life Sciences (CNC-IBILI) Research Unit, University of Coimbra, Coimbra, Portugal

Arnaud Francois Neurosciences Axis, Centre de recherche du CHU de Québec, Québec, QC, Canada; Faculty of Pharmacy, Université Laval, Québec, QC, Canada

K. Michael Gibson Section of Experimental and Systems Pharmacology, College of Pharmacy, Washington State University, Spokane, WA, USA

Maria Grazia Giovannini Department of Health Sciences, Section of Clinical Pharmacology and Oncology, University of Florence, Florence, Italy

Connie G. Glasgow Cardiovascular and Pulmonary Branch, National Heart, Lung, and Blood Institute, National Institutes of Health, Bethesda, MD, USA

Carme Gubern Cerebrovascular Research Group, Biomedical Research Institute (IdiBGi), Girona, Spain

Michael N. Hall Biozentrum, University of Basel, Basel, Switzerland

David C. Hughes Department of Neurobiology, Physiology and Behavior, Functional Molecular Biology Lab, University of California Davis, Davis, CA, USA

Ken-ichi Inoue Second Department of Surgery, Dokkyo Medical University, Mibu, Japan

Thierry Janet EA3808 Molecular Targets and Therapeutics of Alzheimer's Disease, Poitiers University Hospital, University of Poitiers, Poitiers, France

Elisabet Kadar Department of Biology, University of Girona, Girona, Spain

Mohit Kapoor Department of Genetics and Development, Krembil Research Institute, University Health Network, Toronto, Ontario, Canada; Department of Surgery, University of Toronto, Toronto, Ontario, Canada; Arthritis Program, University Health Network, Toronto, Ontario, Canada

Elif Tugce Karoglu Interdisciplinary Graduate Program in Neuroscience, Bilkent University, Ankara, Turkey

Hiroshi Kato Division of Rheumatology, Department of Medicine, State University of New York Upstate Medical University, Syracuse, New York, NY, USA

Goran B.G. Klintmalm Texas A&M University College of Medicine, Dallas, TX, USA; Simmons Transplant Institute, Department of Surgery, Transplant Surgery, Baylor University Medical Center, Dallas, TX, USA

Keiichi Kubota Second Department of Surgery, Dokkyo Medical University, Mibu, Japan

Rakesh C. Kukreja Pauley Heart Center, Division of Cardiology, Department of Internal Medicine, Virginia Commonwealth University Medical Center, Richmond, VA, USA

Daniele Lana Department of Health Sciences, Section of Clinical Pharmacology and Oncology, University of Florence, Florence, Italy

Antonio Leo Science of Health Department, School of Medicine, University of Catanzaro, Italy

Lucia Lisi Institute of Pharmacology, Catholic University Medical School, Rome, Italy

Xiangwei Liu Shanghai Institute of Cardiovascular Diseases, Zhongshan Hospital, Fudan University, Shanghai, PR China; Center for Cardiovascular Research and Alternative Medicine, School of Pharmacy, University of Wyoming College of Health Sciences, Laramie, WY, USA

Kenneth Maiese Cellular and Molecular Signaling, Newark, NJ, USA

Francesco Massari Medical Oncology, Azienda Ospedaliera Universitaria Integrata, University of Verona, Verona, Italy

Greg J. McKenna Simmons Transplant Institute, Baylor University Medical Center, Dallas, TX, USA; Texas A&M University College of Medicine, Dallas, TX, USA

Joel Moss Cardiovascular and Pulmonary Branch, National Heart, Lung, and Blood Institute, National Institutes of Health, Bethesda, MD, USA

Chayan Munshi Environmental Toxicology Laboratory, Department of Zoology, Centre for Advanced Studies, Visva-Bharati (A Central University), Santiniketan, India

Akihiro Nakamura Department of Genetics and Development, Krembil Research Institute, University Health Network, Toronto, Ontario, Canada

Pierluigi Navarra Institute of Pharmacology, Catholic University Medical School, Rome, Italy

Marc Paccalin EA3808 Molecular Targets and Therapeutics of Alzheimer's Disease, Poitiers University Hospital, University of Poitiers, Poitiers, France; Geriatrics Department, Poitiers University Hospital, Poitiers, France; Centre Mémoire de Ressources et de Recherche, France; INSERM CIC-P 1402, Poitiers University Hospital, Poitiers, France

Gustavo Pacheco-Rodriguez Cardiovascular and Pulmonary Branch, National Heart, Lung, and Blood Institute, National Institutes of Health, Bethesda, MD, USA

Guylène Page EA3808 Molecular Targets and Therapeutics of Alzheimer's Disease, Poitiers University Hospital, University of Poitiers, Poitiers, France

Filipe Palavra Laboratory of Pharmacology and Experimental Therapeutics, Institute for Biomedical Imaging and Life Sciences (IBILI), Faculty of Medicine, University of Coimbra, Coimbra, Portugal; Center for Neuroscience and Cell Biology, Institute for Biomedical Imaging and Life Sciences (CNC-IBILI) Consortium, University of Coimbra, Coimbra, Portugal

Andras Perl Division of Rheumatology, Department of Medicine, State University of New York Upstate Medical University, Syracuse, New York, NY, USA

Meagan R. Pitcher Center for Translational Psychiatry, Department of Psychiatry and Behavioral Sciences, Medical School, The University of Texas Health Science Center at Houston, Houston, TX, USA

João Quevedo Center for Translational Psychiatry, Department of Psychiatry and Behavioral Sciences, Medical School, The University of Texas Health Science Center at Houston, Houston, TX, USA; Laboratório de Neurociências, Programa de Pós-Graduação em Ciências da Saúde, Unidade Acadêmica de Ciências da Saúde, Universidade do Extremo Sul Catarinense, Criciúma, SC, Brazil

Flávio Reis Laboratory of Pharmacology and Experimental Therapeutics, Institute for Biomedical Imaging and Life Sciences (IBILI), Faculty of Medicine, University of Coimbra, Coimbra, Portugal; Center for Neuroscience and Cell Biology, Institute for Biomedical Imaging and Life Sciences (CNC.IBILI) Consortium, University of Coimbra, Coimbra, Portugal

Jun Ren Shanghai Institute of Cardiovascular Diseases, Zhongshan Hospital, Fudan University, Shanghai, PR China; Center for Cardiovascular Research and Alternative Medicine, School of Pharmacy, University of Wyoming College of Health Sciences, Laramie, WY, USA

Gislaine Z. Réus Center for Translational Psychiatry, Department of Psychiatry and Behavioral Sciences, Medical School, The University of Texas Health Science Center at Houston, Houston, TX, USA; Laboratório de Neurociências, Programa de Pós-Graduação em Ciências da Saúde, Unidade Acadêmica de Ciências da Saúde, Universidade do Extremo Sul Catarinense, Criciúma, SC, Brazil

Dorit Ron Department of Neurology, University of California, San Francisco, CA, USA

Cinzia Dello Russo Institute of Pharmacology, Catholic University Medical School, Rome, Italy

Emilio Russo Science of Health Department, School of Medicine, University of Catanzaro, Italy

Matteo Santoni Clinica di Oncologia Medica, AOU Ospedali Riuniti, Polytechnic University of the Marche Region, Ancona, Italy

Norisuke Shibuya Second Department of Surgery, Dokkyo Medical University, Mibu, Japan

Soumya S.J. Inter-University Centre for Genomics and Gene Technology, University of Kerala, Thiruvananthapuram, Kerala, India

Wendy K. Steagall Cardiovascular and Pulmonary Branch, National Heart, Lung, and Blood Institute, National Institutes of Health, Bethesda, MD, USA

Sudhakaran P.R. Inter-University Centre for Genomics and Gene Technology, University of Kerala, Thiruvananthapuram, Kerala, India; Department of Biochemistry, University of Kerala, Thiruvananthapuram, Kerala, India; Department of Computational Biology and Bioinformatics, University of Kerala, Thiruvananthapuram, Kerala, India

Marta M. Swierczynska Biozentrum, University of Basel, Basel, Switzerland

Julie Verite EA3808 Molecular Targets and Therapeutics of Alzheimer's Disease, Poitiers University Hospital, University of Poitiers, Poitiers, France

Kara R. Vogel Section of Experimental and Systems Pharmacology, College of Pharmacy, Washington State University, Spokane, WA, USA

Marita A. Wallace Department of Neurobiology, Physiology and Behavior, Functional Molecular Biology Lab, University of California Davis, Davis, CA, USA

Xiaoqiu Wang Center for Animal Biotechnology and Genomics, Texas A&M University, College Station, TX, USA; Department of Animal Science, Texas A&M University, College Station, TX, USA

Guoyao Wu Center for Animal Biotechnology and Genomics, Texas A&M University, College Station, TX, USA; Department of Animal Science, Texas A&M University, College Station, TX, USA

Liang-Jun Yan Department of Pharmaceutical Sciences, UNT System College of Pharmacy, University of North Texas Health Science Center, Fort Worth, TX, USA

About the Editor

Kenneth Maiese is an internationally recognized physician researcher, healthcare executive, and editor with broad research, clinical, and leadership experience in both academia and industry. He was born and raised in New Jersey and was the Valedictorian of his high school class at Pennsauken High School. He graduated from the University of Pennsylvania Summa cum Laude with Distinction and was a Teagle Scholar, Grupe Scholar, and Joseph Collins Scholar at Weill Medical College of Cornell University. Dr Maiese subsequently trained as a physician-scientist at Cornell, the National Institutes of Health (NIH), and as a senior executive at the Harvard School of Public Health. He has extensive experience in academic medicine, healthcare delivery, and drug development, holding positions as tenured Professor and Chair, and Chief of Service of the Department of Neurology and Neurosciences of Rutgers University, Global Head of Translational Medicine and External Innovation, Board Member of the Cancer Institute of New Jersey, Steering Committee Member for the Foundation for the National Institutes of Health, tenured Professor in Neurology, Anatomy & Cell Biology, Molecular Medicine, the Barbara Ann Karmanos Cancer Institute, and the National Institute of Health Center at Wayne State University, and Founding Editor and Editor-in-Chief of multiple international journals. Early in his career, Dr Maiese received outstanding investigator awards, was named a Johnson & Johnson Distinguished Investigator, received the Albrecht Fleckenstein Memorial Award for Distinguished Achievement in Basic Research, was elected to America's Top Physicians and The Best of U.S. Physicians, and his highly cited work has received the distinction of "High Impact Research and Potential Public Health Benefit" by the National Institutes of Health. He has been fortunate to have benefited from continuous funding from sources that include the American Diabetes Association, the American Heart Association, the Bugher Foundation, Johnson and Johnson Focused Giving Award, and the National Institutes of Health. His service on Executive Councils for Graduate Schools has fostered innovative graduate student training programs and he was elected as board and advisor member to develop and execute inaugural MD/PhD Degree Programs and new Translational Medicine Programs. He chairs national grant committees and is a chartered panel member or consultant for numerous national and international foundations as well as multiple study sections and special emphasis panels for the National Institutes of Health. Dr Maiese is frequently honored as the chairperson and/or the plenary speaker for a number of international symposiums, organizations, and presentations to federal policy-makers.

Preface

Throughout the world, the age of the population is increasing at a significant rate. In developed countries over the course of the last half-century, the number of individuals that are over the age of 65 has approximately doubled. This increase in life expectancy and age has been accompanied by a 1% decrease in the age-adjusted death rate over the years 2000 through 2011. Developing nations also share in the increased life expectancy of the world's population. In these countries, the number of elderly people will rise from 5% of the population to approximately 10% of the population over future decades.

This positive news for the increased lifespan of individuals in developed nations as well as developing countries is somewhat buffered by the parallel rise in non-communicable diseases (NCDs). Improvements in healthcare have fostered increased longevity of the global population. For example, of the five leading causes of death that are cardiac disease, cancer, chronic lower respiratory disease, stroke, and traumatic accidents, stroke is no longer ranked as the third leading cause of death. Multiple factors most likely have led to this lower rank for stroke that involve heightened public awareness for the rapid treatment of stroke, a reduction in tobacco consumption, and improved care for disorders involving diabetes mellitus (DM), hypertension, and low-density lipoprotein cholesterol. Active treatment with recombinant tissue plasminogen activator has also led to a reduction in mortality and morbidity in stroke patients. Yet, therapeutic treatments to resolve disability and death from stroke continue to remain limited for the majority of patients.

Overall, a rise in the prevalence of NCDs continues with the advancing age of the world's population. According to World Health Organization (WHO) records, greater than 60% of the annual 57 million global deaths are due to NCDs. Furthermore, almost 80% of these NCDs occur in low- and middle-income countries. Disorders such as DM will continue to grow according to WHO estimates, such that DM will be the seventh leading cause of death by the year 2030. As of 2013, approximately 350 million individuals are estimated to suffer from DM and millions of individuals are estimated to be currently undiagnosed with DM.

The costs of caring for individuals with DM are extremely high. In the United States (US) alone, the costs of care for individuals with DM in 2012 equaled 8915 US dollars per person and almost 17% of the country's gross domestic product. Although early diagnosis of individuals with DM may be critical to prevent the progression of this disease, the presence of impaired glucose tolerance in the young raises additional concerns for the future development of DM. Obesity, now increasing in incidence, is another risk factor for the development of DM since it leads to cellular oxidative stress, insulin resistance, and lipid-induced impairment of pancreatic β cells.

DM can affect multiple systems of the body and can impact cardiovascular disease, neurodegenerative disorders, the musculoskeletal system, cancer, and immune function. Acute and chronic neurodegenerative disorders lead to disability and death in more than 30 million individuals worldwide. It is predicted that the number of individuals with dementia will continue to grow, doubling every 20 years, and reach almost 115 million individuals by 2050. Most of the increase is expected to occur in developing countries, with Alzheimer's disease being considered a significant public health concern. In regards to costs, care for those suffering from dementia equals more than 600 billion US dollars annually. With chronic neurodegenerative disorders such as Parkinson's disease, approximately 50,000 new cases are present in the US alone each year and this number is predicted to double by 2030. During this same period, other disorders such as epilepsy are predicted to affect over 50 million people and peripheral nerve disease is estimated to be present in almost 300 million individuals. With cancer, almost 50% of cancer survivors are 70 years or older, indicating the increased prevalence with advancing age. In contrast, 5% of individuals with cancer are younger than 40 years of age. The number of cancer survivors will continue to grow and reach at least 19 million by 2024 in the US alone. For cardiovascular disease, this disorder remains the leading cause of death and increases with aging, but can easily affect all ages of the world's population. Elevated cholesterol and hypertension are significant risk factors for cardiovascular disease and contribute to approximately 13% of all deaths.

In light of the limited courses of action to treat many NCDs that currently have only symptomatic or a handful of therapeutic options, development of novel therapeutic strategies to address the growing prevalence of NCDs with the accompanying advancing age of the global population is considered to be highly warranted to fill this void. The mechanistic target of rapamycin (mTOR) is one such strategic pathway that has gained high visibility as a critical target for multiple disease entities. mTOR is also known as the mammalian target of rapamycin and the FK506-binding protein 12-rapamycin complex-associated protein 1. mTOR is a 289-kDa serine-threonine protein kinase, present throughout the body, and oversees multiple functions that involve gene transcription, protein formation, oxidative stress, stem cell development, cell metabolism, cell survival, cell senescence, immunity, aging, tissue regeneration and repair, cancer, and pathways of programmed cell death that include autophagy and apoptosis.

Molecules to Medicine with mTOR: Translating Critical Pathways into Novel Therapeutic Strategies is a unique resource that employs a translational medicine approach to present novel work and new insights for the role of mTOR during normal physiology, in disease states, and for the development and refinement of therapeutic strategies applicable to multiple systems of the body. Prominent internationally recognized experts explore the complexity of mTOR signaling and the critical need for therapeutic targeting of this pathway, but maintain a clear and insightful presentation of clinical and basic research perspectives for a broad audience that encompasses healthcare providers, scientists, drug developers, and students.

Molecules to Medicine with mTOR: Translating Critical Pathways into Novel Therapeutic Strategies begins with the presentation of mTOR in cellular development, proliferation, and survival in Section I. Topics of cellular biology and new therapeutic strategies with mTOR that involve the maintenance of proliferation of stem cells, conceptus development, implantation and placentation, myogenesis, cartilage homeostasis, and cell survival during xenobiotic cell injury are presented. Section II extends these topics with the discussion of mTOR in the nervous system and the role of mTOR in congenital disease. In Section III, higher cognitive function and aging are explored with the examination of the role of mTOR in memory formation as well as cognitive impairment, aging processes of the brain, depression, addiction, and behavior modification. The cardiovascular system is discussed in Section IV, highlighting the role of mTOR in nutrient sensing, new vessel growth, cardiac ischemic disease, and the maintenance of cardiac function during metabolic disorders. In Section V, the vital role of mTOR in adaptive and innate immune function, organ transplantation, inflammatory pathways in the nervous system, and glial function are presented. Section VI further expands the analysis of mTOR in the endocrine system and provides scrutiny of the influence of mTOR in endocrine disorders, metabolic disease, diabetogenesis, and integrated pathways that involve both renal and nervous system impairment. The last segment of the book, Section VII, provides a vital perspective for the role of mTOR as a proliferative agent to foster tumor cell growth in a variety of settings and to affect cancer cell metabolic pathways. *Molecules to Medicine with mTOR: Translating Critical Pathways into Novel Therapeutic Strategies* provides an innovative examination of the pivotal function mTOR holds in the human body and highlights the compelling need to critically consider mTOR signaling pathways in the translation of cellular biology to effective and safe clinical treatments for the growing prevalence of multiple disorders that mirror the increasing age and lifespan of the global population.

Kenneth Maiese, Editor

mTOR IN CELLULAR DEVELOPMENT, PROLIFERATION, AND SURVIVAL

Novel Stem Cell Strategies with mTOR

Kenneth Maiese

Cellular and Molecular Signaling, Newark, NJ, USA

1.1 LIFE EXPECTANCY AND THE SIGNIFICANCE OF GLOBAL NONCOMMUNICABLE DISEASES

In developed nations, the age of the population has significantly risen. Life expectancy is approaching almost 80 years for all individuals and has been marked by a 1% decrease in the age-adjusted death rate from the years 2000 through 2011 [1]. The number of individuals over the age of 65 also has doubled during the prior 50 years [2]. If one examines large developing nations, such as China and India, it is expected that the proportion of elderly individuals will increase from the present levels of approximately 5% to almost 10% over the next several decades. Improvements in treatment resources and preventive care have contributed to the increased lifespan of the global population.

However, accompanied with the increase in life expectancy has been a rise in chronic disorders and noncommunicable diseases (NCDs; Table 1.1). Through data obtained by the World Health Organization, greater than 60% of the 57 million annual global deaths result from NCDs. Approximately 80% of these NCDs occur in low- and middle-income countries [3]. NCDs affect approximately 30% of the population under 60 in low- and middle-income countries. In high-income countries, 13% of the population under 60 is affected.

The five leading causes of death are cardiac disease, cancer, chronic lower respiratory disease, stroke, and traumatic accidents [4]. Elevated cholesterol and hypertension are significant risk factors for cardiovascular disease. For example, hypertension contributes to approximately 13% of all deaths [5].

Disorders such as elevated cholesterol and hypertension can also lead to an increase in acute neurodegenerative disorders such as stroke, the fourth leading cause

of death (Table 1.1) [6,7]. Other NCDs, such as diabetes mellitus (DM), can also contribute to acute neurodegenerative diseases that include cerebrovascular disease [8,9]. Stroke leads to multiple complications that affect both the livelihood and minimal daily function of an individual [10,11]. As a leading cause of death [4], stroke affects approximately 15 million individuals every year. In the United States, approximately 800,000 strokes occur per year at an annual cost of 75 billion US dollars [6]. It is of interest to note that stroke is no longer ranked as the third leading cause of death. Factors related to improved management of hypertension and low-density lipoprotein cholesterol disorders, reduction in tobacco consumption, heightened public education and awareness for the need for rapid treatment of cerebrovascular disorders, and more focused management on metabolic disorders such as DM may have contributed to this lower ranking for stroke [6,7]. In addition, treatment with recombinant tissue plasminogen activator has led to a reduction in mortality and morbidity in patients presenting with stroke [12,13]. Yet, overall therapeutic strategies for patients presenting with stroke remain limited for the majority of patients. Treatment with recombinant tissue plasminogen activator is applicable to only a subgroup of patients that requires a narrow therapeutic window.

Together, acute and chronic neurodegenerative disorders lead to disability and death in more than 30 million individuals worldwide annually (Table 1.1) [14]. Acute neurodegenerative disorders can place a severe burden on the world's population [6,15]. Acute traumatic brain injury (TBI) leads to severe neurological disability [16,17]. A twofold detrimental effect on the clinical outcome of patients can occur with TBI. TBI can result in acute injury to the nervous system as well as subsequent chronic impairment [18–20]. In the United States, approximately 50,000 individuals

TABLE 1.1   Global Noncommunicable Diseases

NCD	Clinical presentation
Neurodegenerative disorders	Acute and chronic neurodegenerative disorders result in disability and death in over 30 million individuals worldwide
	TBI results in acute injury to the nervous system as well as subsequent chronic impairment
	More than 5 million individuals have AD and 3.5 million receive treatment at an annual cost of 4 billion US dollars
	Almost 4% of individuals over age 60 suffer from PD globally and this is expected to double by the year 2030
	As the fourth leading cause of death, stroke affects approximately 15 million individuals every year at an annual cost of 75 billion US dollars
	In the year 2030, epilepsy is predicted to affect over 50 million people and neuropathies are estimated to afflict almost 300 million individuals
Cardiovascular disease	As the leading cause of death, cardiac disease affects all ages in the global population
	Elevated cholesterol and hypertension are significant risk factors for cardiovascular disease
	Hypertension contributes to approximately 13% of all deaths and results in acute neurodegenerative disorders such as stroke
Diabetes mellitus	DM affects all systems of the body and leads to pathology in the cardiovascular system, nervous system, immune system, and musculoskeletal system
	DM is increasing throughout the world and approximately 350 million individuals are currently afflicted with this disorder
	Over 8 million individuals may be undiagnosed for metabolic disease
	Overall care for patients with DM consumes 17% of the gross domestic product in the United States
Cancer	The most common cancers among male survivors are prostate, colon and rectal, and melanoma
	The most common cancers among female survivors are breast, uterine corpus, and colon and rectal
	Approximately one half of cancer survivors are at least 70 years or older. Approximately 5% of individuals with cancer are younger than 40 years of age
	In the United States, it is estimated that the number of cancer survivors will increase to 19 million individuals by the year 2024

die every year as a result of TBI and more than 100,000 individuals must live with chronic disability [21]. During severe trauma, approximately 50% of individuals eventually die.

With the aging population, a rise in the incidence of chronic neurodegenerative disorders is also expected to ensue [22,23]. One such example is Alzheimer's disease (AD). The familial cases of AD consist of less than 2% of all presentations [24] and usually occur prior to age 55 [25]. Familial cases of AD occur as an autosomal dominant form of a mutated amyloid precursor protein (APP) gene as well as mutations in the presenilin 1 or 2 genes [26]. Familial AD is present in approximately 200 families worldwide and results from variable single-gene mutations on chromosomes 1, 14, and 21. Mutations on chromosome 1 lead to altered presenilin 2, mutations on chromosome 14 result in altered presenilin 1, and mutations on chromosome 21 lead to altered APP. In contrast, 10% of the global population over the age of 65 is affected with sporadic AD [27]. At present, more than 5 million individuals are diagnosed with AD and 3.5 million are under treatment at an annual cost of almost 4 billion US dollars. It is estimated that the number of those suffering from AD will increase significantly and rise to more than 30 million individuals over the next 15–20 years [14,27,28]. If one considers other chronic progressive neurodegenerative disorders such as Parkinson's disease (PD) [29,30], almost 50,000 new cases present in the United States alone each year. Approximately 1–4% of individuals over age 60 suffer from PD globally and this number of affected individuals may double by the year 2030. By the year 2030, epilepsy is predicted to affect over 50 million people, neuropathies are estimated to afflict almost 300 million individuals, and neurological injuries may change the lives of 243 million individuals [31]. However, similar to acute neurodegenerative disorders, the availability of treatments that can prevent the initiation or block the progression of chronic neurodegenerative disorders is limited.

DM also is a significant NCD for the world's population (Table 1.1) [32]. The incidence of DM is increasing throughout the world and approximately 350 million individuals currently have DM [33–37]. Another 8 million individuals are believed to suffer from metabolic disorders and are currently undiagnosed [38–40]. The costs to care for individuals with DM are high. Approximately 9000 US dollars are spent in the United States for each individual with DM per year and overall care for patients with DM consumes 17% of the gross domestic product [41]. Early diagnosis and proper care of individuals with DM can impact care for this disease by modulating epigenetic changes in age-related genes involved with DM and other degenerative disorders [42–46].

Impaired glucose tolerance in the young raises additional concerns for the risk of developing DM [34]. Obesity is another risk factor for DM [33−37]. Obesity and excess body fat can precipitate pancreatic β-cell injury [47], cellular inflammation [8], impaired growth factor release [48−51], changes in protein tyrosine phosphatase signaling [37,52], oxidative stress [35,53], and insulin resistance [8,32,54,55].

DM can involve any system of the body [8,9]. DM leads to vascular disease resulting in endothelial cell senescence [56], injury to endothelial cells [51,57−62], cardiovascular disease [63−71], platelet dysfunction [72,73], atherosclerosis [11,74−76], loss of endothelial progenitor cells [52,77−81], dysfunctional maintenance and mobilization of endothelial progenitor cells [79,80], and impaired angiogenesis [62,69,82]. DM also can have negative effects on the immune system [33,70,83−88], renal function [89−93], hepatic metabolism [35,54,94−98], and musculoskeletal integrity [53,74,99−101]. DM affects all components of the central and peripheral nervous systems. DM can lead to retinal disease [39,102−104], peripheral nerve disorders [95,105], memory loss [106−108], psychiatric disorders [109,110], stroke [22,46,72,76,111,112], dementia such as AD [50,106,113,114], and impairment of neuronal longevity [50].

Interestingly, tight glucose control does not always result in the resolution of complications from DM [34,115]. Use of diet control treatments may be effective to prevent hyperglycemia events, but can potentially decrease organ mass through processes that involve autophagy [97]. These observations point to the need for additional strategies for the treatment of DM and its complications.

1.2 mTOR STRUCTURE AND SIGNALING

It is clear that NCDs are becoming a growing concern for the global population given the concurrent rise in longevity and life expectancy. Increased lifespan is accompanied by a greater exposure to several NCDs that involve multiple systems of the body as well as processes closely tied to cellular metabolism. Given the need for novel treatments to effectively treat NCDs that currently have limited therapeutic options, the mechanistic target of rapamycin (mTOR) is increasingly being recognized as a critical target for drug discovery development against NCDs that involve cardiovascular disease, neurological disorders, metabolic disease, and cancer.

mTOR is also known as the mammalian target of rapamycin and the FK-506-binding protein 12-rapamycin complex-associated protein 1 (Table 1.2). It is a 289-kDa serine/threonine protein kinase

TABLE 1.2   mTOR Structure and Signaling

mTOR	Highlights
Structure and identification	mTOR is a 289-kDa serine/threonine protein kinase that is encoded by a single gene *FRAP1*
	TOR was identified through the use of rapamycin-resistant TOR mutants in *S. cerevisiae*. The genes *TOR1* and *TOR2* yield two protein isoforms in yeast, known as TOR1 and TOR2
	mTOR is a central component of the protein complexes mTORC1 and mTORC2
	mTORC1 consists of Raptor, the proline-rich Akt substrate 40 kDa (PRAS40), Deptor, and mLST8/GβL
	mTORC2 consists of Rictor, mLST8, Deptor, the mSIN1, and the protein observed with Rictor-1 (Protor-1)
Post-translational phosphorylation	Post-translational phosphorylation of mTOR occurs through multiple avenues
	The C-terminal domain with sequence homology to the catalytic domain of the PI 3-K family has several phosphorylation sites that regulate mTOR
	Within the HEAT domain, serine^{1261} can be phosphorylated in mTORC1 and mTORC2 by insulin signaling through the PI 3-K pathway to increase mTOR activity
Cellular signaling	Rapamycin blocks mTORC1 activity by binding to immunophilin FKBP12 to prevent the phosphorylation of mTOR
	Chronic administration of rapamycin can also inhibit mTORC2 activity
	PRAS40 inhibits mTOR activity and blocks binding of mTORC1 to Raptor
	Deptor blocks mTORC1 activity by binding to the FAT domain of mTOR
	mLST8 promotes mTOR kinase activity through p70S6K and eIF4E-4EBP1 that bind to Raptor
	mTOR signaling relies upon the pathways of PI 3-K, Akt, and AMPK
	In contrast to the inhibition of mTORC1, mTORC2 is activated by the TSC1/TSC2

Akt, protein kinase B; AMPK, AMP activated protein kinase; eIF4E-4EBP1, eukaryotic initiation factor 4E-binding protein 1; FAT domain, FKBP12-rapamycin-associated protein (FRAP), ataxia-telangiectasia (ATM), and the transactivation/transformation domain-associated protein domain; HEAT, Huntingtin, Elongation factor 3, a subunit of protein phosphatase-2A, and TOR1; mTOR, mechanistic target of rapamycin; mTORC1, mTOR complex 1; mTORC2, mTOR complex 2; PI 3-K, phosphoinositide 3-kinase; p70S6K, p70 ribosomal S6 kinase; TOR, target of rapamycin.

that is encoded by a single gene, *FRAP1* [116,117]. Initially, the target of rapamycin (TOR) was identified through the use of rapamycin-resistant TOR mutants in *Saccharomyces cerevisiae* with the genes *TOR1* and *TOR2* that yield two protein isoforms in yeast known as Tor1 and Tor2 [118]. The carboxyterminal domain of mTOR has four domains for catalytic activity that include FAT, FKBP12-rapamycin-binding domain (FRB), the catalytic PI3/PI4-related kinase domain, and FATC [116]. The FAT domain, which consists of FKBP12-rapamycin-associated protein (FRAP), ataxia-telangiectasia (ATM), and the transactivation/transformation domain-associated protein domain, resides adjacent to the FRB that allows association between mTOR and FKBP12 when bound to rapamycin. The PI3/PI4-related kinase domain and the small FATC domain follow these two domains. The N-terminal portion of mTOR contains at least a 20 HEAT (Huntingtin, Elongation factor 3, a subunit of Protein phosphatase-2A, and TOR1) repeat. This area promotes binding with the regulatory proteins Raptor (regulatory-associated protein of mTOR) and Rictor (rapamycin-insensitive companion of mTOR) [119]. Post-translational phosphorylation of mTOR can occur through a number of processes. The C-terminal domain with sequence homology to the catalytic domain of the phosphoinositide 3-kinase (PI 3-K) family [120] contains several phosphorylation sites that regulate mTOR. Within the HEAT domain, serine^{1261} can be phosphorylated in mTOR complex 1 (mTORC1) and mTOR complex 2 (mTORC2) by insulin signaling through the PI 3-K pathway to increase mTOR activity [120]. Phosphorylation of serine^{1261} also results in mTOR serine^{2481} autophosphorylation [120].

mTOR is a central component of the protein complexes mTORC1 and mTORC2 (Figure 1.1) [27,29,121,122]. Rapamycin is a macrolide antibiotic derived from *Streptomyces hygroscopicus* that inhibits both TOR activity and mTOR activity [123]. mTORC1 is more sensitive to the inhibitory effects of rapamycin than mTORC2 [124]. Rapamycin blocks mTORC1 activity by binding to immunophilin FK-506-binding protein 12 (FKBP12) that attaches to FRB at the C-terminal of mTOR to prevent the phosphorylation of mTOR [125]. Chronic administration of rapamycin can inhibit mTORC2 activity that may occur from the disruption of the assembly of mTORC2 [124].

mTORC1 consists of Raptor, the proline-rich Akt substrate 40 kDa (PRAS40), Deptor (DEP domain-containing mTOR interacting protein), and mLST8/GβL (mammalian lethal with Sec13 protein 8, abbreviated as mLST8 or G protein beta subunit-like (GβL)) (Table 1.2) [116]. Phosphorylation of Raptor through several pathways that can include the protein Ras homolog enriched in brain (Rheb) activates mTORC1

and allows it to bind to its complex constituents [126]. Rheb phosphorylates Raptor residue serine^{863} and other residues that include serine^{859}, serine^{855}, serine^{877}, serine^{696}, and threonine^{706}. mTORC1 activity is limited if serine^{863} remains unphosphorylated [127]. Once mTOR is active, mTOR also can modulate Raptor activity that can be blocked by rapamycin [127]. PRAS40 inhibits mTOR activity and can prevent the binding of mTORC1 to Raptor [128]. Phosphorylation of PRAS40 by protein kinase B (Akt) frees PRAS40 from Raptor and allows PRAS40 to be sequestered by the cytoplasmic docking protein 14-3-3 to activate mTORC1 [128–132]. Deptor inhibits mTORC1 activity by binding to the FAT domain of mTOR. Without Deptor, the activity of Akt, mTORC1, and mTORC2 are enhanced [133]. In contrast to PRAS40 and Deptor, mLST8 promotes mTOR kinase activity through p70 ribosomal S6 kinase (p70S6K) and the eukaryotic initiation factor 4E (eIF4E)-binding protein 1 (4EBP1) that bind to Raptor [134]. PRAS40 can block mTORC1 activity by preventing the association of p70S6K and 4EBP1 with Raptor [14,128]. In addition, mLST8 promotes mTOR kinase activity and is known to control insulin signaling through the transcription factor FoxO3 [135], be necessary for Akt and protein kinase C-α (PKCα) phosphorylation [135], and is required for the association between Rictor and mTOR [135].

mTOR signaling is closely associated with the pathways of PI 3-K, Akt, and AMP activated protein kinase (AMPK) (Table 1.2) [64,136]. Akt regulates the activity of the hamartin (tuberous sclerosis 1)/tuberin (tuberous sclerosis 2) (TSC1/TSC2) complex, an inhibitor of mTORC1 [137–140]. Although several regulatory phosphorylation sites are known to exist for TSC1, control of the TSC1/TSC2 complex is primarily mediated though the phosphorylation of TSC2 by Akt, extracellular signal-regulated kinases, activating protein p90 ribosomal S6 kinase 1, AMPK, and glycogen synthase kinase-3β. TSC2 functions as a GTPase-activating protein (GAP) converting a small G protein Rheb (Rheb-GTP) to the inactive GDP-bound form (Rheb-GDP). Once Rheb-GTP is active, Rheb-GTP associates with Raptor to regulate the binding of 4EBP1 to mTORC1 and increase mTORC1 activity [141]. Akt phosphorylates TSC2 on multiple sites that leads to the destabilization of TSC2 and disruption of its interaction with TSC1. Phosphorylation of TSC2 at serine^{939}, serine^{981}, and threonine^{1462} results in the binding of TSC2 to protein 14-3-3, disruption of the TSC1/TSC2 complex, and subsequent activation of Rheb and mTORC1 [142]. During some paradigms of cellular protection, a limited activity of TSC2, as well as AMPK [143], appears necessary since complete knockdown of TSC2 can prevent cellular protection [144].

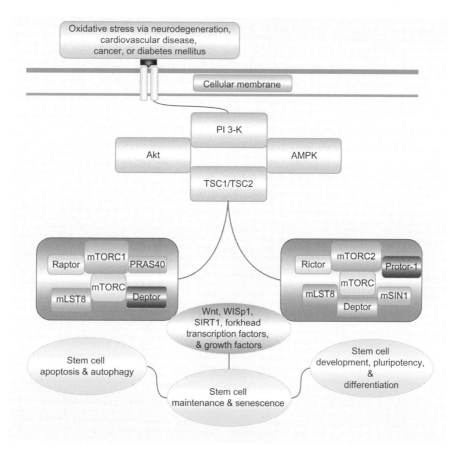

FIGURE 1.1 mTOR oversees stem cell development, pluripotency, differentiation, maintenance, and programmed cell death in clinical disease. Multiple disease entities that involve neurodegenerative disorders, cardiovascular disease, DM, and cancer can be the result of oxidative-stress-mediated cell injury. Through oxidative stress and the generation of ROS, the pathways of mTOR and PI 3-K, protein kinase B (Akt), AMPK, and the TSC1/TSC2 complex become significant determinants of stem cell survival. mTOR is a vital component of mTORC1 and mTORC2. mTORC1 consists of Raptor, the proline-rich Akt substrate 40 kDa (PRAS40), Deptor, and mLST8/GβL. mTORC2 is composed of Rictor, mLST8, Deptor, the mSIN1, and the protein observed with Rictor-1 (Protor-1). Intimately tied to these pathways are Wnt, WISP1, SIRT1, forkhead transcription factors, and growth factors. Ultimately, mTOR signaling governs stem cell apoptosis, autophagy, maintenance, cell senescence, development, pluripotency, and differentiation.

AMPK also controls the activity of the TSC1/TSC2 complex and blocks mTORC1 function. AMPK phosphorylates TSC2 to increase GAP activity to turn Rheb-GTP into Rheb-GDP and thus inhibits the activity of mTORC1 [145]. AMPK also controls TSC1/TSC2 activity through RTP801 (REDD1/product of the *Ddit4* gene) [146]. AMPK activity can increase REDD1 expression during hypoxia to suppress mTORC1 activity by releasing TSC2 from its inhibitory binding to protein 14-3-3 [146]. In addition, AMPK modulates the sirtuin silent mating type information regulation 2 homolog 1 (*S. cerevisiae*) (SIRT1) activity that can be vital for stem cell survival and proliferation [22]. AMPK increases the cellular $NAD^+/NADH$ ratio that leads to deacetylation of the SIRT1 targets peroxisome proliferator-activated receptor-gamma coactivator 1 and the forkhead transcription factors FoxO1 [68] and FoxO3a [147]. AMPK increases nicotinamide phosphoribosyltransferase activity that catalyzes the conversion of nicotinamide to nicotinamide mononucleotide [85], increases nicotinamide adenine dinucleotide (NAD^+) levels, decreases levels of the SIRT1 inhibitor nicotinamide, and promotes SIRT1 transcription [9,148,149]. SIRT1 expression in combination with AMPK activation promotes the induction of autophagy that can protect endothelial cells exposed to oxidized low-density lipoproteins [150]. SIRT1 may function similar to AMPK as an inhibitor of mTOR to block its activity [35].

In regards to mTORC2, mTORC2 consists of Rictor, mLST8, Deptor, the mammalian stress-activated protein kinase interacting protein (mSIN1), and the protein observed with Rictor-1 (Protor-1) [6,119,151–154]. Rictor [155] and mSIN1 [156] can activate Akt at serine[473] and facilitate threonine[308] phosphorylation by phosphoinositide-dependent kinase 1 to promote cell survival [14,151,156]. The kinase domain of mTOR has been shown to phosphorylate mSIN1, preventing the lysosomal degradation of mSIN1 [157]. Protor-1 is a Rictor-binding subunit of mTORC2 and can activate serum and glucocorticoid induced protein kinase 1 (SGK1), since loss of Protor-1 reduces the hydrophobic motif phosphorylation of SGK1 and the substrate N-Myc down-regulated gene 1 in the kidney [158]. mTORC2 also phosphorylates and activates SGK1, a member of the protein kinase A/ protein kinase G/protein kinase C family of protein kinases [159]. mTORC2 controls cytoskeleton remodeling through PKCα and oversees cell migration through the Rac guanine nucleotide exchange factors P-Rex1 and P-Rex2 and through Rho signaling [160]. In contrast to the inhibition of mTORC1, mTORC2 is activated by the TSC1/TSC2 complex through the N-terminal region of TSC2 and the C-terminal region of Rictor [161]. Absence of the TSC1/TSC2 complex leads to the loss of mTORC2 kinase activity *in vitro* [161].

1.3 mTOR AND OXIDATIVE CELL STRESS

mTOR has a critical role in cell signaling pathways and regulates multiple cellular processes that include cellular proliferation, cellular senescence, metabolism, gene transcription, cellular survival, and cellular death (Figure 1.1). Targeting mTOR and its signaling pathways that control stem cell maintenance may prove essential for the development of new strategies for NCDs. Stem cells have the potential for treating multiple disorders. As a result, the ability of mTOR to oversee processes of cell injury and cell death becomes a vital consideration in the development of stem-cell-mediated therapies.

Oxidative stress can markedly impact cellular survival and longevity [11,53,162–167]. During oxidative stress, reactive oxygen species (ROS) are generated that include nitrogen-based free radical species such as nitric oxide and peroxynitrite as well as superoxide free radicals, hydrogen peroxide, and singlet oxygen [163]. Mitochondria also lead to the generation of ROS. Mitochondria produce adenosine triphosphate (ATP) through the oxidation of glucose, pyruvate, and NAD^+ that are present in the cytosol. NAD^+ and flavin adenine dinucleotide (FAD) are reduced to NADH and $FADH_2$ in the tricarboxylic acid cycle. This redox energy from NADH and $FADH_2$ is transferred to oxygen through the electron transport chain. Protons are then transferred from complexes I, III, and IV in the inner membrane to the intermembrane space with a subsequent proton gradient that is formed across the inner membrane. Complex V (ATP synthase) accumulates the energy from this gradient to produce ATP from adenosine diphosphate and inorganic phosphate (P_i).

As a result of this aerobic production for ATP, ROS are generated. At reduced levels, production of ROS may be important for tolerance against hypoxia and ischemia. Moderate levels of ROS also may be required for the regulation of inflammatory cell activation [168]. However, increased levels of ROS can lead to cellular injury and result in mitochondrial and other organelle injury, DNA damage, protein misfolding, and neuronal synaptic dysfunction [17,22,164,169,170]. Protective pathways serve to insulate against damage from ROS and involve vitamins B, C, D, and K [76,85,171–173], glutathione peroxidase [173,174], and superoxide dismutase [163,169,175–182].

1.4 mTOR, STEM CELL SURVIVAL, APOPTOSIS, AND AUTOPHAGY

Closely associated with oxidative stress cell injury are the pathways of programmed cell death that involve apoptosis and autophagy (Figure 1.1) [183–186]. Apoptosis has an early phase that involves the loss of plasma membrane lipid asymmetry and a later phase that leads to genomic DNA degradation [187–189]. The loss of asymmetry of membrane phosphatidylserine (PS) distribution is an early component of apoptosis that can be reversible [32,81,190]. However, if not reversed, cell death ensues and genomic DNA degradation prevails [187,188,191–195]. During the later phase of apoptotic cell injury, cellular DNA is destroyed, which is usually not a reversible process [39,179,196,197]. Externalization of PS residues and the onset of genomic DNA degradation are the result of a series of activation of nucleases and proteases that occur during apoptosis [148,198]. The early phase of apoptosis with membrane PS externalization activates inflammatory cells to engulf and remove injured cells [191,193,199,200]. Inhibition of membrane PS externalization is necessary to prevent the loss of functional cells that are temporarily injured and expressing membrane PS residues.

Under most circumstances, activation of mTOR and its related pathways of PI 3-K and Akt can prevent apoptotic cell death in stem cells. Loss of mTOR activity, such as with rapamycin, leads to endothelial progenitor cell apoptotic death. These detrimental effects during mTOR diminished activity may be related to inhibition of growth factor signaling [201]. Growth factors, such as erythropoietin (EPO) [49,202], are protective against apoptosis through mTOR activity against multiple insults. These include sepsis-associated encephalopathy [203], oxidative stress [129], cerebral microglial injury [204], and beta-amyloid (Aβ) toxicity [205]. EPO relies upon mTOR for cell development, differentiation, and survival [11,35].

EPO requires mTOR for the differentiation of neural precursor cells to achieve a neuronal phenotype [206] and for the protection of retinal progenitor cells from oxidative stress [207]. EPO controls mTOR and its downstream signaling pathways that involve PRAS40 to increase neuronal survival during oxygen-glucose deprivation [129]. EPO, through mTOR and Wnt signaling, maintains microglial survival during oxidative stress [204]. EPO can block Aβ toxicity through Wnt signaling and mTOR pathways and prevent caspase activation and apoptosis [205]. During hypoxia-reoxygenation stress, EPO increases mTOR activity to protect hippocampus-derived neuronal cells [208]. In the mTOR pathway, AMPK may also affect the biological function of EPO. The ability of EPO to oversee neuroinflammation may be linked to AMPK activity [209]. EPO may require a specific level of AMPK activity to alleviate detrimental effects of oxidative stress [210]. In addition, the concentration and activity of EPO may influence the protective actions of mTOR and signaling pathways associated with AMPK. High concentrations of EPO may promote cellular damage and lessen the activity of mTOR [211].

Other growth factors in addition to EPO also rely upon mTOR to maintain stem cell integrity. In experimental models with mice, the growth factors epidermal growth factor (EGF) and fibroblast growth factor (FGF) are protective of stem cells and neurons [212–216]. EGF and FGF use mTOR to maintain the proliferation of neural stem and progenitor cells [217]. EGF requires PI 3-K, Akt, and mTOR pathways to block cell injury during metabolic stress [218] and to prevent memory impairment [219]. Brain-derived neurotrophic factor (BDNF) uses mTOR activation for memory consolidation [220]. Yet, under some conditions, growth factor protection that is dependent upon autophagy may require blockade of mTOR activity. Cortical neurons are protected by BDNF through the induction of autophagy and the inhibition of mTOR during oxygen deprivation [221].

Autophagy is another pathway of programmed cell death that is dependent upon mTOR signaling to influence stem cell survival. Autophagy is a process that recycles components in the cell cytoplasm to remove nonfunctional organelles for disposal and tissue remodeling [15,185,222,223]. Autophagy has classifications that include microautophagy, chaperone-mediated autophagy, and macroautophagy [32]. Macroautophagy is the primary category of autophagy. This process consists of the sequestration of cytoplasmic proteins and organelles into autophagosomes that combine with lysosomes for degradation and recycling [32,66,80,223,224]. Microautophagy results in the invagination of lysosomal membranes for the sequestration and digestion of cytoplasmic components. Chaperone-mediated autophagy employs cytosolic chaperones for the transport of cytoplasmic components across lysosomal membranes [14].

The proteins TOR and mTOR are directly tied with genes that control autophagy [27,29,225]. At least 33 autophagic related genes (Atg) have been identified in yeast that can affect multiple disorders including DM, vascular disease, cancer, and neurodegenerative disorders [10,185,222,223,226–230]. Of these autophagic related genes, Atg1 and Atg13 (also known as Apg13) are associated with PI 3-K, Akt, and the TOR pathways. Dephosphorylation of Atg13 that can occur during starvation leads to the activation of Atg1 and the induction of autophagy. Phosphorylation of Atg13 through a TOR-dependent pathway releases Atg13 from Atg1 and reduces Atg1 activity. In mammals, the homologs of Atg1 are UNC-51 like kinase 1 (ULK1) and ULK2 [29]. Mammalian Atg13 binds to ULK1, ULK2, and FIP200 (focal adhesion kinase family interacting protein of 200 kDa) to activate ULKs, foster phosphorylation of FIP200 by ULKs, and lead to the activation of autophagy [231]. mTOR activity blocks the induction of autophagy by phosphorylating Atg13 and ULKs to inhibit the ULK−Atg13−FIP200 complex.

Under some conditions, autophagy serves as a vital component to promote stem cell survival that involves both SIRT1 and mTOR pathways. In embryonic stem cells, SIRT1 is protective [232] and appears to have an inverse relationship with mTOR [22]. SIRT1 can depress mTOR-mediated pathways as well as promote autophagy to preserve the integrity of human embryonic stem cells exposed to oxidative stress [233]. SIRT1 can foster inhibition of mTOR signaling to promote neuronal growth [234]. SIRT1 activity is necessary to promote autophagy to maintain proteostasis, produce energy during nutrient deprivation, and maintain muscle stem cell activation [235]. In endothelial cells exposed to oxidized low-density lipoproteins that can lead to atherosclerosis, SIRT1 up-regulation in combination with AMPK activity leads to autophagy that is necessary for cellular protection [150].

However, activation of autophagy during mTOR inhibition can also lead to stem cell demise in some circumstances. Autophagy may foster cellular senescence [236]. Endothelial progenitor cells that regenerate vascular endothelium become dysfunctional with the activation of autophagy during exposure to elevated glucose [80].

Wnt signaling pathways with mTOR can also modulate programmed cell death to play a role in stem cell survival. Wnt proteins are cysteine-rich glycosylated proteins that regulate stem cell proliferation and tumor cell growth [237–243]. During the activation of autophagy, breast cancer stem cells have been shown to succumb to apoptosis and the inhibition of Wnt signaling [228]. In addition, growth factors, such as EPO, may require Wnt signaling for the preservation of mesenchymal stem cells [244] and to limit "proapoptotic" forkhead transcription factor activity [245,246].

A downstream target in the Wnt pathway is Wnt1 inducible signaling pathway protein 1 (WISP1), which is also known as CCN4 [9]. WISP1 is a member of the six secreted extracellular matrix-associated CCN family of proteins that are involved in cellular survival and stem cell proliferation [247]. WISP1 is present in the brain, heart, kidney, lung, pancreas, placenta, epithelium, ovaries, small intestine, and spleen [247]. WISP1 is closely tied to mTOR and can significantly influence stem cell survival and proliferation. WISP1 can control induced pluripotent stem cell reprogramming [248,249]. WISP1 is also one of several genes that are overexpressed during pancreatic regeneration [250] and can support vascular regeneration during saphenous vein crush injury [251]. However, WISP1 oversees cellular senescence [252] and does not appear to foster excessive cellular proliferation under circumstances involving aging vascular cells [253]. WISP1 can also be differentially regulated in stem cells. WISP1 expression is increased during stem cell migration [254] and is repressed during hepatic differentiation in adipose-derived stem cells [238]. WISP1 also relies upon pathways of mTOR to prevent injury to cells [6,29]. WISP1 activates mTOR, inhibits PRAS40 [131], and limits TSC2 activity [144] to increase microglial cell survival during oxidative stress and Aβ toxicity. WISP1 also controls the post-translational phosphorylation of AMPK [35,64,136,255]. WISP1 regulates AMPK activation by differentially decreasing phosphorylation of TSC2 at serine[1387], a target of AMPK, and increasing phosphorylation of TSC2 at threone[1462], a target of Akt [144]. This enables WISP1 to provide a minimal but not excessive level of TSC2 and AMPK activity to promote cellular survival during toxic insults [144].

1.5 mTOR, STEM CELL PROLIFERATION, AND STEM CELL DIFFERENTIATION

mTOR can oversee the proliferation of stem cells in multiple systems of the body (Figure 1.1) [29]. Embryonic stem cell proliferation is decreased during the deletion of the C-terminal six amino acids of mTOR that leads to inhibition of kinase activity [256]. Deletion of the *mTOR* gene results in limited trophoblast growth, faulty implantation, and the inability to establish embryonic stem cells [257]. mTOR can also control undifferentiated stem cell growth in human embryonic stem cells as well as lead to activation of cell differentiation. Under some conditions, mTOR

activity leads to mesenchymal stem cell senescence [258]. Loss of mTOR activity can result in cell pluripotency, cell proliferation, and inhibition of mesoderm and endoderm activities in embryonic stem cells [259]. In contrast, activation of mTOR and its downstream signal transduction pathways can lead to cell differentiation. Increased mTOR and p70S6K activity results in embryonic stem cell differentiation [260].

1.6 mTOR, STEM CELLS, AND METABOLIC DISEASE

mTOR pathways are critical components for insulin signaling (Figure 1.1) [64]. mTOR inhibition, such as with rapamycin administration, results in reduced β-cell function and mass, insulin resistance, decreased insulin secretion, and DM [261]. Blockade of mTOR activity with rapamycin also reduces food intake and prevents fat-diet-induced obesity in mice. However, rapamycin administration can impair glucose uptake through the blockade of insulin-generated Akt activation and alteration in the translocation of glucose transporters to the plasma membrane as has been demonstrated in skeletal muscle [99]. Furthermore, loss of p70S6K activity results in hypoinsulinemia, insulin insensitivity to glucose secretion, glucose intolerance, and decreased pancreatic β-cell size [262]. Activation of the mTOR pathways p70S6K and 4EBP1 in pancreatic β-cells in mice leads to improved insulin secretion and resistance to β-cell streptozotocin toxicity and obesity [263].

Yet, activation of mTOR does not always lead to improved cellular metabolism and may require limited activity through the inhibition of this pathway. mTOR can function in a negative feedback loop and produce glucose intolerance by inhibiting the insulin receptor substrate 1 (IRS-1). mTOR signaling through the TSC1/TSC2 complex can inactivate IRS-1 and phosphorylate p70S6K to block IRS-1 activity [264]. With this pathway and mTOR activation, mTOR leads to inhibitory phosphorylation of IRS-1, impaired Akt signaling, and insulin resistance in studies with high-fat-fed obese rats [265]. Consumption of high-fat diets can also activate the renin—angiotensin—aldosterone system with increased circulating angiotensin II that activates mTOR and p70S6K to phosphorylate IRS-1 and foster insulin resistance [266].

Additional studies also suggest that limited activity of mTOR may be beneficial during DM. In studies that have examined caloric restriction in male mice, genes with the greatest statistical change after caloric restriction involved those associated with sirtuin activation and mTOR inhibition [267]. SIRT1 activation can prevent endothelial senescence during hyperglycemia [268], limit endothelial atherosclerotic lesions during elevated lipid states [269], and prevent endothelial cell apoptosis during experimental DM [193,270]. SIRT1 and mTOR signaling appear to have an inverse relationship in the control of cellular metabolism. Hepatic SIRT1 deficiency yields hepatic glucose overproduction, hyperglycemia, initiation of oxidative stress, and inhibition of the gene encoding Rictor that results in impaired mTORC2 and Akt signaling [271]. SIRT1 can function as a negative regulator of unfolded protein response signaling and inhibit mTOR in DM to attenuate hepatic steatosis, ameliorate insulin resistance, and restore glucose homeostasis [272]. During experimental DM models with high glucose exposure to mesangial cells, activation of SIRT1 with mTOR inhibition is required to promote mesangial cell proliferation [273]. SIRT1 is necessary for the protection of endothelial progenitor cells during high glucose exposure [274] and decreased SIRT1 levels have been observed in endothelial progenitor cells in patients with DM [78]. SIRT1 also has been shown to foster angiogenesis for vascular repair mechanisms during DM [275]. SIRT1 may provide progenitor cell protection through the maintenance of mitochondrial pathways. Autophagy and mitochondrial impairment occur in endothelial progenitor cells during exposure to elevated glucose exposure, suggesting a mechanism for dysfunction of endothelial progenitor cells during DM [80]. SIRT1 has been shown to prevent mitochondrial depolarization, cytochrome c release, and Bad, caspase 1, and caspase 3 activation [193,233,270,276].

1.7 mTOR, STEM CELLS, AND THE NERVOUS SYSTEM

In the nervous system, a loss of mTORC1 activity in neural stem cells results in reduced lineage expansion, blocked differentiation, and failed neuronal production (Figure 1.1) [277]. mTOR is necessary for insulin-induced neuronal differentiation in neuronal progenitor cells [278]. mTOR also governs the timing and control of neurogenesis. Inhibition of mTOR through the RTP801/REDD1 pathway delays neuronal differentiation. Over time, the expression of RTP801/REDD1 is altered with maturation. Levels of RTP801/REDD1 in newborn and mature neurons become diminished and mTOR activity is increased to promote the maturation of neurons [279]. Interestingly, with aging, the loss of mTOR activity may affect neural stem cell proliferation. mTOR signaling in the aged brain is reduced and is accompanied by a reduction in the proliferation of active neural stem cells [280]. mTOR activity is also

necessary for the expression of the neuronal phenotype of postmortem neuronal precursors [206].

Growth factors that employ mTOR can also direct stem cell development. EPO promotes Wnt signaling to maintain the survival of mesenchymal stem cells [244]. EPO also requires mTOR to regulate bone homeostasis with osteoblastogenesis and osteoclastogenesis [281]. Differentiation of neural precursor cells that may be used for the treatment of neurodegenerative disorders is dependent upon both EPO and mTOR [206].

Importantly, the degree of mTOR activity may affect different populations of stem cells. Inhibition of mTOR activity can result in stem cell differentiation into astrocytic cells [217]. Combined Akt and mTORC1 inhibition can influence stem cell proliferation and has been shown to lead to reduced neuronal stem cell self-renewal and earlier neuronal and astroglial differentiation [282]. Non-neuronal neighboring cells may represent another factor to influence the growth of neuronal stem cells. Endothelial cells can foster mTOR activity and lead to the expansion of long-term glioblastoma stem-like cells [283].

1.8 mTOR, STEM CELLS, AND THE VASCULAR SYSTEM

Protein kinase signaling plays an important role in stem cell development in the cardiovascular system (Figure 1.1). In particular, mTOR is one of several components important for the proliferation of human embryonic stem-cell-derived cardiomyocytes [284]. mTOR can regulate the proliferation of hematopoietic stem cells and progenitor cells [285]. Loss of mTOR results in endothelial cell death [201]. Similar to neural stem cells that are dependent upon growth factors and mTOR for survival, failed endothelial progenitor cell development may be the result of decreased growth factor signaling and loss of mTOR activity [201]. Insulin-like growth factor relies upon mTOR and Akt for the development of cardiac progenitor cells [286]. mTOR may be required for endothelial progenitor cells to initiate angiogenesis that can subsequently offer neuroprotection during cerebral ischemia [287]. In some studies, mTOR activity has been demonstrated to have a dual role. mTOR not only leads to stem cell proliferation, but also promotes apoptotic cell death, suggesting that a defined degree of mTOR activity is necessary to maximize stem cell viability [288]. For the maintenance of hematopoietic stem cells with blocked differentiation, reduction in mTOR signaling with decreased phosphorylation of p70S6K is required [289].

1.9 mTOR, STEM CELLS, AND TUMORIGENESIS

Strongly associated with proliferative pathways, mTOR also may lead to detrimental effects and promote tumor cell growth through cancer stem cell proliferation. mTOR activity is associated with neurofibromatosis type 1, tuberous sclerosis, Lhermitte−Duclos disease, and glioblastoma multiforme [29]. As a result, the US Food and Drug Administration (FDA) has approved rapamycin (sirolimus) and several rapamycin derivative compounds, known as "rapalogs," that inhibit mTOR for the treatment of subependymal giant cell astrocytoma associated with tuberous sclerosis, renal cancer, and neuroendocrine pancreatic tumors [116,139].

In regards to stem cell proliferation that can foster tumor growth, neuronal activity that promotes high-grade glioma proliferation in neural precursor and oligodendroglial precursor cells has recently been linked to mTOR activity (Figure 1.1) [290,291]. Blockade of mTOR activity can prevent the conversion of astrocytoma cells to oligodendroglioma cells that foster the development of glioblastoma multiforme [292]. Inhibition of mTOR signaling may limit the population of cancer stem cells that can lead to disease recurrence and therapeutic resistance [293]. For example, activation of mTOR has been correlated with chemotherapy resistance of breast cancer cells with stem cell characteristics [294]. Furthermore, in nasopharyngeal carcinoma, cancer stem cell properties are reduced and invasion potential is restrained with the inhibition of mTOR signaling [295].

However, the role of mTOR during tumor growth is complex since mTOR signaling is linked to other proliferative pathways. An example of this involves the Wnt signaling pathway. During injury paradigms, Wnt signaling can be cytoprotective [239,241,296−298]. Activation of Wnt pathways can block inflammatory cell loss during neurodegenerative disorders [187,192,205,299], limit cerebral ischemia [300,301], protect cells during experimental DM [57,58,302], foster neurogenesis [303] and stem cell differentiation [304], and promote wound healing [305]. Growth factors also require Wnt to prevent cerebral endothelial cell injury [58], limit apoptosis during forkhead transcription factor activation [57,306], preserve mesenchymal stem cells [244], maintain immune cells in the nervous system [204], and block Aβ toxicity in cerebral microglia [205]. However, Wnt signaling may lead to glioma proliferation [307,308], metastatic disease [309−312], and malignant melanoma [313]. Prolonged exposure of growth factors such as EPO that rely upon Wnt signaling can have detrimental effects including

the proliferation of cancer [314−316] and blood−brain barrier injury [317]. In addition, downstream pathways of Wnt, such as WISP1, can promote distant metastatic disease [318]. Variants of WISP1 are aggressive in promoting cell growth [319]. In contrast, nonvariant WISP1 expression can block tumor cell invasion, motility, and metastatic disease [320]. Under these scenarios of tumor cell growth with Wnt signaling, mTOR activity may represent a "biological checkpoint" that can limit excessive cell growth that is promoted by Wnt signaling. For example, mTOR activation maintains cell senescence and prevents tumor cell growth that is normally fostered by Wnt [321].

1.10 FUTURE CONSIDERATIONS

The 289-kDa serine/threonine protein kinase mTOR oversees stem cell development, maintenance, proliferation, and differentiation in multiple systems of the body. In human embryonic stem cells, mTOR controls undifferentiated stem cell growth as well as the processes that lead to stem cell differentiation. During metabolic disorders such as DM, activation of mTOR may be necessary for proper pancreatic β-cell function and insulin sensitivity. However, a limited level of mTOR activity with SIRT1 activation is required to restore glucose homeostasis, foster mesangial cell proliferation, promote endothelial progenitor cell number, and allow for vascular repair mechanisms during DM. In the nervous system, mTOR activity is necessary for lineage expansion, cell differentiation, and neuronal cell production. During aging, mTOR activity is diminished and leads to a reduction in the proliferation of active neural stem cells. In the cardiovascular system, mTOR regulates proliferation of hematopoietic stem cells, cardiac progenitor cells, and endothelial progenitor cells that are necessary for angiogenesis. mTOR also plays a prominent role in tumorigenesis through cancer stem cell maintenance, proliferation, and invasion. The function of mTOR in cancer is complex, since at times mTOR can serve to block tumor cell growth.

In the development of stem cell strategies for NCDs, several considerations for mTOR need to be addressed to move forward. First, what are the pathways of programmed cell death that would be most conducive to promote stem cell maintenance and proliferation? Under a number of conditions, mTOR activation can prevent apoptotic cell death and may assist growth factor protection of stem cells through the blockade of apoptosis. However, some populations of stem cells, such as cortical neurons, use autophagic pathways for cell protection that inhibit mTOR.

In combination with SIRT1 activation and inhibition of mTOR, autophagy can also preserve human embryonic stem cells exposed to oxidative stress and promote neuronal cell growth. However, the pathways of autophagy associated with mTOR are not straightforward. Activation of autophagy with mTOR inhibition can also result in stem cell demise. Autophagy may lead to cellular senescence and result in endothelial progenitor cell dysfunction. Further necessary analysis may involve the investigation of the role of pathways such as Wnt signaling and WISP1 that also employ mTOR signaling and can promote stem cell survival, but under some circumstances do not promote excessive cellular proliferation.

Another challenge for targeting mTOR involves the ability to direct stem cell differentiation. How can mTOR activity be finely controlled and balanced to oversee stem cell differentiation and maintain pluripotency? Absent or limited mTOR activity leads to cell pluripotency and cell proliferation. In contrast, activation of mTOR leads to stem cell differentiation, stem cell senescence, and potentially stem cell depletion.

Cellular metabolism for stem cell survival also comes into play with mTOR. mTOR activation can improve insulin secretion, reduce insulin resistance, and promote β-cell function and mass. Yet, mTOR activity as part of a feedback loop also has been shown to lead to inhibitory phosphorylation of IRS-1, impair Akt signaling, and promote insulin resistance in studies with high-fat diets. Furthermore, during mTOR inhibition in DM, SIRT1 can attenuate hepatic steatosis, ameliorate insulin resistance, restore glucose homeostasis, and promote mesangial cell proliferation. Control of mTOR activity and its function in feedback mechanisms must be considered during both physiological glucose homeostasis and during disorders of metabolism to effectively maintain stem cell survival and proliferation.

Similar considerations also must be addressed for mTOR and its impact on other biological systems and disorders as well. In the nervous system, mTOR activation increases lineage expansion, leads to cell differentiation, increases the neuronal cell population, and controls the timing of neurogenesis. However, mTOR activity in neighboring cells, such as endothelial cells, can also influence the neuronal population and lead to detrimental effects that result in the expansion of long-term glioblastoma stem-like cells. mTOR activation in the cardiovascular system is also required for endothelial progenitor cell and cardiac progenitor cell development. However, mTOR activity in the cardiovascular system can have a dual role that not only leads to cell proliferation, but also results in apoptotic cell death, implying that a defined degree of mTOR activity is necessary to maximize stem cell viability. As a proliferative agent, mTOR can also promote tumor cell growth through cancer stem cells. mTOR inhibition may be necessary to block cancer stem cell growth that can lead to disease recurrence and therapeutic resistance. However, mTOR holds a complicated role in tumor cell growth, since it is linked to other proliferative cell pathways necessitating careful governance of the cellular growth properties for mTOR. mTOR may function as a "biological checkpoint" by maintaining cell senescence and blocking tumor cell growth that is normally fostered through proliferative pathways such as Wnt signaling.

Acknowledgments

This research was supported by the following grants to Kenneth Maiese: American Diabetes Association, American Heart Association, NIH NIEHS, NIH NIA, NIH NINDS, and NIH ARRA.

References

[1] Minino AM. Death in the United States, 2011. NCHS Data Brief 2013;115:1–8.

[2] Hayutin A. Global demographic shifts create challenges and opportunities. PREA Quarterly 2007;(Fall):46–53.

[3] World Health Organization. Description of the global burden of NCDs, their risk factors and determinants. Global status report on noncommunicable diseases 2010; 2011. p. 1–176.

[4] Minino AM, Murphy SL. Death in the United States, 2010. NCHS Data Brief 2012;99:1–8.

[5] Sivaraman V, Yellon DM. Pharmacologic therapy that simulates conditioning for cardiac ischemic/reperfusion injury. J Cardiovasc Pharmacol Ther 2014;19(1):83–96.

[6] Maiese K. Cutting through the complexities of mTOR for the treatment of stroke. Curr Neurovasc Res 2014;11(2):177–86.

[7] Pergola PE, White CL, Szychowski JM, Talbert R, Brutto OD, Castellanos M, et al. Achieved blood pressures in the secondary prevention of small subcortical strokes (SPS3) study: challenges and lessons learned. Am J Hypertens 2014;27(8):1052–60.

[8] Esser N, Paquot N, Scheen AJ. Anti-inflammatory agents to treat or prevent type 2 diabetes, metabolic syndrome and cardiovascular disease. Expert Opin Investig Drugs 2015;24(3):283–307.

[9] Maiese K. New insights for oxidative stress and diabetes mellitus. Oxid Med Cell Longev 2015 [Article ID 875961].

[10] Chen W, Sun Y, Liu K, Sun X. Autophagy: a double-edged sword for neuronal survival after cerebral ischemia. Neural Regen Res 2014;9(12):1210–16.

[11] Maiese K. mTOR: driving apoptosis and autophagy for neurocardiac complications of diabetes mellitus. World J Diabetes 2015;6(2):217–24.

[12] Chen H, Zhu G, Liu N, Zhang W. Low-dose tissue plasminogen activator is as effective as standard tissue plasminogen activator administration for the treatment of acute ischemic stroke. Curr Neurovasc Res 2014;11(1):62–7.

[13] Pineda D, Ampurdanes C, Medina MG, Serratosa J, Tusell JM, Saura J, et al. Tissue plasminogen activator induces microglial

inflammation via a noncatalytic molecular mechanism involving activation of mitogen-activated protein kinases and Akt signaling pathways and AnnexinA2 and Galectin-1 receptors. Glia 2012;60(4):526−40.

[14] Maiese K. Driving neural regeneration through the mammalian target of rapamycin. Neural Regen Res 2014;9(15): 1413−17.

[15] Nakka VP, Prakash-Babu P, Vemuganti R. Crosstalk between endoplasmic reticulum stress, oxidative stress, and autophagy: potential therapeutic targets for acute CNS injuries. Mol Neurobiol 2014.

[16] Chong ZZ, Li F, Maiese K. Oxidative stress in the brain: novel cellular targets that govern survival during neurodegenerative disease. Prog Neurobiol 2005;75(3):207−46.

[17] Harish G, Mahadevan A, Pruthi N, Sreenivasamurthy SK, Puttamallesh VN, Keshava Prasad TS, et al. Characterization of traumatic brain injury in human brains reveals distinct cellular and molecular changes in contusion and pericontusion. J Neurochem 2015.

[18] Lee J, Gu X, Wei L, Wei Z, Dix TA, Yu SP. Therapeutic effects of pharmacologically induced hypothermia against traumatic brain injury in mice. J Neurotrauma 2014;31:1417−30.

[19] Maiese K. Coma and impaired consciousness. The Merck Manual. 19th Professional Edition; 2011.

[20] Wang JW, Wang HD, Cong ZX, Zhou XM, Xu JG, Jia Y, et al. Puerarin ameliorates oxidative stress in a rodent model of traumatic brain injury. J Surg Res 2014;186(1):328−37.

[21] Kumar PR, Essa MM, Al-Adawi S, Dradekh G, Memon MA, Akbar M, et al. Omega-3 fatty acids could alleviate the risks of traumatic brain injury—a mini review. J Tradit Complement Med 2014;4(2):89−92.

[22] Maiese K. SIRT1 and stem cells: in the forefront with cardiovascular disease, neurodegeneration and cancer. World J Stem Cells 2015;7(2):235−42.

[23] Martin A, Tegla CA, Cudrici CD, Kruszewski AM, Azimzadeh P, Boodhoo D, et al. Role of SIRT1 in autoimmune demyelination and neurodegeneration. Immunol Res 2015;61(3):187−97.

[24] Maiese K, Chong ZZ, Hou J, Shang YC. New strategies for Alzheimer's disease and cognitive impairment. Oxid Med Cell Longev 2009;2(5):279−89.

[25] Agis-Torres A, Solhuber M, Fernandez M, Sanchez-Montero JM. Multi-target-directed ligands and other therapeutic strategies in the search of a real solution for Alzheimer's disease. Curr Neuropharmacol 2014;12(1):2−36.

[26] Filley CM, Rollins YD, Anderson CA, Arciniegas DB, Howard KL, Murrell JR, et al. The genetics of very early onset Alzheimer disease. Cogn Behav Neurol 2007;20(3):149−56.

[27] Maiese K. Taking aim at Alzheimer's disease through the mammalian target of rapamycin. Ann Med 2014;46(8): 587−96.

[28] Schluesener JK, Zhu X, Schluesener HJ, Wang GW, Ao P. Key network approach reveals new insight into Alzheimer's disease. IET Syst Biol 2014;8(4):169−75.

[29] Chong ZZ, Shang YC, Wang S, Maiese K. Shedding new light on neurodegenerative diseases through the mammalian target of rapamycin. Prog Neurobiol 2012;99(2):128−48.

[30] Mishra AK, Ur Rasheed MS, Shukla S, Tripathi MK, Dixit A, Singh MP. Aberrant autophagy and parkinsonism: does correction rescue from disease progression? Mol Neurobiol 2014;51:893−908.

[31] Organization WH. Neurological disorders: public health challenges. WHO Library Cataloguing-in Publication Data; 2006. p. 1−232.

[32] Maiese K. Programming apoptosis and autophagy with novel approaches for diabetes mellitus. Curr Neurovasc Res 2015;12 (2):173−88.

[33] Jia G, Aroor AR, Martinez-Lemus LA, Sowers JR. Invited review: over-nutrition, mTOR signaling and cardiovascular diseases. Am J Physiol Regul Integr Comp Physiol 2014;307: R1198−206.

[34] Maiese K, Chong ZZ, Shang YC, Hou J. Novel avenues of drug discovery and biomarkers for diabetes mellitus. J Clin Pharmacol 2011;51(2):128−52.

[35] Maiese K, Chong ZZ, Shang YC, Wang S. Novel directions for diabetes mellitus drug discovery. Expert Opin Drug Discov 2013;8(1):35−48.

[36] Rutter MK, Massaro JM, Hoffmann U, O'Donnell CJ, Fox CS. Fasting glucose, obesity, and coronary artery calcification in community-based people without diabetes. Diabetes Care 2012;35:1944−50.

[37] Xu E, Schwab M, Marette A. Role of protein tyrosine phosphatases in the modulation of insulin signaling and their implication in the pathogenesis of obesity-linked insulin resistance. Rev Endocr Metab Disord 2014;15(1):79−97.

[38] Harris MI, Eastman RC. Early detection of undiagnosed diabetes mellitus: a US perspective. Diabetes Metab Res Rev 2000;16 (4):230−6.

[39] Maiese K. Novel applications of trophic factors, Wnt and WISP for neuronal repair and regeneration in metabolic disease. Neural Regen Res 2015;10(4):518−28.

[40] Maiese K, Chong ZZ, Shang YC. Mechanistic insights into diabetes mellitus and oxidative stress. Curr Med Chem 2007;14(16):1729−38.

[41] Centers for Medicare and Medicaid Services. National health expenditure projections 2012−2022. Available from: <http://www.cms.gov>; 2013 [accessed 09.01.15].

[42] Harrison IF, Dexter DT. Epigenetic targeting of histone deacetylase: therapeutic potential in Parkinson's disease? Pharmacol Ther 2013;140(1):34−52.

[43] Maiese K. Epigenetics in the cerebrovascular system: changing the code without altering the sequence. Curr Neurovasc Res 2014;11(1):1−3.

[44] Teschendorff AE, West J, Beck S. Age-associated epigenetic drift: implications, and a case of epigenetic thrift? Hum Mol Genet 2013;22(R1):R7−15.

[45] West J, Beck S, Wang X, Teschendorff AE. An integrative network algorithm identifies age-associated differential methylation interactome hotspots targeting stem-cell differentiation pathways. Sci Rep 2013;3:1630.

[46] Xiao FH, He YH, Li QG, Wu H, Luo LH, Kong QP. A genome-wide scan reveals important roles of DNA methylation in human longevity by regulating age-related disease genes. PLoS One 2015;10(3):e0120388.

[47] Shao S, Yang Y, Yuan G, Zhang M, Yu X. Signaling molecules involved in lipid-induced pancreatic Beta-cell dysfunction. DNA Cell Biol 2013;32(2):41−9.

[48] Maiese K, Chong ZZ, Shang YC, Wang S. Erythropoietin: new directions for the nervous system. Int J Mol Sci 2012;13 (9):11102−29.

[49] Maiese K, Li F, Chong ZZ. New avenues of exploration for erythropoietin. JAMA 2005;293(1):90−5.

[50] White MF. IRS2 integrates insulin/IGF1 signalling with metabolism, neurodegeneration and longevity. Diabetes Obes Metab 2014;16(Suppl. 1):14−15.

[51] Zhang Y, Wang L, Dey S, Alnaeeli M, Suresh S, Rogers H, et al. Erythropoietin action in stress response, tissue maintenance and metabolism. Int J Mol Sci 2014;15(6):10296−333.

[52] Chong ZZ, Maiese K. The Src homology 2 domain tyrosine phosphatases SHP-1 and SHP-2: diversified control of cell growth, inflammation, and injury. Histol Histopathol 2007;22 (11):1251–67.

[53] Liu Y, Palanivel R, Rai E, Park M, Gabor TV, Scheid MP, et al. Adiponectin stimulates autophagy and reduces oxidative stress to enhance insulin sensitivity during high fat diet feeding in mice. Diabetes 2014;64(1):36–48.

[54] Caron AZ, He X, Mottawea W, Seifert EL, Jardine K, Dewar-Darch D, et al. The SIRT1 deacetylase protects mice against the symptoms of metabolic syndrome. FASEB J 2014;28 (3):1306–16.

[55] Maiese K. Paring down obesity and metabolic disease by targeting inflammation and oxidative stress. Curr Neurovasc Res 2015;12(2):107–8.

[56] Arunachalam G, Samuel SM, Marei I, Ding H, Triggle CR. Metformin modulates hyperglycaemia-induced endothelial senescence and apoptosis through SIRT1. Br J Pharmacol 2014;171(2):523–35.

[57] Chong ZZ, Hou J, Shang YC, Wang S, Maiese K. EPO relies upon novel signaling of Wnt1 that requires Akt1, FoxO3a, GSK-3beta, and beta-catenin to foster vascular integrity during experimental diabetes. Curr Neurovasc Res 2011;8(2):103–20.

[58] Chong ZZ, Shang YC, Maiese K. Vascular injury during elevated glucose can be mitigated by erythropoietin and Wnt signaling. Curr Neurovasc Res 2007;4(3):194–204.

[59] Hou J, Chong ZZ, Shang YC, Maiese K. FoxO3a governs early and late apoptotic endothelial programs during elevated glucose through mitochondrial and caspase signaling. Mol Cell Endocrinol 2010;321(2):194–206.

[60] Liu Q, Li J, Cheng R, Chen Y, Lee K, Hu Y, et al. Nitrosative stress plays an important role in wnt pathway activation in diabetic retinopathy. Antioxid Redox Signal 2013;18(10):1141–53.

[61] Schaffer SW, Jong CJ, Mozaffari M. Role of oxidative stress in diabetes-mediated vascular dysfunction: unifying hypothesis of diabetes revisited. Vascul Pharmacol 2012;57(5–6):139–49.

[62] Wang L, Di L, Noguchi CT. Erythropoietin, a novel versatile player regulating energy metabolism beyond the erythroid system. Int J Biol Sci 2014;10(8):921–39.

[63] Aragno M, Mastrocola R, Ghe C, Arnoletti E, Bassino E, Alloatti G, et al. Obestatin induced recovery of myocardial dysfunction in type 1 diabetic rats: underlying mechanisms. Cardiovasc Diabetol 2012;11:129.

[64] Chong ZZ, Maiese K. Mammalian target of rapamycin signaling in diabetic cardiovascular disease. Cardiovasc Diabetol 2012;11(1):45.

[65] Das A, Durrant D, Koka S, Salloum FN, Xi L, Kukreja RC. Mammalian target of rapamycin (mTOR) inhibition with rapamycin improves cardiac function in type 2 diabetic mice: potential role of attenuated oxidative stress and altered contractile protein expression. J Biol Chem 2014;289(7):4145–60.

[66] He C, Zhu H, Li H, Zou MH, Xie Z. Dissociation of Bcl-2-Beclin1 complex by activated AMPK enhances cardiac autophagy and protects against cardiomyocyte apoptosis in diabetes. Diabetes 2013;62(4):1270–81.

[67] Ling S, Birnbaum Y, Nanhwan MK, Thomas B, Bajaj M, Li Y, et al. Dickkopf-1 (DKK1) phosphatase and tensin homolog on chromosome 10 (PTEN) crosstalk via microRNA interference in the diabetic heart. Basic Res Cardiol 2013;108(3):352.

[68] Maiese K, Chong ZZ, Shang YC, Hou J. FoxO proteins: cunning concepts and considerations for the cardiovascular system. Clin Sci (Lond) 2009;116(3):191–203.

[69] Maiese K, Hou J, Chong ZZ, Shang YC. A fork in the path: developing therapeutic inroads with FoxO proteins. Oxid Med Cell Longev 2009;2(3):119–29.

[70] Portbury AL, Ronnebaum SM, Zungu M, Patterson C, Willis MS. Back to your heart: ubiquitin proteasome system-regulated signal transduction. J Mol Cell Cardiol 2012;52(3):526–37.

[71] Zhang C, Zhang L, Chen S, Feng B, Lu X, Bai Y, et al. The prevention of diabetic cardiomyopathy by non-mitogenic acidic fibroblast growth factor is probably mediated by the suppression of oxidative stress and damage. PLoS One 2013;8(12):e82287.

[72] Alexandru N, Popov D, Georgescu A. Platelet dysfunction in vascular pathologies and how can it be treated. Thromb Res 2012;129(2):116–26.

[73] Razmara M, Hjemdahl P, Ostenson CG, Li N. Platelet hyperpro-coagulant activity in Type 2 diabetes mellitus: attenuation by glycoprotein IIb/IIIa inhibition. J Thromb Haemost. 2008;6 (12):2186–92.

[74] Hu P, Lai D, Lu P, Gao J, He H. ERK and Akt signaling pathways are involved in advanced glycation end product-induced autophagy in rat vascular smooth muscle cells. Int J Mol Med 2012;29(4):613–18.

[75] Wang F, Ma X, Zhou M, Pan X, Ni J, Gao M, et al. Serum pigment epithelium-derived factor levels are independently correlated with the presence of coronary artery disease. Cardiovasc Diabetol 2013;12:56.

[76] Xu YJ, Tappia PS, Neki NS, Dhalla NS. Prevention of diabetes-induced cardiovascular complications upon treatment with antioxidants. Heart Fail Rev 2014;19(1):113–21.

[77] Albiero M, Poncina N, Tjwa M, Ciciliot S, Menegazzo L, Ceolotto G, et al. Diabetes causes bone marrow autonomic neuropathy and impairs stem cell mobilization via dysregulated p66Shc and Sirt1. Diabetes 2014;63(4):1353–65.

[78] Balestrieri ML, Servillo L, Esposito A, D'Onofrio N, Giovane A, Casale R, et al. Poor glycaemic control in type 2 diabetes patients reduces endothelial progenitor cell number by influencing SIRT1 signalling via platelet-activating factor receptor activation. Diabetologia 2013;56(1):162–72.

[79] Barthelmes D, Zhu L, Shen W, Gillies MC, Irhimeh MR. Differential gene expression in Lin-/VEGF-R2+ bone marrow-derived endothelial progenitor cells isolated from diabetic mice. Cardiovasc Diabetol 2014;13(1):42.

[80] Kim KA, Shin YJ, Akram M, Kim ES, Choi KW, Suh H, et al. High glucose condition induces autophagy in endothelial progenitor cells contributing to angiogenic impairment. Biol Pharm Bull 2014;37(7):1248–52.

[81] Weinberg E, Maymon T, Weinreb M. AGEs induce caspase-mediated apoptosis of rat BMSCs via TNFalpha production and oxidative stress. J Mol Endocrinol 2014;52(1):67–76.

[82] Chen JX, Tuo Q, Liao DF, Zeng H. Inhibition of protein tyrosine phosphatase improves angiogenesis via enhancing ang-1/tie-2 signaling in diabetes. Exp Diabetes Res 2012;2012:836759.

[83] da Rosa LC, Chiuso-Minicucci F, Zorzella-Pezavento SF, Franca TG, Ishikawa LL, Colavite PM, et al. Bacille Calmette-Guerin/DNAhsp65 prime-boost is protective against diabetes in non-obese diabetic mice but not in the streptozotocin model of type 1 diabetes. Clin Exp Immunol 2013;173(3):430–7.

[84] Hamed S, Bennett CL, Demiot C, Ullmann Y, Teot L, Desmouliere A. Erythropoietin, a novel repurposed drug: an innovative treatment for wound healing in patients with diabetes mellitus. Wound Repair Regen 2014;22(1):23–33.

[85] Maiese K, Chong ZZ, Hou J, Shang YC. The vitamin nicotinamide: translating nutrition into clinical care. Molecules 2009;14 (9):3446–85.

[86] Maiese K, Chong ZZ, Shang YC, Wang S. Translating cell survival and cell longevity into treatment strategies with SIRT1. Rom J Morphol Embryol 2011;52(4):1173–85.

[87] Puthanveetil P, Wan A, Rodrigues B. FoxO1 is crucial for sustaining cardiomyocyte metabolism and cell survival. Cardiovasc Res 2013;97(3):393–403.

[88] Zhao Y, Scott NA, Fynch S, Elkerbout L, Wong WW, Mason KD, et al. Autoreactive T cells induce necrosis and not BCL-2-regulated or death receptor-mediated apoptosis or RIPK3-dependent necroptosis of transplanted islets in a mouse model of type 1 diabetes. Diabetologia 2015;58:140–8.

[89] Hao J, Li F, Liu W, Liu Q, Liu S, Li H, et al. Phosphorylation of PRAS40-Thr246 involves in renal lipid accumulation of diabetes. J Cell Physiol 2013;229:1069–77.

[90] Hao J, Zhu L, Li F, Liu Q, Zhao X, Liu S, et al. Phospho-mTOR: a novel target in regulation of renal lipid metabolism abnormality of diabetes. Exp Cell Res 2013;319(14):2296–306.

[91] Nakazawa J, Isshiki K, Sugimoto T, Araki S, Kume S, Yokomaku Y, et al. Renoprotective effects of asialoerythropoietin in diabetic mice against ischaemia-reperfusion-induced acute kidney injury. Nephrology (Carlton) 2010;15(1):93–101.

[92] Pandya KG, Budhram R, Clark GJ, Lau-Cam CA. Taurine can enhance the protective actions of metformin against diabetes-induced alterations adversely affecting renal function. Adv Exp Med Biol 2015;803:227–50.

[93] Perez-Gallardo RV, Noriega-Cisneros R, Esquivel-Gutierrez E, Calderon-Cortes E, Cortes-Rojo C, Manzo-Avalos S, et al. Effects of diabetes on oxidative and nitrosative stress in kidney mitochondria from aged rats. J Bioenerg Biomembr 2014;46:511–18.

[94] Castano D, Larequi E, Belza I, Astudillo AM, Martinez-Anso E, Balsinde J, et al. Cardiotrophin-1 eliminates hepatic steatosis in obese mice by mechanisms involving AMPK activation. J Hepatol 2014;60(5):1017–25.

[95] Gomes MB, Negrato CA. Alpha-lipoic acid as a pleiotropic compound with potential therapeutic use in diabetes and other chronic diseases. Diabetol Metab Syndr 2014;6(1):80.

[96] Gong H, Pang J, Han Y, Dai Y, Dai D, Cai J, et al. Age-dependent tissue expression patterns of Sirt1 in senescence-accelerated mice. Mol Med Rep 2014;10(6):3296–302.

[97] Lee JH, Lee JH, Jin M, Han SD, Chon GR, Kim IH, et al. Diet control to achieve euglycemia induces significant loss of heart and liver weight via increased autophagy compared with ad libitum diet in diabetic rats. Exp Mol Med 2014;46:e111.

[98] Malla R, Wang Y, Chan WK, Tiwari AK, Faridi JS. Genetic ablation of PRAS40 improves glucose homeostasis via linking the AKT and mTOR pathways. Biochem Pharmacol 2015;96:65–75.

[99] Deblon N, Bourgoin L, Veyrat-Durebex C, Peyrou M, Vinciguerra M, Caillon A, et al. Chronic mTOR inhibition by rapamycin induces muscle insulin resistance despite weight loss in rats. Br J Pharmacol 2012;165(7):2325–40.

[100] Gao J, Li J, An Y, Liu X, Qian Q, Wu Y, et al. Increasing effect of Tangzhiqing formula on IRS-1-dependent PI3K/AKT signaling in muscle. BMC Complement Altern Med 2014;14:198.

[101] Zhang T, Fang M, Fu ZM, Du HC, Yuan H, Xia GY, et al. Expression of PI3-K, PKB and GSK-3 beta in the skeletal muscle tissue of gestational diabetes mellitus. Asian Pac J Trop Med 2014;7(4):309–12.

[102] Busch S, Kannt A, Kolibabka M, Schlotterer A, Wang Q, Lin J, et al. Systemic treatment with erythropoietin protects the neurovascular unit in a rat model of retinal neurodegeneration. PLoS One 2014;9(7):e102013.

[103] Fu D, Wu M, Zhang J, Du M, Yang S, Hammad SM, et al. Mechanisms of modified LDL-induced pericyte loss and retinal injury in diabetic retinopathy. Diabetologia 2012; 55(11):3128–40.

[104] Lee K, Hu Y, Ding L, Chen Y, Takahashi Y, Mott R, et al. Therapeutic potential of a monoclonal antibody blocking the Wnt pathway in diabetic retinopathy. Diabetes. 2012; 61(11):2948–57.

[105] Gomez-Brouchet A, Blaes N, Mouledous L, Fourcade O, Tack I, Frances B, et al. Beneficial effects of levobupivacaine regional anaesthesia on postoperative opioid induced hyperalgesia in diabetic mice. J Transl Med 2015;13(1):208.

[106] Du LL, Chai DM, Zhao LN, Li XH, Zhang FC, Zhang HB, et al. AMPK activation ameliorates alzheimer's disease-like pathology and spatial memory impairment in a streptozotocin-induced alzheimer's disease model in rats. J Alzheimers Dis 2014;43:775–84.

[107] Mao XY, Cao DF, Li X, Yin JY, Wang ZB, Zhang Y, et al. Huperzine A ameliorates cognitive deficits in streptozotocin-induced diabetic rats. Int J Mol Sci 2014;15(5):7667–83.

[108] Zhao Z, Huang G, Wang B, Zhong Y. Inhibition of NF-kappaB activation by Pyrrolidine dithiocarbamate partially attenuates hippocampal MMP-9 activation and improves cognitive deficits in streptozotocin-induced diabetic rats. Behav Brain Res 2013;238:44–7.

[109] Aksu I, Ates M, Baykara B, Kiray M, Sisman AR, Buyuk E, et al. Anxiety correlates to decreased blood and prefrontal cortex IGF-1 levels in streptozotocin induced diabetes. Neurosci Lett 2012;531(2):176–81.

[110] Reagan LP. Diabetes as a chronic metabolic stressor: causes, consequences and clinical complications. Exp Neurol 2012;233 (1):68–78.

[111] Jiang T, Yu JT, Zhu XC, Wang HF, Tan MS, Cao L, et al. Acute metformin preconditioning confers neuroprotection against focal cerebral ischaemia by pre-activation of AMPK-dependent autophagy. Br J Pharmacol 2014;171 (13):3146–57.

[112] Maiese K, Chong ZZ, Hou J, Shang YC. Erythropoietin and oxidative stress. Curr Neurovasc Res 2008;5(2):125–42.

[113] Kapogiannis D, Boxer A, Schwartz JB, Abner EL, Biragyn A, Masharani U, et al. Dysfunctionally phosphorylated type 1 insulin receptor substrate in neural-derived blood exosomes of preclinical Alzheimer's disease. FASEB J 2015;29:589–96.

[114] Maiese K, Chong ZZ, Shang YC. Raves and risks for erythropoietin. Cytokine Growth Factor Rev 2008;19(2):145–55.

[115] Coca SG, Ismail-Beigi F, Haq N, Krumholz HM, Parikh CR. Role of intensive glucose control in development of renal end points in type 2 diabetes mellitus: systematic review and meta-analysis intensive glucose control in type 2 diabetes. Arch Intern Med 2012;172(10):761–9.

[116] Maiese K, Chong ZZ, Shang YC, Wang S. mTOR: on target for novel therapeutic strategies in the nervous system. Trends Mol Med 2013;19(1):51–60.

[117] Neasta J, Barak S, Hamida SB, Ron D. mTOR complex 1: a key player in neuroadaptations induced by drugs of abuse. J Neurochem 2014;130(2):172–84.

[118] Heitman J, Movva NR, Hall MN. Targets for cell cycle arrest by the immunosuppressant rapamycin in yeast. Science 1991;253(5022):905–9.

[119] Chong ZZ, Shang YC, Zhang L, Wang S, Maiese K. Mammalian target of rapamycin: hitting the bull's-eye for neurological disorders. Oxid Med Cell Longev 2010;3(6): 374–91.

[120] Acosta-Jaquez HA, Keller JA, Foster KG, Ekim B, Soliman GA, Feener EP, et al. Site-specific mTOR phosphorylation promotes mTORC1-mediated signaling and cell growth. Mol Cell Biol 2009;29(15):4308–24.

[121] Gulhati P, Bowen KA, Liu J, Stevens PD, Rychahou PG, Chen M, et al. mTORC1 and mTORC2 regulate EMT, motility, and metastasis of colorectal cancer via RhoA and Rac1 signaling pathways. Cancer Res 2011;71(9):3246−56.

[122] Zoncu R, Efeyan A, Sabatini DM. mTOR: from growth signal integration to cancer, diabetes and ageing. Nat Rev Mol Cell Biol 2011;12(1):21−35.

[123] Maiese K. Stem cell guidance through the mechanistic target of rapamycin. World J Stem Cells 2015;7(7):999−1009.

[124] Sarbassov DD, Ali SM, Sengupta S, Sheen JH, Hsu PP, Bagley AF, et al. Prolonged rapamycin treatment inhibits mTORC2 assembly and Akt/PKB. Mol Cell 2006;22(2):159−68.

[125] Chen J, Zheng XF, Brown EJ, Schreiber SL. Identification of an 11-kDa FKBP12-rapamycin-binding domain within the 289-kDa FKBP12-rapamycin-associated protein and characterization of a critical serine residue. Proc Natl Acad Sci USA 1995;92(11): 4947−51.

[126] Bai X, Ma D, Liu A, Shen X, Wang QJ, Liu Y, et al. Rheb activates mTOR by antagonizing its endogenous inhibitor, FKBP38. Science 2007;318(5852):977−80.

[127] Wang L, Lawrence Jr JC, Sturgill TW, Harris TE. Mammalian target of rapamycin complex 1 (mTORC1) activity is associated with phosphorylation of raptor by mTOR. J Biol Chem 2009;284(22):14693−7.

[128] Wang H, Zhang Q, Wen Q, Zheng Y, Philip L, Jiang H, et al. Proline-rich Akt substrate of 40kDa (PRAS40): a novel downstream target of PI3k/Akt signaling pathway. Cell Signal 2012;24(1):17−24.

[129] Chong ZZ, Shang YC, Wang S, Maiese K. PRAS40 is an integral regulatory component of erythropoietin mTOR signaling and cytoprotection. PLoS One 2012;7(9):e45456.

[130] Fonseca BD, Smith EM, Lee VH, MacKintosh C, Proud CG. PRAS40 is a target for mammalian target of rapamycin complex 1 and is required for signaling downstream of this complex. J Biol Chem 2007;282(34):24514−24.

[131] Shang YC, Chong ZZ, Wang S, Maiese K. WNT1 inducible signaling pathway protein 1 (WISP1) targets PRAS40 to govern beta-amyloid apoptotic injury of microglia. Curr Neurovasc Res 2012;9(4):239−49.

[132] Xiong X, Xie R, Zhang H, Gu L, Xie W, Cheng M, et al. PRAS40 plays a pivotal role in protecting against stroke by linking the Akt and mTOR pathways. Neurobiol Dis 2014;66:43−52.

[133] Peterson TR, Laplante M, Thoreen CC, Sancak Y, Kang SA, Kuehl WM, et al. DEPTOR is an mTOR inhibitor frequently overexpressed in multiple myeloma cells and required for their survival. Cell 2009;137(5):873−86.

[134] Kim DH, Sarbassov DD, Ali SM, Latek RR, Guntur KV, Erdjument-Bromage H, et al. GbetaL, a positive regulator of the rapamycin-sensitive pathway required for the nutrient-sensitive interaction between raptor and mTOR. Mol Cell 2003;11(4):895−904.

[135] Guertin DA, Stevens DM, Thoreen CC, Burds AA, Kalaany NY, Moffat J, et al. Ablation in mice of the mTORC components raptor, rictor, or mLST8 reveals that mTORC2 is required for signaling to Akt-FOXO and PKCalpha, but not S6K1. Dev Cell 2006;11(6):859−71.

[136] Martinez de Morentin PB, Martinez-Sanchez N, Roa J, Ferno J, Nogueiras R, Tena-Sempere M, et al. Hypothalamic mTOR: the rookie energy sensor. Curr Mol Med 2014;14(1):3−21.

[137] Chong ZZ, Shang YC, Wang S, Maiese K. A critical kinase cascade in neurological disorders: PI 3-K, Akt, and mTOR. Future Neurol 2012;7(6):733−48.

[138] Janku F, Wheler JJ, Westin SN, Moulder SL, Naing A, Tsimberidou AM, et al. PI3K/AKT/mTOR inhibitors in patients with breast and gynecologic malignancies harboring PIK3CA mutations. J Clin Oncol 2012;30(8):777−82.

[139] Maiese K. Therapeutic targets for cancer: current concepts with PI 3-K, Akt, & mTOR. Indian J Med Res 2013; 137(2):243−6.

[140] Morgan-Warren PJ, Berry M, Ahmed Z, Scott RA, Logan A. Exploiting mTOR signaling: a novel translatable treatment strategy for traumatic optic neuropathy? Invest Ophthalmol Vis Sci 2013;54(10):6903−16.

[141] Sato T, Nakashima A, Guo L, Tamanoi F. Specific activation of mTORC1 by Rheb G-protein in vitro involves enhanced recruitment of its substrate protein. J Biol Chem 2009; 284(19):12783−91.

[142] Cai SL, Tee AR, Short JD, Bergeron JM, Kim J, Shen J, et al. Activity of TSC2 is inhibited by AKT-mediated phosphorylation and membrane partitioning. J Cell Biol 2006;173(2): 279−89.

[143] Hahn-Windgassen A, Nogueira V, Chen CC, Skeen JE, Sonenberg N, Hay N. Akt activates the mammalian target of rapamycin by regulating cellular ATP level and AMPK activity. J Biol Chem 2005;280(37):32081−9.

[144] Shang YC, Chong ZZ, Wang S, Maiese K. Tuberous sclerosis protein 2 (TSC2) modulates CCN4 cytoprotection during apoptotic amyloid toxicity in microglia. Curr Neurovasc Res 2013;10(1):29−38.

[145] Inoki K, Zhu T, Guan KL. TSC2 mediates cellular energy response to control cell growth and survival. Cell 2003; 115(5):577−90.

[146] DeYoung MP, Horak P, Sofer A, Sgroi D, Ellisen LW. Hypoxia regulates TSC1/2-mTOR signaling and tumor suppression through REDD1-mediated 14-3-3 shuttling. Genes Dev 2008; 22(2):239−51.

[147] Canto C, Auwerx J. Caloric restriction, SIRT1 and longevity. Trends Endocrinol Metab 2009;20(7):325−31.

[148] Chong ZZ, Shang YC, Wang S, Maiese K. SIRT1: new avenues of discovery for disorders of oxidative stress. Expert Opin Ther Targets 2012;16(2):167−78.

[149] Wang Y, Liang Y, Vanhoutte PM. SIRT1 and AMPK in regulating mammalian senescence: a critical review and a working model. FEBS Lett 2011;585(7):986−94.

[150] Jin X, Chen M, Yi L, Chang H, Zhang T, Wang L, et al. Delphinidin-3-glucoside protects human umbilical vein endothelial cells against oxidized low-density lipoprotein-induced injury by autophagy upregulation via the AMPK/SIRT1 signaling pathway. Mol Nutr Food Res 2014;58(10):1941−51.

[151] Glidden EJ, Gray LG, Vemuru S, Li D, Harris TE, Mayo MW. Multiple site acetylation of rictor stimulates mammalian target of rapamycin complex 2 (mTORC2)-dependent phosphorylation of Akt protein. J Biol Chem 2012;287(1):581−8.

[152] James MF, Stivison E, Beauchamp R, Han S, Li H, Wallace MR, et al. Regulation of mTOR complex 2 signaling in neurofibromatosis 2-deficient target cell types. Mol Cancer Res 2012;10(5):649−59.

[153] Kamarudin MN, Mohd Raflee NA, Syed Hussein SS, Lo JY, Supriady H, Abdul Kadir H. (R)-(+)-alpha-Lipoic acid protected NG108-15 cells against H2O2-induced cell death through PI3K-Akt/GSK-3beta pathway and suppression of NF-kappabeta-cytokines. Drug Des Dev Ther 2014;8:1765−80.

[154] Tang Z, Baykal AT, Gao H, Quezada HC, Zhang H, Bereczki E, et al. mTor is a signaling hub in cell survival: a mass-spectrometry-based proteomics investigation. J Proteome Res 2014;13(5):2433−44.

[155] Sarbassov DD, Guertin DA, Ali SM, Sabatini DM. Phosphorylation and regulation of Akt/PKB by the rictor-mTOR complex. Science 2005;307(5712):1098−101.

[156] Frias MA, Thoreen CC, Jaffe JD, Schroder W, Sculley T, Carr SA, et al. mSin1 is necessary for Akt/PKB phosphorylation, and its isoforms define three distinct mTORC2s. Curr Biol 2006;16(18):1865–70.

[157] Chen CH, Sarbassov dos D. The mTOR (mammalian target of rapamycin) kinase maintains integrity of mTOR complex 2. J Biol Chem 2011;286(46):40386–94.

[158] Pearce LR, Sommer EM, Sakamoto K, Wullschleger S, Alessi DR. Protor-1 is required for efficient mTORC2-mediated activation of SGK1 in the kidney. Biochem J 2011;436(1):169–79.

[159] Garcia-Martinez JM, Alessi DR. mTOR complex 2 (mTORC2) controls hydrophobic motif phosphorylation and activation of serum- and glucocorticoid-induced protein kinase 1 (SGK1). Biochem J 2008;416(3):375–85.

[160] Jacinto E, Loewith R, Schmidt A, Lin S, Ruegg MA, Hall A, et al. Mammalian TOR complex 2 controls the actin cytoskeleton and is rapamycin insensitive. Nat Cell Biol 2004;6(11):1122–8.

[161] Huang J, Dibble CC, Matsuzaki M, Manning BD. The TSC1-TSC2 complex is required for proper activation of mTOR complex 2. Mol Cell Biol 2008;28(12):4104–15.

[162] Chong ZZ, Li F, Maiese K. Stress in the brain: novel cellular mechanisms of injury linked to Alzheimer's disease. Brain Res Brain Res Rev 2005;49(1):1–21.

[163] Kwon SH, Hong SI, Ma SX, Lee SY, Jang CG. 3′,4′,7-Trihydroxyflavone prevents apoptotic cell death in neuronal cells from hydrogen peroxide-induced oxidative stress. Food Chem Toxicol 2015;80:41–51.

[164] Maiese K, Chong ZZ, Wang S, Shang YC. Oxidant stress and signal transduction in the nervous system with the PI 3-K, akt, and mTOR cascade. Int J Mol Sci 2013;13(11):13830–66.

[165] Mhillaj E, Morgese MG, Trabace L. Early life and oxidative stress in psychiatric disorders: what can we learn from animal models? Curr Pharm Des 2015;21(11):1396–403.

[166] Ozel Turkcu U, Solak Tekin N, Gokdogan Edgunlu T, Karakas Celik S, Oner S. The association of FOXO3A gene polymorphisms with serum FOXO3A levels and oxidative stress markers in vitiligo patients. Gene 2014;536(1):129–34.

[167] Zolotukhin P, Kozlova Y, Dovzhik A, Kovalenko K, Kutsyn K, Aleksandrova A, et al. Oxidative status interactome map: towards novel approaches in experiment planning, data analysis, diagnostics and therapy. Mol Biosyst 2013;9(8):2085–96.

[168] Maiese K. Mitochondria: "mood altering organelles" that impact disease throughout the nervous system. Curr Neurovasc Res 2015;12(4):309–11.

[169] Palma HE, Wolkmer P, Gallio M, Correa MM, Schmatz R, Thome GR, et al. Oxidative stress parameters in blood, liver, and kidney of diabetic rats treated with curcumin and/or insulin. Mol Cell Biochem 2014;386(1–2):199–210.

[170] Zeldich E, Chen CD, Colvin TA, Bove-Fenderson EA, Liang J, Tucker Zhou TB, et al. The neuroprotective effect of Klotho is mediated via regulation of members of the redox system. J Biol Chem 2014;289(35):24700–15.

[171] Bowes Rickman C, Farsiu S, Toth CA, Klingeborn M. Dry age-related macular degeneration: mechanisms, therapeutic targets, and imaging. Invest Ophthalmol Vis Sci 2013;54(14):ORSF68–80.

[172] Miret JA, Munne-Bosch S. Plant amino acid-derived vitamins: biosynthesis and function. Amino Acids 2014;46(4):809–24.

[173] Yousef JM, Mohamed AM. Prophylactic role of B vitamins against bulk and zinc oxide nano-particles toxicity induced oxidative DNA damage and apoptosis in rat livers. Pak J Pharm Sci 2015;28(1):175–84.

[174] Rjiba-Touati K, Ayed-Boussema I, Guedri Y, Achour A, Bacha H, Abid-Essefi S. Effect of recombinant human erythropoietin on mitomycin C-induced oxidative stress and genotoxicity in rat kidney and heart tissues. Hum Exp Toxicol 2015.

[175] Aksu U, Yanar K, Terzioglu D, Erkol T, Ece E, Aydin S, et al. Effect of tempol on redox homeostasis and stress tolerance in mimetically aged drosophila. Arch Insect Biochem Physiol 2014;87:13–25.

[176] Chen Q, Yu W, Shi J, Shen J, Hu Y, Gong J, et al. The effect of extracorporeal membrane oxygenation therapy on systemic oxidative stress injury in a porcine model. Artif Organs 2014;38:426–31.

[177] Gezginci-Oktayoglu S, Sacan O, Bolkent S, Ipci Y, Kabasakal L, Sener G, et al. Chard (Beta vulgaris L. var. cicla) extract ameliorates hyperglycemia by increasing GLUT2 through Akt2 and antioxidant defense in the liver of rats. Acta Histochem 2014;116(1):32–9.

[178] Guo R, Li W, Liu B, Li S, Zhang B, Xu Y. Resveratrol protects vascular smooth muscle cells against high glucose-induced oxidative stress and cell proliferation in vitro. Med Sci Monit Basic Res 2014;20:82–92.

[179] Kim S, Kang IH, Nam JB, Cho Y, Chung DY, Kim SH, et al. Ameliorating the effect of astragaloside IV on learning and memory deficit after chronic cerebral hypoperfusion in rats. Molecules 2015;20(2):1904–21.

[180] Srivastava A, Shivanandappa T. Prevention of hexachlorocyclohexane-induced neuronal oxidative stress by natural antioxidants. Nutr Neurosci 2014;17(4):164–71.

[181] Toblli JE, Cao G, Angerosa M, Rivero M. Long-term phosphodiesterase 5 inhibitor administration reduces inflammatory markers and heat-shock proteins in cavernous tissue of Zucker diabetic fatty rat (ZDF/fa/fa). Int J Impot Res 2015;27(5):182–90.

[182] Turunc Bayrakdar E, Uyanikgil Y, Kanit L, Koylu E, Yalcin A. Nicotinamide treatment reduces the levels of oxidative stress, apoptosis, and PARP-1 activity in Abeta(1-42)-induced rat model of Alzheimer's disease. Free Radic Res 2014;48(2):146–58.

[183] Deruy E, Gosselin K, Vercamer C, Martien S, Bouali F, Slomianny C, et al. MnSOD upregulation induces autophagic programmed cell death in senescent keratinocytes. PLoS One 2010;5(9):e12712.

[184] Lemasters JJ, Nieminen AL, Qian T, Trost LC, Elmore SP, Nishimura Y, et al. The mitochondrial permeability transition in cell death: a common mechanism in necrosis, apoptosis and autophagy. Biochim Biophys Acta 1998;1366(1–2):177–96.

[185] Maiese K, Chong ZZ, Shang YC, Wang S. Targeting disease through novel pathways of apoptosis and autophagy. Expert Opin Ther Targets 2012;16(12):1203–14.

[186] Nagley P, Higgins GC, Atkin JD, Beart PM. Multifaceted deaths orchestrated by mitochondria in neurones. Biochim Biophys Acta 2010;1802:167–85.

[187] Shang YC, Chong ZZ, Hou J, Maiese K. Wnt1, FoxO3a, and NF-kappaB oversee microglial integrity and activation during oxidant stress. Cell Signal 2010;22(9):1317–29.

[188] Viola G, Bortolozzi R, Hamel E, Moro S, Brun P, Castagliuolo I, et al. MG-2477, a new tubulin inhibitor, induces autophagy through inhibition of the Akt/mTOR pathway and delayed apoptosis in A549 cells. Biochem Pharmacol 2012;83(1):16–26.

[189] Wong DZ, Kadir HA, Lee CL, Goh BH. Neuroprotective properties of Loranthus parasiticus aqueous fraction against oxidative stress-induced damage in NG108-15 cells. J Nat Med 2012;66(3):544–51.

[190] Yang Y, Li H, Hou S, Hu B, Liu J, Wang J. The noncoding RNA expression profile and the effect of lncRNA AK126698 on cisplatin resistance in non-small-cell lung cancer cell. PLoS One 2013;8(5):e65309.

[191] Bailey TJ, Fossum SL, Fimbel SM, Montgomery JE, Hyde DR. The inhibitor of phagocytosis, O-phospho-L-serine, suppresses Muller glia proliferation and cone cell regeneration in the light-damaged zebrafish retina. Exp Eye Res 2010; 91(5):601–12.

[192] Chong ZZ, Li F, Maiese K. Cellular demise and inflammatory microglial activation during beta-amyloid toxicity are governed by Wnt1 and canonical signaling pathways. Cell Signal 2007;19(6):1150–62.

[193] Hou J, Chong ZZ, Shang YC, Maiese K. Early apoptotic vascular signaling is determined by Sirt1 through nuclear shuttling, forkhead trafficking, bad, and mitochondrial caspase activation. Curr Neurovasc Res 2010;7(2):95–112.

[194] Lin SH, Maiese K. The metabotropic glutamate receptor system protects against ischemic free radical programmed cell death in rat brain endothelial cells. J Cereb Blood Flow Metab 2001;21(3):262–75.

[195] Popescu NI, Lupu C, Lupu F. Extracellular protein disulfide isomerase regulates coagulation on endothelial cells through modulation of phosphatidylserine exposure. Blood 2010;116 (6):993–1001.

[196] Xin YJ, Yuan B, Yu B, Wang YQ, Wu JJ, Zhou WH, et al. Tet1-mediated DNA demethylation regulates neuronal cell death induced by oxidative stress. Sci Rep 2015;5:7645.

[197] Yu T, Li L, Chen T, Liu Z, Liu H, Li Z. Erythropoietin attenuates advanced glycation endproducts-induced toxicity of schwann cells *in vitro*. Neurochem Res 2015;40(4): 698–712.

[198] Troy CM, Akpan N, Jean YY. Regulation of caspases in the nervous system implications for functions in health and disease. Prog Mol Biol Transl Sci 2011;(99):265–305.

[199] Shang YC, Chong ZZ, Hou J, Maiese K. FoxO3a governs early microglial proliferation and employs mitochondrial depolarization with caspase 3, 8, and 9 cleavage during oxidant induced apoptosis. Curr Neurovasc Res 2009;6(4): 223–38.

[200] Wei L, Sun C, Lei M, Li G, Yi L, Luo F, et al. Activation of Wnt/beta-catenin pathway by exogenous Wnt1 protects SH-SY5Y cells against 6-hydroxydopamine toxicity. J Mol Neurosci 2013;49(1):105–15.

[201] Miriuka SG, Rao V, Peterson M, Tumiati L, Delgado DH, Mohan R, et al. mTOR inhibition induces endothelial progenitor cell death. Am J Transplant 2006;6(9):2069–79.

[202] Maiese K, Chong ZZ, Hou J, Shang YC. Oxidative stress: biomarkers and novel therapeutic pathways. Exp Gerontol 2010;45(3):217–34.

[203] Wang GB, Ni YL, Zhou XP, Zhang WF. The AKT/mTOR pathway mediates neuronal protective effects of erythropoietin in sepsis. Mol Cell Biochem 2014;385(1–2):125–32.

[204] Shang YC, Chong ZZ, Wang S, Maiese K. Erythropoietin and Wnt1 govern pathways of mTOR, Apaf-1, and XIAP in inflammatory microglia. Curr Neurovasc Res 2011; 8(4):270–85.

[205] Shang YC, Chong ZZ, Wang S, Maiese K. Prevention of beta-amyloid degeneration of microglia by erythropoietin depends on Wnt1, the PI 3-K/mTOR pathway, Bad, and Bcl-xL. Aging (Albany NY) 2012;4(3):187–201.

[206] Marfia G, Madaschi L, Marra F, Menarini M, Bottai D, Formenti A, et al. Adult neural precursors isolated from post mortem brain yield mostly neurons: an erythropoietin-dependent process. Neurobiol Dis 2011;43(1):86–98.

[207] Sanghera KP, Mathalone N, Baigi R, Panov E, Wang D, Zhao X, et al. The PI3K/Akt/mTOR pathway mediates retinal progenitor cell survival under hypoxic and superoxide stress. Mol Cell Neurosci 2011;47(2):145–53.

[208] Ryou MG, Choudhury GR, Li W, Winters A, Yuan F, Liu R, et al. Methylene blue-induced neuronal protective mechanism against hypoxia-reoxygenation stress. Neuroscience 2015;301:193–203.

[209] Tsai CF, Kuo YH, Yeh WL, Wu CY, Lin HY, Lai SW, et al. Regulatory effects of caffeic acid phenethyl ester on neuroinflammation in microglial cells. Int J Mol Sci 2015;16 (3):5572–89.

[210] Wang L, Di L, Noguchi CT. AMPK is involved in mediation of erythropoietin influence on metabolic activity and reactive oxygen species production in white adipocytes. Int J Biochem Cell Biol 2014;54:1–9.

[211] Andreucci M, Fuiano G, Presta P, Lucisano G, Leone F, Fuiano L, et al. Downregulation of cell survival signalling pathways and increased cell damage in hydrogen peroxide-treated human renal proximal tubular cells by alpha-erythropoietin. Cell Prolif 2009;42(4):554–61.

[212] Chang ZY, Yeh MK, Chiang CH, Chen YH, Lu DW. Erythropoietin protects adult retinal ganglion cells against NMDA-, trophic factor withdrawal-, and TNF-alpha-induced damage. PLoS One 2013;8(1):e55291.

[213] Czubak P, Bojarska-Junak A, Tabarkiewicz J, Putowski L. A modified method of insulin producing cells' generation from bone marrow-derived mesenchymal stem cells. J Diabetes Res 2014;2014:628591.

[214] Maiese K. Protein kinase C modulates the protective ability of peptide growth factors during anoxia. J Auton Nerv Syst 1994;49(Suppl.):S187–93.

[215] Maiese K, Boccone L. Neuroprotection by peptide growth factors against anoxia and nitric oxide toxicity requires modulation of protein kinase C. J Cereb Blood Flow Metab 1995;15(3):440–9.

[216] Maiese K, Boniece I, DeMeo D, Wagner JA. Peptide growth factors protect against ischemia in culture by preventing nitric oxide toxicity. J Neurosci 1993;13(7):3034–40.

[217] Sato A, Sunayama J, Matsuda K, Tachibana K, Sakurada K, Tomiyama A, et al. Regulation of neural stem/progenitor cell maintenance by PI3K and mTOR. Neurosci Lett 2010; 470(2):115–20.

[218] Kimura R, Okouchi M, Kato T, Imaeda K, Okayama N, Asai K, et al. Epidermal growth factor receptor transactivation is necessary for glucagon-like peptide-1 to protect PC12 cells from apoptosis. Neuroendocrinology 2013;97(4):300–8.

[219] Ramanan VK, Nho K, Shen L, Risacher SL, Kim S, McDonald BC, et al. FASTKD2 is associated with memory and hippocampal structure in older adults. Mol Psychiatry 2015; 20(10):1197–204.

[220] Slipczuk L, Bekinschtein P, Katche C, Cammarota M, Izquierdo I, Medina JH. BDNF activates mTOR to regulate GluR1 expression required for memory formation. PLoS One 2009;4(6):e6007.

[221] Chen A, Xiong LJ, Tong Y, Mao M. Neuroprotective effect of brain-derived neurotrophic factor mediated by autophagy through the PI3K/Akt/mTOR pathway. Mol Med Rep 2013;8 (4):1011–16.

[222] Francois A, Terro F, Quellard N, Fernandez B, Chassaing D, Janet T, et al. Impairment of autophagy in the central nervous system during lipopolysaccharide-induced inflammatory stress in mice. Mol Brain 2014;7(1):56.

[223] Vakifahmetoglu-Norberg H, Xia HG, Yuan J. Pharmacologic agents targeting autophagy. J Clin Invest 2015;125(1):5–13.

[224] Lim YM, Lim H, Hur KY, Quan W, Lee HY, Cheon H, et al. Systemic autophagy insufficiency compromises adaptation to metabolic stress and facilitates progression from obesity to diabetes. Nat Commun 2014;5:4934.

[225] Peng N, Meng N, Wang S, Zhao F, Zhao J, Su L, et al. An activator of mTOR inhibits oxLDL-induced autophagy and apoptosis in vascular endothelial cells and restricts atherosclerosis in apolipoprotein E(−/−) mice. Sci Rep 2014;4:5519.

[226] Chen Y, Liu XR, Yin YQ, Lee CJ, Wang FT, Liu HQ, et al. Unravelling the multifaceted roles of Atg proteins to improve cancer therapy. Cell Prolif 2014;47(2):105−12.

[227] Francois A, Rioux-Bilan A, Quellard N, Fernandez B, Janet T, Chassaing D, et al. Longitudinal follow-up of autophagy and inflammation in brain of APPswePS1dE9 transgenic mice. J Neuroinflamm 2014;11(1):139.

[228] Fu Y, Chang H, Peng X, Bai Q, Yi L, Zhou Y, et al. Resveratrol inhibits breast cancer stem-like cells and induces autophagy via suppressing Wnt/beta-catenin signaling pathway. PLoS One 2014;9(7):e102535.

[229] Geng Y, Ju Y, Ren F, Qiu Y, Tomita Y, Tomoeda M, et al. Insulin receptor substrate 1/2 (IRS1/2) regulates Wnt/beta-catenin signaling through blocking autophagic degradation of dishevelled2. J Biol Chem 2014;289(16):11230−41.

[230] Wu H, Lu MH, Wang W, Zhang MY, Zhu QQ, Xia YY, et al. Lamotrigine reduces beta-site abetapp-cleaving enzyme 1 protein levels through induction of autophagy. J Alzheimers Dis 2015;46(4):863−76.

[231] Jung CH, Jun CB, Ro SH, Kim YM, Otto NM, Cao J, et al. ULK-Atg13-FIP200 complexes mediate mTOR signaling to the autophagy machinery. Mol Biol Cell 2009;20(7):1992−2003.

[232] Chen J, Xavier S, Moskowitz-Kassai E, Chen R, Lu CY, Sanduski K, et al. Cathepsin cleavage of sirtuin 1 in endothelial progenitor cells mediates stress-induced premature senescence. Am J Pathol 2012;180(3):973−83.

[233] Ou X, Lee MR, Huang X, Messina-Graham S, Broxmeyer HE. SIRT1 positively regulates autophagy and mitochondria function in embryonic stem cells under oxidative stress. Stem Cells 2014;32(5):1183−94.

[234] Guo W, Qian L, Zhang J, Zhang W, Morrison A, Hayes P, et al. Sirt1 overexpression in neurons promotes neurite outgrowth and cell survival through inhibition of the mTOR signaling. J Neurosci Res 2011;89(11):1723−36.

[235] Tang AH, Rando TA. Induction of autophagy supports the bioenergetic demands of quiescent muscle stem cell activation. EMBO J 2014;33(23):2782−97.

[236] Chun P. Role of sirtuins in chronic obstructive pulmonary disease. Arch Pharm Res 2014;38:1−10.

[237] Berwick DC, Harvey K. The regulation and deregulation of Wnt signaling by PARK genes in health and disease. J Mol Cell Biol 2014;6(1):3−12.

[238] Heo J, Ahn EK, Jeong HG, Kim YH, Leem SH, Lee SJ, et al. Transcriptional characterization of Wnt pathway during sequential hepatic differentiation of human embryonic stem cells and adipose tissue-derived stem cells. Biochem Biophys Res Commun 2013;434(2):235−40.

[239] Li F, Chong ZZ, Maiese K. Winding through the WNT pathway during cellular development and demise. Histol Histopathol 2006;21(1):103−24.

[240] Maiese K, Li F, Chong ZZ, Shang YC. The Wnt signaling pathway: aging gracefully as a protectionist? Pharmacol Ther 2008;118(1):58−81.

[241] Thorfve A, Lindahl C, Xia W, Igawa K, Lindahl A, Thomsen P, et al. Hydroxyapatite coating affects the Wnt signaling pathway during peri-implant healing in vivo. Acta Biomater 2014;10(3):1451−62.

[242] Wexler EM, Rosen E, Lu D, Osborn GE, Martin E, Raybould H, et al. Genome-wide analysis of a Wnt1-regulated transcriptional network implicates neurodegenerative pathways. Sci Signal 2011;4(193):ra65.

[243] Zeljko M, Pecina-Slaus N, Martic TN, Kusec V, Beros V, Tomas D. Molecular alterations of E-cadherin and beta-catenin in brain metastases. Front Biosci (Elite Ed) 2011;3:616−24.

[244] Danielyan L, Schafer R, Schulz A, Ladewig T, Lourhmati A, Buadze M, et al. Survival, neuron-like differentiation and functionality of mesenchymal stem cells in neurotoxic environment: the critical role of erythropoietin. Cell Death Differ 2009;16(12):1599−614.

[245] Chong ZZ, Maiese K. Erythropoietin involves the phosphatidylinositol 3-kinase pathway, 14-3-3 protein and FOXO3a nuclear trafficking to preserve endothelial cell integrity. Br J Pharmacol 2007;150(7):839−50.

[246] Maiese K. FoxO proteins in the nervous system. Anal Cell Pathol (Amst) 2015;2015:569392.

[247] Maiese K. WISP1: clinical insights for a proliferative and restorative member of the CCN family. Curr Neurovasc Res 2014;11(4):378−89.

[248] Jung DW, Kim WH, Williams DR. Reprogram or reboot: small molecule approaches for the production of induced pluripotent stem cells and direct cell reprogramming. ACS Chem Biol 2014;9(1):80−95.

[249] Yang CS, Lopez CG, Rana TM. Discovery of nonsteroidal anti-inflammatory drug and anticancer drug enhancing reprogramming and induced pluripotent stem cell generation. Stem Cells 2011;29(10):1528−36.

[250] Lim HW, Lee JE, Shin SJ, Lee YE, Oh SH, Park JY, et al. Identification of differentially expressed mRNA during pancreas regeneration of rat by mRNA differential display. Biochem Biophys Res Commun 2002;299(5):806−12.

[251] Price RM, Tulsyan N, Dermody JJ, Schwalb M, Soteropoulos P, Castronuovo Jr JJ. Gene expression after crush injury of human saphenous vein: using microarrays to define the transcriptional profile. J Am Coll Surg 2004;199(3):411−18.

[252] Du J, Klein JD, Hassounah F, Zhang J, Zhang C, Wang XH. Aging increases CCN1 expression leading to muscle senescence. Am J Physiol Cell Physiol 2014;306(1):C28−36.

[253] Marchand A, Atassi F, Gaaya A, Leprince P, Le Feuvre C, Soubrier F, et al. The Wnt/beta-catenin pathway is activated during advanced arterial aging in humans. Aging Cell 2011;10(2):220−32.

[254] Lough D, Dai H, Yang M, Reichensperger J, Cox L, Harrison C, et al. Stimulation of the follicular bulge LGR5+ and LGR6+ stem cells with the gut-derived human alpha defensin 5 results in decreased bacterial presence, enhanced wound healing, and hair growth from tissues devoid of adnexal structures. Plast Reconstr Surg 2013;132(5):1159−71.

[255] Kopp C, Hosseini A, Singh SP, Regenhard P, Khalilvandi-Behroozyar H, Sauerwein H, et al. Nicotinic acid increases adiponectin secretion from differentiated bovine preadipocytes through G-protein coupled receptor signaling. Int J Mol Sci 2014;15(11):21401−18.

[256] Murakami M, Ichisaka T, Maeda M, Oshiro N, Hara K, Edenhofer F, et al. mTOR is essential for growth and proliferation in early mouse embryos and embryonic stem cells. Mol Cell Biol 2004;24(15):6710−18.

[257] Gangloff YG, Mueller M, Dann SG, Svoboda P, Sticker M, Spetz JF, et al. Disruption of the mouse mTOR gene leads to early postimplantation lethality and prohibits embryonic stem cell development. Mol Cell Biol 2004;24(21):9508−16.

[258] Zhang D, Yan B, Yu S, Zhang C, Wang B, Wang Y, et al. Coenzyme Q10 inhibits the aging of mesenchymal stem cells induced by D-galactose through Akt/mTOR signaling. Oxid Med Cell Longev 2015;2015:867293.

[259] Zhou J, Su P, Wang L, Chen J, Zimmermann M, Genbacev O, et al. mTOR supports long-term self-renewal and suppresses

mesoderm and endoderm activities of human embryonic stem cells. Proc Natl Acad Sci USA 2009;106(19):7840–5.

[260] Easley CA, Ben-Yehudah A, Redinger CJ, Oliver SL, Varum ST, Eisinger VM, et al. mTOR-mediated activation of p70 S6K induces differentiation of pluripotent human embryonic stem cells. Cell Reprogram 2010;12(3):263–73.

[261] Fraenkel M, Ketzinel-Gilad M, Ariav Y, Pappo O, Karaca M, Castel J, et al. mTOR inhibition by rapamycin prevents beta-cell adaptation to hyperglycemia and exacerbates the metabolic state in type 2 diabetes. Diabetes 2008;57(4):945–57.

[262] Pende M, Kozma SC, Jaquet M, Oorschot V, Burcelin R, Le Marchand-Brustel Y, et al. Hypoinsulinaemia, glucose intolerance and diminished beta-cell size in S6K1-deficient mice. Nature 2000;408(6815):994–7.

[263] Hamada S, Hara K, Hamada T, Yasuda H, Moriyama H, Nakayama R, et al. Upregulation of the mammalian target of rapamycin complex 1 pathway by Ras homolog enriched in brain in pancreatic beta-cells leads to increased beta-cell mass and prevention of hyperglycemia. Diabetes 2009;58(6):1321–32.

[264] Harrington LS, Findlay GM, Gray A, Tolkacheva T, Wigfield S, Rebholz H, et al. The TSC1-2 tumor suppressor controls insulin-PI3K signaling via regulation of IRS proteins. J Cell Biol 2004;166(2):213–23.

[265] Khamzina L, Veilleux A, Bergeron S, Marette A. Increased activation of the mammalian target of rapamycin pathway in liver and skeletal muscle of obese rats: possible involvement in obesity-linked insulin resistance. Endocrinology 2005;146(3):1473–81.

[266] Kim JA, Jang HJ, Martinez-Lemus LA, Sowers JR. Activation of mTOR/p70S6 kinase by ANG II inhibits insulin-stimulated endothelial nitric oxide synthase and vasodilation. Am J Physiol Endocrinol Metab 2012;302(2):E201–8.

[267] Estep 3rd PW, Warner JB, Bulyk ML. Short-term calorie restriction in male mice feminizes gene expression and alters key regulators of conserved aging regulatory pathways. PLoS One 2009;4(4):e5242.

[268] Orimo M, Minamino T, Miyauchi H, Tateno K, Okada S, Moriya J, et al. Protective role of SIRT1 in diabetic vascular dysfunction. Arterioscler Thromb Vasc Biol 2009;29(6):889–94.

[269] Zhang QJ, Wang Z, Chen HZ, Zhou S, Zheng W, Liu G, et al. Endothelium-specific overexpression of class III deacetylase SIRT1 decreases atherosclerosis in apolipoprotein E-deficient mice. Cardiovasc Res 2008;80(2):191–9.

[270] Hou J, Wang S, Shang YC, Chong ZZ, Maiese K. Erythropoietin employs cell longevity pathways of SIRT1 to foster endothelial vascular integrity during oxidant stress. Curr Neurovasc Res 2011;8(3):220–35.

[271] Wang RH, Kim HS, Xiao C, Xu X, Gavrilova O, Deng CX. Hepatic Sirt1 deficiency in mice impairs mTorc2/Akt signaling and results in hyperglycemia, oxidative damage, and insulin resistance. J Clin Invest 2011;121(11):4477–90.

[272] Li Y, Xu S, Giles A, Nakamura K, Lee JW, Hou X, et al. Hepatic overexpression of SIRT1 in mice attenuates endoplasmic reticulum stress and insulin resistance in the liver. FASEB J 2011;25:1664–79.

[273] Zhang S, Cai G, Fu B, Feng Z, Ding R, Bai X, et al. SIRT1 is required for the effects of rapamycin on high glucose-inducing mesangial cells senescence. Mech Ageing Dev 2012;133(6):387–400.

[274] Balestrieri ML, Rienzo M, Felice F, Rossiello R, Grimaldi V, Milone L, et al. High glucose downregulates endothelial progenitor cell number via SIRT1. Biochim Biophys Acta 2008;1784(6):936–45.

[275] Yuen DA, Zhang Y, Thai K, Spring C, Chan L, Guo X, et al. Angiogenic dysfunction in bone marrow-derived early outgrowth cells from diabetic animals is attenuated by SIRT1 activation. Stem Cells Transl Med 2012;1(12):921–6.

[276] Sun Q, Jia N, Wang W, Jin H, Xu J, Hu H. Activation of SIRT1 by curcumin blocks the neurotoxicity of amyloid-beta25-35 in rat cortical neurons. Biochem Biophys Res Commun 2014;448(1):89–94.

[277] Hartman NW, Lin TV, Zhang L, Paquelet GE, Feliciano DM, Bordey A. mTORC1 targets the translational repressor 4E-BP2, but not S6 kinase 1/2, to regulate neural stem cell self-renewal in vivo. Cell Rep 2013;5(2):433–44.

[278] Han J, Wang B, Xiao Z, Gao Y, Zhao Y, Zhang J, et al. Mammalian target of rapamycin (mTOR) is involved in the neuronal differentiation of neural progenitors induced by insulin. Mol Cell Neurosci 2008;39(1):118–24.

[279] Malagelada C, Lopez-Toledano MA, Willett RT, Jin ZH, Shelanski ML, Greene LA. RTP801/REDD1 regulates the timing of cortical neurogenesis and neuron migration. J Neurosci 2011;31(9):3186–96.

[280] Romine J, Gao X, Xu XM, So KF, Chen J. The proliferation of amplifying neural progenitor cells is impaired in the aging brain and restored by the mTOR pathway activation. Neurobiol Aging 2015;36(4):1716–26.

[281] Kim J, Jung Y, Sun H, Joseph J, Mishra A, Shiozawa Y, et al. Erythropoietin mediated bone formation is regulated by mTOR signaling. J Cell Biochem 2012;113(1):220–8.

[282] Magri L, Cambiaghi M, Cominelli M, Alfaro-Cervello C, Cursi M, Pala M, et al. Sustained activation of mTOR pathway in embryonic neural stem cells leads to development of tuberous sclerosis complex-associated lesions. Cell Stem Cell 2011;9(5):447–62.

[283] Galan-Moya EM, Le Guelte A, Lima Fernandes E, Thirant C, Dwyer J, Bidere N, et al. Secreted factors from brain endothelial cells maintain glioblastoma stem-like cell expansion through the mTOR pathway. EMBO Rep 2011;12(5):470–6.

[284] Foldes G, Mioulane M, Wright JS, Liu AQ, Novak P, Merkely B, et al. Modulation of human embryonic stem cell-derived cardiomyocyte growth: a testbed for studying human cardiac hypertrophy? J Mol Cell Cardiol 2011;50(2):367–76.

[285] Benyoucef A, Calvo J, Renou L, Arcangeli ML, van den Heuvel A, Amsellem S, et al. The SCL/TAL1 transcription factor represses the stress protein DDiT4/REDD1 in human hematopoietic stem/progenitor cells. Stem Cells 2015; 33(7):2268–79.

[286] Engels MC, Rajarajan K, Feistritzer R, Sharma A, Nielsen UB, Schalij MJ, et al. Insulin-like growth factor promotes cardiac lineage induction in vitro by selective expansion of early mesoderm. Stem Cells 2014;32(6):1493–502.

[287] Moon HE, Byun K, Park HW, Kim JH, Hur J, Park JS, et al. COMP-Ang1 potentiates EPC treatment of ischemic brain injury by enhancing angiogenesis through activating AKT-mTOR pathway and promoting vascular migration through activating Tie2-FAK pathway. Exp Neurobiol 2015;24(1):55–70.

[288] Foldes G, Mioulane M, Kodagoda T, Lendvai Z, Iqbal A, Ali NN, et al. Immunosuppressive agents modulate function, growth, and survival of cardiomyocytes and endothelial cells derived from human embryonic stem cells. Stem Cells Dev 2014;23(5):467–76.

[289] Iriuchishima H, Takubo K, Matsuoka S, Onoyama I, Nakayama KI, Nojima Y, et al. Ex vivo maintenance of hematopoietic stem cells by quiescence induction through Fbxw7α overexpression. Blood 2011;117(8):2373–7.

[290] Maiese K. Neuronal activity, mitogens, and mTOR: overcoming the hurdles for the treatment of glioblastoma multiforme. J Transl Sci 2015;1(1):2.

[291] Venkatesh HS, Johung TB, Caretti V, Noll A, Tang Y, Nagaraja S, et al. Neuronal activity promotes glioma growth through neuroligin-3 secretion. Cell 2015;161(4):803–16.

[292] Hu X, Pandolfi PP, Li Y, Koutcher JA, Rosenblum M, Holland EC. mTOR promotes survival and astrocytic characteristics

induced by Pten/AKT signaling in glioblastoma. Neoplasia 2005;7(4):356−68.

[293] Kolev VN, Wright QG, Vidal CM, Ring JE, Shapiro IM, Ricono J, et al. PI3K/mTOR dual inhibitor VS-5584 preferentially targets cancer stem cells. Cancer Res 2015;75(2):446−55.

[294] Karthik GM, Ma R, Lovrot J, Kis LL, Lindh C, Blomquist L, et al. mTOR inhibitors counteract tamoxifen-induced activation of breast cancer stem cells. Cancer Lett 2015;367(1):76−87.

[295] Yang C, Zhang Y, Zhang Y, Zhang Z, Peng J, Li Z, et al. Downregulation of cancer stem cell properties via mTOR signaling pathway inhibition by rapamycin in nasopharyngeal carcinoma. Int J Oncol 2015;47(3):909−17.

[296] Maiese K, Chong ZZ, Shang YC, Hou J. Rogue proliferation versus restorative protection: where do we draw the line for Wnt and forkhead signaling? Expert Opin Ther Targets 2008;12(7):905−16.

[297] Wang S, Sun Z, Zhang X, Li Z, Wu M, Zhao W, et al. Wnt1 positively regulates CD36 expression via TCF4 and PPAR-gamma in macrophages. Cell Physiol Biochem 2015;35(4):1289−302.

[298] Xu D, Zhao W, Pan G, Qian M, Zhu X, Liu W, et al. Expression of Nemo-like kinase after spinal cord injury in rats. J Mol Neurosci 2014;52(3):410−18.

[299] Marchetti B, Pluchino S. Wnt your brain be inflamed? Yes, it Wnt!. Trends Mol Med 2013;19(3):144−56.

[300] Chong ZZ, Shang YC, Hou J, Maiese K. Wnt1 neuroprotection translates into improved neurological function during oxidant stress and cerebral ischemia through AKT1 and mitochondrial apoptotic pathways. Oxid Med Cell Longev 2010;3(2):153−65.

[301] Xing Y, Zhang X, Zhao K, Cui L, Wang L, Dong L, et al. Beneficial effects of sulindac in focal cerebral ischemia: a positive role in Wnt/beta-catenin pathway. Brain Res 2012;1482:71−80.

[302] Pandey S, Chandravati. Targeting Wnt-Frizzled signaling in cardiovascular diseases. Mol Biol Rep 2013;40(10):6011−18.

[303] L'Episcopo F, Tirolo C, Testa N, Caniglia S, Morale MC, Deleidi M, et al. Plasticity of subventricular zone neuroprogenitors in MPTP (1-methyl-4-phenyl-1,2,3,6-tetrahydropyridine) mouse model of parkinson's disease involves cross talk between inflammatory and Wnt/beta-Catenin signaling pathways: functional consequences for neuroprotection and repair. J Neurosci 2012;32(6):2062−85.

[304] Shah N, Morsi Y, Manasseh R. From mechanical stimulation to biological pathways in the regulation of stem cell fate. Cell Biochem Funct 2014;32(4):309−25.

[305] Liu J, Wang Y, Pan Q, Su Y, Zhang Z, Han J, et al. Wnt/beta-catenin pathway forms a negative feedback loop during TGF-beta1 induced human normal skin fibroblast-to-myofibroblast transition. J Dermatol Sci 2012;65(1):38−49.

[306] Maiese K. FoxO transcription factors and regenerative pathways in diabetes mellitus. Curr Neurovasc Res 2015;12 (4):404−13.

[307] Liu Y, Yan W, Zhang W, Chen L, You G, Bao Z, et al. MiR-218 reverses high invasiveness of glioblastoma cells by targeting the oncogenic transcription factor LEF1. Oncol Rep 2012;28 (3):1013−21.

[308] Tu Y, Gao X, Li G, Fu H, Cui D, Liu H, et al. MicroRNA-218 inhibits glioma invasion, migration, proliferation, and cancer stem-like cell self-renewal by targeting the polycomb group gene Bmi1. Cancer Res 2013;73(19):6046−55.

[309] James RG, Davidson KC, Bosch KA, Biechele TL, Robin NC, Taylor RJ, et al. WIKI4, a novel inhibitor of tankyrase and Wnt/ss-catenin signaling. PLoS One 2012;7(12):e50457.

[310] Kafka A, Basic-Kinda S, Pecina-Slaus N. The cellular story of dishevelleds. Croat Med J 2014;55(5):459−67.

[311] Klinke II DJ. Induction of Wnt-inducible signaling protein-1 correlates with invasive breast cancer oncogenesis and reduced type 1 cell-mediated cytotoxic immunity: a retrospective study. PLoS Comput Biol 2014;10(1):e1003409.

[312] Knoblich K, Wang HX, Sharma C, Fletcher AL, Turley SJ, Hemler ME. Tetraspanin TSPAN12 regulates tumor growth and metastasis and inhibits beta-catenin degradation. Cell Mol Life Sci 2014;71(7):1305−14.

[313] Uzdensky AB, Demyanenko SV, Bibov MY. Signal transduction in human cutaneous melanoma and target drugs. Curr Cancer Drug Targets 2013;13(8):843−66.

[314] Hedley BD, Allan AL, Xenocostas A. The role of erythropoietin and erythropoiesis-stimulating agents in tumor progression. Clin Cancer Res 2011;17(20):6373−80.

[315] Maiese K, Li F, Chong ZZ. Erythropoietin and cancer. JAMA 2005;293(15):1858−9.

[316] Zhang C, Li Z, Cao Q, Qin C, Cai H, Zhou H, et al. Association of erythropoietin gene rs576236 polymorphism and risk of adrenal tumors in a Chinese population. J Biomed Res 2014;28(6):456−61.

[317] Ogunshola OO, Moransard M, Gassmann M. Constitutive excessive erythrocytosis causes inflammation and increased vascular permeability in aged mouse brain. Brain Res 2013;1531:48−57.

[318] Ono M, Inkson CA, Sonn R, Kilts TM, de Castro LF, Maeda A, et al. WISP1/CCN4: a potential target for inhibiting prostate cancer growth and spread to bone. PLoS One 2013;8(8):e71709.

[319] Tanaka S, Sugimachi K, Kameyama T, Maehara S, Shirabe K, Shimada M, et al. Human WISP1v, a member of the CCN family, is associated with invasive cholangiocarcinoma. Hepatology 2003;37(5):1122−9.

[320] Soon LL, Yie TA, Shvarts A, Levine AJ, Su F, Tchou-Wong KM. Overexpression of WISP-1 down-regulated motility and invasion of lung cancer cells through inhibition of Rac activation. J Biol Chem 2003;278(13):11465−70.

[321] Castilho RM, Squarize CH, Chodosh LA, Williams BO, Gutkind JS. mTOR mediates Wnt-induced epidermal stem cell exhaustion and aging. Cell Stem Cell 2009;5(3):279−89.

2

mTOR: The Master Regulator of Conceptus Development in Response to Uterine Histotroph During Pregnancy in Ungulates

Xiaoqiu Wang[1,2], *Guoyao Wu*[1,2] *and Fuller W. Bazer*[1,2]

[1]Center for Animal Biotechnology and Genomics, Texas A&M University, College Station, TX, USA [2]Department of Animal Science, Texas A&M University, College Station, TX, USA

2.1 INTRODUCTION

In all mammalian species, the greatest constraint to reproductive performance is embryonic mortality, which, in most cases, claims 20—40% of embryos [1]. Of these embryonic losses, two-thirds occurs during the peri-implantation period of pregnancy due to factors that include asynchrony between conceptus (embryo/fetus and its associated extraembryonic membranes) and uterine signals that regulate conceptus elongation (ungulates), expansion (horse), or invasion (rodents and primates) and uterine receptivity to implantation, resulting in defects in conceptus development and implantation [1,2]. Even for those conceptuses who survive and are eventually born, inappropriate hormonal regulation, metabolism, and cell signaling such as via nutrient-sensing pathways most likely will predispose them to fetal origins of adult disease, the phenomenon known as fetal programming [3—5]. Namely, alterations in fetal nutrition and endocrine status may result in developmental adaptations that permanently change the structure, physiology, and metabolism of the offspring [6—8], thereby predisposing individuals to metabolic, endocrine, and cardiovascular disease in adult life [3,9].

During pregnancy, the uterine epithelia secrete or selectively transport molecules (e.g., nutrients, enzymes, and growth factors) into the uterine lumen that are collectively known as histotroph, which is required for growth and development of the conceptus during the periods of implantation and placentation. Animal studies have shown that an inappropriate nutrient supply reduces utero—placental blood flows and stunts fetal growth, resulting in intrauterine growth restriction (IUGR) [6—8,10—12]. Successful establishment and maintenance of pregnancy requires appropriate development of the conceptus for pregnancy recognition signaling [e.g., interferon tau (IFNT) in ruminants, estrogen in pigs, prolactin, placental lactogen or prolactin-like hormones in rodents, and chorionic gonadotrophin in primates] required for maintenance of the corpus luteum (CL) that secretes progesterone to maintain pregnancy [1,2]. Progesterone is required for an intrauterine environment that supports implantation, placentation, and uterine functions essential for birth of healthy offspring [1,2]. There is growing evidence that the mechanistic target of rapamycin (mTOR) cell signaling pathway is the master regulator of cell growth intensively involved in conceptus development, implantation, and placentation in response to components of uterine histotroph during pregnancy. Therefore, this review focuses on the relationship between selected components of uterine histotroph and mTOR as related to development of conceptuses during preangncy, particularly in ungulates.

2.2 DEVELOPMENTAL EVENTS OF UNGULATE CONCEPTUS

Despite differences in duration of the preimplantation period, as well as type of implantation (noninvasive vs invasive) and placentation (epitheliochorial in pigs, synepitheliochorial in ruminants; endotheliochorial in rodents; and hemochorial in primates), the early stages of embryonic development and phases of blastocyst implantation are common across mammalian species [13−15]. For early development of a zygote (a fertilized ovum), successive cleavage events during stage 1 results in formation of a morula (32- to 64-cell embryo). Compaction is stage 2 of development wherein the morula forms a blastocyst characterized by two distinct cell populations, that is, inner cell mass (ICM) which develops into the embryo/fetus, and trophectoderm (Tr) which gives rise to the placenta. Briefly, the innermost cells of the morula develop gap junctions that increase intercellular communication and coordination, whereas the outer cells of the morula begin compaction by forming cell−cell adhesions known as tight junctions that allow them to become polarized and differentiated for transport and accumulation of water and nutrients in a central cavity called the blastocele [16−20]. The blastocele is surrounded by Tr cells while the ICM is localized to a single pole of the blastocyst. For instance, in sheep, the blastocyst at this stage contains approximately 3000 blastomeres, is 150−200 μm in diameter and is located within the uterine lumen [17,19].

Within the uterine lumen, the events of blastocyst development prior to implantation are divided into five phases (see Figure 2.1), that include: (1) shedding of the zona pellucida (ZP); (2) precontact and blastocyst orientation; (3) apposition of Tr and uterine luminal epithelium (LE); (4) adhesion of Tr to uterine LE; and (5) invasion of the blastocyst into the uterine endometrium which is unique to species such as rodents and primates that have an invasive implantation. Invasive implantation involves migration of the blastocyst through the uterine LE and into the uterine stroma which then becomes decidualized [21−23]. Briefly, during phase 1 the blastocyst continues to grow by mitotic divisions and the blastocele increases in volume and pressure. The ZP breaks down in response to proteolytic enzymes from the Tr, and the blastocyst (ICM and Tr) becomes free to expand further within the uterine lumen. Thereafter, the floating blastocyst, now termed the conceptus, undergoes expansion (horse), elongation from a spherical to tubular and filamentous form (ungulates), or remains spherical prior to implantation (rodents and primates). During phase 2, the preattachment period, the conceptus migrates and undergoes orientation without definitive cellular contacts between conceptus Tr and uterine LE, and initiates pregnancy recognition signaling. IFNT is the pregnancy recognition signal produced and secreted by ruminant conceptuses, whereas estrogen, placental lactogen or prolactin-like hormones, and chorionic gonadotrophin are the pregnancy recognition signals in pigs, rodents, and primates, respectively [24−28]. Apposition is phase 3 wherein the conceptus Tr associates closely with uterine LE for unstable adhesion and, particularly in ungulates, develops finger-like villi or papillae which extend into the superficial ducts of the uterine glands for subsequent stable adhesion and to absorb histotroph. Phase 4 is the adhesion phase during which the Tr becomes firmly adhered to uterine LE and, in ungulates, superficial glandular epithelium (sGE). In sheep and other ruminants, this is the period of interdigitation of conceptus Tr and uterine LE in both caruncular and intercaruncular areas of the endometrium in preparation for development of cotyledons on the chorion and caruncles on the uterine endometrium to form placentomes [23,29]. Also, mononuclear Tr cells differentiate into trophoblast giant binucleate cells that produce placental lactogen, progesterone, and other molecules. In pigs, this is the period of initiation of the epitheliochorial type of placentation in which two apposed cell layers (Tr and uterine LE) form an anatomical interaction involving interdigitated microvilli and increased folding of the endometrium [29,30]. Phase 5 is unique to species with invasive implantation whereby the conceptus (rodents and primates), instead of undergoing expansion or elongation, invades through uterine LE, stroma, and even maternal endothelial cells to achieve close contact between Tr and maternal blood in order to increase nutrient and gas exchange.

With respect to sheep and pigs, the conceptus undergoes a rapid transition from spherical (0.4 mm in diameter at days 10−11) to tubular (1.0 mm in diameter × 33 mm in length at days 12−13) to filamentous forms (68−190 mm in length between days 14 and 16 of pregnancy in sheep; and 700−1000 mm in length between days 14 and 16 of pregnancy in pigs) [1,26]. This process is highly correlated with the composition of histotroph. In fact, it is during this period of morphological and functional transition in development that 30−40% of ungulate conceptuses die as many fail to elongate and achieve sufficient contact between conceptus Tr and uterine LE/sGE for uptake of nutrients and other components of histotroph, resulting in inappropriate outside-in or inside-out signal transduction, and failure of maternal recognition of pregnancy signaling.

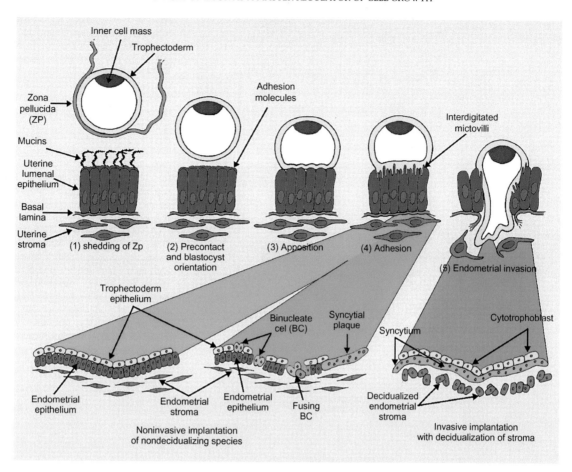

FIGURE 2.1   The phases of blastocyst implantation in mammals.

2.3 ROLE OF mTOR AS A MASTER REGULATOR OF CELL GROWTH

The mTOR, also known as FK506 binding protein 12-rapamycin associated protein 1 is a serine/threonine protein kinase that regulates a variety of biological processes in response to multiple environmental cues, including, but not restricted to nutrients, growth factors, hormones, as well as diverse forms of stress [31–33]. mTOR was first identified and named mammalian target of rapamycin in 1994 [34,35] as the mammalian homolog of TOR discovered during genetic studies of yeast. However, the early studies related to mTOR/TOR started several decades ago with the compound rapamycin, which is a potent antifungal macrolide originally isolated from the soils of Rapa Nui, commonly known as Easter Island [36]. This compound stirred up clinical and research interests due to its antiproliferative and immunosuppressive properties in not only prokaryotes, but also eukaryotes [37–41]. mTOR, the direct target of rapamycin [34,35,42], controls many biological processes, including cell proliferation via regulation of protein,

lipid, and nucleotide synthesis [43,44], ribosome and lysosome biosynthesis, expression of metabolism-regulated genes, autophagy, and cytoskeletal reorganization [45]. Unlike *Saccharomyces cerevisiae* that encodes two different TOR proteins (TOR1 and TOR2) [46], most eukaryotes and all mammals have only one gene that encodes TOR/mTOR. However, all eukaryotes have two TOR/mTOR-containing complexes, that is, the MTOR complex 1 (mTORC1) and the mTOR complex 2 (mTORC2) (see Figure 2.2) [44,47,48]. In mammals, mTOR regulates protein synthesis and, therefore, cell proliferation via mTORC1, which consists of mTOR itself and other components, including: regulatory-associated protein of mTOR (RAPTOR), mammalian lethal with SEC13 protein 8 (MLST8), proline-rich Akt/PKB substrate 40 kDa (PRAS40), and DEP domain-containing mTOR-interacting protein (DEPTOR) [31,45,49–51]. mTORC1 regulates cell proliferation in part by phosphorylating ribosomal protein S6 kinase 1 (RPS6K1) and the eIF-4E-binding protein 1 (EIF4EBP1), known regulators of protein synthesis. mTOR also controls cell survival and spatial aspects of growth, such as cytoskeletal

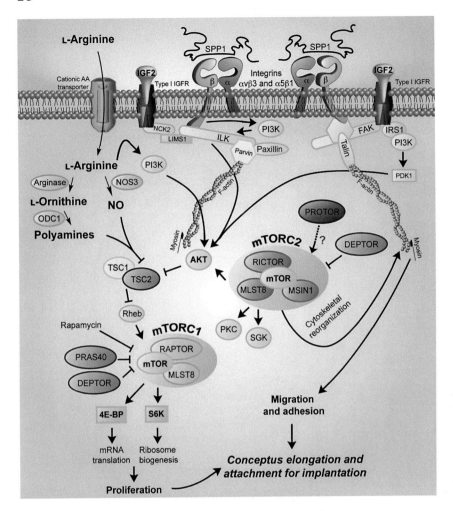

FIGURE 2.2　Model for induction of cell signaling for proliferation, migration, adhesion, and cytoskeletal remodeling of conceptuses via mTORC1 and mTORC2 signaling cascade. AKT1, proto-oncogenic protein kinase 1; FAK, focal adhesion kinase; PDK1, phosphoinositide-dependent protein kinase 1; mTOR, mechanistic target of rapamycin; RAPTOR, regulatory-associated protein of mTOR; RICTOR, rapamycin-insensitive companion of mTOR; IGF2, insulin-like growth factor 2; type I IGF2, type I insulin-like growth factor receptor; ILK, integrin-linked kinases; IRS1, insulin receptor substrate 1; PKC, protein kinase C; SGK, serum/glucocorticoid-regulated kinase; MLST8, mammalian lethal with SEC13 protein 8; PRAS40, proline-rich Akt/PKB substrate 40 kDa; DEPTOR, DEP domain-containing mTOR-interacting protein; MSIN1, mammalian stress-activated MAP kinase interacting protein 1; PROTOR, protein observed with RICTOR; NCK2, noncatalytic region of tyrosine kinase, beta; NO, nitric oxide; NOS3, nitric oxide synthase 3; ODC1, ornithine decarboxylase; PI3K, phosphatidylinositol 3-kinase; LIMS1, LIM and senescent cell antigen-like domains 1; S6K, S6 kinase; SPP1, secreted phosphoprotein 1.

organization and Akt/PKB phosphorylation [52], through mTORC2 that contains mTOR itself and the following components: rapamycin-insensitive companion of mTOR (RICTOR), mammalian stress-activated MAP kinase interacting protein 1 (MSIN1), MLST8, DEPTOR, and protein observed with RICTOR (PROTOR) [51,53–55]. RICTOR, MSIN1, MLST8, and mTOR are the essential core components of the complexes required for maintaining structural integrity. However, DEPTOR and PROTOR are not required for mTORC2 activity, but function as regulatory proteins whereby DEPTOR seems to negatively regulate mTORC2, whereas the function of PROTOR is unclear [56]. mTORC2 controls actin cytoskeleton organization and Akt/PKB phosphorylation [52], which further regulates cell migration, growth, survival, and metabolism.

2.4 HISTOTROPH AND THE mTOR CELL SIGNALING PATHWAY DURING CONCEPTUS DEVELOPMENT

In mammalian species, histotroph includes secretions produced and/or selectively transported by uterine epithelia, particularly uterine sGE and mid- (mGE) to deep- (dGE) glandular epithelia into the uterine lumen. Histotroph contains molecules that include nutrient transport proteins, ions, mitogens, cytokines, lymphokines, enzymes, hormones, growth factors, adhesion proteins, proteases and protease inhibitors, amino acids, glucose, fructose, vitamins, and other substances [2,57,58]. During pregnancy, the composition of histotroph is essential to the development of the conceptuses in pigs, ruminants, rodents, and primates, regardless of differences in type of implantation and placentation. For example, in primates that have invasive implantation, production of histotroph by uterine GE is a primary source of nutrition for conceptus development during at least the first trimester of pregnancy, before hematotrophic nutrition is fully established [59].

Studies with the uterine gland knockout (UGKO) ewe substantiated the essential role for histotroph from uterine glands to regulate luteolysis during normal estrous cycles, early conceptus development and initiation of pregnancy recognition signaling [60–62]. This UGKO ewe model is produced by the administration of a synthetic, nonmetabolizable progestin to neonatal ewe lambs during the critical period of endometrial gland morphogenesis from birth to postnatal day 56 [63]. This early exposure to a progestin permanently ablates development of uterine GE without apparent defects on development of the myometrium or other Müllerian-duct-derived structures in the female reproductive tract or function of the hypothalamic–pituitary–ovarian axis [63,64]. Consequently, the adult UGKO ewes are unable to experience normal estrous cycles or support conceptus elongation due to the absence of histotroph from uterine glands. The UGKO ewes also fail as surrogates to accommodate hatched blastocysts transferred from normal synchronous ewes, in terms of normal elongation and pregnancy recognition signaling [62]. However, uterine histotroph may not be critical for early embryonic development from the zygote to hatched blastocyst stage as large spherical and sometimes early tubular conceptuses can be found in the uterine flushings of bred UGKO ewes [61,62].

Uterine flushings from UGKO ewes on day 14 contain many components such as galectin-15, glycosylated cell adhesion molecule 1 (GLYCAM1), and secreted phophoprotein 1 (SPP1, also known as osteopontin), but other unidentified components are absent or reduced significantly which affects the fertility of these ewes [61,65,66]. mTOR, as the master regulator of cell growth, seems to be heavily involved in such important biological processes. In the following sections, the relationships between identified components in histotroph and mTOR cell signaling pathways are summarized.

2.4.1 Arginine and mTOR

Arginine is a conditionally essential amino acid for adult mammals, including sheep and pigs [67,68]. It is one of the most abundant amino acids deposited in fetal tissue proteins. Concentrations of arginine in histotroph increase by 8–13-fold during the peri-implantation period of pregnancy [3,69–71]. Besides being a building block for protein synthesis, arginine is the common substrate for production of nitric oxide (NO) via NO synthases (NOSs) and for the biosynthesis of polyamines via arginase followed by ornithine decarboxylase 1 (ODC1) [11,72–75]. In addition, supplementation of the diet with arginine increases embryonic and conceptus survival and growth rate of conceptuses in gilts, sheep, and rats [76,77]. This evidence suggests that the abundance of arginine available via histotroph or hematotrophic exchange at the utero–placental interface is important for fetal growth and development during pregnancy.

Arginine transport into cells is mediated primarily by the Na^+-independent system y^+ for cationic amino acids [78]. The cationic amino acid transporter (CAT) system for arginine includes three members, CAT1, CAT2, and CAT3, which are encoded by members of the solute carrier family 7 (SLC7) genes known as *SLC7A1*, *SLC7A2*, and *SLC7A3* genes, respectively. During the peri-implantation period of pregnancy in sheep, expression of both *SLC7A1* and *SLC7A2* mRNAs is enhanced in uterine LE and sGE so that more maternal-derived arginine is transported into uterine lumen, whereas only *SLC7A1* mRNA increases in Tr and extraembryonic endoderm (En) of conceptuses for transport of arginine into those tissues [73]. Moreover, *SLC7A3* mRNA is constitutively and weakly expressed in uterine LE, sGE, and conceptus Tr and En [73]. Therefore, an in utero morpholino antisense oligonucleotide (MAO) loss-of-function approach was used to knockdown translation of *SLC7A1* mRNA, the major arginine transporter in ovine conceptus Tr [73], thereby successfully depriving the conceptus of arginine and resulting in retarded development of conceptuses and a significant decrease in IFNT production [79].

In vitro studies with an established ovine trophectoderm (oTr1) cell line isolated from day 15 ovine conceptuses demonstrated that arginine induces both proliferation and migration of oTr cells through stimulation of the tuberous sclerosis 2 (TSC2)-mTOR-RPS6K-RPS6 signaling cascade [80–82] and mTORC2-mediated cytoskeletal reorganization (X. Wang, G. Wu, G.A. Johnson, and F.W. Bazer, unpublished results), respectively. Consistent with activation of these cell signaling molecules, arginine increases protein synthesis and reduces protein degradation in oTr1 cells [81]. Further studies with oTr1 cells using inhibitors of NOS such as L-NG-nitroarginine methyl ester and ODC1 difluoromethylornithine, as well as donors of NO such as S-nitrosothiol and 1-[N-(2-Aminoethyl)-N-(2-ammonioethyl)amino]diazen-1-ium-1,2-diolate and polyamines (putrescine) revealed that activation of the TSC2-mTORC1 cell signaling pathway is induced by NO and polyamines synthesized from arginine, but also by arginine itself [82]. Therefore, arginine per se is a growth factor that cooperatively stimulates proliferation of oTr cells via the TSC2-mTORC1 cell signaling pathway [82].

2.4.2 Nitric Oxide and mTOR

NO, a product of arginine catabolism, is an important cell signaling gas molecule involved in many physiological processes in mammals. It is generated from conversion of arginine to citrulline by eNOS (NOS3) and/or iNOS (NOS2) in Tr cells and activates guanylate cyclase to produce cyclic guanosine monophosphate that stimulates migration of Tr cells perhaps by modifying the extracellular matrix (ECM), inducing vasodilation of maternal blood vessels [83], or regulating cellular energy metabolism [84]. During elongation and implantation of ovine conceptuses, NOS3 is the major isoform that converts arginine to NO and citrulline [72]. Therefore, *in vivo* knockdown of *NOS3* mRNA reduces NO production by conceptus Tr sufficiently to retard development of the elongating conceptuses; however, the conceptuses produce normal amounts of IFNT [85]. Furthermore, results of *in vitro* studies revealed that NO stimulates proliferation of oTr1 cells via activation of the TSC2-mTORC1 cell signaling cascade, that is, increased abundance of phosphorylated TSC2, mTOR, RPS6K, and EIF4EBP1 [81,82]. However, NO is not required for IFNT production by oTr1 cells [82], which is consistent with results of *in vivo* knockdown of NOS3.

During the peri-implantation period of ovine conceptus development, there is a significant increase in expression of secreted phosphoprotein 1 (SPP1, also known as osteopontin) by uterine GE [86] and NO induces expression of SPP1 that increases cell adhesion and invasion in cultured cells [87,88]. Moreover, hepatocyte-growth factor-induced motility of human trophoblast cells is activated by NO signaling through phosphatidylinositol bisphosphate-3 kinase (PI3K), serine/threonine kinase (AKT), and mTOR [89,90]. Expression of NOS2 is highest in peri-implantation mouse blastocysts [91]. There are increases in NOS3 and NOS2 activities in ovine placentomes between days 30 and 60 of gestation that are sustained to day 140 of gestation to increase placental NO synthesis that is coordinated with increases in placental vascular growth and utero–placental blood flows [91].

2.4.3 Polyamines and mTOR

Polyamines (putrescine, spermidine, and spermine) are products of arginine catabolism required for DNA regulation and protein synthesis [92−94], scavenging reactive oxygen species [95], cell proliferation [81,96,97], and differentiation of tissues [96,98], thereby supporting placental development and embryogenesis in mammals [89,90,99−101]. Depletion of cellular polyamines prevents translation of mRNAs and growth of mammalian cells [102]. During the peri-implantation period of pregnancy in ungulates, increases in expression of ODC1 [72], as well as production of intracellular polyamines, are highly correlated with elongation and implantation of conceptuses [103]. ODC1 is the key enzyme for classical *de novo* biosynthesis of polyamines from arginine, proline, and ornithine [104]; however, it is also known that arginine can be converted to agmatine via arginine decarboxylase (ADC) and agmatine can then be converted to

polyamines via agmatinase (AGMAT) [103,105]. We discovered that the alternative ADC/AGMAT pathway is functional in ovine conceptuses for polyamine biosynthesis and compensates for loss of ODC1 activity following *in vivo* MAO knockdown of translation of *ODC1* mRNA, thereby maintaining the healthy phenotype of conceptuses. Interestingly, not all of the conceptuses that lost ODC1 activity were able to compensate via activation of the alternative ADC/AGMAT pathway in order to maintain physiological levels of polyamines. In fact, for MAO-ODC1 ovine conceptuses with a deficiency in polyamines, conceptus development including elongation was completely retarded. Therefore, the majority of polyamine synthesis may normally be via the conventional ODC1-dependent pathway; however, the ADC/AGMAT-dependent pathway is, at least in sheep, a complimentary or compensatory pathway for production of polyamines for supporting survival and development of conceptuses. In addition, activation of such a compensatory pathway may be associated with genotype of conceptus, perhaps sexual dimorphism or polymorphism, but that has not been investigated.

Cell signaling pathways induced by polyamines include tyrosine- and mitogen-activated protein kinase (MAPK) and proto-oncogenes, c-myc, c-jun, and c-fos [90]. Of interest in reproductive tissues, polyamines also activate mTORC1 cell signaling to stimulate protein synthesis and increase proliferation of both porcine and ovine Tr cells [82,106]. Furthermore, polyamines increase production of IFNT, the pregnancy recognition signal in ruminants, via activation of mTOR cell signaling in oTr1 cells [82]. In rodents, polyamines also stimulate Tr cell motility via activation of mTOR activity, which allows blastocysts to adhere to uterine LE and undergo superficial implantation [107]. Knockout of the *ODC1* gene in mice is not lethal until the gastrulation stage of embryogenesis [108]. There is a requirement for polyamines later in embryogenesis as ODC1-null embryos at the late morula to early blastocyst stages do not survive *in vitro* due to apoptotic cell loss in the ICM, but this condition can be rescued by providing putrescine (a precursor of spermidine and spermine) in drinking water of the dam up to the early implantation stage, but not beyond that stage of pregnancy [108].

2.4.4 Leucine and mTOR

Unlike arginine, leucine is not synthesized *de novo* in any animal cells. However, like arginine, leucine is highly abundant in histotroph during pregnancy and it is an important signaling molecule for conceptus development. In mice, leucine is required for expanded blastocysts to exhibit motility and outgrowth of Tr, which is essential for implantation [107,109–111]. Of note, leucine is the first amino acid that was found to activate mTOR in mammalian cells, and this mechanism helps explain the initial observation in the early 1970s that leucine stimulates protein synthesis and inhibits proteolysis in skeletal muscle of rats. It is now known that leucine induces expression of genes such as insulin-like growth factor 2 (IGF2) [112–114]. Previous studies with leucine and amino acid transporters suggest that uptake of leucine is via transporters, such as SLC3A1, and that leucine stimulates mTOR cell signaling to induce blastocyst motility and differentiation [107,115–117]. Leucine may also function to drive arginine transport as a counter transporter for uptake of arginine via the Na^+-dependent system SLC3A1.

2.4.5 Glutamine and mTOR

Glutamine is a major physiological precursor of ornithine and arginine, and an essential substrate for the synthesis of purine and pyrimidine nucleotides for cell division, amino sugars, and nicotinamide adenine dinucleotide in mammals [70,71,118–120]. Also, glutamine serves as an energy source for rapidly dividing cells. In sheep, the maternal to fetal flux of glutamine is the greatest among all amino acids, particularly between days 13 and 16 of pregnancy when conceptuses are undergoing rapid elongation [69]. Transfer of glutamine from ewes to their fetuses is also greatest among all amino acids measured during mid-gestation [69,121].

Physiological levels of both leucine and glutamine induce cell proliferation via stimulation of mTOR and RPS6K activities [122]. Interestingly, the actions of glutamine require the presence of physiological concentrations of glucose or fructose [a precursor of fructose-6-phosphate (fructose-6-P) and thus glucosamine-6-phosphate (GlcN-6-P)], supporting the view that the hexosamine pathway plays a cell signaling role in conceptus growth and development [123]. Further study demonstrated that glutamine induces proliferation of oTr1 cells via hexosamine-mediated activation of the TSC2-mTOR cell signaling pathway (X. Wang, G. Wu, and F.W. Bazer, unpublished results). Moreover, glutamine is required to facilitate the uptake of leucine, thereby leading to activation of mTORC1 [124].

2.4.6 Fructose and mTOR

During pregnancy, placentae of cetaceans (e.g., whales and porpoises) and ungulates (e.g., pigs and ruminants) are fructogenic as glucose that is not metabolized via glycolysis, pentose cycle, and glycogenesis is converted into fructose by conceptus Tr cells and stored in allantoic fluid [125–127]. In fact, fructose is the most abundant hexose sugar in fetal fluids of cetaceans and ungulates [128–131]. In ewes, for example, the concentration of fructose is between 11.1 and 33.3 mM in allantoic fluid during pregnancy, whereas the maximum concentration of glucose is only 1.1 mM [125]. Fructose is also present, but as a relatively minor sugar compared with glucose, in fetal blood and fetal fluids of humans and other mammals (e.g., dog, cat, guinea pig, rabbit, rat, and ferret) [126,132,133]. In general, high concentrations of fructose are found in the fetal fluids of mammals having epitheliochorial and synepitheliochorial placentae [126], such as pigs and sheep, which are invaluable animal models for studying IUGR in humans [134–136]. The placentae of ungulates contain little or no glycogen. In contrast, rodents and humans with endotheliochorial and hemochorial placentae, respectively, metabolize glucose primarily via glycolysis to form pyruvate and lactate. In all mammals, rates of placental glucose utilization via the pentose cycle may be particularly high to support DNA synthesis and antioxidative reactions.

Studies with pregnant ewes have demonstrated that: (1) injection of glucose into ewes causes a rapid increase in glucose followed by a protracted increase in fructose in fetal blood; (2) the placenta is the site of conversion of glucose to fructose; (3) production of fructose by the placenta is independent of concentrations of glucose in fetal blood; and (4) glucose can move from blood of the conceptus to maternal blood, whereas fructose derived from glucose in the conceptus is not transported into maternal blood [130,137–139]. Studies with pigs confirmed that the placenta is the site of conversion of glucose to fructose [127]. However, the role of fructose during pregnancy in ungulates is largely ignored because fructose is not metabolized via the glycolytic pathway or the Krebs cycle as an energy source.

Proposed roles of fructose are associated with pentose cycle and hexosamine biosynthetic pathways. Studies with fetal pigs and lambs showed that ^{14}C-labeled fructose could be metabolized for synthesis of ribose sugars (precursor for nucleic acids) and generation of reducing equivalents in the form of nicotinamide adenine dinucleotide phosphate, which is related to redox status, as well as synthesis of lipids [140,141]. However, whether this is quantitatively and physiologically significant is unclear, as the purity of the labeled fructose was not verified in the early studies. In porcine placentae obtained on days 25 and 60 of gestation, and in porcine Tr cells, metabolism of fructose (1, 5, and 30 mM) via the pentose cycle, as assessed using $[1-^{14}C]$fructose and $[6-^{14}C]$fructose, was negligible, and there was no detectable production of pyruvate or lactate from fructose (G. Lin, X. Wang, F.W. Bazer, and G. Wu, unpublished data). Recently, we discovered that fructose is actively involved in stimulating cell proliferation and mRNA translation via activation of mTOR cell signaling, and synthesis of glycosaminoglycans, especially hyaluronic acid, via the hexosamine metabolic pathway [123]. Specifically, fructose stimulates phosphorylation of the downstream effectors of mTOR, that is, RPS6K1 and EIF4EBP1, as well as the upstream regulators, that is, PI3K and AKT [123], which are involved with nutrient and energy sensing and protein synthesis and responsiveness to insulin, growth factors, serum, phosphatidic acid, amino acids, and oxidative stress [43].

Once transported into cells, fructose as well as glucose can be metabolized to fructose-6-P, which is further utilized for synthesis of GlcN-6-P via glutamine:fructose-6-phosphate transaminase 1 (GFPT1). Inhibition of GFPT1 activity by azaserine abrogates mTOR activation as well as proliferation of pTr cells [123]. Uridine diphosphate N-acetylglucosamine (UDP-GlcNAc), the product of GlcN-6-P, is associated with intracellular signaling as a substrate for O-linked β-N-acetylglucosamine transferase (OGT), nuclear pore formation and nuclear signaling and the glucose-sensing mechanism, as well as insulin sensitivity of cells [142]. UDP-GlcNAc is the donor substrate for glycosylation of proteins and lipids. Unlike N-linked glycosylation, protein O-GlcNAcylation is an O-linked glycosylation involving attachment of GlcNAc to serine/threonine residues catalyzed by OGT without further extension of GlcNAc, whose removal is catalyzed by O-GlcNAcase (OGA). O-GlcNAcylation is dynamic and reciprocal to phosphorylation at the same or adjacent serine/threonine residues, and often mutually inhibitory; however, recent results have also shown the existence of cooperativeness [143,144]. We further discovered the functional role of fructose to induce proliferation and adhesion of oTr1 cells via activation of AKT-TSC2-mTOR signaling cascade (X. Wang, K.A. Dunlap, G. Wu, and F.W. Bazer, unpublished results). The phosphorylation for activation of this cascade is mediated by O-GlcNAcylation from UDP-GlcNAc, the end product of hexosamine biosynthesis (see Figure 2.3). These results indicate the functional role of fructose in promoting embryonic/fetal growth and development during pregnancy, and also provide insights into understanding of the relationship between excessive fructose intake and metabolic disorders in human medicine.

2.4.7 SPP1 and mTOR

SPP1 is a multifunctional ECM protein that binds to cell surface integrin receptors via its Arg-Gly-Asp amino acid sequence to regulate cell proliferation, migration, adhesion, differentiation, survival, and immune functions in many physiological systems [86,145–151]. During the

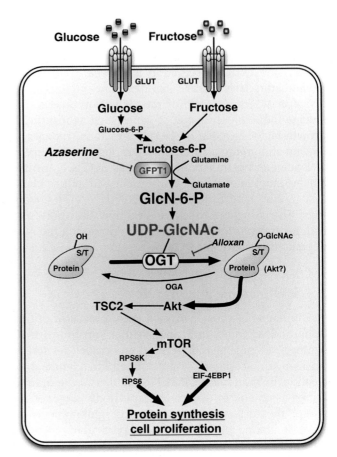

FIGURE 2.3 Schematic diagram of the GFPT1-OGT-mediated Akt-TSC2-mTOR signaling cascade affected by glucose and fructose in oTr1 cells. Fructose stimulates GFPT1 for hexosamine biosynthesis to provide UDP-GlcNAc for O-GlcNAcylation, thereby activating Akt-TSC2-mTOR signaling cascade for stimulation of oTr1 cell proliferation and growth. Akt, proto-oncogenic protein kinase Akt; TSC2, tuberous sclerosis 2; mTOR, mechanistic target of rapamycin; RPS6, ribosomal protein S6; RPS6K, ribosomal protein S6K; EIF4EBP1, eukaryotic translation initiation factor 4E-binding protein 1; GlcN-6-P, glucosamine-6-phosphate; GFPT1, glutamine-fructose-6-phosphate transaminase 1; OGT, O-linked N-acetylglucosamine transferase; OGA, O-GlcNAcase; UDP-GlcNAc, UDP-N-acetylglucosamine.

peri-implantation period of pregnancy, attachment of Tr to uterine LE is facilitated by a mosaic of interactions between integrins and ECM proteins, particularly SPP1, that contribute to stable adhesion of Tr to uterine LE for implantation [152–154]. In most mammals studied, secretion of SPP1 by uterine GE increases significantly during pregnancy [86]. For example, in sheep, implanting conceptuses secrete IFNT to prolong the lifespan of CL which secretes progesterone to induce SPP1 synthesis and secretion from uterine GE into the uterine lumen. Using our oTr1 cell line, we demonstrated that SPP1 binds to αvβ3 and αvβ5 integrin heterodimers to initiate focal adhesion assembly, a prerequisite for adhesion and migration of cells via activation of: (1) P70S6K via crosstalk between mTOR and MAPK pathways; (2) mTOR, PI3K, MAPK3/MAPK1 (Erk1/2), and MAPK14 (p38) signaling to stimulate cell migration; and (3) focal adhesion assembly and myosin II motor activity to induce adhesion of Tr cells [86,146–148,151]. These cell signaling pathways, acting in concert, mediate adhesion, migration, and cytoskeletal remodeling of Tr cells required for expansion and elongation of conceptuses and attachment to uterine LE for implantation. However, SPP1 alone fails to stimulate proliferation of oTr1 cells. Interestingly, further investigations related to individual and combined effects of arginine and recombinant SPP1 have shown their multiple cooperative effects on adhesion, migration, and proliferation of oTr1 cells [155]. At physiological concentrations, arginine (0.2 mM) significantly increased oTr1 cell proliferation, but SPP1 had no effect. However, the combination of arginine and SPP1 significantly increased cell proliferation through activation of PDK1-AKT-TSC2-mTORC1 cell signaling. Arginine is the driving force for cell proliferation and SPP1, by its ability to increase cell spreading and trigger focal adhesion assembly, facilitates cell proliferation and cytoskeletal reorganization. The synergistic effects of the combination of arginine and SPP1 on cell adhesion and migration were achieved via mTORC2-mediated cytoskeleton reorganization, including microtubule α-tubulin, microfilament F-actin, and intermediate filament cytokeratin (X. Wang, G.A. Johnson, G. Wu, and F.W. Bazer, unpublished results). These results suggest that SPP1 together with arginine, activate mTORC1 and mTORC2 in Tr cells to allow blastocysts to make the critical transition from spherical to tubular and filamentous conceptuses, which is a prerequisite for signaling pregnancy recognition, implantation, and placentation required for a successful outcome of pregnancy in ungulates and other mammals.

2.5 SUMMARY

Histotroph includes molecules secreted and/or transported into the uterine lumen and then into the fetal–placental vascular system to support key events involved in the development and survival of conceptuses during pregnancy in mammals. The increase in the abundance of these components reflects activation of pregnancy-associated mechanisms for transport of nutrients into the uterine lumen and then into the fetal–placental blood and fluids. This review provides a framework for studies of constituents, including hexose sugars (i.e., glucose and fructose), ECM proteins (e.g., SPP1), amino acids (e.g., arginine, leucine, and glutamine), and their metabolites (e.g., NO and polyamines) that independently and cooperatively activate nutrient-sensing cell signaling pathways, including mTOR, the master regulator of cell growth (proliferative growth via mTORC1 and spatial growth via mTORC2), for growth, development, and survival of conceptuses, as well as for optimization of culture media for *in vitro* studies of conceptus development. Given the importance of the peri-implantation period of pregnancy in setting the stage for implantation and placentation, the effects of select components of histotroph on conceptus development have long-term consequences for the health and wellbeing of the fetus and into its adulthood, the developmental origins of health and disease concept (DOHAD). Although mechanisms responsible for differential growth and development of the conceptus resulting in DOHAD phenomena remain unclear, they likely involve epigenetic events involving methylation of DNA as well as histone modifications. Moreover, for assisted reproductive technologies used in human medicine and animal agriculture, gametes and embryos may be prone to epigenetic alterations when exposed to specific culture media with an imbalanced mixture of nutrients during *in vitro* procedures. Therefore, future research on the effects of selected components of histotroph on conceptus development, as well as relationships between histotroph-induced mTOR regulation and epigenetic alterations during conceptus development, is of great significance for human medicine and animal agriculture. Understanding such relationships between cell signaling pathways and developmental events is critical for enhancing conceptus development, implantation, and placentation, thereby increasing the probability for birth of healthy offspring.

Acknowledgments

Research in our laboratories was supported by National Research Initiative Competitive Grants from the Animal Reproduction Program (2008-35203-19120, 2009-35206-05211, 2011-67015-20067, and 2011-67015-20028) and Animal Growth & Nutrient Utilization Program (2008-35206-18764) of the USDA National Institute of Food and Agriculture, and Texas A&M AgriLife Research (H-8200). The important contributions of our graduate students and colleagues in this research are gratefully acknowledged.

References

[1] Bazer FW, First NL. Pregnancy and parturition. J Anim Sci 1983;57(Suppl. 2):425–60.

[2] Spencer TE, Bazer FW. Uterine and placental factors regulating conceptus growth in domestic animals. J Anim Sci 2004;82 (E-Suppl):E4–13.

[3] Wu G, Bazer FW, Cudd TA, Meininger CJ, Spencer TE. Maternal nutrition and fetal development. J Nutr 2004;134:2169–72.

[4] Bell AW, Ehrhardt RA. Regulation of placental nutrient transport and implications for fetal growth. Nutr Res Rev 2002;15:211–30.

[5] Barker DJ, Clark PM. Fetal undernutrition and disease in later life. Rev Reprod 1997;2:105—12.

[6] Wang J, Chen L, Li D, Yin Y, Wang X, Li P, et al. Intrauterine growth restriction affects the proteomes of the small intestine, liver, and skeletal muscle in newborn pigs. J Nutr 2008;138:60—6.

[7] Wang X, Lin G, Liu C, Feng C, Zhou H, Wang T, et al. Temporal proteomic analysis reveals defects in small-intestinal development of porcine fetuses with intrauterine growth restriction. J Nutr Biochem 2014;25:785—95.

[8] Wang X, Wu W, Lin G, Li D, Wu G, Wang J. Temporal proteomic analysis reveals continuous impairment of intestinal development in neonatal piglets with intrauterine growth restriction. J Proteome Res 2010;9:924—35.

[9] Waterland RA, Jirtle RL. Early nutrition, epigenetic changes at transposons and imprinted genes, and enhanced susceptibility to adult chronic diseases. Nutrition 2004;20:63—8.

[10] Marsal K. Intrauterine growth restriction. Curr Opin Obstet Gynecol 2002;14:127—35.

[11] Wu G, Pond WG, Flynn SP, Ott TL, Bazer FW. Maternal dietary protein deficiency decreases nitric oxide synthase and ornithine decarboxylase activities in placenta and endometrium of pigs during early gestation. J Nutr 1998;128:2395—402.

[12] Pham TD, MacLennan NK, Chiu CT, Laksana GS, Hsu JL, Lane RH. Uteroplacental insufficiency increases apoptosis and alters p53 gene methylation in the full-term IUGR rat kidney. Am J Physiol Regul Integr Comp Physiol 2003;285:R962—70.

[13] Spencer TE, Johnson GA, Bazer FW, Burghardt RC. Fetal—maternal interactions during the establishment of pregnancy in ruminants. Soc Reprod Fertil Suppl 2007;64:379—96.

[14] Bazer FW, Burghardt RC, Johnson GA, Spencer TE, Wu G. Interferons and progesterone for establishment and maintenance of pregnancy: interactions among novel cell signaling pathways. Reprod Biol 2008;8:179—211.

[15] Bazer FW, Spencer TE, Johnson GA. Interferons and uterine receptivity. Semin Reprod Med 2009;27:90—102.

[16] Barcroft LC, Hay-Schmidt A, Caveney A, Gilfoyle E, Overstrom EW, Hyttel P, et al. Trophectoderm differentiation in the bovine embryo: characterization of a polarized epithelium. J Reprod Fertil 1998;114:327—39.

[17] Bindon BM. Systematic study of preimplantation stages of pregnancy in the sheep. Aust J Biol Sci 1971;24:131—47.

[18] Ziomek CA, Johnson MH. Cell surface interaction induces polarization of mouse 8-cell blastomeres at compaction. Cell 1980;21:935—42.

[19] Wintenberger-Torres S, Flechon JE. Ultrastructural evolution of the trophoblast cells of the pre-implantation sheep blastocyst from day 8 to day 18. J Anat 1974;118:143—53.

[20] Rowson LE, Moor RM. Development of the sheep conceptus during the first fourteen days. J Anat 1966;100:777—85.

[21] Guillomot M. Cellular interactions during implantation in domestic ruminants. J Reprod Fertil Suppl 1995;49:39—51.

[22] Guillomot M, Flechon JE, Wintenberger-Torres S. Conceptus attachment in the ewe: an ultrastructural study. Placenta 1981;2:169—82.

[23] Bazer FW, Spencer TE, Johnson GA, Burghardt RC, Wu G. Comparative aspects of implantation. Reproduction 2009;138:195—209.

[24] Bazer FW. Pregnancy recognition signaling mechanisms in ruminants and pigs. J Anim Sci Biotechnol 2013;4:23.

[25] Bazer FW, Spencer TE, Ott TL. Interferon tau: a novel pregnancy recognition signal. Am J Reprod Immunol 1997;37:412—20.

[26] Bazer FW, Ying W, Wang X, Dunlap KA, Zhou B, Johnson GA, et al. The many faces of interferon tau. Amino Acids 2015;47:449—60.

[27] Bazer FW, Thatcher WW. Theory of maternal recognition of pregnancy in swine based on estrogen controlled endocrine versus exocrine secretion of prostaglandin F2alpha by the uterine endometrium. Prostaglandins 1977;14:397—400.

[28] Bazer FW, Vallet JL, Roberts RM, Sharp DC, Thatcher WW. Role of conceptus secretory products in establishment of pregnancy. J Reprod Fertil 1986;76:841—50.

[29] Bazer FW, Song G, Kim J, Dunlap KA, Satterfield MC, Johnson GA, et al. Uterine biology in pigs and sheep. J Anim Sci Biotechnol 2012;3:23.

[30] Spencer TE, Johnson GA, Bazer FW, Burghardt RC. Implantation mechanisms: insights from the sheep. Reproduction 2004;128:657—68.

[31] Laplante M, Sabatini DM. mTOR signaling in growth control and disease. Cell 2012;149:274—93.

[32] Kim SG, Buel GR, Blenis J. Nutrient regulation of the mTOR complex 1 signaling pathway. Mol Cells 2013;35:463—73.

[33] Chantranupong L, Wolfson RL, Sabatini DM. Nutrient-sensing mechanisms across evolution. Cell 2015;161:67—83.

[34] Brown EJ, Albers MW, Shin TB, Ichikawa K, Keith CT, Lane WS, et al. A mammalian protein targeted by G1-arresting rapamycin-receptor complex. Nature 1994;369:756—8.

[35] Sabatini DM, Erdjument-Bromage H, Lui M, Tempst P, Snyder SH. RAFT1: a mammalian protein that binds to FKBP12 in a rapamycin-dependent fashion and is homologous to yeast TORs. Cell 1994;78:35—43.

[36] Vezina C, Kudelski A, Sehgal SN. Rapamycin (AY-22,989), a new antifungal antibiotic. I. Taxonomy of the producing streptomycete and isolation of the active principle. J Antibiot (Tokyo) 1975;28:721—6.

[37] Segall JE, Block SM, Berg HC. Temporal comparisons in bacterial chemotaxis. Proc Natl Acad Sci USA 1986;83:8987—91.

[38] Visner GA, Lu F, Zhou H, Liu J, Kazemfar K, Agarwal A. Rapamycin induces heme oxygenase-1 in human pulmonary vascular cells: implications in the antiproliferative response to rapamycin. Circulation 2003;107:911—16.

[39] Nair RV, Huang X, Shorthouse R, Adams B, Brazelton T, Braun-Dullaeus R, et al. Antiproliferative effect of rapamycin on growth factor-stimulated human adult lung fibroblasts in vitro may explain its superior efficacy for prevention and treatment of allograft obliterative airway disease in vivo. Transplant Proc 1997;29:614—15.

[40] Janes MR, Fruman DA. Immune regulation by rapamycin: moving beyond T cells. Sci Signal 2009;2:pe25.

[41] Martel RR, Klicius J, Galet S. Inhibition of the immune response by rapamycin, a new antifungal antibiotic. Can J Physiol Pharmacol 1977;55:48—51.

[42] Sabers CJ, Martin MM, Brunn GJ, Williams JM, Dumont FJ, Wiederrecht G, et al. Isolation of a protein target of the FKBP12-rapamycin complex in mammalian cells. J Biol Chem 1995;270:815—22.

[43] Hay N, Sonenberg N. Upstream and downstream of mTOR. Genes Dev 2004;18:1926—45.

[44] Wullschleger S, Loewith R, Hall MN. TOR signaling in growth and metabolism. Cell 2006;124:471—84.

[45] Kim DH, Sarbassov DD, Ali SM, King JE, Latek RR, Erdjument-Bromage H, et al. mTOR interacts with raptor to form a nutrient-sensitive complex that signals to the cell growth machinery. Cell 2002;110:163—75.

[46] Helliwell SB, Wagner P, Kunz J, Deuter-Reinhard M, Henriquez R, Hall MN. TOR1 and TOR2 are structurally and functionally similar but not identical phosphatidylinositol kinase homologues in yeast. Mol Biol Cell 1994;5:105—18.

[47] Guertin DA, Stevens DM, Thoreen CC, Burds AA, Kalaany NY, Moffat J, et al. Ablation in mice of the mTORC components raptor, rictor, or mLST8 reveals that mTORC2 is required for signaling to Akt-FOXO and PKCalpha, but not S6K1. Dev Cell 2006;11:859—71.

[48] Liao XH, Majithia A, Huang X, Kimmel AR. Growth control via TOR kinase signaling, an intracellular sensor of amino acid and energy availability, with crosstalk potential to proline metabolism. Amino Acids 2008;35:761—70.

[49] Hara K, Maruki Y, Long X, Yoshino K, Oshiro N, Hidayat S, et al. Raptor, a binding partner of target of rapamycin (TOR), mediates TOR action. Cell 2002;110:177—89.

[50] Kim DH, Sarbassov DD, Ali SM, Latek RR, Guntur KV, Erdjument-Bromage H, et al. GbetaL, a positive regulator of the rapamycin-sensitive pathway required for the nutrient-sensitive interaction between raptor and mTOR. Mol Cell 2003;11: 895—904.

[51] Loewith R, Jacinto E, Wullschleger S, Lorberg A, Crespo JL, Bonenfant D, et al. Two TOR complexes, only one of which is rapamycin sensitive, have distinct roles in cell growth control. Mol Cell 2002;10:457—68.

[52] Sarbassov DD, Ali SM, Sabatini DM. Growing roles for the mTOR pathway. Curr Opin Cell Biol 2005;17:596—603.

[53] Yang Q, Inoki K, Ikenoue T, Guan KL. Identification of Sin1 as an essential TORC2 component required for complex formation and kinase activity. Genes Dev 2006;20:2820—32.

[54] Laplante M, Sabatini DM. mTOR signaling at a glance. J Cell Sci 2009;122:3589—94.

[55] Pearce LR, Huang X, Boudeau J, Pawlowski R, Wullschleger S, Deak M, et al. Identification of Protor as a novel Rictor-binding component of mTOR complex-2. Biochem J 2007;405:513—22.

[56] Sparks CA, Guertin DA. Targeting mTOR: prospects for mTOR complex 2 inhibitors in cancer therapy. Oncogene 2010;29: 3733—44.

[57] Bazer FW. Uterine protein secretions: relationship to development of the conceptus. J Anim Sci 1975;41:1376—82.

[58] Bazer FW, Spencer TE, Johnson GA, Burghardt RC. Uterine receptivity to implantation of blastocysts in mammals. Front Biosci (Schol Ed) 2011;3:745—67.

[59] Burton GJ, Watson AL, Hempstock J, Skepper JN, Jauniaux E. Uterine glands provide histiotrophic nutrition for the human fetus during the first trimester of pregnancy. J Clin Endocrinol Metab 2002;87:2954—9.

[60] Allison Gray C, Bartol FF, Taylor KM, Wiley AA, Ramsey WS, Ott TL, et al. Ovine uterine gland knock-out model: effects of gland ablation on the estrous cycle. Biol Reprod 2000;62: 448—56.

[61] Gray CA, Burghardt RC, Johnson GA, Bazer FW, Spencer TE. Evidence that absence of endometrial gland secretions in uterine gland knockout ewes compromises conceptus survival and elongation. Reproduction 2002;124:289—300.

[62] Gray CA, Taylor KM, Ramsey WS, Hill JR, Bazer FW, Bartol FF, et al. Endometrial glands are required for preimplantation conceptus elongation and survival. Biol Reprod 2001;64:1608—13.

[63] Gray CA, Taylor KM, Bazer FW, Spencer TE. Mechanisms regulating norgestomet inhibition of endometrial gland morphogenesis in the neonatal ovine uterus. Mol Reprod Dev 2000;57:67—78.

[64] Gray CA, Bazer FW, Spencer TE. Effects of neonatal progestin exposure on female reproductive tract structure and function in the adult ewe. Biol Reprod 2001;64:797—804.

[65] Gray CA, Adelson DL, Bazer FW, Burghardt RC, Meeusen EN, Spencer TE. Discovery and characterization of an epithelial-specific galectin in the endometrium that forms crystals in the trophectoderm. Proc Natl Acad Sci USA 2004;101:7982—7.

[66] Gray CA, Abbey CA, Beremand PD, Choi Y, Farmer JL, Adelson DL, et al. Identification of endometrial genes regulated by early pregnancy, progesterone, and interferon tau in the ovine uterus. Biol Reprod 2006;74:383—94.

[67] Wu G. Functional amino acids in nutrition and health. Amino Acids 2013;45:407—11.

[68] Wu G, Bazer FW, Dai Z, Li D, Wang J, Wu Z. Amino acid nutrition in animals: protein synthesis and beyond. Annu Rev Anim Biosci 2014;2:387—417.

[69] Gao H, Wu G, Spencer TE, Johnson GA, Li X, Bazer FW. Select nutrients in the ovine uterine lumen. I. Amino acids, glucose, and ions in uterine lumenal flushings of cyclic and pregnant ewes. Biol Reprod 2009;80:86—93.

[70] Meier P, Teng C, Battaglia FC, Meschia G. The rate of amino acid nitrogen and total nitrogen accumulation in the fetal lamb. Proc Soc Exp Biol Med 1981;167:463—8.

[71] Wu G, Ott TL, Knabe DA, Bazer FW. Amino acid composition of the fetal pig. J Nutr 1999;129:1031—8.

[72] Gao H, Wu G, Spencer TE, Johnson GA, Bazer FW. Select nutrients in the ovine uterine lumen. V. Nitric oxide synthase, GTP cyclohydrolase, and ornithine decarboxylase in ovine uteri and peri-implantation conceptuses. Biol Reprod 2009;81:67—76.

[73] Gao H, Wu G, Spencer TE, Johnson GA, Bazer FW. Select nutrients in the ovine uterine lumen. III. Cationic amino acid transporters in the ovine uterus and peri-implantation conceptuses. Biol Reprod 2009;80:602—9.

[74] Conrad KP, Vill M, McGuire PG, Dail WG, Davis AK. Expression of nitric oxide synthase by syncytiotrophoblast in human placental villi. FASEB J 1993;7:1269—76.

[75] Dye JF, Vause S, Johnston T, Clark P, Firth JA, D'Souza SW, et al. Characterization of cationic amino acid transporters and expression of endothelial nitric oxide synthase in human placental microvascular endothelial cells. FASEB J 2004;18: 125—7.

[76] Zeng X, Wang F, Fan X, Yang W, Zhou B, Li P, et al. Dietary arginine supplementation during early pregnancy enhances embryonic survival in rats. J Nutr 2008;138:1421—5.

[77] Mateo RD, Wu G, Bazer FW, Park JC, Shinzato I, Kim SW. Dietary L-arginine supplementation enhances the reproductive performance of gilts. J Nutr 2007;137:652—6.

[78] Wu G, Morris Jr SM. Arginine metabolism: nitric oxide and beyond. Biochem J 1998;336(Pt 1):1—17.

[79] Wang X, Frank JW, Little DR, Dunlap KA, Satterfield MC, Burghardt RC, et al. Functional role of arginine during the peri-implantation period of pregnancy. I. Consequences of loss of function of arginine transporter SLC7A1 mRNA in ovine conceptus trophectoderm. FASEB J 2014;28:2852—63.

[80] Kim J, Burghardt RC, Wu G, Johnson GA, Spencer TE, Bazer FW. Select nutrients in the ovine uterine lumen. IX. Differential effects of arginine, leucine, glutamine, and glucose on interferon tau, ornithine decarboxylase, and nitric oxide synthase in the ovine conceptus. Biol Reprod 2011;84:1139—47.

[81] Kim JY, Burghardt RC, Wu G, Johnson GA, Spencer TE, Bazer FW. Select nutrients in the ovine uterine lumen. VIII. Arginine stimulates proliferation of ovine trophectoderm cells through MTOR-RPS6K-RPS6 signaling cascade and synthesis of nitric oxide and polyamines. Biol Reprod 2011;84:70—8.

[82] Wang X, Burghardt RC, Romero JJ, Hansen TR, Wu G, Bazer FW. Functional roles of arginine during the peri-implantation period of pregnancy. III. Arginine stimulates proliferation and interferon tau production by ovine trophectoderm cells via nitric oxide and polyamine-TSC2-MTOR signaling pathways. Biol Reprod 2015;92:75.

[83] Guo H, Marroquin CE, Wai PY, Kuo PC. Nitric oxide-dependent osteopontin expression induces metastatic behavior in HepG2 cells. Dig Dis Sci 2005;50:1288—98.

[84] Dai Z, Wu Z, Yang Y, Wang J, Satterfield MC, Meininger CJ, et al. Nitric oxide and energy metabolism in mammals. Biofactors 2013;39:383—91.

[85] Wang X, Frank JW, Xu J, Dunlap KA, Satterfield MC, Burghardt RC, et al. Functional role of arginine during the peri-implantation period of pregnancy. II. Consequences of loss of function of nitric oxide synthase NOS3 mRNA in ovine conceptus trophectoderm. Biol Reprod 2014;91:59.

[86] Johnson GA, Burghardt RC, Bazer FW, Spencer TE. Osteopontin: roles in implantation and placentation. Biol Reprod 2003;69:1458—71.

[87] Saxena D, Purohit SB, Kumer GP, Laloraya M. Increased appearance of inducible nitric oxide synthase in the uterus and embryo at implantation. Nitric Oxide 2000;4:384—91.

[88] Cartwright JE, Tse WK, Whitley GS. Hepatocyte growth factor induced human trophoblast motility involves phosphatidylinositol-3-kinase, mitogen-activated protein kinase, and inducible nitric oxide synthase. Exp Cell Res 2002;279:219—26.

[89] Kwon H, Wu G, Meininger CJ, Bazer FW, Spencer TE. Developmental changes in nitric oxide synthesis in the ovine placenta. Biol Reprod 2004;70:679—86.

[90] Kwon H, Spencer TE, Bazer FW, Wu G. Developmental changes of amino acids in ovine fetal fluids. Biol Reprod 2003;68:1813—20.

[91] Reynolds LP, Borowicz PP, Vonnahme KA, Johnson ML, Grazul-Bilska AT, Wallace JM, et al. Animal models of placental angiogenesis. Placenta 2005;26:689—708.

[92] Heby O. DNA methylation and polyamines in embryonic development and cancer. Int J Dev Biol 1995;39:737—57.

[93] Igarashi K, Kashiwagi K. Modulation of cellular function by polyamines. Int J Biochem Cell Biol 2010;42:39—51.

[94] Wallace HM, Fraser AV, Hughes A. A perspective of polyamine metabolism. Biochem J 2003;376:1—14.

[95] Chattopadhyay MK, Tabor CW, Tabor H. Polyamines protect Escherichia coli cells from the toxic effect of oxygen. Proc Natl Acad Sci USA 2003;100:2261—5.

[96] Maeda T, Wakasawa T, Shima Y, Tsuboi I, Aizawa S, Tamai I. Role of polyamines derived from arginine in differentiation and proliferation of human blood cells. Biol Pharm Bull 2006;29:234—9.

[97] Nishimura K, Murozumi K, Shirahata A, Park MH, Kashiwagi K, Igarashi K. Independent roles of eIF5A and polyamines in cell proliferation. Biochem J 2005;385:779—85.

[98] Igarashi K, Kashiwagi K. Polyamines: mysterious modulators of cellular functions. Biochem Biophys Res Commun 2000;271:559—64.

[99] Lefevre PL, Palin MF, Beaudry D, Dobias-Goff M, Desmarais JA, Llerena VE, et al. Uterine signaling at the emergence of the embryo from obligate diapause. Am J Physiol Endocrinol Metab 2011;300:E800—8.

[100] Lefevre PL, Palin MF, Chen G, Turecki G, Murphy BD. Polyamines are implicated in the emergence of the embryo from obligate diapause. Endocrinology 2011;152:1627—39.

[101] Lefevre PL, Palin MF, Murphy BD. Polyamines on the reproductive landscape. Endocr Rev 2011;32:694—712.

[102] Mandal S, Mandal A, Johansson HE, Orjalo AV, Park MH. Depletion of cellular polyamines, spermidine and spermine, causes a total arrest in translation and growth in mammalian cells. Proc Natl Acad Sci USA 2013;110:2169—74.

[103] Wang X, Ying W, Dunlap KA, Lin G, Satterfield MC, Burghardt RC, et al. Arginine decarboxylase and agmatinase: an alternative pathway for de novo biosynthesis of polyamines for development of mammalian conceptuses. Biol Reprod 2014;90:84.

[104] Mehrotra PK, Kitchlu S, Farheen S. Effect of inhibitors of enzymes involved in polyamine biosynthesis pathway on pregnancy in mouse and hamster. Contraception 1998;57:55—60.

[105] Bernstein HG, Derst C, Stich C, Pruss H, Peters D, Krauss M, et al. The agmatine-degrading enzyme agmatinase: a key to agmatine signaling in rat and human brain? Amino Acids 2011;40:453—65.

[106] Kong X, Wang X, Yin Y, Li X, Gao H, Bazer FW, et al. Putrescine stimulates the mTOR signaling pathway and protein synthesis in porcine trophectoderm cells. Biol Reprod 2014;91:106.

[107] Martin PM, Sutherland AE, Van Winkle LJ. Amino acid transport regulates blastocyst implantation. Biol Reprod 2003;69:1101—8.

[108] Pendeville H, Carpino N, Marine JC, Takahashi Y, Muller M, Martial JA, et al. The ornithine decarboxylase gene is essential for cell survival during early murine development. Mol Cell Biol 2001;21:6549—58.

[109] Gwatkin RB. Nutritional requirements for post-blastocyst development in the mouse. Amino acids and protein in the uterus during implantation. Int J Fertil 1969;14:101—5.

[110] Martin PM, Sutherland AE. Exogenous amino acids regulate trophectoderm differentiation in the mouse blastocyst through an mTOR-dependent pathway. Dev Biol 2001;240:182—93.

[111] Gwatkin RBL. Amino acid requirements for attachment and outgrowth of the mouse blastocyst in vitro. J Cell Physiol 1966;68:335—43.

[112] Nielsen FC, Ostergaard L, Nielsen J, Christiansen J. Growth-dependent translation of IGF-II mRNA by a rapamycin-sensitive pathway. Nature 1995;377:358—62.

[113] Kimball SR, Shantz LM, Horetsky RL, Jefferson LS. Leucine regulates translation of specific mRNAs in L6 myoblasts through mTOR-mediated changes in availability of eIF4E and phosphorylation of ribosomal protein S6. J Biol Chem 1999;274:11647—52.

[114] Murakami M, Ichisaka T, Maeda M, Oshiro N, Hara K, Edenhofer F, et al. mTOR is essential for growth and proliferation in early mouse embryos and embryonic stem cells. Mol Cell Biol 2004;24:6710—18.

[115] Finch AM, Yang LG, Nwagwu MO, Page KR, McArdle HJ, Ashworth CJ. Placental transport of leucine in a porcine model of low birth weight. Reproduction 2004;128:229—35.

[116] Gao H, Wu G, Spencer TE, Johnson GA, Bazer FW. Select nutrients in the ovine uterine lumen. IV. Expression of neutral and acidic amino acid transporters in ovine uteri and peri-implantation conceptuses. Biol Reprod 2009;80:1196—208.

[117] Van Winkle LJ, Tesch JK, Shah A, Campione AL. System B0,+ amino acid transport regulates the penetration stage of blastocyst implantation with possible long-term developmental consequences through adulthood. Hum Reprod Update 2006;12:145—57.

[118] Wu G, Bazer FW, Davis TA, Kim SW, Li P, Marc Rhoads J, et al. Arginine metabolism and nutrition in growth, health and disease. Amino Acids 2009;37:153—68.

[119] Wu G, Knabe DA, Yan W, Flynn NE. Glutamine and glucose metabolism in enterocytes of the neonatal pig. Am J Physiol 1995;268:R334—42.

[120] Krebs HA, Baverel G, Lund P. Effect of bicarbonate on glutamine metabolism. Int J Biochem 1980;12:69–73.

[121] Bell AW, Kennaugh JM, Battaglia FC, Meschia G. Uptake of amino acids and ammonia at mid-gestation by the fetal lamb. Q J Exp Physiol 1989;74:635–43.

[122] Kim J, Song G, Wu G, Gao H, Johnson GA, Bazer FW. Arginine, leucine, and glutamine stimulate proliferation of porcine trophectoderm cells through the MTOR-RPS6K-RPS6-EIF4EBP1 signal transduction pathway. Biol Reprod 2013;88:113.

[123] Kim J, Song G, Wu G, Bazer FW. Functional roles of fructose. Proc Natl Acad Sci USA 2012;109:E1619–28.

[124] Nicklin P, Bergman P, Zhang B, Triantafellow E, Wang H, Nyfeler B, et al. Bidirectional transport of amino acids regulates mTOR and autophagy. Cell 2009;136:521–34.

[125] Bazer FW, Spencer TE, Thatcher WW. Growth and development of the ovine conceptus. J Anim Sci 2012;90:159–70.

[126] Goodwin RF. Division of the common mammals into two groups according to the concentration of fructose in the blood of the foetus. J Physiol 1956;132:146–56.

[127] White CE, Piper EL, Noland PR. Conversion of glucose to fructose in the fetal pig. J Anim Sci 1979;48:585–90.

[128] Bacon JS, Bell DJ. The identification of fructose as a constituent of the foetal blood of the sheep. Biochem J 1946;40:xlii.

[129] Bacon JS, Bell DJ. Fructose and glucose in the blood of the foetal sheep. Biochem J 1948;42:397–405.

[130] Barklay H, Haas P, et al. The sugar of the foetal blood, the amniotic and allantoic fluids. J Physiol 1949;109:98–102.

[131] Hitchcock MW. Fructose in the sheep foetus. J Physiol 1949; 108:117–26.

[132] Jauniaux E, Hempstock J, Teng C, Battaglia FC, Burton GJ. Polyol concentrations in the fluid compartments of the human conceptus during the first trimester of pregnancy: maintenance of redox potential in a low oxygen environment. J Clin Endocrinol Metab 2005;90:1171–5.

[133] Karvonen MJ. Absence of fructose from human cord blood. Acta Paediatr 1949;37:68–72.

[134] Nathanielsz PW. Animal models that elucidate basic principles of the developmental origins of adult diseases. ILAR J 2006; 47:73–82.

[135] Roberts RM, Smith GW, Bazer FW, Cibelli J, Seidel GE, Bauman DE, et al. Farm animal research in crisis. Science 2009;324:468–9.

[136] Reynolds LP, Borowicz PP, Caton JS, Vonnahme KA, Luther JS, Buchanan DS, et al. Uteroplacental vascular development and placental function: an update. Int J Dev Biol 2010; 54:355–66.

[137] Alexander DP, Andrews RD, Huggett AS, Nixon DA, Widdas WF. The placental transfer of sugars in the sheep: studies with radioactive sugar. J Physiol 1955;129:352–66.

[138] Alexander DP, Huggett AS, Nixon DA, Widdas WF. The placental transfer of sugars in the sheep: the influence of concentration gradient upon the rates of hexose formation as shown in umbilical perfusion of the placenta. J Physiol 1955;129:367–83.

[139] Huggett AS, Warren FL, Warren NV. The origin of the blood fructose of the foetal sheep. J Physiol 1951;113:258–75.

[140] Scott TW, Setchell BP, Bassett JM. Characterization and metabolism of ovine foetal lipids. Biochem J 1967;104:1040–7.

[141] White CE, Piper EL, Noland PR, Daniels LB. Fructose utilization for nucleic acid synthesis in the fetal pig. J Anim Sci 1982;55:73–6.

[142] McClain DA, Lubas WA, Cooksey RC, Hazel M, Parker GJ, Love DC, et al. Altered glycan-dependent signaling induces insulin resistance and hyperleptinemia. Proc Natl Acad Sci USA 2002;99:10695–9.

[143] Heath JM, Sun Y, Yuan K, Bradley WE, Litovsky S, Dell'Italia LJ, et al. Activation of AKT by O-linked N-acetylglucosamine induces vascular calcification in diabetes mellitus. Circ Res 2014;114:1094–102.

[144] Hart GW, Slawson C, Ramirez-Correa G, Lagerlof O. Cross talk between O-GlcNAcylation and phosphorylation: roles in signaling, transcription, and chronic disease. Annu Rev Biochem 2011;80:825–58.

[145] Denhardt DT, Guo X. Osteopontin: a protein with diverse functions. FASEB J 1993;7:1475–82.

[146] Johnson GA, Burghardt RC, Spencer TE, Newton GR, Ott TL, Bazer FW. Ovine osteopontin: II. Osteopontin and alpha(v)beta(3) integrin expression in the uterus and conceptus during the periimplantation period. Biol Reprod 1999;61:892–9.

[147] Johnson GA, Spencer TE, Burghardt RC, Bazer FW. Ovine osteopontin: I. Cloning and expression of messenger ribonucleic acid in the uterus during the periimplantation period. Biol Reprod 1999;61:884–91.

[148] Johnson GA, Spencer TE, Burghardt RC, Taylor KM, Gray CA, Bazer FW. Progesterone modulation of osteopontin gene expression in the ovine uterus. Biol Reprod 2000;62:1315–21.

[149] Senger DR, Perruzzi CA, Papadopoulos-Sergiou A, Van de Water L. Adhesive properties of osteopontin: regulation by a naturally occurring thrombin-cleavage in close proximity to the GRGDS cell-binding domain. Mol Biol Cell 1994;5:565–74.

[150] Sodek J, Ganss B, McKee MD. Osteopontin. Crit Rev Oral Biol Med 2000;11:279–303.

[151] Kim J, Erikson DW, Burghardt RC, Spencer TE, Wu G, Bayless KJ, et al. Secreted phosphoprotein 1 binds integrins to initiate multiple cell signaling pathways, including FRAP1/mTOR, to support attachment and force-generated migration of trophectoderm cells. Matrix Biol 2010;29:369–82.

[152] Aplin JD, Kimber SJ. Trophoblast-uterine interactions at implantation. Reprod Biol Endocrinol 2004;2:48.

[153] Burghardt RC, Johnson GA, Jaeger LA, Ka H, Garlow JE, Spencer TE, et al. Integrins and extracellular matrix proteins at the maternal–fetal interface in domestic animals. Cells Tissues Organs 2002;172:202–17.

[154] Lessey BA, Castelbaum AJ. Integrins and implantation in the human. Rev Endocr Metab Disord 2002;3:107–17.

[155] Wang X, Johnson GA, Burghardt RC, Wu G, Bazer FW. Uterine histotroph and conceptus development. I. Cooperative effects of arginine and secreted phosphoprotein 1 on proliferation of ovine trophectoderm cells via activation of the PDK1-Akt/PKB-TSC2-MTORC1 signaling cascade. Biol Reprod 2015;92:51.

3

mTORC1 in the Control of Myogenesis and Adult Skeletal Muscle Mass

Marita A. Wallace, David C. Hughes and Keith Baar

Department of Neurobiology, Physiology and Behavior, Functional Molecular Biology Lab,
University of California Davis, Davis, CA, USA

3.1 SKELETAL MUSCLE— AN INTRODUCTION

Skeletal muscle is the most abundant tissue in the human body, comprising about 40% of total body mass [1]. The primary functions of skeletal muscle include the control of movement, posture, and breathing. Beyond these mechanical roles skeletal muscle also controls whole-body metabolism [1]. Skeletal muscle is comprised of bundles of muscle fibers (fascicles), nerves, and blood vessels, which are enclosed in a connective tissue sheath called the epimysium. Each fascicle is surrounded by another connective tissue sheath called the perimysium, whereas each muscle fiber within the fascicle is tightly packed and separated from other fibers by the endomyosium. Muscle fibers are long cylindrical and multinucleated cells that lie parallel to each other. The nuclei of muscle fibers are located against the inside of the plasma membrane in muscle that is called the sarcolemma. Centrally located within the muscle fibers are the myofibrils, which are long bundles of contractile and regulatory proteins, myofilaments. The myofibrils are surrounded by the sarcoplasm, which contains important organelles such as the mitochondria (oxidative metabolism), sarcoplasmic reticulum (calcium storage for muscle contraction), and the transverse tubules (pathway for action potentials). The apparatus that contracts the muscle cell consists of two types of myofilaments; a thick filament containing the motor protein myosin and a thin filament composed mainly of the protein actin. Myosin and actin filaments are organized longitudinally in the smallest contractile unit of skeletal muscle, the

sarcomere. It is this alignment that gives skeletal muscle its striated appearance.

Skeletal muscle cells are composed of a heterogeneous collection of muscle fiber types and this allows for the wide variety of capabilities and functions that this organ displays. In human skeletal muscle the three fiber types are referred to as type I, type IIa, and type IIx and are named after the respective myosin heavy-chain (MHC) isoform they contain [2]. In rodent skeletal muscle, another fast fiber type exists, type IIb; however, this protein is not expressed in human skeletal muscle. Each muscle fiber type is characterized not only by specific contractile properties, but by unique metabolic properties as well. Due to their slow contraction speed, low fatigability, high mitochondrial content and oxidative capacity, type I fibers are often called slow oxidative fibers. These fibers are adapted for endurance exercise and are very rare in mice. In contrast, type IIx fibers are called fast glycolytic fibers since they are adapted for short and powerful activities due to their fast contraction speed, low mitochondrial content, and high glycolytic capacity. Type IIa fibers, or fast oxidative fibers, express fast MHC but still have a high oxidative capacity. In fact, in human mixed muscles type IIa fibers can have more mitochondria than type I fibers [3].

Skeletal muscle is a highly adaptive tissue that can alter its size and biochemical makeup in response to changes in its environment [4]. Skeletal muscle size and function are tightly regulated by the balance between the processes controlling muscle protein synthesis (MPS) and degradation. When adult skeletal muscle grows, it does so by increasing the size of each

Molecules to Medicine with mTOR
DOI: http://dx.doi.org/10.1016/B978-0-12-802733-2.00025-6

cell much more than the number of cells [5]. This growth is positively influenced by increased loading (resistance exercise) and changes in nutrition and hormone levels. On the other hand, adult muscle loss (atrophy) or wasting is observed in disease states, disuse, and aging [6,7]. Importantly, muscle atrophy is a strong predictor of morbidity and mortality in cardiovascular, musculoskeletal, nervous, renal, and respiratory diseases, as well as cancer [6,8–10]. Independent of disease, muscle loss occurs as we age (sarcopenia), reducing functional independence and quality of life as well as increasing the risk of falls and fractures [11]. As in disease states, muscle strength is one of the best predictors of longevity with aging [12]. Therefore, the maintenance of skeletal muscle mass is essential for improving longevity as well as the quality of life. Extensive research over the past 15 years has improved our understanding of the molecular signaling pathways involved in controlling skeletal muscle development, growth, and maintenance, as well as those involved in muscle loss and disease. As in all cell types, the activity of the mechanistic (or mammalian) target of rapamycin (mTOR) signaling pathway plays an important role in these processes in skeletal muscle and will be the focus of this chapter.

3.2 THE mTOR SIGNALING PATHWAY IN SKELETAL MUSCLE

mTOR is a serine/threonine (Ser/Thr) kinase that senses mechanical stress, hormones, and cellular nutrients and in turn regulates a wide array of cellular processes including cell growth (proliferation and hypertrophy), metabolism, and autophagy. The mTOR kinase is part of two distinctive complexes, mTOR complex 1 (mTORC1) and 2 (mTORC2), which are generally referred to as being rapamyacin-sensitive and rapamyacin-insensitive, respectively. The mTOR complexes both contain: (1) the mTOR subunit; (2) the DEP domain containing mTOR-interacting protein (DEPTOR); (3) the mammalian lethal with sec-13 protein 8 [mLST8; also known as the G-protein beta subunit-like protein (GβL)]; and (4) the Tti1/Tel2 complex. Unique to mTORC1 is the regulatory-associated protein of mammalian target of rapamycin (raptor) and the proline-rich Akt substrate 40 kDa (PRAS40). In contrast, mTORC2 contains the rapamycin-insensitive companion of mTOR (rictor), protein observed with rictor 1 and 2 (protor1/2), and mammalian stress-activated map kinase-interacting protein 1 (mSIN1). The unique proteins in each complex serve to localize mTOR within the cell, and identify downstream targets.

In skeletal muscle, growth factors and mechanical load (resistance exercise) activate mTORC1 through stimulation of its upstream activator Rheb (the ras homolog enriched in brain). In both cases, this is achieved by the movement of the tuberous sclerosis complex 2 (TSC2) away from Rheb; even though the upstream kinase responsible for phosphorylating and moving TSC2 is different [13]. Amino acids (AAs), on the other hand, activate mTORC1 by moving mTOR to Rheb. Since AAs and resistance exercise activate mTORC1 in different ways, they have an additive effect when they are combined [14]. The control of skeletal muscle growth/hypertrophy via regulation of protein synthesis is the best-known downstream process regulated by mTORC1 activation in skeletal muscle. mTORC1 regulates protein synthesis by controlling the translational activity and capacity of muscle by governing the activity and number of ribosomes, both of which are important for muscle growth [15]. mTOR, primarily through its role in the mTORC1 complex, plays an essential role in regulating adult muscle hypertrophy. However, mTORC1 also plays an important role in embryonic muscle development and regeneration. The following sections will highlight the current knowledge of the role of mTORC1 in the development of muscle from the embryo through to muscle growth and the maintenance of healthy muscle in adult life.

3.3 SKELETAL MUSCLE MYOGENESIS— AN OVERVIEW

Myogenesis, the process of generating skeletal muscle, occurs during embryonic development and is recapitulated in adult skeletal muscle following severe injury. During embryonic muscle development, skeletal muscle precursor cells travel from the somite to the limb buds to form the early myoblasts [16]. The migration of these cells and formation of myoblasts are dependent upon the expression of the tyrosine kinase receptor (RTK), c-met, which is in turn regulated by the transcription factor, paired box protein 3 (Pax-3) [17]. Following the formation of myoblasts, these early muscle cells are stimulated to grow and divide (proliferate) by several cell cycle regulatory factors. Myoblast proliferation is dependent on the expression of muscle regulatory factors (MRFs): myogenic determination protein 1 (MyoD) and myogenic factor 5 (Myf-5) [18–21]. Once sufficient myoblasts have formed, they exit the cell cycle and begin to fuse. Myoblast fusion occurs in two distinct phases. During the initial phase, embryonic myoblasts fuse to form primary (nascent) myotubes [22]. The formation of these primary myotubes requires myogenin and the myocyte enhancer factor 2 and ends when the myotubes become innervated and begin contracting [23–27]. At the onset of contraction, fetal myoblasts fuse to each other or with the primary myotubes to form the secondary fibers that predominate in mature skeletal muscle [22]. MRF4 (also known as Myf-6) and the nuclear factor 1X [28] are required during secondary myogenesis, with MRF4 becoming the prominent MRF expressed in adult skeletal muscle [29–31]. Following the completion of myogenesis, muscle fiber number is largely set varying only ~5% throughout adult life [5].

Myogenesis in adult skeletal muscle (regeneration after injury) depends on the activation of muscle-specific stem cells called satellite cells that have the potential to differentiate into new fibers [32]. This type of myogenesis occurs when mature muscle is injured and satellite cells are activated, proliferate, and fuse to each other or existing fibers to repair the tissue and reestablish homeostasis. In adult muscle, satellite cells lie outside the sarcolemma but inside the basal (external) lamina [33–37]. These skeletal muscle precursor cells can be identified due to the expression of a number of distinctive genetic markers such as Pax-3 and Pax-7 [38,39]. Following muscle damage, there is an accumulation of leukocytes and macrophages, which aid in the removal of cellular debris [40,41]. Macrophages also play an important role in muscle regeneration through the production of proinflammatory cytokines, such as interleukin-1 (IL-1), interleukin-6 (IL-6), and tumor necrosis factor-α, which are involved in satellite cell activation [42]. Upon activation, satellite cells re-enter the cell cycle, proliferate to fill the wound site, then differentiate and fuse with the damaged fibers or form new fibers [43].

3.3.1 The Role of mTOR in Myogenesis

mTOR is an essential component of the myogenic process both *in vivo* and *in vitro*. Rapamyacin inhibits muscle cell differentiation *in vitro* in rat L6 [44] and mouse C2C12 myoblasts [45–47], as well as inhibiting skeletal muscle regeneration in rodents [48,49]. To demonstrate the specific requirement of mTORC1, a rapamycin-resistant mutant of mTOR (RR-mTOR) has been used both in C2C12 myoblasts and mice. In the presence of rapamycin, muscle cells expressing the RR-mTOR mutant were able to differentiate and whole muscles regenerated normally, indicating that the effect of rapamycin was specific to its ability to inhibit mTOR [47,49,50]. Complete knockout of mTOR in mouse muscle results in premature death from a severe myopathy, decreased mitochondrial biogenesis, and reduced oxidative capacity [51].

Erbay and Chen [47] were the first to demonstrate that rapamycin inhibited myogenesis at multiple stages, suggesting that mTORC1 was required for multiple processes during muscle differentiation. Since then, mTOR has been shown to regulate myogenesis at two main stages and these are controlled in both a kinase-independent and -dependent manner. In C2C12 cells, myoblasts fuse to form nascent myotubes when expressing a kinase-inactive mTOR (KI-mTOR); however, further maturation does not occur [52]. Similarly, in mice expressing RR/KI-mTOR, primary myofiber formation occurs during regeneration even in the presence of rapamycin. However, in both cases further maturation of these fibers is arrested by rapamycin. In contrast, muscle regeneration occurs normally in the presence of rapamycin in the RR-mTOR mice [49]. These data suggest that initial fusion of myoblasts requires mTOR, but not its kinase activity, whereas maturation of muscle fibers requires an active form of mTORC1 (see Figure 3.1). What is known about each of these processes will be further discussed below.

3.3.1.1 Kinase-Independent mTOR Regulation of Myogenesis

Recent work suggests that insulin-like growth factor 2 (IGF-2) is the key intermediary of kinase-independent mTOR signaling during myoblast fusion. In this scenario, mTORC1 relocalization by AAs [53,54] drives IGF-2 transcription in both C2C12 myoblasts [54] and in the early stage of skeletal muscle regeneration in mice [49]. There are two mechanisms by which mTOR controls the transcription of IGF-2 in a kinase-independent manner, first by activating a muscle-specific enhancer [54] and second, by reducing the post-transcriptional inhibition of IGF-2 by microRNA-125b [55]. The resulting increased expression of IGF-2 activates the phosphatidylinositide 3-kinase (PI3K)/Akt pathway [54,56], which is essential for fusion. Whether there are other ways that fusion is controlled by the mTOR kinase-independent mechanism remains to be determined.

3.3.1.2 Kinase-Dependent mTOR Regulation of Myogenesis

The final stage of fusion and further maturation of myotubes/myofibers is mediated by secreted proteins, which are regulated in an mTOR kinase-dependent manner. Late in fusion, follistatin is secreted from myocytes and is a powerful stimulator of muscle cell differentiation and maturation [57,58]. The expression of follistatin is negatively regulated by histone deacetylases (HDACs) and positively regulated by the activation of the nitric oxide/cGMP signaling [59–61]. The kinase activity of mTOR enhances production of follistatin by regulating the expression of miR-1 in a MyoD-dependent manner [62]. The mTOR-dependent rise in miR-1 suppresses HDAC4, resulting in enhanced transcription of follistatin [62]. The ability of mTOR to control microRNAs might not be limited to mIR-1 [63].

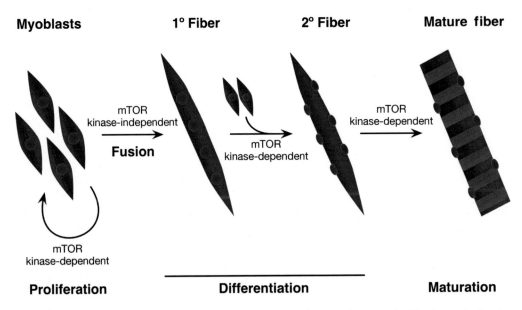

FIGURE 3.1 **Regulation of myogenesis by mTOR.** mTOR signaling controls several stages of skeletal muscle development. The initial proliferation of myoblasts is regulated by the kinase activity of mTOR, however the fusion of myoblasts to form primary (nascent) myofibers is regulated by mTOR in a kinase-independent manner. The final stages of myogenesis, secondary fusion, and the maturation of myofibers, again requires the kinase activity of mTOR.

Sun et al. [62] have shown that a large number of other microRNAs are differentially expressed in a rapamycin-sensitive fashion during myoblast differentiation. How exactly mTOR controls MyoD, and potentially other MRFs, and the subsequent increase in muscle-specific micro-RNAs will most likely be the focus of intense future research.

3.3.1.3 The Role of mTORC1 in Myogenesis

The specific role of mTORC1 in myogenesis has been tested using a muscle-specific knockout of raptor. As with the mTOR-specific knockout animals, these mice develop muscular dystrophy and decreased mitochondria mass and function [64]. Surprisingly though, the *in vitro* studies suggest the opposite affect. *In vitro*, knockdown of raptor augmented myogenic differentiation, whereas overexpression of raptor inhibited myogenic differentiation [65,66]. The only common finding between the knockout in mice and the knockdown in C2C12 cells was the activation of Akt and a shift towards slow myotubes/muscle fibers. It is important to note that the promoter used to generate the raptor knockout in mice is only active in differentiated myocytes [63]. Therefore, further investigation is required using a promoter that deletes raptor prior to myogenic differentiation to determine the role of mTORC1 in early myogenesis. The negative impact of mTORC1 observed *in vitro* appears to act through two main targets, IRS1 and S6K1. Overexpression of raptor in differentiating C2C12 myoblasts results in the disruption

of the insulin receptor substrate 1 (IRS1) and PI3K/Akt pathway through feedback inhibition [66], which is most likely the result of hyperactivated S6K1. The decrease in IRS1/PI3K signaling prevents Akt activation which is required for differentiation [65,67]. These data suggest that both Akt and mTORC1 activation are required for fusion. Developmentally, this is achieved through an mTORC1 kinase-independent activation of IGF-2. By not requiring mTORC1 kinase activity, this prevents S6K1 inhibition of IRS1/PI3K activation. The autocrine/paracrine effect of IRS2 is to activate Akt through a receptor/IRS1/PI3K cascade. Downstream of Akt (see below) mTORC1 kinase activity is activated allowing muscle cell maturation.

3.4 REGULATION OF ADULT SKELETAL MUSCLE MASS

In adult skeletal muscle, muscle cell (fiber) size is tightly regulated by the balance between anabolic and catabolic signaling pathways [6]. The balance of MPS and breakdown (MPB) in healthy adults fluctuates between periods of net positive (e.g., following feeding or mechanical loading) and net negative balance (e.g., fasting or inactivity), that generally result in very small changes in muscle mass over time [68]. In adult muscle, skeletal muscle hypertrophy requires a systematic and progressive change in activity and feeding that over time will increase muscle fiber cross-sectional area with a limited effect on fiber number. The key physiological process involved in muscle hypertrophy is an increase in protein balance; where the rate of MPS exceeds the rate of MPB. Current research suggests that the rate of MPS is more dynamic than MPB, and therefore MPS (translational activity and translational capacity) largely controls muscle fiber size [69]. However, since muscle mass is the largest energy consumer in the body [70–72], a greater muscle mass requires more calories to support the body, even at rest. As a result, signaling to mTORC1 in muscle is controlled not only by positive signals, growth factors, cytokines, mechanical load, and nutrients, but also by negative signals, cellular energy stores, metabolic stress, and oxygen levels. How these stimuli regulate mTORC1 in muscle and how this results in changes in MPS will be discussed in detail in the following sections.

3.4.1 mTORC1 Regulation of Muscle Mass in Response to Nutrients

Since skeletal muscle is the primary storage compartment for AAs, nutrient supply from the diet is a fundamental determinant of muscle cell size. When essential AAs are low, skeletal muscle is broken down to release the AAs necessary to maintain proteostasis throughout the body. This ability to liberate AAs is essential for survival during periods of starvation or nutrient deficiency [73]. In contrast, when AA levels are high, protein synthesis increases so that more AAs can be stored in the proteins that make up the muscle. At the heart of this balance is the mTORC1 signaling pathway [74]. In nutrient-rich states, the activation of mTORC1 stimulates cell growth through the synthesis of protein, lipid, and nucleotides while at the same time inhibiting catabolic processes such as autophagy. Conversely, when nutrients are low mTORC1 is inhibited, slowing down anabolic processes and activating catabolic processes. The availability of glucose and lipids as well as fluctuations in cellular energy (oxygen and adenosine monophosphate (AMP)/adenosine diphosphate (ADP)/adenosine triphosphate (ATP) levels) is communicated to mTORC1 in part by the serine/threonine kinase, AMP-activated protein kinase (AMPK) [74], but also via direct inhibition through AXIN [75]. AMPK is a cellular energy sensor that is activated by an increase in cellular AMP or ADP and serves to block anabolic processes while turning on catabolic processes [74]. In dividing cells, AMPK directly phosphorylates TSC2, preventing its movement away from its target Rheb and decreasing mTORC1 activation [74]. AMPK can also inhibit mTORC1

activity by phosphorylating raptor and increasing its association with the chaperone protein 14-3-3 [74]. However, whether AMPK can shut down mTORC1 in human skeletal muscle remains controversial [76]. In contrast, in a nutrient-rich state AAs are sensed within the cell resulting in the activation of mTORC1 [74]. Even though the complete pathway from AA to mTORC1 remains to be determined, several intermediates that sense AA levels have been identified; the best known of these are the Rag GTPases [74]. In the following sections we will discuss in further detail how the macronutrient intake (protein, carbohydrates, lipids) and more specifically their smaller subunits [AA, glucose, fatty acids (FA)] are involved in the regulation of mTORC1 activity and how this in turn regulates MPS and muscle mass.

3.4.1.1 Protein—Amino Acids

As well as acting as substrate for energy metabolism via anaplerosis and gluconeogenesis, AAs are the building blocks of proteins and play an essential role in the regulation of protein turnover [77]. There are 21 AAs that are required for MPS; 9 of these are referred to as essential AAs (EAA) as they cannot be synthesized in humans and must be supplied in the diet. The anabolic effects of AA feeding on MPS and muscle mass are the result of the EAAs and in particular the branched chain AA leucine that serves as the "trigger" that stimulates MPS [77–82]. Even though leucine triggers the necessary signaling events, all of the AAs are necessary to increase MPS in human muscle [83]. Improvements in muscle mass and strength have been demonstrated in both young and old subjects following chronic AA treatment [84–86]. However, this is believed to be the result of a synergistic effect of the AAs with activity [87] that both signal through the mTORC1 pathway.

The first evidence that growth factors and AA signaled differently was generated in cells lacking TSC2 that were resistant to growth factor withdrawal but still sensitive to AA withdrawal [88,89]. Since then, AAs have been shown to translocate mTORC1 towards membranes where its activator Rheb is located [90,91]. A Raptor—Rheb fusion protein that results in colocalization of mTORC1 and Rheb produces a constitutively active mTORC1 and is insensitive to AA withdrawal [92]. Further, a Raptor—Rheb fusion protein that additionally removes Rheb's ability to localize to membranes, still translocated to membranes in response to AAs and stimulated mTORC1 normally [92]. Together, these studies highlight that AA-mediated activation of mTORC1 requires relocalization of mTORC1 to membrane components that contain Rheb.

The localization and interaction of mTORC1 and Rheb in response to AAs are facilitated by members of the Ras-family of GTP-binding proteins, the Rag

GTPases [90,93]. In mammals the four Rag proteins form heterodimers consisting of one RagA or RagB with one RagC of RagD [90]. AAs stimulate mTORC1 interaction with the Rags through GTP loading RagA/B, which can then directly bind with the mTORC1 subunit raptor [90]. The AA-mediated regulation of Rag activity is currently understood to be under the control of two protein complexes known as the Ragulator and GATOR1 and 2 [92,94]. The Ragulator complex colocalizes with Rheb at the lysosomal membrane and in response to AAs stimulates guanine nucleotide exchange factor (GEF) activity towards RagA/B [92,95]. This activity of Ragulator is controlled by the multiprotein vacuolar H + -adenosine triphosphatase ATPase (v-ATPase) complex [75]. Low AA levels within the lysosomal lumen are detected by v-ATPase and a resulting conformational change alters its interaction with the Ragulator, which in turn blocks GEF activity toward RagA/B [96]. Inhibition of v-ATPase, with either concanamycin A or salicylihalamide A, blocks AA-mediated mTORC1 localization with Rheb and subsequent activation [96]. On the other hand, the GATOR1 complex expresses GTPase activating activity toward the Rag proteins. When intracellular levels of AA are low, the GATOR1 complex holds the RagA/B proteins in an inactive GDP bound state. When AA levels rise, a second GATOR complex

(GATOR2) moves GATOR1 away from RagA/B [94]. Without GATOR1, the Ragulator serves as a GEF towards RagA/B and activates the heterodimer [95]. Beyond Rag signaling, the class III PI3K (Vps34) can also sense AA and in some cell types activates mTORC1 by targeting phospholipase D1 (PLD1) to mTORC1 [97–100]. Overall these studies highlight the important role that the Ragulator, v-ATPase, GATOR, and Vsp34 play in sensing AA levels and recruiting mTORC1 to the lysosome where it can be directly activated by Rheb (see Figure 3.2). However, the majority of these discoveries have been made in HeLa and HEK293 cells. Whether similar processes are involved in AA-mediated activation of mTORC1, MPS, and growth in muscle has yet to be determined.

It is clear that the increase in MPS in response to elevated AAs is dependent upon mTORC1. In animal models, rapamycin inhibits the AA-dependent rise in MPS and mTORC1 activity [82,101–103]. In human studies, ingestion of EAA is associated with an increase in MPS and activation of mTORC1 [104–106] that are completely blocked by rapamycin [107]. As mentioned earlier, very little is known about how AAs activate mTORC1 in adult skeletal muscle. One study has shown an increase in RagB expression in human skeletal muscle following EAA ingestion [108].

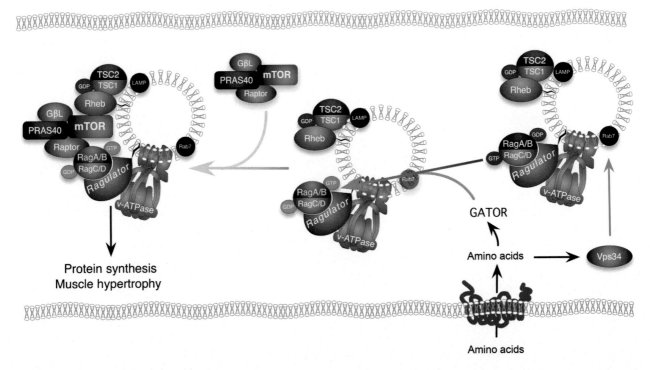

FIGURE 3.2 **AA activation of mTORC1.** AA transporters allow the movement of AA into muscle and an increase in intracellular AAs is sensed by the Vsp34, GATOR, Ragulator, and v-ATPase, resulting in the activation of mTORC1 and subsequently MPS and skeletal muscle hypertrophy. AAs activate the Rag GTPase hetrodimers to translocate mTORC1 in a Ragulator-dependent fashion to membranes that contain Rheb. The Ragulator colocalizes with Rheb and stimulates GEF activity towards RagA/B. However, without inhibition of TSC2, the activation of mTORC1 is limited.

In contrast, in infant pigs AA injection had no effect on RagB expression or the association of RagB with raptor [109]. Beyond the Rag signaling identified in other models, in human skeletal muscle, enhanced mTORC1 activity following AA ingestion is associated with increased expression of several AA transporters [110]. An increase in AA transporters may well be a more potent signal to increase the sensitivity of skeletal muscle to AA availability and this will be an important area of research for the future.

3.4.1.2 Carbohydrates and Fats—Glucose and FA

Carbohydrates and fats in the diet are the main macronutrients that provide the energy to drive cellular processes within skeletal muscle such as growth and contraction. Most of the anabolic processes within muscle require energy in the form of ATP and these are provided from the breakdown of glucose and FA. As mentioned above, AMPK is the most widely studied energy sensor in the cell. However, in AMPK-null cells low-energy conditions still lead to the suppression of mTORC1 activity, highlighting that there are alternative mechanisms that regulate mTORC1 activity in response to the energy status of the cell [111]. One such mechanism involves the protein AXIN that is recruited to the v-ATPase in low-glucose conditions and from there can coordinate the inactivation of the Ragulator and the local activation of AMPK [75]. This serves to directly switch mTORC1 off while simultaneously turning on AMPK. This could explain why it takes more protein to activate mTORC1 and protein synthesis in skeletal muscle during periods of caloric restriction [112]. However, our understanding of the role of glucose and FA in regulating mTORC1 activity in healthy skeletal muscle mass is quite limited and will require further research in the future.

3.4.2 mTORC1 Regulation of Muscle Mass in Response to Growth Factors

Growth factors are naturally occurring proteins or steroid hormones that regulate cell growth and differentiation. Because of their central role in growth,

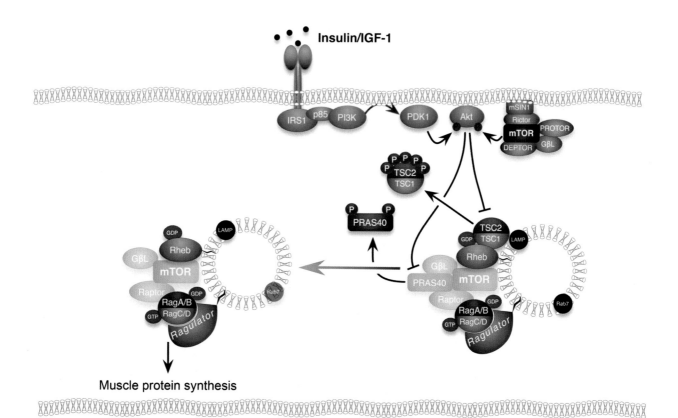

FIGURE 3.3 **Insulin/IGF-1 stimulation of mTORC1.** In this canonical growth factor signaling pathway insulin/IGF-1 binding to its receptor results in the phosphorylation and recruitment of IRS1. IRS1 recruits phosphoinositol-3 kinase (p85 and PI3K) to the cell membrane, which increases PIP3 levels in the membrane resulting in the recruitment of PDK1 and Akt/PKB to the membrane where it is activated by phosphorylation by PDK1 and mTORC2. PKB phosphorylates and moves the TSC1/2 away from Rheb, and phosphorylates PRAS40 which removes it from mTOR. The removal of TSC1/2 and PRAS40 activates the mTORC1 associated with Rheb to increase MPS.

growth factors such as insulin, IGF-1, and testosterone are direct regulators of mTORC1 activity [113–117]. When nutrient levels are abundant, growth factors such as insulin and IGF-1 are preserved in the circulation and promote cell growth and survival via the PI3K/protein kinase B (PKB; also known as Akt) signaling pathway (see Figure 3.3). The binding of insulin or IGF-1 to their RTK results in the recruitment of IRS1/2, which in turn brings PI3K to the plasma membrane [118]. PI3K phosphorylates the third position of the membrane phospholipid phosphoinositol (4,5)-bisphosphate (PIP2) resulting in the production of phosphoinositol (3,4,5)-trisphosphate (PIP3). PIP3 then serves a docking point for proteins containing pleckstrin homology domains such as Akt and the phosphoinositol-dependent kinase-1 (PDK1) at the membrane [119]. The activation of Akt is achieved by the direct phosphorylation by PDK1 at Thr308 and mTORC2 at Ser473 and this activates mTORC1 by phosphorylating and removing TSC2 from Rheb [120,121] and phosphorylating and removing PRAS40 from mTOR. In the following subsections the role of insulin, IGF-1, and testosterone in MPS and muscle hypertrophy and the potential role of mTORC1 in response to these specific growth factors will be highlighted.

3.4.2.1 Insulin

Even though insulin's role in regulating protein synthesis and its ability to stimulate mTORC1 activity have been extensively studied in a range of cell types, its ability to stimulate MPS is still debated. The debate centers on a series of contradictory studies over the past 20 years. These studies differ in the animal/cell model used, the method of insulin administration or removal and/or the level/availability of AAs. In cell culture, insulin treatment of L6 rat myoblasts [122] and nutrient-deprived C2C12 myotubes [123] increases protein synthesis in association with an increase in mTORC1 signaling. In the C2C12 myotubes, the activation of mTORC1 was shown to be the result of canonical PI3K/Akt signaling [123]. Further, using either a PI3K (wortmannin) or mTORC1 (rapamycin) inhibitor prevented the increase in protein synthesis [122]. Overall, these cell culture studies support the role of insulin in promoting MPS by stimulating mTORC1 activity via the PI3K/Akt signaling cascade.

The story with regards to insulin's action in human and animal studies is not as straightforward. A severe reduction in circulating insulin, as observed in untreated type 1 diabetes mellitus in humans and induced diabetes in animals, results in muscle wasting [124,125]. Restoring insulin levels prevents muscle protein loss and increases MPS [124,126–128]. However, in nondiabetic people/animals insulin infusion (to levels above the postabsorptive state) either systemically or via a nearby artery has been shown to have no effect on MPS [129–134]. One possible explanation for this finding is the reduction in plasma AAs observed with insulin infusion [135,136]. In animals [137–140] and humans [141–144] an increase in MPS in response to insulin infusion occurs when the AA pool was maintained [143,144]. Accordingly, when AAs are increased by infusing them together with insulin there is a more consistently observed stimulation of MPS [145,146]. This is not surprising considering the impact that AAs have on mTORC1 location and the additive effect of AA and growth factors discussed above. More recently, the important role of insulin-stimulated vasodilation and the resulting increased delivery of AAs have also been highlighted in the activation of mTORC1 in human skeletal muscle [147].

The activity of mTORC1 targets in animals in response to insulin is consistently increased. Primarily, the phosphorylation of S6K1 [113,136,148] and 4E-BP1 [136,138,148], and the release of eIF-4E from the inactive 4E-BP1/eIF-4E complex [138,148] are all stimulated by insulin. Further, rapamycin and the PI3K inhibitor wortmannin prevent S6K1 activation and MPS in response to insulin [113,148]. The Akt inhibitor (MK-2206) also prevents the insulin-induced activation of S6K1 and 4E-BP1 in skeletal muscle, suggesting that insulin increases mTORC1 activity and protein synthesis through the canonical growth factor pathway. Even though mTORC1 is activated by insulin in muscle, the role of blood flow, AA availability, glucose transport, and metabolic shifts on mTORC1 and MPS needs to be further clarified.

3.4.2.2 Insulin-Like Growth Factor 1

IGF-1 plays an essential role in mediating skeletal muscle development and growth throughout the lifespan [149]. IGF-1 is primarily produced by the liver; however, it is also produced locally in skeletal muscle in response to resistance exercise and muscle injury [150]. IGF-1 produces its effects on skeletal muscle by binding primarily to its receptor (IGF-1R) largely in a paracrine/autocrine fashion. Systemic overexpression of IGF-1 at levels 1.5 times the normal value resulted in a modest increase in skeletal muscle mass [151], whereas skeletal muscle-specific overexpression of IGF-1 resulted in a doubling of muscle mass [152,153]. Transgenic overexpression of IGF-1 has also been shown to attenuate age-related loss of skeletal muscle mass and strength in mice [153]. In cell culture, treatment of muscle cells with IGF-1 also results in an increase in protein synthesis and subsequently myotube hypertrophy [114,154,155]. IGF-1 infusion in humans promotes muscle protein anabolism by directly stimulating MPS and as with insulin this can be augmented by AAs [156].

Mechanistically, Rommel et al. [114] were the first to show that C2C12 cells treated with IGF-1 hypertrophied concomitant with increased phosphorylation of Akt, S6K1, and 4E-BP1. Rapamycin treatment completely blocked the activation of S6K1 and 4E-BP1 and partly inhibited myotube hypertrophy in these cells, confirming the role of mTORC1 in mediating IGF-1's actions on MPS. However, inhibition of PI3K completely blocks myotube hypertrophy as well as phosphorylation of Akt, S6K1, and 4E-BP1. Together, these data highlight that mTORC1 mediates IGF-1-stimulated MPS without affecting MPB, whereas PI3K/Akt both activates MPS (through mTORC1) and blocks MPB (in an mTORC1-independent manner) resulting in hypertrophy.

3.4.2.3 Testosterone

Testosterone is the primary steroid hormone in males but is also found at lower levels in females. Since the isolation and production of testosterone, researchers have been able to demonstrate that administration of this hormone and its derivatives increases MPS, muscle mass, and strength in humans [157–161]. Conversely, reducing circulating testosterone is associated with a reduction in muscle mass [157,162–164]. The use of testosterone and its derivatives is highly effective in rescuing the loss of muscle mass in muscle-wasting conditions in humans [157–159,165,166] and animals [167,168]. Although the effect of testosterone loss on muscle mass and physiology has long been appreciated, the signaling underlying testosterone-mediated MPS and muscle hypertrophy is less clear. Reducing testosterone is associated with a decline in both circulating and intramuscular IGF-1 [169,170], which can be reversed with testosterone treatment [170], suggesting that testosterone may signal through IGF-1 to mTORC1.

Testosterone treatment in C2C12, L6, and primary muscle cells results in myotube hypertrophy concomitant with mTORC1 activation [117,171,172] and hypertrophy is blocked by rapamycin [171,172]. Additionally, the loss of testosterone following castration in mice reduces the phosphorylation of mTOR, S6K1, and 4E-BP1, whereas treatment with the testosterone derivative nandrolone increases the activity of these proteins beyond control levels [117]. Unlike insulin/IGF-1, the effect of testosterones on Akt is equivocal. Testosterone loss has been shown to both increase and decrease Akt phosphorylation [173,174]. However, since testosterone treatment increases IGF-1 levels it is possible that a systemic effect of testosterone is missed in some models [117,175]. Interestingly, in culture it appears that high concentrations of testosterone are needed to activate Akt even

though hypertrophy can occur at low levels [171,117]. In spite of this discrepancy, testosterone-induced myotube hypertrophy and mTORC1 activation are completely blocked by PI3K inhibition [171,172,176]. Therefore, testosterone appears to stimulate muscle hypertrophy and activate mTORC1 through PI3K kinase/Akt.

3.4.3 mTORC1 Regulation of Muscle Mass in Response to Mechanical Load/Exercise

Exercise is one of the most potent regulators of skeletal muscle mass. Exercise regulates MPS and the magnitude of this effect is dependent upon the training status, type of exercise, intensity, and duration of the exercise bout. For example, an acute bout of resistance exercise in untrained people can stimulate MPS 40–150% above resting levels and the chronic application of resistance training is highly associated with skeletal muscle hypertrophy [177,178]. Over the last decade or so there has been a vast array of studies describing the role of mTORC1 in the regulation of MPS and skeletal muscle mass in response to different modes of mechanical load/exercise. The molecular mechanisms that regulate mTORC1 activity in skeletal muscle in response to the different types of mechanical stimuli are starting to emerge. The following sections will highlight the current evidence for whether mTORC1 activity is regulated in response to different mechanical stimuli and the known upstream effectors or sensors of mechanical stimuli that regulate mTORC1 activity.

3.4.3.1 Resistance Exercise/Training

Resistance exercise acutely increases MPS and repeated bouts over 8–12 weeks result in increased skeletal muscle mass and strength [179–184]. Several human studies have shown that MPS rates can be elevated for 48 h following an acute resistance exercise bout [185–190]. The first study to show an association between an increase in mTORC1 activity and enhanced skeletal muscle mass compared the effect of 6 weeks of electrically stimulated contractions in the hindlimb of rats with mTORC1 activation 6 h following the first exercise bout. The strong correlation between these two measures ($r^2 = 0.998$) suggested that mTORC1 played a role in load-induced muscle growth [191]. Similarly, the increase in muscle mass and strength following 14 weeks of high-intensity training in humans correlated with mTORC1 activity 30 min following the first resistance exercise bout [192]. These two studies highlight that the activation of mTORC1 signaling in response to resistance exercise is similar in humans and rodent skeletal muscle and suggest that mTORC1 activation plays a key role in skeletal muscle hypertrophy. Since the initial reports, further support for the prominent role of mTORC1 was provided by studies using chronic mechanical overload induced by removing large weightbearing muscles to drive hypertrophy in the remaining synergists of rodents. Not only is the resulting skeletal muscle hypertrophy associated with an increase in mTORC1 signaling [5,193–197], Bodine et al. [198] showed that rapamycin could completely block overload-induced hypertrophy. Further, using mice with a skeletal muscle-specific RR-mTOR, Hornberger's group showed that overload-induced muscle fiber hypertrophy occurred normally in these mice even in the presence of rapamycin [5]. Lastly, muscle that lacks raptor does not hypertrophy following either 7 or 28 days of overload [199]. Collectively, these data show that mTORC1 is required for overload-induced muscle hypertrophy. However, whether mTORC1 is essential for resistance exercise-mediated muscle hypertrophy (that is more intermittent in nature) has yet to be determined.

As would be expected, human studies have shown that increases in MPS are associated with increased mTORC1 activity [200–205]. Rodent studies have highlighted that mTORC1 signaling is elevated in response to concentric, eccentric, and isometric contractions [206–209], as well as in isolated muscles subjected to passive stretching *ex vivo* [210–212]. These increases in mTORC1 activity in response to acute mechanical loading have also been shown to be necessary for the increase in MPS. Rapamycin treatment prior to *in vivo* resistance exercise in rats and *ex vivo* passive stretching in the EDL muscle of mice has been shown to completely block MPS and mTORC1 signaling [210,213]. Furthermore, rapamycin blocks the initial (first 1–2 h) increase in mTORC1 activity and MPS in humans following resistance exercise [214]. In summary, studies in humans and rodents clearly show that resistance exercise is sufficient to activate mTORC1 and this is necessary for the increase in muscle hypertrophy in response to resistance training. However, whether rapamycin completely blocks an intermittent program of resistance exercise (as performed in humans) has yet to be determined.

3.4.3.2 Regulation of mTORC1 Activity in Response to Mechanical Stimuli

Given that mTORC1 activity is required for the increase in MPS and skeletal muscle hypertrophy in response to resistance exercise, researchers have endeavored to determine how mTORC1 senses mechanical load. Traditionally it has been thought that mTORC1 activity in response to mechanical load is sensed and regulated by the IGF/PI3K/Akt signaling cassette and that Akt phosphorylation of TSC2 activates mTORC1. However, recent evidence suggests mTORC1 activation following loading occurs independent of canonical PI3K/Akt signaling. The following paragraphs will present the current knowledge relating to the upstream regulators of mTORC1 activity in response to mechanical load.

As mentioned above, IGF-1 increases MPS and activates mTORC1 through the PI3K/Akt signaling pathway. IGF-1 expression is elevated in response to mechanical loading [215–217] and originally it was suggested that the increase in IGF-1 expression stimulated mTORC1 and MPS via PI3K/Akt. However, a number of recent studies have challenged this theory. First, mTORC1 activity can be increased independently of PI3K/Akt signaling following mechanical loading [206,210,218,219]. Second, the association between IRS1 and PI3K is decreased (suggesting lower PI3K activity) following resistance exercise [220]. Third, knockin/out studies of members of the IGF-/PI3K/Akt pathway in rodents have demonstrated that mTORC1 activation by loading is not affected by the absence of this pathway. For example, skeletal muscle-specific expression on a dominant-negative IGF-1 receptor had no effect on load-induced muscle hypertrophy and mTORC1 signaling [195]. Additionally, inhibition of Akt through the use of wortmanin or knockout of Akt1 does not affect load-induced mTORC1 activation [194,210]. Finally, muscle-specific deletion of the phosphatase and tensin homolog protein, which should increase PI3K/Akt activity, had no effect on mTORC1 activity or muscle hypertrophy in response to chronic loading [220]. Together, these studies suggest that the IGF-1/PI3K/Akt signaling is not required for mTORC1 signaling in response to mechanical load.

If PI3K/Akt signaling is not required following resistance exercise, then how is mTORC1 activated? Jacobs et al. [13] showed that, like growth factors, resistance exercise still removed TSC2 from Rheb and increased mTORC1 association. Furthermore, the movement of TSC2 was associated with its hyperphosphorylation; however, this hyperphosphorylation occurred in the absence of Akt phosphorylation. The authors showed that TSC2 phosphorylation occurred at RxRxx* sites within TSC2, but have yet to identify the kinase responsible (see Figure 3.4). Future research is therefore required to determine the kinase that regulates TSC2 phosphorylation in response to mechanical stimuli in skeletal muscle.

A potential allosteric activator of mTORC1 is the glycerophospholipid phosphatidic acid (PA) that is believed to bind to the FKBP12-rapamycin-binding domain of mTOR [212]. Exogenous elevations in PA result in increased mTORC1 activity in skeletal

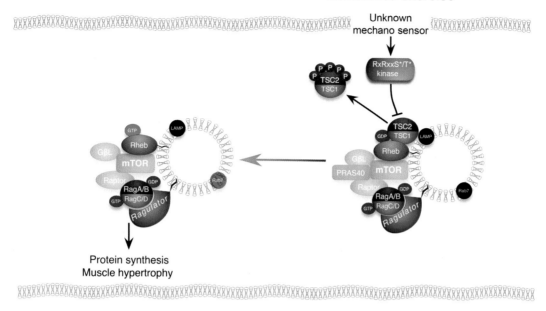

FIGURE 3.4 **Resistance exercise/mechanical load stimulation of mTORC1.** The activation of mTORC1 by resistance exercise is believed to be sensed by an unknown membrane mechanosensor. Like insulin/IGF-1, resistance exercise stimulates mTORC1 activity, MPS, and skeletal muscle hypertrophy via the removal of TSC1/2 from Rheb. However, unlike insulin/IGF-1 signaling, resistance exercise phosphorylates TSC1/2 by an unknown RxRxxS*/T* kinase.

muscle, whereas inhibiting the synthesis of PA prevents the increase in mTORC1 activity in response to loading [212,219]. The level of PA within a cell is regulated by enzymes that alter PA synthesis or degradation. In skeletal muscle, two enzymes that synthesize PA, PLD, and the diacylglycerol kinases (DGK), regulate PA levels in the cell. In mouse skeletal muscle, PLD1 activity increased in response to passive stretching. Additionally, the activation of PLD1 was associated with an increase in PA levels and mTORC1 activity [211]. Although the use of the PLD inhibitor, 1-butanol, blunted mTORC1 activation following passive stretching and eccentric contractions in mouse skeletal muscle [211,219], a more specific PLD inhibitor, 5-fluoro-2-indolyl des-chlorohalopemide, had no effect on either PA levels or mTORC1 signaling [212]. In contrast, stretch increases DGKζ levels and this corresponded with an increase in activity of DGKζ and mTORC1 signaling [212]. Transient overexpression of DGKζ in skeletal muscle also increases mTORC1 activity and induces muscle hypertrophy, whereas the stretch-induced increase in PA and mTORC1 activity is reduced in DGKζ knockout in mice [212]. Overall, these studies propose that DGKζ and PA are involved in mTORC1 activation, increased MPS, and skeletal muscle hypertrophy as a result of resistance exercise.

3.4.4 Coordination of mTORC1 Signaling by Multiple Stimuli in Skeletal Muscle

As discussed above, MPS is controlled by several stimuli including nutrients, growth factors/hormones, and mechanical load/resistance exercise that all converge on mTORC1. Below we discuss how these stimuli can work together to alter mTORC1 activity and increase in MPS and muscle hypertrophy.

3.4.4.1 Amino Acids and Growth Factors

The infusion of AAs together with insulin or IGF-1 in human subjects clearly shows that MPS is greater than either insulin or AA alone [145,146,221]. Similarly, in pigs the MPS effects of insulin and AA are additive until the maximal rate of MPS is reached [139]. In a rat model of type 1 diabetes, that typically has ~80% lower circulating insulin levels, mTORC1 activity is significantly reduced following oral leucine administration when compared to control rats [82]. Together, these studies suggest that insulin can augment AA effects on MPS and that insulin levels must at least be maintained at fasting values to sustain AA-mediated increases in mTORC1 activity. Given that insulin activates mTORC1 through the movement of TSC2 away from Rheb and AAs activate mTORC1 through a Rag-dependent movement of mTOR to Rheb, the molecular biology supports a clear mechanism underlying this additive effect.

The anabolic effects of exogenous testosterone and AA on their own have been heavily investigated, however the possible combined effect of these on MPS is limited to two human studies. The first in young healthy men where oxandrolone and AA alone increased MPS 44% and 94%, respectively, and the combination of oxandrolone and AA had an additive effect increasing MPS 120% [222]. By contrast, in healthy older men prolonged testosterone administration only increases MPS in the fasted state and no additive was observed with AA [223]. Further research is clearly needed to determine the interaction and mechanism (mTORC1 activity) underlying the effect of exogenous testosterone and AA on MPS.

3.4.4.2 Amino Acids and Mechanical Load

Following resistance exercise in the fasted state, muscle protein balance remains negative despite a substantial rise in MPS due to a simultaneous rise in MPB. However, when AAs are provided, either orally or intravenously, muscle protein balance becomes positive due to both an enhanced increase in MPS and a reduction in the rise in MPB [224,225]. As MPS rates remain elevated up to 48 h following an acute bout of resistance exercise it has been proposed that AA ingestion during this period may further augment the increase in MPS following exercise [188]. As discussed earlier, EAAs are the only AAs that appear to be required to stimulate MPS and EAA can increase MPS in a dose–response manner not only at rest but also show an additive effect following exercise [226–228]. Moore and colleagues determined that the optimal amount of protein needed in a young man after an exercise session was 0.25 g/kg body weight. Cermak et al. [14] best illustrated the additive effect of AA and resistance exercise in a meta-analysis that concluded that in both young and old subjects AA significantly increased muscle growth compared with exercise alone. Taken together, these studies highlight that the combination of resistance exercise and ingestion of AAs, specifically EAAs, following exercise results in an overall increase in muscle protein balance due to an augmented increase in MPS and a reduction in MPB.

As with growth factors, the fact that resistance exercise removes TSC2 from Rheb and AAs increase the association of Rheb with mTORC1 means that the molecular mechanism underlying the additive effect is

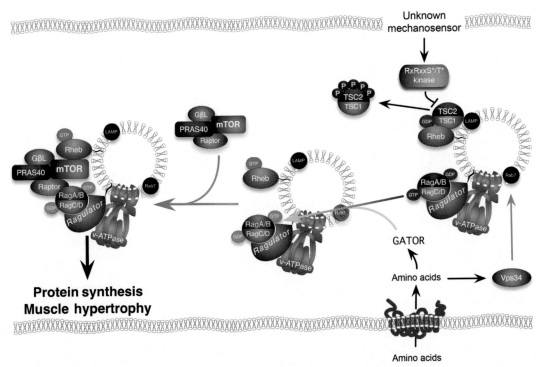

FIGURE 3.5 **Combination of AAs and resistance exercise on mTORC1 activity.** AAs and resistance exercise converge on mTORC1 via different mechanisms, by delivering mTORC1 to Rheb and removing TSC1/2 and activating Rheb, respectively. Therefore, the combination of AA and resistance exercise results in an additive effect on mTORC1 activity, MPS, and skeletal muscle hypertrophy.

clear (see Figure 3.5). As expected, the activation of mTORC1 in novice weightlifters is increased by AA ingestion [186,203,229,230]. By contrast, resistance-trained individuals do not show elevated activation of mTORC1 and MPS following AA ingestion, suggesting that the training status of an individual can alter the ability to stimulate mTORC1 through either the mechanical or the AA pathways [228]. Together, the current research strongly suggests that AA ingestion following resistance exercise can stimulate mTORC1 and MPS more than when these stimuli are given on their own. However, it should be noted that the magnitude of the additive effect of AA ingestion and resistance exercise on mTORC1 activity and MPS may be affected by many factors such as the type, quantity, and timing of protein ingested, and the training status of the individual.

3.4.4.3 Growth Factors and Mechanical Load

Studies into the requirement of insulin for mechanical load-induced muscle hypertrophy suggest that only minimal insulin is needed to support the anabolic effects of mechanical stimuli. Skeletal muscle hypertrophy following overload and MPS following resistance exercise in mildly diabetic rats (moderate reduction in insulin levels) are normal [231,232]. In liver IGF-1-deficient mice ($\sim 80\%$ reduction in IGF-1 serum levels), 16 weeks of resistance exercise resulted in similar increases in muscle strength when compared to control mice [233]. These animals do produce IGF-1 locally, suggesting that hypertrophy may be achieved by local IGF-1 production rather than systemic IGF-1 levels. However, this hypothesis is completely refuted by the findings that: (1) in mice that express a skeletal muscle-specific dominant-negative IGF-1 receptor, where neither insulin nor IGF-1 can activate mTORC1, load-induced mTORC1 activation and muscle hypertrophy occur normally [195] and (2) in the 48 h following resistance exercise there is no increase in either the activation of the IGF-1 receptor or PI3K. Since it appears that both growth factors, through PI3K/Akt, and resistance exercise, through a yet to be identified RxRxxS*/T* kinase, activate mTORC1 by shuttling TSC2 away from Rheb, it makes sense that growth factors and resistance exercise do not have an additive effect on mTORC1 or muscle hypertrophy.

Resistance exercise is known to increase the concentration of serum testosterone and the scale of this rise is dependent on volume, intensity, and rest intervals in-between sets [234]. This postexercise increase in testosterone was believed to play a role in hypertrophy; however, recent work has shown that the exercise-induced increases in testosterone in humans are not important for the increase in MPS, hypertrophy, or strength following resistance training [235,236]. Normal physiological exogenous testosterone levels do contribute to baseline muscle mass. Furthermore, supraphysiological doses of testosterone, when combined with resistance training, produce greater increases in muscle mass and strength than either supraphysiological testosterone treatment or resistance training alone [237]. In fact, studies comparing women and men or men on androgen inhibitors have clearly shown that load-induced increases in muscle mass and strength occur independent of testosterone levels but that the absolute muscle mass is lower with less testosterone [179,238]. Together, these studies suggest that testosterone and resistance exercise have an additive effect on muscle mass, suggesting that they increase muscle mass through parallel pathways. The role that mTORC1 signaling may play in mediating muscle growth signals from the combination of testosterone treatment and resistance exercise combined is currently unknown.

3.5 OUTSTANDING QUESTIONS

Even though we have clearly learned a great deal about how mTORC1 regulates muscle development, growth, and regeneration, there are still a myriad of outstanding questions.

1. During myogenesis, mTORC1 exerts both kinase-dependent and -independent effects. How mTOR in a kinase-independent manner regulates MyoD and the role of mIR-1 and other microRNAs is an outstanding question.
2. There are emerging suggestions that the proportion of carbohydrates and fats in the diet may affect mTORC1 activity and potentially skeletal muscle mass. The effect of high-carbohydrate/low-fat or high-fat/low carbohydrate (ketogenic) diets on skeletal muscle mTORC1 activity, as well as MPS and muscle mass, needs to be determined.
3. The role of Akt in PI3K-induced activation of mTORC1 has clearly been established. However, the identity of the RxRxxS*/T* kinase that is activated by resistance exercise and phosphorylates and moves TSC2 away from Rheb remains unknown and this is one of the most important questions in the field.
4. Even though it is clear that mTORC1 is required for the acute rise in MPS after resistance exercise, whether rapamycin inhibits MPS at later time points and muscle growth as a result of an intermittent exercise program remains to be determined.

5. The ability of testosterone to increase muscle mass has been recognized for almost 100 years. However, how testosterone activates mTORC1 and drives an increase in MPS and muscle size remains an important outstanding question. Furthermore, unlike insulin/IGF-1, testosterone is additive with resistance exercise. This suggests that testosterone might have mTORC1-independent effects that differ from other growth factors.

3.6 SUMMARY

mTORC1 is clearly essential in muscle development and growth from the embryo through to muscle regeneration and hypertrophy in adult skeletal muscle. mTORC1 activity and its subsequent regulation of MPS and muscle mass can be stimulated through a variety of signaling cascades that coordinate growth signals from nutrients, growth factors, and mechanical load/resistance exercise with the metabolic state of the muscle. Even though this chapter has focused on the role of mTORC1 in myogenesis and MPS and muscle hypertrophy in adult skeletal mass, mTORC1 regulates many other processes in skeletal muscle such as metabolism, autophagy, and protein degradation. We know a lot about the role of mTORC1 in skeletal muscle; however, many important questions still remain. As rapamycin is proposed as a treatment to increase lifespan in model organisms [239–241], we need to understand the impact this will have on the function of muscle since muscle strength is predictive of longevity in humans.

References

[1] Collins CA, Partridge TA. Self-renewal of the adult skeletal muscle satellite cell. Cell Cycle 2005;4:1338.

[2] Talmadge RJ. Myosin heavy chain isoform expression following reduced neuromuscular activity: potential regulatory mechanisms. Muscle Nerve 2000;23:661–79.

[3] Russell AP, Feilchenfeldt J, Schreiber S, Praz M, Crettenand A, Gobelet C, et al. Endurance training in humans leads to fiber type-specific increases in levels of peroxisome proliferator-activated receptor-gamma coactivator-1 and peroxisome proliferator-activated receptor-alpha in skeletal muscle. Diabetes 2003;52:2874–81.

[4] Fluck M, Hoppeler H. Molecular basis of skeletal muscle plasticity—from gene to form and function. Rev Physiol Biochem Pharmacol 2003;146:159–216.

[5] Goodman CA, Frey JW, Mabrey DM, Jacobs BL, Lincoln HC, You JS, et al. The role of skeletal muscle mTOR in the regulation of mechanical load-induced growth. J Physiol 2011;589:5485–501.

[6] Glass DJ. Signalling pathways that mediate skeletal muscle hypertrophy and atrophy. Nat Cell Biol 2003;5:87–90.

[7] Glass DJ. Skeletal muscle hypertrophy and atrophy signaling pathways. Int J Biochem Cell Biol 2005;37:1974–84.

[8] Jagoe RT, Goldberg AL. What do we really know about the ubiquitin-proteasome pathway in muscle atrophy? Curr Opin Clin Nutr Metab Care 2001;4:183–90.

[9] Sartorelli V, Fulco M. Molecular and cellular determinants of skeletal muscle atrophy and hypertrophy. Sci STKE 2004;2004:re11.

[10] Goldspink DF, Garlick PJ, McNurlan MA. Protein turnover measured *in vivo* and *in vitro* in muscles undergoing compensatory growth and subsequent denervation atrophy. Biochem J 1983;210:89–98.

[11] Mahoney J, Sager M, Dunham NC, Johnson J. Risk of falls after hospital discharge. J Am Geriatr Soc 1994;42:269–74.

[12] Srikanthan P, Karlamangla AS. Muscle mass index as a predictor of longevity in older adults. Am J Med 2014;127:547–53.

[13] Jacobs BL, You JS, Frey JW, Goodman CA, Gundermann DM, Hornberger TA. Eccentric contractions increase the phosphorylation of tuberous sclerosis complex-2 (TSC2) and alter the targeting of TSC2 and the mechanistic target of rapamycin to the lysosome. J Physiol 2013;591:4611–20.

[14] Cermak NM, Res PT, de Groot LC, Saris WH, van Loon LJ. Protein supplementation augments the adaptive response of skeletal muscle to resistance-type exercise training: a meta-analysis. Am J Clin Nutr 2012;96:1454–64.

[15] Mahoney SJ, Dempsey JM, Blenis J. Cell signaling in protein synthesis ribosome biogenesis and translation initiation and elongation. Prog Mol Biol Transl Sci 2009;90:53–107.

[16] Muralikrishna B, Dhawan J, Rangaraj N, Parnaik VK. Distinct changes in intranuclear lamin A/C organization during myoblast differentiation. J Cell Sci 2001;114:4001–11.

[17] Bladt F, Riethmacher D, Isenmann S, Aguzzi A, Birchmeier C. Essential role for the c-met receptor in the migration of myogenic precursor cells into the limb bud. Nature 1995;376:768–71.

[18] Tapscott SJ, Davis RL, Thayer MJ, Cheng PF, Weintraub H, Lassar AB. MyoD1: a nuclear phosphoprotein requiring a Myc homology region to convert fibroblasts to myoblasts. Science 1988;242:405–11.

[19] Braun T, Rudnicki MA, Arnold HH, Jaenisch R. Targeted inactivation of the muscle regulatory gene Myf-5 results in abnormal rib development and perinatal death. Cell 1992;71:369–82.

[20] Kablar B, Krastel K, Ying C, Asakura A, Tapscott SJ, Rudnicki MA. MyoD and Myf-5 differentially regulate the development of limb versus trunk skeletal muscle. Development 1997;124:4729–38.

[21] Rudnicki MA, Braun T, Hinuma S, Jaenisch R. Inactivation of MyoD in mice leads to up-regulation of the myogenic HLH gene Myf-5 and results in apparently normal muscle development. Cell 1992;71:383–90.

[22] Zhang M, McLennan IS. During secondary myotube formation, primary myotubes preferentially absorb new nuclei at their ends. Dev Dyn 1995;204:168–77.

[23] Hasty P, Bradley A, Morris JH, Edmondson DG, Venuti JM, Olson EN, et al. Muscle deficiency and neonatal death in mice with a targeted mutation in the myogenin gene. Nature 1993;364:501–6.

[24] Nabeshima Y, Hanaoka K, Hayasaka M, Esumi E, Li S, Nonaka I. Myogenin gene disruption results in perinatal lethality because of severe muscle defect. Nature 1993;364:532–5.

[25] Venuti JM. Myogenin is required for late but not early aspects of myogenesis during mouse development. J Cell Biol 1995;128:563–76.

[26] Gossett LA, Kelvin DJ, Sternberg EA, Olson EN. A new myocyte-specific enhancer-binding factor that recognizes a conserved element associated with multiple muscle-specific genes. Mol Cell Biol 1989;9:5022–33.

[27] Molkentin JD, Olson EN. Combinatorial control of muscle development by basic helix-loop-helix and MADS-box transcription factors. Proc Natl Acad Sci USA 1996;93:9366–73.

[28] Messina G, Biressi S, Monteverde S, Magli A, Cassano M, Perani L, et al. Nfix regulates fetal-specific transcription in developing skeletal muscle. Cell 2010;140:554–66.

[29] Hinterberger TJ, Sassoon DA, Rhodes SJ, Konieczny SF. Expression of the muscle regulatory factor MRF 4 during somite and skeletal myofiber development. Dev Biol 1991;147:144–56.

[30] Zhang W, Behringer RR, Olson EN. Inactivation of the myogenic bHLH gene MRF4 results in up-regulation of myogenin and rib anomalies. Genes Dev 1995;9:1388–99.

[31] Olson EN, Arnold HH, Rigby PWJ, Wold BJ. Know your neighbors: three minireview phenotypes in null mutants of the myogenic bHLH Gene MRF4. Cell 1996;85:1–4.

[32] Charge SB, Rudnicki MA. Cellular and molecular regulation of muscle regeneration. Physiol Rev 2004;84:209–38.

[33] Cameron-Smith D. Exercise and skeletal muscle gene expression. Clin Exp Pharmacol Physiol 2002;29:209–13.

[34] Wozniak AC, Kong J, Bock E, Pilipowicz O, Anderson JE. Signaling satellite-cell activation in skeletal muscle: markers, models, stretch, and potential alternate pathways. Muscle Nerve 2005;31:283–300.

[35] Mauro A. Satellite cell of skeletal muscle fibers. J Cell Biol 1961;9:493–5.

[36] Bischoff R, Heintz C. Enhancement of skeletal muscle regeneration. Dev Dyn 1994;201:41–54.

[37] Yablonka-Reuveni Z. Development and postnatal regulation of adult myoblasts. Microsc Res Tech 1995;30:366–80.

[38] Relaix F, Montarras D, Zaffran S, Gayraud-Morel B, Rocancourt D, Tajbakhsh S, et al. Pax3 and Pax7 have distinct and overlapping functions in adult muscle progenitor cells. J Cell Biol 2006;172:91–102.

[39] Seale P, Sabourin LA, Girgis-Gabardo A, Mansouri A, Gruss P, Rudnicki MA. Pax7 is required for the specification of myogenic satellite cells. Cell 2000;102:777–86.

[40] Tidball JG. Inflammatory processes in muscle injury and repair. Am J Physiol Regul Integr Comp Physiol 2005;288:R345–53.

[41] Toumi H, Best TM. The inflammatory response: friend or enemy for muscle injury? Br J Sports Med 2003;37:284–6.

[42] Mantovani A, Sica A, Sozzani S, Allavena P, Vecchi A, Locati M. The chemokine system in diverse forms of macrophage activation and polarization. Trends Immunol 2004;25:677–86.

[43] Hawke TJ, Garry DJ. Myogenic satellite cells: physiology to molecular biology. J Appl Physiol 2001;91:534–51.

[44] Coolican SA, Samuel DS, Ewton DZ, McWade FJ, Florini JR. The mitogenic and myogenic actions of insulin-like growth factors utilize distinct signaling pathways. J Biol Chem 1997;272:6653–62.

[45] Cuenda A, Cohen P. Stress-activated protein kinase-2/p38 and a rapamycin-sensitive pathway are required for C2C12 myogenesis. J Biol Chem 1999;274:4341–6.

[46] Conejo R, Valverde AM, Benito M, Lorenzo M. Insulin produces myogenesis in C2C12 myoblasts by induction of NF-kappaB and downregulation of AP-1 activities. J Cell Physiol 2001;186:82–94.

[47] Erbay E, Chen J. The mammalian target of rapamycin regulates C2C12 myogenesis via a kinase-independent mechanism. J Biol Chem 2001;276:36079–82.

[48] Pallafacchina G, Calabria E, Serrano AL, Kalhovde JM, Schiaffino S. A protein kinase B-dependent and rapamycin-sensitive pathway controls skeletal muscle growth but not fiber type specification. Proc Natl Acad Sci USA 2002;99:9213–18.

[49] Ge Y, Wu AL, Warnes C, Liu J, Zhang C, Kawasome H, et al. mTOR regulates skeletal muscle regeneration *in vivo* through kinase-dependent and kinase-independent mechanisms. Am J Physiol Cell Physiol 2009;297:C1434–44.

[50] Shu L, Zhang X, Houghton PJ. Myogenic differentiation is dependent on both the kinase function and the N-terminal sequence of mammalian target of rapamycin. J Biol Chem 2002;277:16726–32.

[51] Risson V, Mazelin L, Roceri M, Sanchez H, Moncollin V, Corneloup C, et al. Muscle inactivation of mTOR causes metabolic and dystrophin defects leading to severe myopathy. J Cell Biol 2009;187:859–74.

[52] Park IH, Chen J. Mammalian target of rapamycin (mTOR) signaling is required for a late-stage fusion process during skeletal myotube maturation. J Biol Chem 2005;280:32009–17.

[53] Yoon MS, Chen J. Distinct amino acid-sensing mTOR pathways regulate skeletal myogenesis. Mol Biol Cell 2013;24:3754–63.

[54] Erbay E, Park IH, Nuzzi PD, Schoenherr CJ, Chen J. IGF-II transcription in skeletal myogenesis is controlled by mTOR and nutrients. J Cell Biol 2003;163:931–6.

[55] Ge Y, Sun Y, Chen J. IGF-II is regulated by microRNA-125b in skeletal myogenesis. J Cell Biol 2011;192:69–81.

[56] Jiang BH, Aoki M, Zheng JZ, Li J, Vogt PK. Myogenic signaling of phosphatidylinositol 3-kinase requires the serine-threonine kinase Akt/protein kinase B. Proc Natl Acad Sci USA 1999;96:2077–81.

[57] Amthor H, Nicholas G, McKinnell I, Kemp CF, Sharma M, Kambadur R, et al. Follistatin complexes Myostatin and antagonises Myostatin-mediated inhibition of myogenesis. Dev Biol 2004;270:19–30.

[58] Lee SJ, McPherron AC. Regulation of myostatin activity and muscle growth. Proc Natl Acad Sci USA 2001;98:9306–11.

[59] Iezzi S, Di Padova M, Serra C, Caretti G, Simone C, Maklan E, et al. Deacetylase inhibitors increase muscle cell size by promoting myoblast recruitment and fusion through induction of follistatin. Dev Cell 2004;6:673–84.

[60] Minetti GC, Colussi C, Adami R, Serra C, Mozzetta C, Parente V, et al. Functional and morphological recovery of dystrophic muscles in mice treated with deacetylase inhibitors. Nat Med 2006;12:1147–50.

[61] Pisconti A, Brunelli S, Di Padova M, De Palma C, Deponti D, Baesso S, et al. Follistatin induction by nitric oxide through cyclic GMP: a tightly regulated signaling pathway that controls myoblast fusion. J Cell Biol 2006;172:233–44.

[62] Sun Y, Ge Y, Drnevich J, Zhao Y, Band M, Chen J. Mammalian target of rapamycin regulates miRNA-1 and follistatin in skeletal myogenesis. J Cell Biol 2010;189:1157–69.

[63] Ge Y, Chen J. Mammalian target of rapamycin (mTOR) signaling network in skeletal myogenesis. J Biol Chem 2012;287:43928–35.

[64] Bentzinger CF, Romanino K, Cloetta D, Lin S, Mascarenhas JB, Oliveri F, et al. Skeletal muscle-specific ablation of raptor, but not of rictor, causes metabolic changes and results in muscle dystrophy. Cell Metab 2008;8:411–24.

[65] Jaafar R, Zeiller C, Pirola L, Di Grazia A, Naro F, Vidal H, et al. Phospholipase D regulates myogenic differentiation through the activation of both mTORC1 and mTORC2 complexes. J Biol Chem 2011;286:22609–21.

[66] Ge Y, Yoon MS, Chen J. Raptor and Rheb negatively regulate skeletal myogenesis through suppression of insulin receptor substrate 1 (IRS1). J Biol Chem 2011;286:35675–82.

[67] Julien LA, Carriere A, Moreau J, Roux PP. mTORC1-activated S6K1 phosphorylates Rictor on threonine 1135 and regulates mTORC2 signaling. Mol Cell Biol 2010;30:908–21.

[68] Phillips S, Baar K, Lewis N. Nutrition for weight and resistance training. In: Lanham-New SA, Stear S, Shirreffs S, Collins A, editors. Nutrition society textbook on sport and exercise nutrition. Oxford: Wiley-Blackwell; 2011. p. 120–33.

[69] Greenhaff PL, Karagounis LG, Peirce N, Simpson EJ, Hazell M, Layfield R, et al. Disassociation between the effects of amino acids and insulin on signaling, ubiquitin ligases, and protein turnover in human muscle. Am J Physiol Endocrinol Metab 2008;295:E595—604.

[70] Benedict FG. The factors affecting normal basal metabolism. Proc Natl Acad Sci USA 1915;1:105—9.

[71] Sollner-Webb B, Tower J. Transcription of cloned eukaryotic ribosomal RNA genes. Annu Rev Biochem 1986;55:801—30.

[72] Huang J, Manning BD. The TSC1-TSC2 complex: a molecular switchboard controlling cell growth. Biochem J 2008;412:179—90.

[73] Wagenmakers AJ. Muscle amino acid metabolism at rest and during exercise: role in human physiology and metabolism. Exerc Sport Sci Rev 1998;26:287—314.

[74] Jewell JL, Guan KL. Nutrient signaling to mTOR and cell growth. Trends Biochem Sci 2013;38:233—42.

[75] Zhang CS, Jiang B, Li M, Zhu M, Peng Y, Zhang YL, et al. The lysosomal v-ATPase-Regulator complex is a common activator for AMPK and mTORC1, acting as a switch between catabolism and anabolism. Cell Metab 2014;20:526—40.

[76] Lundberg TR, Fernandez-Gonzalo R, Tesch PA. Exercise-induced AMPK activation does not interfere with muscle hypertrophy in response to resistance training in men. J Appl Physiol (1985) 2014;116:611—20.

[77] Kimball SR, Jefferson LS. New functions for amino acids: effects on gene transcription and translation. Am J Clin Nutr 2006;83:500S—7S.

[78] Atherton PJ, Smith K, Etheridge T, Rankin D, Rennie MJ. Distinct anabolic signalling responses to amino acids in C2C12 skeletal muscle cells. Amino Acids 2010;38:1533—9.

[79] Freudenberg A, Petzke KJ, Klaus S. Comparison of high-protein diets and leucine supplementation in the prevention of metabolic syndrome and related disorders in mice. J Nutr Biochem 2012;23:1524—30.

[80] Tipton KD, Gurkin BE, Matin S, Wolfe RR. Nonessential amino acids are not necessary to stimulate net muscle protein synthesis in healthy volunteers. J Nutr Biochem 1999;10:89—95.

[81] Volpi E, Kobayashi H, Sheffield-Moore M, Mittendorfer B, Wolfe RR. Essential amino acids are primarily responsible for the amino acid stimulation of muscle protein anabolism in healthy elderly adults. Am J Clin Nutr 2003;78:250—8.

[82] Anthony JC, Yoshizawa F, Anthony TG, Vary TC, Jefferson LS, Kimball SR. Leucine stimulates translation initiation in skeletal muscle of postabsorptive rats via a rapamycin-sensitive pathway. J Nutr 2000;130:2413—19.

[83] Churchward-Venne TA, Burd NA, Mitchell CJ, West DW, Philp A, Marcotte GR, et al. Supplementation of a suboptimal protein dose with leucine or essential amino acids: effects on myofibrillar protein synthesis at rest and following resistance exercise in men. J Physiol 2012;590:2751—65.

[84] Borsheim E, Bui QU, Tissier S, Kobayashi H, Ferrando AA, Wolfe RR. Effect of amino acid supplementation on muscle mass, strength and physical function in elderly. Clin Nutr 2008;27:189—95.

[85] Dillon EL, Sheffield-Moore M, Paddon-Jones D, Gilkison C, Sanford AP, Casperson SL, et al. Amino acid supplementation increases lean body mass, basal muscle protein synthesis, and insulin-like growth factor-I expression in older women. J Clin Endocrinol Metab 2009;94:1630—7.

[86] Tieland M, van de Rest O, Dirks ML, van der Zwaluw N, Mensink M, van Loon LJ, et al. Protein supplementation improves physical performance in frail elderly people: a randomized, double-blind, placebo-controlled trial. J Am Med Dir Assoc 2012;13:720—6.

[87] Rennie MJ, Bohe J, Smith K, Wackerhage H, Greenhaff P. Branched-chain amino acids as fuels and anabolic signals in human muscle. J Nutr 2006;136:264S—8S.

[88] Roccio M, Bos JL, Zwartkruis FJ. Regulation of the small GTPase Rheb by amino acids. Oncogene 2006;25:657—64.

[89] Smith EM, Finn SG, Tee AR, Browne GJ, Proud CG. The tuberous sclerosis protein TSC2 is not required for the regulation of the mammalian target of rapamycin by amino acids and certain cellular stresses. J Biol Chem 2005;280:18717—27.

[90] Sancak Y, Peterson TR, Shaul YD, Lindquist RA, Thoreen CC, Bar-Peled L, et al. The Rag GTPases bind raptor and mediate amino acid signaling to mTORC1. Science 2008;320:1496—501.

[91] Long X, Ortiz-Vega S, Lin Y, Avruch J. Rheb binding to mammalian target of rapamycin (mTOR) is regulated by amino acid sufficiency. J Biol Chem 2005;280:23433—6.

[92] Sancak Y, Bar-Peled L, Zoncu R, Markhard AL, Nada S, Sabatini DM. Ragulator-Rag complex targets mTORC1 to the lysosomal surface and is necessary for its activation by amino acids. Cell 2010;141:290—303.

[93] Kim E, Goraksha-Hicks P, Li L, Neufeld TP, Guan KL. Regulation of TORC1 by Rag GTPases in nutrient response. Nat Cell Biol 2008;10:935—45.

[94] Bar-Peled L, Chantranupong L, Cherniack AD, Chen WW, Ottina KA, Grabiner BC, et al. A tumor suppressor complex with GAP activity for the Rag GTPases that signal amino acid sufficiency to mTORC1. Science 2013;340:1100—6.

[95] Bar-Peled L, Schweitzer LD, Zoncu R, Sabatini DM. Ragulator is a GEF for the rag GTPases that signal amino acid levels to mTORC1. Cell 2012;150:1196—208.

[96] Zoncu R, Bar-Peled L, Efeyan A, Wang S, Sancak Y, Sabatini DM. mTORC1 senses lysosomal amino acids through an inside-out mechanism that requires the vacuolar H(+)-ATPase. Science 2011;334:678—83.

[97] Byfield MP, Murray JT, Backer JM. hVps34 is a nutrient-regulated lipid kinase required for activation of p70 S6 kinase. J Biol Chem 2005;280:33076—82.

[98] Nobukuni T, Joaquin M, Roccio M, Dann SG, Kim SY, Gulati P, et al. Amino acids mediate mTOR/raptor signaling through activation of class 3 phosphatidylinositol 3OH-kinase. Proc Natl Acad Sci USA 2005;102:14238—43.

[99] Xu L, Salloum D, Medlin PS, Saqcena M, Yellen P, Perrella B, et al. Phospholipase D mediates nutrient input to mammalian target of rapamycin complex 1 (mTORC1). J Biol Chem 2011;286:25477—86.

[100] Yoon MS, Du G, Backer JM, Frohman MA, Chen J. Class III PI-3-kinase activates phospholipase D in an amino acid-sensing mTORC1 pathway. J Cell Biol 2011;195:435—47.

[101] Kimball SR, Jefferson LS, Nguyen HV, Suryawan A, Bush JA, Davis TA. Feeding stimulates protein synthesis in muscle and liver of neonatal pigs through an mTOR-dependent process. Am J Physiol Endocrinol Metab 2000;279:E1080—7.

[102] Suryawan A, Jeyapalan AS, Orellana RA, Wilson FA, Nguyen HV, Davis TA. Leucine stimulates protein synthesis in skeletal muscle of neonatal pigs by enhancing mTORC1 activation. Am J Physiol Endocrinol Metab 2008;295:E868—75.

[103] Vary TC, Anthony JC, Jefferson LS, Kimball SR, Lynch CJ. Rapamycin blunts nutrient stimulation of eIF4G, but not PKCepsilon phosphorylation, in skeletal muscle. Am J Physiol Endocrinol Metab 2007;293:E188—96.

[104] Fujita S, Dreyer HC, Drummond MJ, Glynn EL, Cadenas JG, Yoshizawa F, et al. Nutrient signalling in the regulation of human muscle protein synthesis. J Physiol 2007;582:813—23.

[105] Glynn EL, Fry CS, Drummond MJ, Timmerman KL, Dhanani S, Volpi E, et al. Excess leucine intake enhances muscle anabolic signaling but not net protein anabolism in young men and women. J Nutr 2010;140:1970—6.

[106] Cuthbertson D, Smith K, Babraj J, Leese G, Waddell T, Atherton P, et al. Anabolic signaling deficits underlie amino acid resistance of wasting, aging muscle. FASEB J 2005;19: 422–4.

[107] Dickinson JM, Fry CS, Drummond MJ, Gundermann DM, Walker DK, Glynn EL, et al. Mammalian target of rapamycin complex 1 activation is required for the stimulation of human skeletal muscle protein synthesis by essential amino acids. J Nutr 2011;141:856–62.

[108] Carlin MB, Tanner RE, Agergaard J, Jalili T, McClain DA, Drummond MJ. Skeletal muscle Ras-related GTP binding B mRNA and protein expression is increased after essential amino acid ingestion in healthy humans. J Nutr 2014;144: 1409–14.

[109] Suryawan A, Davis TA. The abundance and activation of mTORC1 regulators in skeletal muscle of neonatal pigs are modulated by insulin, amino acids, and age. J Appl Physiol (1985) 2010;109:1448–54.

[110] Drummond MJ, Glynn EL, Fry CS, Timmerman KL, Volpi E, Rasmussen BB. An increase in essential amino acid availability upregulates amino acid transporter expression in human skeletal muscle. Am J Physiol Endocrinol Metab 2010;298: E1011–18.

[111] Kalender A, Selvaraj A, Kim SY, Gulati P, Brule S, Viollet B, et al. Metformin, independent of AMPK, inhibits mTORC1 in a rag GTPase-dependent manner. Cell Metabolism 2010;11: 390–401.

[112] Areta JL, Burke LM, Camera DM, West DW, Crawshay S, Moore DR, et al. Reduced resting skeletal muscle protein synthesis is rescued by resistance exercise and protein ingestion following short-term energy deficit. Am J Physiol Endocrinol Metab 2014;306:E989–97.

[113] Dardevet D, Sornet C, Vary T, Grizard J. Phosphatidylinositol 3-kinase and p70 s6 kinase participate in the regulation of protein turnover in skeletal muscle by insulin and insulin-like growth factor I. Endocrinology 1996;137:4087–94.

[114] Rommel C, Bodine SC, Clarke BA, Rossman R, Nunez L, Stitt TN, et al. Mediation of IGF-1-induced skeletal myotube hypertrophy by PI(3)K/Akt/mTOR and PI(3)K/Akt/GSK3 pathways. Nat Cell Biol 2001;3:1009–13.

[115] Glass DJ. PI3 kinase regulation of skeletal muscle hypertrophy and atrophy. Curr Top Microbiol Immunol 2010;346:267–78.

[116] Schiaffino S, Mammucari C. Regulation of skeletal muscle growth by the IGF1-Akt/PKB pathway: insights from genetic models. Skelet Muscle 2011;1:4.

[117] White JP, Gao S, Puppa MJ, Sato S, Welle SL, Carson JA. Testosterone regulation of Akt/mTORC1/FoxO3a signaling in skeletal muscle. Mol Cell Endocrinol 2013;365:174–86.

[118] Shepherd PR, Withers DJ, Siddle K. Phosphoinositide 3-kinase: the key switch mechanism in insulin signalling. Biochem J 1998;333(Pt 3):471–90.

[119] Alessi DR, Cohen P. Mechanism of activation and function of protein kinase B. Curr Opin Genet Dev 1998;8:55–62.

[120] Alessi DR, James SR, Downes CP, Holmes AB, Gaffney PR, Reese CB, et al. Characterization of a 3-phosphoinositide-dependent protein kinase which phosphorylates and activates protein kinase Balpha. Curr Biol 1997;7:261–9.

[121] Sarbassov DD, Guertin DA, Ali SM, Sabatini DM. Phosphorylation and regulation of Akt/PKB by the rictor-mTOR complex. Science 2005;307:1098–101.

[122] Kimball SR, Horetsky RL, Jefferson LS. Signal transduction pathways involved in the regulation of protein synthesis by insulin in L6 myoblasts. Am J Physiol 1998;274:C221–8.

[123] Shen WH, Boyle DW, Wisniowski P, Bade A, Liechty EA. Insulin and IGF-I stimulate the formation of the eukaryotic initiation factor 4F complex and protein synthesis in C2C12 myotubes independent of availability of external amino acids. J Endocrinol 2005;185:275–89.

[124] Charlton M, Nair KS. Protein metabolism in insulin-dependent diabetes mellitus. J Nutr 1998;128:323S–7S.

[125] Price SR, Bailey JL, Wang X, Jurkovitz C, England BK, Ding X, et al. Muscle wasting in insulinopenic rats results from activation of the ATP-dependent, ubiquitin-proteasome proteolytic pathway by a mechanism including gene transcription. J Clin Invest 1996;98:1703–8.

[126] Flaim KE, Copenhaver ME, Jefferson LS. Effects of diabetes on protein synthesis in fast- and slow-twitch rat skeletal muscle. Am J Physiol 1980;239:E88–95.

[127] Grzelkowska K, Dardevet D, Balage M, Grizard J. Involvement of the rapamycin-sensitive pathway in the insulin regulation of muscle protein synthesis in streptozotocin-diabetic rats. J Endocrinol 1999;160:137–45.

[128] Pain VM, Garlick PJ. Effect of streptozotocin diabetes and insulin treatment on the rate of protein synthesis in tissues of the rat in vivo. J Biol Chem 1974;249:4510–14.

[129] Denne SC, Liechty EA, Liu YM, Brechtel G, Baron AD. Proteolysis in skeletal muscle and whole body in response to euglycemic hyperinsulinemia in normal adults. Am J Physiol 1991;261:E809–14.

[130] Heslin MJ, Newman E, Wolf RF, Pisters PW, Brennan MF. Effect of hyperinsulinemia on whole body and skeletal muscle leucine carbon kinetics in humans. Am J Physiol 1992;262: E911–18.

[131] Moller-Loswick AC, Zachrisson H, Hyltander A, Korner U, Matthews DE, Lundholm K. Insulin selectively attenuates breakdown of nonmyofibrillar proteins in peripheral tissues of normal men. Am J Physiol 1994;266:E645–52.

[132] Gelfand RA, Barrett EJ. Effect of physiologic hyperinsulinemia on skeletal muscle protein synthesis and breakdown in man. J Clin Invest 1987;80:1–6.

[133] Louard RJ, Fryburg DA, Gelfand RA, Barrett EJ. Insulin sensitivity of protein and glucose metabolism in human forearm skeletal muscle. J Clin Invest 1992;90:2348–54.

[134] Tessari P, Inchiostro S, Biolo G, Vincenti E, Sabadin L. Effects of acute systemic hyperinsulinemia on forearm muscle proteolysis in healthy man. J Clin Invest 1991;88:27–33.

[135] Fukagawa NK, Minaker KL, Young VR, Rowe JW. Insulin dose-dependent reductions in plasma amino acids in man. Am J Physiol 1986;250:E13–17.

[136] Urschel KL, Escobar J, McCutcheon LJ, Geor RJ. Insulin infusion stimulates whole-body protein synthesis and activates the upstream and downstream effectors of mechanistic target of rapamycin signaling in the gluteus medius muscle of mature horses. Domest Anim Endocrinol 2014;47:92–100.

[137] Kimball SR, Jefferson LS, Fadden P, Haystead TA, Lawrence Jr JC. Insulin and diabetes cause reciprocal changes in the association of eIF-4E and PHAS-I in rat skeletal muscle. Am J Physiol 1996;270:C705–9.

[138] Kimball SR, Jurasinski CV, Lawrence Jr JC, Jefferson LS. Insulin stimulates protein synthesis in skeletal muscle by enhancing the association of eIF-4E and eIF-4G. Am J Physiol 1997;272:C754–9.

[139] O'Connor PM, Bush JA, Suryawan A, Nguyen HV, Davis TA. Insulin and amino acids independently stimulate skeletal muscle protein synthesis in neonatal pigs. Am J Physiol Endocrinol Metab 2003;284:E110–19.

[140] Suryawan A, O'Connor PM, Bush JA, Nguyen HV, Davis TA. Differential regulation of protein synthesis by amino acids and insulin in peripheral and visceral tissues of neonatal pigs. Amino Acids 2009;37:97–104.

[141] Bennet WM, Rennie MJ. Protein anabolic actions of insulin in the human body. Diabet Med 1991;8:199–207.

[142] Biolo G, Declan Fleming RY, Wolfe RR. Physiologic hyperinsulinemia stimulates protein synthesis and enhances transport of selected amino acids in human skeletal muscle. J Clin Invest 1995;95:811–19.

[143] Fujita S, Rasmussen BB, Cadenas JG, Grady JJ, Volpi E. Effect of insulin on human skeletal muscle protein synthesis is modulated by insulin-induced changes in muscle blood flow and amino acid availability. Am J Physiol Endocrinol Metab 2006;291:E745–54.

[144] Rasmussen BB, Fujita S, Wolfe RR, Mittendorfer B, Roy M, Rowe VL, et al. Insulin resistance of muscle protein metabolism in aging. FASEB J 2006;20:768–9.

[145] Bennet WM, Connacher AA, Scrimgeour CM, Jung RT, Rennie MJ. Euglycemic hyperinsulinemia augments amino acid uptake by human leg tissues during hyperaminoacidemia. Am J Physiol 1990;259:E185–94.

[146] Hillier TA, Fryburg DA, Jahn LA, Barrett EJ. Extreme hyperinsulinemia unmasks insulin's effect to stimulate protein synthesis in the human forearm. Am J Physiol 1998;274:E1067–74.

[147] Timmerman KL, Lee JL, Dreyer HC, Dhanani S, Glynn EL, Fry CS, et al. Insulin stimulates human skeletal muscle protein synthesis via an indirect mechanism involving endothelial-dependent vasodilation and mammalian target of rapamycin complex 1 signaling. J Clin Endocrinol Metab 2010;95:3848–57.

[148] Azpiazu I, Saltiel AR, DePaoli-Roach AA, Lawrence JC. Regulation of both glycogen synthase and PHAS-I by insulin in rat skeletal muscle involves mitogen-activated protein kinase-independent and rapamycin-sensitive pathways. J Biol Chem 1996;271:5033–9.

[149] Clemmons DR. Role of IGF-I in skeletal muscle mass maintenance. Trends Endocrinol Metab 2009;20:349–56.

[150] Frost RA, Lang CH. Multifaceted role of insulin-like growth factors and mammalian target of rapamycin in skeletal muscle. Endocrinol Metab Clin North Am 2012;41:297–322, vi.

[151] Mathews LS, Hammer RE, Behringer RR, D'Ercole AJ, Bell GI, Brinster RL, et al. Growth enhancement of transgenic mice expressing human insulin-like growth factor I. Endocrinology 1988;123:2827–33.

[152] Coleman ME, DeMayo F, Yin KC, Lee HM, Geske R, Montgomery C, et al. Myogenic vector expression of insulin-like growth factor I stimulates muscle cell differentiation and myofiber hypertrophy in transgenic mice. J Biol Chem 1995;270:12109–16.

[153] Barton-Davis ER, Shoturma DI, Musaro A, Rosenthal N, Sweeney HL. Viral mediated expression of insulin-like growth factor I blocks the aging-related loss of skeletal muscle function. Proc Natl Acad Sci USA 1998;95:15603–7.

[154] Vandenburgh HH, Karlisch P, Shansky J, Feldstein R. Insulin and IGF-I induce pronounced hypertrophy of skeletal myofibers in tissue culture. Am J Physiol 1991;260:C475–84.

[155] Musaro A, Rosenthal N. Maturation of the myogenic program is induced by postmitotic expression of insulin-like growth factor I. Mol Cell Biol 1999;19:3115–24.

[156] Fryburg DA, Jahn LA, Hill SA, Oliveras DM, Barrett EJ. Insulin and insulin-like growth factor-I enhance human skeletal muscle protein anabolism during hyperaminoacidemia by different mechanisms. J Clin Invest 1995;96:1722–9.

[157] Katznelson L, Finkelstein JS, Schoenfeld DA, Rosenthal DI, Anderson EJ, Klibanski A. Increase in bone density and lean body mass during testosterone administration in men with acquired hypogonadism. J Clin Endocrinol Metab 1996;81:4358–65.

[158] Bhasin S, Storer TW, Berman N, Yarasheski KE, Clevenger B, Phillips J, et al. Testosterone replacement increases fat-free mass and muscle size in hypogonadal men. J Clin Endocrinol Metab 1997;82:407–13.

[159] Brodsky IG, Balagopal P, Nair KS. Effects of testosterone replacement on muscle mass and muscle protein synthesis in hypogonadal men—a clinical research center study. J Clin Endocrinol Metab 1996;81:3469–75.

[160] Urban RJ, Bodenburg YH, Gilkison C, Foxworth J, Coggan AR, Wolfe RR, et al. Testosterone administration to elderly men increases skeletal muscle strength and protein synthesis. Am J Physiol 1995;269:E820–6.

[161] Bhasin S, Woodhouse L, Casaburi R, Singh AB, Bhasin D, Berman N, et al. Testosterone dose–response relationships in healthy young men. Am J Physiol Endocrinol Metab 2001;281:E1172–81.

[162] Rogozkin V. Metabolic effects of anabolic steroid on skeletal muscle. Med Sci Sports 1979;11:160–3.

[163] Antonio J, Wilson JD, George FW. Effects of castration and androgen treatment on androgen-receptor levels in rat skeletal muscles. J Appl Physiol (1985) 1999;87:2016–19.

[164] Axell AM, MacLean HE, Plant DR, Harcourt LJ, Davis JA, Jimenez M, et al. Continuous testosterone administration prevents skeletal muscle atrophy and enhances resistance to fatigue in orchidectomized male mice. Am J Physiol Endocrinol Metab 2006;291:E506–16.

[165] Tenover JS. Effects of testosterone supplementation in the aging male. J Clin Endocrinol Metab 1992;75:1092–8.

[166] Kenny AM, Prestwood KM, Gruman CA, Marcello KM, Raisz LG. Effects of transdermal testosterone on bone and muscle in older men with low bioavailable testosterone levels. J Gerontol A Biol Sci Med Sci 2001;56:M266–72.

[167] van Balkom RH, Dekhuijzen PN, van der Heijden HF, Folgering HT, Fransen JA, van Herwaarden CL. Effects of anabolic steroids on diaphragm impairment induced by methylprednisolone in emphysematous hamsters. Eur Respir J 1999;13:1062–9.

[168] Van Balkom RH, Dekhuijzen PN, Folgering HT, Veerkamp JH, Van Moerkerk HT, Fransen JA, et al. Anabolic steroids in part reverse glucocorticoid-induced alterations in rat diaphragm. J Appl Physiol (1985) 1998;84:1492–9.

[169] Grinspoon S, Corcoran C, Lee K, Burrows B, Hubbard J, Katznelson L, et al. Loss of lean body and muscle mass correlates with androgen levels in hypogonadal men with acquired immunodeficiency syndrome and wasting. J Clin Endocrinol Metab 1996;81:4051–8.

[170] Mauras N, Hayes V, Welch S, Rini A, Helgeson K, Dokler M, et al. Testosterone deficiency in young men: marked alterations in whole body protein kinetics, strength, and adiposity. J Clin Endocrinol Metab 1998;83:1886–92.

[171] Wu Y, Bauman WA, Blitzer RD, Cardozo C. Testosterone-induced hypertrophy of L6 myoblasts is dependent upon Erk and mTOR. Biochem Biophys Res Commun 2010;400:679–83.

[172] Basualto-Alarcon C, Jorquera G, Altamirano F, Jaimovich E, Estrada M. Testosterone signals through mTOR and androgen receptor to induce muscle hypertrophy. Med Sci Sports Exerc 2013;45:1712–20.

[173] Ibebunjo C, Eash JK, Li C, Ma Q, Glass DJ. Voluntary running, skeletal muscle gene expression, and signaling inversely regulated by orchidectomy and testosterone replacement. Am J Physiol Endocrinol Metab 2011;300:E327–40.

[174] Haren MT, Siddiqui AM, Armbrecht HJ, Kevorkian RT, Kim MJ, Haas MJ, et al. Testosterone modulates gene expression pathways regulating nutrient accumulation, glucose metabolism and protein turnover in mouse skeletal muscle. Int J Androl 2011;34:55–68.

[175] Yin HN, Chai JK, Yu YM, Shen CA, Wu YQ, Yao YM, et al. Regulation of signaling pathways downstream of IGF-I/insulin by androgen in skeletal muscle of glucocorticoid-treated rats. J Trauma 2009;66:1083−90.

[176] Deane CS, Hughes DC, Sculthorpe N, Lewis MP, Stewart CE, Sharples AP. Impaired hypertrophy in myoblasts is improved with testosterone administration. J Steroid Biochem Mol Biol 2013;138:152−61.

[177] Burd NA, Tang JE, Moore DR, Phillips SM. Exercise training and protein metabolism: influences of contraction, protein intake, and sex-based differences. J Appl Physiol (1985) 2009;106:1692−701.

[178] Tesch PA. Skeletal muscle adaptations consequent to long-term heavy resistance exercise. Med Sci Sports Exerc 1988;20: S132−4.

[179] Cureton KJ, Collins MA, Hill DW, McElhannon Jr FM. Muscle hypertrophy in men and women. Med Sci Sports Exerc 1988;20:338−44.

[180] Kraemer WJ, Nindl BC, Ratamess NA, Gotshalk LA, Volek JS, Fleck SJ, et al. Changes in muscle hypertrophy in women with periodized resistance training. Med Sci Sports Exerc 2004;36:697−708.

[181] Leger B, Cartoni R, Praz M, Lamon S, Deriaz O, Crettenand A, et al. Akt signalling through GSK-3beta, mTOR and Foxo1 is involved in human skeletal muscle hypertrophy and atrophy. J Physiol 2006;576:923−33.

[182] McCall GE, Byrnes WC, Dickinson A, Pattany PM, Fleck SJ. Muscle fiber hypertrophy, hyperplasia, and capillary density in college men after resistance training. J Appl Physiol (1985) 1996;81:2004−12.

[183] Staron RS, Malicky ES, Leonardi MJ, Falkel JE, Hagerman FC, Dudley GA. Muscle hypertrophy and fast fiber type conversions in heavy resistance-trained women. Eur J Appl Physiol Occup Physiol 1990;60:71−9.

[184] Thalacker-Mercer A, Stec M, Cui X, Cross J, Windham S, Bamman M. Cluster analysis reveals differential transcript profiles associated with resistance training-induced human skeletal muscle hypertrophy. Physiol Genom 2013;45:499−507.

[185] Chesley A, MacDougall JD, Tarnopolsky MA, Atkinson SA, Smith K. Changes in human muscle protein synthesis after resistance exercise. J Appl Physiol (1985) 1992;73:1383−8.

[186] Dreyer HC, Drummond MJ, Pennings B, Fujita S, Glynn EL, Chinkes DL, et al. Leucine-enriched essential amino acid and carbohydrate ingestion following resistance exercise enhances mTOR signaling and protein synthesis in human muscle. Am J Physiol Endocrinol Metab 2008;294:E392−400.

[187] MacDougall JD, Gibala MJ, Tarnopolsky MA, MacDonald JR, Interisano SA, Yarasheski KE. The time course for elevated muscle protein synthesis following heavy resistance exercise. Can J Appl Physiol 1995;20:480−6.

[188] Phillips SM, Tipton KD, Aarsland A, Wolf SE, Wolfe RR. Mixed muscle protein synthesis and breakdown after resistance exercise in humans. Am J Physiol 1997;273:E99−107.

[189] Trappe TA, White F, Lambert CP, Cesar D, Hellerstein M, Evans WJ. Effect of ibuprofen and acetaminophen on postexercise muscle protein synthesis. Am J Physiol Endocrinol Metab 2002;282:E551−6.

[190] Welle S, Bhatt K, Thornton CA. Stimulation of myofibrillar synthesis by exercise is mediated by more efficient translation of mRNA. J Appl Physiol (1985) 1999;86:1220−5.

[191] Baar K, Esser K. Phosphorylation of p70(S6k) correlates with increased skeletal muscle mass following resistance exercise. Am J Physiol 1999;276:C120−7.

[192] Terzis G, Georgiadis G, Stratakos G, Vogiatzis I, Kavouras S, Manta P, et al. Resistance exercise-induced increase in muscle mass correlates with p70S6 kinase phosphorylation in human subjects. Eur J Appl Physiol 2008;102:145−52.

[193] Hornberger TA, McLoughlin TJ, Leszczynski JK, Armstrong DD, Jameson RR, Bowen PE, et al. Selenoprotein-deficient transgenic mice exhibit enhanced exercise-induced muscle growth. J Nutr 2003;133:3091−7.

[194] Miyazaki M, McCarthy JJ, Fedele MJ, Esser KA. Early activation of mTORC1 signalling in response to mechanical overload is independent of phosphoinositide 3-kinase/Akt signalling. J Physiol 2011;589:1831−46.

[195] Spangenburg EE, Le Roith D, Ward CW, Bodine SC. A functional insulin-like growth factor receptor is not necessary for load-induced skeletal muscle hypertrophy. J Physiol 2008;586:283−91.

[196] Thomson DM, Gordon SE. Impaired overload-induced muscle growth is associated with diminished translational signalling in aged rat fast-twitch skeletal muscle. J Physiol 2006;574:291−305.

[197] Thomson DM, Brown JD, Fillmore N, Ellsworth SK, Jacobs DL, Winder WW, et al. AMP-activated protein kinase response to contractions and treatment with the AMPK activator AICAR in young adult and old skeletal muscle. J Physiol 2009;587:2077−86.

[198] Bodine SC, Stitt TN, Gonzalez M, Kline WO, Stover GL, Bauerlein R, et al. Akt/mTOR pathway is a crucial regulator of skeletal muscle hypertrophy and can prevent muscle atrophy in vivo. Nat Cell Biol 2001;3:1014−19.

[199] Bentzinger CF, Lin S, Romanino K, Castets P, Guridi M, Summermatter S, et al. Differential response of skeletal muscles to mTORC1 signaling during atrophy and hypertrophy. Skelet Muscle 2013;3:6.

[200] Cuthbertson DJ, Babraj J, Smith K, Wilkes E, Fedele MJ, Esser K, et al. Anabolic signaling and protein synthesis in human skeletal muscle after dynamic shortening or lengthening exercise. Am J Physiol Endocrinol Metab 2006;290:E731−8.

[201] Dreyer HC, Fujita S, Cadenas JG, Chinkes DL, Volpi E, Rasmussen BB. Resistance exercise increases AMPK activity and reduces 4E-BP1 phosphorylation and protein synthesis in human skeletal muscle. J Physiol 2006;576:613−24.

[202] Eliasson J, Elfegoun T, Nilsson J, Kohnke R, Ekblom B, Blomstrand E. Maximal lengthening contractions increase p70 S6 kinase phosphorylation in human skeletal muscle in the absence of nutritional supply. Am J Physiol Endocrinol Metab 2006;291:E1197−205.

[203] Glover EI, Oates BR, Tang JE, Moore DR, Tarnopolsky MA, Phillips SM. Resistance exercise decreases eIF2Bepsilon phosphorylation and potentiates the feeding-induced stimulation of p70S6K1 and rpS6 in young men. Am J Physiol Regul Integr Comp Physiol 2008;295:R604−10.

[204] Holm L, van Hall G, Rose AJ, Miller BF, Doessing S, Richter EA, et al. Contraction intensity and feeding affect collagen and myofibrillar protein synthesis rates differently in human skeletal muscle. Am J Physiol Endocrinol Metab 2010;298:E257−69.

[205] Witard OC, Tieland M, Beelen M, Tipton KD, van Loon LJ, Koopman R. Resistance exercise increases postprandial muscle protein synthesis in humans. Med Sci Sports Exerc 2009;41: 144−54.

[206] Parkington JD, Siebert AP, LeBrasseur NK, Fielding RA. Differential activation of mTOR signaling by contractile activity in skeletal muscle. Am J Physiol Regul Integr Comp Physiol 2003;285:R1086−90.

[207] Burry M, Hawkins D, Spangenburg EE. Lengthening contractions differentially affect p70s6k phosphorylation compared to isometric contractions in rat skeletal muscle. Eur J Appl Physiol 2007;100:409−15.

[208] Witkowski S, Lovering RM, Spangenburg EE. High-frequency electrically stimulated skeletal muscle contractions increase p70s6k phosphorylation independent of known IGF-I sensitive signaling pathways. FEBS Lett 2010;584:2891–5.

[209] Ogasawara R, Sato K, Higashida K, Nakazato K, Fujita S. Ursolic acid stimulates mTORC1 signaling after resistance exercise in rat skeletal muscle. Am J Physiol Endocrinol Metab 2013;305:E760–5.

[210] Hornberger TA, Stuppard R, Conley KE, Fedele MJ, Fiorotto ML, Chin ER, et al. Mechanical stimuli regulate rapamycin-sensitive signalling by a phosphoinositide 3-kinase-, protein kinase B- and growth factor-independent mechanism. Biochem J 2004;380:795–804.

[211] Hornberger TA, Chu WK, Mak YW, Hsiung JW, Huang SA, Chien S. The role of phospholipase D and phosphatidic acid in the mechanical activation of mTOR signaling in skeletal muscle. Proc Natl Acad Sci USA 2006;103:4741–6.

[212] You JS, Lincoln HC, Kim CR, Frey JW, Goodman CA, Zhong XP, et al. The role of diacylglycerol kinase zeta and phosphatidic acid in the mechanical activation of mammalian target of rapamycin (mTOR) signaling and skeletal muscle hypertrophy. J Biol Chem 2014;289:1551–63.

[213] Kubica N, Bolster DR, Farrell PA, Kimball SR, Jefferson LS. Resistance exercise increases muscle protein synthesis and translation of eukaryotic initiation factor 2Bepsilon mRNA in a mammalian target of rapamycin-dependent manner. J Biol Chem 2005;280:7570–80.

[214] Drummond MJ, Fry CS, Glynn EL, Dreyer HC, Dhanani S, Timmerman KL, et al. Rapamycin administration in humans blocks the contraction-induced increase in skeletal muscle protein synthesis. J Physiol 2009;587:1535–46.

[215] DeVol DL, Rotwein P, Sadow JL, Novakofski J, Bechtel PJ. Activation of insulin-like growth factor gene expression during work-induced skeletal muscle growth. Am J Physiol 1990;259:E89–95.

[216] Czerwinski SM, Martin JM, Bechtel PJ. Modulation of IGF mRNA abundance during stretch-induced skeletal muscle hypertrophy and regression. J Appl Physiol (1985) 1994;76:2026–30.

[217] Adams GR, Haddad F. The relationships among IGF-1, DNA content, and protein accumulation during skeletal muscle hypertrophy. J Appl Physiol (1985) 1996;81:2509–16.

[218] Hornberger TA, Chien S. Mechanical stimuli and nutrients regulate rapamycin-sensitive signaling through distinct mechanisms in skeletal muscle. J Cell Biochem 2006;97:1207–16.

[219] O'Neil TK, Duffy LR, Frey JW, Hornberger TA. The role of phosphoinositide 3-kinase and phosphatidic acid in the regulation of mammalian target of rapamycin following eccentric contractions. J Physiol 2009;587:3691–701.

[220] Hamilton DL, Philp A, MacKenzie MG, Baar K. A limited role for PI(3,4,5)P3 regulation in controlling skeletal muscle mass in response to resistance exercise. PLoS One 2010;5:e11624.

[221] Nygren J, Nair KS. Differential regulation of protein dynamics in splanchnic and skeletal muscle beds by insulin and amino acids in healthy human subjects. Diabetes 2003;52:1377–85.

[222] Sheffield-Moore M, Wolfe RR, Gore DC, Wolf SE, Ferrer DM, Ferrando AA. Combined effects of hyperaminoacidemia and oxandrolone on skeletal muscle protein synthesis. Am J Physiol Endocrinol Metab 2000;278:E273–9.

[223] Ferrando AA, Sheffield-Moore M, Paddon-Jones D, Wolfe RR, Urban RJ. Differential anabolic effects of testosterone and amino acid feeding in older men. J Clin Endocrinol Metab 2003;88:358–62.

[224] Biolo G, Tipton KD, Klein S, Wolfe RR. An abundant supply of amino acids enhances the metabolic effect of exercise on muscle protein. Am J Physiol 1997;273:E122–9.

[225] Tipton KD, Ferrando AA, Phillips SM, Doyle Jr D, Wolfe RR. Postexercise net protein synthesis in human muscle from orally administered amino acids. Am J Physiol 1999;276:E628–34.

[226] Borsheim E, Tipton KD, Wolf SE, Wolfe RR. Essential amino acids and muscle protein recovery from resistance exercise. Am J Physiol Endocrinol Metab 2002;283:E648–57.

[227] Miller SL, Tipton KD, Chinkes DL, Wolf SE, Wolfe RR. Independent and combined effects of amino acids and glucose after resistance exercise. Med Sci Sports Exerc 2003;35:449–55.

[228] Moore DR, Robinson MJ, Fry JL, Tang JE, Glover EI, Wilkinson SB, et al. Ingested protein dose response of muscle and albumin protein synthesis after resistance exercise in young men. Am J Clin Nutr 2009;89:161–8.

[229] Koopman R, Pennings B, Zorenc AH, van Loon LJ. Protein ingestion further augments S6K1 phosphorylation in skeletal muscle following resistance type exercise in males. J Nutr 2007;137:1880–6.

[230] Karlsson HK, Nilsson PA, Nilsson J, Chibalin AV, Zierath JR, Blomstrand E. Branched-chain amino acids increase p70S6k phosphorylation in human skeletal muscle after resistance exercise. Am J Physiol Endocrinol Metab 2004;287:E1–7.

[231] Goldberg AL. Role of insulin in work-induced growth of skeletal muscle. Endocrinology 1968;83:1071–3.

[232] Farrell PA, Fedele MJ, Hernandez J, Fluckey JD, Miller III JL, Lang CH, et al. Hypertrophy of skeletal muscle in diabetic rats in response to chronic resistance exercise. J Appl Physiol (1985) 1999;87:1075–82.

[233] Matheny RW, Merritt E, Zannikos SV, Farrar RP, Adamo ML. Serum IGF-I-deficiency does not prevent compensatory skeletal muscle hypertrophy in resistance exercise. Exp Biol Med 2009;234:164–70.

[234] Kraemer WJ, Ratamess NA. Hormonal responses and adaptations to resistance exercise and training. Sports Med 2005;35:339–61.

[235] West DW, Burd NA, Tang JE, Moore DR, Staples AW, Holwerda AM, et al. Elevations in ostensibly anabolic hormones with resistance exercise enhance neither training-induced muscle hypertrophy nor strength of the elbow flexors. J Appl Physiol (1985) 2010;108:60–7.

[236] West DW, Kujbida GW, Moore DR, Atherton P, Burd NA, Padzik JP, et al. Resistance exercise-induced increases in putative anabolic hormones do not enhance muscle protein synthesis or intracellular signalling in young men. J Physiol 2009;587:5239–47.

[237] Bhasin S, Storer TW, Berman N, Callegari C, Clevenger B, Phillips J, et al. The effects of supraphysiologic doses of testosterone on muscle size and strength in normal men. N Engl J Med 1996;335:1–7.

[238] Hanson ED, Sheaff AK, Sood S, Ma L, Francis JD, Goldberg AP, et al. Strength training induces muscle hypertrophy and functional gains in black prostate cancer patients despite androgen deprivation therapy. J Gerontol A Biol Sci Med Sci 2013;68:490–8.

[239] Cox LS, Mattison JA. Increasing longevity through caloric restriction or rapamycin feeding in mammals: common mechanisms for common outcomes? Aging Cell 2009;8:607–13.

[240] Harrison DE, Strong R, Sharp ZD, Nelson JF, Astle CM, Flurkey K, et al. Rapamycin fed late in life extends lifespan in genetically heterogeneous mice. Nature 2009;460:392–5.

[241] Miller RA, Harrison DE, Astle CM, Baur JA, Boyd AR, de Cabo R, et al. Rapamycin, but not resveratrol or simvastatin, extends life span of genetically heterogeneous mice. J Gerontol A Biol Sci Med Sci 2011;66:191–201.

4

mTOR: A Critical Mediator of Articular Cartilage Homeostasis

Akihiro Nakamura[1] *and Mohit Kapoor*[1,2,3]

[1]Department of Genetics and Development, Krembil Research Institute, University Health Network, Toronto, Ontario, Canada [2]Department of Surgery, University of Toronto, Toronto, Ontario, Canada [3]Arthritis Program, University Health Network, Toronto, Ontario, Canada

4.1 INTRODUCTION

Articular cartilage consists of specialized cells called chondrocytes that are essential for its maintenance and overall integrity, the degeneration of which is a hallmark of degenerative joint diseases such as osteoarthritis (OA). In response to injury, aging, or long-term stressful stimuli, chondrocytes—and therefore articular cartilage—become less able to withstand harsh conditions over time. As a result, chondrocytes are unable to maintain a balance between catabolic and anabolic activities in the articular cartilage—ultimately leading to cartilage degeneration. Mechanisms that trigger chondrocytes to lose their ability to maintain articular cartilage homeostasis as well as their own survival are not well understood. Therefore, therapies that can promote chondrocyte survival and stop or delay the process of cartilage degeneration are not available clinically.

Recent studies suggest that the mammalian (or mechanistic) target of rapamycin (mTOR) is crucial for maintenance of articular cartilage homeostasis and could be a potential therapeutic target in delaying the degenerative process. mTOR signaling is implicated in cell growth, proliferation, as well as its ability to negatively regulate autophagy—a cell survival and adaptation process. Since articular cartilage is a postmitotic tissue, it depends on processes such as autophagy to remove damaged organelles, misfolded or aggregated proteins, as well as eliminating pathogens in response to stress. Specifically, autophagy is now emerging as a crucial homeostatic process responsible for the maintenance of chondrocyte survival; suppression of autophagy results in decreased chondroprotection, enhanced apoptosis, and cartilage degeneration. One of the mechanisms through which autophagy is suppressed in the articular chondrocytes is through increased mTOR signaling. In OA cartilage, enhanced mTOR signaling is associated with the suppression of key autophagy mediators and increased chondrocyte cell death [1–3]. Animal studies using mTOR inhibitors or genetic deletion show protection from cartilage-degenerative mechanisms in mouse models of OA, suggesting the therapeutic potential of inhibiting the mTOR signaling pathway to achieve articular cartilage homeostasis [1,3]. In this chapter, we will discuss the most uptodate knowledge of mTOR in chondrocyte biology and articular cartilage homeostasis. We will also discuss the potential of mTOR as a therapeutic target in delaying or stopping the cartilage-degenerative process.

4.2 MECHANISTIC TARGET OF RAPAMYCIN

4.2.1 mTOR Structure

The mTOR protein is a 289-kDa serine-threonine kinase that is a member of the phospho-inositide-3-kinase (PI3K)-related family. This conserved protein

57

integrates with diverse upstream signals to regulate growth-related processes including mRNA translation, ribosome biogenesis, autophagy, and metabolism. Deregulation in the mTOR pathway is implicated in a variety of common diseases such as cancer, diabetes, neurodegeneration, etc.

mTOR interacts with several proteins to form two distinct signaling complexes: mTOR complex 1 (mTORC1) and complex 2 (mTORC2). These mTOR complexes show different sensitivities to upstream inputs resulting in differential downstream outputs. mTORC1 consists of six proteins: mTOR, raptor (regulatory-associated protein of mTOR), pras40 (proline-rich Akt substrate 40 kDa), deptor (DEP domain containing mTOR-interacting protein), mLST8 (mammalian lethal with sec-13 protein 8), and the Tti1/Tel2 complex. Although the exact function of most of the mTOR-interacting proteins in mTORC1 still remains to be fully understood, it has been proposed that raptor affects mTORC1 activity by regulating assembly of the complex and by recruiting substrates for mTOR [4]. It was suggested that mLST8 is essential for an interaction between raptor and mTOR, but the role of mLST8 in the function of mTORC1 is still unclear, as deletion of this protein does not affect mTORC1 activity *in vivo* based on recent findings [5]. Pras40 has been identified to be a direct negative regulator of mTORC1 function [6]. The major function of mTORC1 is to regulate cell growth, proliferation, and survival by sensing mitogen, energy, and nutrient signals [6]. It phosphorylates S6-kinase-1 and eukaryotic translation initiation factor 4E binding protein 1 (4EBP1; also known as EIF4EBP1) to regulate protein synthesis and cell growth [7].

On the other hand, mTORC2 consists of seven proteins: mTOR, Rictor (rapamycin-insensitive companion of mTOR), mSin1 (mammalian stress-activated protein kinase-interacting protein 1), protor1/2 (protein observed with Rictor 1/2), deptor, mLST8, and the Tti1/Tel2 complex. mTORC2 also includes mTOR, deptor, and mLST8; but instead of raptor, mTORC2 contains two special subunits—Rictor and mSin1. There is some evidence that Rictor and mSIN1 stabilize each other, establishing the structural foundation of mTORC2 [8,9]. Rictor also interacts with Protor1, but the physiological function of this interaction is not clear [10,11]. Similar to its role in mTORC1, deptor negatively regulates mTORC2 activity. It has been reported that mTORC2 modulates cell survival in response to growth factors by phosphorylating its downstream kinases [12]. Although it was originally thought to be rapamycin-insensitive because acute treatment with rapamycin did not perturb mTORC2 signaling [13], long-term treatment with rapamycin was shown to affect mTORC2 function [14].

4.2.2 mTOR Signaling Pathway

The mTORC1 pathway integrates inputs from several intracellular and extracellular factors including growth factors, stress, energy status, oxygen, and amino acids to control several biological processes; including protein, lipid synthesis, and autophagy. The key upstream regulator of mTORC1 is the tuberous sclerosis complex (TSC) 1/2. TSC1/2 functions as a GTPase-activating protein for the Ras homolog enriched in brain (Rheb) GTPase. The GTP-bound form of Rheb directly interacts with mTORC1 and strongly stimulates its kinase activity. In this way, mTORC1 is negatively regulated by TSC1/2 through converting Rheb into its inactive GDP-bound state [15]. TSC1/2 transmits many of the upstream signals that activate mTORC1, including growth factors such as insulin and insulin-like growth factor (IGF-1) that stimulate the PI3K/Akt pathway—one of the major upstream signals that modulate mTORC1. Akt phosphorylates mTORC1, causing it to dissociate from Pras40, a raptor binding protein that potently inhibits mTORC1 kinase activity [6]. It has also been reported that proinflammatory cytokines, such as tumor necrosis factor-α (TNF-α), activate mTORC1 through a mechanism conceptually similar to that of growth factors such as IκB kinase β (IKKβ) [16], which inhibit TSC1/2 by phosphorylating TSC1 [16]. This finding provides critical insight into the possible mechanisms associated with inflammation-mediated cellular growth and proliferation. Li et al. [17] supported this theory by showing that inhibition of PI3K was able to prevent TNF-α-induced chondrocyte cell death.

In addition to growth factors and proinflammatory cytokines, several stress signals, such as DNA damage, low energy, and low oxygen (hypoxia), also regulate the activity of mTORC1 though TSC1/2. DNA damage leads to the activation of p53, a tumor suppressor protein that not only inhibits mTOR [18] but also up-regulates the expression of TSC2 and PTEN (phosphatase and tensin homolog deleted on chromosome 10), resulting in the down-regulation of the PI3K-mTORC1 axis [19]. In conditions of low energy and hypoxia, adenosine monophosphate-activated protein kinase (AMPK) phosphorylates TSC2, leading to the up-regulation of TSC1/2 and down-regulation of mTORC1 [20]. Similar to the function of Akt, AMPK also directly phosphorylates mTORC1 [21].

In addition to the PI3K/Akt and AMPK signaling pathways, Wnt/β-catenin signaling, a major pathway regulating cell growth, proliferation, polarity, differentiation, and development, also activates mTORC1 through the inhibition of TSC1/2. This signaling pathway inhibits glycogen synthase kinase 3β, which normally phosphorylates and promotes TSC2 activity [22]. Although these upstream pathways are crucial for the activation of mTORC1, downstream pathways also play important roles in regulating endogenous cell mechanisms. The regulation of protein and lipid synthesis as well as the production of ATP are positively controlled by mTORC1 [23–25]. On the other hand, mTORC1 also functions as a negative regulator of promoting cell survival through the mechanism of autophagy.

4.2.3 Autophagy

Autophagy is a self-degradative process that is essential for cell survival, differentiation, development, and homeostasis [26]. It plays a housekeeping role in removing misfolded or aggregated proteins; clearing damaged organelles, such as mitochondria, endoplasmic reticulum and peroxisomes; as well as eliminating intracellular pathogens in response to stressors such as nutrient deficiencies and hypoxia [27]. Further, autophagy is also involved in promoting cellular senescence, cell surface antigen presentation, protection against genome instability, and preventing necrosis. Deregulation in the autophagy process is implicated in various diseases such as cancer, neurodegeneration, cardiomyopathy, diabetes, liver disease, infections, and autoimmune diseases [28–33].

The autophagy signaling pathway is controlled by a number of autophagy-related genes (Atg) through formation of an autophagosome, a cytoplasmic structure for delivering damaged or obsolete organelles to the lysosome [34,35]. Studies in yeast have identified around 40 Atg proteins, among which 15 play a fundamental role in the formation of the autophagosomal membrane in any type of autophagy [36]. Specifically, Atg1, Atg6, Atg8 (ULK1, Beclin1, and LC3 in mammals, respectively), and Atg5 are the four major positive regulators of autophagy. Unc-51-like kinase 1 (ULK1), one of the Atg1 homologs in mammals, is necessary for the autophagosome to fuse with Atg13 and the focal adhesion kinase family-interacting protein of 200 kDa (FIP200) [37]. ULK1/Atg13/FIP200 is a kinase complex required to initiate autophagy [38,39]. Atg13 is phosphorylated by mTORC1, and FIP200 is crucial for the stability and phosphorylation of ULK1 [39,40]. Beclin1 is also essential for initiating autophagosomal function by forming the Vsp–Beclin1 complex (Beclin1 complex I) of class III PI3K-kinase [41]. Following the induction of autophagy, the formation of an isolation membrane is initiated to engulf cytoplasmic components such as mitochondria and the endoplasmic reticulum.

The Atg12–Atg5–Atg16 complex is essential for autophagosome formation [42,43]. Subsequently, LC3I (localized in the cytosol) is converted to LC3II (localized in the elongated isolation membrane) for maturation of the autophagosome [44]. Finally, the isolated membrane is enclosed by a double or multilayered membrane and subsequently the autophagosome fuses with the lysosome, a multifunctional organelle that can degrade most cellular components [45,46], to form an autolysosome for degradation of the captured materials. Beclin1 complex II (Vps34, p150, Beclin1, and UVRAG) promotes the fusion of autophagosomes with lysosomes [47].

One of the major roles of mTOR is to negatively regulate autophagy. mTORC1 inhibits autophagy through phosphorylation of ULK1 [48]. In addition, mTORC1 directly phosphorylates and subsequently suppresses the formation of ULK1/Atg13/FIP200 [38,39]. Studies show that mTORC1 down-regulates the biogenesis of lysosomes. Upon inhibition of mTORC1, autophagolysosomes are formed by the fusion of autophagosomes with lysosomes to engulf cytoplasmic proteins and organelles—leading to the degeneration of cell components for the organism and cellular adaption to stressors [49]. This evidence clearly suggests that mTOR negatively regulates the formation of an autophagosome by multiple target inhibition, resulting in decreased autophagy. In addition to this process, mTORC1 also regulates death-associated protein 1, a suppressor of autophagy [50].

4.2.4 Cartilage Growth and Development: Role of mTOR

Articular cartilage consists of a single cellular component, the chondrocyte, which is embedded in a complex matrix network. In the embryo, chondrocytes arise from mesenchymal stem cells during chondrogenesis, the earliest phase of skeletal development [51] and begins with proliferation and condensation of limb mesenchymal cells. These cells then exhibit chondrogenic differentiation and subsequently produce a cartilaginous extracellular matrix (ECM) rich in collagen type II to form the cartilage anlage, the original structure of the future long bone [52,53]. The cartilage model is then replaced by bone from the center to the distal ends via the process of endochondral ossification [54] (see Figure 4.1). On the terminal surface zone of the cartilage anlage, flattened cells remain in the resting zone [55,56]. These flattened cells later become articular chondrocytes that reside at the surface of diarthrodial joints and produce hyaline cartilage [57].

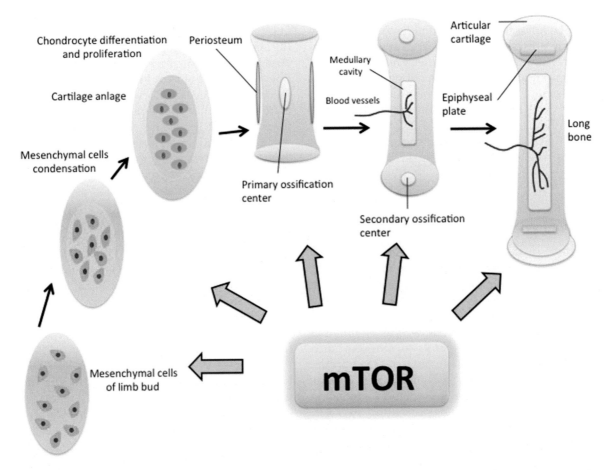

FIGURE 4.1 **mTOR is involved in normal cartilage growth and development.** Long bones are formed by the process of endochondral ossification. Mesenchymal stem cells in the limb bud condense and then differentiate to chondrocytes. Cartilage anlage, a future long bone model, is formed and subsequently bone formation is initiated. mTOR has been shown to play a key role in cartilage, bone, and overall skeletal development.

For normal ossification and cartilage developmental process, contribution of some of the key mediators including Sox9, p57, aggrecan, collagen type II, collagen type X, etc., are critical. During chondrogenesis, chondrocytes express the chondrogenic transcription factor Sox9 that is required for not only proliferation and differentiation of chondrocytes but also regulation of the expression of genes encoding aggrecan and collagen type II, two major cartilage ECM molecules [58–60]. p57, a cell cycle inhibitor is essential for the onset of hypertrophic differentiation and collagen type X, a marker of terminally differentiated hypertrophic chondrocytes that plays a key role in driving the ossification and cartilage growth and developmental process [61]. A variety of signaling molecules including Indian hedgehog, fibroblast growth factors, Akt, p38, Wnt/β-catenin, peroxisome proliferator-activated receptor-gamma (PPARγ), etc., have also been shown to regulate the cartilage developmental process [52,62–67]. The overall process of cartilage development and endochondral ossification is tightly regulated; any disturbances in this regulated process result in musculoskeletal and developmental defects.

Studies show that mTOR may play a crucial role in optimal skeletal growth and development. Chen and Long [68] revealed that mTORC1 and Raptor are essential regulators of normal cartilage and skeletal development in mice. They showed that mTORC1 signaling increases the cell size of chondrocytes and the amount of ECM proteins (type II collagen) produced by chondrocytes at all stages of chondrocyte maturation. They also investigated the role of mTORC2 and its contribution to skeletal development and homeostasis [69]. Loss of Rictor resulted in shorter and narrower skeletal elements, resulting in the reduction of cartilage anlage in mouse embryos as well as delayed chondrocyte hypertrophy. These results indicate that both mTORC1 and mTORC2 are required for normal cartilage development. It has also been shown that inhibition of mTOR by direct infusion of rapamycin into the proximal tibial growth plates in rabbits resulted in a decrease in the size of the growth plates and an abnormally shorter length of long-bones [70]. Similarly, an intraperitoneal injection of rapamycin impairs longitudinal growth in fast-growing rats—associated with altered chondrocyte proliferation, maturation, hypertrophy, and cartilage resorption [71]. Further, Holstein et al. [72] observed that rapamycin delays fracture healing by inhibiting cell proliferation and neovascularization in the callus. In addition, Guan et al. [73] reported that mechanical stress also plays an important role in mTOR activation, which is essential for cell proliferation, chondrogenesis, and cartilage growth during bone development.

mTOR does not independently regulate cartilage growth and development. Intracellular signaling pathways, both upstream and downstream, have also been reported as key regulators in the process of chondrocyte hypertrophy, chondrocyte and osteoblast interactions, and degradation of major cartilage ECM proteins. Starkman et al. [74] observed that the synthesis of IGF-stimulated aggrecans, one of the major ECM components in cartilage, was reduced by inhibition of mTOR and p70S6, both of which are downstream of the PI3K/Akt signaling pathway. Inhibition of mTOR by rapamycin is also reported to reduce transforming growth factor β stimulated-phosphorylation of p70S6 kinase, and the expression of tissue inhibitor of metalloproteinases-3 through the PI3K/Akt/mTOR pathway, leading to inhibition of protein synthesis associated with maintenance of joint tissue integrity [75]. Gelse et al. [76] showed that inhibition of mTOR by rapamycin and inhibition of PI3K by wortmannin were able to decrease the level of hypoxia-inducible factor 1 alpha (HIF-1α), a crucial factor involved in regulating chondrogenesis, joint formation, and stabilization of chondrocytic phenotypes. It has also been reported that expression and activation of AMPK is reduced in degenerated cartilage as well as in chondrocytes [77,78]. Activated AMPK inhibits mTOR and directly phosphorylates the most upstream autophagy inducer, ULK1 [79]. The Wnt/β-catenin signaling pathway also controls multiple developmental processes in skeletal and joint development [80] and is associated with the pathogenesis of OA [81]. Since β-catenin activates mTORC1 through inhibition of TSC1/2, genetic inactivation of the β-catenin in mouse mesenchymal cells *in vivo* results in the loss of osteoblasts, the primary bone cells in bone tissue derived through the processes of intramembranous and endochondral bone formation [82]. These studies suggest that various signaling pathways associated with mTOR play essential roles in cartilage and overall skeletal growth and development.

4.2.5 Articular Cartilage Degeneration: Role of mTOR/Autophagy Signaling

Articular cartilage is a thin layer of connective tissue that consists of structural macromolecules including collagens, proteoglycans, noncollagenous proteins, and glycoproteins without any blood vessels, or nerves [83–85]. Collagen type II represents 90–95% of the collagen in the ECM. Another major ECM molecule present in the articular cartilage is aggrecan. Lack of synthesis of these matrix molecules inhibits maintenance of the cartilage matrix. Distinct zones of articular cartilage include the superficial zone, middle zone, deep zone, and calcified zone [86]. Chondrocytes are the only cell type present in the articular cartilage and their shape, size, and number vary with the region of the articular cartilage they are present in. The superficial zone consists of flattened chondrocytes and the middle zone consists of spherical chondrocytes. Chondrocytes in the deep zone exhibit columnar orientation, whereas chondrocytes in the calcified zone are hypertropic [87].

Chondrocytes are essential in maintaining the integrity of the articular cartilage as well as the balance between the production of catabolic, inflammatory, and anabolic mediators. Cytokines including interleukin (IL)-1β, TNF-α, and IL-6 are major proinflammatory cytokines known to stimulate chondrocytes and induce the production of key catabolic mediators such as metalloproteinases (MMPs) and a disintegrin-like MMP with thrombospondin type 1 motifs (ADAMTS) [88–94]. Articular cartilage homeostasis is maintained by a fine balance between catabolic and anabolic activity within the ECM of articular cartilage. If the balance leans towards enhanced catabolic activity, cartilage integrity is lost—leading to cartilage degeneration.

In adult articular cartilage, the rate of cell turnover is low, resulting in increased susceptibility to chondrocyte cell death and cartilage degeneration over time [95]. Also, articular cartilage is a postmitotic tissue and chondrocytes within rely on endogenous mechanisms such as autophagy to remove dysfunctional organelles and macromolecules. Recent studies suggest that autophagy and its regulation by mTOR are crucial for determining the fate of chondrocytes (survival and cell death) within the articular cartilage as well as the balance between catabolic and anabolic activity. Autophagy also has protective functions against mitochondrial damage and dysfunction [96]. Thus, aberrant mTOR/autophagy signaling has been shown to disrupt cartilage homeostasis, resulting in articular cartilage degeneration (Figure 4.2).

We recently identified that expression of mTOR is enhanced in human OA cartilage as well as surgically induced OA in mouse cartilage (compared to normal

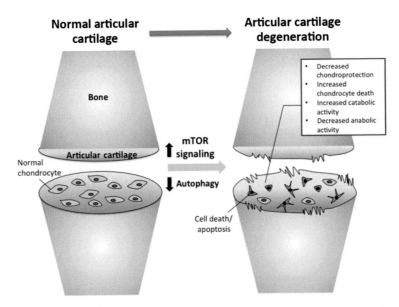

FIGURE 4.2 **Enhanced mTOR signaling and subsequent suppression of autophagy contribute to articular cartilage degeneration.** Enhanced mTOR signaling and suppression of autophagy are associated with reduced chondrocyte survival, enhanced cell death, and catabolic activity resulting in articular cartilage degeneration.

human and sham-surgery control mice, respectively) is associated with reduced autophagy, enhanced chondrocyte apoptosis as well as increased expression of cartilage catabolic mediators [3]. Interestingly, the expression of mTOR is not only up-regulated in articular cartilage, but also in peripheral blood mononuclear cells (PBMCs) in patients with OA [97]. This study also showed a positive correlation between the expression of mTOR in PBMCs and in articular cartilage from patients with end-stage OA. It has also been reported that genetic deletion of the von Hippel-Lindau gene, a tumor suppressor gene, enhances HIF-2α, leading to up-regulation of mTOR, MMP-13, and chondrocyte apoptosis, as well as down-regulation in the expression of autophagic markers such as LC3 and Beclin1 during OA development [98]. Sasaki et al. [99] observed that administration of rapamycin showed increased autophagy by enhancing the expression of LC3II and Beclin1 in human chondrocytes treated with IL-1β, accompanied by significantly decreased expression of MMP-13 and ADAMTS-5, as well as increased expression of collagen type II and aggrecan.

Since aging is one of the major risk factors in degenerative joint diseases such as OA, decreased autophagy in articular cartilage with age could be a major explanation for age-related cartilage degeneration accompanied by increased chondrocyte apoptosis [95,100]. Chang et al. [101] also showed that autophagy is up-regulated in chondrocytes from young patients without any cartilage-degenerative diseases compared to those from elderly patients. It has also been reported

that age-related human OA and mice with surgically induced OA showed a significant reduction in the expression of ULK1, LC3, and Beclin1 in articular cartilage, which was associated with histological changes in age-related cartilage degeneration and increased chondrocyte apoptosis. These results suggest that autophagic activity declines with age, leading to the intracellular accumulation of damaged organelles, decreased chondroprotection and chondrocyte death.

Given that increased mTOR signaling is associated with enhanced articular cartilage-degenerative mechanisms, inhibiting mTOR could be a potential therapeutic strategy for stopping cartilage degeneration. Carames et al. [102] tested the effect of rapamycin by intraperitoneal injections in C57Bl/6J mice subjected to surgical induction of OA. The effect of rapamycin on the pathway was observed by inhibition of phosphorylation of the ribosomal protein S6 and up-regulation of LC3II expression as an indicator of increased autophagy. Rapamycin showed a decrease in the severity of OA-like changes such as cartilage degeneration, loss of surface lamina and fibrillation. This study also showed that rapamycin decreased synovial inflammation, levels of ADAMTS-5, an enzyme that degrades aggrecan, while preserving articular cartilage cellularity. Takayama et al. [103] injected rapamycin intra-articularly and observed delayed cartilage degeneration in murine OA models. Another group created gelatin hydrogels incorporating rapamycin micelles to achieve controlled intra-articular administration of rapamycin in the intra-articular joint space [104].

Similar to systemic administration, a local intra-articular injection was also able to slow down the process of cartilage degeneration in murine OA models. Since short-term administration of glucocorticoid drugs, such as prednisone and dexamethasone, is effective for noninfectious inflammatory arthritis, Liu et al. [105] assessed autophagy activity as well as chondrocyte apoptosis under treatment with glucocorticoids. Although long-term treatment with glucocorticoids increased chondrocyte apoptosis resulting in inhibition of chondrocyte cell growth as well as decreased production of ECM [106], rapamycin enhanced autophagy to protect chondrocytes from apoptosis. Rosenthal et al. [107] focused on the roles of autophagy in the formation of articular cartilage vesicles (ACVs); chondrocyte-derived extracellular organelles responsible for protein secretion, intracellular communication, and pathologic classification. They observed that the porcine chondrocytes treated with rapamycin increased the number of ACVs, the expression of autophagy markers and the formation of autophagosomes, although the chondrocytes from human OA did not show the effect.

In addition to the pharmacological inhibition of mTOR; genetic deficiency of mTOR in cartilage has provided useful information on the role of mTOR in articular cartilage homeostasis. We created cartilage-specific mTOR knockout (KO) mice and showed that

cartilage-specific ablation of mTOR results in increased autophagy and significant protection from cartilage degeneration in mouse models of OA [3]. We also observed that loss of mTOR was associated with a significant decrease in the expression of MMP-13 as well as chondrocyte apoptosis. The exact signaling pathway through which mTOR operates within the articular cartilage is still unknown; however, studies do suggest that mTOR controls autophagy in part by modulating ULK1 expression in the articular cartilage. It has been shown that mTOR KO mice as well as rapamycin treatment in human OA chondrocytes show up-regulation in the expression of ULK1. Further, silencing ULK1 in rapamycin-treated OA chondrocytes with ULK1 siRNA decreases autophagy and increases the expression of OA catabolic factors, including MMP-13 and chemokine (C-C motif) ligand 2 (CCL2) (Figure 4.3).

Another key mediator that has been shown to negatively regulate mTOR in the articular cartilage is PPARγ, a ligand-activated transcription factor required for lipid homeostasis that exhibits anti-inflammatory, anticatabolic, and antifibrotic properties [108]. We recently created cartilage-specific PPARγ KO mice and observed that when subjected to OA surgery, these mice exhibited accelerated chondrocyte cell death and cartilage degeneration associated with enhanced mTOR expression and reduced autophagy

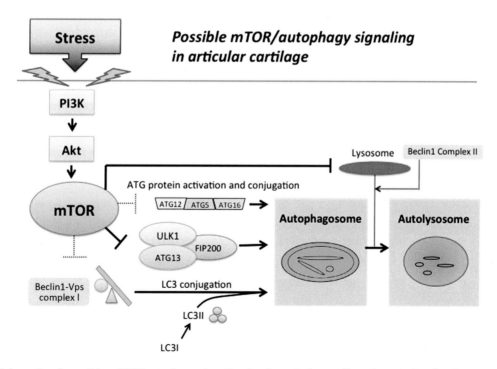

FIGURE 4.3 **Schematic of possible mTOR/autophagy signaling in the articular cartilage.** In articular chondrocytes, stressors such as inflammation or mechanical stress activate mTOR signaling, resulting in the suppression of the ULK1/Atg13/FIP200 complex, along with decreased formation and activation of autophagosomes.

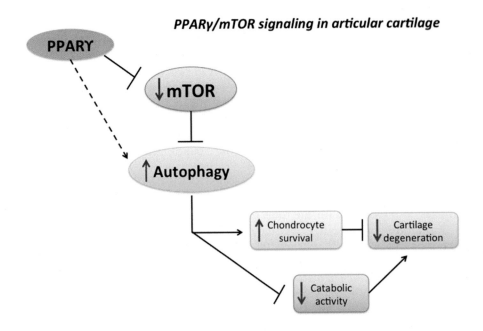

FIGURE 4.4   **Schematic of PPARγ/mTOR signaling pathway in articular cartilage.** In articular cartilage, PPARγ suppresses mTOR signaling and enhances autophagy in articular cartilage resulting in increased chondroprotection, decreased catabolic activity, and protection from cartilage degeneration.

signaling [2]. Furthermore, results showed that restoration of PPARγ expression in PPARγ KO cells *in vitro* decreased mTOR expression and increased autophagy signaling. Further, genetically deleting mTOR in PPARγ KO mice (PPARγ/mTOR double KO mice) was able to increase autophagy, decrease catabolic activity and reverse severe OA phenotypes *in vivo*, suggesting that PPARγ modulates mTOR/autophagy signaling in the articular cartilage (Figure 4.4). The above studies do suggest that targeting mTOR could be of therapeutic importance in stopping cartilage degeneration. However, inhibiting mTOR may not be effective in stopping cartilage degeneration as some studies suggest that inhibiting mTOR may also increase MMP secretion through the activated PI3K/Akt/NFκB pathway [109].

Along with mTOR, Akt activates nuclear factor kappa-light-chain-enhancer of activated B cells (NFκB), which is a key regulator of MMP production [110]. Since mTOR is also controlled by the AMPK pathway, inhibiting only PI3K is not sufficient to inhibit cartilage degeneration. Therefore, it has been proposed that the use of PI3K and mTOR dual inhibitors could be a more effective therapeutic strategy to suppress cartilage degeneration [111]. Some dual inhibitors are currently being tested for efficacy in clinical trials in the field of cancer [112].

It has also been reported that glucosamine, a common supplement in the treatment regime of OA,

could activate autophagy by inhibiting the Akt/FoxO/mTOR pathway [113]. Furthermore, Huang et al. [114] investigated the effects of *n*-3 polyunsaturated fatty acids (PUFAs) and essential fatty acids for mammals on mTOR activity associated with cartilage degeneration. They observed that administration of *n*-3 PUFAs significantly down-regulated mTOR, accompanied by decreased expression of MMP-13 and ADAMTS-5, resulting in inhibition of cartilage degeneration.

4.3 CONCLUSION

Studies suggest that mTOR is crucial for normal cartilage development and maintenance of articular cartilage homeostasis. Current preclinical studies suggest that targeting mTOR could be of therapeutic importance in delaying/stopping articular cartilage degeneration. However, it should be noted that degenerative joint diseases such as OA exhibit degenerative structural and biomechanical changes in not only articular cartilage, but also in adjacent joint tissues including the synovium, subchondral bone, muscles, etc. Since the exact role of mTOR and its signaling in other joint structures is largely unknown; future studies should be directed to determine how mTOR modulation/inhibition affects the homeostasis of these key joint structures to be able to fully characterize the potential of mTOR inhibition in protecting not only cartilage, but other joint tissues affected showing structural alterations in OA and related joint diseases.

Since mTOR is a complex, multifactorial mediator involved in normal biological processes, targeting PPARγ or ULK1 may be an alternative therapeutic strategy to achieve articular cartilage homeostasis and protect articular cartilage from degeneration.

References

[1] Carames B, Taniguchi N, Seino D, Blanco FJ, D'Lima D, Lotz M. Mechanical injury suppresses autophagy regulators and pharmacologic activation of autophagy results in chondroprotection. Arthritis Rheum 2012;64:1182–92.

[2] Vasheghani F, Zhang Y, Li YH, Blati M, Fahmi H, Lussier B, et al. PPARgamma deficiency results in severe, accelerated osteoarthritis associated with aberrant mTOR signalling in the articular cartilage. Ann Rheum Dis 2015;74:569–78.

[3] Zhang Y, Vasheghani F, Li YH, Blati M, Simeone K, Fahmi H, et al. Cartilage-specific deletion of mTOR upregulates autophagy and protects mice from osteoarthritis. Ann Rheum Dis 2014;74:1432–40.

[4] Hara K, Maruki Y, Long X, Yoshino K, Oshiro N, Hidayat S, et al. Raptor, a binding partner of target of rapamycin (TOR), mediates TOR action. Cell 2002;110:177–89.

[5] Guertin DA, Stevens DM, Thoreen CC, Burds AA, Kalaany NY, Moffat J, et al. Ablation in mice of the mTORC components raptor, rictor, or mLST8 reveals that mTORC2 is required for signaling to Akt-FOXO and PKCalpha, but not S6K1. Dev Cell 2006;11:859–71.

[6] Sancak Y, Thoreen CC, Peterson TR, Lindquist RA, Kang SA, Spooner E, et al. PRAS40 is an insulin-regulated inhibitor of the mTORC1 protein kinase. Mol Cell 2007;25:903–15.

[7] Thoreen CC, Chantranupong L, Keys HR, Wang T, Gray NS, Sabatini DM. A unifying model for mTORC1-mediated regulation of mRNA translation. Nature 2012;485:109–13.

[8] Frias MA, Thoreen CC, Jaffe JD, Schroder W, Sculley T, Carr SA, et al. mSin1 is necessary for Akt/PKB phosphorylation, and its isoforms define three distinct mTORC2s. Curr Biol 2006;16:1865–70.

[9] Jacinto E, Facchinetti V, Liu D, Soto N, Wei S, Jung SY, et al. SIN1/MIP1 maintains rictor-mTOR complex integrity and regulates Akt phosphorylation and substrate specificity. Cell 2006;127:125–37.

[10] Thedieck K, Polak P, Kim ML, Molle KD, Cohen A, Jeno P, et al. PRAS40 and PRR5-like protein are new mTOR interactors that regulate apoptosis. PLoS one 2007;2:e1217.

[11] Woo SY, Kim DH, Jun CB, Kim YM, Haar EV, Lee SI, et al. PRR5, a novel component of mTOR complex 2, regulates platelet-derived growth factor receptor beta expression and signaling. J Biol Chem 2007;282:25604–12.

[12] Sarbassov DD, Guertin DA, Ali SM, Sabatini DM. Phosphorylation and regulation of Akt/PKB by the rictor-mTOR complex. Science 2005;307:1098–101.

[13] Jacinto E, Loewith R, Schmidt A, Lin S, Ruegg MA, Hall A, et al. Mammalian TOR complex 2 controls the actin cytoskeleton and is rapamycin insensitive. Nat Cell Biol 2004;6:1122–8.

[14] Sarbassov DD, Ali SM, Sengupta S, Sheen JH, Hsu PP, Bagley AF, et al. Prolonged rapamycin treatment inhibits mTORC2 assembly and Akt/PKB. Mol Cell 2006;22:159–68.

[15] Inoki K, Li Y, Xu T, Guan KL. Rheb GTPase is a direct target of TSC2 GAP activity and regulates mTOR signaling. Genes Dev 2003;17:1829–34.

[16] Lee DF, Kuo HP, Chen CT, Hsu JM, Chou CK, Wei Y, et al. IKK beta suppression of TSC1 links inflammation and tumor angiogenesis via the mTOR pathway. Cell 2007;130:440–55.

[17] Li D, Wu Z, Duan Y, Hao D, Zhang X, Luo H, et al. TNFalpha-mediated apoptosis in human osteoarthritic chondrocytes sensitized by PI3K-NF-kappaB inhibitor, not mTOR inhibitor. Rheumatol Int 2012;32:2017–22.

[18] Metcalfe SM, Canman CE, Milner J, Morris RE, Goldman S, Kastan MB. Rapamycin and p53 act on different pathways to induce G1 arrest in mammalian cells. Oncogene 1997;15:1635–42.

[19] Feng Z, Zhang H, Levine AJ, Jin S. The coordinate regulation of the p53 and mTOR pathways in cells. Proc Natl Acad Sci USA 2005;102:8204–9.

[20] Inoki K, Zhu T, Guan KL. TSC2 mediates cellular energy response to control cell growth and survival. Cell 2003;115:577–90.

[21] Gwinn DM, Shackelford DB, Egan DF, Mihaylova MM, Mery A, Vasquez DS, et al. AMPK phosphorylation of raptor mediates a metabolic checkpoint. Mol Cell 2008;30:214–26.

[22] Inoki K, Ouyang H, Zhu T, Lindvall C, Wang Y, Zhang X, et al. TSC2 integrates Wnt and energy signals via a coordinated phosphorylation by AMPK and GSK3 to regulate cell growth. Cell 2006;126:955–68.

[23] Thoreen CC, Kang SA, Chang JW, Liu Q, Zhang J, Gao Y, et al. An ATP-competitive mammalian target of rapamycin inhibitor reveals rapamycin-resistant functions of mTORC1. J Biol Chem 2009;284:8023–32.

[24] Yu K, Toral-Barza L, Shi C, Zhang WG, Lucas J, Shor B, et al. Biochemical, cellular, and in vivo activity of novel ATP-competitive and selective inhibitors of the mammalian target of rapamycin. Cancer Res 2009;69:6232–40.

[25] Laplante M, Sabatini DM. An emerging role of mTOR in lipid biosynthesis. Curr Biol 2009;19:R1046–52.

[26] Mizushima N. Physiological functions of autophagy. Curr Top Microbiol Immunol 2009;335:71–84.

[27] Levine B, Kroemer G. Autophagy in the pathogenesis of disease. Cell 2008;132:27–42.

[28] Hara T, Nakamura K, Matsui M, Yamamoto A, Nakahara Y, Suzuki-Migishima R, et al. Suppression of basal autophagy in neural cells causes neurodegenerative disease in mice. Nature 2006;441:885–9.

[29] Glick D, Barth S, Macleod KF. Autophagy: cellular and molecular mechanisms. J Pathol 2010;221:3–12.

[30] Rautou PE, Mansouri A, Lebrec D, Durand F, Valla D, Moreau R. Autophagy in liver diseases. J Hepatol 2010;53:1123–34.

[31] Takemura G, Miyata S, Kawase Y, Okada H, Maruyama R, Fujiwara H. Autophagic degeneration and death of cardiomyocytes in heart failure. Autophagy 2006;2:212–14.

[32] Deretic V, Saitoh T, Akira S. Autophagy in infection, inflammation and immunity. Nat Rev Immunol 2013;13:722–37.

[33] Clarke AJ, Ellinghaus U, Cortini A, Stranks A, Simon AK, Botto M, et al. Autophagy is activated in systemic lupus erythematosus and required for plasmablast development. Ann Rheum Dis 2015;74:912–20.

[34] Xie Z, Klionsky DJ. Autophagosome formation: core machinery and adaptations. Nat Cell Biol 2007;9:1102–9.

[35] Zhou H, Huang S. The complexes of mammalian target of rapamycin. Curr Protein Pept Sci 2010;11:409–24.

[36] Mochida K, Oikawa Y, Kimura Y, Kirisako H, Hirano H, Ohsumi Y, et al. Receptor-mediated selective autophagy degrades the endoplasmic reticulum and the nucleus. Nature 2015;522:359–62.

[37] Jung CH, Jun CB, Ro SH, Kim YM, Otto NM, Cao J, et al. ULK-Atg13-FIP200 complexes mediate mTOR signaling to the autophagy machinery. Mol Biol Cell 2009;20:1992–2003.

[38] Ganley IG, Lamdu H, Wang J, Ding X, Chen S, Jiang X. ULK1. ATG13.FIP200 complex mediates mTOR signaling and is essential for autophagy. J Biol Chem 2009;284:12297—305.

[39] Hosokawa N, Hara T, Kaizuka T, Kishi C, Takamura A, Miura Y, et al. Nutrient-dependent mTORC1 association with the ULK1-Atg13-FIP200 complex required for autophagy. Mol Biol Cell 2009;20:1981—91.

[40] Hara T, Takamura A, Kishi C, Iemura S, Natsume T, Guan JL, et al. FIP200, a ULK-interacting protein, is required for autophagosome formation in mammalian cells. J Cell Biol 2008;181:497—510.

[41] Simonsen A, Tooze SA. Coordination of membrane events during autophagy by multiple class III PI3-kinase complexes. J Cell Biol 2009;186:773—82.

[42] Mizushima N, Yamamoto A, Hatano M, Kobayashi Y, Kabeya Y, Suzuki K, et al. Dissection of autophagosome formation using Apg5-deficient mouse embryonic stem cells. J Cell Biol 2001;152:657—68.

[43] Walczak M, Martens S. Dissecting the role of the Atg12-Atg5-Atg16 complex during autophagosome formation. Autophagy 2013;9:424—5.

[44] Kabeya Y, Mizushima N, Ueno T, Yamamoto A, Kirisako T, Noda T, et al. LC3, a mammalian homologue of yeast Apg8p, is localized in autophagosome membranes after processing. EMBO J 2000;19:5720—8.

[45] Levine B, Klionsky DJ. Development by self-digestion: molecular mechanisms and biological functions of autophagy. Dev Cell 2004;6:463—77.

[46] Levine B. Eating oneself and uninvited guests: autophagy-related pathways in cellular defense. Cell 2005;120:159—62.

[47] Itakura E, Kishi C, Inoue K, Mizushima N. Beclin 1 forms two distinct phosphatidylinositol 3-kinase complexes with mammalian Atg14 and UVRAG. Mol Biol Cell 2008;19:5360—72.

[48] Chan EY, Tooze SA. Evolution of Atg1 function and regulation. Autophagy 2009;5:758—65.

[49] Abeliovich H, Dunn WA, Kim Jr J, Klionsky DJ. Dissection of autophagosome biogenesis into distinct nucleation and expansion steps. J Cell Biol 2000;151:1025—34.

[50] Koren I, Reem E, Kimchi A. DAP1, a novel substrate of mTOR, negatively regulates autophagy. Curr Biol 2010;20:1093—8.

[51] Goldring MB, Tsuchimochi K, Ijiri K. The control of chondrogenesis. J Cell Biochem 2006;97:33—44.

[52] Kronenberg HM. Developmental regulation of the growth plate. Nature 2003;423:332—6.

[53] Pizette S, Niswander L. Early steps in limb patterning and chondrogenesis. Novartis Found Symp 2001;232:23—36.

[54] Bobick BE, Chen FH, Le AM, Tuan RS. Regulation of the chondrogenic phenotype in culture. Birth Defects Res C Embryo Today 2009;87:351—71.

[55] DeLise AM, Fischer L, Tuan RS. Cellular interactions and signaling in cartilage development. Osteoarthr Cartil 2000;8:309—34.

[56] Shum L, Coleman CM, Hatakeyama Y, Tuan RS. Morphogenesis and dysmorphogenesis of the appendicular skeleton. Birth Defects Res C Embryo Today 2003;69:102—22.

[57] Pitsillides AA, Beier F. Cartilage biology in osteoarthritis—lessons from developmental biology. Nat Rev Rheumatol 2011;7:654—63.

[58] Horton WA, Royce PM, Steinman B, editors. Extracellular matrix and heritable disorder of connective tissue. New York, NY: Alan R. LIss; 1993.

[59] Doege KJ, Kreis T, Vale R, editors. Guidebook to the extracellular matrix, anchor and adhesion. Oxford: A Sambrook and Tooze Publication at Oxford University Press; 1999.

[60] Ruoslahti E, Yamaguchi Y. Proteoglycans as modulators of growth factor activities. Cell 1991;64:867—9.

[61] Schmid TM, Linsenmayer TF. Immunohistochemical localization of short chain cartilage collagen (type X) in avian tissues. J Cell Biol 1985;100:598—605.

[62] Naski MC, Colvin JS, Coffin JD, Ornitz DM. Repression of hedgehog signaling and BMP4 expression in growth plate cartilage by fibroblast growth factor receptor 3. Development 1998;125:4977—88.

[63] St-Jacques B, Hammerschmidt M, McMahon AP. Indian hedgehog signaling regulates proliferation and differentiation of chondrocytes and is essential for bone formation. Genes Dev 1999;13:2072—86.

[64] Andrade AC, Nilsson O, Barnes KM, Baron J. Wnt gene expression in the post-natal growth plate: regulation with chondrocyte differentiation. Bone 2007;40:1361—9.

[65] Ulici V, Hoenselaar KD, Agoston H, McErlain DD, Umoh J, Chakrabarti S, et al. The role of Akt1 in terminal stages of endochondral bone formation: angiogenesis and ossification. Bone 2009;45:1133—45.

[66] Zhang R, Murakami S, Coustry F, Wang Y, de Crombrugghe B. Constitutive activation of MKK6 in chondrocytes of transgenic mice inhibits proliferation and delays endochondral bone formation. Proc Natl Acad Sci USA 2006;103:365—70.

[67] Monemdjou R, Vasheghani F, Fahmi H, Perez G, Blati M, Taniguchi N, et al. Association of cartilage-specific deletion of peroxisome proliferator-activated receptor gamma with abnormal endochondral ossification and impaired cartilage growth and development in a murine model. Arthritis Rheum 2012;64:1551—61.

[68] Chen J, Long F. mTORC1 signaling controls mammalian skeletal growth through stimulation of protein synthesis. Development (Cambridge, England) 2014;141:2848—54.

[69] Chen J, Holguin N, Shi Y, Silva MJ, Long F. mTORC2 signaling promotes skeletal growth and bone formation in mice. J Bone Miner Res 2015;30:369—78.

[70] Phornphutkul C, Lee M, Voigt C, Wu KY, Ehrlich MG, Gruppuso PA, et al. The effect of rapamycin on bone growth in rabbits. J Orthop Res 2009;27:1157—61.

[71] Alvarez-Garcia O, Carbajo-Perez E, Garcia E, Gil H, Molinos I, Rodriguez J, et al. Rapamycin retards growth and causes marked alterations in the growth plate of young rats. Pediatr Nephrol (Berlin, Germany) 2007;22:954—61.

[72] Holstein JH, Klein M, Garcia P, Histing T, Culemann U, Pizanis A, et al. Rapamycin affects early fracture healing in mice. Br J Pharmacol 2008;154:1055—62.

[73] Guan Y, Yang X, Yang W, Charbonneau C, Chen Q. Mechanical activation of mammalian target of rapamycin pathway is required for cartilage development. FASEB J 2014;28:4470—81.

[74] Starkman BG, Cravero JD, Delcarlo M, Loeser RF. IGF-I stimulation of proteoglycan synthesis by chondrocytes requires activation of the PI 3-kinase pathway but not ERK MAPK. Biochem J 2005;389:723—9.

[75] Qureshi HY, Ahmad R, Sylvester J, Zafarullah M. Requirement of phosphatidylinositol 3-kinase/Akt signaling pathway for regulation of tissue inhibitor of metalloproteinases-3 gene expression by TGF-beta in human chondrocytes. Cell Signal 2007;19:1643—51.

[76] Gelse K, Muhle C, Knaup K, Swoboda B, Wiesener M, Hennig F, et al. Chondrogenic differentiation of growth factor-stimulated precursor cells in cartilage repair tissue is associated with increased HIF-1alpha activity. Osteoarthr Cartil 2008;16:1457—65.

[77] Terkeltaub R, Yang B, Lotz M, Liu-Bryan R. Chondrocyte AMP-activated protein kinase activity suppresses matrix degradation responses to proinflammatory cytokines interleukin-1beta and tumor necrosis factor alpha. Arthritis Rheum 2011;63:1928—37.

[78] Bohensky J, Leshinsky S, Srinivas V, Shapiro IM. Chondrocyte autophagy is stimulated by HIF-1 dependent AMPK activation and mTOR suppression. Pediatr Nephrol (Berlin, Germany) 2010;25:633–42.

[79] Viollet B, Horman S, Leclerc J, Lantier L, Foretz M, Billaud M, et al. AMPK inhibition in health and disease. Crit Rev Biochem Mol Biol 2010;45:276–95.

[80] Yang Y. Wnts and wing: Wnt signaling in vertebrate limb development and musculoskeletal morphogenesis. Birth Defects Res C Embryo Today 2003;69:305–17.

[81] Zhu M, Tang D, Wu Q, Hao S, Chen M, Xie C, et al. Activation of beta-catenin signaling in articular chondrocytes leads to osteoarthritis-like phenotype in adult beta-catenin conditional activation mice. J Bone Miner Res 2009;24:12–21.

[82] Day TF, Guo X, Garrett-Beal L, Yang Y. Wnt/beta-catenin signaling in mesenchymal progenitors controls osteoblast and chondrocyte differentiation during vertebrate skeletogenesis. Dev Cell 2005;8:739–50.

[83] Woo SL, Buckwalter JA. AAOS/NIH/ORS workshop. Injury and repair of the musculoskeletal soft tissues. Savannah, Georgia, June 18–20, 1987. J Orthop Res 1988;6:907–31.

[84] Buckwalter JA, Mankin HJ. Articular cartilage: tissue design and chondrocyte–matrix interactions. Instr Course Lect 1998;47:477–86.

[85] Maroudas A. Physiochemical properties of articular cartilage. In: Freeman MAR, editor. Adult articular cartilage. Kent: Cambridge University Press; 1979. p. 215–90.

[86] Sophia Fox AJ, Bedi A, Rodeo SA. The basic science of articular cartilage: structure, composition, and function. Sports Health 2009;1:461–8.

[87] Guilak F, Mow VC. The mechanical environment of the chondrocyte: a biphasic finite element model of cell–matrix interactions in articular cartilage. J Biomech 2000;33:1663–73.

[88] Billinghurst RC, Dahlberg L, Ionescu M, Reiner A, Bourne R, Rorabeck C, et al. Enhanced cleavage of type II collagen by collagenases in osteoarthritic articular cartilage. J Clin Invest 1997;99:1534–45.

[89] Mitchell PG, Magna HA, Reeves LM, Lopresti-Morrow LL, Yocum SA, Rosner PJ, et al. Cloning, expression, and type II collagenolytic activity of matrix metalloproteinase-13 from human osteoarthritic cartilage. J Clin Invest 1996;97:761–8.

[90] Glasson SS, Askew R, Sheppard B, Carito B, Blanchet T, Ma HL, et al. Deletion of active ADAMTS5 prevents cartilage degradation in a murine model of osteoarthritis. Nature 2005;434:644–8.

[91] Song RH, Tortorella MD, Malfait AM, Alston JT, Yang Z, Arner EC, et al. Aggrecan degradation in human articular cartilage explants is mediated by both ADAMTS-4 and ADAMTS-5. Arthritis Rheum 2007;56:575–85.

[92] Kapoor M, Martel-Pelletier J, Lajeunesse D, Pelletier JP, Fahmi H. Role of proinflammatory cytokines in the pathophysiology of osteoarthritis. Nat Rev Rheumatol 2011;7:33–42.

[93] Ollivierre F, Gubler U, Towle CA, Laurencin C, Treadwell BV. Expression of IL-1 genes in human and bovine chondrocytes: a mechanism for autocrine control of cartilage matrix degradation. Biochem Biophys Res Commun 1986;141:904–11.

[94] Morris EA, Treadwell BV. Effect of interleukin 1 on articular cartilage from young and aged horses and comparison with metabolism of osteoarthritic cartilage. Am J Vet Res 1994;55:138–46.

[95] Carames B, Taniguchi N, Otsuki S, Blanco FJ, Lotz M. Autophagy is a protective mechanism in normal cartilage, and its aging-related loss is linked with cell death and osteoarthritis. Arthritis Rheum 2010;62:791–801.

[96] Lopez de Figueroa P, Lotz MK, Blanco FJ, Carames B. Autophagy activation and protection from mitochondrial dysfunction in human chondrocytes. Arthritis Rheumatol (Hoboken, NJ) 2015;67:966–76.

[97] Tchetina EV, Poole AR, Zaitseva EM, Sharapova EP, Kashevarova NG, Taskina EA, et al. Differences in mammalian target of rapamycin gene expression in the peripheral blood and articular cartilages of osteoarthritic patients and disease activity. Arthritis 2013;2013:461–86.

[98] Weng T, Xie Y, Yi L, Huang J, Luo F, Du X, et al. Loss of Vhl in cartilage accelerated the progression of age-associated and surgically induced murine osteoarthritis. Osteoarthr Cartil 2014;22:1197–205.

[99] Sasaki H, Takayama K, Matsushita T, Ishida K, Kubo S, Matsumoto T, et al. Autophagy modulates osteoarthritis-related gene expression in human chondrocytes. Arthritis Rheum 2012;64:1920–8.

[100] Almonte-Becerril M, Navarro-Garcia F, Gonzalez-Robles A, Vega-Lopez MA, Lavalle C, Kouri JB. Cell death of chondrocytes is a combination between apoptosis and autophagy during the pathogenesis of Osteoarthritis within an experimental model. Apoptosis 2010;15:631–8.

[101] Chang J, Wang W, Zhang H, Hu Y, Wang M, Yin Z. The dual role of autophagy in chondrocyte responses in the pathogenesis of articular cartilage degeneration in osteoarthritis. Int J Mol Med 2013;32:1311–18.

[102] Carames B, Hasegawa A, Taniguchi N, Miyaki S, Blanco FJ, Lotz M. Autophagy activation by rapamycin reduces severity of experimental osteoarthritis. Ann Rheum Dis 2012;71:575–81.

[103] Takayama K, Kawakami Y, Kobayashi M, Greco N, Cummins JH, Matsushita T, et al. Local intra-articular injection of rapamycin delays articular cartilage degeneration in a murine model of osteoarthritis. Arthritis Res Ther 2014;16:482.

[104] Matsuzaki T, Matsushita T, Tabata Y, Saito T, Matsumoto T, Nagai K, et al. Intra-articular administration of gelatin hydrogels incorporating rapamycin-micelles reduces the development of experimental osteoarthritis in a murine model. Biomaterials 2014;35:9904–11.

[105] Liu N, Wang W, Zhao Z, Zhang T, Song Y. Autophagy in human articular chondrocytes is cytoprotective following glucocorticoid stimulation. Mol Med Rep 2014;9:2166–72.

[106] Song YW, Zhang T, Wang WB. Gluocorticoid could influence extracellular matrix synthesis through Sox9 via p38 MAPK pathway. Rheumatol Int 2012;32:3669–73.

[107] Rosenthal AK, Gohr CM, Mitton-Fitzgerald E, Grewal R, Ninomiya J, Coyne CB, et al. Autophagy modulates articular cartilage vesicle formation in primary articular chondrocytes. J Biol Chem 2015;290:13028–38.

[108] Boileau C, Martel-Pelletier J, Fahmi H, Mineau F, Boily M, Pelletier JP. The peroxisome proliferator-activated receptor gamma agonist pioglitazone reduces the development of cartilage lesions in an experimental dog model of osteoarthritis: *in vivo* protective effects mediated through the inhibition of key signaling and catabolic pathways. Arthritis Rheum 2007;56:2288–98.

[109] Osman B, Akool E-S, Doller A, Muller R, Pfeilschifter J, Eberhardt W. Differential modulation of the cytokine-induced MMP-9/TIMP-1 protease-antiprotease system by the mTOR inhibitor rapamycin. Biochem Pharmacol 2011;81:134–43.

[110] Lin TH, Tang CH, Wu K, Fong YC, Yang RS, Fu WM. 15-Deoxy-delta(12,14)-prostaglandin-J2 and ciglitazone inhibit TNF-alpha-induced matrix metalloproteinase 13 production via the antagonism of NF-kappaB activation in human synovial fibroblasts. J Cell Physiol 2011;226:3242–50.

[111] Chen J, Crawford R, Xiao Y. Vertical inhibition of the PI3K/Akt/mTOR pathway for the treatment of osteoarthritis. J Cell Biochem 2013;114:245–9.

[112] Britten CD, Adjei AA, Millham R, Houk BE, Borzillo G, Pierce K, et al. Phase I study of PF-04691502, a small-molecule, oral, dual inhibitor of PI3K and mTOR, in patients with advanced cancer. Invest New Drugs 2014;32:510–17.

[113] Carames B, Kiosses WB, Akasaki Y, Brinson DC, Eap W, Koziol J, et al. Glucosamine activates autophagy *in vitro* and *in vivo*. Arthritis Rheum 2013;65:1843–52.

[114] Huang MJ, Wang L, Jin DD, Zhang ZM, Chen TY, Jia CH, et al. Enhancement of the synthesis of n-3 PUFAs in fat-1 transgenic mice inhibits mTORC1 signalling and delays surgically induced osteoarthritis in comparison with wild-type mice. Ann Rheum Dis 2014;73:1719–27.

5

The Role of mTOR, Autophagy, Apoptosis, and Oxidative Stress During Toxic Metal Injury

Sarmishtha Chatterjee, Chayan Munshi and Shelley Bhattacharya

Environmental Toxicology Laboratory, Department of Zoology, Centre for Advanced Studies, Visva-Bharati (A Central University), Santiniketan, India

5.1 INTRODUCTION

Target of rapamycin (TOR) discovered in 1991, is a highly conserved protein kinase, essential for both fundamental and clinical biology [1]. The story of TOR-signaling begins with a remarkable drug, rapamycin [2]. Rapamycin is a lipophlic macrolide as well as a natural secondary metabolite, produced by *Streptomyces hygroscopicus*. *S. hygroscopicus* is a bacterium, isolated from soil in Easter Island, called Rapa-Nui in 1965, thus the name rapamycin [1,3] was coined. Intracellular rapamycin receptor in all eukaryotes is a small ubiquitous protein, called FK506 binding protein 12 kDa (FKBP12) [4,5]. Rapamycin–FKBP12 complex interacts with evolutionarily conserved TOR proteins to potently inhibit downstream effectors. A single mammalian TOR (mTOR) protein was cloned from several species and alternatively termed as mTOR, FKBP12 and rapamycin associated protein (FRAP), rapamycin and FKBP12 target, sirolimus effector protein, or rapamycin target [6–11]. mTOR is 289 kDa, sharing ~45% identity with the *Streptomyces cerevisiae* TOR proteins and ~56% identity with *Drosophila melanogaster* TOR as well as the human, rat, and mouse mTOR proteins share >95% identity at the amino acid level [9,12]. The TOR proteins function as Ser/Thr protein kinases assigned to a protein family termed the phosphatidylinositol kinase-related kinases (PIKKs) [13,14].

mTOR is a multidomain protein kinase that interacts with other proteins to form two main types of complex, mTOR complexes 1 and 2 (mTORC1 and mTORC2) [15]. mTORC1 associates with raptor, which binds to proteins that are direct substrates for mTORC1, and mTORC2 binds to rictor. Signaling through mTORC1 is much better understood than signaling through mTORC2. mTORC1 is an important node in cellular regulation impacting on cell growth that is linked to aging [15,16]. Structurally, mTOR contains 2549 amino acids and the region of first 1200 N-terminal amino acids contains up to 20 tandem repeated HEAT (a protein–protein interaction structure of two tandem antiparallel α-helices found in Huntingtin Elongation factor 3, PR65/A subunit of protein phosphatase 2A, and TOR) motifs [17]. HEAT repeat region is followed by a FAT [FRAP, ATM, and TRRAP (PIKK family members)] domain and FKPB12–rapamycin binding domain, which serves as a docking site for the rapamycin–FKBP12 complex. Downstream lies a catalytic kinase domain and a FATC (FAT carboxyterminal) domain, located at the C-terminus of the protein [18]. The interactions between FAT and FATC might contribute to the catalytic kinase activity of mTOR [19]. The domain structure of mTOR is sketched in Figure 5.1.

5.2 CELLULAR FUNCTIONS OF mTOR

mTOR is involved in many different cell functions to regulate cellular protein anabolism and catabolism [20,21]. If the cell has enough available amino acids, mTOR is free to signal to other molecules in the cell to build new proteins. On the other hand, if the cell is running low on nutrients, existing proteins and other cell components break down to free the building blocks which can be reused. The process by which the cell breaks down its own components is called autophagy [22], programmed cell death type II, which basically means "eating of the self" [23–26]. The degraded part of the cell is engulfed by a double membrane called phagophore, to separate from the rest of the cell, resulting in a membrane-enclosed bubble of cytosol, called autophagosome. Autophagosome eventually fuses with lysosomes, forming autophagolysosomes as the garbage disposal of the cell that can break down all sorts of cellular components. Key regulators of autophagy include the class I phosphoinositide-3-kinase (PI3K) [27] and adenosine-mono-phosphate activated protein kinase (AMPK) [28] in response to TOR kinase (Figure 5.2).

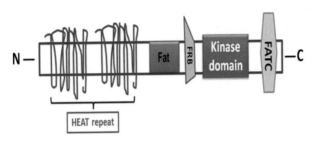

FIGURE 5.1   The domain structure of mTOR.

Activated PI3K signals TOR kinase, especially mTORC1 via protein kinase B (Akt kinase, belonging to the AGC protein kinase family) and modulates autophagy [22,29]. AMPK regulates autophagy through the control of mTORC1. AMPK indirectly inhibited TORC1 by phosphorylation of tuberous sclerosis complex-2 (TSC2) via deactivation of the Rheb GTPase (Ras homology enriched in brain-GTP binding protein) [30–32]. AMPK activation also induces phosphorylation of p53 on Ser15 domain. Phosphorylated p53 inhibits mTORC1 activity and regulates its downstream targets including autophagy [33–35].

In addition, mTOR has profound effects on the control of apoptosis [36]. Apoptosis, programmed cell death type I, plays an important role in various physiological and pathophysiological circumstances [37]. The intrinsic pathway of apoptosis deals with mitochondrial pathway, is initiated by a large variety of upstream stimuli and tightly modulated by various factors including, pro- and antiapoptotic proteins of the Bcl-2 family as well as PI3K/Akt/mTOR pathway [38–40]. The Bcl-2 family of proteins regulates apoptosis by controlling mitochondrial permeability. The proapoptotic Bcl-2 proteins [Bcl-2 associated X-protein (BAX) and proapoptotic BH3 only Bcl-2 family protein (BIM)] may reside in the cytosol but translocate to

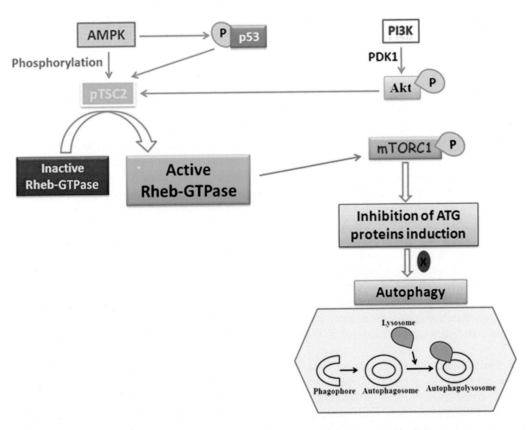

FIGURE 5.2   Role of mTOR in autophagy.

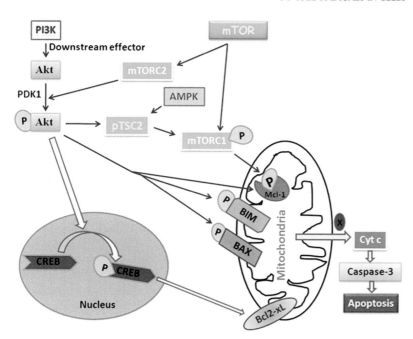

FIGURE 5.3   Role of mTOR in apoptosis.

mitochondria following death signaling and promote the release of cytochrome c (Cyt c) [40]. The PI3K signaling network diversifies into many distinct downstream branches, one of which leads to the phosphorylation of mTORC1 [41–45] on Thr 2446 and Ser 2448 sites by Akt kinase via TSC2 phosphorylation. Akt acts as an antiapoptotic factor that directly or indirectly antagonizes cell death signal transduction via the mitochondrial pathway [46]. Akt directly interferes with the phosphorylation of apoptosis-regulatory proteins which in turn results in a shift to pro- and antiapoptotic stimuli [47–49]. The multidomain Bcl-2 protein BAX and BIM are phosphorylated at serine residues Ser 184 and Ser87, respectively, in an Akt-dependent manner [49,50]. Akt-mediated direct and indirect (by mTORC1) phosphorylation of some antiapoptotic factors, such as myeloid leukemia cell sequence 1 (Mcl-1) located in mitochondrial matrix and vice versa, sometimes diminishes the antiapoptotic properties by decreasing their protein stability. Thus degradation of these proteins via the proteasomal machinery results in a reduction of Mcl-1 protein expression [48,51] which signals the release of Cyt c from mitochondria and induces the intrinsic apoptotic pathway.

On the other hand, mTORC2 elicits its activity as a potent antiapoptotic signal through Mcl-1 [52] and Akt phosphorylation [45,53]. mTORC2 phosphorylates Akt on Ser 473 and Thr 308 region by protein-dependent kinase 1 (PDK), however mTORC2 recognizes Ser 473 region in the hydrophobic motif of Akt as a kinase [53,54] and thereby converts inactive Akt to active Akt. Phosphorylated Akt/active Akt migrates to cytosol,

mitochondria, and nucleus. Nuclear Akt then fulfills the antiapoptotic role [55] through phosphorylation of transcription factors such as cAMP response element-binding (CREB) protein [40,56,57]. Phosphorylated CREB binds to DNA as a dimer via leucine zipper motif and recognizes cAMP-response elements (CREs). CRE then stimulates the transcription of genes and results in the expression of the antiapoptotic factor, Bcl-2 (Bcl2-xL) [58]. The regulation of apoptosis by PI3K/Akt/mTOR pathway is schematically depicted in Figure 5.3.

5.3 FREE RADICALS IN CELLS

Living cells are always subjected to the hazardous effects of exogenously or endogenously produced highly reactive oxidizing molecules and easily acquire electrons from oxidized molecules with which they remain in contact, such as all cellular biomolecules, generating chain reactions and ultimately leading to cell structure damage [59]. Products of oxidative metabolism that provoke oxidative injury are collectively called reactive oxygen species (ROS), which includes superoxide, H_2O_2, and the hydroxyl radical. They are formed either by the loss of a single electron from a nonradical or by the gain of a single electron by a nonradical. Lipid peroxidation is one of the well-studied consequences of ROS. Lipid peroxidation initiators, such as ROS, contribute to the signal transduction cascade that controls cell death and proliferation [60]. The free radicals can trigger cellular damage by covalent binding to macromolecules and enhance lipid peroxidation that depletes the stress of antioxidant enzymes. Inhibition of antioxidant enzymes leads to oxidative stress [61,62]. ROS has a major biological impact due to its endogenous production and high concentration in the cells. It is well known that the principal source of ROS in the cell is the mitochondrial respiratory chain [59,63]. Metal-mediated formation of free radicals causes various modifications to DNA bases, enhances lipid peroxidation, and alters calcium and sulfhydryl homeostasis [64]. Lipid peroxides, formed by the attack of radicals on polyunsaturated fatty acid residues of phospholipids, can further react with redox metals finally producing mutagenic and carcinogenic malondialdehyde, 4-hydroxynonenal, and other exocyclic DNA adducts (etheno and/or propano adducts). While iron (Fe), copper (Cu), chromium (Cr), and cobalt (Co) undergo redox-cycling reactions, for mercury (Hg), cadmium (Cd), and nickel (Ni), the primary route for their toxicity is depletion of glutathione and bonding to sulfhydryl groups of proteins [64,65]. Metal toxicity therefore generates oxidative stress in cell.

5.4 STRESS VIS-À-VIS mTOR

The TOR pathway is an evolutionarily conserved signaling module regulating cell growth in response to a variety of internal and external stimuli, generating oxidative stress [66,67]. mTOR acts in two categories in the surveillance of stress conditions, illustrated in Figure 5.4. The oxidative stress-input to mTOR

generally lies upstream of the TSC complex. ROS induces transcriptional alteration of cellular proteins and impinges the mTOR network signaling [66]. ROS in turn inhibits mTORC1 by direct phosphorylation of TSC2 and indirect activation of AMPK. AMPK results ROS generation, phosphorylates, and thereby activates TSC2, resulting in mTORC1 inhibition [31]. On the contrary, oxidative stress can persuade mTORC1 to play a distinct role in the last stage of autophagy [68,69]. Generally, mTORC1 forms a complex with autophagy-related proteins (ATG) named ATG1, ATG 13, and ATG101, which in turn phosphorylates ATG1 and ATG13 to inhibit the ATG activity in autophagy [28,70]. Induction of stress can reactivate mTORC1 and thereby breaks the mTORC1—ATG protein complex to free ATG1 [69]. ATG1 then leads to the formation of protolysosomal extension by lysosomal-associated membrane protein 1 and microtubule-associated protein 1A/1B light chain 3 (LC3) from autophagolysosome and ultimately mature into functional lysosome by autophagic lysosome reformation [68,69,71].

Toxicants can trigger cellular pathways classified broadly as death and survival signals. Each organ has its own critical threshold towards a toxicant, regulated by an intricate signaling system. Depending on the concentration and duration of exposure to toxicants, cells can utilize a coordinated, preprogrammed signaling pathway to maintain the normal homeostasis of the organ. Considering the complex and interactive mechanisms involved in maintenance of homeostasis of the organism, the present review is further aimed to

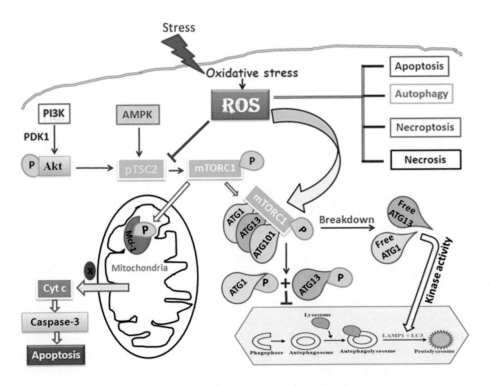

FIGURE 5.4 Oxidative stress and mTOR.

monitor the role of mTOR during cellular injury imposed by toxic metals. Generally, oxidative stress is considered as the first response to toxic metals and a volley of interactions follows to maintain the integrity of the cellular and molecular homeostasis through several coordinated signaling pathways. Toxic metals are usually those which are highly reactive to the lipid membranes of the cells, resulting in formation of injurious free oxygen or hydroxyl radicals. An attempt has been made to weigh the deleteriousness of metals of environmental concern affecting biological systems and the interrelationship of mTOR with different types of cell death.

5.5 ARSENIC (AS) TOXICITY

The main cause of toxic effects of metals is generation of ROS. It has been demonstrated that As is a potential inducer of oxidative stress as evidenced by DNA damage, by superoxide and hydrogen peroxide [72], and by disruption of mitosis while promoting apoptosis [73,74]. It has been amply demonstrated that inorganic As is more toxic than organo-As compounds [75]. Moreover, arsenite induces apoptosis much more strongly than arsenate [76,77] and there is a definite role of free radicals in As toxicity [78].

Various studies on the role of As in cellular apoptosis demonstrate 44.5% apoptosis at a concentration of $200\,\mu M$ arsenite of the JB6C141 cell types [79]; 10 and $20\,\mu M$ As_2O_3 initiated apoptosis in a neural cell line and $40\,\mu M$ caused death in these cells [80]. Paradoxically, As has both carcinogenic [81,82] and effective chemotherapeutic activity [83–85] in acute promyelocytic leukemia (APL), inducing apoptosis within a tumor cell population where a high cumulative dose (340–430 mg) of As_2O_3 was found to cause a complete remission of APL [86]. Thus, normal organs and tissues are liable to As toxicity evidenced by triggering of signaling pathways, such as caspase-dependent or -independent, mitogen-activated protein kinase (MAPK), c-Jun N-terminal kinase (JNK), extracellular-signal-regulated kinase (ERK), and p38 pathways.

Arsenic-induced apoptosis at pharmacological concentrations was mediated by caspase-dependent pathways through mitochondrial checkpoints at $10\,\mu M$ As_2O_3 where the activity of ERK and JNK is dependent on caspase activity, while p38 and MAPK were found to be operative in inducing apoptosis at $40\,\mu M$ of As_2O_3 [87]. As-induced apoptosis of chronic lymphocytic leukemia cells involves inactivation of the kinase AKT and a blockade of the transcription factor NF-κB, as well as up-regulation of PTEN and down-regulation of X-linked inhibitor of apoptosis protein [88]. As can paradoxically increase mTOR activation

and engagement of downstream mTOR effectors in BCR-ABL (tyrosine kinase inhibitor) expressing cells [89] or acute myeloid leukemia cells [90]. The combinations of As with mTORC1 inhibitor (mTORC1i) of rapamycin result in increased apoptosis and enhanced suppressive effects on primary leukemic progenitors [91]. Moreover, the combination of As and mTORC1i may overcome the feedback loop by decreasing activation of the MAPK and AKT signaling pathways in breast cancer [92]. Since carcinogenesis and apoptosis are both tightly linked to the PI3K/Akt/mTOR pathway, it could be expected that As^{3+} would differentially target the pathway to accomplish these opposite effects on cell fate [93].

As_2O_3 has been reported to induce autophagic cell death without activation of caspase-dependent apoptotic cell death [94–96]. Arsenic was reported to induce autophagy at a concentration of $2\,\mu M$ in malignant glioma cell lines [94] and at $6\,\mu M$ in lymphoblastoid cell lines [97]. Autophagy plays complex roles in As-induced death of human promyelocytic leukemia 60 cells; it inhibits As-induced apoptosis in the initiation stage, but amplifies As-mediated apoptotic program on persistent activation of autophagy [98]. Furthermore, carcinogenicity of As can fabricate overproduction of interleukin 6, which sufficiently blocks autophagy by supporting Beclin-1/Mcl-1/mTOR interaction [99].

5.6 CADMIUM (CD) TOXICITY

Cadmium (Cd), the toxic transitional metal, is mainly released from smelting and refining of metals and cigarette smoking, resulting in the pollution of water, air, and soil [65,100]. Cd compounds are used as stabilizers, color pigments, rechargeable batteries, alloys, and can be found in some fertilizers. Cd production, consumption, and emissions to the environment increased worldwide dramatically during the twentieth century [101]. Cd can be absorbed and accumulated in plants and animals and thereby can be accumulated in the human body either through direct exposure to Cd-contaminated environment or by the food chain. Such accumulation contributes to carcinogenesis, immunodepression, and neurodegeneration [100,102,103]. A very long biological half-life of Cd may be responsible for the accumulation in many human organs, including kidney [104], liver [105], lung [106], testis [107], bone [108], and blood system [109]. Cd has a high blood–brain barrier permeability; therefore chronic Cd exposure affects the nervous system, including learning disabilities and hyperactivity in children [110,111], olfactory dysfunction, neurobehavioral defects in attention, psychomotor speed, and memory in industrial workers [112] and also Parkinson's disease by acute Cd poisoining [113].

Cd toxicity is amplified as a consequence of the long biological half-life of the metal and has been associated with blockage of oxidative phosphorylation, glutathione depletion, inhibition of antioxidant enzymatic activity, production of oxidative stress, DNA damage, reduction of protein synthesis, and cell death [65,114]. Cd-induced toxicity is closely associated with the production of ROS [115−117]. Cd accelerates the generation of ROS by depleting cellular glutathione and antioxidant enzymes, such as superoxide dismutase and catalase [115,118] and by inhibiting the electron transfer chain in the mitochondria [119]. It is surmised that the effect of Cd toxicity depends on its concentration and the duration of exposure inducing both autophagic- and apoptotic-related pathways via ROS generation [65,120,121].

However, the mechanisms underlying Cd-induced autophagy are not yet completely understood. Cd (0−10 μM) has been found to induce autophagy through ROS-dependent activation of the serine/threonine kinase 11/liver kinase B1—AMPK—mTOR signaling in skin epidermal cells in a concentration- and time-dependent manner [122]. Cd could also provoke oxidative stress generating ROS inside JB6 cells, which in turn signals energy-sensing LKB1-AMP-activated protein kinase pathway after 12 h of Cd treatment (10 μM) and the effect is further augmented at 24 h after the treatment. AMPK, a heterotrimeric protein complex, is downstream of LKB1 in a signaling pathway that regulates energy homeostasis [123,124]. LKB/AMPK signaling can act as the upstream of mTOR signaling [125,126]. Cd-induced intracellular level of ROS signals LKB1, which in turn phosphorylates AMPK and thus inhibits mTORC1 triggering autophagy in JB6 cells in a significant manner [123,124].

It is commonly accepted that Cd induces apoptotic cell death in cellular systems [127,128]. However, Cd induces apoptosis by either caspase-dependent [129,130] or caspase-independent mechanisms [131,132] depending on the target cell type. Cd-induced apoptotic cell death in JB6 cells was found to be dependent on Ca^{2+}- and H_2O_2-mediated JNK and p53 signaling [133]. Interestingly, autophagic cell death was also induced by Cd (0−10 μM) by the ROS-dependent LKB1-AMPK [122]. Moreover, 10 μM of Cd not only causes apoptosis in skin epidermal cells (JB6) but also leads to necrotic cell death [133] in a time-dependent manner from 1 to 24 h. Generation of ROS due to Cd toxicity elevates the intracellular calcium (Ca^{2+}) level. The increase of free calcium ions is believed to be one of the main mechanisms involved in Cd-induced apoptosis by stimulating the generation of H_2O_2 [133]. Ca^{2+} regulates JNK signaling and also effects H_2O_2 generation. In turn H_2O_2 generated

by Cd regulates both activator protein-1 (AP-1) and p53 pathways in JB6 cells. Collectively, Cd promotes apoptosis mainly through Ca^{2+}-JNK-growth arrest and DNA-damaged inducible protein (GADD45α) as well as a caspase-independent pathway, H_2O_2-AP-1-p53-apoptosis-inducing-actor (AIF) signaling cascades [133]. Ca^{2+} also leads to the activation of MAPKs in Cd-exposed mesangial cells [120] and in tributyltin-treated human T-cell lines [134]. It has been reported [100,135] that Cd at 0−120 mmol/L and 10 and/or 20 μM, respectively, induced neuronal apoptosis by the activation of MAPK and mTOR as well as JNK and PTEN-Akt/mTOR network. Cd toxicity activates ROS production mediated by NADPH oxidase, which in turn activates MAPK [136−138]. Ca^{2+}-dependent protein kinase and PI3K also stimulate MAPK during Cd toxicity [138]. Phospho-p38-MAPK signals GADD45α and induces apoptosis by caspase-3 activation [136,139,140]. JNK and PTEN-Akt/mTOR network diminishes antiapoptotic properties by decreasing their protein stability, which eventually releases Cyt c from mitochondria and induces caspase-dependent apoptosis [51,135]. The effects of Cd toxicity are schematically summarized in Figure 5.5.

5.7 LEAD (PB) TOXICITY

Lead (Pb), a well-known heavy metal, can cause pathophysiological changes in several organ systems including the central nervous system (CNS), cardiovascular, hematopoietic, reproductive, gastrointestinal, and renal systems [141,142]. Pb had modest early uses in ancient medicines and cosmetics, however, in the modern era it has had industrial uses in building materials, paints, and gasoline [143]. Pb has become ubiquitous in the environment. Pb exposure mainly occurs through the respiratory and gastrointestinal systems and absorbed Pb is stored in soft tissues. Inhaled Pb (~30−40%) enters the bloodstream [144]; once absorbed, 99% of Pb is retained in the blood for approximately 30−35 days and over the following 4−6 weeks it is dispersed and accumulated in other tissues [145]. Pb serves no useful purpose in the human body and its unfortunate presence in the body can lead to toxic effects, regardless of the exposure pathway adopted [146]. The effects of Pb on Ca^{2+} fluxes and Ca^{2+}-regulated events are the major mechanisms of lead neurotoxicity [142,147,148]. Pb can cause abnormal hyperphosphorylation of tubulin-associated unit (tau) protein and accumulation of α-synuclein, inducing hippocampal injury affecting the ability to learning and causing memory damage. Pathological hyperphosphorylation and aggregation of tau deposits as neurofibrillary tangles in the brains of those with Alzheimer's disease and other related neurodegenerative disorders are called tauopathies [142,149]. α-Synuclein,

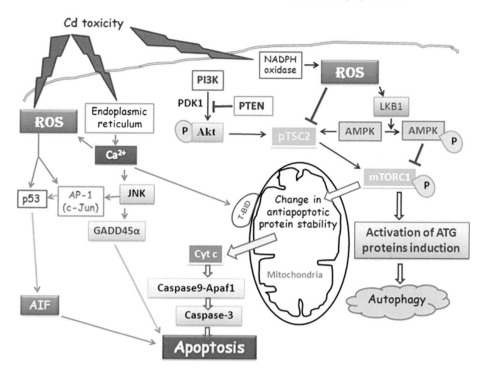

FIGURE 5.5   Toxic signals of Cd.

expressed in the CNS neurons and localized around synaptic vesicles in presynaptic terminals can cause Parkinson's disease or a related disorder, dementia with Lewy bodies through genetic alteration or point mutation [142,150]. The deleterious effects of Pb exposures can involve endoplasmic reticulum (ER) stress in cardiofibroblasts which consecutively generate ROS and a direct depletion of the antioxidant reserves [145,148]. Significant effects have been found on various fundamental cellular processes like protein folding and maturation, ionic transportation, enzyme regulation leading to cell stress [151], and finally cell death, such as autophagy and apoptosis [142,148].

ER has an essential role in multiple cellular processes, such as the folding of secretory membrane proteins, calcium homeostasis, and lipid biosynthesis. When ER trans-membrane sensors detect the accumulation of unfolded proteins, the unfolded protein response (UPR) is initiated to cope with the resulting ER stress [152]. Under stress conditions, a cell performs several adaptive alterations such as autophagy and apoptosis. Pb toxicity can induce robust UPR and ER stress which can mediate autophagy and/or apoptosis to protect the internal system of the cell [142,148]. The regulation of autophagy and to some extent apoptosis is mediated by mTORC1 [32,36]. mTORC1 activity is critical for de novo protein synthesis; the reduced mTORC1 activity may decrease the rate of synthesis of new proteins, which may partly attenuate ER stress on protein folding and processing reactions [148]. Recently it has been reported that Pb

toxicity inhibits the mTORC1 pathway [142,148]. Depending on the concentration and duration of Pb exposure, the internal cellular system then decides to undertake the cellular homeostatic pathway, autophagy/apoptosis, mediated by the ATG proteins. Inhibition of mTORC1 easily shifts the cell towards the induction of autophagy via ATG1–ATG13–ATG101 complex and Beclin-1 (ATG6)–ATG14 complex [23,25]. On the other hand, Beclin-1 plays a significant role in the network of cellular homeostasis, focusing on the cross-regulation between autophagy and apoptosis [153,154]. Pb toxicity can induce apoptosis to promote the activation of caspase-3 via Beclin-1 [142]. It cleaves Beclin-1 from the Beclin-1–ATG14 complex, triggers proapoptotic proteins in mitochondria to release Cyt c which consequently induces apoptosis [153]. The effects of Pb toxicity are portrayed in Figure 5.6.

5.8 MERCURY (HG) TOXICITY

Hg is a well-established environmental toxicant. Hg is known to cause several physiological and biochemical disturbances in mammals. Most of these biochemical reactions are exerted through formation of covalent bonds between sulfur and Hg. Three forms of Hg exist: elemental, inorganic, and organic. Each of them has its own profile of toxicity. Elemental Hg vapor is highly lipid-soluble allowing it to readily cross cellular membranes. It can also be oxidized to mercuric salts from monovalent Hg

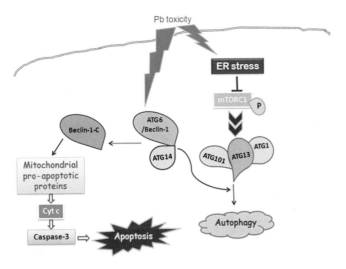

FIGURE 5.6   Toxic signals of Pb.

pathways. Oxidative stress and PI3K, Akt, and mTOR pathways are intimately related to determine the cell fate [166]. The significant biological role of PI3K, Akt, and mTOR is involved in the modulation of apoptosis, autophagy, disorders of cellular metabolism, acute nervous system injury, and chronic neurodegeneration [20,21,166−168]. Moreover, MTOR activates CREB1 through suppression of autophagy, sensitizing subsequent apoptosis [169]. Different metals seem to prefer different signaling pathways and display diverse potential to trigger mTOR pathway cascades [49,135,148]. mTOR signaling pathways integrate both intracellular and extracellular signals and serves as a central regulator of metal toxicity in cellular adaptation. In conclusion, this review elucidates an important role for mTOR modulation of oxidative stress and cell death signaling due to toxic metal injury.

compounds. Thus, when ingested, Hg will be more rapidly absorbed and produce greater toxicity [155].

The distribution kinetics of radioactive Hg in different hepatocellular fractions revealed that Hg treatment increases nuclear and liposomal protein content significantly [156]. On the other hand, generation of an acute-phase response was observed in rabbit treated with mercuric chloride (HgCl$_2$) and remarkable changes in phospholipid profile and acetylcholinesterase activity, [157] while in rat Hg has been found to induce C reactive protein along with a metallothionein-like protein [158]. HgCl$_2$ is also known to damage DNA in fibroblasts of rat and mouse embryos [159,160] and exposure to a nonlethal dose of Hg induces oxidative-stress-mediated apoptosis in rat liver [87,161]. Five micromolar HgCl$_2$ was shown to damage cellular DNA by a nonapoptotic mechanism in the U-937 cell line [162]. Interestingly, recent research [35,163−165] revealed that 5 μM HgCl$_2$ drives autophagy and apoptosis in different cell types of rat liver. Covalent-conjugation of autophagic proteins (ATG5-ATG12-LC3B) stimulates cell death in rat hepatocytes, modulated by Keap1-p62, ERK-p38, and DRAM-p53 regulator proteins through inhibition of mTOR [35,164]. Concurrently, autophagic proteins ATG12 and LC3B exerted a significant role in triggering apoptosis in 5 μM HgCl$_2$-treated rat oval cells [165].

5.9 CONCLUDING REMARKS

The last decade has seen a rapid rise in interest in mTOR signaling cascades. Metal affronted oxidative stress determines the down-regulation and up-regulation of mTOR signaling to maintain cellular homeostasis by different programmed cell death

5.10 FUTURE PERSPECTIVES

Metal-induced oxidative stress plays a significant role in the modulation of apoptosis and autophagy. In this aspect, the PI3K, Akt, and mTOR cascade offers new inroads to the development of novel therapies. Nowadays, with increasing use of nanoparticles in the cosmetics and food industries, attention needs to be focused on metal-nanoparticle-induced oxidative stress and PI3K/Akt/mTOR signaling pathway. It appears to be an important mechanism in manifesting pathways leading to several cardiovascular diseases. New research is focusing on mTOR signaling components in the maintenance of insulin signaling to prevent diabetes mellitus. However, mTOR signaling can also trigger cellular proliferative pathways that can promote aggressive tumor growth. On the other hand, prevention of PI3K and Akt activation can block medulloblastoma growth. Toxicology studies have unraveled that the PI3K/Akt/mTOR pathway is commonly targeted by toxins, especially metals, to produce several hazardous conditions. The PI3K/Akt/mTOR targeted therapy shows therapeutic efficacy, albeit with a variety of toxic side effects. A thorough understanding of the role of the mTOR pathway, with a systemic perspective and computational modeling appears to be the future center of interest.

Acknowledgments

SC is grateful to the National Academy of Sciences, India (NASI), for the award of Research Associate and SB acknowledges NASI for the award of Senior Scientist Platinum Jubilee Fellowship. The authors are grateful to Shuvasree Sarkar, UGC-BSR SRF, for critically going through the manuscript.

References

[1] Loewith R, Hall MN. Target of rapamycin (TOR) in nutrient signaling and growth control. Genetics 2011;189:1177−201.

[2] Benjamin D, Colombi M, Moroni C, Hall MN. Rapamycin passes the torch: a new generation of mTOR inhibitors. Nat Rev Drug Discov 2011;10:868−80.

[3] Vézina C, Kudelski A, Sehgal SN. Rapamycin (AY-22,989), a new antifungal antibiotic. I. Taxonomy of the producing streptomycete and isolation of the active principle. J Antibiot 1975;28:721−6.

[4] Harding MW, Galat A, Uehling DE, Schreiber SL. A receptor for the immunosuppressant FK506 is a cis-trans peptidyl-prolyl isomerase. Nature 1989;341:758−60.

[5] Siekierka JJ, Wiederrecht G, Greulich H, Boulton D, Hung SH, et al. The cytosolic-binding protein for the immunosuppressant FK-506 is both a ubiquitous and highly conserved peptidyl-prolyl cis-trans isomerase. J Biol Chem 1990;265:21011−15.

[6] Brown EJ, Albers MW, Shin TB, Ichikawa K, Keith CT, et al. A mammalian protein targeted by G1-arresting rapamycin-receptor complex. Nature 1994;369:756−8.

[7] Chen Y, Chen H, Rhoad AE, Warner L, Caggiano TJ, et al. A putative sirolimus (rapamycin) effector protein. Biochem Biophys Res Commun 1994;203:1−7.

[8] Chiu MI, Katz H, Berlin V. RAPT1, a mammalian homolog of yeast Tor, interacts with the FKBP12/rapamycin complex. Proc Natl Acad Sci USA 1994;91:12574−8.

[9] Raught B, Gingras AC, Sonenberg N. The target of rapamycin (TOR) proteins. Proc Natl Acad Sci USA 2001;98:7037−44.

[10] Sabatini DM, Erdjument-Bromage H, Lui M, Tempst P, Snyder SH. RAFT1: a mammalian protein that binds to FKBP12 in a rapamycin-dependent fashion and is homologous to yeast TORs. Cell 1994;78:35−43.

[11] Sabers CJ, Martin MM, Brunn GJ, Williams JM, Dumont FJ, et al. Isolation of a protein target of the FKBP12−rapamycin complex in mammalian cells. J Biol Chem 1995;270:815−22.

[12] Abraham RT, Wiederrecht GJ. Immunopharmacology of rapamycin. Annu Rev Immunol 1996;14:483−510.

[13] Hoekstra MF. Responses to DNA damage and regulation of cell cycle checkpoints by the ATM protein kinase family. Curr Opin Genet Dev 1997;7:170−5.

[14] Hunter T. When is a lipid kinase not a lipid kinase—when it is a protein-kinase. Cell 1995;83:1−4.

[15] Hands SL, Proud CG, Wyttenbach A. mTOR's role in ageing: protein synthesis or autophagy? Aging 2009;1:586−97.

[16] Blagosklonny MV, Hall MN. Growth and aging: a common molecular mechanism. Aging 2009;1:357−62.

[17] Tchevkina E, Komelkov A. Protein phosphorylation as a key mechanism of mTORC1/2 signaling pathways. Intech Open Sci 2012. Available from: <http://dx.doi.org/10.5772/48274> [accessed in April 2015].

[18] Gingras AC, Raught B, Sonenberg N. Regulation of translation initiation by FRAP/mTOR. Genes Dev 2001;15:807−26.

[19] Perry J, Kleckner N. The ATRs, ATMs, and TORs are giant HEAT repeat proteins. Cell 2003;112:151−5.

[20] Laplante M, Sabatini DM. mTOR signaling in growth control and disease. Cell 2012;149:274−93.

[21] Sengupta S, Peterson TR, Sabatini DM. Regulation of the mTOR complex 1 pathway by nutrients, growth factors, and stress. Mol Cell 2010;40:310−22.

[22] Bergmann A. Autophagy and cell death: no longer at odds. Cell 2007;131:1032−3.

[23] Barth S, Glick D, Macleod KF. Autophagy: assays and artifacts. J Pathol 2010;221:117−24.

[24] Glick D, Barth S, Macleod KF. Autophagy: cellular and molecular mechanisms. J Pathol 2010;221:3−12.

[25] Klionsky DJ, Abeliovich H, Agostinis P, Agrawal DK, Aliev G, et al. Guidelines for the use and interpretation of assays for monitoring autophagy in higher eukaryotes. Autophagy 2008;4:151−75.

[26] Mizushima N, Ohsumi Y, Yoshimori T. Autophagosome formation in mammalian cells. Cell Struct Funct 2002;27:421−9.

[27] Blommaart EF, Krause U, Schellens JP, Vreeling-Sindelarova H, Meijer AJ. The phosphatidylinositol 3-kinase inhibitors wortmannin and LY294002 inhibit autophagy in isolated rat hepatocytes. Eur J Biochem 1997;243:240−6.

[28] Roach PJ. AMPK→ULK1→autophagy. Mol Cell Biol 2011;31:3082−4.

[29] Mizushima N, Levine B, Cuervo AM, Klionsky DJ. Autophagy fights disease through cellular self digestion. Nature 2008;451:1069−74.

[30] Gwinn DM, Shackelford DB, Egan DF, Mihaylova MM, Mery A, et al. AMPK phosphorylation of raptor mediates a metabolic checkpoint. Mol Cell 2008;30:214−26.

[31] Inoki K, Li Y, Xu T, Guan KL. Rheb GTPase is a direct target of TSC2 GAP activity and regulates mTOR signaling. Gene Dev 2003;17:1829−34.

[32] Martin TD, Chen XW, Kaplan REW, Saltiel AR, Walker CL, et al. Ral and Rheb GTPase activating proteins integrate mTOR and GTPase signaling in aging, autophagy, and tumor cell invasion. Mol Cell 2014;53:209−20.

[33] Budanov AV, Karin M. p53 target genes sestrin1 and sestrin2 connect genotoxic stress and mTOR signaling. Cell 2008;134:451−60.

[34] Feng Z, Zhang H, Levine AJ, Jin S. The coordinate regulation of the p53 and mTOR pathways in cells. Proc Natl Acad Sci USA 2005;102:8204−9.

[35] Chatterjee S, Ray A, Mukherjee S, Agarwal S, Kundu R, et al. Low concentration of mercury induces autophagic cell death in rat hepatocytes. Toxicol Ind Health 2012;30:611−20.

[36] Castedo M, Ferri KF, Kroemer G. Mammalian target of rapamycin (mTOR): pro- and anti-apoptotic. Cell Death Differ 2002;9:99−100.

[37] Lockshin RA, Zakeri Z. Cell death in health and disease. J Cell Mol Med 2007;11:1214−24.

[38] Engelman JA. Targeting PI3K signaling in cancer: opportunities, challenges and limitations. Nat Rev Cancer 2009;9:550−62.

[39] Fulda S, Galluzzi L, Kroemer G. Targeting mitochondria for cancer therapy. Nat Rev Drug Discov 2010;9:447−64.

[40] Fulda S. Shifting the balance of mitochondrial apoptosis: therapeutic perspectives. Front Oncol 2012;2:1−4.

[41] Scott PH, Brunn GJ, Kohn AD, Roth RA, Lawrence JC. Evidence of insulin-stimulated phosphorylation and activation of the mammalian target of rapamycin mediated by a protein kinase B signaling pathway. Proc Natl Acad Sci USA 1998;95:7772−7.

[42] Nave BT, Ouwens M, Withers DJ, Alessi DR, Shepherd PR. Mammalian target of rapamycin is a direct target for protein kinase B: identification of a convergence point for opposing effects of insulin and amino-acid deficiency on protein translation. J Biochem 1999;344:427−31.

[43] Sekulic A, Hudson CC, Homme JL, Yin P, Otterness DM, et al. A direct linkage between the phosphoinositide 3-kinase−AKT signaling pathway and the mammalian target of rapamycin in mitogen-stimulated and transformed cells. Cancer Res 2000;60:3504−13.

[44] Reynolds TH, Bodine SC, Lawrence JC. Control of Ser2448 phosphorylation in the mammalian target of rapamycin by insulin and skeletal muscle load. J Biol Chem 2002;277:17657−62.

[45] Hay N, Sonenberg N. Upstream and downstream of mTOR. Genes Dev 2004;18:1926−45.

[46] Shaw RJ, Cantley LC. Ras, PI(3)K and mTOR signaling controls tumour cell growth. Nature 2006;441:424−30.

[47] Gardai SJ, Hildeman DA, Frankel SK, Whitlock BB, Frasch SC, et al. Phosphorylation of Bax Ser184 by Akt regulates its activity and apoptosis in neutrophils. J Biol Chem 2004;279: 21085−95.

[48] Dan HC, Sun M, Kaneko S, Feldman RI, Nicosia SV, et al. Akt phosphorylation and stabilization of X-linked inhibitor of apoptosis protein (XIAP). J Biol Chem 2004;279:5405−12.

[49] Qi XJ, Wildey GM, Howe PH. Evidence that Ser87 of BimEL is phosphorylated by Akt and regulates BimEL apoptotic function. J Biol Chem 2006;281:813−23.

[50] Yamaguchi H, Wang HG. The protein kinase PKB/Akt regulates cell survival and apoptosis by inhibiting Bax conformational change. Oncogene 2001;20:7779−86.

[51] Maurer U, Charvet C, Wagman AS, Dejardin E, Green DR. Glycogen synthase kinase-3 regulates mitochondrial outer membrane permeabilization and apoptosis by destabilization of MCL-1. Mol Cell 2006;21:749−60.

[52] Mills JR, Hippo Y, Robert F, Chen SMH, Malina A, et al. mTORC1 promotes survival through translational control of Mcl-1. Proc Natl Acad Sci USA 2008;105:10853−8.

[53] Martelli AM, Evangelisti C, Chiarini F, Blalock WL, Papa V, et al. The Phosphatidylinositol 3-Kinase/Akt/Mammalian target of rapamycin signaling network as a new target for acute myelogenous leukemia therapy. Cancer Ther 2007;5:309−30.

[54] Martelli AM, Evangelisti C, Chiarini F, Grimaldi C, Cappellini A, et al. The emerging role of the phosphatidylinositol 3-kinase/ Akt/mammalian target of rapamycin signaling network in normal myelopoiesis and leukemogenesis. Biochim Biophys Acta 2010;1803:991−1002.

[55] Martelli AM, Faenza I, Billi AM, Manzoli L, Evangelisti C, et al. Intranuclear 3′-phosphoinositide metabolism and Akt signaling: new mechanisms for tumorigenesis and protection against apoptosis? Cell Signal 2006;18:1101−7.

[56] Du K, Montminy M. CREB is a regulatory target for the protein kinase Akt/PKB. J Biol Chem 1998;273:32377−9.

[57] Wang L, Piguet AC, Schmidt K, Tordjmann T, Dufour JF. Activation of CREB by tauroursodeoxycholic acid protects cholangiocytes from apoptosis induced by mTOR inhibition. Hepatology 2005;41:1241−51.

[58] Wilson BE, Mochon E, Boxer LM. Induction of bcl-2 expression by phosphorylated CREB proteins during B-cell activation and rescue from apoptosis. Mol Cell Biol 1996;16:5546−56.

[59] Filomeni G, Zio DD, Cecconi F. Oxidative stress and autophagy: the clash between damage and metabolic needs. Cell Death Differ 2015;22:377−88.

[60] Marnett LJ. Lipid peroxidation—DNA damage by malondialdehyde. Mutat Res 1999;424:83−95.

[61] Kyle ME, Micaadei S, Nakae D, Farber JL. Superoxide dismutase and catalase protect cultured hepatocytes from the cytotoxicity of acetaminophen. Biochem Biophys Res Commun 1987;149:889−96.

[62] Kurose I, Higuchi H, Miura S, Saito H, Watanabe N, et al. Oxidative stress mediated apoptosis of hepatocytes exposed to acute ethanol intoxication. Hepatology 1997;25:368−78.

[63] Lee J, Giordano S, Zhang J. Autophagy, mitochondria and oxidative stress: cross-talk and redox signaling. J Biochem 2012;441:523−40.

[64] Valko M, Morris H, Cronin MT. Metals, toxicity and oxidative stress. Curr Med Chem 2005;12:1161−208.

[65] Chatterjee S, Sarkar S, Bhattacharya S. Toxic metals and autophagy. Chem Res Toxicol 2014;27:1887−900.

[66] Reiling JH, Sabatini DM. Stress and mTOR signaling. Oncogene 2006;25:6373−83.

[67] Faghiri Z, Bazana NG. PI3K/Akt and mTOR/p70S6K pathways mediate neuroprotectin D1-induced retinal pigment epithelial cell survival during oxidative stress-induced apoptosis. Exp Eye Res 2010;90:718−25.

[68] Yu L, McPhee CK, Zheng L, Mardones GA, Rong Y, et al. Autophagy termination and lysosome reformation regulated by mTOR. Nature 2010;465:942−6.

[69] Hayat MA. Autophagy: cancer, other pathologies, inflammation, immunity, infection, and aging. Autophagy 2015;5:1−48. Available from: <http://dx.doi.org/10.1016/B978-0-12-801033-4.00001-1>.

[70] Jung CH, Jun CB, Ro SH, Kim YM, Otto NM, et al. ULK-Atg13-FIP200 complexes mediate mTOR signaling to the autophagy machinery. Mol Cell Biol 2009;7:1992−2003.

[71] Chen G, Yu L. Autophagic lysosome reformation. Exp Cell Res 2012;319:142−6.

[72] Kitchin KT. Recent advances in arsenic carcinogenesis: modes of action, animal model systems, and methylated arsenic metabolites. Toxicol Appl Pharmacol 2001;172:249−61.

[73] States JC, Reiners JJ, Pounds JG, Kaplan DJ, Beauerle BD, et al. Arsenite disrupts mitosis and induces apoptosis in SV40-transformed human skin fibroblasts. Toxicol Appl Pharmacol 2002;180:83−91.

[74] Datta S, Saha DR, Ghosh D, Majumdar T, Bhattacharya S, et al. Sub-lethal concentration of arsenic interferes with the proliferation of hepatocytes and induces in vivo apoptosis in *Clarias batrachus* L. Comp Biochem Physiol C 2007;145(Suppl. 3):339−49.

[75] Kaise T, Yamauchi H, Horiuchi Y, Tani T, Watanabe S, et al. A comparative study on acute toxicity of methylarsonic acid, dimethylarsinic acid and trimethylarsine oxide in mice. Appl Organomet Chem 1989;3:273−7.

[76] Ma DC, Sun YH, Chang KZ, Ma XF, Huang SL, et al. Selective induction of apoptosis of NB4 cells from G2-M phase by sodium arsenite at lower doses. Eur J Haematol 1998;61:27−35.

[77] Zeng H. Arsenic suppresses necrosis induced by selenite in human leukemia HL-60 cells. Biol Trace Elem Res 2001;83:1−15.

[78] Bashir S, Sharma Y, Irshad M, Nag TC, Tiwari M, et al. Arsenic induced apoptosis in rat liver following repeated 60 days exposure. Toxicology 2006;217:63−70.

[79] Chen NY, Ma WY, Yang CS, Dong Z. Inhibition of arsenite-induced apoptosis and AP-1 activity by epigallocatechin-3-gallate and theoflavins. J Environ Pathol Toxicol Oncol 2000;19:287−95.

[80] Milton AG, Zalewski PD, Ratnaike RN. Zinc protects against arsenic-induced apoptosis in a neuronal cell line, measured by DEVD-caspase activity. Biometals 2004;17:707−13.

[81] Lu T, Liu J, LeCluyse EL, Zhou YS, Cheng ML, et al. Application of cDNA microarray to the study of arsenic-induced liver diseases in the population of Guizhou, China. Toxicol Sci 2001;59:185−92.

[82] Schuliga M, Chouchane S, Snow ET. Up regulation of glutathione-related genes and enzyme activities in cultured human cells by sublethal concentrations of inorganic arsenic. Toxicol Sci 2002;70:183−92.

[83] Look AT. Arsenic and apoptosis in the treatment of acute promyelocytic leukemia. J Natl Cancer Inst 1998;90:86−8.

[84] Kann S, Estes C, Reichard JF, Huang MY, Sartor MA, et al. Butylhydroquinone protects cells genetically deficient in glutathione biosynthesis from arsenite-induced apoptosis without significantly changing their prooxidant status. Toxicol Sci 2005;87(Suppl. 2):365−84.

[85] Liu ZM, Huang HS. As$_2$O$_3$-induced c-Src/EGFR/ERK signaling is via Sp1 binding sites to stimulate p21WAF1/CIP1 expression in human epidermoid carcinoma A431 cells. Cell Signal 2005;18:244−55.

[86] Huang SY, Yang CH, Chen YC. As$_2$O$_3$ therapy for acute promyelocytic leukemia: an unusual salvage therapy. Leuk Lymphoma 2000;38:283–93.

[87] Ray A, Roy S, Agarwal S, Bhattacharya S. As$_2$O$_3$ toxicity in rat hepatocytes: manifestation of caspase mediated apoptosis. Toxicol Ind Health 2008;24:643–53.

[88] Redondo-Muñoz J, Escobar-Díaz E, Hernández del Cerro M, Pandiella A, Terol MJ, et al. Induction of B-chronic lymphocytic leukemia cell apoptosis by arsenic trioxide involves suppression of the PI3K/Akt survival pathway via JNK activation and PTEN upregulation. Clin Cancer Res 2010;16:4382–91.

[89] Yoon P, Giafis N, Smith J, Mears H, Katsoulidis E, et al. Activation of mammalian target of rapamycin and the p70 S6 kinase by arsenic trioxide in BCR-ABL-expressing cells. Mol Cancer Ther 2006;5:2815–23.

[90] Altman JK, Yoon P, Katsoulidis E, Kroczynska B, Sassano A, et al. Regulatory effects of mammalian target of rapamycin-mediated signals in the generation of arsenic trioxide responses. J Biol Chem 2008;283:1992–2001.

[91] Goussetis DJ, Platanias LC. Arsenic trioxide and the PI3K/AKT pathway in chronic lymphocytic leukemia. Clin Cancer Res 2010. Available from: http://dx.doi.org/10.1158/1078-0432.CCR-10-0072.

[92] Guilbert C, Annis MG, Dong Z, Siegel PM, Miller Jr WH, et al. Arsenic trioxide overcomes rapamycin-induced feedback activation of AKT and ERK signaling to enhance the anti-tumor effects in breast cancer. PLOS One 2013;8:1–12.

[93] Li T, Wang G. Computer-aided targeting of the PI3K/Akt/mTOR Pathway: toxicity reduction and therapeutic opportunities. Int J Mol Sci 2014;15:18856–91.

[94] Kanzawa T, Kondo Y, Ito H, Kondo S, Germano I. Induction of autophagic cell death in malignant glioma cells by arsenic trioxide. Cancer Res 2003;63:2103–8.

[95] Kanzawa T, Zhang L, Xiao L, Germano IM, Kondo Y, et al. Arsenic trioxide induces autophagic cell death in malignant glioma cells by upregulation of mitochondrial cell death protein BNIP3. Oncogene 2005;24:980–91.

[96] Tsai SC, Yang JS, Peng SF, Lu CC, Chiang JH, et al. Bufalin increases sensitivity to AKT/mTOR-induced autophagic cell death in SK-HEP-1 human hepatocellular carcinoma cells. Int J Oncol 2012;41:1431–42.

[97] Bolt M, Byrd RM, Klimecki WT. Autophagy is a biological target of arsenic. In: Jean JS, Bundschuh J, Bhattacharya P, editors. Arsenic in geosphere and human diseases. Boca Raton, FL: CRC Press; 2010. p. 291–2.

[98] Yang YP, Liang ZQ, Gao B, Jia YL, Qin ZH. Dynamic effects of autophagy on arsenic trioxide-induced death of human leukemia cell line HL60 cells. Acta Pharmacol Sin 2008;29:123–34.

[99] Qi Y, Zhang M, Li H, Frank JA, Dai L, et al. Autophagy inhibition by sustained overproduction of IL6 contributes to arsenic carcinogenesis. Cancer Res 2014;74:3740–52.

[100] Chen L, Liu L, Luo Y, Huang S. MAPK and mTOR pathways are involved in cadmium-induced neuronal apoptosis. J Neurochem 2008;105:251–61.

[101] Järup L. Hazards of heavy metal contamination. Br Med Bull 2003;68:167–82.

[102] Lopez E, Figueroa S, Oset-Gasque M, Gonzalez MP. Apoptosis and necrosis: two distinct events induced by cadmium in cortical neurons in culture. Br J Pharmacol 2003;138:901–11.

[103] Kim S, Moon C, Eun S, Ryu P, Jo S. Identification of ASK1, MKK4, JNK, c-Jun, and caspase-3 as a signaling cascade involved in cadmium-induced neuronal cell apoptosis. Biochem Biophys Res Commun 2005;328:326–34.

[104] Johri N, Jacquillet G, Unwin R. Heavy metal poisoning: the effects of cadmium on the kidney. Biometals 2010;23:783–92.

[105] Koyu A, Gokcimen A, Ozguner F, Bayram DS, Kocak A. Evaluation of the effects of cadmium on rat liver. Mol Cell Biochem 2006;284:81–5.

[106] Jiang G, Xu L, Song S, Zhu C, Wu Q, et al. Effects of long-term low-dose cadmium exposure on genomic DNA methylation in human embryo lung fibroblast cells. Toxicology 2008;244:49–55.

[107] Thompson J, Bannigan J. Cadmium: toxic effects on the reproductive system and the embryo. Reprod Toxicol 2008;25:304–15.

[108] Akesson A, Bjellerup P, Lundh T, Lidfeldt J, Nerbrand C, et al. Cadmium-induced effects on bone in a population-based study of women. Environ Health Perspect 2006;114:830–4.

[109] Kocak M, Akcil E. The effects of chronic cadmium toxicity on the hemostatic system. Pathophysiol Haemost Thromb 2006;35:411–16.

[110] Pihl R, Parkes M. Hair element content in learning disabled children. Science 1977;198:204–6.

[111] Wright RO, Amarasiriwardena C, Woolf AD, Jim R, Bellinger DC. Neuropsychological correlates of hair arsenic, manganese and cadmium levels in school-age children residing near a hazardous waste site. Neurotoxicology 2006;27:210–16.

[112] Järup L, Persson B, Edling C, Elinder CG. Renal function impairment in workers previously exposed to cadmium. Nephrone 1993;64:75–81.

[113] Okuda B, Iwamoto Y, Tachibana H, Sugita M. Parkinsonism after acute cadmium poisoning. Clin Neurol Neurosurg 1997;99:263–5.

[114] Templeton DM, Liu Y. Multiple roles of cadmium in cell death and survival. Chem Biol Interact 2010;188:267–75.

[115] Bagchi D, Joshi SS, Bagchi M, Balmoori J, Benner EJ, et al. Cadmium- and chromium-induced oxidative stress, DNA damage, and apoptotic cell death in cultured human chronic myelogenous leukemic K562 cells, promyelocytic leukemic HL-60 cells, and normal human peripheral blood mononuclear cells. J Biochem Mol Toxicol 2000;14:33–41.

[116] Szuster-Ciesielska A, Stachura A, Slotwinska M, Kaminska T, Sniezko R, et al. The inhibitory effect of zinc on cadmium-induced cell apoptosis and reactive oxygen species (ROS) production in cell cultures. Toxicology 2000;145:159–71.

[117] Son YO, Jang YS, Heo JS, Chung WT, Choi KC, et al. Apoptosis-inducing factor plays a critical role in caspase-independent, pyknotic cell death in hydrogen peroxide-exposed cells. Apoptosis 2009;14:796–808.

[118] Shaikh ZA, Vu TT, Zaman K. Oxidative stress as a mechanism of chronic cadmium-induced hepatotoxicity and renal toxicity and protection by antioxidants. Toxicol Appl Pharmacol 1999;154:256–63.

[119] Wang Y, Fang J, Leonard SS, Rao KM. Cadmium inhibits the electron transfer chain and induces reactive oxygen species. Free Radic Biol Med 2004;36:1434–43.

[120] Wang SH, Shih YL, Kuo TC, Ko WC, Shih CM. Cadmium toxicity toward autophagy through ROS-activated GSK-3beta in mesangial cells. Toxicol Sci 2009;108:124–31.

[121] Chiarelli R, Agnello M, Roccheri MC. Seaurchin embryos as a model system for studying autophagy induced by cadmium stress. Autophagy 2011;7:1028–34.

[122] Son YO, Wang X, Hitron JA, Zhang Z, Cheng S, et al. Cadmium induces autophagy through ROS-dependent activation of the LKB1-AMPK signaling in skin epidermal cells. Toxicol Appl Pharmacol 2011;255:287–96.

[123] Hardie DG, Scott JW, Pan DA, Hudson ER. Management of cellular energy by the AMP-activated protein kinase system. FEBS Lett 2003;546:113–20.

[124] Shaw RJ, Kosmatka M, Bardeesy N, Hurley RL, Witters LA, et al. The tumor suppressor LKB1 kinase directly activates

AMP-activated kinase and regulates apoptosis in response to energy stress. Proc Natl Acad Sci USA 2004;101:3329–35.

[125] Ruggero D, Pandolfi PP. Does the ribosome translate cancer? Nat Rev Cancer 2003;3:179–92.

[126] Sabatini DM. mTOR and cancer: insights into a complex relationship. Nat Rev Cancer 2006;6:729–34.

[127] Fujimaki H, Ishido M, Nohara K. Induction of apoptosis in mouse thymocytes by cadmium. Toxicol Lett 2000;115:99–105.

[128] Watjen W, Haase H, Biagioli M, Beyersmann D. Induction of apoptosis in mammalian cells by cadmium and zinc. Environ Health Perspect 2002;110:865–7.

[129] Hossain S, Liu HN, Nguyen M, Shore G, Almazan G. Cadmium exposure induces mitochondria-dependent apoptosis in oligoden-drocytes. Neurotoxicology 2009;30:544–54.

[130] Wang SH, Shih YL, Ko WC, Wei YH, Shih CM. Cadmium-induced autophagy and apoptosis are mediated by a calcium signaling pathway. Cell Mol Life Sci 2008;65:3640–52.

[131] Mao WP, Ye JL, Guan ZB, Zhao JM, Zhang C, et al. Cadmium induces apoptosis in human embryonic kidney (HEK) 293 cells by caspase-dependent and—independent pathways acting on mitochondria. Toxicol In Vitro 2007;21:343–54.

[132] Shih CM, Wu JS, Ko WC, Wang LF, Wei YH, et al. Mitochondria-mediated caspase-independent apoptosis induced by cadmium in normal human lung cells. J Cell Biochem 2003;89:335–47.

[133] Son YO, Lee JC, Hitron JA, Pan J, Zhang Z, et al. Cadmium induces intracellular Ca^{2+}- and H_2O_2-dependent apoptosis through JNK- and p53-mediated pathways in skin epidermal cell line. Toxicol Sci 2010;113:127–37.

[134] Yu ZP, Matsuoka M, Wispriyono B, Iryo Y, Igisu H. Activation of mitogen-activated protein kinases by tributyltin in CCRF-CEM cells: role of intracellular Ca(2þ). Toxicol Appl Pharmacol 2000;168:200–7.

[135] Chen S, Gu C, Xu C, Zhang J, Xu Y, et al. Celastrol prevents cadmium-induced cell death via targeting JNK and PTEN-Akt/mTOR network. J Neurochem 2014;128:256–66.

[136] Wada T, Penninger JM. Mitogen-activated protein kinases in apoptosis regulation. Oncogene 2004;23:2838–49.

[137] Liu Q, Hofmann PA. Protein phosphatase 2A-mediated crosstalk between p38 MAPK and ERK in apoptosis of cardiac myocytes. Am J Physiol Heart Circ Physiol 2004;286:H2204–12.

[138] Yeh JH, Huang CC, Yeh MY, Wang JS, Lee JK, et al. Cadmium-induced cytosolic Ca $^{2+}$ elevation and subsequent apoptosis in renal tubular cells. Basic Clin Pharmacol Toxicol 2009;104:345–51.

[139] Khreiss T, József L, Hossain S, Chan JSD, Potempa LA, et al. Loss of pentameric pymmetry of C-reactive Protein is associated with delayed apoptosis of human neutrophils. J Biol Chem 2002;277:40775–81.

[140] Sarkar D, Su ZZ, Lebedeva IV, Sauane M, Gopalkrishnan RV, Valerie K, et al. mda- 7 (IL-24) mediates selective apoptosis in human melanoma cells by inducing the coordinated over expression of the GADD family of genes by means of p38 MAPK. Proc Natl Acad Sci USA 2002;99:10054–9.

[141] Kosnett MJ, Wedeen RP, Rothenberg SJ, Hipkins KL, Materna BL, et al. Recommendations for medical management of adult lead exposure. Environ Health Perspect 2007;115:463–71.

[142] Zhang J, Cai T, Zhao F, Yao T, Chen Y, et al. The role of α-synuclein and tau hyperphosphorylation-mediated autophagy and apoptosis in lead-induced learning and memory injury. Int J Biol Sci 2012;8:935–44.

[143] Mudipalli A. Lead hepatotoxicity & potential health effects. Indian J Med Res 2007;126:518–27.

[144] Philip AT, Gerson B. Lead poisoning—Part I. Incidence, etiology, and toxicokinetics. Clin Lab Med 1994;14:423–44.

[145] Patrick L. Lead toxicity, a review of the literature. Part 1: exposure, evaluation and treatment. Altern Med Rev 2006;11:2–22.

[146] Yedjou CG, Milner JN, Howard CB, Tchounwou PB. Basic apoptotic mechanisms of lead toxicity in human leukemia (Hl-60) cells. Int J Environ Res Public Health 2010;7:2008–17.

[147] Bressler J, Kim KA, Chakraborti T, Goldstein G. Molecular mechanisms of lead neurotoxicity. Neurochem Res 1999;24:595–600.

[148] Sui L, Zhang P, Yun KL, Zhang HC, Liu L, et al. Lead toxicity induces autophagy to protect against cell death through mTORC1 pathway in cardiofibroblasts. Biosci Rep 2015. Available from: http://dx.doi.org/10.1042/BSR20140164.

[149] Maccioni RB, Cambiazo V. Role of microtubule-associated proteins in the control of microtubule assembly. Physiol Rev 1995;75:835–64.

[150] Gupta A, Dawson VL, Dawson TM. What causes cell death in Parkinson's disease? Ann Neurol 2008;2:3–15.

[151] Garza A, Vega R, Soto E. Cellular mechanisms of lead neurotoxicity. Med Sci Monit 2006;12:57–65.

[152] Braakman I, Bulleid NJ. Protein folding and modification in the mammalian endoplasmic reticulum. Annu Rev Biochem 2011;80:71–99.

[153] Wirawan E, VandeWalle L, Kersse K, Cornelis S, Claerhout S, et al. Caspase-mediated cleavage of Beclin-1 inactivates Beclin-1-induced autophagy and enhances apoptosis by promoting the release of proapoptotic factors from mitochondria. Cell Death Dis 2010. Available from: http://dx.doi.org/10.1038/cddis.2009.16.

[154] Kang R, Zeh HJ, Lotze MT, Tang D. The Beclin 1 network regulates autophagy and apoptosis. Cell Death Differ 2011;18:571–80.

[155] Broussard LA, Hammett-Stabler CA, Winecker RE, Ropero-Miller JD. The toxicology of mercury. Lab Med 2002;33:614–25.

[156] Bose S, Ghosh P, Ghosh S, Chaudhury S, Bhattacharya S. Time dependent distribution [^{203}Hg] mercuric nitrate in the subcellular fraction of rat and fish liver. Biomed Environ Sci 1993;6:195–206.

[157] Ghosh N, Bhattacharya S. Acute phase response of rabbit to $HgCl_2$ and $CdCl_2$. Biomed Environ Sci 1993;6:1–7.

[158] Vinaya Kumar S, Maitra S, Bhattacharya S. In vitro binding of inorganic mercury to the plasma membrane of rat platelet affects Na^+-K^+-ATPase activity and platelet aggregation. BioMetals 2002;15:51–7.

[159] Cantoni O, Christie NT, Robison SH, Costa M. Characterization of DNA lesions produced by $HgCl_2$ in cell culture systems. Chem Biol Interact 1984;49:209–24.

[160] Zasukhina GD, Vasilyeva IM, Sdirkova NI, Krasovsky GN, Vasyukovich LY, et al. Mutagenic effect of thallium and mercury salts on rodent cells with different repair activities. Mutat Res 1983;124:163–73.

[161] Patnaik BB, Ray A, Agarwal S, Bhattacharya S. Induction of oxidative stress by non-lethal dose of mercury in rat liver: possible relationships between apoptosis and necrosis. J Environ Biol 2010;31:413–16.

[162] Ben-Ozer E, Rosenspire A, McCabe M, Worth R, Kindzelskii A, Warra N, et al. Mercuric chloride damages cellular DNA by non-apoptotic mechanism. Mutat Res 2000;470:19–27.

[163] Chatterjee S, Banerjee PP, Chattopadhyay A, Bhattacharya S. Low concentration of $HgCl_2$ drives rat hepatocytes to

autophagy/apoptosis/necroptosis in a time-dependent manner. Toxicol Environ Chem 2013;95:1192−207.

[164] Chatterjee S, Nandi P, Chattopadhyay A, Bhattacharya S. Regulation of autophagy in rat hepatocytes treated *in vitro* with low concentration of mercury. Toxicol Environ Chem 2013;95:504−14.

[165] Chatterjee S, Munshi C, Chattopadhyay A, Bhattacharya S. Mercuric chloride effects on adult rat oval cells-induced apoptosis. Toxicol Environ Chem 2014;95:1722−38.

[166] Maiese K, Chong ZZ, Wang S, Shang YC. Oxidant stress and signal transduction in the nervous system with the PI 3-K, Akt, and mTOR cascade. Int J Mol Sci 2012;13:13830−66.

[167] Qin AP, Liu CF, Qin YY, Hong LZ, Xu M, Yang L, et al. Autophagy was activated in injured astrocytes and mildly decreased cell survival following glucose and oxygen deprivation and focal cerebral ischemia. Autophagy 2010;6:738−53.

[168] Chong ZZ, Maiese K. Mammalian target of rapamycin signaling in diabetic cardiovascular disease. Cardiovasc Diabetol 2012;11. Available from: http://dx.doi.org/10.1186/1475-2840-11-45.

[169] Wang Y, Hua Z, Liua Z, Chena R, Penga H, Guoa J, et al. MTOR inhibition attenuates DNA damage and apoptosis through autophagy-mediated suppression of CREB1. Autophagy 2013;12:2069−86.

mTOR IN GENETIC DISORDERS AND NEURODEGENERATIVE DISEASE

6

The mTOR Signaling Pathway in Neurodegenerative Diseases

Arnaud Francois[1,2], Julie Verite[3], Agnès Rioux Bilan[3], Thierry Janet[3], Frédéric Calon[1,2], Bernard Fauconneau[3], Marc Paccalin[3,4,5,6] and Guylène Page[3]

[1]Neurosciences Axis, Centre de recherche du CHU de Québec, Québec, QC, Canada [2]Faculty of Pharmacy, Université Laval, Québec, QC, Canada [3]EA3808 Molecular Targets and Therapeutics of Alzheimer's Disease, Poitiers University Hospital, University of Poitiers, Poitiers, France [4]Geriatrics Department, Poitiers University Hospital, Poitiers, France [5]Centre Mémoire de Ressources et de Recherche, Poitiers, France [6]INSERM CIC-P 1402, Poitiers University Hospital, Poitiers, France

6.1 INTRODUCTION

The mammalian target of rapamycin (mTOR) is a highly conserved serine-threonine kinase composed of two protein complexes, mTORC1 and mTORC2. mTORC1 is sensitive to rapamycin and implicated in the regulation of autophagy, lysosome biogenesis, energy metabolism, protein and lipid synthesis. The second complex, mTORC2, is involved in the cellular cytoskeleton organization, cell survival, and metabolism [1–3]. mTOR complexes play a central role in cells, and are regulated by multiple signaling pathways as described previously and illustrated in Figure 6.1. Briefly, mTORC1 can be activated by growth factors, insulin, and insulin-like growth factor (IGF) through the phosphoinositide 3 kinase (PI3K)-Akt-tuberous sclerosis complexes 1 and 2 (TSC1, TSC2)-Ras homolog enriched in brain (Rheb) axis [4]. Interestingly, Akt can directly phosphorylate proline-rich Akt substrate 40 kDa (PRAS40), leading to mTORC1 activation [5]. Furthermore, mTORC1 is also under inhibitory control of TSC2 through the signaling pathway double-stranded RNA-dependent protein kinase (PKR)-p53-regulated in development and DNA damage response 1 (REDD1) [6]. Cellular energy and more precisely the increasing of the adenosine monophosphate (AMP)/adenosine triphosphate (ATP) ratio (cellular energy

deficiency) activates AMP-activated protein kinase (AMPK)/REDD1/TSC2 inhibiting mTORC1 [7]. Moreover, AMPK can phosphorylate regulatory-associated protein of mTOR (Raptor), resulting to mTORC1 inhibition [8]. Upon activation, mTORC1 phosphorylates its two major downstream targets, p70 ribosome S6 kinase (p70S6K) and eukaryotic initiation factor 4E-binding protein 1, leading to the release of eukaryotic initiation factor 4E (eIF4E), which is involved in protein synthesis. In comparison to mTORC1 regulation, mTORC2 regulation is poorly understood. Only growth factors stimulate mTORC2 kinase activity, which is mediated through PI3K-dependent mTORC2–ribosome association [4,9].

For several years, many researches into mTOR demonstrated its key role in a wide range of disorders, including diabetes, obesity, neoplasia, seizures, autism spectrum disorders, and neurodegeneration [3]. In the central nervous system (CNS), mTOR regulates synaptic remodeling and long-term potentiation [10–13]. In pathogenic conditions, overactivation of the mTOR pathway is implicated in the formation of erroneous connections between neurons in epilepsy [14,15] and after trauma [16,17]. Moreover, chronically mTOR activation in anorexigenic neurons in the hypothalamus is associated with resistance to the effects of leptin and

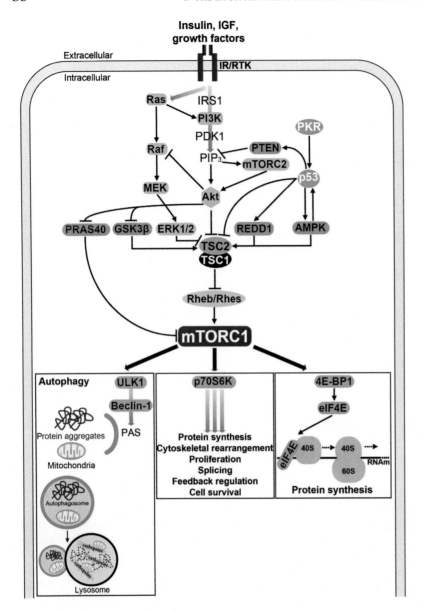

FIGURE 6.1 Scheme of the mTOR signaling pathway. The mTOR pathway could be activated by multiple extracellular signals (insulin, IGF, growth factors) through IR and/or RTK receptor. Different signaling pathways modulate mTORC1 activation. In green, are proteins which activate mTORC1 and, in orange, proteins that decrease its activation. mTORC1 modulates numerous cellular pathways, some of these mentioned in this chapter and resumed in this scheme are: autophagy by the activation of ULK1, protein synthesis by p70S6K and 4E-BP1 activation, cytoskeletal rearrangement, proliferation, splicing, feedback regulation, and cell survival by p70S6K activation. 4E-BP, eIF4E-binding proteins; AMPK, AMP-activated protein kinase; eIF4E, eukaryotic initiation factor 4E; GSK3β, glycogen-synthase kinase-3β; IGF-1, insulin-like growth factor-1; IR, insulin receptor; mTOR, mammalian target of rapamycin; mTORC1/2, mammalian target of rapamycin complex 1/2; p70S6K, p70 ribosome S6 kinase; PI3-K, phosphoinositide 3 kinase; PKR, double-stranded RNA-dependent protein kinase; PRAS40, proline-rich Akt substrate 40 kDa; PTEN, phosphatase and tensin homolog; REDD1, regulation of DNA damage response 1; Rheb, Ras homolog enriched in brain; Rhes, Ras homolog enriched in striatum; RTK, receptor tyrosine kinase.

could influence food intake [18,19]. In Down's syndrome and neurodegenerative disorders, the mTOR signaling axis appears deregulated and contributes to the progression of disease. However, the exact contribution of mTOR signaling in these disorders is complex and poorly understood. In addition, no therapy is yet available to stop and/or reverse the progression of these diseases. The identification of new drug targets remains a huge priority in research on neurodegenerative disorders and mTOR could be an interesting therapeutic target [20].

In this chapter we review recent researches of the mTOR role in major neurodegenerative disorders, including Alzheimer's disease (AD), Parkinson's disease (PD), Huntington's disease (HD), amyotrophic lateral sclerosis (ALS), and multiple sclerosis (MS).

6.2 mTOR IN AD

AD is an age-related and progressive neurodegenerative disorder, characterized clinically by progressive cognitive impairments and pathological signature including β amyloid (Aβ) deposits in the form of extracellular plaques and tau protein aggregates in the form of intracellular neurofibrillary tangles (NFTs) [21,22]. AD is considered to be the main neurodegenerative disorder, affecting over 20 million people worldwide currently, with about 135 million people expected to be diagnosed by 2050. The familial AD form represents less than 2% of all presentations of AD and concerns autosomal dominant mutations in amyloid precursor protein (APP), presenilins 1 or 2 genes [23]. Most AD patients have a sporadic form with a multifactorial etiology. In this form, the first risk factor is aging and the second is the presence of apolipoprotein ε4 allele [23]. Other risk factors increase the susceptibility to develop AD, including cerebrovascular diseases, diabetes, obesity, susceptibility genes, and head trauma [24—28]. Numerous alterations are reported as neuritic plaques and NFTs, oxidative stress, cellular metabolism dysfunction, impairment of proteolysis mechanism and autophagy/proteasome, blood—brain barrier failure, xenobiotic injury, glutamate excitotoxicity, mitochondrial damage, acetylcholine loss, and inflammation dysregulation [29—32]. These alterations were linked to hyper- or hypoactivation of a plethora of cellular signaling pathways such as PI3K/Akt/ mTOR, PKR, intracellular calcium, c-Jun N-terminal kinases, nicotinic acetylcholine receptor signaling, Wnt, AMPK, and silent mating-type information regulator 2 homolog 1 (Sirt1) [29—37].

Analysis of postmortem AD brain revealed hyperactivation of the axis PI3K-Akt-mTOR signaling pathway and of downstream effectors such as p70S6K, 4E-BP1, and eukaryotic elongation factor 2 [38—42].

An interesting study demonstrated that only mTORC1 is activated in hippocampus of AD patients at the severe stage [40]. Recently, a postmortem brain analysis in inferior parietal lobules from non-AD, preclinical AD (PCAD), mild cognitive impairment (MCI), and AD patients showed an mTOR hyperactivation in MCI and AD patients but not in PCAD [43].

During AD, a brain insulin-resistance and a decrease in glucose consumption are observed in AD brain [44—46] as well as a hyperactivation of the insulin receptor (IR)-PI3K-Akt axis [41,47]. Interestingly, Aβ monomers and soluble oligomers can bind directly to IRs [48,49], induce internalization in neurons, and remove IR from dendrites [49,50], mimicking what is described in the AD brain [51].

Concerning the level of mTOR activation in the animal model of AD, contradictory results were observed depending on transgenic AD mouse models and age. Indeed, we previously showed decreased mTOR activation and its downstream p70S6K in two different lines of APP/PS1 transgenic mouse model at 6 months old in the hippocampus and at 12 months old in the cortex [52,53]. Other authors demonstrated an mTOR pathway hyperactivation in 3xTg-AD mouse models [54] and in hippocampus of non-transgenic mice injected with Aβ oligomers [55]. The Aβ effect on mTOR pathway hyperactivity could be dependent on the phosphorylation of the PRAS40 [55] that positively regulates mTOR activity [56,5].

For in vitro studies, similar heterogeneous data are reported. In mixed primary cell culture and in neuroblastoma cells, we showed a reduction in mTOR activation and p70S6K using Aβ1-42 [57,52]. This inhibition of mTOR was also demonstrated in peripheral blood mononuclear cells (PBMCs) of AD patients and was

under the control of the axis PKR-p53-TSC1/TSC2-REDD1 [58,6]. Other studies demonstrated hyperactivation of mTOR using Aβ25-35 [59,60] or stably expressed the 7PA2 familial AD mutation in CHO cells [54].

Taken together these in vivo and in vitro data suggest a dose-dependent effect of Aβ on mTOR signaling, with low and more physiological levels of Aβ causing an increase in mTOR signaling, while significantly higher Aβ concentrations decreases mTOR signaling [61]. Moreover, these opposite results sustain a neuronal dichotomy: apoptotic neurons are characterized by an inhibition of the mTOR signaling pathway while hyperactivated neurons with activated Akt, p70S6K could produce NFTs [62,39].

An indirect link between the mTOR signaling pathway and Aβ was recently revealed through the mTOR enhancer Rheb reported as being decreased in AD brain. In addition, the Rheb reduction is correlated to β-site APP-cleaving enzyme 1 (BACE1) sustaining a facilitation of the Aβ production [63]. BACE1 is the rate-limiting and principal enzyme responsible for Aβ generation in neurons [64]. In the same meticulous study, in vitro and in vivo analysis revealed that Rheb regulates the BACE1 degradation through the proteasome and autophagy, surprisingly in an mTOR-independent manner and by a direct interaction between Rheb and BACE 1 [63].

The other major AD hallmark, tau hyperphosphorylation is positively correlated with the mTOR hyperactivity and its downstream targets in AD brain [42]. In the Drosophila tauopathy model, tau hyperphosphorylation promotes mTOR hyperactivation, promoting neurodegeneration and can be reversed by genetic/ pharmacologic inhibition of mTOR activity [65,66]. In addition, mTOR activation could be able to mediate localization and secretion of tau [67], facilitate its hyperphosphorylation directly and indirectly by inhibiting the protein phosphatase (PP2A) [68,69] and through the p70S6K activation that improved the mRNA tau translation [70]. The tau hyperphosphorylation can be prevented in vitro and in vivo by the inhibitory action of the biguanide metformin on the mTOR activity [71].

Although generally not considered pathognomonic features of AD, the inflammation is present in the AD brain, contributing to its progression [72]. The brain immune cells, microglia, seem to become ineffective with the AD progression to clear Aβ [73] as macrophages [74,75]. T cells are found in brain closed to Aβ deposits and activated microglia [76—78]. Interestingly, the mTOR pathway controls the differentiation of different T-helper cell subsets [79] and the production of cytokines. More precisely, mTORC1 limits the production of proinflammatory cytokines such as interleukin (IL)-12, IL-23, IL-6, and tumor necrosis factor-α through reducing the activity of the transcription factor NF-κB and promotes the expression of anti-inflammatory cytokines such as IL-10 or tumor growth factor-β and of type I interferons [80−82]. Previously we reported a significant decrease in mTOR and p70S6K activation in PBMCs from AD patients and negatively correlated with the Mini Mental State Examination scores [52,83,54]. More recently, a study confirmed our observation and showed a poorer response of lymphocytes from AD patients to the mTOR inhibitor, rapamycin, than healthy elderly control subjects [85]. The mTOR pathway alteration in immune cells certainly contributes to the impaired inflammatory reaction observed in AD.

Another essential role of mTOR is to control the cell proteolysis system, autophagy, which is reported to be impaired in AD [86]. In AD brain and in transgenic mouse models of AD, an accumulation of autophagic vesicles (AVs) in dystrophic neurites was observed [87,88], contributing to the pathology progression by increasing the Aβ production directly in AVs [89−91]. Multiple signaling pathways regulating autophagy were reported to be compromised during AD: mTOR, AMPK, glycogen-synthase kinase-3β (GSK3β) activation, and calcium (Ca^{2+}) signaling [41,92−94]. Other alterations, such as a drastic reduction of beclin-1 brain expression (autophagic protein initiator) [95,96] and a defective lysosomal acidification [97−99], contribute to the proteolysis impairments and the AVs accumulation. The autophagy molecular understanding raises a great deal of attention but its restoration in AD gives some ambiguous results. Indeed, the restoration of autophagy initiation by beclin-1 re-expression in brains of APP J20 transgenic mice [95] and by the rapamycin treatment in 3xTg-AD mice [54] and in J20 transgenic mice [100] can decrease amyloid deposits and tau levels in 3xTg-AD mice. On the other hand, the autophagy stimulation by rapamycin in cell lines expressing the APP mutation *Sweedish* increases the production of Aβ. The correction of lysosomal acidification by L803-mts, a GSK3β inhibitor [101] or by nicotinamide, a NAD^+ precursor [102], improves the Aβ clearance and decreases tau levels. Despite apparent inconsistency, one may propose that the relationships between autophagy and Aβ may change with the progression of the disease. At early stages of Aβ accumulation, induction of autophagy may facilitate its clearance. As the disease progresses, deficiencies in the clearance of autophagic vacuoles may occur and thus further increase autophagy, which may exacerbate the AD phenotype [61].

Recently, another level of complexity emerged in AD and involves mTOR, the relationship between autophagy and inflammation. Indeed, its deregulation was previously showed in autoimmune pathologies, such as Crohn's disease, psoriasis, lupus, arthrosis, and MS [103−107]. We reported recently the implication of this relationship in AD [53,57,108]. Indeed, we revealed that the cytokine IL-1β is an autophagic inducer in primary cellular tricultures with neurons, astrocytes, and microglia, and showed a strong decrease in mTOR and p70S6K activation associated with an accumulation of LC3-II and p62 (two autophagic markers) [57]. Surprisingly, IL-1β treatment alone led to an accumulation of autophagolysosomes (fusion between AVs and lysosomes without rapid resorption) specifically in microglia, which sustained an impairment of autophagy in our *in vitro* AD model [57]. Further experiments *in vivo* in acute and chronic inflammatory mouse models (lipopolysaccharide injection) and in transgenic APPswePS1dE9 mice confirmed the role of IL-1β in the control of brain autophagy. More precisely, we showed a strong correlation between levels of Beclin-1 and IL-1β, a reduction of mTOR signaling pathway activation, associated with AVs accumulation specifically for the APPswePS1dE9 mice [53,108].

The mTOR signaling pathway plays a pivotal role in AD. This deregulated signaling pathway can link all major AD hallmarks: Aβ accumulation, tau hyperphosphorylation, inflammation activation, autophagy impairment, cerebral atrophy, and cognitive deficits.

6.3 THERAPEUTIC STRATEGY FOCUSING mTOR IN AD

Taking into account the translational research showing the malfunction of the mTOR pathway in AD, many authors have therefore chosen mTOR as a therapeutic target.

With regard to the therapeutic aspect, several molecules have been tested to reduce the mTOR activation and stimulate the autophagy. The first molecule already quoted above is rapamycin that mainly inhibits mTORC1 [109]. Different studies demonstrated a beneficial effect of rapamycin treatment in cellular and AD transgenic mouse models, preventing the AD-like cognitive deficits and reducing Aβ and tau pathology [54,100,110−113]. In these studies, the authors maintain that the main therapeutic effect of rapamycin was the autophagy induction. However, as we discuss above, autophagy induction in AD must be modulated carefully and depending on the AD progression. Therapeutic modulation of autophagy in AD may, therefore, require targeting late steps in the autophagic pathway [114].

Other beneficial effects of rapamycin could be taken into consideration; mTOR pathway regulates many different cell functions such as translation, inflammation, cell growth, differentiation, and metabolism. Yates et al. showed that *in vitro* rapamycin treatment of lymphocytes from AD patients modified the expression of 1127 genes, 1033 were up-regulated and 94 genes down-regulated [85]. However, chronic administration of rapamycin at advanced clinical stages may result in deleterious systemic effects due to chronic inhibition of mTOR [115]. It is known that chronic rapamycin treatment also inhibits mTORC2 [116], which is involved in organization of the actin cytoskeleton, phosphorylates Akt with both activates mTORC1 [2]. Rapamycin side effects are known: immunosuppressant, edema, aphthous ulcers, mucositis, rash, hair and nail disorders, metabolic changes including hyperlipidemia, glucose intolerance, decreased insulin sensitivity, and gastrointestinal events [117], and its long-term utility in humans will be limited, while significant promising effects have been shown in animal models.

Besides rapamycin, other molecules which inhibit the mTOR signaling pathway and modulate autophagy have been used in different cellular and animal models of AD. Curcumin can inhibit mTOR and slow down aging [118]. However, other authors showed that curcumin can induce senescence of cells building the vasculature [119]. In oleuropein aglycone-fed mice, brains displayed an astonishingly intense autophagic reaction, as shown by the increases in autophagic marker expression and lysosomal activity. Data obtained with cultured cells confirmed the latter evidence, suggesting mTOR regulation by oleuropein aglycone [120]. Other autophagy inducers by inhibition of mTOR such as cystatin, latrepirdin (Dimebon®), carbamazepin, arctigenin, lamotrigin, wogonin, and ginsenoside showed beneficial results in cellular and animal models of AD [121–128]. Interestingly, GSK-3β inhibitors, by restoring lysosomal acidification and mTOR activity, ameliorate β-amyloid pathology [101].

All these preclinical studies should be rapidly strengthened to bring these molecules to the clinic.

Another way to induce autophagy by inhibiting the mTOR pathway is to stimulate AMPK. The natural polyphenol resveratrol (trans-3,4',5-trihydroxystilbene), largely known for its multiple neuroprotective effects, can activate AMPK in vitro and in vivo models of AD, preventing amyloid toxicity [129,130].

Of the molecules capable of inhibiting mTOR, there is also the peroxisome proliferator-activated receptor (PPAR)-gamma agonists. Authors showed that troglitazone significantly reduced phosphorylation of tau in CHOtau4R in vitro models and pioglitazone rescued cerebrovascular function and counteracted several AD markers detrimental to neuronal function in aged APP mice [131,132]. Recently, a meta-analysis revealed that the efficacy of pioglitazone, well-tolerated in AD and MCI patients, seems to be promising, particularly for patients with comorbid diabetes, however this needs to be further confirmed by well-designed trials with large sample sizes [133].

Among treatments for type 2 diabetes, biguanide metformin also acts on tau phosphorylation through mTORC1 inhibition, PP2A activation, and AMPK activation [117,71]. However, recent studies revealed side effects of biguanide metformin in in vitro, ex vivo, and in vivo AD models with oxidative stress, mitochondrial damage, and alterations in memory function, leading to cell death [134,135].

Interestingly, a very recent study revealed the beneficial effect of activating the μ-opioid receptor (OPRM1) with morphine in in vitro primary cortical neurons exposed to Aβ1-42 oligomers [136]. The morphine could restore the mTOR activation which is inhibited by Aβ1-42 exposure and prevent neuronal loss. A specific antagonist of OPRM1, the CTAP, abolished the beneficial effect of morphine in an mTOR-dependent manner [136].

As seen in this synthesis of the molecules used in in vitro, ex vivo, and in vivo AD models to prevent the activation of the mTOR signaling pathway, thereby stimulating autophagy, clinical passage is still very far off without long-term analysis of these molecules and knowledge of their toxicity. Rapamycin analogs that specifically will target mTORC1 would be important to develop and there is hope for some PPAR agonists for patients who also have type 2 diabetes.

6.4 mTOR IN PD

PD is a heterogeneous and progressive neurodegenerative disorder that affects 2% of the population over 60. It is the second most common neurodegenerative disease after AD [137]. Patients suffering from PD display symptoms of motoric instabilities with resting tremor, rigidity, bradykinesia, and postural instability. Non-motoric symptoms include cognitive impairments, depression, and sleep disorders [138–140]. It perturbs both dopaminergic (substantia nigra pars compacta and striatum) and non-dopaminergic (locus coeruleus, raphe nuclei, nucleus basalis of Meynert, hypothalamus, pedunculopontine nucleus) neuronal systems [138–140]. The other hallmark of PD is the presence of α-synuclein-positive intracytoplasmic inclusions identified in Lewy bodies [141]. The etiology of PD remains uncertain, and as for AD, the major risk factor is age. Different environmental factors were reported to increase risk to develop PD as pesticide exposure (particularly Paraquat and rotenone), prior head injury, rural living, β-blocker use, agricultural occupation, and well-water drinking [142,143]. However, a familial autosomal dominant and recessive form of Parkinsonism exists and genetic analysis identified the SNCA gene that encodes α-synuclein, the major Lewy bodies component of the first Parkinson gene [144]. More recently other genes were reported to be associated to autosomal dominant form as LRRK2, VPS35, eIF4G1, DNAJC13, and CHCHD2 [145–148] and to autosomal recessive form as Parkin, PINK1, and DJ-1 frequently associated with early onset (age less than 40 years) [149].

As for AD, multiple cellular alterations were reported, protein aggregation, perturbation of intracellular protein and membrane trafficking, impairments of the proteolysis system (proteasome and autophagy), mitochondrial damage, and inflammatory deregulations [140,150–154]. An attenuation of the PI3K-Akt-mTOR activation is well established in Parkinson patients and in experimental PD models [155–158]. Postmortem analysis revealed an increase in the REDD1 expression, an mTOR inhibitor in dopaminergic neurons [159]. Recently, a study showed that elevated levels of REDD1 found in PD brain could be due to the loss of Parkin or by its activity mutated in PD and contributing to disease progression [160]. REDD1 was highly induced in a cellular model of PD in which death of neuronal catecholaminergic PC12 cells was triggered by the PD mimetic 6-OHDA (6-hydroxydopamine) and involved the repression of Akt and mTOR activity [161].

In other *in vitro* PD models, mTOR blocking is generally observed. For example, in neuroblastoma cells, Deguil et al. showed that MPP$^+$ (1-methyl-4-phenylpyridinium), metabolite of the 1-methyl-4-phenyl-1,2,3,6-tetrapyridine (MPTP causes cell loss in substantia nigra *in vivo* by inhibiting complex I of the mitochondrial respiratory chain), reduced the expression of phosphorylated mTOR and its downstream targets p70S6K, eIF4E, and 4E-BP1, contributing to translational control deregulation of protein synthesis [162]. And in PC12 cells, Rodriguez-Blanco et al. observed that MPP$^+$ induced the inhibition of the Akt/mTOR pathway and of the autophagic process due to the accumulation of reactive oxygen species [163].

In mice, MPTP treatment induced similar results observed *in vitro*, a reduction of the mTOR and p70S6K activation in striatal and frontal cortical regions [164], associated with learning and memory impairments [165].

Furthermore, mTOR signaling is characterized by the phosphorylation of 4E-BP, leading to release of eIF4E and allowing initiation of translation to occur. But other, different kinases are implicated in 4E-BP phosphorylation.

The L-3,4-Dihydroxyphenylalanine (L-DOPA) remains the most effective symptomatic treatment of Parkinson's disease (PD). However, long-term and chronic administration of L-DOPA produced, dyskinesia symptoms (LID: L-DOPA–induced dyskinesia). Although mTOR activation is reduced in PD brain and in PD mouse model, surprisingly, this adverse effect is mediated by the striatal activation of mTORC1 as shown in 6-OHDA mouse model [166]. The authors demonstrated that the regulation was exerted by extracellular signal-regulated kinases (ERKs) known to inhibit TSC2 (mTOR inhibitor) and sustained a sequential activation of Rheb and mTOR or through

phosphorylation and activation of the mTORC1 component, Raptor [166]. Another analysis in 6-OHDA mice, where the expression of the GTPase Ras homolog enriched in striatum, Rhes (reported as an mTOR activator like Rheb) was deleted, showed resistance to develop LID, associated with a striatal mTORC1 activation reduction without modification of the ERK signaling pathway and associated with a motor deficit improvement [167].

Concerning autophagy, recent genetic analysis revealed that both autosomal dominant and autosomal recessive gene mutations in familial PD are involved in the regulation of autophagy [168]. Indeed, UCH-L1 and LRRK2 mutations induced a chaperone-mediated autophagy blockage [169,170]. In addition, dominant mutations in LRRK2 showed that both human LRRK2 and the *Drosophila* ortholog of LRRK2, phosphorylate 4E-BP [171] and consequently stimulate eIF4E-mediated protein translation both *in vivo* and *in vitro*. However, LRRK2 mutation attenuates resistance to oxidative stress and survival of dopaminergic neurons [171]. PINK1, Parkin, and DJ-1 mutations were reported to disturb the mitochondrial renewal through the mitophagy (selected mitochondria autophagy) and promoting oxidative stress [172–175]. However, the exact contribution of DJ-1 remains to be determined [176]. The A53T α-synuclein mutant protein form might interfere with the autophagy, leading to neurodegeneration [177]. Jiang et al. demonstrated, on SH-SY5Y, that A53T α-synuclein overexpression impairs autophagy and upregulates mTOR/p70S6K signaling, the classical suppressive pathway of autophagy [178]. Interestingly, autophagy seems responsible for the α-synuclein aggregation and its inhibition by the Bafilomycin A1 *in vitro* and *in vivo* PD models showed an inhibition of the α-synuclein degradation, a secretion of α-synuclein small oligomers leading to detrimental microenvironmental responses increasing the inflammatory brain reaction [179–181,153].

6.5 THERAPEUTIC STRATEGY FOCUSING mTOR IN PD

The pharmacologic mTOR inhibitor, rapamycin, which protects neurons from death in both cellular and animal toxin models of PD (e.g., PC12 incubated with MPP$^+$ and mice treated with MPTP) was used in several studies. This protective action could be ascribed to blocked translation of REDD1 [161]. By inhibiting TOR signaling, rapamycin has been shown to lead to 4E-BP hypophosphorylation *in vitro* and *in vivo*. Exposing Parkin/PINK1 mutant animals to rapamycin during development caused an increase in hypophosphorylated 4E-BP and, remarkably, was sufficient to suppress all pathologic phenotypes, including degeneration of dopaminergic neurons in *Drosophila*, muscle degeneration, mitochondrial defects, and locomotor ability [182]. As described previously, surprisingly, striatal mTORC1 activation is induced by chronic L-DOPA treatment and responsible for LID. The administration of rapamycin or its analog CCI-779 (temsirolimus), which is used in the treatment of cancer, in the 6-OHDA mouse and rat model of PD, reduced the mTOR pathway activation in striatal medium spiny neurons without affecting the therapeutic efficiency of L-DOPA [183,166].

Most compounds that induce autophagy, promote clearance of aggregated protein and induce neuroprotection, down-regulating the PI3K/AKT/mTOR pathway or inhibiting mTOR [158].

Curcumin, reported to inhibit mTOR [118], could efficiently reduce the accumulation of A53T α-synuclein mutant protein through down-regulation of mTOR/p70S6K signaling and recovery of autophagy [178].

As reported in AD, resveratrol, known to affect AMPK activation and indirectly modulate mTOR activation, has a beneficial effect on SH-SY5Y exposed to rotenone by preventing cell death by an autophagy-dependent manner. The authors maintain a preservation of the mitochondrial renewal was particularly affected by the rotenone exposure [184]. An interesting flavonol, kaempferol, seems to prevent cell death in SH-SY5Y, primary neuronal cell culture and brain rat slice exposed to rotenone by the preservation of autophagy and reduction of oxidative stress [185]. Other independent mTOR autophagy modulators, such as isorhynchophylline [186–188] and trehalose, exposed on PC12, HEK293, and SH-SY5Y cells transfected with the mutated A53T and A30P α-synuclein gene demonstrated the enhancement of autophagy, improving the clearance of α-synuclein mutant proteins. Moreover, the trehalose was tested in an adenoassociated virus (AAV) A53T α-synuclein-injected rat model at 2 months old and treated with trehalose in drinking water for 6 weeks [189]. Results showed a clearance of α-synuclein aggregation, the protection of nigrostriatal degeneration, the enhancement of autophagy, and prevention of behavioral deficits [189].

Contrary to what is seen in AD, few molecules inhibiting mTOR pathway were tested in PD (rapamycin and resveratrol). In addition, the stimulation of autophagy seems not to induce adverse effects. The enhancement of the autophagy could be a suitable therapeutic strategy. The combination of an mTOR-dependent and mTOR-independent autophagy inducer would be promising [188].

6.6 mTOR IN HD

Contrary to other neurodegenerative disorders, HD is exclusively an autosomal dominant neurodegenerative disorder caused by an expanded polyglutamine (polymorphic CAG) repeat in the huntingtin gene IT15 located on chromosome 4 [190,191]. The polymorphic CAG repeat varys from 11 to 36 in the normal population and the repeat increase from 42 to over 66 copies, with a correlation of age of onset, the longest segments being detected in juvenile HD cases. This mutated gene encodes for a huntingtin mutant protein (mHtt) containing an extend N-terminus polyglutamine (polyQ) with a strand of variable length leading to a toxic gain-of-function [192]. HD patients have a progressive motor dysfunction. Initially, patients manifest uncontrolled, choreic "dance-like" movements [193]. This hyperkinetic phase is followed by a hypokinetic phase where purposeful movement is difficult [194]. Surprisingly, cognitive defects emerge years before diagnosis of HD [195] and are developed in parallel with motor decline. Decreased learning and retrieval of new information, impaired attention, mental flexibility, planning, visuospatial functions, and emotion recognition are the principal cognitive defects reported [196–200]. HD typically starts between 30 and 40 years with a gradual deterioration period of 10–20 years, until death. However, HD patients may differ dramatically in age of onset and disease manifestations, despite similar CAG repeat lengths [193]. Sequence variations in the PPARGC1A gene encoding PGC-1α (involved in mitochondrial function), as well as polymorphisms in PGC1α's downstream targets and subtypes of NMDA receptor genes (GRIN2A and GRIN2B), can exert modifying effects on the age of onset in HD [201,202].

The neuropathology of HD shows principally a selective loss of efferent medium spiny neurons of the striatum (caudate and putamen), of the basal ganglia and widespread degeneration in the brain involving cortical structures [203–206]. Major cellular dysfunctions were reported during HD. Contrary to other neurodegenerative disorders, HD is characterized by a reduction of around 50% of the brain-derived neurotrophic factor concentration essential to maintain neuronal integrity [207,208]. As described previously for AD and PD, HD harbors protein aggregates (mHtt aggregates), mitochondria, proteolysis, vesicle transport dysfunctions, oxidative stress, and neuronal excitotoxicity [193,150–152].

The alteration of mTOR signaling pathway activation during HD seems to be in opposition, overactivated in presymptomatic stage and hypoactivated during the disease. In the postmortem HD brain, a decrease of mTOR activation was reported and immunohistochemistry analysis revealed a sequestration of mTOR in aggregates [209,210]. These observations were reproduced *in vitro* in COS-7 cells and in a mouse model of HD expressing mHtt. The mTOR activation alteration is associated with the reduction in p70S6K and 4E-BP1 activity in these HD models sustaining mRNA translation impairments [209]. More recently, an interesting analysis showed *in vitro* that normal Htt promoted mTOR activation in amino-acid deprivation conditions through its interaction with the GTPase, Rheb, known to stimulate mTOR activity [211]. In the same study, the authors demonstrated that mHtt have a strong affinity for Rheb regardless of amino acid conditions and consequently activate more mTOR than normal Htt. Moreover, striatum-specific deletion of TSC1, a negative regulator of mTORC1, in a TSC1flox$^{/+}$/N171HD mouse HD model by AAV-Cre stereotaxic injection before the symptom apparition, accelerated the onset of HD pathology [211]. Another recent study demonstrated that mTORC1 was hypoactivated in an N171-82Q HD mouse model and that its reactivation by the AAV. caRheb stereotaxic striatal injection increased the mHtt clearance through autophagy, improved motor deficits, and preserved medium spiny neurons [210]. Although Rheb overexpression in striatum of HD mouse model provided positive results, this GTPase is not reduced in the striatum of HD patients, contrary to the GTPase protein Rhes. Its overexpression in N171-82Q HD mouse by AAV-Rhes stereotaxic injection gave similar results to those obtained with Rheb striatal overexpression [210]. Moreover Rhes was initially reported to participate in HD by its capacity to bind mHtt and act as a SUMO (Small Ubiquitin-like MOdifier) E3 ligase to stimulate sumoylation of mHtt, a post-translation modification known to augment mHtt toxicity [212]. Recently, a new role for Rhes was revealed, this GTPase, despite its activator function of mTOR, can surprisingly modulate the autophagy induction in an mTOR-independent manner. Indeed, *in vitro* analysis in PC12 cells showed that Rhes activates autophagy by competitively displacing the inhibitory binding of Bcl-2 to Beclin-1 [213]. Interestingly, overexpression of normal Htt or mHtt in HEK2933 cells abolished the LC3-II increase induced by Rhes overexpression in the same cells, and demonstrated competition between Bcl-2 and Htt to link Rhes and consequently to induced mTOR-independent autophagy [213].

Furthermore, normal Htt is able to promote autophagy induction by direct interaction and activation of ULK1, an mTOR downstream effector inhibited by mTOR activation. Indeed, Htt may promote ULK1 activation during selective autophagy by directly competing for the inhibitory binding of mTORC1 to ULK1 [214], allowing facilitation of the complex formation of ULK1-Atg13-FIP200 required to initiate mTOR-dependent autophagy vesicle formation [215]. In addition, Htt facilitates the p62 recognition of ubiquitinylated cargo giving a scaffold role to Htt [215]. Concerning mHtt, it is interesting to note that deleting the polyQ tract in Htt enhances neuronal autophagic activity and longevity in mice [216]; it is attractive to propose that polyQ expansion compromises the role of Htt in selective autophagy [215].

6.7 THERAPEUTIC STRATEGY FOCUSING mTOR IN HD

Rapamycin was the first molecule to be tested in the HD model in order to promote mHtt aggregates by autophagy. Its beneficial effect was initially demonstrated *in vitro* in PC12 cells overexpressing mHtt [217], and then tested *in vivo* in fly and in Ross/Borchelt and R6/2 HD mouse models and revealed a decrease in aggregate mHtt associated with motor improvement [209]. However, the HD mouse models used in this study developed weight loss (as reported in HD patients) amplified by the rapamycin treatment (known to induce weight loss in humans) and consequently did not modify the mice lifespan between treated and placebo mice [209].

Lithium is known to induce autophagy by increasing the production of inositol and IP3 levels by inhibiting the inositol monophosphatase and inositol transporters [218,219]. However, lithium induces the activation of GSK3β, reported to be able to increase mTOR activation and potentially decrease autophagy induction [209]. Sarkar et al. used a combination of rapamycin and lithium in PC12 cells overexpressing mHtt and showed a synergistic effect of this combination allowing an increase in the clearance of mHtt by autophagy [209]. In this study, the authors revealed that GSK3β seems to be necessary in the autophagic process and its direct inhibition had an adverse effect on the aggregate clearance [209].

As for AD and PD, the potential beneficial effect of polyphenols were tested. A study tested the effect of two polyphenols, fisetin and resveratrol, in PC12 cells overexpressing mHtt, in fly overexpressing mHtt, and in a R6/2 HD mouse model [220]. The authors demonstrated an increase in ERK signaling pathway (known to inhibit mTOR through TSC2 activation) and showed neuroprotection associated with a motor deficit improvement in R6/2 mice [220]. Resveratrol is also reported to increase AMPK signaling and consequently the mTOR signaling pathway [221]. However, another resveratrol trial in N171-82Q HD transgenic mice failed to show any improvement in either motor performance or survival [222]. The neferine, isolated from the lotus seed embryo of *Nelumbo nucifera* and used as therapeutic agent in PC12 cells overexpressing mHtt, revealed a reduction in the level and the toxicity of mHtt through an activation of AMPK and a reduction of the mTOR signaling pathway causing an autophagy induction [223].

The modulation of autophagy in an mTOR-independent manner is the second major strategy employed to promote the clearance of protein aggregates.

Trehalose, a disaccharide, was used to stimulate autophagy and demonstrated beneficial effect *in vitro* to clear mHtt aggregates and in the R6/2 mouse model it was able to improve motor deficit and decreased mHtt brain accumulation by its disaggregating property [224,188]. Another interesting study used the 10-[4'-(N-diethylamino)butyl]-2-chlorophenoxazine as an autophagy inducer and was reported to be an Akt inhibitor and consequently decreased the mTOR activation in Rh1, Rh18, and Rh30 cells [225]. However, in primary neurons transfected with the mHtt, the 10-NPC failed to inhibit Akt but surprisingly remained capable of inducing autophagy and clearing the mHtt [226].

Rilmenidine is reported to induce autophagy in an mTOR-independent manner [227]. This drug is an α_2 adrenergic agonism which acts on α_2-adrenoceptors and imidazoline I_1 receptors in the brain and in the periphery and is used as an antihypertensive agent to treat hypertension [228]. The treatment of HD transgenic mouse model N171-82Q HD showed an improvement in motor deficits and an increase in the autophagic flux associated with a decrease in the mHtt brain level. However, this antihypertensive agent failed to improve the weight loss and lifespan of the N171-82Q mice [228].

In HD, despite the mTOR activation reduction, therapeutic agents inhibiting its activity give interesting results by improving mHtt clearance and motor deficit recovery. The autophagy seems to be a very efficient therapeutic strategy, the combination of mTOR-dependent and mTOR-independent autophagy inducer would be promising. Interestingly, GTPase Rheb and Rhes can modulate mTOR activation and autophagy for Rhes and are potentially interesting therapeutic targets.

6.8 mTOR IN ALS

ALS, also called Lou Gehrig's disease, is a rapidly progressive and fatal neurological disease, caused by the degeneration, in the CNS, of motor neurons controlling voluntary muscles.

The symptoms include muscle weakness and atrophy, spasticity, paralysis, and sometimes dementia [229]. The disease usually begins between 35 and 50 years of age and most patients die from respiratory failure, usually within 3–5 years from the onset of symptoms but about 10% survive for 10 or more years [230].

In 90–95% of ALS cases, the patient has no family history of ALS and the sporadic form occurs without any clearly associated risk factors. In the familial genetic form, mutations in the superoxide dismutase 1 (SOD1) gene are the most common cause (20%) of this disease [231]. SOD1 is an antioxidant enzyme, which protects neurons from free radicals. The mutant SOD1-mediated toxicity is caused by a gain of toxic function rather than a loss of activity [232]. This enzyme induces abnormal proteins sharing an aberrant conformation [233] and abnormal aggregation. Chromogranins may promote secretion of SOD1 mutants and extracellular mutant SOD1 can trigger microgliosis and apoptosis [234]. Other mutations in transactive response DNA-binding protein gene [235] or in the FUS gene (gene coding for protein called "fused in sarcoma") [236] are also associated with genetic ALS.

The mechanism of the disease is not clear but could involve mitochondrial dysfunction [229], protein aggregation [237], or alterations in the immune system [238].

In the transgenic mouse model of ALS, SOD1G93A mice, at about 15 weeks of age, the phosphorylated mTOR immunopositive motor neurons are significantly decreased and a reduction of PI3K-Akt-mTOR-p70S6k activation in the spinal cord is observed in parallel with the motor symptoms from 3 months old [239–241]. In addition, autophagic markers, p62 and LC3-II, were progressively accumulated, suggesting autophagy impairments in these transgenic mice [239,240]. On the other hand, cortical neurons from SOD1G93A mice at 1 month old displayed a hyperexcitability as previously reported in presymptomatic ALS patients [242,243], associated with an increase in the p70S6K activation without mTOR modification [244].

The inactivation of mTOR in ALS would be explained by the sequestration of mTOR into mutant SOD1 aggregates. A similar sequestration was described in cell models, transgenic mice, and the brains of patients with HD, into polyglutamine-containing aggregates [245]. The appearance of mutant SOD1 aggregates in motor neurons of patients or murine models of ALS [246,247] allows proposing that this protective pathway exists in ALS.

6.9 THERAPEUTIC STRATEGY FOCUSING mTOR IN ALS

In SOD1G93A transgenic mice, rapamycin treatment by daily intraperitoneal injection starting 64 days after birth induced an earlier onset of the disease and shorter lifespan. In addition, rapamycin accelerated motor neuron degeneration, apoptotic pathway activation in the spinal cord, and enhanced autophagy, but without SOD1 aggregation reduction in the motor neurons [240]. Another team confirmed that treatment of ALS mice by rapamycin induced an exacerbation of ALS disease [248]. On the other hand, treatment of the same mice with oxotremorine, a stimulator of mTOR, and salubrinal, an inhibitor of ER stress, greatly slowed motor neuron degeneration and extended survival of ALS mice [248]. In the same study, the authors demonstrated that the excitability of motor neurons modulated by glutamate receptor could play a pivotal role in disease progression and engage an adaptive cellular stress response pathway that suppresses mutant SOD1 aggregation and enhances mTOR-mediated protein synthesis to sustain motor neuron survival [248].

Despite the negative effect of rapamycin, the induction of autophagy independently of mTOR improved the elimination of aggregated mutant SOD1 and slowed ALS progression [249]. Interestingly, a clinical trial using a treatment with lithium (usually used for psychiatric medication) and known to be an autophagy inducer [218] during 15 months in ALS patients demonstrated a slowdown of disease progression [250]. A meticulous analysis in mutant SOD1 mice showed that lithium acts as an autophagy inducer, leading to delayed neuronal cell death, decreased reactive gliosis, rescued spinal cord mitochondria, and produced a marked regression of SOD1 aggregates. Resveratrol had the equivalent effect on primary mouse neurons overexpressing the same mutant of SOD1 sustaining no significant mTOR modulation by resveratrol in the context of ALS [251]. Food restriction, another autophagy inducer, had a potential protective effect on the spinal cord of the same ALS mice only at the onset stage but not at the pre-end stage [252]. Trehalose, another autophagy-inducer, decreased protein aggregation and enhanced survival of mouse motor neuron-like NSC-34 cell line transiently transfected with human SOD1G93A, linked to ALS [253].

ALS displayed contradictory facts: the necessity to enhance mTOR activation and simultaneously accelerate the autophagic process in order to limit the SOD1 aggregation. In addition, the excitability of motor neurons appears to be a critical issue for ALS initiation [254]. The use of an independent autophagy inducer associated with mTOR enhancer could be an interesting therapeutic strategy for ALS.

6.10 mTOR IN MS

MS is a common, presumed autoimmune, disorder of the CNS, estimated to affect over 2.5 million people worldwide [255]. In fact, MS is the second most common cause of disability in young adults, just after trauma. The age of symptom onset is mostly between 20 and 40 years, but the disease can occur at any age. Interestingly, women are affected more frequently than men [256].

MS is characterized by well-demarked inflammation, breakdown of myelin sheaths disseminated throughout the white matter, microglia activation, proliferation of astrocytes resulting in gliosis, a loss of oligodendrocytes (OLs), and variable grades of axonal degeneration [257–260]. All of these lesions are caused by lymphocytes, activated microglia, and macrophages, leading to clinical deficits such as weakness in one or more limbs, sensory disturbances, optic neuritis, ataxia, bladder dysfunction, fatigue, and cognitive deficits [261,262]. MS typically starts as an episodic, relapsing–remitting disease with complete or partial recovery between relapses [263,264]. Over time, it evolves in many of the afflicted individuals into a secondary progressive phase characterized by irreversible deterioration of both motor and cognitive functions [263,264]. Furthermore, the etiology of MS is still largely unknown but multiple lines of evidence suggest that the interaction between genetic and environmental factors underlies the risk of developing MS [265], explaining in part why the prevalence of MS varies with geography and ethnicity. Current anti-inflammatory, immunomodulatory, or immunosuppressive treatments, also including newly available drugs, are partly effective in the early stages of the disease, but display very limited benefits in patients affected by progressive MS [266]. Contrary to other pathologies developed in this chapter, the mTOR pathway deregulation was not directly demonstrated in this pathology.

6.11 THERAPEUTIC STRATEGY FOCUSING mTOR IN MS

The immunosuppressive property of rapamycin demonstrated an interesting therapeutic potential of mTOR and its indirect contribution in MS.

Indeed, rapamycin has been shown to prevent the induction and progression of MS in different MS animal models. Rapamycin treatment in experimental autoimmune encephalomyelitis (RR-EAE) has been able to suppress effector T-cell function and simultaneous increase of the percentage of T regulator (Treg) cells [266,267]. In another interesting study, oral administration of rapamycin for 28 days after the clinical onset of disease in a protracted relapsing EAE model induced in Dark Agouti rats showed a significantly milder course of the disease, associated with a reduction in the histopathological signs of EAE [266,268]. More recently, Lisi et al. tested the rapamycin effect when administered at the peak of disease in a chronic model of EAE and showed clinical and histological amelioration of Ch-EAE signs [269]. The inhibition of mTOR signaling by Rheb knockout specifically in T cells reduced the production of Th1 and Th17 lymphocyte effector (implicated in the inflammation deregulation and breakdown of myelin sheaths in MS) associated with lower leukocyte infiltration in the spinal cord and resistance to the EAE development [270].

Taken together, these studies indicate an important role of mTOR in the induction and maintenance of EAE clinical course, specifically in RR-EAE models, by its implication in the regulation of the immune system and inflammatory processes.

Another promising therapeutic strategy in MS is to prevent the loss of OLs, progenitors (OLPs) responsible for axon myelination [271]. Recently, the involvement of the PI3K/Akt/mTOR pathway in OL differentiation and CNS myelination was demonstrated [272]. In this study authors used therapeutic administration of ER β-ligand 2,3-bis(4-hydroxyphenyl)-propionitrile (DPN) through subcutaneous injection in EAE mouse model. They revealed a PI3K/Akt/mTOR signaling pathway activation, a remyelination of axon associated with a restoration of the axon conduction after 4–5 weeks the post-EAE treated with DPN. The activation of the mTOR pathway in OLPs can be modulated by the repression of phosphatase and tensin homolog (PTEN) whose activation leads to repressing the Akt-mTOR axis [273]. The use of bisperoxovanadium (bpV), a PTEN inhibitor, in rat and human OLPs cultured with dorsal root ganglion neurons induced the proliferation of OLPs in a concentration-dependent manner and time-dependent activation of the mTOR signaling pathway [274]. Interestingly, in this study the authors showed that the bpV treatment supplemented with IGF-1 can potentiate the IGF-1 action, reported to be a critical signal for OL development and for myelin formation [274].

Despite the non-direct clues to the mTOR pathway dysregulation in MS, it appears that it is an interesting therapeutic target to slow down disease progression and to lessen the inflammatory reaction. The combination of different molecules (e.g., bpV, IGF-1) gives promising preclinical results in in vivo MS models and must be developed [274].

6.12 CONCLUSION

All the data reported on the mTOR pathway in these various neurodegenerative diseases show the complexity of this signaling pathway (summarized in Table 6.1) in the same process, that is, neuronal loss. Moreover, its essential role in cellular life makes it difficult to modulate its activation without adverse consequences. However, we can retain that mTOR is activated in AD and hypoactivated in PD and in HD (Table 6.1). For AD, studies raise the kinetics of evolution of the activation of mTOR: an increase in pro-DNF neurons and inhibition of mTOR in neurons during apoptosis. For PD, overexpression of REDD1 as mTOR inhibitor is the major factor responsible for the inhibition of mTOR. Surprisingly, in presymptomatic HD, neurons harbor an mTOR hyperactivation that is inversed with disease onset. For these neurodegenerative diseases, the beneficial effects of mTOR inhibition are widely demonstrated in experimental models of these diseases, especially as this inhibition would stimulate autophagy required for the degradation of aberrant proteins. The combination of therapeutic agents targeting the autophagy enhancement in mTOR-dependent and mTOR-independent manners seems promising. However, the autophagy stimulation must proceed carefully and could be dependent on the disease severity, particularly in AD and not really explored in other diseases.

For ALS, mTOR activation is inhibited and its reactivation is beneficial and could be associated with a therapeutic strategy to stimulate autophagy in an mTOR-independent manner (Table 6.1).

Besides, MS mTOR seems to be very difficult to use in correcting this pathologic process by targeting the mTOR pathway because it is involved in regulating Treg cells but also in the survival of OL and OLPs and regeneration of axonal myelination (Table 6.1).

Finally, the specific molecules inhibiting mTOR remain limited and rapamycin harbors side effects already known and will not be an adequate therapeutic agent for chronic use. However, other molecular platforms should be considered to modulate mTOR activity, such as GTPase, Rhes, and Rheb, which are particularly implied in AD, PD, and HD.

TABLE 6.1   Overview of the mTOR Signaling Pathway Alterations Observed in Neurodegenerative Disorders

Neurodegenerative disorders	mTOR signaling pathway alterations		Molecular actors
AD	↑↓Akt	Apoptotic neurons are characterized by an inhibition of the mTOR signaling pathway while hyperactivated neurons with activated Akt, p70S6K could produce neurofibrillary tangles	Aβ, tau hyperphosphorylation, inflammation
	↑↓mTORC1		
	↑↓p70S6K		
	↓AMPK	Decrease mTORC1 activity and autophagy	
	↑PRAS40	Promote mTORC1 hyperactivity	
	↑GSK3β	Decrease mTORC1 activity and autophagy	
	↓Rheb	Decrease the BACE1 degradation (promote Aβ production)	
PD	↓PI3K	Reduction of the mTOR pathway is link to the increase in the REDD1 activity in PD brain	Loss of Parkin or mutated Parkin
	↓Akt		
	↓mTORC1		
	↑REDD1		
	↑mTORC1	The L-dopa treatment induced chronic hyperactivation of mTORC1 through ERK and Rhes activation responsible of LID	L-dopa treatment
	↑ERK		
	↑Rhes		
HD	↑mTORC1	mTOR overactivation in presymptomatic stage, Rhes promote the mHtt toxicity and modulate autophagy in mTOR-independent manner	mHtt
	↑Rheb/Rhes		
	↓mTORC1	mTOR pathway hypoactivation in HD brain	mHtt aggregated mTOR
	↓p70S6K		
ALS	↓PI3K	Reduction of PI3K-Akt-mTOR-p70S6k activation in spinal cord is observed in parallels of the motor symptoms	SOD1 aggregates
	↓Akt		
	↓mTORC1	The reduction of mTOR activation as a therapeutic strategy accelerates the disease progression	
	↓p70S6K		
MS	/	The reduction of mTOR pathway as a therapeutic strategy slows down the disease progression and lessens the inflammatory reaction	/

AD, Alzheimer's disease; PD, Parkinson's disease; HD, Huntington's disease; ALS, amyotrophic lateral sclerosis; MS, multiple sclerosis; mTOR, mammalian target of rapamycin; Rheb, Ras homolog enriched in brain; ERK, extracellular signal-regulated kinase; GSK-3β, glycogen-synthase kinase-3β; AMPK, adenosine monophosphate-activated protein kinase; PRAS40, proline-rich Akt substrate 40 kDa; mHtt, huntingtin mutant protein.

References

[1] Ma XM, Blenis J. Molecular mechanisms of mTOR-mediated translational control. Nat Rev Mol Cell Biol 2009;10:307—18.

[2] Laplante M, Sabatini DM. mTOR signaling in growth control and disease. Cell 2012;149:274—93.

[3] Child ND, Benarroch EE. mTOR: its role in the nervous system and involvement in neurologic disease. Neurology 2014;83:1562—72.

[4] Shimobayashi M, Hall MN. Making new contacts: the mTOR network in metabolism and signalling crosstalk. Nat Rev Mol Cell Biol 2014;15:155—62.

[5] Wang L, Harris TE, Roth RA, Lawrence Jr JC. PRAS40 regulates mTORC1 kinase activity by functioning as a direct inhibitor of substrate binding. J Biol Chem 2007;282:20036—44.

[6] Morel M, Couturier J, Pontcharraud R, Gil R, Fauconneau B, Paccalin M, et al. Evidence of molecular links between PKR and mTOR signalling pathways in Abeta neurotoxicity: role of p53, Redd1 and TSC2. Neurobiol Dis 2009;36:151—61.

[7] Sofer A, Lei K, Johannessen CM, Ellisen LW. Regulation of mTOR and cell growth in response to energy stress by REDD1. Mol Cell Biol 2005;25:5834—45.

[8] Gwinn DM, Shackelford DB, Egan DF, Mihaylova MM, Mery A, Vasquez DS, et al. AMPK phosphorylation of raptor mediates a metabolic checkpoint. Mol Cell 2008;30:214—26.

[9] Zinzalla V, Stracka D, Oppliger W, Hall MN. Activation of mTORC2 by association with the ribosome. Cell 2011;144:757—68.

[10] Hoeffer CA, Klann E. mTOR signaling: at the crossroads of plasticity, memory and disease. Trends Neurosci 2010;33:67—75.

[11] Jaworski J, Spangler S, Seeburg DP, Hoogenraad CC, Sheng M. Control of dendritic arborization by the phosphoinositide-3-kinase-Akt-mammalian target of rapamycin pathway. J Neurosci 2005;25:11300—12.

[12] Ma T, Hoeffer CA, Capetillo-Zarate E, Yu F, Wong H, Lin MT, et al. Dysregulation of the mTOR pathway mediates impairment of synaptic plasticity in a mouse model of Alzheimer's disease. PLoS One 2010;5.

[13] Tang SJ, Reis G, Kang H, Gingras AC, Sonenberg N, Schuman EM. A rapamycin-sensitive signaling pathway contributes to long-term synaptic plasticity in the hippocampus. Proc Natl Acad Sci USA 2002;99:467–72.

[14] Liu J, Reeves C, Michalak Z, Coppola A, Diehl B, Sisodiya SM, et al. Evidence for mTOR pathway activation in a spectrum of epilepsy-associated pathologies. Acta Neuropathol Commun 2014;2:71.

[15] Zeng LH, Rensing NR, Wong M. The mammalian target of rapamycin signaling pathway mediates epileptogenesis in a model of temporal lobe epilepsy. J Neurosci 2009;29:6964–72.

[16] Chen S, Atkins CM, Liu CL, Alonso OF, Dietrich WD, Hu BR. Alterations in mammalian target of rapamycin signaling pathways after traumatic brain injury. J Cereb Blood Flow Metab 2007;27:939–49.

[17] Erlich S, Alexandrovich A, Shohami E, Pinkas-Kramarski R. Rapamycin is a neuroprotective treatment for traumatic brain injury. Neurobiol Dis 2007;26:86–93.

[18] Martinez de Morentin PB, Martinez-Sanchez N, Roa J, Ferno J, Nogueiras R, Tena-Sempere M, et al. Hypothalamic mTOR: the rookie energy sensor. Curr Mol Med 2014;14:3–21.

[19] Yang SB, Tien AC, Boddupalli G, Xu AW, Jan YN, Jan LY. Rapamycin ameliorates age-dependent obesity associated with increased mTOR signaling in hypothalamic POMC neurons. Neuron 2012;75:425–36.

[20] Maiese K, Chong ZZ, Shang YC, Wang S. mTOR: on target for novel therapeutic strategies in the nervous system. Trends Mol Med 2013;19:51–60.

[21] Serrano-Pozo A, Frosch MP, Masliah E, Hyman BT. Neuropathological alterations in Alzheimer disease. Cold Spring Harb Perspect Med 2011;1:a006189.

[22] Nelson PT, Alafuzoff I, Bigio EH, Bouras C, Braak H, Cairns NJ, et al. Correlation of Alzheimer disease neuropathologic changes with cognitive status: a review of the literature. J Neuropathol Exp Neurol 2012;71:362–81.

[23] Mayeux R, Stern Y. Epidemiology of Alzheimer disease. Cold Spring Harb Perspect Med 2012;2(8):pii: a006239.

[24] Reitz C, Bos MJ, Hofman A, Koudstaal PJ, Breteler MM. Prestroke cognitive performance, incident stroke, and risk of dementia: the Rotterdam study. Stroke 2008;39:36–41.

[25] Luchsinger JA, Tang MX, Shea S, Mayeux R. Hyperinsulinemia and risk of Alzheimer disease. Neurology 2004;63:1187–92.

[26] Profenno LA, Porsteinsson AP, Faraone SV. Meta-analysis of Alzheimer's disease risk with obesity, diabetes, and related disorders. Biol Psychiatry 2010;67:505–12.

[27] Lambert JC, Ibrahim-Verbaas CA, Harold D, Naj AC, Sims R, Bellenguez C, et al. Meta-analysis of 74,046 individuals identifies 11 new susceptibility loci for Alzheimer's disease. Nat Genet 2013;45:1452–8.

[28] Fleminger S, Oliver DL, Lovestone S, Rabe-Hesketh S, Giora A. Head injury as a risk factor for Alzheimer's disease: the evidence 10 years on; a partial replication. J Neurol Neurosurg Psychiatry 2003;74:857–62.

[29] De-Paula VJ, Radanovic M, Diniz BS, Forlenza OV. Alzheimer's disease. Subcell Biochem 2012;65:329–52.

[30] Bloom GS. Amyloid-beta and tau: the trigger and bullet in Alzheimer disease pathogenesis. JAMA Neurol 2014;71:505–8.

[31] Spires-Jones TL, Hyman BT. The intersection of amyloid beta and tau at synapses in Alzheimer's disease. Neuron 2014;82: 756–71.

[32] Nunomura A, Castellani RJ, Zhu X, Moreira PI, Perry G, Smith MA. Involvement of oxidative stress in Alzheimer disease. J Neuropathol Exp Neurol 2006;65:631–41.

[33] Hermes M, Eichhoff G, Garaschuk O. Intracellular calcium signalling in Alzheimer's disease. J Cell Mol Med 2010;14:30–41.

[34] Morel M, Couturier J, Lafay-Chebassier C, Paccalin M, Page G. PKR, the double stranded RNA-dependent protein kinase as a critical target in Alzheimer's disease. J Cell Mol Med 2009;13:1476–88.

[35] Buckingham SD, Jones AK, Brown LA, Sattelle DB. Nicotinic acetylcholine receptor signalling: roles in Alzheimer's disease and amyloid neuroprotection. Pharmacol Rev 2009;61:39–61.

[36] Borsello T, Forloni G. JNK signalling: a possible target to prevent neurodegeneration. Curr Pharm Des 2007;13:1875–86.

[37] Chong ZZ, Shang YC, Wang S, Maiese K. A critical kinase cascade in neurological disorders: PI 3-K, Akt, and mTOR. Future neurology 2012;7:733–48.

[38] An WL, Cowburn RF, Li L, Braak H, Alafuzoff I, Iqbal K, et al. Up-regulation of phosphorylated/activated p70 S6 kinase and its relationship to neurofibrillary pathology in Alzheimer's disease. Am J Pathol 2003;163:591–607.

[39] Pei JJ, Hugon J. mTOR-dependent signalling in Alzheimer's disease. J Cell Mol Med 2008;12:2525–32.

[40] Sun YX, Ji X, Mao X, Xie L, Jia J, Galvan V, et al. Differential activation of mTOR complex 1 signaling in human brain with mild to severe Alzheimer's disease. J Alzheimer's Dis 2014;38:437–44.

[41] Griffin RJ, Moloney A, Kelliher M, Johnston JA, Ravid R, Dockery P, et al. Activation of Akt/PKB, increased phosphorylation of Akt substrates and loss and altered distribution of Akt and PTEN are features of Alzheimer's disease pathology. J Neurochem 2005;93:105–17.

[42] Li X, Alafuzoff I, Soininen H, Winblad B, Pei JJ. Levels of mTOR and its downstream targets 4E-BP1, eEF2, and eEF2 kinase in relationships with tau in Alzheimer's disease brain. FEBS J 2005;272:4211–20.

[43] Tramutola A, Triplett JC, Di Domenico F, Niedowicz DM, Murphy MP, Coccia R, et al. Alteration of mTOR signaling occurs early in the progression of Alzheimer disease (AD): analysis of brain from subjects with pre-clinical AD, amnestic mild cognitive impairment and late-stage AD. J Neurochem 2015;133 (5):739–49.

[44] Talbot K, Wang HY, Kazi H, Han LY, Bakshi KP, Stucky A, et al. Demonstrated brain insulin resistance in Alzheimer's disease patients is associated with IGF-1 resistance, IRS-1 dysregulation, and cognitive decline. J Clin Invest 2012;122:1316–38.

[45] Craft S. Insulin resistance syndrome and Alzheimer's disease: age- and obesity-related effects on memory, amyloid, and inflammation. Neurobiol Aging 2005;26(Suppl 1):65–9.

[46] Schubert D. Glucose metabolism and Alzheimer's disease. Ageing Res Rev 2005;4:240–57.

[47] Pei JJ, Khatoon S, An WL, Nordlinder M, Tanaka T, Braak H, et al. Role of protein kinase B in Alzheimer's neurofibrillary pathology. Acta Neuropathol 2003;105:381–92.

[48] Townsend M, Mehta T, Selkoe DJ. Soluble Abeta inhibits specific signal transduction cascades common to the insulin receptor pathway. J Biol Chem 2007;282:33305–12.

[49] Zhao WQ, De Felice FG, Fernandez S, Chen H, Lambert MP, Quon MJ, et al. Amyloid beta oligomers induce impairment of neuronal insulin receptors. FASEB J 2008;22:246–60.

[50] De Felice FG, Vieira MN, Bomfim TR, Decker H, Velasco PT, Lambert MP, et al. Protection of synapses against Alzheimer's-linked toxins: insulin signaling prevents the pathogenic binding of Abeta oligomers. Proc Natl Acad Sci USA 2009;106:1971–6.

[51] Moloney AM, Griffin RJ, Timmons S, O'Connor R, Ravid R, O'Neill C. Defects in IGF-1 receptor, insulin receptor and IRS-1/2 in Alzheimer's disease indicate possible resistance to IGF-1 and insulin signalling. Neurobiol Aging 2010;31:224–43.

[52] Lafay-Chebassier C, Paccalin M, Page G, Barc-Pain S, Perault-Pochat MC, Gil R, et al. mTOR/p70S6k signalling alteration by Abeta exposure as well as in APP-PS1 transgenic models and in patients with Alzheimer's disease. J Neurochem 2005;94:215–25.

[53] Francois A, Rioux Bilan A, Quellard N, Fernandez B, Janet T, Chassaing D, et al. Longitudinal follow-up of autophagy and inflammation in brain of APPswePS1dE9 transgenic mice. J Neuroinflammation 2014;11:139.

[54] Caccamo A, Majumder S, Richardson A, Strong R, Oddo S. Molecular interplay between mammalian target of rapamycin (mTOR), amyloid-beta, and Tau: effects on cognitive impairments. J Biol Chem 2010;285:13107–20.

[55] Caccamo A, Maldonado MA, Majumder S, Medina DX, Holbein W, Magri A, et al. Naturally secreted amyloid-beta increases mammalian target of rapamycin (mTOR) activity via a PRAS40-mediated mechanism. J Biol Chem 2011;286:8924–32.

[56] Sancak Y, Thoreen CC, Peterson TR, Lindquist RA, Kang SA, Spooner E, et al. PRAS40 is an insulin-regulated inhibitor of the mTORC1 protein kinase. Mol Cell 2007;25:903–15.

[57] Francois A, Terro F, Janet T, Rioux Bilan A, Paccalin M, Page G. Involvement of interleukin-1beta in the autophagic process of microglia: relevance to Alzheimer's disease. J Neuroinflammation 2013;10:151.

[58] Damjanac M, Page G, Ragot S, Laborie G, Gil R, Hugon J, et al. PKR, a cognitive decline biomarker, can regulate translation via two consecutive molecular targets p53 and Redd1 in lymphocytes of AD patients. J Cell Mol Med 2009;13:1823–32.

[59] Zhang F, Beharry ZM, Harris TE, Lilly MB, Smith CD, Mahajan S, et al. PIM1 protein kinase regulates PRAS40 phosphorylation and mTOR activity in FDCP1 cells. Cancer Biol Ther 2009;8:846–53.

[60] Bhaskar K, Miller M, Chludzinski A, Herrup K, Zagorski M, Lamb BT. The PI3K-Akt-mTOR pathway regulates Abeta oligomer induced neuronal cell cycle events. Mol Neurodegener 2009;4:14.

[61] Oddo S. The role of mTOR signaling in Alzheimer disease. Front Biosci 2012;4:941–52.

[62] Damjanac M, Rioux Bilan A, Paccalin M, Pontcharraud R, Fauconneau B, Hugon J, et al. Dissociation of Akt/PKB and ribosomal S6 kinase signaling markers in a transgenic mouse model of Alzheimer's disease. Neurobiol Dis 2008;29:354–67.

[63] Shahani N, Pryor W, Swarnkar S, Kholodilov N, Thinakaran G, Burke RE, et al. Rheb GTPase regulates beta-secretase levels and amyloid beta generation. J Biol Chem 2014;289:5799–808.

[64] Vassar R, Kandalepas PC. The beta-secretase enzyme BACE1 as a therapeutic target for Alzheimer's disease. Alzheimer's Res Ther 2011;3:20.

[65] Khurana V, Lu Y, Steinhilb ML, Oldham S, Shulman JM, Feany MB. TOR-mediated cell-cycle activation causes neurodegeneration in a Drosophila tauopathy model. Curr Biol 2006;16:230–41.

[66] Wittmann CW, Wszolek MF, Shulman JM, Salvaterra PM, Lewis J, Hutton M, et al. Tauopathy in Drosophila: neurodegeneration without neurofibrillary tangles. Science 2001;293:711–14.

[67] Tang Z, Ioja E, Bereczki E, Hultenby K, Li C, Guan Z, et al. mTor mediates tau localization and secretion: Implication for Alzheimer's disease. Biochim Biophys Acta 2015;1853:1646–57.

[68] Meske V, Albert F, Ohm TG. Coupling of mammalian target of rapamycin with phosphoinositide 3-kinase signaling pathway regulates protein phosphatase 2A- and glycogen synthase kinase-3-dependent phosphorylation of Tau. J Biol Chem 2008;283:100–9.

[69] Tang Z, Bereczki E, Zhang H, Wang S, Li C, Ji X, et al. Mammalian target of rapamycin (mTor) mediates tau protein dyshomeostasis: implication for Alzheimer disease. J Biol Chem 2013;288:15556–70.

[70] Morita T, Sobue K. Specification of neuronal polarity regulated by local translation of CRMP2 and Tau via the mTOR-p70S6K pathway. J Biol Chem 2009;284:27734–45.

[71] Kickstein E, Krauss S, Thornhill P, Rutschow D, Zeller R, Sharkey J, et al. Biguanide metformin acts on tau phosphorylation via mTOR/protein phosphatase 2A (PP2A) signaling. Proc Natl Acad Sci USA 2010;107:21830–5.

[72] Meraz-Rios MA, Toral-Rios D, Franco-Bocanegra D, Villeda-Hernandez J, Campos-Pena V. Inflammatory process in Alzheimer's disease. Front Integr Neurosci 2013;7:59.

[73] Njie EG, Boelen E, Stassen FR, Steinbusch HW, Borchelt DR, Streit WJ. Ex vivo cultures of microglia from young and aged rodent brain reveal age-related changes in microglial function. Neurobiol Aging 2012;33:195.e1–195.e12.

[74] Fiala M, Lin J, Ringman J, Kermani-Arab V, Tsao G, Patel A, et al. Ineffective phagocytosis of amyloid-beta by macrophages of Alzheimer's disease patients. J Alzheimer's Dis 2005;7:221–32 [discussion 55–62].

[75] Schwartz M, Shechter R. Systemic inflammatory cells fight off neurodegenerative disease. Nat Rev Neurol 2010;6:405–10.

[76] Togo T, Akiyama H, Iseki E, Kondo H, Ikeda K, Kato M, et al. Occurrence of T cells in the brain of Alzheimer's disease and other neurological diseases. J Neuroimmunol 2002;124:83–92.

[77] Browne TC, McQuillan K, McManus RM, O'Reilly JA, Mills KH, Lynch MA. IFN-gamma production by amyloid beta-specific Th1 cells promotes microglial activation and increases plaque burden in a mouse model of Alzheimer's disease. J Immunol 2013;190:2241–51.

[78] McManus RM, Higgins SC, Mills KH, Lynch MA. Respiratory infection promotes T cell infiltration and amyloid-beta deposition in APP/PS1 mice. Neurobiol Aging 2014;35:109–21.

[79] Delgoffe GM, Kole TP, Zheng Y, Zarek PE, Matthews KL, Xiao B, et al. The mTOR kinase differentially regulates effector and regulatory T cell lineage commitment. Immunity 2009;30:832–44.

[80] Katholnig K, Linke M, Pham H, Hengstschlager M, Weichhart T. Immune responses of macrophages and dendritic cells regulated by mTOR signalling. Biochem Soc Trans 2013;41:927–33.

[81] Thomson AW, Turnquist HR, Raimondi G. Immunoregulatory functions of mTOR inhibition. Nat Rev Immunol 2009;9:324–37.

[82] Powell JD, Pollizzi KN, Heikamp EB, Horton MR. Regulation of immune responses by mTOR. Annu Rev Immunol 2012;30:39–68.

[83] Paccalin M, Pain-Barc S, Pluchon C, Paul C, Bazin H, Gil R, et al. The relation between p70S6k expression in lymphocytes and the decline of cognitive test scores in patients with Alzheimer disease. Arch Intern Med 2005;165:2428–9.

[84] Paccalin M, Pain-Barc S, Pluchon C, Paul C, Besson MN, Carret-Rebillat AS, et al. Activated mTOR and PKR kinases in lymphocytes correlate with memory and cognitive decline in Alzheimer's disease. Dement Geriatr Cogn Disord 2006;22:320–6.

[85] Yates SC, Zafar A, Hubbard P, Nagy S, Durant S, Bicknell R, et al. Dysfunction of the mTOR pathway is a risk factor for Alzheimer's disease. Acta Neuropathol Commun 2013;1:3.

[86] Nixon RA. The role of autophagy in neurodegenerative disease. Nat Med 2013;19:983–97.

[87] Ihara Y, Morishima-Kawashima M, Nixon R. The ubiquitin-proteasome system and the autophagic-lysosomal system in Alzheimer disease. Cold Spring Harb Perspect Med 2012;2.

[88] Nixon RA. Autophagy, amyloidogenesis and Alzheimer disease. J Cell Sci 2007;120:4081–91.

[89] Yu WH, Kumar A, Peterhoff C, Shapiro Kulnane L, Uchiyama Y, Lamb BT, et al. Autophagic vacuoles are enriched in amyloid precursor protein-secretase activities: implications for beta-amyloid peptide over-production and localization in Alzheimer's disease. Int J Biochem Cell Biol 2004;36:2531–40.

[90] Yu WH, Cuervo AM, Kumar A, Peterhoff CM, Schmidt SD, Lee JH, et al. Macroautophagy—a novel beta-amyloid peptide-generating pathway activated in Alzheimer's disease. J Cell Biol 2005;171:87–98.

[91] Nilsson P, Saido TC. Dual roles for autophagy: degradation and secretion of Alzheimer's disease Abeta peptide. BioEssays 2014;36:570–8.

[92] Woods NK, Padmanabhan J. Neuronal calcium signaling and Alzheimer's disease. Adv Exp Med Biol 2012;740:1193–217.

[93] Ma T. GSK3 in Alzheimer's disease: mind the isoforms. J Alzheimer's Dis 2014;39:707–10.

[94] Salminen A, Kaarniranta K, Haapasalo A, Soininen H, Hiltunen M. AMP-activated protein kinase: a potential player in Alzheimer's disease. J Neurochem 2011;118:460–74.

[95] Pickford F, Masliah E, Britschgi M, Lucin K, Narasimhan R, Jaeger PA, et al. The autophagy-related protein beclin 1 shows reduced expression in early Alzheimer disease and regulates amyloid beta accumulation in mice. J Clin Invest 2008;118:2190–9.

[96] Jaeger PA, Pickford F, Sun CH, Lucin KM, Masliah E, Wyss-Coray T. Regulation of amyloid precursor protein processing by the Beclin 1 complex. PloS One 2010;5:e11102.

[97] Wolfe DM, Lee JH, Kumar A, Lee S, Orenstein SJ, Nixon RA. Autophagy failure in Alzheimer's disease and the role of defective lysosomal acidification. Eur J Neurosci 2013;37:1949–61.

[98] McBrayer M, Nixon RA. Lysosome and calcium dysregulation in Alzheimer's disease: partners in crime. Biochem Soc Trans 2013;41:1495–502.

[99] Hirakura Y, Lin MC, Kagan BL. Alzheimer amyloid abeta1-42 channels: effects of solvent, pH, and Congo red. J Neurosci Res 1999;57:458–66.

[100] Spilman P, Podlutskaya N, Hart MJ, Debnath J, Gorostiza O, Bredesen D, et al. Inhibition of mTOR by rapamycin abolishes cognitive deficits and reduces amyloid-beta levels in a mouse model of Alzheimer's disease. PloS One 2010;5:e9979.

[101] Avrahami L, Farfara D, Shaham-Kol M, Vassar R, Frenkel D, Eldar-Finkelman H. Inhibition of glycogen synthase kinase-3 ameliorates beta-amyloid pathology and restores lysosomal acidification and mammalian target of rapamycin activity in the Alzheimer disease mouse model: in vivo and in vitro studies. J Biol Chem 2013;288:1295–306.

[102] Liu D, Pitta M, Jiang H, Lee JH, Zhang G, Chen X, et al. Nicotinamide forestalls pathology and cognitive decline in Alzheimer mice: evidence for improved neuronal bioenergetics and autophagy procession. Neurobiol Aging 2013;34:1564–80.

[103] Baumgart DC, Sandborn WJ. Crohn's disease. Lancet 2012;380:1590–605.

[104] Douroudis K, Kingo K, Traks T, Reimann E, Raud K, Ratsep R, et al. Polymorphisms in the ATG16L1 gene are associated with psoriasis vulgaris. Acta Derm Venereol 2012;92:85–7.

[105] Gros F, Arnold J, Page N, Decossas M, Korganow AS, Martin T, et al. Macroautophagy is deregulated in murine and human lupus T lymphocytes. Autophagy 2012;8:1113–23.

[106] Lin NY, Beyer C, Giessl A, Kireva T, Scholtysek C, Uderhardt S, et al. Autophagy regulates TNFalpha-mediated joint destruction in experimental arthritis. Ann Rheum Dis 2013;72:761–8.

[107] Alirezaei M, Fox HS, Flynn CT, Moore CS, Hebb AL, Frausto RF, et al. Elevated ATG5 expression in autoimmune demyelination and multiple sclerosis. Autophagy 2009;5:152–8.

[108] Francois A, Terro F, Quellard N, Fernandez B, Chassaing D, Janet T, et al. Impairment of autophagy in the central nervous system during lipopolysaccharide-induced inflammatory stress in mice. Mol Brain 2014;7:56.

[109] Bove J, Martinez-Vicente M, Vila M. Fighting neurodegeneration with rapamycin: mechanistic insights. Nat Rev Neurosci 2011;12:437–52.

[110] Caccamo A, Magri A, Medina DX, Wisely EV, Lopez-Aranda MF, Silva AJ, et al. mTOR regulates tau phosphorylation and degradation: implications for Alzheimer's disease and other tauopathies. Aging Cell 2013;12:370–80.

[111] Majumder S, Caccamo A, Medina DX, Benavides AD, Javors MA, Kraig E, et al. Lifelong rapamycin administration ameliorates age-dependent cognitive deficits by reducing IL-1beta and enhancing NMDA signaling. Aging Cell 2012;11:326–35.

[112] Jiang T, Yu JT, Zhu XC, Tan MS, Wang HF, Cao L, et al. Temsirolimus promotes autophagic clearance of amyloid-beta and provides protective effects in cellular and animal models of Alzheimer's disease. Pharmacol Res 2014;81:54–63.

[113] Cai Z, Zhao B, Li K, Zhang L, Li C, Quazi SH, et al. Mammalian target of rapamycin: a valid therapeutic target through the autophagy pathway for Alzheimer's disease? J Neurosci Res 2012;90:1105–18.

[114] Boland B, Kumar A, Lee S, Platt FM, Wegiel J, Yu WH, et al. Autophagy induction and autophagosome clearance in neurons: relationship to autophagic pathology in Alzheimer's disease. J Neurosci 2008;28:6926–37.

[115] Aso E, Ferrer I. It may be possible to delay the onset of neurodegenerative diseases with an immunosuppressive drug (rapamycin). Exp Opin Biol Ther 2013;13:1215–19.

[116] Rosner M, Hengstschlager M. Cytoplasmic and nuclear distribution of the protein complexes mTORC1 and mTORC2: rapamycin triggers dephosphorylation and delocalization of the mTORC2 components rictor and sin1. Hum Mol Genet 2008;17:2934–48.

[117] Lamming DW, Ye L, Sabatini DM, Baur JA. Rapalogs and mTOR inhibitors as anti-aging therapeutics. J Clin Invest 2013;123:980–9.

[118] Sikora E, Bielak-Zmijewska A, Mosieniak G, Piwocka K. The promise of slow down ageing may come from curcumin. Curr Pharm Design 2010;16:884–92.

[119] Grabowska W, Kucharewicz K, Wnuk M, Lewinska A, Suszek M, Przybylska D, et al. Curcumin induces senescence of primary human cells building the vasculature in a DNA damage and ATM-independent manner. Age 2015;37:9744.

[120] Grossi C, Rigacci S, Ambrosini S, Ed Dami T, Luccarini I, Traini C, et al. The polyphenol oleuropein aglycone protects TgCRND8 mice against Ass plaque pathology. PLoS One 2013;8:e71702.

[121] Tizon B, Sahoo S, Yu H, Gauthier S, Kumar AR, Mohan P, et al. Induction of autophagy by cystatin C: a mechanism that protects murine primary cortical neurons and neuronal cell lines. PLoS One 2010;5:e9819.

[122] Steele JW, Lachenmayer ML, Ju S, Stock A, Liken J, Kim SH, et al. Latrepirdine improves cognition and arrests progression of neuropathology in an Alzheimer's mouse model. Mol Psychiatry 2013;18:889–97.

[123] Steele JW, Gandy S. Latrepirdine (Dimebon(R)), a potential Alzheimer therapeutic, regulates autophagy and neuropathology in an Alzheimer mouse model. Autophagy 2013;9:617–18.

[124] Li L, Zhang S, Zhang X, Li T, Tang Y, Liu H, et al. Autophagy enhancer carbamazepine alleviates memory deficits and cerebral amyloid-beta pathology in a mouse model of Alzheimer's disease. Curr Alzheimer Res 2013;10:433–41.

[125] Zhu Z, Yan J, Jiang W, Yao XG, Chen J, Chen L, et al. Arctigenin effectively ameliorates memory impairment in Alzheimer's disease model mice targeting both beta-amyloid production and clearance. J Neurosci 2013;33:13138–49.

[126] Wu H, Lu MH, Wang W, Zhang MY, Zhu QQ, Xia YY, et al. Lamotrigine reduces beta-site AbetaPP-cleaving enzyme 1 protein levels through induction of autophagy. J Alzheimer's Dis 2015.

[127] Zhu Y, Wang J. Wogonin increases beta-amyloid clearance and inhibits tau phosphorylation via inhibition of mammalian target of rapamycin: potential drug to treat Alzheimer's disease. Neurol Sci 2015.

[128] Guo J, Chang L, Zhang X, Pei S, Yu M, Gao J. Ginsenoside compound K promotes beta-amyloid peptide clearance in primary astrocytes via autophagy enhancement. Exp Ther Med 2014;8:1271–4.

[129] Vingtdeux V, Giliberto L, Zhao H, Chandakkar P, Wu Q, Simon JE, et al. AMP-activated protein kinase signaling activation by resveratrol modulates amyloid-beta peptide metabolism. J Biol Chem 2010;285:9100–13.

[130] Vingtdeux V, Chandakkar P, Zhao H, d'Abramo C, Davies P, Marambaud P. Novel synthetic small-molecule activators of AMPK as enhancers of autophagy and amyloid-beta peptide degradation. FASEB J 2011;25:219–31.

[131] d'Abramo C, Ricciarelli R, Pronzato MA, Davies P. Troglitazone, a peroxisome proliferator-activated receptor-gamma agonist, decreases tau phosphorylation in CHOtau4R cells. J Neurochem 2006;98:1068–77.

[132] Nicolakakis N, Aboulkassim T, Ongali B, Lecrux C, Fernandes P, Rosa-Neto P, et al. Complete rescue of cerebrovascular function in aged Alzheimer's disease transgenic mice by antioxidants and pioglitazone, a peroxisome proliferator-activated receptor gamma agonist. J Neurosci 2008;28:9287–96.

[133] Liu J, Wang LN, Jia JP. Peroxisome proliferator-activated receptor-gamma agonists for Alzheimer's disease and amnestic mild cognitive impairment: a systematic review and meta-analysis. Drugs Aging 2015;32:57–65.

[134] Picone P, Nuzzo D, Caruana L, Messina E, Barera A, Vasto S, et al. Metformin increases APP expression and processing via oxidative stress, mitochondrial dysfunction and NF-kappaB activation: use of insulin to attenuate metformin's effect. Biochim Biophys Acta 2015;1853:1046–59.

[135] DiTacchio KA, Heinemann SF, Dziewczapolski G. Metformin treatment alters memory function in a mouse model of Alzheimer's disease. J Alzheimer's Dis 2015;44:43–8.

[136] Wang Y, Wang YX, Liu T, Law PY, Loh HH, Qiu Y, et al. mu-Opioid receptor attenuates Abeta oligomers-induced neurotoxicity through mTOR signaling. CNS Neurosci Ther 2015;21:8–14.

[137] Martin I, Dawson VL, Dawson TM. The impact of genetic research on our understanding of Parkinson's disease. Progress Brain Res 2010;183:21–41.

[138] Goetz CG. The history of Parkinson's disease: early clinical descriptions and neurological therapies. Cold Spring Harb Perspect Med 2011;1:a008862.

[139] Gibb WR, Lees AJ. The relevance of the Lewy body to the pathogenesis of idiopathic Parkinson's disease. J Neurol Neurosurg Psychiatr 1988;51:745–52.

[140] Kalia LV, Lang AE. Parkinson's disease. Lancet 2015;386 (9996):896–912.

[141] Jankovic J. Parkinson's disease: clinical features and diagnosis. J Neurol Neurosurg Psychiatr 2008;79:368–76.

[142] Habibi E, Masoudi-Nejad A, Abdolmaleky HM, Haggarty SJ. Emerging roles of epigenetic mechanisms in Parkinson's disease. Funct Integr Genomics 2011;11:523–37.

[143] Noyce AJ, Bestwick JP, Silveira-Moriyama L, Hawkes CH, Giovannoni G, Lees AJ, et al. Meta-analysis of early nonmotor features and risk factors for Parkinson disease. Ann Neurol 2012;72:893–901.

[144] Polymeropoulos MH, Lavedan C, Leroy E, Ide SE, Dehejia A, Dutra A, et al. Mutation in the alpha-synuclein gene identified in families with Parkinson's disease. Science 1997;276: 2045–7.

[145] Vilarino-Guell C, Wider C, Ross OA, Dachsel JC, Kachergus JM, Lincoln SJ, et al. VPS35 mutations in Parkinson disease. Am J Hum Genet 2011;89:162–7.

[146] Chartier-Harlin MC, Dachsel JC, Vilarino-Guell C, Lincoln SJ, Lepretre F, Hulihan MM, et al. Translation initiator EIF4G1 mutations in familial Parkinson disease. Am J Hum Genet 2011;89:398–406.

[147] Vilarino-Guell C, Rajput A, Milnerwood AJ, Shah B, Szu-Tu C, Trinh J, et al. DNAJC13 mutations in Parkinson disease. Hum Mol Genet 2014;23:1794–801.

[148] Funayama M, Ohe K, Amo T, Furuya N, Yamaguchi J, Saiki S, et al. CHCHD2 mutations in autosomal dominant late-onset Parkinson's disease: a genome-wide linkage and sequencing study. Lancet Neurol 2015;14:274–82.

[149] Schrag A, Schott JM. Epidemiological, clinical, and genetic characteristics of early-onset parkinsonism. Lancet Neurol 2006;5:355–63.

[150] Keating DJ. Mitochondrial dysfunction, oxidative stress, regulation of exocytosis and their relevance to neurodegenerative diseases. J Neurochem 2008;104:298–305.

[151] Lin MT, Beal MF. Mitochondrial dysfunction and oxidative stress in neurodegenerative diseases. Nature 2006;443:787–95.

[152] Dantuma NP, Bott LC. The ubiquitin-proteasome system in neurodegenerative diseases: precipitating factor, yet part of the solution. Front Mol Neurosci 2014;7:70.

[153] Menzies FM, Fleming A, Rubinsztein DC. Compromised autophagy and neurodegenerative diseases. Nat Rev Neurosci 2015;16:345–57.

[154] Sankowski R, Mader S, Valdes-Ferrer SI. Systemic inflammation and the brain: novel roles of genetic, molecular, and environmental cues as drivers of neurodegeneration. Front Cell Neurosci 2015;9:28.

[155] Xu Y, Liu C, Chen S, Ye Y, Guo M, Ren Q, et al. Activation of AMPK and inactivation of Akt result in suppression of mTOR-mediated S6K1 and 4E-BP1 pathways leading to neuronal cell death in in vitro models of Parkinson's disease. Cell Signal 2014;26:1680–9.

[156] Inoki K, Corradetti MN, Guan KL. Dysregulation of the TSC-mTOR pathway in human disease. Nat Genet 2005;37:19–24.

[157] Elstner M, Morris CM, Heim K, Bender A, Mehta D, Jaros E, et al. Expression analysis of dopaminergic neurons in Parkinson's disease and aging links transcriptional dysregulation of energy metabolism to cell death. Acta Neuropathol 2011;122:75–86.

[158] Heras-Sandoval D, Perez-Rojas JM, Hernandez-Damian J, Pedraza-Chaverri J. The role of PI3K/AKT/mTOR pathway in the modulation of autophagy and the clearance of protein aggregates in neurodegeneration. Cell Signal 2014;26:2694–701.

[159] Malagelada C, Ryu EJ, Biswas SC, Jackson-Lewis V, Greene LA. RTP801 is elevated in Parkinson brain substantia nigral neurons and mediates death in cellular models of Parkinson's disease by a mechanism involving mammalian target of rapamycin inactivation. J Neurosci 2006;26:9996–10005.

[160] Romani-Aumedes J, Canal M, Martin-Flores N, Sun X, Perez-Fernandez V, Wewering S, et al. Parkin loss of function contributes to RTP801 elevation and neurodegeneration in Parkinson's disease. Cell Death Dis 2014;5:e1364.

[161] Malagelada C, Jin ZH, Jackson-Lewis V, Przedborski S, Greene LA. Rapamycin protects against neuron death in in vitro and in vivo models of Parkinson's disease. J Neurosci 2010;30:1166–75.

[162] Deguil J, Jailloux D, Page G, Fauconneau B, Houeto JL, Philippe M, et al. Neuroprotective effects of pituitary adenylate cyclase-activating polypeptide (PACAP) in MPP+-induced alteration of translational control in Neuro-2a neuroblastoma cells. J Neurosci Res 2007;85:2017–25.

[163] Rodriguez-Blanco J, Martin V, Garcia-Santos G, Herrera F, Casado-Zapico S, Antolin I, et al. Cooperative action of JNK and AKT/mTOR in 1-methyl-4-phenylpyridinium-induced autophagy of neuronal PC12 cells. J Neurosci Res 2012;90:1850–60.

[164] Deguil J, Chavant F, Lafay-Chebassier C, Perault-Pochat MC, Fauconneau B, Pain S. Time course of MPTP toxicity on translational control protein expression in mice brain. Toxicol Lett 2010;196:51–5.

[165] Deguil J, Chavant F, Lafay-Chebassier C, Perault-Pochat MC, Fauconneau B, Pain S. Neuroprotective effect of PACAP on translational control alteration and cognitive decline in MPTP parkinsonian mice. Neurotoxicity Res 2010;17:142–55.

[166] Santini E, Heiman M, Greengard P, Valjent E, Fisone G. Inhibition of mTOR signaling in Parkinson's disease prevents L-DOPA-induced dyskinesia. Sci Signal 2009;2:ra36.

[167] Subramaniam S, Napolitano F, Mealer RG, Kim S, Errico F, Barrow R, et al. Rhes, a striatal-enriched small G protein, mediates mTOR signaling and L-DOPA-induced dyskinesia. Nat Neurosci 2012;15:191–3.

[168] Zhang H, Duan C, Yang H. Defective autophagy in Parkinson's disease: lessons from genetics. Mol Neurobiol 2015;51:89–104.

[169] Cartier AE, Ubhi K, Spencer B, Vazquez-Roque RA, Kosberg KA, Fourgeaud L, et al. Differential effects of UCHL1 modulation on alpha-synuclein in PD-like models of alpha-synucleinopathy. PLoS One 2012;7:e34713.

[170] Orenstein SJ, Kuo SH, Tasset I, Arias E, Koga H, Fernandez-Carasa I, et al. Interplay of LRRK2 with chaperone-mediated autophagy. Nat Neurosci 2013;16:394–406.

[171] Imai Y, Gehrke S, Wang HQ, Takahashi R, Hasegawa K, Oota E, et al. Phosphorylation of 4E-BP by LRRK2 affects the maintenance of dopaminergic neurons in Drosophila. EMBO J 2008;27:2432–43.

[172] Chu CT, Plowey ED, Dagda RK, Hickey RW, Cherra III SJ, Clark RS. Autophagy in neurite injury and neurodegeneration: *in vitro* and *in vivo* models. Methods Enzymol 2009;453:217–49.

[173] Vives-Bauza C, Zhou C, Huang Y, Cui M, de Vries RL, Kim J, et al. PINK1-dependent recruitment of Parkin to mitochondria in mitophagy. Proc Natl Acad Sci USA 2010;107:378–83.

[174] Rakovic A, Grunewald A, Kottwitz J, Bruggemann N, Pramstaller PP, Lohmann K, et al. Mutations in PINK1 and Parkin impair ubiquitination of mitofusins in human fibroblasts. PLoS One 2011;6:e16746.

[175] Thomas KJ, McCoy MK, Blackinton J, Beilina A, van der Brug M, Sandebring A, et al. DJ-1 acts in parallel to the PINK1/parkin pathway to control mitochondrial function and autophagy. Hum Mol Genet 2011;20:40–50.

[176] Trempe JF, Fon EA. Structure and function of Parkin, PINK1, and DJ-1, the three musketeers of neuroprotection. Front Neurol 2013;4:38.

[177] Crews L, Spencer B, Desplats P, Patrick C, Paulino A, Rockenstein E, et al. Selective molecular alterations in the autophagy pathway in patients with Lewy body disease and in models of alpha-synucleinopathy. PLoS One 2010;5:e9313.

[178] Jiang TF, Zhang YJ, Zhou HY, Wang HM, Tian LP, Liu J, et al. Curcumin ameliorates the neurodegenerative pathology in A53T alpha-synuclein cell model of Parkinson's disease through the downregulation of mTOR/p70S6K signaling and

[179] Poehler AM, Xiang W, Spitzer P, May VE, Meixner H, Rockenstein E, et al. Autophagy modulates SNCA/alpha-synuclein release, thereby generating a hostile microenvironment. Autophagy 2014;10:2171–92.

[180] Klucken J, Poehler AM, Ebrahimi-Fakhari D, Schneider J, Nuber S, Rockenstein E, et al. Alpha-synuclein aggregation involves a bafilomycin A 1-sensitive autophagy pathway. Autophagy 2012;8:754–66.

[181] Klucken J, Shin Y, Masliah E, Hyman BT, McLean PJ. Hsp70 reduces alpha-synuclein aggregation and toxicity. J Biol Chem 2004;279:25497–502.

[182] Tain LS, Mortiboys H, Tao RN, Ziviani E, Bandmann O, Whitworth AJ. Rapamycin activation of 4E-BP prevents parkinsonian dopaminergic neuron loss. Nat Neurosci 2009;12:1129–35.

[183] Decressac M, Bjorklund A. mTOR inhibition alleviates L-DOPA-induced dyskinesia in parkinsonian rats. J Parkinson's Dis 2013;3:13–17.

[184] Wu Y, Li X, Zhu JX, Xie W, Le W, Fan Z, et al. Resveratrol-activated AMPK/SIRT1/autophagy in cellular models of Parkinson's disease. Neuro Signals 2011;19:163–74.

[185] Filomeni G, Graziani I, De Zio D, Dini L, Centonze D, Rotilio G, et al. Neuroprotection of kaempferol by autophagy in models of rotenone-mediated acute toxicity: possible implications for Parkinson's disease. Neurobiol Aging 2012;33:767–85.

[186] Lu JH, Tan JQ, Durairajan SS, Liu LF, Zhang ZH, Ma L, et al. Isorhynchophylline, a natural alkaloid, promotes the degradation of alpha-synuclein in neuronal cells via inducing autophagy. Autophagy 2012;8:98–108.

[187] Yu WB, Jiang T, Lan DM, Lu JH, Yue ZY, Wang J, et al. Trehalose inhibits fibrillation of A53T mutant alpha-synuclein and disaggregates existing fibrils. Arch Biochem Biophys 2012;523:144–50.

[188] Sarkar S, Davies JE, Huang Z, Tunnacliffe A, Rubinsztein DC. Trehalose, a novel mTOR-independent autophagy enhancer, accelerates the clearance of mutant huntingtin and alpha-synuclein. J Biol Chem 2007;282:5641–52.

[189] He Q, Koprich JB, Wang Y, Yu WB, Xiao BG, Brotchie JM, et al. Treatment with trehalose prevents behavioral and neurochemical deficits produced in an AAV alpha-synuclein rat model of parkinson's disease. Mol Neurobiol 2015.

[190] Gusella JF, Wexler NS, Conneally PM, Naylor SL, Anderson MA, Tanzi RE, et al. A polymorphic DNA marker genetically linked to Huntington's disease. Nature 1983;306:234–8.

[191] A novel gene containing a trinucleotide repeat that is expanded and unstable on Huntington's disease chromosomes. The Huntington's Disease Collaborative Research Group. Cell 1993; 72:971–83.

[192] Walker FO. Huntington's disease. Lancet 2007;369:218–28.

[193] Zuccato C, Valenza M, Cattaneo E. Molecular mechanisms and potential therapeutical targets in Huntington's disease. Physiol Rev 2010;90:905–81.

[194] Berardelli A, Noth J, Thompson PD, Bollen EL, Curra A, Deuschl G, et al. Pathophysiology of chorea and bradykinesia in Huntington's disease. Mov Disord 1999;14:398–403.

[195] Stout JC, Paulsen JS, Queller S, Solomon AC, Whitlock KB, Campbell JC, et al. Neurocognitive signs in prodromal Huntington disease. Neuropsychology 2011;25:1–14.

[196] Stout JC, Jones R, Labuschagne I, O'Regan AM, Say MJ, Dumas EM, et al. Evaluation of longitudinal 12 and 24 month cognitive outcomes in premanifest and early Huntington's disease. J Neurol Neurosurg Psychiatry 2012;83:687–94.

[197] Snowden JS, Craufurd D, Thompson J, Neary D. Psychomotor, executive, and memory function in preclinical Huntington's disease. J Clin Exp Neuropsychol 2002;24:133–45.

[198] Peavy GM, Jacobson MW, Goldstein JL, Hamilton JM, Kane A, Gamst AC, et al. Cognitive and functional decline in Huntington's disease: dementia criteria revisited. Mov Disord 2010;25:1163–9.

[199] Solomon AC, Stout JC, Johnson SA, Langbehn DR, Aylward EH, Brandt J, et al. Verbal episodic memory declines prior to diagnosis in Huntington's disease. Neuropsychologia 2007;45: 1767–76.

[200] Martin JB, Gusella JF. Huntington's disease. Pathogenesis and management. N Engl J Med 1986;315:1267–76.

[201] Taherzadeh-Fard E, Saft C, Akkad DA, Wieczorek S, Haghikia A, Chan A, et al. PGC-1alpha downstream transcription factors NRF-1 and TFAM are genetic modifiers of Huntington disease. Mol Neurodegener 2011;6:32.

[202] Saft C, Epplen JT, Wieczorek S, Landwehrmeyer GB, Roos RA, de Yebenes JG, et al. NMDA receptor gene variations as modifiers in Huntington disease: a replication study. PLoS Curr 2011;3:RRN1247.

[203] Reiner A, Albin RL, Anderson KD, D'Amato CJ, Penney JB, Young AB. Differential loss of striatal projection neurons in Huntington disease. Proc Natl Acad Sci USA 1988;85:5733–7.

[204] Vonsattel JP, DiFiglia M. Huntington disease. J Neuropathol Exp Neurol 1998;57:369–84.

[205] Rosas HD, Koroshetz WJ, Chen YI, Skeuse C, Vangel M, Cudkowicz ME, et al. Evidence for more widespread cerebral pathology in early HD: an MRI-based morphometric analysis. Neurology 2003;60:1615–20.

[206] Rosas HD, Salat DH, Lee SY, Zaleta AK, Pappu V, Fischl B, et al. Cerebral cortex and the clinical expression of Huntington's disease: complexity and heterogeneity. Brain 2008;131:1057–68.

[207] Baquet ZC, Gorski JA, Jones KR. Early striatal dendrite deficits followed by neuron loss with advanced age in the absence of anterograde cortical brain-derived neurotrophic factor. J Neurosci 2004;24:4250–8.

[208] Zuccato C, Cattaneo E. Role of brain-derived neurotrophic factor in Huntington's disease. Prog Neurobiol 2007;81:294–330.

[209] Ravikumar B, Vacher C, Berger Z, Davies JE, Luo S, Oroz LG, et al. Inhibition of mTOR induces autophagy and reduces toxicity of polyglutamine expansions in fly and mouse models of Huntington disease. Nat Genet 2004;36:585–95.

[210] Lee JH, Tecedor L, Chen YH, Monteys AM, Sowada MJ, Thompson LM, et al. Reinstating aberrant mTORC1 activity in Huntington's disease mice improves disease phenotypes. Neuron 2015;85:303–15.

[211] Pryor WM, Biagioli M, Shahani N, Swarnkar S, Huang WC, Page DT, et al. Huntingtin promotes mTORC1 signaling in the pathogenesis of Huntington's disease. Sci Signal 2014;7: ra103.

[212] Subramaniam S, Sixt KM, Barrow R, Snyder SH. Rhes, a striatal specific protein, mediates mutant-huntingtin cytotoxicity. Science 2009;324:1327–30.

[213] Mealer RG, Murray AJ, Shahani N, Subramaniam S, Snyder SH. Rhes, a striatal-selective protein implicated in Huntington disease, binds beclin-1 and activates autophagy. J Biol Chem 2014;289:3547–54.

[214] Rui YN, Xu Z, Patel B, Chen Z, Chen D, Tito A, et al. Huntingtin functions as a scaffold for selective macroautophagy. Nat Cell Biol 2015;17:262–75.

[215] Ganley IG, Lam du H, Wang J, Ding X, Chen S, Jiang X. ULK1.ATG13.FIP200 complex mediates mTOR signaling and is essential for autophagy. J Biol Chem 2009;284:12297–305.

[216] Zheng S, Clabough EB, Sarkar S, Futter M, Rubinsztein DC, Zeitlin SO. Deletion of the huntingtin polyglutamine stretch enhances neuronal autophagy and longevity in mice. PLoS Genet 2010;6:e1000838.

[217] Ravikumar B, Duden R, Rubinsztein DC. Aggregate-prone proteins with polyglutamine and polyalanine expansions are degraded by autophagy. Hum Mol Genet 2002;11:1107–17.

[218] Sarkar S, Floto RA, Berger Z, Imarisio S, Cordenier A, Pasco M, et al. Lithium induces autophagy by inhibiting inositol monophosphatase. J Cell Biol 2005;170:1101–11.

[219] Phiel CJ, Klein PS. Molecular targets of lithium action. Annu Rev Pharmacol Toxicol 2001;41:789–813.

[220] Maher P, Dargusch R, Bodai L, Gerard PE, Purcell JM, Marsh JL. ERK activation by the polyphenols fisetin and resveratrol provides neuroprotection in multiple models of Huntington's disease. Hum Mol Genet 2011;20:261–70.

[221] Dasgupta B, Milbrandt J. Resveratrol stimulates AMP kinase activity in neurons. Proc Natl Acad Sci USA 2007;104:7217–22.

[222] Ho DJ, Calingasan NY, Wille E, Dumont M, Beal MF. Resveratrol protects against peripheral deficits in a mouse model of Huntington's disease. Exp Neurol 2010;225:74–84.

[223] Wong VK, Wu AG, Wang JR, Liu L, Law BY. Neferine attenuates the protein level and toxicity of mutant huntingtin in PC-12 cells via induction of autophagy. Molecules 2015;20: 3496–514.

[224] Tanaka M, Machida Y, Niu S, Ikeda T, Jana NR, Doi H, et al. Trehalose alleviates polyglutamine-mediated pathology in a mouse model of Huntington disease. Nat Med 2004;10: 148–54.

[225] Thimmaiah KN, Easton JB, Germain GS, Morton CL, Kamath S, Buolamwini JK, et al. Identification of N10-substituted phenoxazines as potent and specific inhibitors of Akt signaling. J Biol Chem 2005;280:31924–35.

[226] Tsvetkov AS, Miller J, Arrasate M, Wong JS, Pleiss MA, Finkbeiner S. A small-molecule scaffold induces autophagy in primary neurons and protects against toxicity in a Huntington disease model. Proc Natl Acad Sci USA 2010; 107:16982–7.

[227] Harron DW. Distinctive features of rilmenidine possibly related to its selectivity for imidazoline receptors. Am J Hypertens 1992;5:91S–98SS.

[228] Rose C, Menzies FM, Renna M, Acevedo-Arozena A, Corrochano S, Sadiq O, et al. Rilmenidine attenuates toxicity of polyglutamine expansions in a mouse model of Huntington's disease. Hum Mol Genet 2010;19:2144–53.

[229] Martin LJ. Mitochondrial pathobiology in ALS. J Bioenerg Biomembr 2011;43:569–79.

[230] Blackhall LJ. Amyotrophic lateral sclerosis and palliative care: where we are, and the road ahead. Muscle Nerve 2012;45:311–18.

[231] Carri MT, Cozzolino M. SOD1 and mitochondria in ALS: a dangerous liaison. J Bioenerg Biomembr 2011;43:593–9.

[232] Reaume AG, Elliott JL, Hoffman EK, Kowall NW, Ferrante RJ, Siwek DF, et al. Motor neurons in Cu/Zn superoxide dismutase-deficient mice develop normally but exhibit enhanced cell death after axonal injury. Nat Genet 1996;13:43–7.

[233] Bosco DA, Morfini G, Karabacak NM, Song Y, Gros-Louis F, Pasinelli P, et al. Wild-type and mutant SOD1 share an aberrant conformation and a common pathogenic pathway in ALS. Nat Neurosci 2010;13:1396–403.

[234] Urushitani M, Sik A, Sakurai T, Nukina N, Takahashi R, Julien JP. Chromogranin-mediated secretion of mutant superoxide dismutase proteins linked to amyotrophic lateral sclerosis. Nat Neurosci 2006;9:108–18.

[235] Sreedharan J, Blair IP, Tripathi VB, Hu X, Vance C, Rogelj B, et al. TDP-43 mutations in familial and sporadic amyotrophic lateral sclerosis. Science 2008;319:1668–72.

[236] Vance C, Rogelj B, Hortobagyi T, De Vos KJ, Nishimura AL, Sreedharan J, et al. Mutations in FUS, an RNA processing protein, cause familial amyotrophic lateral sclerosis type 6. Science 2009;323:1208–11.

[237] Shaw PJ. Molecular and cellular pathways of neurodegeneration in motor neurone disease. J Neurol Neurosurg Psychiatry 2005;76:1046–57.

[238] Mantovani S, Garbelli S, Pasini A, Alimonti D, Perotti C, Melazzini M, et al. Immune system alterations in sporadic amyotrophic lateral sclerosis patients suggest an ongoing neuroinflammatory process. J Neuroimmunol 2009;210:73–9.

[239] Morimoto N, Nagai M, Ohta Y, Miyazaki K, Kurata T, Morimoto M, et al. Increased autophagy in transgenic mice with a G93A mutant SOD1 gene. Brain Res 2007;1167:112–17.

[240] Zhang X, Li L, Chen S, Yang D, Wang Y, Wang Z, et al. Rapamycin treatment augments motor neuron degeneration in SOD1(G93A) mouse model of amyotrophic lateral sclerosis. Autophagy 2011;7:412–25.

[241] Nagano I, Murakami T, Manabe Y, Abe K. Early decrease of survival factors and DNA repair enzyme in spinal motor neurons of presymptomatic transgenic mice that express a mutant SOD1 gene. Life Sci 2002;72:541–8.

[242] Turner MR, Osei-Lah AD, Hammers A, Al-Chalabi A, Shaw CE, Andersen PM, et al. Abnormal cortical excitability in sporadic but not homozygous D90A SOD1 ALS. J Neurol Neurosurg Psychiatry 2005;76:1279–85.

[243] Vucic S, Kiernan MC. Abnormalities in cortical and peripheral excitability in flail arm variant amyotrophic lateral sclerosis. J Neurol Neurosurg Psychiatry 2007;78:849–52.

[244] Carunchio I, Curcio L, Pieri M, Pica F, Caioli S, Viscomi MT, et al. Increased levels of p70S6 phosphorylation in the G93A mouse model of amyotrophic lateral sclerosis and in valine-exposed cortical neurons in culture. Exp Neurol 2010;226: 218–30.

[245] Ravikumar B, Berger Z, Vacher C, O'Kane CJ, Rubinsztein DC. Rapamycin pre-treatment protects against apoptosis. Hum Mol Genet 2006;15:1209–16.

[246] Kato S, Takikawa M, Nakashima K, Hirano A, Cleveland DW, Kusaka H, et al. New consensus research on neuropathological aspects of familial amyotrophic lateral sclerosis with superoxide dismutase 1 (SOD1) gene mutations: inclusions containing SOD1 in neurons and astrocytes. Amyotroph Lateral Scler Other Motor Neuron Disord 2000;1:163–84.

[247] Watanabe M, Dykes-Hoberg M, Culotta VC, Price DL, Wong PC, Rothstein JD. Histological evidence of protein aggregation in mutant SOD1 transgenic mice and in amyotrophic lateral sclerosis neural tissues. Neurobiol Dis 2001;8:933–41.

[248] Saxena S, Roselli F, Singh K, Leptien K, Julien JP, Gros-Louis F, et al. Neuroprotection through excitability and mTOR required in ALS motoneurons to delay disease and extend survival. Neuron 2013;80:80–96.

[249] Hetz C, Thielen P, Matus S, Nassif M, Court F, Kiffin R, et al. XBP-1 deficiency in the nervous system protects against amyotrophic lateral sclerosis by increasing autophagy. Genes Dev 2009;23:2294–306.

[250] Fornai F, Longone P, Cafaro L, Kastsiuchenka O, Ferrucci M, Manca ML, et al. Lithium delays progression of amyotrophic lateral sclerosis. Proc Natl Acad Sci USA 2008;105:2052–7.

[251] Kim D, Nguyen MD, Dobbin MM, Fischer A, Sananbenesi F, Rodgers JT, et al. SIRT1 deacetylase protects against neurodegeneration in models for Alzheimer's disease and amyotrophic lateral sclerosis. EMBO J 2007;26:3169–79.

[252] Zhang K, Shi P, An T, Wang Q, Wang J, Li Z, et al. Food restriction-induced autophagy modulates degradation of mutant SOD1 in an amyotrophic lateral sclerosis mouse model. Brain Res 2013;1519:112–19.

[253] Gomes C, Escrevente C, Costa J. Mutant superoxide dismutase 1 overexpression in NSC-34 cells: effect of trehalose on aggregation, TDP-43 localization and levels of co-expressed glycoproteins. Neurosci Lett 2010;475:145–9.

[254] Mattson MP. Excitation BolsTORs motor neurons in ALS mice. Neuron 2013;80:1–3.

[255] Cross AH, Naismith RT. Established and novel disease-modifying treatments in multiple sclerosis. J Intern Med 2014;275:350–63.

[256] Koch-Henriksen N, Sorensen PS. The changing demographic pattern of multiple sclerosis epidemiology. Lancet Neurol 2010;9:520–32.

[257] Kamm CP, Uitdehaag BM, Polman CH. Multiple sclerosis: current knowledge and future outlook. Eur Neurol 2014;72: 132–41.

[258] Mahad DH, Trapp BD, Lassmann H. Pathological mechanisms in progressive multiple sclerosis. Lancet Neurol 2015;14:183–93.

[259] Kutzelnigg A, Lucchinetti CF, Stadelmann C, Bruck W, Rauschka H, Bergmann M, et al. Cortical demyelination and diffuse white matter injury in multiple sclerosis. Brain 2005; 128:2705–12.

[260] Frischer JM, Bramow S, Dal-Bianco A, Lucchinetti CF, Rauschka H, Schmidbauer M, et al. The relation between inflammation and neurodegeneration in multiple sclerosis brains. Brain 2009;132:1175–89.

[261] Pagnini F, Bosma CM, Phillips D, Langer E. Symptom changes in multiple sclerosis following psychological interventions: a systematic review. BMC Neurol 2014;14:222.

[262] Paparrigopoulos T, Ferentinos P, Kouzoupis A, Koutsis G, Papadimitriou GN. The neuropsychiatry of multiple sclerosis: focus on disorders of mood, affect and behaviour. Int Rev Psychiatry 2010;22:14–21.

[263] Lublin FD, Reingold SC. Defining the clinical course of multiple sclerosis: results of an international survey. National Multiple Sclerosis Society (USA) Advisory Committee on Clinical Trials of New Agents in Multiple Sclerosis. Neurology 1996;46:907–11.

[264] Lublin FD, Reingold SC, Cohen JA, Cutter GR, Sorensen PS, Thompson AJ, et al. Defining the clinical course of multiple sclerosis: the 2013 revisions. Neurology 2014;83:278–86.

[265] Didonna A, Oksenberg JR. Genetic determinants of risk and progression in multiple sclerosis. Clin Chim Acta 2015;449: 16–22.

[266] Dello Russo C, Lisi L, Feinstein DL, Navarra P. mTOR kinase, a key player in the regulation of glial functions: relevance for the therapy of multiple sclerosis. Glia 2013;61:301–11.

[267] Esposito M, Ruffini F, Bellone M, Gagliani N, Battaglia M, Martino G, et al. Rapamycin inhibits relapsing experimental autoimmune encephalomyelitis by both effector and regulatory T cells modulation. J Neuroimmunol 2010;220: 52–63.

[268] Donia M, Mangano K, Amoroso A, Mazzarino MC, Imbesi R, Castrogiovanni P, et al. Treatment with rapamycin ameliorates clinical and histological signs of protracted relapsing experimental allergic encephalomyelitis in Dark Agouti rats and induces expansion of peripheral CD4 + CD25 + Foxp3 + regulatory T cells. J Autoimmun 2009;33:135–40.

[269] Lisi L, Navarra P, Cirocchi R, Sharp A, Stigliano E, Feinstein DL, et al. Rapamycin reduces clinical signs and neuropathic pain in a chronic model of experimental autoimmune encephalomyelitis. J Neuroimmunol 2012;243:43–51.

[270] Delgoffe GM, Pollizzi KN, Waickman AT, Heikamp E, Meyers DJ, Horton MR, et al. The kinase mTOR regulates the differentiation of helper T cells through the selective activation of signaling by mTORC1 and mTORC2. Nat Immunol 2011;12:295–303.

[271] Miron VE, Kuhlmann T, Antel JP. Cells of the oligodendroglial lineage, myelination, and remyelination. Biochim Biophys Acta 2011;1812:184–93.

[272] Kumar S, Patel R, Moore S, Crawford DK, Suwanna N, Mangiardi M, et al. Estrogen receptor beta ligand therapy activates PI3K/Akt/mTOR signaling in oligodendrocytes and promotes remyelination in a mouse model of multiple sclerosis. Neurobiol Dis 2013;56:131–44.

[273] Song MS, Salmena L, Pandolfi PP. The functions and regulation of the PTEN tumour suppressor. Nat Rev Mol Cell Biol 2012;13:283–96.

[274] De Paula ML, Cui QL, Hossain S, Antel J, Almazan G. The PTEN inhibitor bisperoxovanadium enhances myelination by amplifying IGF-1 signaling in rat and human oligodendrocyte progenitors. Glia 2014;62:64–77.

7

mTOR: Exploring a New Potential Therapeutic Target for Stroke

Mar Castellanos[1], Carme Gubern[2] and Elisabet Kadar[3]

[1]Department of Neurology, University Hospital A Coruña, A Coruña, Spain [2]Cerebrovascular Research Group, Biomedical Research Institute (IdiBGi), Girona, Spain [3]Department of Biology, University of Girona, Girona, Spain

7.1 INTRODUCTION

Stroke is a cerebrovascular disease that occurs when the blood flow to an area of the brain is interrupted. The two major categories of stroke are ischemic (due to the occlusion of the blood vessel resulting in a lack of blood and hence oxygen to an area of the brain) and hemorrhagic (due to a rupture of a blood vessel or an abnormal vascular structure that causes bleeding into an area of the brain). About 85% of strokes are ischemic and 15% are hemorrhagic.

Stroke is the third most common cause of death in developed countries, exceeded only by coronary heart disease and cancer, and is the leading cause of serious, long-term disability. Worldwide, about 15 million people suffer a stroke every year: one-third die and one-third are left permanently disabled [1]. Approximately, 795,000 new or recurrent strokes occur per year in the United States [2]. In Europe, the incidence of stroke varies from 101.1 to 239.3 per 100,000 in men and 63.0 to 158.7 per 100,000 in women [3]. Moreover, according to the World Health Organization data, the number of stroke events is likely to increase from 1.1 million per year in 2000 to more than 1.5 million per year in 2025, solely because of the increase in the mean age of the population [4,5].

At this time, the burden of stroke is already huge, both for the stroke patient and for society as a whole. The estimated direct and indirect cost of stroke for 2010 was $73.7 billion in the United States and €64.1 billion in Europe [6]. Due to the expected increase in stroke incidence in the coming years, these costs could well increase by about 10–15% if we do not succeed in preventing more strokes and improving the effectiveness of stroke management in the acute phase. So far, only recombinant tissue plasminogen activator (rt-PA) has been approved for the treatment of acute ischemic stroke [7,8]. This drug is administered with the objective of recanalizing the occluded artery and so achieving reperfusion of the ischemic tissue. By doing so, we try to avoid ischemic lesion growth over the penumbral tissue, a hypoperfused but still viable region that surrounds the ischemic core. The recruitment of the penumbral tissue towards definitive infarcted tissue depends on several factors, such as the collateral circulation, but also the action of the different mediators that are released as a consequence of the ischemic process in a phenomenon known as the ischemic cascade. The inhibition of some of these mediators has been the basis for the development of therapeutic neuroprotective strategies that have been shown to reduce ischemic damage very efficaciously in experimental models of cerebral ischemia, but unfortunately not so efficaciously in human stroke. Different factors have been acknowledged as being responsible for this failure [9] and recommendations have been made in order to try to make a more accurate translation into clinical practice of the positive results of neuroprotective drugs when tested in animals [10]. The continuous improvement in our understanding of the pathophysiological mechanisms participating in the ischemic cascade also opens the way to the development of new neuroprotective pathways in which the same molecule acts at different levels of the cascade (both early and later stages of the ischemic process). These pleiotropic properties confer not only neuroprotective but also neurorepair properties on these molecules, which is thought to be critical in increasing the possibilities of these molecules really being effective in decreasing the very deleterious effects of cerebral ischemia. Among these molecules, the mammalian target of rapamycin (mTOR) has emerged in the last few years as a new potential therapeutic target of different cerebral diseases, including stroke.

This chapter will first briefly review the early and later events of the ischemic cascade in order to understand how the complex mTOR signaling network might participate in this cascade and have a possible role as a therapeutic target in ischemic stroke.

7.2 THE ISCHEMIC CASCADE

Cerebral ischemia results in a cascade of molecular events that are triggered as a consequence of the decrease in the cerebral blood flow (CBF) and the subsequent energetic failure (Figure 7.1). This energy failure leads to metabolic disturbances with changes in the oxygen levels and in glucose metabolism and depletion of energy metabolites including adenosine triphosphate (ATP), phosphocreatine, lactate, and N-acetyl aspartate. The reduction in ATP results in the depolarization of the cellular membranes and in secondary permeability abnormalities, with an intracellular increase in $[Na^+]$, $[Ca^{2+}]$, and $[Cl^-]$, and an extracellular increase in $[K^+]$, which in turn causes the extracellular release of glutamate and other excitatory and inhibitory amino acids such as glycine and γ-aminobutyric acid (GABA), respectively. Glutamate is the main excitatory neurotransmitter in the mammalian brain and is an important mediator of neural plasticity, intracellular communication, and growth and differentiation [11]. However, higher extracellular levels are responsible for the excitotoxic damage occurring in the very early phases of the ischemic process. Glutamate transport is the primary mechanism for maintaining extracellular glutamate concentrations below excitotoxic levels [12,13]. So far, five highaffinity, sodium-dependent plasma membrane glutamate transporter subtypes, which are responsible for this transport, have been identified [14]. Among them, the astroglial excitatory amino acid transporter (EAAT)-2 or GLT-1, also found in some neurons, is responsible for up to 90% of all glutamate transport in adult tissue [15]. The human EAAT2 gene promoter has been isolated and characterized, and an elevated expression of this gene has been found in astrocytes [16]. Bioinformatic analysis of the promoter region has revealed several potential regulatory transcription factor-binding elements, including Sp1, nuclear factor κB, N-myc, and nuclear factor of activated T cells, which may contribute to EAAT2 expression and its regulation. During the ischemic process, the massive release of glutamate has been shown to occur as a result of the impairment of the glutamate uptake by astrocytes due to energetic failure [17] and a polymorphism in the promoter region of the glutamate transporter EAAT2, which is associated with the increased plasma glutamate concentrations and worse neurological outcome in patients with acute ischemic stroke, have been reported [18].

Once released, glutamate activates ionotropic receptors (N-methyl-D-aspartate (NMDA), α-amino-3-hydroxy-5-methyl-4-isoxazole propionate (AMPA), and kainite receptors) and this activation results in an increase in cell permeability to Na^+, K^+, and Ca^{2+}. Among these ions, calcium is a key player since it activates several potentially harmful reactions. The intracellular increase in calcium triggers the inflammatory cascade, which starts with the local expression of inflammatory cytokines, including tumor necrosis factor (TNF)-α and interleukin (IL)-1β, which in turn stimulates the release of other cytokines, mainly IL-6 and IL-8, and chemotactic factors such as leukocyte adhesion molecules including selectins (P-selectin, E-selectin, and L-selectin), the intercellular adhesion molecule 1, the vascular cellular adhesion molecule 1, and the platelet endothelial cellular adhesion molecule [19]. The leukocytes reaching the ischemic zone as a result of the release of chemotactic factors interact with the endothelial cells of the capillary walls, resulting in occlusion of the arteries, and leading to the "non-reflow" phenomenon that impedes the complete recovery of the CBF within the ischemic zone after the recanalization of the occluded artery [20]. Moreover, leukocytes also stimulate the release of vasoconstrictor substances, with secondary damage to the vascular reactivity [21], and of proteolytic enzymes that break down the endothelial wall and permit the leakage of water and erythrocytes, which may result in brain edema and hemorrhagic transformation of the ischemic lesion, respectively [22,23].

The increase in intracellular calcium is also particularly deleterious as it participates in the formation of free radicals and also mediates apoptosis or delayed cellular death [24].

Free radical formation has been shown to contribute to ischemic damage, particularly when ischemia is followed by reperfusion [25]. In fact, reoxygenation during reperfusion provides oxygen to sustain neuronal viability but also provides oxygen as a substrate for numerous enzymatic oxidation reactions that produce reactive oxidants. Different sources of free radical formation have been identified during cerebral ischemia.

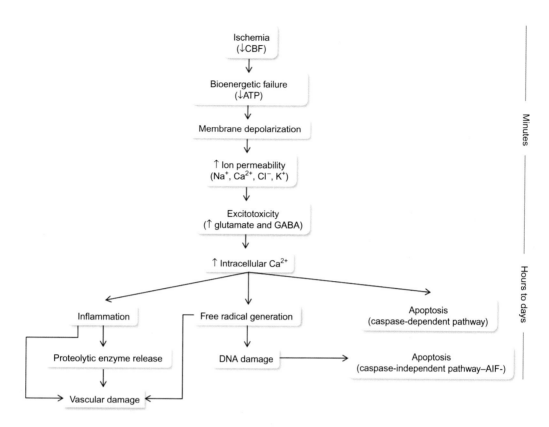

AIF: Apoptosis inducing factor.

FIGURE 7.1 Schematic representation of the main events of the ischemic cascade. Cerebral ischemia leads to a bioenergetic failure with depletion of energy metabolites, mainly ATP, which results in the depolarization of the cellular membranes and in secondary permeability abnormalities with an intracellular increase in [Na$^+$], [Ca^{2+}], and [Cl$^-$], and an extracellular increase in [K$^+$], which in turn causes the extracellular release of glutamate and other excitatory and inhibitory amino acids, such as glycine and GABA, respectively. The increase in the glutamate levels results in an increase in cell permeability to Na$^+$, K$^+$, and Ca^{2+}. The intracellular increase in calcium triggers the inflammatory cascade, which in turn stimulates the release of proteolytic enzymes causing secondary vascular damage. The intracellular calcium also participates in the generation of free radicals with secondary DNA damage and in the apoptotic process through the activation of the caspase-dependent pathway. AIF, apoptosis-inducing factor; ATP, adenosine triphosphate.

A major source of these free radicals, particularly $\cdot O_2^-$ and H_2O_2, is the oxidation of arachidonic acid by cyclo-oxygenase and lipoxygenase [26]. Another source of free radicals is xanthine oxidase, an enzyme that degrades hypoxanthine and xanthine, releasing $\cdot O_2^-$ and H_2O_2 in a process mediated by calcium. Moreover, intracellular calcium also participates in the formation of free radicals through the activation of the neuronal nitric oxide (NO) synthase, which promotes nitric oxide formation and the subsequent synthesis of the highly toxic peroxynitrite radical which is subsequently degraded generating $\cdot OH$ [27].

Besides cellular death due to necrosis mediated by the aforementioned mechanisms, growing evidence suggests that ischemia may additionally induce programmed cell death in the brain through different pathways. The increased intracellular calcium levels activate the mitochondrial apoptotic pathway which is

characterized by changes in Bcl-2 family proteins, cytochrome c release, and caspase-like enzyme activation. The process starts with the activation of calpains secondary to calcium increase which results in the cleaving of the Bcl-2 interacting domain (BID) to its truncated active form (tBID). tBID targets the outer mitochondrial membrane and induces conformational changes in other proapoptotic proteins including Bak, Bax, Bad, and BCl-XS, which can form heterodimers with tBID, as well as antiapoptotic proteins such as Bcl-2 and Bcl-XL. The mechanism by which it is thought that proapoptotic proteins induces the release of apoptogenic factors is the opening of mitochondrial transition pores through which usually sequestered proapoptotic proteins (cytochrome c and the serine protease HtrA2/Omi) are released from the intermembrane space into the cytosol and activate the caspase-dependent mitochondrial apoptotic pathway. It is important to mention that apoptotic and necrotic cellular death are interrelated. In fact, some caspases, such as caspase-3, have been reported to cleave substrate proteins such as poly (ADP-ribose) polymerase (PARP). The inactivation of PARP after this cleavage leads to DNA injury and subsequently to apoptotic cell death, but excessive activation of PARP results in a decrease in ATP, which in turn leads to cellular energy failure and necrotic cell death. Moreover, another group of proapoptotic proteins, which are independent of the caspase pathway, are also released through the transition pores. These include the apoptosis-inducing factor (AIF), endonuclease G, and Bcl-2/adenovirus E1B 19 kDA-interacting protein (BNIP3). Among them, it is known that AIF is released from the mitochondria as a result of the ATP decrease secondary to ischemia. Once it is released, it causes DNA fragmentation and peripheral condensation of nuclear chromatin in a manner that is distinct from the DNA abnormalities that are secondary to caspase-mediated DNA fragmentation [28,29].

At a later stage in the evolution of the ischemic cascade, and with the objective of repairing the damaged tissue, different processes which increase brain plasticity take over. In fact, a slow but consistent recovery can be observed in clinical practice over a period of weeks or months after stroke. Whereas the recovery in the first few days is probably the result of edema resolution and/or reperfusion of the ischemic penumbra, a large part of the recovery afterwards is due mainly to neurorepair processes which are closely related to brain plasticity. Brain plasticity is the ability of some regions of the brain to take on functions which were previously performed by damaged areas. Neuroplasticity includes structural and functional modifications of neurons, astrocytes, and vascular vessels, which are the structures that are damaged during ischemia, and so neurorepair includes neurogenesis and gliagenesis with the necessary synaptogenesis to make the new cells functional, as well as angiogenesis. Neurogenesis is the generation of neurons from neural stem cells. For a long time it was thought that neurogenesis stopped shortly after birth. Now we know that neural stem cells exist not only in the developing nervous system but also in the nervous systems of all adult mammalian organisms, including humans. The mechanisms that regulate ischemia-induced neurogenesis are only partially understood but it is known that neurogenesis after stroke is mediated by molecules such as growth factors including the epidermal growth factor, the basic fibroblast growth factor, the brain-derived neurotrophic factor, and the vascular endothelial growth factor (VEGF), which are released in response to ischemia, erythropoietin, and glutamatergic mechanisms (the blockade of NMDA and AMPA receptors at the time of global ischemia prevent the increase of neurogenesis and, in focal ischemia, this blockade is prevented only by the NMDA receptor) [30–32]. Adult hippocampal neurogenesis is closely associated with angiogenesis from endothelial cell precursors as adequate blood supply is necessary for the survival and development of the new neurons.

Angiogenesis usually occurs in the human brain after stroke but may have to be further stimulated to increase the number of surviving new neurons, a process that seems to be mediated by the recruitment of endothelial progenitor cells (EPCs) to the ischemic area [33]. In clinical practice, the increase in circulating EPCs during the first week after acute ischemic stroke has been shown to be associated with good functional outcome and reduced infarct growth, an effect that seems to be related to an increase in the levels of VEGF [34].

7.3 THERAPEUTIC STRATEGIES FOR ISCHEMIC STROKE

As mentioned above, thrombolytic therapy with rt-PA is the only approved reperfusion therapy for the treatment of ischemic stroke when administered in the first 4.5 h of the evolution of the symptoms [7,8]. However, in spite of its efficacy, rt-PA is still given to a very small proportion of patients (<5% worldwide), which is in part due to the very short therapeutic window for its administration. Moreover, the frequency of recanalization achieved after rt-PA administration is low (<40%) [35] and there are many contraindications to the administration of the drug, which makes it necessary to develop other therapeutic approaches to increase survival and reduce disability in stroke patients. The recently published positive results of endovascular trials [36–40] will undoubtedly improve the scenario but due to the selection criteria for endovascular treatment there will still be patients for whom the endovascular approach will not be an option.

Based on the previously presented knowledge of the different molecules of the ischemic cascade participating as mediators of ischemic damage, a large number of pharmacological interventions directed at avoiding the action of these mediators have been tested in the last few years. These interventions have included the administration of agents to block the events that occur very shortly after the beginning of the ischemia, such as antiexcitotoxics, anti-inflammatories, free radical scavengers, and ion channel modulators, and other agents, such as membrane stabilizers and antiapoptotic drugs, that act at later stages of the ischemic cascade. This neuroprotective approach, whose main objective is to limit the cerebral infarct volume and avoid vulnerable cell death of the ischemic penumbra, proved to be very effective in reducing infarct volume in the experimental models of stroke. However, none of the phase 2b/3 clinical trials carried out so far in patients with ischemic stroke has proved the efficacy of these drugs (Table 7.1).

Different reasons may explain, at least in part, the lack of efficacy achieved in attempting to translate the clear neuroprotective results found in animals to humans [41]. These include the varying therapeutic windows and doses of the administered drugs used in animals and humans, the very different end-points established to evaluate the efficacy of the drugs, sample sizes in clinical trials that are often too small to be able to detect differences in the prespecified endpoints, the heterogeneity of human stroke as compared with the very homogeneous experimental animal stroke models, and the fact that there was no clear knowledge

TABLE 7.1   Phase 2b/3 and Phase 3 Trials in Neuroprotection in Patients with Acute Ischemic Stroke

Drug	Mechanism of action	Result	Cause
EARLY ISCHEMIC INJURY			
CGS1975 (Selfotel)	NMDA antagonist	Negative	Safety
CNS1102 (Cerestat, Aptiganel)	NMDA antagonist	Negative	Safety
Eliprodil	NMDA antagonist	Negative	Safety
Magnesium	NMDA antagonist	Negative	Efficacy
GV150526 (Gavestinel)	Glycine antagonist	Neutral	Efficacy
YM872	AMPA antagonist	Neutral	Efficacy
Fosphenytoin	Indirect glutamate inhibitor	Neutral	Efficacy
BMS-204352	Potassium channel opener[a]	Neutral	Neutral
Lubeluzole	Na+ channel antagonist/ Nitric oxide inhibitor	Negative	Efficacy
	Serotonin agonists	Negative	Efficacy
ONO-2506	Modulator of glutamate transporter uptake	Negative	Efficacy
Nalmefene (Cervene)	Narcotic receptor antagonist (modulation of other non-NMDA receptors)	Neutral	Efficacy
Clomethiazole	GABA agonist	Negative	Efficacy
Nimodipine	Calcium channel blocker	Neutral	Efficacy
Tirilazad	Free radical scavenger	Negative	Efficacy
Ebselen	Free radical scavenger	Negative	Safety
NXY-059	Free radical scavenger	Negative	Efficacy
Enlimobab	Anti-ICAM-1 antibody	Negative	Efficacy
HU23F2G (LeukArrest)	IgG1 human CD18 antibody	Negative	Neutral
Uric acid	Antioxidant	Positive trend	Sample size
LATER ISCHEMIC INJURY			
Ganglioside GM1	Membrane stabilizer	Negative	Neutral
Cerebrolysin	Membrane stabilizer	Positive trend	Not confirmed results
Citicoline	Membrane stabilizer	Negative	Efficacy
bFGF[b] (Fiblast)	Trophic factor/antiapoptosis	Negative	Efficacy

[a]Decreases calcium influx and glutamate release.
[b]bFGF: basic fibroblast growth factor.

about the existence of the penumbra, the target to be protected by the drug, in most of the human trials. Importantly, the mechanism of action of many of the tested neuroprotective drugs was the inhibition of the events at very early stages of the ischemic cascade and, due to the large therapeutic windows of some of the trials, it is likely that these drugs did not have any effect since the mechanism they were trying to block had already taken place. Moreover, due to the complexity of the ischemic cascade, it may be unrealistic to expect a significant neuroprotective effect just by blocking one of the pathways that intervene in the cascade when many different pathways are involved both spatially and temporally in the pathophysiology of stroke. Given this situation, one of the keys to success in inducing neuroprotection in stroke could be the simultaneous modulation of many pathophysiological pathways by a combination of drugs acting at different times or the administration of just one pharmacological agent with pleiotropic effects, which allows the drugs to act at different levels and at different times. Precisely due to these pleiotropic effects, mTOR is emerging as a new potential target to be considered in stroke.

7.4 mTOR PROTEIN

The mTOR (also known as the mechanistic target of rapamycin and FK506-binding protein 12-rapamycin complex-associated protein 1) is a 289-kDa serine/threonine protein kinase, ubiquitously expressed throughout the body, which modulates metabolism, cellular survival, gene transcription, and cytoskeletal components. mTOR is activated through phosphorylation of its specific residues in response to growth factors, nutrients, mitogens, and hormones, whereas the deficiency of growth factors, amino acids, cellular nutrients, and oxygen can down-regulate mTOR activity [42–44]. Serine2448 is a target of Akt and p70 ribosomal S6 kinase (p70S6K), threonine2446 is a target of adenosine monophosphate (AMP)-activated protein kinase (AMPK) and p70S6K, and serine2481 is a site that is rapamycin-insensitive and an autocatalytic site of mTOR. In addition, dual phosphorylation on serine2159 and threonine2164 enhances mTOR activity and promotes autophosphorylation on serine2481 [42].

Structurally, mTOR contains 2549 amino acids and multiple domains (Figure 7.2A). mTOR is considered a member of the phosphatidylinositol 3-kinase (PI3K)-kinase-related kinase superfamily since its C-terminal domain shares strong homology to the catalytic domain of PI3K [45,46]. The C-terminal domain contains the FAT domain (FKBP12-rapamycin-associated protein, ataxia-telangiectasia and transactivation/transformation domain-associated protein), the FKBP12-rapamycin-binding domain (FRB), adjacent to the FAT domain and site of interaction between mTOR and FKBP protein bound to rapamycin, the kinase domain, which contains the phosphorylation sites that regulate mTOR activity, and a FATC (FAT carboxy-terminal) domain. The FAT and FATC domains are always found in combination, so it has been hypothesized that the interactions between FAT and FATC might contribute to the catalytic kinase activity of mTOR via unknown mechanisms. The N-terminal domain of mTOR contains at least 20 HEAT (Huntingtin, Elongation factor 3, a subunit of protein phosphatase-2A and TOR1) repeats, which provide protein interaction between mTOR complex with regulatory-associated protein with mTOR (Raptor) or rapamycin-insensitive companion of mTOR (Rictor). It is the association with either Raptor or Rictor that determines whether mTOR is a component of mTOR complex 1 (mTORC1) or mTOR complex 2 (mTORC2) [42–44,47].

7.5 TOR COMPLEXES AND DOWNSTREAM SIGNALING: mTORC1 AND mTORC2

mTORC1 and mTORC2 multiprotein complexes are responsible for the different physiological functions and, in addition to their different responsiveness to rapamycin treatment, they differ in some components and in their signaling pathways, with mTORC1 being the better characterized of the two complexes (Figure 7.2B).

mTORC1 is considered to be mostly involved in the regulation of the translation initiation machinery influencing cell growth, proliferation, and survival, while mTORC2 participates in cytoskeleton organization, cell migration, modulation of cell cycle progression, and control of cell survival. The identification of mTORC1 and mTORC2 was based on their components and their sensitivity to rapamycin, a macrolide antibiotic from *Streptomyces hygroscopicus* that specifically inhibits mTOR activation. Originally, mTORC1 was considered to be rapamycin-sensitive and mTORC2 to be rapamycin-insensitive but it is now known that prolonged exposure of rapamycin also results in the inhibition of mTORC2. Rapamycin inhibits mTORC1 through the binding to immunophilin FK-506-binding protein 12 (FKBP12) that attaches to the FRB domain at the C-terminal of mTOR to prevent its phosphorylation, whereas it is considered that the inhibition of mTORC2 may involve disrupting the assembly and the integrity of this complex. Both mTOR complexes share the catalytic mTOR subunit, the mammalian lethal with sec-13 protein 8 (mLST8), the DEP domain containing mTOR interacting protein (Deptor),

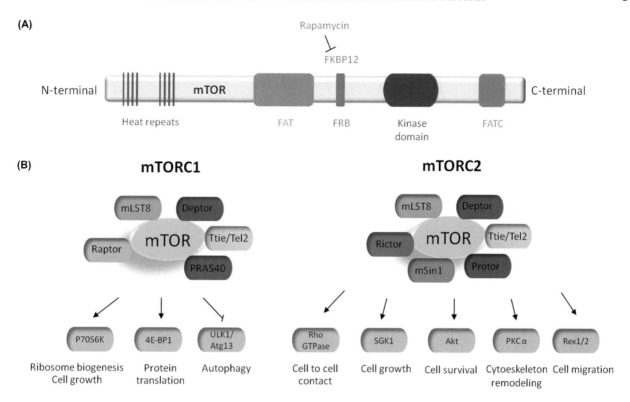

FIGURE 7.2   mTOR domain structure, complexes and downstream signaling. mTOR is a 289-kDa serine/threonine protein kinase, containing multiple domains which include the FAT domain, the FKBP12-rapamycin binding domain (FRB), the kinase domain, the FATC domain and at least 20 HEAT repeats, which provide protein interaction between mTOR complex with Raptor or Rictor (A). mTOR complexes include mTORC1 and mTORC2 which differ in some components and physiological functions as is shown in the figure. Both mTOR complexes share the catalytic mTOR subunit, mLST8, Deptor and the Tti1/Tel2 complex. In contrast, Raptor and PRAS40 are specific to mTORC1, while Rictor, mSin1, and Protor are specific to mTORC2. Downstream targets of mTORC1 include p70S6K1 and 4E-BP1, while Rho GTPases, Akt, SGK1, PKCα, and P-Rex1/P-Rex2 are downstream targets of mTORC2 (B). mTORC, mammalian target of rapamycin complex.

and the Tti1/Tel2 complex. In contrast, Raptor and proline-rich Akt substrate 40 kDa (PRAS40) are specific to mTORC1, while Rictor, mammalian stress-activated map kinase interacting protein 1 (mSin1), and protein observed with rictor 1 and 2 (Protor) are specific of mTORC2 [48,49]. mLST8 interacts with the kinase domain of mTOR, promotes the stabilization of the association between Raptor and mTOR, and is necessary for the Rictor–mTOR interaction and for the stability and activity of mTORC2 complex [42,43]. Deptor is a negative regulator of mTORC1 and mTORC2 through its binding to the FAT domain of mTOR. In the absence of Deptor, the activity of protein kinase B (Akt), mTORC1, and mTORC2 increases [44]. The Tti1/Tel2 complex is important for mTOR stability and for the activity of both mTORC1 and mTORC2, probably through maintenance of the complex assembly. In addition, Tti1 is important due to the association of Tel2 with mTOR [50].

mTORC1 activity is controlled by phosphorylation of Raptor through a number of pathways that involve the protein Ras homolog enriched in brain (Rheb). Rheb phosphorylates Raptor at serine863, among others

residues, and mTORC1 activity can be limited if serine863 remains unphosphorylated [44]. Raptor is a 150-kDa protein that functions to recruit mTOR substrates to the mTORC1 complex, binds to the N-terminal HEAT of mTOR, and regulates the subcellular localization of mTORC1 [42,48]. PRAS40 is a negative regulator of mTORC1. PRAS40 competitively inhibits the binding of mTORC1 to Raptor. The phosphorylation of PRAS40 by Akt at threonine246 or by mTORC1, at serine183, serine212, or serine221, leads to its dissociation from mTORC1, promotes its binding to the cytoplasmic docking protein 14-3-3 and, in this way, facilitates the activation of mTORC1 [44]. Two of the best-characterized downstream targets of mTORC1 are ribosomal protein S6 kinase (p70S6K1) and eukaryotic translation initiation factor 4 epsilon (eIF4E) binding protein 1 (4E-BP1), which are both critical regulators of translation initiation. The phosphorylation of p70S6K fosters mRNA biogenesis, translation of ribosomal proteins, and cell growth. In contrast, the phosphorylation of 4E-BP1 results in its inactivation. Hypophosphorylated 4E-BP1 is active and binds competitively with eukaryotic translation initiation factor 4 gamma (eIF4G) to eIF4E, which regulates translation initiation by interacting with the 5′-mRNA cap structure. The phosphorylation of 4E-BP1 by mTORC1 results in its dissociation from eIF4E, allowing eIF4G to interact with eIF4E and promote protein translation. Binding of 4E-BP1 and p70S6K to Raptor can be prevented during activation of PRAS40 [42,44]. Another downstream target of mTORC1 is ULK1/Atg13 (unc-51-like kinase 1/mammalian autophagy-related gene 13), a kinase complex suppressed by mTORC1 that is required to initiate autophagy, as will be described in Section 7.9 [49].

With regards to the specific components of mTORC2, Rictor and mSin1 can form the structural basis of this complex. In fact, Rictor and mSin1 have been shown to stabilize each other to form mTORC2 and both are required for Akt phosphorylation [43,44]. Rictor is beneficial to the assembly and promotes the activity of this complex. It has been reported that newly synthesized Rictor is susceptible to rapamycin, suggesting that only preformed mTORC2 is resistant to rapamycin, perhaps through steric occlusion, blocking access to the FRB. Although the role of mSin1 is not well understood, it is necessary for the assembly of mTORC2 and for its capacity to phosphorylate Akt. As the deletion of mSin1 is embryonically lethal, it is considered to retain an essential function [43,48]. mTOR phosphorylates mSin1, preventing its lysosomal degradation [44]. Protor binds Rictor independently of mTOR and does not appear to be required for mTORC2-mediated Akt activation. However, Protor may function to activate serum and glucocorticoid-induced protein kinase 1 (SGK1) [44]. In addition to Akt and SGK1, other targets of mTORC2 are protein kinase C alpha (PKCα), P-Rex1, P-Rex2, and Rho GTPases. mTORC2 promotes cell survival through the activation of Akt and cytoskeleton remodeling through PKCα. mTORC2 utilizes Rictor to activate and phosphorylate Akt at serine473, facilitating threonine308 phosphorylation by phosphoinositide-dependent kinase 1. SGK1 is also phosphorylated and activated by mTORC2. SGK1 is a member of the protein kinase A/protein kinase G/protein kinase C (AGC) family of protein kinases and is activated by growth factors to control ion transport and growth. Finally, mTORC2 modulates cell migration, through the activation of P-Rex1 and P-Rex2 by Akt phosphorylation, with mTORC2 acting as a catalytic complex and for cell-to-cell contact using Rho signaling [42,44].

7.6 UPSTREAM REGULATION OF mTOR SIGNALING IN CEREBRAL ISCHEMIA

mTORC1 senses diverse upstream signals (growth factors, stress, energy status, oxygen, and amino acids)

[49]. In contrast, less is known about the mTORC2 pathway. mTORC2 signaling is insensitive to nutrients but does respond to growth factors, such as insulin, through a poorly defined mechanism(s) that requires PI3K [49]. Cerebral ischemia is associated with energy stress with a reduction in ATP levels acting as the trigger of the ischemic cascade. It has been reported that ATP is one of the most important inputs regulating mTOR activity [51] and it has been demonstrated that, upon hypoxia, TOR signaling is inhibited, leading to the down-regulation of protein synthesis [49].

Hypoxia and energy stress inhibit mTORC1 signaling via multiple signaling pathways mediated through the activation of the heterodimer consisting of tuberous sclerosis 1 (TSC1; also known as hamartin) and TSC2 (also known as tuberin). The TSC1/TSC2 complex functions as a GTPase-activating protein (GAP) for the Rheb GTPase, converting active Rheb-GTP to inactive Rheb-GDP. The GTP-bound form of Rheb directly interacts with mTORC1 and strongly stimulates its kinase activity, while TSC1/TSC2 negatively regulates mTORC1 by converting Rheb into its inactive GDP-bound state [49,52] (Figure 7.3). Only TSC2 contains GAP function, however, both TSC1 and TSC2 are required for the functionality of this heterodimer [52]. AMPK is up-regulated under energy stress conditions in response to nutrient deprivation or hypoxia when the intracellular ATP level decreases and AMP increases [53]. AMPK inhibits mTORC1 in two ways. Firstly, AMPK phosphorylates TSC2 on serine1387 and threonine1227 increasing its GAP activity toward Rheb and, secondly, AMPK directly inhibits mTORC1 by phosphorylating Raptor at two highly conserved residues, serine722 and serine792, leading to 14-3-3 protein binding and the allosteric inhibition of mTORC1 [49,52,54]. The phosphorylation of the activation loop threonine172 is absolutely necessary for AMPK activation. Under conditions of energy stress, the serine/threonine liver kinase B1 (LKB1) has been identified as a major kinase phosphorylating the AMPK activation loop at the threonine172 residue and the loss of LKB1 impairs cardiac function during ischemic conditions, illustrating the importance of AMPK signaling in the mTOR pathway for the vascular system [44,55]. In addition, AMPK can also control TSC1/TSC2 activity through stress-induced protein REDD1 (Regulated in Development and DNA damage responses, also known as DDIT4 or RTP801) promoting the increase of its expression [44]. Furthermore, REDD1 also acts independently of the LKB1-AMPK signaling branch to down-regulate mTORC1. The expression of REDD1 is up-regulated upon hypoxia by the transcription factor hypoxia-inducible factor 1 (HIF-1). HIF-1 is a heterodimeric protein complex composed of HIF-1a and HIF-1b subunits. The HIF-1a subunit is rapidly degraded under normoxic conditions, whereas it is stabilized and

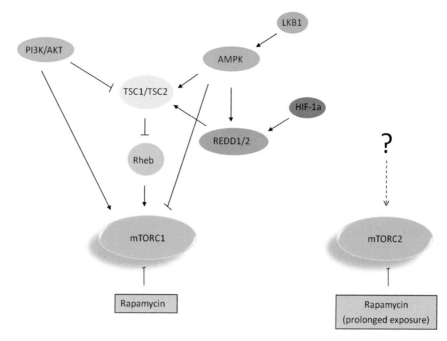

FIGURE 7.3 mTOR signaling in ischemia. During ischemia mTORC1 is inhibited via different signaling pathways mediated through the activation of the TSC1/TSC2 heterodimer. This complex functions as a GTPase-activating protein (GAP) and negatively regulates mTORC1 by converting active Rheb (GTP-bound form that strongly stimulates mTORC1 kinase activity) into its inactive GDP-bound state. Under ischemic conditions, AMPK is up-regulated and inhibits mTORC1 in two ways. Firstly, AMPK phosphorylates TSC2, increasing its GAP activity toward Rheb and, secondly, AMPK directly inhibits mTORC1. AMPK can also control TSC1/TSC2 activity through REDD1 promoting the increase in its expression. LKB1 has been identified as a major kinase activating AMPK in ischemic conditions. REDD1 also acts independently of the LKB1-AMPK signaling branch to down-regulate mTORC1. The expression of REDD1 is up-regulated upon hypoxia by HIF-1a. Decreased mTOR activation is also associated with the PI3K/Akt signaling pathway which is significantly decreased in the context of cerebral ischemia. The upstream signals that regulate mTORC2 signaling in ischemia are unknown. Inactivation of mTORC1 is also pharmacologically induced by rapamycin. It is now known that prolonged exposure of rapamycin also results in the inhibition of mTORC2. mTORC, mammalian target of rapamycin complex; TSC, tuberous sclerosis; Rheb, Ras homolog enriched in brain; AMPK, adenosine monophosphate-activated protein kinase; LKB1, liver kinase B1; HIF, hypoxia-inducible factor.

accumulated upon hypoxia and drives the expression of several genes, including *REDD1* [56,57]. Induction of REDD1 can activate the TSC1/TSC2 complex by competing with TSC2 for 14-3-3 protein binding. Thus, increased REDD1 levels that occur following exposure to hypoxia prevent the inhibitory binding of 14-3-3 protein to TSC2 [58,59]. Taking all this information into account it can be concluded that the inhibition of mTOR activity by REDD1 activation may be mediated either through AMPK-independent or -dependent mechanisms in the context of cerebral ischemia. Finally, the PI3K/Akt signaling pathway is also involved in mTOR signal inhibition in ischemia. The serine473 phosphorylation of Akt is significantly decreased in ischemic stress, which is related to decreased mTOR phosphorylation at serine2448 [52,60].

In the following sections we will summarize the role of mTOR in the main events of the ischemic cascade and discuss how, based on its capability to modulate these different processes, mTOR may acquire the capability of limiting ischemic neuronal death and promoting the neurological recovery (Figure 7.4). However,

further investigation will be necessary to increase knowledge about the intricate cellular signaling pathways of mTOR in cerebral ischemia in order to develop mTOR-reliant clinical strategies for stroke.

7.7 mTOR AND EXCITOTOXIC DAMAGE

Under physiological conditions glutamate may control neuronal synaptic signaling through the activation of the mTORC1−p7026K pathway [44,52]. However, as explained in Section 7.1, higher extracellular levels of glutamate are responsible for the excitotoxic damage occurring in the very early phases of the ischemic process through the activation of ionotropic receptors (NMDA, AMPA, and kainate receptors), which results in an increase in Ca^{2+} cell permeability, a key player in several potentially harmful reactions [12,13,19]. Astroglial EAAT2 (also known as GLT-1) is the main glutamate transporter responsible for maintaining extracellular glutamate concentrations below excitotoxic levels but the detailed mechanism underlying

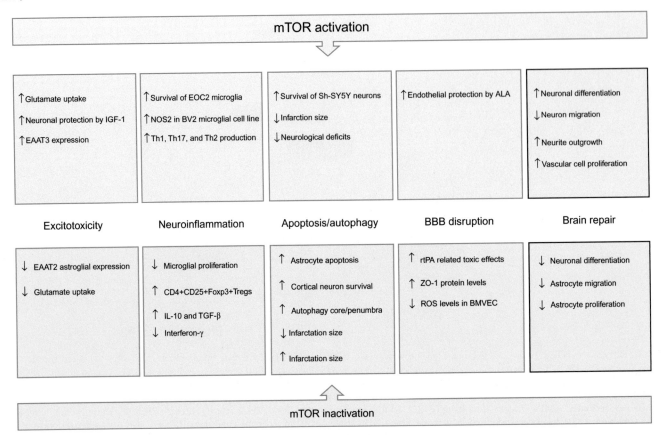

FIGURE 7.4   mTOR role in excitotoxicity, neuroinflammation, apoptosis/autophagy, BBB disruption, and brain repair mechanisms induced after ischemia. The most relevant molecular, cellular, or physiological effects associated with mTOR activation or mTOR inactivation (usually by rapamycin) related to the bibliographic review in excitotoxicity, neuroinflammation, apoptosis/autophagy, BBB disruption, and brain repair mechanisms that occurred after ischemia are summarized in the figure. All the presented data have been reviewed in the cerebral ischemic context and only studies using brain OGD/R *in vitro* models or ischemic injury *in vivo* models such as MCAO have been included. All of the cited information is referenced in the text of the corresponding section. mTOR, mammalian target of rapamycin; BBB, blood−brain barrier; OGD, oxygen-glucose deprivation.

EAAT2 regulation has not been fully elucidated [12,13,15]. It has recently been demonstrated that mTOR is a downstream target of PI3/Akt in the pathway regulating EAAT2 expression in astrocytes. This finding is in agreement with both the impaired EAAT2 expression and glutamate uptake capacity, and the inhibition of mTOR signaling during ischemia [61]. mTOR may also contribute to protection against neuroexcitotoxicity as it induces the expression of another glutamate transporter, EAAT3, which is responsible for the cellular uptake of glutamate in neurons [62]. It is interesting to note that the Leucine-rich repeat in Flightless-1 interacting protein 1 has recently been identified as a positive regulator of EAAT2 expression in astrocytes through the induction of Akt and mTOR phosphorylation at serine473 and serine2448, respectively [63]. Furthermore, Wang et al. have shown that the neuroprotective effect of insulin-like growth factor-1, a multifunctional protein involved in systemic body growth and specifically in neuronal polarity and axonal guidance, against NMDA-induced excitotoxicity in cultured

hippocampal neurons is mediated via the inhibition of the NR2B subunit of NMDA receptor and the activation of PI3K-AKT-mTOR [64]. Therefore, reverting the mTOR inhibition induced by ischemia through activators of this protein, such as 3-benzyl-5-((2-nitrophenoxy) methyl)-dihydrofuran-2(3H)-one (3BDO) [65], could be a therapeutic option to mitigate excitotoxic damage.

7.8 mTOR AND NEUROINFLAMMATION

The inflammatory response is a key element after cerebral ischemia that results in cell damage and death in the subacute phase. As has been explained above, cerebral ischemia results in a cascade of molecular events that are triggered as a consequence of a decrease in the CBF and the subsequent energetic failure which leads to an intracellular increase in calcium triggering the inflammatory cascade. The damaged brain tissue secretes cytokines and chemokines which activate microglia (the innate immune cells residing in the brain) and recruit inflammatory cells to injured area. Activated microglia contribute to initiate and maintain brain inflammation releasing potentially cytotoxic proinflammatory mediators, such as NO. The migrating inflammatory cells include neutrophils and macrophages that secrete additional cytokines and oxygen species, resulting in further tissue damage, oxidative stress, and the activation of matrix metalloproteinases (MMPs) leading to disruption of the blood−brain barrier (BBB) and edema, which is often fatal in ischemic stroke patients [66,67]. T cells also infiltrate the ischemic brain and have detrimental effects after stroke. However, different types of T cells play different roles. They include T helper cells (CD4+) with three major subtypes (Th1, Th2, and Th17), T cytotoxic cells (CD8+) and regulatory T cells (Treg) (also CD4+) [68]. Gu et al. have recently shown reduced infarction in Th1-deficient mice and increased infarction in Th2-deficient mice [69]. Furthermore, deficits of either CD4+ or CD8+ effector T cells result in smaller infarct sizes, while deficits of Treg cells lead to enlarged and delayed infarct sizes [70,71]. Thymus-derived CD4+CD25+Foxp3+ Treg cells in particular are known to be neuroprotective in acute experimental stroke by modulating the function of effector T cells and secreting anti-inflammatory molecules, such as IL-10 and transforming growth factor (TGF)-β [72]. These anti-inflammatory cytokines inhibit the expression of proinflammatory cytokines such as IL-1 β and TNF-α (the most prominent after stroke) and reduce inflammation. As a result of this, much of the research into neuroprotection following stroke has concentrated on mitigating, or modulating, the effects of inflammation. However, the neuroinflammatory response after stroke may not be entirely detrimental. It has been observed that in addition to the deleterious neuroinflammatory response, it also has beneficial effects of helping to clear the large amount of debris caused by brain cell necrotic death after ischemic stroke. This dual role could explain why anti-inflammatory drugs after stroke have not yet shown the expected results in clinical trials [73,74]. In this context, what might the role of mTOR be? Could the inhibition or activation of mTOR improve neurological disability by modulating the immune response after stroke? Recent studies have shown that mTOR is emerging as a critical regulator of immune function. Researchers have proposed that mTOR activity might play a central role in integrating microenvironmental cues to instruct immune cell response, including microglia and T cells [75]. mTOR has been found to enhance the survival of EOC2 microglia during oxygen-glucose deprivation (OGD) and to increase NO synthase 2 expression during hypoxia in the BV2 microglial cell line. mTOR activation has also been detected in response to bacterial lipopolysaccharide (LPS) and cytokines in primary cultures of rat cortical microglia. Additionally, mTOR inhibition by RAD001 reduces microglial proliferation and intracellular levels of cyclooxygenase in response to proinflammatory cytokines suggesting that mTOR controls microglial proinflammatory activation [76]. In the T-cell subset, inhibition of mTOR (both mTORC1 and mTORC2) by rapamycin promotes the generation of CD4 + CD25 + Foxp3 + Tregs, whereas mTORC1 activation promotes Th1 and Th17 production and mTORC2 activation induces Th2 generation. Interestingly, Qin Lu et al. have recently demonstrated that the mTOR signaling pathway was activated at 30 min and returned to its basal level 1 day after intracranial hemorrhage (ICH) and that there were increased levels of phosphorylated mTOR protein predominantly located around the hematoma. Furthermore, rapamycin treatment significantly improved the neurobehavioral deficit after ICH and modulated the immune response, increasing the number of Tregs and the levels of IL-10 and TGF-β, and reducing interferon-γ (associated with the onset and progression of different neural pathology and inducing expression of IL-1β and TNF-α) both in peripheral blood and brain [77]. All this evidence points to mTOR being a novel therapeutic target for the modulation of neuroinflammatory response after stroke, although further investigations will be required to determinate the exact role of this process (deleterious and beneficial effects) during ischemia development.

7.9 mTOR AND CELL VIABILITY: APOPTOSIS AND AUTOPHAGY

During stroke, the decrease in the CBF and the subsequent energetic failure lead to tissue damage due to mechanisms such as those mentioned above and cell death by necrosis or apoptosis. The level of ATP within the cell is the primary factor in determining which mechanism of cell death occurs. Cells within the core infarct (without ATP stores) typically die by necrosis, whereas those in the penumbra die by apoptosis [66]. Necrosis refers to any death associated with cellular lysis that provokes an inflammatory response. Apoptosis refers to programmed cell death with a particular morphology that includes shrinkage and blebbing of cells, rounding and fragmentation of nuclei with condensation, and the margination of chromatin, shrinkage, and phagocytosis of cell fragments without accompanying inflammatory responses (in most cases) [78]. Apoptosis is an ATP-dependent process that can be blocked, allowing damaged tissue to be rescued. Thus, an attractive line of investigation is targeting of delayed cell death pathways in the penumbra. The mTOR pathway, as the primary regulator of the cellular response to changes in energy status as seen following ischemia and reperfusion, appears to be an interesting target to regulate apoptotic neural cell death. However, neuroprotective effects of mTOR are poorly understood and contradictory results have been published. Many studies have demonstrated that the mTOR activation may be necessary to prevent apoptotic cell death during ischemia. In this way, astrocyte OGD-induced apoptosis is increased by treatment with rapamycin [79] and the loss of PRAS40, which may increase mTOR activity, prevents SH-SY5Y neuronal cell line injury induced by OGD [80]. Additionally, different neuroprotectants, such as melatonin [81] and estradiol [82], prevent cerebral ischemic injury through the activation of mTOR/p70S6K signaling pathway and ischemic post-conditioning (IPOC), which refers to brief reperfusion interruptions that reduce infarction after cerebral ischemia and mitigate neurological deficits, provide long-term protection by enhancing Akt and mTOR activity during the acute post-stroke phase [83]. Xiaoxing et al. have recently demonstrated that lentiviral overexpression of the PRAS40 protein reduces infarction size in rats by promoting phosphorylation of Akt, FKHR (FOXO1), PRAS40, and mTOR proteins. Furthermore, PRAS40 gene knockout mice have larger infarctions than wild-type animals after cerebral ischemia suggesting new insights for the role of PRAS40 in the mTOR pathway and its neuroprotective effects [84]. On the other hand, rapamycin can prevent the activation of both mTORC1 and mTORC2 in cortical neurons and improve cell survival following OGD [85].

An emerging stroke literature indicates that in addition to apoptotic cell death, autophagy is seen after stroke [86]. Autophagy is a cellular process that degrades and recycles dysfunctional cellular components, such as misfolded or aggregated proteins or damaged organelles (endoplasmic reticulum (ER), mitochondria, peroxisomes, Golgi apparatus), and that is usually active at basal levels in most cell types and plays a housekeeping role in maintaining the integrity of cells. However, autophagy is strongly induced by cellular stressors including nutrient deprivation, ischemia, and reactive oxygen species (ROS). There are three defined types of autophagy: macroautophagy, microautophagy, and chaperone-mediated autophagy, all of which promote proteolytic degradation of cytosolic components at the lysosome. During macroautophagy, targeted cytoplasmic constituents are included in an autophagosome which fuses with a lysosome degrading and recycling the autolysosome components. In contrast, cytosolic components are directly taken up by the lysosome itself in microautophagy or translocated across the lysosomal membrane in a complex with chaperone proteins in chaperone-mediated

autophagy [87]. In the last few years, normal basal autophagy, which is higher in brain than other tissue, has been accepted and there is an increasing interest especially in the regulation of macroautophagy in neuropathological conditions. Clearance of mutant huntingtin, mutant forms of alpha-synuclein and different forms of tau are dependent on autophagy in Huntington's, familial Parkinson's and Alzheimer's diseases, respectively, and given its role in eliminating "toxic assets" and promoting cell viability, it has been accepted as being beneficial. The role of autophagy in ischemia remains controversial. Increasingly data are suggesting that the activation of autophagy in ischemic brain may also contribute to neuroprotection but other evidence shows that autophagy is involved in a neural death pathway following cerebral ischemia. However, it is clear that determining the signaling pathways that lead to autophagy activation during ischemia may contribute to the development of new therapeutic strategies [86]. TOR acts to inhibit autophagy under nutrient availability and good energy status, whereas TOR kinase is repressed by signals that sense nutrient deprivation, including hypoxia. HIF-1a regulates mTORC1 via REDD1 and REDD2 proteins, as previously described. REDD1 negatively regulates mTORC1 via the TSC1—TSC2 complex and stimulates autophagy to ER stress, which is associated with unfolded protein response and inflammatory signaling. Mitochondrial autophagy induced after dysfunction by oxidative stress also seems to be dependent upon mTOR, where the LKB1-AMPK-TSC1/TSC2 branch may play a role in mTORC1 inhibition [88]. In addition, mTOR has been identified as an upstream kinase of the ULK family, the mammalian homolog of the Atg family, which is widely associated with autophagy pathways in yeast. ULK complexes seem to play a key role in autophagosome formation and the phosphorylation of ULK1 complex by mTOR would interfere with ULK1 recognition of its substrate, thereby inhibiting its function and autophagy. Thus, inhibition of mTORC1 by rapamycin leads to dephosphorylation of ULK1, ULK2, and Atg13 complex in human cells and induces autophagy [89]. A diminished mTOR signaling activity is in line with the elevated autophagy after focal ischemia in ischemic/core and penumbra regions, starting from 6 to 12 h and lasting for up 3 days after ischemia in mice [90]. In a rat model of neonatal hypoxia-ischemia, Carloni et al. have shown that rapamycin increased autophagy, reduced necrotic cell death, and decreased brain injury [91]. Buckley et al. have also demonstrated that rapamycin up-regulation of autophagy reduces infarct size and improves outcomes in both permanent MCAO and embolic MCAO murine models of stroke (without reperfusion and a slow reperfusion similar to that seen in most human stroke, respectively) [92]. On the other hand, rapamycin partially reverses the reduction of infarct size and brain edema and the inhibition of LC3/Beclin-1 induction (two protein markers of autophagy) induced by IPOC, a treatment where inhibition of the autophagy pathway plays a key role in neuroprotection against focal cerebral ischemia [93].

Finally, there is evidence of a crossroad between apoptosis and autophagy, particularly with regard to the antiapoptotic protein Bcl-2 which is also antiautophagic [94]. This antiautophagic effect is mediated through the ability of Bcl-2 to bind with the Beclin-1 protein and prevent autophagosome formation. Stroke injury increases the expression of Bcl-2 and treatment with rapamycin can help to overcome the Bcl-2-initiated blockade of autophagy and improve stroke outcome [95]. In summary, targeting signaling pathways related to apoptosis and autophagy might be a promising option in the treatment of cerebral ischemia, but the exact role of mTOR in these processes and its potential applications have yet to be determined.

7.10 mTOR AND THE BBB

The BBB is formed by the brain microvascular endothelial cells that are closely linked by tight junctions, lack fenestrations, and exhibit low pinocytic activity. The endothelial cells are in close contact with perivascular astrocytes, pericytes, and neurons comprising a neurovascular unit that shows selective permeability for substances to brain and creates a tightly regulated microenvironment in the brain, which is essential for the proper functioning of neurons [96,97]. During ischemic stroke, BBB disruption is a common feature but increasingly evidence indicates that the integrity of this structural and functional barrier is also compromised in other neurological conditions such as epilepsia, multiple sclerosis, Alzheimer's, and Parkinson diseases. Previous studies have found that it is critical to maintain the structure and integrity of brain endothelial cells during ischemia in order to protect the integrity of BBB [98]. Brain endothelial cells are damaged after brain ischemia and reperfusion-induced injury by several mechanisms that include ROS delivery, activation of inflammatory cascades and MMPs. Activated MMPs degrade BBB components usually inducing alterations in tight junctions, protein expression that directly affects BBB permeability and allowing inflammatory cells to infiltrate the brain parenchyma and intensify the injury cascade [66,99]. Recently, the activation of the AKt/mTOR kinase signaling pathway has been related to the beneficial protective effects against brain ischemia of alpha-lipoic acid (ALA), an endogenous short-chain fatty acid that

reduces infarct size. Thus, 1 mM ALA pre- and post-treatment administration of ALA reduced the OGD and reperfusion (OGD/R)-induced lactate dehydrogenase release in both the cell line (bEnd3) and primary cultures of mouse brain microvascular endothelial cells (BMVEC). Furthermore, rapamycin administration eliminates the protective effects of the ALA pre- and post-treatments against OGD/R-induced BMVEC injury blocking the phosphorylation of mTOR and its downstream targets, but not that of Akt [70]. Additionally, it has been suggested that the toxic effects of delayed rt-PA treatment, which exacerbates BBB leakage, may be linked to impaired mTOR signaling in response to hypoxia reoxygenation-associated stress. After 3 or 6 h OGD/R, rt-PA activates mTOR and its target protein kinase, p70S6K but when rt-PA treatment is delayed until 6 h OGD, cell viability is compromised and phosphorylation of p70S6K at 24 h reoxygenation is sharply decreased [100]. It is interesting to notice that in other epithelial models, such as porcine proximal tubular epithelial cells, rapamycin has a restorative function decreasing the paracellular permeability via the modulation of the tight junction protein claudin-1 [101] and that in a rat model of temporal lobe epilepsy, rapamycin-treated rats showed a reduced fluorescein permeability indicating that antiepileptogenic effects in rapamycin may be mediated via the restoration and strengthening of the BBB [102]. On the other hand, recent studies have shown that the induction of autophagy may contribute to increased permeability through occludin degradation in brain endothelial cells under ischemia, providing a new mechanism of BBB disruption in ischemic stroke. In particular, 3-methyladenine, an autophagy inhibitor, reversed the decreased cell viability and occludin levels suggesting that autophagy is involved in ischemic BBB damage [103]. However, contradictory results concerning autophagy-BBB disruption and the role of mTOR in this process again appear since Li et al. demonstrate that pretreatment rapamycin significantly reverses the decreased level of tight junction protein zonula occludens-1 (ZO-1) induced by OGD/R and promotes the distribution of ZO-1 in cell membranes [104]. The hemorrhagic transformation (HT), the most extensively documented consequence of delayed-rt-PA treatment in ischemic stroke patients, is a cerebrovascular injury in which blood vessels are more severely disrupted and BBB leakage may be sustained for weeks to months after insult. HT is closely associated with the activation of MMPs and the role of ROS in the activation of MMPs and the impact of activated MMPs on BBB injury are well documented [97,105]. Thus, neutralization of ROS might be a potential target for minimizing rt-PA's harmful effects on the BBB after ischemia-reperfusion. In support of this, rapamycin attenuates the high levels of ROS induced by OGD/R in BMVEC [104]. Therefore, it is clear that the integrity of BBB is significantly damaged, amplifying ischemic neuronal damage but studies continue to attempt to identify the exact cellular and molecular mechanism of ischemic BBB disruption and the role of mTOR in this process to provide new indicators for therapeutic approaches that minimize brain damage under ischemic stroke.

7.11 mTOR AND BRAIN REPAIR

The brain repair processes activated in the evolution of the ischemic cascade are neurogenesis and gliogenesis, with the necessary synaptogenesis to make the new cells functional, as well as angiogenesis. Adult neurogenesis persists in two regions, the subventricular zone (SVZ) of the lateral ventricles and the subgranular zone of the hippocampal dentate gyrus (DG). In the DG and SVZ, large numbers of neurons are generated from neural stem cells (NSCs) but most of these die and never become integrated into the neuronal circuitry. Neurogenesis is enhanced by increased survival and neuronal specification of stem cells or increased survival of newborn neurons, but little is known about the mechanisms and molecules that make stem cells undergo death in the naive or injured brain. Under pathological conditions such as stroke, the proliferation of NSCs increases and newborn neurons migrate to the site of damage [106]. mTOR is one of the intracellular regulators of the neuroprogenitor—neuron transition and has an essential role in promoting both neuroprogenitor proliferation and differentiation, and in regulating the timing of the transition between these two events. In addition, neuron migration also appears to be under the control of mTOR signaling. The inhibition of REDD1, which should enhance mTOR signaling, accelerates the timing of neuronal differentiation both *in vitro* and *in vivo* [107]. Furthermore, the inhibition of mTOR signaling by rapamycin interferes with insulin-induced neuronal differentiation of cultured rat neuroprogenitors, suppresses neurogenesis and increases the mitotic index in the embryonic chick neural tube. In addition, the inhibition of mTOR activity during kainate-induced epilepsy decreases neuronal cell death, neurogenesis, mossy fiber sprouting, and the development of spontaneous epilepsy and neurosteroid dehydroepiandrosterone promotes survival and dendritic growth of newborn neurons through a sigma(1) receptor-mediated modulation of PI3K-Akt-mTOR-p70S6k signaling [42,107,108]. However, humans and animals with disrupted TSC1/TSC2 complex show defective neuron migration, which indicates that proper neuron migration requires a TSC-dependent inhibition on mTOR signaling. Therefore, it

seems that proper neuronal development depends on suitable regulation of mTOR activity [107]. Corsini et al. have identified the CD95/CD95L system, composed by the "death receptor" CD95 (Fas/APO-1) and its ligand CD95L, as an inducer of injury-induced neurogenesis in the adult brain via activation of the Src/PI3K/AKT/mTOR signaling pathway, which results in a global increase in protein translation associated with neuronal differentiation. Specifically, CD95 stimulation in NSCs induces differentiation and survival of NSCs and, following global ischemia, CD95-mediated brain repair rescued behavioral impairment. It is not surprising that an inducer of apoptosis after brain injury and in neurodegenerative diseases, such as the CD95/CD95L system, promotes neurogenesis. Apoptosis and neurogenesis go hand in hand as it is necessary for damaged cells to be replaced by newborn ones [106]. In addition to neuronal survival, the mTOR pathway is also involved in dendrite development and spine morphogenesis and it has also been identified as being responsible for axon regeneration by promoting protein synthesis in injured neurons and so is crucial for the regulation of synaptic plasticity in adult brain [109–111]. Specifically, the PI-3K/Akt/mTOR pathway has been identified as one of the molecular mechanisms responsible for the beneficial effect of simvastatin, a cholesterol-lowering drug, on neurite outgrowth of primary cortical neurons subjected to OGD [110]. With regards to gliogenesis, the process by which glial cells (astrocytes, oligodendrocytes, Schwann cells, and microglia) are generated through the production of glial progenitor cells and their differentiation into mature glia, it is known that mTORC2-mediated Akt activation is a key driver of neuroglial progenitor/stem cell proliferation and gliogenesis [112] and mTORC1 signaling is important for gliogenesis during late cortical development [113]. It is interesting to note that it has recently been demonstrated that mTOR blockade by rapamycin attenuates astrocyte migration, proliferation, and the production of inflammation mediators after OGD/R [114]. Finally, neovascularization is a physiological response to ischemia that produces collateral vessels to resolve ischemic symptoms or signs. Therapeutic neovascularization has become one of the most important techniques to salvage tissue in critically ischemic patients [115]. It is known that hypoxia enhances vascular cell proliferation and angiogenesis via mTOR-dependent signaling through the activation of HIF-1a and Akt [52,109,116]. Furthermore, the VEGF receptor-2 tyrosine kinase, a crucial regulator of angiogenesis to compensate postischemic cardiovascular dysfunction, acts via the Akt/mTOR/STAT3 signaling cascade [117]. Moon et al. have recently shown that cartilage oligomeric matrix protein (COMP)-Ang1, a soluble, stable and potent Ang1 variant, potentiates the angiogenesis of EPCs and that transplanted EPCs and

COMP-Ang1 were incorporated into the blood vessels and decreased the infarct volume in the rat ischemic brain through the activation of the Akt/mTOR pathway [115].

7.12 CONCLUSIONS

In summary, in this chapter we have reviewed the multiple pathways that mTOR may be modulating after cerebral ischemia. Due to its pleiotropic properties, which allow mTOR to act at different levels as well as at different times of the ischemic cascade, this molecule seems to offer a promising therapeutic strategy to be evaluated as a neuroprotector in ischemic stroke.

References

[1] Lloyd-Jones D, Adams R, Carnethon M, De Simone G, Ferguson TB, Flegal K, American Heart Association Statistics Committee and Stroke Statistics Subcommittee, et al. Heart disease and stroke statistics—2009 update: a report from the American Heart Association Statistics Committee and Stroke Statistics Subcommittee. Circulation 2009;119:e21–8.

[2] Roger VL, Go AS, Lloyd-Jones DM, Adams RJ, Berry JD, Brown TM, American Heart Association Statistics Committee and Stroke Statistics Subcommittee, et al. AHA Heart Disease and Stroke Statistics 2011 update: a report from the American Heart Association. Circulation 2011;123:e18–e209.

[3] EROS Investigators. Incidence of stroke in Europe at the beginning of the 21st century. Stroke 2009;40:1557–63.

[4] Truelsen T, Piechowski-Jóźwiak B, Bonita R, Mather C, Bogousslavsky J, Boysen G. Stroke incidence and prevalence in Europe: a review of the available data. Eur J Neurol 2006;13: 581–98.

[5] Donnan GA, Fisher M, Macleod M, Davis SM. Stroke. Lancet 2008;371:1612–23.

[6] Gustavsson A, Svensson M, Jacobi F, Allgulander C, Alonso J, Beghi E, CDBE2010Study Group, et al. Cost of disorders of the brain in Europe 2010. Eur Neuropsychopharmacol 2011;21: 718–79.

[7] ESO guidelines for management of ischemic stroke—guideline update January (2009). Available from: <http://www.eso-stroke. org/eso-stroke/education/guidelines.html>.

[8] Jauch EC, Saver JL, Adams Jr HP, Bruno A, Connors JJ, Demaerschalk BM, American Heart Association Stroke Council; Council on Cardiovascular Nursing; Council on Peripheral Vascular Disease; Council on Clinical Cardiology, et al. Guidelines for the early management of patients with acute ischemic stroke: a guideline for healthcare professionals from the American Heart Association/American Stroke Association. Stroke 2013;44:870–947.

[9] Gladstone DJ, Black SE, Hakim AM, Heart and Stroke Foundation of Ontario Centre of Excellence in Stroke Recovery. Toward wisdom from failure: lessons from neuroprotective stroke trials and new therapeutic directions. Stroke 2002;33:2123–36.

[10] Saver JL, Albers GW, Dunn B, Johnston KC, Fisher M, STIR VI Consortium. Stroke therapy academic industry roundtable (STAIR) recommendations for extended window acute stroke therapy trials. Stroke 2009;40:2594–600.

[11] Mehta SL, Manhas N, Raghubir R. Molecular targets in cerebral ischemia for developing novel therapeutics. Brain Res Rev 2007; 54:34–66.

[12] Gegelashvili G, Robinson MB, Trotti D, Rauen T. Regulation of glutamate transporters in health and disease. Prog Brain Res 2001;132:267–86.

[13] Maragakis NJ, Rothstein JD. Glutamate transporters: animal models to neurologic diseases. Neurobiol Dis 2004;15:461–73.

[14] Seal RP, Amara SG. Excitatory amino acid transporters: a family influx. Annu Rev Pharmacol Toxicol 1999;39:431–56.

[15] Robinson MB. The family of sodium-dependent glutamate transporters: a focus on the GLT-1/EAAT2 subtype. Neurochem Int 1999;33:479–91.

[16] Su ZZ, Leszczyniecka M, Kang DC, Sarkar D, Chao W, Volsky DJ, et al. Insights into glutamate transport regulation in human astrocytes: cloning of the promoter for excitatory amino acid transporter 2 (EAAT2). Proc Natl Acad Sci USA 2003;100:1955–60.

[17] Rossi DJ, Oshima T, Atwell D. Gluatamate release in severe brain ischaemia is mainly by reversed uptake. Nature 2000;403: 316–21.

[18] Mallolas J, Hurtado O, Castellanos M, Blanco M, Sobrino T, Serena J, et al. A polymorphism in the *EATT2* promoter is associated with higher glutamate concentrations and higher frequency of progressing stroke. J Exp Med 2006;203:711–17.

[19] Feuerstein GZ, Wang X, Barone FC. Inflammatory mediators and brain injury. The role of cytokines and chemokines in stroke and CNS diseases. In: Ginsberg MD, Bogousslavsky J, editors. Cerebrovascular disease: pathophysiology, diagnosis, and management, vol 1. Boston: Blackwell Science; 1998. p. 507–31.

[20] Ames A, Wright LW, Kowade M, Thurston JM, Majors G. Cerebral ischemia. II. The noreflow phenomenon. Am J Pathol 1968;52:437–53.

[21] Härtl R, Schürer L, Schmid-Schönbein GW, del Zopo G. Experimental antileukocyte interventions in cerebral ischemia. J Cereb Blood Flow Metab 1996;16:1108–19.

[22] Hamann GF, Okada Y, Fitridge R, del Zopo GJ. Microvascular basal lamina antigens disappear during cerebral ischemia and reperfusion. Stroke 1995;26:2120–6.

[23] Hamann GF, Okada Y, del Zoppo GJ. Hemorrhagic transformation and microvascular integrity during focal cerebral ischemia/reperfusion. J Cereb Blood Flow Metab 1996;16:1373–8.

[24] Kogure T, Kogure K. Molecular and biochemical events within the brain subjected to cerebral ischemia. Clin Neurosci 1997;4:179–83.

[25] Kontos HA. Oxygen radicals in cerebral vascular injury. Circ Res 1985;57:508–16.

[26] Kukreja R, Kontos H, Hess M, Ellis E. PGH synthase and lipoxygenase generate superoxide in the prseence of NADA and NADPH. Circ Res 1986;59:612–19.

[27] Dawson TM, Dawson VL, Snyder SH. A novel neuronal messenger molecule in brain: the free radical, nitric oxide. Ann Neurol 1992;32:297–311.

[28] Broughton BRS, Reutens DC, Sobey CG. Apoptotic mechanisms after cerebral ischemia. Stroke 2009;40:e331–9.

[29] Culmsee C, Zhu C, Landshamer S, Becattini B, Wagner E, Pellecchia M, et al. Apoptosis-inducing factor triggered by poly (ADP-ribose) polymerase and BID mediates neuronal cell death after oxygen-glucose deprivation and focal cerebral ischemia. J Neurosci 2005;25:10262–72.

[30] Sun Y, Jin K, Xie L, Childs J, Mao XO, Logvinova A, et al. VEGF-induced neuroprotection, neurogenesis, and angiogenesis after focal cerebral ischemia. J Clin Invest 2003;111:1843–51.

[31] Shingo T, Sorokan ST, Shimazaki T, Weiss S. Erythropoietin regulates the *in vitro* and *in vivo* production of neuronal progenitors by mammalian forebrain neural stem cells. J Neurosci 2001;21:9733–43.

[32] Bernabeu R, Sharp FR. NMDA and AMPA/kainate glutamate receptors modulate dentate neurogenesis and CA3 synapsin-I in normal and ischemic hippocampus. J Cereb Blood Flow Metab 2000;20:1669—80.

[33] Zhang ZG, Zhang L, Jiang Q, Chopp M. Bone marrow-derived endothelial progenitor cells participate in cerebral neovascularization after focal cerebral ischemia in the adult mouse. Circ Res 2002;90:284—2888.

[34] Sobrino T, Hurtado O, Moro MA, Rodríguez-Yáñez M, Castellanos M, Brea D, et al. The increase of circulating endothelial progenitor cells after acute ischemic stroke is associated with good outcome. Stroke 2007;38:2759—66.

[35] Alexandrov AV, Grotta JC. Arterial reocclusion in stroke patients treated with intravenous tissue plasminogen activator. Neurology 2002;59:862—7.

[36] Berkhemer OA, Fransen PS, Beumer D, van den Berg LA, Lingsma HF, Yoo AJ, MR CLEAN Investigators, et al. A randomized trial of intraarterial treatment for acute ischemic stroke. N Engl J Med 2015;372:11—20.

[37] Goyal M, Demchuk AM, Menon BK, Eesa M, Rempel JL, Thornton J, ESCAPE Trial Investigators, et al. Randomized assessment of rapid endovascular treatment of ischemic stroke. N Engl J Med 2015;372:1019—30.

[38] Campbell BC, Mitchell PJ, Kleinig TJ, Dewey HM, Churilov L, Yassi N, EXTEND-IA Investigators, et al. Endovascular therapy for ischemic stroke with perfusion-imaging selection. N Engl J Med 2015;372:1009—18.

[39] Saver JL, Goyal M, Bonafe A, Diener HC, Levy EI, Pereira VM, SWIFT PRIME Investigators, et al. Stent-retriever thrombectomy after intravenous t-PA vs. t-PA alone in stroke. N Engl J Med 2015;372:2285—95.

[40] Jovin TG, Chamorro A, Cobo E, de Miquel MA, Molina CA, Rovira A, REVASCAT Trial Investigators, et al. Thrombectomy within 8 h after symptom onset in ischemic stroke. N Engl J Med 2015;372:2296—306.

[41] Gladstone DJ, Black SE, Hakim AM, Heart and Stroke Foundation of Ontario Centre of Excellence in Stroke Recovery. Toward wisdom from failure: lessons from neuroprotective stroke trials and new therapeutic directions. Stroke 2002;33:2123—36.

[42] Chong ZZ, Shang YC, Wang S, Maiese K. A critical kinase cascade in neurological disorders: Pi 3-k, akt, and mtor. Future Neurol 2012;7:733—48.

[43] Chong ZZ, Shang YC, Zhang L, Wang S, Maiese K. Mammalian target of rapamycin: hitting the bull's-eye for neurological disorders. Oxid Med Cell Longev 2010;3:374—91.

[44] Maiese K. Cutting through the complexities of mtor for the treatment of stroke. Curr Neurovasc Res 2014;11:177—86.

[45] Keith CT, Schreiber SL. Pik-related kinases: DNA repair, recombination, and cell cycle checkpoints. Science 1995;270:50—1.

[46] Kunz J, Henriquez R, Schneider U, Deuter-Reinhard M, Movva NR, Hall MN. Target of rapamycin in yeast, tor2, is an essential phosphatidylinositol kinase homolog required for g1 progression. Cell 1993;73:585—96.

[47] Yang Q, Guan KL. Expanding mTOR signaling. Cell Res 2007;17:666—81.

[48] Perluigi M, Di Domenico F, Butterfield DA. mTOR signaling in aging and neurodegeneration: at the crossroad between metabolism dysfunction and impairment of autophagy. Neurobiol Dis 2015. pii: S0969-9961(15)00086-8. Available from:<http://dx.doi.org/10.1016/j.nbd.2015.03.014>. [Epub ahead of print].

[49] Laplante M, Sabatini DM. mTOR signaling in growth control and disease. Cell 2012;149:274—93.

[50] Kaizuka T, Hara T, Oshiro N, Kikkawa U, Yonezawa K, Takehana K, et al. Tti1 and tel2 are critical factors in mammalian target of rapamycin complex assembly. J Biol Chem 2010;285:20109—16.

[51] Dennis PB, Jaeschke A, Saitoh M, Fowler B, Kozma SC, Thomas G. Mammalian TOR: a homeostatic ATP sensor. Science 2001;294:1102—5.

[52] Lipton JO, Sahin M. The neurology of mTOR. Neuron 2014;84:275—91.

[53] Kahn BB, Alquier T, Carling D, Hardie DG. Amp-activated protein kinase: ancient energy gauge provides clues to modern understanding of metabolism. Cell Metab 2005;1:15—25.

[54] Tchevkina E., Komelkov A. Protein phosphorylation as a key mechanism of mTORC1/2 signaling pathways, protein phosphorylation in human health. In: Huang C, editor. Tech; 2012. Available from: <http://www.intechopen.com/books/protein-phosphorylation-in-human-health/protein-phosphorylation-as-a-key-mechanism-of-mtorc1-2-signaling-pathways>

[55] Hardie DG. AMP-activated/SNF1 protein kinases: conserved guardians of cellular energy. Nat Rev Mol Cell Biol 2007;8:774—85.

[56] Martin DE, Hall MN. The expanding TOR signaling network. Curr Opin Cell Biol 2005;17:158—66.

[57] Wullschleger S, Loewith R, Hall MN. TOR signaling in growth and metabolism. Cell 2006;124:471—84.

[58] Brugarolas J, Lei K, Hurley RL, Manning BD, Reiling JH, Hafen E, et al. Regulation of mtor function in response to hypoxia by REDD1 and the TSC1/TSC2 tumor suppressor complex. Genes Dev 2004;18:2893—904.

[59] DeYoung MP, Horak P, Sofer A, Sgroi D, Ellisen LW. Hypoxia regulates TSC1/2-mTOR signaling and tumor suppression through REDD1-mediated 14-3-3 shuttling. Genes Dev 2008;22:239—51.

[60] Hwang SK, Kim HH. The functions of mtor in ischemic diseases. BMB Rep 2011;44:506—11.

[61] Wu X, Kihara T, Akaike A, Niidome T, Sugimoto H. Pi3k/akt/mtor signaling regulates glutamate transporter 1 in astrocytes. Biochem Biophys Res Commun 2010;393:514—18.

[62] Almilaji A, Pakladok T, Guo A, Munoz C, Foller M, Lang F. Regulation of the glutamate transporter eaat3 by mammalian target of rapamycin mTOR. Biochem Biophys Res Commun 2012;421:159—63.

[63] Gubern C, Camos S, Hurtado O, Rodriguez R, Romera VG, Sobrado M, et al. Characterization of Gcf2/Lrrfip1 in experimental cerebral ischemia and its role as a modulator of Akt, mTOR and β-catenin signaling pathways. Neuroscience 2014;268:48—65.

[64] Wang Y, Wang W, Li D, Li M, Wang P, Wen J, et al. IGF-1 alleviates NMDA-induced excitotoxicity in cultured hippocampal neurons against autophagy via the NR2B/PI3K-AKT-mTOR pathway. J Cell Physiol 2014;229:1618—29.

[65] Ge D, Han L, Huang S, Peng N, Wang P, Jiang Z, et al. Identification of a novel mtor activator and discovery of a competing endogenous rna regulating autophagy in vascular endothelial cells. Autophagy 2014;10:957—71.

[66] Majid A. Neuroprotection in stroke: past, present, and future. ISRN Neurol 2014;2014:515716.

[67] Ouyang YB. Inflammation and stroke. Neurosci Lett 2013;548:1—3.

[68] Iadecola C, Anrather J. The immunology of stroke: from mechanisms to translation. Nat Med 2011;17:796—808.

[69] Gu L, Xiong X, Zhang H, Xu B, Steinberg GK, Zhao H. Distinctive effects of T cell subsets in neuronal injury induced by cocultured splenocytes in vitro and by in vivo stroke in mice. Stroke 2012;43:1941—6.

[70] Xie R, Li X, Ling Y, Shen C, Wu X, Xu W, et al. Alpha-lipoic acid pre- and post-treatments provide protection against in vitro ischemia-reperfusion injury in cerebral endothelial cells via akt/mtor signaling. Brain Res 2012;1482:81—90.

[71] Iadecola C, Anrather J. Stroke research at a crossroad: asking the brain for directions. Nat Neurosci 2011;14:1363–8.

[72] Liesz A, Suri-Payer E, Veltkamp C, Doerr H, Sommer C, Rivest S, et al. Regulatory T cells are key cerebroprotective immunomodulators in acute experimental stroke. Nat Med 2009;15:192–9.

[73] Ceulemans AG, Zgavc T, Kooijman R, Hachimi-Idrissi S, Sarre S, Michotte Y. The dual role of the neuroinflammatory response after ischemic stroke: modulatory effects of hypothermia. J Neuroinflammation 2010;7:74.

[74] Ginsberg MD. Neuroprotection for ischemic stroke: past, present and future. Neuropharmacology 2008;55:363–89.

[75] Powell JD, Pollizzi KN, Heikamp EB, Horton MR. Regulation of immune responses by mtor. Annu Rev Immunol 2012;30:39–68.

[76] Dello Russo C, Lisi L, Tringali G, Navarra P. Involvement of mtor kinase in cytokine-dependent microglial activation and cell proliferation. Biochem Pharmacol 2009;78:1242–51.

[77] Lu Q, Gao L, Huang L, Ruan L, Yang J, Huang W, et al. Inhibition of mammalian target of rapamycin improves neurobehavioral deficit and modulates immune response after intracerebral hemorrhage in rat. J Neuroinflammation 2014;11:44.

[78] Fulda S, Gorman AM, Hori O, Samali A. Cellular stress responses: cell survival and cell death. Int J Cell Biol 2010;2010:214074.

[79] Pastor MD, Garcia-Yebenes I, Fradejas N, Perez-Ortiz JM, Mora-Lee S, Tranque P, et al. mTOR/s6 kinase pathway contributes to astrocyte survival during ischemia. J Biol Chem 2009;284:22067–78.

[80] Chong ZZ, Shang YC, Wang S, Maiese K. PRAS40 is an integral regulatory component of erythropoietin mtor signaling and cytoprotection. PLoS One 2012;7:e45456.

[81] Koh PO. Melatonin prevents ischemic brain injury through activation of the mtor/p70s6 kinase signaling pathway. Neurosci Lett 2008;444:74–8.

[82] Koh PO, Cho JH, Won CK, Lee HJ, Sung JH, Kim MO. Estradiol attenuates the focal cerebral ischemic injury through mTOR/p70S6 kinase signaling pathway. Neurosci Lett 2008;436:62–6.

[83] Xie R, Wang P, Ji X, Zhao H. Ischemic post-conditioning facilitates brain recovery after stroke by promoting Akt/mTOR activity in nude rats. J Neurochem 2013;127:723–32.

[84] Xiong X, Xie R, Zhang H, Gu L, Xie W, Cheng M, et al. PRAS40 plays a pivotal role in protecting against stroke by linking the Akt and mTOR pathways. Neurobiol Dis 2014;66:43–52.

[85] Fletcher L, Evans TM, Watts LT, Jimenez DF, Digicaylioglu M. Rapamycin treatment improves neuron viability in an *in vitro* model of stroke. PLoS One 2013;8:e68281.

[86] Gabryel B, Kost A, Kasprowska D. Neuronal autophagy in cerebral ischemia—a potential target for neuroprotective strategies? Pharmacol Rep 2012;64:1–15.

[87] Glick D, Barth S, Macleod KF. Autophagy: cellular and molecular mechanisms. J Pathol 2010;221:3–12.

[88] Jung CH, Ro SH, Cao J, Otto NM, Kim DH. mTOR regulation of autophagy. FEBS Lett 2010;584:1287–95.

[89] Chang YY, Neufeld TP. An Atg1/Atg13 complex with multiple roles in TOR-mediated autophagy regulation. Mol Biol Cell 2009;20:2004–14.

[90] Li WL, Yu SP, Chen D, Yu SS, Jiang YJ, Genetta T, et al. The regulatory role of NF-kappa B in autophagy-like cell death after focal cerebral ischemia in mice. Neuroscience 2013;244:16–30.

[91] Carloni S, Buonocore G, Balduini W. Protective role of autophagy in neonatal hypoxia-ischemia induced brain injury. Neurobiol Dis 2008;32:329–39.

[92] Buckley KM, Hess DL, Sazonova IY, Periyasamy-Thandavan S, Barrett JR, Kirks R, et al. Rapamycin up-regulation of

autophagy reduces infarct size and improves outcomes in both permanent mcal, and embolic MCAO, murine models of stroke. Exp Transl Stroke Med 2014;6:8.

[93] Gao L, Jiang T, Guo J, Liu Y, Cui G, Gu L, et al. Inhibition of autophagy contributes to ischemic postconditioning-induced neuroprotection against focal cerebral ischemia in rats. PLoS One 2012;7:e46092.

[94] Zhao GX, Pan H, Ouyang DY, He XH. The critical molecular interconnections in regulating apoptosis and autophagy. Ann Med 2015;47:305–15.

[95] Buckley KM, Hoda N, Herberg S, Barrett JR, Periyasamy-Thandavan S, Kondrikova G, et al. Induction of autophagy with rapamycin overcomes Bcl-2's deleterious effects on stroke outcome. FASEB J 2013;27:lb514.

[96] Hawkins BT, Davis TP. The blood–brain barrier/neurovascular unit in health and disease. Pharmacol Rev 2005;57:173–85.

[97] Jin R, Yang G, Li G. Molecular insights and therapeutic targets for blood–brain barrier disruption in ischemic stroke: critical role of matrix metalloproteinases and tissue-type plasminogen activator. Neurobiol Dis 2010;38:376–85.

[98] Abbott NJ, Patabendige AA, Dolman DE, Yusof SR, Begley DJ. Structure and function of the blood–brain barrier. Neurobiol Dis 2010;37:13–25.

[99] Shlosberg D, Benifla M, Kaufer D, Friedman A. Blood–brain barrier breakdown as a therapeutic target in traumatic brain injury. Nat Rev Neurol 2010;6:393–403.

[100] Ryou MG, Choudhury GR, Winters A, Xie L, Mallet RT, Yang SH. Pyruvate minimizes rtpa toxicity from *in vitro* oxygen-glucose deprivation and reoxygenation. Brain Res 2013;1530:66–75.

[101] Martin-Martin N, Ryan G, McMorrow T, Ryan MP. Sirolimus and cyclosporine a alter barrier function in renal proximal tubular cells through stimulation of Erk1/2 signaling and claudin-1 expression. Am J Physiol Renal Physiol 2010;298:F672–82.

[102] van Vliet EA, Forte G, Holtman L, den Burger JC, Sinjewel A, de Vries HE, et al. Inhibition of mammalian target of rapamycin reduces epileptogenesis and blood–brain barrier leakage but not microglia activation. Epilepsia 2012;53:1254–63.

[103] Kim K, Shin Y, Kim E, Akram M, Noh D, Kim E, et al. Role of autophagy in blood–brain barrier disruption and tight junction degradation under ischemic condition. Stroke 2015;46: ATP250.

[104] Li H, Gao A, Feng D, Wang Y, Zhang L, Cui Y, et al. Evaluation of the protective potential of brain microvascular endothelial cell autophagy on blood–brain barrier integrity during experimental cerebral ischemia-reperfusion injury. Transl Stroke Res 2014;5:618–26.

[105] Lin CC, Hsieh HL, Shih RH, Chi PL, Cheng SE, Chen JC, et al. NADPH oxidase 2-derived reactive oxygen species signal contributes to bradykinin-induced matrix metalloproteinase-9 expression and cell migration in brain astrocytes. Cell Commun Signal 2012;10:35.

[106] Corsini NS, Sancho-Martinez I, Laudenklos S, Glagow D, Kumar S, Letellier E, et al. The death receptor CD95 activates adult neural stem cells for working memory formation and brain repair. Cell Stem Cell 2009;5:178–90.

[107] Malagelada C, Lopez-Toledano MA, Willett RT, Jin ZH, Shelanski ML, Greene LA. Rtp801/REDD1 regulates the timing of cortical neurogenesis and neuron migration. J Neurosci 2011;31:3186–96.

[108] Li L, Xu B, Zhu Y, Chen L, Sokabe M. DHEA prevents abeta25-35-impaired survival of newborn neurons in the dentate gyrus through a modulation of pi3k-Akt-mTOR signaling. Neuropharmacology 2010;59:323–33.

II. mTOR IN GENETIC DISORDERS AND NEURODEGENERATIVE DISEASE

[109] Pignataro G, Capone D, Polichetti G, Vinciguerra A, Gentile A, Di Renzo G, et al. Neuroprotective, immunosuppressant and antineoplastic properties of mtor inhibitors: current and emerging therapeutic options. Curr Opin Pharmacol 2011;11:378−94.

[110] Wu H, Mahmood A, Qu C, Xiong Y, Chopp M. Simvastatin attenuates axonal injury after experimental traumatic brain injury and promotes neurite outgrowth of primary cortical neurons. Brain Res 2011;1486:121−30.

[111] Swiech L, Perycz M, Malik A, Jaworski J. Role of mTOR in physiology and pathology of the nervous system. Biochim Biophys Acta 2008;1784:116−32.

[112] Lee da Y, Yeh TH, Emnett RJ, White CR, Gutmann DH. Neurofibromatosis-1 regulates neuroglial progenitor proliferation and glial differentiation in a brain region-specific manner. Genes Dev 2010;24:2317−29.

[113] Cloetta D, Thomanetz V, Baranek C, Lustenberger RM, Lin S, Oliveri F, et al. Inactivation of mTORC1 in the developing brain causes microcephaly and affects gliogenesis. J Neurosci 2013;33:7799−810.

[114] Li CY, Li X, Liu SF, Qu WS, Wang W, Tian DS. Inhibition of mtor pathway restrains astrocyte proliferation, migration and production of inflammatory mediators after oxygen-glucose deprivation and reoxygenation. Neurochem Int 2015;83-84: 9−18.

[115] Moon HE, Byun K, Park HW, Kim JH, Hur J, Park JS, et al. Comp-Ang1 potentiates epc treatment of ischemic brain injury by enhancing angiogenesis through activating Akt-mTOR pathway and promoting vascular migration through activating tie2-FAK pathway. Exp Neurobiol 2015;24: 55−70.

[116] Humar R, Kiefer FN, Berns H, Resink TJ, Battegay EJ. Hypoxia enhances vascular cell proliferation and angiogenesis in vitro via rapamycin (mTOR)-dependent signaling. FASEB J 2002; 16:771−80.

[117] Fan W, Li C, Qin X, Wang S, Da H, Cheng K, et al. Adipose stromal cell and sarpogrelate orchestrate the recovery of inflammation-induced angiogenesis in aged hindlimb ischemic mice. Aging Cell 2013;12:32−41.

mTOR Signaling in Epilepsy and Epileptogenesis: Preclinical and Clinical Studies

Antonio Leo[1], Andrew Constanti[2], Antonietta Coppola[3], Rita Citraro[1], Giovambattista De Sarro[1] and Emilio Russo[1]

[1]Science of Health Department, School of Medicine, University of Catanzaro, Italy [2]Department of Pharmacology, UCL School of Pharmacy, London, United Kingdom [3]Epilepsy Centre; Department of Neuroscience, Reproductive and Odontostomatological Sciences, Federico II University, Naples, Italy

8.1 INTRODUCTION

The mammalian/mechanistic target of rapamycin (mTOR), also known as FK506 binding protein 12-rapamycin associated protein 1, is an evolutionarily conserved serine-threonine kinase (289 kDa) belonging to the phosphoinositide 3-kinase (PI3K)-related kinase family [1,2]. Rapamycin (also called sirolimus) is a macrolide antibiotic molecule extracted from the mycelium of the Easter Island (Rapa Nui) soil-dwelling bacterium *Streptomyces hygroscopicus*. In fact, the name rapamycin was created from the following etymology: (Rapamycin (Rapa Nui = Easter Island)). This macrolide, through its specific ability to inhibit mTOR, is mainly used as an antifungal, immunosuppressant, and anticancer agent [3,4]. mTOR is activated by phosphorylation in response to many modulators such as growth factors, mitogens, and hormones. The molecular factors and downstream target signaling molecules associated with the mTOR pathway are numerous and complex. The signal transduction mechanisms linked to mTOR have been studied extensively and have been related to a huge spectrum of fundamental cellular biochemical and physiological processes, such as metabolism, cell growth, proliferation, differentiation, longevity, apoptosis, and autophagy [5,6]. In mammalian cells, the functions of mTOR are mediated through two mTOR heteromeric and functionally distinct protein complexes, mTORC1 and mTORC2, although their precise function still remains unclear. These complexes are formed by mTOR and several proteins, two of which are common to both mTORC1/2: mammalian lethal with Sec13 protein 8 (mLST8, also known as GβL) is a positive regulator of both complexes and DEPTOR (DEP-domain containing mTOR-interacting protein) a physiological negative regulator of both mTOR complexes. mTORC1 has two other associated proteins; the positive regulator named RAPTOR (regulatory-associated protein of mTOR) and PRAS40 (proline-rich AKT substrate of 40 kDa) or AKTS1 (AKT1 substrate), which shows a suppressive action. mTORC2 shows three components in common with mTORC1 and three specific proteins: RICTOR (rapamycin-insensitive companion of mTOR), which plays an important role also for the interaction between mTORC2 and tuberous sclerosis complex 2 (TSC2), a direct activator of this complex [7]; mSIN-1 (mammalian stress-activated protein kinase interacting protein) fundamental for its ability to phosphorylate AKT and other functions [8], and PROTOR-1 (protein observed with RICTOR-1), which is required for efficient mTORC2-mediated activation of SGK-1. Moreover, PROTOR-1 is able to bind RICTOR [9]. In particular, the mTORC1 rapamycin-sensitive complex is fundamental in controlling a wide variety of cellular processes including transcription, translation, autophagy, cell cycle, and microtubule dynamics. Conversely, mTORC2 regulates both development of the cytoskeleton and also controls cell survival. This complex shows a different sensitivity to rapamycin, which inhibits mTOR

DOI: http://dx.doi.org/10.1016/B978-0-12-802733-2.00006-2

through binding with FKBP12 (FK 506-binding protein of 12 kDa). In particular, it has been demonstrated that rapamycin, at least in some cell lines, only inhibits the assembly and function of mTORC2 after long treatment [10].

Dysregulation of the mTOR pathway has been frequently correlated with several disease conditions such as tumorigenesis, type 2 diabetes, inflammation, and neurological diseases [1,6,11]. The role of mTOR in the brain is crucial, since it is implicated in several physiological and pathological processes including protein translation, protein synthesis in axons and dendrites of neurons, autophagy, and microtubule dynamics [12,13]. The mTOR pathway is a key regulator during the development of the cerebral cortex. In fact, during neuronal development, it has been suggested that mTOR could control protein expression and other cellular mechanisms including differentiation, axon growth, navigation, and synaptogenesis, which plays an important role in neuronal excitability [14,15]. In the adult central nervous system (CNS), mTOR is involved in crucial roles including learning and memory. In fact, studies have demonstrated the ability of mTOR to promote protein synthesis at the synaptic level, fundamental both for synaptic plasticity and for memory formation. In particular, mTOR plays an important role in the consolidation of memory through the long-term potentiation (LTP) mechanism [16,17]. Other studies reported a linkage between enhancing mTORC1 activity and memory improvement. Rapamycin seems to disrupt this process in several behavioral models. Moreover, rapamycin administration has also been shown to decrease both the spatial hippocampus memory and the consolidation memory in many brain regions [18–21]. However, mTORC1 hyperactivity has been associated with memory deficits in human patients and experimental models of tuberous sclerosis (TS) [22]. In a mouse model of TS, rapamycin rescued memory performance, which could be, at least in part, related to an abnormal mTOR activity [23]. Puighermanal et al. [24] have also shown that a cannabinoid can modulate hippocampal long-term memory through mTOR signaling. Rapamycin or the protein synthesis inhibitor anisomycin abolished the amnesic-like effects of Δ^9-THC. In addition to its role in memory, mTOR acts in the hypothalamus as an energy sensor to control both food intake and body energy balance. Leucine, by inducing mTOR activity, decreases food intake and body weight, whereas rapamycin, along with leucine, blocked these effects [25]. In the hypothalamus, mTOR is also involved in the regulation of puberty onset and luteinizing hormone secretion, whereas rapamycin caused inhibition of the gonadotrophic axis at puberty [26]. Other relevant functions of mTOR in CNS are

sleep and circadian rhythms. The importance of mTOR signaling in CNS physiology is underscored by the myriad disorders in which mTOR pathway disruption is implicated, such as tumors, autism, mood disorders, neurodegenerative diseases, as well as epilepsy [27,28]. The possible link between mTOR and neurodegenerative diseases could also be autophagy. Autophagy is an evolutionarily conserved catabolic process that degrades bulk cytosol in lysosomal compartments, enabling amino acids and fatty acids to be recycled. mTOR is one of the fundamental regulators of autophagy [29]. A typical hallmark of neurodegenerative diseases, such as Alzheimer's, Parkinson's and Huntington's diseases, is the aberrant accumulation of protein aggregates in the brain [30,31]. The clearance of these proteins would seem to be increased by mTOR inhibition [32,33]. However, many other neurodegenerative disorders have been linked to dysregulation of the mTOR pathway. Pharmacological manipulation of mTOR signaling is thus proving to be a promising therapeutic branch for the treatment of several neurological disorders [27].

The aim of this chapter is to review the current knowledge on the role of mTOR signaling in epilepsy and epileptogenesis. In fact, it has been widely commented that another potential field of application of the pharmacological manipulation of mTOR signaling is indeed epilepsy and epileptogenesis, which share some interesting commonalities with many neurological disorders. In particular, many neurological and psychiatric disorders present as a common comorbidity in epileptic patients [1,27,34,35].

Epilepsy is a common and debilitating disorder of the nervous system defined as a brain state that promotes periodic, unprovoked seizures, which are unpredictable and sometimes progressively severe. The term "seizures" refers to an abnormal, paroxysmal change in the electrical activity of the brain. Epileptogenesis (latency period) is a dynamic process that progressively alters neuronal excitability; this term indicates a cascade of events that appears in a specific critical period of time which starts after or during the occurrence of an insult, such as traumatic brain injury (TBI), infection, or genetic predisposition, and finishes at the onset of repetitive spontaneous seizures [36,37]. Currently, epileptic syndromes are treated with pharmacological compounds (antiepileptic drugs: AEDs) that are solely symptomatic therapies; in fact, they primarily suppress seizures but do not necessarily possess antiepileptogenic or disease-modifying properties. Unfortunately, the AEDs currently available have not improved the outcome of refractory epilepsy, which is one of the major problems in epilepsy management [38]. Therefore, one of the major unmet needs in the field of epilepsy is the identification of novel disease-modifying therapies that can completely prevent epilepsy (antiepileptogenic drugs) or slow its progression. Therefore, this crucial silent period offers a range of opportunity in which an appropriate treatment might prevent or modify the epileptogenic process and thus significantly affect the life of a potential epileptic patient. The silent period also offers the possibility to discover early biomarkers of epileptogenesis, which are crucial for early diagnosis, and perhaps also to direct new therapies. Such markers may well prove to be neuroinflammatory in nature, since neuroinflammation appears to be a prominent feature of several epilepsy syndromes in human patients and animal models [39]. A rational antiepileptogenic strategy would thus seem to be to target primary cell signaling pathways that initially trigger the downstream mechanisms causing epileptogenesis.

Considering the mTOR involvement in major multiple cellular functions which can influence neuronal excitability, it is not surprising that the mTOR signaling cascade could be responsible for or participate in the development of spontaneous seizures, and that this pathway could represent an important target for both epileptogenesis and seizure pharmacotherapy [40–43]. In fact, many preclinical and some clinical data have underscored the mTOR pathway as fundamental in both genetic and acquired epilepsy syndromes [44]. Some genetic disorders strongly support the primary hypothesis that mTOR plays a fundamental role in the pathophysiology of epilepsy. In fact, excessive activation of mTOR signaling as a consequence of loss-of-function mutations of genes encoding for natural mTOR inhibitors such as *TSC1* and *TSC2* coding for the proteins hamartin and tuberin, respectively, phosphatase and tensin homolog (PTEN) and STE20-related kinase adaptor alpha are linked both to the development of cortical malformations and epilepsy. These malformations or "mTORopathies"-related epilepsies include: focal CD, hemimegalencephaly, ganglioglioma, and TSC and are characterized by disorganized cortical lamination, cytomegaly, and intractable epilepsy. The term "mTORopathies" was created to describe neurological disorders characterized by altered cortical architecture, abnormal neuronal morphology, and intractable epilepsy as a consequence of excessive mTOR signaling. Often, patients with focal cortical malformations and intractable seizures require epilepsy surgery [45–47]. At the same time, hyperactivation of the mTOR pathway has also been observed in animal models of acquired epilepsy, such as models of temporal lobe epilepsy (TLE), models of TBI, and models of infantile spasms. Rapamycin and other mTOR inhibitors inhibiting mTOR signaling decrease seizures, delay seizure development, or prevent epileptogenesis in many experimental models of mTOR hyperactivation. To date, the mechanisms by which mTOR inhibition gives rise to the inhibition of seizure activity in several experimental models is unfortunately unclear. Nevertheless, the use of selective mTOR inhibitors can represent an important new therapeutic strategy for managing or eventually preventing epilepsy due to these disorders [48].

8.2 PRECLINICAL EVIDENCE ON THE ROLE OF THE mTOR PATHWAY IN EPILEPTOGENESIS

Many experimental models of genetic and acquired epilepsy, in which mTOR hyperactivation was evidenced, are responsive to mTOR inhibitors [1,41,49]. This evidence supports the hypothesis that dysregulation of the mTOR pathway is a key condition for the development of epileptogenesis and epilepsy. In agreement with this, mTOR inhibitors might be useful to manage some epileptic syndromes and epileptogenesis. In the following sections, all relevant current data in epileptogenesis models relating to mTOR are reviewed (Table 8.1).

8.2.1 TSC Models

Within the genetic epilepsy syndromes, one particular condition has drawn particular attention, since it is strongly linked with dysregulation of the mTOR pathway. TSC is an inherited autosomal disorder resulting from a mutation of one of two tumor suppressor genes: *TSC1* and *TSC2*. The protein products of these two genes, hamartin and tuberin, form a functional protein complex that inhibits the mTOR pathway, which controls cell growth and proliferation. In TSC, benign tumors may develop in multiple organs of the body such as skin, liver, heart, kidney, lung, and the brain, in which it is often associated with the development of subependymal giant cell astrocytoma (SEGA) among other tumors [72]. The majority of children and adolescents affected by TSC show CNS complications such as epilepsy/seizures, cognitive impairment, challenging behavioral problems, and autism-like symptoms causing significant morbidity and mortality. This disease represents one of the most common genetic causes of intractable epilepsy ($\sim 0.01\%$) [73], which is expressed in ~ 80–90% of affected patients. Epilepsy onset occurs during the first few months of life. Different types of seizures are linked with TSC including focal, complex focal, infantile spasms, tonic, clonic, tonic–clonic, atonic, myoclonic, and atypical absences [74,75]. Seizures related to TSC are currently clinically

TABLE 8.1   Preclinical Evidence on the Role of mTOR Inhibitors in Preventing Epileptogenesis

Experimental model	Type of epilepsy	mTOR inhibitor	Dose	Protocol of administration	Response	Hypothesized mechanism(s)	Reference(s)
Amygdala stimulation-induced SE	Acquired epilepsy (TLE)	Rapamycin	6 mg/kg daily	Treatment (2 weeks) after amygdala stimulation	Did not stop the epileptogenic process and no decrease in disease severity	Antiepileptogenic effects of mTOR inhibitors could be restricted to certain animal models, treatment timing or experimental conditions	Sliwa et al. [50]
Electrical stimulation of the angular bundle (SE)	Acquired epilepsy (TLE)	Rapamycin	6 mg/kg/day IP	Treatment started 4 h after the induction of SE	Reduced and in some rats also prevented seizure onset	Decreased both BBB leakage and seizure development in TLE rat model, but it did not reduce the inflammatory response after SE induction	van Vliet et al. [51]
Kainate-induced SE	Acquired epilepsy (TLE)	Rapamycin	6 mg/kg/day IP	Treatment started prior to kainate-induced SE	Suppressed the development of seizures	Blocked cell death, neurogenesis, mossy fiber sprouting, and the development of spontaneous epilepsy	Zeng et al. [42]; Tang et al. [52]
Kainate-induced SE	Acquired epilepsy (TLE)	Rapamycin	6 mg/kg/day IP	Treatment started within 1 h of kainate injection	Enhanced mTOR pathway activation, higher than with kainate alone	Pro-epileptogenic effects	Zeng et al. [53]
Kainate-induced SE	Acquired epilepsy (TLE)	Rapamycin	6 mg/kg/day IP	Treatment started at a longer time period from kainate	Antiepileptogenic effects	Antiepileptogenic effects	Zeng et al. [53]
NS-Pten KO	Genetic epilepsy (cortical dysplasia)	Rapamycin	10 mg/kg IP	Treatment (2 weeks) started at the 4th and 5th weeks of life	Reduced the severity and the duration of the seizure activity, even 3 weeks after cessation of treatment	Hypothesized changes in subcellular structures and, possibly, in processes that are involved in synaptic plasticity and membrane excitability	Ljungberg et al. [54]
NSE-Pten KO	Genetic epilepsy (cortical dysplasia)	Rapamycin	10 mg/kg IP	Treatment started in mice of 5–6 weeks old	Reduced seizures and behavior abnormalities (comorbidity)	Inhibition of anatomical, cellular, and behavioral abnormalities related with mTOR pathway hyperactivation	Zhou et al. [55]
				One treatment per day for five consecutive days per week			
Pilocarpine-induced SE	Acquired epilepsy (TLE)	Rapamycin	3 mg/kg/day IP	Treatment started 24 h after pilocarpine administration for 2 months	Antiepileptogenic effects	Inhibited mossy fiber sprouting. Influenced GABAergic synaptic reorganization	Buckmaster and Wen [56]
Pilocarpine-induced SE	Acquired epilepsy (TLE)	Rapamycin	10 mg/kg IP	Treatment started 24 h later and continued for 2 months	Blocked mossy fiber sprouting. No effects on seizure frequency	Antiepileptogenic effects of mTOR inhibitors could be restricted to certain animal models, treatment timing or experimental conditions	Heng et al. [57]; Buckmaster and Lew [58]; Buckmaster [59]
Pilocarpine-induced SE	Acquired epilepsy (TLE)	Rapamycin	6 mg/kg/day IP	Treatment started before pilocarpine administration	Suppressed seizures	5-HT6 receptor could mediate epilepsy by enhancing of mTOR pathway signaling	Wang et al. [60]

(Continued)

TABLE 8.1 (Continued)

Experimental model	Type of epilepsy	mTOR inhibitor	Dose	Protocol of administration	Response	Hypothesized mechanism(s)	Reference(s)
PTEN KO mice	Genetic epilepsy	Rapamycin	6 mg/kg/day IP	Treatment started 2–5 days post tamoxifen injection	Reduced seizures	Inhibited mossy fiber sprouting	Pun et al. [61]
TBI	Acquired epilepsy	Rapamycin	6 mg/kg/day IP	Treatment started 1 h after injury and continued for 1 month	Prevented the development of post-traumatic epilepsy	Decreased neuronal degeneration and mossy fiber sprouting, although this effect did not directly correlate with inhibition of epileptogenesis	Guo et al. [62]; Berdichevsky et al. [63]
$Tsc1^{GFAP}$CKO mice	Genetic epilepsy (tuberous sclerosis)	Rapamycin	3 mg/kg IP	Treatment started at P14	Prevented the onset of seizures and premature death	Inhibited astrogliosis, neuronal disorganization and increased brain size. Changes in molecular targets of mTOR, which are implicated in membrane excitability	Zeng et al. [64]; Wong et al. [65]
$Tsc1^{null-neuron}$ mice	Genetic epilepsy (tuberous sclerosis)	Rapamycin and everolimus	6 mg/kg IP	Treatment started at P7-P9, every other day, up to 92 days (P100)	Lack of spontaneous seizures during treatment	Improved neurofilament abnormalities, myelination, and cell enlargement. Reduced levels of phospho-S6	Meikle et al. [66]
$Tsc2^{GFAP1}$CKO	Genetic epilepsy (tuberous sclerosis)	Rapamycin	3 mg/kg IP	Treatment started at P14, 5 days/week, up to the end of experiments	Rescued the animals from epilepsy and increased their survival	Inhibited astrogliosis, neuronal disorganization and increased brain size. Reduced levels of phospho-S6	Goto et al. [67]; Zeng et al. [68]
WAG/Rij rat	Genetic epilepsy (absence epilepsy)	Rapamycin	1 mg/kg OS	Treatment started at P45 and continued for 17 weeks	Decreased the development of absence seizures	The antiepileptogenic mechanism(s) in the WAG/Rij rat absence model remain unclear. However, it was suggested that antiepileptogenic and antiepileptic effects could be mediated by inhibition of the release of inflammatory cytokines	Russo et al. [69–71]

BBB, blood–brain barrier; CKO, conditional knockout; GABA, gamma-aminobutyric acid; GFAP, glial fibrillary acid protein; IP, intraperitoneally; OS, orally; KO, knockout; mTOR, mammalian target of rapamycin; NS-Pten KO, neuron subset-specific Pten knockout; NSE, neuron-specific enolase; P, postnatal day; PTEN, phosphatase and tensin homolog; SE, status epilepticus; TBI, traumatic brain injury; TLE, temporal lobe epilepsy; TSC1, tuberous sclerosis complex 1; TSC2, tuberous sclerosis complex 2; WAG/Rij rat, Wistar Albino Glaxo/Rij rat.

treated by AEDs, corticosteroids, vagus nerve stimulators, and ketogenic diet. However, in some cases these treatments are ineffective. In fact, a retrospective study conducted between 2002 and 2008, in patients with TSC and correlated epilepsy, has demonstrated that ~62.5% of these individuals developed refractory seizures [76]. Intermittent calorie-restricted diets seem to induce a set of biochemical and metabolic changes, such as reduced glucose levels, reduced inflammatory markers, increased sirtuins (enzymes linked with lifespan extension in mammals), increased AMP-activated protein kinase (AMPK) signaling, inhibition of mTOR signaling, and an increase in autophagy. Studies in experimental models of epilepsy suggest that these biochemical and metabolic changes might reduce epileptogenesis [77]; interestingly, in agreement, the ketogenic diet was previously found to inhibit the mTOR pathway in rats [78].

Among current AEDs, vigabatrin has been shown to have unique efficacy against seizures in TSC, especially infantile spasms, a typically devastating type of childhood seizure. To date, the exact mechanism by which vigabatrin is effective in TSC remains unclear. However, Zhang et al. [79] demonstrated in a knockout mouse model of TSC: $Tsc1^{flox/flox}$-glial fibrillary acidic protein (GFAP)-Cre knockout ($Tsc1^{GFAP}$CKO), that vigabatrin, at doses of 100 and 200 mg/kg, in addition to its already-proven mechanism of action to increase brain γ-aminobutyric acid (GABA) levels by inhibition of γ-aminobutyrate transaminase [80], partially inhibits the mTOR pathway activity and glial proliferation in vivo, as well as reducing mTOR pathway activation in cultured astrocytes from both knockout and control mice. Therefore, vigabatrin seems to also directly act on the mTOR pathway; however, an indirect action cannot yet be excluded. Furthermore, it was reported that a preventative antiepileptic treatment of infants with TSC (and at high risk of epilepsy), with vigabatrin, but also levetiracetam, valproic acid, and topiramate, markedly improved their risk of developing mental retardation and reduced the incidence of drug-resistant seizures [81,82]. However, also in TSC, it is clear that in the presence of pharmacoresistant epilepsy, surgery still plays an important role, as well as in other common forms of epilepsy [76]. Clinical and preclinical evidence confirms the use of mTOR inhibitors such as rapamycin and everolimus in many TSC-related disease manifestations, in which it was discovered that the mTOR signaling pathway is hyperactivated. Among these TSC-associated manifestations are included: SEGAs, renal angiomyolipoma, skin manifestations, and epilepsy [49]. In particular, the first discovery for a possible role of mTOR in TSC goes back to 2002, when Onda and Crino [83] explained that in a model for human tuberous giant cells, a complete loss of TSC2 expression and hyperactivation of the mTOR pathway was present during cortical development. Thereafter, many studies with TSC1 and TSC2 knockout and knockdown mouse models have supported important information on the TSC-mTOR signaling cascade on the neuronal morphology, axonogenesis, dendritic arborization, regulation of neurotransmitter-receptor expression, neuronal myelination, neuronal autophagy, cortical architecture, and astrocyte proliferation. Moreover, it was also demonstrated that hyperactivation of mTOR signaling could be involved in neuronal hyperexcitability. Animal models of TSC are crucial to study the link between mTOR, TSC, and epilepsy [84,85]. Regarding autophagy, in knockout mice and in human TSC patients, it has been shown that an impairment of this process, after disinhibition of mTOR, is responsible for epileptogenesis [86,87].

Further evidence on the role of mTOR in TSC-associated epilepsy originates from the use, in these models, of mTOR inhibitors. Zeng and Xu [64] were the first to demonstrate the antiepileptogenic effect of mTOR inhibitors (rapamycin; 3 mg/kg) in $Tsc1^{GFAP}$CKO mice, where the $Tsc1$ gene was conditionally inactivated in glial cells. Rapamycin, in $TSC1^{GFAP}$ mice, after an early treatment, was able to prevent both the onset of neurological abnormalities and seizures, which appeared between one and two months of age. In particular, the early treatment (P14), with rapamycin inhibited the progressive astrogliosis and it preserved the organization of hippocampal neurons as well as the levels of astrocyte-specific glutamate transporters, which could be considered a possible mechanism of seizure generation and epileptogenesis. Moreover, it has also been noted that S6 phosphorylation, a marker of mTOR pathway activation, was decreased after early treatment with rapamycin (3 mg/kg for 5 weeks) [64,65]. In a knockout mouse model of TSC in which $Tsc1$ was ablated in most neurons during cortical development, rapamycin, 6 mg/kg intraperitoneally (IP) every other day, and its derivate everolimus, 3 or 6 mg/kg IP every other day (treatment beginning at postnatal day (P) 7–9 every other day, for up to 92 days (P100)), were able to reverse the animal phenotype, rescuing the mutants from epilepsy [66]. Early treatment with rapamycin (3 mg/kg/day IP) in other mouse models of TSC, also rescued the mutants from epilepsy and increased their survival. In particular, these effects have also been observed in $Tsc2^{GFAP1}$CKO mice. In comparison to $Tsc1^{GFAP1}$CKO mice, the $Tsc2^{GFAP1}$CKO mice show higher levels of mTOR activation, which are reversed by mTOR inhibitors [67,68].

Fu and Cawthon [88] hypothesized that defects of GABAergic interneurons are linked both to the onset of epilepsy and autism-like symptoms in TSC patients. To test this hypothesis, they developed a knockout mouse model in which the deletion of the $Tsc1$ gene was generated in GABAergic interneuron progenitor cells. These mice had reduced growth and survival. Moreover, cortical and hippocampal GABAergic interneurons in these animals were enlarged and displayed hyperactivation of the mTOR pathway. The consequences of these molecular changes were linked to a reduced seizure threshold. Furthermore, the authors found that the $Tsc1$ gene plays an important role in GABAergic interneuron development, migration, and function.

8.2.2 CD Models

Human CD (also known as malformation of cortical development) is another recognized type of "mTORopathy" characterized by intractable epilepsy. Epilepsy surgery remains a valuable treatment option to achieve seizure freedom in this condition. Since CD has been linked to mutations of genes encoding for mTOR regulators [89], mTOR inhibitors through their antiepileptogenic mechanisms might be useful for the treatment of CD-related epilepsy [89–92]. Recently, Lee et al. [93] have also observed that the mTOR pathway is deregulated by the up-regulated miRNAs, such as hsa-miR-21 and hsa-miR-155, in CD. It was demonstrated that brain-specific deletion of the mouse homolog PTEN was able to mimic several features of human phakomatosis (a group of rare hereditary neurocutaneous disorders), including cortical and hippocampal disorganization, aberrant mossy fiber sprouting and seizures. The tumor suppressor PTEN is a negative regulator of PI3K (phosphatidylinositol-4,5-bisphosphate 3-kinase) signaling and is needed for maintenance of mammalian neuronal size [94−96]. Dysregulation of the PI3K-Akt-mTOR pathway is responsible for many human cancers, as well as various brain disorders, including macrocephaly, mental retardation, seizures, Lhermitte−Duclos disease (a rare tumor of the cerebellum), and autism [54,97,98]. Furthermore, PTEN deficiency *in vivo* is linked with the excessive growth, migration, and proliferation of dysplastic cells in CD [99].

With the exception of TSC, the molecular basis of several types of CD remains completely unclear. To better understand the role of mTOR in CD and epilepsy, Ljungberg and Sunnen [54] have characterized neuron subset-specific *Pten* knockout (NS-*Pten* KO) mice as an experimental model of CD. These mutant PTEN mice have been previously described as $Pten^{loxP/loxP}$; Gfap-Cre mice [94,95] and exhibit *Pten*-selective inactivation in postnatal granule neurons of the dentate gyrus and cerebellum. This animal model, as well as CD patients, shows neuronal hypertrophy as a consequence of enhanced mTOR activity. Furthermore, these mutant mice express abnormal electroencephalographic activity with spontaneous seizures that are evident at 4−6 weeks of age. Short-term treatment (2 weeks) with rapamycin (10 mg/kg IP five times per week), when the epileptiform activity and other related pathological disorders are not completely developed, suppresses the severity of seizures [54]. In additional work, a PTEN mutant model was created in which *Pten* and *Tsc1* deletion was directed to a limited number of postnatal, post-mitotic hippocampal and cortical neurons. Rapamycin (1 mg/kg IP) in these young adult (5−6 weeks old) mice was able to reduce both seizure frequency and duration. Rapamycin also ameliorated the anxiety-like behavior of the mice in this model. It is well known that epilepsy is often associated with psychiatric comorbidity, such as depression and anxiety [100]. This latter point represents an unmet need in epilepsy and particular attention should be paid on the mTOR pathway considering its involvement also in major depression. Likewise, the mTOR inhibitors in some studies also ameliorated the major structural disorders of these animal models such as macrocephaly and neuronal hypertrophy [55,69,101].

8.2.3 The WAG/Rij Rat Model of Absence Epilepsy

Russo et al. [70] have recently demonstrated the involvement of the mTOR pathway in another animal model of genetic epilepsy. In fact, the WAG/Rij rat represents a well-validated animal model of *absence*-type (non-convulsive) epilepsy, epileptogenesis, and mild-depression comorbidity [102−104]. In particular, the authors investigated the effects of rapamycin after chronic, subchronic, and acute administrations in the WAG/Rij rat model. In addition, considering the now-recognized role of proneuroinflammatory cytokines in the development of epilepsy [105], the link between antiepileptic effects of rapamycin and neuroinflammation was also studied in parallel. To this purpose, the proinflammatory bacterial endotoxin lipopolysaccharide (LPS) was administered before rapamycin. In fact, according to Kovacs et al. [106], LPS seems to worsen seizures through generation of abnormal neuroinflammatory responses in WAG/Rij rats. Russo et al. [70] found that the chronic oral treatment with rapamycin (1 mg/kg/day; OS) in WAG/Rij rats started at P45 (prior to seizure development), and continued for 17 weeks, was able to decrease the development of absence seizures. Moreover, this chronic rapamycin treatment also showed prodepressant effects in the WAG/Rij rats and also in Wistar rats (non-epileptic controls) at the age of 6 months, whereas at the age of 10 months, no differences were noted.

The antiepileptogenic effects of rapamycin, together with other findings, have confirmed the involvement of the mTOR pathway in both the epileptogenic and depressive-like processes in the WAG/Rij model. In agreement with this hypothesis, it was also established that WAG/Rij rats, in comparison to Wistar rats, have higher levels of total mTOR in several brain areas, including the cortex, hippocampus, and thalamus. Moreover, WAG/Rij rats, in comparison to Wistar rats, also showed an age-related decline in neurogenesis as a consequence of the higher levels of mTOR activation [70,71,107]. In an additional study by the same research group, it was reported that in the early phases of the inflammatory response induced by LPS (3 μg/rat in 5 μl saline), (following Toll-like receptor 4 [TLR4] stimulation and activation of innate immunity), there is an interaction between AMPK and AKT through the mTOR pathway. Such mechanisms could be responsible for LPS proepileptic and prodepressant effects in the WAG/Rij rat model. Rapamycin treatment, (0.5 mg/kg IP) followed 30 min later by LPS infusion in the WAG/Rij rat brain, regulated the LPS-induced neuroinflammatory processes and the authors suggest that this action could be nuclear factor-kappa B (NF-κB)-dependent. Moreover, it was proposed that both the antiepileptic and antiepileptogenic effects of mTOR inhibitors could be partly mediated by the inhibition/delay in the release of CNS proinflammatory cytokines [69].

8.2.4 TLE Models

Pun et al. [61], through the use of PTEN KO mice, observed the involvement of the mTOR pathway in giving rise to abnormal hippocampal circuits, which are a critical step in the development of human TLE. In fact, experimental PTEN ablation in vivo mimics several pathological markers of TLE, including neuronal hypertrophy, increased dendritic spine density, and spontaneous seizures. During epileptogenesis, adult-generated dentate granule cells (DGCs) form aberrant neuronal connections with neighboring DGCs increasing neuronal excitability in the hippocampus. In particular, Sutula and Dudek [108] detected that mossy fiber sprouting occurred when granule cell axons arising in the inner molecular layer of the dentate gyrus formed recurrent excitatory connections, which potentially enhanced susceptibility to seizures. The authors' conclusion was that PTEN deletion among hippocampal granule cells was sufficient to develop spontaneous seizures in a few weeks, and that mTOR signaling played a fundamental role in this process. Therefore, hyperactivation of the mTOR pathway as a result of PTEN deletion is a possible mechanism of epileptogenesis also in TLE. Moreover, rapamycin administration (6 mg/kg/day IP, 5 days/week) was effective in inhibiting epileptogenesis and the presence of abnormal granule cells in the PTEN animal model [61]. Accordingly, it was demonstrated that mTOR inhibitors rescued fiber sprouting by promoting the survival of the somatostatin/green fluorescent protein (GFP)-positive interneurons after pilocarpine-induced *status epilepticus* (SE) in mice. In particular, epileptic mice treated with vehicle had an increased GFP-positive axon length per dentate gyrus, as a consequence of GABAergic axon sprouting. On the contrary, epileptic mice treated with rapamycin (3 mg/kg/day IP, starting 24 h after pilocarpine treatment for 2 months) and control mice had similar axon length. Therefore, according to the authors, GABAergic synaptic reorganization after epileptogenic treatments could be achieved by treatment with mTOR inhibitors [56].

These and others studies support the role of mossy fiber sprouting in the epileptogenic process. Some of them report that rapamycin is also able to decrease both epileptiform activity and mossy fiber sprouting in mouse models of TLE, such as pilocarpine and kainate-post SE spontaneous seizures. Zeng et al. [42] demonstrated that the mTOR pathway is markedly enhanced, in a biphasic manner, after kainate-induced SE. In particular, the first enhancement of mTOR activation was observed in both hippocampus and cortex, a few hours after seizure activity, whereas the second enhancement appeared several days later, but only in the hippocampus. The exact mechanism by which kainate induces this mTOR enhancement is unclear, above all for the second peak of mTOR activation. However, it was hypothesized that excessive release of glutamate could be involved in this process. Furthermore, rapamycin (6 mg/kg/day IP) administered prior to kainate-induced SE, blocked cell death, neurogenesis, mossy fiber sprouting, and the development of spontaneous epilepsy in this mouse model of TLE. In contrast, when rapamycin (6 mg/kg/day IP) was administered 24 h *after* kainate, it inhibited mossy fiber sprouting and reduced seizure frequency, but less efficaciously than with pretreatment. Similar results were obtained in the pilocarpine-induced SE model. Thus, the authors claimed that the mTOR pathway mediates mechanisms of epileptogenesis in kainate and pilocarpine rat models and rapamycin could have antiepileptogenic effects in these models [42,52,108]. However, paradoxical effects of mTOR inhibition have also been reported. In fact, mTOR activation can have both proapoptotic and antiapoptotic effects, depending on different phases of the cell cycle [2]. Zeng et al. [53] observed that rapamycin administration (6 mg/kg IP) within 1 h of kainate injection in rats induced *enhancement* of the mTOR pathway, higher than with kainate alone, whereas when rapamycin was administered at a longer time period from kainate injection, the expected inhibition of the mTOR signaling cascade was observed. Therefore, mTOR would seem to act as a master switch that regulates, under different situations, neuronal death and epileptogenesis.

More recently, it was demonstrated that the role of mossy fiber sprouting in TLE is controversial and not completely understood. In mice treated with pilocarpine to induce SE, a high dose of rapamycin (10 mg/kg/day IP) was administered, beginning 24 h later and continuing for 2 months, to block mossy fiber sprouting and to investigate seizure frequency. In comparison to the vehicle, rapamycin completely blocked mossy fiber sprouting, whereas it did not affect seizure frequency. Therefore, mossy fiber sprouting would not appear to be necessary for epileptogenesis in this model of TLE. Similarly, it was reported that post-treatment with rapamycin (6 mg/kg daily IP for 2 weeks) after amygdala electrical stimulation-induced SE, did not stop the epileptogenic process and did not decrease disease severity. These data suggest that the antiepileptogenic effects of mTOR inhibition could be restricted to certain animal models, treatment timing, or experimental conditions [50,57].

In an interesting recent study, van Vliet et al. [51] treated rats daily with rapamycin for 7 days

(6 mg/kg/day, IP) starting 4 h after the induction of SE evoked by electrical stimulation of the angular bundle, and reported that the pharmacological properties of mTOR inhibitors were not apparently a consequence of the suppression of brain inflammation, as indicated by normal microglial activation in comparison to a vehicle group. In particular, the authors found that chronic treatment with rapamycin was able to reduce, and in some rats prevent, seizure onset (see above). Likewise, rapamycin reduced other potential features related to epilepsy and epileptogenesis, such as mossy fiber sprouting, neuronal loss, and blood–brain barrier (BBB) leakage. At odds with this, the authors did not find a change in microglial activation and inflammation as a consequence of treatment. The effects after a single treatment with rapamycin on inflammation were studied through changes in plasma levels of IL-1β, IL-6, and C-reactive protein. Therefore, blood samples were collected 6 weeks after SE at various intervals. In particular, the plasma levels of these inflammatory proteins did not differ between control, vehicle, and rapamycin-treated rats. Moreover, it was also reported by immunostaining procedures (CD11b/c and CD68 macrophage marker staining), that rapamycin did not contain monocyte and microglial activation in the brain. Accordingly, rapamycin did not reduce the number of reactive astrocytes. Finally, the relationship between seizure incidence, microglial activation, and BBB permeability (fluorescein permeability index) was also examined, but no correlation was found between seizure incidence and microglial activation, whereas a correlation did exist between seizures and BBB permeability. In fact, rapamycin treatment decreased both BBB leakage and seizure development in this TLE rat model, but it did not reduce the inflammatory response after SE induction. According to the authors, their study does not support the role of neuroinflammation in the epileptogenic processes, whereas preventing BBB leakage could have an important role in the antiepileptogenic effect of rapamycin [51]. A major limitation on these conclusions is the measure of inflammatory parameters in the blood, where they appeared not to be altered by SE itself. Therefore, no possible effects of rapamycin on circulating cytokines could have been measured.

8.2.5 Other Models of Epileptogenesis

TBI is a major cause of death, mental diseases, and disability. Among the consequences, TBI post-traumatic epilepsy is very common and is a cause of significant morbidity and mortality in TBI patients [109]. Studies have reported that mTOR inhibitors might have antiepileptogenic effects in the development of post-traumatic epilepsy in an animal model of TBI. Rapamycin was tested in an experimental model of controlled cortical impact (CCI) injury, a well-validated model of TBI, in which it has also been proven that an aberrant activation of mTORC1 occurs. Rapamycin administration (6 mg/kg/day IP), started 1 h after CCI and continued for 1 month, reduced mTORC1 hyperactivation and decreased neurodegeneration. Furthermore, quite importantly, it also prevented the development of post-traumatic epilepsy [62].

mTOR involvement in TBI has also been demonstrated in a rat hippocampal organotypic culture model of post-traumatic epilepsy. Ictal activity was measured both by lactate production and by multiple electrode array recordings. Moreover, cell death was identified by lactate dehydrogenase release measurements and Nissl staining. The conclusions of the study were mTOR activation in post-traumatic epileptogenesis is strongly PI3K-dependent and dual inhibitors of PI3K and mTOR (e.g., NVP-BEZ235) could be more effective in comparison to rapamycin alone to prevent epilepsy development. Moreover, in this model, rapamycin showed antiepileptogenic properties even after a transient inhibition of mTOR. In fact, an early termination of treatment did not increase epileptic activity or cell death [63].

Wang et al. [60] studied the role of 5-HT6 serotonin receptors in epilepsy, demonstrating that these receptors are overexpressed in samples of human brain tissue obtained from patients with intractable TLE and in a pilocarpine TLE rat model. Moreover, a hyperactivation of mTOR signaling was also found in the TLE model. According to the authors, there was a link between mTOR and 5-HT6 receptor expression, which could be responsible for the development of seizures. In particular, 5-HT6 receptor activation could mediate epilepsy by enhancing the mTOR pathway. In fact, pretreatment of rats with SB-399885 (10 mg/kg IP), an antagonist of 5-HT6 receptors, before pilocarpine administration, suppressed seizures and mTOR activation. At the same time, rapamycin pretreatment (6 mg/kg IP) also suppressed seizures. They concluded that the 5-HT6/mTOR pathway could be a promising potential therapeutic target to treat epilepsy.

8.3 PRECLINICAL EVIDENCE ON THE ROLE OF mTOR IN SEIZURE GENERATION AND THE EFFECTIVENESS OF mTOR INHIBITORS

The evidence reported above and many other reports have clearly documented the role of mTOR signaling in epileptogenesis as well as the potential therapeutic antiepileptogenic effects of mTOR inhibitors. However, studies have also investigated the potential *antiseizure* role of mTOR inhibitors. To date, the majority of antiseizure drugs act on the neuronal excitability controlling both receptors and ion channels. However, in comparison to the commonly used antiseizure drugs, mTOR inhibitors, at least directly, do not influence neuronal excitability. In fact, in vitro studies have not revealed any direct effect of rapamycin on neuronal bioelectrical activity [41,110,111]. Nevertheless, the mTOR pathway can influence neuronal excitability indirectly through mechanisms controlling synaptic structure and plasticity. In fact, Ras-PI3K-Akt-mTOR and Ras-MAPK signaling pathways play an important role in the regulation of dendrite arborization and spine formation, which are critical for the functioning of neurons and neuronal networks. Moreover, synaptic plasticity may influence neuronal excitability and might be related to mechanisms of epileptogenesis [14,15,61]. Furthermore, the mTOR pathway probably also affects neuronal excitability by modulating the expression of ion channels and receptors [17,65,112,113].

To test the antiseizure properties of mTOR inhibitors, both chronic animal models and acute seizure models have been used (Table 8.2). Hartman et al. [116], in NIH Swiss mice, screened rapamycin effects in acute seizure tests similar to those currently used to screen the therapeutic potential of future AEDs. In this study, both the short-term and the long-term effects of rapamycin (4.5 mg/kg IP) were evaluated. In particular, in the short-term treatment, a single dose of rapamycin protected mice only against tonic hindlimb extension in the maximal electroshock seizure threshold test, whereas in the 6-Hz stimulation model and in the pentylenetetrazole (PTZ)-induced seizure model, it had no effect. Regarding kainate-induced seizures, the short-term treatment with this mTOR inhibitor showed mixed effects. In fact, no differences were reported in overall seizure scores or maximum seizure scores, whereas it was reported that rapamycin shortened the time to generate two seizures, but without influencing seizure activity over the first 2 h of exposure.

Long-term repetitive treatment with rapamycin (three daily doses) protected solely against kainate-induced seizures, but only at late times after seizure onset. According to several studies, the potential antiseizure mechanisms of mTOR inhibitors could also be mediated by reduced neuronal excitability and/or neurotransmitter release. Furthermore, mTOR is necessary for long-term hippocampal synaptic plasticity [116,121].

Similarly, Chachua et al. [115] in several acute seizure models and treatment paradigms compared rapamycin effects on immature (P15) and young adult rats (P55-60). Rapamycin was administered at doses of 3 mg/kg IP in P15 rats and 3 or 6 mg/kg in P55-60 rats. In P15 rats, the higher dose was not used, because rapamycin at 3 mg/kg markedly influenced the body weight. In particular, a single dose of rapamycin had anticonvulsant effects in the PTZ-model only in immature rats but did not have effects in adult rats. Long-term pretreatment had a *pro*-convulsant effect on KA-induced seizures. Regarding the flurothyl seizure model, only a 4-hour pretreatment was anticonvulsant both in adult and immature rats. In immature rats, short-term pretreatment did not show effects against N-methyl-D-aspartic acid (NMDA) or kainate-induced seizures. Moreover, the lack of rapamycin effects, above all in immature rats, was correlated with decreased neuropeptide Y (NPY) expression in the cortex and hippocampus, suggesting a possible link between mTOR signaling and the NPY system. In these studies, rapamycin showed variable efficacy on acute seizures, which were age-, time-, treatment paradigm-, and model-dependent. In conclusion from their results, the authors suggested that rapamycin was a poor anticonvulsant that may have beneficial effects only against epileptogenesis.

Another study has demonstrated that the antiseizure role of mTOR inhibitors is still poorly understood and controversial. In particular, in immature and adult Sprague-Dawley rats, pretreated with rapamycin (5 mg/kg/day IP for 1—3 consecutive days) prior to induction of seizures by PTZ (35—60 mg/kg, IP), pilocarpine (255—300 mg/kg, IP), or kainate (10—20 mg/kg, IP), it did not have antiseizure activity in adult rats. However, in immature rats, rapamycin decreased the minimal dose of PTZ necessary to induce seizures being, therefore, *pro*-convulsant. Moreover, in this study, it was found that rapamycin treatment downregulated expression of the potassium chloride cotransporter (KCC2) both in the thalamus and hippocampus. This down-regulation could, by inducing a depolarizing shift of the chloride equilibrium potential, increase susceptibility to pilocarpine-induced seizures in immature rats [112,115,116]. In fact, it was reported that the levels of KCC2 mRNA encoding the cotransporter were inversely related with neuronal excitability [122].

Further preclinical studies on chronic epilepsy models (genetic and acquired epilepsy) have demonstrated that mTOR inhibitors can effectively manage seizures dependent on mTOR pathway dysregulation [41]. In $Tsc1^{GFAP}$CKO mice with conditional inactivation of the $Tsc1$ gene primarily in glia, the development of seizures typically starts between 1 and 2 months of age, and a late treatment with rapamycin (3 mg/kg IP) after onset of neurological abnormalities was performed.

TABLE 8.2 Preclinical Evidence on the Role of mTOR Inhibitors in Preventing Seizures

Experimental model	Type of epilepsy	mTOR inhibitor	Dose	Protocol of administration	Response	Hypothesized mechanism(s)	Reference(s)
Acute biallelic deletion of *Tsc1* in adult mice.	Genetic epilepsy	Rapamycin	5 and 10 mg/kg/day IP	Treatment started after seizure onset	5 mg/kg only abolished seizures 10 mg/kg abolished seizures and prolonged median survival	Not to work as a classical anticonvulsant. Effectively reduced pS6 levels	Abs et al. [114]
Acute seizure models in Sprague-Dawley rats	Acquired epilepsy	Rapamycin	3 or 6 mg/kg IP	Used different treatment paradigm in P15 (immature) and P55-60 (young adult) rats	Shown variable efficacy on acute seizures, which are age-, time-, treatment paradigm-, and model-dependent	The lack of effects, above all in immature rats, has been correlated with decreased NPY expression in the cortex and hippocampus	Chachua et al. [115]
Acute seizure models in Sprague-Dawley rats	Acquired epilepsy	Rapamycin	5 mg/kg IP	Treatment started, in immature and mature rats, prior to induction of seizures by PTZ, pilocarpine, or kainate	Shown variable efficacy on acute seizures, which are age-, time-, treatment paradigm-, and model-dependent	The anticonvulsant role is poorly understood and controversial. However, it was reported that the treatment down-regulates KCC2 expression in CNS, which could increase susceptibility to pilocarpine-induced seizures in immature rats	Huang et al. [112]
Acute seizure tests in NIH Swiss mice	Acquired epilepsy	Rapamycin	4.5 mg/kg IP	Short-term treatment (single dose) 3 h before seizure onset and long-term treatment (three daily doses) before seizure onset	Showed variable efficacy on acute seizures, which were age-, time-, treatment paradigm-, and model-dependent	The link between mTOR activity and excessive neuronal activity during seizures is not clear. However, it has been hypothesized that a reduction in neuronal excitability and/or neurotransmitter release may occur with rapamycin	Hartman et al. [116]
Multiple-hit rat model of infantile spasms	Epileptic encephalopathy	Rapamycin	Different doses	Different treatment started 24 h after	Reduced spasms in a dose-related way. No effects on other seizures type	Unclear	Raffo et al. [117]
NS-*Pten* KO (cortical dysplasia)	Genetic epilepsy	Rapamycin	10 mg/kg IP	Intermittent treatments (over a period of 5 months)	Epileptiform activity remained suppressed and survival increased after additional intermittent treatments	The exactly mechanism by which mTOR inhibitors suppress seizures is still unclear. However, hypothesized changes in subcellular structures and, possibly, in processes that are involved in synaptic plasticity and membrane excitability may occur	Ljungberg et al. [54]; Sunnen et al. [118]
NS-*Pten* KO mice (cortical dysplasia)	Genetic epilepsy	Rapamycin	10 mg/kg/day IP	Treatment started at P9	Suppressed both mTOR hyperactivation and seizures, even after they were established	Attenuated epileptiform activity and reduced astrogliosis and microgliosis in this model	Nguyen et al. [119]

(Continued)

II. mTOR IN GENETIC DISORDERS AND NEURODEGENERATIVE DISEASE

TABLE 8.2 (Continued)

Experimental model	Type of epilepsy	mTOR inhibitor	Dose	Protocol of administration	Response	Hypothesized mechanism(s)	Reference(s)
PTEN mutant model	Genetic epilepsy	Temsirolimus	7.5 mg/kg	Treatment started from 6 to 16 weeks, when mutant mice were symptomatic	Decreased seizures and mortality as well as with rapamycin	Unclear	Kwon et al. [99]
Rat model of cryptogenic infantile spasms	Epileptic encephalopathy	Rapamycin	3 mg/kg IP	Pretreatment	No effects in this model		Chachua et al. [120]
$Tsc1^{GFAP}$CKO mice (tuberous sclerosis)	Genetic epilepsy	Rapamycin	3 mg/Kg IP	Late treatment started at 6 weeks of age after onset of neurological abnormalities	Decreased seizures dramatically, even though to a lesser degree in comparison to early treatment seizures	Inhibited astrogliosis, neuronal disorganization and increased brain size. Changes in molecular targets of mTOR, which are implicated in membrane excitability	Zeng et al. [64]
WAG/Rij rat (absence epilepsy)	Genetic epilepsy	Rapamycin	0.5, 1, and 3 mg/kg IP and 0.1–3 mg/kg OS	Acute treatment and subchronic (7-day) treatment after seizures onset	Decreased the development of absence seizures	The antiepileptogenic mechanism(s) in WAG/Rij rat absence model remain unclear. However, it was suggested that antiepileptogenic and antiepileptic effects could be mediated by inhibition of the release of inflammatory cytokines	Russo et al. [69–71]

CKO, conditional knockout; CNS, central nervous system; GFAP, glial fibrillary acid protein; IP, intraperitoneally; OS, orally; KCC2, potassium chloride cotransporter 2; KO, knockout; mTOR, mammalian target of rapamycin; NPY, neuropeptide Y; NS-Pten KO, neuron subset-specific Pten knockout; P, postnatal day; pS6, phospho-S6; PTEN, phosphatase and tensin homolog; PTZ, pentylenetetrazole; TSC 1, tuberous sclerosis complex 1; TSC 2, tuberous sclerosis complex 2; WAG/Rij rat = Wistar Albino Glaxo/Rij rat.

Rapamycin treatment dramatically decreased seizures after a few days of treatment, even though to a lesser degree in comparison to early treatment, Moreover, after 3 weeks of treatment, a significant number of mice became seizure-free. Therefore, rapamycin was effective in preventing seizures and prolonging survival in this model [64].

In order to emphasize the role of the mTORC1 pathway in epilepsy development, an interesting study was conducted by Abs and Goorden [114] after biallelic *Tsc1* gene deletion in adult *Tsc1* heterozygous and wild-type mice. Biallelic *Tsc1* gene deletion led, both in knockout and wild-type mice, to the development of seizures as a consequence of mTORC1 hyperactivation, enhanced neuronal excitability of CA1 pyramidal cells, and reduced threshold for protein-synthesis-dependent LTP in the Shaffer collaterals. Both processes may contribute to epileptic discharges. Treatment with rapamycin (5 mg/kg IP per day) after

seizure onset, in both animal groups, decreased seizures, but did not prolong the median survival. On the contrary, rapamycin treatment (10 mg/kg IP per day) reduced both seizure frequency and increased survival. Furthermore, the authors also measured rapamycin blood levels in mice. In particular, rapamycin blood levels were higher than in the brain; however, the clearance of rapamycin from blood was faster than in the brain. According to the authors, this observation could explain why intermittent treatment in some experimental models, such as the multiple-hit model, decreased seizures [114]. In conclusion, this result suggests that mTOR could be a promising target for managing seizures not only in the presence of major brain pathology, but also in other types of epilepsies that result from increased mTOR hyperactivation.

Nguyen et al. [119] observed that rapamycin treatment started at postnatal week 9 (10 mg/kg IP 5 days/week) in NS-*Pten* KO mice, a validated model of CD, was able to suppress both mTOR hyperactivation and seizures, even after they were established (4–6 weeks). Furthermore, rapamycin also reduced astrogliosis and microgliosis. In agreement with other studies, the authors proposed that the mTOR signaling pathway plays a critical role also in the maintenance of epilepsy. Moreover, the authors suggested that mTOR inhibitors could be candidates to treat CD and correlated epilepsy [119].

In NS-*Pten* KO mice, it was demonstrated that a single treatment with rapamycin (10 mg/kg IP) was able to reduce both epileptiform activity and mossy fiber sprouting for several weeks before epilepsy recurred. On the contrary, epileptiform activity remained suppressed and survival increased after additional intermittent treatments (over a period of 5 months) with rapamycin, whereas without mTOR inhibition, epilepsy worsened and survival decreased [54,118].

Russo et al. [70], in WAG/Rij rats, observed antiepileptogenic effects (see Section 6.2.3) as well as antiepileptic effects after rapamycin treatment. In particular, it was reported that both the acute treatment (0.5, 1, and 3 mg/kg) and subchronic (7-day) treatment (0.1–3 mg/kg) with rapamycin, respectively, IP and orally administered, surprisingly prevented the LPS-induced increase in absence seizures. Moreover, the subchronic treatment produced contrasting antidepressant properties between WAG/Rij rats and Wistar rats (used as control), whereas acute treatment did not show significant differences on immobility time in both strains.

In the PTEN mutant mouse model, treatment with rapamycin ester (named temsirolimus; 7.5 mg/kg) started from 6 to 16 weeks, when mice were symptomatic, decreased seizures and mortality as well as with rapamycin [99]. Similarly, late treatment with rapamycin was still capable of reducing both the epileptic activity and cell death in the organotypic hippocampal culture model of post-traumatic epilepsy [63] (see Section 8.2.4).

Infantile spasms (West syndrome (WS); epileptic spasms), is a devastating and common epileptic encephalopathy of infancy. WS occurs between 4 and 6 months of age and consists of a triad of cluster epileptic spasms, hypsarrhythmic electroencephalography (EEG) (a disorganized and asynchronous high-amplitude background with superimposed spikes and slow waves) and mental retardation. The pharmacotherapies of WS include adrenocorticotropic hormone (ACTH), corticosteroids, and/or vigabatrin. In a rat model of cryptogenic infantile spasms, in which seizures were triggered with NMDA between P10 and 15, a pretreatment with vigabatrin (250 mg/kg IP) decreased the spasms, whereas rapamycin (3 mg/kg

IP) did not show protective effects in this model. Moreover, it was also observed that both ACTH (0.3 mg/kg subcutaneously) after chronic treatment and methylprednisolone (60 mg/kg IPi) after chronic pretreatment, decreased spasms [120].

However, it was reported (see above) that hyperactivation of the mTOR signaling cascade is implicated in the pathogenesis of disorders correlated with WS. The therapeutic properties of rapamycin, for infantile spasms, were tested in the multiple-hit rat model of ACTH-refractory symptomatic infantile spasms. Rapamycin, administered at different doses and protocols, after the onset of spasms, was able to reduce spasms in a dose-related way. Furthermore, the drug also improved visuospatial learning in the barn maze. However, rapamycin did not have effects on other seizure types such as myoclonic seizures, body jumps, or tonic seizures. According to the authors, these different effects on spasms and other epileptic events could be linked to different pathogenetic mechanisms. Nevertheless, this study provides some evidence that rapamycin could be an effective new option for the treatment of infantile spasms [117].

8.4 CLINICAL EVIDENCE FOR THE EFFECTIVENESS OF mTOR INHIBITORS IN TSC-RELATED EPILEPSY

Epilepsy represents one of the oldest neurological disorders. Nowadays, there is a compelling need to discover new pharmacological therapies able to treat this serious CNS disorder. In particular, the major goal of research will be to identify disease-modifying drugs, which could modulate the etiopathogenetic mechanisms underlying the disease. To date, various preclinical evidence (see Section 8.3) and available clinical studies, particularly in genetic epilepsy, suggest that mTOR represents one of the most promising therapeutic options both to treat and to prevent epilepsy [40,48,123]. Nowadays, the importance of mTOR inhibitors in the management of epilepsy is quickly growing and in most epilepsy congresses such the Twelfth Eilat Conference on New Antiepileptic Drugs (AEDs)-EILAT XII, sessions were dedicated to the use everolimus in epilepsy pharmacotherapy [124] (Table 8.3).

Everolimus, a rapamycin analog, is a selective inhibitor of the mTOR pathway. Similarly to rapamycin, everolimus acts specifically on the mTORC1 complex [134]. The US Food and Drug Administration approved everolimus for the treatment of patients with TSC associated with inoperable SEGAs and renal angiomyolipomas [131,135,136]. However, it was also observed that everolimus could be a disease-modifying drug, effective in treating other

TABLE 8.3	Clinical Evidence for mTOR Inhibitors in Preventing TSC-Related Epilepsy

Disease	Study type	Drug	Dose/duration of treatment	Patient(s)	Outcome	Reference(s)
SEGAs associated with TSC	EXIST-1	Everolimus	At the moment of report, the children were still on treatment	Eight children under 3 years of age	One patient showed cessation of seizures and two other patients reported a significant reduction in the number of seizures	Kotulska et al. [125]
Sturge–Weber syndrome-associated epilepsy	Open-label, phase 2, controlled study	Everolimus	Treatment started at a dosage of 5 mg/m^2/day, dosed once per day in the morning	Male or female between the ages of 2 and 18 years	This study is currently recruiting participants	ClinicalTrials.gov [126]
TSC and refractory epilepsy	Case report	Everolimus	4.5 mg/m^2/day for up to 12 months	Young boy (10 years old)	Completely abolished epileptic seizures	Perek-Polnik et al. [127]
TSC and refractory epilepsy	Case report	Rapamycin	Treatment started at 0.05 mg/kg/day and then it was increased to 0.15 mg/kg/day for up to 10 months	Young girl (10 years old)	Reduced seizures. No effects on cortical tubers	Muncy et al. [128]
TSC and refractory epilepsy	Case report	Everolimus	The dosage was gradually augmented to reach a plateau serum concentration of 5–10 ng/ml. The observation period was continued for 9 months	Four girls and three boys, mean age 5 years	Two of these patients did not respond to everolimus, whereas in 4 patients there was a reduction in seizure frequency. One of these patients stopped treatment	Wiegand et al. [129]
TSC and refractory epilepsy	Open label	Rapamycin	Treatment was a 1–5 mg daily dose. The median duration of treatment was 18 months	Seven patients, median age 6 years	In one of these patients, a 90% reduction of seizure frequency was observed, whereas in four patients the reduction achieved was between 50 and 90%. The last two patients had a reduction <50%. The treatment also improved facial angiofibromas and cognition	Cardamone et al. [130]
TSC and refractory epilepsy	Prospective, multicenter, open-label, phase 1/2 clinical trial	Everolimus	The dosage was gradually augmented to reach a plateau serum concentration of 5–15 ng/ml. Treatment was for 12 weeks	Patients ≥ 2 years of age	The majority of these patients showed an improvement in seizure control	Franz et al. [131]; Krueger et al. [132]
TSC and related epilepsy	Prospective open-label phase 1–2 study	Everolimus	The dosage was gradually augmented to reach a plateau serum concentration of 5–15 ng/ml. Treatment was for 6 months	16 patients	Nine patients showed a decrease in seizure frequency, 6 did not show significant reduction, whereas in 1 patient, there was an *increased* seizure frequency	Krueger et al. [74]
TSC-associated refractory seizures	A three-arm, randomized, double-blind, placebo-controlled study	Everolimus	Everolimus titrated from 3 to 7 ng/ml and also from 9 to 15 ng/ml and placebo	Male or female between the ages of 2 and 65 years	This study is currently recruiting participants	Novartis Pharmaceuticals [133]

EXIST-1, EXamining everolimus In a Study of TSC; SEGA, subependymal giant cell astrocytoma; TSC, tuberous sclerosis complex.

characteristics of TSC, such as epilepsy. In fact, in a young boy (10 years old) affected by TSC and pharmacoresistant epilepsy, where tumor resection was excluded, everolimus administration (4.5 mg/m^2/day for 12 months) completely stopped epileptic seizures. On the contrary, during the treatment with conventional AEDs: oxcarbazepine (38 mg/kg/day), topiramate (10 mg/kg/day), and levetiracetam (50 mg/kg/day), focal epilepsy was reported. Dramatic seizure reduction was also achieved with rapamycin administration for 10 months, in a 10-year-old girl with TSC and refractory epilepsy. The starting dose was 0.05 mg/kg/day, which was then increased to 0.15 mg/kg/day. This treatment did not, however, have any effects on cortical tubers [127,128].

The efficacy of everolimus was also studied in four girls and three boys (mean age 5 years), with refractory epilepsy and TSC. The period of observation lasted for 9 months, whereas the everolimus dosage was gradually augmented to reach a plateau serum concentration of 5–10 ng/ml. It was reported that the serum concentration fluctuated during the treatment, and required frequent changes in everolimus dosage. Only two of these patients failed to respond to everolimus, whereas in four patients, there was a reduction in seizure frequency. One of these patients stopped treatment because of side effects (rash) induced by the drug [129].

Recently, an open-label case series in seven patients (median age 6 years), with TSC and refractory epilepsy, described the efficacy of rapamycin (1–5 mg daily dose), which was reported to have only minimal adverse effects. In one of these patients, a 90% reduction of seizure frequency was observed, whereas in four patients, the reduction achieved was between 50 and 90%. The last two patients had a reduction of less than 50%. The median duration of treatment with rapamycin was 18 months (6–36 months). Moreover, this treatment was reported to improve other characteristics of TSC, such as facial angiofibromas and cognition [130].

Krueger et al. [74], in a prospective open-label phase 1–2 study designed to assess the efficacy of everolimus in patients with SEGA (associated with TSC) (primary end points), reported a reduction in seizure frequency (secondary end points), after 6 months of treatment with everolimus, which was administered orally at a dose of 3.0 mg/m^2 of body surface and successively adjusted to obtain a blood concentration of 5–15 ng/ml. In particular, of 16 patients with TSC, nine showed a decrease in seizure frequency, six did not show significant reduction, whereas in one there was an *increased* seizure frequency. The effect of everolimus, after 12 weeks of treatment in the management of pharmacoresistant epilepsy in patients ≥2 years affected by SEGAs, was also investigated in a prospective, multicenter, open-label, phase 1/2 clinical trial. The majority of these patients showed an improvement in seizure control and in quality of life following treatment with everolimus, which was titrated to a serum concentration of 5–15 ng/ml [132]. In agreement with everolimus' efficacy in patients with SEGAs, a multicenter randomized placebo-controlled phase 3 trial was performed with the aim of supporting the use of everolimus in patients with SEGAs and TSC-related epilepsy. Everolimus was titrated to reach serum levels of 5–15 ng/ml in patients where the median age was 9.5 years, whereas the median duration of treatment was 41.9 weeks. In particular, no patients discontinued everolimus treatment due to adverse events. Therefore, these findings strongly suggest everolimus as an effective treatment for SEGA-related epilepsy. Moreover, everolimus could be a disease-modifying treatment for other characteristics of TS [125,131].

Domanska-Pakiela et al. [137] also reported, in a prospective study of five neonatal patients with TSC, that the epilepsy onset was preceded, 1–8 days before, by epileptiform discharges. Therefore, EEG recordings could have predictive value during infancy in patients with TSC. Still in progress, some clinical trials including a multicenter placebo-controlled study in patients with TSC. The aim of this study will be to evaluate the efficacy and safety of everolimus in TSC patients with refractory partial-onset seizures [75].

8.5 CONCLUSIONS AND PERSPECTIVES

Epilepsy represents one of the oldest and most prevalent neurological disorders. It is also one of the most challenging areas for clinicians aiming to achieve an effective long-term therapeutic management of epilepsy patients. To date, the mTOR pathway represents, perhaps, the most promising molecular target for producing a better understanding and treatment of this disease. The first favorable point in this regard is the availability of a number of already-marketed drugs acting on this target; such drugs (i.e., rapamycin, everolimus, etc.) are already used clinically for other conditions (e.g., as immunosuppressants in kidney transplantation, in treatment of advanced renal carcinoma and breast cancer), therefore, their safety profile has already been studied. Following this convenience, clinical trials and small case report series/studies in epilepsy are already available. Preclinical data strongly support mTOR inhibitor efficacy in treating some genetic forms of epilepsy (e.g., TSC) but also in the prevention of epilepsy *development* (antiepileptogenic effect). Some controversial results with mTOR inhibitors also exist and, while it is clear that the mTOR pathway might be beneficially targeted, some drawbacks of "general" mTOR inhibition must still be considered and more studies are warranted (ideally with more selective drugs), before full clinical translation of mTOR targeting can result. In particular, the time-window, duration, and dosages to be used in the prevention of epileptogenesis are still far from being clearly identified. On the other hand, mTOR inhibitors possess only a very low efficacy in preclinical animal models of seizures/epilepsy and therefore, their effectiveness in *stopping* seizures is probably very limited and/or in any case, it might require time before some kind of effectiveness can be observed; generally, a few days of treatment are necessary to observe an antiseizure effect.

Despite several drugs acting on mTOR that are already currently available (both considering marketed and experimental drugs) [1,138], none of these agents has selectivity for the CNS, which obviously underlies the appearance of peripheral side effects that might limit dosages and therefore efficacy. The development of more specific and selective drugs acting in the CNS is therefore a highly relevant field of research, which still lacks positive results. As mentioned above, clinical trials are currently undergoing for both everolimus and rapamycin in epilepsy. The results of these trials will indeed shed light on the possibility to use such molecules in TSC patients; however, this will only be an initial step, which will need further clinical trials to understand the possible use of mTOR inhibitors in other clinical situations with patients at risk of epilepsy development (e.g., TBI). Of note, everolimus has only a limited ability to cross the BBB and to arrive in the brain [139]; therefore, while the actual preliminary clinical data available are encouraging, a higher effectiveness might be found using other molecules with a higher ability to act on the brain and therefore reduce peripheral side effects, which represents, at the moment, the major limitation in the choice of the dose.

In conclusion, as in the case of many other diseases, we feel that the mTOR pathway represents a great opportunity for future novel drug development in epilepsy therapeutics; the actual available data with mTOR inhibitors, both clinical and preclinical are intriguing and highly support future experiments. Unmet needs are represented by the lack of brain-selective inhibitor molecules with a better safety profile and clinical trials in other epilepsy syndromes.

References

[1] Russo E, Citraro R, Constanti A, De Sarro G. The mTOR signaling pathway in the brain: focus on epilepsy and epileptogenesis. Mol Neurobiol 2012;46:662–81.

[2] Asnaghi L, Bruno P, Priulla M, Nicolin A. mTOR: a protein kinase switching between life and death. Pharmacol Res 2004;50:545–9.

[3] Vezina C, Kudelski A, Sehgal SN. Rapamycin (AY-22,989), a new antifungal antibiotic. I. Taxonomy of the producing streptomycete and isolation of the active principle. J Antibiot (Tokyo) 1975;28:721–6.

[4] Sofroniadou S, Goldsmith D. Mammalian target of rapamycin (mTOR) inhibitors: potential uses and a review of haematological adverse effects. Drug Saf 2011;34:97–115.

[5] Weichhart T. Mammalian target of rapamycin: a signaling kinase for every aspect of cellular life. Methods Mol Biol 2012;821:1–14.

[6] Tsang CK, Qi H, Liu LF, Zheng XF. Targeting mammalian target of rapamycin (mTOR) for health and diseases. Drug Discov Today 2007;12:112–24.

[7] Huang J, Wu S, Wu CL, Manning BD. Signaling events downstream of mammalian target of rapamycin complex 2 are attenuated in cells and tumors deficient for the tuberous sclerosis complex tumor suppressors. Cancer Res 2009;69:6107–14.

[8] Frias MA, Thoreen CC, Jaffe JD, Schroder W, Sculley T, Carr SA, et al. mSin1 is necessary for Akt/PKB phosphorylation, and its isoforms define three distinct mTORC2s. Curr Biol 2006;16:1865–70.

[9] Pearce LR, Huang X, Boudeau J, Pawlowski R, Wullschleger S, Deak M, et al. Identification of Protor as a novel Rictor-binding component of mTOR complex-2. Biochem J 2007;405:513–22.

[10] Sarbassov DD, Ali SM, Sengupta S, Sheen JH, Hsu PP, Bagley AF, et al. Prolonged rapamycin treatment inhibits mTORC2 assembly and Akt/PKB. Mol Cell 2006;22:159–68.

[11] Chong ZZ, Yao Q, Li HH. The rationale of targeting mammalian target of rapamycin for ischemic stroke. Cell Signal 2013;25:1598–607.

[12] Sandsmark DK, Pelletier C, Weber JD, Gutmann DH. Mammalian target of rapamycin: master regulator of cell growth in the nervous system. Histol Histopathol 2007;22:895–903.

[13] Wong M. Mammalian target of rapamycin (mTOR) pathways in neurological diseases. Biomed J 2013;36:40–50.

[14] Jaworski J, Spangler S, Seeburg DP, Hoogenraad CC, Sheng M. Control of dendritic arborization by the phosphoinositide-3'-kinase-Akt-mammalian target of rapamycin pathway. J Neurosci 2005;25:11300–12.

[15] Kumar V, Zhang MX, Swank MW, Kunz J, Wu GY. Regulation of dendritic morphogenesis by Ras-PI3K-Akt-mTOR and Ras-MAPK signaling pathways. J Neurosci 2005;25:11288–99.

[16] Hoeffer CA, Klann E. mTOR signaling: at the crossroads of plasticity, memory and disease. Trends Neurosci 2010;33:67–75.

[17] Raab-Graham KF, Haddick PC, Jan YN, Jan LY. Activity- and mTOR-dependent suppression of Kv1.1 channel mRNA translation in dendrites. Science 2006;314:144–8.

[18] Dash PK, Orsi SA, Moore AN. Spatial memory formation and memory-enhancing effect of glucose involves activation of the tuberous sclerosis complex-mammalian target of rapamycin pathway. J Neurosci 2006;26:8048–56.

[19] Hoeffer CA, Tang W, Wong H, Santillan A, Patterson RJ, Martinez LA, et al. Removal of FKBP12 enhances mTOR-Raptor interactions, LTP, memory, and perseverative/repetitive behavior. Neuron 2008;60:832–45.

[20] Parsons RG, Gafford GM, Helmstetter FJ. Translational control via the mammalian target of rapamycin pathway is critical for the formation and stability of long-term fear memory in amygdala neurons. J Neurosci 2006;26:12977–83.

[21] Tischmeyer W, Schicknick H, Kraus M, Seidenbecher CI, Staak S, Scheich H, et al. Rapamycin-sensitive signalling in long-term consolidation of auditory cortex-dependent memory. Eur J Neurosci 2003;18:942–50.

[22] Ehninger D, de Vries PJ, Silva AJ. From mTOR to cognition: molecular and cellular mechanisms of cognitive impairments in tuberous sclerosis. J Intellect Disabil Res 2009;53:838–51.

[23] Ehninger D, Han S, Shilyansky C, Zhou Y, Li W, Kwiatkowski DJ, et al. Reversal of learning deficits in a Tsc2+/− mouse model of tuberous sclerosis. Nat Med 2008;14:843–8.

[24] Puighermanal E, Marsicano G, Busquets-Garcia A, Lutz B, Maldonado R, Ozaita A. Cannabinoid modulation of hippocampal long-term memory is mediated by mTOR signaling. Nat Neurosci 2009;12:1152–8.

[25] Cota D, Proulx K, Smith KA, Kozma SC, Thomas G, Woods SC, et al. Hypothalamic mTOR signaling regulates food intake. Science 2006;312:927–30.

[26] Roa J, Garcia-Galiano D, Varela L, Sanchez-Garrido MA, Pineda R, Castellano JM, et al. The mammalian target of rapamycin as novel central regulator of puberty onset via modulation of hypothalamic Kiss1 system. Endocrinology 2009;150:5016–26.

[27] Lipton JO, Sahin M. The neurology of mTOR. Neuron 2014;84:275–91.

[28] Chong ZZ, Shang YC, Zhang L, Wang S, Maiese K. Mammalian target of rapamycin: hitting the bull's-eye for neurological disorders. Oxid Med Cell Longev 2010;3:374–91.

[29] Nyfeler B, Bergman P, Wilson CJ, Murphy LO. Quantitative visualization of autophagy induction by mTOR inhibitors. Methods Mol Biol 2012;821:239–50.

[30] Arrasate M, Finkbeiner S. Protein aggregates in Huntington's disease. Exp Neurol 2012;238:1–11.

[31] Irvine GB, El-Agnaf OM, Shankar GM, Walsh DM. Protein aggregation in the brain: the molecular basis for Alzheimer's and Parkinson's diseases. Mol Med 2008;14:451–64.

[32] Ravikumar B, Vacher C, Berger Z, Davies JE, Luo S, Oroz LG, et al. Inhibition of mTOR induces autophagy and reduces toxicity of polyglutamine expansions in fly and mouse models of Huntington disease. Nat Genet 2004;36:585–95.

[33] Berger Z, Ravikumar B, Menzies FM, Oroz LG, Underwood BR, Pangalos MN, et al. Rapamycin alleviates toxicity of different aggregate-prone proteins. Hum Mol Genet 2006;15:433–42.

[34] Lacey CJ, Salzberg MR, D'Souza WJ. Risk factors for depression in community-treated epilepsy: systematic review. Epilepsy Behav 2015;43C:1–7.

[35] Citraro R, Leo A, de Fazio P, De Sarro G, Russo E. Antidepressants but not antipsychotics have antiepileptogenic effects with limited effects on comorbid depressive-like behavior in the WAG/Rij rat model of absence epilepsy. Br J Pharmacol 2015.

[36] Goldberg EM, Coulter DA. Mechanisms of epileptogenesis: a convergence on neural circuit dysfunction. Nat Rev Neurosci 2013;14:337–49.

[37] Pitkanen A, Lukasiuk K. Mechanisms of epileptogenesis and potential treatment targets. Lancet Neurol 2011;10:173–86.

[38] Kwan P, Brodie MJ. Refractory epilepsy: mechanisms and solutions. Expert Rev Neurother 2006;6:397–406.

[39] Pernot F, Heinrich C, Barbier L, Peinnequin A, Carpentier P, Dhote F, et al. Inflammatory changes during epileptogenesis and spontaneous seizures in a mouse model of mesiotemporal lobe epilepsy. Epilepsia 2011;52:2315–25.

[40] Meng XF, Yu JT, Song JH, Chi S, Tan L. Role of the mTOR signaling pathway in epilepsy. J Neurol Sci 2013;332:4–15.

[41] Wong M. A critical review of mTOR inhibitors and epilepsy: from basic science to clinical trials. Expert Rev Neurother 2013;13:657–69.

[42] Zeng LH, Rensing NR, Wong M. The mammalian target of rapamycin signaling pathway mediates epileptogenesis in a model of temporal lobe epilepsy. J Neurosci 2009;29:6964–72.

[43] White HS, Loscher W. Searching for the ideal antiepileptogenic agent in experimental models: single treatment versus combinatorial treatment strategies. Neurotherapeutics 2014;11:373–84.

[44] McDaniel SS, Wong M. Therapeutic role of mammalian target of rapamycin (mTOR) inhibition in preventing epileptogenesis. Neurosci Lett 2011;497:231–9.

[45] Crino PB. mTOR: a pathogenic signaling pathway in developmental brain malformations. Trends Mol Med 2011;17:734–42.

[46] Crino PB. Focal brain malformations: seizures, signaling, sequencing. Epilepsia 2009;50(Suppl. 9):3–8.

[47] Wong M, Crino PB. mTOR and epileptogenesis in developmental brain malformations. In: Noebels JL, Avoli M, Rogawski MA, Olsen RW, Delgado-Escueta AV, editors. Jasper's basic mechanisms of the epilepsies. 4th ed. Bethesda, MD: Oxford University Press, USA; 2012.

[48] Ostendorf AP, Wong M. mTOR inhibition in epilepsy: rationale and clinical perspectives. CNS Drugs 2015;29:91–9.

[49] Curatolo P, Moavero R. mTOR inhibitors as a new therapeutic option for epilepsy. Expert Rev Neurother 2013;13:627–38.

[50] Sliwa A, Plucinska G, Bednarczyk J, Lukasiuk K. Post-treatment with rapamycin does not prevent epileptogenesis in the amygdala stimulation model of temporal lobe epilepsy. Neurosci Lett 2012;509:105–9.

[51] van Vliet EA, Forte G, Holtman L, den Burger JC, Sinjewel A, de Vries HE, et al. Inhibition of mammalian target of rapamycin reduces epileptogenesis and blood-brain barrier leakage but not microglia activation. Epilepsia 2012;53:1254–63.

[52] Tang H, Long H, Zeng C, Li Y, Bi F, Wang J, et al. Rapamycin suppresses the recurrent excitatory circuits of dentate gyrus in a mouse model of temporal lobe epilepsy. Biochem Biophys Res Commun 2012;420:199–204.

[53] Zeng LH, McDaniel S, Rensing NR, Wong M. Regulation of cell death and epileptogenesis by the mammalian target of rapamycin (mTOR): a double-edged sword? Cell Cycle 2010;9:2281–5.

[54] Ljungberg MC, Sunnen CN, Lugo JN, Anderson AE, D'Arcangelo G. Rapamycin suppresses seizures and neuronal hypertrophy in a mouse model of cortical dysplasia. Dis Model Mech 2009;2:389–98.

[55] Zhou J, Blundell J, Ogawa S, Kwon CH, Zhang W, Sinton C, et al. Pharmacological inhibition of mTORC1 suppresses anatomical, cellular, and behavioral abnormalities in neural-specific Pten knock-out mice. J Neurosci 2009;29:1773–83.

[56] Buckmaster PS, Wen X. Rapamycin suppresses axon sprouting by somatostatin interneurons in a mouse model of temporal lobe epilepsy. Epilepsia 2011;52:2057–64.

[57] Heng K, Haney MM, Buckmaster PS. High-dose rapamycin blocks mossy fiber sprouting but not seizures in a mouse model of temporal lobe epilepsy. Epilepsia 2013;54:1535–41.

[58] Buckmaster PS, Lew FH. Rapamycin suppresses mossy fiber sprouting but not seizure frequency in a mouse model of temporal lobe epilepsy. J Neurosci 2011;31:2337–47.

[59] Buckmaster PS. Does mossy fiber sprouting give rise to the epileptic state? Adv Exp Med Biol 2014;813:161–8.

[60] Wang L, Lv Y, Deng W, Peng X, Xiao Z, Xi Z, et al. 5-HT6 receptor recruitment of mTOR modulates seizure activity in epilepsy. Mol Neurobiol 2014.

[61] Pun RY, Rolle IJ, Lasarge CL, Hosford BE, Rosen JM, Uhl JD, et al. Excessive activation of mTOR in postnatally generated granule cells is sufficient to cause epilepsy. Neuron 2012;75:1022–34.

[62] Guo D, Zeng L, Brody DL, Wong M. Rapamycin attenuates the development of posttraumatic epilepsy in a mouse model of traumatic brain injury. PLoS One 2013;8:e64078.

[63] Berdichevsky Y, Dryer AM, Saponjian Y, Mahoney MM, Pimentel CA, Lucini CA, et al. PI3K-Akt signaling activates mTOR-mediated epileptogenesis in organotypic hippocampal culture model of post-traumatic epilepsy. J Neurosci 2013;33:9056–67.

[64] Zeng LH, Xu L, Gutmann DH, Wong M. Rapamycin prevents epilepsy in a mouse model of tuberous sclerosis complex. Ann Neurol 2008;63:444–53.

[65] Wong M, Ess KC, Uhlmann EJ, Jansen LA, Li W, Crino PB, et al. Impaired glial glutamate transport in a mouse tuberous sclerosis epilepsy model. Ann Neurol 2003;54:251–6.

[66] Meikle L, Pollizzi K, Egnor A, Kramvis I, Lane H, Sahin M, et al. Response of a neuronal model of tuberous sclerosis to mammalian target of rapamycin (mTOR) inhibitors: effects on mTORC1 and Akt signaling lead to improved survival and function. J Neurosci 2008;28:5422–32.

[67] Goto J, Talos DM, Klein P, Qin W, Chekaluk YI, Anderl S, et al. Regulable neural progenitor-specific Tsc1 loss yields giant cells with organellar dysfunction in a model of tuberous sclerosis complex. Proc Natl Acad Sci USA 2011;108:E1070–79.

[68] Zeng LH, Rensing NR, Zhang B, Gutmann DH, Gambello MJ, Wong M. Tsc2 gene inactivation causes a more severe epilepsy phenotype than Tsc1 inactivation in a mouse model of tuberous sclerosis complex. Hum Mol Genet 2011;20:445–54.

[69] Russo E, Andreozzi F, Iuliano R, Dattilo V, Procopio T, Fiume G, et al. Early molecular and behavioral response to lipopolysaccharide in the WAG/Rij rat model of absence epilepsy and depressive-like behavior, involves interplay between AMPK, AKT/mTOR pathways and neuroinflammatory cytokine release. Brain Behav Immun 2014;42:157–68.

[70] Russo E, Citraro R, Donato G, Camastra C, Iuliano R, Cuzzocrea S, et al. mTOR inhibition modulates epileptogenesis, seizures and depressive behavior in a genetic rat model of absence epilepsy. Neuropharmacology 2013;69:25–36.

[71] Russo E, Follesa P, Citraro R, Camastra C, Donato A, Isola D, et al. The mTOR signaling pathway and neuronal stem/progenitor cell proliferation in the hippocampus are altered during the development of absence epilepsy in a genetic animal model. Neurol Sci 2014;35:1793–9.

[72] Curatolo P, Maria BL. Tuberous sclerosis. Handb Clin Neurol 2013;111:323–31.

[73] Poduri A, Lowenstein D. Epilepsy genetics—past, present, and future. Curr Opin Genet Dev 2011;21:325–32.

[74] Krueger DA, Care MM, Holland K, Agricola K, Tudor C, Mangeshkar P, et al. Everolimus for subependymal giant-cell astrocytomas in tuberous sclerosis. N Engl J Med 2010;363:1801–11.

[75] Curatolo P. Mechanistic Target of Rapamycin (mTOR) in tuberous sclerosis complex-associated epilepsy. Pediatr Neurol 2015;52:281–9.

[76] Chu-Shore CJ, Major P, Camposano S, Muzykewicz D, Thiele EA. The natural history of epilepsy in tuberous sclerosis complex. Epilepsia 2010;51:1236–41.

[77] Ruppe V, Dilsiz P, Reiss CS, Carlson C, Devinsky O, Zagzag D, et al. Developmental brain abnormalities in tuberous sclerosis complex: a comparative tissue analysis of cortical tubers and perituberal cortex. Epilepsia 2014;55:539–50.

[78] McDaniel SS, Rensing NR, Thio LL, Yamada KA, Wong M. The ketogenic diet inhibits the mammalian target of rapamycin (mTOR) pathway. Epilepsia 2011;52:e7–11.

[79] Zhang B, McDaniel SS, Rensing NR, Wong M. Vigabatrin inhibits seizures and mTOR pathway activation in a mouse model of tuberous sclerosis complex. PLoS One 2013;8:e57445.

[80] Ben-Menachem E. Mechanism of action of vigabatrin: correcting misperceptions. Acta Neurol Scand Suppl 2011;5–15.

[81] Jozwiak S, Kotulska K, Domanska-Pakiela D, Lojszczyk B, Syczewska M, Chmielewski D, et al. Antiepileptic treatment before the onset of seizures reduces epilepsy severity and risk of mental retardation in infants with tuberous sclerosis complex. Eur J Paediatr Neurol 2011;15:424–31.

[82] Bombardieri R, Pinci M, Moavero R, Cerminara C, Curatolo P. Early control of seizures improves long-term outcome in children with tuberous sclerosis complex. Eur J Paediatr Neurol 2010;14:146–9.

[83] Onda H, Crino PB, Zhang H, Murphey RD, Rastelli L, Gould Rothberg BE, et al. Tsc2 null murine neuroepithelial cells are a model for human tuber giant cells, and show activation of an mTOR pathway. Mol Cell Neurosci 2002;21:561–74.

[84] Feliciano DM, Lin TV, Hartman NW, Bartley CM, Kubera C, Hsieh L, et al. A circuitry and biochemical basis for tuberous sclerosis symptoms: from epilepsy to neurocognitive deficits. Int J Dev Neurosci 2013;31:667–78.

[85] Wong M. A tuber-ful animal model of tuberous sclerosis at last? Epilepsy Curr 2012;12:15–16.

[86] McMahon J, Huang X, Yang J, Komatsu M, Yue Z, Qian J, et al. Impaired autophagy in neurons after disinhibition of mammalian target of rapamycin and its contribution to epileptogenesis. J Neurosci 2012;32:15704–14.

[87] Di Nardo A, Wertz MH, Kwiatkowski E, Tsai PT, Leech JD, Greene-Colozzi E, et al. Neuronal Tsc1/2 complex controls autophagy through AMPK-dependent regulation of ULK1. Hum Mol Genet 2014;23:3865–74.

[88] Fu C, Cawthon B, Clinkscales W, Bruce A, Winzenburger P, Ess KC. GABAergic interneuron development and function is modulated by the Tsc1 gene. Cereb Cortex 2012;22:2111–19.

[89] Galanopoulou AS, Gorter JA, Cepeda C. Finding a better drug for epilepsy: the mTOR pathway as an antiepileptogenic target. Epilepsia 2012;53:1119–30.

[90] Fauser S, Essang C, Altenmuller DM, Staack AM, Steinhoff BJ, Strobl K, et al. Long-term seizure outcome in 211 patients with focal cortical dysplasia. Epilepsia 2015;56:66–76.

[91] Liu J, Reeves C, Michalak Z, Coppola A, Diehl B, Sisodiya SM, et al. Evidence for mTOR pathway activation in a spectrum of epilepsy-associated pathologies. Acta Neuropathol Commun 2014;2:71.

[92] Lim JS, Kim WI, Kang HC, Kim SH, Park AH, Park EK, et al. Brain somatic mutations in MTOR cause focal cortical dysplasia type II leading to intractable epilepsy. Nat Med 2015;21:395–400.

[93] Lee JY, Park AK, Lee ES, Park WY, Park SH, Choi JW, et al. miRNA expression analysis in cortical dysplasia: regulation of mTOR and LIS1 pathway. Epilepsy Res 2014;108:433–41.

[94] Backman SA, Stambolic V, Suzuki A, Haight J, Elia A, Pretorius J, et al. Deletion of Pten in mouse brain causes seizures, ataxia and defects in soma size resembling Lhermitte–Duclos disease. Nat Genet 2001;29:396–403.

[95] Kwon CH, Zhu X, Zhang J, Knoop LL, Tharp R, Smeyne RJ, et al. Pten regulates neuronal soma size: a mouse model of Lhermitte–Duclos disease. Nat Genet 2001;29:404–11.

[96] Eng C. PTEN: one gene, many syndromes. Hum Mutat 2003;22:183–98.

[97] Zhou XP, Waite KA, Pilarski R, Hampel H, Fernandez MJ, Bos C, et al. Germline PTEN promoter mutations and deletions in Cowden/Bannayan–Riley–Ruvalcaba syndrome result in aberrant PTEN protein and dysregulation of the phosphoinositol-3-kinase/Akt pathway. Am J Hum Genet 2003;73:404–11.

[98] Zhou XP, Marsh DJ, Morrison CD, Chaudhury AR, Maxwell M, Reifenberger G, et al. Germline inactivation of PTEN and dysregulation of the phosphoinositol-3-kinase/Akt pathway cause human Lhermitte–Duclos disease in adults. Am J Hum Genet 2003;73:1191–8.

[99] Kwon CH, Zhu X, Zhang J, Baker SJ. mTor is required for hypertrophy of Pten-deficient neuronal soma in vivo. Proc Natl Acad Sci USA 2003;100:12923−8.

[100] Brooks-Kayal AR, Bath KG, Berg AT, Galanopoulou AS, Holmes GL, Jensen FE, et al. Issues related to symptomatic and disease-modifying treatments affecting cognitive and neuropsychiatric comorbidities of epilepsy. Epilepsia 2013;54 (Suppl. 4):44−60.

[101] Jernigan CS, Goswami DB, Austin MC, Iyo AH, Chandran A, Stockmeier CA, et al. The mTOR signaling pathway in the prefrontal cortex is compromised in major depressive disorder. Prog Neuropsychopharmacol Biol Psychiatry 2011;35:1774−9.

[102] Citraro R, Leo A, Aiello R, Pugliese M, Russo E, De Sarro G. Comparative analysis of the treatment of chronic antipsychotic drugs on epileptic susceptibility in genetically epilepsy-prone rats. Neurotherapeutics 2015;12:250−62.

[103] Sarkisova K, van Luijtelaar G. The WAG/Rij strain: a genetic animal model of absence epilepsy with comorbidity of depression [corrected]. Prog Neuropsychopharmacol Biol Psychiatry 2011;35:854−76.

[104] Blumenfeld H, Klein JP, Schridde U, Vestal M, Rice T, Khera DS, et al. Early treatment suppresses the development of spike-wave epilepsy in a rat model. Epilepsia 2008;49:400−9.

[105] Vezzani A, Friedman A, Dingledine RJ. The role of inflammation in epileptogenesis. Neuropharmacology 2013;69:16−24.

[106] Kovacs Z, Czurko A, Kekesi KA, Juhasz G. Intracerebroventricularly administered lipopolysaccharide enhances spike-wave discharges in freely moving WAG/Rij rats. Brain Res Bull 2011;85:410−16.

[107] Gurol G, Demiralp DO, Yilmaz AK, Akman O, Ates N, Karson A. Comparative proteomic approach in rat model of absence epilepsy. J Mol Neurosci 2015;55:632−43.

[108] Sutula TP, Dudek FE. Unmasking recurrent excitation generated by mossy fiber sprouting in the epileptic dentate gyrus: an emergent property of a complex system. Prog Brain Res 2007;163:541−63.

[109] Pitkanen A, Bolkvadze T, Immonen R. Anti-epileptogenesis in rodent post-traumatic epilepsy models. Neurosci Lett 2011;497:163−71.

[110] Ruegg S, Baybis M, Juul H, Dichter M, Crino PB. Effects of rapamycin on gene expression, morphology, and electrophysiological properties of rat hippocampal neurons. Epilepsy Res 2007;77:85−92.

[111] Daoud D, Scheld HH, Speckmann EJ, Gorji A. Rapamycin: brain excitability studied in vitro. Epilepsia 2007;48:834−6.

[112] Huang X, McMahon J, Yang J, Shin D, Huang Y. Rapamycin down-regulates KCC2 expression and increases seizure susceptibility to convulsants in immature rats. Neuroscience 2012;219:33−47.

[113] Lozovaya N, Gataullina S, Tsintsadze T, Tsintsadze V, Pallesi-Pocachard E, Minlebaev M, et al. Selective suppression of excessive GluN2C expression rescues early epilepsy in a tuberous sclerosis murine model. Nat Commun 2014;5:4563.

[114] Abs E, Goorden SM, Schreiber J, Overwater IE, Hoogeveen-Westerveld M, Bruinsma CF, et al. TORC1-dependent epilepsy caused by acute biallelic Tsc1 deletion in adult mice. Ann Neurol 2013;74:569−79.

[115] Chachua T, Poon KL, Yum MS, Nesheiwat L, DeSantis K, Veliskova J, et al. Rapamycin has age-, treatment paradigm-, and model-specific anticonvulsant effects and modulates neuropeptide Y expression in rats. Epilepsia 2012;53:2015−25.

[116] Hartman AL, Santos P, Dolce A, Hardwick JM. The mTOR inhibitor rapamycin has limited acute anticonvulsant effects in mice. PLoS One 2012;7:e45156.

[117] Raffo E, Coppola A, Ono T, Briggs SW, Galanopoulou AS. A pulse rapamycin therapy for infantile spasms and associated cognitive decline. Neurobiol Dis 2011;43:322−9.

[118] Sunnen CN, Brewster AL, Lugo JN, Vanegas F, Turcios E, Mukhi S, et al. Inhibition of the mammalian target of rapamycin blocks epilepsy progression in NS-Pten conditional knockout mice. Epilepsia 2011;52:2065−75.

[119] Nguyen LH, Brewster AL, Clark ME, Regnier-Golanov A, Sunnen CN, Patil VV, et al. mTOR inhibition suppresses established epilepsy in a mouse model of cortical dysplasia. Epilepsia 2015.

[120] Chachua T, Yum MS, Veliskova J, Velisek L. Validation of the rat model of cryptogenic infantile spasms. Epilepsia 2011;52: 1666−77.

[121] Tang SJ, Reis G, Kang H, Gingras AC, Sonenberg N, Schuman EM. A rapamycin-sensitive signaling pathway contributes to long-term synaptic plasticity in the hippocampus. Proc Natl Acad Sci USA 2002;99:467−72.

[122] Rivera C, Li H, Thomas-Crusells J, Lahtinen H, Viitanen T, Nanobashvili A, et al. BDNF-induced TrkB activation downregulates the K+-Cl- cotransporter KCC2 and impairs neuronal Cl-extrusion. J Cell Biol 2002;159:747−52.

[123] Zeng LH, Rensing NR, Wong M. Developing Antiepileptogenic drugs for acquired epilepsy: targeting the mammalian target of rapamycin (mTOR) pathway. Mol Cell Pharmacol 2009;1:124−9.

[124] Bialer M, Johannessen SI, Levy RH, Perucca E, Tomson T, White HS. Progress report on new antiepileptic drugs: a summary of the Twelfth Eilat Conference (EILAT XII). Epilepsy Res 2015;111:85−141.

[125] Kotulska K, Chmielewski D, Borkowska J, Jurkiewicz E, Kuczynski D, Kmiec T, et al. Long-term effect of everolimus on epilepsy and growth in children under 3 years of age treated for subependymal giant cell astrocytoma associated with tuberous sclerosis complex. Eur J Paediatr Neurol 2013;17:479−85.

[126] ClinicalTrials.gov. Baylor College of Medicine adjunctive everolimus (RAD 001) therapy for epilepsy in children with Sturge−Weber syndrome (SWS); NCT01997255; https://clinicaltrials.gov/ct2/show/NCT01997255.

[127] Perek-Polnik M, Jozwiak S, Jurkiewicz E, Perek D, Kotulska K. Effective everolimus treatment of inoperable, life-threatening subependymal giant cell astrocytoma and intractable epilepsy in a patient with tuberous sclerosis complex. Eur J Paediatr Neurol 2012;16:83−5.

[128] Muncy J, Butler IJ, Koenig MK. Rapamycin reduces seizure frequency in tuberous sclerosis complex. J Child Neurol 2009;24:477.

[129] Wiegand G, May TW, Ostertag P, Boor R, Stephani U, Franz DN. Everolimus in tuberous sclerosis patients with intractable epilepsy: a treatment option? Eur J Paediatr Neurol 2013;17:631−8.

[130] Cardamone M, Flanagan D, Mowat D, Kennedy SE, Chopra M, Lawson JA. Mammalian target of rapamycin inhibitors for intractable epilepsy and subependymal giant cell astrocytomas in tuberous sclerosis complex. J Pediatr 2014;164:1195−200.

[131] Franz DN, Belousova E, Sparagana S, Bebin EM, Frost M, Kuperman R, et al. Efficacy and safety of everolimus for subependymal giant cell astrocytomas associated with tuberous sclerosis complex (EXIST-1): a multicentre, randomised, placebo-controlled phase 3 trial. Lancet 2013;381:125−32.

[132] Krueger DA, Wilfong AA, Holland-Bouley K, Anderson AE, Agricola K, Tudor C, et al. Everolimus treatment of refractory epilepsy in tuberous sclerosis complex. Ann Neurol 2013;74:679—87.

[133] Novartis Pharmaceuticals. A placebo-controlled study of efficacy and safety of 2 trough-ranges of everolimus as adjunctive therapy in patients with tuberous sclerosis complex (TSC) and refractory partial-onset seizures (EXIST-3); NCT01713946; https://clinicaltrials.gov/ct2/show/NCT01713946?term= everolimus+epilepsy&rank=3.

[134] Hasskarl J. Everolimus. Recent Results Cancer Res 2014;201: 373—92.

[135] Gipson TT, Jennett H, Wachtel L, Gregory M, Poretti A, Johnston MV. Everolimus and intensive behavioral therapy in an adolescent with tuberous sclerosis complex and severe behavior. Epilepsy Behav Case Rep 2013;1:122—5.

[136] Kwiatkowski DJ, Palmer MR, Jozwiak S, Bissler J, Franz D, Segal S, et al. Response to everolimus is seen in TSC-associated SEGAs and angiomyolipomas independent of mutation type and site in TSC1 and TSC2. Eur J Hum Genet 2015.

[137] Domanska-Pakiela D, Kaczorowska M, Jurkiewicz E, Kotulska K, Dunin-Wasowicz D, Jozwiak S. EEG abnormalities preceding the epilepsy onset in tuberous sclerosis complex patients—a prospective study of 5 patients. Eur J Paediatr Neurol 2014;18:458—68.

[138] Zhang YJ, Duan Y, Zheng XF. Targeting the mTOR kinase domain: the second generation of mTOR inhibitors. Drug Discov Today 2011;16:325—31.

[139] Gottschalk S, Cummins CL, Leibfritz D, Christians U, Benet LZ, Serkova NJ. Age and sex differences in the effects of the immunosuppressants cyclosporine, sirolimus and everolimus on rat brain metabolism. Neurotoxicology 2011;32:50—7.

mTOR, Autophagy, Aminoacidopathies, and Human Genetic Disorders

Garrett R. Ainslie, K. Michael Gibson and Kara R. Vogel

Section of Experimental and Systems Pharmacology, College of Pharmacy, Washington State University, Spokane, WA, USA

9.1 INTRODUCTION TO MECHANISTIC TARGET OF RAPAMYCIN (mTOR) SIGNALING

9.1.1 Mechanistic Target of Rapamycin

The 289-kDA serine/threonine kinase mechanistic target of rapamycin (mTOR; FRAP, RAFT, mammalian target of rapamycin) regulates metabolism, protein synthesis, cellular growth, tissue repair and regeneration, cytoskeletal organization, stem cell regeneration or quiescence, cellular survival, and aging. mTOR is encoded by the single gene, *FRAP1* (FK506-Binding Protein 12-Rapamycin Complex-Associated Protein 1) located on chromosome 1p36. *FRAP* encodes the single high-molecular weight mTOR protein composed of 2549 amino acids and several conserved structural domains including: 20 tandem HEAT (Huntingtin, elongation factor 3, a subunit of protein phosphatase 2A, PI3 kinase TOR) repeats of the N-terminus, the C-terminal kinase domain, two focal adhesion targeting (FAT) domains, and the rapamycin interactive FRB domain immediately upstream of the C-terminus [1] (Figure 9.1). The product forms two major protein complexes mTORC1 (mTOR, regulatory-associated protein of mTOR known as Raptor), mammalian lethal with SEC13 protein 8 (MLST8, also GβL), 40-kDa pro-rich protein kinase C (PKC, also Akt) substrate (PRAS40), DEPTOR, and mTORC2 (mTOR, the rapamycin-insensitive companion of mTOR, known as RICTOR), GβL, mammalian stress-activated protein kinase interacting protein 1 (mSIN1), Protor 1/2, DEPTOR, and TTI1 and TEL2, each with differing kinase specificity [2,3] (Figure 9.2).

9.1.2 mTOR Signaling and Nutrition (Amino Acids, Sugars, Growth Factors)

Cellular and organismal growth occur through highly regulated anabolic processes including: DNA synthesis, mRNA translation, protein and lipid synthesis, and organellular biogenesis. Under nutrient-rich conditions mTOR integrates signals from stress, growth factors, and energy status, cues from adenosine triphosphate (ATP) and amino acid levels to balance these anabolic processes vis-a-vis catabolism and autophagy inhibition [4].

9.1.3 Upstream mTOR

mTOR activity is controlled by extracellular molecules in the brain including glutamate, the ubiquitous excitatory amino acid, insulin-like growth factor 1, vascular endothelial growth factor, brain-derived neurotrophic factor, ciliary neurotrophic factor, and axonal guidance molecules. The extracellular cues signal to mTOR via the phosphatidylinositol-4, 5-bisphosphate 3-kinase (PI3K) and/or the mitogen-activated protein kinase (MAPK) pathways. Growth factors stimulate mTOR through the PI3K/Akt signaling pathway (Figure 9.2).

Akt is a serine/threonine kinase, also known as protein kinase B, downstream of PI3K which phosphorylates two adjacent sites on the mTOR protein (Thr 2446 and Ser 2448). The tuberous sclerosis complex (TSC), composed of the tumor suppressors TSC1 (hamartin) and TSC2 (tuberin), is a GTPase-activating protein (GAP) which functions downstream of Akt. Phosphorylation of TSC2 by Akt inhibits TSC2

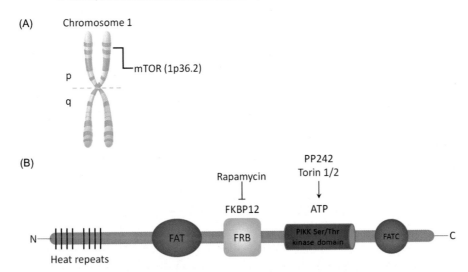

FIGURE 9.1 The mammalian target of rapamycin (mTOR). (A) FK506-binding protein 12-rapamycin complex-associated protein 1 (FRAP1) is located on chromosome 1p36. (B) The mTOR protein, composed of 2549 amino acids and several conserved structural domains, comprises 20 tandem HEAT repeats on the N-terminus, a C-terminal kinase domain, two focal adhesion targeting (FAT) domains and the FKBP-rapamycin-binding (FRB) domain. Rapamycin acts via inhibition of the FRB domain, whereas torin 1/2 and PP242 are inhibitors of ATP binding.

allowing guanosine triphosphate (GTP)-bound Rheb (Ras homolog enriched in brain) to activate mTOR. The non-linear mTORC1/Akt interrelation involves a regulatory positive feedback as well as a negative feedback loop which blocks the PI3K activation of Akt.

Adenosine monophosphate (AMP)-activated protein kinase (AMPK) also inhibits mTORC1 activity through phosphorylation of distinct residues of TSC 2 and can further down-regulate mTORC1 through kinase activity with Raptor [5] (Figure 9.2).

9.1.4 Amino Acid Regulation of mTOR

The Rag family of GTPases form heterodimers of RagA or RagB with RagC or RagD, which represent a key link between mTOR and amino acids [6]. In the absence of amino acids, Rag heterodimers assume the inactive conformation. The Ragulator, a lysosome-bound pentomeric protein complex maintains inactive Rag. Recent data suggest that the Ragulator tethers Rag GTPases to the lysosomes, thereby localizing mTOR to this catabolic organelle. The active form of the Rag occurs by the GAPs action of GATOR1 (DEPDC5, NPRL2, and NPRL3) and GATOR2 (MIOS, WDR24, WDR59, SEH1L, and SEC13) [7]. Two models of amino acid sensing by mTOR have emerged: leucyl-tRNA synthetase cytoplasmic activation of mTOR and the inside-to-outside lysosomal model in which amino acids accumulated inside the lysosome signal through the v-ATPase to Ragulator to the activated the RagA/B-GTP-RagC/D-GDP complex. Potentially, multiple signals act at the lysosome to coordinate mTOR in response to amino acids [8]. In particular, leucine, glutamine, and arginine have been implicated in mTORC1 regulation (Figure 9.3).

The branched-chain amino acid (BCAA) leucine plays a unique role in metabolic regulation, which surpasses the basic role of acting as a substrate in protein synthesis, but also regulates protein synthesis alone and in synergy with insulin in adipose tissue and skeletal muscle. In adipocytes, leucine, followed by leucine analogs, was the most potent amino acid linked to rapamycin-sensitive 4E binding protein 1 (4EBP1) phosphorylation [9]. Leucine maintains glucose homeostasis through the glucose—alanine cycle, promotes gluconeogenesis in a hypocaloric state, and can contribute to the anaplerotic supply of ketoisocaproate through acetyl coenzyme A to the citric acid cycle (TCA cycle). Furthermore, leucine can induce insulin release associated with mitochondrial biogenesis, mitochondrial mass, and oxygen consumption, emphasizing the importance of leucine-driven metabolism and energy utilization in myocytes and adipocytes [10]. Some of these actions are inhibited by rapamycin, suggesting that leucine acts on the mTOR pathway, while other effects are mTOR-independent. Extensive research is focused on the role of various amino acids in mTOR signaling, the tissue specificity for their individual roles, and the influence of the different transporters and molecular machines such as the Rag GTPases, Ragulator, and the vacuolar H^+ adenosine triphosphatase (v-ATPase) in the variable signaling potency and efficacy of different amino acids. The Guan group has

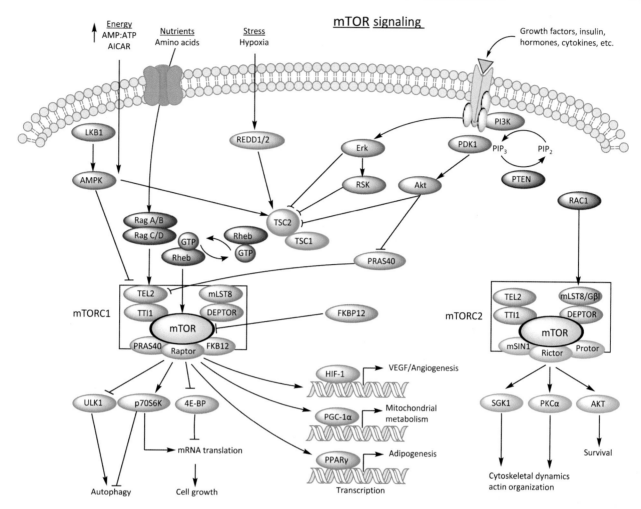

FIGURE 9.2   The mTOR signaling pathway. Hormones, growth factors, and various cytokines trigger signaling events to mTOR via the phosphatidylinositol-4, 5-bisphosphate 3-kinase (PI3K) and/or the mitogen-activated protein kinase (MAPK) pathways. Growth factors stimulate mTOR through the PI3K/protein kinase B (PKB, Akt) signaling pathway. Akt is a serine/threonine kinase, also known as protein kinase B, downstream of PI3K which phosphorylates two adjacent sites on the mTOR protein (Thr 2446 and Ser 2448). The tuberous sclerosis complex, composed of the tumor suppressors TSC1 and TSC2, is a GTPase-activating protein (GAP) which functions downstream of Akt. Phosphorylation of TSC2 by Akt inhibits TSC2 allowing GTP-bound Ras homolog enriched in brain (Rheb) to activate mTOR. The non-linear mTORC1/Akt interrelation involves a regulatory positive feedback as well as a negative feedback loop which blocks the PI3K activation of Akt. Hypoxic conditions may activate the TSC2 via DNA-damage-inducible transcript (DDIT4, REDD1/2). Under conditions of high adenosine triphosphate (ATP), serine/threonine kinase 1 (STK11, or liver kinase B1, LKB1) sequentially activates AMP-activated protein kinase (AMPK) then TSC2. Other nutrients, such as amino acids, activate Ras-related GTP-binding proteins A/B/C/D (Rag A/B/C/D), which are positive modulators of mTORC1 activity. Promotion of mTORC1 to its active state results in the suppression of unc-51 autophagy activating kinase 1 (ULK1) and eukaryotic translation initiation factor 4E binding protein (4EBP), while phosphorylating S6K. The activation of mTORC1 also results in the promotion of key transcription factors, including: peroxisome proliferator activated receptor gamma (PPARγ), PPARγ cofactor 1 alpha (PGC-1α) and hypoxia inducible factor (HIF-1). mTORC2 acts on serum/glucocorticoid regulated kinase 1 (SGK1), protein kinase C alpha (PKCα) and Akt cascades. Although pathways relating to mTORC2 are less understood relative to mTORC1, certain growth factors are believed to promote mTORC2 activity, and there is also evidence that Ras-related C3 botulinum toxin substrate 1 (RAC1) can have an equivalent effect. mTOR, mechanistic target of rapamycin; DEPTOR, DEP domain containing mTOR-interacting protein; MAPKAP1, also mSIN1, MAPK associated protein 1; mLST8, mTOR associated protein, LST8 homolog; PTEN, phosphatase and tensin homolog; PRAS40, proline rich AKT1 substrate; PDK1, pyruvate dehydrogenase kinase 1; RICTOR, RAPTOR independent complex of mTOR complex 2; RSK, ribosomal protein S6 kinase; TEL2, telomere maintenance 2; TTI1, tubulin tyrosine ligase; TSC, tuberous sclerosis complex; mTORC, mechanistic target of rapamycin complex.

recently shown that while leucine/mTOR signaling depends upon the Rag GTPases, glutamine stimulates mTOR in a Rag GTPase-independent manner. Furthermore, glutamine recruitment of mTORC1 to the lysosome did not require the Ragulator, but requires v-ATPase and the adenosine diphosphate ribosylation factor 1 (Arf1) GTPase [11]. Further evidence of differential amino acid sensing by mTOR came from the

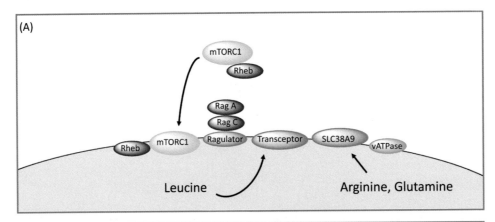

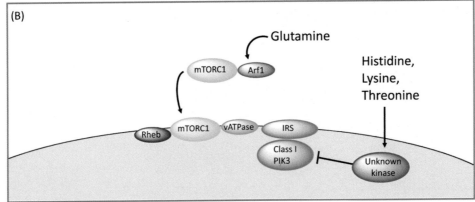

FIGURE 9.3   The role of various amino acids in autophagy. (A) The Ras-related GTP-binding protein (Rag) family of GTPases form hetero-dimers of RaGA and RaGC. The lysosome-bound pentomeric protein Ragulator complex maintains inactive RaGs. Amino acids, leucine, and/or arginine/glutamine accumulate inside the lysosome and signal to Ragulator via solute carrier family 38, member 9 (SLC38A9) transport, respectively. (B) Glutamine recruitment of mTORC1 to the lysosome does not require Ragulator. In this mechanism the v-ATPase and the Arf1 GTPase are requisite. Negative regulation of mTOR occurs through an intraperoxisomal kinase activation by histidine, lysine, and threonine which acts to suppress the insulin-controlled IRS-phosphoinositide 3-kinase (PI3K) pathway. Arf1, ADP-ribosylation factor 1; IRS, isoleucyl-tRNA sythetase; Rheb, GTP-bound Ras homolog enriched in brain; v-ATPase, vacuolar [H^+] ATPase; GTP, guanosine triphosphate; mTORC, mechanistic target of rapamycin complex.

Sabatini group. In their recently published mechanistic work involving arginine signaling to mTOR, these investigators documented that the lysosomal solute carrier 38A9 interacts with Rag GTPases and Ragulator, transports arginine with low affinity and is required for arginine activation of the mTORC1 pathway [12]. Through various pathways now under investigation, amino acid stimulation increases S6K1 and 4EBP1 phosphorylation, whereupon amino acid depletion rapidly dephosphorylates the two downstream targets of mTOR S6K1 and 4EBP1 [7].

9.1.5 Downstream mTORC1 Signaling (Cell Growth, Translation Initiation, Inhibition of Autophagy)

Rheb activation of mTOR in response to nutrients, growth factors, and amino acids leads to the downstream phosphorylation of the translational control targets, the 40S ribosomal S6 kinase 1 (S6K1), and the eukaryotic translation initiation factor 4E (eIF4E)-binding protein 1 (4EBP1). mTORC1 initiates cap-dependent translation through the oppression of 4EBP tight binding to eIF4E. In a second regulatory act, mTORC1 augments translation through S6K1, a critical effector of mTOR. Phosphorylation of the rapamycin-sensitive residue T389 (a common residue for mTOR activity assay) of S6K1 is mediated by mTOR, an interaction with mTORC1 through the TOR signaling motifs of Raptor [13]. S6K targets the well-described translation initiation factor eukaryotic initiation factor 4B, allowing its recruitment to the translation preinitiation complex and subsequent eIF4A RNA-helicase enhancement. Phosphorylation of S6K1 by mTOR allows phosphoinositide-dependent kinase-1 to phosphorylate and activate S6K1. mTOR targeting of translational machinery gives it a prominent regulatory role in protein synthesis and cellular growth. The active S6K1 also phosphorylates and inhibits insulin receptor substrate 1 (IRS1), with an important role in negative feedback on insulin/PI3K/AKT/mTOR signaling [14]. The mTORC1/S6K signaling axis also contributes to regulatory outcomes in transcription, translation, insulin resistance, cellular growth, metabolism, and autophagy. Alterations in mTOR signaling can result in a number of pathological outcomes including obesity, metabolic disease, cancer, epilepsy, and diabetes [15].

9.1.6 Autophagy

The process of autophagy, first observed by Porter and Ashford in the 1960s, is an important function in cellular homeostasis in which cytoplasmic contents are delivered to the lysosome for recycling. Christian de Duve, who received the Nobel Prize for the discovery of lysosomes and peroxisomes, later coined the phrase autophagy, Greek for "self-eating." The three major types are macroautophagy, the major process, microautophagy, and chaperone-mediated autophagy. In microautophagy, engulfment of the substrate occurs by invagination of the lysosomal membrane, most often via non-selective processes. The process proceeds by several steps: (i) induction at the phagophore assembly site (PAS); (ii) elongation of the phagophore and subsequent engulfment of the cellular debris; (iii) fusion of the autophagosome with a lysosome to form a phagolysosome; and (iv) degradation of contents by lysosomal hydrolases and recycling of basic materials [16] (Figure 9.4). In addition to general autophagy, a number of selective autophagy processes exist in which autophagic adaptor proteins mediate recruitment of specific molecules or organelles to the elongating phagophore. These include: mitophagy, pexophagy, nucleophagy, selective autophagy of endoplasmic reticulum, ribophagy, glycophagy, aggrephagy, and xenophagy.

mTOR negatively regulates autophagy at the induction stage. Autophagy induction occurs during conditions of low glucose, growth factors, and amino acids, and during starvation or catabolic state, resulting in cellular recycling of aggregates and macromolecules, all of which prolongs cellular survival [17]. AMPK, protein kinase A, and the mTOR signaling pathways are the major regulatory pathways involved in balancing cellular energy levels, metabolic homeostasis, and autophagy to sustain cellular viability.

9.1.7 Apoptosis

Apoptosis (Greek, for "falling off" or "falling apart") is a genetically programmed process of cellular demise which turns on in response to intra- and extracellular signaling. Upon activation in type 1 cell death, signaling cascades activate catabolic enzymes, which degrade the cellular structures, shrink the cell, induce pyknosis (the condensation of the nuclear chromatin), which is also a characteristic morphological feature of apoptosis, and karyorrhexis (nuclear fragmentation). Autophagy and apoptosis represent two processes wherein aged, damaged, or superfluous organelles or cells are eradicated. The interrelationship between the two processes is complex in that signaling inputs, stressors, nutrients, and growth factors alter the thresholds, leading to variable cellular outcomes: autophagic cytoprotection or autophagic cellular death (type II cell death). For example, autophagy can protect the cell from apoptosis during metabolic stressors such as low nutrient conditions [18]. Atg5 knockout experiments in mice and autophagy inhibition experiments *in vitro* have linked caspase 3 levels and programmed cell death to autophagy [19]. However, autophagy inhibition does not necessarily activate apoptosis. Autophagy can trigger necrosis or apoptosis

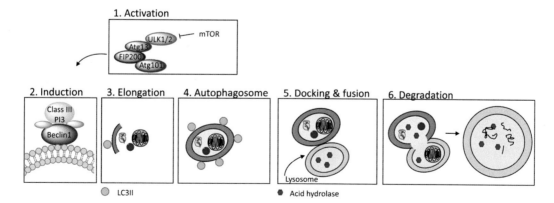

FIGURE 9.4 The autophagic procession. Autophagy proceeds by several steps: (1) Activation of the ULK-complex regulated by mTOR, (2) induction at the phagophore assembly site (PAS) which entails activation of a complex with the biomarker beclin-1, (3) elongation of the phagophore and subsequent engulfment (4) of the cellular debris, (5) fusion of the autophagosome with a lysosome to form a phagolysosome, and (6) degradation of contents by lysosomal hydrolases and recycling of basic materials. Atg, autophagy-related protein; LC3-I/II, microtubule-associated protein 1 light chain 3; ULK1, unc-51 like autophagy activating kinase 1; RB1CC1, FIP200, RB1-inducible coiled-coil 1.

as a primary response to stress signals. In many cases autophagy precedes apoptosis, potentially to assist catabolically or as an ATP source to facilitate apoptotic processes. On the other hand, in some cell types, autophagic cell death resulted with caspase inhibition. Massive autophagy thus has the capacity for cytotoxicity [18,20].

9.1.8 Pharmacology of mTOR: Rapamycin and Second-Generation Rapalogs

As mTOR functions at the center of the metabolic hub for catabolic and anabolic balance with downstream implications on cell survival, mTOR influences the health of cells, in particular with regard to metabolic disease and epilepsy, and the nervous system function with pharmacological applications in carcinogenesis and cancer treatment, diabetes, aging, and neurological disorders. mTOR kinase was first discovered in rapamycin-resistant yeast, TOR mutants, in 1991. Rapamycin (sirolimus) is a non-antibiotic macrolide immunosuppressant isolated from the Easter Island soil bacterium *Streptomyces hygroscopicus* by Suren Sehgal in the 1970s. It was originally developed as an antifungal until its potent immunosuppressive and antiproliferative properties were unveiled. The first Food and Drug Administration (FDA) approval for rapamycin came in 1999 for its use in organ transplantation immunosuppression [21].

Rapalogs are relatively hydrophobic and readily cross the plasma membrane. In both yeast and mammals, rapamycin complexes with the 12-kDa FK-Binding Protein to inhibit Raptor-bound mTOR kinase activity, thus blocking mTORC1 but not mTORC2. The conjugation is non-obligate, and free rapamycin can itself inhibit mTORC1. Recent work has shown that long-term treatment with rapamycin can exert inhibitory effects on the assembly of mTORC2 in Hela and PC3 cells [22]. mTORC2 promotes cellular survival through Akt/PKB activation. Akt in turn phosphorylates mTORC2 on SIN1 T86 resulting in a positive feedback loop [23]. The next generation of rapalogs were intended to (i) reduce the toxicological profile which includes risk of nephrotoxicity, hepatotoxicity, lung, pancreas, and systemic toxicity and (ii) improve the potency through mTORC1/2 coinhibition. Rapamycin analogs, temsirolimus, everolimus, and ridaforolimus (Figure 9.5) have improved hydrophilicity and pharmacokinetic profiles. Second-generation inhibitors of mTOR, termed ATP-competitive mTOR kinase inhibitors block the kinase functions of both mTORC1 and mTORC2. Among these are AZD-8055, OSI-027, and INK128 in clinical trials, or the second-generation rapalogs in preclinical use Torin 1, Torin 2, and WYE-132 which show improved potency, but the issue of toxicity remains [24]. Dual Akt/mTOR inhibitors (i.e., BEZ235 or PI103) also show toxicity. Despite these limitations,

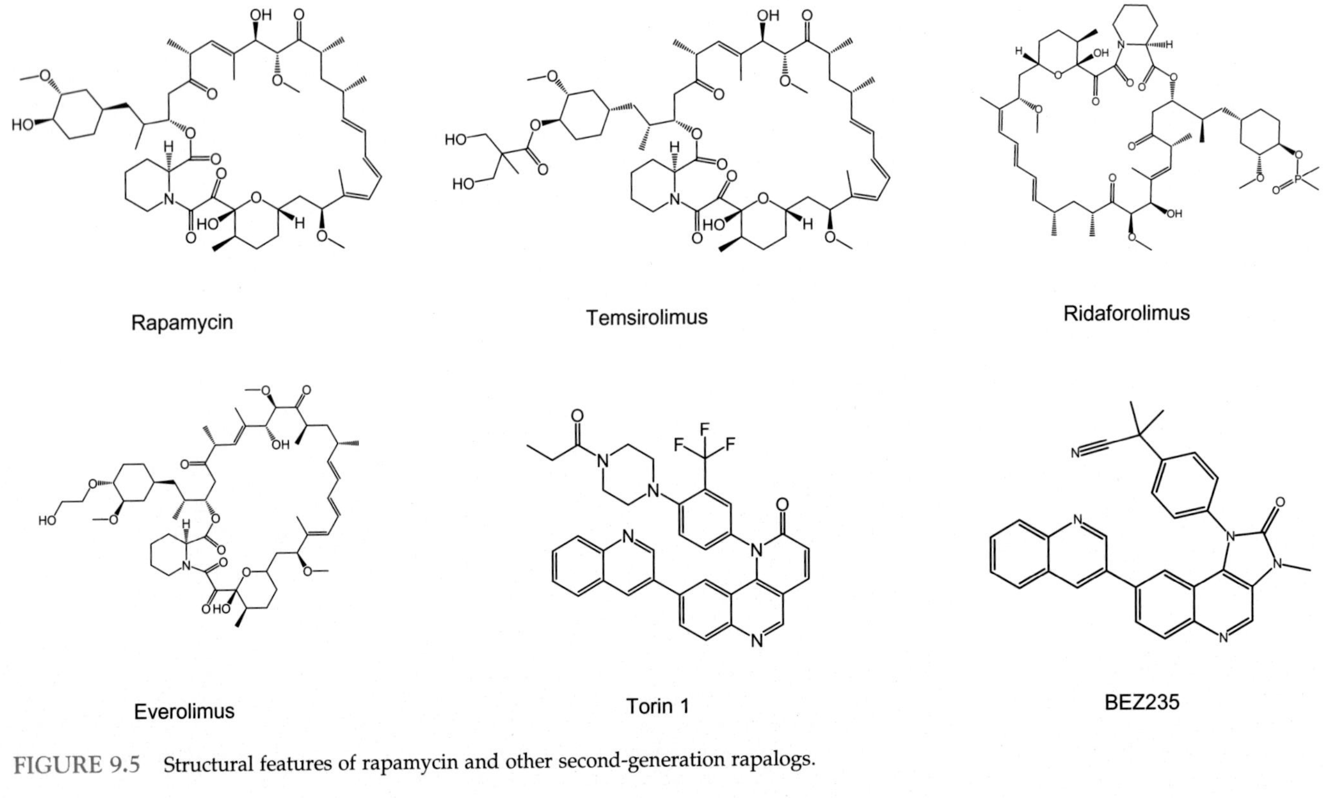

FIGURE 9.5   Structural features of rapamycin and other second-generation rapalogs.

TABLE 9.1 Selected Second-Generation "Rapalogs" in Clinical Trials for the Treatment or Management of Heritable Metabolic Disease

	Specificity	Disease
Sirolimus	Inhibitor of FKBP12 IC50 = 0.1 nM (HEK293 cells)	• Tuberous sclerosis • Neurofibromatoses • Angiofibroma • Neurofibroma associated with: • Neurofibromatosis type 1

Everolimus (RAD01)	Inhibitor of FKBP12
	
	IC50 = 1.6–2.4 nM (cell-free assay)
		• Tuberous sclerosis
		• TSC-related autism
		• Plexiform neurofibroma associated with:
		  • Neurofibromatosis type 1
		  • Neurofibromatosis type 2

Temsirolimus	IC50 = 1.76 μM (cell-free assay)	• Diabetes mellitus, types 1 and 2 • Hypoglycemia • Pompe disease • Fanconi anemia • Cystinosis (kidney transplants)
Zotarolimus	Inhibitor of FKBP12 IC50 = 2.8 nM	• Diabetes mellitus, type 1

FKBP12, 12-kDa FK-binding protein.

clinical trials for rapamycin, rapamycin analogs, and second-generation rapalogs continue in the areas of oncology, transplantation, aging, and coronary artery disease, diabetes, and epilepsy (Table 9.1).

9.2 THE ROLE OF mTOR IN HERITABLE DISEASES

9.2.1 Neurodegenerative Diseases and Autophagy

As part of their pathogenesis, a number of neurological diseases result from genetic predispositions in an autophagic pathway [25]. As normal homeostasis in the central nervous system (CNS) depends on autophagy for the recycling of cellular proteins and neurotrophic factors, it is no surprise that defects in autophagy are a hallmark in neurodegenerative diseases, such as Alzheimer's disease (AD), Parkinson's disease (PD), Huntington's disease (HD), and amyotrophic lateral sclerosis (ALS) [26].

9.2.2 Alzheimer's Disease

Alzheimer's disease (AD) is the most common progressive neurodegenerative disease, affecting

approximately 20% of women and 10% of men above the age of 65 years and a total of 30 million people worldwide. Progression of this disease occurs with the accumulation of misfolded proteins, changes in neuroinflammation, increases in oxidative stress and the loss of synaptic contacts and progressive neuronal cell death [27]. Although most cases (95%) have an unexplained etiology, heritable forms of Alzheimer's disease exist and typically involve mutations in the genes encoding for amyloid precursor protein (APP), presenilin-1, and presenilin-2. APP is critical for the anchoring of actin in the cytoskeleton, while the latter two genes regulate the formation of amyloid-β, a neurotoxic protein which forms plaques and protein aggregates [28].

Genome-wide association studies have shown autophagy to be transcriptionally down-regulated during aging in Alzheimer's patients, contradictory to the normal aging processes [29]. Lysosomal pathway abnormalities precede formation of the neurofibrillary tangles and senile plaques, with increased lysosomes, lysosomal hydrolases (including cathepsins), and AD progression marked by lysosomal dysfunction and build-up of $A\beta_{1-42}$ and vacuolar structures. In a mouse model of AD, modulation of autophagic vacuole proliferation occurred via mTOR pathways [30,31].

9.2.3 Parkinson's Disease

Parkinson's Disease (PD), an incurable progressive neurodegenerative disorder, is the second most common neurodegenerative disease, affecting 2% of the population over the age of 60 years [28]. Clinical features of PD include resting tremor, rigid motor functions, bradykinesia, and postural instability with cognitive impairments. The molecular mechanisms driving this disorder remain unclear; however, a great number of mutations have been identified. The most common PD mutations alter the leucine-rich region kinase 2 (LRRK2, also known as PARK8) leading to the familial autosomal dominant form of the disease. The G2019S mutation in LRRK2 increases kinase activity (gain-of-function mutation) activating the calcium-dependent AMPK effector of autophagy [32].

Mitochondrial dysfunction is an early sign of PD, regulated by the associated genes DNA phosphatase/kinase 1 (PNK1) and Parkin. These two genes mediate degradation of damaged mitochondria via mitophagy and result in the accumulation of dysfunctional mitochondria [33]. Although PD is typically considered a mitochondrial disorder, the presence of Lewy bodies implies defective protein handling (i.e., proteolysis). The normal degradation of α-synuclein, a cytoplasmic protein, occurs by chaperon-mediated autophagy, a process defective in PD patients. Rapamycin has been shown to normalize autophagic processes in PD disease models (*in vitro* and murine) [34]. These findings underscore the fact that PD is not merely a mitochondrial disease.

9.2.4 Huntington's Disease

HD affects the basal ganglia and other portions of the brain responsible for movement, resulting in uncontrollable spasms termed "chorea." HD is an autosomal dominant disorder caused by a mutation to the huntington gene (HTT, chromosome 4p16.3). This gene includes a stretch of CAG (glutamine) nucleotide repeats, and the age of onset inversely correlates with a greater repeat number. The normal role of the huntingtin protein, although not well understood, is believed to have anti-apoptotic properties through inhibition of pro-caspase 9 processing [35].

Autophagic processes impede the progression of HD, and *in vitro* experiments have shown that they act to compensate for huntingtin protein accumulation. The accumulation of two proteins mutant huntingtin protein and α-synuclein are the major mechanism of neurotoxicity in HD, with consequences on transcriptional regulation and mitochondrial dysfunction. Both mTOR-dependent (rapamycin) and mTOR-independent (trehalose) inducers of autophagy improve the clearance of these protein aggregates *in vitro* and in a murine model of HD [36,37].

9.2.5 Amyotropic Lateral Sclerosis

Amyotropic lateral sclerosis (ALS) affects motor neurons and promotes myasthenia, muscle atrophy, and paralysis. The typical age of onset is between 35 and 50 years of age with death expected 3–5 years after onset [38]. The vast majority of cases (90–95%) are sporadic and unexplained; however, the familial form is typically due to mutations in superoxide dismutase (SOD1). SOD1 is an antioxidant and neuroprotective enzyme with involvement in apoptosis. Some hypothesized causes for the pathophysiology of ALS include protein aggregation, lactate dyscrasia, and mitochondrial pathobiology [39,28].

There is an extensive body of work in various model organisms evaluating the molecular mechanisms involved in ALS, from *Drosophila* to zebrafish and mouse models. A reverse genetic screen was conducted in a *Drosophila* model to identify potential mutations resulting in an ALS phenotype. These studies carried out by Delvadigamani et al. identified known mutations from human ALS (TDP-43, Alsin2, and SOD1) and observed a decrease in *Drosophila* Tor

S6K activity. Furthermore, rapamycin treatment mitigates the phenotype (i.e., loss of locomotor function and neuronal axon defects) in zebrafish and mouse models of ALS [40,41]. A second report defined the role of aggrephagy in ALS. Scotter et al. demonstrated that the ubiquitin proteasome system (UPS) is involved in the degradation of transactive response DNA-binding protein 43 (TDP-43). Inhibition of the UPS resulted in TDP-43 aggregates, but dual inhibition of autophagy and aggrephagy most closely reflected the human condition [42]. Lastly, an mTOR-independent autophagy inducer, trehalose, improved motor neuron survival in an ALS mouse model [43].

9.3 AMINOACIDOPATHIES

9.3.1 Phenylketonuria

Phenylketonuria (PKU) is an autosomal recessive disorder which results from accumulation of phenylalanine. PKU is most often associated with mutation in the gene *PAH* which is responsible for transcription of phenylalanine hydrolase (Figure 9.6). The PAH gene is located on the long (q) arm of chromosome 12. Mutations to the PAH gene result in a deficiency in hepatic phenylalanine hydroxylase [44] and ensuing accumulation of phenylalanine. Phenylalanine hydroxylase, in concert with the cofactor tetrahydrobiopterin (BH_4), converts

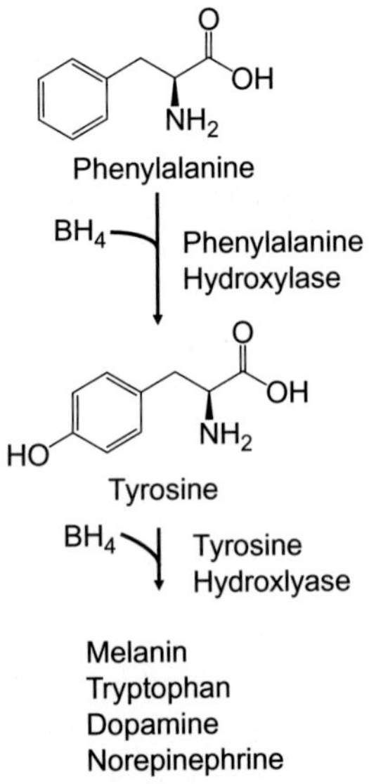

FIGURE 9.6 *Phenylalanine metabolism. Conversion of the essential amino acid phenylalanine to tyrosine requires catalysis by phenylalanine hydroxylase and the cofactor tetrahydrobiopterin (BH_4). Tyrosine is further metabolized by tyrosine hydroxylase, also requiring BH_4, and allowing the metabolism of critical downstream neurotransmitters, amino acids, and hormones. Genetic alterations in the Pah locus or enzymes of BH_4 recycling or synthesis can result in phenylketonuria or BH_4 deficiency.*

phenylalanine to tyrosine. Tyrosine, in turn, is used to synthesize tryptophan, neurotransmitters (dopamine and norepinephrine), and melanin (Figure 9.6). Treatment involves strict lifelong dietary restriction of phenylalanine, and without treatment, children can develop a permanent intellectual disability, often accompanied by seizures, developmental delays, and psychological disorders. Phenylalanine is transported into the brain by the large neutral amino acid (LNAA) carrier, L-amino acid transporter 1 (LAT1). Several pathomechanisms may contribute to phenylalanine toxicity and neuropsychological dysfunction: (i) high concentrations of phenylalanine inhibit the LAT1-mediated transport of LNAAs, decreasing their brain penetration; (ii) reduced pyruvate kinase activity; and (iii) inhibition of mTORC1 activity [45].

More recently the dysregulation of the mTOR pathway has been examined in PKU. Sanayamma and colleagues used K562-D cells, a cell line sensitive to proliferation in the presence of phenylalanine, to survey the impact of phenylalanine on mTOR signaling. K562 cells were cultured for 4 days with phenylalanine (5 mM) and BCAAs (leucine, isoleucine, and valine) at 1 mM. Of the BCAAs, only valine improved cell proliferation versus phenylalanine alone. Valine did not facilitate cell proliferation by itself, yet it did reduce the phenylalanine-dependent growth inhibition at several tested doses (0.2−2 mM valine). Finally, Western blot analysis indicated that phenylalanine reduced 70S6K phosphorylation and valine restored mTOR activity to control levels [45].

9.3.2 BH₄ Deficiency

Deficiency of BH₄, an essential cofactor for phenylalanine, tyrosine, and tryptophan hydroxylases, as well as nitric oxide synthases, can cause several metabolic disorders, including PKU and PD [46,47]. *De novo* synthesis of BH₄ from GTP proceeds through sequential enzymatic reactions mediated by GTP cyclohydrolase I (GTPCH), 6-pyruvoyl-tetrahydropterine synthase (PTPS), and sepiapterin reductase (SR). Heterogeneous genetic lesions have been identified in BH₄ patients, including autosomal recessive variants in both synthesis and recycling enzymes: GTPCH, PTPS, SR, and quinoid dihydropteridin reductase [48,49] (Figure 9.6).

The mTOR signaling pathway has been studied in a BH₄-deficient $Spr^{-/-}$ mouse devoid of the gene encoding SR [50]. In the Spr null mouse, mTORC1 signaling was inactivated in favor of autophagy (as measured by conversion of LC3-I to LC3-II and confirmed by electron microscopy). Implementation of a tyrosine supplemented diet initiated at birth rescued the animal growth and mTORC1 inhibition (measured by

phosphorylation of S6K at Thr389) by 25 days of life, and also inactivated autophagy [50,51]. Relative to $Spr^{+/+}$ mice, phosphorylation of Akt and AMPK was examined with no difference between the two genotypes. Further elucidation of this mechanism was conducted using NIH3T3 mouse fibroblasts. Under high phenylalanine and low tyrosine conditions, fluorescent imaging studies showed ectopic expression of RhebQ64L, which induced phosphorylation of HA-S6K and attenuated LC3 formation under tyrosine-depleted conditions. Finally, this pathway was studied and confirmed using the lymphocytes collected from untreated and treated BH₄-deficient PKU patients.

9.3.3 Tyrosinemia Type I

Tyrosinemia is a hereditary inborn error of metabolism whereby the body cannot effectively break down tyrosine, resulting in hepatic and renal toxicity and neurological impairments. Three types of tyrosinemia (types I–III) occur in humans, each with distinct pathologies resulting from defects in different enzymes. The most severe form is tyrosinemia type I which results from inactive fumaryl acetate hydrolase, accumulation of fumarylacetoacetate (FAA), the product of tyrosine decarboxylation succinylacetoacetate (a mitochondrial toxin which inhibits phosphorylation within the Krebs cycle), as well as succinylacetone [52].

In vitro studies performed by Jorquera and Tanguay in CHO-V79 and human HepG2 cells, showed that FAA-mediated cell death was induced primarily by apoptosis through activation of cyclin B-p34cdc2 kinase [53]. Early activation of caspase 1 was followed by mitochondrial cytochrome c release and caspase 3 activation measured by colorimetric assay and mitochondrial membrane potential DiOC₆ staining. Parallel work in a Fah and 4-hydroxyphenylpyruvate dioxygenase (Hpd) double-mutant mouse ($Fah^{-/-}$ $Hpd^{-/-}$) showed that cytochrome c was released from mitochondria following the administration of HGA [54]. The authors also studied caspase inhibitors (YVAD, caspase 1; and DEVD, caspase 1 and 3) in hepatocyte culture and in the mouse model. Caspase inhibition prevented apoptosis in these systems; further molecule studies are necessary to determine any prospective involvement/aberrations in mTOR signaling that could be related to these mitochondrial or apoptotic outcomes in tyrosinemia.

9.3.4 Histidinemia

Histidinemia, also known as histidinuria, was first described in 1961 by H. Ghadimi in two young children [55]. This disease results from a defect in histidase (also called histidinase or histidine ammonia-lysase EC 4.3.1.3), which catalyzes the irreversible non-oxidative deamination of histidine to urocanic acid, primarily in the liver and skin. Diagnostic confirmation presents marked elevation of urinary excretion of histidine and its transaminated byproducts and/or blood histidine levels about 4–14 times above normal [56,57]. Histidinemia is one of the most common inborn errors of human metabolism occurring in the range of 1:8000 in Japan and 1:37,000 in Sweden. The impact of this biochemical defect varies considerably from relatively benign to severe developmental delays and deficits. Treatments include dietary restriction of histidine, without any improvements.

The role of mTOR in the pathophysiology of this aminoacidopathy has not been studied in a disease-specific model. However, some evidence links the histidase product urocanic acid to a protective mechanism against ultraviolet B (UVB)-induced apoptosis in keratinocytes, as examined in histidinemic mice [58]. Separate *in vivo* studies of the inducible mTOR-deficient K5-CreER(T2); mTOR(fl/fl) mouse model found that mTOR deletion attenuated UVB-induced S6K phosphorylation and Ser473 phosphorylation of the mTORC2 target Akt, effectively increasing apoptosis. Both mTORC1 and mTORC2 play unique and complimentary roles in UVB-induced signaling in the skin [59].

Histidine on the other hand has been more specifically implicated in the mTOR pathway. In a 2008 publication [60], the role of essential amino acids on protein synthesis in the mTOR pathway (Figure 9.3) was examined in mammary epithelial cells. Culture conditions with excessive leucine, isoleucine, and valine increased S6K1 phosphorylation relative to cells deprived of essential amino acids; however, lysine, histidine, and threonine together inhibited S6K1 phosphorylation in bovine and murine mammary epithelial cells. Enhancing the levels of each amino acid to five-fold above normal media concentrations resulted in no observable effect, implying a synergistic process. However, a negative synergy occurred with the collective increase in leucine, histidine, and threonine lowering mTOR Ser 2448 phosphorylation.

A secondary connection of histidine to the mTOR pathway involves the histidine transporter (SLC15A1). SLC15A1, a lysosomal transporter of histidine and oligopeptides, is critical in the mTOR-mediated inflammatory response. A histidine transporter-deficient mouse (Slc15a4$^{-/-}$) develops histidine accumulation and diminished pH regulation in lysosomes followed by changes in vacuolar H$^+$-ATPase and mTOR activities [61]. Further elucidation of the molecular mechanisms of histidine/mTOR signaling in cytoprotection and inflammatory response would advance the understanding of the regulatory role of amino acids in mTOR signaling and prospectively open up new interventions for histidinemia.

9.3.5 Maple Syrup Urine Disease

Maple syrup urine disease (MSUD) or branched-chain ketoacid dehydrogenase (BCKDH) deficiency is a large neutral aminoacidopathy in which BCAAs, leucine, valine, and isoleucine accumulate. The most common defect in this rare disorder (incidence 1:180,000) occurs by a mutation on chromosome 19 encoding for the E1α subunit of BCKDH. MSUD is an autosomal recessive disorder, and dietary restriction of leucine, valine, and isoleucine at an early age can decrease the detrimental neurological outcomes. Despite strict dietary adherence, many MSUD patients experience long-term deficits. Liver transplantation offers potential correction of the underlying deficit, and more than 100 liver transplantations have been performed for classical MSUD [62].

Mills et al. sought to address the molecular mechanisms of MSUD using primary microglial cells, the brain's resident macrophage, in which the effects of BCAAs were examined. Microglial cells exposed to media containing high levels of BCAA favored the M2 (cell proliferation, repair) versus M1 (inflammation) state and increased interleukin-10 and phagocytic activity and oxidative stress conditions [63,64]. These experiments did not study the potential mediation of these BCAA effects by the mTOR pathway; nonetheless emerging studies have shown TSC/mTOR regulation of the innate inflammatory response and other regulatory roles of mTOR signaling in the immune system [65–67]. Interestingly, mTOR pathways have not been studied in an MSUD disease model despite the prominent pathophysiological role of leucine in MSUD and the role of leucine in mTOR activity mediated by amino acids.

9.3.6 Succinic Semialdehyde Dehydrogenase Deficiency

Succinic semialdehyde dehydrogenase (SSADH), encoded on chromosome 6p22.3 (ALDH5A1), is a mitochondrial enzyme responsible for conversion of succinic semialdehyde to succinic acid. Succinic semialdehyde is produced from gamma-aminobutyric acid (GABA) (elevated in SSADH) and converted to succinic acid which is shunted into the TCA cycle (Figure 9.7). Excess succinic semialdehyde is converted to γ-hydroxybutyric acid (GHB) a neuromodulator and agonist of the GHB and GABA$_B$ receptors. Additional downstream products of GABA metabolism

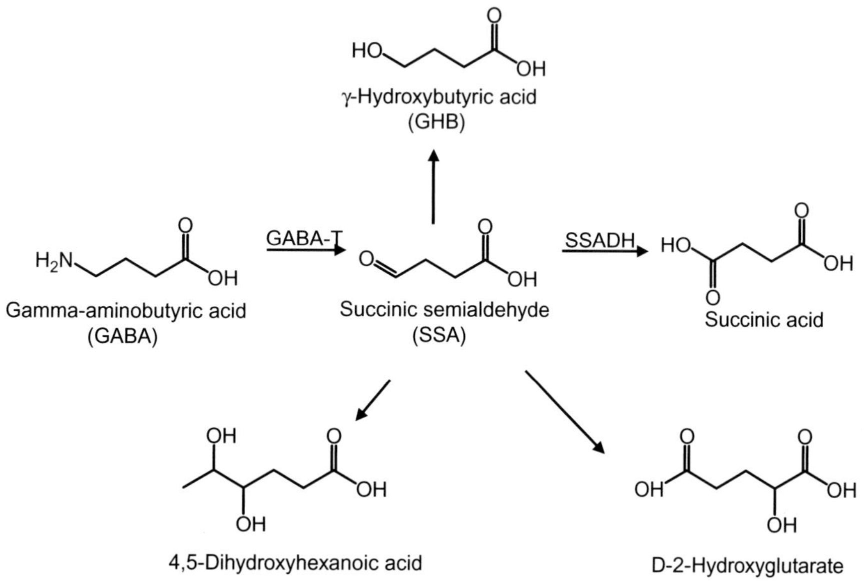

FIGURE 9.7 Gamma-aminobutyric acid (GABA) metabolism. In congenital disorders of GABA transaminase (GABA-T) and of succinic semialdehyde dehydrogenase (SSADH), GABA accumulates. Other metabolites accumulate in SSADH as potential contributors to the heterogeneity of disease severity observed including: succinic semialdehyde (SSA), the neuromodulator and potential neurotransmitter γ-hydroxybutyric acid (GHB), 4,5-dihydroxyhexanoic acid, and D-2-hydroxyglutarate. Studies in yeast and mice have implicated GABA in mTOR-dependent autophagy inhibition.

are also elevated including 4,5-dihydroxyhexanoic acid and D-2-hydroxygluterate. The clinical phenotype of SSADH deficiency includes delays in speech, mental and motor development, as well as the development of seizures, ataxia, and hypotonia. Currently, SSADH deficiency is managed using symptomatic agents (e.g., vigabatrin and sodium valproate to reduce GHB) [68]. Vigabatrin, an irreversible inhibitor of gamma-aminobutyric acid transaminase (GABA-T), further increases GABA but decreases GHB levels. Deleterious side effects (including visual field loss) have rendered vigabatrin treatment unfavorable. Researchers have begun to investigate the potential involvement of mTOR in the retinal toxicities associated with vigabatrin administration. Mice treated with vigabatrin demonstrated a dose-dependent increase in GABA and corresponding increase in mitochondrial size and number. This phenotype was partially corrected with Torin 1. Pilot studies also suggest that vigabatrin application elevated p2448 mTOR and caspase 3 [69].

Mechanistic studies by the Subramani group investigated the effects of GABA elevation in UGA2 mutant yeast (the GABA shunt is conserved), in which threefold elevations of GABA partially inhibited pexophagy, prompting further mechanistic studies. Exogenous application of 10 mM GABA to yeast media decreased pexophagy (measured by Western blot analysis of Pot1-GFP) and mitophagy (measured by Western blot of Om45-GFP). Importantly, these effects of GABA on autophagy could be attenuated by

rapamycin, correcting pexophagy and mitophagy readout levels, and also normalizing redox stress. Further examination in the murine model for SSADH deficiency (aldh5a1$^{-/-}$) showed elevated brain and liver mitochondrial numbers, increased mitochondrial size in the liver and higher pS6. Mitochondrial number normalized with administration of rapamycin (5 mg/kg body weight) as determined by transmission electron microscopy and SOD1 activity assay [70]. These studies cumulatively highlight the importance of GABA in mTOR signaling and implicate alterations of autophagy in SSADH and perhaps other GABAergic disorders.

9.3.7 GABA-T Deficiency

GABA-T, a mitochondrial enzyme encoded by the gene ABAT located on chromosome 16p13.2, converts GABA to SSA. GABA-T deficiency is regarded as even more rare than SSADH deficiency, and had been identified in only a few patients, perhaps in part due to its similar clinical presentation to severe neonatal or infantile epileptic encephalopathy. Analogous to SSADH-deficient patients, individuals lacking sufficient GABA-T activity manifest elevations in GABA with similar pathological implications [71]. GABA is an agonist of metabotropic GABA$_B$ receptors which in turn modulate mTOR [71a]. In a recent study of GABA-T deficient patients and mice, ABAT mutations led to mtDNA depletion implicating a role for GABA-T (and GABA) in the mitochondrial nucleotide salvage pathway [72].

9.4 OTHER AMINOACIDOPATHIES

9.4.1 Cystathioninuria

Cystathioninuria or cystathionase deficiency is an autosomal recessive disease. It can be clinically identified with the measurement of high amounts of urinary cystathionine, an intermediate in cysteine and α-ketobutyric acid production. Cystathioninuria may associate with developmental and intellectual delays. There are no therapies specific for cysathioninuria; however, a high-cysteine diet given to cystathioninuric mice ($Cth^{-/-}$) appeared beneficial [73]. The untreated $Cth^{-/-}$ mouse exhibits acute myopathy along with elevated expression of asparagine synthetase. Additionally, hepatic and skeletal muscle glutathione pools were depleted compared to wild-type mice. Oxidative stress sensitivity was exacerbated when mice or hepatocytes were exposed to low-cysteine conditions. Mice fed a high-cysteine diet were less sensitive to oxidative stress and had lessened muscle atrophy. The authors concluded that the levels of cysteine are important in the pathogenesis of this disease as the mouse model progressed. Skeletal myofibers accumulated microtubule-associated protein 1A/1B-light chain 3 (LC-III) and p62, and mice declined in health until death by severe paralysis by 4 weeks of age. These two markers are indicative of aberrant autophagy [73].

9.4.2 Carnitine Deficiency

Carnitine is important in the transfer of long-chain fatty acids into the mitochondria. Carnitine deficiency is an autosomal recessive disorder of carnitine transport function. Symptoms may present as episodes of hypoketotic hypoglycemia, hepatomegaly, elevated liver transaminases, and hyperammonemia [74]. Carnitine supplementation is the approved treatment for this disorder, and in some cases other dietary adjustments are made to include riboflavin, glycine, and biotin [75].

Secondary to the neurological abnormalities, cardiomyopathy develops to a significant extent in these patients. This pathology has been linked to mTOR pathways in a carnitine-deficient mouse model ($Acsl1^{T-/-}$). These mice exhibit abnormal cardiac mitochondria, which are half the normal size, containing twice the mitochondrial DNA as wild-type littermates [76]. Phosphorylation of S6 kinase was also observed to be fivefold increased in the mutant versus control mice and cardiac hypertrophy occurred

without lipid accumulation. The authors postulated that the explanation for mTOR involvement in this disease relates to the metabolic shift from fatty acid metabolism to catabolism of glucose and amino acids which are known to modulate mTOR directly. This hypothesis has been verified in cancer cells which frequently favor glycolysis.

9.5 HERITABLE AUTISM SPECTRUM DISORDERS

The prevalence of metabolic abnormalities in heritable autism spectrum disorders (ASDs) is unknown, although several metabolic deficiencies have been associated with autistic symptoms, including PKU, histidinemia, adenylosuccinate lyase deficiency, dihydropyrimidine dehydrogenase deficiency, 5′-nucleotidase superactivity, and phosphoribosylpyrophosphate synthetase deficiency [77]. Autism is also a prominent manifestation in the heritable epilepsy disorders, TSC, and in the GABA disorder, SSADH deficiency. Several ASDs with causal mutations in known genetic loci have evidence of mTOR and autophagy involvement including Rett syndrome, neurofibromatosis 1, and fragile X syndrome (Figure 9.8).

9.5.1 Rett Syndrome

Rett syndrome is a neurological impairment caused by mutation in the gene methyl-CpG-binding protein 2 (MECP2). MECP2 is an X-linked gene that results in lethal neonatal encephalopathy in males, but not in females who mature to develop Rett syndrome. Symptoms first manifest approximately a year after birth and progressively worsen resulting in loss of motor and language skills, hand stereotypies, seizures, and ASD [78]. Speech, physical, and occupational therapies are the only available treatments for the majority of symptoms [79,80].

Mouse models have facilitated the characterization of loss-of-function mutations in MECP2 [81]. These models accurately modeled the delayed onset of neurological symptoms in females (6 weeks of age in mice) which also included behavioral phenotypes. Further work with the model has provided some insight into the role MECP2 plays in neurodegeneration; however, the molecular mechanisms remain unclear. Hypothesizing that aberrations in neuronal protein synthesis contribute to ASD, Ricciardi et al. investigated the AKT/mTOR pathway in a Rett syndrome murine model [82]. Immunohistochemistry showed a reduction of ribosomal protein S6 in Mecp2 mutant mice at 8 weeks. Closer examination during

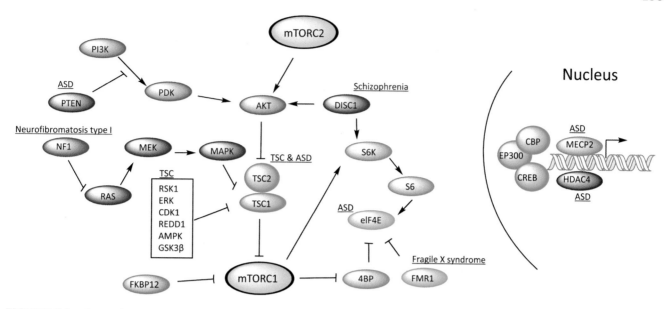

FIGURE 9.8 Genetic loci involving mTOR signaling in neurological diseases and epilepsies. Mutations in phosphatase and tensin homolog (PTEN) can result in autism spectrum disorders (ASDs). Under normal conditions PTEN facilitates phosphorylation of PDK by PI3K and ultimately initiates AKT. Additional causal genetic alterations in ASDs occur in the nuclear transcriptional regulators methyl-CpG-binding protein 2 (MECP2), implicated in most cases of Rett syndrome, and histone deacetylase 4 (HDAC4). Autosomal dominant inheritance of a pathological mutation in tuberous sclerosis complex 1/2 (TSC 1/2) results in tuberous sclerosis complex, a disorder characterized by tubers, epilepsy in the majority of cases, and not infrequently ASDs. Several other negative regulators of TSC have been implicated in tuberous sclerosis, including: ribosomal protein S6 kinase 1 (RSK1), extracellular signal-related kinase (ERK), cyclin-dependent kinase 1 (CDK1), DNA-damage-inducible transcript (DDIT4, also REDD1), adenosine monophosphate-activated protein kinase (AMPK), and glycogen synthase kinase 3 beta (GSK3β). Eukaryotic translation initiation factor 4E (eIF4E, another gene with connections to the occurrence of ASDs) aids in the initiation of translation by recruiting ribosomes to messenger RNA. The fragile X mental retardation (FMR1, causative in fragile X disorder) protein confers eIF4E to the active state. Ultimately, Ras blocks TSC via sequential phosphorylation of mitogen-activated protein kinases, MEK and MAPK. The Ras pathway is inactivated by the neurofibromin gene (NF1) product, whereas mutations in NF1 result in neurofibromatosis type I. Disrupted-in-schizophrenia I (DISC1) acts on AKT and S6 kinase 1 (S6K1), and mutations in DISC1 have been implicated in the psychological disorder schizophrenia and TSC.

development in various brain regions revealed a decrease in protein synthesis, attributed to aberrant Akt/mTOR signaling. Western blot analysis of the p70-S6K revealed that phosphorylation suppressed mTOR signaling as well as Akt in the brain.

9.5.2 Neurofibromatosis Type 1

Neurofibromatosis type 1 is a commonly occurring (~1:2500 individuals) autosomal dominant disorder caused by a mutation to the neurofibromin gene (NF1) located on chromosome 17q11.2 [83]. This disorder is distinguished by café-au-lait spots, neurofibromas, and Losch nodules of the iris. Other clinical presentations vary considerably, but include dysmorphic features, cardiovascular malformation, high expression of benign lesions, and delayed cognitive development [84]. The intellectual impairment in NF1 is accompanied by social, emotional, and developmental deficits typical of ASD. The NF1 gene product, neurofibromin, is a GTPase acting in the Ras signaling pathway (Figure 9.8). Nonetheless, clinical studies involving Ras inhibitors have shown limited effects and have led to

the examination of alternative pathways involved with the NF1 mutation [85]. Primary astrocytes cultured from Nf1$^{flox/flox}$ mice were shown to have upregulated mRNA translation. Using a methionine incorporation assay as a measure of protein production, ~12-fold more methionine was detected in mutants relative to control after 45 min. Cultured primary Nf1 mutant astrocytes were treated with rapamycin (5 μM), which normalized methionine incorporation to control levels. Western blot analysis also showed high phosphorylation of S6 which was corrected by both rapamycin and LY294002 (a PI3K inhibitor). Following these in vitro mechanistic studies, these investigators confirmed that rapamycin could normalize ribosomal S6 aggregates in vivo and promote normal astrocyte proliferation, indicating that rapalogs should be further examined to treat neurofibromatosis. Additional studies using Pten, Tsc1, and Tsc2 knockout mice, each over expressing Rheb, have also been conducted using rapamycin intervention. Unlike the elevated phospho-S6 observed in Nf1 mutants, these models were shown to have elevated phospho-histone-H3, which normalized with rapamycin [86].

9.5.3 Fragile X Syndrome

Fragile X syndrome is the leading cause of heritable autism. Fragile X mutations can be classified as a premutation if there are between 50 and 230 CGG repeats in the promoter region of the fragile X mental retardation-1 gene (*FMR1*) and a full mutation if there are greater than 200 repeats [86] (Figure 9.8). The loss of function of FMR1 activates PI3K via PI3K enhancer resulting in mTOR activation. Current findings suggest that elevations in mGluR1 and mGluR2 (both glutamate receptors) lead to reduced localization of AMPA receptors into the postsynaptic membrane impairing plasticity [87]. Research findings have supported this hypothesis and sparked interest in utilizing glutamate, GABA, and/or mTOR signaling pathways as therapeutic targets [88,89]. Unexpected and convoluted results have been derived from *Drosophila* providing key evolutionary differences in the role of FMR1. However, Fmr1 knockout zebrafish have demonstrated key behavioral measurements in concordance with the human phenotype and have potential as a disease model [90].

9.5.4 Summary of Heritable Autism Disorders

With the prevalence of ASDs reaching 1 in 68 children (1 in 42 boys and 1 in 189 girls) scientists will continue to explore the molecular complexities and involvement of mTOR signaling and autophagy in autism [91].

9.5.5 mTORopathy: Tuberous Sclerosis and Genetic Epilepsy

Tuberous sclerosis is a phakomatosis (or neurocutaneous disorder) named for the cortical tubers which constitute one of three types of TSC-associated benign brain tumors. Tuberous sclerosis has an incidence of 1:6000 with one-third familial and two-thirds of cases arising from *de novo* mutations. The familial form of the disease inherits in an autosomal dominant pattern with causal mutations found in the genes *TSC1* (chromosome 9q34; less severe) and *TSC2* (chromosome 16p13.3; most severe) [92]. The protein products hamartin (TSC1) and tuberin (TSC2) account for 10−15% and 75−80% of sporadic cases, respectively (Figure 9.8). Hamartin functions in cellular adhesion and tuberin in the cell cycle, while together they form the TSC 1/2 complex which negatively regulates Rheb, an mTOR activity enhancer. The disease is multisystemic with variable penetrance characterized by hamartomas (non-malignant excess of tissue) of the brain, skin, liver, kidney, eyes, lungs, and heart [93]. The three types of brain lesions are cortical tubers,

which are believed to cause TSC-associated seizures, subependymal nodules which develop near the cerebral ventricles calcifying within the early months of life, and subependymal giant cell astrocytomas (SEGAs) which occur in 15% of cases. The major clinical features of tuberous sclerosis can also include facial angiofibromas, non-traumatic ungula, hypomelanotic macules, shagreen patches, retinal nodular hamartomas or achromic patches, cardiac rhabdomyoma, lymphangiomyomatosis, renal angiomyolipoma, bone cysts, or hamartomatous rectal polyps [92]. Twenty to sixty percent of TSC cases are associated with ASDs which accounts for 1−4% of autism cases, reduced intellectual quotient (30% very low), and psychiatric disorders. Epilepsy occurs in about 80% of TSC patients, the most common CNS manifestation, as well as infantile spasms 50% [94,95].

The variable presentation and progression of TSC cause challenges for the care and management of patients. Multiple antiepileptic agents have been employed in TSC with continued cases of refractory epilepsy. Vigabatrin, an irreversible GABA transaminase inhibitor, improves TSC-associated infantile spasms 95% of the time, an outcome likely related to augmenting GABA levels in a setting of deficient GABAergic inhibition, scarcity of GABA interneurons, and changes to GABA receptors found in TSC [92]. A systematic review of outcomes in epilepsy surgery in TSC found 101 patients that were seizure-free (57%) and improved frequency by more than 90% in another 32 patients (18%) [96], and brain resection surgery is effective in 40% of children who continue to manifest seizures [97].

Studies in both conventional and conditional knockout animal models of TSC have shed light on the molecular pathogenesis of the disease. Tang and others used an autistic mouse (TSC2 deficient, $Tsc2^{+/-}$) to study mTOR [98]. In this mouse model, mTOR is constitutively overactivated and results in postnatal spine pruning defects. In this model, rapamycin corrected the pruning effects as well as the ASD-like behavior effects. Using an Atg7^{CKO} neuronal autophagy-deficient (Atg7 is required for autophagosome formation) mouse and a Tsc2$^{+/-}$:Atg7^{CKO} double-mutant, rapamycin was incapable of correcting the ASD phenotype. The increase in spine density was accompanied by elevated mTOR phosphorylation, ribosomal protein S6, and LC3-II levels, which has been observed in both mouse and human [98]. This work indicated that autophagy is essential in dendritic spine pruning and divergent mTOR and autophagic function in ASD. The authors suggested that further evaluation of the downstream targets of mTOR in ASD patients could provide a therapeutic target for certain ASD cases.

As traditional antiepileptic drugs tend to focus on end-stage mechanisms to balance excitability/inhibitory neurotransmission, recent studies in TSC have focused on modifying epileptogenic processes and signaling mechanisms upstream. As an mTORopathy with hyperactive mTOR signaling, mTOR modulation with rapamycin and related pharmaceuticals represent a rational study point. Early treatment with rapamycin in TSC mice appears to prevent the underlying molecular and histological manifestations leading to epilepsy, and later treatment in TSC2$^{+/-}$ (haploinsufficient) mice can reverse some clinical features of the disease [99,100]. In TSC mouse models mTOR inhibition has shown the potential to improve cognitive, behavioral, developmental, lifespan outcomes, and seizures. A number of clinical trials are underway to test the utility of rapamycin in children and adult patients [101].

While the US FDA has not approved rapamycin/ rapalog treatment for TSC, in 2010 the FDA approved everolimus (Afinitor®), an orally administered mTOR inhibitor for SEGAs and angiomyelolipoma kidney tumors associated with TSC. A growing number of clinical trials poised to assess treatment outcomes with rapamycin and everolimus in TSC are underway including: outcomes on SEGA (NCT02338609), LAM (Phase III NCT00790400), cutaneous manifestations (Phase I complete NCT01031901) and facial angiofibromas (Phase I NCT01853423), brain mTOR activity and hyperexcitability (Phase II NCT02451696), neurocognition (Phase II NCT01289912; Phase II NCT01954693), autism (Phase2/ 3 NCT01730209; Phase II NCT01929642), epilepsy (Phase I NCT01070316; Phase III NCT01713946), as well as long-term outcomes on height, weight, and development (Phase IV NCT02338609).

Clinical trials in TSC have extended the traditional application of rapalogs outside of transplantation immunosuppression. In a study of 23 medically refractory epilepsy patients with TSC treated for 12 weeks with everolimus, 17 of 20 experienced seizure reductions by a median reduction of 73%, as well as improvements in behavior and quality of life [102]. A number of questions have arisen such as, what is the mechanism of mTOR's effective action, whether this is one of antiseizure or antiepileptogenesis, and will mTOR inhibitors also prove advantageous in the treatment of epilepsy outside of TSC-related epilepsy [99]?

9.5.6 Focal Cortical Dysplasia

Focal cortical dysplasia (FCD), a common form of epilepsy in children, accounts for 25% of cases of medically refractory epilepsy [103]. *De novo* brain somatic mutations causing gain-of-function in mTOR were discovered after deep sequencing of brain tissue from 73 patients with FCD type II. Rapamycin intervention in mice with FCD suppressed the cytomegalic neurons and epileptic seizures associated with the focal cortical expression of mutant mTOR [104].

9.5.7 Temporal Lobe Epilepsy

Temporal lobe epilepsy, regarded as polygenic, heterogeneous, and complex, constitutes the most common partial epilepsy, with few genes implicated. However, in familial temporal lobe epilepsy, a number of documented mutations occur within the genes for epitempin or leucine-rich glioma-inactivated 1 and the voltage-gated sodium channel 1(SCN1 A/B) [105]. The prognosis for refractory cases includes worsening memory, mood, and quality of life. Animal studies have begun to address the potential for mTOR therapy in temporal lobe epilepsy, and have started to address the mechanistic question of antiepileptogenesis versus antiseizure.

In a kainite-induced epileptic rodent model of acquired temporal lobe epilepsy, rapamycin administration delayed the development of epilepsy [106]. In a pilocarpine-induced temporal lobe epilepsy murine model, seizure frequency remained high with rapamycin treatment that was undertaken 1 day after development of status epilepticus. Although rapamycin initiation occurred late in epileptic development, a 42% improvement in mossy fiber sprouting occurred with rapamycin-treated mice compared to vehicle, and the epilepsy-related hypertrophy of the dentate gyrus was also reduced. However, granule cell proliferation, hilar neuron loss, and ectopic granule cells remained unchanged with this treatment in the pilocarpine model [107]. Mossy fiber sprouting, previously viewed as early and irreversible, may in fact be progressive and reinforce seizures [99]. The clinical implications for rapamycin in epilepsy treatment have shown some promising results in preclinical models of temporal lobe epilepsy. Further research will help elucidate the full potential of mTOR inhibition in epilepsy, and address the question of whether rapalogs function best in the presymptomatic stages or as antiseizure drugs.

9.6 LYSOSOMAL STORAGE DISORDERS

9.6.1 Pompe Disease (Glycogen Storage Disease Type II)

Pompe disease is an autosomal recessive lysosomal storage disorder associated with deficiency in acid alpha-glucosidase (GAA), leading to accumulation of glycogen in lysosomal bodies [108]. The most notable symptoms are deterioration in cardiac and skeletal muscle, and fatality by cardiorespiratory failure. Enzyme replacement therapy alleviates cardiac but not skeletal symptoms of the disease. The selective nature of recombinant therapy is in part due to the accumulation of autophagic vesicles, which can capture and sequester recombinant GAA. Both patients with Pompe disease and GAA-deficient mice exhibit analogous phenotypes of constitutively activated autophagy in skeletal muscle. Nishiyama et al. examined these traits in tissue samples from patients, and then studied the markers of the Akt/mTOR pathway in cultured patient fibroblasts. These investigators observed that patient fibroblasts had decreased phosphorylation of AKT and p70 S6 kinase. Treatment with insulin, which increases of AKT and p70 S6 kinase phosphorylation, resulted in suppression of autophagy, which improved the localization of recombinant GAA via decreased lysosomal accumulation. The same group further demonstrated that the Akt/mTOR pathway is responsible for up-regulation of autophagy in Pompe disease via examination of the phosphorylation status of Akt and S6K [109].

9.6.2 Niemann—Pick C Disease

Niemann—Pick type C (NP-C) disease is an autosomal recessive and lethal disease caused by defects in cholesterol transport proteins, NPC1 and NPC2 [110]. These defects result in the intralysosomal and -endosomal accumulation of cholesterol and glycosphingolipids with an increase in basal autophagy [110,111]. The symptoms may vary but typically include hepatosplenomegaly and delay of motor development. Neurological symptoms include abnormal upward saccades, poor head control, diminished motor skills, and loss of hearing and development of tremor. Only recently have treatments for NP-C become available, namely Miglustat, a reversible inhibitor of glycosphingolipid synthesis. Other drugs being studied as potential treatments include cholesterol-lowering agents, vorinostat, and β-cyclodextrin [112].

Rapamycin has previously been shown to decrease low-density lipoprotein-receptor expression, and therefore was tested as a potential NP-C therapy in human fibroblasts [113,110]. The latter study confirmed the hypothesis that rapamycin could lower cholesterol in cultured fibroblasts and enhanced the effects of vorinostat and β-cyclodextrin in combination. The observed reduction in autophagy by rapamycin is proposed to be due to decreased Beclin-1 (part of the multiprotein complex involved in the autophagic nucleation step) expression instead of the Akt/mTOR pathway based on work *in vitro* and in the mouse [114].

9.6.3 G_{M1}-Gangliosidosis

The lysosomal storage disorder G_{M1}-gangliosidosis is caused by mutations in the *GLB1* gene (3p21.33), encoding for the enzyme β-galactosidase (β-gal). This disease is associated with skeletal dysplasia, visceromegaly, and progressive neurodegeneration. Severe forms demonstrate these clinical symptoms between 1 and 3 years of age [115]. Less severe presentations may develop as late as 30 years of age. Enhanced autophagy, as in other lysosomal storage diseases, is a conserved feature in G_{M1}-gangliosidosis. The mechanism of elevated autophagy has been studied in the β-gal$^{-/-}$ mouse model. Similar to the Pacheco study (see NP-C disease) overactivation of autophagy occurs via Beclin-1 and not the Akt/mTOR pathway.

9.7 HERITABLE MITOCHONDRIAL DISEASES

Mitochondria are the cellular machinery responsible for regulating reactive oxygen species, initiating apoptosis, and producing energy in the form of ATP. Mitochondria also have numerous metabolic responsibilities involving pyruvate, amino acids, fatty acids, and steroids. Numerous genetic alterations can result in mitochondrial diseases, the target tissue of these defects will dictate the symptoms, and, in light of the link between many of these metabolic processes and mTOR, studies of rapamycin and/or autophagic processes have been undertaken in Leigh's syndrome, Kearns—Sayre syndrome, and complex I deficiency (Table 9.2).

9.7.1 Leigh's Syndrome

Leigh's syndrome, also called juvenile subacute necrotizing encephalomyelopathy, is a rare inherited metabolic disorder and neurodegenerative disease (affecting 1:40,000 live births). Typically this disease arises in infancy or early childhood. There have been many mutations of mitochondrial and nuclear DNA reported to impact oxidative phosphorylation. In addition to these genetic factors, dysfunction in the pyruvate dehydrogenase complex has been associated with Leigh's syndrome. Typically Leigh's syndrome is inherited in an autosomal recessive manner, but can be X-linked recessive via mitochondrial inheritance. The phenotype is one of progressive neurological deterioration which may include loss of motor skills, anorexia, vomiting, irritability, and seizure. The disease progresses to general weakness, hypotonia, and lactic acidosis which can impair respiratory and renal function [116]. Clinically, this is the most commonly observed pediatric mitochondrial disease. The prognosis

TABLE 9.2 Heritable Mitochondrial Diseases, Genetics and Symptoms

	Cause	Symptoms
Alpers disease	POLG gene mutation	Seizures, dementia, spasticity, blindness, liver dysfunction, and cerebral degeneration
Barth syndrome (lethal infantile cardiomyopathy)	X-linked recessive	Skeletal myopathy, cardiomyopathy, short stature, and neutropenia
Carnitine-acyl-carnitine deficiency	Autosomal recessive	Seizures, apnea, bradycardia, vomiting, lethargy, coma, enlarged liver, limb weakness, myoglobin in the urine, Reye-like symptoms triggered by fasting
Carnitine deficiency syndrome	Autosomal recessive	Cardiomyopathy, failure to thrive, and altered consciousness or coma, sometimes hypertonia
Creatine deficiency syndromes	GAMT and AGAT: autosomal recessive; SLC6A8: X-linked. E40 + E9	General: mental retardation, seizures, speech delay. Additional possible symptoms: GAMT, behavioral disorder, including autistic behaviors; movement disorders SLC6A8, growth retardation; (males) mild to severe mental retardation; (females) learning; and behavior problems
Coenzyme Q10 deficiency	Probably autosomal recessive	Encephalomyopathy, developmental delay, exercise intolerance, ragged-red fibers, and recurrent myoglobin in the urine
Complex I deficiency	Autosomal	Macrocephaly with progressive leukodystrophy, non-specific encephalopathy, hypertrophic cardiomyopathy, myopathy, liver disease
Complex II deficiency	Probably autosomal recessive	Encephalomyopathy, failure to thrive, developmental delay, hyoptonia, lethargy, respiratory failure, ataxia, myoclonus. Lactic acidosis common
Complex III deficiency	Probably autosomal recessive	Fatal infantile encephalomyopathy, congenital lactic acidosis, hypotonia, dystrophic posturing, seizures, and coma
Complex IV deficiency	Autosomal recessive	Myopathy and hypotonia
Complex V deficiency	MTATP6 gene mutation	Slow, progressive myopathy
CPEO	Mitochondrial DNA point mutations: A3243G (most common)	Similar to those of KSS plus: visual myopathy, retinitis pigmentosa, dysfunction of the central nervous system
CPT I deficiency	Autosomal recessive	Enlarged liver, recurrent Reye-like episodes triggered by fasting or illnesses
CPT II deficiency	Autosomal recessive	Myopathic: exercise intolerance, fasting intolerance, muscle pain, muscle stiffness, and myoglobin in the urine
Kearns–Sayre syndrome	Due to mitochondria DNA deletions	Progressive paralysis of certain eye muscles, atypical retinitis pigmentosa, progressive degeneration, pigmentary degeneration of the retina, cardiomyopathy
Long-chain acyl-CoA dehydrongenase deficiency	Autosomal recessive	Failure to thrive, enlarged liver and heart, metabolic encephalopathy, hypotonia
LCHAD	Autosomal recessive	Encephalopathy, liver dysfunction, cardiomyopathy, and myopathy. Also pigmentary retinopathy and peripheral neuropathy
Leigh disease	Autosomal recessive	Seizures, hypotonia, fatigue, nystagmus, poor motor function, ataxia
Luft disease	Unknown inheritance	Hypermetabolism, fever, heat intolerance, profuse perspiration, polyphagia, polydipsia, ragged-red fibers, and resting tachycardia
MAD/glutaric aciduria type II	Defects of the flavoproteins responsible for transferring electrons (ETF or ETF-dehydrogenase)	

(Continued)

TABLE 9.2   (Continued)

	Cause	Symptoms
Medium-chain acyl-CoA dehydrongenase deficiency	Autosomal recessive	Episodes of encephalopathy, enlarged and fatty degeneration of the liver, low blood carnitine
Mitochondrial encephalomyopathy lactic acidosis and stroke-like episodes	Mitochondrial DNA point mutations: A3243G (most common)	Short stature, seizures, stroke-like episodes with focused neurological deficits, recurrent headaches, cognitive regression, ragged-red fibers
Myoclonic epilepsy and ragged-red fiber disease	Mitochondrial DNA point mutations: A8344G, T8356C	Myoclonus, epilepsy, progressive ataxia, muscle weakness/degeneration, deafness, dementia
Mitochondrial recessive ataxia syndrome	POLG mutation, recessive inheritance	Encephalopathy, balance problems, ataxia, epilepsy, cognitive impairment, psychiatric symptoms, eye movement disorders, involuntary movements, peripheral neuropathy
Myoneurogastrointestinal disorder and encephalopathy	Autosomal recessive	Progressive external ophthalmoplegia, limb weakness, peripheral neuropathy, digestive tract disorders, leukodystrophy, lactic acidosis, ragged-red fibers
Pyruvate caboxylase deficiency	Varies	Seizures and spasticity. Lactic acidosis, hypoglycemia, severe retardation, failure to thrive
Pyruvate dehydrogenase deficiency	X-linked recessive	Lactic acidosis, ataxia, pyruvic acidosis, spinal and cerebellar degeneration
Short-chain acyl-CoA dehydrogenase deficiency	Autosomal recessive	Failure to thrive, developmental delay, hypoglycemia
SCHAD	Autosomal recessive	Encephalopathy, liver disease, cardiomyopathy

POLG, polymerase gamma; GAMT, guanidinoacetate N-Methyltransferase; AGAT, arginine:glycine amidinotransferase; LCHAD, long-chain 3-hydroxyacyl-CoA dehydrogenase; SCHAD, short-chain 3-hydroxyacyl-CoA.

is poor, typically resulting in death by 3 years of age [117–119]. Despite progress in identifying key genetic mutations and pathogenesis, little is understood about the biochemical mechanism of Leigh's syndrome; treatments are palliative. One investigational drug, EPI-743, has increased the lifespan of Leigh's syndrome patients and reversed disease progression [120]. EPI-743 counters oxidative stress via the target enzyme NADPH quinone oxidoreducatase 1 [121].

mTOR intervention has been considered as a treatment approach to Leigh syndrome. The $Ndufs4^{-/-}$ (the gene responsible for the assembly and stability of complex I of the mitochondrial electron transport chain) mouse resembles the human disorder phenotypically and survives to approximately 50 days of age [122]. Daily administration of rapamycin (8 mg/kg body weight, intraperitoneally, starting at day 10 of life) significantly elongated the lifespan of mutant mice and attenuated lesions observed in the vehicle-treated group (e.g., astrocyte activation and glial reactivity). The trademark metabolite profiles in mutant mice included decreased free amino acids, free fatty acids, and increased oxidative stress with reduced levels of GABA and dopamine. Rapamycin intervention resulted in a shift to amino acid catabolism from glycolysis, decreasing the accumulation of glycolytic intermediates. These findings in the mouse model suggest that mTOR pathways should be further investigated in mitochondrial diseases [123].

9.7.2 Nicotinamide Adenine Dinucleotide Dehydrogenase Deficiency (Complex I Deficiency)

The gene coding for nicotinamide adenine dinucleotide (NADH) dehydrogenase (complex I) is located on chromosome 11p11.11. The protein, NADH dehydrogenase, is a part of the mitochondrial respiratory chain, which catalyzes transfer of electrons from NADH to ubiquinone. Complex I deficiency is often linked to other mitochondrial diseases, such as Leigh's syndrome. Mutations in this pathway have also been identified in other neurological diseases, including bipolar disorder, PD and schizophrenia [124]. The occurrence of schizophrenia in complex 1 deficiency, related to the gene disrupted-in-schizophrenia I (DISC1), results in an increase in pAkt and pS6. This has been shown in a DISC1 knockdown cell line in which rapamycin intervention attenuated biochemical and behavioral outcomes [125].

9.7.3 Kearns—Sayre Syndrome

Kearns—Sayre syndrome (KSS) is a rare neuromuscular disease progressing to chronic external ophthalmoplegia, atypical retinitis pigmentosa, and pigmentary degeneration of the retina. KSS is caused by a deletion in mitochondrial DNA resulting in increased protein damage, inhibition of the UPS, decreased amino acid recycling, and autophagy excitation in key cell types [126]. KSS can occur spontaneously in early development, or through germline mutations in mitochondrial polymerase gamma [127]. Approaches to mitigate the clinical manifestations of KSS include: peroxisome proliferator activated receptor (PPAR) agonists, the AMPK agonist, resveratrol, Sirt1 agonists, antioxidant supplementation, such as with L-carnitine, coenzyme Q_{10}, $MitoQ_{10}$, N-acetyl cysteine, and vitamins C, E, K1, and B [128]. Rapalog administration has yet to be evaluated. Attempts to identify the mechanism of pathogenesis have unveiled the importance of autophagy in KSS. In one study, six cell types with mutations in mitochondrial DNA were isolated from KSS patients, and subjected to microarray analysis. These findings suggested inhibition of transcripts involved in ubiquitin-mediated proteasome activity, amino acid degradation, macroautophagy, and induction of the AMP kinase pathway [126].

9.8 DEFECTS IN AUTOPHAGY

9.8.1 Crohn's Disease

Crohn's disease (CD) is a debilitating idiopathic autoimmune condition localized mainly to the gastrointestinal tract. Treatment ranges from antidiarrheal agents, steroids, immunosuppression (including rapamycin), to surgical resection [129]. The cause of CD is unknown, but there are numerous genetic and environmental factors which increase the risk for developing CD. Mutations have been observed in many genes, such as those involved in immune response and in autophagy, namely ATG16L1 and NOD2 (nucleotide-binding oligomerization domain 2) [130,131]. ATG16L1 is essential for autophagy [131] and NOD2, one of the earliest disovered mutations, is required for caspase recruitment.

9.8.2 Hereditary Spastic Paraparesis

Hereditary spastic paraparesis is caused by heterogeneous mutations leading to dysfunction of the pyramidal tract and developing into spasticity of the lower limbs (this form is denoted as the pure form). Inheritance can vary from X-linked, autosomal dominant, or autosomal recessive [132]. The most common gene mutation, SPG3A, presents prior to age 10 and encodes a GTPase and Golgi body transmembrane protein [133]. Mutations in at least 59 different genes have

been confirmed including the gene encoding spastin, a mitochondrial protein and signaling molecule [134]. Mutations in genes necessary for autophagy have recently been identified.

The role of autophagy in this disease first appeared when Oz-Levi et al. identified a mutation in the gene TECPR2 on the q arm of chromosome 14. The effect of deletions of TECPR2 was examined in vitro by transfecting COS-7, HEK293, and HeLa cells with the normal and mutated forms of the gene. This gene had previously been shown to positively regulate autophagosome sequestration through interactions with Atg8 homologs and was confirmed in these cell models. Decreased accumulation of LC3-II was observed in cultured patient skin fibroblasts [135]. Subsequent work examining mutation in autophagy genes, by Vantaggiato et al., identified mutations in a different subset of the disease occurring in zinc finger FYVE containing 26 (ZFYVE26). Patient skin fibroblasts and lymphocytes had reduced colocalization of MAP1LC3B compared to controls in both basal conditions and autophagy-inducing conditions. This work suggested that ZFYVE26 is occurring in a subset of hereditary spastic paraparesis, and that this gene is necessary for autophagosome maturation. Only a small subset of mutations identified to date are involved in autophagy in hereditary spastic paraparesis.

9.8.3 Vici Syndrome

Vici syndrome is a recessive disorder first described in 1988 which affects multiple systems resulting in agenesis of the corpus callosum, cataracts, cardiomyopathy, immunodeficiency, and hyperglycemia. Unlike hereditary spastic paraparesis, Vici syndrome is the result of mutation in a single gene, EPG5. EPG5 encodes for the epg-5 protein, an essential mediator in starvation-induced autophagy [136].

Researchers from the United Kingdom sought to further examine the molecular mechanisms involved in Vici syndrome. They utilized immunofluorescence on two patient skeletal muscle samples and observed an association between muscle atrophy and increased autophagy-associated proteins Nbr1 and p62/SQSTM1 [137]. Cultured patient skin fibroblasts were subsequently treated with rapamycin. There was no response to rapamycin, indicating that the deficiency is downstream of mTOR. This supports the hypothesis that epg-5 acts in late autolysosome formation. An Epg5 knockout mouse model with consistent pathology including muscle denervation, atrophy or myofibers, progressive lower-limb paralysis, and truncated survival, also experiences defective maturation of autophagosomes to autolysosomes [138].

9.9 OTHER HERITABLE DISORDERS

9.9.1 Cystic Fibrosis

Cystic fibrosis (CF) is acquired by a mutation to the CF transmembrane conductance regulator protein. This protein facilities transmembrane anion transport across epithelial cell membranes. The most marked symptom is the accumulation of mucus in the lungs and subsequent infection. These lung infections can be moderately managed with antibiotics, but chronic inflammation has detrimental effects. Chronic inflammation results in increased ROS, decreased autophagy, and decreased clearance of bacterial or fungal infections [139]. Approaches to induce inhibition of mTOR using rapamycin have been considered, but there remains the contradiction of employing an immunosuppressant to treat an infection. Therefore, in CF, a selective mTOR-independent autophagy inducer is required. To address this clinical need, a therapeutic peptide has been derived from beclin-1 (tat-beclin-1) [140]. This peptide was shown to induce autophagy *in vivo*, but has not yet been tested in any model of CF [141].

9.9.2 Diabetes Mellitus

Over 350 million people are affected by diabetes mellitus (DM) worldwide [142]. DM is classified as being either a non-insulin-dependent (type 1) or insulin-dependent (type 2) disease. Type 1 DM is an autoimmune response to pancreatic β cells, resulting in the loss of insulin production and regulation. Type 2 DM typically develops after age 40 and is a progressive deterioration of glucose tolerance accompanied by β-cell compensation with hyperplasia, preceded by a decrease in β-cell mass and function. Genome-wide association studies continue in order to advance understanding of the heritability and genes implicated in type 2 diabetes. Congenital diabetes can occur by defective insulin secretion or by mutations in the monogenic forms of the disease, termed MODY, with more than 11 types involving genes such as hepatic nuclear factor 4α, hepatic nuclear factor 1α, glucokinase, insulin promoter factor 1, hepatic nuclear factor 1β, and others. DM is a disease with non-specific effects on the entire body, including damage to the neuroglia vascular unit, endothelial cell injury, and loss of angiogenesis [143]. Additionally DM is a disease of chronic oxidative stress, which can result in mitochondrial DNA damage and impaired cell function and ultimately apoptosis and autophagy [144].

To combat these systems in an environment high in oxidative stress, several approaches have been employed, including protein tyrosine phosphatases, antioxidants, and growth factors. One intriguing growth factor with promise is erythropoietin (EPO). EPO has shown efficacy in models of diabetic retinal degeneration, and is protective toward endothelial cells in *in vitro* models of DM [145,146]. EPO has non-specific effects in the body, including its regulation of mTOR via PI3-K/Akt pathways. EPO regulates mTOR activity through post-translational phosphorylation of PRAS40, an inhibitory element, resulting in PI3K-dependent subcellular binding of PRAS40 to the docking protein 14-3-3. Changes in the phosphorylation status of PRAS40 were shown to be independent of other protective pathways of EPO involving extracellular signal-related kinase (ERK 1/2) or signal transducer and activator of transcription (STAT). Regulation of autophagy has been mechanistically studied in an $Agt5^{-/-}$ mouse for diabetes-induced glomerulosclerosis, confirming the promise of manipulating autophagy pathways as a therapeutic approach for DM [147].

Additional studies implicate mTOR in DM. One study used the Zucker diabetic fatty (ZDF) rat model to investigate the effects of diabetes on brain protein expression, autophagy, and antioxidant defense in DM. The ZDF rat manifests elevated protein aggregates and glycosylation, and increased mTOR phosphorylation and S6 ribosomal protein were associated with the protein aggregates. Proteolytic pathways were inhibited as suggested by increases in LC3-II/LC3-I levels in the brain [148].

9.10 CONCLUSIONS AND SUMMARY

mTOR integrates numerous cues to orchestrate outcomes on growth and metabolism, with developmental, neurological, and psychological consequences resulting from genetic mutations within the *FRAP1* gene, in mTOR's upstream effectors, as well as in loci related to input cues such as amino acids and insulin. In the two and a half decades since the discovery of TOR in yeast, advances in mTOR's normophysiological and pathological role have inspired new clinical trials in a number of complex diseases (Table 9.1). Many intricacies of mTOR await preclinical investigations, but manipulation of the mTOR and autophagy pathways represent a promising area of pharmacological research.

References

[1] Hay N, Sonenberg N. Upstream and downstream of mTOR. Genes Dev 2004;18:1926—45.
[2] Fleming A, Noda T, Yoshimori T, Rubinsztein DC. Chemical modulators of autophagy as biological probes and potential therapeutics. Nat Chem Biol 2011;7:9—17.

[3] Lipton JO, Sahin M. The neurology of mTOR. Neuron 2014;84:275–91.

[4] Jung CH, Ro S-H, Cao J, Otto NM, Kim D-H. mTOR regulation of autophagy. FEBS Lett 2010;584:1287–95.

[5] Inoki K, Kim J, Guan K-L. AMPK and mTOR in cellular energy homeostasis and drug targets. Annu Rev Pharmacol Toxicol 2012;52:381–400.

[6] Zoncu R, Efeyan A, Sabatini DM. mTOR: from growth signal integration to cancer, diabetes and ageing. Nat Rev Mol Cell Biol 2011;12:21–35.

[7] Kim E. Mechanisms of amino acid sensing in mTOR signaling pathway. Nutr Res Pract 2009;3:64–71.

[8] Jewell JL, Russell RC, Guan K-L. Amino acid signalling upstream of mTOR. Nat Rev Mol Cell Biol 2013;14:133–9.

[9] Lynch CJ. Role of leucine in the regulation of mTOR by amino acids: revelations from structure-activity studies. J Nutr 2001;131:861S–5S.

[10] Li F, Yin Y, Tan B, Kong X, Wu G. Leucine nutrition in animals and humans: mTOR signaling and beyond. Amino Acids 2011;41:1185–93.

[11] Jewell JL, Kim YC, Russell RC, Yu F-X, Park HW, Plouffe SW. Differential regulation of mTORC1 by leucine and glutamine. Science 2015;347:194–8.

[12] Wang S, Tsun Z-Y, Wolfson RL, Shen K, Wyant GA, Plovanich ME. Metabolism. Lysosomal amino acid transporter SLC38A9 signals arginine sufficiency to mTORC1. Science (New York, N.Y.) 2015;347:188–94.

[13] Hanrahan J, Blenis J. Rheb activation of mTOR and S6K1 signaling. Methods Enzymol 2006;407:542–55.

[14] Ma XM, Blenis J. Molecular mechanisms of mTOR-mediated translational control. Nat Rev Mol Cell Biol 2009;10:307–18.

[15] Dann SG, Selvaraj A, Thomas G. mTOR complex 1-S6K1 signaling: at the crossroads of obesity, diabetes and cancer. Trends Mol Med 2007;13:252–9.

[16] Mariño G, Niso-Santano M, Baehrecke EH, Kroemer G. Self-consumption: the interplay of autophagy and apoptosis. Nat Rev Mol Cell Biol 2014;15:81–94.

[17] Mizumura K, Choi AMK, Ryter SW. Emerging role of selective autophagy in human diseases. Front Pharmacol 2014;5. Available from: http://www.ncbi.nlm.nih.gov/pmc/articles/PMC4220655/ [Accessed 8.06.2015].

[18] Maiuri MC, Zalckvar E, Kimchi A, Kroemer G. Self-eating and self-killing: crosstalk between autophagy and apoptosis. Nat Rev Mol Cell Biol 2007;8:741–52.

[19] Ryter SW, Mizumura K, Choi AMK. The impact of autophagy on cell death modalities. Int J Cell Biol 2014;2014:502676.

[20] Kang R, Zeh HJ, Lotze MT, Tang D. The Beclin 1 network regulates autophagy and apoptosis. Cell Death Differ 2011;18:571–80.

[21] Powell JD, Pollizzi KN, Heikamp EB, Horton MR. Regulation of immune responses by mTOR. Annu Rev Immunol 2012;30:39–68.

[22] Sarbassov DD, Ali SM, Sengupta S, Sheen J-H, Hsu PP, Bagley AF. Prolonged rapamycin treatment inhibits mTORC2 assembly and Akt/PKB. Mol Cell 2006;22:159–68.

[23] Humphrey SJ, Yang G, Yang P, Fazakerley DJ, Stöckli J, Yang JY. Dynamic adipocyte phosphoproteome reveals that Akt directly regulates mTORC2. Cell Metab 2013;17:1009–20.

[24] Schenone S, Brullo C, Musumeci F, Radi M, Botta M. ATP-competitive inhibitors of mTOR: an update. Curr Med Chem 2011;18:2995–3014.

[25] Menzies FM, Fleming A, Rubinsztein DC. Compromised autophagy and neurodegenerative diseases. Nat Rev Neurosci 2015;16:345–57.

[26] Ghavami S, Shojaei S, Yeganeh B, Ande SR, Jangamreddy JR, Mehrpour M. Autophagy and apoptosis dysfunction in neurodegenerative disorders. Prog Neurobiol 2014;112:24–49.

[27] Querfurth HW, LaFerla FM. Alzheimer's disease. N Engl J Med 2010;362:329–44.

[28] Martin LJ. Mitochondrial pathobiology in Parkinson's disease and amyotrophic lateral sclerosis. J Alzheimer's Dis 2010;20 (Suppl. 2):S335–56.

[29] Lipinski MM, Zheng B, Lu T, Yan Z, Py BF, Ng A. Genome-wide analysis reveals mechanisms modulating autophagy in normal brain aging and in Alzheimer's disease. Proc Natl Acad Sci USA 2010;107:14164–9.

[30] Cataldo AM, Peterhoff CM, Schmidt SD, Terio NB, Duff K, Beard M. Presenilin mutations in familial Alzheimer disease and transgenic mouse models accelerate neuronal lysosomal pathology. J Neuropathol Exp Neurol 2004;63:821–30.

[31] Yu WH, Cuervo AM, Kumar A, Peterhoff CM, Schmidt SD, Lee J-H. Macroautophagy—a novel β-amyloid peptide-generating pathway activated in Alzheimer's disease. J Cell Biol 2005;171:87–98.

[32] Gómez-Suaga P, Luzón-Toro B, Churamani D, Zhang L, Bloor-Young D, Patel S. Leucine-rich repeat kinase 2 regulates autophagy through a calcium-dependent pathway involving NAADP. Hum Mol Genet 2011;21(3):511–25. Available from: http://dx.doi.org/10.1093/hmg/ddr481.

[33] Geisler S, Holmström KM, Treis A, Skujat D, Weber SS, Fiesel FC. The PINK1/Parkin-mediated mitophagy is compromised by PD-associated mutations. Autophagy 2010;6:871–8.

[34] Dehay B, Bové J, Rodríguez-Muela N, Perier C, Recasens A, Boya P. Pathogenic lysosomal depletion in Parkinson's disease. J Neurosci 2010;30:12535–44.

[35] La Spada AR, Weydt P, Pineda VV. Huntington's Disease Pathogenesis: Mechanisms and Pathways. In: Lo DC, Hughes RE, editors. Neurobiology of huntington's disease: applications to drug discovery. Boca Raton (FL): CRC Press; 2011. Available from: http://www.ncbi.nlm.nih.gov/books/NBK55992/ [Accessed 22.06.2015].

[36] Sarkar S, Davies JE, Huang Z, Tunnacliffe A, Rubinsztein DC. Trehalose, a novel mTOR-independent autophagy enhancer, accelerates the clearance of mutant huntingtin and alpha-synuclein. J Biol Chem 2007;282:5641–52.

[37] Sarkar S, Rubinsztein DC. Huntington's disease: degradation of mutant huntingtin by autophagy. FEBS J 2008;275:4263–70.

[38] Blackhall LJ. Amyotrophic lateral sclerosis and palliative care: where we are, and the road ahead. Muscle Nerve 2012;45:311–18.

[39] Vadakkadath Meethal S, Atwood CS. Lactate dyscrasia: a novel explanation for amyotrophic lateral sclerosis. Neurobiol Aging 2012;33:569–81.

[40] Lattante S, de Calbiac H, Le Ber I, Brice A, Ciura S, Kabashi E. Sqstm1 knock-down causes a locomotor phenotype ameliorated by rapamycin in a zebrafish model of ALS/FTLD. Hum Mol Genet 2015;24:1682–90.

[41] Staats KA, Hernandez S, Schönefeldt S, Bento-Abreu A, Dooley J, Van Damme P. Rapamycin increases survival in ALS mice lacking mature lymphocytes. Mol Neurodegener 2013;8:31.

[42] Scotter EL, Vance C, Nishimura AL, Lee Y-B, Chen H-J, Urwin H. Differential roles of the ubiquitin proteasome system and autophagy in the clearance of soluble and aggregated TDP-43 species. J Cell Sci 2014;127:1263–78.

[43] Zhang X, Chen S, Song L, Tang Y, Shen Y, Jia L. MTOR-independent, autophagic enhancer trehalose prolongs motor neuron survival and ameliorates the autophagic flux defect in a mouse model of amyotrophic lateral sclerosis. Autophagy 2014;10:588–602.

[44] Blau N, van Spronsen FJ, Levy HL. Phenylketonuria. Lancet 2010;376:1417–27.

[45] Sanayama Y, Matsumoto A, Shimojo N, Kohno Y, Nakaya H. Phenylalanine sensitive K562-D cells for the analysis of the biochemical impact of excess amino acid. Sci Rep 2014;4:6941.

[46] Kure S, Hou DC, Ohura T, Iwamoto H, Suzuki S, Sugiyama N. Tetrahydrobiopterin-responsive phenylalanine hydroxylase deficiency. J Pediatr 1999;135:375–8.

[47] Choi HJ, Lee SY, Cho Y, No H, Kim SW, Hwang O. Tetrahydrobiopterin causes mitochondrial dysfunction in dopaminergic cells: implications for Parkinson's disease. Neurochem Int 2006;48:255–62.

[48] Thöny B, Blau N. Mutations in the BH4-metabolizing genes GTP cyclohydrolase I, 6-pyruvoyl-tetrahydropterin synthase, sepiapterin reductase, carbinolamine-4a-dehydratase, and dihydropteridine reductase. Hum Mutat 2006;27:870–8.

[49] Thöny B, Blau N. Mutations in the GTP cyclohydrolase I and 6-pyruvoyl-tetrahydropterin synthase genes. Hum Mutat 1997;10:11–20.

[50] Kwak SS, Jeong M, Choi JH, Kim D, Min H, Yoon Y. Amelioration of behavioral abnormalities in BH(4)-deficient mice by dietary supplementation of tyrosine. PloS One 2013;8:e60803.

[51] Kwak SS, Suk J, Choi JH, Yang S, Kim JW, Sohn S. Autophagy induction by tetrahydrobiopterin deficiency. Autophagy 2011;7:1323–34.

[52] Russo PA, Mitchell GA, Tanguay RM. Tyrosinemia: a review. Pediatr Dev Pathol 2014;4:212–21.

[53] Jorquera R, Tanguay RM. Cyclin B-dependent kinase and caspase-1 activation precedes mitochondrial dysfunction in fumarylacetoacetate-induced apoptosis. FASEB J 1999;13:2284–98.

[54] Kubo S, Sun M, Miyahara M, Umeyama K, Urakami K, Yamamoto T. Hepatocyte injury in tyrosinemia type 1 is induced by fumarylacetoacetate and is inhibited by caspase inhibitors. Proc Natl Acad Sci USA 1998;95:9552–7.

[55] Ghadimi H, Partington MW, Hunter A. A familial disturbance of histidine metabolism. N Engl J Med 1961;265:221–4.

[56] Taylor RG, Levy HL, McInnes RR. Histidase and histidinemia. Clinical and molecular considerations. Mol Biol Med 1991;8:101–16.

[57] Virmani K, Widhalm K. Histidinemia: a biochemical variant or a disease? J Am Coll Nutr 1993;12:115–24.

[58] Barresi C, Stremnitzer C, Mlitz V, Kezic S, Kammeyer A, Ghannadan M. Increased sensitivity of histidinemic mice to UVB radiation suggests a crucial role of endogenous urocanic acid in photoprotection. J Invest Dermatol 2011;131:188–94.

[59] Carr TD, DiGiovanni J, Lynch CJ, Shantz LM. Inhibition of mTOR suppresses UVB-induced keratinocyte proliferation and survival. Cancer Prev Res (Phila) 2012;5:1394–404.

[60] Prizant RL, Barash I. Negative effects of the amino acids Lys, His, and Thr on S6K1 phosphorylation in mammary epithelial cells. J Cell Biochem 2008;105:1038–47.

[61] Kobayashi T, Shimabukuro-Demoto S, Yoshida-Sugitani R, Furuyama-Tanaka K, Karyu H, Sugiura Y. The histidine transporter SLC15A4 coordinates mTOR-dependent inflammatory responses and pathogenic antibody production. Immunity 2014;41:375–88.

[62] Soltys KA, Mazariegos GV, Strauss KA. Living related transplantation for MSUD—caution, or a new path forward? Pediatr Transplant 2015;19:247–8.

[63] Mills CD. M1 and M2 macrophages: oracles of health and disease. Crit Rev Immunol 2012;32:463–88.

[64] De Simone R, Vissicchio F, Mingarelli C, De Nuccio C, Visentin S, Ajmone-Cat MA. Branched-chain amino acids influence the immune properties of microglial cells and their responsiveness to pro-inflammatory signals. Biochim Biophys Acta 2013;1832:650–9.

[65] Weichhart T, Costantino G, Poglitsch M, Rosner M, Zeyda M, Stuhlmeier KM. The TSC-mTOR signaling pathway regulates the innate inflammatory response. Immunity 2008;29:565–77.

[66] Xie L, Sun F, Wang J, Mao X, Xie L, Yang S-H. mTOR signaling inhibition modulates macrophage/microglia-mediated neuroinflammation and secondary injury via regulatory T cells after focal ischemia. J Immunol (Baltimore, Md.: 1950) 2014;192:6009–19.

[67] Hotamisligil GS, Erbay E. Nutrient sensing and inflammation in metabolic diseases. Nat Rev Immunol 2008;8:923–34.

[68] Knerr I, Pearl PL, Bottiglieri T, Snead OC, Jakobs C, Gibson KM. Therapeutic concepts in succinate semialdehyde dehydrogenase (SSADH; ALDH5a1) deficiency (gamma-hydroxybutyric aciduria). Hypotheses evolved from 25 years of patient evaluation, studies in Aldh5a1$^{-/-}$ mice and characterization of gamma-hydroxybutyric acid pharmacology. J Inherit Metab Dis 2007;30:279–94.

[69] Vogel KR, Ainslie GR, Jansen EEW, Salomons GS, Gibson KM. Torin 1 partially corrects vigabatrin-induced mitochondrial increase in mouse. Ann Clin Transl Neurol 2015;2:699–706.

[70] Lakhani R, Vogel KR, Till A, Liu J, Burnett SF, Gibson KM. Defects in GABA metabolism affect selective autophagy pathways and are alleviated by mTOR inhibition. EMBO Mol Med 2014;6:551–66.

[71] Pearl PL, Parviz M, Vogel K, Schreiber J, Theodore WH, Gibson KM. Inherited disorders of gamma-aminobutyric acid metabolism and advances in ALDH5A1 mutation identification. Dev Med Child Neurol 2014;57:611–17. Available from: http://dx.doi.org/10.1111/dmcn.12668.

[71a] Workman ER, Niere F, Raab-Graham KF. mTORC1-dependent protein synthesis underlying rapid antidepressant effect requires GABABR signaling. Neuropharmacology 2013;73:192–203.

[72] Besse A, Wu P, Bruni F, Donti T, Graham BH, Craigen WJ. The GABA transaminase, ABAT, is essential for mitochondrial nucleoside metabolism. Cell Metab 2015;21:417–27.

[73] Ishii I, Akahoshi N, Yamada H, Nakano S, Izumi T, Suematsu M. Cystathionine gamma-lyase-deficient mice require dietary cysteine to protect against acute lethal myopathy and oxidative injury. J Biol Chem 2010;285:26358–68.

[74] Magoulas PL, El-Hattab AW. Systemic primary carnitine deficiency: an overview of clinical manifestations, diagnosis, and management. Orphanet J Rare Dis 2012;7:68.

[75] Carnitine deficiency treatment and management. 2014. Available from: http://emedicine.medscape.com/article/942233-treatment [Accessed 14.06.2014].

[76] Ellis JM, Mentock SM, Depetrillo MA, Koves TR, Sen S, Watkins SM. Mouse cardiac acyl coenzyme a synthetase 1 deficiency impairs Fatty Acid oxidation and induces cardiac hypertrophy. Mol Cell Biol 2011;31:1252–62.

[77] Page T. Metabolic approaches to the treatment of autism spectrum disorders. J Autism Dev Disord 2000;30:463–9.

[78] Moog U, Smeets EEJ, van Roozendaal KEP, Schoenmakers S, Herbergs J, Schoonbrood-Lenssen AMJ. Neurodevelopmental disorders in males related to the gene causing Rett syndrome in females (MECP2). Eur J Paediatr Neurol 2003;7:5–12.

[79] Pohodich AE, Zoghbi HY. Rett syndrome: disruption of epigenetic control of postnatal neurological functions. Hum Mol Genet 2015;24:R10–16. Available from: http://dx.doi.org/10.1093/hmg/ddv217.

[80] Zoghbi HY, Bear MF. Synaptic dysfunction in neurodevelopmental disorders associated with autism and intellectual disabilities. Cold Spring Harb Perspect Biol 2012;4:a009886.

[81] Guy J, Hendrich B, Holmes M, Martin JE, Bird A. A mouse Mecp2-null mutation causes neurological symptoms that mimic Rett syndrome. Nat Genet 2001;27:322−6.

[82] Ricciardi S, Boggio EM, Grosso S, Lonetti G, Forlani G, Stefanelli G. Reduced AKT/mTOR signaling and protein synthesis dysregulation in a Rett syndrome animal model. Hum Mol Genet 2011;20:1182−96.

[83] Shen MH, Harper PS, Upadhyaya M. Molecular genetics of neurofibromatosis type 1 (NF1). J Med Genet 1996;33:2−17.

[84] Mautner V-F, Kluwe L, Friedrich RE, Roehl AC, Bammert S, Högel J. Clinical characterisation of 29 neurofibromatosis type-1 patients with molecularly ascertained 1.4 Mb type-1 NF1 deletions. J Med Genet 2010;47:623−30.

[85] Dasgupta B, Yi Y, Chen DY, Weber JD, Gutmann DH. Proteomic analysis reveals hyperactivation of the mammalian target of rapamycin pathway in neurofibromatosis 1-associated human and mouse brain tumors. Cancer Res 2005;65: 2755−60.

[86] Gipson TT, Johnston MV. Plasticity and mTOR: towards restoration of impaired synaptic plasticity in mTOR-related neurogenetic disorders. Neural Plast 2012;2012. Available from: http://www.ncbi.nlm.nih.gov/pmc/articles/PMC3350854/ [Accessed 16.06. 2015].

[87] Gross C, Hoffmann A, Bassell GJ, Berry-Kravis EM. Therapeutic strategies in fragile X syndrome: from bench to bedside and back. Neurotherapeutics 2015;2012:1−10. Available from: http://dx.doi.org/10.1007/s13311-015-0355-9.

[88] Busquets-Garcia A, Maldonado R, Ozaita A. New insights into the molecular pathophysiology of fragile X syndrome and therapeutic perspectives from the animal model. Int J Biochem Cell Biol 2014;53:121−6.

[89] Sharma A, Hoeffer CA, Takayasu Y, Miyawaki T, McBride SM, Klann E. Dysregulation of mTOR signaling in fragile X syndrome. J Neurosci 2010;30:694−702.

[90] Kim L, He L, Maaswinkel H, Zhu L, Sirotkin H, Weng W. Anxiety, hyperactivity and stereotypy in a zebrafish model of fragile X syndrome and autism spectrum disorder. Prog Neuropsychopharmacol Biol Psychiatry 2014;55:40−9.

[91] Newsletter Signup (megamenu)|Autism Speaks. 2015. Available from: https://www.autismspeaks.org/newsletter-signup-megamenu?wmode=transparent [Accessed 15.06.2015].

[92] Curatolo P, Bombardieri R, Jozwiak S. Tuberous sclerosis. Lancet 2008;372:657−68.

[93] Curatolo P, Maria BL. Tuberous sclerosis. Handb Clin Neurol 2013;111:323−31.

[94] Ehninger D, Silva AJ. Rapamycin for treating tuberous sclerosis and autism spectrum disorders. Trends Mol Med 2011;17:78−87.

[95] Curatolo P. Mechanistic target of rapamycin (mTOR) in tuberous sclerosis complex-associated epilepsy. Pediatr Neurol 2015;52:281−9.

[96] Jansen FE, van Huffelen AC, Algra A, van Nieuwenhuizen O. Epilepsy surgery in tuberous sclerosis: a systematic review. Epilepsia 2007;48:1477−84.

[97] Fallah A, Guyatt GH, Snead OC, Ebrahim S, Ibrahim GM, Mansouri A. Predictors of seizure outcomes in children with tuberous sclerosis complex and intractable epilepsy undergoing resective epilepsy surgery: an individual participant data meta-analysis. PLoS One 2013;8:e53565.

[98] Tang G, Gudsnuk K, Kuo S-H, Cotrina ML, Rosoklija G, Sosunov A. Loss of mTOR-dependent macroautophagy causes autistic-like synaptic pruning deficits. Neuron 2014;83: 1131−43.

[99] Wong M. Rapamycin for treatment of epilepsy: antiseizure, antiepileptogenic, both, or neither? Epilepsy Curr 2011;11:66−8.

[100] Ehninger D, Han S, Shilyansky C, Zhou Y, Li W, Kwiatkowski DJ. Reversal of learning deficits in a Tsc2+/− mouse model of tuberous sclerosis. Nat Med 2008;14:843−8.

[101] Jülich K, Sahin M. Mechanism-based treatment in tuberous sclerosis complex. Pediatr Neurol 2014;50:290−6.

[102] Krueger DA, Wilfong AA, Holland-Bouley K, Anderson AE, Agricola K, Tudor C. Everolimus treatment of refractory epilepsy in tuberous sclerosis complex. Ann Neurol 2013;74:679−87.

[103] Wong M. Mammalian target of rapamycin (mTOR) activation in focal cortical dysplasia and related focal cortical malformations. Exp Neurol 2013;244:22−6.

[104] Lim JS, Kim W, Kang H-C, Kim SH, Park AH, Park EK. Brain somatic mutations in MTOR cause focal cortical dysplasia type II leading to intractable epilepsy. Nat Med 2015;21:395−400.

[105] Salzmann A, Malafosse A. Genetics of temporal lobe epilepsy: a review. Epilepsy Res Treat 2012;2012:e863702.

[106] Zeng L-H, Rensing NR, Wong M. The mammalian target of rapamycin (mTOR) signaling pathway mediates epileptogenesis in a model of temporal lobe epilepsy. J Neurosci 2009;29:6964−72.

[107] Buckmaster PS, Lew FH. Rapamycin suppresses mossy fiber sprouting but not seizure frequency in a mouse model of temporal lobe epilepsy. J Neurosci 2011;31:2337−47.

[108] Hers HG. Alpha-glucosidase deficiency in generalized glycogen storage disease (Pompe's disease). Biochem J 1963;86:11−16.

[109] Nishiyama Y, Shimada Y, Yokoi T, Kobayashi H, Higuchi T, Eto Y. Akt inactivation induces endoplasmic reticulum stress-independent autophagy in fibroblasts from patients with Pompe disease. Mol Genet Metab 2012;107:490−5.

[110] Wehrmann ZT, Hulett TW, Huegel KL, Vaughan KT, Wiest O, Helquist P. Quantitative comparison of the efficacy of various compounds in lowering intracellular cholesterol levels in Niemann−Pick type C fibroblasts. PLoS One 2012;7:e48561.

[111] Alobaidy H. Recent advances in the diagnosis and treatment of Niemann−Pick disease type C in children: a guide to early diagnosis for the general pediatrician. Int J Pediatr 2015;2015:816593.

[112] Tang Y, Li H, Liu J-P. Niemann−Pick disease type C: from molecule to clinic. Clin Exp Pharmacol Physiol 2010;37:132−40.

[113] Sharpe LJ, Brown AJ. Rapamycin down-regulates LDL-receptor expression independently of SREBP-2. Biochem Biophys Res Commun 2008;373:670−4.

[114] Pacheco CD, Kunkel R, Lieberman AP. Autophagy in Niemann−Pick C disease is dependent upon Beclin-1 and responsive to lipid trafficking defects. Hum Mol Genet 2007;16:1495−503.

[115] Regier DS, Tifft CJ. GLB1-related disorders. In: Pagon RA, Adam MP, Ardinger HH, Wallace SE, Amemiya A, Bean LJ, editors. GeneReviews®. Seattle (WA): University of Washington; 1993. Available from: http://www.ncbi.nlm.nih.gov/books/NBK164500/ [Accessed 10.06.2015].

[116] Rahman S, Blok RB, Dahl HH, Danks DM, Kirby DM, Chow CW. Leigh syndrome: clinical features and biochemical and DNA abnormalities. Ann Neurol 1996;39:343−51.

[117] Ma Y-Y, Wu T-F, Liu Y-P, Wang Q, Song J-Q, Li X-Y. Genetic and biochemical findings in Chinese children with Leigh syndrome. J Clin Neurosci 2013;20:1591−4.

[118] Montpetit VJ, Andermann F, Carpenter S, Fawcett JS, Zborowska-Sluis D, Giberson HR. Subacute necrotizing encephalomyelopathy. A review and a study of two families. Brain 1971;94:1−30.

[119] Sofou K, De Coo IFM, Isohanni P, Ostergaard E, Naess K, De Meirleir L. A multicenter study on Leigh syndrome: disease

course and predictors of survival. Orphanet J Rare Dis 2014;9:52.

[120] Martinelli D, Catteruccia M, Piemonte F, Pastore A, Tozzi G, Dionisi-Vici C. EPI-743 reverses the progression of the pediatric mitochondrial disease—genetically defined Leigh syndrome. Mol Genet Metab 2012;107:383—8.

[121] Sadun AA, Chicani CF, Ross-Cisneros FN, Barboni P, Thoolen M, Shrader WD. Effect of EPI-743 on the clinical course of the mitochondrial disease Leber hereditary optic neuropathy. Arch Neurol 2012;69:331—8.

[122] Johnson SC, Yanos ME, Kayser E-B, Quintana A, Sangesland M, Castanza A. mTOR inhibition alleviates mitochondrial disease in a mouse model of Leigh syndrome. Science 2013;342:1524—8.

[123] Geach T. Neurometabolic disease: treating mitochondrial diseases with mTOR inhibitors—a potential treatment for Leigh syndrome? Nat Rev Neurol 2014;10:2.

[124] Chou AP, Li S, Fitzmaurice AG, Bronstein JM. Mechanisms of rotenone-induced proteasome inhibition. Neurotoxicology 2010;31:367—72.

[125] Kim JY, Duan X, Liu CY, Jang M-H, Guo JU, Pow-anpongkul N. DISC1 regulates new neuron development in the adult brain via modulation of AKT-mTOR signaling through KIAA1212. Neuron 2009;63:761—73.

[126] Alemi M, Prigione A, Wong A, Schoenfeld R, DiMauro S, Hirano M. Mitochondrial DNA deletions inhibit proteasomal activity and stimulate an autophagic transcript. Free Radic Biol Med 2007;42:32—43.

[127] Kaukonen J, Juselius JK, Tiranti V, Kyttälä A, Zeviani M, Comi GP. Role of adenine nucleotide translocator 1 in mtDNA maintenance. Science 2000;289:782—5.

[128] Valero T. Editorial (thematic issue: mitochondrial biogenesis: pharmacological approaches). Curr Pharm Des 2014;20:5507—9.

[129] Baumgart DC, Sandborn WJ. Crohn's disease. Lancet 2012;380:1590—605.

[130] Brinar M, Vermeire S, Cleynen I, Lemmens B, Sagaert X, Henckaerts L. Genetic variants in autophagy-related genes and granuloma formation in a cohort of surgically treated Crohn's disease patients. J Crohn's Colitis 2012;6:43—50.

[131] Prescott NJ, Fisher SA, Franke A, Hampe J, Onnie CM, Soars D. A nonsynonymous SNP in ATG16L1 predisposes to ileal Crohn's disease and is independent of CARD15 and IBD5. Gastroenterology 2007;132:1665—71.

[132] Depienne C, Stevanin G, Brice A, Durr A. Hereditary spastic paraplegias: an update. Curr Opin Neurol 2007;20:674—80.

[133] Namekawa M, Ribai P, Nelson I, Forlani S, Fellmann F, Goizet C. SPG3A is the most frequent cause of hereditary spastic paraplegia with onset before age 10 years. Neurology 2006;66:112—14.

[134] Klebe S, Stevanin G, Depienne C. Clinical and genetic heterogeneity in hereditary spastic paraplegias: from SPG1 to SPG72 and still counting. Rev Neurol (Paris) 2015;171

(6—7):505—30. Available from: http://dx.doi.org/10.1016/j.neurol.2015.02.017.

[135] Oz-Levi D, Ben-Zeev B, Ruzzo EK, Hitomi Y, Gelman A, Pelak K. Mutation in TECPR2 reveals a role for autophagy in hereditary spastic paraparesis. Am J Hum Genet 2012;91:1065—72.

[136] Tian Y, Li Z, Hu W, Ren H, Tian E, Zhao Y. C. elegans screen identifies autophagy genes specific to multicellular organisms. Cell 2010;141:1042—55.

[137] Cullup T, Kho AL, Dionisi-Vici C, Brandmeier B, Smith F, Urry Z. Recessive mutations in EPG5 cause Vici syndrome, a multisystem disorder with defective autophagy. Nat Genet 2013;45:83—7.

[138] Zhao H, Zhao YG, Wang X, Xu L, Miao L, Feng D. Mice deficient in Epg5 exhibit selective neuronal vulnerability to degeneration. J Cell Biol 2013;200:731—41.

[139] Luciani A, Villella VR, Esposito S, Brunetti-Pierri N, Medina D, Settembre C. Defective CFTR induces aggresome formation and lung inflammation in cystic fibrosis through ROS-mediated autophagy inhibition. Nat Cell Biol 2010;12:863—75.

[140] Shoji-Kawata S, Sumpter R, Leveno M, Campbell GR, Zou Z, Kinch L. Identification of a candidate therapeutic autophagy-inducing peptide. Nature 2013;494:201—6.

[141] Junkins RD, McCormick C, Lin T-J. The emerging potential of autophagy-based therapies in the treatment of cystic fibrosis lung infections. Autophagy 2014;10:538—47.

[142] Maiese K. mTOR: driving apoptosis and autophagy for neurocardiac complications of diabetes mellitus. World J Diabetes 2015;6:217—24.

[143] Association AD. Diagnosis and classification of diabetes mellitus. Diabetes Care 2014;37:S81—90.

[144] Fu D, Wu M, Zhang J, Du M, Yang S, Hammad SM. Mechanisms of modified LDL-induced pericyte loss and retinal injury in diabetic retinopathy. Diabetologia 2012;55:3128—40.

[145] Chong ZZ, Shang YC, Wang S, Maiese K. PRAS40 is an integral regulatory component of erythropoietin mTOR signaling and cytoprotection. PloS One 2012;7:e45456.

[146] Chong ZZ, Hou J, Shang YC, Wang S, Maiese K. EPO relies upon novel signaling of Wnt1 that requires Akt1, FoxO3a, GSK-3β, and β-catenin to foster vascular integrity during experimental diabetes. Curr Neurovasc Res 2011;8:103—20.

[147] Lenoir O, Jasiek M, Hénique C, Guyonnet L, Hartleben B, Bork T. Endothelial cell and podocyte autophagy synergistically protect from diabetes-induced glomerulosclerosis. Autophagy 2015;11(7):1130—45. Available from: http://dx.doi.org/10.1080/15548627.2015.1049799.

[148] Talaei F, Van Praag VM, Shishavan MH, Landheer SW, Buikema H, Henning RH. Increased protein aggregation in Zucker diabetic fatty rat brain: identification of key mechanistic targets and the therapeutic application of hydrogen sulfide. BMC Cell Biol 2014;15:1.

mTOR IN MEMORY, BEHAVIOR, AND AGING

10

mTOR Involvement in the Mechanisms of Memory: An Overview of Animal Studies

Maria Grazia Giovannini and Daniele Lana

Department of Health Sciences, Section of Clinical Pharmacology and Oncology,
University of Florence, Florence, Italy

Memory, the ability of an animal to adapt its behavior in response to environmental stimuli, is not a unitary function, but depends on the structural and functional plasticity of several brain regions, different cell types, multiple neurotransmitter systems, and transduction pathways [1,2].

Learning experiences start cascades of events in the brain which can result in short-term memories (STM) that last minutes to hours, or long-term memories (LTM) that last days, weeks, and even a lifetime [3]. A major question that the neurobiologists of memory are still facing is whether STM and LTM are interdependent or independent phenomena. Some cellular mechanisms that underlie the development of STM overlap with those of LTM, but other mechanisms are independent and peculiar of either form of memory [1,4]. According to the current hypotheses, formation of STM and LTM depends upon biochemical processes that act in parallel and on different timescales [1,4]. Indeed, pharmacological treatments block STM independently from LTM [4–6]. Nevertheless, to better answer this question too, it would be necessary to demonstrate that STM can be suppressed without affecting LTM.

A unique characteristic that distinguishes LTM from STM is the need for a consolidation period. During consolidation, it has been demonstrated that structural and functional modifications of synapses take place [7], the most important of which is protein synthesis, unique for the development of LTM [8–12]. Protein translation in the synaptic-dendritic compartment represents a mechanism that can produce rapid changes of protein content in response to synaptic activity [13–15].

To better understand memory, it is of the utmost importance to clarify how and where in the brain experiences are encoded into lasting memories. Among the various brain structures involved in memory, there is a general consensus that the cornu ammonis area 1 (CA1) region of the hippocampus plays a major role in the formation of new memories about experienced events (episodic or autobiographical memory) [16], and in spatial memory and navigation, particularly in memory encoding [16–21], both in animals and humans. It should be mentioned that the hippocampus is strictly connected with the entorhinal cortex to form the hippocampal formation and additional output pathways project to other cortical areas including the prefrontal cortex. The entorhinal cortex, in turn, is reciprocally connected with many other parts of the cerebral cortex. It is currently believed that memory consolidation takes place through the bidirectional circuitry between the hippocampus and cortical areas [22,23].

Hippocampal long-term potentiation (LTP) is a use-dependent increase of synaptic efficiency that is being widely studied as a model of the synaptic changes that underlie learning and memory (for review, see Refs [24,25]). Indeed, both LTP and memory are defined as persistent cellular modifications that result from transient stimuli. The biochemical changes that underlie LTP correlate closely with those involved in learning [25]. LTP, like memory, can be divided into early LTP (E-LTP), which does not require ongoing protein synthesis, and late LTP (L-LTP), sensitive to blockers of either transcription or translation [26]. E-LTP is therefore assimilated to STM, whereas L-LTP is assimilated to LTM. Several neurotransmitter systems, including the glutamatergic [27], cholinergic [28], and adrenergic [29–31] systems participate in the molecular events that lead to induction and maintenance of L-LTP and memory. These and other *in vitro* paradigms that model synaptic plasticity, such as long-term depression (LTD), are very important to shed light on the mechanisms through which neurons alter their connectivity and give rise to memories.

In this chapter, on the bases of the current literature, we shall describe the current view on the molecular mechanisms of synaptic plasticity and memory, with particular emphasis on the involvement of mammalian target of rapamycin (mTOR) in hippocampal LTM formation.

10.1 mTOR AS A REGULATOR OF TRANSLATION IN THE BRAIN

The mammalian target of rapamycin, recently renamed "mechanistic" target of rapamycin [32], is an evolutionary conserved, ubiquitous high-molecular weight serine-threonine protein kinase that belongs to the phosphatidylinositol 3-kinase-related kinase protein family [33]. mTOR is a cytoplasmic protein that regulates cell growth, proliferation, and survival [34] by increasing protein translation [35]. mTOR is expressed not only in peripheral tissues but also in the central nervous system, particularly in neurons, but also in glia cells [36,37]. In neurons, mTOR is localized in the cytoplasm and dendrites, where its function is mediated through two large biochemical complexes defined as mammalian target of rapamycin complex 1 (mTORC1) and mammalian target of rapamycin complex 2 (mTORC2). The two complexes have different protein composition (extensively reviewed elsewhere, see Refs [38,39]). The mTORC1 is characterized by the presence of proline-rich Akt substrate of 40 kDa (PRAS40) and by the regulator-associated protein of the mammalian target of rapamycin (Raptor) [39], essential for mTORC1 activity. The mTORC2 comprehends the rapamycin-insensitive companion of mTOR (Rictor), the mammalian stress-activated MAP kinase-interacting protein 1, both essential for mTORC2 function, and proteins observed with Rictor 1 and 2 (Protor 1 and 2) [40–43]. Only mTORC1 is extremely sensitive to acute inhibition by rapamycin (with an IC50 in the nanomolar range), although prolonged treatment with high doses of rapamycin can also inhibit mTORC2 activity [44].

Activation of mTOR by growth factors [45,46] and glutamate receptors is well documented [47]. Recently it has been demonstrated that mTOR can be activated also by G-protein-coupled receptors [48,49], such as muscarinic acetylcholine receptors in a neuroblastoma cell line *in vitro* [50]. Furthermore, mTOR activation is downstream of nicotinic acetylcholine receptors in cultured nonsmall-cell lung carcinoma cells [51]. The above results indicate that in the brain mTOR can be downstream not only of the excitatory glutamate input through *N*-methyl-D-aspartate receptor (NMDAR), but also of other neurotransmitter systems, such as ACh through cholinergic receptors.

In the central nervous system, activation of mTOR, through regulation of protein translation, results in different outcomes, depending upon age. During embryogenesis, mTOR is essential for early brain development [52]. During senescence, mTOR seems to participate in the regulation of autophagy to prevent the accumulation of toxic aggregates of proteins that are currently considered to be the cause of several brain neurodegenerative disorders [53,54].

In the adult brain, mTOR modulates translation in specific cellular departments, including dendrites and dendritic spines [55] directly regulating peripheral protein synthesis, independently from the nuclear transcription machinery.

The mTOR pathway regulates synaptic plasticity coordinating, through numerous potential intracellular mechanisms, the timing and subcellular localization of translation machinery for the synthesis of new proteins [56]. Downstream effectors of mTOR include p70S6 Kinase (p70S6K), and eukaryotic Elongation Factor 1A and 2 (eEF1A and eEF2), which are mostly involved in ribosome recruitment to mRNA, and the eukaryotic initiation factor 4E binding proteins (4E-BPs), which regulate both the initiation and elongation phases of translation [45]. One of the mechanisms through which mTOR activation regulates protein translation may come from the effects of downstream effectors of mTORC1, such as 4E-BP1, on protein translation. These effects are not limited simply to switching "on" or "off" protein synthesis; they can also alter the range and the type of nascent proteins by mediating a modulation between cap-dependent and cap-independent translation [57]. A question then arises as to how mRNAs are selected or "tagged" [58] at the synapses to be translated. A major class of mRNAs, regulated downstream of mTORC1, are 5′ terminal oligoyrimidine tract mRNAs (TOP mRNAs) [59]. TOP mRNAs are transported into dendrites and axons where they represent one of the most abundant mRNA species [60,61], and are locally translated [62] through mTORC1 activity [63]. However, the importance of local TOP mRNAs translation in neurons is still largely speculative.

mTOR may modulate synaptic plasticity not only promoting the synthesis of proteins, but also suppressing translation of selected mRNAs. Kv1.1 is a voltage-gated potassium channel present in axons and in the somatodendritic compartment, whose activation lowers neuronal activity. Calcium-stimulation of mTORC1 activity reduces levels of Kv1.1 in cultured neurons. On the contrary, it has been demonstrated that treatment with rapamycin increases the Kv1.1 protein expression in hippocampal dendrites, without altering its axonal expression. These data suggest that mTORC1 supports synaptic activity not only through increase, but also through reduction, of local protein synthesis, suppressing translation of selected mRNAs [64].

10.2 mTOR INVOLVEMENT IN LTP AND LTD

There are several mechanisms through which mTOR may modulate synaptic plasticity and LTM formation [14]. The involvement of mTOR in synaptic plasticity was first demonstrated when it was shown that rapamycin [45,65,66] impairs long-term facilitation in aplysia [67] and crayfish [68], and prevents L-LTP formation in the rat [69]. Later, mTOR-dependent activation of dendritic p70S6K was shown to be necessary for the induction phase of protein-synthesis-dependent synaptic plasticity [70]. Further studies [62,71] showed an interesting interplay between extracellular regulated kinase (ERK) and mTOR pathways at CA3–CA1 synapses: high-frequency stimulation (HFS) induces LTP that depends upon translation of proteins regulated by mTOR, whereas ERK is necessary for HFS stimulation of mTOR through activation of 3-phosphoinositide-dependent protein kinase-1 and protein kinase B Akt upstream of mTOR. Furthermore, Tsokas and colleagues demonstrated that stimuli that induce L-LTP lead to mTORC1 activation that in turn translates TOP mRNAs [71] as well as mRNAs that have well-defined roles in synaptic plasticity and memory mechanisms, such as NMDAR 1, calcium-calmodulin protein kinase II alpha, postsynaptic density protein 95, activity-regulated cytoskeleton-associated protein, and protein kinase M zeta [72–75].

LTD, a different form of hippocampal synaptic plasticity mediated by metabotropic glutamate receptors (mGluR), has also been reported to be sensitive to rapamycin [76–79]. However, not all laboratories have found that rapamycin impairs LTD mediated by mGluR [80], suggesting that inhibition of mTOR by rapamycin is not the limiting step in LTD formation.

An interesting question is how the activation of mTOR signaling, at the crossroads of different intracellular transduction pathways activated by glutamate, can generate either LTP or LTD, which have completely different physiological outcomes. This specificity may be due to the activation of different pools of mTOR

sequestered into highly organized regions close to the plasma membrane nearby different glutamate receptors, or through specific trafficking of mRNAs to different neuronal compartments [15].

Due to the lack of selective inhibitors of mTORC2 activation, this complex has been so far less investigated, therefore its precise function in the brain is still elusive. The downstream targets of mTORC2 include Akt, protein kinase C (PKC) and serum and glucocorticoid-regulated kinase 1. Recently Thomanetz and colleagues produced a mouse strain in which Rictor was conditionally deleted in the developing and the adult central nervous system [81]. In this transgenic mouse strain it was found that the brain was smaller than in wild-type (wt) siblings and the morphology of neurons was strongly affected. This phenotype may depend on the dysregulation of the actin cytoskeleton and on impaired activation of several PKC isoforms, affected by the conditional deletion of mTORC2. PKC isoforms and their targets are implicated in neurological diseases including spinocerebellar ataxia and have been shown to affect learning and memory, suggesting that aberrations of mTORC2 signaling might be involved in neurodegenerative diseases of the brain [82].

Huang and colleagues [83] demonstrated that conditional deletion of Rictor in the postnatal murine forebrain greatly reduces mTORC2 activity and selectively impairs both hippocampal L-LTP and long-term fear memories retention. Moreover, A-443654, a compound that activates mTORC2, converts E-LTP into L-LTP and enhances LTM. Thus, these data suggest not only that mTORC2 has a necessary role in synaptic plasticity and memory [83], but also that mTORC2 could be a therapeutic target for the treatment of cognitive dysfunctions.

A comprehensive overview of the literature on the effects of mTOR in LTD and LTP mechanisms is reported in Table 10.1.

10.3 mTOR INVOLVEMENT IN LTM

The first experiments that demonstrated the participation of mTOR in memory development *in vivo* were those of Parsons et al. [91] and Bekinschtein et al. [92], who showed that acquisition or consolidation of fear memories in the hippocampus or amygdala of the rat requires mTOR activity. Formation of long-term fear memory is impaired when rapamycin is injected bilaterally into the amygdala during training [91]. Rapamycin injection disrupts memory performance 24 h, but not 3 h, after training [89], demonstrating that the animal can form the fear-associated STM. Thus, rapamycin specifically impairs the consolidation of the

TABLE 10.1 Overview of the Literature on the Effects of mTOR in Synaptic Plasticity *In Vitro*

Experimental paradigm	Species	Drug	Concentration	Effect	Ref.
fEPSP recording sensory neurons cultures	Aplysia	Rapamycin	20 nM	No effect on LTF at 24 h Complete inhibition at 72 h	[67]
EJPs recording opener muscle	Crayfish	Rapamycin	100 nM	LTF inhibition	[68]
fEPSP recording hippocampal slices	BALB/c57 mouse SD rat	Rapamycin	200 nM	L-LTP inhibition E-LTP unaffected Inhibition of BDNF-induced synaptic potentiation	[69]
fEPSP recording hippocampal slices	Wistar rats (1 month)	Rapamycin	1 μM	Inhibition of L-LTP only during the induction paradigm, but not after establishment of L-LTP Prevention of conversion from STP to LTP	[70]
fEPSP recording hippocampal slices	C57BL/6 mice	Rapamycin	20 nM	Prevention of mGluR-LTD induced by DHPG	[76]
fEPSP recording hippocampal slices	SD rats	Rapamycin	1 μM or 100 nM	Inhibition of LTP maintenance induced by strong HFS At 100 nM partial inhibition of LTP maintenance	[71]
fEPSP recording hippocampal slices	4E-BP2 KO mice	–		Induction of L-LTP with E-LTP paradigm L-LTP paradigm evoke short-lasting and lower initial potentiation LTP	[84]
fEPSP recording hippocampal slices	Tsc2 +/− mice	Rapamycin	20 nM	More robust mGluR-LTD, insensitive to rapamycin	[85]
fEPSPs recording hippocampal slices	SD rats	Rapamycin	200 nM	LTP not affected 10 min after tetanic stimulations	[72]
fEPSPs recording hippocampal slices	Tsc2 +/− rats	–		Paired-pulse plasticity enhancement LTP and LTD reduction	[86]
fEPSPs recording hippocampal slices	SD rats (18–30 days)	Rapamycin	200 nM	Impairment of LTP induction 1 h after tetanization	[73]
fEPSP recording hippocampal slices	S6K1/S6K2 KO mice	–		L-LTP not affected E-LTP compromised in S6K1-deficient mice	[87]
fEPSP recording hippocampal slices	Tsc2 +/− mice	–		Abnormal LTP	[88]
fEPSP recording hippocampal slices	FKBP12 cKO mice	Rapamycin	20 nM	Enhanced, rapamycin-insensitive L-LTP	[89]
fEPSP recording hippocampal slices	FMRP KO mice	Rapamycin	20 nM	Enhanced, rapamycin-insensitive mGluR-LTD in FMRP KO mice	[77]
fEPSP recording hippocampal slices	Tsc2 +/− mice	Rapamycin	20 nM	Decreased mGluR-LTD in Tsc2 +/− mice Enhancement of mGluR-LTD by rapamycin	[80]
fEPSP recording organotypic hippocampal slices	SD rats	Rapamycin	200 nM	mGluR-LTD blockade	[78]
fEPSP recording hippocampal slices	Cyfip1 heterozygous mice	Rapamycin	20 nM	Reduced LTP in wt and Cyfip1 heterozygous mice mGluR-LTD not impaired Cyfip1 heterozygous mice	[79]
fEPSP recording hippocampal slices	ΔRG transgenic mice	–		Normal hippocampal E-LTP, L-LTP, and NMDAR-LTD impaired mGluR-LTD	[90]
fEPSP recording hippocampal slices	Tcs1 +/− and Tsc2 −/− cKO mice	–		mGluR-LTD impairment	
fEPSP recording hippocampal slices	Rictor fb KO and wt mice	A-443654 (Akt inhibitor)	0.5 μM	A-443654 converted E-LTP into L-LTP in wt mice but not in A-443654 Impaired L-LTP in Rictor fb KO	[83]

Note: fEPSP, field excitatory postsynaptic potentials; EJPs, excitatory junctional potentials; LTF, long-term facilitation; mGluR, metabotropic glutamate receptors; HFS, high-frequency stimulation; KO, knockout; cKO, conditional knockout; wt, wild type; SD, Sprague Dawley; C57BL, C57 Black; FMRP, fragile X mental retardation 1; Cyfip1, cytoplasmic FMR1-interacting protein 1; A-443645, compound A; BDNF, brain-derived neurotrophic factor; mGluR, metabotropic glutamate receptor; DHPG, (*S*)-3,5-dihydroxyphenylglycine; NMDAR, *N*-methyl-D-aspartate receptor.

memory to long-term storage. Later, it has been repeatedly demonstrated that rapamycin blocks consolidation or reconsolidation of contextual fear memory [93–96]. Furthermore, central administration of rapamycin *in vivo* disrupts the consolidation of different types of memories that depend on other brain regions. Rapamycin has been shown to disrupt hippocampus-dependent spatial memory [97], auditory-cortex-dependent memory [98,99], gustatory-cortex-dependent memory for taste aversion [100], and prefrontal-cortex-dependent fear memory [101].

Nevertheless, while it has been repeatedly demonstrated that rapamycin blocks consolidation or reconsolidation of contextual fear memory [91–96], its effects on cue-based fear memory are far less consistent [91,94,95]. Rapamycin directly infused into the amygdala impairs cue-based fear memory consolidation and reconsolidation [91]; surprisingly, although systemic rapamycin administration blocks post-training mTOR activation in the amygdala, it does not impair consolidation and reconsolidation of fear-potentiated startle to an odor-conditioned stimulus, a cued-based fear memory [95]. It is therefore not clear whether systemic rapamycin treatment would attenuate auditory fear memory, a cue-based fear memory.

Chronic administration of rapamycin in the diet to mice enhances spatial learning in young mice and restores memory of an aversive event in old mice. The authors also showed that rapamycin has anxiolytic and antidepressant-like effects at all ages tested [102].

It has been shown in multiple species and in a variety of behavioral paradigms that a consolidated memory becomes sensitive to disruption following reactivation [103,104]. This time-lapse of vulnerability requires *de novo* protein synthesis to reconsolidate the engram, empirically defining the reconsolidation phase of memory ([104]; for review see Ref. [105]). Translational control by mTOR appears essential in reconsolidation of the engram following retrieval. Indeed, following retrieval, animals exhibit increased phosphorylation of p70S6K [94], and rapamycin administration after reactivation disrupts subsequent retention [91,93–96,106]. Recently, it has also been demonstrated that blockade of mTOR with rapamycin administered systemically immediately or 12 h after training or reactivation impairs both consolidation and reconsolidation of an auditory fear memory [107,108]. These data demonstrate that biphasic mTOR activation and protein translation are essential for both consolidation and reconsolidation mechanisms that contribute to the formation, restabilization, and persistence of long-term auditory fear memories. These findings are particularly interesting in view of a possible treatment model for reducing the emotional strength of established traumatic memories, such as post-traumatic stress disorder and specific phobias, through pharmacologic blockade of mTOR using rapamycin or analogs (rapalogs) upon reactivation of the emotionally arousing memory [107,108].

Recent data indicating that mTORC1 contributes to cocaine-dependent place preference and cue-induced reinstatement memory processes [109,110] raise the possibility that mTORC1 is involved in the reconsolidation of memories associated with drugs of abuse, particularly alcohol. It has been demonstrated that reconsolidation of alcohol-related memories triggered by the most behaviorally relevant cues for relapse, odor and taste, activates mTORC1 in amygdala and cortical areas of rats, and causes synaptic protein synthesis [111]. Furthermore, systemic or local amygdala inhibition of mTORC1 using rapamycin during reconsolidation disrupts alcohol-associated memories, leading to longlasting suppression of relapse. These findings provide evidence that the mTORC1 pathway and its downstream effectors are crucial in alcohol-related memory reconsolidation [111]. Since disruption of the reconsolidation of memories associated with drugs of abuse has been proposed as a potential strategy to prevent relapse, these data indicate the mTORC1 pathway as a possible target in the therapy of alcoholism relapse.

Interestingly, it was reported that in cultured neurons and hippocampal slices from transgenic mouse model of Alzheimer's disease (AD) and in hippocampal slices from wt mice exposed to exogenous Aβ1-42, the mTOR signaling pathway is inhibited, and this effect correlates with impairment in synaptic plasticity [112]. On the contrary, up-regulation of mTOR signaling by both pharmacological and genetic methods prevents Aβ1-42-induced synaptic impairment, indicating that dysregulation of the mTOR pathway could play a role in the synaptic dysfunction that characterizes AD [112]. These data confirm the idea that mTOR prevents the accumulation of toxic protein aggregates regulating autophagy [53,54], which is a major degradation pathway for organelles and aggregated proteins [113]. mTOR regulates autophagy through activation of mTORC2 which phosphorylates and activates Akt [114] that in turn positively regulates mTORC1, stimulating its function, and inhibiting autophagy. While excessive autophagy can lead to cell death, increased autophagy facilitates the clearance of aggregation-prone proteins such as Aβ1-42 [115], thus promoting neuronal survival in several neurodegenerative disorder models. The role of the mTOR pathway and of autophagy in AD is still unclear. Nevertheless, it has been shown that long-term mTOR inhibition by rapamycin in mice overexpressing human amyloid precursor protein V717F (PDAPP), a transgenic mouse model of AD, decreases the levels of Aβ1-42 and prevents

AD-like cognitive deficits in the Morris water maze (MWM) test [116]. These effects were paralleled by activation of autophagy in transgenic PDAPP mice, but not in wt mice. These data suggest that inhibition of mTOR by rapamycin can lower Aβ1-42 levels and slow the progression of cognitive impairments, possibly through stimulation of autophagy [117].

Finally, recent reports indicate that LTM deficits can be associated with hyperactivation of the mTOR-p70S6K signaling pathway and with an imbalance in protein synthesis [118]. Puighermanal and colleagues demonstrated that stimulation of the cannabinoid receptor type 1 in mouse hippocampus *in vivo* by the agonist tetrahydrocannabinol (THC) overactivates the mTOR pathway, activating both p70S6K and 4E-BP, as well as other downstream components of the translational apparatus and factors that participate in the initiation step of translation [119]. Contrary to what may be expected, the subsequent protein translation increase seems to be responsible for the memory impairments caused by cannabinoid consumption [119]. Indeed, inhibition of the mTOR/p70S6K pathway with systemic administration of rapamycin prevents the phosphorylation of p70S6K after administration of a cannabinoid agonist, recovering also the ensuing memory deficits [119]. It seems therefore that for proper memory storage a precise control of mTOR activity, as well as of its downstream effectors and of the translational machinery, is necessary [120]. Indeed, not only reduced but also enhanced activation of the mTOR signaling cascade have been recently associated with memory disruption [121]. In the same direction lead the findings showing that, although basal mTOR activity seems to be necessary for memory consolidation, an increase in mTOR signaling can disrupt memory processing. Indeed, mutant mice with constant activation of the hippocampal mTOR pathway display memory deficits [88,122]. Patients with tuberous sclerosis and animal models of this genetic disease, such as heterozygous mice for tuberous sclerosis complex 1 (Tsc1) and tuberous sclerosis complex 2 (Tsc2), that cause reduction of Tsc1–Tsc2 activation and increase of mTORC1 activity, are associated with memory deficits [88,122]. Interestingly, administration of rapamycin rescues memory deficits [88], suggesting that memory impairments in these transgenic mice are due, at least in part, to abnormal/increased mTOR signaling. This is also the case for the fragile X mental retardation protein (Knockout) KO mice, an animal model of fragile X syndrome [77], and for the 12 kDa FK506-binding protein (FKBP12) KO mice, which show enhanced mTOR and p70S6K phosphorylation in the hippocampus. These mutant mouse strains display enhanced associative contextual fear memory and anomalous performances in the novel object recognition task [89].

The mechanisms through which dysregulation/hyperactivation of mTOR activity may lead to memory deficits is still not known. It is suggested that mTOR carefully modulates translation during a distinct temporal window, and failure to properly deactivate mTOR, once activated, can be as detrimental as blocking its activation. On a parallel level, it can be mentioned that hyperactivation of mTOR plays an important role in cancer development and genetic disorders [123]. It has been shown that rapamycin extends lifespan [124], and rapalogs may be useful in the treatment of cancer, cardiovascular disease, and autoimmunity [125]. Nevertheless, the requirement of proper mTORC1 activation for synaptic plasticity and LTM consolidation should be taken into account before using these drugs, although therapeutic doses of rapamycin may be below the threshold to disrupt memory formation [116,126]. Indeed, patients treated with a rapalog, everolimus, as immunosuppressant did not experience cognitive impairments during treatment [127].

A comprehensive overview of the literature on the effects of mTOR in memory mechanisms is reported in Table 10.2.

10.4 INTERPLAY BETWEEN ACH AND THE ERK AND mTOR PATHWAYS IN INHIBITORY AVOIDANCE MEMORY ENCODING

The effects of ACh on learning and memory in the hippocampus appear to be mediated by muscarinic ACh receptors (mAChRs) [131,132], and nicotinic ACh receptors (nAChRs) [133–135]. Several years ago Feig and Lipton [136] demonstrated that the activation of mAChRs stimulates new protein synthesis in hippocampal CA1 dendrites, indicating a role of local protein synthesis in mAChR-dependent synaptic plasticity. Recent results demonstrated that the mTOR pathway is downstream of mAChRs and nAChRs [51]. Specifically, the M2 subtype of mAChRs utilizes a mitogen-activated protein kinase (MAPK)-dependent mechanism to activate mTOR, whereas the M3 subtype utilizes either MAPK-dependent or -independent mechanisms, depending on the cellular context [50]. Furthermore, a fine regulation of mTOR and p70S6K by the M4 subtype of mAChRs in PC12 cells has been demonstrated [137], and the cholinergic agonist carbachol significantly increases mTOR and p70S6K activation in CA1 pyramidal neurons *in vitro* [6]. All the above data link the cholinergic system to mTOR activation and to local protein synthesis through multiple MAPK- and/or PKC-dependent mechanisms. Emphasis has recently being given to the role of the α7 nAChR subtype in cognitive functions and

TABLE 10.2 Overview of the Literature on the Effects of mTOR in Memory Mechanisms *In Vivo*

Experimental paradigm	Species	Drug	Dose	Administration route	Administration time	Effect	Ref.
FMs discrimination test	Mongolian gerbils	Rapamycin	60 nM, 1 µl/side	Intra-auditory cortex	1. Shortly after FM training 2. 1 day before the initial training	1. No impairment of newly acquired memory trace for 24 h, deficits at 48 h 2. Learning not affected	[99]
1. MWM 2. FC	4E-BP2 KO mice	—				1. Impaired spatial learning and memory 2. Conditioned fear-associative memory deficits	[84]
1. CTA 2. LI plus CTA	Wistar rats	—				Phosphorylation of S6K1, eEF2, correlates with taste memory consolidation	[128]
MWM	LE rats	Rapamycin	0.9 ng, 1 µl/side	Intradorsal hippocamps	Immediately post-training	Impaired long-term spatial memory at 48 h	[97]
1. ORT 2. Radial-arm maze (RAM)	SD rats	LY294002 (PI3K inhibitor)	5 mM, 7.5 µl	i.c.v.	1. 10 min before the start of the sample phase in ORT 2. 20 min before the first trial of the day in RAM	1. Deficits in long-term ORT memory 2. No impairment in RAM spatial learning	[129]
Contextual and auditory conditioned stimuli FC	LE rats	Rapamycin	5 µg/µl, 0.5 µl/side	Intra-amigdala	Immediately after training	Impairment in both context and auditory conditioned stimuli fear conditioning	[91]
– Open field – Rotarod – Light/dark tests – T-maze – CTA – Step-through PA – Elevated plus maze – Social dominance tube test	4E-BP2 KO mice	—				– Altered behavior in the open field test, rotarod test, light/dark test, T-maze and CTA test – Normal behavior in PA test, elevated plus maze and social dominance Tube test	[130]
Step-down IA	Wistar rats	Rapamycin	60 nM, 0.5 µl/side	Intrahippocampal CA1	15 min before training	Impairment in inhibitory avoidance LTM without affecting STM	[92]
– MWM – FC – Social Interaction – Nest Building	Tsc1 +/− mice	—				– Spatial learning deficits in the MWM – Impaired context conditioning – Reduced social functioning	[122]
– Contextual FC – CTA – MWM	S6K1 and S6K2 KO mice	—				– S6K1 KO mice: Deficit in contextual fear memory and CTA Impaired spatial memory – S6K2 KO mice: Decreased contextual fear memory 7 days after training Reduction in LI of CTA	[87]

(Continued)

TABLE 10.2 (Continued)

Experimental paradigm	Species	Drug	Dose	Administration route	Administration time	Effect	Ref.
– Contextual FC – Locomotor apparatus test – Dark/light box – Elevated plus maze	C57BL/6 mice	Rapamycin	20–40 mg/kg	IP	Contextual FC 1. 30 min before acquisition 2. Immediately after reactivation of memory at 24 h (4 times) 3. Immediately after reactivation of memory at 48 h Other tests 4. Injection before test	Normal spatial memory 1. Rapamycin impairs long-term contextual fear memory (24 h) 2. Rapamycin inhibits reconsolidation for 21 days 3. Rapamycin reduces recall for at least 7 days 4. Rapamycin does not alter activity or anxiety-like behavior	[93]
Pavlovian olfactory conditioning (massed and spaced memory protocol)	Drosophila Fmr1 mutants	–				Defects in 1-day memory after spaced but not after massed training	[118]
– FC – MWM	Tsc2 +/− mice	Rapamycin	– FC: 5 mg/kg – MWM: 5 mg/kg – or 1 mg/kg	IP	– Fear conditioning: injection for 5 days before test and 3 h before test – MWM: daily 3 h before training	Rapamycin rescues behavioral deficits in FC and MWM	[88]
– ORT – MWM – Step-down IA	Wistar rats	Rapamycin	20 pmol, 0.8 µl/side	Intradorsal hippocampal CA1	1. Immediately, 180 min, 540 min after training 2. After reactivation	1. Rapamycin immediately or 180 min after training impairs LTM retention at 24 h 2. Rapamycin after memory reactivation hinders persistence of memory trace	[106]
– Contextual and cued FC – MWM – Y-maze reversal task – ORT – Marble burying	FKBP12 cKO mice	–				– Enhanced contextual fear memory – Autistic/obsessive-compulsive-like perseveration in MWM, Y-maze reversal task, ORT, and marble burying test	[89]
FMs discrimination	Mongolian gerbils	Rapamycin	60 nM, 1 µl/side	Intra-auditory cortex	1. 1 day before initial conditioning 2. shortly after conditioning	1. Rapamycin prevents the FMs discrimination increment 2. No impairment	[98]

Behavioral task	Animal	Drug	Dose	Route/region	Timing	Effect	Reference
– Trace FC – ORT	SD rats	Rapamycin	7 ng, 1 μl/side	Intramedial prefrontal cortex	Immediately after training	– Impairment of long-term (3 days and 6 days) trace fear memory – No effect on short-term trace fear memory and ORT memory	[101]
– CTA – LI plus CTA	Wistar rats	Rapamycin	10 μM, 1 μl/side	Intragustatory cortex	1. 25 min before pre-exposure in CTA 2. 100 min after pre-exposure in LI	Higher aversion rates in both CTA and LI paradigms	[100]
– ORT – Context-recognition	CD-1 mice, CB1 KO mice	Rapamycin	1 mg/kg	IP	3 h before behavioral tasks	Nonamnesic doses of rapamycin abrogate the amnesic-like effects of THC	[119]
MWM	3xTg-AD mice	Rapamycin	2.24 mg/kg bw/day (14 mg/kg chow, 5 g chow/day)	Orally rapamycin-supplementedchow	Daily	Rescue of early learning and memory deficits	[126]
Olfactory and context FC	SD rats	Rapamycin	40 mg/kg	IP	1. Immediately after training 2. Immediately after reactivation	1. Impaired consolidation of context but not odor cued fear memory 2. Impaired reconsolidation of context but not odor cued fear memory	[95]
Induction of reinstatement of cocaine seeking	SD rats	Rapamycin	50 μg, 0.5 μl/side	Intranucleus accumbes core	30 min before reinstatement test	Decrease of cue-induced reinstatement of cocaine seeking	[109]
Contextual FC	Tsc2 +/− mice	CDPPB (mGluR5 agonist)	10 mg/kg	IP	30 min prior to training	Positive modulation of mGluR5 reverses synaptic and behavioral deficits in Tsc2 +/− mice	[80]
MWM	Aged PDAPP Tg mice	Rapamycin	2.24 mg/kg bw/day (14 mg/kg chow, 5 g chow/day)	Orally rapamycin-supplemented chow	Daily	Improvement of learning deficits in PDAPP Tg mice	[116]
Contextual FC	LE rats	Rapamycin	1. 1, 2, or 5 μg/μl, 1 μl/side 2. 5 μg/μl, 1 μl/side	Intradorsal hippocampus	1. Immediately after training 2. 30 min prior to the retrieval session	1. Impairment of long-term contextual fear memory (24 h) 2. Impairment of reconsolidation of contextual fear memory	[94]
Place-conditioning and assessment of locomotor activity	Aged C57BL/6J mice	Rapamycin	10 mg/kg	IP	1 h prior to test	– No effect on test – Impairment of place preference and long-term locomotor sensitization	[110]
– PA – MWM – Tail suspension – Elevated plus maze	Male and female C57BL/6 mice (lifespan)	Rapamycin	2.24 mg/kg bw/day (14 mg/kg chow, 5 g chow/day)	Rapamycin supplemented chow	Daily	– Enhancement of cognitive function in adult mice – Blockade of cognitive decline in older animals – Decrease of anxiety and depressive-like behavior at all ages tested	[102]

(Continued)

TABLE 10.2 (Continued)

Experimental paradigm	Species	Drug	Dose	Administration route	Administration time	Effect	Ref.
ORT	Wistar rats	Rapamycin	600 nM, 0.5 µl/side	1. Intrabasolateral amigdala 2. Intradorsal hippocampus	15 min before, immediately after, or 6 h after training or reactivation	1. Impaired formation of LTM. Impaired LTM before or after reactivation 2. Impaired formation of LTM. Impaired LTM before reactivation	[96]
– Operant alcohol self-administration – Home-cage alcohol consumption	LE rats	Rapamycin	1. 20 mg/kg 2. 50 µg, 0.5 µl/side	1. IP 2. Intra-amigdala central nucleus	Immediately after the memory reactivation session	Rapamycin attenuates alcohol seeking relapse after reactivation and attenuates consumption on subsequent days	[111]
– Contextual and auditory FC – MWM	Rictor fb KO mice	A-443654 (mTORC2 activator)	2.5 mg/kg	IP	Immediately after training	– Long-term, but not short-term, fear memory and spatial memory is impaired in mTORC2 KO mice – A-443654 enhances LTM in wt mice but not in mTORC2 KO mice	[83]
Negatively-reinforced olfactory learning (massed or spaced memory protocol)	TORC2 KO Drosophila	–				Long-term spaced (but not massed) memory is impaired	
Step-down IA	Wistar rats	Rapamycin	1.5 nmol, 5 µl	ICV	30 min before acquisition	Impaired long-term but not short-term IA memory	[6]
Auditory FC	Aged C57BL/6 mice	Rapamycin	40 mg/kg	IP	Immediately or 12 h after training or reactivation	– Impairment of both consolidation and reconsolidation of memory – Impairment of reconsolidation, dependent upon reactivation of the memory trace	[107]

Note: MWM, Morris water maze; FMs, frequency-modulated tones; FC, contextual and cued fear conditioning; CTA, conditioned taste aversion; LI, latent inhibition; ORT, object recognition test; RAM, radial-arm maze; PA, passive avoidance; IA, inhibitory avoidance; LE, Long Evans; SD, Sprague Dawley; KO, knockout; bw, body weight; wt, wild type; S6K1, p70S6K1; S6K2, p70S6K2; LT294002, 2-morpholin-4-yl-8-phenylchromen-4-one; CDPPB, 3-cyano-N-(1,3-diphenyl-1H-pyrazol-5-yl)benzamide; ICV, intracerebroventricular; IP, intraperitoneal.

pathophysiological mechanisms of dementia. The α7 nicotinic receptor, in addition to its ionotropic activity, is associated with metabotropic activity coupled to Ca^{2+}-regulated signaling [138], and it has been demonstrated that mTOR activation is downstream of nicotinic receptors in cultured nonsmall-cell lung carcinoma cells [51].

In a recent study from our laboratory, we showed that mTOR and p70S6K are massively activated in most of CA1 pyramidal neurons soon after acquisition of an inhibitory avoidance (IA) LTM [6]. A fairly rapid and transient inactivation of mTORC1 and, consequently, of p70S6K by rapamycin impairs formation of LTM with no effect on STM, demonstrating that mTORC1 activation is necessary for LTM. These data are consistent with those reported by Hoeffer and colleagues who demonstrated that rapamycin disrupts fear-associated LTM formation 24 h, but not 3 h, after acquisition.

An intriguing result is the observation that administration of the mAChR antagonist scopolamine increases activation of mTOR and p70S6K 1 h after acquisition of the IA. Antagonism by scopolamine on presynaptic inhibitory M2/M4 muscarinic autoreceptors [139–142] massively increases ACh release from septo-hippocampal cholinergic terminals [143]. ACh, in the presence of the nonselective muscarinic antagonist scopolamine, binds to postsynaptic nAChRs with consequent activation of downstream effectors such as mTOR. The link of nAChR to the mTOR pathway and the involvement of nAChR to mediate ACh postsynaptic responses in the hippocampus are substantiated by several studies [51,134,144–146]. An alternative explanation for the increased activation of mTOR following scopolamine administration is that the large and longlasting increase of ACh release [143] may in time overcome the postsynaptic antagonistic effect of the competitive antagonist scopolamine on the postsynaptic mAChRs, with consequent activation of intracellular pathways that lead to activation of mTOR. Therefore, ACh function on any given circuit and intracellular pathway may depend on the differential expression of postsynaptic mAChRs and nAChRs and upon the temporal dynamics of ACh levels in the synaptic cleft.

We showed that administration of muscarinic plus nicotinic antagonists *in vivo* block the scopolamine-induced increase of mTOR activation 1 h after acquisition of IA [6]. mTOR activation at a longer time (4 h) after administration of the two drugs is sufficient to develop LTM [6]. We envisage that activation of mTOR at later times is sufficient to activate downstream effectors leading to LTM formation, or that other neurotransmitter systems [46,147] and/or other intracellular signaling pathways are involved in IA

LTM formation [148]. A further explanation may come from data that demonstrate that mTORC1 activity regulates only a small component of total protein synthesis [149,150]. Additional mTORC1-independent regulatory signals are required to induce LTM since stimulation of mTORC1 probably generates a set of proteins that are important, but not sufficient, for neuronal plasticity or memory [32].

An interesting interconnection between the cholinergic system and mTOR comes from the work by Tsokas and colleagues [71]. They demonstrated that mTORC1 activity stimulates protein encoding in dendrites through TOP mRNAs. eEF1A, one of the TOP mRNAs translated during plasticity [71], may play an important role in the regulation of the expression of one muscarinic acetylcholine receptor subtype [151].

A mechanistic model that may help explaining the integrated role of cholinergic activation and the downstream effectors ERK and mTOR in the formation of hippocampal IA STM and LTM is the following: the cholinergic septo-hippocampal pathway, activated during acquisition of an IA memory [132], releases ACh from its synaptic terminals which impinges on and activates both muscarinic and nicotinic pre- and postsynaptic ACh receptors. Postsynaptic mAChRs indirectly activate the intracellular pathways of ERK and mTOR, responsible, with different contributions and different downstream effectors, for IA STM and LTM formation [6,132]. Inhibitory muscarinic presynaptic receptors, blocked by muscarinic antagonists, massively increase ACh release [143] that impinges on postsynaptic nAChRs. Activation of postsynaptic nAChRs indirectly leads to activation of mTOR and formation of the mTORC1 complex that increases, through p70S6K activation, local protein synthesis that is necessary for IA LTM memory [6]. A crosstalk between ERK and mTOR at different levels of the signaling flow has been described. Indeed, ERK was found to phosphorylate and inhibit the function of Tsc2, albeit through different mechanisms and at different phosphorylation sites [152]. For instance, a recent study [62] showed an interesting interplay between ERK and mTOR pathways at CA3–CA1 synapses: ERK is required for the HFS-induced activation of the mTOR pathway in the hippocampus.

Further studies demonstrate a complex interplay among the cholinergic system, ERK, and mTOR. For instance, Connor and colleagues showed in mouse CA1 region that coactivation of mAChRs and β-adrenergic receptors facilitates the conversion of short-term potentiation (STP) to LTP through an ERK- and mTOR-dependent mechanism which requires translation initiation [153]. mTOR is known to phosphorylate p70S6K on Thr389, while ERK is able to phosphorylate p70S6K on Thr421/Ser424 [154]. It was also shown that the mAChRs agonist oxotremorine induces phosphorylation of p70S6K at Thr389, which is not dependent upon activation of mTOR but possibly upon the ERK pathway activation [155]. A previous study showed that mTOR could phosphorylate p70S6K on Thr421/Ser424, a specific site of ERK and inversely, ERK phosphorylation p70S6K on Thr389 [154], making the story even more complex. It is also currently accepted that ERK and mTORC1 synergistically regulate eukaryotic translation initiation factor 4E and translation initiation in LTM and synaptic plasticity [156,157].

It seems therefore that independent/concurrent/synergic recruitment and activation of ERK and mTOR signaling cascades may be a conserved mechanism for the precise regulation of translation downstream of various neuromodulatory receptors.

10.5 CONCLUSIONS

The mechanisms that are involved in the formation of LTM are the focus of intense investigations, and although a few answers are now available, we are still far from having a complete understanding of the process. It is still a matter of investigation whether STM and LTM mechanisms are processed in series, or in parallel. The most common view at present is that different molecular mechanisms may be needed to form STM and LTM, whereas some mechanisms are involved in both. *De novo* protein synthesis is required to stabilize a STM into a LTM, and in this chapter we have summarized our present knowledge on the complex involvement of the mTOR pathway in the mechanisms of synaptic plasticity and memory. How mTOR activation enhances protein translation in the somatodendritic compartment is well understood. Nevertheless the exact mechanisms downstream of mTOR activation involved in synaptic plasticity and memory formation have not been completely unraveled. The precise involvement of the two complexes, mTORC1 and mTORC2, in memory is not clear. Also, how dysregulation of mTORC1 complex may result in impairment or improvement of memories, depending upon the systems involved is still a matter of debate. The use of animal models, and especially of transgenic/KO mouse strains will help researchers to understand the tuning of these fine mechanisms in further detail, to shed light not only on basic mechanisms but also, hopefully, to find in the future pharmacological treatments able to treat memory deficits.

References

[1] Izquierdo I, Barros DM, Mello e Souza T, de Souza MM, Izquierdo LA, Medina JH. Mechanisms for memory types differ. Nature 1998;393:635–6.

[2] Giovannini MG, Lana D, Pepeu G. The integrated role of ACh, ERK and mTOR in the mechanisms of hippocampal inhibitory avoidance memory. Neurobiol Learn Mem 2015;119:18–33.

[3] McGaugh JL. Time-dependent processes in memory storage. Science 1966;153:1351–8.

[4] Izquierdo LA, Barros DM, Vianna MR, Coitinho A, deDavid e Silva T, Choi H, et al. Molecular pharmacological dissection of short- and long-term memory. Cell Molec Neurobiol 2002;22:269–87.

[5] Vianna MR, Izquierdo LA, Barros DM, Medina JH, Izquierdo I. Intrahippocampal infusion of an inhibitor of protein kinase A separates short- from long-term memory. Behav Pharmacol 1999;10:223–7.

[6] Lana D, Cerbai F, Di Russo J, Boscaro F, Giannetti A, Petkova-Kirova P, et al. Hippocampal long term memory: effect of the cholinergic system on local protein synthesis. Neurobiol Learn Mem 2013;106:246–57.

[7] Igaz LM, Vianna MRM, Medina JH, Izquierdo I. Two time periods of hippocampal mRNA synthesis are required for memory consolidation of fear-motivated learning. J Neurosci 2002;22:6781–9.

[8] Davis HP, Squire LR. Protein synthesis and memory: a review. Psychol Bull 1984;96:518–59.

[9] Freeman FM, Rose SP, Scholey AB. Two time windows of anisomycin-induced amnesia for passive avoidance training in the day-old chick. Neurobiol Learn Mem 1995;63:291–5.

[10] Tiunova A, Anokhin K, Rose SP, Mileusnic R. Involvement of glutamate receptors, protein kinases, and protein synthesis in memory for visual discrimination in the young chick. Neurobiol Learn Mem 1996;65:233–4.

[11] Bourtchouladze R, Abel T, Berman N, Gordon R, Lapidus K, Kandel ER. Different training procedures recruit either one or two critical periods for contextual memory consolidation, each of which requires protein synthesis and PKA. Learn Mem 1998;5:365–74.

[12] Schafe GE, Nadel NV, Sullivan GM, Harris A, LeDoux JE. Memory consolidation for contextual and auditory fear conditioning is dependent on protein synthesis, PKA, and MAP kinase. Learn Mem 1999;6:97–110.

[13] Bailey CH, Kandel ER, Si K. The persistence of long-term memory: a molecular approach to self-sustaining changes in learning-induced synaptic growth. Neuron 2004;44:49–57.

[14] Kelleher III RJ, Govindarajan A, Jung HY, Kang H, Tonegawa S. Translational control by MAPK signaling in long-term synaptic plasticity and memory. Cell 2004;116:467–79.

[15] Hoeffer CA, Klann E. mTOR signaling: at the crossroads of plasticity, memory and disease. Trends Neurosci 2010;33:67–75.

[16] Squire LR. Memory and the hippocampus: a synthesis from findings with rats, monkeys, and humans. Psychol Rev 1992;99:195–231.

[17] Eichenbaum H. The hippocampus and declarative memory: cognitive mechanisms and neural codes. Behav Brain Res 2001;127:199–207.

[18] Hasselmo ME, Wyble BP, Wallenstein GV. Encoding and retrieval of episodic memories: role of cholinergic and GABAergic modulation in the hippocampus. Hippocampus 1996;6:693–708.

[19] Vinogradova OS. Hippocampus as comparator: role of the two input and two output systems of the hippocampus in selection and registration of information. Hippocampus 2001;11:578–98.

[20] Lisman JE, Grace AA. The hippocampal-VTA loop: controlling the entry of information into long-term memory. Neuron 2005;46:703–13.

[21] Moser EI, Kropff E, Moser MB. Place cells, grid cells, and the brain's spatial representation system. Ann Rev Neurosci 2008;31:69–89.

[22] Lavenex P, Amaral DG. Hippocampal-neocortical interaction: a hierarchy of associativity. Hippocampus 2000;10:420–30.

[23] Lavenex PB, Amara DG, Lavenex P. Hippocampal lesion prevents spatialrelational learning in adult macaque monkeys. J Neurosci 2006;26:4546–58.

[24] Malenka RC, Nicoll RA. Long-term potentiation—a decade of progress? Science 1999;285:1870–4.

[25] Bliss TV, Collingridge GL. A synaptic model of memory: long-term potentiation in the hippocampus. Nature 1993;361:31–9.

[26] Kandel ER. The molecular biology of memory storage: a dialogue between genes and synapses. Science 2001;294:1030–8.

[27] Collingridge G. Synaptic plasticity. The role of NMDA receptors in learning and memory. Nature 1987;330:604–5.

[28] Blitzer RD, Gil O, Landau EM. Cholinergic stimulation enhances long-term potentiation in the CA1 region of rat hippocampus. Neurosci Lett 1990;119:207–10.

[29] Watabe AM, Zaki PA, O'Dell TJ. Coactivation of beta-adrenergic and cholinergic receptors enhances the induction of long-term potentiation and synergistically activates mitogen-activated protein kinase in the hippocampal CA1 region. J Neurosci 2000;20:5924–31.

[30] Giovannini MG, Blitzer RD, Wong T, Asoma K, Tsokas P, Morrison JH, et al. Mitogen-activated protein kinase regulates early phosphorylation and delayed expression of Ca2 + /calmodulin-dependent protein kinase II in long-term potentiation. J Neurosci 2001;21:7053–62.

[31] Winder DG, Martin KC, Muzzio IA, Rohrer D, Chruscinski A, Kobilka B, et al. ERK plays a regulatory role in induction of LTP by theta frequency stimulation and its modulation by beta-adrenergic receptors. Neuron 1999;24:715–26.

[32] Graber TE, McCamphill PK, Sossin WS. A recollection of mTOR signaling in learning and memory. Learn Mem 2013;20:518–30.

[33] Jung CH, Ro SH, Cao J, Otto NM, Kim DH. mTOR regulation of autophagy. FEBS Lett 2010;584:1287–95.

[34] Martin DE, Hall MN. The expanding TOR signaling network. Curr Opin Cell Biol 2005;17:158–66.

[35] Abelaira HM, Reus GZ, Neotti MV, Quevedo J. The role of mTOR in depression and antidepressant responses. Life Sci 2014;101:10–14.

[36] Dello Russo C, Lisi L, Feinstein DL, Navarra P. mTOR kinase, a key player in the regulation of glial functions: relevance for the therapy of multiple sclerosis. Glia 2013;61:301–11.

[37] Chong ZZ, Li F, Maiese K. The pro-survival pathways of mTOR and protein kinase B target glycogen synthase kinase-3beta and nuclear factor-kappaB to foster endogenous microglial cell protection. Int J Mol Med 2007;19:263–72.

[38] Dibble CC, Manning BD. Signal integration by mTORC1 coordinates nutrient input with biosynthetic output. Nat Cell Biol 2013;15:555–64.

[39] Laplante M, Sabatini DM. mTOR signaling in growth control and disease. Cell 2012;149:274–93.

[40] Jacinto E, Loewith R, Schmidt A, Lin S, Rüegg MA, Hall A, et al. Mammalian TOR complex 2 controls the actin cytoskeleton and is rapamycin insensitive. Nat Cell Biol 2004;6:1122–8.

[41] Jacinto E, Facchinetti V, Liu D, Soto N, Wei S, Jung SY, et al. SIN1/MIP1 maintains rictor-mTOR complex integrity and regulates Akt phosphorylation and substrate specificity. Cell 2006;127:125–37.

[42] Pearce LR, Huang X, Boudeau J, Pawłowski R, Wullschleger S, Deak M, et al. Identification of Protor as a novel Rictor-binding component of mTOR complex-2. Biochem J 2007;405:513–22.

[43] Sarbassov DD, Ali SM, Kim DH, Guertin DA, Latek RR, Erdjument-Bromage H, et al. Rictor, a novel binding partner of mTOR, defines a rapamycin-insensitive and raptor-independent pathway that regulates the cytoskeleton. Curr Biol 2004;14:1296–302.

[44] Sarbassov DD, Ali SM, Sengupta S, Sheen JH, Hsu PP, Bagley AF, et al. Prolonged rapamycin treatment inhibits mTORC2 assembly and Akt/PKB. Mol Cell 2006;22:159–68.

[45] Hay N, Sonenberg N. Upstream and downstream of mTOR. Gen Develop 2004;18:1926–45.

[46] Slipczuk L, Bekinschtein P, Katche C, Cammarota M, Izquierdo I, Medina JH. BDNF activates mTOR to regulate GluR1 expression required for memory formation. PLoS One 2009;4:e6007.

[47] Lenz G, Avruch J. Glutamatergic regulation of the p70S6 kinase in primary mouse neurons. J Biol Chem 2005;280:38121–4.

[48] Wang L, Proud CG. Regulation of the phosphorylation of elongation factor 2 by MEK-dependent signalling in adult rat cardiomyocytes. FEBS Lett 2002;531:285–9.

[49] Arvisais EW, Romanelli A, Hou X, Davis JS. AKT-independent phosphorylation of TSC2 and activation of mTOR and ribosomal protein S6 kinase signaling by prostaglandin F2alpha. J Biol Chem 2006;281:26904–13.

[50] Slack BE, Blusztajn JK. Differential regulation of mTOR-dependent S6 phosphorylation by muscarinic acetylcholine receptor subtypes. J Cell Biochem 2008;104:1818–31.

[51] Zheng Y, Ritzenthaler JD, Roman J, Han S. Nicotine stimulates human lung cancer cell growth by inducing fibronectin expression. Am J Resp Cell Mol Biol 2007;37:681–90.

[52] Hentges KE, Sirry B, Gingeras AC, Sarbassov D, Sonenberg N, Sabatini D, et al. FRAP/mTOR is required for proliferation and patterning during embryonic development in the mouse. Proc Natl Acad Sci USA 2001;98:13796–801.

[53] O'Neill C. PI3-kinase/Akt/mTOR signaling: impaired on/off switches in aging, cognitive decline and Alzheimer's disease. Exp Geront 2013;48:647–53.

[54] Lipton JO, Sahin M. The neurology of mTOR. Neuron 2014;84:275–91.

[55] Sutton MA, Schuman EM. Local translational control in dendrites and its role in long-term synaptic plasticity. J Neurobiol 2005;64:116–31.

[56] Russo E, Citraro R, Constanti A, De Sarro G. The mTOR signaling pathway in the brain: focus on epilepsy and epileptogenesis. Mol Neurobiol 2012;46:662–81.

[57] Bove J, Martinez-Vicente M, Vila M. Fighting neurodegeneration with rapamycin: mechanistic insights. Nat Rev Neurosci 2011;12:437–52.

[58] Frey U, Morris RG. Synaptic tagging and long-term potentiation. Nature 1997;385:533–6.

[59] Thoreen CC, Chantranupong L, Keys HR, Wang T, Gray NS, Sabatini DM. A unifying model for mTORC1-mediated regulation of mRNA translation. Nature 2012;485:109–13.

[60] Moccia R, Chen D, Lyles V, Kapuya EEY, Kalachikov S, Spahn CM, et al. An unbiased cDNA library prepared from isolated Aplysia sensory neuron processes is enriched for cytoskeletal and translational mRNAs. J Neurosci 2003;23:9409–17.

[61] Poon MM, Choi SH, Jamieson CA, Geschwind DH, Martin KC. Identification of process-localized mRNAs from cultured rodent hippocampal neurons. J Neurosci 2006;26:13390–9.

[62] Tsokas P, Ma T, Iyengar R, Landau EM, Blitzer RD. Mitogen-activated protein kinase upregulates the dendritic translation machinery in long-term potentiation by controlling the

mammalian target of rapamycin pathway. J Neurosci 2007;27:5885–94.

[63] Gobert D, Topolnik L, Azzi M, Huang L, Badeaux F, Desgroseillers L, et al. Forskolin induction of late-LTP and up-regulation of 5′ TOP mRNAs translation via mTOR, ERK, and PI3K in hippocampal pyramidal cells. J Neurochem 2008;106:1160–74.

[64] Raab-Graham KF, Haddick PC, Jan YN, Jan LY. Activity- and mTOR-dependent suppression of Kv1.1 channel mRNA translation in dendrites. Science 2006;314:144–8.

[65] Takei N, Kawamura M, Hara K, Yonezawa K, Nawa H. Brain-derived neurotrophic factor enhances neuronal translation by activating multiple initiation processes—comparison with the effects of insulin. J Biol Chem 2001;276:42818–25.

[66] Takei N, Inamura N, Kawamura M, Namba H, Hara K, Yonezawa K, et al. Brain-derived neurotrophic factor induces mammalian target of rapamycin-dependent local activation of translation machinery and protein synthesis in neuronal dendrites. J Neurosci 2004;24:9760–9.

[67] Casadio A, Martin KC, Giustetto M, Zhu HX, Chen M, Bartsch D, et al. A transient, neuron-wide form of CREB-mediated long-term facilitation can be stabilized at specific synapses by local protein synthesis. Cell 1999;99:221–37.

[68] Beaumont V, Zhong N, Fletcher R, Froemke RC, Zucker RS. Phosphorylation and local presynaptic protein synthesis in calcium and calcineurin-dependent induction of crayfish long-term facilitation. Neuron 2001;32:489–501.

[69] Tang SJ, Reis G, Kang H, Gingras AC, Sonenberg N, Schuman EM. A rapamycin-sensitive signaling pathway contributes to long-term synaptic plasticity in the hippocampus. Proc Natl Acad Sci USA 2002;99:467–72.

[70] Cammalleri M, Lutjens R, Berton F, King AR, Simpson C, Francesconi W, et al. Time-restricted role for dendritic activation of the mTOR-p70(S6K) pathway in the induction of late-phase long-term potentiation in the CA1. Proc Natl Acad Sci USA 2003;100:14368–73.

[71] Tsokas P, Grace EA, Chan P, Ma T, Sealfon SC, Iyengar R, et al. Local protein synthesis mediates a rapid increase in dendritic elongation factor 1A after induction of late long-term potentiation. J Neurosci 2005;25:5833–43.

[72] Gong R, Park CS, Abbassi NR, Tang SJ. Roles of glutamate receptors and the mammalian target of rapamycin (mTOR) signaling pathway in activity-dependent dendritic protein synthesis in hippocampal neurons. J Biol Chem 2006;281:18802–15.

[73] Kelly MT, Crary JF, Sacktor TC. Regulation of protein kinase M zeta synthesis by multiple kinases in long-term potentiation. J Neurosci 2007;27:3439–44.

[74] Lee CC, Huang CC, Wu MY, Hsu KS. Insulin stimulates post-synaptic density-95 protein translation via the phosphoinositide 3-kinase-Akt-mammalian target of rapamycin signaling pathway. J Biol Chem 2005;280:18543–50.

[75] Schratt GM, Nigh EA, Chen WG, Hu L, Greenberg ME. BDNF regulates the translation of a select group of mRNAs by a mammalian target of rapamycin phosphatidylinositol 3-kinase-dependent pathway during neuronal development. J Neurosci 2004;24:7366–77.

[76] Hou L, Klann E. Activation of the phosphoinositide 3-kinase-Akt-mammalian target of rapamycin signaling pathway is required for metabotropic glutamate receptor-dependent long-term depression. J Neurosci 2004;24:6352–61.

[77] Sharma A, Hoeffer CA, Takayasu Y, Miyawaki T, McBride SM, Klann E, et al. Dysregulation of mTOR signaling in fragile X syndrome. J Neurosci 2010;30:694–702.

[78] Lebeau G, Miller LC, Tartas M, McAdam R, Laplante I, Badeaux F, et al. Staufen 2 regulates mGluR long-term depression and Map1b mRNA distribution in hippocampal neurons. Learn Mem 2011;18:314–26.

[79] Bozdagi O, Sakurai T, Dorr N, Pilorge M, Takahashi N, Buxbaum JD. Haplo insufficiency of Cyfip1 produces fragile X-like phenotypes in mice. PLoS One 2012;7:e42422.

[80] Auerbach BD, Osterweil EK, Bear MF. Mutations causing syndromic autism define an axis of synaptic pathophysiology. Nature 2011;480:63–8.

[81] Thomanetz V, Angliker N, Cloëtta D, Lustenberger RM, Schweighauser M, Oliveri F, et al. Ablation of the mTORC2 component rictor in brain or Purkinje cells affects size and neuron morphology. J Cell Biol 2013;201:293–308.

[82] Angliker N, Rüegg MA. In vivo evidence for mTORC2-mediated actin cytoskeleton rearrangement in neurons. BioArchitecture 2013;3:113–18.

[83] Huang W, Zhu PJ, Zhang S, Zhou H, Stoica L, Galiano M, et al. mTORC2 controls actin polymerization required for consolidation of long-term memory. Nat Neurosci 2013;16:441–8.

[84] Banko JL, Poulin F, Hou L, DeMaria CT, Sonenberg N, Klann E. The translation repressor 4E-BP2 is critical for eIF4F complex formation, synaptic plasticity, and memory in the hippocampus. J Neurosci 2005;25:9581–90.

[85] Banko JL, Hou L, Poulin F, Sonenberg N, Klann E. Regulation of eukaryotic initiation factor 4E by converging signaling pathways during metabotropic glutamate receptor-dependent long-term depression. J Neurosci 2006;26:2167–73.

[86] von der Brelie C, Waltereit R, Zhang L, Beck H, Kirschstein T. Impaired synaptic plasticity in a rat model of tuberous sclerosis. Eur J Neurosci 2006;23:686–92.

[87] Antion MD, Merhav M, Hoeffer CA, Reis G, Kozma SC, Thomas G, et al. Removal of S6K1 and S6K2 leads to divergent alterations in learning, memory, and synaptic plasticity. Learn Mem 2008;15:29–38.

[88] Ehninger D, Han S, Shilyansky C, Zhou Y, Li WD, Kwiatkowski DJ, et al. Reversal of learning deficits in a Tsc2 (+/−) mouse model of tuberous sclerosis. Nat Med 2008;14:843–8.

[89] Hoeffer CA, Tang W, Wong H, Santillan A, Patterson RJ, Martinez LA, et al. Removal of FKBP12 enhances mTOR-Raptor interactions, LTP, memory, and perseverative/repetitive behavior. Neuron 2008;60:832–45.

[90] Chévere-Torres I, Kaphzan H, Bhattacharya A, Kang A, Maki JM, Gambello MJ, et al. Metabotropic glutamate receptor-dependent long-term depression is impaired due to elevated ERK signaling in the ΔRG mouse model of tuberous sclerosis complex. Neurobiol Dis 2012;45:1101–10.

[91] Parsons RG, Gafford GM, Helmstetter FJ. Translational control via the mammalian target of rapamycin pathway is critical for the formation and stability of long-term fear memory in amygdala neurons. J Neurosci 2006;26:12977–83.

[92] Bekinschtein P, Katche C, Slipczuk LN, Igaz LM, Cammarota M, Izquierdo I, et al. mTOR signaling in the hippocampus is necessary for memory formation. Neurobiol Learn Mem 2007;87:303–7.

[93] Blundell J, Kouser M, Powell CM. Systemic inhibition of mammalian target of rapamycin inhibits fear memory reconsolidation. Neurobiol Learn Mem 2008;90:28–35.

[94] Gafford GM, Parsons RG, Helmstetter FJ. Consolidation and reconsolidation of contextual fear memory requires mammalian target of rapamycin-dependent translation in the dorsal hippocampus. Neuroscience 2011;182:98–104.

[95] Glover EM, Ressler KJ, Davis M. Differing effects of systemically administered rapamycin on consolidation and reconsolidation of context vs. cued fear memories. Learn Mem 2010;17:577–81.

[96] Jobim PFC, Pedroso TR, Werenicz A, Christoff RR, Maurmann N, Reolon GK, et al. Impairment of object recognition memory by rapamycin inhibition of mTOR in the amygdala or hippocampus around the time of learning or reactivation. Behav Brain Res 2012;228:151—8.

[97] Dash PK, Orsi SA, Moore AN. Spatial memory formation and memory-enhancing effect of glucose involves activation of the tuberous sclerosis complex-Mammalian target of rapamycin pathway. J Neurosci 2006;26:8048—56.

[98] Schicknick H, Schott BH, Budinger E, Smalla KH, Riedel A, Seidenbecher CI, et al. Dopaminergic modulation of auditory cortex-dependent memory consolidation through mTOR. Cer Cort 2008;18:2646—58.

[99] Tischmeyer W, Schicknick H, Kraus M, Seidenbecher CI, Staak S, Scheich H, et al. Rapamycin-sensitive signalling in long-term consolidation of auditory cortex-dependent memory. Eur J Neurosci 2003;18:942—50.

[100] Belelovsky K, Kaphzan H, Elkobi A, Rosenblum K. Biphasic activation of the mTOR pathway in the gustatory cortex is correlated with and necessary for taste learning. J Neurosci 2009;29:7424—31.

[101] Sui L, Wang J, Li BM. Role of the phosphoinositide 3-kinase-Akt-mammalian target of the rapamycin signaling pathway in long-term potentiation and trace fear conditioning memory in rat medial prefrontal cortex. Learn Mem 2008;15:762—76.

[102] Halloran J, Hussong SA, Burbank R, Podlutskaya N, Fischer KE, Sloane LB, et al. Chronic inhibition of mammalian target of rapamycin by rapamycin modulates cognitive and non-cognitive components of behavior throughout lifespan in mice. Neuroscience 2012;223:102—13.

[103] Misanin JR, Miller RR, Lewis DJ. Retrograde amnesia produced by electroconvulsive shock after reactivation of a consolidated memory trace. Science 1968;160:554—5.

[104] Nader K, Schafe GE, Le Doux JE. Fear memories require protein synthesis in the amygdala for reconsolidation after retrieval. Nature 2000;406:722—6.

[105] Tronson NC, Taylor JR. Molecular mechanisms of memory reconsolidation. Nat Rev Neurosci 2007;8:262—75.

[106] Myskiw JC, Rossato JI, Bevilaqua LR, Medina JH, Izquierdo I, Cammarota M. On the participation of mTOR in recognition memory. Neurobiol Learn Mem 2008;89:338—51.

[107] Mac Callum PE, Hebert M, Adamec RE, Blundell J. Systemic inhibition of mTOR kinase via rapamycin disrupts consolidation and reconsolidation of auditory fear memory. Neurobiol Learn Mem 2014;112:176—85.

[108] Benjamin D, Colombi M, Moroni C, Hall MN. Rapamycin passes the torch: a new generation of mTOR inhibitors. Nat Rev Drug Discov 2011;10:868—80.

[109] Wang X, Luo YX, He YY, Li FQ, Shi HS, Xue LF, et al. Nucleus accumbens core mammalian target of rapamycin signaling pathway is critical for cue-induced reinstatement of cocaine seeking in rats. J Neurosci 2010;30:12632—41.

[110] Bailey J, Ma D, Szumlinski KK. Rapamycin attenuates the expression of cocaine-induced place preference and behavioral sensitization. Addict Biol 2012;17:248—58.

[111] Barak S, Liu F, Ben Hamida S, Yowell QV, Neasta J, Kharazia V, et al. Disruption of alcohol-related memories by mTORC1 inhibition prevents relapse. Nat Neurosci 2013;16:1111—17.

[112] Ma T, Hoeffer CA, Capetillo-Zarate E, Yu F, Wong H, Lin MT, et al. Dysregulation of the mTOR pathway mediates impairment of synaptic plasticity in a mouse model of Alzheimer's disease. PLoS One 2010;5:e12945.

[113] Rubinsztein DC, Gestwicki JE, Murphy LO, Klionsky DJ. Potential therapeutic applications of autophagy. Nat Rev Drug Discov 2007;6:304—12.

[114] Zeng Z, Sarbassov dos D, Samudio IJ, Yee KW, Munsell MF, Ellen Jackson C, et al. Rapamycin derivatives reduce mTORC2 signaling and inhibit AKT activation in AML. Blood 2007;109:3509—12.

[115] Jaeger PA, Wyss-Coray T. All-you-can-eat: autophagy in neurodegeneration and neuroprotection. Mol Neurodegener 2009;4:16.

[116] Spilman P, Podlutskaya N, Hart MJ, Debnath J, Gorostiza O, Bredesen D, et al. Inhibition of mTOR by rapamycin abolishes cognitive deficits and reduces amyloid-beta levels in a mouse model of Alzheimer's disease. PLoS One 2010;5:e9979.

[117] Perluigi M, Di Domenico F, Butterfield DA. mTOR signaling in aging and neurodegeneration: at the crossroad between metabolism dysfunction and impairment of autophagy. Neurobiol Dis 2015, pii: S0969-9961(15)00086-8.

[118] Bolduc FV, Bell K, Cox H, Broadie KS, Tully T. Excess protein synthesis in Drosophila Fragile X mutants impairs long-term memory. Nat Neurosci 2008;11:1143—5.

[119] Puighermanal E, Marsicano G, Busquets-Garcia A, Lutz B, Maldonado R, Ozaita A. Cannabinoid modulation of hippocampal long-term memory is mediated by mTOR signaling. Nat Neurosci 2009;12:1152—8.

[120] Puighermanal E, Busquets-Garcia A, Maldonado R, Ozaita A. Cellular and intracellular mechanisms involved in the cognitive impairment of cannabinoids. Phil Trans R Soc Lond Ser B Biol Sci 2012;367:3254—63.

[121] Troca-Marín JA, Alves-Sampaio A, Montesinos ML. Deregulated mTOR-mediated translation in intellectual disability. Progr Neurobiol 2012;96:268—82.

[122] Goorden SM, van Woerden GM, van der Weerd L, Cheadle JP, Elgersma Y. Cognitive deficits in Tsc1 +/− mice in the absence of cerebral lesions and seizures. Ann Neurol 2007;62:648—55.

[123] Rosner M, Hanneder M, Siegel N, Valli A, Fuchs C, Hengstschläger M. The mTOR pathway and its role in human genetic diseases. Mut Res 2008;659:284—92.

[124] Harrison DE, Strong R, Sharp ZD, Nelson JF, Astle CM, Flurkey K, et al. Rapamycin fed late in life extends lifespan in genetically heterogeneous mice. Nature 2009;460:392—5.

[125] Wullschleger S, Loewith R, Hall MN. TOR signaling in growth and metabolism. Cell 2006;124:471—84.

[126] Caccamo A, Majumder S, Richardson A, Strong R, Oddo S. Molecular interplaybetween mammalian target of rapamycin (mTOR), amyloid-beta, and Tau: effects on cognitive impairments. J Biol Chem 2010;285:13107—20.

[127] Lang UE, Heger J, Willbring M, Domula M, Matschke K, Tugtekin SM. Immunosuppression using the mammalian target of rapamycin (mTOR) inhibitor everolimus: pilot study shows significant cognitive and affective improvement. Transplant Proc 2009;41:4285—8.

[128] Belelovsky K, Elkobi A, Kaphzan H, Nairn AC, Rosenblum K. A molecular switch for translational control in taste memory consolidation. Eur J Neurosci 2005;22:2560—8.

[129] Horwood JM, Dufour F, Laroche S, Davis S. Signalling mechanisms mediated by the phosphoinositide 3-kinase/Akt cascade in synaptic plasticity and memory in the rat. Eur J Neurosci 2006;23:3375—84.

[130] Banko JL, Merhav M, Stern E, Sonenberg N, Rosenblum K, Klann E. Behavioral alterations in mice lacking the translation repressor 4E-BP2. Neurobiol Learn Mem 2007;87:248—56.

[131] Barros DM, Pereira P, Medina JH, Izquierdo I. Modulation of working memory and of long- but not short-term memory by cholinergic mechanisms in the basolateral amygdala. Behav Pharmacol 2002;13:163—7.

[132] Giovannini MG, Pazzagli M, Malmberg-Aiello P, Della CL, Rakovska AD, Cerbai F, et al. Inhibition of acetylcholine-induced activation of extracellular regulated protein kinase prevents the

encoding of an inhibitory avoidance response in the rat. Neuroscience 2005;136:15—32.

[133] Decker MW, Brioni JD, Bannon AW, Arneric SP. Diversity of neuronal nicotinic acetylcholine receptors: lessons from behavior and implications for CNS therapeutics. Life Sci 1995;56:545—70.

[134] Marti BD, Ramirez MR, Dos Reis EA, Izquierdo I. Participation of hippocampal nicotinic receptors in acquisition, consolidation and retrieval of memory for one trial inhibitory avoidance in rats. Neuroscience 2004;126:651—6.

[135] Mitsushima D, Sano A, Takahashi T. A cholinergic trigger drives learning-induced plasticity at hippocampal synapses. Nat Commun 2013;4:2760—9.

[136] Feig S, Lipton P. Pairing the cholinergic agonist carbachol with patterned Schaffer collateral stimulation initiates protein synthesis in hippocampal CA1 pyramidal cell dendrites via a muscarinic, NMDA-dependent mechanism. J Neurosci 1993;13:1010—21.

[137] Chan GP, Wu EH, Wong YH. Regulation of mTOR and p70 S6 kinase by the muscarinic M4 receptor in PC12 cells. Cell Biol Int 2009;33:230—8.

[138] Berg DK, Conroy WG. Nicotinic alpha 7 receptors: synaptic options and downstream signaling in neurons. J Neurobiol 2002;53:512—23.

[139] Quirion R, Wilson A, Rowe W, Aubert I, Richard J, Doods H, et al. Facilitation of acetylcholine release and cognitive performance by an M2-muscarinic receptor antagonist in aged memory-impaired rats. J Neurosci 1995;15:1455—62.

[140] Raiteri M, Leardi R, Marchi M. Heterogeneity of presynaptic muscarinic receptors regulating neurotransmitter release in the rat brain. J Pharmacol Exp Ther 1984;228:209—14.

[141] Douglas CL, Baghdoyan HA, Lydic R. Prefrontal cortex acetylcholine release, EEG slow waves, and spindles are modulated by M2 autoreceptors in C57BL/6J mouse. J Neurophysiol 2002;87:2817—22.

[142] Zhang W, Basile A, Gomeza J, Volpicelli L, Levey AI, Wess J. Characterization of central inhibitory muscarinic autoreceptors by the use of muscarinic acetylcholine receptor knock-out mice. J Neurosci 2002;22:1709—17.

[143] Scali C, Vannucchi MG, Pepeu G, Casamenti F. Peripherally injected scopolamine differentially modulates acetylcholine release *in vivo* in the young and aged rats. Neurosci Lett 1995;197:171—4.

[144] Sun X, Ritzenthaler JD, Zhong X, Zheng Y, Roman J, Han S. Nicotine stimulates PPARbeta/delta expression in human lung carcinoma cells through activation of PI3K/mTOR and suppression of AP-2alpha. Cancer Res 2009;69:6445—53.

[145] Bell KA, Shim H, Chen CK, McQuiston AR. Nicotinic excitatory postsynaptic potentials in hippocampal CA1 interneurons are predominantly mediated by nicotinic receptors that contain alpha4 and beta2 subunits. Neuropharmacology 2011;61:1379—88.

[146] Gu Z, Yakel JL. Timing-dependent septal cholinergic induction of dynamic hippocampal synaptic plasticity. Neuron 2011;71:155—65.

[147] Izquierdo I, Medina JH, Izquierdo LA, Barros DM, de Souza MM, Mello e Souza T. Short- and long-term memory are differentially regulated by monoaminergic systems in the rat brain. Neurobiol Learn Mem 1998;69:219—24.

[148] Khakpai F, Nasehi M, Haeri-Rohani A, Eidi A, Zarrindast MR. Scopolamine induced memory impairment; possible involvement of NMDA receptor mechanisms of dorsal hippocampus and/or septum. Behav Brain Res 2012;231:1—10.

[149] Yanow SK, Manseau F, Hislop J, Castellucci VF, Sossin WS. Biochemical pathways by which serotonin regulates translation in the nervous system of Aplysia. J Neurochem, 1998;70:572—83.

[150] Choo AY, Yoon SO, Kim SG, Roux PP, Blenis J. Rapamycin differentially inhibits S6Ks and 4E-BP1 to mediate cell-type-specific repression of mRNA translation. Proc Natl Acad Sci USA 2008;105:17414—19.

[151] McClatchy DB, Fang G, Levey AI. Elongation factor 1A family regulates the recycling of the M4 muscarinic acetylcholine receptor. Neurochem Res 2006;31:975—88.

[152] Corradetti MN, Guan KL. Upstream of the mammalian target of rapamycin: do all roads pass through mTOR? Oncogene 2006;25:6347—60.

[153] Connor SA, Maity S, Roy B, Ali DW, Nguyen PV. Conversion of short-term potentiation to long-term potentiation in mouse CA1 by coactivation of β-adrenergic and muscarinic receptors. Learn Mem 2012;19:535—42.

[154] Lafay-Chebassier C, Perault-Pochat MC, Page G, Bilan AR, Damjanac M, Pain S, et al. The immunosuppressant rapamycin exacerbates neurotoxicity of A beta peptide. J Neurosci Res 2006;84:1323—34.

[155] Deguil J, Perault-Pochat MC, Chavant F, Lafay-Chebassier C, Fauconneau B, Pain S. Activation of the protein p70S6K via ERK phosphorylation by cholinergic muscarinic receptors stimulation in human neuroblastoma cells and in mice brain. Toxicol Lett 2008;182:91—6.

[156] Panja D, Dagyte G, Bidinosti M, Wibrand K, Kristiansen AM, Sonenberg N, et al. Novel translational control in arc-dependent long term potentiation consolidation *in vivo*. J Biol Chem 2009;284:31498—511.

[157] Gal-Ben-Ari S, Rosenblum K. Molecular mechanisms underlying memory consolidation of taste information in the cortex. Front Behav Neurosci 2012;5:87.

11

Mammalian Target of Rapamycin (mTOR), Aging, Neuroscience, and Their Association with Aging-Related Diseases

Ergul Dilan Celebi-Birand[1], Elif Tugce Karoglu[1], Fusun Doldur-Balli[2] and Michelle M. Adams[1,3]

[1]Interdisciplinary Graduate Program in Neuroscience, Bilkent University, Ankara, Turkey [2]Department of Molecular Biology and Genetics, Bilkent University, Ankara, Turkey [3]Department of Psychology, Bilkent University, Ankara, Turkey

11.1 INTRODUCTION

According to the World Health Organization's report, the world population is rapidly aging. The average life expectancy is increasing worldwide and the ratio of the older population (i.e., persons 65 years or older) to the overall population has gotten larger. The number of people over 60 years is expected to increase from 605 million, an estimation made in 2000, to two billion by 2050, and the percentage of the aged population will have doubled from 11 to 22% [1]. With this rise in average lifespan there is a surge in the amount of age-related cognitive deficits as well as age-associated diseases. This can interfere with an independent lifestyle for the elderly, as well as put a burden on potential caregivers. For instance, the prevalence of age-related dementia, which is a progressive brain disorder, increases with age. There are no definitive therapies against dementia, and informal and formal care cause a burden on society, families, and caregivers for the elderly [2]. Thus, finding appropriate interventions that alter the course of aging-related disorders is quite important.

Aging is defined as a normal physiological process in which gradual deterioration in body functions occurs [3]. Gradual declines in sensory, motor, and cognitive functions are accompanied with brain aging [4]. With the growing number of older people, aging-associated cognitive decline, which is related to cognitive impairments, has become an important issue. A significant loss of learning and memory capacity, decline in visual and hearing abilities, some gastrointestinal tract problems, motor deficits, deterioration of balance control, and fall-related injuries are several clinical problems in the elderly population associated with the aging of the nervous system [5]. Beyond these clinical manifestations, age-related cognitive decline is associated with cognitive dysfunctions, reduced self-sufficiency and loss of independence, which reduce quality of life for older adults [6]. Cognitive capacity decreases during the process of aging even in the absence of a diagnosed pathology. It is a result of multifactorial changes which are combinations of genetic and environmental etiologies [7]. Dissecting the effects of individual factors and revealing the molecular, cellular, and cognitive mechanisms are important in terms of devising restorative interventions for age-related cognitive decline.

In this chapter, we will first provide data on the course of normal versus pathological aging. We will then consider the role of mammalian target of rapamycin (mTOR) signaling in aging, neurological and non-neurological diseases. Taken together, the available data suggest that mTOR signaling may

Molecules to Medicine with mTOR
DOI: http://dx.doi.org/10.1016/B978-0-12-802733-2.00007-4

185

underlie the cellular and molecular changes with aging. The potential for factors such as the mTOR pathway to impact those aging-related behavioral and molecular changes, as well as possible interventions, also are considered.

11.2 COGNITIVE ALTERATIONS ASSOCIATED WITH NORMAL AGING

Normal aging is associated with behavioral and cognitive alterations. However, these changes in age-related cognitive capacities are a multidimensional process; decline across one cognitive domain may not be in parallel with or indicative of a falling-off that would occur in another domain. Therefore, some cognitive abilities may decrease at older age but some cognitive skills are preserved.

The pattern of age-related cognitive decline and alterations differs between pathological and normal aging. Therefore, during the analysis of age-related cognitive decline, considering these two directions of normal versus pathological aging is important (Table 11.1). Cognitive decline and alterations in behavior can occur in attention, learning, memory, and executive functions, as well as language abilities. However, the most common and significant declines are usually observed in the learning and memory domains, whose deficiency affects and reduces the well-being of older adults [8]. Therefore, focusing on specific components of these two cognitive domains including episodic memory, semantic memory, and spatial learning-memory will be important for helping elderly individuals maintain their independence.

Episodic memory is memory for autobiographical events and its retrieval depends on contextual or temporal cues [9] and performance on tasks utilizing this ability declines with normal aging [10,11]. In, addition to poor episodic memory abilities during older ages, functional magnetic resonance imaging studies revealed that older adults show a different pattern of hippocampal formation and parahippocampal activity during the episodic memory tasks when they are compared with the young control group, and these age-related alterations in the hippocampal and parahippocampal activities may play a role in the age-related decline in episodic memory [12] (Figures 11.1 and 11.2). In the pathological aging conditions, people show a larger decline than normal aging groups in terms of episodic memory. Older people with mild cognitive impairment [13] and people with Alzheimer's disease (AD) show poorer performance in episodic memory tasks than their healthy old group counterparts [9]. In terms of neural activity, older people with AD show a more widespread and decreased activity than the healthy older group, with observations of changes in the posterior cingulate and hippocampal formation [14]. These activity alterations could account for the performance differences between normal and pathological aging.

Semantic memory, which is referred to as acquired knowledge and concepts independent from contextual cues like the meaning of the words and geographical facts, is another component of memory [9]. Unlike the episodic component of the memory domain, age-related cognitive decline is not so evident in semantic memory [15]. Experimental evidence suggests that there may be some decline but it is not necessarily a linear one since it is has been shown that 55- to 60-year-old people tend to show better performance in semantic memory tasks than the younger and older controls [11]. Also, in terms of brain activity, neural activity dedifferentiation found in episodic memory tasks is not found in semantic memory tasks between old and young groups, which indicates that the course of activation required for semantic memory is similar between young and old groups [16]. Therefore, based on these findings it could be concluded that semantic memory is preserved during normal aging. On the other hand, in the pathological aging conditions like AD, this preserved pattern is not observed and there is an evident decline in semantic memory [17]. The differences in performance in semantic memory tasks between AD and normal aging are likely explained by the volume loss in left medial perirhinal cortex, entorhinal cortex, and hippocampal formation [18].

Spatial learning and memory are components of the learning and memory domains in which particular locations are associated with distinct stimuli or cues, and these stimuli are used to learn the location of the object of interest. This process requires both learning and memory abilities [19]. In different animal models [20–23] and human studies [24,25], it has been demonstrated that the performance related to spatial learning and memory reduces with normal aging. Older groups tend to show poor spatial learning and memory ability when they are compared with young groups in related tasks including the radial maze, Morris water maze, virtual maze, and so forth. It has been suggested that age-related deficits in spatial memory are associated with hippocampal integrity and during aging as hippocampal function declines, spatial learning and memory deficits begin to occur [20,23]. In terms of comparisons between normal and pathological aging conditions, it has been shown that older people with Lewy body dementia and AD tend to show worse performance in spatial learning and memory-related tasks when compared to healthy old groups, and people with Lewy body dementia tend to show more

TABLE 11.1 Summary of the Molecular, Cellular, and Behavioral Findings Associated with Cognitive Deficits, and Effects of mTOR Manipulations in Normal and Pathological Brain Ageing Conditions

Findings → Conditions ↓	Neural loss	Molecular alterations	Behavioral alterations	Effects of mTOR manipulations
Normal aging	No neural loss	Subtle changes in synaptic integrity	Decline in domains including spatial learning-memory, episodic memory	Partial inhibition by rapamycin • Decreased levels of IL-1β • Increased NMDA signaling • Augmented monoamine levels • Better cognitive performance
Parkinson's disease	Dopaminergic neural loss especially in substantia nigra pars compacta	Deficits in dopaminergic circuitry	• Motor symptoms including tremors and ataxia • Anxiety and depression	Partial inhibition by rapamycin • Suppressed PD-related pathological phenotype, this suppression is mediated by 4EBP activity • Ameliorated PD-related mitochondrial defects • Protective against PD-related neural death PD mimetics • Inhibited mTOR pathway • Neural loss Inhibiton of all actions of mTOR by Torin1 • Not protective against PD • Induced neural death
Alzheimer's disease	Neural loss especially hippocampus and cerebral neocortex	• Amyloid-beta protein accumulation • Presence of neurofibrillary tangles • Hyperphosphorylated tau protein	Deficits across domains including memory, spatial learning, language, attention, visual perception, executive functions, and social function	Partial inhibition by rapamycin • Better spatial learning and memory • Reduced amyloid-beta levels • Induced autophagy in hippocampus
Frontotemporal dementia	Neural loss in frontal and temporal lobes	Proteinopathies including TDP-43	• Abnormal social behavior and apathies • Cognitive deficits in executive function, attention, visuospatial abilities, language, and memory	Partial inhibition by rapamycin • Reduced decline in learning and memory • Reduced neural loss • Decreased levels of TDP-43 products
Huntington's disease	Neural loss especially in hippocampus and basal ganglia	Presence of mutant Huntingtin protein • Decreased uptake of glutamate • Increased NMDA signaling • Decreased levels of BDNF	• Motor dysfunction, muscle rigidity • Deficits in memory and executive functions • Anxiety and depression	Partial inhibition by rapamycin • Protection against neural death and toxicity • Decreased protein aggregation • Reduced motor and behavioral deficits

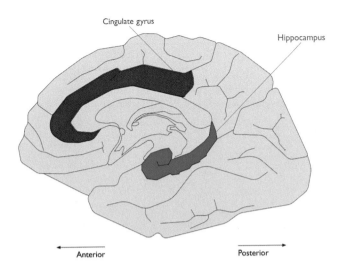

FIGURE 11.1 Human brain medial view. A schematic drawing illustrating the medial view of the human brain. Hippocampus, found in the medial temporal lobe and cingulate gyrus, positioned in parts of the frontal and parietal lobes, are highlighted. In this view, the anterior part of the brain is towards the left and the posterior to the right.

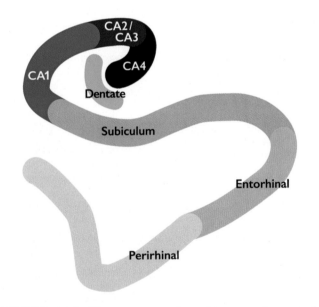

FIGURE 11.2 Hippocampus and parahippocampal gyrus. A schematic diagram showing the hippocampal regions CA1, CA2/CA3, CA4, dentate gyrus, entorhinal and perirhinal cortices (parahippocampal gyrus). CA1, CA3, dentate gyrus, entorhinal and perirhinal cortices are affected regions both in normal and pathological aging conditions, and their aberrant functioning is likely to underlie learning and memory deficits in the elderly.

cognitive deficits in spatial memory than people with AD [26]. Therefore, although some cognitive abilities decline with normal aging, this impairment becomes more severe during pathological aging.

11.3 CELLULAR AND MOLECULAR ALTERATIONS ASSOCIATED WITH NORMAL BRAIN AGING

Determining the cellular and molecular changes that are occurring in the aging brain and underlie cognitive decline is key to determining the targets for possible interventions. Early studies suggested that age-related cognitive decline is a result of significant cell [27,28] and synapse loss [29,30], but subsequent studies using unbiased counting techniques have shown that overall loss of cells [31–33] and synapse [34,35] associated with aging was not observed. Instead, it has been proposed that subtle changes in synaptic integrity rather than large-scale alterations are likely to underlie age-related cognitive decline [7]. For example, there is an age-related decline in axospinous-type synapses in the dentate gyrus of the hippocampus, a structure that is integrally linked to learning and memory [30,36]. In the hippocampus, Smith et al. reported no changes in synaptophysin levels, a key synaptic protein, between young and old rats, however, when the aged animals were separated according to their memory ability, they found decreases in synaptophysin levels for behaviorally impaired old rats in the molecular layer of the dentate gyrus and the lacunosum-moleculare layer of CA3 [37] (Figure 11.2). Thus, unlike pathological aging that is associated with a disease state, it is likely that there are very subtle region- and layer-specific age-related changes in synapses during normal aging, suggesting the importance of examining other aspects of synapses which may also be changing with age.

Multiple factors may be affecting the function of synapses in the aging brain. For example, brain-derived neurotrophic factor (BDNF), which is a neurotrophin family member and a mediator of synaptogenesis, synaptic plasticity, neuronal survival, and differentiation in the mammalian brain [38,39] is another factor that is related to cognitive aging. Age-related learning impairments were reported in a study which used a heterozygous BDNF knockout mouse model [40].

Neurogenesis, which is a process required for maintenance of cellular turnover and cognitive plasticity in adult brains, was shown to decrease dramatically in the subventricular zone (SVZ) of the forebrain lateral ventricles of old rodents. SVZ is the main source of neural stem cells. The major decline was detected during the transition from young age to middle age [41]. Since newborn granule cells are proposed to be involved in pattern separation [42], drastic decreases in neurogenesis at old age might be related to cognitive decline from this aspect [43]. The endocannabinoid system functions in neuroprotection, regulation of neurogenesis, and BDNF expression [43,44]. This system is involved in the aging process and indeed mutant mice lacking cannabinoid 1 (CB1) receptor performed much worse in several learning and memory tests as aging progressed [45,46]. These findings indicated an accelerated cognitive aging phenotype associated with the deletion of the CB1 receptor gene, and further research proposed an age-related function for CB1 receptor on cortical glutamatergic neurons in habituation and acquisition of spatial learning [44].

In terms of gene expression profiles that are altered with increasing age, the most affected gene expression profiles are those whose functions are involved in calcium signaling and the cyclic adenosine monophosphate response element-binding protein pathway [47], synaptic plasticity, mitochondrial function, stress response, DNA repair, and inflammation [48,49]. A gene expression study in which young and old male and female zebrafish brains were compared found that neurogenesis, cell differentiation, and development of brain- and nervous-system-related genes are differentially expressed [50]. Epigenetic regulations depending on histone modifications are also associated with the aging process [51,52]. Since changes in the chromatin structure lead to outcomes at the transcriptional level, they may result in alterations of the hippocampal gene expression and eventually may impair cognitive functions [53]. Alterations in the expression of genes that function in inflammation and apoptosis are accelerated in neurodegenerative diseases [43].

A key molecular target that may contribute to age-related cognitive decline is the glutamate receptor. Glutamate receptors are the primary mediators of excitatory transmission in the central nervous system [54], and play an important role in learning and memory, connecting them to age-related decline [55,56]. In the adult hippocampus, the primary types of glutamate receptors are N-methyl-D-aspartate (NMDA) and alpha-amino-3-hydroxy-5-methyl-4-isoxazole propionic acid (AMPA). Each of the pharmacologically defined classes of ionotropic glutamate receptors are composed of multimeric assemblies of protein subunits [54]. The critical subunits for NMDA receptors are NR1, NR2A, and NR2B, and for AMPA receptors are GluR1 and GluR2. Families of subunits are recognized by sequence homology [54,57,58] and supported by coprecipitation studies revealing distinct subunit assembly patterns [59,60], and common electrophysiological and pharmacological properties demonstrated in expression systems [54,61,62]. Finally, glutamate receptors have been linked to learning and memory such that decreases in levels lead to memory impairments [63] and increases lead to memory enhancement [64].

Glutamate receptors, in particular the NMDA receptor, have been implicated in age-related cognitive decline [65–67]. Age-related changes in glutamate receptor levels have been observed in many studies, although the data are equivocal as to the direction of change. Although age-related decreases in NMDA receptor levels, specifically in NR1 and NR2B, have been described in some studies [68,69], there are studies with contradictory results that show no changes [70]. Additionally, age-related decreases in the AMPA receptor subunits GluR1 and GluR2 levels have been reported in all the subfields of the rat hippocampus [71]. In addition to the molecular alterations, functional changes in excitatory transmission mediated by glutamate receptors have been reported in the aging hippocampus. For example, long-term potentiation (LTP), an NMDA receptor-dependent plasticity mechanism, and a model of cellular learning and memory, has been shown to have a faster rate of decay in aged animals [65]. The threshold for induction of LTP is elevated in aged rats [72], and the late phase of LTP has been found to be significantly reduced in aged mice [23]. Moreover, there is a shift in LTP from being an NMDA-dependent mechanism to one that is dependent on voltage-dependent Ca^{2+} channels in hippocampal CA1 of aged rats [73]. Additionally, there is a significant reduction in both NMDA and non-NMDA glutamate receptor-mediated excitatory postsynaptic potentials in CA1 of 27-month-old rats as compared to 9-month-old rats [74]. Finally, the magnitude of LTP is reduced in the aged animals [68,69]. Thus, these data suggest that glutamate receptor changes, particularly NMDA receptors, may contribute to age-related declines in cognitive function.

In the brain, the astroglia interact very closely with glutamatergic synapses forming what is called the tripartite synapse. It is becoming clear that astroglia play a very important role in synaptic transmission in addition to their role as a trophic support for neurons. They integrate and process synaptic information and control synaptic transmission and plasticity [75]. While the primary focus of aging-related research has been on alterations in excitatory synapses that contribute to age-related cognitive decline, very little work has been done on the contribution of glial changes during aging and our knowledge is rudimentary at best. It is well established that as there are changes in excitatory synapses, there are parallel changes in glial cells. For example during the estrous cycle there are fluctuations in the circulating levels of estrogen. In the rodent hippocampus, it is well-documented that excitatory synapse number increases when estrogen levels are high [76]. Moreover, there is evidence that as the number of these excitatory synapses increases and decreases, the number of astrocytic processes increases and decreases in a concomitant manner [77,78]. In the context of aging, there is evidence suggesting that glial fibrillary acidic protein (GFAP)-positive astrocytes become fewer, smaller and less complex in the CA1 region of the hippocampus [79]. The level of glutamine synthetase, which is an enzyme that is critical for glutamate homeostasis and is maintained by astrocytes to help neurons produce glutamate, is reduced in the hippocampus but the levels of S100β, a calcium-binding protein that regulates astrocytic movement and shape, is increased in the aged brain [80]. Beside these findings, our knowledge of glial modifications in the context of aging is still very limited. The age-related changes in glia are more complex than expected, thus, they should be further investigated in relation to alterations in excitatory synapses. Thus, while some of the observations of cellular and molecular changes in the normal aging brain may be similar to those seen in pathological aging, the normal aging brain changes are more subtle in nature despite resulting in cognitive impairments. Differences in the molecular and cellular changes between normal and pathological aging will be discussed further below (Table 11.1).

11.4 mTOR SIGNALING PATHWAY

mTOR is a serine/threonine kinase, belonging to the phosphatidyl inositol 3′ kinase-related kinase family [81]. It was named after its sensitivity to rapamycin, an antifungal agent that is synthesized by the bacterial species *Streptomyces hygroscopicus* [82]. The interaction of rapamycin is an indirect interaction rather than a direct one. Rapamycin forms a complex with cyclophilin protein FKBP12, and this protein—inhibitor complex then binds to the FKBP12-rapamycin-binding (FRB) domain of mTOR (Figure 11.3). This interaction inhibits access to mTOR's active site [83—85]. mTOR homologs have been identified in several non-vertebrate or non-mammalian organisms including budding yeast [86,87], *C. elegans*, *Drosophila*, and zebrafish [88—90].

mTOR is found in two distinct protein complexes in the cell: mTOR complex 1 (mTORC1) and mTOR complex 2 (mTORC2). mTORC1 is composed of mTOR, mLST8, DEPTOR, Raptor, AKT1S1/PRAS40, and promotes transcription, translation, cell cycle progression, ribosome biogenesis, and cell growth through repression of autophagy [91,92] (Figure 11.3). mTORC2 is composed of mTOR, mLST8, DEPTOR, Rictor, mSin1 (also known as MAPKAP1), and PRR5 (also known as PROTOR-1) [93,85] (Figure 11.3). mTORC2 regulates cell survival, cytoskeleton, cell cycle, and cell metabolism [92]. Early evidence suggested that rapamycin only inhibited the activity of mTORC1 but not mTORC2, however chronic rapamycin treatment has later been shown to inhibit mTORC2 through binding to free mTOR [94]. mTOR interacts with Akt in both complexes. mTORC1 is indirectly activated by active Akt, whereas phosphorylation of Akt at Ser473 residue via mTORC2 upregulates the overall Akt/mTOR pathway activity [95,96].

11.5 mTORC1

Deregulation of mTORC1 has been implicated in aging, cancer, obesity, type 2 diabetes, and neurodegenerative diseases [97]. mTORC1 is activated in response to nutrients, growth factors, oxygen, energy status of the cell, and inflammation [93]. Active mTORC1 in turn regulates eukaryotic translation and protein synthesis through phosphorylation of S6 kinase 1 (S6K1) and eukaryotic translation initiation factor 4E (eIF4E) binding protein 1 (4E-BP1) [98,99]. eIF4E normally binds to the m7GpppN cap located at the 5′-end of mRNAs. Unphosphorylated 4E-BP1 represses translation, and its phosphorylation via mTORC1 which causes dissociation of 4E-BP1 from eIF4E, and enhances eIF4E-mediated translation. S6K1 regulates protein synthesis, cell growth, and proliferation through an mTORC1-dependent mechanism. mTORC1 phosphorylates and activates S6K1 [100,101]. Both S6K and 4E-BP proteins contain a common amino acid motif called TOR signaling (TOS) motif, which has been suggested to be essential for interaction with Raptor in mTORC1 [102,103].

Growth factors such as insulin and insulin-like growth factor 1 (IGF1) activate receptor tyrosine kinases (RTKs). Active RTKs induce production of phosphatidylinositol (3,4,5)-trisphosphate (PIP3) and activate Akt. Akt exerts its effect on mTORC1 through two mechanisms: inactivating the tuberous sclerosis complex (TSC) via phosphorylation, and inhibiting the interaction of the proline-rich Akt substrate 40 kDa (PRAS40) with mTORC1.

When phosphorylated by Akt, PRAS40 can bind to the 14-3-3 protein. Although the downstream events initiated upon this binding are not clear yet, it has been proposed that binding of the 14-3-3 protein to phosphorylated PRAS40 inhibits it from binding to mTOR, thus relieving inhibition on mTOR [104]. The negative regulation of mTORC1 activity by PRAS40 might be due to inhibition of mTORC1-mediated phosphorylation of 4E-BP1 and S6K1 [105,106].

TSC1 and TSC2, proteins upstream of mTOR identified as tumor suppressors, form heterodimers and act in TSC to inhibit mTOR signaling by converting Ras homolog enriched in brain (Rheb) from guanosine triphosphate (GTP)-bound to guanosine diphosphate (GDP)-bound forms. Rheb can only activate mTOR when it is bound to GTP and farnesylated [107], hence TSC indirectly inhibits mTOR signaling through Rheb [108,109]. Akt phosphorylates TSC2, leading to

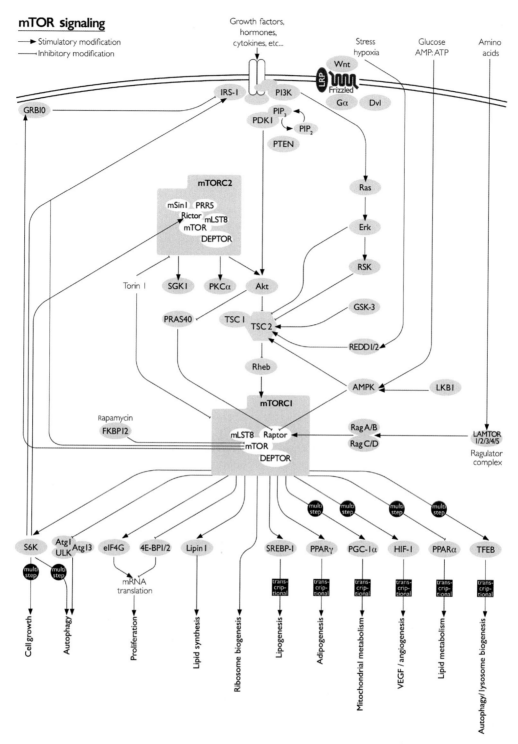

FIGURE 11.3   The mTOR signaling pathway. mTOR is found in two distinct protein complexes: mTORC1 and mTORC2. mTORC1 promotes cell growth through a multistep process mediated by S6K, proliferation by relieving inhibition on eIF4G, lipid synthesis, lipogenesis, and lipid metabolism by activating lipogenic transcription factors Lipin 1, SREBP-1, and PPARα. mTORC1 is also involved in ribosome biogenesis, adipogenesis, angiogenesis, and mitochondrial metabolism, and inhibits autophagic activity through inhibition of Atg1/Atg13/ULK complex. mTORC2's effects on these cellular processes are largely indirect, as mTORC2 exerts most of its effects through mTORC1 and its signaling partners. mTORC1 in turn regulates the activity of mTORC2 through phosphorylation of mSin1 in the mTORC2.

the disruption of the TSC1–TSC2 heterodimer [110] and relieves repression on mTOR. AMP-dependent protein kinase (AMPK), on the other hand, is activated by serine/threonine kinase LKB1/STK11-mediated phosphorylation, and activates TSC2 in turn through phosphorylation [111,112].

Recently, FKBP38 has been identified as another component of mTORC1. FKBP38 binds to FRB domain of mTOR, the binding site for another protein belonging to the same family: fKBP12. However, unlike FKBP12, FKBP38 does not need rapamycin to bind mTOR. It has been also shown that Rheb-GTP binds to FKBP38 with higher affinity than Rheb-GDP, and active Rheb prevents FKBP38 from binding to mTOR and relieving it from inhibition [113,114]. Yet another study suggested that the interaction of Rheb with FKBP38 did not differ between GTP- or GDP-bound forms [115].

An additional response mechanism of mTOR to cytokines and growth factors has been suggested to involve signal transducer and activator of transcription 3 (STAT3). Phosphorylation of STAT3 at Tyr705 by Janus kinase or RTKs has been shown to facilitate dimerization of STAT3 and the translocation of the dimer to the nucleus. Upon nuclear translocation, STAT3 regulates cell growth, proliferation, and cell survival [116]. Recently, phosphorylation of STAT3 at Ser727 by various kinases including mTOR has been shown. Furthermore, rapamycin was found to reduce phosphorylated STAT3 levels [117], and loss of TSC1 and TSC2 has been linked to increased levels of phosphorylated STAT3, a common observation in several human cancers. Elevated STAT3 activity promotes tumorigenesis by positively regulating the expression of genes that are involved in mechanisms crucial for cancer cells such as cell growth, proliferation, survival, angiogenesis, and metastasis [118].

mTORC1 also promotes tumorigenesis through lipid synthesis, a mechanism required for actively proliferating cells. mTORC1 positively regulates sterol-regulatory-element-binding proteins (SREBPs), transcription factors that control lipogenic genes and their expression [119]. While mTORC1 induces nuclear translocation of active SREBPs, inhibition of mTORC1 reduces SREBP expression at transcriptional and translational levels [93]. The nuclear translocation of SREBPs requires Lipin 1, a transcriptional coactivator that is phosphorylated by mTORC1. Phosphorylated Lipin 1 is translocated out of nucleus, and its exclusion from the nucleus relieves SREBPs from suppression. Free SREBPs then bind to target genes involved in lipid synthesis [120].

Since growth factors and nutrients regulate adipose tissue accumulation, it is not surprising that mTORC1 has a role in adipogenesis. Indeed, mTORC1 inhibition has been shown to have severe effects on adipogenesis and maintenance of adipose cells [93]. mTORC1 exerts its positive effects on adipogenesis indirectly, through peroxisome proliferator-activated receptor γ (PPARγ). PPARγ controls expression of genes involved in fatty acid synthesis and processing [121].

A related protein, PPARα, regulates genes that are involved in hepatic ketogenesis and fatty acid oxidation. Evidence suggests that mTORC1 inhibits PPARα activity by recruiting its negative regulator, nuclear receptor corepressor into the nucleus [122]. Interestingly, mTORC1 activity has been shown to be increased in the livers of old mice [123], suggesting a link between metabolism, aging, and mTORC1.

mTORC1 signaling has been shown to influence proinflammatory processes and neo angiogenesis through induction of hypoxia-inducible factor 1 α (HIF1α) [124]. Cancer cells often have to cope with oxygen deprivation, and HIF1α promotes expression of genes that shift cellular energy metabolism from aerobic to anaerobic respiration and regulates glycolysis and glucose transport [125]. mTORC1 positively regulates HIF1α through 4E-BP1-eIF4-dependent translational activation. The HIF1α–MTORC1 axis is also involved in angiogenesis and promotes tumorigenesis [93].

In addition to growth factors, amino acids are crucial for activation of mTORC1. A recent study proposed an inside-out model of amino acid sensing. According to this model, amino acids accumulate in the lysosomal lumen, and this accumulation initiates a signaling cascade that involves vacuolar H^+-adenoside triphosphate ATPase (v-ATPase). The interaction of v-ATPase with Ragulator at the lysosomal surface recruits mTORC1 to the lysosome. Ragulator is a protein complex that is essential for amino-acid-dependent activation of mTORC1 [126]. The Ragulator complex has guanine nucleotide exchange factor activity on Rag GTPases, through which mTORC1 is recruited to the lysosomal surface. Upon amino acid stimulation, RagA or RagB, both Rag GTPases, interacts with Raptor in mTORC1 [127]. This recruitment allows mTORC1 to interact with Rheb since Rheb resides at the lysosomal surface.

mTORC1 regulates cellular homeostasis in several ways. One way is modulation of autophagy. Self-degradation of proteins, and even entire organelles, is achieved through formation of the autophagosome. This process is crucial for clearance of damaged, dangerous, or toxic particles from the cell. mTORC1 is a known suppressor of autophagy. Studies in yeast and mammals suggested that mTORC1 phosphorylates autophagy-related proteins Atg1 and Atg13 found in a complex that is essential for autophagosome formation. Phosphorylation by mTORC1 disrupts the complex and inhibits the activation of the autophagic pathway. In mammals, in addition to Atg1 and Atg13 proteins, an Atg homolog named ULK is negatively regulated by mTORC1, indicating a conserved mechanism for mTORC1-mediated regulation of autophagy [128].

Finally, lysosome biogenesis and its modulation by the mTORC1−TFEB (transcription factor EB) axis is an important regulator of cellular homeostasis. Lysosomes are responsible for lipid homeostasis, clearance of proteins and organelles, energy metabolism, stress response, and pathogen defense [129]. mTORC1 is one of the key molecules regulating TFEB, which in turn promotes formation of autophagosomes and their fusion with lysosomes [130]. Phosphorylation of TFEB by mTORC1 upon activation of the nutrient-sensing pathway, allows 14-3-3 protein to bind TFEB and inhibit it from translocating to the nucleus. As lysosomal dysfunction is implicated in aging and age-related diseases, such as cancer and neurodegeneration, inhibition of mTORC1 is considered as a potential therapeutic approach [93].

11.6 mTORC2

In contrast to substantial evidence emerging on the signaling cascades involving mTORC1, less is known about upstream and downstream components of the mTORC2 signaling cascade. It is known that mTORC2 is insensitive to nutrients or energy status of the cell but sensitive to growth factors and insulin [131]. Insulin has been suggested to facilitate mTORC2 binding to ribosome through a poorly understood PI3K-dependent mechanism [132]. Recent evidence suggests an interplay between mTORC1 and mTORC2 in response to insulin stimulation, and their regulatory role in insulin receptor substrate 1 (IRS-1) degradation through growth factor receptor-bound protein 10-dependent mechanism [133,134]. When active, mTORC2 is involved in cell survival, cytoskeleton organization, and cellular metabolism [97].

The best-characterized role of mTORC2 is phosphorylation of Akt at the Ser473 residue, a post-translational mechanism required for full Akt activation [96]. This mTORC2-dependent phosphorylation of Akt links mTORC2 to the mTORC1 signaling pathway. However, studies in which mTORC2 is knocked out or knocked down suggest that mTORC2 function is not essential for mTORC1 activity [135]. mTORC1 might indirectly inhibit mTORC2 activity through phosphorylation of mSin1 by S6K1 [136]. Other substrates of mTORC2 are identified as serum- and glucocorticoid-induced protein kinase 1, protein kinase C (PKC) isoforms (in particular, PKCα). The former is involved

in ion transport and cell growth, while the latter links mTORC2 to cytoskeleton organization [97].

As we have mentioned earlier, recent studies provide evidence of the inhibition of mTORC2 activity upon long-term rapamycin treatment. A model involving dephosphorylation of mTORC2 components mSin1 and Rictor, and subsequent dissociation of mTORC2 in response to rapamycin was proposed to explain rapamycin's inhibitory effect on mTORC2 [137]. Similar to mTORC1, the TSC1−TSC2 heterodimer was found to be associated with mTORC2. However, this interaction was independent of Rheb, indicating distinct regulatory mechanisms acting on both mTOR complexes. Furthermore, while TSC has an inhibitory effect on mTORC1, it activates mTORC2 [138]. Some recent studies propose the association of mTORC2 with ribosomes, and an endoplasmic reticulum (ER) subcompartment called mitochondrial-associated ER membrane (MAM) in response to growth factor stimulation. It seems that mTORC2 is required for maintenance of MAM, regulation of mitochondrial metabolism and cell survival, although these claims remain to be further explored [132,139,140].

11.7 mTOR SIGNALING AND AGE-RELATED DISEASES

The main focus in the last decade has been on mTOR and its relationship to lifespan and disease prevention. It is clear from a large amount of evidence that mTOR overactivation promotes aging and disease, which in turn reduces lifespan [97]. By contrast, blocking mTOR retards aging and disease progression, and significantly increases lifespan. Less work has been performed about the role of mTOR in brain aging and disease but evidence from the role mTOR signaling plays in non-neurological disorders indicates that it could alter the course of normal and pathological brain aging.

11.7.1 mTOR and Metabolic Diseases

The mTOR signaling pathway is a nutrient-sensing pathway, meaning that it is regulated by nutrients and several other stimuli, and when activated, it influences cellular processes that have vast effects from metabolism to growth and cellular survival to aging [97]. Owing to its close relation with metabolic processes, mTOR is involved in several metabolic diseases, including diabetes mellitus. Diabetic complications, such as insulin resistance, have been linked with overactivation of the mTOR pathway [124]. Overactivated mTOR has been shown to cause insulin resistance, partly due to activation of S6 kinase (S6K) via active mTOR and subsequent degradation of IRSs. This feedback loop renders insulin signaling impaired [134].

In relation to diabetic complications, the mTOR signaling pathway has also been shown to regulate pancreatic β-cell proliferation and total β-cell mass [141]. Progressive pancreatic β-cell loss is a common characteristic of type 1 and type 2 diabetes [142]. Inhibition of mTORC1 via rapamycin treatment only abolished glucose-stimulated β-cell expansion, but not lipid-induced expansion in zebrafish. The latter has been shown to require insulin/IGF1 signaling, and in combination with the mTOR pathway, they exerted their nutrient-sensing effects that promote pancreatic β-cell neogenesis. These findings suggested a distinct role for mTOR signaling in progression of diabetes and diabetic complications [143].

The vascular endothelial dysfunction and its most common consequence, cardiovascular diseases, are associated with advancing age. In one study, arteries of older mice, which displayed endothelial dysfunction, have been shown to harbor increased mTOR signaling. SIRT-1, AMPK-1, and mTOR signaling pathways have been shown to be linked to the vascular aging phenotype and several studies suggest that their convergence and crosstalk rather than their independent activities affect vascular endothelial cells. Inhibition of mTOR by rapamycin or rapamycin analogs such as life-long caloric restriction (CR) prevents increased arterial mTOR signaling, improves nitric oxide amounts and ergo, prevents endothelial dysfunction in old mice [144].

11.7.2 mTOR and Cancer

There is emerging evidence that cancer is an age-related disease, and reduced or delayed carcinogenesis is associated with slow, healthy aging. The incidence of common cancers, such as lung, colon, breast, prostate, pancreatic, thyroid, gastric, and brain, increases with age. Not surprisingly, people who live more than 90 years are protected from cancer. Furthermore, a similar pattern has been shown in several animal models such as mice and rats. In long-lived mice and rats, cancer prevalence is very low despite high levels of oxidative stress [145].

The PI3K/Akt/mTOR pathway is among the most frequently altered pathways in human cancers. This may be due to its control over processes that are crucial for tumorigenesis, such as cell cycle progression, cell survival, cellular metabolism and energetics, and transcription [146]. Activation or aberrant functioning of the mTOR pathway by either activation

of oncogenes such as Ras, Raf, MEK, PI3K, and Akt or loss of tumor suppressors such as PTEN, NF-1, and P53 has been reported by several research groups studying human cancers [147,128,148,149]. Mutations in the genes encoding the proteins involved in this pathway have been implicated in various cancers. Although mutations that have been identified affect upstream elements such as PTEN [150], PIK3CA [151], PIK3R1 [152] and Akt [153], rather than directly targeting mTOR, aberrant functioning of this pathway is clearly involved in cancer development, and particularly in ovarian cancer [146].

There are several aspects of tumorigenicity on which mTOR has direct or indirect effects. The mechanism through which mTOR is acting is translation initiation, which is required for normal cell growth and proliferation. When aberrant, translational control leads to uncontrolled growth of the cell. Eukaryotic translation initiation is regulated by eukaryotic initiation factors eIF4G, eIF4E, eIF4A, and eIF4F. Of these, eIF4E is regulated by inhibitory eIF4E binding proteins (4E-BPs). 4E-BPs mainly act via binding to eIF4E and inhibiting its association with eIF4F, and therefore lowering translation levels. However, when phosphorylated by active mTOR in the mTORC1 complex upon mitogenic stimulation, 4E-BP1 dissociates from eIF4E, allowing it to form the translation initiation complex and proceed with the translation process [154–156].

Growth factors such as insulin and nutrients regulate mTORC1 activity via activation of the PI3K/Akt pathway [128], and the overexpression of growth factor receptors such as insulin-like growth factor receptor and human epidermal growth factor receptor 2 may activate the PI3K/Akt and mTOR pathways [157]. mTORC2, on the other hand, affects tumorigenesis through its role in glucose metabolism, organization of the actin cytoskeleton, and apoptosis in addition to its distinct role in cell growth and proliferation [96,158]. A subunit of mTORC2, Rictor, is overexpressed in some cancer types. Rictor overexpression leads to overactive mTORC2 activity, which has been shown to be associated with higher proliferative and invasive potential of tumors [159,160]. TSC1 or TSC2 mutations, on the other hand, cause tuberous sclerosis, a familial cancer syndrome which leads to benign tumors but may progress to malignant ones [161].

The close relationship between mTOR, aging, and cancer may be based on the phenomenon called gero-conversion. In proliferating cells, cellular growth and cell division are balanced in response to nutrients and growth factors. In aging cells, the mTOR pathway shifts from cellular arrest and quiescence into senescence. In neoplasms and cancers, cellular senescence is overcome and cells start dividing uncontrollably.

Hence, both mechanisms depend partly on how the mTOR pathway acts and how it is regulated [145].

11.7.3 mTOR Signaling and Normal and Pathological Brain Aging

mTOR is expressed at high levels in the brain, mainly in neurons but also in glial cells. It plays an important role in development for neuron survival and growth. In the adult brain, mTOR controls synaptic plasticity and processes that underlie learning and memory [162]. For example, increases in GluR1 levels were shown to be necessary in the dorsal hippocampus of rat for the learning and memory consolidation process around an inhibitory avoidance training and 3 h after the training. This study also demonstrated that the increase in the level of GluR1 was obtained through the BDNF/mTOR signaling pathway [163]. mTOR contributes to changes in LTP by regulating protein synthesis in the adult brain [162]. Thus, in normal conditions mTOR is necessary for brain function. However, during normal and pathological aging, mTOR signaling may underlie some of the deficits in brain areas controlling learning and memory (Table 11.1).

In terms of normal brain aging, manipulations of the mTOR pathway result in ameliorating effects against age-related decline. It has been shown that partial inhibition of the mTOR pathway by rapamycin extends lifespan in many model organisms. However, another crucial issue is whether this lifespan increase is correlated with preserved cognitive functions in older ages which would normally decline. Studies in mouse models indicate that rapamycin treatment starting at younger ages has an enhancing effect on spatial memory and learning in older ages. It was shown that these ameliorating effects of rapamycin are associated with reduced IL-1β, which is a proinflammatory cytokine associated with memory impairments in older adults' levels in the hippocampus, and increased NMDA signaling. However, these enhancing effects on learning and memory are not observed if the rapamycin treatment starts at older ages [164]. Another study indicated that partial and chronic inhibition of mTOR by rapamycin ameliorated both the cognitive and non-cognitive components of behavior in a mouse model. Eight-month-old animals treated with rapamycin showed better performance in terms of spatial learning-memory; and 25-month-old animals treated with rapamycin showed better long-term memory performance, which is related to aversive stimuli when they are compared with non-treated controls. Additionally, non-cognitive components of behavior, such as anxiety and depressive behaviors, which are also observed in older populations, are reduced in rapamycin-treated groups. The mechanism through which this reduction is occurring is thought to be mediated by partial inhibition of mTOR, which results in increased major monoamine levels including epinephrine, norepinephrine, dopamine, and 5-hydroxytryptamine [165]. In conclusion, the data indicate that in the normal aging condition, that is, without pathology, partial inhibition of mTOR could improve cognitive decline and have ameliorating effects on cognitive and non-cognitive components of behavior that decline with normal aging.

To date, more information has been obtained with respect to the mTOR signaling pathway and pathological brain aging. In terms of normal brain aging, several reports suggest that a decreased rate of hippocampal neurogenesis has been associated with cognitive impairment in the elderly, in particular with the learning and memory processes [166]. In the past it was thought that neurons do not regenerate, however, it is clear now that neural stem and/or progenitor cells with the capacity to generate new neurons throughout the lifespan have been found in certain regions of the adult mammalian brain [167]. The fate of hippocampal neural stem cells depends on both the internal signaling pathways and external stimuli, such as growth factors and nutrients [168,169]. Different subtypes of neural stem cells are affected differentially by aging. For instance, the number of quiescent hippocampal neural stem cells has a slower rate of decay than the active hippocampal neural stem cells. Although some reports suggest that the total number of hippocampal neural stem cells does not decline with age but rather the number of quiescent hippocampal neural stem cells increases in response to the aging process [170,171], another study showed a significant decline in the number, and the proliferation rate of hippocampal neural stem cells in aging mice. Interestingly, the mTOR signaling pathway activity was shown to be impaired in these cells, and its restoration stimulated the proliferation of hippocampal neural stem cells in the aged mouse brain. Furthermore, inhibition of the mTOR pathway was shown to be sufficient to induce premature aging in the young mice [172].

11.7.4 Parkinson's Disease

Parkinson's disease (PD) is a neurodegenerative disease which is characterized by progressive dopaminergic neural loss, especially in substantia nigra pars compacta, and it results in deficits in motor functions including tremors and ataxia [173]. Currently, there is no decisive treatment for PD; one reason is that underlying pathology and the etiology of PD are not very well known. However, current studies have revealed associations of neural loss in PD with apoptosis, oxidative stress, DNA damage, mitochondrial dysfunction, aberrant protein aggregation, and endoplasmic reticulum stress [174–176].

Interestingly, studies have also suggested that components of the mTOR pathway could play a role in PD progression and manipulations targeting this pathway may promise therapeutic strategies for PD (Table 11.1). It has been shown that treatment with rapamycin, which is the pharmacological inhibitor of mTOR, suppresses the PD-related pathological phenotype, including muscle degeneration, locomotor activities, and mitochondrial defects, and these effects of rapamycin were mediated by 4E-BP activity in the *Drosophila melanogaster* model of PD [177]. In contrast, it was reported that *in vitro* models of PD, in which PD was induced by specific toxins, inhibited mTOR activity. In this model inhibited mTOR activity, which is related to increased AMPK and decreased Akt activity, was associated with neural loss [178]. Based on these studies, it may be the case that both inhibition and activation of the mTOR pathway could be associated with PD-related neural death. Another study indicated that selective suppression of the mTOR pathway is protective against PD-related neural death, and it was shown that although rapamycin is an inhibitor of mTOR activity, it spares some activity of mTOR, such as regulation of Akt phosphorylation, while it inhibits other mTOR activities like S6K (also known as p70S6K) activation. Also, in this study when all the actions of mTOR were inhibited, dephosphorylation of Akt and neural death were induced in both *in vitro* and mouse models of PD [179]. Therefore for PD, it could be argued that some activities of mTOR are essential for the survival of the neurons and others could be harmful for the neurons during the disease progression. Moreover, complete inhibition of mTOR activity may not be protective against neural death in PD, as the protective effect of rapamycin is associated with its selective suppression of mTOR activity.

11.7.5 Alzheimer's Disease

Another pathology associated with advanced age is Alzheimer's disease (AD). AD is the most common progressive neurodegenerative disease in the aging population; cognitive deficits can be observed across domains like memory, spatial learning, language, attention, visual perception, executive functions, and social function [180]. Progressive amyloid-beta protein accumulation and presence of neurofibrillary tangles with hyperphosphorylated tau protein are the pathological hallmarks

of AD, and they are associated with progressive neural loss [181]. One reason for the accumulation of these harmful proteins that has been proposed is the presence of an altered autophagy response against these toxic substances in the affected cells [182]. Moreover, since there is an established relationship between the mTOR pathway and autophagy, the mTOR pathway has become a target for interventions and manipulations in order to understand its role in AD (Table 11.1). Evidence demonstrates that long-term inhibition of mTOR with rapamycin in the transgenic AD mouse model reduces and prevents AD-like cognitive deficits; rapamycin-treated transgenic mice show better performance in spatial memory and learning tasks. This inhibition lowers the levels of amyloid-beta protein (Aβ42) and prevents or delays the onset of AD. Also, it has been shown that autophagy is induced in the hippocampus of rapamycin-treated transgenic mice. However, these ameliorating effects could not be found in wild-type mice treated with rapamycin, hence the effects of rapamycin may depend on the endogenous Aβ levels in mice [183]. In another study, mTOR was inhibited by rapamycin in 2- and 15-month-old transgenic AD model mice. In 2-month-old mice, this inhibition prevented cognitive decline in learning and memory, increased autophagy, and decreased the levels of Aβ, tau, and tangle accumulation by increasing the soluble Aβ and tau levels, whereas in 15-month-old mice, this inhibition did not change the soluble or non-soluble levels of Aβ and tau and did not ameliorate cognitive deficits [184]. Taken together, inhibition of mTOR via rapamycin could ameliorate the progression of AD and AD-related cognitive deficits, but the extent of these effects may depend on the AD progression.

11.7.6 Frontotemporal Dementia

Frontotemporal dementia is an age-related neurodegenerative disease characterized by neural loss in frontal and temporal lobes (Figure 11.1). Clinical manifestations of frontotemporal dementia in response to neural loss in these specific regions are observed as alterations of certain personality traits. These alterations include abnormal social behavior and apathies, and cognitive deficits in executive function, attention, visuospatial abilities, language, and memory [185]. The molecular and cellular alterations underlying the neural loss are related to proteinopathies, which refers to the presence of malformed or abnormal proteins. One of the proteins associated with frontotemporal dementia is TDP-43. TDP-43 is a DNA/RNA-binding protein, found in the cytoplasmic ubiquitin inclusions of the affected cells, and plays a role as a main component in the disease progression [186].

Since the neurodegeneration process in frontotemporal dementia is associated with proteinopathies, induction of autophagy in the affected cells has become a focus as a possible intervention. Studies conducted in the mouse model of frontotemporal dementia demonstrated that inhibition of the mTOR pathway by rapamycin prevents motor and cognitive deficits including learning and memory declines, reduces the neural loss and the caspase-3 levels, which is the indicator of the neural death, and decreases levels of TDP-43 products [187] (Table 11.1). Therefore, it is likely that the inhibition of mTOR, which might result in autophagy activation, may be a promising therapeutic approach for age-related neurodegenerative diseases with proteinopathies.

11.7.7 Huntington's Disease

Huntington's disease (HD) is an inherited neurodegenerative disease affecting the whole brain, yet some regions, such as the basal ganglia and hippocampus, are more vulnerable. Clinical symptoms include motor dysfunction, muscle rigidity, problems in memory and executive functions, anxiety, and depression. HD is a polyglutamine tract disorder, emerging when the number of CAG repeats in the HTT gene encoding Huntingtin protein exceeds a specific threshold and mutant Huntingtin protein, which has toxic effects, is produced [188]. Alterations due to mutant Huntingtin protein expression are decreased uptake of glutamate and increased NMDA receptor signaling at the striatal excitatory synapses. Decreased levels of BDNF, and altered striatal dopamine and endocannabinoid signaling accompany degeneration in HD [189].

As in the cases of AD and frontotemporal dementia, abnormal protein aggregation has been observed in HD. Thus, activation of autophagy through the mTOR pathway could have therapeutic effects in HD. *In vitro* studies conducted with HD models indicated that inhibition of mTOR with small molecule enhancers of rapamycin increased the clearance of the mutant Huntingtin fragments in the affected cells (Table 11.1). Another study using the *Drosophila melanogaster* model of HD demonstrated that these small molecule enhancers of rapamycin protect against toxicity induced by mutant Huntingtin protein [190]. Other studies in *Drosophila melanogaster* have supportive findings of rapamycin protection against neural death and toxicity associated with HD, and studies with a mouse model of HD showed that rapamycin treatment reduces the HD-related motor and behavioral deficits and decreased protein aggregation [191]. It could be said that interventions like the rapamycin targeting of

mTOR and the autophagy mechanism could be a promising therapeutic approach and area of research for HD.

11.8 INTERVENTIONS FOR ALTERING mTOR ACTIVITY

It has been previously mentioned throughout this chapter that rapamycin treatment, which inhibits mTOR, can alter the course of aging-related diseases and pathologies. Moreover, it appears to have beneficial effects on lifespan and some of the declines associated with aging. There are many studies reporting that rapamycin treatment extends lifespan in various organisms. For instance, rapamycin treatment late in life (at 600 days of age) extended lifespan in both male and female genetically heterogeneous mice [192]. Another study in *Drosophila* suggested that genes encoded by mitochondrial DNA have influence on the mTOR pathway, and that mTOR inhibition via rapamycin increases mitochondrial respiration and decreases H_2O_2 production [193]. Similar studies in yeast suggested reduced TOS due to deletion of *TOR1* gene or rapamycin treatment, and modulation of reactive oxygen species by enhancement of the mitochondrial respiration as a possible mechanism for lifespan extension [194,195].

Rapamycin treatment has also been shown to prevent age-related metabolic diseases including obesity [196,197] and diabetic complications such as retinopathy, nephropathy, and coronary disease in animal models and humans [198,196,199]. Rapamycin mimics starvation or CR, depending on the administered concentration. Therefore, it tricks the organism to think that nutrient supply is growing short. Starvation has several metabolic effects including increased gluconeogenesis, ketogenesis, lipolysis, decreased insulin levels, and it leads to insulin resistance [200].

Given the fact that mTOR signaling is almost universally activated in cancer, rapamycin, rapamycin analogs (i.e., rapalogs), and dietary regimens such as CR have become promising approaches in cancer prevention and treatment [201–203]. Rapalogs, such as temsirolimus and everolimus, are being used as anticancer drugs, the former being the first rapalog approved by FDA for use in humans [204,205]. Although rapalogs are currently being used in numerous clinical trials, their role as anticancer agents remains modest. This modest effect may be due to their lack of toxicity, and hence inability to kill cancer cells at relevant concentrations. Rather, they act to slow down the growth of cancer cells and prove more effective when used in combination with other anticancer drugs. Regardless of their disadvantage as anticancer drugs, they are still effective agents for cancer-preventive and antiaging purposes.

Another intervention that may be beneficial is CR, which is thought to act by blocking the mTOR pathway in a manner similar to rapamycin treatment. Even short-term (i.e., 8 weeks) CR had a protective effect on renal cells of the aged rats. Reduced mitochondrial oxidative damage, retardation of renal senescence, and increased autophagic activity have been accompanied by decreased mTOR levels, and increased SIRT-1 and AMPK levels, in these old rats [206]. Of these mechanisms, autophagic activity has particular significance since its decrease during aging has been considered as the underlying mechanism for the accumulation of abnormal/damaged macromolecules. Degradation of long-lived proteins, unfolded or incorrectly folded proteins, and damaged organelles occur through autophagic pathways. Hence, it is essential for maintenance of cellular and organismal homeostasis [207–209].

It is important to point out that both rapamycin and CR have been shown to affect males and females differentially. In one study, tissue- and sex-specific effects of rapamycin on the proteasome-chaperone network have been shown [80]. For instance, levels of proteasome-related chaperones increased more in females than males. Furthermore, brains of rapamycin-treated females had higher levels of mTOR pathway proteins (p-mTOR, p-Akt, rpS6, and 4E-BP1), than males and other tissues tested. These results suggest that rapamycin increases proteasome activity in the brains of female mice, compared to other tissues and male counterparts.

Another study showed that short-term CR in male mice shifted the gene expression profile toward a more feminine type. In this microarray-based study, CR has been shown to change expression levels of over 3000 genes involved in hormone signaling, aging, cell cycle, and apoptosis [210]. When the overall life expectancy of males and females is taken into account, i.e., that females of most mammals have a longer lifespan than males, these results seem quite explanatory for both the underlying mechanisms for CR-induced longevity and sex-dependent variations in response to CR and CR mimetics.

Taken altogether, rapamycin, rapalogs, and CR have been shown to have positive effects on various age-related diseases, and to extend overall and healthy lifespan in both animals and humans. Thus, the effects of CR and the CR-mimetic rapamycin on lifespan have been the subject of ongoing research. It will be important to determine the best regimen, either dietary or therapeutic drug dose, for the optimal antiaging effects. Moreover, more research needs to be performed to determine the exact effects of CR and CR mimetics on normal brain aging.

11.9 CONCLUSION

Normal aging is accompanied by a range of biological changes, many of which diminish the quality of life and interfere with the ability of the elderly to function unassisted in society. Even in the absence of evident pathology, cognitive function declines during old age and is often reflected as a significant loss of learning and memory ability. Undoubtedly, the biological changes are multifactorial in nature, and are likely caused by a combination of genetic and environmental factors. Nevertheless, understanding the contribution of individual factors to age-related changes in cognitive function would provide important insight into the mechanisms by which such changes occur, as well as suggest strategies for preventing or ameliorating cognitive changes.

The mTOR pathway appears to be a very important target for altering the course of aging-related cognitive decline and diseases that are both neurological and non-neurological in nature. Understanding the exact effects of mTOR on aging as well as the appropriate therapeutic approaches is the subject of ongoing research. Undoubtedly, these data will contribute to our understanding of both normal and pathological aging and how it can be prevented or altered.

Acknowledgment

We would like to thank Can I. Birand for helping with preparation of the figures. The research being done in the laboratory of Michelle M. Adams is currently supported by an Installation Grant from the European Molecular Biology Organization and the Scientific and Technological Research Council of Turkey (TUBITAK) project number 214S236.

References

[1] World Health Organisation. WHO ageing and life-course. Ageing Lifecourse; 2013.
[2] Costa N, Ferlicoq L, Derumeaux-Burel H, Rapp T, Garnault V, Gillette-Guyonnet S, et al. Comparison of informal care time and costs in different age-related dementias: a review. BioMed Res Int 2013;2013:1–15.
[3] López-Otín C, Blasco MA, Partridge L, Serrano M, Kroemer G. The hallmarks of aging. Cell 2013;153:1194–217.
[4] Mora F. Successful brain aging: plasticity, environmental enrichment, and lifestyle. Dialogues Clin Neurosci 2013;15:45–52.
[5] Sarlak G, Jenwitheesuk A, Chetsawang B, Govitrapong P. Effects of melatonin on nervous system aging: neurogenesis and neurodegeneration. J Pharmacol Sci 2013;123:9–24.
[6] Thomas W, Blanchard J. Moving beyond place: aging in community. Generations 2009;33:12–17.
[7] Peters R. Ageing and the brain. Postgrad Med J 2006;82:84–8.
[8] Arslan-Ergul A, Ozdemir AT, Adams MM. Aging, neurogenesis, and caloric restriction in different model organisms. Aging Dis 2013;4:221–32.
[9] Salmon DP, Bondi MW. Neuropsychological assessment of dementia. Annu Rev Psychol 2009;60:257–82.

[10] Mitchell DB, Brown AS, Murphy DR. Dissociations between procedural and episodic memory: effects of time and aging. Psychol Aging 1990;5:264–76.

[11] Nilsson L-G. Memory function in normal aging. Acta Neurol Scand Suppl 2003;179:7–13.

[12] Cabeza R, Daselaar SM, Dolcos F, Prince SE, Budde M, Nyberg L. Task-independent and task-specific age effects on brain activity during working memory, visual attention and episodic retrieval. Cereb Cortex 2004;14:364–75.

[13] Wang Q-S, Zhou J-N. Retrieval and encoding of episodic memory in normal aging and patients with mild cognitive impairment. Brain Res 2002;924:113–15.

[14] Greicius MD, Srivastava G, Reiss AL, Menon V. Default-mode network activity distinguishes Alzheimer's disease from healthy aging: evidence from functional MRI. Proc Natl Acad Sci USA 2004;101:4637–42.

[15] Craik FIM. Memory changes in normal aging. Curr Dir Psychol Sci 1994;3:155–8.

[16] St-Laurent M, Abdi H, Burianová H, Grady CL. Influence of aging on the neural correlates of autobiographical, episodic, and semantic memory retrieval. J Cogn Neurosci 2011;23:4150–63.

[17] Giffard B. The nature of semantic memory deficits in Alzheimer's disease: new insights from hyperpriming effects. Brain 2001;124:1522–32.

[18] Hirni DI, Kivisaari SL, Monsch AU, Taylor KI. Distinct neuroanatomical bases of episodic and semantic memory performance in Alzheimer's disease. Neuropsychologia 2013;51:930–7.

[19] Olton DS, Samuelson RJ. Remembrance of places passed: spatial memory in rats. J Exp Psychol Anim Behav Process 1976;2:97–116.

[20] Fischer W, Björklund A, Chen K, Gage FH. NGF improves spatial memory in aged rodents as a function of age. J Neurosci 1991;11:1889–906.

[21] Herndon JG, Moss MB, Rosene DL, Killiany RJ. Patterns of cognitive decline in aged rhesus monkeys. Behav Brain Res 1997;87:25–34.

[22] Frick KM, Baxter MG, Markowska AL, Olton DS, Price DL. Age-related spatial reference and working memory deficits assessed in the water maze. Neurobiol Aging 1995;16:149–60.

[23] Bach ME, Barad M, Son H, Zhuo M, Lu YF, Shih R, et al. Age-related defects in spatial memory are correlated with defects in the late phase of hippocampal long-term potentiation in vitro and are attenuated by drugs that enhance the cAMP signaling pathway. Proc Natl Acad Sci USA 1999;96:5280–5.

[24] Newman MC, Kaszniak AW. Spatial memory and aging: performance on a human analog of the morris water maze. Aging Neuropsychol Cogn 2010;7:86–93.

[25] Moffat S. Age differences in spatial memory in a virtual environment navigation task. Neurobiol Aging 2001;22:787–96.

[26] Sahgal A, McKeith IG, Galloway PH, Tasker N, Steckler T. Do differences in visuospatial ability between senile dementias of the Alzheimer and Lewy body types reflect differences solely in mnemonic function? J Clin Exp Neuropsychol 1995;17:35–43.

[27] Devaney KO, Johnson HA. Neuron loss in the aging visual cortex of man. J Gerontol 1980;35:836–41.

[28] Henderson G, Tomlinson BE, Gibson PH. Cell counts in human cerebral cortex in normal adults throughout life using an image analysing computer. J Neurol Sci 1980;46:113–36.

[29] Bondareff W. Synaptic atrophy in the senescent hippocampus. Mech Ageing Dev 1979;9:163–71.

[30] Geinisman Y, de Toledo-Morrell L, Morrell F. Loss of perforated synapses in the dentate gyrus: morphological substrate of memory deficit in aged rats. Proc Natl Acad Sci USA 1986;83:3027–31.

[31] Haug H, Eggers R. Morphometry of the human cortex cerebri and corpus striatum during aging. Neurobiol Aging 1991;12:336–8; (discussion 352–5).

[32] Rapp PR, Gallagher M. Preserved neuron number in the hippocampus of aged rats with spatial learning deficits. Proc Natl Acad Sci USA 1996;93:9926–30.

[33] Rasmussen T, Schliemann T, Sørensen JC, Zimmer J, West MJ. Memory impaired aged rats: no loss of principal hippocampal and subicular neurons. Neurobiol Aging 1996;17:143–7.

[34] Newton IG, Forbes ME, Linville MC, Pang H, Tucker EW, Riddle DR, et al. Effects of aging and caloric restriction on dentate gyrus synapses and glutamate receptor subunits. Neurobiol Aging 2008;29:1308–18.

[35] Shi L, Adams MM, Linville MC, Newton IG, Forbes ME, Long AB, et al. Caloric restriction eliminates the aging-related decline in NMDA and AMPA receptor subunits in the rat hippocampus and induces homeostasis. Exp Neurol 2007;206:70–9.

[36] Geinisman Y, Berry RW, Disterhoft JF, Power JM, Van der Zee EA. Associative learning elicits the formation of multiple-synapse boutons. J Neurosci 2001;21:5568–73.

[37] Smith TD, Adams MM, Gallagher M, Morrison JH, Rapp PR. Circuit-specific alterations in hippocampal synaptophysin immunoreactivity predict spatial learning impairment in aged rats. J Neurosci 2000;20:6587–93.

[38] Park H, Poo MM. Neurotrophin regulation of neural circuit development and function. Nat Rev Neurosci 2013;14:7–23.

[39] Cunha C, Brambilla R, Thomas KL. A simple role for BDNF in learning and memory? Front Mol Neurosci 2010;3:1.

[40] Petzold A, Endres T. Chronic BDNF deficiency leads to age-dependent impairment in spatial learning Material & Methods. Neurobiol Learn Mem 2015;120:1–3.

[41] Hamilton LK, Joppé SE, Cochard LM, Fernandes KJL. Aging and neurogenesis in the adult forebrain: what we have learned and where we should go from here. Eur J Neurosci 2013;37:1978–86.

[42] Yassa MA, Stark CEL. Pattern separation in the hippocampus. Trends Neurosci 2011;34:515–25.

[43] Bilkei-Gorzo A. The endocannabinoid system in normal and pathological brain ageing. Philos Trans R Soc B Biol Sci 2012;367:3326–41.

[44] Legler A, Monory K, Lutz B. Age differences in the role of the cannabinoid type 1 receptor on glutamatergic neurons in habituation and spatial memory acquisition. Life Sci 2015;1–7.

[45] Bilkei-Gorzo A, Racz I, Valverde O, Otto M, Michel K, Sastre M, et al. Early age-related cognitive impairment in mice lacking cannabinoid CB1 receptors. Proc Natl Acad Sci USA 2005;102: 15670–5.

[46] Albayram O, Bilkei-Gorzo A, Zimmer A. Loss of CB1 receptors leads to differential age-related changes in reward-driven learning and memory. Front Aging Neurosci 2012;4:1–8.

[47] Burke SN, Barnes CA. Neural plasticity in the ageing brain. Nat Rev Neurosci 2006;7:30–40.

[48] Lu T, Pan Y, Kao S-Y, Li C, Kohane I, Chan J, et al. Gene regulation and DNA damagein the ageing human brain. Nature 2004;429:883–91.

[49] Jiang CH, Tsien JZ, Schultz PG, Hu Y. The effects of aging on gene expression in the hypothalamus and cortex of mice. Proc Natl Acad Sci USA 2001;98:1930–4.

[50] Arslan-Ergul A, Adams MM. Gene expression changes in aging zebrafish (Danio rerio) brains are sexually dimorphic. BMC Neurosci 2014;15:29.

[51] Moroz LL, Kohn AB. Do different neurons age differently? Direct genome-wide analysis of aging in single identified cholinergic neurons. Front Aging Neurosci 2010;2:1–18.

[52] Cheung I, Shulha HP, Jiang Y, Matevossian A, Wang J, Weng Z, et al. Developmental regulation and individual differences of neuronal H3K4me3 epigenomes in the prefrontal cortex. Proc Natl Acad Sci USA 2010;107:8824−9.

[53] Peleg S, Sananbenesi F, Zovoilis A, Burkhardt S, Bahari-Javan S, Agis-Balboa RC, et al. Altered histone acetylation is associated with age-dependent memory impairment in mice. Science 2010;328:753−6.

[54] Hollmann M, Heinemann S. Cloned glutamate receptors. Annu Rev Neurosci 1994;17:31−108.

[55] Morris RG, Anderson E, Lynch GS, Baudry M. Selective impairment of learning and blockade of long-term potentiation by an N-methyl-D-aspartate receptor antagonist, AP5. Nature 1986;5:774−6.

[56] Bliss TV, Collingridge GL. A synaptic model of memory: long-term potentiation in the hippocampus. Nature 1993;361:31−9.

[57] Hollmann M, O'Shea-Greenfield A, Rogers SW, Heinemann S. Cloning by functional expression of a member of the glutamate receptor family. Nature 1989;342:643−8.

[58] Nakanishi S. Molecular diversity of glutamate receptors and implications for brain function. Science 1992;258:597−603.

[59] Brose N, Huntley GW, Stern-Bach Y, Sharma G, Morrison JH, Heinemann SF. Differential assembly of coexpressed glutamate receptor subunits in neurons of rat cerebral cortex. J Biol Chem 1994;269:16780−4.

[60] Puchalski RB, Louis JC, Brose N, Traynelis SF, Egebjerg J, Kukekov V, et al. Selective RNA editing and subunit assembly of native glutamate receptors. Neuron 1994;13:131−47.

[61] Boulter J, Hollmann M, O'Shea-Greenfield A, Hartley M, Deneris E, Maron C, et al. Molecular cloning and functional expression of glutamate receptor subunit genes. Science 1990;249:1033−7.

[62] Keinänen K, Wisden W, Sommer B, Werner P, Herb A, Verdoorn TA, et al. A family of AMPA-selective glutamate receptors. Science 1990;249:556−60.

[63] Tsien JZ, Huerta PT, Tonegawa S. The essential role of hippocampal CA1 NMDA receptor-dependent synaptic plasticity in spatial memory. Cell 1996;87:1327−38.

[64] Tang YP, Shimizu E, Dube GR, Rampon C, Kerchner GA, Zhuo M, et al. Genetic enhancement of learning and memory in mice. Nature 1999;401:63−9.

[65] Barnes CA. Normal aging: regionally specific changes in hippocampal synaptic transmission. Trends Neurosci 1994;17:13−18.

[66] Morrison JH, Gazzaley AH. Age-related alterations of the N-methyl-D-aspartate receptor in the dentate gyrus. Mol Psychiatry 1996;1:356−8.

[67] Morrison JH, Hof PR. Life and death of neurons in the aging brain. Science 1997;278:412−19.

[68] Clayton DA, Mesches MH, Alvarez E, Bickford PC, Browning MD. A hippocampal NR2B deficit can mimic age-related changes in long-term potentiation and spatial learning in the Fischer 344 rat. J Neurosci 2002;22:3628−37.

[69] Eckles-Smith K, Clayton D, Bickford P, Browning MD. Caloric restriction prevents age-related deficits in LTP and in NMDA receptor expression. Brain Res Mol Brain Res 2000;78:154−62.

[70] Adams MM, Shah RA, Janssen WG, Morrison JH. Different modes of hippocampal plasticity in response to estrogen in young and aged female rats. Proc Natl Acad Sci USA 2001;98:8071−6.

[71] Pagliusi SR, Gerrard P, Abdallah M, Talabot D, Catsicas S. Age-related changes in expression of AMPA-selective glutamate receptor subunits: is calcium-permeability altered in hippocampal neurons? Neuroscience 1994;61:429−33.

[72] Barnes CA, Rao G, Houston FP. LTP induction threshold change in old rats at the perforant path—granule cell synapse. Neurobiol Aging 21:613−0.

[73] Shankar S, Teyler TJ, Robbins N. Aging differentially alters forms of long-term potentiation in rat hippocampal area CA1. J Neurophysiol 1998;79:334−41.

[74] Barnes C, Rao G, Shen J. Age-related decrease in the N-Methyl-d-AspartateR-mediated excitatory postsynaptic potential in hippocampal region CA1. Neurobiol Aging 1997;18:445−52.

[75] Perea G, Navarrete M, Araque A. Tripartite synapses: astrocytes process and control synaptic information. Trends Neurosci 2009;32:421−31.

[76] Woolley CS, McEwen BS. Roles of estradiol and progesterone in regulation of hippocampal dendritic spine density during the estrous cycle in the rat. J Comp Neurol 1993;336:293−306.

[77] Klintsova A, Levy WB, Desmond NL. Astrocytic volume fluctuates in the hippocampal CA1 region across the estrous cycle. Brain Res 1995;690:269−74.

[78] Luquin S, Naftolin F, Garcia-Segura LM. Natural fluctuation and gonadal hormone regulation of astrocyte immunoreactivity in dentate gyrus. J Neurobiol 1993;24:913−24.

[79] Cerbai F, Lana D, Nosi D, Petkova-Kirova P, Zecchi S, Brothers HM, et al. The neuron-astrocyte-microglia triad in normal brain ageing and in a model of neuroinflammation in the rat hippocampus. PloS ONE 2012;7:e45250.

[80] Rodriguez K a, Dodds SG, Strong R, Galvan V, Sharp ZD, Buffenstein R. Divergent tissue and sex effects of rapamycin on the proteasome-chaperone network of old mice. Front Mol Neurosci 2014;7:83.

[81] Keith CT, Schreiber SL. PIK-related kinases: DNA repair, recombination, and cell cycle checkpoints. Science 1995;270:50−1.

[82] Vézina C, Kudelski A, Sehgal SN. Rapamycin (AY-22,989), a new antifungal antibiotic. I. Taxonomy of the producing streptomycete and isolation of the active principle. J Antibiot (Tokyo) 1975;28:721−6.

[83] Huang S, Bjornsti MA, Houghton PJ. Rapamycins: mechanism of action and cellular resistance. Cancer Biol Ther 2003;2:222−32.

[84] Yang H, Rudge DG, Koos JD, Vaidialingam B, Yang HJ, Pavletich NP. mTOR kinase structure, mechanism and regulation. Nature 2013;497:217−23.

[85] Hubbard PA, Moody CL, Murali R. Allosteric modulation of Ras and the PI3K/AKT/mTOR pathway: emerging therapeutic opportunities. Front Physiol 2014;5:1−7.

[86] Cafferkey R, Young PR, McLaughlin MM, Bergsma DJ, Koltin Y, Sathe GM, et al. Dominant missense mutations in a novel yeast protein related to mammalian phosphatidylinositol 3-kinase and VPS34 abrogate rapamycin cytotoxicity. Mol Cell Biol 1993;13:6012−23.

[87] Kunz J, Henriquez R, Schneider U, Deuter-Reinhard M, Movva NR, Hall MN. Target of rapamycin in yeast, TOR2, is an essential phosphatidylinositol kinase homolog required for G1 progression. Cell 1993;73:585−96.

[88] Long X, Spycher C, Han ZS, Rose AM, Müller F, Avruch J. TOR deficiency in C. elegans causes developmental arrest and intestinal atrophy by inhibition of mRNA translation. Curr Biol 2002;12:1448−61.

[89] Oldham S. Genetic and biochemical characterization of dTOR, the Drosophila homolog of the target of rapamycin. Genes Dev 2000;14:2689−94.

[90] Fleming A, Rubinsztein DC. Zebrafish as a model to understand autophagy and its role in neurological disease. Biochim Biophys Acta 2011;1812:520−6.

[91] Sarbassov DD, Ali SM, Kim D-H, Guertin DA, Latek RR, Erdjument-Bromage H, et al. Rictor, a novel binding partner of mTOR, defines a rapamycin-insensitive and raptor-independent pathway that regulates the cytoskeleton. Curr Biol 2004;14: 1296−302.

[92] Sabatini DM. mTOR and cancer: insights into a complex relationship. Nat Rev Cancer 2006;6:729−34.

[93] Laplante M, Sabatini DM. Regulation of mTORC1 and its impact on gene expression at a glance. J Cell Sci 2013;126:1713–19.

[94] Sarbassov DD, Ali SM, Sengupta S, Sheen J-H, Hsu PP, Bagley AF, et al. Prolonged rapamycin treatment inhibits mTORC2 assembly and Akt/PKB. Mol Cell 2006;22:159–68.

[95] Inoki K, Li Y, Xu T, Guan K-L. Rheb GTPase is a direct target of TSC2 GAP activity and regulates mTOR signaling. Genes Dev 2003;17:1829–34.

[96] Sarbassov DD, Guertin DA, Ali SM, Sabatini DM. Phosphorylation and regulation of Akt/PKB by the rictor-mTOR complex. Science 2005;307:1098–101.

[97] Laplante M, Sabatini DM. mTOR signaling in growth control and disease. Cell 2012;149:274–93.

[98] Blommaart EFC, Luiken JJFP, Blommaart PJE, van Woerkom GM, Meijer AJ. Phosphorylation of ribosomal protein S6 is inhibitory for autophagy in isolated rat hepatocytes. J Biol Chem 1995;270:2320–6.

[99] Hara K, Yonezawa K, Weng Q-P, Kozlowski MT, Belham C, Avruch J. Amino acid sufficiency and mTOR regulate p70 S6 kinase and eIF-4E BP1 through a common effector mechanism. J Biol Chem 1998;273:14484–94.

[100] Fingar DC, Salama S, Tsou C, Harlow E, Blenis J. Mammalian cell size is controlled by mTOR and its downstream targets S6K1 and 4EBP1/eIF4E. Genes Dev 2002;16:1472–87.

[101] Fingar DC, Richardson CJ, Tee AR, Cheatham L, Tsou C, Blenis J. mTOR controls cell cycle progression through its cell growth effectors S6K1 and 4E-BP1/eukaryotic translation initiation factor 4E. Mol Cell Biol 2004;24:200–16.

[102] Nojima H, Tokunaga C, Eguchi S, Oshiro N, Hidayat S, Yoshino K, et al. The mammalian target of rapamycin (mTOR) partner, raptor, binds the mTOR substrates p70 S6 kinase and 4E-BP1 through their TOR signaling (TOS) motif. J Biol Chem 2003;278:15461–4.

[103] Schalm SS, Fingar DC, Sabatini DM, Blenis J. TOS motif-mediated raptor binding regulates 4E-BP1 multisite phosphorylation and function. Curr Biol 2003;13:797–806.

[104] Kovacina KS, Park GY, Bae SS, Guzzetta AW, Schaefer E, Birnbaum MJ, et al. Identification of a proline-rich Akt substrate as a 14-3-3 binding partner. J Biol Chem 2003;278:10189–94.

[105] Thedieck K, Polak P, Kim ML, Molle KD, Cohen A, Jenö P, et al. PRAS40 and PRR5-like protein are new mTOR interactors that regulate apoptosis. PloS ONE 2007;2:e1217.

[106] Sancak Y, Thoreen CC, Peterson TR, Lindquist RA, Kang SA, Spooner E, et al. PRAS40 is an insulin-regulated inhibitor of the mTORC1 protein kinase. Mol Cell 2007;25:903–15.

[107] Tee AR, Manning BD, Roux PP, Cantley LC, Blenis J. Tuberous sclerosis complex gene products, tuberin and hamartin, control mTOR signaling by acting as a gtpase-activating protein complex toward rheb. Curr Biol 2003;13:1259–68.

[108] Van Slegtenhorst M, de Hoogt R, Hermans C, Nellist M, Janssen B, Verhoef S, et al. Identification of the tuberous sclerosis gene TSC1 on chromosome 9q34. Science 1997;277:805–8.

[109] Zhang Y, Gao X, Saucedo LJ, Ru B, Edgar BA, Pan D. Rheb is a direct target of the tuberous sclerosis tumour suppressor proteins. Nat Cell Biol 2003;5:578–81.

[110] Ma L, Chen Z, Erdjument-Bromage H, Tempst P, Pandolfi PP. Phosphorylation and functional inactivation of TSC2 by Erk implications for tuberous sclerosis and cancer pathogenesis. Cell 2005;121:179–93.

[111] Corradetti MN, Inoki K, Bardeesy N, DePinho RA, Guan K-L. Regulation of the TSC pathway by LKB1: evidence of a molecular link between tuberous sclerosis complex and Peutz-Jeghers syndrome. Genes Dev 2004;18:1533–8.

[112] Hong S-P, Leiper FC, Woods A, Carling D, Carlson M. Activation of yeast Snf1 and mammalian AMP-activated protein kinase by upstream kinases. Proc Natl Acad Sci USA 2003;100:8839–43.

[113] Chiu MI, Katz H, Berlin V. RAPT1, a mammalian homolog of yeast Tor, interacts with the FKBP12/rapamycin complex. Proc Natl Acad Sci USA 1994;91:12574–8.

[114] Chen J, Zheng XF, Brown EJ, Schreiber SL. Identification of an 11-kDa FKBP12-rapamycin-binding domain within the 289-kDa FKBP12-rapamycin-associated protein and characterization of a critical serine residue. Proc Natl Acad Sci USA 1995;92:4947–51.

[115] Wang X, Fonseca BD, Tang H, Liu R, Elia A, Clemens MJ, et al. Re-evaluating the roles of proposed modulators of mammalian target of rapamycin complex 1 (mTORC1) signaling. J Biol Chem 2008;283:30482–92.

[116] Qi Q-R, Yang Z-M. Regulation and function of signal transducer and activator of transcription 3. World J Biol Chem 2014;5:231–9.

[117] Kim J-H, Yoon M-S, Chen J. Signal transducer and activator of transcription 3 (STAT3) mediates amino acid inhibition of insulin signaling through serine 727 phosphorylation. J Biol Chem 2009;284:35425–32.

[118] Kortylewski M, Jove R, Yu H. Targeting STAT3 affects melanoma on multiple fronts. Cancer Metastasis Rev 2005;24:315–27.

[119] Horton JD, Goldstein JL, Brown MS. SREBPs: activators of the complete program of cholesterol and fatty acid synthesis in the liver. J Clin Invest 2002;109:1125–31.

[120] Peterson TR, Sengupta SS, Harris TE, Carmack AE, Kang SA, Balderas E, et al. mTOR complex 1 regulates lipin 1 localization to control the SREBP pathway. Cell 2011;146:408–20.

[121] Rosen ED, MacDougald OA. Adipocyte differentiation from the inside out. Nat Rev Mol Cell Biol 2006;7:885–96.

[122] Lefebvre P, Chinetti G, Fruchart J-C, Staels B. Sorting out the roles of PPAR alpha in energy metabolism and vascular homeostasis. J Clin Invest 2006;116:571–80.

[123] Sengupta S, Peterson TR, Laplante M, Oh S, Sabatini DM. mTORC1 controls fasting-induced ketogenesis and its modulation by ageing. Nature 2010;468:1100–4.

[124] Blagosklonny MV. TOR-centric view on insulin resistance and diabetic complications: perspective for endocrinologists and gerontologists. Cell Death Dis 2013;4:e964.

[125] Majmundar AJ, Wong WJ, Simon MC. Hypoxia-inducible factors and the response to hypoxic stress. Mol Cell 2010;40:294–309.

[126] Bar-Peled L, Schweitzer LD, Zoncu R, Sabatini DM. Ragulator is a GEF for the rag GTPases that signal amino acid levels to mTORC1. Cell 2012;150:1196–208.

[127] Sancak Y, Peterson TR, Shaul YD, Lindquist RA, Thoreen CC, Bar-Peled L, et al. The Rag GTPases bind raptor and mediate amino acid signaling to mTORC1. Science 2008;320:1496–501.

[128] Zoncu R, Efeyan A, Sabatini DM. mTOR: from growth signal integration to cancer, diabetes and ageing. Nat Rev Mol Cell Biol 2011;12:21–35.

[129] Singh R, Cuervo AM. Autophagy in the cellular energetic balance. Cell Metab 2011;13:495–504.

[130] Settembre C, Zoncu R, Medina DL, Vetrini F, Erdin S, Erdin S, et al. A lysosome-to-nucleus signalling mechanism senses and regulates the lysosome via mTOR and TFEB. EMBO J 2012;31:1095–108.

[131] Dalle Pezze P, Sonntag AG, Thien A, Prentzell MT, Godel M, Fischer S, et al. A dynamic network model of mTOR signaling reveals TSC-independent mTORC2 regulation. Sci Signal 2012;5:ra25.

[132] Zinzalla V, Stracka D, Oppliger W, Hall MN. Activation of mTORC2 by association with the ribosome. Cell 2011;144:757–68.

[133] Destefano MA, Jacinto E. Regulation of insulin receptor substrate-1 by mTORC2 (mammalian target of rapamycin complex 2). Biochem Soc Trans 2013;41:896–901.

[134] Yu Y, Yoon S-O, Poulogiannis G, Yang Q, Ma XM, Villen J, et al. Phosphoproteomic analysis identifies Grb10 as an mTORC1 substrate that negatively regulates insulin signaling. Science 2011;332:1322–6.

III. mTOR IN MEMORY, BEHAVIOR, AND AGING

[135] Guertin DA, Stevens DM, Thoreen CC, Burds AA, Kalaany NY, Moffat J, et al. Ablation in mice of the mTORC components raptor, rictor, or mLST8 reveals that mTORC2 is required for signaling to Akt-FOXO and PKCalpha, but not S6K1. Dev Cell 2006;11:859–71.

[136] Liu P, Guo J, Gan W, Wei W. Dual phosphorylation of Sin1 at T86 and T398 negatively regulates mTORC2 complex integrity and activity. Protein Cell 2014;5:171–7.

[137] Rosner M, Hengstschläger M. Cytoplasmic and nuclear distribution of the protein complexes mTORC1 and mTORC2: rapamycin triggers dephosphorylation and delocalization of the mTORC2 components rictor and sin1. Hum Mol Genet 2008;17:2934–48.

[138] Huang J, Dibble CC, Matsuzaki M, Manning BD. The TSC1-TSC2 complex is required for proper activation of mTOR complex 2. Mol Cell Biol 2008;28:4104–15.

[139] Oh WJ, Wu C, Kim SJ, Facchinetti V, Julien L-A, Finlan M, et al. mTORC2 can associate with ribosomes to promote cotranslational phosphorylation and stability of nascent Akt polypeptide. EMBO J 2010;29:3939–51.

[140] Betz C, Stracka D, Prescianotto-Baschong C, Frieden M, Demaurex N, Hall MN. Feature article: mTOR complex 2-Akt signaling at mitochondria-associated endoplasmic reticulum membranes (MAM) regulates mitochondrial physiology. Proc Natl Acad Sci USA 2013;110:12526–34.

[141] Balcazar N, Sathyamurthy A, Elghazi L, Gould A, Weiss A, Shiojima I, et al. mTORC1 activation regulates beta-cell mass and proliferation by modulation of cyclin D2 synthesis and stability. J Biol Chem 2009;284:7832–42.

[142] Cnop M, Welsh N, Jonas J-C, Jorns A, Lenzen S, Eizirik DL. Mechanisms of pancreatic-cell death in type 1 and type 2 diabetes: many differences, few similarities. Diabetes 2005;54: S97–107.

[143] Maddison LA, Chen W. Nutrient excess stimulates β-cell neogenesis in zebrafish. Diabetes 2012;61:2517–24.

[144] Donato AJ, Morgan RG, Walker AE, Lesniewski LA. Cellular and molecular biology of aging endothelial cells. J Mol Cell Cardiol 2015.

[145] Blagosklonny MV. Molecular damage in cancer: an argument for mTOR-driven aging. Aging 2011;3:1130–41.

[146] Mabuchi S, Kuroda H, Takahashi R, Sasano T. The PI3K/ AKT/mTOR pathway as a therapeutic target in ovarian cancer. Gynecol Oncol 2015;137:173–9.

[147] Zhou H, Huang S. mTOR signaling in cancer cell motility and tumor metastasis. Crit Rev Eukaryot Gene Expr 2010;20:1–16.

[148] Feng Z, Zhang H, Levine AJ, Jin S. The coordinate regulation of the p53 and mTOR pathways in cells. Proc Natl Acad Sci USA 2005;102:8204–9.

[149] Levine AJ, Feng Z, Mak TW, You H, Jin S. Coordination and communication between the p53 and IGF-1-AKT-TOR signal transduction pathways. Genes Dev 2006;20:267–75.

[150] Obata K, Morland SJ, Watson RH, Hitchcock A, Chenevix-Trench G, Thomas EJ, et al. Frequent PTEN/MMAC mutations in endometrioid but not serous or mucinous epithelial ovarian tumors. Cancer Res 1998;58:2095–7.

[151] Levine DA, Bogomolniy F, Yee CJ, Lash A, Barakat RR, Borgen PI, et al. Frequent mutation of the PIK3CA gene in ovarian and breast cancers. Clin Cancer Res 2005;11:2875–8.

[152] Philp AJ, Campbell IG, Leet C, Vincan E, Rockman SP, Whitehead RH, et al. The phosphatidylinositol 3′-kinase p85alpha gene is an oncogene in human ovarian and colon tumors. Cancer Res 2001;61:7426–9.

[153] Carpten JD, Faber AL, Horn C, Donoho GP, Briggs SL, Robbins CM, et al. A transforming mutation in the pleckstrin homology domain of AKT1 in cancer. Nature 2007;448:439–44.

[154] Pause A, Belsham GJ, Gingras AC, Donzé O, Lin TA, Lawrence JC, et al. Insulin-dependent stimulation of protein synthesis by phosphorylation of a regulator of 5′-cap function. Nature 1994;371:762–7.

[155] Hay N, Sonenberg N. Upstream and downstream of mTOR. Genes Dev 2004;18:1926–45.

[156] Gingras AC, Raught B, Gygi SP, Niedzwiecka A, Miron M, Burley SK, et al. Hierarchical phosphorylation of the translation inhibitor 4E-BP1. Genes Dev 2001;15:2852–64.

[157] Vivanco I, Sawyers CL. The phosphatidylinositol 3-Kinase AKT pathway in human cancer. Nat Rev Cancer 2002;2:489–501.

[158] Jacinto E, Loewith R, Schmidt A, Lin S, Rüegg MA, Hall A, et al. Mammalian TOR complex 2 controls the actin cytoskeleton and is rapamycin insensitive. Nat Cell Biol 2004;6:1122–8.

[159] Masri J, Bernath A, Martin J, Jo OD, Vartanian R, Funk A, et al. mTORC2 activity is elevated in gliomas and promotes growth and cell motility via overexpression of rictor. Cancer Res 2007;67:11712–20.

[160] Hietakangas V, Cohen SM. TOR complex 2 is needed for cell cycle progression and anchorage-independent growth of MCF7 and PC3 tumor cells. BMC Cancer 2008;8:282.

[161] Inoki K, Corradetti MN, Guan K-L. Dysregulation of the TSC-mTOR pathway in human disease. Nat Genet 2005;37:19–24.

[162] Hoeffer CA, Klann E. mTOR signaling: at the crossroads of plasticity, memory and disease. Trends Neurosci 2010;33:67–75.

[163] Slipczuk L, Bekinschtein P, Katche C, Cammarota M, Izquierdo I, Medina JH. BDNF activates mTOR to regulate GluR1 expression required for memory formation. PLoS One 2009;4:e6007.

[164] Majumder S, Caccamo A, Medina DX, Benavides AD, Javors MA, Kraig E, et al. Lifelong rapamycin administration ameliorates age-dependent cognitive deficits by reducing IL-1β and enhancing NMDA signaling. Aging Cell 2012;11:326–35.

[165] Halloran J, Hussong SA, Burbank R, Podlutskaya N, Fischer KE, Sloane LB, et al. Chronic inhibition of mammalian target of rapamycin by rapamycin modulates cognitive and non-cognitive components of behavior throughout lifespan in mice. Neuroscience 2012;223:102–13.

[166] Costa V, Lugert S, Jagasia R. Role of adult hippocampal neurogenesis in cognition in physiology and disease: pharmacological targets and biomarkers. Handb Exp Pharmacol 2015;228:99–155.

[167] Gage FH. Stem cells of the central nervous system. Curr Opin Neurobiol 1998;8:671–6.

[168] Couillard-Despres S, Iglseder B, Aigner L. Neurogenesis, cellular plasticity and cognition: the impact of stem cells in the adult and aging brain—a mini-review. Gerontology 2011;57: 559–64.

[169] Kempermann G, Kuhn HG, Gage FH. Genetic influence on neurogenesis in the dentate gyrus of adult mice. Proc Natl Acad Sci USA 1997;94:10409–14.

[170] Lazarov O, Mattson MP, Peterson DA, Pimplikar SW, van Praag H. When neurogenesis encounters aging and disease. Trends Neurosci 2010;33:569–79.

[171] Lugert S, Basak O, Knuckles P, Haussler U, Fabel K, Götz M, et al. Quiescent and active hippocampal neural stem cells with distinct morphologies respond selectively to physiological and pathological stimuli and aging. Cell Stem Cell 2010;6:445–56.

[172] Romine J, Gao X, Xu X-M, So KF, Chen J. The proliferation of amplifying neural progenitor cells is impaired in the aging brain and restored by the mTOR pathway activation. Neurobiol Aging 2015;36:1716–26.

[173] Jankovic J. Parkinson's disease: clinical features and diagnosis. J Neurol Neurosurg Psychiatry 2008;79:368–76.

[174] Hetz C, Mollereau B. Disturbance of endoplasmic reticulum proteostasis in neurodegenerative diseases. Nat Rev Neurosci 2014;15:233–49.

[175] Guo C, Sun L, Chen X, Zhang D. Oxidative stress, mitochondrial damage and neurodegenerative diseases. Neural Regen Res 2013;8:2003–14.

[176] Brasnjevic I, Hof PR, Steinbusch HWM, Schmitz C. Accumulation of nuclear DNA damage or neuron loss: molecular basis for a new approach to understanding selective neuronal vulnerability in neurodegenerative diseases. DNA Repair (Amst) 2008;7:1087–97.

[177] Tain LS, Mortiboys H, Tao RN, Ziviani E, Bandmann O, Whitworth AJ. Rapamycin activation of 4E-BP prevents parkinsonian dopaminergic neuron loss. Nat Neurosci 2009;12:1129–35.

[178] Xu Y, Liu C, Chen S, Ye Y, Guo M, Ren Q, et al. Activation of AMPK and inactivation of Akt result in suppression of mTOR-mediated S6K1 and 4E-BP1 pathways leading to neuronal cell death in in vitro models of Parkinson's disease. Cell Signal 2014;26:1680–9.

[179] Malagelada C, Jin ZH, Jackson-Lewis V, Przedborski S, Greene LA. Rapamycin protects against neuron death in in vitro and in vivo models of Parkinson's disease. J Neurosci 2010;30:1166–75.

[180] McKhann G, Drachman D, Folstein M, Katzman R, Price D, Stadlan EM. Clinical diagnosis of Alzheimer's disease: report of the NINCDS-ADRDA Work Group under the auspices of Department of Health and Human Services Task Force on Alzheimer's Disease. Neurology 1984;34:939–44.

[181] Crews L, Masliah E. Molecular mechanisms of neurodegeneration in Alzheimer's disease. Hum Mol Genet 2010;19:R12–20.

[182] Cuervo AM, Dice JF. Age-related decline in chaperone-mediated autophagy. J Biol Chem 2000;275:31505–13.

[183] Spilman P, Podlutskaya N, Hart MJ, Debnath J, Gorostiza O, Bredesen D, et al. Inhibition of mTOR by rapamycin abolishes cognitive deficits and reduces amyloid-β levels in a mouse model of Alzheimer's disease. PLoS One 2010;5:e9979.

[184] Majumder S, Richardson A, Strong R, Oddo S. Inducing autophagy by rapamycin before, but not after, the formation of plaques and tangles ameliorates cognitive deficits. PLoS One 2011;6:e25416.

[185] Boxer AL, Miller BL. Clinical features of frontotemporal dementia. Alzheimer Dis Assoc Disord 2005;19(Suppl. 1):S3–6.

[186] Chen-Plotkin AS, Lee VM-Y, Trojanowski JQ. TAR DNA-binding protein 43 in neurodegenerative disease. Nat Rev Neurol 2010;6:211–20.

[187] Wang I-F, Guo B-S, Liu Y-C, Wu C-C, Yang C-H, Tsai K-J, et al. Autophagy activators rescue and alleviate pathogenesis of a mouse model with proteinopathies of the TAR DNA-binding protein 43. Proc Natl Acad Sci USA 2012;109:15024–9.

[188] Trushina E, McMurray CT. Oxidative stress and mitochondrial dysfunction in neurodegenerative diseases. Neuroscience 2007;145:1233–48.

[189] Sepers MD, Raymond LA. Mechanisms of synaptic dysfunction and excitotoxicity in Huntington's disease. Drug Discov Today 2014;19:990–6.

[190] Floto RA, Sarkar S, Perlstein EO, Kampmann B, Schreiber SL, Rubinsztein DC. Erratum: small molecule enhancers of rapamycin-induced TOR inhibition promote autophagy, reduce toxicity in Huntington's disease models and enhance killing of mycobacteria by macrophages (autophagy). Autophagy 2007;3:620–2.

[191] Ravikumar B, Vacher C, Berger Z, Davies JE, Luo S, Oroz LG, et al. Inhibition of mTOR induces autophagy and reduces toxicity of polyglutamine expansions in fly and mouse models of Huntington disease. Nat Genet 2004;36:585–95.

[192] Harrison DE, Strong R, Sharp ZD, Nelson JF, Astle CM, Flurkey K, et al. Rapamycin fed late in life extends lifespan in genetically heterogeneous mice. Nature 2009;460:392–5.

[193] Villa-Cuesta E, Holmbeck MA, Rand DM. Rapamycin increases mitochondrial efficiency by mtDNA-dependent reprogramming of mitochondrial metabolism in Drosophila. J Cell Sci 2014;127:2282–90.

[194] Bonawitz ND, Chatenay-Lapointe M, Pan Y, Shadel GS. Reduced TOR signaling extends chronological life span via increased respiration and upregulation of mitochondrial gene expression. Cell Metab 2007;5:265–77.

[195] Pan Y, Nishida Y, Wang M, Verdin E. Metabolic regulation, mitochondria and the life-prolonging effect of rapamycin: a mini-review. Gerontology 2012;58:524–30.

[196] Kolosova NG, Muraleva NA, Zhdankina AA, Stefanova NA, Fursova AZ, Blagosklonny MV. Prevention of age-related macular degeneration-like retinopathy by rapamycin in rats. Am J Pathol 2012;181:472–7.

[197] Yang S-B, Tien A-C, Boddupalli G, Xu AW, Jan YN, Jan LY. Rapamycin ameliorates age-dependent obesity associated with increased mTOR signaling in hypothalamic POMC neurons. Neuron 2012;75:425–36.

[198] Lloberas N, Cruzado JM, Franquesa M, Herrero-Fresneda I, Torras J, Alperovich G, et al. Mammalian target of rapamycin pathway blockade slows progression of diabetic kidney disease in rats. J Am Soc Nephrol 2006;17:1395–404.

[199] Keogh A, Richardson M, Ruygrok P, Spratt P, Galbraith A, O'Driscoll G, et al. Sirolimus in de novo heart transplant recipients reduces acute rejection and prevents coronary artery disease at 2 years: a randomized clinical trial. Circulation 2004;110:2694–700.

[200] Blagosklonny MV. Rapalogs in cancer prevention: anti-aging or anticancer? Cancer Biol Ther 2012;13:1349–54.

[201] Garber K. Rapamycin's resurrection: a new way to target the cancer cell cycle. J Natl Cancer Inst 2001;93:1517–19.

[202] Bjornsti M-A, Houghton PJ. The TOR pathway: a target for cancer therapy. Nat Rev Cancer 2004;4:335–48.

[203] Kwitkowski VE, Prowell TM, Ibrahim A, Farrell AT, Justice R, Mitchell SS, et al. FDA approval summary: temsirolimus as treatment for advanced renal cell carcinoma. Oncologist 2010;15:428–35.

[204] Hudes Carducci M, Tomczak P, Dutcher J, Figlin R, Kapoor A, Staroslawska E, The Global ARCC, G, et al. Temsirolimus, interferon alpha or both for advanced renal cell carcinoma. N Engl J Med 2007;356:2271–81.

[205] Mabuchi S, Altomare DA, Connolly DC, Klein-Szanto A, Litwin S, Hoelzle MK, et al. RAD001 (everolimus) delays tumor onset and progression in a transgenic mouse model of ovarian cancer. Cancer Res 2007;67:2408–13.

[206] Ning YC, Cai GY, Zhuo L, Gao JJ, Dong D, Cui S, et al. Short-term calorie restriction protects against renal senescence of aged rats by increasing autophagic activity and reducing oxidative damage. Mech Ageing Dev 2013;134:570–9.

[207] Cuervo AM, Bergamini E, Brunk UT, Dröge W, Ffrench M, Terman A. Autophagy and aging: the importance of maintaining "clean" cells. Autophagy 2005;1:131–40.

[208] Levine B, Klionsky DJ. Development by self-digestion: molecular mechanisms and biological functions of autophagy. Dev Cell 2004;6:463–77.

[209] Rajawat YS, Hilioti Z, Bossis I. Aging: central role for autophagy and the lysosomal degradative system. Ageing Res Rev 2009;8:199–213.

[210] Estep PW, Warner JB, Bulyk ML. Short-term calorie restriction in male mice feminizes gene expression and alters key regulators of conserved aging regulatory pathways. PLoS One 2009;4:e5242.

12

The Role of mTOR in Mood Disorders Pathophysiology and Treatment

Gislaine Z. Réus[1,2], Meagan R. Pitcher[1], Camila O. Arent[2] and João Quevedo[1,2]

[1]Center for Translational Psychiatry, Department of Psychiatry and Behavioral Sciences, Medical School, The University of Texas Health Science Center at Houston, Houston, TX, USA [2]Laboratório de Neurociências, Programa de Pós-Graduação em Ciências da Saúde, Unidade Acadêmica de Ciências da Saúde, Universidade do Extremo Sul Catarinense, Criciúma, SC, Brazil

12.1 INTRODUCTION

Bipolar disorder (BD) is a severe mood disorder characterized by recurrent episodes of mania followed by depression. Although the clinical characteristic for the diagnosis of BD is the presence of manic symptoms, depression represents the predominant mood state in people with BD type I and BD type II. Moreover, BD is associated with high morbidity, mortality, and risk of suicide [1]. It has been shown that people suffering with BD die 10–20 years earlier than the general population [2]. Though a large selection of treatment options, including lithium, antipsychotics, anticonvulsants, and psychotherapies, have been used [3–6] around 45% of people with BD do not respond to treatment. Furthermore, negative outcomes continue to occur among individuals suffering from BD [7,8].

Major depressive disorder (MDD) is a serious public health problem and one of the most common psychiatric disorders, with a lifetime prevalence of 17% in the United States and an estimated that 121 million people affected worldwide [9,10]. Almost one million lives are lost yearly due to suicide, which translates to 3000 suicide deaths every day [11]. Moreover, people suffering from severe depression have high rates of morbidity with profound economic and social consequences [12]. Although the treatment of depression is generally safe and effective, it is far from ideal. For example, standard antidepressants usually require approximately 1 month or more for antidepressant effects to manifest [13]. It is important to note that the high risk of suicide and other deliberate acts of self-harm during the first month of treatment is not uncommon, and when it does occur has been postulated to be due to a mismatch in symptom improvement. That is, physical energy improves first, while resolution of depressive mood and negative thoughts is more gradual. For these reasons, more studies investigating the neurobiology of depression, as well as new treatments with more effective and faster time to action are needed.

Despite significant progress in mood disorder research over the past decades, pathophysiological mechanisms remain largely unknown, thereby limiting therapeutic development. The complexity and heterogeneity of mood disorders involve multiple causes, such as inflammation, genetic associations, and environmental factors. This heterogeneity could be related to the poor effectiveness, variability of response to antidepressant and mood stabilizer drugs, delay in clinical response, and side effects in people being treated for mood disorders.

12.2 MAMMALIAN TARGET OF RAPAMYCIN (mTOR) IN BD PATHOPHYSIOLOGY AND TREATMENT

People with BD experience recurrent affective episodes of depression and mania. BD I features

Molecules to Medicine with mTOR
DOI: http://dx.doi.org/10.1016/B978-0-12-802733-2.00015-3

depression and mania mood states, while BD II features depression and hypomania mood states. People with BD face high morbidity and mortality and high rates of suicidality [14]. A roadblock to therapeutic development in BD is the lack of understanding of the underlying pathophysiologic pathways. The standard-of-care treatment for BD is lithium. Although lithium is known to have activity in several BD-relevant pathways, the mechanism by which it improves BD symptoms remains unknown [15]. Exploration of mammalian Target of Rapamycin (mTOR) pathway involvement in BD could open new opportunities for drug development.

Three independent studies implicate the 17q25.3 chromosomal region in the heritability of BD [16–18]. A detailed analysis of this region revealed associations between single nucleotide polymorphisms (SNPs) in this region and BD. In this region there were 62 Raptor (RPTOR) SNPs associated with BD [19]. RPTOR is the component of the mTORC1 protein complex that makes is responsive to rapamycin. An intronic RPTOR SNP had the strongest odds ratio association with BD [19]. This study indicates that there may be a genetic association between mTORC1, or mTORC1 regulation, and BD [19] (Table 12.1). Gene variants for protein kinase B (AKT1), an interactor in the mTOR pathway, also appear to be associated with BD diagnosis in two human studies [26,27].

Na^+/K^+ APTase dysregulation has been investigated in mouse and rat models for mania behavior [28–30] and linked to BD in humans by several lines of evidence. In humans with BD there is reduction in Na^+/K^+ APTase expression in the postmortem brain [31,32], abnormal regulation of endogenous Na^+/K^+ APTase inhibition molecules [33–37] and associations with Na^+/K^+ APTase allele variants [38]. In one rodent model of BD mania, investigators can induce mania-like behavior in rats by injecting the Na^+/K^+ APTase inhibitor ouabain directly into the brain [23,24]. There is apparent mTOR involvement in two studies utilizing the rat brain Na^+/K^+ APTase inhibition model. Rats with Na^+/K^+ APTase inhibition have increased mTOR and AKT1 activation in the brain during the time they exhibit mania-like behavior [23,24] (Table 12.1).

A 2015 study performed by Machado-Vieira and colleagues examined gene expression of mTOR and AKT1 in non-medicated BD depression subjects who were early in their illness course [39] (Table 12.2). The subject group consisted of 25 BD I and II individuals experiencing a depressive episode, by Hamilton Depression Rating Score (HAMD), who were within 5 years of their first affective episode [39]. Gene expression was assessed from RNA isolated from blood samples in a qualitative real-time polymerase chain reaction [39]. Compared to matched healthy controls, the BD subjects had decreased expression of AKT1 and mTOR [39]. After a course of lithium treatment, AKT1 and mTOR remained unchanged from the untreated state in the BD subjects [39]. Because blood samples were used, the effect of lithium treatment on mTOR activity in the brain may not be represented by this result. Further, because the study only examined subjects early in the course of disease, it remains unknown if mTOR activity has a relationship with BD disease progression or onset.

The consolidated results of the rodent and human studies of mTOR expression level may indicate that mTOR activity changes congruent to mood state. In rodents, mania is associated with increased mTOR activity, whereas in humans BD depression is associated with decreased mTOR activity. However, a direct association between the two mood poles and mTOR

TABLE 12.1 Summary of mTOR Alterations Involved with Pathophysiology of Mood Disorders

Subject	Alteration	References
MDD—human	Altered mTOR signaling, and reduced mTOR expression	Feyissa et al. [20]; Jernigan et al. [21]
MDD—animal	Activated/phosphorylated mTOR signaling	Chandran et al. [22]
BD—human	Genetic and regulation of mTORC1; decreased expression of mTOR	Rajkumar et al. [19]
BD—animal	Na^+/K^+ APTase inhibition increases mTOR	Kim et al. [23]; Yu et al. [24]; Machado-Vieira et al. [25]

TABLE 12.2 Summary of mTOR Alterations Involved with Treatment of Mood Disorders

Subject	Alteration	References
MDD—human	Ketamine antidepressant effect is mediated by mTOR activation	Hashimoto [40]
MDD—animal model	Ketamine; scopolamine, imipramine, fluoxetine, ECT, and sleep deprivation antidepressant effects it seems to be mediated by mTOR	Li et al. [41]; Monteggia and Kavalali [42]; Jeon et al. [43]; Warren et al. [44]; Elfving and Wegener [45]; Vecsey et al. [46]
BD—human	Lithium treatment did not alter mTOR	Machado-Vieira et al. [25]
BD—animal model	Lithium increases mTORC1 activity	Inoki et al. [47]

Note: BD, bipolar disorder; ECT, electroconvulsive therapy; MDD, major depressive disorder; mTOR, mammalian target of rapamycin.

has not yet been shown in humans. It would be interesting if mTOR pathway activity unified the opposing poles of BD behavior. This would require a study of mTOR pathway protein expression, preferably from postmortem human brains, of subjects in mania, depression, and euthymic mood states.

Lithium is the standard of care for mania in BD, has an excellent ability to control symptoms, and reduces mortality and morbidity [48,49]. How lithium works to improve BD is unknown but several hypotheses have been proposed [15]. In 1996, Klein and Melton described a novel function of lithium as a potent and specific inhibitor of glycogen synthase kinase 3-beta (GSK-3β) activity [50]. Since this discovery, GSK-3β has been explored as a potentially critical protein integrating BD pathophysiology [25,51,52]. In 2006, Inoki and colleagues observed that treating HEK293 cells with lithium lead to increased ribosomal S6 kinase 1 (S6K) phosphorylation in a dose-dependent manner [47]. In that study, lithium had no significant effect on Akt phosphorylation [47]. S6K lies downstream of mTORC1 in the phosphorylation cascade and the observation that lithium increases mTORC1 activity supports the notion that GSK-3β is an inhibitor of mTOR signaling [47] (Table 12.2). It is interesting to note that mania behavior in the above described Na^+/K^+ APTase inhibition rat model of BD is alleviated by lithium treatment [53]. However, a direct relationship between lithium and mTOR activity as it pertains to improvement in BD symptoms has not been formally explored.

In the last two decades, a considerable number of studies have shown a central role of the glutamatergic system in mood disorders, including MDD and BD [54–57]. Glutamate is the principal mediator of excitatory synaptic transmission in the central nervous system of mammalian. Under normal conditions, glutamate plays a prominent role in synaptic plasticity, learning, and memory. Also it is involved in neural development as well as the phases of cell proliferation and migration [58]. However, under pathological conditions glutamate could lead to excitotoxicity, which in turn could lead to cell death [59]. Glutamate acts on specific receptors located on the surface of the cell membrane, classified according to their pharmacological and functional properties [60]. Its action occurs in three different cellular compartments: presynaptic neurons, postsynaptic neurons, and glia. The function of glutamate is exerted through two types of receptors: ionotropic (iGluR), including N-methyl-D-aspartate (NMDA), α-amino-3-hydroxy-5-methyl-4-isoxazolepropionic acid (AMPA), and kainate, and metabotropic (mGluR). Interestingly, patients with MDD presented a significant increase in serum levels of glutamate compared with healthy controls [61,62]. In addition, there was a related increase in glutamate levels in the postmortem frontal cortex of patients with MDD and BD [63], and abnormalities in the levels of glutamate reuptake proteins in the postmortem prefrontal cortex (PFC) of patients with BD [64].

Besides glutamate, its receptors seem to play a key role in the pathophysiology and treatment of mood disorders. For instance, there were reported changes in binding properties of the NMDA receptor in the brain of suicide victims [65], and alterations on gene expression of NMDA subunits R1 and R2B were correlated to be a risk for BD [66,67]. Furthermore, several preclinical and clinical studies have demonstrated that NMDA receptor antagonists, for example ketamine, have presented antidepressant properties. In this regard, rapid actions of ketamine (0.5 mg/kg) were observed in patients who were resistant to two or more classes of standard antidepressants [68,69] (Table 12.2). Still, previous preclinical studies have shown that ketamine presents antidepressant effects in rodents submitted to stress, including maternal deprivation and chronic mild stress [70–72].

A better understood role for mTOR in BD treatment is the use of ketamine as a treatment in BD depression. In this context, the hypothesis is that BD depression will respond to ketamine treatment due to activation of the mTOR pathway, as observed in cases of unipolar depression. This hypothesis assumes that both unipolar and BD depression have a common underlying mechanism. In three clinical studies, people with BD I or BD II in a depressive episode were treated with ketamine and experienced significant improvements in depression symptoms [73–75]. In the first study, performed by Diazgranados and colleagues, people with fourth edition of the Diagnostic and Statistical Manual of Mental Disorders (DSM-IV) BD I or II were rated on the Montgomery—Asberg Depression Rating Scale (MADRS) at baseline [73]. Subjects then received intravenous 0.5 mg/kg ketamine hydrochloride or saline and in a randomized, placebo-controlled crossover study [73]. A response to ketamine was defined as a 50% improvement from baseline MADRS score [54]. Twelve of seventeen BD subjects who received ketamine experienced an improvement in MADRS score at some point during the drug trial [73]. Summary data demonstrated that subjects treated with ketamine had improvement in depression symptoms at 40 min after treatment, sustainable for up to 3 days after treatment, with three subjects experiencing improvement that lasted 2 weeks [73]. The Diazgranados BD depression study result was later confirmed in a replication study in an identical-design randomized, placebo-controlled crossover study with a new independent sample of fifteen subjects [74]. As observed previously, subjects treated with ketamine had improvement in MADRS depression score at 40 min after treatment that was sustained at 3 days after treatment [74]. In the replication study, subjects treated with ketamine also had reduced suicide ideation, as measured by a linear mixed model of suicide item scores from the MADRS, HAMD, and Beck Depression Inventory [74]. A final clinical study examined sublingual dosing of ketamine in a non-systematic case series design [75]. A weakness of this study is that a clinical questionnaire was not used as an endpoint, instead subjects rated their present mood, 90 min after ketamine dosing, on a scale from one to 10 [75]. Additionally, there was inconsistency in the frequency and duration of ketamine dosing among subjects [75]. Eleven people with DSM-IV BD I or II were given various doses of 10 mg racemic ketamine as sublingual drops [75]. Eight of 11 subjects reported a positive response to ketamine after the first dose [75].

Each of the above described studies of ketamine utility for BD features BD subject groups consisting of less than 20 members and has subject groups containing both BD I and BD II diagnoses. The current evidence for ketamine as a treatment for BD depression is promising, but weakly supported. More studies, with more powerful or prospective design, are required to determine if mTOR activation by ketamine may be a useful therapeutic strategy in BD depression. Given the observation that mTOR expression is decreased in peripheral tissues in one previously described BD study [39], this line of study may be fruitful. Indeed, in future clinical trials peripheral blood may be used alongside mood questionnaire outcomes to determine if ketamine treatment is efficacious.

12.3 mTOR IN MDD PATHOPHYSIOLOGY AND TREATMENT

Although the symptoms of MDD are well characterized, the neurobiology of this disorder is still largely unknown. MDD can be considered a polygenic disease, which is determined by environmental influences in genetically predisposed individuals [76]. During the progress of MDD molecular, cellular, structural, and functional changes are observed in the brain of affected individuals. In fact, there is evidence for the involvement of neurotrophic factors, such as brain-derived neurotrophic factor (BDNF) [77], inflammatory cytokines [78], hypothalamus—pituitary—adrenal axis activation [79], and increased glutamate levels in MDD. More recently, the mTOR pathway has been reported to be important in the pathophysiology of MDD [80].

Neuronal mTOR function is influenced by activity of growth factors, cytokines, and glutamate activity through NMDA and metabotropic receptors (mGluR),

which have been reported to be involved with MDD. In addition, mTOR is associated with local protein synthesis and formation of new synapses [81]. Postmortem studies have been shown a dysregulation in the mTOR signaling as well as a dysfunction in the glutamatergic system in the PFC of depressed individuals [20,82,83] (Table 12.1). mTOR dysregulation in people with MDD is related to a reduction in p70 ribosomal protein S6 kinase (p70S6K), which induces phosphorylation of eukaryotic translation initiation factor 4B and is involved with the initiation of protein translation [21].

In PFC of human postmortem samples from MDD subjects, a decrease in the expression of mTOR and downstream targets for this pathway was observed, suggesting that impairment in the mTOR pathway in MDD could lead to a reduction in protein translation [21,84]. In a placebo-controlled clinical study one group demonstrated that a single subanesthetic dose of ketamine, caused a rapid antidepressant effect within hours in treatment-resistant people with MDD. This study suggested that the ketamine effects were mediated by the mTOR signaling pathway [40]. In fact, Miller et al. [85] also demonstrated that ketamine antidepressant effects are mediated through mTOR activation. In that study the authors showed that genetic deletion of GluN2B, a subunit of NMDA receptor, from principal cortical neurons occluded the effects of ketamine in suppression of depression-like behavior [85]. In addition, ketamine induced a rapid increase in mTOR phosphorylation, which is abolished in GluN2B knockout animals [85]. Ketamine antidepressant effects it seems to be mediated by other pathways that are related to mTOR. For example, a study from our group demonstrated that ketamine treatment was able to reverse an increase in pro-inflammatory cytokines, such as TNF-α, IL-1 and IL-6 in the serum and cerebrospinal fluid, and depressive-like behavior induced by maternal care deprivation in rats [86]. Furthermore, ketamine treatment reduced oxidative damage and increased antioxidant enzymes in the rat brain induced also by maternal care deprivation [87].

Although the association between marked deficits in synaptic proteins and dysregulation of mTOR signaling in MDD is rather new and studies are limited, there are more studies that detected that mTOR signaling pathway has an antidepressant response to various drugs. This effect seems to be more associated with antidepressant NMDA receptor antagonists, such as ketamine. In fact, ketamine stimulates mTOR signaling in the PFC, induces synaptogenesis, and increases BDNF levels in the hippocampus [88—91]. Meanwhile, infusion of rapamycin, an inhibitor of the mTORC1 complex, blocks the antidepressant-like effects of ketamine [91]. The antidepressant effects of ketamine are mediated by NMDA receptor antagonism leading to an increase in synaptic signaling proteins and increased number and function of new spine synapses in the PFC of rats [92]. Moreover, it has been demonstrated that a single dose of ketamine rapidly reversed the chronic stress-induced behavioral and synaptic deficits in an mTOR-dependent manner [41] (Table 12.2). Ketamine stimulates mTOR synaptogenesis and glutamate transmission resulting in an activation of AKT and extracellular signal-regulated kinases signaling, also increasing cAMP response element-binding protein, BDNF and its receptor tropomyosin receptor kinase B levels [93]. It has been also reported in the current literature that there is a reduction in PI3K-Akt-mTOR signaling pathway components in the PFC and amygdala of stressed rats [22,94,95] (Table 12.1). In another study the rat PFC inhibition of calcineurin, a protein that contributes to the regulation of neurotransmission, affected neuronal structure and plasticity, neuronal excitability, and induced depression-like behavior, followed by decrease in mTOR activity. These effects were reversed by the activation of mTOR by NMDA, promoting antidepressant effects [96,97]. It has been demonstrated in rodents that the subchronic administration of rapamycin exhibits an antidepressant-like effect in classical tests of antidepressant responses, such as forced swimming and tail suspension tests. In addition, acute administration of different NMDA receptor antagonists or antagonists of group II of the metabotropic glutamate receptors produce a fast antidepressant effect mediated by mTOR signaling pathway activation [92,98–101].

Recent findings have demonstrated that scopolamine produces rapid antidepressant effects in individuals with depression [102]. Scopolamine is a centrally acting competitive inhibitor of the muscarinic cholinergic receptor site [103,104] and blocks all five receptor subtypes with high affinity. These receptors are located at presynaptic and postsynaptic sites in cholinergic neurons, as well as glutamatergic synapses. Postsynaptic M1 receptors are reported to regulate long-term depression through enhanced internalization of AMPA and/or NMDA receptors [105]. Also, by blocking these receptors, scopolamine inhibits this process and thereby enhances synaptic plasticity and synaptogenesis [106]. Antidepressant mechanisms by which scopolamine acts are currently described as similar to ketamine effects, dependent on mTORC1 signaling and glutamate AMPA transmission [42]. In fact, a recent study by Voleti and colleagues [107] showed that pretreatment with rapamycin or the AMPA receptor antagonist NBQX completely blocked the antidepressant effects of scopolamine in the forced swimming test. These actions were correlated with the number and function of spine synapses in layer V pyramidal neurons in the PFC. Moreover, using microdialysis experiments, the authors also demonstrated that scopolamine rapidly increases extracellular glutamate levels in the medial PFC [107]. Thus, these results suggest an increase in glutamatergic synaptic plasticity possibly mediated by mTOR activation as the underlying cellular substrate for scopolamine action.

Classical antidepressants effects also appear to be mediated by the mTOR signaling pathway [22]. Jeon and colleagues [43] demonstrated that administration of imipramine, a tricyclic antidepressant, inhibited PI3K/AKT/mTOR signaling. Warren and colleagues [44] demonstrated that fluoxetine, an inhibitor of serotonin reuptake, in combination with the psychostimulant methylphenidate, induced an increase in mTOR activity. Thus, further studies are needed to characterize the mTOR signaling pathway in depression and its participation in the mechanisms of antidepressants. A recent study demonstrated that the antidepressant-like effect of ascorbic acid, an essential nutrient in human diets, is dependent on PI3K and mTOR activation and GSK-3β inhibition [108]. Interestingly, Park et al. [91] in an in vitro study using cultured hippocampal neurons that antidepressants, paroxetine, ketamine, tranylcypromine and paroxetine activated mTOR signaling and increased synaptic protein levels and dendritic outgrowth mediated by mTOR signaling. On the other hand, the antidepressant imipramine, fluoxetine and sertraline effects were not mediated by mTOR pathway in the Park study [91]. However, a study with rodents demonstrated that fluoxetine treatment in the postnatal period in combination with sodium butyrate, a histone deacetylase (HDAC) inhibitor prevented a dysregulation of HDAC4 and mTOR as well as anxiety- and depression-like behavior [109]. The authors also demonstrated that treatment with fluoxetine in rodents in adulthood that received fluoxetine in postnatal period was able to reverse the enhanced HDAC4 expression and improve mTOR signaling and associated genes [109].

Electroconvulsive therapy is considered effective of MDD treatment, and this antidepressant strategy is seems to modulate the mTOR signaling pathway. In fact, a study demonstrated that after repeated electroconvulsive seizures sessions, the mTOR pathway was activated in the frontal cortex and the hippocampus of rats, leading to an increase in vascular endothelial growth factor [45].

Recently, it has been proposed that sleep deprivation may represent a powerful non-pharmacological antidepressant treatment that rapidly alleviates depressive symptoms on a night of total sleep deprivation, which is effective in ~60% of depressed people [110]. A study showed that the effects of 5 h sleep deprivation may influence mTORC1 signaling in the

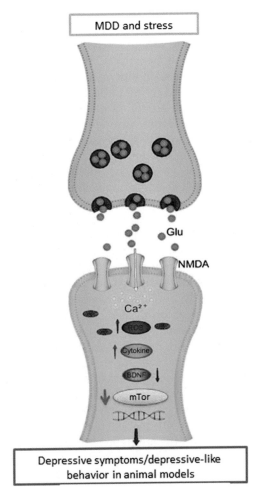
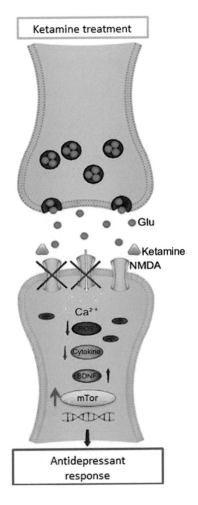

FIGURE 12.1   The role of mTOR and signaling cascades in the pathophysiology and treatment of MDD. In stress or MDD situations there is an increase in Glu levels, which in turn could lead a super activation of NMDA receptors, leading an increase in Ca^{2+} influx. An excitotoxicity induced by Glu is related to oxidative stress, increased levels of pro-inflammatory cytokines, decreased levels of neurotrophic factors, such as BDNF, and decreased mTOR activation and phosphorylation. These intracellular alterations are associated with depressive symptoms in patients with MDD and depressive-like behavior in animal models of depression. On the other hand, treatment with ketamine, an antagonist of NMDA receptor, acts decreasing oxidative stress, inflammatory cytokines, and increasing BDNF levels and mTOR activity and phosphorylation. BDNF, brain-derived neurotrophic factor; Glu, glutamate; MDD, major depressive disorder; mTOR, mammalian target of rapamycin; NMDA, N-methyl-D-aspartate; ROS, reactive oxygen species.

mouse hippocampus [46]. Changes in glutamatergic neurotransmission may contribute to the antidepressant response of sleep deprivation; however, the exact underlying mechanisms are unknown (Figure 12.1).

12.4 CONCLUSIONS

Preclinical and human studies implicate the mTOR signaling pathway in mood disorder pathophysiology. However, more research is required to determine how perturbations in mTOR signaling directly contribute to mood disorder manifestation. These future studies could be carried out in preclinical models or in large human cohorts to identify genetic contributors. An especially

promising finding is that the fast-acting antidepressant effect of ketamine acts through effects on the mTOR pathway. The mTOR pathway should be examined more thoroughly in the future to identify new therapeutic drugs with mood disorder-alleviating effects.

References

[1] Miller C, Bauer MS. Excess mortality in bipolar disorders. Curr Psychiatry Rep 2014;16:499.
[2] Laursen TM, Wahlbeck K, Hallgren J, Westman J, Osby U, Alinaghizadeh H, et al. Life expectancy and death by diseases of the circulatory system in patients with bipolar disorder or schizophrenia in the Nordic countries. PloS One 2013;8: e67133.

[3] Frank E, Kupfer DJ, Thase ME, Mallinger AG, Swartz HA, Fagiolini AM, et al. Two-year outcomes for interpersonal and social rhythm therapy in individuals with bipolar I disorder. Arch Gen Psychiatry 2005;62:996–1004.

[4] Mitchell PB, Hadzi-Pavlovic D. Lithium treatment for bipolar disorder. Bull World Health Organ 2000;78:515–17.

[5] McElroy SL, Keck Jr PE, Pope Jr HG, Hudson JI. Valproate in the treatment of bipolar disorder: literature review and clinical guidelines. J Clin Psychopharmacol 1992;12:42s–52s.

[6] Scherk H, Pajonk FG, Leucht S. Second-generation antipsychotic agents in the treatment of acute mania: a systematic review and meta-analysis of randomized controlled trials. Arch Gen Psychiatry 2007;64:442–55.

[7] Sassi RB, Soares JC. Emerging therapeutic targets in bipolar mood disorder. Expert Opin Ther Target 2001;5:587–99.

[8] Schou M. Perspectives on lithium treatment of bipolar disorder: action, efficacy, effect on suicidal behavior. Bipolar Disord 1999;1:5–10.

[9] Kessler RC, Chiu WT, Demler O, Merikangas KR, Walters EE. Prevalence, severity, and comorbidity of 12-month DSM-IV disorders in the National Comorbidity Survey Replication. Arch Gen Psychiatry 2005;62:617–27.

[10] Sattler R, Rothstein JD. Targeting an old mechanism in a new disease—protection of glutamatergic dysfunction in depression. Biol Psychiatry 2007;61:137–8.

[11] World Health Organization, World suicide prevention day. Available from: http://www.who.int/mediacentre/events/annual/world_suicide_prevention_day/en/; 2012. Accessed 30 April, 2014.

[12] Nemeroff CB, Owens MJ. Treatment of mood disorders. Nat Neurosci 2002;5:1068–70.

[13] Berton O, Nestler EJ. New approaches to antidepressant drug discovery: beyond monoamines. Nat Rev Neurosci 2006;7:137–51.

[14] Malhi GS, Bargh DM, Kuiper S, Coulston CM, Das P. Modeling bipolar disorder suicidality. Bipolar Disord 2013;15:559–74.

[15] Oruch R, Elderbi MA, Khattab HA, Pryme IF, Lund A. Lithium: a review of pharmacology, clinical uses, and toxicity. Eur J Pharmacol 2014;740:464–73.

[16] Dick DM, Foroud T, Flury L, Bowman ES, Miller MJ, Rau NL, et al. Genomewide linkage analyses of bipolar disorder: a new sample of 250 pedigrees from the national institute of mental health genetics initiative. Am J Hum Genet 2003;73:107–14.

[17] Ewald H, Wikman FP, Teruel BM, Buttenschon HN, Torralba M, Als TD, et al. A genome-wide search for risk genes using homozygosity mapping and microarrays with 1,494 single-nucleotide polymorphisms in 22 eastern cuban families with bipolar disorder. Am J Med Genet B Neuropsychiatr Genet 2005;133B:25–30.

[18] Fullston T, Gabb B, Callen D, Ullmann R, Woollatt E, Bain S, et al. Inherited balanced translocation T(9;17)(Q33.2;Q25.3) concomitant with a 16p13.1 duplication in a patient with schizophrenia. Am J Med Genet B Neuropsychiatr Genet 2011;156:204–14.

[19] Rajkumar AP, Christensen JH, Mattheisen M, Jacobsen I, Bache I, Pallesen J, et al. Analysis of T(9;17)(Q33.2;Q25.3) chromosomal breakpoint regions and genetic association reveals novel candidate genes for bipolar disorder. Bipolar Disord 2015;17(2):205–11.

[20] Feyissa AM, Chandran A, Stockmeier CA, Karolewicz B. Reduced levels of NR2A and NR2B subunits of NMDA receptor and PSD-95 in the prefrontal cortex in major depression. Prog Neuropsychopharmacol Biol Psychiatry 2009;33:70–5.

[21] Jernigan CS, Goswami DB, Austin MC, Iyo AH, Chandran A, Stockmeier CA, et al. The mTOR signaling pathway in the prefrontal cortex is compromised in major depressive disorder. Prog Neuropsychopharmacol Biol Psychiatry 2011;35:1774–9.

[22] Chandran A, Iyo AH, Jernigan CS, Legutko B, Austin MC, Karolewicz B. Reduced phosphorylation of the mTOR signaling pathway components in the amygdala of rats exposed to chronic stress. Prog Neuropsychopharmacol Biol Psychiatry 2013;40:240–5.

[23] Kim SH, Yu HS, Park HG, Ha K, Kim YS, Shin SY, et al. Intracerebroventricular administration of ouabain, a Na/K-atpase inhibitor, activates mtor signal pathways and protein translation in the rat frontal cortex. Prog Neuropsychopharmacol Biol Psychiatry 2013;45:73–82.

[24] Yu HS, Kim SH, Park HG, Kim YS, Ahn YM. Activation of Akt signaling in rat brain by intracerebroventricular injection of ouabain: a rat model for mania. Prog Neuropsychopharmacol Biol Psychiatry 2010;34:888–94.

[25] Machado-Vieira R, Manji HK, Zarate Jr CA. The role of lithium in the treatment of bipolar disorder: convergent evidence for neurotrophic effects as a unifying hypothesis. Bipolar Disord 2009;11(Suppl. 2):92–109.

[26] Karege F, Meary A, Perroud N, Jamain S, Leboyer M, Ballmann E, et al. Genetic overlap between schizophrenia and bipolar disorder: a study with Akt1 gene variants and clinical phenotypes. Schizophr Res 2012;135:8–14.

[27] Karege F, Perroud N, Schurhoff F, Meary A, Marillier G, Burkhardt S, et al. Association of Akt1 gene variants and protein expression in both schizophrenia and bipolar disorder. Genes Brain Behav 2010;9:503–11.

[28] Ruktanonchai DJ, El-Mallakh RS, Li R, Levy RS. Persistent hyperactivity following a single intracerebroventricular dose of ouabain. Physiol Behav 1998;63:403–6.

[29] Kirshenbaum GS, Burgess CR, Dery N, Fahnestock M, Peever JH, Roder JC. Attenuation of mania-like behavior in Na(+),K(+)-Atpase Alpha3 mutant mice by prospective therapies for bipolar disorder: melatonin and exercise. Neuroscience 2014;260:195–204.

[30] Kirshenbaum GS, Clapcote SJ, Duffy S, Burgess CR, Petersen J, Jarowek KJ, et al. Mania-like behavior induced by genetic dysfunction of the neuron-specific Na+,K+-Atpase Alpha3 sodium pump. Proc Natl Acad Sci USA 2011;108:18144–9.

[31] Rose AM, Mellett BJ, Valdes Jr R, Kleinman JE, Herman MM, Li R, et al. Alpha 2 isoform of the Na,K-Adenosine triphosphatase is reduced in temporal cortex of bipolar individuals. Biol Psychiatry 1998;44:892–7.

[32] Tochigi M, Iwamoto K, Bundo M, Sasaki T, Kato N, Kato T. Gene expression profiling of major depression and suicide in the prefrontal cortex of postmortem brains. Neurosci Res 2008;60:184–91.

[33] Christo PJ, El-Mallakh RS. Possible role of endogenous ouabain-like compounds in the pathophysiology of bipolar illness. Med Hypotheses 1993;41:378–83.

[34] Croyle ML, Woo AL, Lingrel JB. Extensive random mutagenesis analysis of the Na+/K+-Atpase alpha subunit identifies known and previously unidentified amino acid residues that alter ouabain sensitivity—implications for ouabain binding. Eur J Biochem 1997;248:488–95.

[35] El-Mallakh RS, Stoddard M, Jortani SA, El-Masri MA, Sephton S, Valdes Jr R. Aberrant regulation of endogenous ouabain-like factor in bipolar subjects. Psychiatry Res 2010;178:116–20.

[36] Goldstein I, Lerer E, Laiba E, Mallet J, Mujaheed M, Laurent C, et al. Association between sodium- and potassium-activated adenosine triphosphatase alpha isoforms and bipolar disorders. Biol Psychiatry 2009;65:985–91.

[37] Grider G, El-Mallakh RS, Huff MO, Buss TJ, Miller J, Valdes Jr R. Endogenous digoxin-like immunoreactive factor (Dlif) serum concentrations are decreased in manic bipolar patients compared to normal controls. J Affect Disord 1999;54:261–7.

[38] Mynett-Johnson L, Murphy V, McCormack J, Shields DC, Claffey E, Manley P, et al. Evidence for an allelic association between bipolar disorder and a Na+, K+ adenosine triphosphatase alpha subunit gene (Atp1a3). Biol Psychiatry 1998;44:47–51.

[39] Machado-Vieira R, Zanetti MV, Teixeira AL, Uno M, Valiengo LL, Soeiro-de-Souza MG, et al. Decreased Akt1/Mtor pathway mrna expression in short-term bipolar disorder. Eur Neuropsychopharmacol 2015;25:468–73.

[40] Hashimoto K. Role of the mTOR signaling pathway in the rapid antidepressant action of ketamine. Expert Rev Neurother 2011;11:33–6.

[41] Li N, Liu RJ, Dwyer JM, Banasr M, Lee B, Son H, et al. Glutamate N-methyl-D-aspartate receptor antagonists rapidly reverse behavioral and synaptic deficits caused by chronic stress exposure. Biol Psychiatry 2011;69:754–61.

[42] Monteggia LM, Kavalali ET. Scopolamine and ketamine: evidence of convergence? Biol Psychiatry 2013;74:712–13.

[43] Jeon SH, Kim SH, Kim Y, Kim YS, Lim Y, Lee YH, et al. The tricyclic antidepressant imipramine induces autophagic cell death in U87MG glioma cells. Biochem Biophys Res 2011;23:311–17.

[44] Warren BL, Iñiguez SD, Alcantara LF, Wright KN, Parise EM, Weakley SK, et al. Juvenile administration of concomitant methylphenidate and fluoxetine alters behavioral reactivity to reward-and mood related stimuli and disrupts ventral tegmental area gene expression in adulthood. J Neurosci 2011;31:10347–58.

[45] Elfving B, Wegener G. Electroconvulsive seizures stimulate the vegf pathway via mTORC1. Synapse 2012;66:340–5.

[46] Vecsey CG, Peixoto L, Choi JH, Wimmer M, Jaganath D, Hernandez PJ, et al. Genomic analysis of sleep deprivation reveals translational regulation in the hippocampus. Physiol Genomics 2012;44:981–91.

[47] Inoki K, Ouyang H, Zhu T, Lindvall C, Wang Y, Zhang X, et al. Tsc2 integrates Wnt and energy signals via a coordinated phosphorylation by Ampk and Gsk3 to regulate cell growth. Cell 2006;126:955–68.

[48] Grof P, Muller-Oerlinghausen B. A critical appraisal of lithium's efficacy and effectiveness: the last 60 years. Bipolar Disord 2009;11(Suppl 2):10–19.

[49] Latalova K, Kamaradova D, Prasko J. Suicide in bipolar disorder: a review. Psychiatr Danub 2014;26:108–14.

[50] Klein PS, Melton DA. A molecular mechanism for the effect of lithium on development. Proc Natl Acad Sci USA 1996;93:8455–9.

[51] Yin L, Wang J, Klein PS, Lazar MA. Nuclear receptor rev-erbalpha is a critical lithium-sensitive component of the circadian clock. Science 2006;311:1002–5.

[52] Lavoie J, Hébert M, Beaulieu JM. Looking beyond the role of glycogen synthase kinase-3 genetic expression on electroretinogram response: what about lithium? Biol Psychiatry 2015;77: e15–17.

[53] Li R, El-Mallakh RS, Harrison L, Changaris DG, Levy RS. Lithium prevents ouabain-induced behavioral changes. Toward an animal model for manic depression. Mol Chem Neuropathol 1997;31:65–72.

[54] Jun C, Choi Y, Lim SM, Bae S, Hong YS, Kim JE, et al. Disturbance of the glutamatergic system in mood disorders. Exp Neurobiol 2014;23:28–35.

[55] Soeiro-de-Souza MG, Salvadore G, Moreno RA, Otaduy MC, Chaim KT, Gattaz WF, et al. Bcl-2 rs956572 polymorphism is associated with increased anterior cingulate cortical glutamate in euthymic bipolar I disorder. Neuropsychopharmacology 2013;38:468–75.

[56] Reus GZ, Vieira FG, Abelaira HM, Michels M, Tomaz DB, dos Santos MA, et al. MAPK signaling correlates with the antidepressant effects of ketamine. J Psychiatr Res 2014;55:15–21.

[57] Reus GZ, Stringari RB, Ribeiro KF, Ferraro AK, Vitto MF, Cesconetto P, et al. Ketamine plus imipramine treatment induces antidepressant-like behavior and increases CREB and BDNF protein levels and PKA and PKC phosphorylation in rat brain. Behav Brain Res 2011;221:166–71.

[58] McDonald JW, Johnston MV. Physiological and pathophysiological roles of excitatory amino acids during central nervous system development. Brain Res Brain Res Rev 1990;5:41–70.

[59] Pittenger C, Sanacora G, Krystal JH. The NMDA receptor as a therapeutic target in major depressive disorder. CNS Neurol Disord Drug Targets 2007;6:101–15.

[60] Sanacora G, Zarate CA, Krystal JH, Manji HK. Targeting the glutamatergic system to develop novel, improved therapeutics for mood disorders. Nat Rev Drug Discov 2008;7:426–37.

[61] Kim JS, Schmid-Burgk W, Claus D, Kornhuber HH. Increased serum glutamate in depressed patients. Arch Psychiatr Nervenkr 1982;232:299–304.

[62] Mitani H, Shirayama Y, Yamada T, Maeda K, Ashby Jr CR, Kawahara R. Correlation between plasma levels of glutamate, alanine and serine with severity of depression. Prog Neuropsychopharmacol Biol Psychiatry 2006;30:1155–8.

[63] Hashimoto K, Sawa A, Iyo M. Increased levels of glutamate in brains from patients with mood disorders. Biol Psychiatry 2007;62:1310–16.

[64] Rao JS, Kellom M, Reese EA, Rapoport SI, Kim HW. Dysregulated glutamate and dopamine transporters in postmortem frontal cortex from bipolar and schizophrenic patients. J Affect Disord 2012;136:63–71.

[65] Nowak G, Ordway GA, Paul IA. Alterations in the N-methyl-D-aspartate (NMDA) receptor complex in the frontal cortex of suicide victims. Brain Res 1995;675:157–64.

[66] Mundo E, Tharmalingham S, Neves-Pereira M, Dalton EJ, Macciardi F, Parikh SV, et al. Evidence that the N-methyl-D-aspartate subunit 1 receptor gene (GRIN1) confers susceptibility to bipolar disorder. Mol Psychiatry 2003;8:241–5.

[67] Martucci L, Wong AH, De Luca V, Likhodi O, Wong GW, King N, et al. N-methyl-D-aspartate receptor NR2B subunit gene GRIN2B in schizophrenia and bipolar disorder: polymorphisms and mRNA levels. Schizophr Res 2006;84:214–21.

[68] Zarate Jr CA, Singh JB, Carlson PJ, Brutsche NE, Ameli R, Luckenbaugh DA. A randomized trial of an N-methyl-D-aspartate antagonist in treatment-resistant major depression. Arch Gen Psychiatry 2006;63:856–64.

[69] Fond G, Loundou A, Rabu C, Macgregor A, Lancon C, Brittner M, et al. Ketamine administration in depressive disorders: a systematic review and meta-analysis. Psychopharmacology 2014;231: 3663–76.

[70] Garcia LS, Comim CM, Valvassori SS, Reus GZ, Stertz L, Kapczinski F, et al. Ketamine treatment reverses behavioral and physiological alterations induced by chronic mild stress in rats. Prog Neuropsychopharmacol Biol Psychiatry 2009;33:450–5.

[71] Fraga DB, Réus GZ, Abelaira HM, De Luca RD, Canever L, Pfaffenseller B, et al. Ketamine alters behavior and decreases BDNF levels in the rat brain as a function of time after drug administration. Revista Brasileira Psiquiatria 2013;35:262–6.

[72] Reus GZ, Abelaira HM, dos Santos MA, Carlessi AS, Tomaz DB, Neotti MV, et al. Ketamine and imipramine in the nucleus accumbens regulate histone deacetylation induced by maternal deprivation and are critical for associated behaviors. Behav Brain Res 2013;256:451–6.

[73] Diazgranados N, Ibrahim L, Brutsche NE, Newberg A, Kronstein P, Khalife S, et al. A randomized add-on trial of an N-Methyl-D-aspartate antagonist in treatment-resistant bipolar depression. Arch Gen Psychiatry 2010;67:793–802.

[74] Zarate Jr CA, Brutsche NE, Ibrahim L, Franco-Chaves J, Diazgranados N, Cravchik A, et al. Replication of ketamine's antidepressant efficacy in bipolar depression: a randomized controlled add-on trial. Biol Psychiatry 2012;71:939–46.

[75] Lara DR, Bisol LW, Munari LR. Antidepressant, mood stabilizing and procognitive effects of very low dose sublingual ketamine in refractory unipolar and bipolar depression. Int J Neuropsychopharmacol 2013;16:2111–17.

[76] Charney DS, Manji HK. Life stress, genes, and depression: multiple pathways lead to increased risk and new opportunities for intervention. Sci STKE 2004;2004(225):re5.

[77] Duman RS. Role of neurotrophic factors in the etiology and treatment of mood disorders. Neuromolecular Med 2004;5:11–25.

[78] Maes M, Vandoolaeghe E, Ranjan R, Bosmans E, Bergmans R, Desnyder R. Increased serum interleukin-1-receptor-antagonist concentrations in major depression. J Affect Disord 1995;36:29–36.

[79] Wingenfeld K, Wolf OT. Stress, memory, and the hippocampus. Front Neurol Neurosci 2014;34:109–20.

[80] Niciu MJ, Ionescu DF, Richards EM, Zarate Jr CA. Glutamate and its receptors in the pathophysiology and treatment of major depressive disorder. J Neural Transm 2014;121:907–24.

[81] Hoeffer CA, Klann E. mTOR signaling: at the crossroads of plasticity, memory and disease. Trends Neurosci 2010;33:67–75.

[82] Deschwanden A, Karolewicz B, Feyissa AM, Treyer V, Ametamey SM, Johayem A, et al. Reduced metabotropic glutamate receptor 5 density in major depression determined by [11C] ABP688 PET and postmortem study. Am J Psychiatric 2011;168:727–34.

[83] Karolewicz B, Cetin M, Aricioglu F. Beyond the glutamate N-methyl D-aspartate receptor in major depressive disorder: the mTOR signaling pathway. Bull Clin Psychopharmacol 2011;21:1–6.

[84] Goswami DB, Jernigan CS, Chandran A, Iyo AH, May WL, Austin MC, et al. Gene expression analysis of novel genes in the prefrontal cortex of major depressive disorder subjects. Prog Neuropsychopharmacol Biol Psychiatry 2013;43:126–33.

[85] Miller OH, Yang L, Wang CC, Hargroder EA, Zhang Y, Delpire E, et al. GluN2B-containing NMDA receptors regulate depression-like behavior and are critical for the rapid antidepressant actions of ketamine. Elife 2014;:e03581.

[86] Réus GZ, Carlessi AS, Titus SE, Abelaira HM, Ignácio ZM, da Luz JR, et al. A single dose of S-ketamine induces long-term antidepressant effects and decreases oxidative stress in adulthood rats following maternal deprivation. Dev Neurobiol 2015. [in press].

[87] Réus GZ, Nacif MP, Abelaira HM, Tomaz DB, dos Santos MA, Carlessi AS, et al. Ketamine ameliorates depressive-like behaviors and immune alterations in adult rats following maternal deprivation. Neurosci Lett 2015;584:83–7.

[88] Garcia LB, Comim CM, Valvassori SS, Réus GZ, Barbosa LM, Andreazza AC, et al. Acute administration of ketamine induces antidepressant-like effects in the forced swimming test and increases BDNF levels in the rat hippocampus. Prog Neuropsychopharmacol Biol Psychiatry 2008;32:140–4.

[89] Garcia LB, Comim CM, Valvassori SS, Réus GZ, Andreazza AC, Stertz L, et al. Chronic administration of ketamine elicits antidepressant-like effects in rats without affecting hippocampal brain-derived neurotrophic factor protein levels. Basic Clin Pharmacol Toxicol 2008;103:502–6.

[90] Hoeffer CA, Tang W, Wong H, Santillan A, Patterson RJ, Martinez LA, et al. Removal of FKBP12 enhances mTOR-Raptor interactions, LTP, memory, and perseverative/repetitive behavior. Neuron 2008;60:832–45.

[91] Park SW, Lee JG, Seo MK, Lee CH, Cho HY, Lee BJ, et al. Differential effects of antidepressant drugs on mTOR signalling in rat hippocampal neurons. Int J Neuropsychopharmacol 2014;17:1831–46.

[92] Li N, Lee B, Liu RJ, Banasr M, Dwyer JM, Iwata M, et al. mTOR-dependent synapse formation underlies the rapid antidepressant effects of NMDA antagonists. Science 2010;20:959–64.

[93] Abelaira HM, Réus GZ, Neotti MV, Quevedo J. The role of mTOR in depression and antidepressant responses. Life Sci 2014;101:10–14.

[94] Sui L, Wang J, Li BM. Role of the phosphoinositide 3-kinase-Akt-mammalian target of the rapamycin signaling pathway in long-term potentiation and trace fear conditioning memory in rat medial prefrontal cortex. Learn Mem 2008;15:762–76.

[95] Howell KR, Kutiyanawalla A, Pillai A. Long-term continuous corticosterone treatment decreases VEGF receptor-2 expression in frontal cortex. PLoS One 2011;6:e20198.

[96] Zhu WL, Shi HS, Wang SJ, Wu P, Ding ZB, Lu L. Hippocampal CA3 calcineurin activity participates in depressive-like behavior in rats. J Neurochem 2011;117:1075–86.

[97] Yu JJ, Zhang Y, Wang Y, Wen ZY, Liu XH, Qin J, et al. Inhibition of calcineurin in the prefrontal cortex induced depressive-like behavior through mTOR signaling pathway. Psychopharmacology 2013;225:361–72.

[98] Cleary C, Linde JAS, Hiscock KM, Hadas I, Belmaker RH, Agam G, et al. Antidepressive-like effects of rapamycin in animal models: implications for mTOR inhibition as a new target for treatment of affective disorders. Brain Res Bull 2008;76:469–73.

[99] Maeng S, Zarate Jr CA, Du J, Schloesser RJ, McCammon J, Chen G, et al. Cellular mechanisms underlying the antidepressant effects of ketamine: role of α-amino-3-hydroxy-5-methylisoxazole-4-propionic acid receptors. Biol Psychiatry 2008;63:349–52.

[100] Yoon SC, Seo MS, Kim SH, Jeon WJ, Ahn YM, Kang UG, et al. The effect of MK-801 on mTOR/p70S6K and translation-related proteins in rat frontal cortex. Neurosci Lett 2008;434:23–8.

[101] Koike H, Iijima M, Chaki S. Involvement of the mammalian target of rapamycin signaling in the antidepressant-like effect of group II metabotropic glutamate receptor antagonists. Neuropharmacology 2011;61:1419–23.

[102] Yang C, Hashimoto K. Combination of nitrous oxide with isoflurane or scopolamine for treatment-resistant major depression. Clin Psychopharmacol Neurosci 2015;13:118–20.

[103] Jaffe RJ, Novakovic V, Peselow ED. Scopolamine as an antidepressant: a systematic review. Clin Neuropharmacol 2013;36:24–6.

[104] Silveira MM, Malcolm E, Shoaib M, Winstanley CA. Scopolamine and amphetamine produce similar decision-making deficits on a rat gambling task via independent pathways. Behav Brain Res 2015;281:86–95.

[105] Caruana D, Warburton EC, Bashir ZI. Induction of activity-dependent LTD requires muscarinic receptor activation in medial prefrontal cortex. J Neurosci 2011;31:18464–78.

[106] Duman RS. Pathophysiology of depression and innovative treatments: remodeling glutamatergic synaptic connections. Dialogues Clin Neurosci 2014;16:11–27.

[107] Voleti B, Navarria A, Liu RJ, Banasr M, Li N, Terwilliger R, et al. Scopolamine rapidly increases mammalian target of

rapamycin complex 1 signaling, synaptogenesis, and antidepressant behavioral responses. Biol Psychiatry 2013;74:742—9.

[108] Moretti M, Budni J, Freitas AE, Rosa PB, Rodrigues AL. Antidepressant-like effect of ascorbic acid is associated with the modulation of mammalian target of rapamycin pathway. J Psychiatric Res 2014;48:16—24.

[109] Sarkar A, Chachra P, Kennedy P, Pena CJ, Desouza LA, Nestler EJ, et al. Hippocampal HDAC4 contributes to postnatal fluoxetine-evoked depression-like behavior. Neuropsychopharmacology 2014;39:2221—32.

[110] Hemmeter UM, Hemmeter-Spernal J, Krieg JC. Sleep deprivation in depression. Exp Rev Neurother 2010;10:1101—15.

13

mTOR and Drugs of Abuse

Jacob T. Beckley and Dorit Ron

Department of Neurology, University of California, San Francisco, CA, USA

13.1 INTRODUCTION

Drugs of abuse are a group of psychoactive chemicals that are unique for their self-reinforcing property. The psychoactive symptoms of intoxication differ depending on the type of drug, but one commonality is that drugs of abuse can cause feelings of euphoria. Also, despite distinct pharmacological properties, drugs of abuse are capable of hijacking learning processes, particularly in circuits that encode motivated and goal-directed behavior [1]. Drug-taking is a spectrum of behavioral adaptations, from occasional use to pathological abuse. When at its most pathological, substance-use disorders are characterized by compulsive drug-taking, and decision-making processes that are specifically geared towards taking or obtaining the drug. Relapse is also a devastating symptom of a substance-use disorder, where even after extended periods of withdrawal, certain contexts or experiences trigger drug memories and make abstinent users crave the drug [2].

As defined by the fifth edition of the "Diagnostic and Statistical Manual of Mental Disorders," substance-use disorders are defined by four groups of symptoms: impaired control, social impairment, risky use, and pharmacological criteria. Each grouping has several criteria, with a total of 11 potential symptoms. The severity of the substance-use disorder increases with the presence of more criteria, with a mild disorder being defined as 2–3 symptoms present, a moderate disorder being 4–5 symptoms, and a severe disorder being six or more symptoms. [3]. See Table 13.1 for the specific criteria under each grouping that defines substance-use disorders.

Given that the primary symptoms of substance-use disorders include compulsive drug-taking, compromised decision-making, and relapse [2,4], it is apparent that drugs of abuse severely compromise learning and memory systems in multiple brain circuits. The kinase mechanistic target of rapamycin (mTOR) is critically involved in mediating long-term memory formation through its role in local, activity-dependent protein translation [5], and while research on mTOR and drugs of abuse is in its nascent period, it is evident that mTOR plays a central role in the development of pathological behaviors associated with escalated drug use. This chapter will give details on how mTOR mediates the cellular and behavioral changes that drugs exert on brain circuits.

13.2 DRUGS OF ABUSE

Drugs with abuse potential span a range of unrelated chemicals with distinct pharmacology. For example, psychostimulants, opiates, nicotine, cannabis, and dissociative hallucinogens have a well-defined but distinct pharmacological site of action, whereas alcohol affects several ion channels with a low affinity. Below is a brief explainer for each abused drug class. Two prominent class of drugs, abused inhalants, which include nitrous oxide and volatile organic solvents like toluene, and psychedelics, which include psilocybin and lysergic acid diethylamide, are not described because there are no experiments connecting these types of drugs with mTOR signaling, although this is likely due to a lack of research studies, not because there is no effect of these compounds on mTOR.

13.2.1 Psychostimulants

This class includes drugs such as cocaine, amphetamine, methamphetamine, and 3,4-methylenedioxymethamphetamine (MDMA; also known as ecstasy).

Molecules to Medicine with mTOR
DOI: http://dx.doi.org/10.1016/B978-0-12-802733-2.00005-0

TABLE 13.1 DSM-V Substance-Use Disorders Criteria

Grouping	Criteria
Impaired control	1. Take larger amounts or for a longer duration than intended 2. Persistent desire to reduce use 3. Spend an inordinate amount of time obtaining, using, or recovering from the substance 4. Craving: intense desire to take the substance
Social impairment	1. Failure to fulfill major obligations at work, school, or home 2. Continued use despite recurrent social or interpersonal problems that are caused or worsened by substance use 3. Other activities are reduced or given up because of drug use
Risky use	1. Recurrent use in physically hazardous situations 2. Continued use despite a physical or psychological problem that is caused or exacerbated by the substance
Pharmacological criteria	1. Tolerance: higher dose is required to achieve the desired effect 2. Withdrawal: physical or psychological symptoms that develop when there is no substance in blood or tissue

Note: There are 11 criteria that fall under four groupings, and the more criteria that a patient exhibits, the more severe the disorder [3].

Also included is methylphenidate (Ritalin), which is commonly prescribed for treating attention-deficit hyperactivity disorder (ADHD). These drugs all promote wakefulness and alertness, and can induce euphoria. At low concentrations, these drugs can increase focus and attention, which is why methylphenidate and amphetamine are prescribed for cognitive disorders like ADHD. All of these compounds increase dopamine concentrations in the synaptic cleft [6]. Cocaine and methylphenidate increase dopamine by blocking the dopamine transporter (DAT), which inhibits reuptake. Substituted amphetamines, which include amphetamine, methamphetamine, and MDMA, can reverse the flow of dopamine through the transporter, causing dopamine efflux into the synaptic cleft. All of these compounds act on the norepinephrine and serotonergic transporter as well, with MDMA having a higher affinity for the serotonin transporter than other psychostimulants, which may give rise to some its more psychedelic effects.

13.2.2 Opioids

The members of this drug class, such as morphine, codeine, hydrocodone, and heroin, all similarly act as agonists at opioid receptors, and their positively reinforcing property is mediated primarily by stimulating mu opioid receptors (MORs) [7]. Morphine is the psychoactive compound in opium, which is a resin that is extracted from the plant *Papaver somniferum*. While many opioids are derivatives of morphine and therefore their synthesis starts from the plant extraction process, other opioids, such as the potent fentanyl, are synthetically derived. Opioids are commonly prescribed for their potent analgesic properties, but they are abused for their euphorigenic and sedative effects. Opioids have many side effects, especially gastrointestinal symptoms such as nausea and constipation. Respiratory depression is another serious adverse side effect, particularly after an overdose, and can result in coma or death. With prolonged usage, tolerance and withdrawal symptoms are prevalent.

13.2.3 Nicotine

Nicotine is the psychoactive compound found in tobacco, which is the cured leaf from one of a number of plants of the genus *Nicotiana*, the most common being *Nicotiana tabacum*. Nicotine is a stimulant, but unlike the psychostimulant drug class, its primary mechanism of action is that it is an agonist at acetylcholine nicotinic receptors. Acutely, nicotine promotes alertness and focus, and can facilitate short-term episodic memory and working memory [8]. Long-term chronic use of tobacco products increases the risk of a number of serious health complications, including cardiovascular diseases, such as stroke and cancer. Long-term nicotine use also causes dependence, and serious withdrawal symptoms occur with smoking cessation.

13.2.4 Alcohol

Alcohol, which is short for ethyl alcohol, is the product from the fermentation of sugars. Alcohol has many uses other than recreational consumption, as it is a common solvent, antiseptic, and fuel source. Alcohol is a central nervous system depressant, although lower doses induce feelings of euphoria, disinhibition, and sociability. Higher doses impair sensory, motor, and cognitive functions. Unlike many other drugs of abuse, alcohol does not have a well-defined single site of action that underlies its behavioral and cognitive effects, but rather has a wide range of cellular targets, including actions as an antagonist of the glutamatergic N-methyl-D-aspartate (NMDA) receptor, and an agonist at the γ-aminobutyric acid A (GABA-A) receptor [9]. Long-term, chronic alcohol consumption can lead to profound disturbances in cognition and memory and can induce physical dependence. With the most serious levels of alcohol dependence, alcohol withdrawal syndrome can occur, which includes symptoms such as agitation, seizures, and delirium tremens.

13.2.5 Cannabis

The primary psychoactive compound in the cannabis plant is Δ9-tetrahydrocannabinol (THC), which acts as a partial agonist at the cannabinoid receptor 1 (CB1). THC has a variety of psychoactive effects, including euphoria, impaired short-term memory, and altered perception of time [10]. Depending on the dose, cannabis can be anxiolytic or anxiogenic [11]. THC is also an analgesic, antiemetic, and stimulates appetite. Long-term effects of chronic cannabis use include an increased risk of developing or exacerbating a psychiatric disorder, including depression and anxiety disorders.

13.2.6 Dissociative Hallucinogens

This drug class includes arylcyclohexylamines, of which phencyclidine (PCP) and ketamine are members. Drugs that have dissociative properties cause a separation of conscious feelings from the external environment, or in other words, a dissociation between mind and body. Because of this effect, these drugs are powerful global anesthetics, and ketamine is still used as such in veterinary medicine. The primary mechanism of action for arylcyclohexylamines is that these compounds are NMDA channel blockers. Other effects include disorientation, behavioral disinhibition, irregular breathing, and ataxia [12]. Recent research suggests that low-dose treatment of ketamine has rapid antidepressant effects, and this effect is due to the activation of mTOR in frontal cortical regions [13]. As such ketamine is currently being tested in humans for the treatment of major depression.

Despite diverse pharmacological properties, all of these drugs of abuse can cause feelings of euphoria and affect the brain circuits that underlie motivational learning. In 1988, Di Chiara and Imperato found that drugs with abuse potential commonly acted on the mesolimbic dopamine system, by acutely increasing dopamine released from ventral tegmental area (VTA) neurons into the nucleus accumbens (NAc) [14]. This research clearly illustrated that the mesolimbic pathway, which is critically involved in encoding the value of reinforcing stimuli [15], is a primary contributor to the unique effects of drugs of abuse. Following chronic drug use, there is a compensatory reduction in dopaminergic signaling, and the drugs which elicited a large dopaminergic efflux following acute use now result in less dopamine release, indicative of a drug-tolerant state [16]. Other positively reinforcing stimuli also evoke less dopamine release following compulsive drug-taking, reflecting a state where non-drug stimuli are potentially less rewarding and less able to shape behavior.

Beyond the mesolimbic dopamine system, drugs of abuse profoundly affect cortical brain areas that are normally involved in decision-making processes. The prefrontal cortex (PFC) integrates sensory, motor, and subcortical input and exerts top-down control over neocortical and subcortical regions, including the VTA and NAc [17]. In an addictive state, the ability of the cortex to regulate striatal activity becomes dysfunctional, and there is a concomitant loss in the ability to control drug intake [18,19]. The cortex becomes hyper-responsive to drugs or drug cues, while it is less responsive to other non-drug stimuli, and so decision-making processes are geared towards taking the drug. Essentially, drugs of abuse acutely reinforce their own acquisition, and after chronic use, decision-making processes are biased towards obtaining or taking the drug.

Molecular adaptations induced by drugs of abuse underlie the mechanisms that lead to the development and maintenance of the behavioral phenotypes described above. Specifically, short-term drug exposure results in the activation of G-protein-coupled receptors and/or ion channels, leading to the activation of intracellular signaling cascades that can produce short-terms effects such as channel phosphorylation and localization, and/or long-term changes that are marked by increases in gene expression. These molecular adaptations in turn trigger long-term maladaptations that include stable changes in gene expression, stabilization or weakening of synapses, and morphological changes (for more details, please see reviews by Ron and Jurd [20] and Ahmadiantehrani et al. [21]).

13.3 ANIMAL MODELS OF DRUG ABUSE

There are a number of ways that researchers utilize animal models to study the effects of drugs of abuse. The most straightforward way to model the development of substance-use disorders is to train the animal to make a response to receive the drug, which is also known as the self-administration paradigm. For all drugs except alcohol, this is usually achieved by having a rodent or non-human primate make a response, such as a lever press or a nose poke through a portal, to receive an intravenous injection of the drug. With alcohol, the instrumental responding allows the test subject to receive a small amount of alcohol to consume. Cues such as a light or a tone can be added to indicate the availability of the drug, and then can be exploited to promote drug-seeking. The self-administration paradigm has several phases that can be tested, including acquisition, extinction, and reinstatement. In acquisition, the subject presses a lever a set number of times to receive the drug reward. Over many sessions, the amount of responding increases, as does the drug intake. Extinction is when the responding stimulus, such as the lever, is available but the drug is no longer received after pressing. Reinstatement is an animal model of relapse: after an animal has fully extinguished the instrumental responding behavior, the subject may receive a small priming injection of the drug or the cues could be presented, which will result in the subject pressing the lever once again for the drug even though the drug is not delivered after responding. There are other offshoots of self-administration that utilize operant conditioning in slightly different variants. For example, responding for a drug reinforcer can be put on a progressive ratio of reinforcement, meaning that each successive drug delivery takes more responses to receive the reward. The progressive ratio for drug reward tests the motivation the subject has to receive the reward, and the point at which the subject stops responding because the cost is too high is called the breakpoint.

Other animal models are arguably simpler because they involve non-contingent injections of the drug, and test for select aspects of drug administration. For example, locomotor sensitization, also called behavioral sensitization, occurs due to repeated systemic injections of an abused drug, usually a psychostimulant. Following each successive injection, the hyperlocomotor response to the drug increases. Sensitization is a long-lasting phenomenon, with increased locomotion evident even after long periods of abstinence. This behavior requires the dopaminergic mesolimbic pathway and corticostriatal glutamatergic circuits, and is the behavioral correlate of the modifications that drugs impart on these circuits [22]. While this behavior is almost always only tested in animals, one report showed that locomotor sensitization to amphetamine occurred in humans as well [23].

Another commonly used animal model is conditioned place preference (CPP). In this model, the subject receives a non-contingent administration of the drug in one context, and an inert substance in a second context. On the test day, the subject has access to both contexts, and commonly with drugs with abuse potential, the subject will prefer to be in the drug-paired context, indicating that the drug-paired environment has taken on the value of the reinforcing property of the drug [24]. Similarly to the self-administered paradigm, CPP is a behavior that can be extinguished and

reinstated, allowing experimenters to test different aspects of drug-seeking. CPP is also evident in humans [25], although like locomotor sensitization, it is only evident in subjects who have not had prior experience with the drug. Nevertheless, these paradigms allow researchers to gain valuable insight into how drugs of abuse commandeer normal learning processes, and also whether particular treatments lessen the impact of abused drugs.

13.4 mTOR

mTOR plays a central role in cellular processes that support the formation of long-term memories [26], and as such, is a critical player in the induction and maintenance of aberrant plasticity mechanisms induced by drugs of abuse. The mTOR signaling pathway is activated by a number of neurotransmitter and growth factor receptors, including dopamine D1 receptors [27], serotonergic 5-hydroxytryptamine 6 (5-HT6) receptors [28], MORs [29], tropomyosin receptor kinase B, the receptor for brain-derived neurotrophic factor (BDNF) [30], and Ca^{2+}-permeable α-amino-3-hydroxy-5-methyl-4-isoxazolepropionic acid (AMPA) receptors [31], among others. Interestingly, mTOR is activated due to stimulation of the NMDA receptor, through activation of phosphatidylinositol 3-kinase (PI3K) [32], and also by the inhibition of the NMDA GluN2B receptor subunit [13], which may be reciprocally linked to dendritic translational machinery [33]. Given that drugs of abuse directly target a number of these receptors, such as the dopamine D1 receptors and MORs, it is not surprising that use of these drugs may lead to the direct activation of mTOR.

Downstream of receptor stimulation, two primary means of mTOR activation are through extracellular signal-regulated kinase (Erk), which is a kinase within the mitogen-activated protein kinase pathway, and Akt, also known as protein kinase B, which is activated by PI3K [34]. These kinases inhibit the activity of the tuberous sclerosis complex 1/2 (TSC1/2). The TSC1/2 heterodimer acts as a guanosine triphosphatase activating protein, which promotes the protein Ras homolog enriched in brain (Rheb) in its guanosine diphosphate-bound, inactive state. Rheb in its GTP-bound state activates mTOR, so by phosphorylating and inhibiting TSC1/2, Erk and Akt serve to activate mTOR by disinhibiting its interaction with GTP-Rheb [34,35].

mTOR is a unique kinase because it signals via two distinct multiprotein complexes: mTOR complex 1 (mTORC1) and complex 2 (mTORC2). These complexes are distinguishable in part by the presence of their unique adapter proteins; for example, mTORC1 contains regulatory-associated protein of mTOR and mTORC2 contains rapamycin insensitive companion of mTOR (RICTOR). mTORC2 regulates actin cytoskeleton reorganization [36], and reduction of mTORC2 activity impairs long-term memory formation by disrupting actin polymerization processes [37]. Nearly all research on mTOR and drugs of abuse focuses on mTORC1. Nevertheless, given that mTORC2 regulates the cytoskeleton, a process that is intertwined with synaptic plasticity and memory formation, the role of mTORC2 in mediating the effects of drugs of abuse is an important future line of research, and one study has shown a link between mTORC2 and drug-induced neuroadaptations [38].

The primary targets of mTORC1 kinase activity are ribosomal protein S6 Kinase, 70 kDa (S6K) and the eukaryotic translation initiation factor-4E (eIF4E) binding protein (4E-BP). S6K is activated when phosphorylated by mTORC1, and it phosphorylates several residues on ribosomal protein S6 (S6), which is a component of the 40S ribosomal subunit. Phosphorylation of 4E-BP inhibits its function by causing dissociation from eIF4E. Many researchers measure activity of mTORC1 by examining changes in the phosphorylation state of S6K, 4E-BP, and S6. mTORC1 promotes the initiation and elongation of mRNAs that contain the 5' terminal oligopyrimidine motif, and many members of this class of mRNAs encode components of the ribosomal machinery, suggesting that mTORC1 promotes the translation of proteins through its direct control over the translation of ribosomal proteins [39,40].

The functions of mTORC1 have received more research attention than mTORC2, in part because of the discovery of a selective mTORC1 inhibitor. The inhibitor for which the protein is named, rapamycin, was discovered from the bacteria *Streptomyces hygroscopicus* found on Easter Island [41]. Originally used as an antifungal agent, rapamycin has robust immunosuppressive properties and is marketed as a treatment to prevent organ rejection. Rapamycin acts by binding to the protein FK506-binding protein 1A, 12 kDa (FKBP12), and this complex then binds to the FKBP12-rapamycin-binding domain on mTORC1, which inhibits mTORC1's ability to associate with its substrates [42]. Interestingly, an acute treatment of rapamycin only inhibits mTORC1, while chronic treatment disrupts the function of mTORC2 as well [43]. Nevertheless, rapamycin has been an invaluable pharmacological tool to study the impact of mTORC1 on various cellular and behavioral outcomes, and since there are no known selective inhibitors of mTORC2, research on mTORC1 has received more attention.

One of the primary functions of mTOR in the brain is to regulate translation locally within the dendrites. The placement in the dendrites is critical, for it allows

mTOR to respond directly to incoming synaptic input. This activity-dependent regulation of local protein synthesis allows mTOR to be a master regulator of long-lasting synaptic plasticity processes. It is important to note that mTOR regulates synaptic plasticity processes that require protein synthesis, such as late long-term potentiation (LTP), but not necessarily plasticity that is more transient and only requires trafficking, like early LTP. As such, mTOR is critically important for long-term memory formation because of the maintenance of the plasticity, which requires protein synthesis, but not the induction of synaptic plasticity [44]. In cultured hippocampal neurons, for example, rapamycin does not affect the induction of LTP following tetanic stimulation, but the late-phase LTP is abolished [45]. Furthermore, mTOR heterozygote mice have normal contextual and auditory fear memory, and normal hippocampal cornu ammonis area 1 (CA1) neuron LTP induction, but an administration of a concentration of rapamycin that has no effect in wild-type mice reduces late-phase LTP and impairs both contextual fear conditioning and the reconsolidation of contextual fear conditioning [46]. Reducing mTORC2 activity also selectively impairs the late-phase LTP. Mice with reduced mTORC2 activity, via *RICTOR* knockdown, have reduced actin polymerization, and display impairments in auditory and contextual fear conditioning. These deficiencies are restored by increasing actin polymerization [37]. Together, these findings indicate that both mTORC1 and mTORC2 are critically involved in the maintenance phase of LTP and the formation of long-term memories, but they regulate distinct, yet complementary, cellular components.

13.5 DRUGS OF ABUSE AND mTOR

Substance abuse disorders occur because of aberrant learning and synaptic plasticity within brain circuits involved in motivation, reinforcement learning, and decision-making [1]. One of the primary roles of mTORC1 in the nervous system is to promote activity-dependent protein translation to support the dynamic plasticity processes. Because mTORC1 is critically involved in plasticity mechanisms, it is likely that mTORC1 plays a role in the neuroadaptations induced by drugs of abuse. In fact, for most drug classes, studies have shown that mTORC1 is important for the expression of drug-induced behaviors. Thus, the mTORC1 pathway may be the common link between the molecular adaptations induced by the structurally diverse drugs of abuse and the common behavioral outcomes. Below is a summary of the contributions of mTOR signaling to the effects of drugs of abuse, organized by drug class.

13.5.1 Psychostimulants

Cocaine is the most commonly studied drug of abuse in preclinical research and as such there are more studies on cocaine and mTOR compared to other drugs, from acute effects of cocaine on mTORC1 signaling to the effects of disrupting mTORC1 on cocaine-induced behavioral phenotypes. While the effects of cocaine on the nervous system may not be so similar to how other drugs affect brain circuits, cocaine is a good model for other psychostimulants, such as amphetamine, methamphetamine, methylphenidate, and MDMA.

Acutely, cocaine increases phospho-S6 in the frontal cortex, VTA, and the NAc, indicating that mTORC1 is activated following a cocaine challenge [47]. The mechanism underlying cocaine's ability to activate mTORC1 has not been delineated, but because cocaine's primary mechanism of action is to block the DAT and increase extracellular dopamine, it is likely that dopamine triggers mTORC1 signaling, and D1 receptor activation can lead to the activation of mTORC1 [27,48]. The behavioral relevance of mTORC1 signaling following cocaine treatments has been shown in a number of ways. For example, inhibiting mTORC1 blocks locomotor sensitization to cocaine. It does this at two stages of sensitization: both at the induction stage, tested by injecting rapamycin prior to the daily cocaine injections; and at the expression stage, tested by injecting rapamycin prior to the challenge cocaine injection following repeated cocaine and withdrawal [47,49]. Rapamycin also blocks the expression of CPP induced by cocaine, meaning that inhibition of mTORC1 prior to the place preference test and not during the cocaine-context training session is sufficient to block the cocaine CPP. Cocaine CPP increases mTORC1 activity in the VTA, which is blocked by inhibiting the metabotropic glutamate receptor 1 (mGluR1). Furthermore, blocking mGluR1 or protein translation in the VTA blocks cocaine CPP [50]. Rapamycin also blocks the reconsolidation of CPP memories, which is tested in the following manner: following a CPP test day, rodents are reexposed to the cocaine-paired side, which triggers the cocaine-context paired memory. In this stage of memory reconsolidation, memories are more labile and subject to manipulation. If rapamycin is injected immediately after a reconsolidation session, then cocaine fails to induce CPP on a subsequent test day, even when the test is 14 days after the reconsolidation session, suggesting that mTORC1 is required for the reconsolidation of the cocaine memory [51]. Counterintuitively, reexposure to a cocaine-paired environment appears to reduce mTORC1 signaling, through decreased activation of Akt, in the NAc and hippocampus [52]. However, it is possible that for CPP, rapamycin is acting through brain regions such as the VTA to block the cocaine-induced behavior, and not the NAc or hippocampus.

mTORC1 signaling is also important for the motivational drive to receive a cocaine reward. Rapamycin reduces the breakpoint for responding to cocaine on a progressive ratio schedule of reinforcement [53]. Rats that have a higher breakpoint for cocaine responding display lower levels of *mTOR* gene expression in the NAc and possibly in the dorsal striatum [54]. It is unknown however whether the decreased gene expression is a compensatory response to higher mTOR activity or whether this reflects a reduction in mTOR protein.

mTORC1 is likely important for mediating relapse to cocaine as well. mTORC1 is activated in the NAc shell 24 hours after the final session of prolonged cocaine self-administration [53]. Furthermore, after prolonged withdrawal following extended access to cocaine self-administration, AMPA receptor function is altered in the NAc core, a physiological ramification that likely mediates relapse. Rapamycin reverses the alterations in AMPA receptor function in NAc core neurons, indicating that mTORC1 is likely responsible for altering AMPA physiology in the prolonged withdrawal state [55]. Finally, mTORC1 mediates the reinstatement of cocaine-seeking triggered by cocaine cues. Cue-induced reinstatement increases mTORC1 activity in the NAc shell, and reducing mTORC1 activity with rapamycin blocks reinstatement. The effect is specific for cocaine because sucrose reinstatement is unaffected by rapamycin [56]. Interestingly, NMDA in the NAc promotes reinstatement and increases mTORC1 activity, suggesting the possibility that mTORC1 is activated by glutamate following repeated cocaine use and withdrawal. Overall, each stage of cocaine-taking and -seeking involves mTORC1 signaling, particularly in the NAc, and therefore mTORC1 is likely a central player in the neuroadaptations that occur due to escalated cocaine use.

Other psychostimulants likely alter behavior and trigger neuroadaptations through the activation of mTOR. For example, repeated methylphenidate during adolescence increases *mTOR* gene expression in the VTA [57]. Also, mTORC1 inhibition disrupts a form of methamphetamine-induced CPP. If methamphetamine is administered prior to the CPP test, then the amount of place preference is sensitized. However, mTORC1 inhibition within the NAc blocks this CPP sensitization, without affecting normal levels of methamphetamine CPP [58]. While more experiments are needed to show how other psychostimulants affect mTORC1 signaling, results from these studies suggest that mTORC1 likely plays a role in the development of behaviors associated with the use of psychostimulants other than cocaine.

There is also a link between methamphetamine and mTOR that is outside of the nervous system. Methamphetamine at high doses or taken chronically is toxic and leads to cell death through various mechanisms including autophagy. Autophagy is actively inhibited by mTORC1, and several studies have shown that methamphetamine inhibits mTORC1 activity in various cell culture lines, including PC12 cells [59] and SK-N-SH cells [60]. Furthermore, methamphetamine reduces mTORC1 activity and causes autophagy in human brain microvascular cells, suggesting that methamphetamine may disrupt the blood−brain barrier [61]. Finally, MDMA, which also can cause cell death at high concentrations, induces autophagy in cardiomyocytes and causes left ventricle contractile dysfunction. MDMA triggers an increase in phosphorylation of mTOR at threonine 2446 [62], which is the target of 5′-adenosine monophosphate-activated protein kinase (AMPK) and leads to an inhibition of mTORC1 activity [63]. Whether methamphetamine also inhibits mTOR through AMPK is not clear, and it is unknown whether behaviorally relevant concentrations of methamphetamine or MDMA cause cell death. Nevertheless, the actions of methamphetamine and MDMA to inhibit mTORC1 and trigger autophagy show that these psychostimulants may act quite differently in cells that are outside of the nervous system.

13.5.2 Opioids

In contrast to psychostimulants that directly increase dopamine availability in synapses, opioids modulate dopaminergic activity through opioid receptors [7], and it is also likely that opioids activate mTORC1 through dopamine-independent mechanisms. In fact, activation of either MORs or delta opioid receptors increases mTORC1 activation through the activation of Akt [29,64]. Also, brain regions that show mTORC1 signaling following opioid use may differ from psychostimulants. For example, acquisition of morphine-induced CPP activates mTORC1 in the hippocampus CA3, but not CA1, nor the NAc or VTA. Furthermore, inhibiting mTORC1 or MOR in CA3 blocks morphine CPP and reduces the morphine-induced increase in mTORC1 activation [65]. This study indicates that in the hippocampus MOR stimulation with morphine triggers mTORC1 activation, and also suggests that morphine-induced CPP differs from cocaine-induced CPP in the brain regions that are directing the behavior. Nevertheless, similarly to cocaine, rapamycin blocks the reconsolidation of morphine CPP, indicating that mTORC1 is involved in different aspects of CPP learning and memory [51].

While there is little known about how self-administration of an opioid acts on mTOR signaling, there are some intriguing signs that mTOR is involved in different aspects of tolerance to opioids. For example, chronic morphine treatment decreases the size of

VTA dopamine neurons, leading to increased neuron firing but decreased release of dopamine, which may trigger tolerance to the rewarding effects of morphine. Morphine reduces cell size via decreased mTORC2 activation, as reversing the deficits in mTORC2 signaling reverses the morphine-induced alterations in cell size [38]. Tolerance to the analgesic effects of morphine may also be due to mTOR signaling. For example, repeated intrathecal injections of morphine activate mTORC1 in dorsal horn neurons, which is dependent on activation of MORs and Akt. Activating mTORC1 in the dorsal horn reduces morphine-induced analgesia, while reducing morphine-induced mTORC1 activation reduces development of morphine tolerance and hyperalgesia [66].

Finally, a clinical study showed that rapamycin decreased cue-induced craving in abstinent heroin users, but had no effect on the anxiogenic effect of the heroin cues [67]. So far, this is the only clinical experiment regarding the use of rapamycin as a treatment option for people with substance abuse disorders.

13.5.3 Alcohol

Alcohol is unique among the drugs with abuse potential because it does not have a primary receptor target but rather affects a wide range of channels and receptors. Despite its heterogeneous pharmacology, alcohol affects circuits similarly to other drugs of abuse. Also, similar to cocaine, an acute injection of alcohol increases mTORC1 activity in the NAc, as evidenced by increased phosphorylation of S6K and 4E-BP [68]. A signaling cascade leading to the activation of mTORC1 in the NAc by alcohol has been identified. Specifically, acute systemic administration of alcohol or self-administration of alcohol in rats with a history of excessive alcohol-drinking increases the activity of H-Ras, a central upstream initiator of mTORC1 activity, in the NAc [69]. H-Ras activation leads to the activation of the PI3K [70], which in turn activates Akt [71]. Systemic administration of alcohol or excessive alcohol-drinking activates the PI3K/Akt pathway in the NAc [68] and inhibition of H-Ras, PI3K, or Akt attenuates excessive alcohol-drinking in rodents [69,72]. Downstream of Akt activation is mTORC1 [34] and acute non-contingent administration or a history of alcohol-drinking result in the activation of mTORC1. Importantly, the activation of mTORC1 can be detected even after 24 hours of withdrawal, a time point when alcohol has been fully metabolized [68]. Thus, mTORC1 activation is a more enduring alteration following chronic alcohol consumption. Mechanistically, alcohol-induced mTORC1 signaling promotes an increase in the levels of proteins that are involved in excitatory synaptic transmission. Specifically, alcohol drinking increases the protein levels of the AMPA subunit GluA1 and the scaffolding protein Homer, both proteins whose translation is linked to mTORC1 activation [68,73]. The activation of mTORC1 by alcohol is behaviorally relevant because inhibition of mTORC1 by rapamycin reduces alcohol-induced locomotor sensitization and CPP. Furthermore, a single administration of rapamycin, either systemically or injected directly into the NAc, significantly decreases alcohol-seeking in rodents without affecting water consumption. Rapamycin also does not affect intake of the natural rewarding substances, sucrose or saccharine [68], which suggests that the Akt/mTORC1 pathway is selectively involved in responding to alcohol and not generally involved in responding to any positive reinforcer.

The reactivation of alcohol memories after prolonged withdrawal also involves mTORC1 signaling. Presenting the alcohol odor after a period of abstinence activates mTORC1 in the orbitofrontal cortex, medial PFC, and amygdala. In the orbitofrontal cortex and amygdala, the presentation of the cue is sufficient to increase the expression of GluA1, postsynaptic density protein 95 (PSD-95), Homer, and activity-regulated cytoskeleton-associated protein (Arc), with Arc's increase requiring mTORC1 activation. This suggests that an alcohol cue presentation triggers alterations that promote excitatory synaptic transmission in these brain regions. The reactivation of the alcohol memory is a salient experience, and it increases lever-pressing for alcohol in a subsequent test session. If rapamycin is administered into the amygdala immediately following the cue presentation, during the period of memory reconsolidation, then responding for alcohol is greatly reduced, indicating mTORC1 in the amygdala is important in encoding the reconsolidation of the alcohol memory [74]. Together, these data provide evidence that the mTORC1 signaling pathway is critically involved in the expression of a wide range of behaviors that develop due to alcohol exposure, including alcohol-drinking and the reconsolidation of the memory of alcohol reward.

13.5.4 Nicotine

Overall, there is very little known about how nicotine affects mTOR signaling in the nervous system and whether mTOR is necessary for the development of nicotine-induced behaviors. One study showed that locomotor sensitization to nicotine increases mTORC1 activity selectively in the basolateral amygdala, and that rapamycin injected directly into the basolateral amygdala prior to each nicotine injection blocks locomotor sensitization. Also, rapamycin injected prior to the challenge nicotine injection following a period of withdrawal blocks locomotor sensitization, indicating that rapamycin blocks both the development and expression of nicotine-induced locomotor sensitization [75]. The role of mTORC1 in mediating nicotine's behavioral effects is likely regionally selective, as intra-VTA rapamycin does not block locomotor sensitization to nicotine [76]. Despite these studies showing mTORC1's role in nicotine-induced behavioral sensitization, there have been no further studies on whether mTOR is necessary for other nicotine behaviors, such as CPP, self-administration, or reinstatement.

There has been more research on nicotine and mTORC1 with regard to the carcinogenic effects of nicotine and other tobacco products. For example, nicotine and the tobacco product 4-(methylnitrosamino)-1-(3-pyridyl)-1-butanone (nicotine-derived nitrosamine ketone) increases Akt and mTOR activity in three lung cancer lines. Furthermore, nicotine increases proliferation of the non-small-lung cancer cell line (NSLCL), an effect that is dependent on Akt [77]. Nicotine increases the expression of two transcription factors in NSLCLs, peroxisome proliferator-activated receptor beta/delta subtype and hypoxia-inducible factor 1 alpha, both of which are dependent on mTORC1 activation and are tumorigenic [78,79]. Finally, nicotine exposure increases the expression of survivin, an inhibitor of apoptosis, in normal human bronchial epithelial cells via activation of mTORC1 [80]. While there is little known about whether mTORC1 is important for the rewarding, motivational, or cognitive aspects of nicotine use, it is apparent that nicotine activates mTORC1 in cells outside of the nervous system, such as the bronchioles and lungs, and promotes tumorigenesis while inhibiting apoptosis.

13.5.5 Cannabis

The effects of the psychoactive chemicals from the cannabis plant, namely THC, on the addiction neurocircuitry have been less elucidated compared to other drugs of abuse. However, it has been shown that THC or other CB1 agonists activate mTORC1 in the hippocampus, an effect that is blocked by the CB1 antagonist Rimonabant [81]. THC impairs performance on a novel object-recognition task, a behavioral test that assesses recognition memory. The impairment due to THC is reversed by rapamycin, suggesting that the amnesic effect of THC is due to mTORC1 activation. CB1 receptors are localized presynaptically, and THC's ability to impair performance on the novel object task depends on CB1 localized on GABA neurons and also NMDA receptors, suggesting that the mechanism of action is that THC inhibits GABA release from presynaptic terminals, thereby disinhibiting glutamate inputs and leading to aberrant mTORC1 activation [81].

The mechanism by which THC activates mTORC1 is not clear, however it is known that THC activates mTOR in a number of brain regions, including the hippocampus, striatum, frontal cortex, and amygdala [82]. THC causes a range of symptoms, including hypothermia, hypolocomotion, analgesia, and catalepsy, and may also be anxiolytic at low doses and anxiogenic at higher doses [11]. mTORC1 plays a role in the impairment of recognition memory, and it is also important for the anxiogenic effect [82]. However, many of the psychoactive effects of THC are independent of mTORC1.

Cannabinoids have differential effects on cell viability depending on the specific compound. For example, THC increases cell viability of dissociated striatal cells in culture following NMDA treatment, an effect that is blocked by inhibiting Akt or mTORC1. In these cells, THC increases *BDNF* gene expression, an effect that requires mTORC1. In an animal model of Huntington's disease, *BDNF* mRNA is reduced in the striatum, and this is reversed by overexpression of CB1 receptors, providing a link between CB1, mTORC1, and *BDNF* expression [83]. Conversely, cannabidiol, the other major phytocannabinoid, decreases cell viability in human breast cancer cell lines by inducing apoptosis and autophagy in a CB1-independent manner. In this case, cannabidiol inhibits the activation of mTORC1 [84]. More studies are needed to determine how THC and cannabidiol interact with mTORC1 signaling, and whether mTORC1 plays a role in mediating any of the effects of THC or cannabidiol on brain function.

13.5.6 Dissociative Hallucinogens

This class of drugs may be the least abused of the drug classes that have been summarized, but their interactions with mTOR signaling are intriguing. The most well-known drug of the arylcyclohexylamine class, ketamine, has typically been used as a dissociative anesthetic, and is abused because of its potent hallucinogenic properties. Recently, it was discovered that ketamine has rapid antidepressant properties, an effect that is mediated by mTORC1. For example, ketamine activates mTORC1 in the PFC and triggers an increase in proteins associated with excitatory synaptic transmission, such as PSD-95, GluA1, and Arc. Ketamine also increases dendritic spine density and overall glutamatergic synaptic physiology in the PFC [13]. It is thought that increased excitatory synaptic transmission in the PFC is at least partially responsible for ketamine's antidepressant effects. This is supported by the finding that levels of mTORC1 and downstream target S6K are decreased in the postmortem cortical tissue from depressed patients [85]. The dilemma with using ketamine for therapeutic purposes is that the drug drastically alters consciousness and has severe side effects. However, combining a subthreshold dose of ketamine with lithium, a glycogen synthase kinase 3 beta inhibitor, also robustly activates mTORC1 and increases excitatory synaptic transmission in the PFC [86]. The actions of ketamine are mimicked by genetic deletion of the NMDA GluN2B subunit, suggesting that other drugs that inhibit this receptor subunit may also have antidepressant properties. While counterintuitive because of GluN2B's important role in facilitating excitatory synaptic transmission, deleting GluN2B increases spine density and local dendritic protein translation in PFC neurons [87]. GluN2B may be acting as a sensor of neuronal excitation, and when glutamate levels are too high, it reduces local protein synthesis. It is possible that ketamine, which is a potent NMDA channel blocker, increases protein synthesis by blocking incoming synaptic input.

Activation of mTORC1 in the PFC may have antidepressant properties, but it may affect other aspects of cognition. For example, stimulation of the serotonergic 5-HT6 receptor, which activates mTORC1 in the PFC, impairs novel object recognition, an effect that is reversed by rapamycin. Either neonatal treatment of the arylcyclohexylamine PCP, or social isolation beginning immediately after weaning and lasting for a few weeks, also activates mTORC1 in the PFC and impairs novel object recognition, which again is reversed by inhibiting mTORC1 [28]. This study points to the complex nature of mTORC1's actions in the PFC, which may be antidepressant in some instances, but may also be potentially detrimental.

13.6 CONCLUSIONS AND FUTURE DIRECTIONS

Drugs of abuse are diverse both in terms of their pharmacological site of action and their behavioral effects. However, drugs of abuse share cellular, physiological, and behavioral commonalities that can be modeled in rodents. This review lays out the possibility that the central molecular link between the reinforcing actions of most drugs of abuse is mTORC1 (Figure 13.1). Targeting mTORC1 for the treatment of substance-use disorders is a promising direction because, unlike other molecules that have been targeted, mTOR signaling may be a more general phenomenon for all drugs of abuse. Inhibition of mTORC1 does not affect consumption of natural rewards like sucrose, is not anxiogenic, aversive, or rewarding, and does not alter locomotion in animal models [68]. Furthermore, the fact that a single administration of rapamycin leads to a long-lasting reduction in alcohol-seeking [74] suggests that a short course of treatment may be sufficient to reduce craving or relapse.

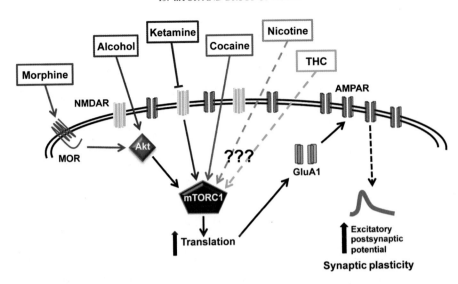

FIGURE 13.1   Activation of mTORC1 by drugs of abuse. Drugs of abuse differentially act on neuronal receptors and channels, but they trigger signaling pathways that converge on mTORC1. Morphine is an agonist at MORs, which can activate mTORC1 via Akt [29], while alcohol likely activates Akt/mTORC1 via H-Ras (H-Ras not shown) [69]. Ketamine activates mTORC1 via the inhibition of NMDA receptors, specifically the GluN2B subunit [13]. Cocaine acutely activates mTORC1 [47], although the mechanism is not known. It is also not known if/how THC or nicotine activate mTORC1, although both THC and nicotine produce behaviors that require activation of mTORC1 [75,81]. MOR, mu opioid receptor; NMDAR, N-methyl-D-aspartate receptor; AMPAR, α-amino-3-hydroxy-5-methyl-4-isoxazolepropionic acid receptor; mTORC1, mechanistic target of rapamycin complex 1; GluA1, glutamatergic AMPA receptor subunit 1.

More research is needed to determine the role of mTOR in all stages of the development of substance-use disorders for each drug class, but mTORC1 inhibitors may be an efficacious therapy to treat this devastating disease.

Acknowledgments

This work was supported by NIH/NIAAA grants P50 AA017072 (DR) and F32 AA023703 (JTB).

References

[1] Luscher C, Malenka RC. Drug-evoked synaptic plasticity in addiction: from molecular changes to circuit remodeling. Neuron 2011;69:650−63.

[2] Koob GF, Volkow ND. Neurocircuitry of addiction. Neuropsychopharmacology 2010;35:217−38.

[3] American Psychiatric Association. Diagnostic and statistical manual of mental disorders. 5th ed. Arlington, VA: American Psychiatric Association; 2013.

[4] Redish AD, Jensen S, Johnson A. A unified framework for addiction: vulnerabilities in the decision process. Behav Brain Sci 2008;31:415−37.

[5] Buffington SA, Huang W, Costa-Mattioli M. Translational control in synaptic plasticity and cognitive dysfunction. Annu Rev Neurosci 2014;37:17−38.

[6] Sulzer D. How addictive drugs disrupt presynaptic dopamine neurotransmission. Neuron 2011;69:628−49.

[7] Ting-A-Kee R, van der Kooy D. The neurobiology of opiate motivation. Cold Spring Harb Perspect Med 2012;10:a012096.

[8] Heishman SJ, Kleykamp BA, Singleton EG. Meta-analysis of the acute effects of nicotine and smoking on human performance. Psychopharmacology 2010;210:453−69.

[9] Vengeliene V, Bilbao A, Molander A, Spanagel R. Neuropharmacology of alcohol addiction. Br J Pharmacol 2008;154:299−315.

[10] Ameri A. The effects of cannabinoids on the brain. Prog Neurobiol 1999;58:315−48.

[11] Tambaro S, Bortolato M. Cannabinoid-related agents in the treatment of anxiety disorders: current knowledge and future perspectives. Recent Patents CNS Drug Discov 2012;7:25−40.

[12] Bey T, Patel A. Phencyclidine intoxication and adverse effects: a clinical and pharmacological review of an illicit drug. Cal J Emerg Med 2007;8:9−14.

[13] Li N, Lee B, Liu R-J, Banasr M, Dwyer JM, Iwata M, et al. mTOR-dependent synapse formation underlies the rapid antidepressant effects of NMDA antagonists. Science 2010;329:959−64.

[14] Di Chiara G, Imperato A. Drugs abused by humans preferentially increase synaptic dopamine concentrations in the mesolimbic system of freely moving rats. Proc Natl Acad Sci USA 1988;85:5274−8.

[15] Baik J-H. Dopamine signaling in reward-related behaviors. Front Neural Circuits 2013;7:152.

[16] Diana M. The dopamine hypothesis of drug addiction and its potential therapeutic value. Front Psychiatry 2011;2:64.

[17] Feil J, Sheppard D, Fitzgerald PB, Yücel M, Lubman DI, Bradshaw JL. Addiction, compulsive drug seeking, and the role of frontostriatal mechanisms in regulating inhibitory control. Neurosci Biobehav Rev 2010;35:248−75.

[18] Kalivas PW. The glutamate homeostasis hypothesis of addiction. Nat Rev Neurosci 2009;10:561−72.

[19] Goldstein RZ, Volkow ND. Dysfunction of the prefrontal cortex in addiction: neuroimaging findings and clinical implications. Nat Rev Neurosci 2011;12:652−69.

[20] Ron D, Jurd R. The "ups and downs" of signaling cascades in addiction. Sci STKE 2005;2005:re14.

[21] Ahmadiantehrani S, Warnault V, Legastelois R, Ron D. From signaling pathways to behavior: the light and dark sides of alcohol. In: Noronha A, Cui C, Harris RA, Crabbe JC, editors. Neurobiology of alcohol dependence. San Diego: Academic Press; 2014. p. 155−71.

[22] Steketee JD, Kalivas PW. Drug wanting: behavioral sensitization and relapse to drug-seeking behavior. Pharmacol Rev 2011;63:348−65.

[23] Strakowski SM, Sax KW, Setters MJ, Keck PE. Enhanced response to repeated d-amphetamine challenge: evidence for behavioral sensitization in humans. Biol Psychiatry 1996;140: 872−80.

[24] Napier TC, Herrold AA, de Wit H. Using conditioned place preference to identify relapse prevention medications. Neurosci Biobehav Rev 2013;37:2081−6.

[25] Childs E, de Wit H. Amphetamine-induced place preference in humans. Biol Psychiatry 2009;65:900−4.

[26] Hoeffer CA, Klann E. mTOR signaling: at the crossroads of plasticity, memory and disease. Trends Neurosci 2010;33:67−75.

[27] Santini E, Heiman M, Greengard P, Valjent E, Fisone G. Inhibition of mTOR signaling in Parkinson's disease prevents L-DOPA-induced dyskinesia. Sci Signal 2009;2:ra36.

[28] Meffre J, Chaumont-Dubel S, Mannoury la Cour C, Loiseau F, Watson DJ, Dekeyne A, et al. 5-HT(6) receptor recruitment of mTOR as a mechanism for perturbed cognition in schizophrenia. EMBO Mol Med 2012;4:1043−56.

[29] Polakiewicz RD, Schieferl SM, Gingras AC, Sonenberg N, Comb MJ. mu-Opioid receptor activates signaling pathways implicated in cell survival and translational control. J Biol Chem 1998;273:23534−41.

[30] Takei N, Inamura N, Kawamura M, Namba H, Hara K, Yonezawa K, et al. Brain-derived neurotrophic factor induces mammalian target of rapamycin-dependent local activation of translation machinery and protein synthesis in neuronal dendrites. J Neurosci 2004;24:9760−9.

[31] Perkinton MS, Sihra TS, Williams RJ. Ca(2+)-permeable AMPA receptors induce phosphorylation of cAMP response element-binding protein through a phosphatidylinositol 3-kinase-dependent stimulation of the mitogen-activated protein kinase signaling cascade in neurons. J Neurosci 1999;19: 5861−74.

[32] Perkinton MS, Ip J, Wood GL, Crossthwaite AJ, Williams RJ. Phosphatidylinositol 3-kinase is a central mediator of NMDA receptor signalling to MAP kinase (Erk1/2), Akt/ PKB and CREB in striatal neurones. J Neurochem 2002;80:239−54.

[33] Sutton MA, Taylor AM, Ito HT, Pham A, Schuman EM. Postsynaptic decoding of neural activity: eEF2 as a biochemical sensor coupling miniature synaptic transmission to local protein synthesis. Neuron 2007;55:648−61.

[34] Mendoza MC, Er EE, Blenis J. The Ras-ERK and PI3K-mTOR pathways: cross-talk and compensation. Trends Biochem Sci 2011;36:320−8.

[35] Ma XM, Blenis J. Molecular mechanisms of mTOR-mediated translational control. Nat Rev Mol Cell Biol 2009;10:307−18.

[36] Oh WJ, Jacinto E. mTOR complex 2 signaling and functions. Cell Cycle 2011;10:2305−16.

[37] Huang W, Zhu PJ, Zhang S, Zhou H, Stoica L, Galiano M, et al. mTORC2 controls actin polymerization required for consolidation of long-term memory. Nat Neurosci 2013;16:441−8.

[38] Mazei-Robison MS, Koo JW, Friedman AK, Lansink CS, Robison AJ, Vinish M, et al. Role for mTOR signaling and

neuronal activity in morphine-induced adaptations in ventral tegmental area dopamine neurons. Neuron 2011;72:977−90.

[39] Lipton JO, Sahin M. The neurology of mTOR. Neuron. 2014;84:275−91.

[40] Costa-Mattioli M, Sossin WS, Klann E, Sonenberg N. Translational control of long-lasting synaptic plasticity and memory. Neuron 2009;61:10−26.

[41] Vézina C, Kudelski A, Sehgal SN. Rapamycin (AY-22,989), a new antifungal antibiotic. I. Taxonomy of the producing streptomycete and isolation of the active principle. J Antibiotics 1975;28:721−6.

[42] Hartford CM, Ratain MJ. Rapamycin: something old, something new, sometimes borrowed and now renewed. Clin Pharm Ther 2007;82:381−8.

[43] Lamming DW, Ye L, Katajisto P, Goncalves MD, Saitoh M, Stevens DM, et al. Rapamycin-induced insulin resistance is mediated by mTORC2 loss and uncoupled from longevity. Science 2012;335:1638−43.

[44] Bekinschtein P, Cammarota M, Igaz LM, Bevilaqua LRM, Izquierdo I, Medina JH. Persistence of long-term memory storage requires a late protein synthesis- and BDNF- dependent phase in the hippocampus. Neuron 2007;53:261−77.

[45] Tang SJ, Reis G, Kang H, Gingras A-C, Sonenberg N, Schuman EM. A rapamycin-sensitive signaling pathway contributes to long-term synaptic plasticity in the hippocampus. Proc Natl Acad Sci USA 2002;99:467−72.

[46] Stoica L, Zhu PJ, Huang W, Zhou H, Kozma SC, Costa-Mattioli M. Selective pharmacogenetic inhibition of mammalian target of Rapamycin complex I (mTORC1) blocks long-term synaptic plasticity and memory storage. Proc Natl Acad Sci USA 2011;108:3791−6.

[47] Wu J, McCallum SE, Glick SD, Huang Y. Inhibition of the mammalian target of rapamycin pathway by rapamycin blocks cocaine-induced locomotor sensitization. Neuroscience 2011;172:104−9.

[48] Gangarossa G, Ceolin L, Paucard A, Lerner-Natoli M, Perroy J, Fagni L, et al. Repeated stimulation of dopamine D1-like receptor and hyperactivation of mTOR signaling lead to generalized seizures, altered dentate gyrus plasticity, and memory deficits. Hippocampus 2014;24:1466−81.

[49] Bailey J, Ma D, Szumlinski KK. Rapamycin attenuates the expression of cocaine-induced place preference and behavioral sensitization. Addict Biol 2012;17:248−58.

[50] Yu F, Zhong P, Liu X, Sun D, Gao H-Q, Liu Q-S. Metabotropic glutamate receptor I (mGluR1) antagonism impairs cocaine-induced conditioned place preference via inhibition of protein synthesis. Neuropsychopharmacology 2013;38:1308−21.

[51] Lin J, Liu L, Wen Q, Zheng C, Gao Y, Peng S, et al. Rapamycin prevents drug seeking via disrupting reconsolidation of reward memory in rats. Int J Neuropsychopharmacol 2014;17:127−36.

[52] Shi X, Miller JS, Harper LJ, Poole RL, Gould TJ, Unterwald EM. Reactivation of cocaine reward memory engages the Akt/ GSK3/mTOR signaling pathway and can be disrupted by GSK3 inhibition. Psychopharmacology 2014;231:3109−18.

[53] James MH, Quinn RK, Ong LK, Levi EM, Charnley JL, Smith DW, et al. mTORC1 inhibition in the nucleus accumbens "protects" against the expression of drug seeking and "relapse" and is associated with reductions in GluA1 AMPAR and CAMKIIα levels. Neuropsychopharmacology 2014;39:1694−702.

[54] Brown AL, Flynn JR, Smith DW, Dayas CV. Down-regulated striatal gene expression for synaptic plasticity-associated proteins in addiction and relapse vulnerable animals. Int J Neuropsychopharmacol 2011;14:1099−110.

[55] Scheyer AF, Wolf ME, Tseng KY. A protein synthesis-dependent mechanism sustains calcium-permeable AMPA

receptor transmission in nucleus accumbens synapses during withdrawal from cocaine self-administration. J Neurosci 2014;34:3095—100.

[56] Wang X, Luo YX, He YY, Li FQ, Shi HS, Xue LF, et al. Nucleus accumbens core mammalian target of rapamycin signaling pathway is critical for cue-induced reinstatement of cocaine seeking in rats. J Neurosci 2010;30:12632—41.

[57] Warren BL, Iñiguez SD, Alcantara LF, Wright KN, Parise EM, Weakley SK, et al. Juvenile administration of concomitant methylphenidate and fluoxetine alters behavioral reactivity to reward- and mood-related stimuli and disrupts ventral tegmental area gene expression in adulthood. J Neurosci 2011;31:10347—58.

[58] Narita M, Akai H, Kita T, Nagumo Y, Sunagawa N, Hara C, et al. Involvement of mitogen-stimulated p70-S6 kinase in the development of sensitization to the methamphetamine-induced rewarding effect in rats. Neuroscience 2005;132:553—60.

[59] Li Y, Hu Z, Chen B, Bu Q, Lu W, Deng Y, et al. Taurine attenuates methamphetamine-induced autophagy and apoptosis in PC12 cells through mTOR signaling pathway. Toxicol Lett 2012;215:1—7.

[60] Kongsuphol P, Mukda S, Nopparat C, Villarroel A, Govitrapong P. Melatonin attenuates methamphetamine-induced deactivation of the mammalian target of rapamycin signaling to induce autophagy in SK-N-SH cells. J Pineal Res 2009;46:199—206.

[61] Ma J, Wan J, Meng J, Banerjee S, Ramakrishnan S, Roy S. Methamphetamine induces autophagy as a pro-survival response against apoptotic endothelial cell death through the Kappa opioid receptor. Cell Death Dis 2014;5:e1099.

[62] Shintani-Ishida K, Saka K, Yamaguchi K, Hayashima M, Nagai H, Takemura G, et al. MDMA induces cardiac contractile dysfunction through autophagy upregulation and lysosome destabilization in rats. Biochim Biophys Acta 2014;1842:691—700.

[63] Cheng SWY, Fryer LGD, Carling D, Shepherd PR. Thr2446 is a novel mammalian target of rapamycin (mTOR) phosphorylation site regulated by nutrient status. J Biol Chem 2004;279:15719—22.

[64] Olianas MC, Dedoni S, Onali P. Signaling pathways mediating phosphorylation and inactivation of glycogen synthase kinase-3β by the recombinant human δ-opioid receptor stably expressed in Chinese hamster ovary cells. Neuropharmacology. 2011;60:1326—36.

[65] Cui Y, Zhang XQ, Cui Y, Xin WJ, Jing J, Liu XG. Activation of phosphatidylinositol 3-kinase/Akt-mammalian target of Rapamycin signaling pathway in the hippocampus is essential for the acquisition of morphine-induced place preference in rats. Neuroscience 2010;171:134—43.

[66] Xu J-T, Zhao J-Y, Zhao X, Ligons D, Tiwari V, Atianjoh FE, et al. Opioid receptor-triggered spinal mTORC1 activation contributes to morphine tolerance and hyperalgesia. J Clin Invest 2014;124:592—603.

[67] Shi J, Jun W, Zhao L-Y, Xue Y-X, Zhang X-Y, Kosten TR, et al. Effect of rapamycin on cue-induced drug craving in abstinent heroin addicts. Eur J Pharmacol 2009;615:108—12.

[68] Neasta J, Ben Hamida S, Yowell Q, Carnicella S, Ron D. Role for mammalian target of rapamycin complex 1 signaling in neuroadaptations underlying alcohol-related disorders. Proc Natl Acad Sci USA 2010;107:20093—8.

[69] Ben Hamida S, Neasta J, Lasek AW, Kharazia V, Zou M, Carnicella S, et al. The small G protein H-Ras in the mesolimbic system is a molecular gateway to alcohol-seeking and excessive drinking behaviors. J Neurosci 2012;32:15849—58.

[70] Castellano E, Downward J. Role of RAS in the regulation of PI 3-kinase. Curr Top Microbiol Immunol 2010;346:143—69.

[71] Burgering BM, Coffer PJ. Protein kinase B (c-Akt) in phosphatidylinositol-3-OH kinase signal transduction. Nature 1995;376:599—602.

[72] Neasta J, Ben Hamida S, Yowell QV, Carnicella S, Ron D. AKT signaling pathway in the nucleus accumbens mediates excessive alcohol drinking behaviors. Biol Psychiatry 2011;70:575—82.

[73] Slipczuk L, Bekinschtein P, Katche C, Cammarota M, Izquierdo I, Medina JH. BDNF activates mTOR to regulate GluR1 expression required for memory formation. PLoS One 2009;4:e6007.

[74] Barak S, Liu F, Ben Hamida S, Yowell QV, Neasta J, Kharazia V, et al. Disruption of alcohol-related memories by mTORC1 inhibition prevents relapse. Nat Neurosci 2013;16:1111—17.

[75] Gao Y, Peng S, Wen Q, Zheng C, Lin J, Tan Y, et al. The mammalian target of rapamycin pathway in the basolateral amygdala is critical for nicotine-induced behavioural sensitization. Int J Neuropsychopharmacol 2014;17:1881—94.

[76] Addy NA, Fornasiero EF, Stevens TR, Taylor JR, Picciotto MR. Role of calcineurin in nicotine-mediated locomotor sensitization. J Neurosci 2007;27:8571—80.

[77] Tsurutani J, Castillo SS, Brognard J, Granville CA, Zhang C, Gills JJ, et al. Tobacco components stimulate Akt-dependent proliferation and NFkappaB-dependent survival in lung cancer cells. Carcinogenesis 2005;26:1182—95.

[78] Sun X, Ritzenthaler JD, Zhong X, Zheng Y, Roman J, Han S. Nicotine stimulates PPARbeta/delta expression in human lung carcinoma cells through activation of PI3K/mTOR and suppression of AP-2alpha. Cancer Res 2009;69:6445—53.

[79] Zhang Q, Tang X, Zhang Z-F, Velikina R, Shi S, Le AD. Nicotine induces hypoxia-inducible factor-1alpha expression in human lung cancer cells via nicotinic acetylcholine receptor-mediated signaling pathways. Clin Cancer Res 2007;13:4686—94.

[80] Jin Q, Menter DG, Mao L, Hong WK, Lee H-Y. Survivin expression in normal human bronchial epithelial cells: an early and critical step in tumorigenesis induced by tobacco exposure. Carcinogenesis 2008;29:1614—22.

[81] Puighermanal E, Marsicano G, Busquets-Garcia A, Lutz B, Maldonado R, Ozaita A. Cannabinoid modulation of hippocampal long-term memory is mediated by mTOR signaling. Nat Neurosci 2009;12:1152—8.

[82] Puighermanal E, Busquets-Garcia A, Gomis-González M, Marsicano G, Maldonado R, Ozaita A. Dissociation of the pharmacological effects of THC by mTOR blockade. Neuropsychopharmacology 2013;38:1334—43.

[83] Blázquez C, Chiarlone A, Bellocchio L, Resel E, Pruunsild P, García-Rincón D, et al. The CB1 cannabinoid receptor signals striatal neuroprotection via a PI3K/Akt/mTORC1/BDNF pathway. Cell Death Differ 2015;22:1618—29.

[84] Shrivastava A, Kuzontkoski PM, Groopman JE, Prasad A. Cannabidiol induces programmed cell death in breast cancer cells by coordinating the cross-talk between apoptosis and autophagy. Mol Cancer Ther 2011;10:1161—72.

[85] Jernigan CS, Goswami DB, Austin MC, Iyo AH, Chandran A, Stockmeier CA, et al. The mTOR signaling pathway in the prefrontal cortex is compromised in major depressive disorder. Prog Neuropsychopharmacol Biol Psychiatry 2011;35:1774—9.

[86] Liu R-J, Fuchikami M, Dwyer JM, Lepack AE, Duman RS, Aghajanian GK. GSK-3 inhibition potentiates the synaptogenic and antidepressant-like effects of subthreshold doses of ketamine. Neuropsychopharmacology 2013;38:2268—77.

[87] Miller OH, Yang L, Wang C-C, Hargroder EA, Zhang Y, Delpire E, et al. GluN2B-containing NMDA receptors regulate depression-like behavior and are critical for the rapid antidepressant actions of ketamine. Elife 2014;3:e03581.

mTOR IN CARDIOVASCULAR DEVELOPMENT AND DISEASE

mTOR as a Modulator of Metabolite Sensing Relevant to Angiogenesis

Soumya S.J.[1], Athira A.P.[2], Binu S.[2] and Sudhakaran P.R.[1,2,3]

[1]Inter-University Centre for Genomics and Gene Technology, University of Kerala, Thiruvananthapuram, Kerala, India
[2]Department of Biochemistry, University of Kerala, Thiruvananthapuram, Kerala, India [3]Department of Computational Biology and Bioinformatics, University of Kerala, Thiruvananthapuram, Kerala, India

14.1 INTRODUCTION

Angiogenesis is the process of formation of new blood vessels by sprouting of preexisting capillaries and occurs throughout postnatal life in both physiological and pathological conditions. Being an important process for the supply of oxygen and nutrients through establishing blood supply to target tissues, angiogenesis is tightly regulated by the action of a number of proangiogenic and antiangiogenic factors. Physiological angiogenesis occurs during embryogenesis, in the female reproductive system during the development of follicles, corpus luteum formation and embryo implantation [1,2]. Excessive angiogenesis occurs during disease conditions such as hemangiomas, psoriasis, obesity, and cancer, and insufficient angiogenesis occurs during wound healing and in ischemic conditions. In response to angiogenic stimuli, otherwise quiescent endothelial cells (ECs) are activated and undergo a sequence of events which includes the secretion of matrix metalloproteases (MMPs) and other proteases that degrade the extracellular matrix (ECM), cell migration into the newly created space, proliferation, establishing cell—cell and cell—matrix interactions and neovessel formation [3]. Multiple secreted factors, such as growth factors, cytokines, chemokines, guidance molecules, and their cognate receptors on the cell surface play a critical role in coordinating these processes. Changes in the extracellular microenvironment and oxygen tension and metabolite status within the cell can be linked with the angiogenic state of the cell. Hypoxic conditions can alter the oxidative metabolism resulting in accumulation of metabolites such as lactate and changes in adenosine triphosphate (ATP). Further, certain metabolites can accumulate consequent on their altered metabolism associated with certain pathological conditions. Cells sense these conditions through sensing mechanisms such as hypoxia-inducible factor (HIF) and metabolic regulators. These cause an increase in the expression of angiogenic factors particularly vascular endothelial growth factor (VEGF) thereby inducing angiogenic response. Among the various metabolic regulators in the cell, such as peroxisome proliferator-activated receptor (PPAR), peroxisome proliferator-activated receptor-gamma coactivator 1-α (PGC1α), 5′ adenosine monophosphate (AMP)-activated protein kinase (AMPK) and Forkhead box O (FOXO), mammalian target of rapamycin (mTOR) plays an important role in linking the metabolic microenvironment of the cell with its angiogenic potential. Detailed analyses of the structure and the mechanism of action and regulation of the mTOR protein kinase are given elsewhere in this volume. The following sections give a brief account of the recent advances in molecular and cellular mechanisms of angiogenesis and review the role of mTOR in linking metabolite status of the cell with angiogenesis by integrating various intracellular signaling pathways relevant to angiogenesis.

14.2 CELLULAR AND MOLECULAR MECHANISMS OF ANGIOGENESIS

Growth of blood vessels in postnatal life for organ and tissue development as well as during tissue repair in adult organs takes place through the process of angiogenesis. It involves an activation phase characterized by initiation and progression of neovascularization and resolution phase characterized by termination and stabilization of the vessels. During the last two and a half decades, intense research using different model systems has provided great insights into the molecular and cellular mechanisms of angiogenesis [4–7]. The characteristic features of angiogenesis comprise: (i) degradation of the basement membrane by proteases secreted by the ECs of existing vasculature, in response to the angiogenic stimuli, proliferation, and formation of tiny endothelial sprouts; (ii) directed migration of EC at the sprout tip towards the angiogenic stimuli and the proliferation of EC below the sprout; (iii) expression cell adhesion molecules such as selectins, CD31, integrins leading to establishment of cell–cell contact and intercellular junctions; (iv) formation of lumen; (v) stabilization of the vessel by organizing pericytes and vascular smooth muscle cells as supporting cells and deposition of basement membrane; and (vi) anastomosis with the existing vessels and establishing blood flow. The principal early event is the "activation" of quiescent EC mediated by the VEGF family of growth factors, particularly VEGF A, and their cognate receptors which have been recognized as master regulators of angiogenesis.

Apart from up-regulating proteases in response to VEGF, a small fraction of the EC that recognize the VEGF gradient acquire a specific tip cell phenotype which is characterized by the formation of numerous filopodia that extend towards the direction in which EC migrates. The ECs that trail tip cells, called "stalk cells," are less motile but support sprouting and vessel extension. Tip cells, being the leading ECs at the tips of vascular sprouts which regulate EC differentiation during angiogenesis, are guided towards avascular regions, where it can sense angiogenic factors using filopodia. Vascular endothelial growth factor receptor-2 (VEGFR2) and VEGFR3 expressed by tip cell filopodia can sense VEGF and activated tip cell expresses Delta-like ligand 4 (Dll4) which is a ligand for Notch receptor. Tip cell and stalk cell together form a vascular sprout and Dll4 binds to Notch receptors on stalk cells to regulate angiogenesis [8,9]. The stimulation of the Dll4/Notch signaling pathway in the tip cell suppresses the VEGF response in the following EC and induces its differentiation to form stalk cells of the lumen of sprouting vessel [10]. ECs behind the stalk cells are differentiated to "phalanx cells," which form the innermost layer of the blood vessel. Endothelial tip cells of two sprouts come together to form a new blood vessel. Tip cells also modulate angiogenesis by

secreting angiogenic factors such as proteases, chemokines, Apelin, angiopoietin-2 (Ang2), and platelet-derived growth factor (PDGF) [11–13]. Excessive tip cell formation results in hyperdense non-functional vasculature. Notch regulates tip cell activity by increasing the expression of VEGFR1, which is a negative regulator of tip cell differentiation [14].

Once initiated, EC sprouting continues in a highly directional manner until tip cells connect with adjacent vessels and undergo anastomosis, which leads to the fusion of the contacting vessels. Proangiogenic growth factors and various molecules, such as semaphorins and ephrins, mediate invasive tip cell sprouting. During sprout elongation, "stalk cells" maintain connectivity with parental vessels and initiate partitioning defective-3-mediated vascular lumen morphogenesis. Expression of VEGFR1, activation of Notch, roundabout homolog 4 (Robo 4) and Wnt signaling in stalk cells repress tip cell behavior to maintain the hierarchical organization of sprouting ECs. Shuffling and exchange of positions between tip cells and stalk cells also occur during angiogenic sprouting. Upon contact with other vessels, tip cell behavior is repressed and vessels fuse by the process of anastomosis, which is assisted by associated myeloid cells. On contact with other ECs, tip cells lose their motile phenotype, generate tight endothelial cell–cell junctions and fuse with recipient vessels to form a continuous unobstructed lumen, which allows blood flow. Pericytes and vascular smooth muscle cells are recruited to developing vasculature by PDGF-β and transforming growth factor-β [15,16]. EC sprouting behavior is suppressed by the deposition of basement membrane at the abluminal surface and strengthening of cell–cell junctions which reestablish a mature quiescent phenotype.

14.3 SIGNALING PATHWAYS IN ANGIOGENESIS

Angiogenesis is regulated by growth factors such as VEGF, fibroblast growth factor (FGF), PDGF, hepatocyte growth factor, Ang1, Ang2, placental growth factor (PlGF), epidermal growth factor, and ephrins. Angiogenic growth factors bind to receptors on ECs and ligand–receptor binding induces several signal transduction pathways resulting in EC proliferation, migration, and neovascularization. Ligand–receptor complexes, such as VEGF/VEGF receptors, Angiopoietin/Tie2 receptors, Dll4/Notch receptors, Ephrin/Eph receptors and Slit/Roundabout (Slit/Robo) receptors are mainly involved in inducing signaling events relevant to angiogenesis [17].

14.3.1 VEGF/VEGFR Signaling

VEGF is the most important proangiogenic growth factor which can potentiate microvascular hyperpermeability accompanied with angiogenesis. The VEGF family comprises seven members, of which VEGF-A represents the most potent mitogenic and chemoattractant signal for EC. VEGF binds to VEGF-receptor and activates multiple downstream signaling molecules, such as mitogen-activated protein kinase (p38MAPK), Phosphatidylinositol-3-kinases (PI3K), protein kinase B (PKB) (Akt), phospholipase Cγ (PLCγ), and small GTPase that can stimulate EC functions relevant to angiogenesis [18,19]. Three VEGF tyrosine kinase receptors have been identified: the fms-like tyrosine kinase Flt-1 (VEGFR-1/Flt-1), the kinase domain region, also referred to as fetal liver kinase (VEGFR-2/KDR/Flk-1), and Flt-4 (VEGFR-3). VEGFR-2 appears to be the most important receptor in VEGF-induced mitogenesis and permeability. VEGF/VEGFR2 binding induces the production of platelet-activating factor (PAF) by ECs, stimulates their mitosis and migration, and increases vascular permeability. PAF stimulates the migration of ECs and also induces the production of FGF and angiogenic chemokines. VEGF/VEGFR binding results in several intracellular downstream signaling events leading to EC proliferation, survival, migration, and permeability. Binding of VEGF to VEGFR results in receptor dimerization and activation leading to autophosphorylation of tyrosine residues in the cytoplasmic domains of VEGF-receptor creating a binding site for the VEGFR-associated protein, Shc-related adaptor protein (Sck), and PLCγ1 [20]. PLCγ1 activates PKC-Ras-ERK signaling, translocating extracellular signal-regulated kinase (ERK) to the nucleus and activating transcription factors such as c-jun and c-fos promoting gene transcription relevant to EC proliferation. VEGF/VEGFR binding also activates p38MAPK and focal adhesion kinase leading to actin reorganization and EC migration. It also activates the PI3K signaling pathway which mediates the effect of VEGF on ECs. ERK/MAPK pathways could induce angiogenesis but not vascular permeability, whereas activation of the PI3K pathway was required for both angiogenesis and vascular permeability [21]. VEGF produced by ECs can also exert an autocrine effect [22].

14.3.2 Angiopoietin/Tie Signaling

Angiopoietin/Tie signaling maintains vascular quiescence and homeostasis of vessels by promoting EC survival and vascular maturation. Ang/Tie signaling comprises Ang1, Ang2, Tie1, and Tie2. Ang1/Tie2 stimulates deposition of basement membrane and thereby vessel maturation through the activation of the ERK/MAPK pathway and PI3K-Akt pathway [23]. Conversely, Ang2 competes with Ang1 for binding to Tie2 and destabilizes vasculature [24]. But the inhibitory effect of Ang2/Tie2 on vessel leakage has also been reported [25]. It appears that the activity of Ang2 may depend on the activation state of endothelium. The function of the Tie1 receptor is to modulate the activity of Tie2 [26].

14.3.3 Ephrin/Eph Signaling

Ephrin/Eph signaling complex comprises Ephrin ligands A and B, Eph receptors A and B. Ephrin/Eph signaling inhibits endothelial sprouting and angiogenesis induced by VEGF. Conflicting role of Ephrin/Eph signaling on angiogenesis was also reported. Loss of Ephrin B2 caused inhibition of tip cell filopodial extension and endothelial sprouting, whereas in mouse xenograft models of cancer, EphB4 reduced angiogenesis [27,28].

14.3.4 Slit/Robo Signaling

Neuronal signaling system Slit ligand/Roundabout receptor signaling also regulates angiogenesis by participating in blood vessel guidance. Slit ligands 1, 2, and 3 and Robo receptors 1, 2, 3, and 4 are present. Among these, Robo 1 and 4 are involved in angiogenesis [29]. Slit 2 on tumor cells interacts with Robo1 on ECs to induce angiogenesis in a PI3K-dependent manner as well as by activating mTORC2-Akt-Rac signaling pathway [30]. Robo4 is reported to possess conflicting roles on angiogenesis; some reports suggest the inhibitory effect of Robo4 on angiogenesis. Robo4 possesses proangiogenic effect also as Robo4-Fc competes with ligand for Robo 4 and inhibits angiogenesis [31]. The angiostatic effect of Robo4 was evidenced by stimulation of VEGF-mediated angiogenesis in Robo4 knockout mice [32]. In a mammary fat pad transplant assay, inactivation of Robo4 caused an increase in the production of VEGF. Robo4 also inhibits angiogenesis by inhibiting Src signaling downstream of VEGF [33].

14.3.5 Dll4/Notch Signaling

As indicated above, Dll4/Notch signaling regulates vascular sprouting by controlling the fate of tip (leading)/stalk (following) cells. As indicated before, tip cells are responsible for branching and stalk cells are responsible for tube formation. Tip cells can sense angiogenic signals in the microenvironment through filopodia.

VEGF binds to VEGFR2 in the tip cell filopodia leading to the up-regulation of Dll4 via the PI3K-Akt pathway [34]. Dll4 binds to Notch receptor and signals following ECs to become stalk cells [35]. *Ex vivo* studies showed that VEGF treatment up-regulated Dll4 in mouse retina [36]. Inhibition of Dll4/Notch signaling leads to the inhibition of angiogenesis. In xenograft mouse tumor models, treatment with anti-Dll4 inhibited angiogenesis induced by VEGF [37].

14.4 PI3K-Akt SIGNALING DOWNSTREAM OF VEGF

The PI3K-Akt signaling pathway and MEK-ERK signaling pathway are two major mitogenic signaling pathways. mTOR integrates these signaling pathways to regulate the mitogenic action of various growth factors. The PI3K-Akt signaling pathway, an intracellular signaling pathway activated by receptor tyrosine kinases, is important in regulating angiogenesis. Phosphatidylinositol 3-kinases (PI3K) are a family of enzymes that phosphorylate the $3'$-OH of the inositol ring of phosphatidylinositol. Phosphatidylinositol 3,4,5-triphosphate (PIP3) is an important lipid second messenger generated by PI3K, which plays a key role in several signal transduction pathways. PIP3 activates the serine/threonine kinases, phosphatidylinositol-dependent kinase-1 (PDK1) and Akt. The phosphatase and tensin homolog (PTEN) gene encodes a phosphatase that opposes the action of PI3K, thereby reducing the level of activated Akt by dephosphorylating PIP3 to phosphatidylinositol 4,5-diphosphate (PIP2) [38].

Akt is a serine/threonine kinase (also known as PKB) which plays a critical role in regulating diverse cellular functions including EC survival, EC migration, gene transcription, and protein synthesis relevant to angiogenesis. Akt controls protein synthesis and cell growth by leading to the phosphorylation of mTOR. mTOR phosphorylates p70-S6 kinase-1 (S6K1) and 4E-binding protein 1 (4E-BP1) regulating protein synthesis relevant to angiogenesis. Akt mediates cell survival by directly inhibiting proapoptotic proteins like Bcl-2-associated death promoter and caspase-9. Endothelial nitric oxide synthase (eNOS) activated by Akt generates nitric oxide, resulting in an increase in vascular permeability and cellular migration. Akt also inhibits glycogen synthase kinase 3β (GSK3β) and increases β-catenin activity which translocates to the nucleus and increases transcription of genes involved in angiogenesis [38] (Figure 14.1).

The paracrine loop between endothelial—non-ECs through various growth factors develops an angiogenic environment which activates ECs for neovascularization. The PI3K-Akt-mTOR signaling pathway is known to play a key role in angiogenesis and VEGF, in part,

signals through the PI3K-Akt-mTOR pathway to regulate angiogenesis. It was reported that the metabolite status of the cell also modulates the production of VEGF through the activation of the PI3K-Akt-mTOR signaling pathway. There exists an autocrine feedforward loop in tumor cells where tumor-derived VEGF induces production of VEGF via VEGFR2-dependent activation of mTOR. Nutrient sensing kinase mTOR can sense the metabolite status of the cell and coordinate with the production of angiogenic growth factors. mTOR is activated by proangiogenic growth factors, whereas it is inhibited by nutrient shortage. mTOR is also involved in hypoxia-mediated angiogenesis by inducing the translation of HIF1α. Therefore, understanding the nutrient sensing and integration of signaling pathways by mTOR may lead to the development of strategies targeting mTOR for antiangiogenic treatment.

14.5 mTOR-A METABOLITE SENSOR

It has become increasingly evident that mTOR functions as a central regulator of cell growth by regulating translation, transcription, ribosome biogenesis, nutrient transport, and autophagy in response to nutrient availability. As described elsewhere in this volume in detail, mTOR is a serine/threonine kinase and downstream target of the PI3K-Akt signaling pathway. It is a member of the phosphatidylinositol 3-kinase-related kinase family. It contains a number of HEAT repeats within its N-terminus (It contains four proteins that give rise to the acronym HEAT (**H**untingtin, **e**longation factor 3 (EF3), protein phosph**a**tase 2A (PP2**A**), and the yeast kinase **T**OR1), a FKBP-12-rapamycin-associated protein (FRAP), ataxia—telangiectasia mutated (ATM), transformation/transcription domain-associated protein (TRRAP) domain (FRAP-ATM-TRRAP [FAT]) and an FAT C-terminal domain. HEAT and FAT domains are distinct functional motifs that mediate protein interactions. In addition to this, mTOR contains a FKBP12-rapamycin-binding domain that is the minimal region required to bind FKBP12/rapamycin and lies N-terminal to the kinase domain. Recent evidence has also indicated the presence of a nuclear shuttling signal within mTOR involving Leu545 and Leu547, although the mechanism behind mTOR nuclear transport is yet to be elucidated [39].

mTOR is present in two distinct complexes: mTOR complex 1 (mTORC1) and mTOR complex 2 (mTORC2). mTORC1 is a heterotrimeric protein kinase consisting of mTOR catalytic subunit and associated proteins, raptor (regulatory associated protein of mTOR) and GβL (G protein β-subunit like protein). The core component of mTORC1 is raptor that binds

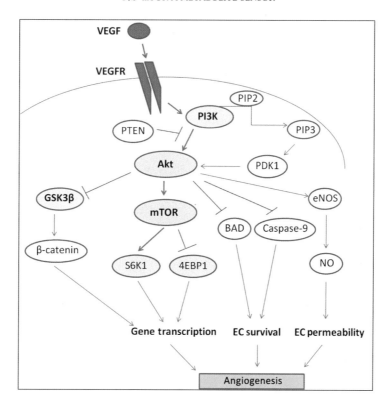

FIGURE 14.1 Activation of mTOR by VEGF signaling. VEGF mediates its effect via VEGFR2, which activates the Akt-mTOR signaling pathway through PI3K that converts PIP2 to PIP3, which further activates PDK1. Activated PDK1 activates Akt, leading to the activation of mTOR and S6Kinase which finally promotes angiogenesis. PTEN opposes the action of PI3K, by converting PIP3 to PIP2 thereby reducing the level of activated Akt. Activated Akt up-regulates the production of eNOS causing an increase in the production of nitric oxide, which is a key signaling molecule involved in EC migration and permeability. Akt inhibits BAD and caspase-9, key molecules involved in apoptosis and it also inhibits GSK3β, which is a key molecule involved in the phosphorylation of β catenin. mTOR, mammalian target of rapamycin; VEGF, vascular endothelial growth factor; VEGFR2, vascular endothelial growth factor receptor 2; PTEN, phosphatase and tensin homolog; PIP2, phosphatidylinositol 4,5-diphosphate; PIP3, phosphatidylinositol 3,4,5-triphosphate; EC, endothelial cell; BAD, Bcl-2-associated death promoter; GSK3β, glycogen synthase kinase 3β. (⟶ Activation, ⊣ Inhibition).

to HEAT repeats. It also contains LST8 (lethal with sec13 protein 8) which acts as a TOR interactor and stabilizes mTOR−Raptor interactions [39]. Eukaryotic initiation factor 4E-BP1, 2, and 3 and the p70 S6 kinases (S6K1 and S6K2) act as downstream signaling molecules of mTORC1 in regulating cell growth and proliferation. The targets of S6K1 include ribosomal protein S6, S6K1 Aly/REF-like Target, eukaryotic translation initiation factor 4B, and eukaryotic elongation factor 2 kinase [39]. S6K and 4E-BP proteins contain a TOR signaling motif to facilitate raptor binding. 4E-BP1 also contains an RAIP motif which is necessary for efficient phosphorylation of substrate by mTORC1. mTORC2 is a heterotetrameric protein kinase consisting of mTOR, GβL, and instead of raptor, it contains two other proteins; rapamycin insensitive companion of mTOR (rictor) and MAPK associated protein 1 (mSin1) [39].

mTORC1 is rapamycin-sensitive and converges multiple signals reflecting the availability of growth factors, nutrients, or energy to promote either cellular growth or catabolic processes during stress [40]. TSC1−TSC2 (tuberous sclerosis complex) through its GAP (GTPase activating protein) activity towards the small G protein Rheb (Ras homolog enriched in brain) is a critical negative regulator of mTORC1. Growth factors signal to mTORC1 via Akt, which inactivates TSC2 to activate mTORC1. Low ATP levels lead to the AMPK-dependent activation of TSC2 and phosphorylation of raptor to reduce mTORC1 signaling. Amino acid availability is signaled to mTORC1 through signaling involving the Ras-related GTPase (Rag) and Ragulator complex protein LAMTOR1. mTORC1 modulates biological processes by acting through downstream targets such as 4E-BP1 and p70S6 Kinase. mTORC2, which is not directly susceptible to rapamycin, promotes cellular survival by activating Akt. mTORC2 also regulates organization of actin cytoskeleton by activating protein kinases Cα and regulates ion transport and growth via serum and glucocorticoid-regulated kinase 1 (SGK1) phosphorylation. Abnormal mTOR signaling occurs during cancer, cardiovascular disease, and metabolic disorders [40].

mTOR regulates cellular processes such as cell cycle progression and hypoxia by signaling through substrates other than S6K and 4E-BP. mTOR-SGK1-p27 signaling is an important modulator of cell cycle progression. mTOR acts as a positive regulator of HIF-1α and thereby induces expression of genes involved in angiogenesis, erythropoiesis, and glucose metabolism. The PI3K/Akt/mTOR signaling pathway up-regulated the expression of HIF-1α in prostate cancer cells [41], while treatment with rapamycin down-regulated the expression of HIF-1α in both normoxic and hypoxic prostate cancer cells [42].

14.6 mTOR INTEGRATING HYPOXIA AND ANGIOGENESIS

Hypoxic regions are hypovascular and hypoxia demands angiogenesis to meet the metabolic requirements of cells by up-regulating the expression of angiogenic growth factors, particularly VEGF through HIF1α [43]. During hypoxia, HIF accumulates and is transported to the nucleus where it induces the expression of growth factors which in a paracrine manner induce EC proliferation and migration. PI3K and Ras pathways can also increase the expression of HIF. ECs also possess oxygen-sensing mechanisms including nicotinamide adenine dinucleotide (NAD+) phosphate oxidases, eNOS, and heme oxygenases [44]. ECs also express oxygen sensors that interact with HIF and this is an important molecular mechanism for detecting changes in oxygen tension. HIF-1α can heterodimerize with HIF-1β/aryl hydrocarbon nuclear translocator subunit to form transcriptionally active complex which can induce expression of genes involved in angiogenesis [45]. The activity of HIF is regulated by O$_2$ sensors such as prolyl hydroxylase domain (PHD) proteins and factor inhibiting HIFs [46]. PHD and HIF are expressed in ECs and O$_2$ sensing by PHD/HIF system modulates angiogenesis [47]. Of the three isoforms, HIF-1α regulates EC proliferation, migration, sprouting, and stabilization of vessels by recruiting pericytes and SMCs, whereas HIF-2α regulates vascular integrity, assembly, and morphogenesis. HIF-1α also stimulates the expression of VEGF, VEGFR2, and intracellular adhesion molecule [48–50]. Embryonic expression of a dominant-negative HIF that inhibits both HIF-1α and HIF-2α leads to defective vascular remodeling and failed vascular sprouting in the yolk sac and embryo [51]. HIF promotes differentiation of vascular progenitors to ECs to induce angiogenesis [52]. Transplantation of HIF-1α-overexpressing endothelial progenitor cells improves revascularization of ischemic hindlimbs [53].

There is increasing evidence indicating a role for mTOR in sensing and integrating hypoxia-responsive pathways in modulating angiogenesis. mTOR can sense variation in oxygen tension and modulate angiogenesis by inducing the translation of HIF. mTORC1 promotes glucose uptake and glycolysis by inducing the expression of HIF1α. But mTORC1 activates HIF1α while it is inhibited by hypoxia as a part of the feedback mechanism. Hypoxia down-regulates mTORC1 signaling and this is regulated by HIF by up-regulating the expression of Redd1(regulated in development and DNA damage responses 1) and Redd2 which further inhibits the phosphorylation of 4E-BP1 and S6K1 [54,55]. Inhibition of mTORC1 by Redd1 occurs even under constitutive Akt activation and thus overrides hypoxic stress and repress mitogenic activation of mTORC1 [56]. Bnip3 (Bcl-2/adenovirus E1B 19-kDa interacting protein 3) is also involved in hypoxic signaling through mTORC1. HIF-1α induces the expression of Bnip3 during hypoxia [57]. Bnip3 interacts with Rheb and inhibits Rheb's ability to activate mTORC1 and subsequently causes S6K dephosphorylation. Both mTORC1 and mTORC2 are necessary for hypoxia-induced EC proliferation. An additional regulatory mechanism of mTOR signaling via cytoplasmic/nuclear shuttling was reported. In rat aortic ECs, induction of hypoxia promotes phosphorylation of mTOR and causes nuclear translocation of phosphorylated mTOR to regulate angiogenesis.

14.7 mTOR INTEGRATING INFLAMMATION AND ANGIOGENESIS

Angiogenesis and inflammation are codependent and, during inflammation, angiogenesis is initiated by the activation of different cell populations, which release a variety of angiogenic factors. Chronic inflammatory diseases are known to be associated with aberrant angiogenesis, such as appearance of newly formed blood vessels in granulation tissue. Inflammation and angiogenesis are capable of potentiating each other and there is growing evidence that the angiogenesis that accompanies chronic inflammation tends to prolong and intensify the inflammatory response. Antiangiogenic drugs can be used for the treatment of inflammatory diseases. Inflammatory mediators are capable of inhibiting or promoting angiogenesis. During inflammation, infiltrating and resident inflammatory cells are capable of producing angiogenic regulators including VEGF and cytokines. Angiogenic regulators exhibit proangiogenic or antiangiogenic properties that may influence the intensity of the angiogenic response elicited during inflammation. Chemokines induce VEGF-A gene expression, suggesting that the chemokine works in concert with VEGF to promote angiogenesis during inflammation. Lipid inflammatory mediators, metabolic products of cyclooxygenase (COX) and lipoxygenase (LOX) also regulate angiogenesis by regulating the production of VEGF. Inflammatory cytokine tumor necrosis factor-α (TNF-α) has been implicated in the priming of endothelial "tip cells" for migration induced by VEGF.

mTOR acts as a regulator linking inflammation to angiogenesis. IκB kinase β, downstream of TNF-α signaling, inactivates TSC1 leading to activation of mTOR thereby enhancing angiogenesis. mTOR promoted macrophage induced tumor angiogenesis through downstream effector signal transducer and activator of transcription 3 (STAT3) and inhibition of mTOR resulted in macrophage mediated antitumor effect. By activating the Janus kinase (JAK)—STAT signaling pathway, mTOR modulates secretion of matrix remodeling enzymes such as MMP2, MMP9, urokinase plasminogen activator (uPA), and plasminogen activator inhibitor-1 [58], which as indicated before play a critical role in angiogenesis.

14.8 mTOR INTEGRATING METABOLITE STATUS AND ANGIOGENESIS

Metabolites regulate the process of angiogenesis and ECs are equipped with mechanisms to sense cellular metabolite concentrations. mTOR acts as a major cellular metabolite sensor and the ability of mTOR to increase the expression of genes involved in neovascularization likely contributes to modulation of angiogenic process.

14.8.1 Metabolite Status of the Cell and Angiogenesis

Metabolite alterations occur in ECs in avascular tissues to survive stress and hypoxic conditions and to induce angiogenesis by modulating the expression of VEGF. Cells in avascular tissue experience insufficiency of oxygen and nutrients and associated metabolic alteration and activate oxygen and metabolite-sensing mechanisms. Metabolites of carbohydrate, lipid, and amino acid contribute to the regulation of angiogenesis. Angiogenesis involves communication between ECs and multiple neighboring cells such as fibroblasts, adipocytes, smooth muscle cells, inflammatory cells, and pericytes [59—62]. Complex interactions occur between ECs and supporting cells in the microenvironment via angiogenic regulators. Metabolites accumulated in the host cells stimulate angiogenesis by modulating the production of angiogenic growth factors which act on ECs in a paracrine manner. In addition to this, metabolite

status of the ECs also contributes to autocrine signaling leading to angiogenesis. Accumulation of unusual metabolites also occurs during pathological conditions. ECs rely on glycolysis for production of ATP by converting glucose into lactate under physiological conditions [63]. However, ECs retain the ability to switch to oxidative metabolism of glucose, amino acids, and fatty acids in case of reduced glycolytic rates [64]. Results reported from our group and elsewhere showed that lactate, citrate, sarcosine, and metabolites of arachidonic acid can influence the expression and angiogenic potential of VEGF [65—68] (Figure 14.2).

Lactate is an important metabolic product of glycolysis, regulating angiogenesis. Under hypoxic conditions, cells switch over to anaerobic metabolic pathways, with a concomitant increase in lactate. The level of lactate increases significantly in tissues during wound healing and tumor development when compared to normal tissue. Studies reported from our laboratory and others showed the proangiogenic effect of lactate by up-regulating VEGF [65]. Lactate also inactivates PHDs and thereby triggers HIF-driven VEGF expression. Lactate modulates cellular NAD+ levels and causes a reduction in the level of PAR [poly(ADP-ribose)] modification of VEGF, thereby enhancing its biological activity [65]. Citrate is a key metabolite involved in mitochondrial energy metabolism and acts as a carbon source for the synthesis of fatty acid and cholesterol. Citrate is synthesized from acetyl CoA formed from glucose under normoxic conditions, but in hypoxic conditions it is formed by rerouting glutamine metabolism in mitochondria. Citrate is reported to have altered metabolism in tumor conditions. In prostate cancer, cells are characterized by high levels of citrate. Recently we observed that citrate at high concentrations caused EC proliferation and angiogenesis [66], suggesting that the angiogenic effect may be produced only in conditions of excessive accumulation of citrate [69].

Lipid metabolites regulating angiogenesis include LOX and COX metabolites of arachidonic acid which are accumulated during inflammatory conditions. COX and LOX metabolites are reported to possess opposing effects on angiogenesis by modulating the expression of VEGF, which further depends on the relative levels of metabolites present in the microenvironment. COX metabolite prostaglandin E$_2$ is proangiogenic, whereas prostaglandin D$_2$ is antiangiogenic in nature [70]. Our study also showed similar divergent angiogenic effects of 15-LOX metabolites, 15(S)-HETE [15(S)-hydroxyeicosatetraenoic acid] being proangiogenic and 15(S)-HPETE [15(S)-hydroperoxyeicosatetraenoic acid] being antiangiogenic. 15(S)-HETE induces angiogenesis by up-regulating VEGF through activation of the PI3K-Akt-mTOR signaling pathway and 15(S)-HPETE

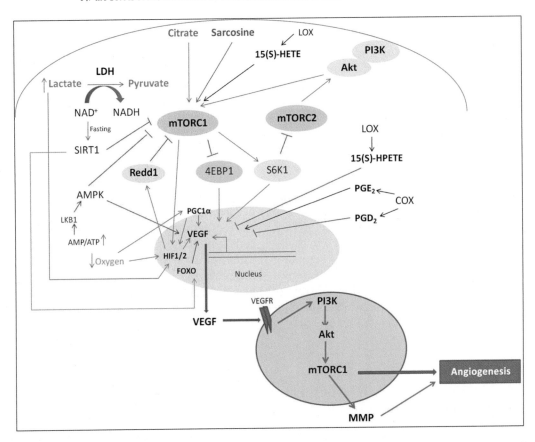

FIGURE 14.2 Integrating metabolite sensing with angiogenesis by mTOR. Metabolites such as citrate, sarcosine, and 15(S)-HETE activate mTORC1 through activating mTOR-upstream signaling molecules PI3K and Akt. Activated mTORC1 activates S6K1 and inhibits 4E-BP1, thereby inducing VEGF translation. Lactate accumulated during hypoxic conditions induces expression of HIF1α, which up-regulates VEGF together with its transcriptional coactivator PGC1α, that is directly activated by mTORC1. HIF1α also exerts a feedback mechanism on mTORC1 by activating Redd1. The LKB1/AMPK pathway senses cellular energy levels and inhibits mTORC1 when ATP is low. SIRT1 senses NAD+ levels and activates the expression of FOXO proteins, which activates VEGF translation. Akt is activated by mTORC2, which is further regulated in a feedback manner by S6Kinase. VEGF exerts a paracrine effect on endothelial cells via VEGFR2 and activates the PI3K-Akt-mTORC1 signaling pathway. mTOR up-regulates MMPs that initiate matrix cleavage. Similar metabolite effects can also occur in endothelial cells causing VEGF synthesis and the secreted VEGF can produce autocrine effects. mTORC1, mammalian target of rapamycin complex 1; mTORC2, mammalian target of rapamycin complex 2; 4E-BP1, 4E-binding protein 1; VEGF, vascular endothelial growth factor; HIF1α, hypoxia-inducible factor 1α; LKB1, liver kinase B1; AMPK, adenosine monophosphate-activated protein kinase; ATP, adenosine triphosphate; SIRT1, sirtuin 1; NAD+, nicotinamide adenine dinucleotide; FOXO, Forkhead box O; VEGFR2, vascular endothelial growth factor receptor 2; MMP, matrix metalloprotease. (⟶ Activation, ⊣ Inhibition).

inhibits angiogenesis by inducing apoptosis of ECs [71,72]. Apart from this, steroids such as 17β-estradiol have been reported to directly regulate VEGF gene transcription in endometrial cells and in Ishikawa adenocarcinoma cells. Furthermore, oxysterol, the ligand for LxR (liver X receptor), can induce the expression of VEGF through PPAR/LxR heterodimerization and activation (reviewed in Ref. [69]).

Regulation of VEGF expression by amino acid has been reported, but the results are conflicting. Drogat et al. reported that glutamine deprivation induces VEGF mRNA expression, but leads to a decrease in VEGF expression at the protein level in A549/8 human carcinoma cells [73]. In contrast, Marjon and colleagues found that glutamine deprivation induced both transcription and translation of VEGF in a human breast adenocarcinoma cell line [74]. In tumor conditions, GCN2/ATF4 (general control nonderepressible 2/activating transcription factor 4) pathway is central for tumors to adapt to amino acid deprivation in the tumor microenvironment, eliciting gene expression that enhances tumor angiogenesis. Sarcosine, an intermediate of glycine metabolism, is a normal physiological metabolite; however, it is accumulated in the tissue and urine of prostate cancer patients. Recently, we observed that sarcosine at high concentrations exerts proangiogenic effect by modulating VEGF expression [67]. Purine and pyrimidine nucleotides are degraded and their metabolites accumulate during ischemia, of which adenosine and nicotinamide have been reported to induce vascular growth; but the mechanism of their action has not been examined.

Metabolites regulate the expression of VEGF by activating various intracellular signaling pathways, especially the PI3K-Akt-mTOR signaling pathway. It was observed that accumulation of citrate and sarcosine in ECs promotes activation of the PI3K-Akt-mTOR pathway and increased VEGF production. 15(S)-HETE induces angiogenesis by up-regulating VEGF through activation of the PI3K-Akt-mTOR signaling pathway as evidenced by the reversal of its effect by rapamycin. Up-regulation of VEGF by sarcosine is also mediated through the PI3K/Akt/mTOR pathway.

Cells are equipped with metabolite-sensing mechanisms. Metabolic regulators such as PGC-1α, LKB1 (liver kinase B1), PPAR-ϒ, AMPK, SIRT1 (sirtuin 1), and FOXO can sense the metabolite status of both endothelial and non-ECs and modulate angiogenesis based on alterations in the level of metabolites [75]. These metabolic regulators can modulate signaling events and transcription factors relevant to angiogenesis. Metabolic regulators often stimulate angiogenesis in nutrient-deprived conditions. In ischemic tissues, PGC1α senses oxygen and nutrient scarcity and thereby stimulates angiogenesis by up-regulating VEGF through interaction with estrogen-related receptor-α [76]. AMP senses energy deprivation and increased cellular levels of AMP activate AMPK and thereby induce angiogenesis by up-regulating VEGF expression [77]. LKB1 also activates AMPK and induces VEGF-driven angiogenesis. During fasting, NAD + activates Sirt-1 which limits the antiangiogenic effect of FOXO transcription factors and modulates angiogenesis [75]. Identifying the molecular basis of the vascular—metabolic interface will help improve the efficacy of antiangiogenic treatment.

While the basic steps of angiogenesis indicated above operate in different tissues the vascular network of each organ will be established through important tissue-specific mechanisms. Angiogenesis and vascular density in an actively metabolizing tissue that requires more oxygen and nutrients are relatively high. As indicated above, signaling molecule mTOR can sense the nutrient status in growing vessels and link these signals to modulate cellular responses during neovascularization. mTOR appears to act as the molecular link between the metabolite status of the cell and angiogenesis by integrating metabolite-sensing mechanisms to modulate the expression of angiogenic growth factors.

14.8.2 Metabolite Sensing by mTOR

Metabolite sensors help in maintaining cellular homeostasis by regulating metabolic processes. mTOR, especially mTORC1, is sensitive to the metabolite status of the cell. mTORC1 integrates many signaling pathways and responds to the availability of nutrients. mTOR can sense the scarcity of nutrients and directs cells to store energy by reducing metabolic activities.

14.8.3 Adenosine Triphosphate

mTOR is sensitive to the level of ATP in the cell. Low levels of ATP inactivate mTOR to reduce energy-requiring metabolic pathways in cells. In response to energy stress, AMPK is activated and inhibits mTORC1 by distinct TSC2-dependent and -independent mechanisms. The energy-sensing AMPK inhibits mTOR signal transduction by phosphorylating and activating TSC2 [78]. In TSC2-independent mechanisms, AMPK phosphorylates Raptor and induces binding of 14-3-3 to Raptor thereby inhibiting mTORC1. Similar to hypoxic conditions, Redd1 also inhibits mTORC1 signaling in response to glucose withdrawal or energy depletion. Folliculin (FLCN) and FLCN-interacting proteins (FNIP1 and FNIP2) were reported to be involved in nutrient sensing through AMPK and mTOR [79,80], where FLCN and FNIP1 are phosphorylated in an AMPK- and mTOR-dependent manner.

14.8.4 Glucose

AMPK can influence glucose homeostasis during energy deprivation by inhibiting mTORC1. AMPK also promotes uptake of glucose and amino acids which in turn results in the activation of mTORC1. mTORC1 further promotes glucose uptake and hyperactivation of mTORC1 under glucose—excess conditions have a negative feedback effect on glucose catabolism. Prolonged activation of S6K1 caused by mTOR hyperactivation, phosphorylates and inactivates the insulin receptor substrate [81], leading to insulin desensitization and inhibition of Akt activity. This feedback loop reduces glucose uptake and glycogen synthesis, and increases gluconeogenesis as a result of mTORC1 hyperactivation.

14.8.5 Protein

mTOR regulates protein synthesis by phosphorylating translation factors such as 4E-BP1 and S6K1. During energy stress, AMPK-mTORC1 crosstalk regulates protein synthesis. AMPK attenuates mTORC1 signaling through phosphorylation and activation of TSC2 [82], a negative regulator of mTORC1. AMPK also directly phosphorylates Raptor, which induces 14-3-3 binding to raptor and inhibits mTORC1 activity.

It therefore appears that AMPK activation on one side causes inhibition of mTORC1 but can induce up-regulation of VEGF [77], suggesting that AMPK can modulate angiogenesis through mTORC1-dependent (inhibitory) and -independent (stimulatory) mechanisms, highlighting the importance of metabolic context of the cell in the control of angiogenesis.

14.8.6 Lipid

There is increasing evidence in support of a role for mTOR in the regulation of lipid biosynthesis. Inhibition of mTORC1 inhibits sterol regulatory element-binding transcription factor 1 (SREBP1/2), resulting in a significant decrease in lipogenic gene expression [83]. mTORC1 also promotes the function of SREBP through phosphorylating Lipin-1, a phosphatidic acid phosphatase, and preventing its nuclear entry and suppression of SREBP1/2 [84]. Furthermore, mTORC1 regulates lipid synthesis by up-regulating PPARγ, a key regulator of lipid uptake and adipogenesis [85].

Though detailed mechanisms of sensing of metabolites by mTOR have not been elucidated, certain models have been proposed to explain amino acid and phosphatidic acid sensing by mTOR [86]. Mediators involved in metabolite sensing by mTOR are Vps34 (vacuolar protein sorting 34), MAP4K3 (mitogen activated protein kinase kinase kinase kinase3), Rag (Ras-related GTP binding protein) GTPases, and RalA. mTOR-dependent metabolite sensing is linked to Vps34, which activates mTOR downstream signaling molecule S6K1. Starvation of amino acid or glucose inhibits Vps34 apparently in a Ca^{2+}/calmodulin-dependent manner. Vps34 coordinates nutrient signals upstream of mTORC1 as evidenced by the absence of the effect of rapamycin on the activity of Vps34. MAP4K3 is a nutrient-sensitive kinase within the mTOR pathway. Amino acids activate MAP4K3, which further activates S6K1 and 4E-BP1 in a rapamycin-sensitive manner. Knockdown of MAP4K3 inhibits S6K1 activation induced by amino acids. In addition to these, Rag GTPases also act as mediators of amino acid signaling to mTORC1. Rag GTPases are mTORC1 associated proteins and interacts with Raptor in a rapamycin sensitive manner. Rag GTPases modulate the cellular localization of mTOR in response to amino acids and act similar to TSC1–TSC2 /Rheb input to mTORC1 [87,88]. RalA is another GTPase which is linked to metabolite sensing by mTORC1 and activation of RalA is nutrient-dependent. Amino acids and glucose activate RalA by inducing its binding to GTP. Inhibition of RalA also inhibited the phosphorylation of S6K1 and 4E-BP1 following treatment with amino acid or glucose in vitro, confirming the involvement of RalA in metabolite sensing of mTORC1. Knockdown studies suggested that RalA acts downstream of Rheb [89].

14.9 TRANSCRIPTIONAL CONTROL OF ANGIOGENIC MEDIATORS BY mTOR

mTOR appears to act as a critical switch for endothelial catabolism and anabolism and a central regulator of EC integrity and angiogenesis. The mTOR pathway integrates the nutrient status, hypoxia, and cellular energy status to control cell growth and angiogenesis. mTOR activates downstream targets S6K1 and 4E-BP1 to modulate protein synthesis relevant to angiogenesis. mTOR activity is tightly regulated by TSC1–TSC2 heterodimer and Rheb. TSC1–TSC2 complex inhibits cell growth and proliferation by inhibiting mTOR signaling. Through numerous phosphorylation events, the TSC1–TSC2 complex acts as a sensor and integrator of cell growth conditions relaying signals from diverse cellular pathways to modulate mTORC1 activity. Angiogenic growth factors, intracellular signaling molecules, and intercellular adhesion molecules signal through the PI3K-Akt-mTOR signaling pathway in ECs to regulate angiogenesis. In addition to this, proangiogenic signals, either nutrient or cytokine in origin, activate the PI3K-Akt-mTOR signaling pathway to up-regulate angiogenic growth factors. This has been amply evidenced by the demonstration of the role of mTOR in VEGF expression in different conditions [66,67].

Anti-mTOR and rapamycin analogs inhibit angiogenesis in breast cancer models both in vivo and in vitro by down-regulating VEGF translation [90]. Also, in mice bearing human tumor xenografts, mTOR inhibitor RAD001 significantly reduced tumor angiogenesis [91]. mTOR induces the translation of HIF1α mRNA, thereby enhancing the production of VEGF. PI3K-Akt-mTOR-S6K1 signaling is activated in Kaposi's sarcoma conditions and treatment with rapamycin inhibited the formation of Kaposi's sarcoma lesions through its antiangiogenic action. Rapamycin also inhibited metastatic tumor growth and angiogenesis in mouse models in vivo by decreasing the production of VEGF as well as by inhibiting the response of ECs to VEGF. Elevated levels of HIF1α due to the loss of von Hippel–Lindau disease are associated with renal cell carcinoma. In a xenograft model using human renal cell carcinoma, rapamycin inhibited the translation of HIF1α which correlated with down-regulation of VEGF and inhibition of angiogenesis. mTOR-specific inhibitor CCI-779 inhibited the expression of HIF1α and VEGF in xenograft models of human rhabdomyosarcoma in mice [92]. Thoridazine inhibited angiogenesis and tumor growth in human ovarian cancer xenografts in nude mice by targeting the PI3K-Akt-mTOR pathway [93]. Akt-mTOR-S6K1 signaling also promotes progression of hepatocellular carcinoma by inducing angiogenesis.

Akt regulates angiogenesis by activating mTOR signaling through its inhibitory effects on TSC1/TSC2 complex. Metabolites activate Akt in a PI3K-dependent manner and activated Akt phosphorylates and activates mTOR kinase. As indicated before, mTOR is generally inhibited by TSC2 which is a GTPase-activating protein regulating the GTPase function of Rheb. TSC1 stabilizes TSC2 and prevents its ubiquitinylation. Kinases such as Akt and ERK phosphorylate and inactivate TSC2 and activate mTOR. mTOR further phosphorylates downstream signaling molecules, S6K1, and translation initiation factor 4E-BP and induces gene transcription and synthesis of proteins relevant to angiogenesis [94].

The signaling between Akt and mTOR involves a regulatory feedback mechanism in addition to the PI3K-Akt-mTOR-S6K1-4E-BP1 downstream signaling pathway. Reports suggest that mTORC1-downstream target S6Kinase can inhibit the activation of Akt by inhibiting the rictor subunit of mTORC2 complex. Short-term exposure of ECs to rapamycin inhibits angiogenesis by targeting downstream signaling molecules of mTORC1, such as S6Kinase and 4E-BP1, but did not affect Akt. However, long-term exposure of ECs to rapamycin affects cell survival by inhibiting the activation of Akt. Akt is phosphorylated and activated by PDK1 and putative PDK2 in a PI3K-dependent manner. In this context, mTORC2 acts as PDK2 and activates Akt. PDK1 phosphorylates Akt at Thr308 leading to partial activation of Akt and phosphorylation of Akt at Ser473 by mTORC2 stimulates full enzymatic activity. mTORC2 is generally insensitive to rapamycin, but long-term exposure of cells to rapamycin inhibits mTORC2 by affecting the association of its subunits mTOR and rictor and thereby inhibits the phosphorylation of Akt. The proline-rich Akt substrate of 40 kDa (PRAS40) acts as modulator of the Akt and mTOR-mediated signaling pathways. Phosphorylation of PRAS40 by Akt results in dissociation of PRAS40 from mTORC1, thus relieving its inhibitory effect on mTORC1 activity [94].

As indicated earlier, the initial stage of angiogenesis involves the degradation of ECM by proteases such as MMPs, ADAMTS (A Disintegrin And Metalloproteinase with Thrombospondin Motifs) and uPA. It was reported that the production of MMPs is regulated by mTOR as rapamycin inhibits the production of MMP9 in tumor conditions. mTOR, specifically mTORC2, regulates MMP9 activity and invasion of glioma tumor cells through the Raf1-MEK-ERK signaling pathway [95] (Figure 14.3).

14.9.1 MicroRNAs and mTOR

MicroRNAs (miRNA) have emerged as modulators of biological pathways that are essential in cancer initiation, development, and progression. miRNAs target the mTOR pathway in several ways, either by interacting directly with mTOR itself or targeting key genes within the pathway, which ultimately affects mTOR function. These genes include upstream regulators of mTOR, such as PI3K and Akt, and negative regulators, such as PTEN. miRNAs regulate mTOR activity relevant to angiogenesis. The antiangiogenic effect of miR-100 is mediated by repressing mTOR signaling in endothelial and vascular smooth muscle cells [96]. miR-18a overexpression in tumor xenografts from human gastric cancer showed reduced phosphorylation of S6K1 and 4E-BP1 and associated down-regulation of VEGF and HIF-1α [97]. miR-382 activates the PI3K-Akt-mTOR signaling pathway by targeting PTEN during hypoxic conditions [98].

14.10 mTOR AS A TARGET AGAINST ANGIOGENESIS

Inhibition of the PI3K-Akt-mTOR signaling pathway would be a promising antiangiogenic approach in pathological conditions such as tumors. mTOR inhibitors can inhibit angiogenesis by down-regulating the production of angiogenic growth factors by endothelial and non-ECs. Rapamycin, specific inhibitor of mTOR, produces antiangiogenic effects by down-regulating the production of VEGF and by inhibiting the translation of HIF1α. It also decreases cell survival by inhibiting Akt. These properties make rapamycin a powerful anticancer agent. The effect of rapamycin depends on the duration of the treatment because of the differential effect of rapamycin on mTORC1 and mTORC2. Rapamycin analogs, such as temsirolimus, everolimus, and AP23574 are being developed as antiangiogenic agents [99,100]. Naturally occurring substances which inhibit angiogenesis have been found to affect mTOR (Athira et al., unpublished data). The possible beneficial effect of combination therapy of mTOR inhibitors and VEGF inhibitors cannot be excluded, particularly in cases of anti-VEGF-resistant cancers.

14.11 CONCLUSION

Although mTOR was first identified as the target of rapamycin, recent studies indicated the role of mTOR in hypoxia and metabolite sensing. It has become immensely evident that mTOR plays a significant role in integrating hypoxia and metabolite status of the cellular microenvironment with angiogenic pathways. But it is not fully clear how mTOR senses metabolite and energy status. mTOR modulates

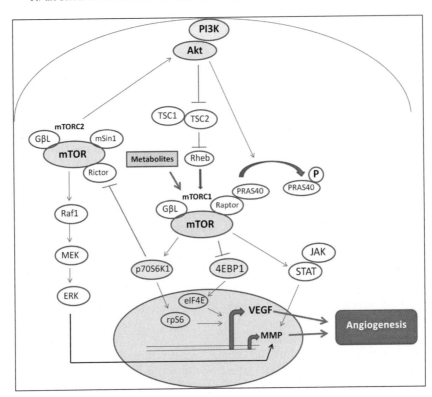

FIGURE 14.3 Transcriptional regulation of angiogenesis mediators by mTOR. Activation of Akt through PI3K leads to the activation of the mTOR signaling pathway by inhibiting TSC1—TSC2 complex. Inhibition of TSC1—TSC2 complex inhibits Rheb, which further activates mTORC1 complex and P70S6K1. Activation of these signaling molecules leads to the activation of ribosomal protein S6 which further up-regulates VEGF production and promotes angiogenesis. PRAS40 acts as modulator of the Akt-mTOR signaling pathways. Akt promotes angiogenesis by dissociating PRAS40 from mTORC1 and relieving its inhibition on mTORC1 and eIF4E. In a feedback manner S6Kinase, the downstream signaling molecule of mTORC1, inhibits Akt by inhibiting the rictor subunit of mTORC2 complex. mTORC2 activates the Raf 1-MEK-ERK signaling pathway, and mTORC1 activates STAT in the JAK-STAT pathway, resulting in the up-regulation of MMP expression required to initiate angiogenesis. mTOR, mammalian target of rapamycin; TSC1 and 2, tuberous sclerosis complexes 1 and 2; mTORC1, mammalian target of rapamycin complex 1; mTORC2, mammalian target of rapamycin complex 2; VEGF, vascular endothelial growth factor; PRAS40, proline-rich Akt substrate of 40 kDa; STAT, signal transducer and activator of transcription; JAK, Janus kinase; MMP, matrix metallo-protease. (⟶ Activation, —⊣ Inhibition).

angiogenesis by phosphorylating target proteins involved in regulating gene expression and synthesis of proteins relevant to angiogenesis and metabolism. Since mTOR acts as a converging point for various signaling pathways and regulatory molecules relevant to hypoxia, nutrient sensing and angiogenesis and multiple feedback mechanisms fine tune the activity of mTOR, targeting mTOR would be a useful strategy for antiangiogenic therapy in cancer, particularly in cases where resistance to anti-VEGF therapy develops.

Acknowledgments

Financial assistance received from KSCSTE, Government of Kerala (Athira A. P., Binu S., and P. R. Sudhakaran), UGC, New Delhi (Soumya S. J. and P. R. Sudhakaran) and DST, Government of India (Binu S. and P. R. Sudhakaran) is gratefully acknowledged.

References

[1] Modlich U, Kaup FJ, Augustin HG. Cyclic angiogenesis and blood vessel regression in the ovary: blood vessel regression during luteolysis involves endothelial cell detachment and vessel occlusion. Lab Invest 1996;74:771—80.

[2] Folkman J. Angiogenesis in cancer, vascular, rheumatoid and other disease. Nat Med 1995;1:27—31.

[3] Karamysheva AF. Mechanisms of angiogenesis. Biochem (Mosc) 2008;73:751—62.

[4] Eilken HM, Adams RH. Turning on the angiogenic microswitch. Nat Med 2010;16:853—4.

[5] Carmeliet P, Jain RK. Molecular mechanisms and clinical applications of angiogenesis. Nature 2011;473:298—307.

[6] Chung AS, Ferrara N. Developmental and Pathological Angiogenesis. Ann Rev of Cel and Dev Biol 2011;27:563—84.

[7] Herbert SP, Stainier DYR. Molecular control of endothelial cell behaviour during blood vessel morphogenesis. Nat Rev Mol Cell Biol 2011;12:551—64.

[8] Gerhardt H, Golding M, Fruttiger M, et al. VEGF guides angiogenic sprouting utilizing endothelial tip cell filopodia. J Cel Biol 2003;161:1163—77.

[9] Benedito R, Rocha SF, Woeste M, Zamykal M, Radtke F, Casanovas O, et al. Notch-dependent VEGFR3 upregulation allows angiogenesis without VEGF-VEGFR2 signalling. Nature 2012;484:110−14.

[10] Siemerink MJ, Klaassen I, Van Noorden CJ, Schlingemann RO. Endothelial tip cells in ocular angiogenesis: potential target for anti-angiogenesis therapy. J Histochem Cytochem 2013;61:101−15.

[11] Strasser GA, Kaminker JS, Tessier-Lavigne M. Microarray analysis of retinal endothelial tip cells identifies CXCR4 as a mediator of tip cell morphology and branching. Blood 2010;115:5102−10.

[12] Harrington LS, Sainson RCA, Williams CK, Taylor JM, Shi W, Li JL, et al. Regulation of multiple angiogenic pathways by Dll4 and Notch in human umbilical vein endothelial cells. Microvasc Res 2008;75:144−54.

[13] Del Toro R, Prahst C, Mathivet T, Siegfried G, Kaminker JS, Larrivee B, et al. Identification and functional analysis of endothelial tip cell-enriched genes. Blood 2010;116:4025−33.

[14] Krueger J, Liu D, Scholz K, Zimmer A, Shi Y, Klein C, et al. Flt1 acts as a negative regulator of tip cell formation and branching morphogenesis in the zebrafish embryo. Development 2011;138:2111−20.

[15] Gaengel K, Genove G, Armulik A, Betsholtz C. Endothelial mural cell signaling in vascular development and angiogenesis. Arterioscler Thromb Vasc Biol 2009;29:630−8.

[16] Jain RK. Molecular regulation of vessel maturation. Nat Med 2003;9:685−93.

[17] Ziyad S, Iruela-Arispe ML. Molecular mechanisms of tumor angiogenesis. Genes Cancer 2011;2:1085−96.

[18] Wang F, Yamauchi M, Muramatsu M, Osawa T, Tsuchida R, Shibuya M. RACK1 regulates VEGF/Flt1-mediated cell migration via activation of a PI3K/Akt pathway. J Biol Chem 2011;286:9097−106.

[19] Shibuya M. Vascular Endothelial Growth Factor (VEGF) and its receptor (VEGFR) signaling in angiogenesis: a crucial target for anti- and pro-angiogenic therapies. Gene Canc 2011;2:1097−105.

[20] Hoeben A, Landuyt B, Highley MS, Wildiers H, Van Oosterom AT, De Bruijn EA. Vascular endothelial growth factor and angiogenesis. Pharmacol Rev 2004;56:549−80.

[21] Serban D, Leng J, Cheresh D. H-ras regulates angiogenesis and vascular permeability by activation of distinct downstream effectors. Circ Res 2008;102:1350−8.

[22] Byrne AM, Bouchier-Hayes DJ, Harmey JH. Angiogenic and cell survival functions of vascular endothelial growth factor (VEGF). J Cell Mol Med 2005;9:777−94.

[23] Saharinen P, Eklund L, Miettinen J. Angiopoietins assemble distinct Tie2 signalling complexes in endothelial cell−cell and cell−matrix contacts. Nat Cell Biol 2008;10:527−37.

[24] Scharpfenecker M, Fiedler U, Reiss Y, Augustin HG. The Tie-2 ligand angiopoietin-2 destabilizes quiescent endothelium through an internal autocrine loop mechanism. J Cell Sci 2005;118:771−80.

[25] Daly C, Pasnikowski E, Burova E. Angiopoietin-2 functions as an autocrine protective factor in stressed endothelial cells. Proc Natl Acad Sci USA 2006;103:15491−6.

[26] Yuan HT, Venkatesha S, Chan B. Activation of the orphan endothelial receptor Tie1 modifies Tie2-mediated intracellular signaling and cell survival. FASEB J 2007;21:3171−83.

[27] Wang Y, Nakayama M, Pitulescu ME. Ephrin-B2 controls VEGF-induced angiogenesis and lymphangiogenesis. Nature 2010;465:483−6.

[28] Kertesz N, Krasnoperov V, Reddy R. The soluble extracellular domain of EphB4 (sEphB4) antagonizes EphB4-EphrinB2 interaction, modulates angiogenesis, and inhibits tumor growth. Blood 2006;107:2330−8.

[29] Huminiecki L, Gorn M, Suchting S, Poulsom R, Bicknell R. Magic roundabout is a new member of the roundabout receptor family that is endothelial specific and expressed at sites of active angiogenesis. Genomics 2002;79:547−52.

[30] Wang B, Xiao Y, Ding BB. Induction of tumor angiogenesis by Slit-Robo signaling and inhibition of cancer growth by blocking Robo activity. Cancer Cell 2003;4:19−29.

[31] Suchting S, Heal P, Tahtis K, Stewart LM, Bicknell R. Soluble Robo4 receptor inhibits in vivo angiogenesis and endothelial cell migration. FASEB J 2005;19:121−3.

[32] Jones CA, London NR, Chen H. Robo4 stabilizes the vascular network by inhibiting pathologic angiogenesis and endothelial hyperpermeability. Nat Med 2008;14:448−53.

[33] Marlow R, Binnewies M, Sorensen LK. Vascular Robo4 restricts proangiogenic VEGF signaling in breast. Proc Natl Acad Sci USA 2010;107:10520−5.

[34] Liu ZJ, Shirakawa T, Li Y. Regulation of Notch1 and Dll4 by vascular endothelial growth factor in arterial endothelial cells: implications for modulating arteriogenesis and angiogenesis. Mol Cell Biol 2003;23:14−25.

[35] Hellstrom M, Phng LK, Hofmann JJ. Dll4 signalling through Notch1 regulates formation of tip cells during angiogenesis. Nature 2007;445:776−80.

[36] Lobov IB, Renard RA, Papadopoulos N. Delta-like ligand 4 (Dll4) is induced by VEGF as a negative regulator of angiogenic sprouting. Proc Natl Acad Sci USA 2007;104:3219−24.

[37] Ridgway J, Zhang G, Wu Y. Inhibition of Dll4 signalling inhibits tumour growth by deregulating angiogenesis. Nature 2006;444:1083−7.

[38] Karar J, Maity A. PI3K/AKT/mTOR pathway in angiogenesis. Front Mol Neurosci 2011;4:51.

[39] Dunlop EA, Tee AR. Mammalian target of rapamycin complex 1: signalling inputs, substrates and feedback mechanisms. Cell Signal 2009;21:827−35.

[40] Dann SG, Selvaraj A, Thomas G. mTOR Complex1−S6K1 signaling: at the crossroads of obesity, diabetes and cancer. Trends Mol Med 2007;13:252−9.

[41] Zhong H, Chiles K, Feldser D, Laughner E, Hanrahan C, Georgescu MM, et al. Modulation of hypoxia-inducible factor 1alpha expression by the epidermal growth factor/phosphatidylinositol 3-kinase/PTEN/AKT/FRAP pathway in human prostate cancer cells: implications for tumor angiogenesis and therapeutics. Cancer Res 2000;60:1541−5.

[42] Hudson CC, Liu M, Chiang GG, Otterness DM, Loomis DC, Kaper F, et al. Regulation of hypoxia-inducible factor 1alpha expression and function by the mammalian target of rapamycin. Mol Cell Biol 2002;22:7004−14.

[43] Krock BL, Skuli N, Simon MC. Hypoxia-induced angiogenesis: good and evil. Genes Cancer 2011;1117−33.

[44] Ward JP. Oxygen sensors in context. Biochim Biophys Acta 2008;1777:1−14.

[45] Semenza GL. Targeting HIF-1 for cancer therapy. Nat Rev Cancer 2003;3:721−32.

[46] Kaelin WG, Ratcliffe PJ. Oxygen sensing by metazoans: the central role of the HIF hydroxylase pathway. Mol Cell 2008;30:393−402.

[47] Ramirez Bergeron DL, Runge A, Adelman DM, Gohil M, Simon MC. HIF-dependent hematopoietic factors regulate the development of the embryonic vasculature. Dev Cell 2006;11:81−92.

[48] Grunewald M, Avraham I, Dor Y, Bachar-Lustig E, Itin A, Jung S, et al. VEGF-induced adult neovascularization: recruitment, retention, and role of accessory cells. Cell 2006;124:175−89.

[49] Elvert G, Kappel A, Heidenreich R, Englmeier U, Lanz S, Acker T, et al. Cooperative interaction of hypoxia-inducible factor-2alpha (HIF-2alpha) and Ets-1 in the transcriptional activation of vascular endothelial growth factor receptor-2 (Flk-1). J Biol Chem 2003;278:7520—30.

[50] Lee SP, Youn SW, Cho HJ, Li L, Kim TY, Yook HS, et al. Integrin-linked kinase, a hypoxia-responsive molecule, controls postnatal vasculogenesis by recruitment of endothelial progenitor cells to ischemic tissue. Circulation 2006;114:150—9.

[51] Licht AH, Müller-Holtkamp F, Flamme I, Breier G. Inhibition of hypoxia-inducible factor activity in endothelial cells disrupts embryonic cardiovascular development. Blood 2006;107:584—90.

[52] Tillmanns J, Rota M, Hosoda T, Misao Y, Esposito G, Gonzalez A, et al. Formation of large coronary arteries by cardiac progenitor cells. Proc Natl Acad Sci USA 2008;105:1668—73.

[53] Jiang M, Wang B, Wang C, He B, Fan H, Shao Q, et al. In vivo enhancement of angiogenesis by adenoviral transfer of HIF-1alpha-modified endothelial progenitor cells (Ad-HIF-1alpha-modified EPC for angiogenesis). Int J Biochem Cell Biol 2008;40:2284—95.

[54] Brugarolas J, Lei K, Hurley RL, Manning BD, Reiling JH, Hafen E, et al. Regulation of mTOR function in response to hypoxia by REDD1 and the TSC1/TSC2 tumor suppressor complex. Genes Dev 2004;18:2893—904.

[55] Arsham AM, Howell JJ, Simon MC. A novel hypoxia-inducible factor-independent hypoxic response regulating mammalian target of rapamycin and its targets. J Biol Chem 2003;278:29655—60.

[56] DeYoung MP, Horak P, Sofer A, Sgroi D, Ellisen LW. Hypoxia regulates TSC1/2-mTOR signaling and tumor suppression through REDD1-mediated 14-3-3 shuttling. Genes Dev 2008;22:239—51.

[57] Sowter HM, Ratcliffe PJ, Watson P, Greenberg AH, Harris AL. HIF-1-dependent regulation of hypoxic induction of the cell death factors BNIP3 and NIX in human tumors. Cancer Res 2001;61:6669—73.

[58] Busch S, Renaud SJ, Schleussner E, Graham CH, Markert UR. mTOR mediates human trophoblast invasion through regulation of matrix-remodeling enzymes and is associated with serine phosphorylation of STAT3. Exp Cell Res 2009;315:1724—33.

[59] Gomes FG, Nedel F, Alves AM, Nör JE, Tarquinio SB. Tumor angiogenesis and lymphangiogenesis: tumor/endothelial crosstalk and cellular/microenvironmental signaling mechanisms. Life Sci 2013;92:101—7.

[60] Mor F, Quintana FJ, Cohen IR. Angiogenesis-inflammation cross-talk: vascular endothelial growth factor is secreted by activated T cells and induces Th1 polarization. J Immunol 2004;172:4618—23.

[61] Sozzani S, Rusnati M, Riboldi E, Mitola S, Presta M. Dendritic cell-endothelial cell cross-talk in angiogenesis. Trends Immunol 2007;28:385—92.

[62] Cao Y. Angiogenesis modulates adipogenesis and obesity. J Clin Invest 2007;117:2362—8.

[63] De Bock K, Georgiadou M, Carmeliet P. Role of endothelial cell metabolism in vessel sprouting. Cell Metab 2013;18:634—47.

[64] Dranka BP, Hill BG, Darley-Usmar VM. Mitochondrial reserve capacity in endothelial cells: the impact of nitric oxide and reactive oxygen species. Free Radical Biol Med 2010;48:905—14.

[65] Kumar VB, Viji RI, Kiran MS, Sudhakaran PR. Endothelial cell response to lactate: implication of PAR modification of VEGF. J Cell Physiol 2007;211:477—85.

[66] Binu S, Soumya SJ, Sudhakaran PR. Metabolite control of angiogenesis: angiogenic effect of citrate. J Physiol Biochem 2013;69:383—95.

[67] Sudhakaran PR, Binu S, Soumya SJ. Effect of sarcosine on endothelial function relevant to angiogenesis. J Cancer Res Ther 2014;10:603—10.

[68] Soumya SJ, Binu S, Helen A, Anil Kumar K, Reddanna P, Sudhakaran PR. Effect of 15-lipoxygenase metabolites on angiogenesis: 15(S)-HPETE is angiostatic and 15(S)-HETE is angiogenic. Inflamm Res 2012;61:707—18.

[69] Kumar VB, Binu S, Soumya SJ, Haritha K, Sudhakaran PR. Regulation of vascular endothelial growth factor by metabolic context of the cell. Glycoconj J 2014;31:427—34.

[70] Viji RI, Kumar VB, Kiran MS, Sudhakaran PR. Modulation of cyclooxygenase in endothelial cells by fibronectin: relevance to angiogenesis. J Cell Biochem 2008;105:158—66.

[71] Soumya SJ, Binu S, Helen A, Reddanna P, Sudhakaran PR. 15(S)-HETE-induced angiogenesis in adipose tissue is mediated through activation of PI3K/Akt/mTOR signaling pathway. Biochem Cell Biol 2013;91:498—505.

[72] Soumya SJ, Binu S, Helen A, Reddanna P, Sudhakaran PR. 15-LOX metabolites and angiogenesis: angiostatic effect of 15(S)-HPETE involves induction of apoptosis in adipose endothelial cells. Peer J 2014;2:e635.

[73] Drogat B, Bouchecareilh M, North S, Petibois C, Deleris G, Chevet E, et al. Acute L-glutamine deprivation compromises VEGF-A upregulation in A549/8 human carcinoma cells. J Cell Physiol 2007;212:463—72.

[74] Marjon PL, Bobrovnikova-Marjon EV, Abcouwer SF. Expression of the pro-angiogenic factors vascular endothelial growth factor and interleukin-8/CXCL8 by human breast carcinomas is responsive to nutrient deprivation and endoplasmic reticulum stress. Mol Cancer 2004;3:4.

[75] Fraisl P, Mazzone M, Schmidt T, Carmeliet P. Regulation of angiogenesis by oxygen and metabolism. Dev Cell 2009;16:167—79.

[76] Arany Z, Foo SY, Ma Y, Ruas JL, Bommi-Reddy A, Girnun G, et al. HIF-independent regulation of VEGF and angiogenesis by the transcriptional coactivator PGC-1alpha. Nature 2008;451:1008—12.

[77] Ouchi N, Shibata R, Walsh K. AMP-activated protein kinase signaling stimulates VEGF expression and angiogenesis in skeletal muscle. Circ Res 2005;96:838—46.

[78] Corradetti MN, Inoki K, Bardeesy N, DePinho RA, Guan KL. Regulation of the TSC pathway by LKB1: evidence of a molecular link between tuberous sclerosis complex and Peutz-Jeghers syndrome. Genes Dev 2004;18:1533—8.

[79] Hasumi H, Baba M, Hong SB, Hasumi Y, Huang Y, Yao M, et al. Identification and characterization of a novel folliculin-interacting protein FNIP2. Gene 2008;415:60—7.

[80] Takagi Y, Kobayashi T, Shiono M, Wang L, Piao X, Sun G, et al. Interaction of folliculin (Birt-Hogg-Dubé gene product) with a novel Fnip1-like (FnipL/Fnip2) protein. Oncogene 2008;27:5339—47.

[81] Tremblay F, Brule S, Hee Um, S, Li Y, Masuda K, Roden M, et al. Identification of IRS-1 Ser-1101 as a target of S6K1 in nutrient- and obesity induced insulin resistance. Proc Natl Acad Sci USA 2007;104:14056—61.

[82] Inoki K, Zhu T, Guan KL. TSC2 mediates cellular energy response to control cell growth and survival. Cell 2003;115:577—90.

[83] Wang BT, Ducker GS, Barczak AJ, Barbeau R, Erle DJ, Shokat KM. The mammalian target of rapamycin regulates cholesterol biosynthetic gene expression and exhibits a rapamycin-resistant transcriptional profile. Proc Natl Acad Sci USA 2011;108:15201—6.

[84] Peterson TR, Sengupta SS, Harris TE, Carmack AE, Kang SA, Balderas E, et al. mTOR complex 1 regulates lipin 1 localization to control the SREBP pathway. Cell 2011;146:408—20.

[85] Zhang HH, Huang J, Duvel K, Boback B, Wu S, Squillace RM, et al. Insulin stimulates adipogenesis through the Akt-TSC2-mTORC1 pathway. PLoS One 2009;4:e6189.

[86] Goldberg EL, Smithey MJ, Lutes LK, Uhrlaub JL, Nikolich-Zugich J. Immune memory-boosting dose of rapamycin impairs macrophage vesicle acidification and curtails glycolysis in effector CD8 cells, impairing defense against acute infections. J Immunol 2014;193:757−63.

[87] Jewell JL, Guan KL. Nutrient signaling to mTOR and cell growth. Trends Biochem Sci 2013;38:233−42.

[88] Kim SG, Buel GR, Blenis J. Nutrient regulation of the mTOR complex 1 signaling pathway. Mol Cells 2013;35:463−73.

[89] Maehama T, Tanaka M, Nishina H, Murakami M, Kanaho Y, Hanada K. RalA functions as an indispensable signal mediator for the nutrient-sensing system. J Biol Chem 2008;283:35053−9.

[90] Falcon BL, Barr S, Gokhale PC, Chou J, Fogarty J, Depeille P, et al. Reduced VEGF production, angiogenesis, and vascular regrowth contribute to the antitumor properties of dual mTORC1/mTORC2 inhibitors. Cancer Res 2011;71:1573−83.

[91] Huynh H, Pierce Chow KH, Soo KC, Toh HC, Choo SP, Foo KF, et al. RAD001 (everolimus) inhibits tumour growth in xenograft models of human hepatocellular carcinoma. J Cell Mol Med 2009;13:1371−80.

[92] Emstoff MS. mTOR pathway and mTOR inhibitors in cancer therapy. Br J Clin Pharmacol 2011;71:970.

[93] Park MS, Dong SM, Kim BR, Seo SH, Kang S, Lee EJ, et al. Thioridazine inhibits angiogenesis and tumor growth by targeting the VEGFR-2/PI3K/mTOR pathway in ovarian cancer xenografts. Oncotarget 2014;5:4929−34.

[94] Shahbazian D, Roux PP, Mieulet V, Cohen MS, Raught B, Taunton J, et al. The mTOR/PI3K and MAPK pathways converge on eIF4B to control its phosphorylation and activity. EMBO J 2006;25:2781−91.

[95] Das G, Shiras A, Shanmuganandam K, Shastry P. Rictor regulates MMP-9 activity and invasion through Raf-1-MEK-ERK signaling pathway in glioma cells. Mol Carcinog 2011;50:412−23.

[96] Grundmann S, Hans FP, Kinniry S, Heinke J, Helbing T, Bluhm F, et al. MicroRNA-100 regulates neovascularization by suppression of mammalian target of rapamycin in endothelial and vascular smooth muscle cells. Circulation 2011;123:999−1009.

[97] Zheng Y, Li S, Ding Y, Wang Q, Luo H, Shi Q, et al. The role of miR-18a in gastric cancer angiogenesis. Hepatogastroenterology 2013;60:1809−13.

[98] Seok JK, Lee SH, Kim MJ, Lee YM. MicroRNA-382 induced by HIF-1α is an angiogenic miR targeting the tumor suppressor phosphatase and tensin homolog. Nucleic Acids Res 2014;42:8062−72.

[99] Mead H, Zeremski M, Guba M. mTOR signaling in angiogenesis. In: Polunovsky VA, Houghton PJ, editors. Cancer Drug Discovery and Development. USA: Springer; 2010. p. 49−74. ISBN: 978-1-60327-270-4, 978-1-60327-271-1.

[100] Polunovsky VA, Houghton PJ. In: Polunovsky VA, Houghton PJ, editors. mTOR Pathway and mTOR Inhibitors in Cancer Therapy. USA: Springer; 2010ISBN: 978-1-60327-270-4, 978-1-60327-271-1.

15

Role of mTOR Signaling in Cardioprotection

Anindita Das and Rakesh C. Kukreja

Pauley Heart Center, Division of Cardiology, Department of Internal Medicine, Virginia Commonwealth University Medical Center, Richmond, VA, USA

15.1 INTRODUCTION

The antifungal drug rapamycin (Sirolimus®) was first isolated from the soil bacterium *Streptomyces hygroscopicus* on Easter Island (Rapa Nui) in 1975 [1,2]. Subsequently, in the 1990s, the immunosuppressive property of rapamycin was recognized; it has since been primarily used in the treatment of organ rejection in transplant recipients [3–5]. The antiproliferative effect of rapamycin, first discovered in yeast system, is based on its ability to bind to its intracellular receptor, the peptidyl-prolyl *cis/trans* isomerase FKBP12 (FK506 binding protein) [6]. In the early 1990s, genetic screens in budding yeast identified TOR1 and TOR2 (target of rapamycin) as mediators of antiproliferative effects of rapamycin on yeast [7–9]. Shortly afterwards in 1994 and 1995, mammalian target of rapamycin (mTOR, also known as mechanistic TOR) was purified and characterized for the first time as the physical target of rapamycin [10–12]. The rapamycin–FKBP12 complex binds and specifically acts as an allosteric inhibitor of mTOR, an atypical serine/threonine kinase [13]. Recently, a great deal of attention has focused on the potential use of rapamycin for coating drug-eluting stents to prevent restenosis following coronary angioplasty [14]. Restenosis, a reduction in lumen diameter after angioplasty and stent implantation, is the result of arterial damage with subsequent neointimal tissue proliferation. Due to its antiproliferative property, rapamycin prevents intimal growth of graft coronary arteries and reduces the incidence of vasculopathy [15].

15.2 mTOR SIGNALING PATHWAY

mTOR belongs to the phosphatidylinositol 3-kinase (PI3K)-related kinase family with a molecular weight of ~ 289 kDa. It interacts with other molecular components to form two physically and functionally distinct complexes, mTOR complex 1 (mTORC1) and mTOR complex 2 (mTORC2) [16]. Both complexes share the catalytic mTOR subunit, mLST8 (mammalian lethal with sec-13 protein 8), Deptor (DEP domain-containing mTOR-interacting protein), and Tti1-Tel2 (Tel two interacting protein 1/Tel2) complex [17–19]. Other components of mTORC1 are Raptor (regulatory-associated protein of mTOR) and PRAS40 (proline-rich AKT substrate) [20,21]. Alternatively, mTORC2 contains Rictor (rapamycin-insensitive companion of mTOR) and mSin1 (mammalian stress-activated MAP kinase-interacting protein 1) [22,23]. Raptor and Rictor function as scaffold proteins for integrity of the complexes; regulate their subcellular localization and interactions with substrates and regulator [24–26].

Over the past two decades, remarkable progress has been made to characterize the function and regulation of mTORC1, yet the mechanisms of mTORC2 activation are less well elucidated. mTORC1 is a signal integrator responding to multiple signals from growth factors, stressors such as hypoxia, nutrients, and energy status to control cellular homeostasis, growth, proliferation, and protein and lipid synthesis (Figure 15.1) [24,26]. mTORC1 activates the ribosomal proteins S6 kinase 1 and 2 (S6K1/2) by phosphorylating their hydrophobic motif (HM), on Thr389 and Thr 388, respectively, which promotes mRNA biogenesis, as well as translational

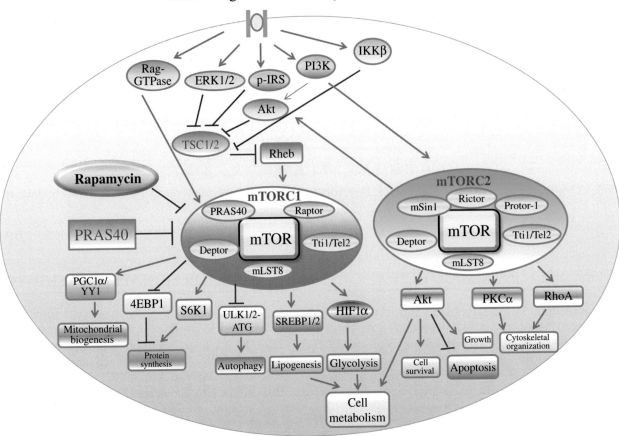

FIGURE 15.1   The mammalian target of rapamycin (mTOR) signaling pathway. eIF4E-Bp1, eukaryotic translation initiation factor 4E-binding protein 1; ATG, autophagy-related gene; DEPTOR, DEP domain-containing mTOR-interacting protein; ERK1/2, extracellular signal regulated kinase 1/2; HIF-1α, hypoxia-inducible factor-1α; IKKβ, inhibitor of NF-κB kinase-β; p-IRS, phosphoinsulin receptor substrate; mLST8, mammalian lethal with sec-13 protein 8; PGC1α, peroxisome proliferator-activated receptor γ coactivator-1α; PI3K, phosphoinositide 3-kinase; PKCα, protein kinase C α; PRAS40, proline-rich Akt substrate 40; Protor-1, protein observed with rictor-1; Raptor; regulatory-associated protein in mTOR; Rheb, Ras homolog enriched in brain; RhoA, Ras homolog gene family, member A; S6K1, S6 kinase 1; SREBP1/2, sterol regulatory element-binding protein 1/2; TSC1/2, tuberous sclerosis protein 1/2; Tt1, Tel two interacting protein 1; ULK1/2, unc-51-like kinase 1/2; and YY1, Ying-Yang 1.

initiation and elongation of protein synthesis [25,27]. In contrast, mTORC1 phosphorylates and inhibits eukaryotic translation initiation factor 4E (eIF4E)-binding protein 1 (4E-BP1), which triggers the initiation of cap-dependent protein translation processes [25–27]. mTORC1 also activates the transcription factors sterol regulatory element-binding protein 1 and 2 (SREBP1/2), which control the expression of lipogenic genes and promote the synthesis of cellular membrane lipids [25,28]. In addition, mTORC1 increases mitochondrial DNA content and the expression of genes involved in oxidative metabolism through the peroxisome proliferator-activated receptor γ (PPAR-γ)-mediated activation of transcription factor Ying-Yang 1, which ultimately promotes mitochondrial biogenesis and oxidative metabolism [24,29]. Furthermore, mTORC1 increases glycolytic flux by activating the transcription

and the translation of hypoxia-inducible factor 1α [30]. Overall, mTORC1 has an evolutionarily conserved role in promoting anabolic cell growth, but it also inhibits the catabolic process of autophagy, the central degradation process of proteins and organelles in cells [24,31]. mTORC1 directly phosphorylates and suppresses ULK1/Atg13/FIP200 (unc-51-like kinase 1/mammalian autophagy-related gene 13/focal adhesion kinase family-interacting protein of 200 kDa), a kinase complex which promotes autophagosome formation [32,33]. mTORC1 also suppresses autophagy by regulating death-associated protein 1 and another crucial autophagic protein, Atg7 [34].

The knowledge of mTORC2 regulation and function has lagged greatly behind that of mTORC1. mTORC2 signaling is insensitive to nutrient derivation, but it responds to growth factors such as insulin via PI3 kinase-dependent mechanisms (Figure 15.1) [35]. mTORC2 phosphorylates the turn motif and the HM of several AGC (protein kinase A, G, and C) kinases, including PKB, SGK1 (serum/glucocorticoid-induced kinase 1), PKCs (protein kinase C), and Rho1 (GDP–GTP exchange protein-2 [Rom2]), resulting in their stabilization and activation [36]. Rictor enables mTORC2 to directly phosphorylate Akt at its HM Ser473 and facilitates Thr308 phosphorylation by PDK1 (phosphoinositide-dependent kinase 1) as part of the insulin signaling cascade [37]. Akt regulates essential cellular processes such as survival, growth, proliferation, metabolism, and apoptosis through the phosphorylation of several effectors [24].

Rictor of mTORC2 is directly phosphorylated by S6K1 downstream of mTORC1 and this phosphorylation event exerts a negative regulatory effect on the mTORC2-dependent phosphorylation of Akt-S473 within cells [38]. In contrast, upon stimulation with growth factors, mTORC2 activates Akt which in turn enhances mTORC1 activity through inactivation of TSC1/2 (tuberous sclerosis complex). The TSC2 is inactivated by Akt-dependent phosphorylation, which destabilizes TSC2 and disrupts its interaction with TSC1 [39,40]. TSC1/2 complex inhibits mTORC1 through its GTPase-activating protein activity toward Ras homolog enriched in brain (Rheb) [41]. These studies underscore the interconnection between mTORC1 and mTORC2 [42]. Rapamycin preferentially inhibits mTORC1, however recent studies show that chronic treatment with the drug also inhibits mTORC2 [43,44].

15.3 mTOR IN MYOCARDIAL ISCHEMIA/REPERFUSION INJURY

Cardiovascular disease (CVD) produces immense health and economic burdens in the United States and globally. Despite advances in the treatment of CVD, it remains the leading global cause of death, accounting for 17.3 million deaths per year, a number that is expected to grow to >23.6 million by 2030 [45]. CVDs are a group of disorders of the heart and blood vessels. They include coronary heart disease, cerebrovascular disease or heart stroke, heart attack (myocardial infarction), heart failure, dilated cardiomyopathy, hypertrophic cardiomyopathy, angina, arrhythmia, atherosclerosis, hypertension, rheumatic heart disease, peripheral vascular disease, congenital heart disease, and so on. Acute myocardial infarction (AMI), one of the most catastrophic acute cardiac disorders, is the major cause of the detrimental effects of coronary heart disease on the myocardium [46]. AMI, caused by acute myocardial ischemia due to insufficient coronary blood flow to supply to meet the oxygen demand in the heart,

is a major cause of hundreds of thousands of deaths and disability each year worldwide. The most effective therapeutic intervention for reducing acute myocardial ischemic (AMI) injury by limiting the myocardial infarct size as well as preserving left ventricular (LV) function is early and effective myocardial reperfusion [47]. Paradoxically, the reintroduction of oxygen-rich blood during reperfusion to the ischemic tissue also has detrimental effects, a phenomenon known as myocardial reperfusion injury [48,49]. Thus, the prevention of postischemic cardiac cell death may represent the best therapeutic modality to limit myocardial injury.

Although reperfusion is mandatory to salvage ischemic myocardium from infarction [50,51], an irreversible ischemic myocardial cell injury is developed after prolonged coronary occlusion (after 2 h of ischemia) [52,53]. Therefore, in the last three decades, intensive research has focused on identifying the remarkable cardioprotective effect of innate adaptive responses to ischemia/reperfusion (I/R) insult by different conditioning strategies which can provide therapeutic paradigms for cardioprotection. The prime cardioprotective paradigm of ischemic preconditioning (IPC) was established by Murry et al. in 1986 [54]. Protection using IPC was accomplished by brief intermittent episodes of ischemia and reperfusion, resulting in substantial myocardial protection from prolonged periods of ischemia. The identification of the mechanism of this phenomenon has provided a conceptual framework for developing novel therapeutic strategies to mimic the cardioprotecive effects of IPC with pharmacological agents. IPC has been successfully translated to humans with ischemic heart disease, but because of its nature ("pre") can only be used in elective settings, such as percutaneous coronary interventions and coronary artery bypass grafting, and not in AMI [55]. In 2003, Zhao et al. identified another cardioprotective phenomenon called postconditioning (PostC), which attenuates reperfusion injury [56]. PostC differs from IPC in that it is instituted at the immediate onset of arterial reperfusion after a sustained episode of coronary occlusion. PostC has been successfully translated to humans with ischemic heart disease, and as an intervention that is performed at immediate reperfusion. Furthermore, it can also be used in patients undergoing interventional reperfusion of AMI [55].

Accumulating evidence indicates the important regulatory role of mTOR in myocardial I/R injury. Previous studies report that ischemic [57,58] or pharmacological preconditioning/postconditioning [59,60] initiate PI3K-Akt signaling cascade and confer cardioprotection through recruitment of mTOR-p70S6K. In addition, mTOR inhibition with rapamycin abolished early [59] and delayed [58] preconditioning-induced cardioprotection as well as protection at reperfusion [57,60]. On the

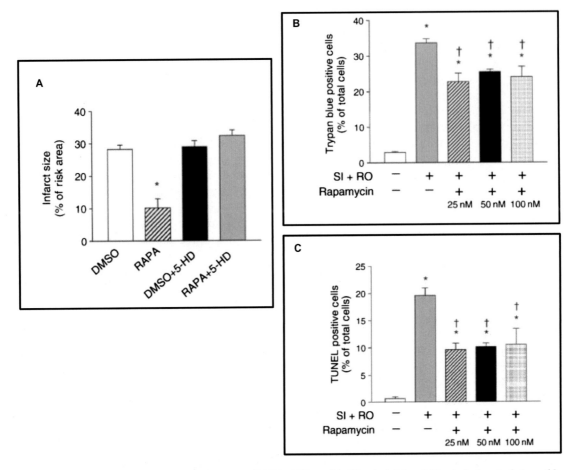

FIGURE 15.2 (A). Acute preconditioning with rapamycin (RAPA, 0.25 mg/kg, IP, administered 30 min before isolation of heart) reduces infarct size in isolated mice hearts following 20 min global ischemia and 30 min reperfusion. Note that RAPA-induced reduction of infarct size is blocked by 5-HD (5-hydroxydecanoate; 100 μM, infused 10 min before ischemia) (*P < 0.001 vs. DMSO control, DMSO + 5-HD, RAPA + 5-HD; n = 6 mice/group). (B, C) Effect of rapamycin on cardiomyocyte (B) necrosis and (C) apoptosis: Pretreatment with rapamycin for 1 h prior to simulated ischemia (SI) for 40 min and reoxygenation (RO) for 1 or 18 h decreases the number of trypan blue-positive cells as well as TUNEL-positive nuclei, indicating improved cell viability and reduced apoptosis with rapamycin (P, 0.001 vs. control; †P < 0.001 vs. SI-RO).

contrary, in 2006, the preconditioning-like cardioprotective effect of rapamycin in isolated mouse heart and cardiomyocytes was first reported [61]. Rapamycin treatment reduced infarct size after I/R injury by attenuating necrosis and apoptosis in cardiomyocytes (Figure 15.2). This study also demonstrated that attenuation of I/R injury with rapamycin was mediated through opening of mitochondrial ATP-sensitive potassium channels (mitoK$_{ATP}$ channel). Later, the signaling pathways by which rapamycin triggers preconditioning-like anti-infarct effect in *ex vivo* and *in vivo* myocardial I/R mouse models was further investigated [62]. Rapamycin triggered unique cardioprotective signaling through activation of Janus kinase 2 (JAK2)-STA3 [62] (Figure 15.3). The JAK-signal transducer and activator of transcription (STAT) pathway is composed of a family of receptor-associated cytosolic tyrosine kinases (JAKs) that phosphorylate a tyrosine residue in cognate of STATs [63].

Previous studies demonstrated that activation of JAK-STAT signaling in response to IPC confers cardioprotection via prosurvival signaling cascades or inhibition of proapoptotic factors [64,65]. Rapamycin-induced cardioprotection against I/R injury was abolished with inhibitor of JAK2 (AG-490) or STAT3 (stattic) as well as *in situ* knockdown of STAT3 [62] (Figure 15.4). Preconditioning with rapamycin also induced phosphorylation of ERK, STAT3-dependent eNOS, and glycogen synthase kinase-3β (GSK-3β) in concert with increased prosurvival Bcl-2 to Bax ratio [62] (Figure 15.5).

15.4 ROLE OF mTORC1 VERSUS mTORC2 IN MYOCARDIAL INFARCTION

Several genetic and pharmacological approaches have now revealed that mTORC1 inhibition is beneficial after myocardial infarction [66–69]. mTORC1 is

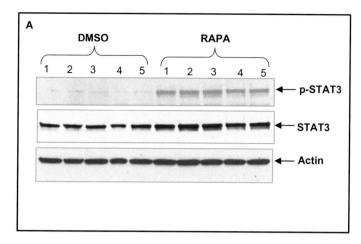

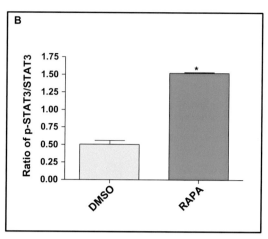

FIGURE 15.3 Rapamycin (RAPA) enhances phosphorylation of STAT3 at tyrosine 705. (A) Representative immunoblots for p-STAT3, total STAT3, and actin expression in whole heart after 1 h of RAPA treatment (0.25 mg/kg, IP). (B) Densitometry analysis of immunoblots for the ratio of p-STAT3/STAT3. *$P < 0.0001$ vs. DMSO control, $n = 5$ per group.

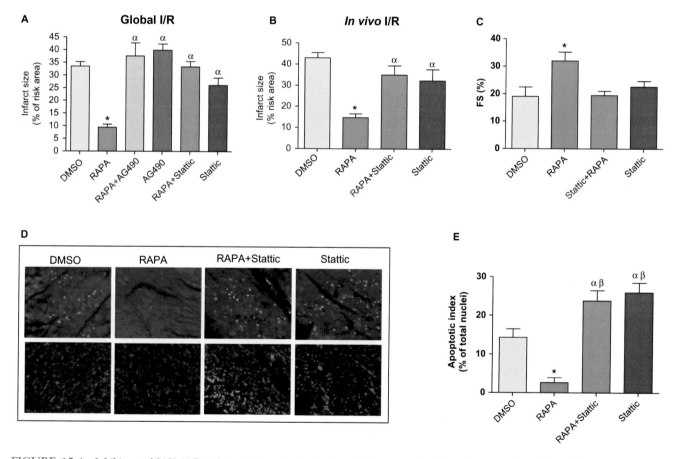

FIGURE 15.4 Inhibitors of JAK (AG-490) and STAT3 (Stattic) abolish infarct-limiting effect of rapamycin (RAPA) following I/R. Stattic (20 mg/kg) or AG-490 (40 mg/kg) was administered intraperitoneally (IP) 30 min before rapamycin (0.25 mg/kg, IP) treatment. After 1 h of rapamycin treatment, the heart is subjected to I/R. (A) Myocardial infarct size following 20 min of no-flow global ischemia and 30 min of reperfusion, *$P < 0.001$ vs. DMSO (solvent of rapamycin); $^{\alpha}P < 0.001$ vs. RAPA, $n = 7$ per group. (B) Myocardial infarct size following in situ I/R by 30 min ligation of left coronary artery and 24 h reperfusion. *$P < 0.001$ vs. DMSO; $^{\alpha}P < 0.01$ vs. RAPA, $n = 6$ per group. (C) Cardiac function (fractional shortening, FS), monitored by echocardiography following in situ I/R, is improved with rapamycin treatment. Stattic abolished the rapamycin-induced functional improvement. *$P < 0.05$ vs. DMSO, $n = 5$ per group. (D) Rapamycin reduces myocardial apoptosis, determined by TUNEL assay after in situ I/R. Stattic treatment blocks rapamycin-induced reduction of apoptosis. Representative images of TUNEL-positive nuclei in green fluorescent color and total nuclei (blue) staining with 4,5-diamino-2-phenylindole (DAPI). (E) Bar diagram showing quantitative data of TUNEL-positive nuclei in the peri-infarct region of myocardium. *$P < 0.01$ vs. DMSO; $^{\alpha}P < 0.05$ vs. RAPA; $^{\beta}P < 0.001$ vs. DMSO; $n = 5$ per group.

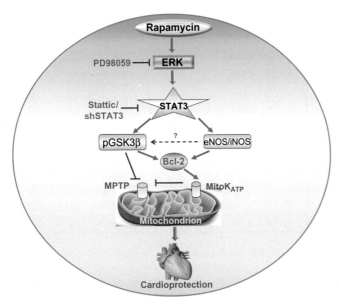

FIGURE 15.5 Proposed signaling pathways of STAT3-mediated cardioprotection by rapamycin against I/R injury.

activated in the remote myocardium during chronic myocardial infarction and contributes to ventricular remodeling [66,68]. Pharmacological mTOR inhibition with everolimus prevents adverse LV remodeling and cardiac dilation, which limits infarct size with improved cardiac function following chronic myocardial infarction [66]. The mTOR inhibition increased autophagy and decreased proteasome activity in the border zone of the infarcted myocardium [66]. Rapamycin and S6K inhibitors alleviate cardiac ischemic remodeling and cardiomyocyte apoptosis through PDK2 phosphorylation of Akt following myocardial infarction [70]. Since Rheb is required for mTORC1 signaling, a recent study reported that the partial cardiac-specific deletion of Rheb1 and pharmacological inhibition of mTORC1 signaling exert cardioprotection against adverse cardiac remodeling in mice [71]. These studies suggest that inhibition of mTORC1 could be a potential therapeutic strategy to limit infarct size as well as attenuate adverse LV remodeling after myocardial infarction.

Selective activation of mTORC2 with concurrent inhibition of mTORC1 signaling led to decreased cardiomyocyte apoptosis and tissue damage after myocardial infarction [68]. This study showed that selective inhibition of mTORC1 with clinically relevant adeno-associated virus serotype 9 gene therapy with PRAS40 (AAV-PRAS40) signaling enhanced cardioprotection (by limiting myocardial infarction, improving cardiac function) and reduced mortality. Decreased mTORC1 signaling slowed degradation of insulin receptor substrate-1, consequently improving insulin signaling

and mTORC2 function in AAV-PRAS40-treated hearts. Conversely, mTORC2 deletion with AAV-sh-Rictor exacerbated the decline in cardiac function and remodeling following a myocardial infarction. Such a combinatorial approach of inhibition of mTORC1 in conjunction with activation of mTORC2 was important for the regulation of ischemic damage and cardiac remodeling after myocardial infarction. A previous study also demonstrated the crucial role of mTORC2 activation in myocardium protection from I/R injury by constitutive phosphorylation of Akt-Ser473 in PH domain leucine-rich repeat protein phosphatase-1, a selective Akt-Ser473 phosphatase, knockout mouse [72].

The pivotal role of mTORC2 has also been shown in IPC [73]. In isolated perfused mouse hearts, IPC induced mTORC2 activity, leading to phosphorylation of Akt at Ser473 and subsequent phosphorylation of GSK-3β, endothelial nitric oxide synthase, p70S6K, and ribosomal S6. The protective effect of IPC was lost by pretreatment with dual mTORC inhibitors but not with rapamycin, an mTORC1 inhibitor, which confirmed the crucial role of mTORC2 activation in cardioprotection [73].

mTOR overexpression was also found to reduce cell death of ventricular cardiomyocytes in response to hypoxia through activation of NF-κB and suppression of inducible death factor BCL-2/adenovirus E1B 19 kDa interacting protein 3 (Bnip3) [74]. Although, the relative activation of mTORC1 and C2 following overexpression of mTOR was not examined in this study, these findings raise the interesting possibility that activation of Akt by mTORC2 may be required for direct triggering of NF-κB signaling for averting cell death of ventricular myocytes during hypoxia [74,75]. Therefore, careful modulation of mTOR activity and a more selective approach towards mTORC1 inhibition may be critical for the treatment of myocardial ischemia.

Interestingly, the function of mTOR is distinct between prolonged ischemia and reperfusion [76]. Since, mTOR is an essential regulator for protein synthesis and cardiac hypertrophy, inhibition of mTOR as well as protein synthesis during ischemia is beneficial in reducing energy consumption and endoplasmic reticulum stress [76,77]. However, the cardioprotective role of mTOR signaling in reperfusion injury is still controversial. Earlier studies suggested that activation of mTOR during the reperfusion phase may induce the cell survival signaling mechanisms and mitochondrial biogenesis, which may lead to recovery from myocardial ischemia during reperfusion [29,60].

Autophagy is an evolutionarily constitutive intracellular catalytic process that targets dysfunctional or damaged cytoplasmic constituents to the lysosome for degradation and recycling [78]. Autophagy is under the control of multiple signaling pathways, including nutrients, stress, hormones, growth factors, and intracellular energy information [79]. Autophagy was also induced by ischemia and further enhanced by reperfusion in the mouse heart, *in vivo* [80]. Autophagy resulting from ischemia was accompanied by activation of AMPK and inhibited by dominant negative AMPK. Importantly, autophagy plays distinct roles during ischemia and reperfusion: autophagy may be protective during ischemia, whereas it may be detrimental during reperfusion [80]. Since mTOR is another key modulator of autophagy, the inhibition of GSK-3β and resultant mTOR-dependent attenuation of autophagy are detrimental during prolonged ischemia, but they are beneficial during reperfusion [76]. Moreover, excessive activation of autophagy during reperfusion leads to cell death [80], although autophagy also plays a beneficial role during myocardial reperfusion in some experimental conditions [81]. Accumulating studies indicate that mTORC1 inhibition is protective during ischemia through the activation of autophagy, reduction of protein synthesis, and subsequent activation of mTORC2 [42,69].

15.5 mTOR IN CARDIAC HYPERTROPHY

Cardiac hypertrophy, characterized by cardiomyocyte enlargement and structural remodeling, leads to myocardial wall thickness and is caused by physiological stimuli such as exercise, or pathological stimuli from pressure or volume overload induced by hypertension, myocardial infarction, or valvular heart disease [82]. Cardiac hypertrophy is, initially, an adaptive mechanism to maintain normal cardiac function. However, with prolonged non-reversible remodeling this leads to dilated cardiomyopathy and heart failure [83]. This hypertrophic response is comprised of an increase in cell size and protein content, as well as complex alterations in gene transcription and translation [84]. Since protein synthesis is regulated by the PI3K-AKT-mTOR signaling pathways, numerous studies were conducted to understand the key role of mTOR in the development of cardiac hypertrophy. Independent studies from different groups indicated that the hypertrophic stimulus with β-adrenergic stimulation [85], angiotensin-II [86], or insulin growth factor-1 [87] induced mTORC1 activity in cardiomyocytes, and the pharmacological inhibition of mTORC1 with rapamycin reversed cardiac hypertrophy [88]. Rapamycin also attenuated transverse aortic constriction (TAC)-induced cardiac hypertrophy as well as physical-exercise-induced physiological hypertrophy by inactivating mTORC1 [77,89–92]. Depression of ubiquitin proteasome system (UPS) activities

contributes to cardiac dysfunction resulting in pressure-overloaded hearts of mice by enhanced cardiomyocyte apoptosis through accumulation of proapoptotic proteins [93]. Akt is known to phosphorylate several proapoptotic substrates for subsequent UPS-mediated elimination [93]. Rapamycin administration augments activation of Akt, increases pressure overload-induced degradation of phosphorylated inhibitor of κB, enhances expression of cellular inhibitor of apoptosis protein 1, and decreases active caspase-3 [94]. Long-term rapamycin treatment (2 weeks) in pressure-overload myocardium-blunted hypertrophy, improved contractile function, and inhibited apoptosis by reducing caspase-3 and calpain activation [94]. These studies indicate that an increase in targeted polyubiquitination (Ub), Akt activation, and survival signaling by inhibition of mTORC1 with rapamycin during hypertrophy can tilt the molecular balance toward mTORC2 survival signaling in the myocardium [94].

mTORC1 activity is strongly stimulated by upstream Rheb in the heart and Rheb-mTORC1 signaling is involved in pathologic heart remodeling [24]. A targeted Rheb1-mTORC1 inhibition with astragaloside IV or cardiac-specific rheb1 deletion effectively alleviated adverse cardiac remodeling after myocardial infarction and hypertrophy in mouse models [71]. As previously underlined, among mTORC1 regulatory binding partners, PRAS40 interacts with Raptor and inhibits Reb1-induced activation of mTORC1 kinase [21,95]. PRAS40 is phosphorylated by Akt in cardiomyocytes, which relieves the inhibitory function of PRAS40 on mTORC1. Cardiomyocyte-specific overexpression of PRAS40 blocks mTORC1 in cardiomyocytes and blunts pathological remodeling after long-term pressure overload and preserves cardiac function [69].

However, other studies have challenged the importance of mTOR-mediated protein synthesis and growth during hypertrophy. The cardiac hypertrophic growth in response to physiological and pathological stimuli was not different in cardiac-specific transgenic mice overexpressing mTOR or mutant mTOR compared with that of non-transgenic littermates [96,97]. Interestingly, mTOR overexpression did not significantly increase mTORC1 signaling during pathological hypertrophy induced by TAC in this study; instead, it actually decreased phosphorylation of S6K [97]. Therefore, the specific activation of mTORC2 by mTOR overexpression may contribute to preserve cardiac function during TAC. Moreover, the deletion of the ribosomal S6 kinases, direct downstream effectors of mTOR, did not attenuate pathological, physiological, or insulin-like growth factor receptor-induced cardiac hypertrophy [98]. Inducible cardiac-specific raptor-deficient mice did not develop adaptive hypertrophy in response to pressure overload, however these mice developed ventricular dilation and cardiac dysfunction associated with autophagy, apoptosis, and mitochondrial and metabolic gene alteration [99]. These studies indicate that mTOR signaling may be essential for adaptive cardiomyocyte growth to pressure overload. Further investigations are needed to better understand the specific role of mTORC1 and mTORC2 in cardiac hypertrophy.

15.6 mTOR IN CARDIAC METABOLISM AND DIABETES

The increasing prevalence of obesity and associated metabolic disorders are well-recognized risk factors to individual and public health in developing countries. Metabolic syndrome (MS) is a constellation of obesity, hypertension, and disorders of lipid and carbohydrate metabolism. MS is a major risk factor for the development of type 2 diabetes, atherosclerosis, and coronary artery disease [100]. Since the mTOR signaling pathway responds to nutrients and growth factor levels, its role in the regulation of metabolism has attracted intense interest during the last decade, together with the notion of intermittent versus persistent signaling to explain its differential effects. A short-term activation of mTORC1 signaling during availability of nutrients within a physiological range is necessary for anabolic metabolism, energy storage, and consumption, as well as normal cell/tissue development [101]. This is supported by a study which reported that skeletal muscle-specific Raptor-deficient mice begin to develop muscular dystrophy in conjunction with decreased PGC-1α-mediated oxidative metabolism [102]. In contrast, a persistent activation of mTORC1 is involved in the presence of nutritional excess including glucose and amino acids, genetic and diet-induced animal models of obesity, and metabolic disorders in the liver [103], skeletal muscle [103,104], adipose tissue [105,106], as well as in the heart [107,108].

15.7 mTOR SIGNALING IN ADIPOGENESIS AND LIPOGENESIS

mTORC1 activity is high in the livers of obese rodents, which leads to degradation of IRS1 and poor insulin signaling, as well as hepatic insulin resistance [109]. Lipogenesis is paradoxically very active in the liver of insulin-resistant rodents [24]. mTORC1 hyperactivation by overfeeding promotes lipogenesis through induction of SREBP-1c (SREBP) cleavage and activation [110,111]. Insulin activates mTORC1 through Akt-mediated phosphorylation and inhibition of TSC2

(tuberous sclerosis complex 2) [39–41]. Rapamycin blocks Akt-induced SREBP-1 expression and nuclear accumulation, the expression of several lipogenic genes, and the synthesis of various classes of lipids [111]. The knockdown of Raptor, but not Rictor, showed similar effects, indicating that SREBP-1 activation mainly depends on mTORC1, but not mTORC2 [28,112,113]. Using the mTORC1 inhibitor, rapamycin, other independent studies also confirmed the significant role of mTORC1 in the regulation of energy production through profound effects on hepatic fatty acid metabolism [114,115]. Additionally, the antidiabetic drug, metformin, which is a known to activate adenosine monophosphate-activated protein kinase (AMPK) and also subsequently inhibits mTORC1, reduced hepatic lipid content by promoting fatty acid oxidation, impairing SREBP-1c expression and cleavage [116]. In addition, using liver-specific Rictor knockout mice, a recent study has established the crucial role of hepatic mTORC2 in lipogenesis through activation of Akt-mTORC1-SREBP-1c [117].

Pharmacological and genetic studies have demonstrated that mTORC1 signaling also plays a fundamental role in lipid storage by stimulating the synthesis of triglycerides in white adipose tissue (WAT) and the differentiation of preadipocytes into white adipocytes through the translational control of the master regulator of adipogenesis, PPAR-γ [118,119]. mTOR inhibition with rapamycin reduced mRNA and protein levels of PPAR-γ and C/EBP-α (CCAAT/enhancer binding protein) and the expression of numerous lipogenic genes [118,120,121]. Constitutive activation of mTORC1 through TSC2 deletion or the deletion of 4E-BP1/2 induces PPAR-γ and C/EBP-α expression and promotes adipogenesis [102,119]. Additionally, S6K1-deficient mice had reduced adipose tissue mass and were protected against diet-induced obesity [106]. Adipocyte-specific deletion of Raptor in mice resulted in lean mice with reduced WAT mass which are resistant to high-fat diet (HFD)-induced obesity [113].

15.8 mTOR IN DIABETIC HEARTS

MS, the combination of hypertension, obesity, dyslipidemia, and insulin resistance, is a precursor of diabetes mellitus [122]. The diabetic heart shows early maladaptation to alteration in lipid and/or glucose metabolism, accompanied by lipid accumulation, and cardiac insulin resistance, which ultimately lead to cardiac dysfunction and heart failure [123,124]. Hyperglycemia and hyperlipidemia as a result of diabetes lead to oxidative stress and inflammation in the cardiovascular system which causes endothelial dysfunction and cardiomyopathy [125,126]. Sustained hyperactive mTORC1 signaling plays a crucial role in connecting metabolic stress and CVDs through stimulation of oxidative stress and inflammatory responses [88,101]. Inhibition of AMPK during metabolic stress leads to activation of mTORC1 [42,101,127]. Enhancing AMPK activation during ischemia reduces infarct size in the diabetic heart following I/R injury size [128]. Moreover, the reduction of AMPK activity by ablation of liver kinase B1 (LKB1, a regulation of AMPK) in the heart leads to energy deprivation and impairs cardiac function during aerobic or ischemic conditions [129].

In HFD-fed obese mice, increased activation of the Akt/mTOR pathway causes vascular senescence and vascular dysfunction, which increase the susceptibility to peripheral and cerebral ischemia [130]. Rapamycin treatment of diet-induced obese mice or of transgenic mice with long-term activation of endothelial Akt inhibited activation of mTOR and Akt, prevented vascular senescence without altering body weight, and reduced the severity of limb necrosis and ischemic stroke [130]. In the HFD-induced obesity and MS mouse model, enhanced cardiac mTORC1 activity suppressed autophagy and increased ischemic injury [67]. Pharmacological and genetic inhibition of mTORC1 restored autophagy and normalized the increase in infarct size observed in HFD mice [67]. This study provided a mechanistic explanation for the susceptibility of ischemic heart with metabolic abnormality and underscores the role of mTORC1 inhibition as an effective therapeutic option to reduce cardiac ischemic injury in the presence of MS. A reduction in myocardial autophagosome formation with associated induction of apoptosis inflammation, mitochondrial dysfunction, and fibrosis are also reported in a swine model of MS [131,132]. In another mouse model of HFD-induced obesity, cardiac autophagosome formation was reduced, with increased mTOR activity and cardiac hypertrophy and contractile dysfunction [133]. Rapamycin treatment restored cardiac function by promoting autophagy through inhibition of mTOR and stimulation of AMPK activity [133]. Interestingly, HFD promoted the initiation of autophagy and accumulation of autophagy, although it disrupted autophagosome maturation probably at the step of autophagosome fusion with lysosomes, autophagic flux [134]. As the major activator of mTOR, the Akt family of serine–threonine kinases is also activated by HFD in the heart. HFD-feeding up-regulated cardiac expression of Akt2, but not Akt1 and Akt3. Akt2 knockout ameliorated HFD-feeding-induced cardiac pathological hypertrophy and contractile anomalies by facilitating cardiac autophagosome maturation processes. Rescuing of the cardiac autophagy by Akt2 knockout inhibited apoptosis and improved mitochondrial performance, intracellular Ca^{2+} homeostasis, and

contractile protein levels [134]. Autophagosome formation and autophagic flux were also inhibited in the heart of the streptozotocin-induced type I mouse model [135]. Therefore, restoring myocardial cellular autophagy may represent a novel therapeutic target for preserving myocardial structure and function in obesity and MS.

Recently, we reported the effect of chronic treatment (28 days) of rapamycin (0.25 mg/kg, intraperitoneally) in identifying novel protein targets involved in preservation of cardiac function in type 2 diabetic mice [136]. The metabolic status was significantly improved in these mice as shown by reductions in plasma glucose, insulin, triglyceride levels, as well as body weight. However, a similar treatment strategy with rapamycin did not affect body weight, metabolic parameters, or cardiac function in non-diabetic mice. Another study showed that long-term (6 months) treatment with rapamycin in female diabetic mice decreased body weight and fat mass without affecting food intake. The markers of fatty acid oxidation and mitochondrial biogenesis were elevated in the gonadal WAT. In addition, circulating non-esterified free fatty acids were reduced, circulating adiponectin was elevated, and insulin sensitivity was improved in these mice [137]. Although earlier studies report that chronic treatment with rapamycin (2 mg/kg/day) impairs whole-body insulin sensitivity by disrupting mTORC2 activity and blocking the ability of mTORC2-Akt to inhibit hepatic gluconeogenesis [43,44]; a well-controlled cardioprotective dose of rapamycin (0.25 mg/kg/day) preferentially inhibited mTORC1, without disrupting mTORC2 in type 2 diabetic mice [136]. Rapamycin inhibited the enhanced phosphorylation of mTOR and S6 ribosomal protein (downstream target of mTORC1) due to diabetes in the heart, without interfering with phosphorylation of Akt (Ser^{473}, target of mTORC2). A proteomic approach identified alteration of four cytoskeletal/contractile proteins in diabetic mouse heart as compared to non-diabetic heart. The expression of the key contractile/cytoskeletal proteins, MHC6α (myosin heavy polypeptide 6α) and MyBP-C (myosin binding protein C), were suppressed, but MLC-2 (myosin light chain 2) was increased in diabetic heart as compared to non-diabetic heart. Chronic treatment with rapamycin reversed the alteration of the cardiac expression of myosin isozymes.

Our proteomic study also identified the alteration of three proteins involved in glucose metabolism [136]. These included Pgm2, glucose phosphomutase which catalyzes the conversion of glucose 1 phosphate to the glycolytic intermediate glucose 6-phosphate. Suppression of Pgm2 in diabetic heart was completely recovered following chronic rapamycin treatment [136]. PDH E1α, pyruvate dehydrogenase E1α, contributes to transforming pyruvate into acetyl-CoA, which can be used in the citric acid cycle to carry out cellular respiration. Chronic inhibition of PDH with the cardiac-specific overexpression of PDH inhibitor triggers an adaptive metabolic response in transgenic mice [138]. The transgenic mice with overexpression of cardiac-specific PDH are resistant to HFD-induced cardiomyocyte lipid accumulation and exhibit normal functional recovery after myocardial I/R injury. The adaptive metabolic reprogramming is involved with the activation of AMPK kinase and induction of the transcriptional coactivator, PGC-1α. Chronic treatment with rapamycin reduced PDH E1α in diabetic heart, with induction of PYGB protein (glycogen phosphorylase), which may improve cardiac metabolism and help recover cardiac function [136].

Increased production of reactive oxygen species (ROS) in diabetic hearts is a major contributing factor in the development and progression of diabetic cardiomyopathy [139,140]. In diabetic hearts, oxidative stress is exacerbated in the presence of fatty acids, which leads to mitochondrial uncoupling [140]. Therapeutic strategies that either reduce ROS production or augment myocardial antioxidant defense mechanisms have been shown to be efficacious in the prevention of diabetes-induced myocardial dysfunction [141−144]. Interestingly, oxidative stress, as measured by glutathione levels and lipid peroxidation, was significantly reduced in rapamycin-treated diabetic hearts [136]. Rapamycin treatment augmented the mitochondrial antioxidant enzyme PRX-5 (peroxiredoxin 5) in diabetic hearts, which may help in alleviating oxidative stress in the diabetic heart [136].

Another powerful approach to overcome ROS-induced toxicity is to limit the cellular availability of transition metals, most notably iron sequestration [145]. An excess of free iron must be detoxified by sequestering in ferritin, the major intracellular iron storage protein, by oxidizing Fe(II) to Fe(III); ferritin heavy chain (FHC) possesses that ferroxidase activity [146]. A decrease in the abundance of FHC increases the levels of iron deposition and oxidative stress, which leads to cardiomyocyte death in failing hearts following ischemia or pressure overload [147]. The expression of FHC is also induced with chronic rapamycin treatment in the hearts of diabetic mice [136]. Therefore, it appears that chronic rapamycin treatment improves metabolic status and preserves cardiac function in diabetic mice, through attenuation of oxidative stress and alteration of antioxidants as well as contractile and glucose metabolic protein expression (Figure 15.6).

A recent study shows that specific mTORC1 inhibition by PRAS40 prevents the development of diabetic cardiomyopathy by improving metabolic profile in HFD-induced type 2 diabetic mice and leptin receptor

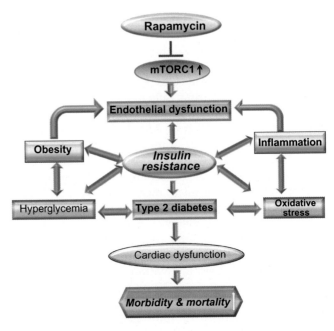

FIGURE 15.6 Proposed scheme of cardioprotection by rapamycin in type 2 diabetes. Rapamycin inhibits mTOR signaling and subsequently prevents endothelial dysfunction, obesity, hyperglycemia, insulin resistance, inflammation, and oxidative stresses. This eventually prevents diabetic-induced cardiac dysfunction in type 2 diabetes.

mutant db/db mice [148]. PRAS40 overexpression prevents cardiac dysfunction by blunting hypertrophic cardiac remodeling. In this context, a selective approach towards mTORC1 inhibition may be a promising therapeutic modality in overcoming metabolic abnormalities and preventing cardiac dysfunction in patients with diabetes.

15.9 mTOR INHIBITION AND REPERFUSION THERAPY IN DIABETIC HEARTS

The heart of a diabetic patient is more susceptible to AMI injury [122] and clinical studies suggest that diabetes is associated with higher mortality after AMI due to more extensive atherosclerotic lesions, hypertrophy, and LV dysfunction [149,150]. Although reperfusion is mandatory to salvage ischemic myocardium from infarction, reperfusion itself contributes to irreversible ischemic myocardial injury by formation of ROS, intracellular Ca^{2+} overload, mitochondrial dysfunction, activation of intracellular proteolysis, and uncoordinated excess contractile activity [49,55]. Currently there are no therapies approved by the FDA that can directly diminish the detrimental effect of myocardial reperfusion injury. A recent, seminal study showed that rapamycin administered at the onset of reperfusion following

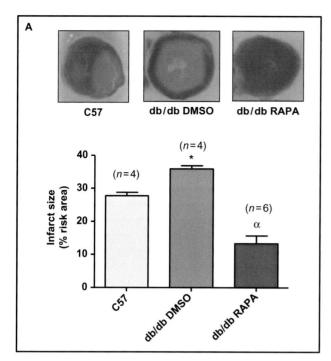

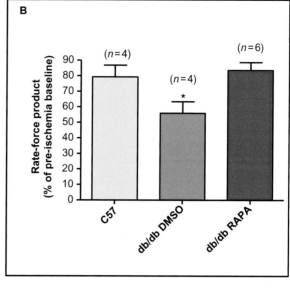

FIGURE 15.7   Reperfusion therapy with rapamycin (RAPA) reduces infarct size in diabetic mouse (db/db) hearts following ischemia/reperfusion (I/R). (A) Myocardial infarct size in C57 and db/db mice following 30 min of no-flow global ischemia and 1 h of reperfusion in Langendorff mode. Rapamycin (100 nM) or DMSO (solvent of rapamycin) is infused at the time of reperfusion (*$P < 0.05$ vs. C57; $^{\alpha}P < 0.001$ vs. other). (B) Product of heart rate and ventricular developed force (% of preischemic baseline) (*$P < 0.05$ vs. other).

global ischemia significantly reduces myocardial infarct size in the hearts of diabetic mice [151] (Figure 15.7). Moreover, rapamycin treatment improved cardiac function with increased coronary flow rates in diabetic heart following I/R injury. Rapamycin treatment during reoxygenation following simulated ischemia also reduced necrosis, apoptosis, and preserved mitochondrial membrane potential in adult primary cardiomyocytes isolated from diabetic mice [151]. Interestingly, rapamycin infusion at the onset of reperfusion inhibited phosphorylation of S6 ribosomal protein (target of mTORC1) in hearts of normoglycemic mice without interfering with Akt phosphorylation at Ser473, a target of mTORC2 [151] (Figure 15.8). Contrary to expectations, the phosphorylation of S6 was inhibited, but Akt phosphorylation was enhanced with rapamycin treatment in diabetic hearts [151]. Thus specific inhibition of mTORC1 activity as well as activation of mTORC2 may provide beneficial effects of cardioprotection by rapamycin treatment at reperfusion in diabetic heart following I/R injury.

The phosphorylation and subsequent activation of STAT3 also contribute to cardioprotection via prosurvival signaling cascade or the inhibition of proapoptotic factors after I/R injury [64,151]. The diabetic heart shows a decreased level of activated STAT3 which is associated with insulin resistance and dilated cardiomyopathy [151–153]. Rapamycin treatment

significantly increased both total and phosphorylated STAT3 in the hearts of diabetic mice [151] (Figure 15.8). Rapamycin treatment at reperfusion exerted a robust infarct-sparing effect in HFD-fed wild-type mice which was associated with improved cardiac function after global I/R injury. Conversely, the infarct-limiting effect of rapamycin was abolished in HFD-fed cardiac-specific STAT3-deficient mice [151] (Figure 15.9). Moreover, the protective effect of rapamycin treatment at reoxygenation in cardiomyocytes isolated from HFD-fed STAT3-deficient mice was abolished after SI/RO. These compelling studies revealed that reperfusion therapy with rapamycin in diabetic hearts provides protection through the STAT3-AKT signaling pathway.

15.10   CONCLUSIONS

It is quite clear from the above-reviewed studies that mTOR signaling is critical in cardiovascular and metabolic diseases. On the other hand, the clinical applications of mTOR inhibitors in these disorders have not yet been established. Further studies are needed to clarify the cardioprotective mechanisms regulated by mTOR signaling in order to approve the clinical use of mTOR inhibitors in patients. Thus, targeting mTOR signaling by rapamycin or its safer analogs

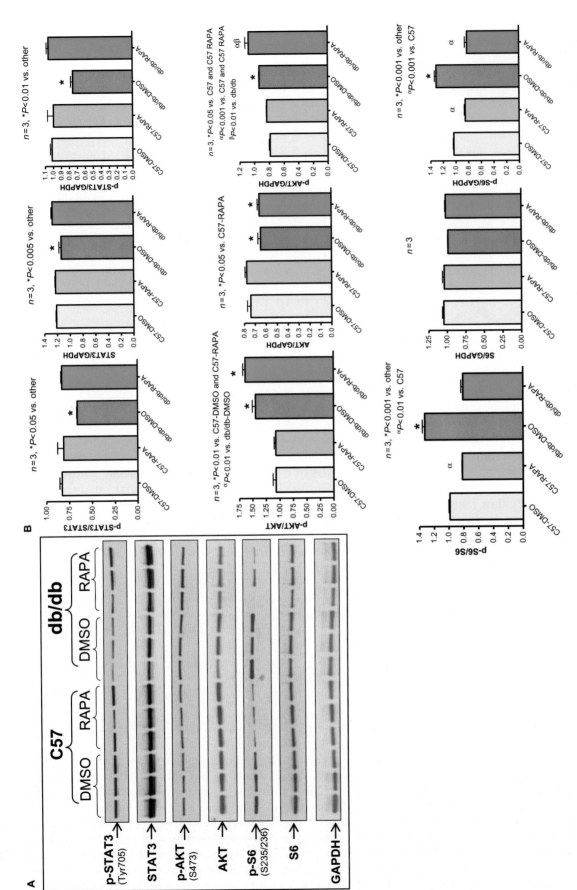

FIGURE 15.8 Reperfusion therapy with rapamycin restores phosphorylation of STAT3 in diabetic heart, but inhibits mTORC1. Rapamycin (100 nM) or DMSO (solvent of rapamycin) is infused at the time of reperfusion following 30 min of no-flow global ischemia in Langendorff mode. (A) Representative immunoblots of phospho-STAT3, STAT3, phospho-AKT, AKT, phospho-S6, S6, and GAPDH in hearts of C57 and db/db mice. (B) Densitometric analysis of the ratios of phosphorylated (p) to total protein, total protein to GAPDH, and phosphorylated proteins to GAPDH.

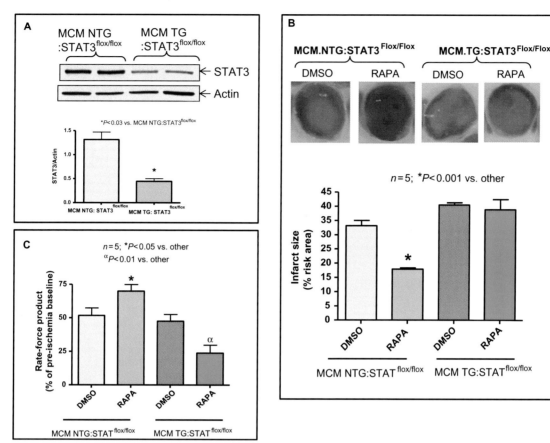

FIGURE 15.9 Cardiac-specific STAT3 deficiency abolishes the infarct-limiting effect of rapamycin (RAPA) against ischemia/reperfusion (I/R) injury in diabetic mice. (A) STAT3-deficient (MCM TG:STAT3$^{flox/flox}$) and WT (MCM NTG:STAT3$^{flox/flox}$) male mice (8–10 weeks) are fed a high-fat diet (HFD) for 16 weeks, after which they receive tamoxifen (20 mg/kg/day IP) for 10 days. Rapamycin (100 nM) or DMSO (solvent of rapamycin) is infused at the time of reperfusion following 30 min of no-flow global ischemia in Langendorff mode. Representative immunoblots of STAT3 and GAPDH in hearts of HFD-fed and tamoxifen-treated STAT3-deficient and WT mice. Densitometric analysis representing fold change in STAT3/GAPDH ratio. (B) Myocardial infarct sizes of WT and STAT3-deficient mice following global I/R as well as infusion of rapamycin (RAPA, 100 nM) or DMSO (solvent of rapamycin) at reperfusion. (C) Product of heart rate and ventricular developed force (% of preischemic baseline).

with less adverse effects could be promising for treatment of cardiovascular and metabolic diseases including I/R injury, hypertrophy, and type 2 diabetes.

Acknowledgments

This work was supported by grants from the National Institutes of Health R37 HL051045, R01 HL079424, R01 HL093685, and R01 HL118808 (R. C. Kukreja) and A. D. Williams' Fund of Virginia Commonwealth University Grant UL1RR031990 (A. Das) and CTSA (UL1TR000058 from the National Center for Advancing Translational Sciences) and the CCTR (Center for Clinical and Translational Research) Endowment Fund (A. Das).

References

[1] Sehgal SN, Baker H, Vezina C. Rapamycin (AY-22,989), a new antifungal antibiotic. II. Fermentation, isolation and characterization. J Antibiot (Tokyo) 1975;28:727–32.

[2] Vezina C, Kudelski A, Sehgal SN. Rapamycin (AY-22,989), a new antifungal antibiotic. I. Taxonomy of the producing streptomycete and isolation of the active principle. J Antibiot (Tokyo) 1975;28:721–6.

[3] Morris RE, Wu J, Shorthouse R. A study of the contrasting effects of cyclosporine, FK 506, and rapamycin on the suppression of allograft rejection. Transplant Proc 1990;22:1638–41.

[4] Morris RE, Meiser BM, Wu J, Shorthouse R, Wang J. Use of rapamycin for the suppression of alloimmune reactions in vivo: schedule dependence, tolerance induction, synergy with cyclosporine and FK 506, and effect on host-versus-graft and graft-versus-host reactions. Transplant Proc 1991;23:521–4.

[5] Stepkowski SM, Chen H, Daloze P, Kahan BD. Rapamycin, a potent immunosuppressive drug for vascularized heart, kidney, and small bowel transplantation in the rat. Transplantation 1991;51:22–6.

[6] Koltin Y, Faucette L, Bergsma DJ, Levy MA, Cafferkey R, Koser PL, et al. Rapamycin sensitivity in Saccharomyces cerevisiae is mediated by a peptidyl-prolyl cis-trans isomerase related to human FK506-binding protein. Mol Cell Biol 1991;11:1718–23.

[7] Cafferkey R, Young PR, McLaughlin MM, Bergsma DJ, Koltin Y, Sathe GM, et al. Dominant missense mutations in a novel yeast protein related to mammalian phosphatidylinositol 3-kinase and VPS34 abrogate rapamycin cytotoxicity. Mol Cell Biol 1993;13:6012–23.

[8] Cafferkey R, McLaughlin MM, Young PR, Johnson RK, Livi GP. Yeast TOR (DRR) proteins: amino-acid sequence alignment and identification of structural motifs. Gene 1994;141:133–6.

[9] Kunz J, Henriquez R, Schneider U, Deuter-Reinhard M, Movva NR, Hall MN. Target of rapamycin in yeast, TOR2, is an essential phosphatidylinositol kinase homolog required for G1 progression. Cell 1993;73:585–96.

[10] Brown EJ, Albers MW, Shin TB, Ichikawa K, Keith CT, Lane WS, et al. A mammalian protein targeted by G1-arresting rapamycin-receptor complex. Nature 1994;369:756–8.

[11] Sabatini DM, Erdjument-Bromage H, Lui M, Tempst P, Snyder SH. RAFT1: a mammalian protein that binds to FKBP12 in a rapamycin-dependent fashion and is homologous to yeast TORs. Cell 1994;78:35–43.

[12] Sabers CJ, Martin MM, Brunn GJ, Williams JM, Dumont FJ, Wiederrecht G, et al. Isolation of a protein target of the FKBP12-rapamycin complex in mammalian cells. J Biol Chem 1995;270:815–22.

[13] Schmelzle T, Hall MN. TOR, a central controller of cell growth. Cell 2000;103:253–62.

[14] Morice MC, Serruys PW, Sousa JE, Fajadet J, Ban HE, Perin M, et al. A randomized comparison of a sirolimus-eluting stent with a standard stent for coronary revascularization. N Engl J Med 2002;346:1773–80.

[15] Delgado JF, Manito N, Segovia J, Almenar L, Arizon JM, Camprecios M, et al. The use of proliferation signal inhibitors in the prevention and treatment of allograft vasculopathy in heart transplantation. Transplant Rev (Orlando.) 2009;23:69–79.

[16] Guertin DA, Sabatini DM. The pharmacology of mTOR inhibition. Sci Signal 2009;2:e24.

[17] Kaizuka T, Hara T, Oshiro N, Kikkawa U, Yonezawa K, Takehana K, et al. Tti1 and Tel2 are critical factors in mammalian target of rapamycin complex assembly. J Biol Chem 2010;285:20109–16.

[18] Kim DH, Sarbassov DD, Ali SM, Latek RR, Guntur KV, Erdjument-Bromage H, et al. GbetaL, a positive regulator of the rapamycin-sensitive pathway required for the nutrient-sensitive interaction between raptor and mTOR. Mol Cell 2003;11:895–904.

[19] Peterson TR, Laplante M, Thoreen CC, Sancak Y, Kang SA, Kuehl WM, et al. DEPTOR is an mTOR inhibitor frequently overexpressed in multiple myeloma cells and required for their survival. Cell 2009;137:873–86.

[20] Kim DH, Sarbassov DD, Ali SM, King JE, Latek RR, Erdjument-Bromage H, et al. mTOR interacts with raptor to form a nutrient-sensitive complex that signals to the cell growth machinery. Cell 2002;110:163–75.

[21] Sancak Y, Thoreen CC, Peterson TR, Lindquist RA, Kang SA, Spooner E, et al. PRAS40 is an insulin-regulated inhibitor of the mTORC1 protein kinase. Mol Cell 2007;25:903–15.

[22] Jacinto E, Facchinetti V, Liu D, Soto N, Wei S, Jung SY, et al. SIN1/MIP1 maintains rictor-mTOR complex integrity and regulates Akt phosphorylation and substrate specificity. Cell 2006;127:125–37.

[23] Sarbassov DD, Ali SM, Kim DH, Guertin DA, Latek RR, Erdjument-Bromage H, et al. Rictor, a novel binding partner of mTOR, defines a rapamycin-insensitive and raptor-independent pathway that regulates the cytoskeleton. Curr Biol 2004;14:1296–302.

[24] Laplante M, Sabatini DM. mTOR signaling in growth control and disease. Cell 2012;149:274–93.

[25] Laplante M, Sabatini DM. Regulation of mTORC1 and its impact on gene expression at a glance. J Cell Sci 2013;126:1713–19.

[26] Wullschleger S, Loewith R, Hall MN. TOR signaling in growth and metabolism. Cell 2006;124:471–84.

[27] Ma XM, Blenis J. Molecular mechanisms of mTOR-mediated translational control. Nat Rev Mol Cell Biol 2009;10:307–18.

[28] Laplante M, Sabatini DM. An emerging role of mTOR in lipid biosynthesis. Curr Biol 2009;19:R1046–52.

[29] Cunningham JT, Rodgers JT, Arlow DH, Vazquez F, Mootha VK, Puigserver P. mTOR controls mitochondrial oxidative function through a YY1-PGC-1alpha transcriptional complex. Nature 2007;450:736–40.

[30] Hudson CC, Liu M, Chiang GG, Otterness DM, Loomis DC, Kaper F, et al. Regulation of hypoxia-inducible factor 1 alpha expression and function by the mammalian target of rapamycin. Mol Cell Biol 2002;22:7004–14.

[31] Sciarretta S, Zhai P, Volpe M, Sadoshima J. Pharmacological modulation of autophagy during cardiac stress. J Cardiovasc Pharmacol 2012;60:235–41.

[32] Egan DF, Shackelford DB, Mihaylova MM, Gelino S, Kohnz RA, Mair W, et al. Phosphorylation of ULK1 (hATG1) by AMP-activated protein kinase connects energy sensing to mitophagy. Science 2011;331:456–61.

[33] Ganley IG, Lam DH, Wang J, Ding X, Chen S, Jiang X. ULK1. ATG13.FIP200 complex mediates mTOR signaling and is essential for autophagy. J Biol Chem 2009;284:12297–305.

[34] Hsu PP, Kang SA, Rameseder J, Zhang Y, Ottina KA, Lim D, et al. The mTOR-regulated phosphoproteome reveals a mechanism of mTORC1-mediated inhibition of growth factor signaling. Science 2011;332:1317–22.

[35] Gan X, Wang J, Su B, Wu D. Evidence for direct activation of mTORC2 kinase activity by phosphatidylinositol 3,4,5-trisphosphate. J Biol Chem 2011;286:10998–1002.

[36] Xie J, Herbert TP. The role of mammalian target of rapamycin (mTOR) in the regulation of pancreatic beta-cell mass: implications in the development of type-2 diabetes. Cell Mol Life Sci 2012;69:1289–304.

[37] Sarbassov DD, Guertin DA, Ali SM, Sabatini DM. Phosphorylation and regulation of Akt/PKB by the rictor-mTOR complex. Science 2005;307:1098–101.

[38] Dibble CC, Asara JM, Manning BD. Characterization of Rictor phosphorylation sites reveals direct regulation of mTOR complex 2 by S6K1. Mol Cell Biol 2009;29:5657–70.

[39] Inoki K, Li Y, Zhu T, Wu J, Guan KL. TSC2 is phosphorylated and inhibited by Akt and suppresses mTOR signalling. Nat Cell Biol 2002;4:648–57.

[40] Manning BD, Tee AR, Logsdon MN, Blenis J, Cantley LC. Identification of the tuberous sclerosis complex-2 tumor suppressor gene product tuberin as a target of the phosphoinositide 3-kinase/akt pathway. Mol Cell 2002;10:151–62.

[41] Dibble CC, Elis W, Menon S, Qin W, Klekota J, Asara JM, et al. TBC1D7 is a third subunit of the TSC1-TSC2 complex upstream of mTORC1. Mol Cell 2012;47:535–46.

[42] Sciarretta S, Volpe M, Sadoshima J. Mammalian target of rapamycin signaling in cardiac physiology and disease. Circ Res 2014;114:549–64.

[43] Lamming DW, Ye L, Katajisto P, Goncalves MD, Saitoh M, Stevens DM, et al. Rapamycin-induced insulin resistance is mediated by mTORC2 loss and uncoupled from longevity. Science 2012;335:1638–43.

[44] Sarbassov DD, Ali SM, Sengupta S, Sheen JH, Hsu PP, Bagley AF, et al. Prolonged rapamycin treatment inhibits mTORC2 assembly and Akt/PKB. Mol Cell 2006;22:159–68.

[45] Mozaffarian D, Benjamin EJ, Go AS, Arnett DK, Blaha MJ, Cushman M, et al. Heart disease and stroke statistics—2015 update: a report from the American Heart Association. Circulation 2015;131:e29–322.

[46] Hausenloy DJ, Yellon DM. Myocardial ischemia-reperfusion injury: a neglected therapeutic target. J Clin Invest 2013;123:92−100.

[47] Hochman JS, Choo H. Limitation of myocardial infarct expansion by reperfusion independent of myocardial salvage. Circulation 1987;75:299−306.

[48] Braunwald E, Kloner RA. Myocardial reperfusion: a double-edged sword? J Clin Invest 1985;76:1713−19.

[49] Yellon DM, Hausenloy DJ. Myocardial reperfusion injury. N Engl J Med 2007;357:1121−35.

[50] Ginks WR, Sybers HD, Maroko PR, Covell JW, Sobel BE, Ross Jr J. Coronary artery reperfusion. II. Reduction of myocardial infarct size at 1 week after the coronary occlusion. J Clin Invest 1972;51:2717−23.

[51] Maroko PR, Libby P, Ginks WR, Bloor CM, Shell WE, Sobel BE, et al. Coronary artery reperfusion. I. Early effects on local myocardial function and the extent of myocardial necrosis. J Clin Invest 1972;51:2710−16.

[52] Reimer KA, Lowe JE, Rasmussen MM, Jennings RB. The wavefront phenomenon of ischemic cell death. 1. Myocardial infarct size vs duration of coronary occlusion in dogs. Circulation 1977;56:786−94.

[53] Reimer KA, Jennings RB. The "wavefront phenomenon" of myocardial ischemic cell death. II. Transmural progression of necrosis within the framework of ischemic bed size (myocardium at risk) and collateral flow. Lab Invest 1979;40:633−44.

[54] Murry CE, Jennings RB, Reimer KA. Preconditioning with ischemia: a delay of lethal cell injury in ischemic myocardium. Circulation 1986;74:1124−36.

[55] Heusch G. Molecular basis of cardioprotection: signal transduction in ischemic pre-, post-, and remote conditioning. Circ Res 2015;116:674−99.

[56] Zhao ZQ, Corvera JS, Halkos ME, Kerendi F, Wang NP, Guyton RA, et al. Inhibition of myocardial injury by ischemic postconditioning during reperfusion: comparison with ischemic preconditioning. Am J Physiol Heart Circ Physiol 2003;285:H579−88.

[57] Hausenloy DJ, Mocanu MM, Yellon DM. Cross-talk between the survival kinases during early reperfusion: its contribution to ischemic preconditioning. Cardiovasc Res 2004;63:305−12.

[58] Kis A, Yellon DM, Baxter GF. Second window of protection following myocardial preconditioning: an essential role for PI3 kinase and p70S6 kinase. J Mol Cell Cardiol 2003;35:1063−71.

[59] Gross ER, Hsu AK, Gross GJ. Opioid-induced cardioprotection occurs via glycogen synthase kinase beta inhibition during reperfusion in intact rat hearts. Circ Res 2004;94:960−6.

[60] Jonassen AK, Sack MN, Mjos OD, Yellon DM. Myocardial protection by insulin at reperfusion requires early administration and is mediated via Akt and p70s6 kinase cell-survival signaling. Circ Res 2001;89:1191−8.

[61] Khan S, Salloum F, Das A, Xi L, Vetrovec GW, Kukreja RC. Rapamycin confers preconditioning-like protection against ischemia-reperfusion injury in isolated mouse heart and cardiomyocytes. J Mol Cell Cardiol 2006;41:256−64.

[62] Das A, Salloum FN, Durrant D, Ockaili R, Kukreja RC. Rapamycin protects against myocardial ischemia-reperfusion injury through JAK2-STAT3 signaling pathway. J Mol Cell Cardiol 2012;53:858−69.

[63] Dawn B, Xuan YT, Guo Y, Rezazadeh A, Stein AB, Hunt G, et al. IL-6 plays an obligatory role in late preconditioning via JAK-STAT signaling and upregulation of iNOS and COX-2. Cardiovasc Res 2004;64:61−71.

[64] Suleman N, Somers S, Smith R, Opie LH, Lecour SC. Dual activation of STAT-3 and Akt is required during the trigger phase of ischaemic preconditioning. Cardiovasc Res 2008;79:127−33.

[65] Xuan YT, Guo Y, Han H, Zhu Y, Bolli R. An essential role of the JAK-STAT pathway in ischemic preconditioning. Proc Natl Acad Sci U S A 2001;98:9050−5.

[66] Buss SJ, Muenz S, Riffel JH, Malekar P, Hagenmueller M, Weiss CS, et al. Beneficial effects of mammalian target of rapamycin inhibition on left ventricular remodeling after myocardial infarction. J Am Coll Cardiol 2009;54:2435−46.

[67] Sciarretta S, Zhai P, Shao D, Maejima Y, Robbins J, Volpe M, et al. Rheb is a critical regulator of autophagy during myocardial ischemia: pathophysiological implications in obesity and metabolic syndrome. Circulation 2012;125:1134−46.

[68] Volkers M, Konstantin MH, Doroudgar S, Toko H, Quijada P, Din S, et al. Mechanistic target of rapamycin complex 2 protects the heart from ischemic damage. Circulation 2013;128:2132−44.

[69] Volkers M, Toko H, Doroudgar S, Din S, Quijada P, Joyo AY, et al. Pathological hypertrophy amelioration by PRAS40-mediated inhibition of mTORC1. Proc Natl Acad Sci USA 2013;110:12661−6.

[70] Di R, Wu X, Chang Z, Zhao X, Feng Q, Lu S, et al. S6K inhibition renders cardiac protection against myocardial infarction through PDK1 phosphorylation of Akt. Biochem J 2012;441:199−207.

[71] Wu X, Cao Y, Nie J, Liu H, Lu S, Hu X, et al. Genetic and pharmacological inhibition of Rheb1-mTORC1 signaling exerts cardioprotection against adverse cardiac remodeling in mice. Am J Pathol 2013;182:2005−14.

[72] Miyamoto S, Purcell NH, Smith JM, Gao T, Whittaker R, Huang K, et al. PHLPP-1 negatively regulates Akt activity and survival in the heart. Circ Res 2010;107:476−84.

[73] Yano T, Ferlito M, Aponte A, Kuno A, Miura T, Murphy E, et al. Pivotal role of mTORC2 and involvement of ribosomal protein S6 in cardioprotective signaling. Circ Res 2014;114:1268−80.

[74] Dhingra R, Gang H, Wang Y, Biala AK, Aviv Y, Margulets V, et al. Bidirectional regulation of nuclear factor-kappaB and mammalian target of rapamycin signaling functionally links Bnip3 gene repression and cell survival of ventricular myocytes. Circ Heart Fail 2013;6:335−43.

[75] Regula KM, Baetz D, Kirshenbaum LA. Nuclear factor-kappaB represses hypoxia-induced mitochondrial defects and cell death of ventricular myocytes. Circulation 2004;110:3795−802.

[76] Zhai P, Sciarretta S, Galeotti J, Volpe M, Sadoshima J. Differential roles of GSK-3beta during myocardial ischemia and ischemia/reperfusion. Circ Res 2011;109:502−11.

[77] McMullen JR, Sherwood MC, Tarnavski O, Zhang L, Dorfman AL, Shioi T, et al. Inhibition of mTOR signaling with rapamycin regresses established cardiac hypertrophy induced by pressure overload. Circulation 2004;109:3050−5.

[78] Mizushima N, Levine B, Cuervo AM, Klionsky DJ. Autophagy fights disease through cellular self-digestion. Nature 2008;451:1069−75.

[79] Glick D, Barth S, Macleod KF. Autophagy: cellular and molecular mechanisms. J Pathol 2010;221:3−12.

[80] Matsui Y, Takagi H, Qu X, Abdellatif M, Sakoda H, Asano T, et al. Distinct roles of autophagy in the heart during ischemia and reperfusion: roles of AMP-activated protein kinase and Beclin 1 in mediating autophagy. Circ Res 2007;100:914−22.

[81] Hamacher-Brady A, Brady NR, Gottlieb RA. The interplay between pro-death and pro-survival signaling pathways in myocardial ischemia/reperfusion injury: apoptosis meets autophagy. Cardiovasc Drugs Ther 2006;20:445−62.

[82] Heineke J, Molkentin JD. Regulation of cardiac hypertrophy by intracellular signalling pathways. Nat Rev Mol Cell Biol 2006;7:589−600.

[83] Nadal-Ginard B, Kajstura J, Leri A, Anversa P. Myocyte death, growth, and regeneration in cardiac hypertrophy and failure. Circ Res 2003;92:139—50.

[84] Dorn GW, Force T. Protein kinase cascades in the regulation of cardiac hypertrophy. J Clin Invest 2005;115:527—37.

[85] Simm A, Schluter K, Diez C, Piper HM, Hoppe J. Activation of p70(S6) kinase by beta-adrenoceptor agonists on adult cardiomyocytes. J Mol Cell Cardiol 1998;30:2059—67.

[86] Sadoshima J, Izumo S. Rapamycin selectively inhibits angiotensin II-induced increase in protein synthesis in cardiac myocytes in vitro. Potential role of 70-kD S6 kinase in angiotensin II-induced cardiac hypertrophy. Circ Res 1995;77:1040—52.

[87] Lavandero S, Foncea R, Perez V, Sapag-Hagar M. Effect of inhibitors of signal transduction on IGF-1-induced protein synthesis associated with hypertrophy in cultured neonatal rat ventricular myocytes. FEBS Lett 1998;422:193—6.

[88] Soesanto W, Lin HY, Hu E, Lefler S, Litwin SE, Sena S, et al. Mammalian target of rapamycin is a critical regulator of cardiac hypertrophy in spontaneously hypertensive rats. Hypertension 2009;54:1321—7.

[89] Gao XM, Wong G, Wang B, Kiriazis H, Moore XL, Su YD, et al. Inhibition of mTOR reduces chronic pressure-overload cardiac hypertrophy and fibrosis. J Hypertens 2006;24:1663—70.

[90] Kemi OJ, Ceci M, Wisloff U, Grimaldi S, Gallo P, Smith GL, et al. Activation or inactivation of cardiac Akt/mTOR signaling diverges physiological from pathological hypertrophy. J Cell Physiol 2008;214:316—21.

[91] Shioi T, McMullen JR, Tarnavski O, Converso K, Sherwood MC, Manning WJ, et al. Rapamycin attenuates load-induced cardiac hypertrophy in mice. Circulation 2003;107:1664—70.

[92] Zhang D, Contu R, Latronico MV, Zhang J, Rizzi R, Catalucci D, et al. MTORC1 regulates cardiac function and myocyte survival through 4E-BP1 inhibition in mice. J Clin Invest 2010;120:2805—16.

[93] Tsukamoto O, Minamino T, Okada K, Shintani Y, Takashima S, Kato H, et al. Depression of proteasome activities during the progression of cardiac dysfunction in pressure-overloaded heart of mice. Biochem Biophys Res Commun 2006;340:1125—33.

[94] Harston RK, McKillop JC, Moschella PC, Van LA, Quinones LS, Baicu CF, et al. Rapamycin treatment augments both protein ubiquitination and Akt activation in pressure-overloaded rat myocardium. Am J Physiol Heart Circ Physiol 2011;300:H1696—706.

[95] Vander HE, Lee SI, Bandhakavi S, Griffin TJ, Kim DH. Insulin signalling to mTOR mediated by the Akt/PKB substrate PRAS40. Nat Cell Biol 2007;9:316—23.

[96] Shen WH, Chen Z, Shi S, Chen H, Zhu W, Penner A, et al. Cardiac restricted overexpression of kinase-dead mammalian target of rapamycin (mTOR) mutant impairs the mTOR-mediated signaling and cardiac function. J Biol Chem 2008;283:13842—9.

[97] Song X, Kusakari Y, Xiao CY, Kinsella SD, Rosenberg MA, Scherrer-Crosbie M, et al. mTOR attenuates the inflammatory response in cardiomyocytes and prevents cardiac dysfunction in pathological hypertrophy. Am J Physiol Cell Physiol 2010;299:C1256—66.

[98] McMullen JR, Shioi T, Zhang L, Tarnavski O, Sherwood MC, Dorfman AL, et al. Deletion of ribosomal S6 kinases does not attenuate pathological, physiological, or insulin-like growth factor 1 receptor-phosphoinositide 3-kinase-induced cardiac hypertrophy. Mol Cell Biol 2004;24:6231—40.

[99] Shende P, Plaisance I, Morandi C, Pellieux C, Berthonneche C, Zorzato F, et al. Cardiac raptor ablation impairs adaptive hypertrophy, alters metabolic gene expression, and causes heart failure in mice. Circulation 2011;123:1073—82.

[100] Eberly LE, Prineas R, Cohen JD, Vazquez G, Zhi X, Neaton JD, et al. Metabolic syndrome: risk factor distribution and 18-year mortality in the multiple risk factor intervention trial. Diabetes Care 2006;29:123—30.

[101] Yang Z, Ming XF. mTOR signalling: the molecular interface connecting metabolic stress, aging and cardiovascular diseases. Obes Rev 2012;13(Suppl. 2):58—68.

[102] Bentzinger CF, Romanino K, Cloetta D, Lin S, Mascarenhas JB, Oliveri F, et al. Skeletal muscle-specific ablation of raptor, but not of rictor, causes metabolic changes and results in muscle dystrophy. Cell Metab 2008;8:411—24.

[103] Khamzina L, Veilleux A, Bergeron S, Marette A. Increased activation of the mammalian target of rapamycin pathway in liver and skeletal muscle of obese rats: possible involvement in obesity-linked insulin resistance. Endocrinology 2005;146:1473—81.

[104] Drake JC, Alway SE, Hollander JM, Williamson DL. AICAR treatment for 14 days normalizes obesity-induced dysregulation of TORC1 signaling and translational capacity in fasted skeletal muscle. Am J Physiol Regul Integr Comp Physiol 2010;299:R1546—54.

[105] Ranieri SC, Fusco S, Panieri E, Labate V, Mele M, Tesori V, et al. Mammalian life-span determinant p66shcA mediates obesity-induced insulin resistance. Proc Natl Acad Sci U S A 2010;107:13420—5.

[106] Um SH, Frigerio F, Watanabe M, Picard F, Joaquin M, Sticker M, et al. Absence of S6K1 protects against age- and diet-induced obesity while enhancing insulin sensitivity. Nature 2004;431:200—5.

[107] Sung MM, Koonen DP, Soltys CL, Jacobs RL, Febbraio M, Dyck JR. Increased CD36 expression in middle-aged mice contributes to obesity-related cardiac hypertrophy in the absence of cardiac dysfunction. J Mol Med (Berl) 2011;89:459—69.

[108] Turdi S, Kandadi MR, Zhao J, Huff AF, Du M, Ren J. Deficiency in AMP-activated protein kinase exaggerates high fat diet-induced cardiac hypertrophy and contractile dysfunction. J Mol Cell Cardiol 2011;50:712—22.

[109] Tremblay F, Brule S, Hee US, Li Y, Masuda K, Roden M, et al. Identification of IRS-1 Ser-1101 as a target of S6K1 in nutrient- and obesity-induced insulin resistance. Proc Natl Acad Sci USA 2007;104:14056—61.

[110] Duvel K, Yecies JL, Menon S, Raman P, Lipovsky AI, Souza AL, et al. Activation of a metabolic gene regulatory network downstream of mTOR complex 1. Mol Cell 2010;39:171—83.

[111] Porstmann T, Santos CR, Griffiths B, Cully M, Wu M, Leevers S, et al. SREBP activity is regulated by mTORC1 and contributes to Akt-dependent cell growth. Cell Metab 2008;8:224—36.

[112] Cybulski N, Polak P, Auwerx J, Ruegg MA, Hall MN. mTOR complex 2 in adipose tissue negatively controls whole-body growth. Proc Natl Acad Sci U S A 2009;106:9902—7.

[113] Polak P, Cybulski N, Feige JN, Auwerx J, Ruegg MA, Hall MN. Adipose-specific knockout of raptor results in lean mice with enhanced mitochondrial respiration. Cell Metab 2008;8:399—410.

[114] Brown NF, Stefanovic-Racic M, Sipula IJ, Perdomo G. The mammalian target of rapamycin regulates lipid metabolism in primary cultures of rat hepatocytes. Metabolism 2007;56:1500—7.

[115] Peng T, Golub TR, Sabatini DM. The immunosuppressant rapamycin mimics a starvation-like signal distinct from amino acid and glucose deprivation. Mol Cell Biol 2002;22:5575—84.

IV. mTOR IN CARDIOVASCULAR DEVELOPMENT AND DISEASE

[116] Zhou G, Myers R, Li Y, Chen Y, Shen X, Fenyk-Melody J, et al. Role of AMP-activated protein kinase in mechanism of metformin action. J Clin Invest 2001;108:1167–74.

[117] Hagiwara A, Cornu M, Cybulski N, Polak P, Betz C, Trapani F, et al. Hepatic mTORC2 activates glycolysis and lipogenesis through Akt, glucokinase, and SREBP1c. Cell Metab 2012;15:725–38.

[118] Kim JE, Chen J. Regulation of peroxisome proliferator-activated receptor-gamma activity by mammalian target of rapamycin and amino acids in adipogenesis. Diabetes 2004;53:2748–56.

[119] Zhang HH, Huang J, Duvel K, Boback B, Wu S, Squillace RM, et al. Insulin stimulates adipogenesis through the Akt-TSC2-mTORC1 pathway. PLoS One 2009;4:e6189.

[120] Cho HJ, Park J, Lee HW, Lee YS, Kim JB. Regulation of adipocyte differentiation and insulin action with rapamycin. Biochem Biophys Res Commun 2004;321:942–8.

[121] Gagnon A, Lau S, Sorisky A. Rapamycin-sensitive phase of 3T3-L1 preadipocyte differentiation after clonal expansion. J Cell Physiol 2001;189:14–22.

[122] Clavijo LC, Pinto TL, Kuchulakanti PK, Torguson R, Chu WW, Satler LF, et al. Metabolic syndrome in patients with acute myocardial infarction is associated with increased infarct size and in-hospital complications. Cardiovasc Revasc Med 2006;7:7–11.

[123] Goldberg IJ, Trent CM, Schulze PC. Lipid metabolism and toxicity in the heart. Cell Metab 2012;15:805–12.

[124] Pulakat L, DeMarco VG, Ardhanari S, Chockalingam A, Gul R, Whaley-Connell A, et al. Adaptive mechanisms to compensate for overnutrition-induced cardiovascular abnormalities. Am J Physiol Regul Integr Comp Physiol 2011;301:R885–95.

[125] Lee S, Zhang H, Chen J, Dellsperger KC, Hill MA, Zhang C. Adiponectin abates diabetes-induced endothelial dysfunction by suppressing oxidative stress, adhesion molecules, and inflammation in type 2 diabetic mice. Am J Physiol Heart Circ Physiol 2012;303:H106–15.

[126] Zhang H, Dellsperger KC, Zhang C. The link between metabolic abnormalities and endothelial dysfunction in type 2 diabetes: an update. Basic Res Cardiol 2012;107:237.

[127] Zhang P, Hu X, Xu X, Fassett J, Zhu G, Viollet B, et al. AMP activated protein kinase-alpha2 deficiency exacerbates pressure-overload-induced left ventricular hypertrophy and dysfunction in mice. Hypertension 2008;52:918–24.

[128] Paiva MA, Rutter-Locher Z, Goncalves LM, Providencia LA, Davidson SM, Yellon DM, et al. Enhancing AMPK activation during ischemia protects the diabetic heart against reperfusion injury. Am J Physiol Heart Circ Physiol 2011;300:H2123–34.

[129] Jessen N, Koh HJ, Folmes CD, Wagg C, Fujii N, Lofgren B, et al. Ablation of LKB1 in the heart leads to energy deprivation and impaired cardiac function. Biochim Biophys Acta 2010;1802:593–600.

[130] Wang CY, Kim HH, Hiroi Y, Sawada N, Salomone S, Benjamin LE, et al. Obesity increases vascular senescence and susceptibility to ischemic injury through chronic activation of Akt and mTOR. Sci Signal 2009;2:ra11.

[131] Li ZL, Woollard JR, Ebrahimi B, Crane JA, Jordan KL, Lerman A, et al. Transition from obesity to metabolic syndrome is associated with altered myocardial autophagy and apoptosis. Arterioscler Thromb Vasc Biol 2012;32:1132–41.

[132] Li ZL, Ebrahimi B, Zhang X, Eirin A, Woollard JR, Tang H, et al. Obesity-metabolic derangement exacerbates cardiomyocyte loss distal to moderate coronary artery stenosis in pigs without affecting global cardiac function. Am J Physiol Heart Circ Physiol 2014;306:H1087–101.

[133] Guo R, Zhang Y, Turdi S, Ren J. Adiponectin knockout accentuates high fat diet-induced obesity and cardiac dysfunction: role of autophagy. Biochim Biophys Acta 2013;1832:1136–48.

[134] Xu X, Hua Y, Nair S, Zhang Y, Ren J. Akt2 knockout preserves cardiac function in high-fat diet-induced obesity by rescuing cardiac autophagosome maturation. J Mol Cell Biol 2013;5:61–3.

[135] Xu X, Kobayashi S, Chen K, Timm D, Volden P, Huang Y, et al. Diminished autophagy limits cardiac injury in mouse models of type 1 diabetes. J Biol Chem 2013;288:18077–92.

[136] Das A, Durrant D, Koka S, Salloum FN, Xi L, Kukreja RC. Mammalian target of rapamycin (mTOR) inhibition with rapamycin improves cardiac function in type 2 diabetic mice: potential role of attenuated oxidative stress and altered contractile protein expression. J Biol Chem 2014;289:4145–60.

[137] Deepa SS, Walsh ME, Hamilton RT, Pulliam D, Shi Y, Hill S, et al. Rapamycin modulates markers of mitochondrial biogenesis and fatty acid oxidation in the adipose tissue of db/db mice. J Biochem Pharmacol Res 2013;1:114–23.

[138] Chambers KT, Leone TC, Sambandam N, Kovacs A, Wagg CS, Lopaschuk GD, et al. Chronic inhibition of pyruvate dehydrogenase in heart triggers an adaptive metabolic response. J Biol Chem 2011;286:11155–62.

[139] Boudina S, Sena S, Theobald H, Sheng X, Wright JJ, Hu XX, et al. Mitochondrial energetics in the heart in obesity-related diabetes: direct evidence for increased uncoupled respiration and activation of uncoupling proteins. Diabetes 2007;56:2457–66.

[140] Boudina S, Abel ED. Diabetic cardiomyopathy, causes and effects. Rev Endocr Metab Disord 2010;11:31–9.

[141] Cai L, Wang Y, Zhou G, Chen T, Song Y, Li X, et al. Attenuation by metallothionein of early cardiac cell death via suppression of mitochondrial oxidative stress results in a prevention of diabetic cardiomyopathy. J Am Coll Cardiol 2006;48:1688–97.

[142] Matsushima S, Kinugawa S, Ide T, Matsusaka H, Inoue N, Ohta Y, et al. Overexpression of glutathione peroxidase attenuates myocardial remodeling and preserves diastolic function in diabetic heart. Am J Physiol Heart Circ Physiol 2006;291:H2237–45.

[143] Matsushima S, Ide T, Yamato M, Matsusaka H, Hattori F, Ikeuchi M, et al. Overexpression of mitochondrial peroxiredoxin-3 prevents left ventricular remodeling and failure after myocardial infarction in mice. Circulation 2006;113:1779–86.

[144] Shen X, Zheng S, Metreveli NS, Epstein PN. Protection of cardiac mitochondria by overexpression of MnSOD reduces diabetic cardiomyopathy. Diabetes 2006;55:798–805.

[145] Pham CG, Bubici C, Zazzeroni F, Papa S, Jones J, Alvarez K, et al. Ferritin heavy chain upregulation by NF-kappaB inhibits TNFalpha-induced apoptosis by suppressing reactive oxygen species. Cell 2004;119:529–42.

[146] Torti FM, Torti SV. Regulation of ferritin genes and protein. Blood 2002;99:3505–16.

[147] Omiya S, Hikoso S, Imanishi Y, Saito A, Yamaguchi O, Takeda T, et al. Downregulation of ferritin heavy chain increases labile iron pool, oxidative stress and cell death in cardiomyocytes. J Mol Cell Cardiol 2009;46:59–66.

[148] Volkers M, Doroudgar S, Nguyen N, Konstandin MH, Quijada P, Din S, et al. PRAS40 prevents development of diabetic cardiomyopathy and improves hepatic insulin sensitivity in obesity. EMBO Mol Med 2014;6:57–65.

[149] Alegria JR, Miller TD, Gibbons RJ, Yi QL, Yusuf S. Infarct size, ejection fraction, and mortality in diabetic patients with acute myocardial infarction treated with thrombolytic therapy. Am Heart J 2007;154:743–50.

[150] Cubbon RM, Wheatcroft SB, Grant PJ, Gale CP, Barth JH, Sapsford RJ, et al. Temporal trends in mortality of patients with diabetes mellitus suffering acute myocardial infarction: a comparison of over 3000 patients between 1995 and 2003. Eur Heart J 2007;28:540—5.

[151] Das A, Salloum FN, Filippone SM, Durrant DE, Rokosh G, Bolli R, et al. Inhibition of mammalian target of rapamycin protects against reperfusion injury in diabetic heart through STAT3 signaling. Basic Res Cardiol 2015;110:31.

[152] Drenger B, Ostrovsky IA, Barak M, Nechemia-Arbely Y, Ziv E, Axelrod JH. Diabetes blockade of sevoflurane postconditioning is not restored by insulin in the rat heart: phosphorylated signal transducer and activator of transcription 3- and phosphatidylinositol 3-kinase-mediated inhibition. Anesthesiology 2011;114:1364—72.

[153] Hilfiker-Kleiner D, Hilfiker A, Fuchs M, Kaminski K, Schaefer A, Schieffer B, et al. Signal transducer and activator of transcription 3 is required for myocardial capillary growth, control of interstitial matrix deposition, and heart protection from ischemic injury. Circ Res 2004;95:187—95.

16

Role of Mammalian Target of Rapamycin (mTOR) in Cardiac Homeostasis in Metabolic Disorders

Xiangwei Liu[1,2] *and Jun Ren*[1,2]

[1]Shanghai Institute of Cardiovascular Diseases, Zhongshan Hospital, Fudan University, Shanghai, PR China
[2]Center for Cardiovascular Research and Alternative Medicine, School of Pharmacy, University of Wyoming College of Health Sciences, Laramie, WY, USA

16.1 INTRODUCTION

The ever-increasing prevalence of obesity, diabetes mellitus, and cardiometabolic disorders in general, has imposed a great challenge to our society in the twenty-first century [1,2]. The complex interplay between metabolic disorders and coronary heart diseases, among many risk factors, is believed to contribute to the pathogenesis of heart diseases in individuals with cardiometabolic diseases [3,4]. Cardiometabolic syndrome is defined as a constellation of metabolic disturbances, including hyperglycemia, dyslipidemia, and hypertension. Ample clinical and experimental studies have revealed a close tie between abdominal obesity and the etiology of metabolic syndrome [5], therefore elevating the unique role for obesity as an independent risk factor for heart diseases. Despite the subtle discrepancies in the definition of metabolic syndrome among various health organizations including the World Health Organization, the European Group for the Study of Insulin Resistance, the National Health and Nutrition Examination, and the National Cholesterol Education Program's Adult Treatment Panel III (NCEP: ATPIII), cardinal players remain somewhat reminiscent of metabolic syndrome including central obesity, glucose intolerance, insulin resistance, dyslipidemia, and hypertension [6–9]. The prevalence of metabolic syndrome is found to be age-dependent, with participants aged 60 or above displaying the highest rate of metabolic syndrome [10]. In addition, the prevalence of metabolic disorders displays a positive correlation with obesity, particularly in the severely obese (body mass index >35) [11–13]. Obesity is associated with not only risk of atherosclerotic vascular diseases but also ventricular hypertrophy, heart failure, arrhythmia, and stroke, possibly through activation of sympathetic tone, adipokine, and profibrotic signaling mechanisms [14]. Moreover, obesity is closely associated with onset and development of type 2 diabetes, which may contribute directly or indirectly to heart dysfunction [13]. More than four decades ago, Rubler and colleagues first reported myopathic changes in patients with diabetes mellitus independent of any preexisting micro- and macrovascular diseases [15]. This postmortem study suggested that patients who suffer from diabetes mellitus may develop diabetic cardiomyopathy, a cardiac muscle disease possibly due to unfavorable metabolic alterations independent of concomitant coronary artery diseases [15,16]. More clinical studies have revealed that diabetic cardiomyopathy may evolve independently to heart failure with either preserved or reduced left ventricular ejection fraction [16]. Although a number of theories have been postulated for the onset and development of obesity-related or diabetic cardiomyopathy including sympathetic overactivation, inflammation, apoptosis, oxidative stress, intracellular Ca^{2+} mishandling, dysregulated autophagy, glucose and lipid toxicity, as well as hyperinsulinemia [16–19], effective clinical management against cardiomyopathy in metabolic disorders, particularly obesity and diabetes mellitus, still remains dismal. Current treatment for myopathy associated with metabolic disorders (mainly obesity and diabetes) is somewhat limited to diuretics, lifestyle modification, weight reduction, and treatment according to heart failure guidelines [16]. A better understanding of cell signaling mechanisms at the convergence of metabolic control and cardiac homeostasis is thus pertinent to the proper clinical management of heart diseases in metabolic disorders. Recent evidence has indicated a rather unique and important role for insulin signaling cascade in the maintenance of cardiac geometry and function [20,21]. Insulin is a well-known regulator of metabolism and heart function with a central role in linking metabolic stress and cardiovascular diseases [21]. Upon activation, insulin receptor is capable of catalyzing tyrosine phosphorylation en route to activation of the phosphatidylinositol 3 (PI-3)-kinase-Akt-mammalian target of rapamycin (also called mechanistic target of rapamycin, mTOR) signaling pathway. Uncontrolled insulin signal has been demonstrated to be responsible for the pathologic hypertrophy and heart failure in insulin-resistant situations such as obesity and type 2 diabetes [22,23]. In this chapter, we review the evidence supporting a unique role for mTOR as an interface connecting metabolic stress and heart diseases. We will discuss the ability of mTOR in integrating diverse environmental signals to control cellular growth and organismal homeostasis. Meanwhile, the therapeutic potentials of targeting mTOR signaling in the management of metabolic-disorder-associated heart diseases are discussed.

mTOR is a highly evolutionarily conserved atypical serine/threonine kinase with an essential role in the regulation of protein synthesis, cell growth and proliferation, metabolism, autophagy, ribosomal and mitochondrial biogenesis, many of which closely interact with others [24—26]. In addition, mTOR serves as a vital downstream signaling regulator for a wide variety of intra- and extracellular signals, such as insulin, growth factors, angiotensin II (Ang II), PI-3 kinase, and protein kinase B (a.k.a., Akt) [27]. In order to exert its multiple biological functions, mTOR needs to interact with a number of cofactors and forms two distinct functional complexes, namely mTOR complex 1 (mTORC1) and mTOR complex 2 (mTORC2) [28,29]. Experimental models with deletion of mTOR or other crucial components of mTOR complexes revealed that mTOR is directly involved in energy metabolism under both physiological and pathophysiological settings to control vital cellular processes necessary for cardiac development, geometry, and contractile function [30—32]. Here we will focus on some of the recent evidence regarding the important role of mTOR signaling in the regulation of cardiac metabolic, structural, and functional homeostasis.

16.2 SIGNALING MECHANISM OF mTOR

mTOR, a serine/threonine kinase, belongs to the family of phosphatidylinositol kinase-related kinase with a molecular weight of 289 kDa. It is evolutionarily conserved [33,34] and was first identified as a physical target of rapamycin to suppress proliferation and immunity of mammalian cells [35]. As a result of the rapamycin binding, mTORC1 activity is suppressed. Nonetheless, mTORC2 is less sensitive to rapamycin [36]. It is well established that mTORC1 is composed of several structural components including mTOR, regulatory-associated protein of mTOR (raptor), mammalian lethal with sec-13 protein 8, DEP domain-containing mTOR-interacting protein, proline-rich protein kinase B (Akt) substrate 40 (PRAS40), and Tel two interacting protein 1/Tel2 [37—39]. On the other hand, mTORC2 contains structural components including mTOR, the scaffold protein rapamycin-insensitive companion of mTOR (Rictor), protein 1, stress-activated map kinase-interacting mammalian lethal with sec-13 protein 8, protein observed with rictor, DEP domain-containing mTOR-interacting protein, and Tel two interacting protein 1/Tel2 [40,41]. It has been established that these protein components help to maintain complex structures, regulate subcellular translocation, and interact with the up- and downstream signals [42]. It is noteworthy that mTORC1 and mTORC2 display rather distinct biological functions while the mechanisms of mTORC2 are less examined compared with those for mTORC1 [43]. Ample evidence has indicated a master regulator role for mTORC1 in cell adaption [26]. mTORC1 is turned on by growth factors, insulin, and nutrient signals, en route to the regulation of protein synthesis, cell growth and proliferation, ribosomal and mitochondrial biogenesis, autophagy, and metabolic homeostasis manifested as decreased fatty acid oxidation and increased glucose utilization [44,45]. Ras homolog enriched in brain (Rheb), a small GTPase protein, is perhaps the most commonly known activator for mTORC1 [46,47]. Rheb is under the negative regulation of tuberose sclerosis complex (TSC) 1/2, a guanosine triphosphatase (GTPase) activating protein to promote switch from active Rheb—GTP (GTP, guanriphosphat osphate) into the inactive Rheb-GDP (GDP, guanosine diphosphate) and thus inhibit mTORC1 activity [48]. Many growth factors, such as insulin and amino acids (AAs), are capable of turning on mTORC1 through a PI-3

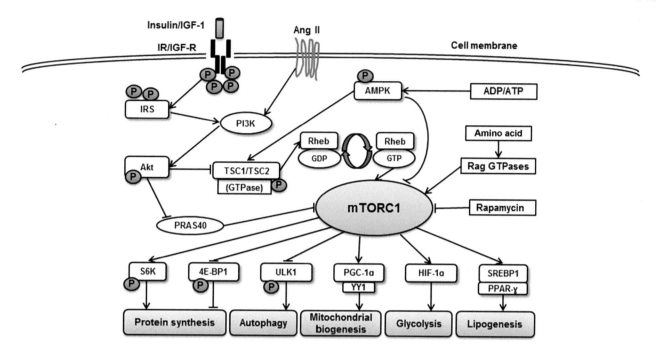

FIGURE 16.1   Model of mTOR signaling pathways regulating cardiac metabolic homeostasis. Growth factors and hormones such as insulin, amino acids, and angiotensin II (Ang II), activates the IRS1/PI3K/Akt pathway, which suppresses the phosphorylation of TSC2, then leads to the increase of Rheb–GTP, therefore stimulating the activity of mTORC1. On the other hand, AMPK, stimulated by a low ATP level, inhibits mTORC1 through an increase in the phosphorylation of TSC2. mTOR can also be induced by amino-acid-dependent Rag GTPase. Rapamycin directly inhibits mTOR. mTORC1 is capable of inducing protein synthesis via S6 kinase and 4E-BP1. mTORC1 may also inhibit autophagy and facilitate mitochondrial biogenesis, glycolysis, and lipogenesis. Arrows represent activation, whereas bars represent inhibition. mTOR, mammalian target of rapamycin; TSC2, tuberose sclerosis complex 2; Rheb, Ras homolog enriched in brain; AMPK, adenosine monophosphate-activated protein kinase; ATP, adenosine triphosphate; mTORC1, mammalian target of rapamycin complex 1.

kinase-Akt-dependent inhibition of TSC1/2 to facilitate the active "on" switch of the Rheb–GTP complex [49–51]. During starvation, the energy fuel sensor adenosine monophosphate-activated protein kinase (AMPK) inhibits mTORC1 through activation of TSC1/2 and subsequently suppression of the Rheb–GTP complex [52–54]. Not surprisingly, mTOR complexes (in particular mTORC1) serve as a central hub for the seemingly reciprocal regulations between Akt and AMPK in cell and energy metabolism [55]. In particular, activation of Akt at Thr[308] in the presence of low AMPK status (or high adenosine triphosphate (ATP) to adenosine monophosphate (AMP) ratio) is known to activate mTORC1 and the energy biosynthetic processes. Depletion of energy as a result of high AMPK activity status leads to inactivation of mTORC1, activation of the transcriptional factor foxo forkhead box, subgroup O (FoxO), and promotes constitution of mTORC2 that would result in phosphorylation of Akt at Ser[473] [55]. In addition, AMPK can directly suppress mTORC1 by phosphorylation of raptor [56–58]. Nutrients and AAs are deemed strong activators of mTORC1, possibly through activation of Akt. AAs promote the interaction of Rag GTPases with mTORC1 to translocate the complex from cytoplasm to the lysosomal membranes where mTOR gets activated by

Rheb [59–61]. A scheme is shown to summarize the cell signaling mechanisms for mTOR (Figure 16.1).

The most well-characterized biological function of mTORC1 is to promote protein synthesis, mainly through S6 kinase (S6K) and eukaryotic translation initiation factor (eIF)-4E-binding protein 1 (4E-BP1) [62–64]. Moreover, the activation of sterol regulatory element binding transcription factor 1 and peroxisome proliferator-activated receptor-γ (PPAR-γ) in response to mTORC1 may trigger lipogenesis [65–67]. In addition, mTOR (likely through mTORC1) controls mitochondrial oxidative function and mitochondrial biogenesis through a complex between the transcriptional factor Yin-Yang 1 (YY1) and peroxisome proliferator-activated receptor coactivator-1α (PGC-1α) [67,68]. More evidence suggested that mTORC1 may activate hypoxia-inducible factor 1 (HIF-1α) to promote glycolysis [69]. Last, but not least, mTOR serves as the most important master regulator for autophagy, an evolutionarily conserved cellular process to degrade and recycle aged or damaged organelles and molecules for regeneration of energy and nutrients [70,71]. Autophagy plays a pivotal role in maintaining cardiac function but also contributes to the pathogenesis of heart disease [72–74]. Aberrant autophagy regulation has been identified in a number of pathological myocardial conditions including heart failure, cardiac hypertrophy, alcoholic cardiomyopathy, ischemic cardiomyopathy, acute myocardial infarction, postinfarction remodeling, and cardiac senescence [73–76]. Importantly, mTORC1 inhibits autophagy through its phosphorylation of unc-51-like autophagy activating kinase 1 and also direct regulation of lysosomal function [53,56,58,76,77].

16.3 CARDIAC METABOLIC HOMEOSTASIS

Heart is a high dynamic energy-intensive organ requiring a continuous energy supply in the form of ATP to support systolic and diastolic work, ion channel activity, and cell integrity [78,79]. Despite the high energy demand, hearts maintain little energy back up, unlike the other major energy consumptive tissues such as skeletal muscles. Therefore, it is not surprising that many pathological factors may alter cardiac energy supply and thus force the heart to undergo metabolic adaption (such as switching from carbohydrates to fatty acids for energy supply). If the stress continues, these pathological influences may affect cardiac function, resulting in the development of cardiac dysfunction and ultimately irreversible heart failure [80,81]. The mammalian heart is an "omnivore," which utilizes multiple substrates, including free fatty acids (FFA), carbohydrates, AAs, and ketone bodies, as its energy source for pumping function [79]. However, the make-up of cardiac energy substrates changes dynamically during embryonic development and with the appearance of pathological stresses [82]. During early embryonic development stage, glycolysis serves as the major energy source [82]. With the later differentiation of embryonic stem cells into cardiomyocytes, a higher energy demand is needed for mitochondrial oxidative phosphorylation. However, the relatively low levels of oxygen and circulating fatty acids fail to surpass glycolysis as the main energy source [82]. Interestingly, the rise in mitochondrial oxidative capacity with development seems to coincide with a drastic drop in the proliferative ability of cardiomyocytes, indicating the possible presence of crosstalk among cell differentiation, proliferation, and metabolic state [79]. Immediately after birth, a dramatic increase in fatty acid oxidation starts to develop in concordance with a loss in glycolysis [83,84]. Typically, in the postnatal heart under normal aerobic conditions, it is perceived that 50–70% of energy may come from fatty acids, while the remaining amount is mostly attributed to glucose and lactate [85]. In response to the changing physiological/environmental stimuli, the normal adult heart maintains metabolic flexibility to allow rapid adjustment to suit energy substrate availability [85,86]. However, cardiac uptake and oxidation are not balanced under pathological conditions such as obesity and type 2 diabetes [87]. For example, a shift occurs in myocardial substrate preference from high-energy-efficient glucose oxidation to less-energy-efficient fatty acid oxidation in chronic diseases such as obesity and heart failure [88]. Diminished metabolic flexibility (such as the increase in myocardial glucose uptake and oxidation) renders the heart susceptible to contractile dysfunction and is considered a hallmark for heart failure [88,89]. It is thus essential to maintain appropriate metabolic balance and adaption (metabolic flexibility) under physiological and pathological situations for a healthy heart.

16.4 mTOR AND CARDIAC METABOLISM IN DEVELOPING HEART

During development, the heart undergoes gradual environmental changes in fuel supply, and the metabolic flexibility is essential to cope with these environmental changes [90]. Studies using murine models with "loss of function" of mTOR indicated that mTOR is indispensable for the developing heart for survival, proliferation, physiological hypertrophy, and energy metabolism of cardiomyocytes. Cardiomyocyte-specific mTOR deletion leads to over 90% embryonic lethality and early mortality in the postnatal stage, accompanied with a 50% loss of cardiomyocytes [91]. As stated earlier, Rheb serves as an important regulator for mTORC1. Cardiac-specific Rheb-deficient mice (Rheb$^{-/-}$) die between postnatal days 8 and 10 due to the insufficient cardiomyocyte hypertrophy and impaired sarcomere maturation. However, genetic ablation of 4E-BP1, the activity of which is regulated by mTORC1, rescued these unfavorable changes. To this end, it may be concluded that Rheb-dependent mTORC1 activation is essential for cardiomyocyte hypertrophic growth during the early postnatal period [92]. Furthermore, muscle-specific mTOR conditional knockout mice (mTOR-mKO) may survive the embryonic and perinatal development. However, these mice exhibit growth retardation, cardiac pathology, and premature death. Cardiac muscles from mTOR-mKO mice displayed cardiomyocyte loss and dilated cardiomyopathy as a result of severe apoptosis and necrosis. These myopathic changes seemed to be associated with dampened fatty acid oxidation and glycolysis. Phosphorylation of S6 kinase, a crucial target downstream of mTORC1 and mTORC2, is reduced [93]. Further study also revealed a key role for mTOR in altered maternal nutrition-induced fetal development and metabolic pattern. Maternal diet-induced obesity leads to cardiac hypertrophy in offspring associated with hyperinsulinemia-induced activation of mTOR, ERK, and oxidative stress [94].

Notably, mTOR-regulated autophagy is essential to the physiological metabolic adaptation of neonatal hearts [95]. Neonates suffer abrupt nutrient deprivation at birth when the transplacental nutrient supply is suddenly cut-off [96]. Autophagy therefore provides an essential adaptive role during this perinatal period to maintain survival during early neonatal nutrient starvation in the heart [95]. A recent study observed normal development, despite mortality on postnatal day 1 in a knock-in mouse model with expression of a constitutively active RagA GTPase to govern mTOR activation by AAs. Within an hour of birth, mTORC1 was overtly suppressed in wild-type neonates coinciding with the dropped plasma levels of AAs. However, such a phenomenon in mTOR inhibition is absent in RagA GTPase neonates despite a comparable drop in blood levels of nutrients [43]. It may be concluded that mTOR inhibition triggered by starvation and nutrient deprivation at birth is vital in neonatal autophagy induction to maintain adequate nutrient homeostasis [43]. Later, autophagy will have to be suppressed once milk-feeding is established. Experimental studies using cardiac-specific deletion of insulin receptor substrate 1 and 2 (IRS1/2) revealed that deletion of these IRS ablates the physiological suppression of autophagy that would otherwise parallel the postnatal increase in circulating insulin, leading to cardiomyocyte loss, heart failure, and premature death. Interestingly, reactivation of mTOR using AA supplementation may ameliorate these detrimental cardiac outcomes resulting from IRS deletion, indicating a pivotal role of suppression of mTOR-dependent autophagy (from circulating insulin levels) in neonatal growth, survival, and cardiac development [95]. A more recent study revealed a role for sestrin in the regulation of Rag GTPases, sestrin overexpression inhibited amino-acid-induced Rag guanine nucleotide exchange and mTORC1 translocation to lysosomes. Mice deficient in sestrins displayed poor postnatal survival along with defective mTORC1 inactivation during neonatal fasting [97].

16.5 mTOR AND CARDIOMETABOLIC SYNDROME

Emerging evidence suggests that overnutrition, obesity, and cardiometabolic syndrome often overlap and are companied with chronic cardiac comorbidities. Metabolic disorders in these comorbidities are characterized as insulin resistance, pronounced inflammation, oxidative stress, mitochondrial dysfunction, and interstitial fibrosis. Levels of autophagy seem to decrease in these metabolic anomalies associated with marked apoptosis [98]. Raptor is a unique component of mTORC1. In an epidemic study involving both subjects of normal lifespan and subjects with a longevity (>95 years) from a homogeneous population of American Japanese ancestry, genetic variation in RAPTOR is found in association with overweight/ obesity [99]. Mice overexpressing a loss-of-function mutant AMPK, a negative upstream regulator of mTOR, were fed a high-fat diet and exhibited accentuated glucose intolerance, dampened fatty acid oxidation, and more severe cardiac hypertrophy with cardiomyocyte contractile dysfunction and intracellular Ca^{2+} handling disorder. These unfavorable effects are likely mediated through reduced phosphorylation of Akt and mTOR [100]. An independent study revealed that Ang II type 2 receptor-mediated crosstalk between Ang II and mTOR ameliorated overnutrition-induced activation of mTOR signaling in cardio-vascular tissue of rats, mice, and humans to confer cardioprotection [101]. In line with these observations, high-fat diet-fed flies exhibit increased levels of triglycerides and alterations in insulin/glucose homeostasis, in a manner reminiscent of mammalians. High-fat diet-feeding also causes cardiac anomalies reminiscent of diabetic cardiomyopathy in mammalians. These metabolic and cardiotoxic phenotypes elicited by a high-fat diet are interrupted by inhibiting insulin-TOR signaling. These investigators further proved that reducing insulin-TOR activity (by expressing TSC1-2, 4E-BP1, or FoXO) in the heart may protect against fat diet-induced cardiac fat accumulation and dysfunction [102]. These findings suggest that the cardiac benefit by targeting mTOR is unlikely due to a secondary systemic effect of mTOR inhibition. It has been well documented that insulin regulates both glucose uptake and postnatal cardiac growth through an mTOR-dependent manner, supporting the notion that mTOR is at the convergence point between nutrient metabolism and cardiac growth [103]. Taken together, mTOR signaling is pivotal to heart dysfunction under obesity and metabolic syndrome. The activation of mTOR should be a central signaling molecule governing cardiac remodeling. Manipulation against mTOR, either genetically or pharmaceutically, should be promising in alleviating heart anomalies in obesity, diabetes, and the overall cardiometabolic diseases [26,104].

As stated earlier, autophagy plays an important role in cardiac anomalies in obesity and other cardiometabolic disorders [105]. Data from our laboratory suggest that Akt2 knockout preserves cardiac function in high-fat diet-induced obesity by rescuing cardiac autophagosome maturation through the inhibition of Akt-mTORC1 activation [104,106]. Administration of rapamycin negates or greatly attenuates fat diet-induced/adiponectin-deficiency-accentuated obesity, cardiac hypertrophy, and contractile dysfunction as well as AMPK dephosphorylation, and p62 buildup [107]. Our observation further revealed that a high-fat diet suppresses autophagy and adiponectin-knockout augments the suppressed autophagy, whereas rapamycin, an inducer of autophagy by inhibiting mTOR, effectively rescues dietary obesity and fat diet-induced cardiac anomalies, supporting a permissive role for the mTOR-regulated autophagy change in obesity-associated heart anomalies [107]. Interestingly, mTOR seems to be responsible for the increased myocardial infarction size and ischemia/reperfusion injury in individuals with uncorrected obesity and metabolic syndrome, due to the uncontrolled mTOR signaling and subsequently inhibition of autophagy. This is supported by the observation that high-fat diet-fed mice exhibits dysregulated cardiac activation of Rheb and mTORC1, particularly during ischemia, along with inhibition of autophagy and pronounced ischemic injury in the heart. Pharmacological and genetic inhibition of mTORC1 restored autophagy and abrogated the increase in infarct size observed in high-fat diet-fed mice. However, such approaches failed to protect against high-fat diet-fed mice with genetic disruption of autophagy [108]. Myocardial infarct size, after 30 min of normothermic global ischemia and 60 min of reperfusion in Langendorff mode, is significantly smaller in rapamycin treated db/db mice accompanied with restored phosphorylation of STAT3 and Akt as well as reduced ribosomal protein S6 phosphorylation. Furthermore, in inducible cardiac-specific STAT3 deletion mice, rapamycin is unable to rescue against I/R injured cardiomyocytes in high-fat diet-induced mice [109]. However, controversial evidence suggested that cardiac-specific transgenic mice overexpressing mTOR exhibit better cardiac function recovery and less necrotic markers, creatine kinase and lactate dehydrogenase, under normal and high-fat diet feedings, suggesting that mTOR may be of benefit against heart injury after I/R in diet-induced obesity mice [110]. Nonetheless, the precise interplay between mTOR and autophagy in the regulation of cardiac homeostasis warrants further scrutiny under metabolic disorders.

16.6 mTOR AND DIABETIC CARDIOMYOPATHY

Although cardiac complications of diabetes mellitus are mainly associated with ischemic heart diseases due to progressive development of atherosclerosis, patients suffer from diabetic cardiomyopathy independent of coronary artery disease or hypertension. This type of muscle-originated pathology in the heart may contribute to the onset and development of heart failure in these patients [16,111]. Adding to the complexity are the etiological differences between type 1 and type 2 diabetes for the pathogenesis of respective cardiomyopathies [16]. As mentioned above, glucose, lactate, and fatty acids provide the bulk of the energy under the physiological setting to maintain cardiac performance. However, in type 1 diabetes with insufficient insulin secretion, glucose transporter 4 and fatty acid translocase/cluster of differentiation 36 cannot be translocated to cell membranes due to the absence of insulin [112]. As a result, glucose and fatty acids are unable to be taken up to promote mitochondrial oxidation, leading to dampened ATP supply and cardiac contractile dysfunction [112]. Data from our lab revealed that high glucose (25.5 mM) of adult rat cardiomyocytes leads to depressed peak shortening, reduced maximal velocity of shortening/relengthening ($\pm dl/dt$) and prolongs time-to-90% relengthening, evaluated using an IonOptix MyoCam system. High glucose also activates Akt, mTOR, and p70s6k and suppresses GSK-3β. These effects can be abolished by insulin-like growth factor 1 (IGF-1). The IGF-1-elicited protective effect against high glucose was nullified by either LY294002 (an Akt inhibitor) or rapamycin. Thus, inhibition of mTOR is sufficient to protect cardiomyocytes against high-glucose toxicity [113]. While circulating insulin level and peripheral insulin sensitivity may regulate mTOR activity, it is much more complicated *in vivo*. Streptozotocin-induced diabetes is associated with reduced myocardial contractility, overall loss of oxidative capacity, a shift toward a slower myosin heavy chain isoform, and decreased mTOR tissue content. All of these changes appear to be reversible with insulin. Supplementation with AA mix in diabetic mice partially restored myocardial and oxidative dysfunction and increased mTOR tissue content. However, the combination of insulin and AAs did not generate any synergistic effects on enzymatic or functional properties, indicating possible involvement of an mTOR-dependent mechanism in AA supplement-induced beneficial effects on oxidative and contractile function in diabetic hearts [114]. Not surprisingly, mTOR is normally inactivated in type 1 diabetes as a result of hypoinsulinemia and dampened insulin signaling. This is in line with the observation of increased autophagy in type 1 diabetes. Activation of mTOR (or inhibition of autophagy) [115], either through insulin or AA supplement, offers a beneficial impact for cardiomyocytes in the setting of type 1 diabetes.

In type 2 diabetes, the distinct feature/core concept has to be dampened insulin sensitivity, usually with hyperinsulinemia at the beginning and hypoinsulinemia at the end, ultimately leading to uncontrolled blood glucose. It has been suggested that mTOR inhibition is beneficial in the setting of type 2 diabetes. This is supported by the evidence that autophagy induction may be a useful therapeutic avenue in the treatment of diabetic cardiomyopathy in type 2 diabetes mellitus [106]. Chronic rapamycin treatment significantly reduces body weight, heart weight, plasma glucose, triglyceride, and insulin levels in db/db mice, accompanied with improved fractional shortening, and reduced oxidative stress. Rapamycin interrupted enhanced phosphorylation of mTOR and S6 kinase, although not Akt, in db/db mouse hearts. Proteomic and Western blot analyses identified significant changes in several cytoskeletal/contractile proteins, glucose metabolism proteins, and antioxidant proteins in db/db hearts following rapamycin treatment [116]. Compelling evidence has indicated that specific mTORC1 inhibition by PRAS40 prevents the onset and development of diabetic cardiomyopathy which is manifested as improved metabolic function, blunted hypertrophic growth, and preserved cardiac function [117]. Moreover, a combination of acute hypoinsulinemia and hypolipidemia triggered by the administration of niacin led to diminished protein synthesis, mTOR and eIF2B activity, correlated with changes in energy status and/or redox potential [118]. These findings favor the presence of an interplay between the metabolic substrate-FFA and protein synthesis. In a separate study, cardiac-specific leptin receptor deletion (induced) triggered left ventricular dysfunction and irreversible lethal heart failure in less than 10 days in mice, associated with marked decreases in cardiac mitochondrial ATP, phosphorylated mTOR, and AMPK. It is possible that the lethal heart failure caused by specific deletion of cardiomyocyte leptin receptors is mediated through AMPK/mTOR-regulated action in cardiomyocyte energy metabolism [119].

Although current findings are still controversial with regards to cardiac mTOR signaling in type 1 and type 2 diabetes mellitus, differences in the levels of insulin and adiposity between these types may be able to provide some pathophysiological insights into the pathogenesis of diabetic cardiomyopathies, which is essential for the development of new treatment strategies.

16.7 MECHANISMS OF mTOR-REGULATED CARDIAC METABOLISM

mTOR directly regulates utilization of energy metabolic substrates. mTOR deletion induces a specific defect in fatty acid utilization but not glucose utilization in the heart. mTOR is believed to regulate metabolic homeostasis by controlling the transcriptional activity of multiple key transcription factors and enzymes. Likewise, mTORC1 stimulates mRNA translation and other anabolic processes, the important energy-consuming processes in a variety of diseases including diabetes, heart disease, and cancer. The notion of a role for mTOR in cardiac energy metabolism regulation received convincing support from experimental studies. For example, cardiac-specific mTOR deletion impairs myocardial palmitate oxidation, whereas increasing glucose oxidation, accompanied with down-regulated expression levels and enzymatic activity of various fatty acid metabolic genes, with unchanged PGC-1α, electron transport chain subunits, mitochondrial DNA, morphology and pyruvate-supported and the mitochondria-uncoupler carbonyl cyanide 4-(trifluoromethoxy) phenylhydrazone-stimulated respirations [67]. In line with these observations from mTOR deletion, ablation of cardiac raptor in adulthood also imposed a switch from fatty acid to glucose oxidation 3 weeks after raptor deletion, when cardiac function is still normal. However, cardiac function deteriorated rapidly after that time point. Likewise, pressure overload resulted in severely dilated cardiomyopathy in these mice at only 1 week without a prior phase of adaptive hypertrophy. It is conceived that blunted protein synthesis capacity may underscore the loss of adaptive cardiomyocyte hypertrophy in the cardiac raptor deletion model. Possible contribution from a shift in metabolic substrate and the loss of mitochondrial content must be considered as they both appeared early on in pressure overload [120]. A role for mTOR in the regulation of cardiac metabolism also received consolidation from its inhibitor rapamycin. Rapamycin is suggested to disrupt the positive transcriptional feedback loop between CCAAT/enhancer-binding protein-α and PPAR-γ, two key transcription factors in fatty acid synthesis and uptake, by directly targeting the transactivation activity of PPAR-γ [121]. Rapamycin also inhibits the accumulation of HIF-1α and HIF-1α-dependent hypoxia or $CoCl_2$-induced transcription. Further evidence pinpoints the oxygen-dependent degradation domain of GAL4-HIF-1α fusion proteins as a critical target for the rapamycin-sensitive, mTOR-dependent signaling cascade leading to HIF-1α stabilization by $CoCl_2$, a process deemed essential for glycolysis [122].

Mitochondria play a key role in cardiac metabolic homeostasis and mitochondrial function is often impaired in individuals with diabetes, obesity, and cardiometabolic syndromes. mTOR may regulate mitochondrial function (both biogenesis and function) through selectively promoting translation of nucleus-encoded mitochondria-related mRNAs. Translation of nucleus-encoded mitochondria-related mRNAs in response to mTOR activation facilitates

ATP production capacity and may be achieved through inhibition of the eIF4E-BPs. Aberrant mTOR signaling has been shown to underlie mitochondrial injury leading to compromised cellular energy metabolism in pathological conditions such as insulin resistance and obesity [123]. Not only can mitochondrial integrity regulate insulin sensitivity, insulin may reciprocally regulate mitochondrial metabolism in cardiomyocytes via regulation of mitochondrial fusion. Three hours of insulin treatment promote mitochondrial protein Opa-1, mitochondrial fusion, increased mitochondrial membrane potential, and elevated both intracellular ATP levels and oxygen consumption in cardiomyocytes, in an Akt-mTOR-NFKB-dependent manner. The metabolic benefits triggered by insulin were negated with silencing of Opa-1 or Mfn2 [124]. In skeletal muscles, rapamycin down-regulated the gene expression of the mitochondrial transcriptional regulators PGC-1α, estrogen-related receptor α, and nuclear respiratory factors, leading to decreased mitochondrial gene expression and oxygen consumption. Further study revealed that the transcription factor YY1 may serve as a common target for mTOR and PGC-1α and is required for rapamycin-dependent repression of these mitochondrial genes [68].

Metabolic regulation of mTOR in the heart is also accomplished by autophagy, a crucial machinery to maintain cellular homeostasis. Defected or uncontrolled autophagy is present in a wide range of pathologies including heart diseases [105,125,126]. Obesity suppresses autophagy through activation of mTORC1, likely due to elevated levels of circulating lipids, AAs, and insulin [44,127,128]. Ample experimental evidence has also suggested that a decreased autophagy level may be maladaptive, whereas autophagy induction, such as using the mTOR inhibitor rapamycin, rescues against the detrimental effects in obese hearts [44,127]. Along the same line, mTOR is up-regulated in conjunction with suppressed autophagy in type 2 diabetes mellitus, via hyperinsulinemia and hyperglycemia. Inhibition of mTOR using rapamycin activates autophagy and provokes cardioprotection against diabetes [127,129,130]. Moreover, defective autophagy may also contribute to enhanced concomitant pathologies such as ischemia/reperfusion injury in obese and type 2 diabetic hearts [109,110]. However, it is noteworthy that decreased mTOR signaling is present in type 1 diabetes, possibly due to compromised insulin signaling. This received convincing support that restoration of mTOR rescues against diabetic cardiomyopathy in type 1 diabetes [116]. In general, autophagy is normally down-regulated in type 1 diabetes. It is believed that restoration of autophagy protects type 1 diabetic hearts although contradictory findings are also present. For example, hyperglycemia drastically suppressed autophagic flux in cardiomyocytes. Unexpectedly, inhibition of autophagy using 3-methyladenine or short-hairpin RNA-mediated silencing of the *Beclin1* or *Atg7* gene, provided an adaptive response to limit high-glucose-induced cardiotoxicity. Conversely, upregulation of autophagy with rapamycin or overexpression of Beclin1 or Atg7 predisposed cardiomyocytes to be more susceptible to high-glucose toxicity [131]. Furthermore, heart damage caused by type 1 diabetes was found to be rescued in an autophagy-deficient model, in which mitophagy was increased. Therefore selective autophagy, such as mitophagy, is expected to play a much more important role in the setting of a "global insufficiency" of autophagy [132]. It is accepted that unregulated mTOR inhibits autophagy and is likely responsible for cardiac anomalies in particular cardiac remodeling in obesity and type 2 diabetes. Nonetheless, the control of mTOR in autophagy seems to be dysregulated in the face of type 1 diabetes. Further study is warranted to better understand mTOR involvement in the regulation of autophagy in cardiac dysfunction in type 1 diabetes mellitus.

16.8 CONCLUSIONS AND FUTURE PERSPECTIVES

Over recent decades, the regulatory mechanisms governing metabolic disorders have drawn much attention for scientists and physicians in various disciplines including endocrinology, cardiology, oncology, and general public health. Here we have reviewed the metabolic contribution of mTOR in the regulation of cardiac homeostasis under cardiometabolic diseases. As a cell stress sensor and master regulator whose activity is regulated by energy metabolic substrates, AAs, insulin, growth factors, and hormones, mTOR and mTOR complex play a cardinal role in the regulation of cellular metabolism and growth in the heart. mTOR signaling is believed to drive the adaption responses to pathological stimuli via protein synthesis, autophagy, ribosomal and mitochondrial biogenesis, as well as cell metabolism. Activation of mTOR promotes cellular anabolism, suppresses catabolism, and promotes cell growth manifested as cell proliferation, differentiation, hypertrophy, and loss of autophagy. This might contribute to cardiac remodeling and the detrimental effects of obesity, type 2 diabetes, and overall cardiometabolic disorders. Tight control of mTOR to limit these unfavorable responses, especially autophagy, is expected to serve as a pharmaceutical target in these comorbidities. The mTOR inhibitor rapamycin and its analogs (rapalogs) have been used in clinical trials to treat many diseases [133]. Nonetheless, rapalogs presented some modest efficacies in many clinical trials. It is speculated that mTORC1-regulated "overwhelming" negative feedback loops, such as activation of receptor tyrosine kinases, PI3 kinase-Akt signaling, and Ras-MAPK signaling may have contributed to the somewhat limited clinical success of these rapamycin analogs. On the contrary, under insufficient nutrition, such as starvation, inactivation of mTOR fosters cells to switch into an energy- and amino-acid-saving mode. Recent advances in identifying novel drugs suppressing mTOR to turn on autophagy have also shed some light and therapeutic promises for the management of obesity, diabetes, and metabolic disorders, where mTOR signaling may be hyperactive.

References

[1] Xu Y, Wang L, He J, Bi Y, Li M, Wang T, China Noncommunicable Disease Surveillance Group, et al. Prevalence and control of diabetes in Chinese adults. JAMA 2013;310:948−59.

[2] Selvin E, Lazo M, Chen Y, Shen L, Rubin J, McEvoy JW, et al. Diabetes mellitus, prediabetes, and incidence of subclinical myocardial damage. Circulation 2014;130:1374−82.

[3] Turer AT, Hill JA, Elmquist JK, Scherer PE. Adipose tissue biology and cardiomyopathy: translational implications. Circ Res 2012;111:1565−77.

[4] Collins S. A heart-adipose tissue connection in the regulation of energy metabolism. Nat Rev Endocrinol 2014;10:157−63.

[5] Despres JP. Abdominal obesity as important component of insulin-resistance syndrome. Nutrition 1993;9:452−9.

[6] Alberti KG, Zimmet PZ. Definition, diagnosis and classification of diabetes mellitus and its complications. Part 1: diagnosis and classification of diabetes mellitus provisional report of a WHO consultation. Diabet Med 1998;15:539−53.

[7] Balkau B, Charles MA. Comment on the provisional report from the WHO consultation. European Group for the Study of Insulin Resistance (EGIR). Diabet Med 1999;16:442−3.

[8] Grundy SM, Cleeman JI, Daniels SR, Donato KA, Eckel RH, Franklin BA, et al. Diagnosis and management of the metabolic syndrome: an American Heart Association/National Heart, Lung, and Blood Institute Scientific Statement. Circulation 2005;112:2735−52.

[9] Xu X, Ren J. Cardiac stem cell regeneration in metabolic syndrome. Curr Pharm Des 2013;19:4888−92.

[10] Ford ES, Giles WH, Dietz WH. Prevalence of the metabolic syndrome among US adults: findings from the third National Health and Nutrition Examination Survey. JAMA 2002;287:356−9.

[11] de Ferranti SD, Gauvreau K, Ludwig DS, Neufeld EJ, Newburger JW, Rifai N. Prevalence of the metabolic syndrome in American adolescents: findings from the Third National Health and Nutrition Examination Survey. Circulation 2004;110:2494−7.

[12] Weiss R, Dziura J, Burgert TS, Tamborlane WV, Taksali SE, Yeckel CW, et al. Obesity and the metabolic syndrome in children and adolescents. N Engl J Med 2004;350:2362−74.

[13] Gallagher EJ, LeRoith D. Obesity and diabetes: the increased risk of cancer and cancer-related mortality. Physiol Rev 2015;95:727−48.

[14] Mahajan R, Lau DH, Sanders P. Impact of obesity on cardiac metabolism, fibrosis, and function. Trends Cardiovasc Med 2015;25:119−26.

[15] Rubler S, Dlugash J, Yuceoglu YZ, Kumral T, Branwood AW, Grishman A. New type of cardiomyopathy associated with diabetic glomerulosclerosis. Am J Cardiol 1972;30:595−602.

[16] Seferovic PM, Paulus WJ. Clinical diabetic cardiomyopathy: a two-faced disease with restrictive and dilated phenotypes. Eur Heart J 2015.

[17] Kubli DA, Gustafsson AB. Unbreak my heart: targeting mitochondrial autophagy in diabetic cardiomyopathy. Antioxid Redox Signal 2015;22:1527−44.

[18] Zhang Y, Babcock SA, Hu N, Maris JR, Wang H, Ren J. Mitochondrial aldehyde dehydrogenase (ALDH2) protects against streptozotocin-induced diabetic cardiomyopathy: role of GSK3beta and mitochondrial function. BMC Med 2012;10:40.

[19] Ren J, Ceylan-Isik AF. Diabetic cardiomyopathy: do women differ from men? Endocrine 2004;25:73−83.

[20] Yao H, Han X. The cardioprotection of the insulin-mediated PI3K/Akt/mTOR signaling pathway. Am J Cardiovasc Drugs 2014;14:433−42.

[21] Bridges D, Saltiel AR. Phosphoinositides: key modulators of energy metabolism. Biochim Biophys Acta 2015;1851:857−66.

[22] Stanley WC, Dabkowski ER, Ribeiro Jr. RF, O'Connell KA. Dietary fat and heart failure: moving from lipotoxicity to lipoprotection. Circ Res 2012;110:764−76.

[23] Giacco F, Brownlee M. Oxidative stress and diabetic complications. Circ Res 2010;107:1058−70.

[24] Le Hir H, Seraphin B. EJCs at the heart of translational control. Cell 2008;133:213−16.

[25] Cetrullo S, D'Adamo S, Tantini B, Borzi RM, Flamigni F. mTOR, AMPK, and Sirt1: key players in metabolic stress management. Crit Rev Eukaryot Gene Expr 2015;25:59−75.

[26] Kim YC, Guan KL. mTOR: a pharmacologic target for autophagy regulation. J Clin Invest 2015;125:25−32.

[27] Jewell JL, Guan KL. Nutrient signaling to mTOR and cell growth. Trends Biochem Sci 2013;38:233−42.

[28] Ma XM, Blenis J. Molecular mechanisms of mTOR-mediated translational control. Nat Rev Mol Cell Biol 2009;10:307−18.

[29] Sanchez Canedo C, Demeulder B, Ginion A, Bayascas JR, Balligand JL, Alessi DR, et al. Activation of the cardiac mTOR/p70(S6K) pathway by leucine requires PDK1 and correlates with PRAS40 phosphorylation. Am J Physiol Endocrinol Metab 2010;298:E761−9.

[30] Anmann T, Varikmaa M, Timohhina N, Tepp K, Shevchuk I, Chekulayev V, et al. Formation of highly organized intracellular structure and energy metabolism in cardiac muscle cells during postnatal development of rat heart. Biochim Biophys Acta 2014;1837:1350−61.

[31] Maillet M, van Berlo JH, Molkentin JD. Molecular basis of physiological heart growth: fundamental concepts and new players. Nat Rev Mol Cell Biol 2013;14:38−48.

[32] Shimobayashi M, Hall MN. Making new contacts: the mTOR network in metabolism and signalling crosstalk. Nat Rev Mol Cell Biol 2014;15:155−62.

[33] Wullschleger S, Loewith R, Hall MN. TOR signaling in growth and metabolism. Cell 2006;124:471−84.

[34] Kapahi P, Chen D, Rogers AN, Katewa SD, Li PW, Thomas EL, et al. With TOR, less is more: a key role for the conserved nutrient-sensing TOR pathway in aging. Cell Metab 2010;11:453−65.

[35] Kim DH, Sarbassov DD, Ali SM, King JE, Latek RR, Erdjument-Bromage H, et al. mTOR interacts with raptor to form a nutrient-sensitive complex that signals to the cell growth machinery. Cell 2002;110:163−75.

[36] Laplante M, Sabatini DM. mTOR signaling in growth control and disease. Cell 2012;149:274–93.

[37] Feldman ME, Apsel B, Uotila A, Loewith R, Knight ZA, Ruggero D, et al. Active-site inhibitors of mTOR target rapamycin-resistant outputs of mTORC1 and mTORC2. PLoS Biol 2009;7:e38.

[38] Zhang D, Contu R, Latronico MV, Zhang J, Rizzi R, Catalucci D, et al. mTORC1 regulates cardiac function and myocyte survival through 4E-BP1 inhibition in mice. J Clin Invest 2010;120:2805–16.

[39] Zhang L, You J, Sidhu J, Tejpal N, Ganachari M, Skelton TS, et al. Abrogation of chronic rejection in rat model system involves modulation of the mTORC1 and mTORC2 pathways. Transplantation 2013;96:782–90.

[40] Zhou Y, Wang D, Gao X, Lew K, Richards AM, Wang P. mTORC2 phosphorylation of Akt1: a possible mechanism for hydrogen sulfide-induced cardioprotection. PLoS One 2014;9:e99665.

[41] Zhao X, Lu S, Nie J, Hu X, Luo W, Wu X, et al. Phosphoinositide-dependent kinase 1 and mTORC2 synergistically maintain postnatal heart growth and heart function in mice. Mol Cell Biol 2014;34:1966–75.

[42] Xie J, Proud CG. Crosstalk between mTOR complexes. Nat Cell Biol 2013;15:1263–5.

[43] Efeyan A, Zoncu R, Chang S, Gumper I, Snitkin H, Wolfson RL, et al. Regulation of mTORC1 by the Rag GTPases is necessary for neonatal autophagy and survival. Nature 2013;493:679–83.

[44] Kubli DA, Gustafsson AB. Cardiomyocyte health: adapting to metabolic changes through autophagy. Trends Endocrinol Metab 2014;25:156–64.

[45] Sciarretta S, Volpe M, Sadoshima J. Mammalian target of rapamycin signaling in cardiac physiology and disease. Circ Res 2014;114:549–64.

[46] Mazhab-Jafari MT, Marshall CB, Stathopulos PB, Kobashigawa Y, Stambolic V, Kay LE, et al. Membrane-dependent modulation of the mTOR activator Rheb: NMR observations of a GTPase tethered to a lipid-bilayer nanodisc. J Am Chem Soc 2013;135:3367–70.

[47] Bai X, Ma D, Liu A, Shen X, Wang QJ, Liu Y, et al. Rheb activates mTOR by antagonizing its endogenous inhibitor, FKBP38. Science 2007;318:977–80.

[48] Garami A, Zwartkruis FJ, Nobukuni T, Joaquin M, Roccio M, Stocker H, et al. Insulin activation of Rheb, a mediator of mTOR/S6K/4E-BP signaling, is inhibited by TSC1 and 2. Mol Cell 2003;11:1457–66.

[49] Vander Haar E, Lee SI, Bandhakavi S, Griffin TJ, Kim DH. Insulin signalling to mTOR mediated by the Akt/PKB substrate PRAS40. Nat Cell Biol 2007;9:316–23.

[50] Shioi T, McMullen JR, Kang PM, Douglas PS, Obata T, Franke TF, et al. Akt/protein kinase B promotes organ growth in transgenic mice. Mol Cell Biol 2002;22:2799–809.

[51] Mammucari C, Schiaffino S, Sandri M. Downstream of Akt: FoxO3 and mTOR in the regulation of autophagy in skeletal muscle. Autophagy 2008;4:524–6.

[52] Zheng Q, Zhao K, Han X, Huff AF, Cui Q, Babcock SA, et al. Inhibition of AMPK accentuates prolonged caloric restriction-induced change in cardiac contractile function through disruption of compensatory autophagy. Biochim Biophys Acta 2015;1852:332–42.

[53] Shang L, Wang X. AMPK and mTOR coordinate the regulation of Ulk1 and mammalian autophagy initiation. Autophagy 2011;7:924–6.

[54] Inoki K, Kim J, Guan KL. AMPK and mTOR in cellular energy homeostasis and drug targets. Annu Rev Pharmacol Toxicol 2012;52:381–400.

[55] Vadlakonda L, Dash A, Pasupuleti M, Anil Kumar K, Reddanna P. The paradox of Akt-mTOR interactions. Front Oncol 2013;3:165.

[56] Egan D, Kim J, Shaw RJ, Guan KL. The autophagy initiating kinase ULK1 is regulated via opposing phosphorylation by AMPK and mTOR. Autophagy 2011;7:643–4.

[57] Ginion A, Auquier J, Benton CR, Mouton C, Vanoverschelde JL, Hue L, et al. Inhibition of the mTOR/p70S6K pathway is not involved in the insulin-sensitizing effect of AMPK on cardiac glucose uptake. Am J Physiol Heart Circ Physiol 2011;301:H469–77.

[58] Kim J, Kundu M, Viollet B, Guan KL. AMPK and mTOR regulate autophagy through direct phosphorylation of Ulk1. Nat Cell Biol 2011;13:132–41.

[59] Shaw RJ. mTOR signaling: RAG GTPases transmit the amino acid signal. Trends Biochem Sci 2008;33:565–8.

[60] Jewell JL, Russell RC, Guan KL. Amino acid signalling upstream of mTOR. Nat Rev Mol Cell Biol 2013;14:133–9.

[61] Neishabouri SH, Hutson SM, Davoodi J. Chronic activation of mTOR complex 1 by branched chain amino acids and organ hypertrophy. Amino Acids 2015;47:1167–82.

[62] Holz MK, Ballif BA, Gygi SP, Blenis J. mTOR and S6K1 mediate assembly of the translation preinitiation complex through dynamic protein interchange and ordered phosphorylation events. Cell 2005;123:569–80.

[63] Ben-Sahra I, Howell JJ, Asara JM, Manning BD. Stimulation of de novo pyrimidine synthesis by growth signaling through mTOR and S6K1. Science 2013;339:1323–8.

[64] Li Q, Ren J. Chronic alcohol consumption alters mammalian target of rapamycin (mTOR), reduces ribosomal p70s6 kinase and p4E-BP1 levels in mouse cerebral cortex. Exp Neurol 2007;204:840–4.

[65] Powers T. Cell growth control: mTOR takes on fat. Mol Cell 2008;31:775–6.

[66] Duvel K, Yecies JL, Menon S, Raman P, Lipovsky AI, Souza AL, et al. Activation of a metabolic gene regulatory network downstream of mTOR complex 1. Mol Cell 2010;39:171–83.

[67] Zhu Y, Soto J, Anderson B, Riehle C, Zhang YC, Wende AR, et al. Regulation of fatty acid metabolism by mTOR in adult murine hearts occurs independently of changes in PGC-1alpha. Am J Physiol Heart Circ Physiol 2013;305:H41–51.

[68] Cunningham JT, Rodgers JT, Arlow DH, Vazquez F, Mootha VK, Puigserver P. mTOR controls mitochondrial oxidative function through a YY1-PGC-1alpha transcriptional complex. Nature 2007;450:736–40.

[69] Cheng SC, Quintin J, Cramer RA, Shepardson KM, Saeed S, Kumar V, et al. mTOR- and HIF-1alpha-mediated aerobic glycolysis as metabolic basis for trained immunity. Science 2014;345:1250684.

[70] Blagosklonny MV. Hypoxia, MTOR and autophagy: converging on senescence or quiescence. Autophagy 2013;9:260–2.

[71] Goswami SK, Das DK. Autophagy in the myocardium: dying for survival? Exp Clin Cardiol 2006;11:183–8.

[72] Ren J, Taegtmeyer H. Too much or not enough of a good thing—the Janus faces of autophagy in cardiac fuel and protein homeostasis. J Mol Cell Cardiol 2015;84:223–6.

[73] McLendon PM, Robbins J. Proteotoxicity and cardiac dysfunction. Circ Res 2015;116:1863–82.

[74] Saito T, Sadoshima J. Molecular mechanisms of mitochondrial autophagy/mitophagy in the heart. Circ Res 2015;116:1477–90.

[75] Ren J, Zhang Y. Emerging potential of therapeutic targeting of autophagy and protein quality control in the management of cardiometabolic diseases. Biochim Biophys Acta 2015;1852:185–7.

[76] Gatica D, Chiong M, Lavandero S, Klionsky DJ. Molecular mechanisms of autophagy in the cardiovascular system. Circ Res 2015;116:456–67.

[77] Nazio F, Strappazzon F, Antonioli M, Bielli P, Cianfanelli V, Bordi M, et al. mTOR inhibits autophagy by controlling ULK1 ubiquitylation, self-association and function through AMBRA1 and TRAF6. Nat Cell Biol 2013;15:406—16.

[78] Ingwall JS, Weiss RG. Is the failing heart energy starved? On using chemical energy to support cardiac function. Circ Res 2004;95:135—45.

[79] Kolwicz Jr. SC, Purohit S, Tian R. Cardiac metabolism and its interactions with contraction, growth, and survival of cardiomyocytes. Circ Res 2013;113:603—16.

[80] Doehner W, Frenneaux M, Anker SD. Metabolic impairment in heart failure: the myocardial and systemic perspective. J Am Coll Cardiol 2014;64:1388—400.

[81] Carley AN, Taegtmeyer H, Lewandowski ED. Matrix revisited: mechanisms linking energy substrate metabolism to the function of the heart. Circ Res 2014;114:717—29.

[82] Xavier-Neto J, Sousa Costa AM, Figueira AC, Caiaffa CD, Amaral FN, Peres LM, et al. Signaling through retinoic acid receptors in cardiac development: doing the right things at the right times. Biochim Biophys Acta 2015;1849:94—111.

[83] Lopaschuk GD, Jaswal JS. Energy metabolic phenotype of the cardiomyocyte during development, differentiation, and postnatal maturation. J Cardiovasc Pharmacol 2010;56: 130—40.

[84] Iruretagoyena JI, Davis W, Bird C, Olsen J, Radue R, Teo Broman A, et al. Metabolic gene profile in early human fetal heart development. Mol Hum Reprod 2014;20:690—700.

[85] Sambandam N, Lopaschuk GD. AMP-activated protein kinase (AMPK) control of fatty acid and glucose metabolism in the ischemic heart. Prog Lipid Res 2003;42:238—56.

[86] Grossman AN, Opie LH, Beshansky JR, Ingwall JS, Rackley CE, Selker HP. Glucose-insulin-potassium revived: current status in acute coronary syndromes and the energy-depleted heart. Circulation 2013;127:1040—8.

[87] Goldberg IJ, Trent CM, Schulze PC. Lipid metabolism and toxicity in the heart. Cell Metab 2012;15:805—12.

[88] Yan J, Young ME, Cui L, Lopaschuk GD, Liao R, Tian R. Increased glucose uptake and oxidation in mouse hearts prevent high fatty acid oxidation but cause cardiac dysfunction in diet-induced obesity. Circulation 2009;119:2818—28.

[89] Wang ZV, Li DL, Hill JA. Heart failure and loss of metabolic control. J Cardiovasc Pharmacol 2014;63:302—13.

[90] Lavrentyev EN, He D, Cook GA. Expression of genes participating in regulation of fatty acid and glucose utilization and energy metabolism in developing rat hearts. Am J Physiol Heart Circ Physiol 2004;287:H2035—42.

[91] Zhu Y, Pires KM, Whitehead KJ, Olsen CD, Wayment B, Zhang YC, et al. Mechanistic target of rapamycin (mTOR) is essential for murine embryonic heart development and growth. PLoS One 2013;8:e54221.

[92] Tamai T, Yamaguchi O, Hikoso S, Takeda T, Taneike M, Oka T, et al. Rheb (Ras homologue enriched in brain)-dependent mammalian target of rapamycin complex 1 (mTORC1) activation becomes indispensable for cardiac hypertrophic growth after early postnatal period. J Biol Chem 2013;288: 10176—87.

[93] Zhang P, Shan T, Liang X, Deng C, Kuang S. Mammalian target of rapamycin is essential for cardiomyocyte survival and heart development in mice. Biochem Biophys Res Commun 2014;452:53—9.

[94] Fernandez-Twinn DS, Blackmore HL, Siggens L, Giussani DA, Cross CM, Foo R, et al. The programming of cardiac hypertrophy in the offspring by maternal obesity is associated with hyperinsulinemia, AKT, ERK, and mTOR activation. Endocrinology 2012;153:5961—71.

[95] Riehle C, Wende AR, Sena S, Pires KM, Pereira RO, Zhu Y, et al. Insulin receptor substrate signaling suppresses neonatal autophagy in the heart. J Clin Invest 2013;123:5319—33.

[96] Kuma A, Hatano M, Matsui M, Yamamoto A, Nakaya H, Yoshimori T, et al. The role of autophagy during the early neonatal starvation period. Nature 2004;432:1032—6.

[97] Peng M, Yin N, Li MO. Sestrins function as guanine nucleotide dissociation inhibitors for Rag GTPases to control mTORC1 signaling. Cell 2014;159:122—33.

[98] Li ZL, Woollard JR, Ebrahimi B, Crane JA, Jordan KL, Lerman A, et al. Transition from obesity to metabolic syndrome is associated with altered myocardial autophagy and apoptosis. Arterioscler Thromb Vasc Biol 2012;32:1132—41.

[99] Morris BJ, Carnes BA, Chen R, Donlon TA, He Q, Grove JS, et al. Genetic variation in the raptor gene is associated with overweight but not hypertension in American men of Japanese ancestry. Am J Hypertens 2015;28:508—17.

[100] Turdi S, Kandadi MR, Zhao J, Huff AF, Du M, Ren J. Deficiency in AMP-activated protein kinase exaggerates high fat diet-induced cardiac hypertrophy and contractile dysfunction. J Mol Cell Cardiol 2011;50:712—22.

[101] Pulakat L, Demarco VG, Whaley-Connell A, Sowers JR. The impact of overnutrition on insulin metabolic signaling in the heart and the kidney. Cardiorenal Med 2011;1:102—12.

[102] Birse RT, Choi J, Reardon K, Rodriguez J, Graham S, Diop S, et al. High-fat-diet-induced obesity and heart dysfunction are regulated by the TOR pathway in Drosophila. Cell Metab 2010;12:533—44.

[103] Sharma S, Guthrie PH, Chan SS, Haq S, Taegtmeyer H. Glucose phosphorylation is required for insulin-dependent mTOR signalling in the heart. Cardiovasc Res 2007;76:71—80.

[104] Zhang Y, Xu X, Ren J. MTOR overactivation and interrupted autophagy flux in obese hearts: a dicey assembly? Autophagy 2013;9:939—41.

[105] Xu X, Ren J. Unmasking the janus faces of autophagy in obesity-associated insulin resistance and cardiac dysfunction. Clin Exp Pharmacol Physiol 2012;39:200—8.

[106] Xu X, Hua Y, Nair S, Zhang Y, Ren J. Akt2 knockout preserves cardiac function in high-fat diet-induced obesity by rescuing cardiac autophagosome maturation. J Mol Cell Biol 2013;5: 61—3.

[107] Guo R, Zhang Y, Turdi S, Ren J. Adiponectin knockout accentuates high fat diet-induced obesity and cardiac dysfunction: role of autophagy. Biochim Biophys Acta 2013;1832:1136—48.

[108] Sciarretta S, Zhai P, Shao D, Maejima Y, Robbins J, Volpe M, et al. Rheb is a critical regulator of autophagy during myocardial ischemia: pathophysiological implications in obesity and metabolic syndrome. Circulation 2012;125:1134—46.

[109] Das A, Salloum FN, Filippone SM, Durrant DE, Rokosh G, Bolli R, et al. Inhibition of mammalian target of rapamycin protects against reperfusion injury in diabetic heart through STAT3 signaling. Basic Res Cardiol 2015;110:31.

[110] Aoyagi T, Higa JK, Aoyagi H, Yorichika N, Shimada BK, Matsui T. Cardiac mTOR rescues the detrimental effects of diet-induced obesity in the heart after ischemia-reperfusion. Am J Physiol Heart Circ Physiol 2015;308:H1530—9.

[111] Westermeier F, Navarro-Marquez M, Lopez-Crisosto C, Bravo-Sagua R, Quiroga C, Bustamante M, et al. Defective insulin signaling and mitochondrial dynamics in diabetic cardiomyopathy. Biochim Biophys Acta 2015;1853:1113—18.

[112] Lam A, Lopaschuk GD. Anti-anginal effects of partial fatty acid oxidation inhibitors. Curr Opin Pharmacol 2007;7:179—85.

[113] Li SY, Fang CX, Aberle II NS, Ren BH, Ceylan-Isik AF, Ren J. Inhibition of PI-3 kinase/Akt/mTOR, but not calcineurin signaling, reverses insulin-like growth factor I-induced protection

against glucose toxicity in cardiomyocyte contractile function. J Endocrinol 2005;186:491—503.

[114] Pellegrino MA, Patrini C, Pasini E, Brocca L, Flati V, Corsetti G, et al. Amino acid supplementation counteracts metabolic and functional damage in the diabetic rat heart. Am J Cardiol 2008;101:49e—56e.

[115] Kobayashi S, Liang Q. Autophagy and mitophagy in diabetic cardiomyopathy. Biochim Biophys Acta 2015;1852:252—61.

[116] Das A, Durrant D, Koka S, Salloum FN, Xi L, Kukreja RC. Mammalian target of rapamycin (mTOR) inhibition with rapamycin improves cardiac function in type 2 diabetic mice: potential role of attenuated oxidative stress and altered contractile protein expression. J Biol Chem 2014;289:4145—60.

[117] Volkers M, Doroudgar S, Nguyen N, Konstandin MH, Quijada P, Din S, et al. PRAS40 prevents development of diabetic cardiomyopathy and improves hepatic insulin sensitivity in obesity. EMBO Mol Med 2014;6:57—65.

[118] Crozier SJ, Anthony JC, Schworer CM, Reiter AK, Anthony TG, Kimball SR, et al. Tissue-specific regulation of protein synthesis by insulin and free fatty acids. Am J Physiol Endocrinol Metab 2003;285:E754—62.

[119] Hall ME, Smith G, Hall JE, Stec DE. Cardiomyocyte-specific deletion of leptin receptors causes lethal heart failure in Cre-recombinase-mediated cardiotoxicity. Am J Physiol Regul Integr Comp Physiol 2012;303:R1241—50.

[120] Shende P, Plaisance I, Morandi C, Pellieux C, Berthonneche C, Zorzato F, et al. Cardiac raptor ablation impairs adaptive hypertrophy, alters metabolic gene expression, and causes heart failure in mice. Circulation 2011;123:1073—82.

[121] Kim JE, Chen J. Regulation of peroxisome proliferator-activated receptor-gamma activity by mammalian target of rapamycin and amino acids in adipogenesis. Diabetes 2004;53:2748—56.

[122] Hudson CC, Liu M, Chiang GG, Otterness DM, Loomis DC, Kaper F, et al. Regulation of hypoxia-inducible factor 1alpha expression and function by the mammalian target of rapamycin. Mol Cell Biol 2002;22:7004—14.

[123] Morita M, Gravel SP, Chenard V, Sikstrom K, Zheng L, Alain T, et al. mTORC1 controls mitochondrial activity and biogenesis through 4E-BP-dependent translational regulation. Cell Metab 2013;18:698—711.

[124] Parra V, Verdejo HE, Iglewski M, Del Campo A, Troncoso R, Jones D, et al. Insulin stimulates mitochondrial fusion and function in cardiomyocytes via the Akt-mTOR-NFkappaB-Opa-1 signaling pathway. Diabetes 2014;63:75—88.

[125] Mellor KM, Reichelt ME, Delbridge LM. Autophagy anomalies in the diabetic myocardium. Autophagy 2011;7:1263—7.

[126] Mellor KM, Bell JR, Ritchie RH, Delbridge LM. Myocardial insulin resistance, metabolic stress and autophagy in diabetes. Clin Exp Pharmacol Physiol 2013;40:56—61.

[127] Ren SY, Xu X. Role of autophagy in metabolic syndrome-associated heart disease. Biochim Biophys Acta 2015;1852:225—31.

[128] Kim KH, Jeong YT, Oh H, Kim SH, Cho JM, Kim YN, et al. Autophagy deficiency leads to protection from obesity and insulin resistance by inducing Fgf21 as a mitokine. Nat Med 2013;19:83—92.

[129] Kobayashi S, Xu X, Chen K, Liang Q. Suppression of autophagy is protective in high glucose-induced cardiomyocyte injury. Autophagy 2012;8:577—92.

[130] Park S, Pak J, Jang I, Cho JW. Inhibition of mTOR affects protein stability of OGT. Biochem Biophys Res Commun 2014;453:208—12.

[131] Kobayashi S, Xu X, Chen K, Liang Q. Suppression of autophagy is protective in high glucose-induced cardiomyocyte injury. Autophagy 2012;8:577—92.

[132] Xu X, Kobayashi S, Chen K, Timm D, Volden P, Huang Y, et al. Diminished autophagy limits cardiac injury in mouse models of type 1 diabetes. J Biol Chem 2013;288:18077—92.

[133] Wataya-Kaneda M. Mammalian target of rapamycin and tuberous sclerosis complex. J Dermatol Sci 2015;79:93—100.

mTOR IN THE IMMUNE SYSTEM AND AUTOIMMUNE DISORDERS

17

Roles of Mechanistic Target of Rapamycin in the Adaptive and Innate Immune Systems

Hiroshi Kato and Andras Perl

Division of Rheumatology, Department of Medicine, State University of New York Upstate Medical University, Syracuse, New York, NY, USA

17.1 INTRODUCTION

Mechanistic target of rapamycin (mTOR) is a serine-threonine kinase which senses a variety of environmental cues, such as oxygen, glucose, amino acids, lipids, growth factors, and cytokines, and integrates these signals to orchestrate cell growth, proliferation, and differentiation. The roles of mTOR in the development and regulation of immune system have been rapidly emerging. Better understanding of this paradigm is expected to link the modulation of metabolic pathways to the restoration of immune-homeostasis, thereby treatment of a broad range of disease entities, including infectious disease, autoimmune disease, allograft rejection, and malignancy. In this chapter, we discuss the roles of the mTOR pathway in the development and function of T cells, B cells, and antigen-presenting cells (APCs), as well as potential application of mTOR blockade to the treatment of autoimmune and inflammatory diseases.

17.2 OVERVIEW OF mTOR AND mTOR COMPLEXES

Rapamycin was originally extracted from *Streptomyces hygroscopicus* found in soil on Easter Island in an effort to find potentially new antibiotics [1]. Rapamycin was found to inhibit the growth of yeast and to have immunosuppressive and antitumor potentials [2]. In yeast, rapamycin forms a complex with an immunophilin, FK-binding protein 12 (FKBP12) [3]. The target of rapamycin (TOR) was found to be a

289-kDa serine-threonine kinase. mTOR was later discovered and found to be an evolutionarily conserved serine-threonine kinase that is a member of the phosphoinositide 3-kinase related kinase family [4,5]. The N terminus of mTOR is characterized by a cluster of huntingtin, elongation factor 3, a subunit of protein phosphatase 2A, and TOR1 (HEAT) repeats that play a role in protein–protein interactions (Figure 17.1A). Next, there is a FRAP, ATM, and TRRAP (FAT) domain, followed by the FKBP12 rapamycin-binding (FRB) domain. Adjacent to the FRB is the kinase domain responsible for the mTOR's serine-threonine kinase activity and also the binding site for mTOR kinase-specific inhibitors. The C terminus consists of the FATS domain that appears to play a role in maintaining structural integrity. mTOR forms two distinct complexes to serve as a sensor and a coordinator of a broad range of environmental cues (Figure 17.1B). mTOR complex 1 (mTORC1) is composed of regulatory-associated protein of mTOR (RAPTOR), mammalian lethal with Sec13 protein 8 (mLST8), the proline-rich Akt substrate 40 kDa (PRAS40), and DEP domain containing mTOR-interacting (DEPTOR) [6]. mTORC2 contains mLST8 and DEPTOR in addition to the scaffolding protein RAPTOR-independent companion of TOR (RICTOR), mSIN1 proteins, and the protein observed with RICTOR (PROTOR) [6]. Rapamycin and other rapalogs bind to FKBP12 and, by binding to the FRB site on mTOR, are believed to block the ability of RAPTOR to bind to mTOR, thereby inhibiting mTORC1 [7]. However, prolonged treatment with rapamycin for some tissues and cell types is known to inhibit mTORC2 [8].

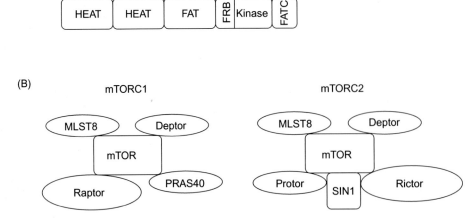

FIGURE 17.1   Structure of mTOR complexes. (A) mTOR is an evolutional conserved 289-kDa serine-threonine kinase that consists of two N-terminal HEAT (huntingtin, elongation factor 3, subunit of PP2A, and TOR) domains, which mediate protein–protein interactions, adjacent to a FAT domain. The 12-kDa FK506-binding protein (FKBP12) binds to the FRB domain to inhibit mTOR. The mTOR kinase catalytic domain lies C-terminal to the FRB site, where mTOR kinase-specific inhibitors bind. The FATS domain maintains structural integrity of the mTOR kinase. (B) mTOR associates with two distinct sets of adapter proteins to form two intracellular signaling complexes with unique substrate specificity. mTORC1 is composed of RAPTOR, mLST8, PRAS40, and DEPTOR, the latter two of which suppress the mTORC1 activity. mTORC2 contains mLST8 and DEPTOR, but is distinguished by the scaffolding protein RICTOR, mSIN1 proteins, and PROTOR.

17.3 UPSTREAM mTOR REGULATORS

mTOR senses a wide range of environmental cues in the form of growth factors, cytokines, nutrients, energy, and stress (Figure 17.2). Whereas we have gained much insight into the mechanisms by which these inputs regulate mTORC1 signaling, what regulates mTORC2 signaling largely remains undefined [9–11]. Ras homolog enriched in brain (RHEB) lies immediately upstream of mTORC1. RHEB is a small GTAase and is a critical regulator of mTORC1 [12–14]. RHEB is negatively regulated by the GTPase-activating protein activity of a complex consisting of tuberous sclerosis complex 1 (TSC1) and TSC2. Akt or ERK1/2 inactivates TSC through phosphorylation, which leads to the activation of RHEB and mTORC1. Growth factors such as insulin or insulin-like growth factors activate phosphatidylinositol 3-kinase (PI3K). PI3K activates 3-phosphoinositide-dependent protein kinase-1 (PDK1) through the production of phosphatidylinositol (3,4,5)-triphosphate. PDK1 in turn activates Akt, leading to mTORC1 activation. Notch 1 regulates early thymocyte differentiation and survival in an Akt–mTOR-pathway-dependent manner [15]. A growing number of cytokines and other immunological signals have been shown to activate mTOR. CD28 is a potent activator of PI3K and thus an important activator of mTORC1 in T cells [16,17]. Alternatively, the coinhibitory molecule PD-1 ligand 1 inhibits mTOR activity by binding to PD-1 on the surface of T cells [18]. IL-2 and IL-4 also activate mTORC1 via

PI3K activation [19,20]. As compared with rapid mTOR activation by the TCR, IL-7 induces delayed yet sustained PI3K–Akt signaling and mTOR activation in a STAT5-dependent manner, which contributes to glucose uptake and trophic effects in T cells [21,22]. In CD8$^+$ T cells, IL-12 and IFN-γ prolong the activation of mTOR upon stimulation [23]. Likewise, IL-1 promotes Th17 differentiation in part by activating mTORC1 [24]. Leptin modulates regulatory T (Treg) function via mTOR activation [25]. Besides the growth factors and cytokines, many other environmental cues regulate mTORC1 via TSC. For example, a low ATP: AMP ratio turns on the AMP-activated kinase (AMPK) [26,27], which in turn phosphorylates TSC. Whereas Akt inactivates TSC, phosphorylation of TSC by AMPK promotes its ability to inhibit mTORC1. Likewise, phosphorylation of TSC by glycogen synthase kinase 3β (GSK3β) inhibits mTOR [28]. Wnt signaling inhibits GSK3β and thus promotes mTORC1 signaling. The hypoxia-induced factor protein regulated in the development of DNA damage response 1 inhibits mTOR by promoting the assembly and activation of TSC in the setting of low oxygen [29]. Although growth-factor- or cytokine-induced activation of RHEB is central to mTORC1 signaling, mTORC1 activation requires the presence of amino acids [30]. Amino acids enhance mTORC1 activity by promoting the interaction between RHEB and mTORC1. Amino acids activate the Rag GTPases, which physically interact with RAPTOR and promote the localization of mTORC1 with RHEB on the surface of late endosomes and

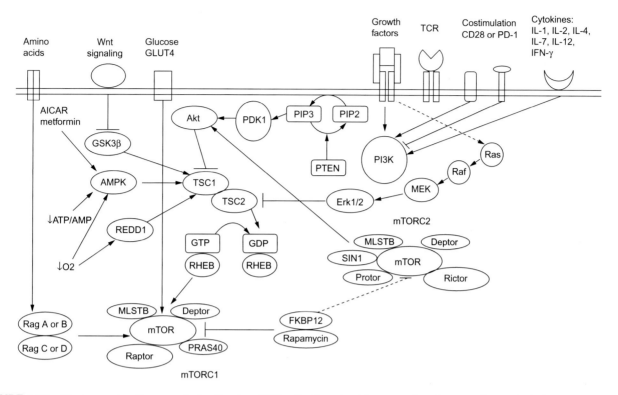

FIGURE 17.2 Upstream signaling cascade leading to mTOR activation. A wide range of environmental cues, including growth factors, TCR and costimulatory signals, cytokines, energy, oxygen, and nutrients regulate mTORC1. Upstream regulation of mTORC2 remains poorly understood. The details of mTORC1 and mTORC2 signaling pathways are discussed in the text.

lysosomes [30]. Branch chain amino acids (BCAA) such as leucine are among the most potent activators of mTORC1. In line with this, Treg can promote infectious tolerance in part by depleting BCAA, culminating in mTOR inhibition and further Treg generation [31]. With regard to the mTORC2 regulators, recent studies shed light on the role of ribosome in linking PI3K to mTOR2 activation [32,33]. Further, a small GTPase, RAC1, directly binds to mTOR to activate mTORC1 and mTORC2 [34].

17.4 DOWNSTREAM mTOR SIGNALING

Phosphorylation of S6K1 and 4E-BP1 serve as makers of mTORC1 activation (Figure 17.3). S6K1 promotes translation of mRNA and is activated by mTORC1, whereas 4E-BP1 is an inhibitor of eIF-4E and its phosphorylation by mTOR leads to its inactivation [35]. Hypoxia inducible factor (HIF) is controlled by mTORC1 in a 4E-BP1-dependent manner [9]. HIF- and mTORC1-activated sterol regulatory element-binding protein 1 (SREBP1) plays critical roles in regulating glucose and lipid metabolism, respectively [36]. SREBP1 up-regulates the genetic programs involved in lipid biosynthesis and pentose phosphate pathway.

Additionally, mTORC1 plays a role in mitochondrial biosynthesis and autophagy. Indeed, mTORC1 blockade leads to an increase in autophagy whereas mitochondrial biosynthesis is enhanced in the setting of mTORC1 activation [37,38]. Phosphorylation of Akt at serine 473 serves as a surrogate maker of mTORC2 activation. This mTORC2-dependent phosphorylation of Akt is distinct from PDK1 dependent phosphorylation of Akt at threonine 308 that is downstream of mTORC1. In fact, upstream Akt activation as determined by threonine 308 phosphorylation is robust in the absence of mTORC2 [39,40]. Alternatively, deletion of RICTOR led to slight inhibition of Akt phosphorylation at threonine 308 [41]. Therefore, the paradigm regarding the mTORC2-dependent and independent Akt activity has been evolving. However, several studies clearly demonstrated that the Akt ability to regulate Forkhead box protein O1 (FOXO1) and FOXO3 is mTORC2-dependent [40]. Akt activation leads to the phosphorylation of these transcription factors, which results in their sequestration in the cytoplasm, thereby inhibiting their activity. Another downstream target of mTORC2 is the serum glucocorticoid-regulated kinase-1 (SGK1). SGK1 is similar to Akt and S6K1 in that it is an AGC (cAMP-dependent, cGMP-dependent, and protein kinase C) kinase [42]. Like

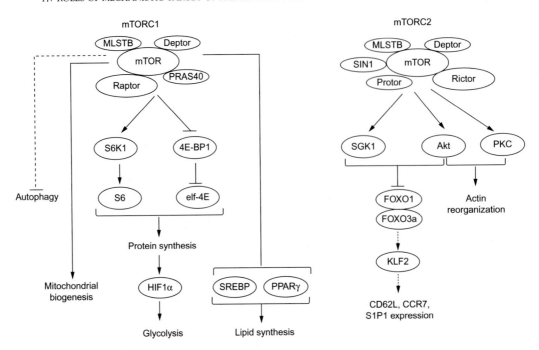

FIGURE 17.3   Downstream of mTORC1 and mTORC2 signaling. Upon mTORC1 activation, mTOR phosphorylates S6K1, leading to the phosphorylation of ribosomal S6 protein, which culminates in enhanced protein synthesis. Phosphorylation of 4E-BP1 by mTOR allows elf-4E to participate in the translation—initiation complexes. mTORC1 activity promotes glycolysis, lipid synthesis, and mitochondrial biogenesis and inhibits autophagy. mTORC2 activation leads to the phosphorylation of Akt at serine 473 and SGK1, leading to the phosphorylation and sequestration of FOXO proteins in the cytoplasm. This prevents the FOXO proteins from activating the transcription of KLF2, which positively regulates CD62L, CCR7, and S1P1.

Akt, SGK1 phosphorylates and inhibits FOXO family members [43]. mTORC2 phosphorylates and activates protein kinase C alpha (PKCα), culminating in actin reorganization [44]. Importantly, mTORC2 promotes Th2 differentiation via activation of PKCθ [41].

17.5 ROLES OF mTOR IN T-CELL HOMEOSTASIS AND ACTIVATION

TSC1 and TSC2 function as an integral complex to strictly control mTORC1 activity. TSC1-mediated control of mTOR signaling promotes quiescence of naïve T cells by controlling cell size, cell cycle entry, and metabolic machinery [45–47]. The abrogation of quiescence predisposes TSC1-deficient T cells to apoptosis, which results in the loss of conventional T cells and iNKT cells. The remaining Tsc (−/−) T cells exist in a unique semiactivated (CD44+CD122−) state *in vivo* and exhibit increased levels of activation, cell cycle entry, and cytokine expression as compared with activated wild-type T cells. Nonetheless, Tsc (−/−) T cells fail to generate effective antibacterial response even when the excessive apoptosis is blocked. Tsc1 (−/−) T cells exhibit increased mTORC1, but diminished mTORC2

activity as compared with wild-type T cells, where treatment with rapamycin partially reverses the T-cell homeostasis and survival [45]. These observations clarify a critical role of TSC1 in establishing naïve T-cell quiescence to facilitate immune-homeostasis and function.

Similar to FK506, immunosuppressive action of rapamycin depends on its ability to bind to immunophilin FKBP12 [48]. However, unlike FK506 and cyclosporine A, rapamycin does not inhibit calcineurin and thus does not inhibit T-cell-receptor (TCR)-induced NF-AT activation. Accordingly, rapamycin does not block the expression of TCR-induced genes, such as IL-2, IFN-γ, TNF-α, as well as genes for NF-AT-dependent proteins such as cbl-b, GRAIL, DGK-α, Egr-2, and Egr-3 [49]. MicroRNAs fine-tune TCR responsiveness by modulating the mTOR activity as evidenced when Dicer-deficient CD4+ T cells exhibited enhanced TCR responsiveness in association with costimulation-independent IL-2 production and increased mTORC1 and mTORC2 activity, where genetic ablation of *Raptor* and *Rictor* partially reversed such abnormalities [50]. Importantly, rapamycin induces T-cell anergy even in the presence of costimulation [17,51–53], where calcineurin inhibitor cyclosporine A opposes the rapamycin-induced anergy [52]. Likewise, induction of T-cell proliferation in the presence of rapamycin was unable to overcome anergy [17]. Alternatively, cell cycle arrest in G1 in the absence of mTOR blockade did not induce anergy [51]. These observations suggest that rapamycin induces anergy not by inhibiting cell proliferation, but by inhibiting mTOR.

17.6 LINKING T-CELL METABOLISM TO FUNCTION

Naïve T cells exploit catabolic metabolism, that is, they generate ATP via tricarboxylic acid (TCA) cycle and oxidative phosphorylation. They also employ autophagy to generate molecules required for protein synthesis and energy. The quiescent state of lymphocytes is actively maintained by numerous regulatory transcription factors, such as Krüppel-like factor (KLF2) and the FOXOs [54–56]. Upon activation, lymphocytes become anabolic and start to utilize aerobic glycolysis to generate ATP; this is known as the Warburg effect and is employed by cancer cells [57]. This seemingly inefficient means to generate energy as compared with TCA cycle and mitochondrial respiration is important to match the increased demand of protein, nucleotide, and lipid biosynthesis in the context of lymphocyte activation [58]. Such a metabolic switch during transition from resting to activated T cells appears to be orchestrated by immunological signals. For example, CD28-induced PI3K activation leads to Akt activation, which in turn promotes the surface expression of glucose transporters [59–61]. Activation of mTORC1 via HIF promotes the expression of proteins involved in glycolysis and glucose uptake, whereas mTORC1-dependent activation of SREBP results in the up-regulation of proteins essential for the pentose phosphate pathway as well as fatty acid and steroid synthesis [36]. Accordingly, blocking these metabolic pathways inhibits T-cell activation. As an example, metformin and AICAR, which mimic energy depletion and activate mTORC1 inhibitor, AMPK, inhibit IL-2 production, and promote T-cell anergy [62–64]. The glucose analog and mTORC1 inhibitor, 2-deoxyglucose, blocks T-cell cytokine production and promotes anergy [64]. As noted earlier, mTORC1 activity requires the presence of amino acids besides the growth factor stimulation. As such, the leucine antagonist N-acetyl-leucine-amide can inhibit mTORC1 even in the presence of potent mTORC1 activators and inhibits T-cell function [65]. A series of these observations collectively support an intriguing notion that modulation of immune-cell metabolism is potentially an effective strategy to control the immune-cell function.

17.7 ROLES OF mTOR IN CD4+ HELPER-T-CELL DIFFERENTIATION

Functional diversity is one of the hallmarks of adaptive immunity, as exemplified by the remarkable ability of CD4+ T helper (Th) cells to differentiate into diverse lineages in response to various environmental cues. In this regard, the cytokine milieu provides potent signals that dictate the T-cell-lineage commitment and functional development. Given the pivotal roles of mTOR in sensing a broad range of environmental cues, it is tempting to speculate that mTOR plays an important role in T-cell-lineage commitment. In fact, CD4+ T-cell-specific deletion of mTOR led to defective Th1, Th2, and Th17 differentiation even under strongly polarizing conditions [66] (Figure 17.4). Alternatively, such cells became Foxp3+ Treg cells. Mechanistically, the inability to differentiate into effector helper T (Th) lineage was associated with decreased STAT4, STAT6, and STAT3 phosphorylation in response to IL-12, IL-4, and IL-6, respectively. Such diminished STAT activation was associated with decreased expression of Th-lineage-defining transcription factors; T-bet, GATA-3, and ROR-γt, respectively. To define the roles of individual mTOR complexes in Th-lineage development, T-cell-specific deletion of RHEB and RICTOR has been employed. Of note, RHEB (−/−) T cells failed to differentiate into Th1 and Th17 cells, associated with defective STAT4 and STAT3 phosphorylation in response to IL-12 and IL-6 respectively, whereas such cells retained the ability to become Th2 cells [39]. However, raptor, but not rheb, links the glucose metabolism to IL-4 and IL-2 receptor expression on CD4+ T cells and subsequent STAT6 and STAT5 phosphorylation, which culminate in Th2 differentiation [67]. Amino acid transporter, ASCT2, couples the TCR and CD28 signals to glutamine uptake and mTORC1 activation and mediates Th1 and Th17 differentiation [68]. The mechanisms by which mTOR instructs STAT activation have largely remained unclear. Some observations suggest interaction between mTOR and the STATs [69–71]. Alternatively, increased activity of SOCS3 was noted in RHEB (−/−) T cells [72,73], which may in part account for the defective Th17 differentiation given that SOCS3 is a potent inhibitor of STAT3. The Th17 cell is a metabolically demanding cell in that it requires mTOR-HIF-1α-dependent glycolytic pathways for its differentiation [74] and that Th17 differentiation is compromised by depletion of specific amino acids [75] or increases in lipid metabolism [76]. In contrast to RHEB (−/−) T cells, RICTOR (−/−) T cells failed to differentiate into Th2 cells [39,41], associated with deceased IL-4-induced STAT6 phosphorylation and

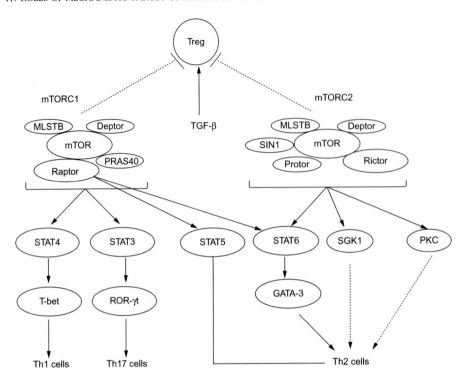

FIGURE 17.4   Roles of mTOR complexes in T-helper-cell lineage development. mTORC1 promotes the phosphorylation of STAT4 and STAT3 in response to IL-12 and IL-6, possibly via suppression of SOCS3, which culminates in Th1 and Th17 differentiation, respectively. In mTORC1 complex, Raptor specifically links the glucose metabolism to the activation of IL-2->STAT5 and IL-4->STAT6 pathways, leading to Th2 differentiation. mTORC2 promotes the phosphorylation of STAT6 in response to IL-4 possibly via suppression of SOCS5, which leads to Th2 differentiation. mTORC2 activates SGK1 and PKC, both of which potentially contribute to Th2 differentiation. Whether a dual blockade of mTORC1 and mTORC2 is required for optimal CD4$^+$CD25$^+$Foxp3$^+$ Treg differentiation remains controversial and may be dictated by the availability TGF-β.

increased SOCS5 activity [39]. Such a defective Th2 differentiation was attributed to decreased PKCθ activity and NF-κB-mediated transcription [41]. SGK1 downstream of mTORC2 promotes Th2 differentiation, whereas it negatively regulates Th1 differentiation in a TCF-1-dependent manner [77].

17.8 ROLES OF mTOR BLOCKADE IN CD4$^+$CD25$^+$FOXP3$^+$ TREG DEVELOPMENT

The genetic studies suggest that both mTORC1 and mTORC2 need to be blocked for effective Treg generation, which points to an intriguing model where Treg differentiation represents a default T-cell-lineage fate in the absence of mTOR activity [39,66] (Figure 17.4). However, two important questions remain to be clarified: (1) Does Treg require blockade of both of these mTOR complexes for its differentiation? (2) Does Treg need mTOR blockade throughout its stage of development?

17.8.1 Does Treg Require Dual mTORC1 and 2 Blockades for Its Differentiation?

Additional evidence supports the aforementioned model that both mTORC1 and mTORC2 need to be blocked for Treg differentiation. Picomolar concentrations of rapamycin that do not inhibit mTORC2 can generate Tregs in RICTOR null T cells, which indicates that selective mTORC1 blockade enhances Treg differentiation in T cells in which mTORC2 is genetically ablated. Additionally, human Treg has diminished mTORC1 and mTORC2 activity, where Treg suppressive function is compromised by enforced expression of Akt [78]. mTORC2 activates Akt and SGK1, which in turn phosphorylate and inhibit the activation of FOXO1 and FOXO3a. Hence, mTORC2 inhibition leads to activation of FOXO1 and FOXO3a, which promotes Foxp3 transcription [79]. However, several lines of evidence challenge the model regarding the need for dual mTORC1 and 2 blockades for Treg differentiation. TSC1 deficiency, which is expected to promote Akt activity, enhanced thymic natural Treg development independent of the Akt–mTOR pathway [80].

Rapamycin, which specifically inhibits mTORC1, but not mTORC2 under normal circumstances, was shown to promote Treg generation in various immune-mediated diseases [81–83]. TGF-β promotes Treg differentiation even when mTORC1 and mTORC2 are active [84,85]. The selective mTORC1 blockade can enhance Treg generation in the presence of a low concentration of TGF-β that is otherwise insufficient to induce Treg differentiation in the absence of rapamycin [86]. mTOR-deficient T cells display constitutively phosphorylated SMAD3, indicative of active TGF-β signaling [66]. This was attributed to increased T-cell responsiveness to TGF-β rather than increased TGF-β in the culture media. In the absence of exogenous TGF-β, neutralization of TGF-β abrogated Treg generation from the mTOR-null T cells, suggesting that TGF-β was contributing to Treg differentiation in the absence of mTOR [66]. A series of these findings support a model where the availability of TGF-β signaling dictates the requirement for mTOR blockade in Treg differentiation (Figure 17.4).

17.8.2 Does Treg Need mTOR Blockade Throughout Its Stage of Development?

The majority of observations point to the indispensable roles of mTOR blockade in Treg development, as exemplified by when constitutively active Akt–mTOR axis abrogated Treg generation in the thymus [87]. Ligation of neuropilin-1 on Treg by semaphorin-4a potentiates the Treg stability and function by inhibiting the Akt–mTOR axis [88]. CD5 promotes differentiation of extrathymic Tregs by blocking mTOR [89]. The PI3K inhibitor phosphatase and tensin homolog (PTEN) is a negative regulator of mTOR. Complete loss of PTEN in T cells impairs peripheral as well as central tolerance [90] and disrupts the Foxp3 induction [91]. Treg-specific deletion of PTEN leads to development of autoimmune lymphoproliferative syndrome associated with excessive germinal center formation and follicular helper-T-cell generation, which was partially blocked by deletion of IFN-γ signaling [92]. Such PTEN-deficient Treg exhibits lineage instability [93] as well as increased mTORC2 activity and glycolysis, but impaired mitochondrial fitness [92]. Insulin suppresses Treg ability to secrete IL-10 in an Akt–mTOR-pathway-dependent manner [94]. The absence of mTOR activity confers Treg selective advantage to proliferate as Treg is more resistant to apoptosis than effector T cells in the presence of rapamycin [95,96]. Such resistance to mTOR inhibition seems to be mediated by up-regulation of Pim-2 kinase [97]. Pim-2 kinase has similar signaling attributes to Akt in terms of its ability to control activation, growth, and survival. However, unlike Akt, the activation of Pim-2 kinase is mTOR-independent

[98]. Upon mTOR blockade, Pim-2 kinase expression increases in part by Foxp3-induced transcription. Along these lines, less mTOR activation has been noted upon Treg stimulation, which may in part be attributed to an increase in PTEN [95]. Accordingly, IL-2 stimulation inhibits mTOR activity, but enhanced STAT5 phosphorylation in Treg. These findings raise a possibility that Treg may employ distinct metabolic pathways for proliferation and function. In this regard, Th17 differentiation is associated with HIF-1α-dependent up-regulation of glycolytic pathways, whereas HIF-1α deficiency or blockade of glycolysis tips the balance from Th17 differentiation towards Treg differentiation [74,99]. Mechanistically, HIF-1α transcriptionally activates ROR-γt and promotes proteosomal degradation of Foxp3 [99].

However, it is important to turn our attention to a few potentially contradictory observations, which may allow us to better define the dynamically changing roles of the mTOR pathway in Treg development. A study showed that transient mTOR blockade allowed Tregs to exit from the anergic state and enter the cell cycle. Such Tregs exhibit high mTOR activity, vigorously proliferate, and demonstrate high suppressive function [25]. In the same study, however, chronic rapamycin treatment or genetic mTOR ablation did not allow Tregs to exit the anergic state and proliferate. Likewise, another group demonstrated that Treg required mTORC1 for proliferation and its immunosuppressive function [100].

We reason that the aforementioned discrepant observations are potentially attributed to the variability of the studied cell population, culture conditions, and the timing, duration, and method of mTOR blockade.

17.9 ROLES OF mTOR IN CD8$^+$ T-CELL DEVELOPMENT

Similar to CD4$^+$ T cells, quiescence of naïve CD8$^+$ T cells is actively maintained by transcriptions factors, such as Schlafen-2, KLF2, KL4, and E74-like factor 4 (ELF4) [101]. Upon TCR engagement, mTOR plays a role in reversing the quiescent state by inhibiting ELF4 and KLF4 expression [102]. As is the case for CD4$^+$ T cells, the activated CD8$^+$ T cells switch from catabolism to anabolism and oxidative glycolysis [103]. In fact, antigen recognition in CD8$^+$ T cells leads to mTOR- and MAP-kinase-signaling-induced ribosomal S6 phosphorylation [104]. IFN-γ and IL-12 lead to prolonged mTORC1-dependent T-bet expression in CD8$^+$ T cells [23]. mTOR positively regulates interferon regulatory factor 4, which confers CD8$^+$ T-cell proliferative potential and effector function by down-regulating the expression of CDK inhibitors and proapoptotic genes and up-regulating T-bet and Blimp-1, respectively [105].

These findings couple TCR engagement and CD8$^+$ effector generation through mTOR activation. Alternatively, senescence of effector memory CD8$^+$ T cells, as exemplified by decreased proliferation, telomerase activity, and mitochondrial function, is reversed by blockade of the p38-MAPK pathway in an mTOR-independent manner [106].

The initial expansion of CD8$^+$ effector T cells is followed by a contraction phase and concomitant development of CD8$^+$ memory T cells [107]. CD8$^+$ memory T cells express IL-7 receptor (CD127) and IL-2 receptor β chain (CD122). The memory development is associated with the re-expression of CD62L and the up-regulation of CCR7, both of which allow T-cell trafficking through the lymph nodes. At the transcriptional level, conversion from effector to memory cell is associated with a shift from T-bet to eomesodermin (Eomes)-driven gene expression [108,109] as well as metabolic switch from anabolism and catabolism. Analogous correlation between the metabolism and expression of lineage-specifying transcription factors is observed in inflammatory CD4$^+$ T cells versus CD4$^+$CD25$^+$Foxp3$^+$ Tregs (Table 17.1). In line with these observations, low-dose rapamycin promotes the effector to memory transition in the setting of lymphocytic choriomeningitis virus infection [110]. More specifically, rapamycin treatment during the expansion phase increased the overall number of memory CD8$^+$ T cells, whereas it selectively expanded the highly functional memory CD8$^+$ T cells when given during the contraction phase. Such effects of rapamycin appeared to be T-cell-intrinsic. In fact, RAPTOR knockdown in the antigen-specific T cells enhanced memory T-cell response, comparably to rapamycin treatment [110]. FKBP12 is an intracellular binding partner of rapamycin and FKBP12—rapamycin complex inhibits the mTOR pathway. Rapamycin failed to promote memory T-cell differentiation in antigen-specific CD8$^+$ T cells in which FKBP was genetically

ablated [110]. Tumor necrosis factor receptor-associated factor 6 (TRAF6)-deficient CD8$^+$ T cells failed to differentiate into long-lived memory T cells, whereas they retained the ability to differentiate to effector cells [111]. TRAF6-deficient CD8$^+$ T cells were defective in AMPK activation with metformin and mitochondrial fatty acid oxidation (FAO) in response to growth factor withdrawal. However, rapamycin restored FAO and promoted memory CD8$^+$ T-cell development not only in TARF6-deficient T cells but also in wild-type T cells [111]. Dual blockade of mTOR and CTLA-4 enhances antitumor and antibacterial CD8$^+$ memory T-cell response, which is associated with increased mitochondrial biogenesis and spare respiratory capacity [112]. Importantly, mTOR blockade affects the expression of relevant transcription factors in CD8$^+$ T cells. IL-7-induced mTOR activation promotes T-bet expression, whereas IL-15 promotes the transition to memory T cells in part through the up-regulation of Eomes. IFN-γ receptor signaling activates mTOR and dampens CD8$^+$ memory T-cell response and its Eomes expression in the presence of weak TCR stimulation [113]. Rapamycin inhibits T-bet expression and promotes Eomes expression independent of IL-15 [114]. IL-7 and IL-15 receptor expression is repressed in cells that express constitutively active Akt, in which survival of memory CD8$^+$ T cells is diminished [115]. In contrast to memory CD8$^+$ T cells in the secondary lymphoid organs, rapamycin blocks the generation of tissue-resident memory CD8$^+$ T cells in the mucosal sites, which confers protection against CD8$^+$ T-cell-mediated lethal intestinal autoimmunity [116]. Unlike the memory development, acquisition of effector function by memory CD8$^+$ T cells, in particular IFN-γ expression, requires rapamycin-insensitive, but Akt—mTORC2-dependent up-regulation of immediate-early glycolytic switch [117].

Exhausted CD8$^+$ T cells in the setting of chronic viral infection exhibit reduced TCR-activation of the

TABLE 17.1 Metabolic Control of T-Cell Development via mTORC1

	Anabolism	Catabolism
mTORC1 activity	High	Low
Major source of energy	Glycolysis	Oxidative phosphorylation
CD4$^+$ T cells	Th1, Th17 cells	Tregs
Transcription factors	T-bet, ROR-γt, HIF-1α	Foxp3, Pim-2 kinase, and FOXO transcription factors
CD8$^+$ T cells	Cytotoxic CD8$^+$ T cells	Memory CD8$^+$ T cells
Transcription factors	T-bet, IRF-4, Blimp-1	Eomes, FOXO, and KLF2 transcription factors

Notes: Effector CD4$^+$ T cells, such as Th1 and Th17 cells, are in an anabolic state as exemplified by increased mTOR and glycolytic activity. Their lineage development is driven by mTOR-dependent expression of HIF-1α as well as T-bet (Th1 cells) and ROR-γt (Th17 cells). In contrast, CD4$^+$CD25$^+$FoxP3$^+$ Tregs are in catabolic state characterized by low mTOR and glycolytic activity. mTOR blockade enhances the expression of Foxp3, Pim-2 kinase, and FOXO transcription factors, culminating in Treg differentiation. Analogous correlation between the metabolism and expression of lineage-defining transcription factors is seen in cytotoxic versus memory CD8$^+$ T cells, where mTOR blockade facilitates the conversion from cytotoxic to memory CD8$^+$ T cells.

Akt—mTOR pathway, resulting in increased transcriptional activity of FOXO1, where PD-L1-blockade restores the mTOR activity [118]. FOXO1 in turn directly activates the transcription of PD-L1, suggesting that a positive feedback loop of PD-L1→FoxO1→PD-L1 might enable CD8$^+$ T cells to adapt to chronic infection.

The ability of mTOR blockade to enhance CD8$^+$ memory T-cell development has important implications for vaccine development. Treatment with rapamycin and metformin can enhance antitumor memory response [23,110,111]. Vaccination given concomitantly with the rapalog temsirolimus resulted in enhanced memory formation as well as antitumor immunity. Treatment with rhesus macaques with sirolimus during vaccinia virus vaccination led to enhanced central and effector memory CD8$^+$ T-cell development [119]. Rapamycin augments the CD8$^+$ T-cell response against a virus-like particle derived from hepatitis B virus core antigen [120]. Finally, in the context of T-cell-mediated antibacterial responses and graft rejection, rapamycin selectively expanded responses to an antigen when it was in the context of a bacterial infection, but failed to do so when the same antigen was presented as part of a transplant [121], suggesting that rapamycin may achieve tolerance to transplanted organs without compromising antimicrobial immunity.

17.10 ROLES OF mTOR IN T-CELL TRAFFICKING

Naïve T cells express CD62L and CCR7, which allow the cells to circulate through secondary lymphoid organs [122]. Once the cells recognize cognate antigen and differentiate into effector cells, CD62L and CCR7 are down-regulated, which allow the effector cells to exit out of the lymphoid organs. Consistent with the ability of mTOR to promote effector generation, the PI3K—mTOR axis plays a role in down-regulating CD62L and CCR7 [123]. In CD8$^+$ T cells that lack the PI3K inhibitor PTEN, where mTOR is hyperactivated, CD62L and CCR7 remain down-regulated. In contrast, rapamycin treatment leads to persistent expression of these molecules. As noted earlier, resolution of viral infection is associated with transition from effector to memory CD8$^+$ T cells. This transition is associated with the re-expression of CD62L and CCR7, which allows memory cells to circulate through secondary lymphoid organs for continuous surveillance.

The expression of CD62L, CCR7, and the memory marker CD127 has been linked to the FOXOs and KLF2 [124,125]. mTORC2 activates Akt, which leads to the inactivation of FOXOs and decreased KLF2 expression. As KLF2 positively regulates the transcription of these trafficking molecules, the expression of CD62L and CCR7 decline upon Akt and mTOR activation. In addition to such an mTORC2-dependent mechanism, Akt can regulate CD8$^+$ T-cell trafficking in an mTORC1-dependent, but FOXOs-independent, manner [126]. Besides CD62L, CCR7, and CD127, the G-protein-couple receptor sphingosine 1-phosphate receptor 1 (S1P1) is also regulated by KLF2 [127]. S1P1 is critical in promoting T-cell egress from lymph nodes. The immunosuppressive agent FTY720 acts on S1P1, promoting sequestration of T cells in the lymph nodes [128]. Of note, S1P1 leads to mTOR activation [129], which enhances Th1 differentiation and antagonizes Treg differentiation [130]. Transgenic expression of S1P1 on Treg impairs its suppressive function, whereas rapamycin opposes S1P1-mediated attenuation of Treg function.

17.11 ROLES OF mTOR IN B-CELL DEVELOPMENT

As compared with T cells, the mTOR paradigm in B-cell differentiation and function remains to be better defined. Nonetheless, a growing body of evidence indicates the critical importance of optimal mTOR activity in B-cell development. As an example, in the mTOR-hypomorph mouse, B-cell development appeared to be more sensitive to low mTOR activity than T-cell development [131]. Such mice exhibited a partial block in the large pre-B to small pre-B stages of development. In the periphery, there was an increase in mature B cells with a concomitant decrease in transitional and marginal zone B cells. The mTOR-hypomorph B cells demonstrated defective proliferation in response to B-cell mitogens. Signaling through the B-cell receptor and CD40 was more compromised than Toll-like receptor (TLR) signaling. Additionally, antibody production to T-cell-dependent and -independent antigens was diminished. B-cell class switching also requires mTORC1, where rapamycin skews antibody response away from high-affinity variant epitopes towards more conserved elements of hemagglutinin in the setting of influenza vaccination [132]. Alternatively, TSC1-deficient B cells demonstrated hyperactive mTORC1 activity [133]. Such mice exhibited B-cell maturation defects, in particular significant reduction in marginal zone B cells, which was partially reversed with rapamycin. Similar to the mTOR-hypomorph mice, humoral response to T-cell-dependent and -independent antigens was impaired.

Analogous to T cells, the quiescent state of B cells is actively maintained by many regulatory transcription factors [134]. B-cell-activating factor (BAFF) mediates its function by activating mTOR and Pim-2 kinase [135].

For example, Pim-2-deficient B cells can survive in the presence of BAFF, which is abrogated by treatment with rapamycin. These two pathways converge and culminate in the expression of antiapoptotic protein Mcl-1, which is critical for executing these survival signals. Upon activation, B cells rapidly increase glucose uptake and glycolysis. mTOR is likely central to this process given that HIF plays a role in the up-regulation of glycolytic pathway for B-cell development [136].

In B cells lacking mSIN1, which is a component of mTORC2, an increase in IL-7 receptor expression and recombinase activating gene (RAG) activity were observed [137]. This was attributed to mTORC2 potential to activate Akt2, which led to inhibition of FOXO1 and thus down-regulation of IL-7 receptor and RAG. In line with this, PI3K inhibited RAG expression via mTORC2 activation [138].

BCR signaling activates the mTOR pathway [139]. However, in the case of T-cell-independent antigens, only prolonged antigen exposure leads to mTOR activation sufficient for B-cell activation. Alternatively, engagement of inhibitory Fcγ receptors inhibits PI3K signaling and thus down-regulates mTOR activity [140]. Collectively, it is plausible that the net balance of mTOR activation via BCR signaling and the mTOR inhibition via Fcγ receptor signaling dictates the overall mTOR activity as well as the B-cell activation status. Besides the BCR signaling, TLR and CD40 signaling also activate the PI3K–mTOR axis in B cells. Rapamycin inhibits anti-CD40-mediated proliferation and the anti-CD40 potential to prevent apoptosis of splenic B cells [141]. LPS-induced B-cell proliferation and differentiation are inhibited by rapamycin [142]. Nonetheless, individual B-cell subsets appear to be differentially controlled by mTOR. For example, marginal zone B cells maintain high levels of mTOR activity in response to nutrients in the absence of mitogens, whereas follicular B cells exhibit a lower level of mTOR activity and are less sensitive to rapamycin.

17.12 ROLES OF mTOR IN INNATE IMMUNITY

The effects of the mTOR pathway on the development and function of innate immune cells, such as dendritic cell (DC)s, appear to be highly context-dependent as reviewed below, such as the types of cells, the immunological milieu in which they are generated, and the outcomes studied.

Plasmacytoid DCs (pDCs) play critical roles in the antiviral immune response through production of type I interferons [143], which is mediated by TLR7 and TLR9 signaling [144]. Rapamycin abrogates the ability of TLR7 and TLR9 agonists to induce IFN-α and

-β [145,146], where mTOR blockade disrupts the complex of TLR9 and MyD88, preventing nuclear translocation of IRF7 [145]. The targeted mTOR blockade in macrophages and DCs by means of microencapsulated rapamycin diminished IFN-α and IFN-β production and attenuated response to yellowfever virus vaccine. Human myeloid DC (mDC) produces type I and III IFNs in an mTOR-dependent manner [147].

In contrast to pDCs, a large body of evidence suggests that mTOR suppresses the expression of inflammatory phenotypes of myeloid phagocytes, such as monocytes, macrophages, and mDCs. Stimulation of these cells with LPS in the presence of rapamycin leads to increased IL-6, IL-12, and IL-23 expression via NF-κB activation, but diminished STAT3-dependent IL-10 production [148]. Deletion of TSC2, that induces mTOR activation, diminishes NF-κB, promotes STAT3 activity, and reverses such a proinflammatory phenotype. IL-10 phosphorylates STAT3 in a PI3k/Akt/mTOR-dependent manner and confers macrophages' anti-inflammatory potential [149]. Likewise, JAK3 inhibits TLR-mediated production of IL-6, IL-12, and TNF-α by human monocytes in a PI3K–Akt pathway-dependent manner [150]. TSC-deficient mDC exhibits down-regulated IRF-4-dependent MHC-II expression, which is restored by rapamycin [151]. Rapamycin-treated monocytes exhibit potent Th1 and Th17 polarizing potential. Accordingly, rapamycin treatment confers protection against *Listeria monocytogenes* infection [148]. The effect of rapamycin on LPS-induced IL-12 production is in part dependent on IL-10-STAT3 pathway, whereas GSK3 regulates IL-12 production independent of IL-10 [152]. Contrary to the aforementioned observations, treatment of monocytes with *Candida albicans* cell wall shifted metabolism from oxidative phosphorylation to glycolysis. This was associated with permissive histone modifications of genes involved in glycolytic pathways and Akt-, mTOR-, and HIF-1α-dependent up-regulation of TNF-α production [153].

Accumulating evidence suggests that the states of DCs dictate the lineage fate of T cells which they interact with. For example, upon activation, DCs express costimulatory molecules and cytokines that induce robust T-cell responses. In contrast, when T cells encounter antigen presented by resting DCs, this interaction leads to antigen-specific tolerance [154]. Rapamycin inhibits macropinocytosis in APCs [155], suggesting that mTOR plays a role in the antigen uptake and processing. Several growth factors and cytokines promote the differentiation and maturation of bone-marrow-derived DCs. Flt3 ligand promotes the generation of conventional DCs, CD8+ DCs, and pDCs [156], which is mTOR-dependent as exemplified by when rapamycin inhibits Flt3 ligand-induced DC maturation [157]. In contrast, deletion of PTEN leads to hyperactivation of mTOR, which results in excessive expansion of the DC compartment, in particular CD8$^+$ DCs. *In vitro* differentiation of bone-marrow-derived DCs in the presence of IL-4 and GM-CSF is inhibited by rapamycin [158]. Such rapamycin-treated DCs down-regulate MHC-II and costimulatory molecules and induce T-cell tolerance (anergy) as opposed to activation [159]. Additionally, such tolerogenic DCs can promote the generation of Tregs [160]. Likewise, ATP-competitive mTORC1 and mTORC2 inhibition, but not mTORC1 inhibition by rapamycin, induces B7H1 on DCs in a STAT3-dependent manner and such DCs induce Tregs in a B7H1-dependent manner [161]. *In vivo*, rapamycin-treated allogeneic DCs can induce tolerance and prevent graft rejection in mouse models of solid organ transplantation [159,162]. Interestingly, whereas rapamycin-treated mature DCs can produce increased amounts of IL-12, they fail to promote Th1 response, but induce Tregs [163].

Natural killer cells (NK cells) up-regulate glycolysis and effector function including IFN-γ expression in an mTOR-dependent manner, where the effector function is dependent on glycolysis [164]. IL-15 drives mTOR activation in NK cells, whereas mTOR confers optimal IL-15 responsiveness, maturation, and effector functions of NK cells [165]. PDK1 confers IL-15 responsiveness of NK cells and is essential for its early development, but is dispensable for its terminal differentiation and maturation [166]. Optimal mTOR activity is critical for invariant NKT (iNKT) cell development, as exemplified when genetic mTORC1 ablation and mTOR hyperactivation due to folliculin-interacting-protein deficiency both led to disrupted iNKT cell development [167,168]. Whereas both mTORC1 and mTORC2 play nonredundant roles in thymic iNKT cell development, mTORC2, but not mTORC1, is essential for NKT17 cell development [169].

The mTOR pathway in other innate immune cells has largely remained unexplored, but a recent study has shown that a component of mTORC2, rictor, negatively regulates Fcε receptor I-induced degranulation of mast cells, correlating with suppression of calcium mobilization and cytoskeletal rearrangement [170].

17.13 TARGETING mTOR IN THE TREATMENT OF IMMUNE-MEDIATED DISEASES

As noted earlier, the principal mechanisms of action of rapamycin include induction of Tregs and T-cell anergy as well as abrogation of effector T-cell differentiation [171]. This is distinct from that of calcineurin inhibitors, which work by blocking TCR signaling and subsequent NF-AT activation. In fact, cyclosporine A opposes rapamycin-induced tolerance [52]. In this regard, sirolimus, rather than FK506, expands liver FOXP3$^+$ Tregs in the setting of liver transplantation [172]. Likewise, a sirolimus-based, but calcineurin inhibitor-free, regimen has been developed to promote stable mixed donor chimerism after nonmyeloablative allogeneic hematopoietic stem cell transplantation for sickle cell disease [173]. As infections are an unavoidable complication of immunosuppression, what renders rapamycin more appealing than other immunosuppressives includes its potential to enhance antiviral memory response [110] as well as Th1- and Th17-driven antimicrobial response presumably via induction of proinflammatory phenotypes in mDCs [148].

mTOR blockade has emerged as a novel therapeutic approach in autoimmune diseases. SLE T cells exhibit increased mTOR activity [174] and mTOR blockade with rapamycin is therapeutic in SLE [175,176]. We have recently reported that mTORC1, but not mTORC2, was increased in lupus T cells and that mTORC1 mediated the expansion of Th17 cells and IL-4-producing CD4$^-$CD8$^-$ cells and contraction of CD4$^+$CD25$^+$FOXP3$^+$ Tregs in SLE [176,177]. A recent study suggests potential efficacy of rapamycin in ameliorating thrombotic microangiopathy in the setting of antiphospholipid syndrome nephropathy [178]. Rapamycin expanded Tregs from patients with type I diabetes *in vitro* [179] and improved β-cell function [180]. Rapamycin may also be therapeutic in steroid-refractory autoimmune hepatitis [181].

While aforementioned observations in clinical and translational studies are all encouraging, numerous questions remain to be answered before applying mTOR blockade to the treatment of immunological disorders. As discussed earlier, mTOR blockade appears to pose reciprocal effects on the adaptive immune system; T and B cells, versus the innate immune system, in particular mDCs. Accordingly, the net outcome of mTOR blockade likely depends on what cell types are mainly driving the inflammatory pathology. Thus, it is imperative to define the effects of mTOR blockade in a disease-entity, disease-stage, and organ-manifestation-specific manner. Whereas Treg is likely an important target of mTOR blockade, the benefits of mTOR blockade in enhancing Treg proliferation and function appear to depend on the timing and duration of mTOR blockade, as exemplified when transient mTOR blockade, but not chronic mTOR blockade or genetic mTOR ablation, allowed Treg to proliferate and acquire suppressive function [25]. Thus, it is tempting to clarify whether transient versus chronic mTOR inhibition confers differential impact on human autoimmune diseases by comparing the benefits of intermittent pulse versus daily oral rapamycin.

References

[1] Vezina C, Kudelski A, Sehgal SN. Rapamycin (AY-22,989), a new antifungal antibiotic. I. Taxonomy of the producing streptomycete and isolation of the active principle. J Antibiot (Tokyo) 1975;28(10):721–6.

[2] Sabatini DM. mTOR and cancer: insights into a complex relationship. Nat Rev Cancer 2006;6(9):729–34.

[3] Harding MW, Galat A, Uehling DE, Schreiber SL. A receptor for the immunosuppressant FK506 is a cis-trans peptidyl-prolyl isomerase. Nature 1989;341(6244):758–60.

[4] Guertin DA, Sabatini DM. The pharmacology of mTOR inhibition. Sci Signal 2009;2(67):24.

[5] Zoncu R, Efeyan A, Sabatini DM. mTOR: from growth signal integration to cancer, diabetes and ageing. Nat Rev Mol Cell Biol 2011;12(1):21–35.

[6] Laplante M, Sabatini DM. mTOR signaling at a glance. J Cell Sci 2009;122(Pt 20):3589–94.

[7] Abraham RT, Wiederrecht GJ. Immunopharmacology of rapamycin. Annu Rev Immunol 1996;14:483–510.

[8] Sarbassov DD, Ali SM, Sengupta S, Sheen JH, Hsu PP, Bagley AF, et al. Prolonged rapamycin treatment inhibits mTORC2 assembly and Akt/PKB. Mol Cell 2006;22(2):159–68.

[9] Sengupta S, Peterson TR, Sabatini DM. Regulation of the mTOR complex 1 pathway by nutrients, growth factors, and stress. Mol Cell 2010;40(2):310–22.

[10] Yecies JL, Manning BD. Transcriptional control of cellular metabolism by mTOR signaling. Cancer Res 2011;71(8):2815–20.

[11] Caron E, Ghosh S, Matsuoka Y, Ashton-Beaucage D, Therrien M, Lemieux S, et al. A comprehensive map of the mTOR signaling network. Mol Syst Biol 2010;6:453.

[12] Yamagata K, Sanders LK, Kaufmann WE, Yee W, Barnes CA, Nathans D, et al. Rheb, a growth factor- and synaptic activity-regulated gene, encodes a novel Ras-related protein. J Biol Chem 1994;269(23):16333–9.

[13] Yee WM, Worley PF. Rheb interacts with Raf-1 kinase and may function to integrate growth factor- and protein kinase A-dependent signals. Mol Cell Biol 1997;17(2):921–33.

[14] Saucedo LJ, Gao X, Chiarelli DA, Li L, Pan D, Edgar BA. Rheb promotes cell growth as a component of the insulin/TOR signalling network. Nat Cell Biol 2003;5(6):566–71.

[15] Ciofani M, Zuniga-Pflucker JC. Notch promotes survival of pre-T cells at the beta-selection checkpoint by regulating cellular metabolism. Nat Immunol 2005;6(9):881–8.

[16] Powell JD, Delgoffe GM. The mammalian target of rapamycin: linking T cell differentiation, function, and metabolism. Immunity 2010;33(3):301–11.

[17] Colombetti S, Basso V, Mueller DL, Mondino A. Prolonged TCR/CD28 engagement drives IL-2-independent T cell clonal expansion through signaling mediated by the mammalian target of rapamycin. J Immunol 2006;176(5):2730–8.

[18] Francisco LM, Salinas VH, Brown KE, Vanguri VK, Freeman GJ, Kuchroo VK, et al. PD-L1 regulates the development, maintenance, and function of induced regulatory T cells. J Exp Med 2009;206(13):3015–29.

[19] Abraham RT. Mammalian target of rapamycin: immunosuppressive drugs uncover a novel pathway of cytokine receptor signaling. Curr Opin Immunol 1998;10(3):330–6.

[20] Stephenson LM, Park DS, Mora AL, Goenka S, Boothby M. Sequence motifs in IL-4R alpha mediating cell-cycle progression of primary lymphocytes. J Immunol 2005;175(8):5178–85.

[21] Wofford JA, Wieman HL, Jacobs SR, Zhao Y, Rathmell JC. IL-7 promotes Glut1 trafficking and glucose uptake via STAT5-mediated activation of Akt to support T-cell survival. Blood 2008;111(4):2101–11.

[22] Rathmell JC, Farkash EA, Gao W, Thompson CB. IL-7 enhances the survival and maintains the size of naive T cells. J Immunol 2001;167(12):6869–76.

[23] Rao RR, Li Q, Odunsi K, Shrikant PA. The mTOR kinase determines effector versus memory CD8+ T cell fate by regulating the expression of transcription factors T-bet and Eomesodermin. Immunity 2010;32(1):67–78.

[24] Gulen MF, Kang Z, Bulek K, Youzhong W, Kim TW, Chen Y, et al. The receptor SIGIRR suppresses Th17 cell proliferation via inhibition of the interleukin-1 receptor pathway and mTOR kinase activation. Immunity 2010;32(1):54–66.

[25] Procaccini C, De Rosa V, Galgani M, Abanni L, Cali G, Porcellini A, et al. An oscillatory switch in mTOR kinase activity sets regulatory T cell responsiveness. Immunity 2010;33(6):929–41.

[26] Tamas P, Hawley SA, Clarke RG, Mustard KJ, Green K, Hardie DG, et al. Regulation of the energy sensor AMP-activated protein kinase by antigen receptor and Ca2+ in T lymphocytes. J Exp Med 2006;203(7):1665–70.

[27] Motoshima H, Goldstein BJ, Igata M, Araki E. AMPK and cell proliferation—AMPK as a therapeutic target for atherosclerosis and cancer. J Physiol 2006;574(Pt 1):63–71.

[28] Inoki K, Ouyang H, Zhu T, Lindvall C, Wang Y, Zhang X, et al. TSC2 integrates Wnt and energy signals via a coordinated phosphorylation by AMPK and GSK3 to regulate cell growth. Cell 2006;126(5):955–68.

[29] Brugarolas J, Lei K, Hurley RL, Manning BD, Reiling JH, Hafen E, et al. Regulation of mTOR function in response to hypoxia by REDD1 and the TSC1/TSC2 tumor suppressor complex. Genes Dev 2004;18(23):2893–904.

[30] Sancak Y, Peterson TR, Shaul YD, Lindquist RA, Thoreen CC, Bar-Peled L, et al. The Rag GTPases bind raptor and mediate amino acid signaling to mTORC1. Science 2008;320(5882):1496–501.

[31] Cobbold SP, Adams E, Farquhar CA, Nolan KF, Howie D, Lui KO, et al. Infectious tolerance via the consumption of essential amino acids and mTOR signaling. Proc Natl Acad Sci USA 2009;106(29):12055–60.

[32] Zinzalla V, Stracka D, Oppliger W, Hall MN. Activation of mTORC2 by association with the ribosome. Cell 2011;144(5):757–68.

[33] Oh WJ, Wu CC, Kim SJ, Facchinetti V, Julien LA, Finlan M, et al. mTORC2 can associate with ribosomes to promote cotranslational phosphorylation and stability of nascent Akt polypeptide. EMBO J 2010;29(23):3939–51.

[34] Saci A, Cantley LC, Carpenter CL. Rac1 regulates the activity of mTORC1 and mTORC2 and controls cellular size. Mol Cell 2011;42(1):50–61.

[35] Ma XM, Blenis J. Molecular mechanisms of mTOR-mediated translational control. Nat Rev Mol Cell Biol 2009;10(5):307–18.

[36] Duvel K, Yecies JL, Menon S, Raman P, Lipovsky AI, Souza AL, et al. Activation of a metabolic gene regulatory network downstream of mTOR complex 1. Mol Cell 2010;39(2):171–83.

[37] Yu L, McPhee CK, Zheng L, Mardones GA, Rong Y, Peng J, et al. Termination of autophagy and reformation of lysosomes regulated by mTOR. Nature 2010;465(7300):942–6.

[38] Cunningham JT, Rodgers JT, Arlow DH, Vazquez F, Mootha VK, Puigserver P. mTOR controls mitochondrial oxidative function through a YY1-PGC-1alpha transcriptional complex. Nature 2007;450(7170):736–40.

[39] Delgoffe GM, Pollizzi KN, Waickman AT, Heikamp E, Meyers DJ, Horton MR, et al. The kinase mTOR regulates the differentiation of helper T cells through the selective activation of signaling by mTORC1 and mTORC2. Nat Immunol 2011;12(4):295–303.

[40] Guertin DA, Stevens DM, Thoreen CC, Burds AA, Kalaany NY, Moffat J, et al. Ablation in mice of the mTORC components

raptor, rictor, or mLST8 reveals that mTORC2 is required for signaling to Akt-FOXO and PKCalpha, but not S6K1. Dev Cell 2006;11(6):859−71.

[41] Lee K, Gudapati P, Dragovic S, Spencer C, Joyce S, Killeen N, et al. Mammalian target of rapamycin protein complex 2 regulates differentiation of Th1 and Th2 cell subsets via distinct signaling pathways. Immunity 2010;32(6):743−53.

[42] Garcia-Martinez JM, Alessi DR. mTOR complex 2 (mTORC2) controls hydrophobic motif phosphorylation and activation of serum- and glucocorticoid-induced protein kinase 1 (SGK1). Biochem J 2008;416(3):375−85.

[43] Brunet A, Park J, Tran H, Hu LS, Hemmings BA, Greenberg ME. Protein kinase SGK mediates survival signals by phosphorylating the forkhead transcription factor FKHRL1 (FOXO3a). Mol Cell Biol 2001;21(3):952−65.

[44] Ikenoue T, Inoki K, Yang Q, Zhou X, Guan KL. Essential function of TORC2 in PKC and Akt turn motif phosphorylation, maturation and signalling. EMBO J 2008;27(14):1919−31.

[45] Yang K, Neale G, Green DR, He W, Chi H. The tumor suppressor Tsc1 enforces quiescence of naive T cells to promote immune homeostasis and function. Nat Immunol 2011;12(9):888−97.

[46] Wu Q, Liu Y, Chen C, Ikenoue T, Qiao Y, Li CS, et al. The tuberous sclerosis complex-mammalian target of rapamycin pathway maintains the quiescence and survival of naive T cells. J Immunol 2011;187(3):1106−12.

[47] O'Brien TF, Gorentla BK, Xie D, Srivatsan S, McLeod IX, He YW, et al. Regulation of T-cell survival and mitochondrial homeostasis by TSC1. Eur J Immunol 2011;41(11):3361−70.

[48] Powell JD, Zheng Y. Dissecting the mechanism of T-cell anergy with immunophilin ligands. Curr Opin Investig Drugs 2006;7(11):1002−7.

[49] Safford M, Collins S, Lutz MA, Allen A, Huang CT, Kowalski J, et al. Egr-2 and Egr-3 are negative regulators of T cell activation. Nat Immunol 2005;6(5):472−80.

[50] Marcais A, Blevins R, Graumann J, Feytout A, Dharmalingam G, Carroll T, et al. microRNA-mediated regulation of mTOR complex components facilitates discrimination between activation and anergy in CD4 T cells. J Exp Med 2014;211(11):2281−95.

[51] Allen A, Zheng Y, Gardner L, Safford M, Horton MR, Powell JD. The novel cyclophilin binding compound, sanglifehrin A, disassociates G1 cell cycle arrest from tolerance induction. J Immunol 2004;172(8):4797−803.

[52] Powell JD, Lerner CG, Schwartz RH. Inhibition of cell cycle progression by rapamycin induces T cell clonal anergy even in the presence of costimulation. J Immunol 1999;162(5):2775−84.

[53] Vanasek TL, Khoruts A, Zell T, Mueller DL. Antagonistic roles for CTLA-4 and the mammalian target of rapamycin in the regulation of clonal anergy: enhanced cell cycle progression promotes recall antigen responsiveness. J Immunol 2001;167(10):5636−44.

[54] Buckley AF, Kuo CT, Leiden JM. Transcription factor LKLF is sufficient to program T cell quiescence via a c-Myc-dependent pathway. Nat Immunol 2001;2(8):698−704.

[55] Berger M, Krebs P, Crozat K, Li X, Croker BA, Siggs OM, et al. An Slfn2 mutation causes lymphoid and myeloid immunodeficiency due to loss of immune cell quiescence. Nat Immunol 2010;11(4):335−43.

[56] Modiano JF, Johnson LD, Bellgrau D. Negative regulators in homeostasis of naive peripheral T cells. Immunol Res 2008;41(2):137−53.

[57] Deberardinis RJ, Sayed N, Ditsworth D, Thompson CB. Brick by brick: metabolism and tumor cell growth. Curr Opin Genet Dev 2008;18(1):54−61.

[58] Jones RG, Thompson CB. Revving the engine: signal transduction fuels T cell activation. Immunity 2007;27(2):173−8.

[59] Frauwirth KA, Riley JL, Harris MH, Parry RV, Rathmell JC, Plas DR, et al. The CD28 signaling pathway regulates glucose metabolism. Immunity 2002;16(6):769−77.

[60] Frauwirth KA, Thompson CB. Activation and inhibition of lymphocytes by costimulation. J Clin Invest 2002;109(3):295−9.

[61] Frauwirth KA, Thompson CB. Regulation of T lymphocyte metabolism. J Immunol 2004;172(8):4661−5.

[62] Jhun BS, Oh YT, Lee JY, Kong Y, Yoon KS, Kim SS, et al. AICAR suppresses IL-2 expression through inhibition of GSK-3 phosphorylation and NF-AT activation in Jurkat T cells. Biochem Biophys Res Commun 2005;332(2):339−46.

[63] Nath N, Giri S, Prasad R, Salem ML, Singh AK, Singh I. 5-Aminoimidazole-4-carboxamide ribonucleoside: a novel immunomodulator with therapeutic efficacy in experimental autoimmune encephalomyelitis. J Immunol 2005;175(1):566−74.

[64] Zheng Y, Delgoffe GM, Meyer CF, Chan W, Powell JD. Anergic T cells are metabolically anergic. J Immunol 2009;183(10):6095−101.

[65] Hidayat S, Yoshino K, Tokunaga C, Hara K, Matsuo M, Yonezawa K. Inhibition of amino acid-mTOR signaling by a leucine derivative induces G1 arrest in Jurkat cells. Biochem Biophys Res Commun 2003;301(2):417−23.

[66] Delgoffe GM, Kole TP, Zheng Y, Zarek PE, Matthews KL, Xiao B, et al. The mTOR kinase differentially regulates effector and regulatory T cell lineage commitment. Immunity 2009;30(6):832−44.

[67] Yang K, Shrestha S, Zeng H, Karmaus PW, Neale G, Vogel P, et al. T cell exit from quiescence and differentiation into Th2 cells depend on Raptor-mTORC1-mediated metabolic reprogramming. Immunity 2013;39(6):1043−56.

[68] Nakaya M, Xiao Y, Zhou X, Chang JH, Chang M, Cheng X, et al. Inflammatory T cell responses rely on amino acid transporter ASCT2 facilitation of glutamine uptake and mTORC1 kinase activation. Immunity 2014;40(5):692−705.

[69] Busch S, Renaud SJ, Schleussner E, Graham CH, Markert UR. mTOR mediates human trophoblast invasion through regulation of matrix-remodeling enzymes and is associated with serine phosphorylation of STAT3. Exp Cell Res 2009;315(10):1724−33.

[70] Zhou J, Wulfkuhle J, Zhang H, Gu P, Yang Y, Deng J, et al. Activation of the PTEN/mTOR/STAT3 pathway in breast cancer stem-like cells is required for viability and maintenance. Proc Natl Acad Sci USA 2007;104(41):16158−63.

[71] Wang B, Xiao Z, Chen B, Han J, Gao Y, Zhang J, et al. Nogo-66 promotes the differentiation of neural progenitors into astroglial lineage cells through mTOR-STAT3 pathway. PLoS One 2008;3(3):e1856.

[72] Yamamoto K, Yamaguchi M, Miyasaka N, Miura O. SOCS-3 inhibits IL-12-induced STAT4 activation by binding through its SH2 domain to the STAT4 docking site in the IL-12 receptor beta2 subunit. Biochem Biophys Res Commun 2003;310(4):1188−93.

[73] Yu CR, Mahdi RM, Ebong S, Vistica BP, Gery I, Egwuagu CE. Suppressor of cytokine signaling 3 regulates proliferation and activation of T-helper cells. J Biol Chem 2003;278(32):29752−9.

[74] Shi LZ, Wang R, Huang G, Vogel P, Neale G, Green DR, et al. HIF1alpha-dependent glycolytic pathway orchestrates a metabolic checkpoint for the differentiation of TH17 and Treg cells. J Exp Med 2011;208(7):1367−76.

[75] Sundrud MS, Koralov SB, Feuerer M, Calado DP, Kozhaya AE, Rhule-Smith A, et al. Halofuginone inhibits TH17 cell differentiation by activating the amino acid starvation response. Science 2009;324(5932):1334−8.

[76] Cui G, Qin X, Wu L, Zhang Y, Sheng X, Yu Q, et al. Liver X receptor (LXR) mediates negative regulation of mouse and human Th17 differentiation. J Clin Invest 2011;121(2):658−70.

[77] Heikamp EB, Patel CH, Collins S, Waickman A, Oh MH, Sun IH, et al. The AGC kinase SGK1 regulates TH1 and TH2 differentiation downstream of the mTORC2 complex. Nat Immunol 2014;15(5):457−64.

[78] Crellin NK, Garcia RV, Levings MK. Altered activation of AKT is required for the suppressive function of human CD4+CD25+ T regulatory cells. Blood 2007;109(5):2014−22.

[79] Merkenschlager M, von Boehmer H. PI3 kinase signalling blocks Foxp3 expression by sequestering Foxo factors. J Exp Med 2010;207(7):1347−50.

[80] Chen H, Zhang L, Zhang H, Xiao Y, Shao L, Li H, et al. Disruption of TSC1/2 signaling complex reveals a checkpoint governing thymic CD4+ CD25+ Foxp3+ regulatory T-cell development in mice. FASEB J 2013;27(10):3979−90.

[81] Battaglia M, Stabilini A, Roncarolo MG. Rapamycin selectively expands CD4+CD25+FoxP3+ regulatory T cells. Blood 2005;105(12):4743−8.

[82] Kang J, Huddleston SJ, Fraser JM, Khoruts A. De novo induction of antigen-specific CD4+CD25+Foxp3+ regulatory T cells in vivo following systemic antigen administration accompanied by blockade of mTOR. J Leukoc Biol 2008;83(5):1230−9.

[83] Kopf H, de la Rosa GM, Howard OM, Chen X. Rapamycin inhibits differentiation of Th17 cells and promotes generation of FoxP3+ T regulatory cells. Int Immunopharmacol 2007;7 (13):1819−24.

[84] Zhou L, Lopes JE, Chong MM, Ivanov II, Min R, Victora GD, et al. TGF-beta-induced Foxp3 inhibits T(H)17 cell differentiation by antagonizing RORgammat function. Nature 2008;453 (7192):236−40.

[85] Marie JC, Letterio JJ, Gavin M, Rudensky AY. TGF-beta1 maintains suppressor function and Foxp3 expression in CD4+CD25+ regulatory T cells. J Exp Med 2005;201(7):1061−7.

[86] Gabrysova L, Christensen JR, Wu X, Kissenpfennig A, Malissen B, O'Garra A. Integrated T-cell receptor and costimulatory signals determine TGF-beta-dependent differentiation and maintenance of Foxp3+ regulatory T cells. Eur J Immunol 2011;41(5):1242−8.

[87] Haxhinasto S, Mathis D, Benoist C. The AKT-mTOR axis regulates de novo differentiation of CD4+Foxp3+ cells. J Exp Med 2008;205(3):565−74.

[88] Delgoffe GM, Woo SR, Turnis ME, Gravano DM, Guy C, Overacre AE, et al. Stability and function of regulatory T cells is maintained by a neuropilin-1-semaphorin-4a axis. Nature 2013;501(7466):252−6.

[89] Henderson JG, Opejin A, Jones A, Gross C, Hawiger D. CD5 instructs extrathymic regulatory T cell development in response to self and tolerizing antigens. Immunity 2015;42(3):471−83.

[90] Suzuki A, Yamaguchi MT, Ohteki T, Sasaki T, Kaisho T, Kimura Y, et al. T cell-specific loss of Pten leads to defects in central and peripheral tolerance. Immunity 2001;14(5):523−34.

[91] Sauer S, Bruno L, Hertweck A, Finlay D, Leleu M, Spivakov M, et al. T cell receptor signaling controls Foxp3 expression via PI3K, Akt, and mTOR. Proc Natl Acad Sci USA 2008;105 (22):7797−802.

[92] Shrestha S, Yang K, Guy C, Vogel P, Neale G, Chi H. Treg cells require the phosphatase PTEN to restrain TH1 and TFH cell responses. Nat Immunol 2015;16(2):178−87.

[93] Huynh A, DuPage M, Priyadharshini B, Sage PT, Quiros J, Borges CM, et al. Control of PI(3) kinase in Treg cells maintains homeostasis and lineage stability. Nat Immunol 2015;16(2):188−96.

[94] Han JM, Patterson SJ, Speck M, Ehses JA, Levings MK. Insulin inhibits IL-10-mediated regulatory T cell function: implications for obesity. J Immunol 2014;192(2):623−9.

[95] Zeiser R, Leveson-Gower DB, Zambricki EA, Kambham N, Beilhack A, Loh J, et al. Differential impact of mammalian target of rapamycin inhibition on CD4+CD25+Foxp3+ regulatory T cells compared with conventional CD4+ T cells. Blood 2008;111(1):453−62.

[96] Strauss L, Czystowska M, Szajnik M, Mandapathil M, Whiteside TL. Differential responses of human regulatory T cells (Treg) and effector T cells to rapamycin. PLoS One 2009;4(6):e5994.

[97] Basu S, Golovina T, Mikheeva T, June CH, Riley JL. Cutting edge: Foxp3-mediated induction of pim 2 allows human T regulatory cells to preferentially expand in rapamycin. J Immunol 2008;180(9):5794−8.

[98] Fox CJ, Hammerman PS, Thompson CB. The Pim kinases control rapamycin-resistant T cell survival and activation. J Exp Med 2005;201(2):259−66.

[99] Dang EV, Barbi J, Yang HY, Jinasena D, Yu H, Zheng Y, et al. Control of T(H)17/T(reg) balance by hypoxia-inducible factor 1. Cell 2011;146(5):772−84.

[100] Zeng H, Yang K, Cloer C, Neale G, Vogel P, Chi H. mTORC1 couples immune signals and metabolic programming to establish T(reg)-cell function. Nature 2013;499(7459):485−90.

[101] Yamada T, Park CS, Mamonkin M, Lacorazza HD. Transcription factor ELF4 controls the proliferation and homing of CD8+ T cells via the Kruppel-like factors KLF4 and KLF2. Nat Immunol 2009;10(6):618−26.

[102] Yamada T, Gierach K, Lee PH, Wang X, Lacorazza HD. Cutting edge: expression of the transcription factor E74-like factor 4 is regulated by the mammalian target of rapamycin pathway in CD8+ T cells. J Immunol 2010;185(7):3824−8.

[103] Pearce EL. Metabolism in T cell activation and differentiation. Curr Opin Immunol 2010;22(3):314−20.

[104] Salmond RJ, Emery J, Okkenhaug K, Zamoyska R. MAPK, phosphatidylinositol 3-kinase, and mammalian target of rapamycin pathways converge at the level of ribosomal protein S6 phosphorylation to control metabolic signaling in CD8 T cells. J Immunol 2009;183(11):7388−97.

[105] Yao S, Buzo BF, Pham D, Jiang L, Taparowsky EJ, Kaplan MH, et al. Interferon regulatory factor 4 sustains CD8(+) T cell expansion and effector differentiation. Immunity 2013;39(5): 833−45.

[106] Henson SM, Lanna A, Riddell NE, Franzese O, Macaulay R, Griffiths SJ, et al. p38 signaling inhibits mTORC1-independent autophagy in senescent human CD8(+) T cells. J Clin Invest 2014;124(9):4004−16.

[107] Finlay D, Cantrell DA. Metabolism, migration and memory in cytotoxic T cells. Nat Rev Immunol 2011;11(2):109−17.

[108] Joshi NS, Kaech SM. Effector CD8 T cell development: a balancing act between memory cell potential and terminal differentiation. J Immunol 2008;180(3):1309−15.

[109] Takemoto N, Intlekofer AM, Northrup JT, Wherry EJ, Reiner SL. Cutting edge: IL-12 inversely regulates T-bet and eomesodermin expression during pathogen-induced CD8+ T cell differentiation. J Immunol 2006;177(11):7515−19.

[110] Araki K, Turner AP, Shaffer VO, Gangappa S, Keller SA, Bachmann MF, et al. mTOR regulates memory CD8 T-cell differentiation. Nature 2009;460(7251):108−12.

[111] Pearce EL, Walsh MC, Cejas PJ, Harms GM, Shen H, Wang LS, et al. Enhancing CD8 T-cell memory by modulating fatty acid metabolism. Nature 2009;460(7251):103−7.

[112] Pedicord VA, Cross JR, Montalvo-Ortiz W, Miller ML, Allison JP. Friends not foes: CTLA-4 blockade and mTOR inhibition cooperate during CD8+ T cell priming to promote memory formation and metabolic readiness. J Immunol 2015;194(5):2089−98.

[113] Stoycheva D, Deiser K, Starck L, Nishanth G, Schluter D, Uckert W, et al. IFN-gamma regulates CD8+ memory T cell differentiation and survival in response to weak, but not strong, TCR signals. J Immunol 2015;194(2):553−9.

[114] Li Q, Rao RR, Araki K, Pollizzi K, Odunsi K, Powell JD, et al. A central role for mTOR kinase in homeostatic proliferation induced CD8+ T cell memory and tumor immunity. Immunity 2011;34(4):541–53.

[115] Hand TW, Cui W, Jung YW, Sefik E, Joshi NS, Chandele A, et al. Differential effects of STAT5 and PI3K/AKT signaling on effector and memory CD8 T-cell survival. Proc Natl Acad Sci USA 2010;107(38):16601–6.

[116] Sowell RT, Rogozinska M, Nelson CE, Vezys V, Marzo AL. Cutting edge: generation of effector cells that localize to mucosal tissues and form resident memory CD8 T cells is controlled by mTOR. J Immunol 2014;193(5):2067–71.

[117] Gubser PM, Bantug GR, Razik L, Fischer M, Dimeloe S, Hoenger G, et al. Rapid effector function of memory CD8+ T cells requires an immediate-early glycolytic switch. Nat Immunol 2013;14(10):1064–72.

[118] Staron MM, Gray SM, Marshall HD, Parish IA, Chen JH, Perry CJ, et al. The transcription factor FoxO1 sustains expression of the inhibitory receptor PD-1 and survival of antiviral CD8(+) T cells during chronic infection. Immunity 2014;41(5):802–14.

[119] Turner AP, Shaffer VO, Araki K, Martens C, Turner PL, Gangappa S, et al. Sirolimus enhances the magnitude and quality of viral-specific CD8+ T-cell responses to vaccinia virus vaccination in rhesus macaques. Am J Transplant 2011;11(3):613–18.

[120] Storni T, Ruedl C, Schwarz K, Schwendener RA, Renner WA, Bachmann MF. Nonmethylated CG motifs packaged into virus-like particles induce protective cytotoxic T cell responses in the absence of systemic side effects. J Immunol 2004;172(3):1777–85.

[121] Ferrer IR, Wagener ME, Robertson JM, Turner AP, Araki K, Ahmed R, et al. Cutting edge: rapamycin augments pathogen-specific but not graft-reactive CD8+ T cell responses. J Immunol 2010;185(4):2004–8.

[122] Finlay D, Cantrell D. Phosphoinositide 3-kinase and the mammalian target of rapamycin pathways control T cell migration. Ann NY Acad Sci 2010;1183:149–57.

[123] Sinclair LV, Finlay D, Feijoo C, Cornish GH, Gray A, Ager A, et al. Phosphatidylinositol-3-OH kinase and nutrient-sensing mTOR pathways control T lymphocyte trafficking. Nat Immunol 2008;9(5):513–21.

[124] Fabre S, Carrette F, Chen J, Lang V, Semichon M, Denoyelle C, et al. FOXO1 regulates L-Selectin and a network of human T cell homing molecules downstream of phosphatidylinositol 3-kinase. J Immunol 2008;181(5):2980–9.

[125] Kerdiles YM, Beisner DR, Tinoco R, Dejean AS, Castrillon DH, DePinho RA, et al. Foxo1 links homing and survival of naive T cells by regulating L-selectin, CCR7 and interleukin 7 receptor. Nat Immunol 2009;10(2):176–84.

[126] Macintyre AN, Finlay D, Preston G, Sinclair LV, Waugh CM, Tamas P, et al. Protein kinase B controls transcriptional programs that direct cytotoxic T cell fate but is dispensable for T cell metabolism. Immunity 2011;34(2):224–36.

[127] Carlson CM, Endrizzi BT, Wu J, Ding X, Weinreich MA, Walsh ER, et al. Kruppel-like factor 2 regulates thymocyte and T-cell migration. Nature 2006;442(7100):299–302.

[128] Matloubian M, Lo CG, Cinamon G, Lesneski MJ, Xu Y, Brinkmann V, et al. Lymphocyte egress from thymus and peripheral lymphoid organs is dependent on S1P receptor 1. Nature 2004;427(6972):355–60.

[129] Liu G, Burns S, Huang G, Boyd K, Proia RL, Flavell RA, et al. The receptor S1P1 overrides regulatory T cell-mediated immune suppression through Akt-mTOR. Nat Immunol 2009;10(7):769–77.

[130] Liu G, Yang K, Burns S, Shrestha S, Chi H. The S1P(1)-mTOR axis directs the reciprocal differentiation of T(H)1 and T(reg) cells. Nat Immunol 2010;11(11):1047–56.

[131] Zhang S, Readinger JA, DuBois W, Janka-Junttila M, Robinson R, Pruitt M, et al. Constitutive reductions in mTOR alter cell size, immune cell development, and antibody production. Blood 2011;117(4):1228–38.

[132] Keating R, Hertz T, Wehenkel M, Harris TL, Edwards BA, McClaren JL, et al. The kinase mTOR modulates the antibody response to provide cross-protective immunity to lethal infection with influenza virus. Nat Immunol 2013;14(12):1266–76.

[133] Benhamron S, Tirosh B. Direct activation of mTOR in B lymphocytes confers impairment in B-cell maturation andloss of marginal zone B cells. Eur J Immunol 2011;41(8):2390–6.

[134] Plas DR, Rathmell JC, Thompson CB. Homeostatic control of lymphocyte survival: potential origins and implications. Nat Immunol 2002;3(6):515–21.

[135] Woodland RT, Fox CJ, Schmidt MR, Hammerman PS, Opferman JT, Korsmeyer SJ, et al. Multiple signaling pathways promote B lymphocyte stimulator dependent B-cell growth and survival. Blood 2008;111(2):750–60.

[136] Kojima H, Kobayashi A, Sakurai D, Kanno Y, Hase H, Takahashi R, et al. Differentiation stage-specific requirement in hypoxia-inducible factor-1alpha-regulated glycolytic pathway during murine B cell development in bone marrow. J Immunol 2010;184(1):154–63.

[137] Lazorchak AS, Liu D, Facchinetti V, Di Lorenzo A, Sessa WC, Schatz DG, et al. Sin1-mTORC2 suppresses rag and il7r gene expression through Akt2 in B cells. Mol Cell 2010;39(3):433–43.

[138] Llorian M, Stamataki Z, Hill S, Turner M, Martensson IL. The PI3K p110delta is required for down-regulation of RAG expression in immature B cells. J Immunol 2007;178(4):1981–5.

[139] Donahue AC, Fruman DA. Proliferation and survival of activated B cells requires sustained antigen receptor engagement and phosphoinositide 3-kinase activation. J Immunol 2003;170(12):5851–60.

[140] Ono M, Bolland S, Tempst P, Ravetch JV. Role of the inositol phosphatase SHIP in negative regulation of the immune system by the receptor Fc(gamma)RIIB. Nature 1996;383(6597):263–6.

[141] Sakata A, Kuwahara K, Ohmura T, Inui S, Sakaguchi N. Involvement of a rapamycin-sensitive pathway in CD40-mediated activation of murine B cells in vitro. Immunol Lett 1999;68(2–3):301–9.

[142] Donahue AC, Fruman DA. Distinct signaling mechanisms activate the target of rapamycin in response to different B-cell stimuli. Eur J Immunol 2007;37(10):2923–36.

[143] Colonna M, Trinchieri G, Liu YJ. Plasmacytoid dendritic cells in immunity. Nat Immunol 2004;5(12):1219–26.

[144] Kawai T, Akira S. Innate immune recognition of viral infection. Nat Immunol 2006;7(2):131–7.

[145] Cao W, Manicassamy S, Tang H, Kasturi SP, Pirani A, Murthy N, et al. Toll-like receptor-mediated induction of type I interferon in plasmacytoid dendritic cells requires the rapamycin-sensitive PI(3)K-mTOR-p70S6K pathway. Nat Immunol 2008;9(10):1157–64.

[146] Colina R, Costa-Mattioli M, Dowling RJ, Jaramillo M, Tai LH, Breitbach CJ, et al. Translational control of the innate immune response through IRF-7. Nature 2008;452(7185):323–8.

[147] Fekete T, Pazmandi K, Szabo A, Bacsi A, Koncz G, Rajnavolgyi E. The antiviral immune response in human conventional dendritic cells is controlled by the mammalian target of rapamycin. J Leukoc Biol 2014;96(4):579–89.

[148] Weichhart T, Costantino G, Poglitsch M, Rosner M, Zeyda M, Stuhlmeier KM, et al. The TSC-mTOR signaling pathway regulates the innate inflammatory response. Immunity 2008;29(4):565–77.

[149] Zhu YP, Brown JR, Sag D, Zhang L, Suttles J. Adenosine 5′-monophosphate-activated protein kinase regulates

IL-10-mediated anti-inflammatory signaling pathways in macrophages. J Immunol 2015;194(2):584—94.

[150] Wang H, Brown J, Gao S, Liang S, Jotwani R, Zhou H, et al. The role of JAK-3 in regulating TLR-mediated inflammatory cytokine production in innate immune cells. J Immunol 2013;191(3):1164—74.

[151] Pan H, O'Brien TF, Wright G, Yang J, Shin J, Wright KL, et al. Critical role of the tumor suppressor tuberous sclerosis complex 1 in dendritic cell activation of CD4 T cells by promoting MHC class II expression via IRF4 and CIITA. J Immunol 2013;191(2):699—707.

[152] Ohtani M, Nagai S, Kondo S, Mizuno S, Nakamura K, Tanabe M, et al. Mammalian target of rapamycin and glycogen synthase kinase 3 differentially regulate lipopolysaccharide-induced interleukin-12 production in dendritic cells. Blood 2008;112 (3):635—43.

[153] Cheng SC, Quintin J, Cramer RA, Shepardson KM, Saeed S, Kumar V, et al. mTOR- and HIF-1alpha-mediated aerobic glycolysis as metabolic basis for trained immunity. Science 2014;345(6204):1250684.

[154] Steinman RM, Nussenzweig MC. Avoiding horror autotoxicus: the importance of dendritic cells in peripheral T cell tolerance. Proc Natl Acad Sci USA 2002;99(1):351—8.

[155] Hackstein H, Taner T, Logar AJ, Thomson AW. Rapamycin inhibits macropinocytosis and mannose receptor-mediated endocytosis by bone marrow-derived dendritic cells. Blood 2002;100(3):1084—7.

[156] McKenna HJ, Stocking KL, Miller RE, Brasel K, De Smedt T, Maraskovsky E, et al. Mice lacking flt3 ligand have deficient hematopoiesis affecting hematopoietic progenitor cells, dendritic cells, and natural killer cells. Blood 2000;95(11):3489—97.

[157] Sathaliyawala T, O'Gorman WE, Greter M, Bogunovic M, Konjufca V, Hou ZE, et al. Mammalian target of rapamycin controls dendritic cell development downstream of Flt3 ligand signaling. Immunity 2010;33(4):597—606.

[158] Hackstein H, Taner T, Zahorchak AF, Morelli AE, Logar AJ, Gessner A, et al. Rapamycin inhibits IL-4-induced dendritic cell maturation in vitro and dendritic cell mobilization and function in vivo. Blood 2003;101(11):4457—63.

[159] Taner T, Hackstein H, Wang Z, Morelli AE, Thomson AW. Rapamycin-treated, alloantigen-pulsed host dendritic cells induce ag-specific T cell regulation and prolong graft survival. Am J Transplant 2005;5(2):228—36.

[160] Turnquist HR, Raimondi G, Zahorchak AF, Fischer RT, Wang Z, Thomson AW. Rapamycin-conditioned dendritic cells are poor stimulators of allogeneic CD4+ T cells, but enrich for antigen-specific Foxp3+ T regulatory cells and promote organ transplant tolerance. J Immunol 2007;178(11):7018—31.

[161] Rosborough BR, Raich-Regue D, Matta BM, Lee K, Gan B, DePinho RA, et al. Murine dendritic cell rapamycin-resistant and rictor-independent mTOR controls IL-10, B7-H1, and regulatory T-cell induction. Blood 2013;121(18):3619—30.

[162] Reichardt W, Durr C, von Elverfeldt D, Juttner E, Gerlach UV, Yamada M, et al. Impact of mammalian target of rapamycin inhibition on lymphoid homing and tolerogenic function of nanoparticle-labeled dendritic cells following allogeneic hematopoietic cell transplantation. J Immunol 2008;181 (7):4770—9.

[163] Turnquist HR, Cardinal J, Macedo C, Rosborough BR, Sumpter TL, Geller DA, et al. mTOR and GSK-3 shape the CD4+ T-cell stimulatory and differentiation capacity of myeloid DCs after exposure to LPS. Blood 2010;115(23):4758—69.

[164] Donnelly RP, Loftus RM, Keating SE, Liou KT, Biron CA, Gardiner CM, et al. mTORC1-dependent metabolic

reprogramming is a prerequisite for NK cell effector function. J Immunol 2014;193(9):4477—84.

[165] Marcais A, Cherfils-Vicini J, Viant C, Degouve S, Viel S, Fenis A, et al. The metabolic checkpoint kinase mTOR is essential for IL-15 signaling during the development and activation of NK cells. Nat Immunol 2014;15(8):749—57.

[166] Yang M, Li D, Chang Z, Yang Z, Tian Z, Dong Z. PDK1 orchestrates early NK cell development through induction of E4BP4 expression and maintenance of IL-15 responsiveness. J Exp Med 2015;212(2):253—65.

[167] Park H, Tsang M, Iritani BM, Bevan MJ. Metabolic regulator Fnip1 is crucial for iNKT lymphocyte development. Proc Natl Acad Sci USA 2014;111(19):7066—71.

[168] Shin J, Wang S, Deng W, Wu J, Gao J, Zhong XP. Mechanistic target of rapamycin complex 1 is critical for invariant natural killer T-cell development and effector function. Proc Natl Acad Sci USA 2014;111(8):E776—83.

[169] Wei J, Yang K, Chi H. Cutting edge: discrete functions of mTOR signaling in invariant NKT cell development and NKT17 fate decision. J Immunol 2014;193(9):4297—301.

[170] Smrz D, Cruse G, Beaven MA, Kirshenbaum A, Metcalfe DD, Gilfillan AM. Rictor negatively regulates high-affinity receptors for IgE-induced mast cell degranulation. J Immunol 2014;193(12):5924—32.

[171] Waickman AT, Powell JD. mTOR, metabolism, and the regulation of T-cell differentiation and function. Immunol Rev 2012;249(1):43—58.

[172] Levitsky J, Mathew JM, Abecassis M, Tambur A, Leventhal J, Chandrasekaran D, et al. Systemic immunoregulatory and proteogenomic effects of tacrolimus to sirolimus conversion in liver transplant recipients. Hepatology 2013;57(1):239—48.

[173] Hsieh MM, Kang EM, Fitzhugh CD, Link MB, Bolan CD, Kurlander R, et al. Allogeneic hematopoietic stem-cell transplantation for sickle cell disease. N Engl J Med 2009;361 (24):2309—17.

[174] Fernandez DR, Telarico T, Bonilla E, Li Q, Banerjee S, Middleton FA, et al. Activation of mammalian target of rapamycin controls the loss of TCRzeta in lupus T cells through HRES-1/Rab4-regulated lysosomal degradation. J Immunol 2009;182(4):2063—73.

[175] Fernandez D, Bonilla E, Mirz N, Niland B, Perl A. Rapamycin reduces disease activity and normalizes T cell activation-induced calcium fluxing in patients with systemic lupus erythematosus. Arthritis Rheum 2006;54(9):2983—8.

[176] Lai ZW, Borsuk R, Shadakshari A, Yu J, Dawood M, Garcia R, et al. mTOR activation triggers IL-4 production and necrotic death of double-negative T cells in patients with systemic lupus eryhthematosus. J Immunol 2013;191(5):2236—46.

[177] Kato H, Perl A. Mechanistic target of rapamycin complex 1 expands Th17 and IL-4+ CD4-CD8- double-negative T cells and contracts regulatory T cells in systemic lupus erythematosus. J Immunol 2014;192(9):4134—44.

[178] Canaud G, Terzi F. Inhibition of the mTORC pathway in the antiphospholipid syndrome. N Engl J Med 2014;371(16):1554—5.

[179] Putnam AL, Brusko TM, Lee MR, Liu W, Szot GL, Ghosh T, et al. Expansion of human regulatory T-cells from patients with type 1 diabetes. Diabetes 2009;58(3):652—62.

[180] Piemonti L, Maffi P, Monti L, Lampasona V, Perseghin G, Magistretti P, et al. Beta cell function during rapamycin monotherapy in long-term type 1 diabetes. Diabetologia 2011;54 (2):433—9.

[181] Chatrath H, Allen L, Boyer TD. Use of sirolimus in the treatment of refractory autoimmune hepatitis. Am J Med 2014;127 (11):1128—31.

18

The Role of mTOR Inhibitors in Solid Organ Transplantation

Greg J. McKenna[1,2] *and Goran B.G. Klintmalm*[2,3]

[1]Simmons Transplant Institute, Baylor University Medical Center, Dallas, TX, USA [2]Texas A&M University College of Medicine, Dallas, TX, USA [3]Simmons Transplant Institute, Department of Surgery, Transplant Surgery, Baylor University Medical Center, Dallas, TX, USA

18.1 INTRODUCTION

Successful organ transplantation requires suppression of the immune system to prevent rejection of the transplanted allograft. The catalyst for organ transplantation came more than 35 years ago in 1978 with the development and use of the first calcineurin inhibitor (CNI), cyclosporine [1], for use as an immunosuppressant. This was the ingredient that made solid organ transplantation feasible, pushing organ rejection rates down below 20% and propelling the 1-year survival rates for liver transplantation from 30% up to 90%. However, these advances related to CNI-based immunosuppression came with a price, in the form of long-term morbidity and adverse effects, particularly nephrotoxicity and malignancies. CNI nephrotoxicity is a common story shared across all organ transplantation, with 97% of kidney allograft biopsies showing nephrotoxicity at 10 years [2,3] and up to 25% of liver transplant recipients [4] developing end-stage renal disease. The search for an immunosuppressant that might avoid CNI nephrotoxicity and malignancies has been the goal ever since.

It was the ongoing search for a less morbid immunosuppressant for organ transplantation that brought sirolimus (rapamycin), the first mammalian target of rapamycin (mTOR) inhibitor, to the forefront, allowing it initially to gain Food and Drug Administration (FDA) approval. Dr Suren Sehgal is considered the father of the mTOR inhibitors, as he and his group at Ayerst Labs in Montreal in 1972 first isolated and investigated a putative antifungal agent from a rare bacteria

(*Streptomyces hygroscopicus*) [5], found in soil samples from Easter Island. Dr Sehgal and his group named the antifungal rapamycin (derived from "Rapa Nui," the native Polynesian name for Easter Island). Dr Sehgal also sent the sample to the National Cancer Institute where the substance was found to have "fantastic activity" against solid tumors, however economic pressures led the management at Ayerst Labs to decide against further pursuing the compound.

In 1987 Dr Seghal, by then at the company Wyeth Ayerst, was still researching the abandoned drug and identified immunosuppressive effects with sirolimus [6,7], which ended its potential as an antifungal or antibiotic. However, with the advent of the CNI tacrolimus being developed as a new immunosuppressant for organ transplantation, researchers noted the structural similarities between sirolimus and tacrolimus [6,8]. This led two separate research groups—Dr Randall Morris's group at Stanford University [9] and Dr Roy Calne's group [10] at Cambridge—to independently study the immunosuppressive properties of sirolimus with an eye towards solid organ transplantation.

Like tacrolimus, sirolimus was found to bind the same key protein known as FKBP1A (FK506 binding protein-1A). However, unlike tacrolimus, this protein-bound complex did not inhibit calcineurin phosphatase [11], and therefore sirolimus was not a CNI. Because sirolimus was not a CNI, it was thought it might avoid the nephrotoxicity associated with CNI use, and animal studies verified that sirolimus lacked the nephrotoxicity seen with CNI use [12]. This made sirolimus a potentially new class of immunosuppression with a side effect profile distinct from the CNI currently in use.

In 1991, Joseph Heitman, Mike Hall, and Rao Movva at Sandoz Pharmaceuticals (now Novartis) in Basel Switzerland used molecular genetics studies in yeast to identify the molecular target of sirolimus, which they termed "TOR" (target of rapamycin). In 1994, the mammalian homolog of these yeast proteins was identified by Dr David M Sabatini and Dr Solomon Snyder at Johns Hopkins University, and also by Dr Robert Abraham and Dr Stuart Schreiber at Harvard University who coined the term "mTOR" for the protein.

The initial research into sirolimus and its interaction with mTOR were focused along its immunosuppressive properties. It was apparent early that sirolimus arrested the cell cycle at the G1/S interface, limiting cell division, particularly in proliferating lymphocytes [13]. Sirolimus did not interfere with the early events after T-cell activation in the manner of the CNI cyclosporine and tacrolimus [14], but rather sirolimus inhibited IL-2 (interleukin 2)-induced binding of transcription factors in the proliferating T-cell and inhibited cell cycle progression [13]. The ability to inhibit growth factor signaling was a characteristic feature of sirolimus, and it impacted both immune and nonimmune cells. The mechanistic pathway for sirolimus was initially studied in the context of its role as an immunosuppressant and mTOR emerged as a powerful non-nephrogenic source of immunosuppression. But it would be almost 15 years before it became apparent that sirolimus' role in immunosuppression was really just a small segment of the complex, multifaceted, multi-input mTOR pathway that we now know maintains cellular homeostasis. Nonetheless the field of organ transplantation is where the mTOR inhibitor was really born, and so we explore that role in transplantation here.

18.2 mTOR INHIBITOR TRANSPLANT IMMUNOLOGY AND PHARMACOLOGY

Successful organ transplantation requires preventing rejection of the allograft by the immune system. The alloimmune response to the donor allograft needs several components to accomplish the T-cell proliferation, T-cell differentiation, B-cell activation, and alloantibody production. Donor antigens are presented to recipient lymphocytes by antigen-presenting cells, which then interact with the T-cell receptor complex. The alloantigen recognition by this receptor complex triggers the first of what is termed the "three-signal hypothesis of lymphocyte activation" [13]. Along with

simultaneous costimulatory molecule binding, which represents the second signal, a cascade of events ultimately leads to the migration from the cytoplasm to the nucleus, of NFAT (nuclear factor of activated T-cell). Once in the nucleus, NFAT promotes the transcription of both IL-2 and the IL-2 receptor (IL-2R) which help to amplify this proliferative response. When IL-2 binds the IL-2R that has migrated to the T-cell surface (the third signal), an intracellular cascade activates the mTOR, initiating the G0 to G1 transition of the cell cycle and which leads to cell division, proliferation, and differentiation of that specific T-cell [11,13,14]. This response is known as T-cell activation, and these activated T cells will target any cells carrying that specific donor antigen which in the case of transplantation leads to allograft rejection. This is a very simplified view of the complex mTOR pathway, but it highlights that segment relevant to the alloimmune response in organ transplantation.

Rapamycin interacts with FKBP1A (the same protein to which tacrolimus also binds in an independent process) to create a FKBP1A—rapamycin complex which binds to the FRB (FKBP—rapamycin-binding) domain of the protein mTOR [15]. By binding the FRB domain, it allosterically inhibits the nearby catalytic site [15] inhibiting the kinase activity and blocks downstream activation of the p70S6 kinase (ribosomal protein S6 kinase) gene. p70S6 kinase controls the translation of cell cycle regulating proteins cyclin-E and cyclin-A dependent kinase which trigger the cell cycle transition from G1 to S phase [16,17]. By inhibiting mTOR, we can prevent cell cycle transition and limit cellular proliferation, particularly the proliferation and activation of lymphocytes [18] that is necessary for the alloimmune response. Rapamycin is the "godfather" of a class of molecules that are termed mTOR inhibitors, because they specifically inhibit the mTOR molecule and halt activation of the pathway.

There are currently two mTOR inhibitors that have been approved for use in solid organ transplantation (Figure 18.1). The first is sirolimus (brand name Rapamycin, Rapamune), a macrolide that was the first mTOR inhibitor isolated. The medication is supplied as 0.5-mg and 1-mg tablets, and due to its long half-life of 62 h, it is typically dosed as a once-daily medication [19]. A standard initial dosing of sirolimus in transplantation is 2 mg once daily, and the daily dose is subsequently titrated to a serum blood level target between 4 and 10 ng/ml. Higher serum levels of sirolimus are not necessary for adequate immunosuppression, and merely lead to more adverse events.

The second mTOR inhibitor approved for use in solid organ transplantation is everolimus [brand name Zortress (USA), and Certain (Europe)], a semisynthetic rapamycin analog having a hydroxyethyl ether

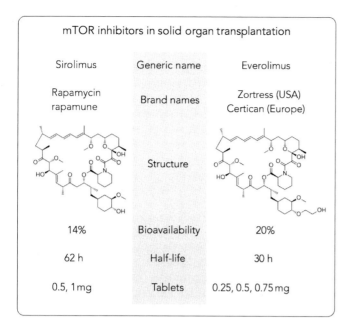

FIGURE 18.1 Comparisons of the two mTOR inhibitors approved for use in solid organ transplantation.

derivative, and was developed for better oral administration [15]. It was synthesized as part of a program designed to identify a rapamycin derivative with improved pharmacokinetics and pharmacodynamics, more consistent bioavailability and improved stability and solubility, making it an improved version more suitable for transplantation indications [20,21]. Everolimus is supplied as 0.25-, 0.5-, and 0.75-mg tablets, and with its shorter half-life of 30 h, it is dosed at a twice-daily medication for transplant. For oncology indications, everolimus (brand name Affinitor) is dosed as a once-daily medication, however because of a drug interaction that exists between cyclosporine and everolimus it is dosed as a twice-daily medication for transplant indications. A standard initial dosing of everolimus is 1.5 mg twice daily, and the dosing is subsequently titrated to a serum blood level target between 3 and 12 ng/ml [22].

Oral bioavailability for mTOR inhibitors is low (sirolimus 14%, everolimus 20%) [2,23]. They are both metabolized by cytochrome p450 3A4 subunit, and are impacted by the pump P-glycoprotein. This is relevant since many of the medications used in transplantation, such as antifungals like fluconazole and voraconazole, can interact with the 3A4 subunit and subsequently increase or decrease the serum drug level of sirolimus leading to either potential rejection or drug toxicity.

mTOR inhibitors are mostly excreted in the bile (sirolimus 91%, everolimus 98%) [19,20], so good biliary excretory function is necessary for drug clearance. While this is not especially relevant to kidney and heart

transplantation, in cases of liver transplantation, allograft dysfunction can alter mTOR inhibitor clearance.

Sirolimus has a labeled indication for use in kidney transplant in conjunction with a CNI, and everolimus has a labeled indication for both liver and kidney transplant in conjunction with a CNI. However, in most circumstances, mTOR inhibitors are used in an off-label fashion. *De novo* therapy occurs when an mTOR inhibitor is started in the immediate post-transplant period. Conversion therapy occurs when a CNI is initiated as the primary immunosuppression for an allograft, and during the maintenance phase, this CNI is discontinued and an mTOR inhibitor started along with an antimetabolite. mTOR inhibitor conversion protocols are used to either avoid an adverse effect from CNI-based immunosuppression, or to take advantage of specific benefit from inhibiting the mTOR pathway beyond immunosuppression.

18.3 mTOR INHIBITOR CLINICAL TRIAL EXPERIENCE

Since the first human trials in 1995 [24], mTOR inhibitors have been extensively evaluated in the transplant population. Both sirolimus and everolimus have had numerous large clinical trials in kidney, liver, and heart transplant populations [25–30]. The results of these trials have been varied with diverse outcomes across the different organs. It is this lack of a common experimental experience that has limited regulatory approval of mTOR inhibitors, and the routine adoption of these medications.

18.4 KIDNEY TRANSPLANTATION: SIROLIMUS

The first two major studies of mTOR inhibitor use in organ transplantation examined the role for sirolimus in kidney transplantation. The first study [27], was a US multicenter trial of 719 *de novo* kidney recipients receiving cyclosporine and steroids, who were randomized to one of three groups to receive either sirolimus 2 mg/day, sirolimus 5 mg/day, or azathioprine 2–3 mg/kg/day. The incidence of composite efficacy failure (death, graft loss, rejection) at 6 months was better in both sirolimus arms (18.7% vs. 16.8% vs. 32.3%, $P = 0.002$) mostly due to improvements in rejection rates, and these improvements persisted to 24 months. The second study [25] was a worldwide multicenter trial of 576 *de novo* kidney recipients receiving cyclosporine and steroids, who were randomized to one of three groups to also receive either sirolimus 2 mg/day, sirolimus 5 mg/day, or placebo.

In this second study, the incidence of composite efficacy failure (death, graft loss, rejection) at 6 months was better in both sirolimus arms (30.0% vs. 25.6% vs. 47.7%, $P = 0.002$) mostly due to improvements in rejection rates, and these improvements persisted to 36 months. In both the first and the second studies, there was no difference in patient and graft survival at either 6, 24, or 36 months. However, in both studies, the glomerular filtration rate (GFR) was unexpectedly lower in the sirolimus groups compared to either azathioprine or placebo.

The third major study [31,32] was a randomized, multicenter, worldwide trial of 525 de novo kidney recipients, comparing a control group regimen of sirolimus, cyclosporine, and steroids to a study cohort of a sirolimus-based regimen where the cyclosporine was slowly withdrawn at 3 months. The hypothesis was that CNI withdrawal would result in improved renal function. In the cyclosporine withdrawal cohort, the sirolimus was dosed to a target trough concentration of 16–24 ng/ml during the first 12 months overlapping the withdrawal period, and then targeted to 12–20 ng/ml after that point. At 12, 24, and 36 months the graft and patient survival were similar in both groups. The incidence of rejection postrandomization was higher in the cyclosporine withdrawal cohort (10.2% vs. 5.6%). Most importantly, the GFR was significantly improved in the cyclosporine withdrawal cohort at months 12, 24, and 36, with a mean GFR improvement of +11.5 ml/min at month 36.

A fourth major trial of sirolimus in renal transplantation examined conversion from CNI to a sirolimus-based regimen. The CONVERT trial [33] was a randomized controlled trial of 830 maintenance kidney recipients (from 6 to 120 months after transplant) who were stratified based on their baseline GFR. Sirolimus was initiated with a single loading dose of 12–20 mg and adjusted to achieve a target sirolimus serum level of 8–20 ng/ml. There was no benefit associated with conversion to sirolimus regarding improved renal function, and there was a higher level of nephrotic range proteinuria following conversion. Because of a higher rate of serious adverse events (particularly pneumonia, rejection, graft loss, and death) in patients in the sirolimus group having significant renal dysfunction, the trial discontinued enrolling patients with a GFR <40 ml/min midstudy.

Based on the results of these randomized trials, which demonstrated sirolimus to be both safe and effective, sirolimus was approved by the FDA for the prevention of organ rejection in kidney transplant recipients in September 1999. This represented the first time an mTOR inhibitor was approved for use in patients by the FDA.

18.5 KIDNEY TRANSPLANTATION: EVEROLIMUS

Following an initial phase I trial that established the safety and tolerability of everolimus in renal transplantation [34], the first set of randomized trials assessed the efficacy of everolimus in conjunction with full-dose cyclosporine. Two similar phase III randomized controlled trials (B201 and B251) both compared the efficacy of two separate everolimus dosing regimens (1.5 and 3 mg/day) along with full-dose cyclosporine and steroids, to a control group of MMF (2 g/day) with cyclosporine and steroids. In the European-based Study B201 of 588 de novo renal recipients [26], there was no statistical difference in graft loss, patient death, or acute rejection between the three groups at 12 months. At 36 months the primary composite endpoint (death, graft loss, rejection) was similar although the graft loss was lowest in everolimus 1.5 mg/day dosing compared to both the everolimus 3 mg/day dosing and the control group (7.2% vs. 16.7% vs. 10.7%, respectively) [35]. In the US-based study B251 [36] of 583 de novo renal recipients, there was no statistical difference in graft loss, patient death, or acute rejection between the three groups at 36 months. The rate of antibody-treated rejection was lower in the everolimus 1.5 mg/day group compared to the control group at both 12 months (7.8% vs. 16.3%) and 36 months (9.8% vs. 18.4%). Although the everolimus-based regimens were found to be safe in terms of adverse effects, both studies showed an increased incidence of renal dysfunction, suggesting that a concomitant CNI reduction might ultimately improve the renal function.

The next set of randomized trials therefore assessed everolimus along with a reduced dose of cyclosporine. Two similar phase III randomized controlled trials (A2306 and A2307) [26] both compared the efficacy of a pair of separate everolimus dosing regimens (1.5 and 3 mg/day) in combination with reduced dosing of cyclosporine and steroids, to a control group of MMF (2 g/day) along with full dosing of cyclosporine and steroids. The main difference between the two studies was the level of reduction of the cyclosporine dosing in the study cohort. The study A2307, which looked at 256 de novo renal recipients, used induction therapy with the anti-IL-2 receptor drug Basiliximab to enable a further reduction of cyclosporine dosing, as compared to study A2306, which examined 237 de novo recipients with a reduction in cyclosporine dosing that did not use induction. Comparing the outcomes of renal function in study A2306 to B201 and B251 [26,36], highlighted an improved serum creatinine, and GFR with the reduced cyclosporine dosing along with everolimus. There was no difference in efficacy between the groups in the trials, but the incidence of rejection was increased in the everolimus 1.5 mg/day group in study A2306 compared to the same group in study A2307 (25.0% vs. 13.7%) which suggested that anti-IL-2 receptor induction therapy was beneficial in reducing the risk of rejection when used with everolimus and reduced cyclosporine. These pairs of studies show that everolimus can allow cyclosporine levels to be reduced by at least 57% at the months without affecting safety or efficacy, to enable renal benefits.

The final pivotal study for everolimus in kidney transplantation was study A2309 [37], a large 24-month phase IIIb trial of 833 *de novo* renal transplant recipients who were randomized to one of two serum levels of everolimus (low: trough 3–8 ng/ml, or high: trough 6–12 ng/ml) with a concomitant 60% reduction in cyclosporine dosing, or to a control group of mycophenolic acid along with standard dosing of cyclosporine. The composite efficacy failure (death, graft loss, rejection) showed no statistical difference as the low everolimus group, high everolimus group, and control group (32.9%, 26.9%, and 27.4%, respectively). With regard to renal function, the mean GFR at 24 months showed no significant difference with 52.5, 49.4, and 50.5 ml/min in the low everolimus, high everolimus, and control groups, respectively. There was a higher proportion of patients that discontinued therapy in the everolimus group compared to controls. Nonetheless, the trial showed that when dosing everolimus to serum trough level of 3–8 ng/ml, you could safely allow a reduction of almost 60% of the CNI.

Despite not demonstrating the superior renal benefits, based on these five studies, everolimus was a safe and effective therapy. In April 2010, the FDA approved everolimus for renal transplantation in adult patients at low-moderate immunologic risk.

Following FDA approval, the idea of eliminating CNI (rather than reducing CNI) using everolimus was examined in the long-term ZEUS study of 300 *de novo* low immunological risk kidney recipients. The patients were randomized at 4.5 months to either remain on cyclosporine, or to undergo a conversion to everolimus with the elimination of cyclosporine [38]. The GFR in the everolimus group was 8.2 ml/min higher at 5-year follow-up, however the rate of acute rejection was also higher in the everolimus arm compared to the cyclosporine arm (13.6% vs. 7.5%, P = 0.09). The majority of these rejection was Banff 1 rejection that was reversible and did not affect long-term outcomes. This trial demonstrated that an improvement in long-term renal function could indeed by safely achieved following a conversion to everolimus from CNI. This trial has led the way to making early CNI elimination with everolimus the standard approach for using mTOR inhibitors in renal transplantation.

18.6 LIVER TRANSPLANTATION: SIROLIMUS

After sirolimus was approved for use in kidney transplantation, there was much initial promise regarding the potential for mTOR inhibitors as immunosuppression in liver transplantation. Few suspected, however, that the liver registration trials for sirolimus would lead to such controversy—one that has had repercussions that impacted investigations into the potential beneficial indications of mTOR inhibitors for more than a decade.

There were two multicenter, randomized controlled phase II trials for sirolimus. The first trial, Wyeth 211 [39] was a study of 112 *de novo* liver recipients taking sirolimus and cyclosporine compared to 52 control patients taking standard tacrolimus. The sirolimus patients received a loading dose of sirolimus 15 mg followed by a daily dose of sirolimus 5 mg daily. Unfortunately the results of this trial were never formally published, which has limited scrutiny. There was numerically more hepatic artery thrombosis (HAT) in the sirolimus arm compared to the control group, however it was not statistically significant (9.0% vs. 3.8%, P = 0.10). There was no statistical difference in graft survival or patient survival and the incidence of rejection was less with sirolimus.

The second randomized controlled trial, Wyeth 220 [40] was supposed to be a study of 300 subjects (150 subjects in each study arm), comparing a regimen of sirolimus and tacrolimus with a control group with standard tacrolimus dosing. The sirolimus patients received a similar loading dose of sirolimus 15 mg followed by a daily dose of sirolimus 5 mg daily. Unfortunately, the study was terminated early after 21 months because of an imbalance in serious adverse events. The prematurely terminated trial had 222 *de novo* liver recipients with 110 patients in the sirolimus arm and 112 patients in the control group. While there was no statistical difference in HAT rate (5.5% vs. 0.9%, P = 0.07) when combined with separate cases of portal vein thrombosis (PVT) there were numerically more episodes of allograft vascular thrombosis in the sirolimus group (8% vs. 3%, P = 0.07). There was a higher incidence of graft loss (26.4% vs. 12.5%, P = 0.009) and patient death (20% vs. 8%, P = 0.01) as well as a higher incidence of sepsis in the sirolimus arm (20.4% vs. 7.2%, P = 0.006). For many years, the specific details of this study remained uncertain as it was more than 11 years after the trial completed that the results were published. Because of the lack of information regarding the two registration trials, and an inability to verify the details, there was much misinformation regarding the use of sirolimus in liver transplantation.

Based on the results of these two registration trials, the FDA placed a "black box warning" on the use of sirolimus in *de novo* liver transplant recipients, and did not approve sirolimus for use in liver transplantation. The black box warning stated "The use of Rapamune in combination with tacrolimus was associated with excess mortality and graft loss in a study of *de novo* transplant patients. Many of these patients had evidence of infection at or near the time of death. In this and another study in *de novo* liver transplant patients, the use of Rapamune in combination with cyclosporine or tacrolimus was associated with an increase in HAT; most cases of HAT occurred within 30 days post transplantation and most led to graft loss or death." While neither trial actually showed a statistically significant difference in HAT incidence, when the FDA combined the results from the two trials, and also included separate cases of PVT, a significant difference in the combined HAT/PVT incidence was seen with *de novo* sirolimus use (7% vs. 2%, $P = 0.02$). The results from the two registration trials became the justification for the warning, and the use of sirolimus in liver transplantation became severely restricted due to concerns of HAT.

In 2009, the FDA issued a second warning for sirolimus in liver transplantation, this time regarding sirolimus conversion, after a registration trial found a higher overall treatment failure rate, increased infections, and "increased mortality in patients converted from a calcineurin-inhibitor to sirolimus" [41]. At the time of the warning, details again were scarce and it would be 3 years after the warning before the details of the Sirolimus Liver Conversion Trial were published to allow examination [29]. This large trial had 607 maintenance liver recipients who were randomized 2:1 to either abrupt conversion from CNI to sirolimus, or continuation on CNI. The sirolimus group were given a loading dose of 15 mg and dosed to a supratherapeutic level of 10–20 ng/ml. There was no difference between the two groups with respect to improved renal function. There were numerically more deaths after sirolimus conversion compared to controls (3.3% vs. 1.4%) although it was not statistically significant, and there were no graft loss in either arm. The incidence of rejection was higher in the sirolimus conversion group (6.4% vs. 1.9%) and there were a significant number of adverse events (likely a function of the supratherapeutic dosing) that lead to the discontinuation of sirolimus in nearly every patient in the trial (98.5%) by 5 years after conversion.

Despite these black box warnings, several experienced clinicians and transplant centers continue to use sirolimus off-label in liver transplant recipients, because of a perceived benefit from the drug, along with questions as to the validity of the results in the various registration trials. Between 1998 and 2008, 8.8% of all patients transplanted used sirolimus as part of maintenance immunosuppression [42]. In more than 15 years since these registration trials were performed, none of the findings have been replicated, and in more than 20 studies since that time that have reported on *de novo* sirolimus use in liver transplantation, all of them have either shown no difference in HAT, or even a reduced incidence of HAT with sirolimus. A large observational study of 252 patients from the combined programs of the University of Alberta and University of Colorado [43], as well as a separate large observational study of *de novo* sirolimus 640 patients from Baylor University Medical Center in Dallas (Figure 18.2) [44,45] have both demonstrated a significantly reduced HAT incidence with *de novo* sirolimus use, in direct contradiction to the black box warning. In both of these large observational studies, loading doses were not given, and supratherapeutic dosing was avoided using a sirolimus dose of 2 mg daily as compared with 5 mg daily in the registration trials.

Over time, it has become apparent that the failure of sirolimus in the two registration trials and the large conversion trial are likely a function of study design and an incomplete understanding of the most appropriate immunosuppressive dosing for sirolimus that is needed in transplantation [46]. In the two registration trials of *de novo* use, and the trial of conversion, a large loading dose of sirolimus was given, and supratherapeutic dosing levels were used. As the dosing regimens have evolved to a more normatherapeutic dosing, HAT has become essentially nonexistent.

18.7 LIVER TRANSPLANTATION: EVEROLIMUS

In pursuing approval from the FDA for use in liver transplantation, everolimus inherited the burdens of the black box warning associated with sirolimus—a "guilt by association." However, the lessons learned from the failed sirolimus registration trials, such as avoidance of supratherapeutic dosing and avoidance of large loading doses, allowed clinicians investigating everolimus to design more successful trials. A variety of randomized controlled trials gave support to the safe use of everolimus in liver transplantation, while reiterating many of the adverse effects common among all mTOR inhibitors.

The first randomized trial of everolimus in liver transplantation was a 2006 study by Levy et al. [47] of 199 patients that assessed the safety and the tolerability of everolimus in liver transplant recipients. The patients were randomized to receive either cyclosporine plus everolimus at one of three doses (0.5, 1, or 2 mg twice

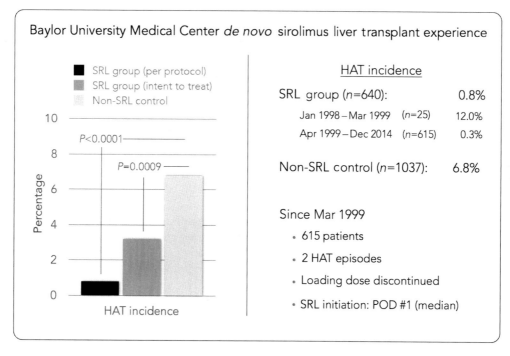

FIGURE 18.2   The Baylor University Medical Center (Dallas) experience of *de novo* sirolimus immunosuppression in 640 liver transplant recipients, from January 1998 to December 2014, showing significantly lower HAT incidence with *de novo* sirolimus use compared to controls (0.3% vs. 6.8%, P<0.0001). The sirolimus group was examined as "intent-to-treat" (identifying all HAT episodes including those where sirolimus was discontinued prior to HAT) or as "per protocol" (identifying all HAT episodes the developed when actively taking sirolimus). Patients from January 1998 to March 1999 received large loading dose with supratherapeutic sirolimus dosing. SRL, sirolimus; HAT, hepatic artery thrombosis; POD, postoperative day.

daily) or to cyclosporine plus placebo. Unfortunately, because the treatment groups were small, they were not powered enough to show a difference in outcomes, nonetheless, there were no differences in the major primary endpoints (death, graft loss, or rejection) between the groups. Because of the black box warning associated with sirolimus, HAT was a significant focus of the trial, however there was no evidence of increased incidence of HAT with everolimus. There were more adverse events in the everolimus groups, particularly the high-dose group, and the discontinuation rate was high, as only 22% of patients were maintained on everolimus until the 3-year follow-up. The study showed that an acceptable level of safety and tolerability could be achieved on everolimus [22].

The second major randomized trial of everolimus in liver transplantation was a 2010 study by Masetti et al. [48,49] of 78 *de novo* liver recipients that were treated for 30 days with cyclosporine, and then randomized to either continue cyclosporine as monotherapy or to convert to everolimus monotherapy (where the everolimus was initiated on post-transplant day 10 and then the cyclosporine was discontinued on day 30). At 1 year post-transplant, the everolimus monotherapy group had a dramatically improved renal function with a GFR 87.7 ml/min vs. 59.9 ml/min in

the cyclosporine monotherapy control. The incidence of chronic kidney disease (CKD) of stage ≥3 in the everolimus group was 15.4% compared to 52.2% in the control group. There was no difference in complications, rejection, or patient survival between groups. This study showed that early conversion to everolimus with discontinuation of CNI was associated with a substantial improvement in renal function, and also served to highlight that the most effective intervention to avoid long-term nephrotoxicity was early intervention.

The third major randomized trial of everolimus in liver transplantation was the 2012 PROTECT trial in 2012 by Fischer et al. [50], which was designed to determine the potential renal benefits of CNI withdrawal using everolimus. The study design of a 4-week delay in introducing everolimus reflected concerns of the potential risk of HAT due to the sirolimus black box warning. In the trial, 375 patients received either tacrolimus or cyclosporine along with basiliximab induction, and at 4 weeks post-transplant they were randomized to either continue the CNI or be converted to everolimus monotherapy (assuming the patient had GFR >50 ml/min). Only 203 patients were randomized as 172 patients failed to advance to randomization due to adverse events

(16%), GFR < 50 ml/min (12%), and graft loss (11%). The incidence of the composite efficacy failure endpoint (death, graft loss, rejection) in the everolimus group was similar to the control group (20.8% vs. 20.4%, respectively), as well as the incidence of biopsy-proven rejection (17.7% vs. 15.3%, respectively). There was an improvement of the calculated GFR [Modification of Diet in Renal Disease (MDRD) method] of +7.8 ml/min in the everolimus group. There were significantly more adverse events in the everolimus group, however the incidence of wound healing impairment was the same and there were no episodes of HAT. The PROTECT trial showed that an early conversion to everolimus could be safely accomplished, with the possibility of improvement in nephrotoxicity.

The final major randomized trial of everolimus in liver transplantation, and the one that led to registration of the drug for liver transplantation by the FDA, was the H2304 trial in 2012 [51]. This was a prospective, multicenter, randomized trial of 719 *de novo* transplant recipients, randomized to one of three groups: an everolimus group with reduced tacrolimus dosing, an everolimus group with tacrolimus elimination, and a control arm with standard tacrolimus dosing. Everolimus was not initiated until 30 days posttransplant, once again, in deference to concerns from the black box warning for sirolimus regarding HAT. The everolimus with tacrolimus elimination arm was halted during the study because of an increased risk of rejection (although those patients still randomized to that group were allowed to continue with therapy). The everolimus group with reduced tacrolimus had significantly less biopsy-proven rejection compared to controls at both 12 months (4.1% vs. 10.7%) and 24 months (6.1% vs. 13.3%) [51,52]. The primary composite endpoint (death, graft loss, rejection) for the everolimus with reduced tacrolimus group was 6.7% vs. 9.7% compared to controls. The change in GFR was superior in the everolimus with reduced tacrolimus group with a difference of +8.5 ml/min compared to controls. There were no differences in HAT or wound healing complications. These improvements in GFR persisted when the study was assessed at 24 months [52] and 36 months [53]. Of note, regarding the patients in the halted arm of everolimus along with tacrolimus elimination, while the group's higher rate of rejection led to discontinuation of this study arm, this group also had a substantial and persistent improvement in GFR that, unlike the other arms, continued to increase over the 36-month study period. The H2304 trial results showed that the use of everolimus in the adult population was safe and effective, even up to a 3-year follow-up [54], and the use of everolimus allowed a CNI reduction, which was beneficial to renal function.

The results from the H2304 registration trial led to FDA approval of everolimus for use in liver transplantation making it the first mTOR inhibitor approved for use in liver transplantation. Everolimus was also the first immunosuppressant approved by the FDA for liver transplantation in over 10 years.

18.8 HEART TRANSPLANTATION: SIROLIMUS

Since the development of sirolimus and the understanding of its non-nephrotoxic nature, a focus in cardiac transplantation has been to use this new class of immunosuppression to minimize CNI nephrotoxicity. However, with further understanding of the antiproliferative effects of sirolimus, a further focus in the field of heart transplantation has been to reduce the chronic allograft vasculopathy related to CNI [55].

The first major randomized trial regarding mTOR inhibitors in cardiac transplantation [56] involved 136 heart transplant recipients and compared a regimen of sirolimus, cyclosporine, and steroids, to a control group of azathioprine, cyclosporine, and steroids to assess safety and efficacy. At 6 months, the incidence of moderate/severe rejection was lower in the sirolimus group compared to controls (32.4% vs. 56.8%, P = 0.01) with comparable survival and improvements in limiting progression of vasculopathy. Unfortunately, there was a 44% rate of discontinuation in the sirolimus group due to adverse effects.

The next trial [57] examined the efficacy of CNI withdrawal using sirolimus in a trial of 34 heart recipients having underlying renal dysfunction. Sirolimus was initiated and the CNI was withdrawn gradually over 12 weeks. The change in GFR from baseline was significantly higher in the sirolimus group, which increased +12.1 ml/min, compared to a reduction in GFR reduction from baseline of the control group of −5.4 ml/min (P = 0.004). This trial showed that sirolimus could be used to safely accomplish CNI withdrawal in heart transplant recipients, with substantial benefit. A third trial [58] employed the same CNI withdrawal protocol to demonstrate that sirolimus attenuated the increase in mean plaque volume compared to controls (0.1 mm vs. 1.28 mm) and that CNI elimination using sirolimus limited cardiac allograft vasculopathy.

Several trials examined the long-term impact of sirolimus in cardiac transplantation with contradictory findings. A 2012 study [59] of 103 cardiac transplant recipients compared sirolimus conversion along with complete CNI withdrawal, to a control group. Conversions were a mean 1.2 years post-transplant. Plaque progression was attenuated in the sirolimus conversion group (0.7 vs. 9.3, $P = 0.0003$) with most attenuation occurring in those converted in the first 2 years. There was increased 5-year survival in the sirolimus conversion group compared to controls (97.4% vs. 81.8%, $P = 0.006$). Conversion to sirolimus attenuated vasculopathy and improved long-term survival.

In contrast to this first long-term trial [59] is a 2015 long-term study [60] of 126 *de novo* cardiac transplant recipients randomized to receive either sirolimus with low-dose tacrolimus, or full-dose tacrolimus with mycophenolate mofetil (MMF). Analysis took place 8 years after transplantation. There were no differences in rejection or freedom from chronic allograft vasculopathy. The renoprotective benefits seen early with sirolimus did not persist to 5 and 8 years post-transplant, and the trend for superior survival at 5 years did not persist to 8 years. A multivariate analysis at 8 years did not demonstrate any benefits with sirolimus and low-dose CNI.

Although sirolimus never received FDA approval for use in heart transplantation, several transplant centers have routinely used mTOR inhibitors, such as sirolimus, for more than a decade, to reduce or discontinue CNI to preserve renal function and limit chronic allograft vasculopathy.

18.9 HEART TRANSPLANTATION: EVEROLIMUS

The immunosuppressive potential for everolimus in heart transplantation was first assessed in study B253 [30,61], a randomized, multicenter, double blind trial of 634 *de novo* heart transplant recipients, randomized to receive either one of two everolimus doses (1.5 or 3 mg) along with cyclosporine and steroids, or to a control group of azathioprine with cyclosporine and steroids. At 6 months when examining the primary composite endpoint (death, graft loss, rejection), in comparison to the control group, there was superior efficacy in both the lower everolimus dose group (36.4% vs. 46.7%, $P = 0.03$) and the higher everolimus dose group (27.0% vs. 46.7%, $P < 0.001$), In addition, compared to controls, the everolimus groups demonstrated reduced vascular intimal thickening, significantly less vasculopathy (33% vs. 58.3%), and also a threefold lower incidence of cytomegalovirus (CMV) infection rate [61]. The results for this study were the first to establish the value of everolimus 1.5 mg dosing in limiting the development of allograft vasculopathy in heart transplant recipients up to 24 months, and it also demonstrated a dose-related relationship to developing adverse effects [62].

Following this pivotal trial was study A2411, a randomized multicenter trial of *de novo* everolimus use in heart transplantation [63]. The study had 176 *de novo* heart transplant recipients randomized to receive either everolimus (trough level 3—8 ng/ml) along with reduced-dose cyclosporine and steroids, or a control group with MMF (3 g/day) along with standard-dose cyclosporine and steroids. At 12 months, the rejection rates were similar between the two groups and there was a threefold reduction in CMV infection, however there was no difference in renal function between the groups.

A third smaller trial of 37 heart recipients [64] was designed to demonstrate successful conversion to a reduced-dose cyclosporine protocol to minimize the nephrotoxicity. The trial showed heart recipients could be successfully converted to everolimus at a mean cyclosporine level of 68.5 ng/ml—a very low CNI level, without any increase in rejection.

A fourth larger trial [65] of 282 thoracic transplant recipients further examined the renal function benefits with a conversion to an everolimus and reduced cyclosporine dosing protocol. Over a 12-month period, the CNI was decreased a mean 57%. The mean GFR increased for those in the everolimus group, but decreased in the controls at 12 months (+4.6 ml/min vs. −0.5 ml/min). Those that switched earlier benefited the most from conversion (+7.8 ml/min for those in the lowest third for time post-transplant). However, the main concern from the study was a higher incidence of serious adverse events with everolimus compared to controls (46.8% vs. 31.0%, $P = 0.02$), particularly infections and pneumonia. Additionally, the 12-month mortality was numerically higher in the everolimus group (7.8% vs. 4.8%, $P = $ NS) and the majority of the deaths in the everolimus group occurred in the first 90 days compared to controls (77% vs. 33%). This large trial brought into question the safety of everolimus for use in heart transplant based on these serious adverse events.

The SCHEDULE trial [66] was a randomized trial of 115 *de novo* heart transplant recipients randomized to either everolimus and low-dose cyclosporine, or standard doses of cyclosporine with MMF. In the study cohort of everolimus and low-dose cyclosporine, the cyclosporine was subsequently withdrawn and the everolimus dosing increased. The GFR at 12 months was significantly higher in this everolimus study cohort compared to controls (79.8 m/min vs. 61.5 ml/min, $P < 0.001$), there was also significantly less intimal thickening and a reduced incidence of chronic allograft vasculopathy. The incidence of rejection was significantly increased in the everolimus study cohort, while the CMV infection was sixfold lower with similar bacterial infection.

Everolimus did not gain FDA approval as an indication for heart transplantation. It received a warning that in *de novo* heart transplant patients, it "resulted in an increased mortality often associated with serious infections within the first 3 months post-transplantation," related to the fourth trial. Nonetheless, despite the warnings, everolimus is used by some centers in an off-label fashion. The SCHEDULE trial is a good example of how centers use everolimus in an off-label fashion successfully, particularly when clinicians are successful in managing side effects of mTOR inhibitors to prevent discontinuation.

18.10 MANAGEMENT OF mTOR INHIBITOR SIDE EFFECTS

In order to achieve the benefits of mTOR inhibitors in transplantation, a clinician must manage the side effect profile (Figure 18.3). The challenge of managing the adverse effects is one of the reasons many clinical trials of sirolimus and everolimus show mixed results; it is difficult to identify the benefits of a drug when it has been discontinued. In some transplant centers, as well as in most clinical trials, more than 50% of recipients have mTOR inhibitor discontinued due to side effects. Ironically, the majority of these adverse effects can be managed and the rate of discontinuation should be less than 5% of patients.

18.10.1 Hyperlipidemia

Hyperlipidemia is the most common side effect associated with mTOR inhibitors. It occurs in almost 75% of patients and leads to significant elevations of total cholesterol, LDL, and triglycerides, which typically generate concerns [67].

The exact mechanism for the increased lipids and triglycerides is not completely clear, as mTOR inhibition affects numerous areas that impact lipid and cholesterol metabolism and any of them maybe the cause. mTOR inhibitors block lipases that impair lipid catabolism and clearance [67]; they alter fatty acid uptake and the synthesis of triglycerides which reduce fat deposition in adipose tissue [68], and they down-regulate genes needed for fatty acid transport and esterification. mTOR inhibitors block endogenous cholesterol synthesis by inhibiting mammalian 3-hydroxy-3-methylglutaryl coenzyme A reductase. In addition, mTOR inhibitors can decrease cholesterol reuptake, and increase cholesterol efflux from macrophages [69,70].

The main concern regarding hyperlipidemia from mTOR inhibitors is that it can lead to increased cardiovascular events, since hyperlipidemia is a well-established cardiac risk factor [71]. However, despite the increase in lipids and triglycerides from mTOR inhibitors, animal studies show the antiproliferative and anti-inflammatory effects of mTOR inhibitors actually reduce atherosclerosis. In addition, a large study

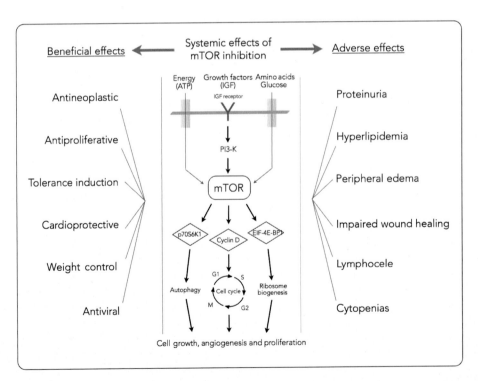

FIGURE 18.3　The beneficial and the adverse systemic effects of inhibition of the multifaceted mTOR pathway. ATP, adenosine triphosphate; IGF, insulin growth factor; PI3-K, phosphoinisitide 3-kinase; mTOR, mammalian target of rapamycin.

of liver transplant recipients receiving *de novo* sirolimus showed no impact on cardiovascular events with sirolimus, despite the fact these patients had higher cholesterol and triglyceride levels and had a significantly higher Framingham risk score [72]. The decreased cholesterol reuptake and increased cholesterol efflux from macrophages as a result of the mTOR inhibitors maybe the reason for the reduced cardiovascular events, since this prevents the macrophages from evolving into foam cells, a necessary component in the development of atheroma.

Management of hyperlipidemia and hypertriglyceridemia from mTOR inhibitors requires different strategies for each. Both require diet, weight, and lifestyle modification. In addition to this, hyperlipidemia requires statins as a first-line therapy for LDL levels >190 mg/dl. Hypertriglyceridemia should be treated with omega-3-acid ethyl esters (fish oil tablets) and refractory cases should add statin therapy.

Given the cardioprotective benefits of mTOR inhibitors, when significant hyperlipidemia occurs, clinicians may reconsider discontinuing the mTOR inhibitor, since decreasing the mTOR inhibitor may paradoxically increase the cardiovascular risk when those cardioprotective benefits are lost. With significant triglyceridemia, despite the cardiovascular risks that are abrogated by the mTOR inhibitor, the risk of pancreatitis persists, and so discontinuation of the mTOR inhibitor maybe considered in those cases that are refractory to statin and fish oil therapy.

18.10.2 Cytopenias

Cytopenias are the second most common side effect from sirolimus, stemming from sirolimus' impact on the cell cycle, and the subsequent reduced cellular proliferation on bone marrow production. Anemia, neutropenia, and thrombocytopenia all frequently occur with mTOR inhibitor use. The anemia due to mTOR inhibitors typically occurs in the initial post-transplant period and resolves as patients get further from the time of transplant. Most anemia is asymptomatic, however sometimes iron supplementation or erythropoietin injections are necessary; transfusions are rare.

In most instances, the neutropenia from mTOR inhibitors is not clinically significant, as there is not an increase in the incidence of bacterial infection with mTOR inhibitor use [44]. The neutropenia (and thrombocytopenia) associated with mTOR inhibitors, can be improved by limiting other medications that also cause cytopenias, including sulfas, H-2 blockers, and proton pump inhibitors. For severe neutropenia, a granulocyte colony-stimulating factor such as filgastrim can be considered.

18.10.3 Oral Ulcers

Oral ulcers are a common side effect of mTOR inhibitors, typically affecting 25% of transplant recipients, with reports of up to 60% [73,74]. A possible pathophysiological mechanism is from the direct toxic effects on the oral mucous membranes by the mTOR inhibitor [75]. These painful ulcers typically present on the lips, cheek, and tongue as gray, and erythematous, superficial, ulcers, usually less than 1 cm in size. The ulcers typically appear within a week of initiation of the mTOR inhibitor, but can occur any time after initiation [75,76]. In the past, these oral ulcerations have been mistakenly attributed to herpes simplex virus [25,46], and they were frequently cited as a reason for discontinuation of the mTOR inhibitor. The ulcers are easily treated with kenalog-in-orabase, a topical highly potent corticosteroid combined with a dental paste, which seals the topical steroid on the ulcer preventing it from being washed away by saliva. Symptoms improve in 1 day, with resolution of the ulcer in 2−3 days. Severe ulcers, and those refractory to kenalog-in-orabase therapy can often limit oral intake and necessitate a reduction in the mTOR inhibitor dosage, and when this fails discontinuation is necessary. Ulcers that persist after discontinuation of mTOR inhibitor need oncological assessment [73].

18.10.4 Wound Healing Issues

By blocking the mTOR pathway, and cell cycle regulation and proliferation, mTOR inhibitors limit the cellular responses that are involved in wound healing. However, in addition to cell cycle regulation, mTOR is involved in several specific processes that impact fibrogenesis (Figure 18.4A), and inhibition of the mTOR pathways leads to: (a) reduced synthesis of type 1 collagen mRNA; (b) reduced fibroblast proliferation resulting from inhibition of the profibrotic growth factors MGP-1, PDGF, and TGF-β; and (c) disrupted fibroblast attachment from altering the α-1β-3 integrin pathway [77−80 The effects on these components of the wound healing pathway can impact and interfere with surgical incision healing and there are reports of increased risks of incisional hernias in patients taking mTOR inhibitors [81,82 particularly when the mTOR inhibitor is combined with an antiproliferative such as mycophenolate.

Despite the clear impact on wound healing processes, nearly all trials of sirolimus in liver transplantation that investigated complications related to wound healing showed no difference in the incidence of infection or hernias [43,83−85]. Only one trial, which used supratherapeutic dosing of sirolimus at 5 mg daily,

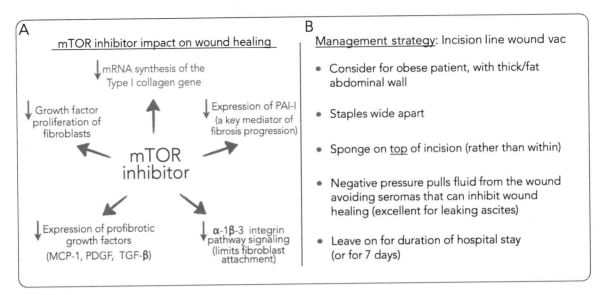

FIGURE 18.4 (A) The mTOR pathway is involved in several specific processes that impact fibrogenesis, and mTOR inhibitors have a significant effect on wound healing. mRNA, messenger ribonucleic acid; MCP-1, monocyte chemotactic protein 1; PDGF, platelet-derived growth factor; TGF-β, transforming growth factor beta; PAI-1, plasminogen activator inhibitor-1. (B) Incision-line wound vacs are an effective strategy to promote incisional healing in high risk patients (e.g., obese, diabetic) on sirolimus.

describes a significantly higher incidence of wound complications.

Wound complications in patients taking mTOR inhibitors are more common in those who are diabetic and obese. In these higher-risk recipients who will be taking an mTOR inhibitor, at the time of transplant, an incision-line "wound vac" can be placed on the incision for up to 1 week postoperatively while the patient is in hospital (Figure 18.4B). This removes any fluid from within the wound that might inhibit healing and it can draw the wound edges together.

Because of the impact that mTOR inhibitors have on the wound healing process, if elective surgery is planned, the mTOR inhibitor should be temporarily converted to a CNI at least 2 weeks prior to surgery and for approximately 4 weeks after surgery to allow wound healing.

18.10.5 Proteinuria

Unlike CNIs, the mTOR inhibitors do not cause any renal tubular dysfunction *per se*, although they are associated with a higher incidence of proteinuria. While the mechanism explaining this proteinuria is not clearly understood, it is considered due to morphological changes that occur within podocytes and remodeling of the glomerular filter apparatus [86]. Some propose that mTOR inhibitors decrease VEGF synthesis and expression which leads to podocyte injury. *In vitro* and *in vivo* studies of podocytes and mTOR

inhibitors show that the mTOR inhibitors reduce levels of the proteins nephrin, podocin, and synaptopodin, three proteins important for maintaining podocyte morphology. While debate has existed as to whether proteinuria is a cause of, or is an effect of, renal dysfunction, evidence is building that proteinuria can induce inflammatory and fibrogenic effects [87] and thus the mTOR inhibitor-induced proteinuria should be treated if possible.

Treatment strategies to manage and limit the proteinuria from mTOR inhibitors include statins and ACE inhibitors [88], both of which regulate inflammatory mediators and can ameliorate the reduced nephrin and podocin proteins, and ultimately stabilize the podocytes [89].

18.10.6 Pneumonitis

Pneumonitis is a rare but serious complication associated with mTOR inhibitors. Its incidence is dose-related, and as such is seen less commonly in transplant patients compared to oncology patients. The incidence is also mTOR inhibitor-related, with a reported incidence of 1–6% with sirolimus, and 0.5–1% with everolimus [90–93].

mTOR inhibitor-associated pneumonitis is a noninfectious, nonmalignant, interstitial lung disease that typically presents with a nonproductive cough, dyspnea, and rarely hemoptysis. A chest X-ray shows a nonspecific "ground-glass" infiltrate and a bronchoscopic alveolar lavage that is negative for infection [94,95]. The pathogenesis of the pneumonitis is not clearly defined although various possibilities include a cell-mediated autoimmune response or a direct alveolar toxic injury.

Treatment for mild symptoms includes a dose reduction of the mTOR inhibitor along with a course of corticosteroids once infection has been ruled out. If significant symptoms exist, the mTOR inhibitor should be discontinued.

18.10.7 Rash

Dermatitis and rash can occur with mTOR inhibitor use, with an incidence of 10–30% [74]. These symptoms typically appear soon after initiation of the mTOR and result from blocking the epidermal growth factor pathway [73]. Some resolve spontaneously and some mild rashes will resolve with topical corticosteroids, or a dose reduction of the mTOR inhibitor. Refractory dermatological complications typically necessitate discontinuation of the mTOR inhibitor.

18.10.8 Peripheral Edema

Peripheral edema, particularly pedal edema, is a common complication of mTOR inhibitors, although the reported incidence depends both on the type of organ transplanted, and particular mTOR inhibitor used. The incidence ranges from 10% of heart transplant recipients [96], to 33% of liver recipients [85,97] and to 30–60% of kidney recipients [98,99]. Peripheral edema maybe less frequent with everolimus compared to sirolimus as matched control retrospective studies show fivefold less edema with everolimus [23,100,101].

The peripheral edema from mTOR inhibitors is due to altered lymphatic drainage. mTOR inhibitors block VEGF and have an antilymphangiogenic impact on the lymphatic endothelial cells [102,103].

The peripheral edema can be controlled with low doses of a loop diuretic (i.e., furosemide), compression stockings, and a reduction of mTOR inhibitor dosing [99]. Any calcium channel blockers such as nifedipine and amlopdine should be changed to a different antihypertensive, as these common post-transplant drugs routinely cause, or worsen, peripheral edema. If the symptoms persist despite these therapies, then the mTOR inhibitor may need to be discontinued.

18.10.9 Lymphocele

A large meta-analysis of randomized controlled trials shows a twofold increase in lymphoceles in kidney transplant recipients with mTOR inhibitor administration [104]. The lymphocele incidence is reported in up to 16% of kidney transplant recipients receiving everolimus and 20% of recipients receiving sirolimus [27,105]. Intervention is needed if there is hydronephrosis, alteration of flow of the vasculature, or clinical evidence of an infection. Treatment includes percutaneous aspiration and, if that fails, a laparoscopic marsupialization, neither of which necessitates discontinuation of the mTOR inhibitor.

18.11 SPECIAL MANAGEMENT CIRCUMSTANCES

Because the mTOR pathway is so multifaceted, and controls so many normal cellular processes that maintain cellular homeostasis, inhibition of this pathway with mTOR inhibitors can also have beneficial effects that can be utilized to yield an advantage for the recipient.

18.11.1 Renal Protection

Chronic renal dysfunction is common in liver recipients and impacts long-term survival. An SRTR database study of 36,849 liver recipients showed 26% had CKD by 10 years post-transplant [49,106]. Numerous factors impact post-transplant renal dysfunction, including pretransplant comorbidities and renal function, however long-term use of CNI has the greatest impact. Long-term maintenance immunosuppression with CNI can cause an irreversible chronic nephrotoxicity from interstitial fibrosis, hyaline thickening, and glomerulosclerosis [107]. Additionally, the diabetes and hypertension that are common side effects of CNI can indirectly worsen underlying renal dysfunction from CNI nephrotoxicity.

To manage renal dysfunction, CNI can be converted to an mTOR inhibitor. Recipients should take an antimetabolite [i.e., MMF 1000 mg twice daily, or mycophenolic acid (Myfortic) 720 mg twice daily] prior to converting, to minimize risk of rejection. Either sirolimus 2 mg daily or everolimus 1.5 mg twice daily is initiated without a loading dose, and then the CNI is discontinued upon initiation of the mTOR inhibitor.

Many studies and trials have examined the role of mTOR inhibitor conversion and CNI elimination in maintaining renal function. The results, however, have been anything but definitive. Early studies were favorable and showed improvement in renal function following conversion [108,109]. However more recent studies, as well as a recent meta-analysis, have questioned the benefits of mTOR conversion [110]. All of these trials of mTOR inhibitor conversion typically were small studies, with short-term outcomes, a heterogeneous time to conversion, and variable preconversion renal function. A recent large, multicenter, registration trial of sirolimus conversion [29] found no benefits, and instead showed an increase in infectious complications following the conversion. Many questions exist regarding the study's design, the supratherapeutic dosing of the sirolimus, as well as the interpretation of the results [46]. Conversely, a second recent large multicenter trial of mTOR conversion [111] showed significant improvement in renal function after an early conversion, albeit with an increased incidence in rejection.

There is a balance that exists between conversion and rejection with regards to the timing of mTOR inhibitor conversion. Very early conversions to an mTOR inhibitor in liver transplantation are associated with a high incidence of rejection, as the CNI-free regimen is often not enough immunosuppression for that point in the post-transplant course. In general a later conversion after 1 year is associated with a much lower incidence of rejection, of less than 2% following conversion. Conversely, the earlier the conversion from CNI to an mTOR inhibitor, the better, in terms of improved renal function. With early removal of CNI, the nephrotoxic effects related to vasoconstriction from CNI can resolve while still reversible. With later conversions, these changes develop into chronic fibrotic changes that are irreversible following removal of CNI (Figure 18.5A−D). By 2 years post-transplant, a change to mTOR inhibitors starts to be less effective, and after this point, it can even be detrimental due to the

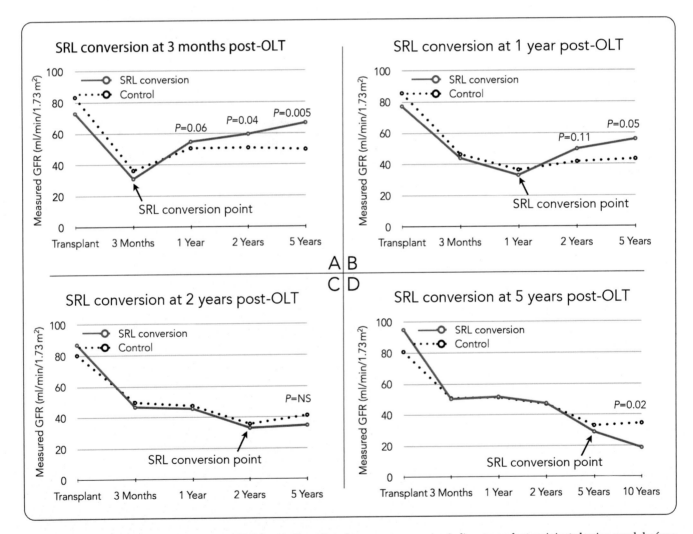

FIGURE 18.5 Outcomes of conversion to a CNI-free sirolimus-based immunosuppression in liver transplant recipients having renal dysfunction (GFR <50 ml/min/1.73 m²), and stratified to protocolized conversion time points from transplantation [112]. (A) $n = 86$ patients converted to sirolimus at 3 months post-transplant showing significant improvement in renal function out to 5 years. (B) $n = 37$ patients converted to sirolimus at 1 year post-transplant showing significant improvement in renal function out to 5 years. (C) $n = 34$ patients converted to sirolimus at 2 years post-transplant showing no difference in renal function out to 5 years. (D) $n = 38$ patients converted to sirolimus at 5 years post-transplant showing worse renal function with sirolimus conversion out to 10 years. SRL, sirolimus; GFR, glomerular filtration rate.

increased proteinuria caused by the mTOR inhibitor [112] and the impact of this proteinuria on a chronically damaged kidney. This variable benefit versus detriment related to the timing of mTOR inhibitor conversion post-transplant is the likely reason for the varied results demonstrated in the conversion trials.

When planning a conversion to an mTOR inhibitor for renal protective benefit in a liver transplant recipient, the aim should be to convert patients in a window that is after 2 months post-transplant to minimize rejection, and prior to 2 years to avoid detrimental late changes once chronicity is established. The protocol is as follows: (i) A 24-h urine should be collected and measured for proteinuria. If the 24-h protein is <300 mg/day, a conversion to an mTOR inhibitor is planned. (ii) An antimetabolite should be initiated prior to conversion, with a dosing related to timing that reflects the rejection risk: MMF 1500 mg twice daily for conversions <2 months post-transplant, MMF 1000 mg twice daily for conversions between 2 months and 1 year post-transplant, and MMF 500 mg twice daily for conversions after 1 year. (iii) Once the antimetabolite is established, an mTOR inhibitor is initiated (either sirolimus 2 mg daily or everolimus 1.5 mg twice daily) and the CNI is discontinued. Target serum trough levels for sirolimus and everolimus are 4–10 ng/ml.

18.11.2 Neurotoxicity

Neurotoxicity from CNI is the most common cause of neurologic complications after liver transplantation. Calcineurin is found in very high concentrations in neural tissues, and disruption to the blood–brain barrier that occurs in cirrhotic patients allows CNI to enter the brain in the post-transplant period where it can interact with calcineurin in the brain [113]. Symptoms include headache, tremors, confusion, dysarthrias, and seizures, and these symptoms are potentially reversible if the insulting agent can be eliminated [113]. Both tacrolimus and cyclosporine can cause neurotoxic symptoms, although they are more common with tacrolimus [114].

When neurotoxic symptoms develop, the first action should be to reduce the CNI. If symptoms do not improve, the CNI can be discontinued and an mTOR inhibitor can be initiated [115]. When conversion to an mTOR inhibitor occurs within the first 2 months (the typical timing for neurotoxic symptoms) there is a high risk of developing rejection. A typical mTOR conversion protocol for neurotoxicity reflects this increased risk of rejection: (i) initiate high-dose antimetabolite, that is, MMF 1500 mg twice daily; (ii) initiate corticosteroids with a slow taper; (iii) start sirolimus

2 mg daily (or everolimus 1.5 mg twice daily), aiming for a trough level of 10 ng/ml. The clinician should have a low threshold of suspicion for rejection in these patients converted to an mTOR inhibitor for neurotoxicity.

18.11.3 Immunoregulatory—Tolerance Induction

One benefit from mTOR inhibitor conversion in transplant recipients that are still being fully understood and delineated is an immunomodulatory benefit. Regulatory T cells, called Tregs, are a phenotype of T cells that suppress the immune response to self and nonself antigens and play an important role in the induction of immunologic tolerance. These Tregs produce IL-17 and secrete effector cytokines such as TNF [116], and strongly inhibit the proliferation of CD4 responder T cells. Tregs play a role in suppression of donor-activated effector T cells and promote tolerance induction. However Tregs also play a role in limiting chronic rejection—a common cause of long-term graft loss in kidney transplants—as chronic rejection is associated with a decrease in Tregs [117].

Sirolimus promotes the conversion of T cells into a suppressor phenotype to become CD4 + CD25 + regulatory T cells [118 While cyclosporine and tacrolimus completely inhibit this conversion from occurring, sirolimus has been shown to significantly increase this proportion of alloantigen-specific regulatory T cells [119], and a conversion from CNI to sirolimus results in a consistent and sustained expansion of the Treg population [120]. An increase in Treg population is associated with long-term graft survival [121].

Conversion to mTOR inhibitors may represent a way to modulate the immune system to one that induces tolerance and limits the development of chronic rejection of allograft. While further investigations are needed, this immunomodulatory benefit of mTOR inhibitors on preventing chronic rejection maybe the most exciting.

18.11.4 Antineoplastic—Hepatocellular Carcinoma

Both cyclosporine and tacrolimus have been shown to promote tumor formation, and there are several reports describing an increased recurrence of hepatocellular carcinoma (HCC) due to CNI, with the tumors developing in a dose-dependent fashion [122–124]. Conversely, mTOR inhibitors are known to be antineoplastic. The serine-threonine kinase mTOR pathway normally regulates a comprehensive pathway of cell growth, proliferation, metabolism, nutrition, angiogenesis, and survival [125,126]. There are several animal models that have shown an association between the development of HCC and mTOR pathway activation [127,128]. The mTOR inhibitors sirolimus and everolimus that are used in liver transplant, both block this tumor pathway and have been shown to inhibit cell growth and proliferation. Recent data show that treatment of advanced HCC with sirolimus or everolimus in high doses (without any surgery or transplantation) has shown partial responses and a few complete responses by RECIST criteria [129,130].

Given the role of mTOR in tumor development, and the known impact of mTOR inhibitors in limiting HCC progression, it follows that using mTOR inhibitors as the prime immunosuppression in the post-transplant setting might improve HCC recurrence. The main debate is that serum drug levels required for antineoplastic activity are higher than the ones used for transplantation, and that using antineoplastic dosing in a transplant recipient might lead to increased wound and anastomotic healing complications and thrombosis of the allograft. Nonetheless there are several studies [83,131,132] that demonstrate improved survival and reduced HCC recurrence when the immunosuppression included sirolimus at normal dosing regimens. A recent meta-analysis of five studies of sirolimus-based immunosuppression in recipients with HCC showed lower HCC recurrence rate (4.9–12.9% vs. 17.3–38.7%, $P < 0.001$), longer recurrence-free survival, and improved overall survival (5-year survival 80% vs. 59–62%, $P < 0.001$) with sirolimus use, compared to patients who did not receive sirolimus post-transplant [133,134]. The results from two future large multicenter randomized trials of mTOR inhibitor immunosuppression in recipients with HCC, such as the SiLVER trial, and the HCC-RAD trial, will be useful in determining whether a benefit exists for mTOR inhibitor use in liver recipients transplanted for HCC [135].

A typical immunosuppression strategy for recipients transplanted for HCC is: (i) sirolimus 2 mg daily or everolimus 1.5 mg twice daily (no loading dose, titrated to level 4–10 ng/ml); (ii) cyclosporine 2–3 mg/kg twice daily titrated to a level of 150 ng/ml; and (iii) steroid taper of 200 mg daily to discontinuation over 10 days.

18.11.5 Antineoplastic—Skin Cancer

Organ transplant recipients have an increased incidence of cancers in the post-transplant setting, as cancers are a function of immunosuppression. Skin cancer is by far the most common cancer found in transplant patients, and it impacts 16–22% of all transplant recipients; 95% of these are nonmelanoma skin cancer. The incidence of squamous cell carcinoma (SCC) is

increased 50–250-fold compared to the general population and the incidence of Kaposi sarcoma is elevated 25–84-fold [136,137]. Given the impact of the mTOR pathway on cell proliferation, angiogenesis, and neoplasms, there is potential to mitigate this large incidence of skin cancer found in transplant patients, through the use of mTOR inhibitors [138].

There have been five prospective studies that have examined the impact of mTOR inhibitors on nonmelanoma skin cancer incidence [139–143]. The studies all demonstrate that sirolimus improved the incidence of nonmelanoma skin cancer. The CONVERT trial of 830 patients [141] showed a significantly lower incidence of skin cancer in patients following conversion to sirolimus (1.2 per 100 person years vs. 4.3, $P < 0.001$). A second large trial demonstrated that a marked reduction of SCC occurred in the sirolimus group after 2 years (HR 0.38, 95% CI 0.17–0.84), particularly in those with a prior history of SCC [142]. This trial demonstrated the benefits of converting patients to sirolimus, particularly those patients who have developed a SCC on CNI immunosuppression. Based on these study results, patients who are at risk for developing SCC, as well as those who have had a prior SCC, should be converted to an mTOR inhibitor.

A typical immunosuppression strategy to convert patients for SCC is: (i) initiate an antimetabolite, that is, MMF 1000 mg twice daily if patient is not already on; (ii) once the antimetabolite is initiated, start sirolimus 2 mg daily or everolimus 1.5 mg twice daily (no loading dose, titrated to level 4–10 ng/ml); (iii) discontinue CNI once sirolimus or everolimus is initiated (no need to overlap).

18.11.6 Hemolytic Uremic Syndrome

Hemolytic uremic syndrome (HUS) is a disease characterized by a triad of hemolytic anemia, thrombocytopenia, and renal insufficiency. The disease is caused by a toxic insult causing endothelial damage, complement dysregulation, platelet activation, and microangiopathy [144]. While rare in the general population, HUS is more common in the transplant population because CNI, particularly tacrolimus, can cause the endothelial damage that triggers HUS [145]. It occurs in up to 5% of kidney transplant patients [146] and 2% of liver transplant patients.

Early recognition of the triad of symptoms is key, particularly in liver transplant recipients who commonly present with a nonpathologic anemia, thrombocytopenia, and renal insufficiency. Treatment involves discontinuation of the CNI and conversion to an mTOR inhibitor. Because these conversions take place in the first few days post-transplant, the clinician should be aware of the high risk of allograft rejection with early conversion. A typical mTOR conversion protocol for HUS reflects this risk of rejection: (i) initiate high-dose antimetabolite, that is, MMF 1500 mg twice daily; (ii) initiate corticosteroids with a slow taper; (iii) start sirolimus 2 mg daily or everolimus 1.5 mg twice daily aiming for a trough of 10 ng/ml; and (iv) consider an antibody-based therapy such as thymoglobulin for additional bridging immunosuppression.

18.11.7 Anti-Cytomegalovirus Effects

CMV is a herpetic virus that commonly occurs in immunocompromised patients. CMV utilizes the host cell's protein synthesis to support viral replication, and CMV is a strong activator of phosphatidylinositol 3-kinase (PI-3K) [147]. PI-3K activates mTOR, which subsequently initiates cell ribosome biogenesis and protein synthesis. *In vitro* data suggest mTOR inhibition impairs viral replication, and mTOR inhibitors limit this CMV-dependent PI-3K activation of the mTOR pathway preventing viral replication.

Several studies have shown that both sirolimus and everolimus can reduce the incidence of CMV infection [30,43,148] in renal, cardiac, and liver recipients. The large experience of sirolimus in *de novo* liver transplant from Baylor University Medical Center shows a two-fold reduction in CMV viremia and a fourfold reduction in invasive CMV disease [72,149]. In patients at high risk for CMV infection based on donor/recipient CMV status (i.e., donor CMV-seropositive/recipient CMV-seronegative) consideration should be given to use sirolimus or everolimus-based immunosuppression to limit CMV infection.

18.11.8 Cardioprotection

A frequent adverse effect of mTOR inhibitors is hyperlipidemia, and a common misperception is that these elevated lipids lead to an increased risk of cardiovascular disease. *In vitro* and *in vivo* studies have demonstrated that inhibition of the mTOR pathway impacts progression of atherosclerosis. The antiproliferative effects of sirolimus can inhibit vascular smooth muscle cell proliferation that causes neointimal thickening. In addition, mTOR inhibitors decrease cholesterol reuptake and increase cholesterol efflux out of macrophages. This helps prevent the macrophages from developing into foam cells, a key component in the development of atherosclerosis (in patients not taking an mTOR inhibitor elevated lipids leads to foam cell formation).

A large study of 401 liver transplant recipients [72] who received *de novo* sirolimus showed no difference in the risk of cardiovascular disease over a 12-year study period. Comparing the patients who received *de novo* sirolimus to the control group, there was no difference in myocardial infarction (1.0% vs. 1.2%, $P = NS$), cerebrovascular accidents (2.9% vs. 2.9%, $P = NS$), and aortic aneurysm (0.2% vs. 0.3%, $P = NS$). However, the patients receiving sirolimus were more commonly male, older, and have more pretransplant diabetes, hypertension, and elevated lipids. This made the sirolimus patients a higher-risk group, and a Framingham Risk Model analysis predicts that the patients in the cohort who received sirolimus should have had nearly double the 10-year risk of cardiovascular disease, despite having numerically less cardiovascular disease. This indicates that sirolimus has cardioprotective benefits for transplant recipients.

18.11.9 Weight Gain

Excessive weight gain is common after liver transplant, and up to 50% of patients become obese after transplant. In particular, patients on cyclosporine and steroids are very prone to weight gain and obesity. This has significant repercussions since obesity is associated with diabetes, hypertension, and cardiovascular risk. The mTOR pathway controls cellular homeostasis, by incorporating nutrient-sensing signals, and it is a critical mediator in regulating satiety and body weight [150,151]. During states of energy deprivation or hypoxia, mTOR is inactivated to limit cellular proliferation and growth. Thus mTOR inhibitors themselves can replicate a physiologic condition of decreased energy stores. In a study of 210 *de novo* liver recipients [152] receiving initial sirolimus immunosuppression (Figure 18.6), the median weight was significantly lower (and essentially unchanged) in the sirolimus cohort compared to a control group at both 2 years (75.3 kg vs. 84.1 kg, $P = 0.05$) and 5 years post-transplant (79.5 kg vs. 88.6 kg, $P = 0.04$). Immunosuppression using mTOR inhibitors may represent a way to limit the typical post-transplant weight gain that is so common in patients receiving CNI and steroids. Conversely, mTOR inhibitors should be avoided in sarcopenic or malnourished transplant recipients in the immediate postoperative periods in order to allow weight gain to occur.

18.11.10 Antiproliferative—Vasculopathy

The mTOR pathway is crucial for cellular proliferation and mTOR inhibitors such as sirolimus and everolimus have antiproliferative effects that can be beneficial. In cardiac transplant, cardiac allograft vasculopathy is a major factor limiting the long-term survival after transplantation. The vasculopathy is difficult to treat and can lead to retransplantation. Allograft vasculopathy is an

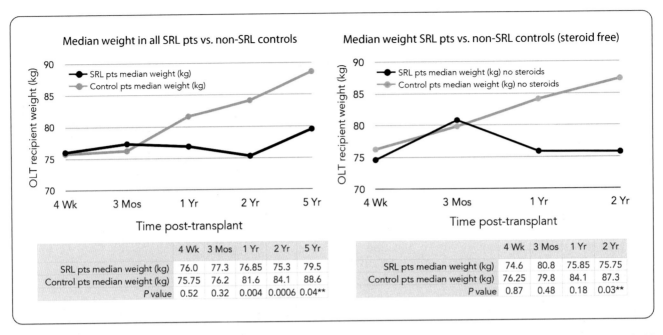

FIGURE 18.6   Impact of *de novo* sirolimus immunosuppression on post-transplant weight gain in 210 liver transplant recipients, stratified by steroid dosing. Sirolimus led to significantly reduced post-transplant weight gain, abrogating the impact of CNI on weight [152]. SRL, sirolimus; OLT, liver transplant recipient.

accelerated form of cardiovascular disease that is characterized by concentric fibrous intimal hyperplasia along the length of the coronary vessels [153] and this hyperplasia represents a target for antiproliferative therapeutics.

Sirolimus and everolimus have both been shown to reduce intimal thickening as seen by intravascular ultrasound (IVUS) in numerous clinical trials [58,59,154]. Rapidly progressive first-year intimal thickening seen on IVUS has been validated as predictive of poor outcome and predictive of developing chronic allograft vasculopathy by 5 years. Raichlin et al. [58] showed sirolimus attenuated any mean plaque volume increase leaving the plaque volume almost unchanged compared to controls (0.1 mm vs. 1.28 mm). Toplinsky et al. [59] showed sirolimus limited plaque progression compared to controls (0.7 vs. 9.3, $P = 0.0003$).

The number of trials, and the consistency of the findings regarding attenuation of chronic allograft vasculopathy by mTOR inhibitors, emphasize that using sirolimus or everolimus as part of the immunosuppression regimen in cardiac transplantation clearly is a valid strategy to minimize the development of vasculopathy in the allograft.

18.11.11 Antiproliferative—Hepatitis C Virus

Because of the mTOR pathway involvement on fibrosis and wound healing, as well as its role in controlling viral replication, mTOR inhibitors can impact hepatitis C virus (HCV) viral activity and the fibrosis progression that leads to HCV recurrence.

Several studies have shown a benefit in mTOR inhibitors limiting fibrosis progression. A large trial of liver recipients using *de novo* sirolimus monitored the progression of fibrosis with protocolized yearly biopsies [44] and the sirolimus cohort showed improved mean fibrosis stage and significantly lower incidence of advanced fibrosis (stage ≥ 2) at both 1 and 2 years post-transplant compared to controls. Another large trial [155] described a reduction in fibrosis scores on serial biopsies at a mean of 16.6 months in the patients taking sirolimus. This second trial also showed a reduced level of viral activity in the sirolimus group. Both trials showed numerically improved survival in the sirolimus groups, although these improvements were not statistically significant. In contrast to the trials that used biopsy data to demonstrate that sirolimus improved fibrosis progression with no difference in survival, is a recent study of the SRTR Database [156]. This registry database showed that there was increased risk of 3-year mortality in HCV patients treated with sirolimus, although that impact was not seen in non-HCV patients treated with sirolimus. Because this study reviewed a large registry database, the study lacks the granularity to elucidate the discrepancies with the biopsy-related studies. It is doubtful, however, whether this discrepancy will need to be resolved in the future given the current success of antiretroviral and other anti-HCV therapies in eliminating HCV.

18.12 SUMMARY

mTOR inhibitors came of age in the field of transplantation. This is where mTOR inhibitors were first approved by the FDA, and it is where the drugs were first used in a clinical setting. mTOR inhibitors are a class of drugs with a significant breadth of complications, largely because the mTOR pathway is so intimately involved in cellular function and homeostasis. These complications have limited the full adoption of these drugs in transplantation, and the key to broader adoption is a better understanding and communication of how to manage these potential complications.

The relationship between transplantation and mTOR inhibitors has been a complicated one, filled with misunderstanding and mistrust. However there has been a cadre of clinicians who have remained faithful to the class, and continued to use these drugs—even off-label when necessary—because of a belief in their potential and of their benefits and unique advantages. The study of mTOR inhibitors in transplantation has fostered new avenues of research in other areas of medicine, making these drugs some of the most exciting and diverse therapies in many other fields. Maybe, as they reach broad acceptance elsewhere, mTOR inhibitors might ultimately find acceptance in transplantation.

References

[1] Calne RY, White DJ, Thiru S, Evans DJ, McMaster P, Dunn DC, et al. Cyclosporin A in patients receiving renal allografts from cadaver donors. Lancet 1978;2:1323–7.

[2] Shihab F, Christians U, Smith L, Wellen JR, Kaplan B. Focus on mTOR inhibitors and tacrolimus in renal transplantation: pharmacokinetics, exposure-response relationships, and clinical outcomes. Transpl Immunol 2014;31:22–32.

[3] Nankivell BJ, Burrows RJ, Fung CL, O'Connell PJ, Allen RD, Chapman JR. The natural history of chronic allograft nephropathy. N Engl J Med 2003;349:2326–33.

[4] Lamattina JC, Foley DP, Mezrich JD, Fernandez LA, Vidyasagar V, D'Alessandro AM, et al. Chronic kidney disease stage progression in liver transplant recipients. Clin J Am Soc Nephrol 2011;6:1851–7.

[5] Vézina C, Kudelski A, Sehgal SN. Rapamycin (AY-22,989) a new antifungal antibiotic. I. Taxonomy of the producing streptomycete and isolation of the active principle. J Antibiot 1975;28:721–6.

[6] Thomson AW, Woo J. Immunosuppressive properties of FK-506 and rapamycin. Lancet 1989;19:443–4.

[7] Martel RR, Klicius J, Galet S. Inhibition of the immune response by rapamycin, a new antifungal antibiotic. Can J Physiol Pharmacol 1977;55:48–51.

[8] Meiser BM, Wang J, Morris RE. Rapamycin: a new and highly active immunosuppressive macrolide with an efficacy superior to cyclosporine. Progress in Immunology. Berlin Heidelberg: Springer; 19891195–8

[9] Morris RE, Meiser BM. Identification of a new pharmacologic action for an old compound. Med Sci Res 1989;17:609–10.

[10] Calne RY, Collier DS, Lim S, Pollard SG, Samaan A, White DJ, et al. Rapamycin for immunosuppression in organ allografting. Lancet 1989;2:227.

[11] Gummert JF, Ikonen T, Morris RE. Newer immunosuppressive drugs: a review. J Am Soc Nephrol 1999;10:1366–80.

[12] Whiting PH, Woo J, Adam BJ, Hasan NU, Davidson RJ, Thomson AW. Toxicity of rapamycin—a comparative and combination study with cyclosporine at immunotherapeutic dosages in the rat. Transplantation 1991;52:203–8.

[13] Corthay A. A three-cell model for activation of naïve T helper cells. Scand J Immunol 2006;64:93–6.

[14] Dumont FJ, Su Q. Mechanism of action of the immunosuppressant rapamycin. Life Sci 1996;58:373–95.

[15] Benjamin D, Columbi M, Moroni C, Hall MN. Rapamycin passes the torch: a new generation of mTOR inhibitors. Nat Rev Drug Discov 2011;10:868–80.

[16] Terada N, Lucas JJ, Szepesi A, Franklin RA, Domenico J, Gelfand EW. Rapamycin blocks cell cycle progression of activated T cells prior to events characteristic of the middle to late G1 phase of the cycle. J Cell Physiol 1993;154:7–15.

[17] Gibbons JJ, Abraham RT. Mammalian target of rapamycin: discovery of rapamycin reveals a signaling pathway important for normal and cancer cell growth. Sem Oncol 2009;36:S3–17.

[18] Li J, Kim SG, Blenis J. Rapamycin: one drug, many effects. Cell Metab 2014;19:373–9.

[19] Mahalati K, Kahan BD. Clinical pharmacokinetics of sirolimus. Clin Pharmacokinet 2001;40:573–85.

[20] Lebwohl D, Anak O, Sahmoud T, Klimovsky J, Elmroth I, Haas T, et al. Development of everolimus, a novel oral mTOR inhibitor, across a spectrum of diseases. Ann NY Acad Sci 2013;1291:14–32.

[21] Kirchner GI, Meier-Wiedenbach I, Manns MP. Clinica pharmacokinetics of everolimus. Clin Pharmacokinet 2004;43:83–95.

[22] Trotter JF, Lizardo-Sanchez L. Everolimus in liver transplantation. Curr Opin Organ Transplant 2014;19:578–82.

[23] Salvadori M, Bertoni E. Long-term outcome of everolimus treatment in transplant patients. Transpl Res Risk Manage 2011;3:77–90.

[24] Kahan BD. Concentration-controlled immunosuppressive regimens using cyclosporine with sirolimus of brequinar in human renal transplantation. Transplant Proc 1995;27:33–6.

[25] MacDonald AS, Rapamune Global Study Group. A worldwide, phase III, randomized, controlled, safety and efficacy study of a sirolimus/cyclosporine regimen for prevention of acute rejection in recipients of primary mismatched renal allografts. Transplantation 2001;71:271–80.

[26] Vitko S, Tedesco H, Eris J, Pascual J, Whelchel J, Magee JC, et al. Everolimus with optimized cyclosporine dosing in renal transplant recipients: 6-month safety and efficacy results of two randomized studies. Am J Transplant 2004;4:626–35.

[27] Kahan BD. Efficacy of sirolimus compared with azathioprine for reduction of acute allograft rejection: a randomized multicenter study. The Rapamune US Study Group. Lancet 2000;356:194–202.

[28] De Simone P, Metselaar HJ, Fischer L, Dumortier J, Boudjema K, Hardwigsen J, et al. Conversion from a calcineurin inhibitor to everolimus therapy in maintenance liver transplant recipients: a prospective, randomized, multicenter trial. Liver Transpl 2009;15:1262–9.

[29] Abdelmalek MF, Humar A, Stickel F, Andreone P, Pascher A, Barroso E, et al. Sirolimus conversion regimen versus continued calcineurin inhibitors in liver allograft recipients: a randomized trial. Am J Transplant 2012;12:694–705.

[30] Eisen HJ, Tuzcu EM, Dorent R, Kobashigawa J, Mancini D, Valantine-von Kaeppler HA, et al. Everolimus for the prevention of allograft rejection and vasculopathy in cardiac-transplant recipients. N Engl J Med 2003;349:847–58.

[31] Oberbauer R, Kreis H, Johnson RW, Mota A, Claesson K, Ruiz JC, et al. Long-term improvement in renal function with sirolimus after early cyclosporine withdrawal in renal transplant recipients: 2-year results of the Rapamune Maintenance Regimen Study. Transplantation 2003;76:364–70.

[32] Russ G, Segoloni G, Oberbauer R, Legendre C, Mota A, Eris J, et al. Superior outcomes in renal transplantation after early cyclosporine withdrawal and sirolimus maintenance therapy, regardless of baseline renal function. Transplantation 2005;80:1204–11.

[33] Schena FP, Pascoe MD, Alberu J, del Carmen Rial M, Oberbauer R, Brennan DC, et al. Conversion from calcineurin inhibitors to sirolimus maintenance therapy in renal allograft recipients: 24-month efficacy and safety results from the CONVERT trial. Transplantation 2009;87:233–42.

[34] Kahan B, Wong RL, Carter C, Katz SH, Von Fellenberg J, Van Buren CT, et al. A phase I study of a 4-week course of SDZ-RAD (RAD) quiescent cyclosporine-prednisone-treated renal transplants recipients. Transplantation 1999;68:1100–6.

[35] Vitko S, Margreiter R, Weimar W, Dantal J, Kuypers D, Winkler M, et al. Three-year efficacy and safety results from a study of everolimus versus mycophenolate mofetil in de novo renal transplant patients. Am J Transplant 2005;5:2521–30.

[36] Lorber MI, Mugaonkar S, Butt KM, Elkhammas E, Rajagopalan PR, Kahan B, et al. Everolimus versus mycophenolate mofetil the prevent of rejection in de novo renal transplant recipients: a 3-year randomized, multiventer, phase III study. Transplantation 2005;80:244–52.

[37] Cibrik D, Silva HT, Vathsala A, Lackova E, Cornu-Artis C, Walker RG, et al. Randomized trial of everolimus-facilitated calcineurin inhibitor inimization over 24 months in renal transplantation. Transplantation 2013;95:933–42.

[38] Budde K, Lehner F, Sommerer C, Reinke P, Arns W, Eisenberger U, et al. Five-year outcomes in kidney transplant patients converted from cyclosporine to everolimus: the randomized ZEUS study. Am J Transpl 2015;15:119–28.

[39] Wiesner R, Klintmalm G, McDiarmid S. Sirolimus immunotherapy results in reduced rates of acute rejection in de novo orthotopic liver transplant recipients. Am J Transplant 2002;2:464.

[40] Asrani SK, Wiesner R, Trotter JF, Klintmalm G, Katz E, Maller E, et al. De novo sirolimus and reduced dose tacrolimus versus standard-dose tacrolimus after liver transplantation: the 2000–2003 phase II prospective randomized trial. Am J Transplant 2014;14:356–66.

[41] Information for Healthcare Professions: Sirolimus (marketed as Rapamune), http://www.fda.gov/Drugs/DrugSafety/PostmarketDrugSafetyInformationforPatientsandProviders/DrugSafetyInformationforHealthcareProfessionals/ucm165015.html.

[42] Scientific Registry of Transplant Recipients. Immunosuppression use for maintenance between discharge and one year, http://www.srtr.org/annual_reports/; 2008.

[43] Molinari M, Berman K, Meeberg G, Shapiro JA, Bigam D, Trotter JF, et al. Multicentric outcome analysis of sirolimus-based immunosuppression in 252 liver transplant recipients. Transpl Int 2010;23:155–68.

[44] McKenna GJ, Trotter JF, Klintmalm E, Onaca N, Ruiz R, Jennings LW, et al. Limiting hepatitis C virus progression in liver transplant recipients using sirolimus-based immunosuppression. Am J Transplant 2011;11:2379–87.

[45] McKenna GJ, Sanchez EQ, Khan T, Nikitin D, Vasani S, Chinnakotla S, et al. Sirolimus and hepatic artery complications: reassessing the black box warning? Am J Transplant 2008;8:306.

[46] McKenna GJ, Trotter JF. Sirolimus conversion for renal dysfunction in liver transplant recipients: the devil really is in the details... Am J Transplant 2012;12:521–2.

[47] Levy G, Schmidli H, Punch J, Tuttle-Newhall E, Mayer D, Neuhaus P, et al. Safety, tolerability, and efficacy of everolimus in de novo liver transplant recipients: 12- and 36-month results. Liver Transpl 2006;12:1640–8.

[48] Masetti M, Montalti R, Rompianesi G, Codeluppi M, Gerring R, Romano A, et al. Early withdrawl of calcineurin inhibitors and everolimus monotherapy in de novo liver transplant recipients preserves renal function. Am J Transplant 2010;10:2252–62.

[49] McKenna G, Trotter JF. Does early (CNI) conversion lead to eternal (renal) salvation? Am J Transplant 2010;10:2189–90.

[50] Fischer L, Klempnauer J, Beckebaum S, Metselaar HJ, Neuhaus P, Schemmer P, et al. A randomized, controlled study to assess the conversion from calcineurin-inhibitors to everolimus after liver transplantation—PROTECT. Am J Transplant 2012;12:1855–65.

[51] De Simone P, Nevens F, De Carlis L, Metselaar HJ, Beckebaum S, Saliba F, et al. Everolimus with reduced tacrolimus improves renal function in de novo liver transplant recipients: a randomized controlled trial. Am J Transplant 2012;12:3008–20.

[52] Saliba F, De Simone P, Nevens F, De Carlis L, Metselaar HJ, Beckebaum S, et al. Renal function at two years in liver transplant patients receiving everolimus: results of a randomized, multicenter study. Am J Transplant 2013;13:1734–45.

[53] Fischer L, Saliba F, Kaiser GM, De Carlis L, Metselaar HJ, De Simone P, et al. Three-year outcomes in de novo liver transplant patients receiving everolimus with reduced tacrolimus: follow-up results from a randomized, multicenter study. Transplantation 2015;99:1455–62.

[54] Ganschow R, Pollok JM, Janofsky M, Junge G. The role of everolimus in liver transplantation. Clin Exp Gastroenterol 2014;7:329–43.

[55] Kushwaha SS. mTOR inhibitors as primary immunosuppression after heart transplant: confounding factors in clinical trials. Am J Transplant 2014;14:1958–9.

[56] Keogh A, Richardson M, Ruygrok P, Spratt P, Galbraith A, O'Driscoll G, et al. Sirolimus in de novo heart transplant recipients reduces acute rejection and prevents coronary artery disease at 2 years: a randomized clinical trial. Circulation 2004;110:2694–700.

[57] Kushwaha SS, Khalpey Z, Frantz RP, Rodeheffer RJ, Clavell AL, Daly RC, et al. Sirolimus in cardiac transplantation: use as a primary immunosuppressant in calcineurin inhibitor-induced nephrotoxicity. J Heart Lung Transplant 2005;24:2129–36.

[58] Raichlin E, Bae JH, Khalpey Z, Edwards BS, Kremers WK, Clavell AL, et al. Conversion to sirolimus as primary immunosuppression attenuates the progression of allograft vasculopathy after cardiac transplantation. Circulation 2007;116:2726–33.

[59] Toplinsky Y, Hasin T, Raichlin E, Boilson BA, Schirger JA, Pereira NL, et al. Sirolimus as primary immunosuppression attenuates allograft vasculopathy with improved late survival and decreased cardiac events after cardiac transplantation. Circulation 2012;125:708–20.

[60] Guethoff S, Stroeh K, Grinninger C, Koenig MA, Kleinert EC, Rieger A, et al. De novo sirolimus with low-dose tacrolimus versus full-dose tacrolimus with mycophenolate mofetil after heart transplantation—8 year results. J Heart Lung Transplant 2015;34:634–42.

[61] Viganò M, Tuzcu EM, Benza R, Boissonnat P, Haverich A, Hill J, et al. Prevention of acute rejection and allograft vasculopathy by everolimus in cardiac transplants recipients: a 24-month analysis. J Heart Lung Transplant 2007;26:584–92.

[62] Gurk-Turner C, Manitpisitkul W, Cooper M. A comprehensive review of everolimus clinical reports: a new mammalian target of rapamycin inhibitor. Transplantation 2012;94:659–68.

[63] Lehmkuhl HB, Arizon J, Vigano M, Almenar L, Gerosa G, Maccherini M, et al. Everolimus with reduced cyclosporine versus MMF with standard cyclosporine in de novo heart transplant recipients. Transplantation 2009;88:115–22.

[64] Schweiger M, Wasler A, Prenner G, Stiegler P, Stadlbauer V, Schwartz M, et al. Everolimus and reduced cyclosporine rough levels in maintenance heart transplant recipients. Transpl Immunol 2006;16:46–51.

[65] Arora S, Gude E, Sigurdardottir V, Mortensen SA, Eiskjaer H, Riise G, et al. Improvement in renal function after everolimus introduction and calcineurin inhibitor reduction in maintenance thoracic transplant recipients: the significance of baseline glomerular filtration rate. J Heart Lung Transplant 2012;31:259–65.

[66] Arora S, Andreassen AK, Andersson B, Gustafsson F, Eiskjaer H, Botker HE, et al. The effect of everolimus initiation and calcineurin inhibitor elimination on cardiac allograft vasculopathy in de novo recipients: one-year results of a Scandinavian Randomized Trial. Am J Transplant 2015;15:1967–75.

[67] Busaidy NL, Farooki A, Dowlati A, Perentesis JP, Dancey JE, Doyle LA, et al. Management of metabolic effects associated with anticancer agents tareting the PI3K-Akt-mTOR pathway. J Clin Oncol 2012;30:2919–28.

[68] Houde VP, Brûlé S, Festuccia WT, Blanchard PG, Bellmann K, Deshaies Y, et al. Chronic rapamycin treatment causes glucose intolerance and hyperlipidemia by upregulating and impairing lipid deposition in adipose tissue. Diabetes 2010;59:1338–48.

[69] Ma KL, Ruan XZ, Powis SH, Moorhead JF, Varghese Z. Anti-atherosclerotic effects of sirolimus on human vascular smooth muscles cells. Am J Physiol Heart Circ Physiol 2007;292:H2721–8.

[70] Mathis AS, Jin X, Friedman GS, Peng F, Carl SM, Knipp GT. The pharmacodynamics effects of sirolimus and sirolimus-calcineurin inhibitor combinations on macrophage scavenger and nuclear hormone receptors. J Pharm Sci 2007;96:209–22.

[71] Lipid Research Clinic. The lipid research clinics coronary primary prevention trial results: I. Reduction in incidence of coronary heart disease. JAMA 1984;251:351–64.

[72] McKenna GJ, Trotter JF, Klintmalm E, Ruiz R, Onaca N, Testa G, et al. Sirolimus and cardiovascular disease risk in liver transplantation. Transplantation 2013;15:215–21.

[73] Campistol JM, de Fijter JW, Flechner SM, Langone A, Morelon E, Stockfleth E. mTOR inhibitor-associated dermatologic and mucosal problems. Clin Transplant 2010;24:149–56.

[74] Kaplan B, Qazi Y, Wellen JR. Strategies for the management of adverse events associated with mTOR inhibitors. Transplant Rev 2014;28:126–33.

[75] Mahé E, Morelon E, Lechaton S, Sand KH, Mansouri R, Ducasse MF, et al. Cutaneous adverse events in renal transplant recipients receiving sirolimus-based therapy. Transplantation 2005;79:476–82.

[76] Sonis S, Treister N, Chawla S, Demetri G, Haluska F. Preliminary characterization of oral lesions associated with inhibitors of mammalian target of rapamycin in cancer patients. Cancer 2010;116:210–15.

[77] Shegogue D, Trohanowksa M. Mammalian target of rapamycin positively regulates collagen type 1 production via a phosphatidylinositol 3-kinase-independent pathway. J Biol Chem 2004;279:23166–75.

[78] Akselband Y, Harding MW, Nelson PA. Rapamycin inhibits spontaneous and fibroblast growth factor beta-stimulated proliferation of endothelial cells and fibroblasts. Transplant Proc 1991;23:2833–6.

[79] Sehgal SN. Rapamune (RAPA, rapamycin, sirolimus): mechanism of action of immunosuppressive effects results from blockade of signal transduction and inhibition of cell cycle progression. Clin Biochem 1998;31:335–40.

[80] Bongelo RGB, Fuhro R, Wang Z, Valeri CR, Andry C, Salant DJ, et al. Rapamycin ameliorates proteinuria-associated tubulointerstitial inflammation and fibrosis in experimental membranous nephropathy. J Am Soc Nephrol 2005;16:2063–72.

[81] Knight RJ, Villa M, Laskey R, Benavides C, Schoenberg L, Welsh M, et al. Risk factors for impaired wound healing in sirolimus-treated renal transplant recipients. Clin Transplant 2007;21:460–5.

[82] Kahn J, Muller H, Iberer F, Kniepeiss D, Duller D, Rehak P, et al. Incisional hernia following liver transplantation: incidence and predisposing factors. Clin Transplant 2007;21:423–6.

[83] Chinnakotla S, Davis GL, Vasani S, Kim P, Tomiyama K, Sanchez E, et al. Impact of sirolimus on the recurrence of hepatocellular carcinoma after liver transplantation. Liver Transpl 2009;15:1834–42.

[84] Dunkleberg JC, Trotter JF, Wachs M, Bak T, Kugelmas M, Steinberg T, et al. Sirolimus as primary immunosuppression in liver transplantation is not associated with hepatic artery or wound complications. Liver Transpl 2003;9:463–8.

[85] Watson CJE, Gimson AES, Alexander GJ, Allison ME, Gibbs P, Smith JC, et al. A randomized controlled trial of late conversion from calcineurin inhibitor (CNI)-based to sirolimus-based immunosuppression in liver transplant recipients with impaired renal function. Liver Transpl 2007;13:1694–702.

[86] Muller-Krebs S, Weber L, Tsobaneli J, Kihm LP, Reiser J, Zeier M, et al. Cellular effects of everolimus and sirolimus on podocytes. PLoS One 2013;8:e80340.

[87] Abbatae M, Zoja C, Remuzzi G. How does proteinuria cause progressive renal damage? J Am Soc Nephrol 2006;17:2974–84.

[88] Blanco S, Vaquero M, Gomez-Guerrero C, Lopez D, Egido J, Romero R. Potential role of angiotensin-converting enzyme inhibitors and statins on early podocyte damage in a model of type 2 diabetes mellitus, obesity, and mild hypertension. Am J Hypertens 2005;18:557–65.

[89] Zoja C, Corna D, Gagliardini E, Conti S, Arnaboldi L, Benigni A, et al. Adding a statin to a combination of ACE inhibitor and ARB normalizes proteinuria in experimental diabetes, which translates into full renoprotection. Am J Physiol Renal Physiol 2010;299:F1203–11.

[90] Morcos A, Nair S, Keane MP, McElvaney NG, McCormick PA. Interstitial pneumonitis is a frequent complication in liver transplant recipients treated with sirolimus. Ir J Med Sci 2012;181:231–5.

[91] Lopez P, Kohler S, Dimri S. Interstitial lung disease associated with mTOR inhibitors in solid organ transplant recipients: results from a large phase III clinical trial program of everolimus and review of the literature. J Transplant 2014. Available from: http://dx.doi.org/10.1155/2014/305931.

[92] Roberts RJ, Wells AC, Unitt E, Griffiths M, Tasker AD, Allison ME, et al. Sirolimus-induced pneumonitis following liver transplantation. Liver Transpl 2007;13:853–6.

[93] Alexandru S, Ortiz A, Baldovi S, Milicua JM, Ruiz-Escribano E, Egido J, et al. Severe everolimus-associated pneumonitis in a renal transplant recipient. Nephrol Dial Transplant 2008;23:3353–5.

[94] Creel PA. Management of mTOR inhibitor side effects. Clin J Oncol Nurs 2009;13(Suppl):19–23.

[95] Albiges L, Chamming's F, Duclos B, Stern M, Motzer RJ, Ravaud A, et al. Incidence and management of mTOR inhibitor-associated pneumonitis in patients with metastatic renal cell carcinoma. Ann Oncol 2012;23:1943–53.

[96] Romagnoli J, Citterio F, Nanni G, Tondolo V, Castagneto M. Severe limb lymphedema in sirolimus-treated patients. Transplant Proc 2005;37:834–6.

[97] Vivarelli M, Dazzi A, Cucchetti A, Gasbarrini A, Zanello A, Di Gioia P, et al. Sirolimus in liver transplant recipients: a large single-center experience. Transplant Proc 2010;42:2579–84.

[98] De Simone P, Carrai P, Precisi A, Petruccelli S, Baldoni L, Balzano E, et al. Conversion to everolimus monotherapy in maintenance liver transplantation: feasibility, safety, and impact on renal function. Transpl Int 2009;22:279–86.

V. mTOR IN THE IMMUNE SYSTEM AND AUTOIMMUNE DISORDERS

[99] Gharbi C, Gueutin V, Izzedine H. Oedema, solid organ transplantation and mammalian target of rapamycin inhibitor/proliferation signal inhibitors (mTOR-I/PSIs). Clin Kidney J 2014;7:115—20.

[100] Mora JA, Almenar L, Martinez-Dolz L, Salvador A. mTOR inhibitors and unilateral edema. Rev Esp Cardiol 2008;61:987—8.

[101] Ribezzo M, Boffini M, Ricci D, Barbero C, Bonato R, Attisani M, et al. Incidence and treatment of lymphedema in heart transplant patients treated with everolimus. Transplant Proc 2014;46:2334—8.

[102] Huber S, Bruns CJ, Schmid G, Hermann PC, Conrad C, Niess H, et al. Inhibition of the mammalian target of rapamycin impedes lymphangiogensis. Kidney Int 2007;71:771—7.

[103] Mäkinen T, Jussila L, Veikkola T, Karpanen T, Kettunen MI, Pulkkanen KJ, et al. Inhibition of lymphangiogenesis with resulting lymphedema in transgenic mice expressing soluble VEGF receptor-3. Nat Med 2001;7:199—205.

[104] Pengel LH, Liu LQ, Morris PJ. Do wound complications or lymphoceles occur more often in solid organ transplant recipients on mTOR inhibitors? A systematic review of randomized controlled trials. Transpl Int 2011;24:1216—30.

[105] Vitko S, Margreiter R, Weimar W, Dantal J, Viljoen HG, Li Y, et al. Everolimus (Certican) 12-month safety and efficacy versus mycophenolate mofetil in de novo renal transplant recipients. Transplantation 2004;78:1532—40.

[106] Ruebner R, Goldberg D, Abt PL, Bahirwani R, Levine M, Sawinski D, et al. Risk of end-stage renal disease among liver transplant recipients with pretransplant renal dysfunction. Am J Transplant 2012;12:2958—65.

[107] Naesens M, Kuypers DRJ, Sarwal M. Calcineurin inhibitor nephrotoxicity. Clin J Am Soc Nephrol 2009;4:481—508.

[108] Fairbanks KD, Eustace JA, Fine D, Thuluvath PJ. Renal function improves in liver transplant recipients when switched from a calcineurin inhibitor to sirolimus. Liver Transpl 2003;9:1079—85.

[109] Cotterell AH, Fisher RA, King AL, Gehr TW, Dawson S, Sterling RK, et al. Calcineurin inhibitor-induced chronic nephrotoxicity in liver transplant patients is reversible using rapamycin as the primary immunosuppressive agent. Clin Transplant 2002;16:49—51.

[110] Asrani SK, Leise MD, West CP, Murad MH, Pederson RA, Erwin PJ, et al. Wieser use of sirolimus in liver transplant recipients with renal insufficiency: a systematic review and meta-analysis. Hepatology 2010;52:1360—70.

[111] Teperman L, Moonka D, Sebastian A, Sher L, Marotta P, Marsh C, et al. Calcineurin inhibitor-free mycophenolate mofetil/sirolimus maintenance in liver transplantation: the randomized spare-the-nephron trial. Liver Transpl 2014;19:675—89.

[112] McKenna GJ, Trotter JF, Klintmalm E, Onaca N, Ruiz R, Campsen J, et al. The impact of timing in sirolimus conversion for renal insufficiency in liver transplant recipients. Liver Transpl 2011;17:S84.

[113] Senzolo M, Ferronato C, Burra P. Neurologic complications after solid organ transplantation. Transpl Int 2009;22:269—78.

[114] Bechstein WO. Neurotoxicity of calcineurin inhibitors: impact and clinical management. Transpl Int 2000;13:313—26.

[115] Forgacs B, Merhav HJ, Lappin J, Miele L. Successful conversion to rapamycin for calcineurin inhibitor related neurotoxicity following liver transplantation. Transplant Proc 2005;37:1912—14.

[116] Boros P, Bromberg JS. Human FOXP3 + regulatory T cells in transplantation. Am J Transplant 2009;9:1719—24.

[117] Louis S, Braudeau C, Giral M, Dupont A, Moizant F, Robillard N, et al. Contrasting CD25hiCD4 + Tcells/FOXP3 patterns in chronic rejection and operational drug-free tolerance. Transplantation 2006;81:398—407.

[118] Kim KW, Chung BH, Kim BM, Cho ML, Yang CW. The effect of mammalian target of rapamycin inhibition on T helper type 17 and regulatory T cell differentiation in vitro and in vivo in kidney transplant recipients. Immunology 2015;144:68—78.

[119] Levitsky J, Mathew JM, Abecassis M, Tambur A, Leventhal J, Chandrasekaran D, et al. Systemic immunoregulatory and proteogenomic effects of tacrolimus to sirolimus conversion in liver transplant recipients. Hepatology 2013;57:239—48.

[120] Gao W, Lu Y, El Essawy B, Oukka M, Kuchroo VK, Strom TB. Contrasting effects of cyclosporine and rapamycin in de novo generation of alloantigen specific regulatory T cells. Am J Transplant 2007;7:1722—32.

[121] Gong N, Chen Z, Wang J, Fang A, Li Y, Xiang Y, et al. Immunoregulatory effects of sirolimus versus tacrolimus treatment in kidney allograft recipients. Cell Immunol 2015; (in press).

[122] Hojo M, Morimoto T, Maluccio M, Asano T, Morimoto K, Lagman M, et al. Cyclosporine induces cancer progression by a cell-autonomous mechanism. Nature 1999;397:530—4.

[123] Vivarelli M, Cucchetti A, La Barba G, Ravaiolo M, Del Gaudio M, Lauro A, et al. Liver transplantations for hepatocellular carcinoma under calcineurin inhibitors: reassessment of risk factors for tumor recurrence. Ann Surg 2008;248:857—62.

[124] Friese CE, Ferrel L, Liu T, Ascher NL, Roberts JP. Effects of systemic cyclosporine on tumor recurrence after liver transplantation in a model of hepatocellular carcinoma. Transplantation 1999;67:510—5113.

[125] Finn RS. Current and future treatment strategies with advanced hepatocellular carcinoma: role of mTOR inhibition. Liver Cancer 2012;1:247—56.

[126] Fasolo A, Sessa C. Targeting mTOR pathways in human malignancies. Curr Pharm Dis 2012;18:2766—77.

[127] Menon S, Yecies JL, Zhang HH, Howell JJ, Nicholatos J, Harputugil E, et al. Chronic activation of mTOR complex 1 is sufficient to cause hepatocellular carcinoma in mice. Sci Signal 2012;5:24—51.

[128] Sahin F, Kannangai R, Adegbola O, Wang J, Su G, Torbenson M. mTOR and F70 S6 kinase expression in primary liver neoplasms. Clin Cancer Res 2004;10:8421—5.

[129] Zhu AX, Abrams TA, Miksad R, Blaszkowsky LS, Meyerhardt JA, Zheng H, et al. Phase 1-2 study of everolimus in advanced hepatocellular carcinoma. Cancer 2011;117:5094—102.

[130] Decaens T, Luciani A, Itti E, Hulin A, Roudot-Thoraval F, Laurent A, et al. Phase II study of sirolimus in treatment-naïve patients with advanced hepatocellular carcinoma. Dig Liver Dis 2012;44:610—16.

[131] Toso C, Merani S, Bigam DL, Shapiro AM, Kneteman NM. Sirolimus-based immunosuppression is associated with increased survival after liver transplantation for hepatocellular carcinoma. Hepatology 2010;51:1237—43.

[132] Zimmerman MA, Trotter JF, Wachs M, Bak T, Campsen J, Skibba A, et al. Sirolimus-based immunosuppression following liver transplantation for hepatocellular carcinoma. Liver Transpl 2008;14:633—8.

[133] Menon KV, Hakeem AR, Heaton ND. Meta-analysis: recurrence and survival following the use of sirolimus in liver transplantation for hepatocellular carcinoma. Aliment Pharmacol Ther 2013;37:411—19.

[134] Liang W, Wang D, Ling X, Kao AA, Kong Y, Shang Y, et al. Sirolimus-based immunosuppression in liver transplantation for hepatocellular carcinoma: a meta-analysis. Liver Transpl 2012;18:62—9.

[135] Schnitzbauer AA, Zuelke C, Graeb C, Rochon J, Bilbao I, Burra P, et al. A prospective randomized open-labeled, trial comparing sirolimus contining versus mTOR-inhibitor free

immunosuppression in patients undergoing liver transplantation for hepatocellular carcinoma. BMC Cancer 2010;10:190.

[136] Schrem H, Barg-Hock H, Strassbourg CP, Schwarz A, Klempnauer J. Aftercare for patients with transplant organs. Dtsch Arztebl Int 2009;106:148−56.

[137] Krynitz B, Edgren G, Lindelof B, Baecklund E, Brattstrom C, Wilczek H, et al. Risk of skin cancer and other malignancies in kidney, liver, heart and lung transplant recipients 1970 to 2008—a Swedish population-based study. Int J Cancer 2013;132:1429−38.

[138] Alter M, Satzger I, Schrem H, Kapp A, Gutzmer R. Nonmelanoma skin cancer is reduced after switch of immunosuppressionto mTOR-inhibitors in organ transplant recipients. J Dtsch Dermatol Ges 2014;12:480−8.

[139] Salgo R, Gossmann J, Schofer H, Kachel HG, Kuck J, Geiger H, et al. Switch to a sirolimus-based immunosuppression in long-term renal transplant recipients: reducedr rate of (pre-) malignancies and nonmelanoma skin cancer in a prospective, randomized, assessor-blinded, controlled clinical trial. Am J Transplant 2010;10:1385−93.

[140] Campbell SB, Walker R, Tai SS, Jiang Q, Russ GR. Randomized controlled trial of sirolimus for renal transplant recipients at high risk for nonmelanoma skin cancer. Am J Transplant 2012;12:1146−56.

[141] Alberu J, Pascoe MD, Campistol JM, Schena FP, Rial Mdel C, Polinsky M, et al. Lower malignancy rates in renal allograft recipients converted to siorlimus-based calcineurin inhibitor-free immunotherapy: 24 month results from the CONVERT trial. Transplantation 2011;92:302−10.

[142] Euvard S, Morelon E, Rostaing L, Goffin E, Brocard A, Tromme I, et al. Sirolimus and secondary skin-cancer prevention in kidney transplantation. N Engl J Med 2012;367: 329−39.

[143] Hoogendijk-van den Akker JM, Harden PN, Holtsma AJ, Proby CM, Wolterbeek R, Bouwes Bavinck JN, et al. Two-year randomized controlled prospective trial converting treatment of stable renal transplant recipients with cutaneous invasive squamous cell carcinomas to sirolimus. J Clin Oncol 2013;31:1317−27.

[144] Noris M, Remuzzi G. Atypical hemolytic-uremic syndrome. N Engl J Med 2009;361:1676−87.

[145] Pham PT, Peng A, Wilkinson AH, Gritsch HA, Lassman C, Pham PC, et al. Cyclosporine and tacrolimus-associated thrombotic microangiopathy. Am J Kidney Dis 2000;36:844−50.

[146] Reynolds JC, Agodoa LY, Yuan CM, Abbott KC. Thrombotic microangiopathy after renal transplantation in the United States. Am J Kidney Dis 2003;42:1058−68.

[147] Kudchodkar SB, Yu Y, Maguire TG. Human cytomegalovirus infection induces rapamycin-insensitive phosphorylation of downstream effects of mTOR kinase. J Virol 2004;78:11030−9.

[148] Haririan A, Morawski K, West M. Sirolimus exposure during the early post-transplant period reduces the risk of CMV infection relative to tacrolimus in renal allograft recipients. Clin Transplant 2007;21:466−71.

[149] McKenna GJ, Sanchez EQ, Khan T, Nikitin S, Vasani S, Chinnakotla S, et al. Protection from cytomegalovirus infectin using sirolimus as initial immunosuppression in liver transplants. Am J Transplant 2008;8:545.

[150] Dias VC, Madsen KL, Mulder KE, Keelan M, Yatscoff RW, Thomson ABR. Oral administration of rapamycin and cyclosporine differentially alter intestinal function in rabbits. Dig Dis Sci 1998;43:2227−36.

[151] Rovira J, Arellano EM, Burke JT, Brault Y, Moya-Rull D, Banon-Maneus E, et al. Effect of mTOR inhibitor on body weight: from an experimental rat model to human transplant patients. Transpl Int 2008;21:992−8.

[152] McKenna GJ, Trotter J, Kintmalm E, Sanchez EQ, Chinnakotla S, Randall HB, et al. The effect of sirolimus on body weight in liver transplantation: can we limit a major comorbidity? Hepatology 2009;50:590A.

[153] Ramzy D, Rao V, Brahm J, Miriuka S, Delgado D, Ross HJ. Cardiac allograft vasculopathy. Can J Surg 2005;48:319−27.

[154] Raichlin E, Khalpey Z, Kremers W, Frantz RP, Rodeheffer RJ, Clavell AL, et al. Replacement of calcineurin-inhibitors with sirolimus as primary immu nosuppression in stable cardiac transplant recipients. Transplantation 2007;84:467−74.

[155] Asthana S, Toso C, Meeberg G, Bigam DL, Mason A, Shapiro AJM, et al. The impact of sirolimus on hepatitis C recurrence after liver transplantation. Can J Gastroenterol 2011;25: 28−34.

[156] Watt K, Dierkhising R, Heimbach J, Charlton M. Impact of sirolimus and tacrolimus on mortality & graft loss in liver transplant recipients with and without hepatitis C virus: an analysis of the scientific Registry of Transplant Recipients Database. Liver Transpl 2012;18:1029−36.

19

mTOR and Neuroinflammation

Filipe Palavra[1,2], António Francisco Ambrósio[1,2] and Flávio Reis[1,2]

[1]Laboratory of Pharmacology and Experimental Therapeutics, Institute for Biomedical Imaging and Life Sciences (IBILI), Faculty of Medicine, University of Coimbra, Coimbra, Portugal
[2]Center for Neuroscience and Cell Biology, Institute for Biomedical Imaging and Life Sciences (CNC.IBILI) Consortium, University of Coimbra, Coimbra, Portugal

19.1 INTRODUCTION

The mammalian target of rapamycin (mTOR) is a ubiquitous protein kinase that regulates several important physiological functions. Given the wide distribution of mTOR in various cell types throughout the body, the impairment of its signaling pathways has already been described in the pathogenesis of a number of systemic disorders, such as cancer, diabetes mellitus, cardiovascular disease, and in a variety of neurological conditions. From tuberous sclerosis complex (a genetic disorder that represents the prototypical disease related to abnormal mTOR signaling) to dementia, passing through epilepsy and disorders belonging to the spectrum of autism, there are already several bridges established between the dysfunction of the nervous system and mTOR [1].

In pathophysiological terms, some of these diseases share common mechanistic aspects, in particular of an inflammatory nature. In fact, inflammation involving the nervous system, so-called neuroinflammation, has been assuming a relevant role in contemporary neuroscience and its understanding has been shedding some light on extremely complex diseases, such as, for example, typical neurodegenerative nosological entities like Alzheimer's and Parkinson's diseases [2].

Neuroinflammation is, therefore, a scientific hot topic and this chapter addresses the relationships that have been established recently between neuroinflammatory processes and mTOR signaling. This chapter is not focused on specific diseases (those will be discussed throughout the book), but generically in the role and consequences of neuroinflammation and,

of course, in the intervention of mTOR pathways in each one of them.

19.2 NEUROINFLAMMATION: AN OVERVIEW

Inflammation is a physiologic response of the innate immune system to a variety of stimuli (ranging from aseptic insults to microbial infections) that aims to protect and defend the body from dangerous variations of homeostasis produced by those initial insults. In general, an acute inflammatory response has a rapid onset and a variable limited duration. Moreover, it is generally accompanied by a systemic response characterized by a rapid alteration in the levels of several plasma proteins, known as the acute-phase response. In a great variety of clinical conditions, persistent immune activation can result in chronic inflammation and this frequently leads to pathologic consequences. In fact, the role of chronic low-grade inflammation in the pathophysiology of several diseases (even in those where, from the outset, inflammatory mechanisms would not be so intuitively linked to the underlying pathology) is gaining a critical relevance. Chronic inflammation is now seen as a key player in the genesis and/or progression of a variety of conditions and several effective therapeutic strategies precisely pass through its control [3].

Peripheral inflammation is able to trigger and potentiate the development of an inflammatory reaction that specifically involves the nervous system and that relies on some cellular and molecular players that are different from those implicated in systemic reactions. The blood—brain barrier (BBB), microglia, astrocytes, and neurons are some of these players that make inflammation in the nervous system a challenging phenomenon, with many features still to be discovered and with a great potential to help understanding the mechanisms underlying several neurological conditions.

The BBB is a highly specialized form of endothelium that was thought to completely separate the central nervous system (CNS) from the peripheral immune system. However, it is currently evident that BBB is not only permeable to proinflammatory molecules coming from peripheral inflammation, but it can also be stimulated to be a source of these mediators, both releasing and transmitting these substances and allowing leukocytes to migrate into the CNS parenchyma [4,5]. Some active transport systems have been characterized facilitating the delivery of cytokines into the brain. These interactions become even more evident in the circumventricular organs, in which an incomplete barrier at the blood—brain interface exists [6]. Proinflammatory cytokines, such as tumor necrosis factor (TNF), interleukin (IL)-6, and IL-1β are known to modulate the permeability of the BBB and one of the mechanisms by which this can be done is by altering the resistance of tight junctions in endothelial cells in brain vasculature [7]. Damaging tight junction proteins and their interaction with the cytoskeleton naturally results in an increase in permeability of the barrier, not only to molecules of considerable size which, in normal conditions, should not contact CNS parenchyma, but also for all types of cells of the immune system, mainly leukocytes. The movement of these cells across the barrier is also regulated to some extent by other humoral factors, such as chemokines. As an example, CCL19 and CCL21 enable T-cell adhesion to the BBB, whereas CXCL12 may contribute to reduce CNS T-cell infiltration [8]. Many of these molecules are produced by astrocytes and since they are a key part in defining the barrier itself, it is very interesting to see how they can contribute to the regulation of an immunological attack to the CNS.

The resident macrophages of the CNS, known as microglial cells, also play a crucial role in neuroinflammation. They are usually in a quiescent and inactive state, but in response to several proinflammatory cytokines and signaling molecules, they can turn into active cells, with phagocytic potential and with the capacity to release more proinflammatory mediators in the process. These cells can remain activated for long periods of time, releasing quantities of cytokines and neurotoxic molecules that may play a role in long-term neurodegeneration [9]. The macrophage activation process occurs by different pathways and normally allows cells to acquire one of two phenotypes: M1 or M2. M1 (classically activated) macrophages are cells with effector capacity, being stimulated by interferon-γ and TNF to produce an aggressive immune response. M2 (alternatively activated) cells are usually stimulated by IL-4 and they are able to antagonize prototypic inflammatory responses and markers, assuming a more regulatory activity [10]. Switching from M2 to M1 state seems to have an effect on the intensity of peripheral immune responses and it is likely that is also true for microglial cells and for neuroinflammation, although this is an area with some controversy and great need for research.

As indicated above, astrocytes, when stimulated, are able to release proinflammatory signaling molecules. This different family of glial cells is very likely to have a role whose relevance to inflammatory mechanisms is far superior to what is currently known. Together with microglial cells, and beyond their role in controlling the BBB, astrocytes can interfere with the regulation of synaptic function and may be involved in the development of neurodegeneration seen in some types of dementia [11]. This dynamic relationship existing between BBB endothelial cells, microglia, and astrocytes is very likely to also involve neurons, in a cellular network in which, in the presence of a stimulus generating an inflammatory reaction, the activity of a cell type directly influences the behavior of the others.

Nevertheless, for this network to be considered as such, it is imperative that cells can communicate effectively and quickly with each other. In this sense, there are molecular components of neuroinflammatory response that deserve, in this section, a particular focus. Cytokines are low-molecular-weight proteins or glycoproteins that assist in regulating the development of an immune response, causing its exacerbation or reduction. Although there are some typical pro- and anti-inflammatory cytokines, this distinction should not be rigidly attributed to specific molecules, since their signaling effects may differ, depending on the location they are in the CNS and on the context of the disease. This is related to an attribute of these molecules known as pleiotropy, but they exhibit more characteristics, such as redundancy, synergy, antagonism, and cascade induction, which permit them to regulate cellular activity in a coordinated and interactive way. As an example, IL-6 activates T cells and stimulates the production of other inflammatory molecules, such as C-reactive protein and fibrinogen [12]. Another common cascade leading to inflammation and degeneration is that of TNF receptor type 1-associated death domain protein and TNF receptor-associated factor 2 protein, which are able to recruit enzymes responsible for the activation of the transcription factor NF-κB and for the induction of c-jun N-terminal kinase pathways, culminating in the modulation of apoptosis and inflammation [13]. A direct proapoptotic effect by TNF signaling is exerted through the Fas-associated protein with death domain and caspase-8, an enzyme strongly linked to neurodegeneration [14]. When bound to the IL-1 receptor complex, IL-1β is also a relevant initiator of several signal transduction cascades in neuroinflammation, such as mitogen-activated protein kinase pathways [15].

Chemokines are a superfamily of small polypeptides, most of which contain 90−130 amino acid residues. They control the adhesion, chemotaxis, and activation of many types of leukocyte populations and subpopulations, thus being the major regulators of leukocyte traffic. Despite having a low physiological concentration inside the CNS, the levels of certain chemokines are strongly up-regulated in chronic neuroinflammation, such as monocyte chemoattractant protein-1 [16]. These molecules are also involved in regulating chemotaxis of astrocytes and microglia and they may also adversely affect neurogenesis and contribute to disrupting neuronal function.

The complement system, activated by classical, alternative, or lectin-binding pathways, is one of the most relevant effectors of the humoral branch of the immune system. Its activity in the CNS was not considered relevant until relationships between glial cells and complement proteins were characterized. C3a, C3b, and C5a, for example, have a role in controlling microglia chemotaxis and phagocytic functions, as well as in synaptic remodeling in neuroinflammation [17]. A recent clinical trial using a monoclonal antibody called eculizumab, a C5 inhibitor, demonstrated a benefit of that treatment in reducing neuroinflammation in neuromyelitis optica, an autoimmune disorder of the CNS, normally associated with the production of antiaquaporin-4 autoantibodies [18]. It is likely that the control of the activity of some proteins belonging to the complement cascade proves to be useful in treating several neurologic conditions characterized by a chronic neuroinflammatory status. Nevertheless, it is essential to increase knowledge about the relationship existing between the complement system and neuroinflammation, in order to identify new efficient therapeutic targets that may be clinically useful.

In recent years, an enzyme that is becoming increasingly associated with neuroinflammation and neurodegeneration, potentially being one of those key targets, is the enzyme cyclooxygenase (COX), which converts arachidonic acid to eicosanoids, such as prostaglandins and thromboxanes. It has two common isoforms, COX-1 and COX-2, with different roles both in normal physiology and pathology. Prostaglandin synthesis due to COX-1 activity is seen in microglia and this is one aspect that may contribute to the release of such molecules after the activation of these cells [19]. The COX-1 pathway is predominantly proinflammatory and its activation has been observed in traumatic brain injury and in neurodegenerative diseases, such as Alzheimer's disease [20,21]. On the other hand, COX-2 has notably been shown to have anti-inflammatory properties and, being expressed mainly in neurons, it seems to be associated with synaptic functioning and memory formation [22]. Cytokine signaling also interferes with these pathways. IL-1β induces COX-2 activity and prostaglandin E2 biosynthesis and the COX pathway itself stimulates the production of IL-6 [23,24], which serves as an example of the many existing feedback loops in the regulation of immune response. COX-2 production also influences leukocyte migration across a damaged BBB [25].

In summary, the concept of neuroinflammation has a wide scope and numerous complexities, not only intrinsic in their nature, but also regarding its connection with systemic inflammation [26]. It is now clear that neuroinflammatory mechanisms cause and accelerate long-term neurologic conditions that have a huge impact in modern societies.

After this overview, we will now focus on describing how the mTOR fits these mechanisms and how the complex signaling pathways mediated by that protein kinase integrate the main consequences of neuroinflammation.

19.3 OVERVIEW OF mTOR SIGNALING: UPSTREAM AND DOWNSTREAM PATHWAYS

The mTOR is an evolutionarily highly conserved serine/threonine protein kinase, member of the phosphoinositide 3-kinase (PI3K)-related kinase family and cell survival pathway. It is a molecular sensor coupled with nutrient abundance and energy that integrates extracellular and intracellular events, with impact on several brain functions and disorders, as recently reviewed [1,27−33]. mTOR is activated by phosphorylation and regulates cell growth, proliferation, metabolism, and survival in response to various growth factors, nutrients, mitogens and hormones, including glucose, amino acids, fatty acids, insulin-like growth factor-1, epidermal growth factor (EGF), vascular endothelial growth factor, insulin, and cytokines, which work as upstream modulators of mTOR kinase [34−36].

mTOR exerts its function through two complexes (mTORC1 and mTORC2), which have different composition [37–39]. The two common proteins shared by both complexes are mLST8 (mammalian lethal with Sec13 protein 8, also known as GβL), which works as a positive regulator; and DEPTOR (DEP-domain containing mTOR-interacting protein), which is a negative regulator. mTORC1 associates with Raptor (regulatory-associated protein of mTOR), a positive regulator also involved in substrate recruitment, and PRAS40 (proline-rich AKT substrate of 40 kDa), the suppressive action component, which itself is inhibited by Akt. mTORC2 associates with Rictor (rapamycin-insensitive companion of mTOR), which plays a role in the activating interaction between mTORC2 and tuberous sclerosis complex 2 (TSC2), which is a direct activator of this complex, with mSIN-1 (mammalian stress-activated protein kinase interacting protein), which is necessary for the assembly of the complex and for its capacity to phosphorylate Akt, and with PROTOR-1 (protein observed with RICTOR-1), which has been shown to bind to RICTOR and seems to play a role in enabling mTORC2 to efficiently activate serum- and glucocorticoid-induced kinase 1 (SGK1) [1,27,37–39].

mTORC1 and mTORC2 complexes also have different sensitivity to rapamycin, which inhibits mTOR phosphorylation and activation by binding to protein FKBP12 (FK 506-binding protein of 12 kDa). Although it was previously thought that mTORC2 is completely rapamycin-insensitive, it is now known that rapamycin is able to inhibit mTORC2 after prolonged exposure, at least in some cell lines [40,41]. In addition, the two complexes have distinct regulation and seem to accomplish distinct functions. Whereas mTORC1 is involved in the regulation of cell growth and proliferation in response to growth factors, mitogens, nutrients, and stress stimuli, mTORC2 has been associated with cell survival and cycle progression, being insensitive to nutrients or cellular energy. In fact, mTORC2 has been less studied, but it is already known that it phosphorylates Akt in response to hormones or growth factors, thus regulating actin cytoskeleton and cell survival [41].

Although the increasing interest in the mTOR signaling pathway has allowed considerable advances in the past few years, much remains to be known regarding the identity and way of action of mTOR upstream modulators and downstream targets/effectors, but exciting aspects with putative/possible physiological and pathophysiological impact were already revealed [42–46]. As mentioned above, mTORC1 senses growth factors, mitogens, nutrients (amino acids and energy), and stress signals, thus regulating key cellular functions, including cell growth, proliferation, and survival [47]. As an example, mTORC1 is indirectly activated by type I insulin-like growth factor receptors which stimulates PI3K to catalyze intracellular phosphatidylinositol-3,4,5-triphosphate synthesis, which activates AKT that regulates mTOR, via tumor suppressor proteins tuberous sclerosis 1 and 2 (TSC1 and 2, also named hamartin and tuberin) [48].

Nutrients (such as amino acids, particularly leucine) are also upstream positive mTOR regulators, hypothetically acting upstream of TSC1/2, even though further studies are still needed to clarify the molecular mechanisms involved [49–52]. mTORC1 also responds to energy and stress stimuli, acting as a sensor to cellular energy status, via the AMP-activated protein kinase (AMPK) and the TSC2 [53]. Under low-energy conditions, AMPK is activated causing mTOR inhibition through phosphorylation of TSC2. In addition, under hypoxic conditions, mTORC1 seems to be negatively regulated via TSC1/2 and/or AMPK-dependent mechanisms [54,55].

Several downstream targets of mTORC1 have been identified, including the recently described S6Ks (p70 ribosomal protein S6 kinase 1/2) and 4E-BPs (eukaryotic initiation factor 4 binding proteins), which are involved in the regulation of cell growth, cell cycle proliferation, and metabolism [37–39]. In addition, mTOR has also been involved in the regulation of other proteins with distinct functions, such as ornithine decarboxylase, a key enzyme in the polyamine biosynthetic pathway, hypoxia-inducible factor 1α, a transcription factor involved in the regulation of important biological processes (such as angiogenesis, inflammation, bioenergetics, proliferation, motility, and apoptosis), lipin, which plays a role in lipid biosynthesis and diacylglycerol production in the brain, as well as STAT3, which is involved in the effects mediated by several cytokines (such as IL-6 and IL-10) [44–46,56–58].

The upstream modulators of mTORC2 are as yet almost unknown, but insulin and related pathways have been suggested as the main activators. mTORC2 activation is mediated by the phosphorylation of Akt and other AGC-family kinases, such as SGK and protein kinases C (PKCs), playing an important role in cell survival and organization of the cytoskeleton [37,59]. Akt appears to be an upstream regulator of mTORC1 while also being a downstream target of mTORC2. In fact, mTOR-dependent phosphorylated Akt regulates different cellular processes, such as cell growth, proliferation, apoptosis, and glucose metabolism [60]. PKCs are also phosphorylated by mTORC2, thus having a role on actin organization and cell motility, as well as Rho GTPases, which are regulated by mTORC2 and seem to play a role in the actin cytoskeleton [41,61–63]. Finally, SGK1 is phosphorylated by mTORC2, thus regulating ion transport and growth [64,65].

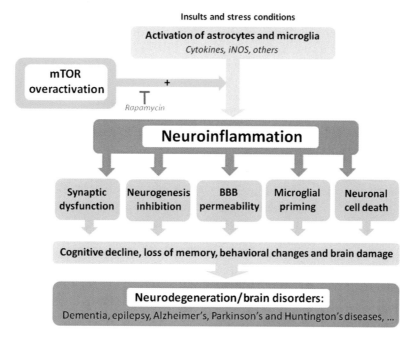

FIGURE 19.1 Schematic diagram of the proposed links between mTOR overactivation and brain disorders, in which neuroinflammation plays a crucial role. BBB, blood−brain barrier; iNOS, inducible nitric oxide synthase; mTOR, mammalian target of rapamycin.

Despite the information already known, further research is still needed to identify the downstream mTOR targets and/or clarify the precise physiological and pathological roles played in conjugation with mTORC1 and mTORC2 signaling. Even so, the already known impact of the mTOR pathway in the brain might be extremely relevant, as it is clearly associated with mechanisms with a crucial impact on the nervous system, including control of protein translation and synthesis, autophagy, and microtubule dynamics. The following sections of this chapter will discuss the putative role of mTOR pathways on neuroinflammation, a particularly relevant aspect with a broad spectrum of intervention and with several ways to disrupt CNS homeostasis, implicated in various neurological diseases (Figure 19.1).

19.4 mTOR IN NEUROINFLAMMATION AND IN ITS CONSEQUENCES

As previously stated, mTOR is a ubiquitous multieffector protein with a large structure (in contains 2549 amino acid residues, weighing approximately 250 kDa), corresponding to a highly conserved homolog of the yeast protein, target of rapamycin (TOR). Rapamycin (also known as sirolimus) is a macrolide firstly isolated from a bacteria strain of the *Streptomyces* genus, which was collected in Easter Island soil and initially described as an antifungal agent [66]. Its immunosuppressant properties were

later discovered and, in 1999, the US Food and Drug Administration approved the drug for preventing renal allograft rejection [67].

Rapamycin does not impair mTOR catalytic activity per se, but it disrupts mTOR−protein complex formation, thus blocking the signaling pathway. More precisely, rapamycin acts as an allosteric inhibitor of mTORC1, which, in a complex with FKBP12, binds to the FRB (FKBP12-rapamycin binding) domain of the kinase [68,69]. This is thought to modify the conformation of the enzyme and, therefore, it weakens the integrity of the kinase complex and prevents the association with its substrates [69]. Since it crosses the BBB and, when acutely administered, it does not affect the activity of mTORC2 or other kinases [68], rapamycin is assumed as an extremely relevant tool for studying mTORC1 function in the brain. In addition, if systemically administered, rapamycin does not induce anxiety or general locomotion impairment in rodents and so it has no negative impact on animal behavioral studies [70]. However, when a prolonged treatment with rapamycin is made, mTORC2 activity is inhibited in non-neuronal tissue (it should be noted that it is not clear whether this occurs in all brain structures) and this effect has been linked to insulin resistance [71].

19.4.1 mTOR and Synaptic Dysfunction

Synaptic dysfunction can be not only a consequence of prolonged neuroinflammation, but also features in the early stages of several pathologies, suggesting that inflammatory mechanisms may be involved in the transition from healthy synapse transmission to a disease state. Synaptic dysfunction can manifest itself in several ways, including complete loss of synaptic function (that is, naturally associated with cell loss, particularly with neuronal scarcity) and impairment of synaptic plasticity.

Evidence associating mTOR signaling to synaptic plasticity has been derived from studies using rapamycin in *Aplysia* and crayfish, which firstly intended to study long-term facilitation (LTF) in these living beings [72,73]. Rapamycin was also shown to be responsible for the blockage of eukaryotic elongation factor 2 kinase phosphorylation during LTF-mediated elongation in *Aplysia* [74]. These findings enhanced the critical role TOR signaling exerts in various phases of long-lasting plastic changes in invertebrates.

By using rapamycin, Tang and Schuman firstly demonstrated the role of mTOR in the late phase of N-methyl-D-aspartate (NMDA) receptor-dependent hippocampal long-term potentiation (LTP) and they also showed that several elements of mTOR machinery colocalized with postsynaptic markers [75]. These data suggested that mTOR inhibition impaired synaptic translation machinery and this was the starting point to several experiments showing that, in the process of establishing protein synthesis-dependent plasticity, mTOR is responsible for regulating the availability of factors involved in translation [76,77].

mTOR exerts another important function in synaptic plasticity, which is long-term depression (LTD) dependent on the metabotropic glutamate receptor (mGluR) [78]. This has been studied in mice modeling fragile X syndrome, which has allowed some understanding of the molecular aspects underlying cognitive dysfunction associated with this genetic disease [79]. But an interesting (and unanswered, so far) question arising from these data is how the activation of NMDA receptors and mGluR can both promote mTOR signaling and generate LTP and LTD, which are completely different (or even antagonistic) physiologic phenomena. Probably, one explanation may go through the activation of specific pools of mTOR contained in regions of the plasma membranes with a high level of organization, in terms of receptors. The specific trafficking of mRNA substrates to different neuronal compartments may also determine synaptic responses to stimulation and, therefore, modify the response of mTOR-dependent protein machinery to available transcripts.

But, if the activity of mTOR underlying neuronal development and synaptic plasticity has been well studied, less is known regarding its potential function in basic synaptic transmission. This issue was addressed in recent studies, using transgenic mice with altered mTORC1 signaling pathways. In FKBP12-deficient mice, in which an increase in mTORC1 activity is expected, no changes in basal synaptic transmission in hippocampal slices were observed [80]. On the other hand, two different studies showed a regulatory activity performed by mTORC1 in several features of synaptic transmission [81,82]. The molecular features justifying these differences are not known, but they can rely on autophagy [81] and/or on synapse formation [82]. Importantly, this seems to be independent of the type of synapse in consideration, since it has been described for glutamatergic, dopaminergic, and GABAergic neurons, reinforcing the universal control mTORC1 may exert in neurotransmission [81,82].

A clinical aspect that stands out from all this is the association of mTOR machinery with learning and memory. Long-lasting forms of synaptic plasticity and memory require protein synthesis and dendritic changes [83]. Since mTORC1 controls synaptic protein translation, it is intuitive to think that the kinase may have an important role in the pathophysiology of neurological conditions characterized by cognitive decline, loss of memory and behavioral changes, such as Alzheimer's, Parkinson's, and Huntington's diseases (Figure 19.1). In fact, that is already under study and there is the hope that the knowledge related to the mTOR signaling pathways could generate new therapeutic targets that can provide effective treatments to patients with these conditions [84–86].

19.4.2 mTOR and Neurogenesis Inhibition

The differentiation of neural progenitor cells to neurons, known as neurogenesis, occurs in certain restricted brain areas, such as in the subgranular layer of the dentate gyrus of the hippocampus, in the subventricular zone of lateral ventricles, and in the amygdala [87]. Some degree of dysfunction (either inhibition or stimulation) of this process can have a role in the pathophysiology of some neurological conditions, mainly in their early and presymptomatic phases. A link with dementia, particularly with Alzheimer's disease, is quite intuitive, but the extent to which neurogenesis affects the development of clinical signs and symptoms is currently unknown [88]. A number of proinflammatory cytokines (like IL-6, TNF, and IL-18) may contribute to the inhibition of neurogenesis, inducing death of neural progenitor cells [89]. These molecules are released by activated microglia and this suggests that microglia-mediated neuroinflammation may have a detrimental effect on neurogenesis.

However, this should not be seen exclusively, because changes in microglia phenotype were showed to play a role in helping the differentiation of neural progenitor cells [90].

A number of studies have implicated the mTOR signaling pathway in the regulation of self-renewal and/or proliferation of neural progenitor cells, namely in rapidly developing embryonic and early postnatal brain [91–93]. In a more recent study, Paliouras and colleagues revealed that, in addition to those early developmental effects, activation of mTOR continues to have a significant role in the mature brain, where its effects are necessary for the maintenance and proliferation of the transiently amplifying (TA) progenitor pool *in vitro* and *in vivo* [94]. In fact, mTORC1 signaling was specifically observed in TA progenitor cells of the subventricular niche, drawing attention to the possible stage-specific modulatory activity of mTOR in this brain region. This pathway seems to be important in response to different proliferative endogenous or exogenous (such as EGF) signals, acting as a mediator of cell expansion in the adult subventricular region [94]. Moreover, downregulation of mTOR signaling and quiescence of the subventricular zone during aging seem to be linked, highlighting the key regulator effect of this pathway on neurogenesis, not only in the adult, but essentially in the aging brain [94].

Thus, after documenting the existence of a relationship between the mTOR signaling pathway and the regulation of neurogenesis and neuroinflammation (common to a wide variety of neurological conditions) one of the most relevant biological aspects that can compromise the viability of neural progenitor cells, it makes sense to think of using mTOR as a therapeutic target to enhance neuronal repair mechanisms. Nevertheless, although attractive, this concept lacks extensive research.

19.4.3 mTOR and Neuronal Death

Neurons can die by two essential mechanisms: necrosis, as occurs in acute ischemia or trauma, and apoptosis, which is programmed cell death, considered as part of normal physiology, but also as a relevant mechanism in chronic neurodegenerative disorders. Neuroinflammation may be responsible for an activation of apoptotic mechanisms, since many proapoptotic pathways are up-regulated by signaling molecules that are produced in excess by cells implicated in inflammation. This suggests that neuroinflammation can directly interfere with neuronal apoptosis and, thus, it can cause acute damage and

accelerate long-term degeneration. The TNF family of cytokines is one of the most involved in activating proapoptotic machinery and molecules such as tumor-necrosis-factor-related apoptotic-ligand have been shown to cause neuronal death [95]. This can also be the outcome of the activation of several receptors, such as TNF receptors 1 and 2 (TNFR1 and TNFR2, respectively) [96].

Nitric oxide (NO) causes neuronal apoptosis by inhibiting mitochondrial metabolism and therefore increasing glutamate release and excitotoxic cell death [97]. This is a relevant aspect in neuroinflammation, since NO is synthesized and released from astrocytes and microglia, mainly through the enzyme inducible NO synthase (iNOS). The expression of iNOS is tightly regulated by complex molecular mechanisms, which involve transcriptional and post-transcriptional aspects. With respect to this, the mTOR pathway has been described as having both stimulatory and inhibitory actions, depending on the cell type and activating stimulus [98–100]. In a study by Lisi and colleagues, using astrocytes in culture, when rapamycin was administered together with a proinflammatory stimulus it induced a transient upregulation of iNOS mRNA, which was accompanied by an increase in the rate of its degradation [101]. Such effects did not result in augmenting either iNOS protein levels or nitrite production to any significant level and combined with data coming from a different study, revealing a marked anti-inflammatory effect of rapamycin on microglial cells [102]. These data suggest that mTOR inhibitors may have a beneficial effect in preventing apoptosis occurring in inflammatory-based CNS diseases.

Necroptosis is a recently described mechanism of regulated necrosis that seems to contribute to neuronal death in acute lesions, such as stroke and trauma. It is mainly regulated by the kinase activity of receptor-interacting protein kinase 1 (RIP1) and RIP3, but Liu and colleagues recently characterized an intervention of mTOR in the complex regulatory mechanisms of necroptosis [103]. In fact, mTOR is a downstream effector of PKB (or Akt) and, using a neuronal cell line, those authors were able to demonstrate that inhibiting Akt/mTOR signaling with Akt inhibitor VIII, rapamycin, or both, it was possible to prevent mitochondrial reactive oxygen species (ROS) production and necroptosis by over 50%, without interfering with RIP1-RIP3 activity [103]. Such information is clinically attractive and the understanding of necroptosis signaling mechanisms is likely to project new therapeutic targets that, by reducing cell death, may improve functional outcomes after traumatic brain injury, stroke, or other acute injuries of the CNS (Figure 19.2).

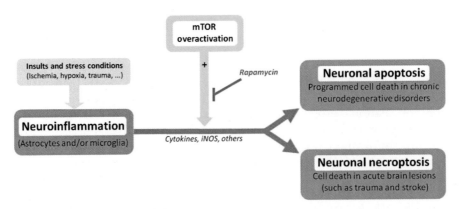

FIGURE 19.2 Schematic diagram of the proposed links between mTOR overactivation and neuroinflammation-evoked neuronal apoptosis and necroptosis. iNOS, inducible nitric oxide synthase; mTOR, mammalian target of rapamycin.

19.4.4 mTOR and Microglial Priming

Acute or chronic inflammatory insults may sensitize immune cells, leading to an exaggerated proinflammatory response when establishing contact with a secondary stimulus. This is known as priming and has long been observed in peripheral immune cells, mainly in macrophages [104]. Microglia share some of these abilities and characteristic responses, which contributes to suggest that these cells play a relevant role in accelerating neurodegeneration in response to peripheral inflammation [105]. Nevertheless, even local stimulation by soluble and fibrillar amyloid β induces microglial priming through NADPH oxidase activation, resulting in an increase in ROS release, which in turn contributes to oxidative and neuronal stress [106].

Recent studies revealed that mTOR controls microglial priming in response to proinflammatory cytokines, also playing a relevant role in maintaining microglial viability and proliferation [102]. In fact, Dello Russo and colleagues detected increased levels of phosphorylated mTOR in primary cultures of microglial cells in response to both lipopolysaccharide (LPS) and a mixture of proinflammatory cytokines [102]. However, after pharmacologically inhibiting the activity of mTOR, microglial responses elicited by cytokines were clearly reduced, while the impact of that inhibition in LPS-triggered responses was lower. All these effects appeared to be selective for cell types, since mTOR inhibitors did not impact on proinflammatory activation and cell viability in cultures of astrocytes [102].

Hypoxia is also an interesting condition to deepen the knowledge related to the role of mTOR in microglial priming. In fact, microglia is activated in hypoxic conditions and animal studies have shown that, after cerebral ischemia, the degree of activation is correlated with neuronal cell death, highlighting the

role inflammatory mediators may have on neuronal survival in hypoxic environments [107]. By using a BV-2 microglial cell line, Lu and colleagues demonstrated that mTOR enhances the expression of iNOS during hypoxia [107]. Since the excessive activity of this enzyme correlates with neuronal damage and with worse clinical outcomes [108], mTOR inhibition emerges as a desirable therapeutic strategy to be used in the context of stroke. In a recent study, rapamycin injection into the lateral ventricles of rats 6 h after focal cerebral ischemia resulted in a significant reduction in lesion volume and in an improvement of motor deficits [109]. These effects may be the result of a positive modulation in poststroke neuroinflammatory responses, consisting of a reduction in deleterious actions of immune cells and of an enhancement in their protective activities [109].

There is also a wealth of body of evidence suggesting that in traumatic brain injury early inflammation contributes to worsen late clinical outcomes [110]. Considering the role microglial priming may have in those early inflammatory events and the regulatory effects of mTOR signaling exerted on them, Erlich and colleagues tested the hypothesis of using rapamycin as a neuroprotective agent in mice, after a brain injury [111]. It was possible to demonstrate that intraperitoneal rapamycin injection, 4 h following closed head injury, significantly improved animals' functional recovery, reduced microglia activation, and increased the number of surviving neurons at the site of the lesion [111].

In animal models of epilepsy, particularly temporal lobe epilepsy, the inhibition of mTOR signaling with rapamycin yielded very positive results, with a significant reduction in the occurrence of seizures or even a prevention of epilepsy development [112]. Even in neuropathological terms, treatment with rapamycin reverses mossy fiber sprouting and neuronal cell loss, which are believed to have a relevant role in epileptogenesis [112]. However, so far, there is no robust evidence to support that this benefit of rapamycin in animal models of epilepsy is due to an effect on glial reactivity. In fact, brain inflammation and gliosis do not seem to effectively contribute to epileptogenesis and it is likely that rapamycin has a preferential effect on BBB permeability or in specific inflammatory pathways that are not currently characterized [113].

All this evidence highlights how challenging the relationship between microglial priming and mTOR signaling pathways can be. However, this complexity can become an opportunity for drug development. As an example, the case of resveratrol, a natural non-flavonoid polyphenolic found in grapes, red wine, mulberries, peanuts, and other plants, should be noted [114]. This compound significantly attenuates microglial cell overactivation by reducing the expression of several neurotoxic proinflammatory mediators and cytokines, through activation of mTOR signaling pathway [115]. In this context, it may be reasonable to consider resveratrol as an anti-inflammatory drug candidate, with a broad spectrum of action, in clinical grounds.

19.4.5 mTOR and Glycogen Synthase Kinase-3

Glycogen synthase kinase-3 (GSK3) is also a serine/threonine protein kinase that has been recently characterized as a mediator of inflammation. The inhibition of this enzyme was shown to be responsible for an increase in IL-10 levels (anti-inflammatory cytokine) and for a significant increase in the production of several proinflammatory cytokines, after Toll-like receptor (TLR) stimulation [116]. Following these observations, various roles of GSK3 in neuroinflammation have been described. It appears to be involved in microglial

migration and priming [117] and in the regulation of the production of TNF-α via NF-κB [118], IL-6 [119], and NO in microglia [120]. GSK3 is also associated with a reduced BBB integrity and with an increase in the traffic of monocytes across the barrier [121].

GSK3 is involved in the activation of myeloid dendritic cells and in regulating their ability to secrete cytokines [122]. During dendritic cell generation, mTOR inhibition leads to a certain degree of resistance to maturation, but if the inhibition is prior to TLR activation, an increase in the production of proinflammatory cytokines is noted [123]. Dendritic cells and the cytokines they produce are able to influence the differentiation of T-helper cells and, thus, the orchestration of all the mechanisms of the immune response [124]. In a recent study by Wang et al. [125], deepening the interactions between mTOR and GSK3, the authors demonstrated that mTORC1 inhibition resulted in the loss of GSK3 (S21/9) phosphorylation and that this was the main effect by which rapamycin was able to influence the levels of pro- and anti-inflammatory cytokine production, using LPS-stimulated cells. These paradoxical effects of mTORC1 inhibition were also shown to be due to its regulation of GSK3, because by using dendritic cells expressing a constitutively active GSK3, authors found an inability of the mTORC1 pathway to inactivate that enzyme, which resulted in a loss of rapamycin capacity to increase the inflammatory response and to suppress IL-10 [125]. In contrast, by directly inhibiting GSK3, the effects of rapamycin on the levels of pro- and anti-inflammatory cytokines could be rescued. Since mTOR has the ability to increase the inflammatory response through its regulation of GSK3, this highlights how relevant it could be for GSK3 targeting in the presence of mTORC1 inhibition to abolish rapamycin's inflammatory properties (Figure 19.3). In fact, in clinical

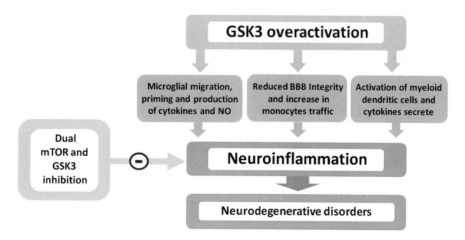

FIGURE 19.3 Schematic diagram of the proposed links between mTOR and GSK3 overactivation in brain disorders with underlying neuroinflammation. BBB: blood—brain barrier; GSK3, glycogen synthase kinase-3; NO, nitric oxide; mTOR, mammalian target of rapamycin.

grounds, there are reports that corroborate the existence of adverse events of an inflammatory nature in patients undergoing treatment with rapamycin [126,127] and, therefore, the inhibition of GSK3 could be an interesting strategy to avoid them. Lithium chloride is a well-known substance that has been shown to inhibit GSK and recent evidence suggests that low, non-toxic concentrations of such a compound have indeed anti-inflammatory effects [128].

The emergence of GSK3 as a potential therapeutic target in conditions wherein neuroinflammation plays an important role, in pathophysiological terms, has already contributed to the development of interesting research, notably in the field of Alzheimer's disease [129]. However, the establishment of any clinically relevant role for these pathways requires further work to be done.

19.5 CONCLUSIONS

Although there are abundant data linking mTOR signaling pathways to neuroinflammation, significant gaps in our knowledge still remain. In the complexity of all the above-mentioned phenomena, mTOR functions as a signal integrator, shaping the response of CNS cell elements to the myriad of activation/inhibition signals to which they are exposed. This makes mTOR a very interesting therapeutic target, making it plausible to think that the pharmacological control of its activity can translate into a clinical benefit that may be common to several neurological conditions. Logically, all this requires extensive research. It is crucial to increase our knowledge in this area to provide patients and physicians with new drugs able to modify the natural history of several life-threatening conditions that are currently real public health problems.

Acknowledgments

The authors' research is supported by the Portuguese Foundation for Science and Technology (FCT) (Strategic Projects Pest-C/SAU/UI3282/2013 and UID/NEU/04539/2013), COMPETE-FEDER and by a research grant from Biogen.

References

[1] Wong M. Mammalian target of rapamycin (mTOR) pathways in neurological diseases. Biomed J 2013;36:40–50.

[2] O'Callaghan JP, Sriram K, Miller DB. Defining "neuroinflammation". Ann N Y Acad Sci 2008;1139:318–30.

[3] Palavra F, Ciampi-Díaz E, Sena A. Cardiometabolic risk, inflammation and neurodegenerative disorders. In: Palavra F, Reis F, Marado D, Sena A, editors. Biomarkers of cardiometabolic risk, inflammation and disease. Switzerland: Springer International Publishing; 2015. p. 133–59. Chapter 7.

[4] Laflamme N, Lacroix S, Rivest S. An essential role of interleukin-1beta in mediating NF-kappaB activity and COX-2 transcription in cells of the blood-brain barrier in response to a systemic and localized inflammation but not during endotoxemia. J Neurosci 1999;19:10923–30.

[5] De Vries HE, Blom-Roosemalen MC, van Oosten M, de Boer AG, van Berkel TJ, Breimer DD, et al. The influence of cytokines on the integrity of the blood-brain barrier in vitro. J Neuroimmunol 1996;64:37–43.

[6] Quan N, Stern EL, Whiteside MB, Herkenham M. Induction of pro-inflammatory cytokine mRNAs in the brain after peripheral injection of subseptic doses of lipopolysaccharide in the rat. J Neuroimmunol 1999;93:72–80.

[7] Wong D, Dorovini-Zis K, Vincent SR. Cytokines, nitric oxide, and cGMP modulate the permeability of an in vitro model of the human blood-brain barrier. Exp Neurol 2004;190:446–55.

[8] Engelhardt B. T cell migration into the central nervous system during health and disease: different molecular keys allow access to different central nervous system compartments. Clin Exp Neuroimmunol 2010;1:79–93.

[9] Liu B, Hong JS. Role of microglia in inflammation-mediated neurodegenerative diseases: mechanisms and strategies for therapeutic intervention. J Pharmacol Exp Ther 2003;304:1–7.

[10] Das A, Sinha M, Datta S, Abas M, Chaffee S, Sen CK, et al. Monocyte and macrophage plasticity in tissue repair and regeneration. Am J Pathol 2015;185:2596–606.

[11] Xing B, Bachstetter AD, Van Eldik LJ. Microglial p38alpha MAPK is critical for LPS-induced neuron degeneration, through a mechanism involving TNFalpha. Mol Neurodegener 2011;6:84.

[12] Mihara M, Hashizume M, Yoshida H, Suzuki M, Shiina M. IL-6/IL-6 receptor system and its role in physiological and pathological conditions. Clin Sci (Lond) 2012;122:143–59.

[13] Hohmann HP, Brockhaus M, Baeuerle PA, Remy R, Kolbeck R, van Loon AP. Expression of the types A and B tumor necrosis factor (TNF) receptors is independently regulated, and both receptors mediate activation of the transcription factor NF-kappa B. TNF alpha is not needed for induction of a biological effect via TNF receptors. J Biol Chem 1990;265:22409–17.

[14] Kischkel FC, Lawrence DA, Chuntherapai A, Schow P, Kim KJ, Ashkenazi A. Apo2L/TRAIL-dependent recruitment of endogenous FADD and caspase-8 to death receptors 4 and 5. Immunity 2000;12:611–20.

[15] Kim SH, Smith CJ, Van Eldik LJ. Importance of MAPK pathways for microglial pro-inflammatory cytokine IL-1 beta production. Neurobiol Aging 2004;25:431–9.

[16] Sokolova A, Hill MD, Rahimi F, Warden LA, Halliday GM, Shepherd CE. Monocyte chemoattractant protein-1 plays a dominant role in the chronic inflammation observed in Alzheimer's disease. Brain Pathol 2009;19:392–8.

[17] Zabel MK, Kirsch WM. From development to dysfunction: microglia and the complement cascade in CNS homeostasis. Ageing Res Rev 2013;12:749–56.

[18] Pittock SJ, Lennon VA, McKeon A, Mandrekar J, Weinshenker BG, Lucchinetti CF, et al. Eculizumab in AQP4-IgG-positive relapsing neuromyelitis optica spectrum disorders: an open-label pilot study. Lancet Neurol 2013;12:554–62.

[19] Hoozemans JJ, Rozemuller AJ, Janssen I, De Groot CJ, Veerhuis R, Eikelenboom P. Cyclooxygenase expression in microglia and neurons in Alzheimer's disease and control brain. Acta Neuropathol 2001;101:2–8.

[20] Schwab JM, Beschorner R, Meyermann R, Gözalan F, Schluesener HJ. Persistent accumulation of cyclooxygenase-1-expressing microglial cells and macrophages and transient upregulation by endothelium in human brain injury. J Neurosurg 2002;96:892–9.

[21] Frautschy SA. Thinking outside the box about COX-1 in Alzheimer's disease. Neurobiol Dis 2010;38:492–4.

[22] Trepanier CH, Milgram NW. Neuroinflammation in Alzheimer's disease: are NSAIDs and selective COX-2 inhibitors the next line of therapy? J Alzheimers Dis 2010;21:1089–99.

[23] Zhao Y, Usatyuk PV, Gorshkova IA, He D, Wang T, Moreno-Vinasco L, et al. Regulation of COX-2 expression and IL-6 release by particulate matter in airway epithelial cells. Am J Respir Cell Mol Biol 2009;40:19–30.

[24] Molina-Holgado E, Ortiz S, Molina-Holgado F, Guaza C. Induction of COX-2 and PGE(2) biosynthesis by IL-1beta is mediated by PKC and mitogen-activated protein kinases in murine astrocytes. Br J Pharmacol 2000;131:152–9.

[25] Choi S-H, Aid S, Choi U, Bosetti F. Cyclooxygenases-1 and -2 differentially modulate leukocyte recruitment into the inflamed brain. Pharmacogenomics J 2010;10:448–57.

[26] Lyman M, Lloyd DG, Ji X, Vizcaychipi MP, Ma D. Neuroinflammation: the role and consequences. Neurosci Res 2014;79:1–12.

[27] Takei N, Nawa H. mTOR signaling and its roles in normal and abnormal brain development. Front Mol Neurosci 2014;7:28.

[28] Graber TE, McCamphill PK, Sossin WS. A recollection of mTOR signaling in learning and memory. Learn Mem 2013;20:518–30.

[29] Meng XF, Yu JT, Song JH, Chi S, Tan L. Role of the mTOR signaling pathway in epilepsy. J Neurol Sci 2013;332:4–15.

[30] Russo E, Citraro R, Constanti A, De Sarro G. mTOR signaling pathway in the brain: focus on epilepsy and epileptogenesis. Mol Neurobiol 2012;46:662–81.

[31] Cho CH. Frontier of epilepsy research—mTOR signaling pathway. Exp Mol Med 2011;43:231–74.

[32] Chong ZZ, Shang YC, Zhang L, Wang S, Maiese K. Mammalian target of rapamycin: hitting the bull's-eye for neurological disorders. Oxid Med Cell Longev 2010;3:374–91.

[33] Hoeffer CA, Klann E. mTOR signaling: at the crossroads of plasticity, memory and disease. Trends Neurosci 2010;33:67–75.

[34] Cornu M, Albert V, Hall MN. mTOR in aging, metabolism, and cancer. Curr Opin Genet Dev 2013;23:53–62.

[35] Zoncu R, Efeyan A, Sabatini DM. mTOR: from growth signal integration to cancer, diabetes and ageing. Nat Rev Mol Cell Biol 2011;12:21–35.

[36] Hands SL, Proud CG, Wyttenbach A. mTOR's role in ageing: protein synthesis or autophagy? Aging (Albany NY) 2009;1:586–97.

[37] Laplante M, Sabatini DM. mTOR signaling in growth control and disease. Cell 2012;149:274–93.

[38] Wang X, Proud CG. mTORC1 signaling: what we still don't know. J Mol Cell Biol 2011;3:206–20.

[39] Loewith R, Jacinto E, Wullschleger S, Lorberg A, Crespo JL, Bonenfant D, et al. Two TOR complexes, only one of which is rapamycin sensitive, have distinct roles in cell growth control. Mol Cell 2002;10:457–68.

[40] Sarbassov DD, Ali SM, Sengupta S, Sheen JH, Hsu PP, Bagley AF, et al. Prolonged rapamycin treatment inhibits mTORC2 assembly and Akt/PKB. Mol Cell 2006;22:159–68.

[41] Jacinto E, Loewith R, Schmidt A, Lin S, Rüegg MA, Hall A, et al. Mammalian TOR complex 2 controls the actin cytoskeleton and is rapamycin insensitive. Nat Cell Biol 2004;6:1122–8.

[42] Zhang YJ, Duan Y, Zheng XF. Targeting the mTOR kinase domain: the second generation of mTOR inhibitors. Drug Discov Today 2011;16:325–31.

[43] Russell RC, Fang C, Guan KL. An emerging role for TOR signaling in mammalian tissue and stem cell physiology. Development 2011;138:3343–56.

[44] Zhou H, Huang S. The complexes of mammalian target of rapamycin. Curr Protein Pept Sci 2010;11:409–24.

[45] Swiech L, Perycz M, Malik A, Jaworski J. Role of mTOR in physiology and pathology of the nervous system. Biochim Biophys Acta 2008;1784:116–32.

[46] Hay N, Sonenberg N. Upstream and downstream of mTOR. Genes Dev 2004;18:1926–45.

[47] Fingar DC, Blenis J. Target of rapamycin (TOR): an integrator of nutrient and growth factor signals and coordinator of cell growth and cell cycle progression. Oncogene 2004;23:3151–71.

[48] Huang S, Houghton PJ. Targeting mTOR signaling for cancer therapy. Curr Opin Pharmacol 2003;3:371–7.

[49] Sancak Y, Peterson TR, Shaul YD, Lindquist RA, Thoreen CC, Bar-Peled L, et al. The Rag GTPases bind raptor and mediate amino acid signaling to mTORC1. Science 2008;320:1496–501.

[50] Kim MS, Kuehn HS, Metcalfe DD, Gilfillan AM. Activation and function of the mTORC1 pathway in mast cells. J Immunol 2008;180:4586–95.

[51] Inoki K, Li Y, Zhu T, Wu J, Guan KL. TSC2 is phosphorylated and inhibited by Akt and suppresses mTOR signalling. Nat Cell Biol 2002;4:648–57.

[52] Gao X, Zhang Y, Arrazola P, Hino O, Kobayashi T, Yeung RS, et al. Tsc tumour suppressor proteins antagonize amino-acid-TOR signalling. Nat Cell Biol 2002;4:699–704.

[53] Inoki K, Zhu T, Guan KL. TSC2 mediates cellular energy response to control cell growth and survival. Cell 2003;115:577–90.

[54] Schneider A, Younis RH, Gutkind JS. Hypoxia-induced energy stress inhibits the mTOR pathway by activating an AMPK/REDD1 signaling axis in head and neck squamous cell carcinoma. Neoplasia 2008;10:1295–302.

[55] Brugarolas J, Lei K, Hurley RL, Manning BD, Reiling JH, Hafen E, et al. Regulation of mTOR function in response to hypoxia by REDD1 and the TSC1/TSC2 tumor suppressor complex. Genes Dev 2004;18:2893–904.

[56] Cole-Edwards KK, Bazan NG. Lipid signaling in experimental epilepsy. Neurochem Res 2005;30:847–53.

[57] Sarbassov DD, Ali SM, Sabatini DM. Growing roles for the mTOR pathway. Curr Opin Cell Biol 2005;17:596–603.

[58] Levy DE, Lee CK. What does Stat3 do? J Clin Invest 2002;109:1143–8.

[59] Parrales A, López E, Lee-Rivera I, López-Colomé AM. ERK1/2-dependent activation of mTOR/mTORC1/p70S6K regulates thrombin-induced RPE cell proliferation. Cell Signal 2013;25:829–38.

[60] Manning BD, Cantley LC. AKT/PKB signaling: navigating downstream. Cell 2007;129:1261–74.

[61] Guertin DA, Stevens DM, Thoreen CC, Burds AA, Kalaany NY, Moffat J, et al. Ablation in mice of the mTORC components raptor, rictor, or mLST8 reveals that mTORC2 is required for signaling to Akt-FOXO and PKCalpha, but not S6K1. Dev Cell 2006;11:859–71.

[62] Aeder SE, Martin PM, Soh JW, Hussaini IM. PKC-eta mediates glioblastoma cell proliferation through the Akt and mTOR signaling pathways. Oncogene 2004;23:9062–9.

[63] Sarbassov DD, Ali SM, Kim DH, Guertin DA, Latek RR, Erdjument-Bromage H, et al. Rictor, a novel binding partner of mTOR, defines a rapamycin-insensitive and raptor-independent pathway that regulates the cytoskeleton. Curr Biol 2004;14:1296–302.

[64] García-Martínez JM, Alessi DR. mTOR complex 2 (mTORC2) controls hydrophobic motif phosphorylation and activation of serum- and glucocorticoid-induced protein kinase 1 (SGK1). Biochem J 2008;416:375–85.

[65] Lang F, Böhmer C, Palmada M, Seebohm G, Strutz-Seebohm N, Vallon V. Patho)physiological significance of the serum and glucocorticoid-inducible kinase isoforms. Physiol Rev 2006;86:1151–78.

[66] Vézina C, Kudelski A, Sehgal SN. Rapamycin (AY-22,989), a new antifungal antibiotic. I. Taxonomy of the producing streptomycete and isolation of the active principle. J Antibiot (Tokyo) 1975;28:721−6.

[67] Hartford CM, Ratain MJ. Rapamycin: something old, something new, sometimes borrowed and now renewed. Clin Pharmacol Ther 2007;82:381−8.

[68] Dowling RJ, Topisirovic I, Fonseca BD, Sonenberg N. Dissecting the role of mTOR: lessons from mTOR inhibitors. Biochim Biophys Acta 2010;1804:433−9.

[69] Yip CK, Murata K, Walz T, Sabatini DM, Kang SA. Structure of the human mTOR complex I and its implications for rapamycin inhibition. Mol Cell 2010;38:768−74.

[70] Lin J, Liu L, Wen Q, Zheng C, Gao Y, Peng S, et al. Rapamycin prevents drug seeking via disrupting reconsolidation of reward memory in rats. Int J Neuropsychopharmacol 2014;17:127−36.

[71] Lamming DW, Ye L, Katajisto P, Gonçalves MD, Saitoh M, Stevens DM, et al. Rapamycin-induced insulin resistance is mediated by mTORC2 loss and uncoupled from longevity. Science 2012;335:1638−43.

[72] Khan A, Pepio AM, Sossin WS. Serotonin activates S6 kinase in a rapamycin-sensitive manner in Aplysia synaptosomes. J Neurosci 2001;21:382−91.

[73] Beaumont V, Zhong N, Fletcher R, Froemke RC, Zucker RS. Phosphorylation and local presynaptic protein synthesis in calcium- and calcineurin-dependent induction of crayfish long-term facilitation. Neuron 2001;32:489−501.

[74] Carrol M, Warren O, Fan X, Sossin WS. 5-HT stimulates eEF2 dephosphorylation in a rapamycin-sensitive manner in Aplysia neurites. J Neurochem 2004;90:1464−76.

[75] Tang SJ, Schuman EM. Protein synthesis in the dendrite. Philos Trans R Soc Lond B Biol Sci 2002;357:521−9.

[76] Tsokas P, Grace EA, Chan P, Ma T, Sealfon SC, Iyengar R, et al. Local protein synthesis mediates a rapid increase in dendritic elongation factor 1A after induction of late long-term potentiation. J Neurosci 2005;25:5833−43.

[77] Cammalleri M, Lütjens R, Berton F, King AR, Simpson C, Francesconi W, et al. Time-restricted role for dendritic activation of the mTOR-p70S6K pathway in the induction of late-phase long-term potentiation in the CA1. Proc Natl Acad Sci USA 2003;100:14368−73.

[78] Huber KM, Roder JC, Bear MF. Chemical induction of mGluR5- and protein synthesis-dependent long-term depression in hippocampal area CA1. J Neurophysiol 2001;86:321−5.

[79] Huber KM, Gallagher SM, Warren ST, Bear MF. Altered synaptic plasticity in a mouse model of fragile X mental retardation. Proc Natl Acad Sci U S A 2002;99:7746−50.

[80] Hoeffer CA, Tang W, Wong H, Santillan A, Patterson RJ, Martinez LA, et al. Removal of FKBP12 enhances mTOR-Raptor interactions, LTP, memory, and perseverative/repetitive behavior. Neuron 2008;60:832−45.

[81] Hernandez D, Torres CA, Setlik W, Cebrián C, Mosharov EV, Tang G, et al. Regulation of presynaptic neurotransmission by macroautophagy. Neuron 2012;74:277−84.

[82] Weston MC, Chen H, Swann JW. Multiple roles for mammalian target of rapamycin signalling in both glutamatergic and GABAergic synaptic transmission. J Neurosci 2012;32:11441−52.

[83] Costa-Mattioli M, Sossin WS, Klann E, Sonenberg N. Translational control of long-lasting synaptic plasticity and memory. Neuron 2009;61:10−26.

[84] Roscic A, Baldo B, Crochemore C, Marcellin D, Paganetti P. Induction of autophagy with catalytic mTOR inhibitors reduces huntingtin aggregates in a neuronal cell model. J Neurochem 2011;119:398−407.

[85] Ma T, Hoeffer CA, Capetillo-Zarate E, Yu F, Wong H, Lin MT, et al. Dysregulation of the mTOR pathway mediates impairment of synaptic plasticity in a mouse model of Alzheimer's disease. PLoS One 2010;5:e12845.

[86] Tain LS, Mortiboys H, Tao RN, Ziviani E, Bandmann O, Whitworth AJ. Rapamycin activation of 4E-BP prevents parkinsonian dopaminergic neural loss. Nat Neurosci 2009;12:1129−35.

[87] Bernier PJ, Bedard A, Vinet J, Levesque M, Parent A. Newly generated neurons in the amygdala and adjoining cortex of adult primates. Proc Natl Acad Sci U S A 2002;99:11464−9.

[88] Martinez-Canabal A. Reconsidering hippocampal neurogenesis in Alzheimer's disease. Front Neurosci 2014;8:147.

[89] Liu YP, Lin HI, Tzeng SF. Tumor necrosis factor-alpha and interleukin-18 modulate neuronal cell fate in embryonic neural progenitor culture. Brain Res 2005;1054:152−8.

[90] Butovsky O, Ziv Y, Schwartz A, Landa G, Talpalar AE, Pluchino S, et al. Microglia activated by IL-4 or IFN-gamma differentially induce neurogenesis and oligodendrogenesis from adult stem/progenitor cells. Mol Cell Neurosci 2006;31:149−60.

[91] Magri L, Cambiaghi M, Cominelli M, Alfaro-Cervello C, Cursi M, Pala M, et al. Sustained activation of mTOR pathway in embryonic neural stem cells leads to development of tuberous sclerosis complex-associated lesions. Cell Stem Cell 2011;9: 447−62.

[92] Raman L, Kong X, Gilley JA, Kernie SG. Chronic hypoxia impairs murine hippocampal development and depletes the postnatal progenitor pool by attenuating mammalian target of rapamycin signalling. Pediatr Res 2011;70:115−20.

[93] Sato A, Sunayama J, Matsuda K, Tachibana K, Sakurada K, Tomiyama A, et al. Regulation of neural stem/progenitor cell maintenance by PI3K and mTOR. Neurosci Lett 2010;470: 115−20.

[94] Paliouras GN, Hamilton LK, Aumont A, Joppé SE, Barnabé-Heider F, Fernandes KJ. Mammalian target of rapamycin signalling is a key regulator of the transit-amplifying progenitor pool in the adult and aging forebrain. J Neurosci 2012;32:15012−26.

[95] Nitsch R, Bechmann I, Deisz RA, Haas D, Lehmann TN, Wendling U, et al. Human brain-cell death induced by tumour-necrosis-factor-related apoptosis-inducing ligand (TRAIL). Lancet 2000;356:827−8.

[96] Micheau O, Tschopp J. Induction of TNF receptor I-mediated apoptosis via two sequential signalling complexes. Cell 2003;114:181−90.

[97] Bal-Price A, Brown GC. Inflammatory neurodegeneration mediated by nitric oxide from activated glia-inhibiting neuronal respiration, causing glutamate release and excitotoxicity. J Neurosci 2001;21:6480−91.

[98] Tsou HK, Su CM, Chen HT, Hsieh MH, Lin CJ, Lu DY, et al. Integrin-linked kinase is involved in TNF-alpha-induced inducible nitric-oxide synthase expression in myoblasts. J Cell Biochem 2010;109:1244−53.

[99] Jin HK, Ahn SH, Yoon JW, Park JW, Lee EK, Yoo JS, et al. Rapamycin down-regulates inducible nitric oxide synthase by inducing proteasomal degradation. Biol Pharm Bull 2009;32:988−92.

[100] Tang CH, Lu DY, Tan TW, Fu WM, Yang RS. Ultrasound induces hypoxia-inducible factor-1 activation and inducible nitric-oxide synthase expression through the integrin/integrin-linked kinase/Akt/mammalian target of rapamycin pathway in osteoblasts. J Biol Chem 2007;282:25406−15.

[101] Lisi L, Navarra P, Feinstein DL, Dello Russo C. The mTOR kinase inhibitor rapamycin decreases iNOS mRNA stability in astrocytes. J Neuroinflammation 2011;8:1.

[102] Dello Russo C, Lisi L, Tringali G, Navarra P. Involvement of mTOR kinase in cytokine-dependent microglial activation and cell proliferation. Biochem Pharmacol 2009;78:1242–51.

[103] Liu Q, Qiu J, Liang M, Golinski J, van Leyen K, Jung JE, et al. Akt and mTOR mediate programmed necrosis in neurons. Cell Death Dis 2014;5:e1084.

[104] Nestel FP, Price KS, Seemayer TA, Lapp WS. Macrophage priming and lipopolysaccharide-triggered release of tumor necrosis factor alpha during graft-versus-host disease. J Exp Med 1992;175:405–13.

[105] Frank MG, Baratta MV, Sprunger DB, Watkins LR, Maier SF. Microglia serve as a neuroimmune substrate for stress-induced potentiation of CNS pro-inflammatory cytokine responses. Brain Behav Immun 2007;21:47–59.

[106] Schilling T, Eder C. Amyloid-beta-induced reactive oxygen species production and priming are differently regulated by ion channels in microglia. J Cell Physiol 2011;226:3295–302.

[107] Lu DY, Liou HC, Tang CH, Fu WM. Hypoxia-induced iNOS expression in microglia is regulated by the PI3-kinase/Akt/mTOR signaling pathway and activation of hypoxia inducible factor-1alpha. Biochem Pharmacol 2006;72:992–1000.

[108] Iadecola C, Zhang F, Casey R, Nagayama M, Ross ME. Delayed reduction of ischemic brain injury and neurological deficits in mice lacking the inducible nitric oxide synthase gene. J Neurosci 1997;17:9157–64.

[109] Xie L, Sun F, Wang J, Mao X, Xie L, Yang SH, et al. mTOR signaling inhibition modulates macrophage/microglia-mediated neuroinflammation and secondary injury via regulatory T cells after focal ischemia. J Immunol 2014;192:6009–19.

[110] Kirino T. Delayed neuronal death. Neuropathology 2000;20 (Suppl):S95–7.

[111] Erlich S, Alexandrovich A, Shohami E, Pinkas-Kramarski R. Rapamycin is a neuroprotective treatment for traumatic brain injury. Neurobiol Dis 2007;26:86–93.

[112] Wong M. Mammalian target of rapamycin (mTOR) inhibition as a potential antiepileptogenic therapy: from tuberous sclerosis to common acquired epilepsies. Epilepsia 2010;51:27–36.

[113] van Vliet EA, Forte G, Holtman L, den Burger JC, Sinjewel A, de Vries HE, et al. Inhibition of mammalian target of rapamycin reduces epileptogenesis and blood–brain barrier leakage but not microglia activation. Epilepsia 2012;53:1254–63.

[114] Shakibaei M, Harikumar KB, Aggarwal BB. Resveratrol addiction: to die or not to die. Mol Nutr Food Res 2009;53:115–28.

[115] Zhong LM, Zong Y, Sun L, Guo JZ, Zhang W, He Y, et al. Resveratrol inhibits inflammatory responses via the mammalian target of rapamycin signaling pathway in cultured LPS-stimulated microglial cells. PLoS One 2012;7:e32195.

[116] Martin M, Rehani K, Jope RS, Michalek SM. Toll-like receptor-mediated cytokine production is differentially regulated by glycogen synthase kinase 3. Nat Immunol 2005;6:777–84.

[117] Yuskaitis CJ, Jope RS. Glycogen synthase kinase-3 regulates microglial migration, inflammation, and inflammation-induced neurotoxicity. Cell Signal 2009;21:264–73.

[118] Wang MJ, Huang HY, Chen WF, Chang HF, Kuo JS. Glycogen synthase kinase-3β inactivation inhibits tumor necrosis factor-α production in microglia by modulating nuclear factor κB and MLK3/JNK signaling cascades. J Neuroinflammation 2010;7:99.

[119] Beurel E, Jope RS. Lipopolysaccharide-induced interleukin-6 production is controlled by glycogen synthase kinase-3 and STAT3 in the brain. J Neuroinflammation 2009;6:9.

[120] Huang WC, Lin YS, Wang CY, Tsai CC, Tseng HC, Chen CL, et al. Glycogen synthase kinase-3 negatively regulates anti-inflammatory interleukin-10 for lipopolysaccharide-induced iNOS/NO biosynthesis and RANTES production in microglial cells. Immunology 2009;128(Suppl. 1):e275–86.

[121] Ramirez SH, Fan S, Zhang M, Papugani A, Reichenbach N, Dykstra H, et al. Inhibition of glycogen synthase kinase 3beta (GSK3beta) decreases inflammatory responses in brain endothelial cells. Am J Pathol 2010;176:881–92.

[122] Liu KJ, Lee YL, Yang YY, Shih NY, Ho CC, Wu YC, et al. Modulation of the development of human monocyte-derived dendritic cells by lithium chloride. J Cell Physiol 2011;226:424–33.

[123] Brown J, Wang H, Hajishengallis GN, Martin M. TLR-signaling networks: an integration of adaptor molecules, kinases, and cross-talk. J Dent Res 2011;90:417–27.

[124] O'Shea JJ, Paul WE. Mechanisms underlying lineage commitment and plasticity of helper CD4 + T cells. Science 2010;327:1098–102.

[125] Wang H, Brown J, Gu Z, Garcia CA, Liang R, Alard P, et al. Convergence of the mammalian target of rapamycin complex 1- and glycogen synthase kinase 3-β-signaling pathways regulates the innate inflammatory response. J Immunol 2011;186:5217–26.

[126] Thaunat O, Beaumont C, Chatenoud L, Lechaton S, Mamzer-Bruneel MF, Varet B, et al. Anemia after late introduction of sirolimus may correlate with biochemical evidence of a chronic inflammatory state. Transplantation 2005;80:1212–19.

[127] Dittrich E, Schmaldienst S, Soleiman A, Hörl WH, Pohanka E. Rapamycin-associated post-transplantation glomerulonephritis and its remission after reintroduction of calcineurin-inhibitor therapy. Transpl Int 2004;17:215–20.

[128] Green HF, Nolan YM. GSK-3 mediates the release of IL-1β, TNF-α and IL-10 from cortical glia. Neurochem Int 2012;61:666–71.

[129] Cai Z, Zhao Y, Zhao B. Roles of glycogen synthase kinase 3 in Alzheimer's disease. Curr Alzheimer Res 2012;9:864–79.

20

mTOR in Multiple Sclerosis: The Emerging Role in the Regulation of Glial Biology

Cinzia Dello Russo, Pierluigi Navarra and Lucia Lisi

Institute of Pharmacology, Catholic University Medical School, Rome, Italy

20.1 INTRODUCTION

Multiple sclerosis (MS) is an autoimmune disorder of the central nervous system (CNS), which represents the most common cause of acquired neurologic disability among young adults affecting approximately 2.3 million people worldwide [1]. The CNS pathology is characterized by scattered areas of demarked inflammation, called lesions or plaques, breakdown of myelin sheaths (demyelination), microglia activation, astrocyte proliferation with subsequent gliosis, and variable degrees of axonal damage [2]. Disease's etiology remains unknown; however both genetic [3] and environmental factors [4] seem to play a role. The disease clinical course is highly variable and largely unpredictable. In most patients (more than 80%), MS begins as an episodic relapsing/remitting disease (RRMS), with complete or partial recovery in between relapses. Over time, the frequency of relapses decreases and the majority of patients (up to 90% within 25 years) enter a progressive phase (secondary progressive MS, SPMS) [5]. In 10–20% of the cases, the relapsing phase is missing and the disease is progressive from the onset (primary progressive MS, PPMS) [6]. Progressive MS is a highly disabling condition. In this stage, neurodegeneration becomes prominent, leading to permanent disability; thus approximately 50% of RRMS patients are confined in the wheelchair within 25 years from clinical onset by severe motor impairment, increasing paralysis and

spasticity [5]. The clinical onset of PPMS is 10 year delayed in comparison to RRMS but the rate of progression is quite similar, thus suggesting common underlying pathological mechanisms [7].

The process responsible for the acute inflammatory lesions within the CNS is probably initiated in the periphery by autoreactive $CD4^+$ lymphocytes (particularly T helper 1, Th1, and Th17) directed against an unknown myelin antigen. Once activated these cells can cross the blood brain barrier (BBB), infiltrate the CNS and upon re-encountering the driver autoantigen promote an inflammatory response [8]. The latter is further sustained by the recruitment of other peripheral immune cells (including myeloid, $CD8^+$ T cells, B cells, and natural killer cells) and parenchymal glia [5]. Although Th1 and Th17 were considered the main immune effector cells in MS, macrophages and microglia assume an important role in sustaining neuroinflammation, thus significantly contributing to disease pathogenesis. Activated microglia and macrophages are frequently observed in active MS plaques [9], but their exact function is still unclear. In MS, proinflammatory, neurotoxic and mylien-attacking microglia and macrophages predominate, even though some cells with anti-inflammatory, neuroprotective and remyelination-promoting proprieties are also present, which suggest a potential dichotomic role of microglia in MS pathogenesis. In general, two main phenotypes occur prominently in inflammatory lesions: the

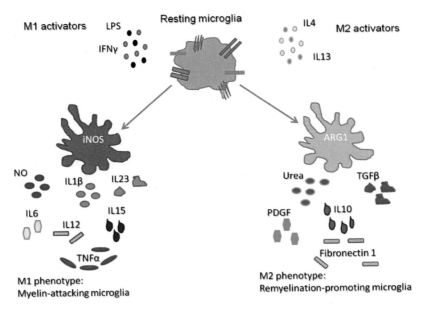

FIGURE 20.1   Schematic of microglia polarization in CNS demyelination diseases, such as MS.

classically activated M1 cells and the alternative activated M2 cells [10]. In this chapter, we will refer to the M1 status of activation for the myelin-attacking microglia and to the M2 phenotype for the cells with remyelination-promoting proprieties (Figure 20.1).

The outcome of inflammatory processes is often dictated by the correct balance between these different statuses of activation (it is not possible to affirm that a particular polarization status is in absolute better than another). Said that, inflammation mostly sustained by M1 microglia is often associated to demyelination and can directly contributes to axonal damage in both MS and EAE [10]. On the other hand, M2 microglial cells probably favors remyelination that often occurs in the early stages of disease, but it is insufficient to repair severe and long-lasting lesions as those predominant during progression [11]. Moreover, chronic activation of microglia results in the long term in disruption of microglial functions, making them less responsive to anti-inflammatory stimuli or less adept at phagocytosis [10]. In addition to microglial activation, astrocytes can be also critically involved in disease pathogenesis. In certain phases of disease, reduction of astrocytic activity seems to be beneficial, yet in others loss or disruption of astrocyte functions appears to contribute or exacerbate the inflammatory processes associated with MS. Astrocytes not only have the ability to enhance immune responses and inhibit myelin repair, but they can also be protective and limit CNS inflammation while supporting oligodendrocyte and axonal regeneration. The extensive distribution of astrocytes throughout the white and grey matter and their intimate relation with blood vessels of the CNS, as well as the

numerous immunity-related actions that these cells are capable of, imply that astrocytes may have a prominent role in neuroinflammatory disorders, such as MS [12]. Chronic inflammation sustained by glial cells and demyelination are the main causes of axonal damage and subsequent neurodegeneration in MS [7].

Ten disease-modifying therapies (DMTs) are now available for the treatment of RRMS, and several oral and biological drugs are in different stages of development [13]. These therapies are partly effective in the early phases of the disease (i.e., significantly reducing the rate and severity of relapses), but none of them significantly alters MS long-term prognosis [14]. These drugs are probably unable to target the inflammatory component, most active during the progressive phase of the disease, i.e., the activation of innate immune system sustained by microglia and dendritic cells [5]. Moreover, in this later stage of disease, inflammation is mostly compartmentalized within the CNS behind an intact BBB, not sufficiently permeable to most available drugs [15]. In this scenario, it might be postulated that more aggressive immunomodulatory or immunosuppressive treatments or a much earlier start of such therapies are required to substantially affect the timing and the severity of progression in MS. In particular, there is need of drugs that have the ability to easily cross the BBB and target both the adaptive and the innate immune system.

In this regard, signaling pathways regulated by the "mechanistic" target of rapamycin (mTOR) kinase have been described as crucial regulators of the immune system, including both the innate and adaptive immune responses [16]. In addition, a direct role

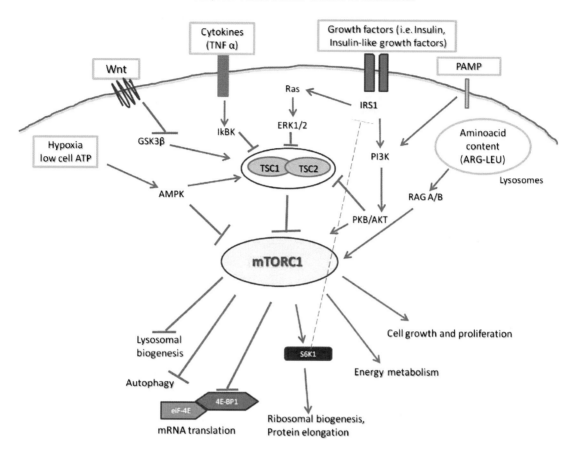

FIGURE 20.2 Schematic representing the main intracellular targets as well as the main cellular processes regulated by mTORC1.

of mTOR in the modulation of glial biology has been increasingly recognized, including microglial and astrocyte immune activation [17], as well as oligodendrocyte development and myelin formation [18,19]. Moreover, pharmacological inhibition of mTOR has been shown to exert beneficial effects in different experimental models of MS, thus supporting the relevance of this pathway in disease pathogenesis. A thorough understanding of the physiological functions regulated by mTOR within the brain, for example, the role in directing microglial polarization or the remyelination process after injury, can contribute to the development of novel effective treatments for MS, which may include selective inhibition or enhancement of mTOR activity in different cell types. Current evidence will be reviewed in this chapter.

20.2 mTOR, THE "MECHANISTIC" TARGET OF RAPAMYCIN

The "mechanistic" target of rapamycin (mTOR) is a conserved serine/threonine protein kinase belonging to the phosphoinositide 3-kinase (PI3K) family, primarily involved in the regulation of cell growth and

metabolism [20]. The mTOR pathway integrates diverse environmental clues with the final effect of promoting anabolic and inhibiting catabolic intracellular processes [21]. In mammals, the kinase is encoded by a single gene and signals in association with different partners through at least two distinct complexes, mTORC1 (Figure 20.2) and mTORC2 (Figure 20.3). They share the kinase catalytic subunit mTOR and are distinguished by two scaffolding subunits, RAPTOR (the regulatory-associated protein of mTOR) and RICTOR (the rapamycin-insensitive companion of mTOR), respectively [22].

The mTOR complex 1 is activated in response to different intracellular and extracellular clues, among which growth factors, cytokines, energy status, oxygen and amino acids (Figure 20.2). Several upstream regulators of its activity converge on the tuberous sclerosis complex, consisting of TSC1 and TSC2 proteins. The TSC1/2 complex normally inhibits mTORC1 activation by favoring the GDP-bound inactive state of Rheb, a positive activator of mTORC1. As reviewed by Laplante and Sabatini [22], growth factors increase mTORC1 activity, by promoting the phosphorylation and degradation of the TSC1/2 complex. This occurs via ligand-dependent activation of receptor tyrosine

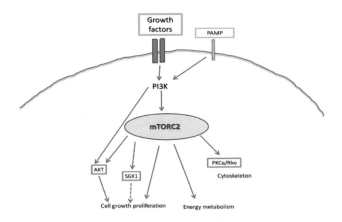

FIGURE 20.3 Schematic representing the main intracellular targets as well as the main cellular processes regulated by mTORC2.

kinases (RTKs) followed by activation of PI3K and RAS. The effector kinases of these pathways, namely the protein kinase B (PKB/AKT) and the extracellular-signal-regulated kinases (ERK) 1 and 2, promote the phosphorylation of TSC1/2. (Figure 20.2). In contrast, when cellular energy is low, the AMP kinase (AMPK) blocks mTORC1 activity by increasing the inhibitory activity of TSC2 [22]. The activation of mTORC1 generally increases the cellular capacity of protein generation. The two main downstream targets of mTORC1, the eukaryotic initiation factor 4E-binding protein 1 (4E-BP1) and the ribosomal S6 protein kinase 1 (S6K1), are key components of the protein translation machinery [23]. In addition, the S6K1 kinase is part of a retro-inhibitory feedback mechanism that controls the magnitude of RTK activation, by phosphorylation and inactivation of IRS1 (Figure 20.2, dotted line). IRS1 is a docking protein, that in its tyrosine-phosphorylated form couples the insulin receptor to its downstream effectors [24]. In addition to protein production, mTORC1 regulates other biosynthetic and metabolic pathways as well as intracellular process (i.e., macroautophagy and lysosome biogenesis), involved in the control of cell growth and metabolism (Figure 20.2) [22].

In contrast to mTORC1, mechanisms leading to mTORC2 activation are less characterized. However, data suggest that mTORC2 is also activated by PI3K, via direct phosphorylation of specific mTORC2 interactors, like RICTOR (Figure 20.3) [25]. In addition, AMPK can directly modulate the activity of mTORC2 by phosphorylation of this complex, and thus blocking its nuclear translocation and regulated gene transcription [26]. mTORC2 participate to the regulation of protein synthesis, by providing a quality control in the process of protein production via phosphorylation of amino acid residues in nascent polypeptide chains, including AKT [27]. Additional functions of mTORC2

include the regulation of the catalytic activity of several kinases belonging to the ACG family, such as AKT, the serum- and glucocorticoid-induced protein kinase (SGK1), protein kinase C (PKC)α and PKCθ. In particular, AKT is a key regulator of cell survival and proliferation, with mTORC2 promoting its phosphorylation at Ser_{473} and maximal activation [28,29]. SGK1 is another important kinase in the control of cell proliferation [30], whereas PKCα, along with other effectors (like paxilin and Rho GTPases) is important in the regulation of the dynamic of actin cytoskeleton [31,32]. Interestingly, mTORC2 dependent phosphorylation of PKCθ contributes to T helper 2 (Th2) differentiation [33].

In conclusion, it has been shown that the two complexes interact at many different levels, which allows the integration of diverse signals and the regulation of appropriate cellular responses [21]. As mentioned above, mTOR is both a downstream effector of AKT (i.e., mTORC1) as well as an important upstream regulator of the kinase synthesis and activity (i.e., mTORC2). Furthermore, the S6K (a downstream target of mTORC1) is a negative modulator of mTORC2 activity, whereas the TSC complex (which normally inhibits mTORC1 activation) exerts a positive effect on mTORC2 functions [34]. This complicated interplay appears particularly relevant in the regulation of the immune system [16]. In fact, numerous immunological signals are able to activate the mTOR kinase, which in turn regulates cell metabolism, lineage differentiation and cell functions within both the innate and adaptive immune system [35,36]. Consistently, deregulation of the mTOR pathway has been invariably associated with different autoimmune disorders, which suggests the potential beneficial effect of mTOR inhibitor drugs [16].

20.3 mTOR INHIBITORS

The mTOR kinase is the main target of the immunosuppressant drug, rapamycin. This drug isolated from *Streptomyces hygroscopicus* from soil samples collected from Easter Island (Rapa Nui, from where the name), was originally found to have antifungal proprieties [37]. However, its immunosuppressive activity became rapidly the most relevant clinical feature. Both in the United States and Europe, the drug has been approved and successfully used to prevent allograft rejection in kidney transplantation in combination with cyclosporine or corticosteroids. However, the list of potential clinical applications of rapamycin has been rapidly expanding [38]. The drug mainly inhibits mTORC1 activity by forming a trimolecular complex with mTOR and the immunophilin, FKBP12 (FK506-binding protein of 12 kDa; also known as PPIase FKBP1A), and disrupting the association with RAPTOR [39,40].

However, not all the functions mediated by mTORC1 are sensitive to rapamycin, the inhibition of cap-dependent translation and the induction of autophagy are in part resistant to rapamycin [41]. On the other hand, studies of Sabatini's group have shown that rapamycin given at higher concentrations and in chronic treatments also interferes with mTORC2 regulatory functions in certain cell types [42].

Due to its lipophilic nature, rapamycin readily crosses the BBB thus exerting direct effects within the CNS [43]. However, reduced solubility and stability in water represented a challenge for intravenous or oral dosage formulations, and contributed to the complexity of rapamycin dosing as well as to large intra- and inter-individual variability in pharmacokinetic profiles. These issues have been overcome either by formulating the drug as nanometer-sized crystals [44], or by the development of novel drugs, the so-called rapalogs. These drugs share the same molecular mechanisms of action of rapamycin, but display improved pharmacokinetic properties. Rapalogs includes CCI-779 (tensirolimus), RAD-001 (everolimus), and AP23573 (ridaforolimus or deforolimus) [34]. Among these mTOR inhibitors, CCI-779 is a pro-drug of rapamycin, with important anti-proliferative properties [45]; whereas RAD-001 is a 40-O-(2-hydroxyethyl) derivative of rapamycin, thus a more water-soluble compound. RAD-001 is currently approved as immunosuppressant for the prevention of solid organ rejection, and for the treatment of benign and malignant cancers [46]. The antiproliferative effects of rapalogs are halted by concomitant and direct activation of AKT. In fact, mTORC1 inhibition is followed by suppression of the negative feedback on PI3K activity mediated by S6K/IRS1 (Figure 20.2), and in addition rapalogs fail to efficiently block the stimulatory effects of mTORC2 on AKT (Figure 20.3) [47].

These limitations have led to the development of a second generation of mTOR inhibitors, with greater potential clinical use in oncology: the ATP-competitive mTOR inhibitors, which block both mTORC1 and mTORC2 activity [48]. Unlike rapamycin, a specific allosteric inhibitor of mTORC1, these ATP-competitive inhibitors target the catalytic site of the enzyme, thus promoting a broader, more potent and sustained inhibition of mTOR, and preventing the activation of PI3K/AKT caused by the de-repression of negative feedbacks [49]. This is due to the effective inhibition of rapamycin-insensitive mTORC2 activity in addition to mTORC1 inhibition, and also to a more comprehensive and sustained mTORC1 inhibition as demonstrated by sustained reduction of 4E-BP1 phosphorylation [50]. Actually, many compounds with different chemical structures show these functions (as we have recently reviewed [34]) and some of them are being tested in clinical trials in the oncology field. Surprisingly, catalytic mTOR inhibition has weaker effect on the growth and function of normal T cells, and appears to be less immunosuppressive than rapamycin [51].

As reported in section 20.2, mTOR inhibition can be also achieved by activation of AMPK. Specific AMPK activators have been developed with significant antiproliferative effects in several cancers [52]. Interestingly, the antidiabetic drug metformin has been found to promote AMPK activation both *in vitro* and *in vivo* [53]. Recent data suggest that AMPK activation and thus mTOR inhibition are important molecular mechanisms of action underlying both the antidiabetic and the antiproliferative actions of this drug [26,54]. Moreover, the antitumor effects of metformin have been recently shown to be immune mediated [55]. Taken together, these data suggest the potential pharmacological role of metformin beyond diabetes.

20.4 mTOR IN THE PERIPHERAL IMMUNE SYSTEM

The mTOR pathway constitutes a central rheostat in the immune system, integrating diverse extra- and intra-cellular signals with the metabolism and the immunological responses of different immune cells [16]. The involvement of mTOR was first studied in the regulation of T-cell biology and it is currently well characterized [56]. Recent work demonstrates that mTOR signaling is relevant both for the maintenance of naïve T-cell quiescence and the development of effector T cells (Teffs) [16]. Numerous immunological signals, including T-cell receptor (TCR), co-stimulatory signals, and inflammatory cytokines, induce mTOR activation which in turn regulates lineage differentiation and immune function [57]. Consistently, mTOR deficient T cells display normal activation and interleukin (IL)-2 production after initial stimulation, but fail to differentiate into Th1, Th2, or Th17 effector cells [58]. Moreover, deletion of Rheb selectively prevents mTORC1 activation, suppressing Th1 and Th17 generation, whereas rictor-deficient T cells develop into Th2 [59]. These data suggest that all Teffs require mTOR activity, but each complex exerts different functions in various subsets. Consistently, mTOR inhibition restricts Teff development, but invariably promotes the expansion of T regulatory cells (Tregs) [60]. Tregs, a specific subset critically involved in the reduction of immune responses and maintenance of immune tolerance, are metabolically distinct from other Teffs. They show a high rate of lipid oxidation and mitochondrial membrane potential, in line with higher levels of AMPK activity and reduced activity of mTORC1 [61]. Metformin (i.e., a further stimulation of AMPK activity), increases Treg number *in vivo* [62]. Finally, the mTOR pathway plays also a key role in the regulation of CD8+ T cells, controlling their activation and the transition to memory cells [61]. Consistently, mTOR inhibition by different means has been associated with T-cell anergy, induction of Tregs, and enhanced memory T-cell responses to microbial pathogens [63]. Additionally, the PI3K/mTOR pathway is required for complete B-cell development, homeostasis and immune response. In fact, rapamycin can suppress human B-cell proliferation induced after CD40 ligation, reducing the number antibodies producing cells [64].

In addition to the extensive literature on the role of mTOR in the regulation of the adaptive immune system, more recent studies have described the central function of mTOR in the regulation of the innate immune responses [35]. In dendritic cells (DCs) and macrophages, two cardinal phagocytes of the innate immune system, stimulation of toll-like receptors (TLRs) by different PAMPs (pathogen associated molecular patterns) increases both mTORC1 and 2 activity via activation of a PI3K mediated signaling pathway (Figures 20.2–20.3) [65]. Consistently, inhibition of mTOR activity by rapamycin interferes with the complete development of DCs and the inflammatory activation of macrophages [35]. In this regard, it should be kept in mind that macrophages exert a complex role in host defense and maintenance of tissue homeostasis, which rely on a delicate balance, as described for microglial cells, between their proinflammatory activity (M1 activated status) and their immunomodulatory (M2) functions. In isolated human monocytes and primary myeloid dendritic cells (mDCs), mTOR activation inhibits the production of pro-inflammatory cytokines (IL-12, IL-23, IL-6, and tumor necrosis factor α, TNFα) while it enhances the release of the antinflammatory cytokine IL-10 [65]. Consistently, rapamycin has been shown to increase TNFα production in peritoneal activated macrophages, possibly via inhibition of IL-10 production [66]. These data suggest that mTOR is involved in the regulation of macrophage polarization, and its activation probably favor an M2 status. In support of this hypothesis, it has been shown that the bone morphogenetic protein-7 mediates monocyte polarization into M2 macrophages, via activation of the PI3K-AKT-mTOR pathway [67]; and rapamycin, by blocking mTOR, unbalances the polarization of human macrophages toward the M1 status [68]. In contrast, constitutive activation of mTORC1 interferes with IL-4-dependent M2 macrophage polarization, and increases inflammatory responses to proinflammatory stimuli [69]. Similarly, other studies showed that mTOR inhibition decreases the global inflammatory response of *in vitro* generated DCs [70,71] and reduces the production of nitric oxide (NO) by macrophages

activated with the bacterial endotoxin lipopolysaccharide (LPS) [72]. These data suggest that the role of mTOR in orchestrating the immune response is complex, depending on the inflammatory environment as well as on the inflammatory stimuli. Further studies to fully elucidate the role of each mTOR complex in the regulation of the innate immune responses are therefore necessary.

20.5 mTOR IN THE REGULATION OF GLIAL INFLAMMATORY RESPONSES

20.5.1 Microglia

Experimental evidence from our group and others support the notion that mTOR is involved in the regulation of microglial inflammatory activation, thus making this kinase a possible target for therapeutic intervention to modulate brain inflammatory responses (for a recent review, see [17]). In primary cultures of rat microglial cells, mTOR can be activated in response to different inflammatory stimuli (i.e., the bacterial endotoxin LPS, or a mixture of pro-inflammatory cytokines), and by exposure to conditioned medium harvested by glioma cells under different conditions [73,74]. Interestingly, pharmacological inhibition of mTOR can interfere with microglial activation, but the final outcome significantly depends on the initial stimulus. In fact, mTOR inhibitors were able to reduce the activity and expression of inducible NO synthase (NOS2) when induced by cytokines; to increase the expression, leaving significantly unaffected the activity, of NOS2 in LPS-treated microglia; and to significantly increase NOS2 expression and activity in glioma activated-microglial cells [73,74]. Thus mTOR inhibitors hold the potential to influence microglial polarization; considering that NOS2 is regarded as a marker of M1-polarization [75]. In our studies, we also observed an involvement of mTOR in the regulation of the cycloxygenase (COX) pathway [73], another M1 polarization marker [75]. In particular in presence of cytokines, RAD001 tended to reduce COX-2 protein levels but when administrated alone significantly augmented the intracellular amount of COX-2, further underlying the dual role of mTOR in the regulation of microglial inflammatory responses. In addition to our data, it has been shown that pharmacological inhibition of mTOR can reduce COX-2 and NOS2 protein levels (without affecting the mRNA steady state level of these genes) in catalase-exposed microglial cells [76]; and in parallel, activation of autophagy suppressed NOS2 and IL6 expression of LPS-stimulated microglia [77]. These data would therefore suggest that mTOR inhibition can restrict M1 polarization of microglia. In contrast, rapamycin enhanced the expression of both COX-2 and the microsomal prostaglandin (PG) E synthase-1 and the release of PGE2 and PGD2 in microglial cells activated by LPS [78]; thus supporting an opposite role of mTOR on microglial polarization. Exposure of primary rat microglial cultures or the BV-2 microglial cell line to hypoxia further confirms the involvement of the PI3K/AKT/mTOR signaling pathway together with the activation of hypoxia inducible factor-1α (HIF-1α) in the regulation of microglial NOS2 expression [79].

Microglial cells, like macrophages, are highly versatile cells: under pathological conditions microglia do not constitute a uniform cell population, but rather comprise a family of cells with diverse phenotypes with different effects, either beneficial or detrimental [80]. Data reviewed in this paragraph strongly suggest that mTOR is a crucial regulator of microglial function; and the ability of mTOR to regulate microglial polarization can be important in inflammatory diseases, like MS. However, actually at least *in vitro* is not yet possible to define the exact role of mTOR in microglial as well as in macrophage polarization since conflicting literature is available. Such discrepancy may be due to the different pro-inflammatory stimuli used to activate the cells *in vitro* and/or the different time of incubation. In fact, in our studies on primary cultures of rat microglial cells, we observed mTOR activation in response to different stimuli, even though the magnitude of activation and the final effect on microglial activation was different [73,74]. In addition, studies from our laboratory and others have shown that mTOR is critically involved in maintaining microglial viability, under basal condition or in presence of different stressors [73,81,82].

In conclusion, it is relevant for MS to underlie that in presence of a more relevant patho-physiological stimulus, like a mixture of pro-inflammatory cytokines, the block of mTOR is an anti-M1 promoter strategy, thus suggesting that mTOR inhibition could restrict the detrimental effect of mylien-attacking microglia. In contrast, data on LSP-microglia showing that mTOR inhibition may increase M1 polarization would suggest a potential harmful effect of these drugs in neuroinflammatory based diseases.

20.5.2 Astrocytes

mTOR is also a key regulator of intracellular processes in astroglial cells, and various mTOR upstream regulators have been reported to play an important role in astrocyte physiology (as we have previously reviewed [17]). Our group first demonstrated that mTOR interferes with astrocyte pro-inflammatory activation [83]. In fact rapamycin transiently increased

NOS2 mRNA levels, whereas reduced levels of NOS2 mRNA were detected after 48 h, suggesting that rapamycin can modify NOS2 mRNA stability. In this regard, we found that rapamycin significantly reduced the half-life of NOS2 mRNA, from 4 h to 50 min. Moreover, rapamycin induced a significant upregulation of tristetraprolin, a protein involved in the regulation of NOS2 mRNA stability [83]. Rapamycin may therefore exert contrasting effects in astrocytes, in which an initial increase in inflammatory gene expression, such as NOS2, is followed by a compensatory down regulation due to a reduced mRNA half-life. Moreover, mTOR is involved in the regulation of IL-6 secretive pathway. In fact in primary cultures of rat astrocytes IL-6 expression requires the activation of p38 MAPK and depends on NF-κB transcriptional activity. Activation of these pathways in astrocytes occurs when the PI3K-mTOR-AKT pathway is inhibited [84], thus suggesting that the mTOR kinase normally restricts its secretion. In this regard, IL-6 is a pleiotropic cytokine involved both in the glial scar formation as well as in the regulation of neuroregenerative processes after brain injury. In addition, astrocyte activation under oxygen-glucose deprivation/reoxygenation, is accompanied by phosphorylation of mTOR and its downstream substrate, the S6K1. In this experimental model, rapamycin significantly inhibited mTOR signal pathway, thus suppressing astrocyte proliferation via modulation of cell cycle progression. Moreover, rapamycin attenuated astrocytic migration and mitigated the production of inflammatory factors, such as TNF-α and NOS2 [85]. These findings therefore support the role of mTOR in mediating astrocyte proinflammatory activation with possible detrimental consequences on neuronal functions.

20.6 mTOR IN THE REGULATION OF THE MYELINATION PROCESS

It is well established that the PI3K/AKT pathway promotes myelination in both the central and the peripheral nervous system (CNS and PNS). The mTOR kinase is both a downstream effector of PI3K/AKT as well as a key modulator of AKT activity via mTORC2 (Figures 20.2–20.3), thus suggesting a role in the regulation of oligodendrocyte (OLG) development and myelin synthesis [19]. In this regard, early studies by Wood and collaborators have shown that rapamycin significantly interferes with the development of galactocerebroside positive immature OLGs and myelinating OLGs *in vitro* [86]. Both mTOR complexes appeared to be involved in the complete maturation of OLGs *in vitro*, albeit with different mechanisms. The inhibitory effects of rapamycin were selective on OLG development, since the drug did not affect neuronal and astrocytic differentiation from neonatal subventricular zone (SVZ)-derived neurospheres. These data were further expanded by analyzing changes in the proteome of oligodendrocyte progenitor cells (OPCs) during their differentiation in the presence of rapamycin [87]. The authors identified mTOR-induced proteins (among which cytoskeletal proteins, myelin proteins, and proteins involved in cholesterol and fatty acid biosynthesis) as well as mTOR-downregulated proteins. Proteomic data were concordant with previously published mRNA microarray data [88]. However, a subset of proteins appeared to be regulated at the translational level, possibly by mTORC1. For these proteins, increased or unchanged mRNA levels in response to rapamycin treatment [88] were followed by reduced protein synthesis [87]. The involvement of mTOR in oligodendrocyte differentiation, albeit in a later stage of the maturation process, was demonstrated also by Guardiola-Diaz and colleagues [89]. This group showed that the activation of mTOR seemed to be critical for cytoskeletal organization and the expression of normal levels of major myelin proteins *in vitro*, as well as for the onset of myelination in the postnatal mouse brain. Consistently, a strong activation of mTOR was found during postnatal rat brain development, particularly in the white matter as myelination peaked in the forebrain [86]. Taken together these data underline the importance of the mTOR pathway in driving the terminal differentiation of OLGs and thus the myelination process *in vivo*.

This is further indicated by work from the group of Macklin who generated transgenic mice overexpressing a constitutive active form of AKT under the PLP-promoter in myelinating cells. These animals showed increased myelination in the CNS, with an active myelination process that lasted through adulthood [90]. However, prolonged myelination eventually becomes pathologic, and myelin defects are observed at the electron microscopy analysis of the optic nerves in 6 month old animals. Moreover, transgenic mice did not survive past 14 months [90]. In line with these findings, cell-specific loss of PTEN (a tumor suppressor protein that restricts the extent of PI3K activation), in OLGs and Schwann cells resulted in hypermyelination. The latter was indeed associated with the development of progressive neuropathy and myelin sheath abnormalities [91]. In the PLP-AKT mice, increased activation of mTOR together with increased phosphorylation levels of mTORC1 major downstream targets, S6K and S6 ribosomal protein (S6RP) was detected, thus suggesting the involvement of mTORC1 downstream AKT activation [92]. The induction of mTORC1 activation was particularly evident in the corpus callosum of transgenic mice, in oligodendrocyte appearing cells. Consistently with previous findings [86,89], increased mTOR activation was also detected in the corpus callosum of wild type animals during brain development, thus further underlying the crucial role of this pathway in the regulation of myelination in the CNS. Of note, peripheral myelination was not influenced by the overexpression of constitutively active AKT under the PLP-promoter in myelinating cells [90]. However, mice lacking mTOR in the Schwann cells display hypomyelinated sciatic nerves [93], which indicates that myelination in PNS is at least in part regulated by the mTOR signaling.

The critical role of mTOR in the regulation of myelin synthesis in the CNS is further demonstrated by studies on conditional knockout (cKO) mice, in which the ablation of the mTOR kinase or specific interactors, like RAPTOR and RICTOR was selectively obtained in the OLG lineage. In particular, myelination onset and myelin thickness are affected in the in the spinal cord of mTOR cKO animals [94]. Consistently, differentiation of OPCs appeared reduced during development and in the adult spinal cord [94]. Moreover, deletion of mTORC1 in OLGs in RAPTOR cKO mice resulted in developmental delay and dysmyelination [95]. On the other hand, selective changes in myelin proteins and RNA expression were observed in different regions of the CNS in the animal model of RICTOR ablation in OLGs. However, these changes did not result in a dysmyelinating phenotype at the ultrastructural level [95]. Consistently, data from Suter's laboratory identified mTORC1 as crucial regulator of CNS myelination, whereas mTORC2 seems to play a minor role [96]. Interestingly, OLG-specific over-activation of mTORC1, via ablation of TSC1, reduces the extent of myelination, probably due to the downregulation of Akt signaling and lipogenic pathways [96]. These data demonstrate the importance of a balanced regulation of mTOR activation and action in OLG, in order to achieve proper myelination in the CNS. Interestingly, rapamycin was able to improve myelination when injected *in vivo* in a model of tuberous sclerosis [97] as well as in explant cultures from neuropathic mice by activating autophagic mechanisms [98]. Taken together, these data suggest that under pathological conditions rapamycin may contribute to re−equilibrate mTOR activity thus favoring the myelination process.

20.7 mTOR INHIBITORS AND NEUROPROTECTION IN EAE

The experimental autoimmune encephalomyelitis (EAE) is a widely used animal model, to study the pathologic mechanism of MS disease and to screen possible therapeutic agents. EAE is a T-cell-mediated autoimmune demyelinating disease that can be induced in several species by immunization with myelin antigens, like myelin basic protein (MBP), proteolipid protein (PLP), myelin oligodendrocyte glycoprotein (MOG), or via adoptive transfer of myelin-reactive T cells. [5]. EAE can follow either relapsing-remitting (RR) or chronic (Ch) disease courses, which may resemble the relapsing-remitting and the primary progressive forms of MS, respectively. There are some differences between Ch−EAE and RR−EAE models that in part depend on the immunization protocol used to induce the pathology. In our experience, using subcutaneous MOG35-55 peptide injection in complete Freund's adjuvant, animals develop a long-lasting disease, showing clinical symptoms 6−8 days after the MOG35-55 booster injection and reaching the peak of disease at 10−14 days. The disease tends to remain stable or progress over the time, in fact, animals barely recover unless effectively treated [99]. Therefore, this model can be used to screen drugs with potential beneficial effects in the progressive forms of MS, which still represent an unmet medical need.

The pharmacological properties of mTOR inhibitors, including the ability to restrict CD4+ effector T-cell differentiation and promote the generation of Treg cells, together with the immune-modulatory actions on glial cells (as elucidated in the previous paragraphs), made these drugs possible candidates for the treatment of EAE as well as MS. First, Martel and collaborators showed that rapamycin completely prevented the development of EAE, and related this pharmacological effect to the inhibition of the peripheral immune system [100]. In EAE, animals immunized with mouse spinal cord homogenates together with the *Mycobacterium tuberculosis*, rapamycin significantly decreases the severity of the disease alone and in combination with 1,25-dihydroxyvitamin D3 [101,102]. Importantly, in a Ch−EAE model rapamycin, in association with IL-2 mixed with a particular IL-2 monoclonal antibody, displayed beneficial effects in the treatment of ongoing disease whereas rapamycin alone, injected 2 days after immunization, delayed the onset of clinical signs, even though all the mice eventually developed severe disease [103]. Furthermore, rapamycin has been shown to prevent the induction and the progression of RR−EAE. This beneficial effect has been associated with suppression of effector T-cell function and simultaneous increase of the percentage of Treg cells [104]. In another study, rapamycin was orally administered for 28 days after the clinical onset of disease at 3 mg/kg in a protracted relapsing EAE model induced in Dark Agouti rats. Animals treated with rapamycin showed a significantly milder course of the disease, associated with a reduction of the histopathological signs of EAE. The authors showed an increased percentage of splenic

Treg cells together with a concomitant reduction of splenic CD8+ T cells [105]. Delgoffe and colleagues generated Rheb cKO in T cells. These animals developed a milder disease after immunization with MOG35-55, due to reduced leukocyte infiltration in the spinal cord and clonal expansion of Th1 and Th17 cells in periphery [59]. These findings demonstrate that the selective deletion of mTORC1 signaling leads to enhanced Th2 differentiation *in vivo*, even under strongly Th1 and Th17 promoting conditions. Procaccini and collaborators used EGFP-Foxp3 mice (mice co-expressing the enhanced green fluorescent protein, EGFP, under the control of the endogenous Foxp3 promoter/enhancer elements, specific of Treg cells) to demonstrate that acute rapamycin administration (12 h before priming with self-antigen) ameliorated the clinical course of RR−EAE, via increasing Treg proliferation [106]. Other EAE-studies were carried out to elucidate if other pathways can synergize with the mTOR pathway in the pathogenesis of EAE. Importantly SIGIRR, a negative regulator of IL-1 receptor and Toll-like receptor signaling, controls Th17 cell expansion and effector function through the IL-1-induced mTOR signaling pathway; in fact, SIGIRR-/- mice show an earlier onset of neurological impairment and markedly increased disease severity compared to wild type animals [107]. Interestingly, another downstream pathway of mTOR activation, the HIF-1α-dependent glycolitic pathway, mediates T-cell differentiation. HIF-1α-induced metabolic reprogramming orchestrates lineage differentiation of T cells and therefore, blocking glycolysis alters Th17 and Treg differentiation and protects mice from EAE [108]. Taken together, these studies support the hypothesis that the mTOR pathway is critically involved in the development of EAE, although most of the mechanistic evaluations were restricted to the peripheral immune system and most of the studies mentioned so far were carried out on RR−EAE models.

More recent evidence from our laboratory and others groups suggest similar beneficial effects of mTOR inhibitors in Ch−EAE models. We observed that rapamycin ameliorates clinical and histological signs of Ch-EAE when administered at the peak of disease (therapeutic approach) [109]. Moreover, T cells harvested from rapamycin treated animals were less responsive to *ex-vivo* administration of 20 μg/ml MOG$_{35−55}$. However, rapamycin treatment did not change the extent of peripheral immune cell infiltrates, nor the astrocyte activation in the brain, despite significantly improving myelination. In this experimental model, the peripheral inflammatory component appears to be modest in the later stages of disease [110], and it is mostly compartmentalized behind an intact BBB [111]. Therefore, we can hypothesize that the observed beneficial effects of rapamycin are the consequence of reduced microglial inflammatory activation and/or a direct protective effect on myelination, due to the ability of the drug to cross the BBB. In this model, rapamycin was also able to reduce signs of neuropathic pain. The latter represents the most prevalent pain syndrome observed in MS patients, and the most difficult to treat [112]. In our studies, rapamycin increased the sensitivity threshold for mechanical allodynia, which is usually reduced at the clinical onset of disease. For this evaluation, rapamycin was injected in a prophylactic manner, and the treatment completely abolished the development of EAE [109]. Consistently, the group of Tiwari-Woodruff using a Ch–EAE model, showed that prophylactic administration of the estrogen receptor (ER) β ligand 2,3-bis(4-hydroxyphenyl)-propionitrile (DPN) decreases clinical severity. DNP is neuroprotective, stimulates endogenous myelination, and improves axon conduction, albeit without altering peripheral cytokine production or reducing CNS inflammation [113]. The protective effects of DPN treatment seemed to be mediated by increased ERβ phosphorylation and PI3K/Akt/mTOR activation [114]. In addition, they investigated the effects of the highly selective ERβ agonist, indazole chloride (Ind-Cl), on functional remyelination in a Ch–EAE model. Therapeutic Ind-Cl administration improved clinical signs of disease and animal motor performance. In contrast to previous findings, the drug decreased Th1 cytokine production in the periphery and reactive astrocytes, activated microglia, and T-cell infiltration in the brain of EAE mice. Moreover, Ind-Cl increased callosal myelination and mature oligodendrocytes, leading to callosal conduction and refractoriness. However, therapeutic Ind-Cl-induced remyelination was independent of its effects on the immune system, as Ind-Cl increased remyelination also in cuprizone demyelinating model. The authors linked the beneficial effects of Ind-CI to direct activation of the mTOR pathway in the OPCs, thus increased OLG survival and differentiation [115].

Although indirectly a possible beneficial effect of mTOR inhibition in EAE can be deduced from studies using metformin. Restricting infiltration of mononuclear cells into the CNS, oral administration of metformin attenuates the EAE disease in both RR–EAE and in Ch–EAE, and inhibits the expression of inflammatory cytokines and their mediators (IL6, TNFα, NOS2) [116]. In addition, metformin attenuates T-cell autoreactivity in the peripheral compartment of EAE Lewis rats and protects OLGs in activated mixed glial cultures [117]. Overall these effects are linked to AMPK activity and lipid alteration: AMPK activity and lipids alterations were restored by metformin treatment in the CNS of treated EAE animals [116]. AMPK activation lead to mTOR inhibition, as described in

Section 20.2, hence metformin effects on EAE models can be also mediated by inhibition of mTOR. These data, thus, demonstrate that mTOR is a key regulator in both RR–EAE and Ch–EAE that deserves more attention to understand potential implication for MS therapy.

20.8 CONCLUSIONS

In summary, the mTOR kinase appears to play a key role in regulating peripheral and central inflammatory processes that could contribute to the pathogenesis of MS. Beneficial effects of mTOR inhibitors are observed in different EAE models (both RR–EAE and Ch–EAE), making the use of mTOR inhibitors a good candidate for testing in neurological disorders and diseases with an inflammatory component. However, two clinical trials using two different rapalogs are listed in the clinicaltrials.gov database, without any report on beneficial effects in the human pathology. In particular, a phase I/II, open-label pilot trial to evaluate the safety of sirolimus in patients with relapsing-remitting MS enrolled 8 patients and terminated in January 2007, and a second multicenter, randomized, double-blind, long-term extension study to determine the safety, tolerability, and preliminary efficacy of CCI-779 in subjects with RR–MS or SPMS enrolled 221 patients and has been completed in November 2005. For both trials no study results were posted in clinicaltrials.gov nor in Pubmed. The critical role played by mTOR in the myelination process as well as in the regulation of neuronal functions (reviewed elsewhere [17]) may limit the use of mTOR inhibitors, or require specific timing to avoid interfering with processes of remyelination and axonal regeneration that occur after damage.

References

[1] Browne P, Chandraratna D, Angood C, Tremlett H, Baker C, Taylor BV, et al. Atlas of multiple sclerosis 2013: a growing global problem with widespread inequity. Neurology 2014;83 (11):1022–4.

[2] Popescu BF, Pirko I, Lucchinetti CF. Pathology of multiple sclerosis: where do we stand? Continuum (Minneap Minn) 2013;19 (4 Multiple Sclerosis):901–21.

[3] Didonna A, Oksenberg JR. Genetic determinants of risk and progression in multiple sclerosis. Clin Chim Acta 2015; pii: S0009-8981(15)00056-X.

[4] Hedström AK, Olsson T, Alfredsson L. The role of environment and lifestyle in determining the risk of multiple sclerosis. Curr Top Behav Neurosci 2015;26:87–104.

[5] Duffy SS, Lees JG, Moalem-Taylor G. The contribution of immune and glial cell types in experimental autoimmune encephalomyelitis and multiple sclerosis. Mult Scler Int 2014;2014:285245.

[6] Loleit V, Biberacher V, Hemmer B. Current and future therapies targeting the immune system in multiple sclerosis. Curr Pharm Biotechnol 2014;15(3):276–96.

[7] Fitzner D, Simons M. Chronic progressive multiple sclerosis - pathogenesis of neurodegeneration and therapeutic strategies. Curr Neuropharmacol 2010;8(3):305–15.

[8] Comabella M, Khoury SJ. Immunopathogenesis of multiple sclerosis. Clin Immunol 2012;142(1):2–8.

[9] Bogie JF, Stinissen P, Hendriks JJ. Macrophage subsets and microglia in multiple sclerosis. Acta Neuropathol 2014;128 (2):191–213.

[10] Cherry JD, Olschowka JA, O'Banion MK. Neuroinflammation and M2 microglia: the good, the bad, and the inflamed. J Neuroinflammation 2014;11:98.

[11] Michailidou I, de Vries HE, Hol EM, van Strien ME. Activation of endogenous neural stem cells for multiple sclerosis therapy. Front Neurosci 2015;8:454.

[12] Miljković D, Timotijević G, Mostarica Stojković M. Astrocytes in the tempest of multiple sclerosis. FEBS Lett 2011;585 (23):3781–8.

[13] Chiurchiù V. Novel targets in multiple sclerosis: to oxidative stress and beyond. Curr Top Med Chem 2014;14(22):2590–9.

[14] Wingerchuk DM, Carter JL. Multiple sclerosis: current and emerging disease-modifying therapies and treatment strategies. Mayo Clin Proc 2014;89(2):225–40.

[15] Lassmann H, van Horssen J, Mahad D. Progressive multiple sclerosis: pathology and pathogenesis. Nat Rev Neurol 2012;8 (11):647–56.

[16] Yang H, Wang X, Zhang Y, Liu H, Liao J, Shao K, et al. Modulation of TSC-mTOR signaling on immune cells in immunity and autoimmunity. J Cell Physiol 2014;229(1):17–26.

[17] Dello Russo C, Lisi L, Feinstein DL, Navarra P. mTOR kinase, a key player in the regulation of glial functions: relevance for the therapy of multiple sclerosis. Glia 2013;61(3):301–11.

[18] Norrmén C, Suter U. Akt/mTOR signalling in myelination. Biochem Soc Trans 2013;41(4):944–50.

[19] Wood TL, Bercury KK, Cifelli SE, Mursch LE, Min J, Dai J, et al. mTOR: a link from the extracellular milieu to transcriptional regulation of oligodendrocyte development. ASN Neuro 2013;5 (1):e00108.

[20] Wullschleger S, Loewith R, Hall MN. TOR signaling in growth and metabolism. Cell 2006;124(3):471–84.

[21] Foster KG, Fingar DC. Mammalian target of rapamycin (mTOR): conducting the cellular signaling symphony. J Biol Chem 2010;285(19):14071–7.

[22] Laplante M, Sabatini DM. mTOR signaling in growth control and disease. Cell 2012;149(2):274–93.

[23] Ma XM, Blenis J. Molecular mechanisms of mTOR-mediated translational control. Nat Rev Mol Cell Biol 2009;10(5):307–18.

[24] Brummer T, Schmitz-Peiffer C, Daly RJ. Docking proteins. FEBS J 2010;277:4356–69.

[25] Sparks CA, Guertin DA. Targeting mTOR: prospects for mTOR complex 2 inhibitors in cancer therapy. Oncogene 2010;29 (26):3733–44.

[26] Vallianou NG, Evangelopoulos A, Kazazis C. Metformin and cancer. Rev Diabet Stud 2013;10(4):228–35.

[27] Oh WJ, Wu CC, Kim SJ, Facchinetti V, Julien LA, Finlan M, et al. mTORC2 can associate with ribosomes to promote cotranslational phosphorylation and stability of nascent Akt polypeptide. EMBO J 2010;29(23):3939–51.

[28] Hresko RC, Mueckler M. mTOR.RICTOR is the Ser473 kinase for Akt/protein kinase B in 3T3-L1 adipocytes. J Biol Chem 2005;280(49):40406–16.

[29] Sarbassov DD, Guertin DA, Ali SM, Sabatini DM. Phosphorylation and regulation of Akt/PKB by the rictor-mTOR complex. Science 2005;307(5712):1098–101.

[30] García-Martínez JM, Alessi DR. mTOR complex 2 (mTORC2) controls hydrophobic motif phosphorylation and activation of serum- and glucocorticoid-induced protein kinase 1 (SGK1). Biochem J 2008;416(3):375–85.

[31] Jacinto E, Loewith R, Schmidt A, Lin S, Rüegg MA, Hall A, et al. Mammalian TOR complex 2 controls the actin cytoskeleton and is rapamycin insensitive. Nat Cell Biol 2004;6(11):1122–228.

[32] Sarbassov DD, Ali SM, Kim DH, Guertin DA, Latek RR, Erdjument-Bromage H, et al. Rictor, a novel binding partner of mTOR, defines a rapamycin-insensitive and raptor-independent pathway that regulates the cytoskeleton. Curr Biol 2004;14 (14):1296–302.

[33] Lee K, Gudapati P, Dragovic S, Spencer C, Joyce S, Killeen N, et al. Mammalian target of rapamycin protein complex 2 regulates differentiation of Th1 and Th2 cell subsets via distinct signaling pathways. Immunity 2010;32(6):743–53.

[34] Lisi L, Aceto P, Navarra P, Dello Russo C. mTOR Kinase: a possible pharmacological target in the management of chronic pain. Biomed Res Int 2015;2015:394257.

[35] Katholnig K, Linke M, Pham H, Hengstschläger M, Weichhart T. Immune responses of macrophages and dendritic cells regulated by mTOR signalling. Biochem Soc Trans 2013;41 (4):927–33.

[36] Liu P, Guo J, Gan W, Wei W. Dual phosphorylation of Sin1 at T86 and T398 negatively regulates mTORC2 complex integrity and activity. Protein Cell 2014;5(3):171–7.

[37] Dennis PB, Fumagalli S, Thomas G. Target of rapamycin (TOR): balancing the opposing forces of protein synthesis and degradation. Curr Opin Genet Dev 1999;9(1):49–54.

[38] Sofroniadou S, Goldsmith D. Mammalian target of rapamycin (mTOR) inhibitors: potential uses and a review of haematological adverse effects. Drug Saf 2011;34(2):97–115.

[39] Hartford CM, Ratain MJ. Rapamycin: something old, something new, sometimes borrowed and now renewed. Clin Pharmacol Ther 2007;82(4):381–8.

[40] Yip CK, Murata K, Walz T, Sabatini DM, Kang SA. Structure of the human mTOR complex I and its implications for rapamycin inhibition. Mol Cell 2010;38(5):768–74.

[41] Thoreen CC, Sabatini DM. Rapamycin inhibits mTORC1, but not completely. Autophagy 2009;5(5):725–6.

[42] Sarbassov DD, Ali SM, Sengupta S, Sheen JH, Hsu PP, Bagley AF, et al. Prolonged rapamycin treatment inhibits mTORC2 assembly and Akt/PKB. Mol Cell 2006;22(2):159–68.

[43] Pong K, Zaleska MM. Therapeutic implications for immunophilin ligands in the treatment of neurodegenerative diseases. Curr Drug Targets CNS Neurol Disord 2003;2(6):349–56.

[44] Shen LJ, Wu FL. Nanomedicines in renal transplant rejection--focus on sirolimus. Int J Nanomedicine 2007;2(1):25–32.

[45] Stenner-Liewen F, Grünwald V, Greil R, Porta C. The clinical potential of temsirolimus in second or later lines of treatment for metastatic renal cell carcinoma. Expert Rev Anticancer Ther 2013;13(9):1021–33.

[46] Hasskarl J. Everolimus. Recent Results Cancer Res 2014;201:373–92.

[47] Huang J, Dibble CC, Matsuzaki M, Manning BD. The TSC1-TSC2 complex is required for proper activation of mTOR complex 2. Mol Cell Biol 2008;28(12):4104–15.

[48] Benjamin D, Colombi M, Moroni C, Hall MN. Rapamycin passes the torch: a new generation of mTOR inhibitors. Nat Rev Drug Discov 2011;10(11):868–80.

[49] Wander SA, Hennessy BT, Slingerland JM. Next-generation mTOR inhibitors in clinical oncology: how pathway complexity informs therapeutic strategy. J Clin Invest 2011;121(4):1231−41.

[50] Schenone S, Brullo C, Musumeci F, Radi M, Botta M. ATP-competitive inhibitors of mTOR: an update. Curr Med Chem 2011;18(20):2995−3014.

[51] Janes MR, Limon JJ, So L, Chen J, Lim RJ, Chavez MA, et al. Effective and selective targeting of leukemia cells using a TORC1/2 kinase inhibitor. Nat Med 2010;16(2):205−13.

[52] Plews RL, Mohd Yusof A, Wang C, Saji M, Zhang X, Chen CS, et al. A novel dual AMPK activator/mTOR inhibitor inhibits thyroid cancer cell growth. J Clin Endocrinol Metab 2015; jc20141777.

[53] Zhou G, Myers R, Li Y, Chen Y, Shen X, Fenyk-Melody J, et al. Role of AMP-activated protein kinase in mechanism of metformin action. J Clin Invest 2001;108(8):1167−74.

[54] Pollak M. Overcoming drug development bottlenecks with repurposing: repurposing biguanides to target energy metabolism for cancer treatment. Nat Med 2014;20(6):591−3.

[55] Eikawa S, Nishida M, Mizukami S, Yamazaki C, Nakayama E, Udono H. Immune-mediated antitumor effect by type 2 diabetes drug, metformin. Proc Natl Acad Sci U S A 2015;112 (6):1809−14.

[56] Chapman NM, Chi H. mTOR links environmental signals to T cell fate decisions. Front Immunol 2015;5:686.

[57] Zeng H, Chi H. mTOR signaling and transcriptional regulation in T lymphocytes. Transcription 2014;5(2):e28263.

[58] Chi H. Regulation and function of mTOR signalling in T cell fate decisions. Nat Rev Immunol 2012;12(5):325−38.

[59] Delgoffe GM, Pollizzi KN, Waickman AT, Heikamp E, Meyers DJ, Horton MR, et al. The kinase mTOR regulates the differentiation of helper T cells through the selective activation of signaling by mTORC1 and mTORC2. Nat Immunol 2011;12(4):295−303.

[60] Inoki K. Role of TSC-mTOR pathway in diabetic nephropathy. Diabetes Res Clin Pract 2008;82(Suppl. 1):S59−62.

[61] MacIver NJ, Michalek RD, Rathmell JC. Metabolic regulation of T lymphocytes. Annu Rev Immunol 2013;31:259−83.

[62] Michalek RD, Gerriets VA, Jacobs SR, Macintyre AN, MacIver NJ, Mason EF, et al. Cutting edge: distinct glycolytic and lipid oxidative metabolic programs are essential for effector and regulatory CD4+ T cell subsets. J Immunol 2011;186(6):3299−303.

[63] Thomson AW, Turnquist HR, Raimondi G. Immunoregulatory functions of Mtor inhibition. Nat Rev Immunol 2009;9 (5):324−37.

[64] Heidt S, Roelen DL, Eijsink C, van Kooten C, Claas FH, Mulder A. Effects of immunosuppressive drugs on purified human B cells: evidence supporting the use of MMF and rapamycin. Transplantation 2008;86(9):1292−300.

[65] Weichhart T, Säemann MD. The PI3K/Akt/mTOR pathway in innate immune cells: emerging therapeutic applications. Ann Rheum Dis 2008;67(Suppl 3):p. iii 70−4.

[66] Baker AK, Wang R, Mackman N, Luyendyk JP. Rapamycin enhances LPS induction of tissue factor and tumor necrosis factor-alpha expression in macrophages by reducing IL-10 expression. Mol Immunol 2009;46(11−12):2249−55.

[67] Rocher C, Singla DK. SMAD-PI3K-Akt-mTOR pathway mediates BMP-7 polarization of monocytes into M2 macrophages. PLoS One 2013;8(12):e84009.

[68] Mercalli A, Calavita I, Dugnani E, Citro A, Cantarelli E, Nano R, et al. Rapamycin unbalances the polarization of human macrophages to M1. Immunology 2013;140(2):179−90.

[69] Byles V, Covarrubias AJ, Ben-Sahra I, Lamming DW, Sabatini DM, Manning BD, et al. The TSC-mTOR pathway regulates macrophage polarization. Nat Commun 2013;4:2834.

[70] Monti P, Mercalli A, Leone BE, Valerio DC, Allavena P, Piemonti L. Rapamycin impairs antigen uptake of human dendritic cells. Transplantation 2003;75(1):137−45.

[71] Fischer R, Turnquist HR, Taner T, Thomson AW. Use of rapamycin in the induction of tolerogenic dendritic cells. Handb Exp Pharmacol 2009;188:215−32.

[72] Weinstein SL, Finn AJ, Davé SH, Meng F, Lowell CA, Sanghera JS, et al. Phosphatidylinositol 3-kinase and mTOR mediate lipopolysaccharide-stimulated nitric oxide production in macrophages via interferon-beta. J Leukoc Biol 2000;67(3):405−14.

[73] Dello Russo C, Lisi L, Tringali G, Navarra P. Involvement of mTOR kinase in cytokine-dependent microglial activation and cell proliferation. Biochem Pharmacol 2009;78(9):1242−51.

[74] Lisi L, Laudati E, Navarra P, Dello Russo C. The mTOR kinase inhibitors polarize glioma-activated microglia to express a M1 phenotype. J Neuroinflammation 2014;11:125.

[75] Orihuela, R, McPherson, CA, Harry, GJ. Microglial M1/M2 polarization and metabolic states. Br J Pharmacol 2015.

[76] Jang BC, Paik JH, Kim SP, Shin DH, Song DK, Park JG, et al. Catalase induced expression of inflammatory mediators via activation of NF-kappaB, PI3K/AKT, p70S6K, and JNKs in BV2 microglia. Cell Signal 2005;17(5):625−33.

[77] Han HE, Kim TK, Son HJ, Park WJ, Han PL. Activation of autophagy pathway suppresses the expression of iNOS, IL6 and cell death of LPS-stimulated microglia cells. Biomol Ther (Seoul) 2013;21(1):21−8.

[78] de Oliveira AC, Candelario-Jalil E, Langbein J, Wendeburg L, Bhatia HS, Schlachetzki JC, et al. Pharmacological inhibition of Akt and downstream pathways modulates the expression of COX-2 and mPGES-1 in activated microglia. J Neuroinflammation 2012;9:2.

[79] Lu DY, Liou HC, Tang CH, Fu WM. Hypoxia-induced iNOS expression in microglia is regulated by the PI3-kinase/Akt/mTOR signaling pathway and activation of hypoxia inducible factor-1alpha. Biochem Pharmacol 2006;72(8):992−1000.

[80] Schwartz M, Butovsky O, Brück W, Hanisch UK. Microglial phenotype: is the commitment reversible? Trends Neurosci 2006;29(2):68−74.

[81] Chong ZZ, Li F, Maiese K. The pro-survival pathways of mTOR and protein kinase B target glycogen synthase kinase-3beta and nuclear factor-kappaB to foster endogenous microglial cell protection. Int J Mol Med 2007;19(2):263−72.

[82] Shang YC, Chong ZZ, Wang S, Maiese K. Erythropoietin and Wnt1 govern pathways of mTOR, Apaf-1, and XIAP in inflammatory microglia. Curr Neurovasc Res 2011;8(4):270−85.

[83] Lisi L, Navarra P, Feinstein DL, Dello Russo C. The mTOR kinase inhibitor rapamycin decreases iNOS mRNA stability in astrocytes. J Neuroinflammation 2011;8(1):1.

[84] Codeluppi S, Fernandez-Zafra T, Sandor K, Kjell J, Liu Q, Abrams M, et al. Interleukin-6 secretion by astrocytes is dynamically regulated by PI3K-mTOR-calcium signaling. PLoS One 2014;9(3):e92649.

[85] Li CY, Li X, Liu SF, Qu WS, Wang W, Tian DS. Inhibition of mTOR pathway restrains astrocyte proliferation, migration and production of inflammatory mediators after oxygen-glucose deprivation and reoxygenation. Neurochem Int 2015; pii: S0197-0186(15)00044-3.

[86] Tyler WA, Gangoli N, Gokina P, Kim HA, Covey M, Levison SW, et al. Activation of the mammalian target of rapamycin (mTOR) is essential for oligodendrocyte differentiation. J Neurosci 2009; 29(19):6367−78.

[87] Tyler WA, Jain MR, Cifelli SE, Li Q, Ku L, Feng Y, et al. Proteomic identification of novel targets regulated by the mammalian target of rapamycin pathway during oligodendrocyte differentiation. Glia 2011;59(11):1754−69.

[88] Dugas JC, Tai YC, Speed TP, Ngai J, Barres BA. Functional genomic analysis of oligodendrocyte differentiation. J Neurosci 2006;26(43):10967–83.

[89] Guardiola-Diaz HM, Ishii A, Bansal R. Erk1/2 MAPK and mTOR signaling sequentially regulates progression through distinct stages of oligodendrocyte differentiation. Glia 2012;60 (3):476–86.

[90] Flores AI, Narayanan SP, Morse EN, Shick HE, Yin X, Kidd G, et al. Constitutively active Akt induces enhanced myelination in the CNS. J Neurosci 2008;28(28):7174–83.

[91] Goebbels S, Oltrogge JH, Kemper R, Heilmann I, Bormuth I, Wolfer S, et al. Elevated phosphatidylinositol 3,4,5-trisphosphate in glia triggers cell-autonomous membrane wrapping and myelination. J Neurosci 2010;30(26):8953–64.

[92] Narayanan SP, Flores AI, Wang F, Macklin WB. Akt signals through the mammalian target of rapamycin pathway to regulate CNS myelination. J Neurosci 2009;29(21):6860–70.

[93] Sherman DL, Krols M, Wu LM, Grove M, Nave KA, Gangloff YG, et al. Arrest of myelination and reduced axon growth when Schwann cells lack mTOR. J Neurosci 2012;32(5):1817–25.

[94] Wahl SE, McLane LE, Bercury KK, Macklin WB, Wood TL. Mammalian target of rapamycin promotes oligodendrocyte differentiation, initiation and extent of CNS myelination. J Neurosci 2014;34(13):4453–65.

[95] Bercury KK, Dai J, Sachs HH, Ahrendsen JT, Wood TL, Macklin WB. Conditional ablation of raptor or rictor has differential impact on oligodendrocyte differentiation and CNS myelination. J Neurosci 2014;34(13):4466–80.

[96] Lebrun-Julien F, Bachmann L, Norrmén C, Trötzmüller M, Köfeler H, Rüegg MA, et al. Balanced mTORC1 activity in oligodendrocytes is required for accurate CNS myelination. J Neurosci 2014;34(25):8432–48.

[97] Meikle L, Pollizzi K, Egnor A, Kramvis I, Lane H, Sahin M, et al. Response of a neuronal model of tuberous sclerosis to mammalian target of rapamycin (mTOR) inhibitors: effects on mTORC1 and Akt signaling lead to improved survival and function. J Neurosci 2008;28(21):5422–32.

[98] Rangaraju S, Verrier JD, Madorsky I, Nicks J, Dunn Jr WA, Notterpek L. Rapamycin activates autophagy and improves myelination in explant cultures from neuropathic mice. J Neurosci 2010;30(34):11388–97.

[99] Feinstein DL, Galea E, Gavrilyuk V, Brosnan CF, Whitacre CC, Dumitrescu-Ozimek L, et al. Peroxisome proliferator-activated receptor-gamma agonists prevent experimental autoimmune encephalomyelitis. Ann Neurol 2002;51(6):694–702.

[100] Martel RR, Klicius J, Galet S. Inhibition of the immune response by rapamycin, a new antifungal antibiotic. Can J Physiol Pharmacol 1977;55(1):48–51.

[101] Branisteanu DD, Mathieu C, Bouillon R. Synergism between sirolimus and 1,25-dihydroxyvitamin D3 in vitro and in vivo. J Neuroimmunol 1997;79(2):138–47.

[102] van Etten E, Branisteanu DD, Verstuyf A, Waer M, Bouillon R, Mathieu C. Analogs of 1,25-dihydroxyvitamin D3 as dose-reducing agents for classical immunosuppressants. Transplantation 2000;69(9):1932–42.

[103] Webster KE, Walters S, Kohler RE, Mrkvan T, Boyman O, Surh CD, et al. In vivo expansion of T reg cells with IL-2-mAb complexes: induction of resistance to EAE and long-term acceptance of islet allografts without immunosuppression. J Exp Med 2009;206(4):751–60.

[104] Esposito M, Ruffini F, Bellone M, Gagliani N, Battaglia M, Martino G, et al. Rapamycin inhibits relapsing experimental autoimmune encephalomyelitis by both effector and regulatory T cells modulation. J Neuroimmunol 2010;220(1–2):52–63.

[105] Donia M, Mangano K, Amoroso A, Mazzarino MC, Imbesi R, Castrogiovanni P, et al. Treatment with rapamycin ameliorates clinical and histological signs of protracted relapsing experimental allergic encephalomyelitis in Dark Agouti rats and induces expansion of peripheral CD4+ CD25+ Foxp3+ regulatory T cells. J Autoimmun 2009;33(2):135–40.

[106] Procaccini C, De Rosa V, Galgani M, Abanni L, Calì G, Porcellini A, et al. An oscillatory switch in mTOR kinase activity sets regulatory T cell responsiveness. Immunity 2010;33 (6):929–41.

[107] Gulen MF, Kang Z, Bulek K, Youzhong W, Kim TW, Chen Y, et al. The receptor SIGIRR suppresses Th17 cell proliferation via inhibition of the interleukin-1 receptor pathway and mTOR kinase activation. Immunity 2010;32(1):54–66.

[108] Shi LZ, Wang R, Huang G, Vogel P, Neale G, Green DR, et al. HIF1alpha-dependent glycolytic pathway orchestrates a metabolic checkpoint for the differentiation of TH17 and Treg cells. J Exp Med 2011;208(7):1367–76.

[109] Lisi L, Navarra P, Cirocchi R, Sharp A, Stigliano E, Feinstein DL, et al. Rapamycin reduces clinical signs and neuropathic pain in a chronic model of experimental autoimmune encephalomyelitis. J Neuroimmunol 2012;243(1–2):43–51.

[110] Juedes AE, Hjelmström P, Bergman CM, Neild AL, Ruddle NH. Kinetics and cellular origin of cytokines in the central nervous system: insight into mechanisms of myelin oligodendrocyte glycoprotein-induced experimental autoimmune encephalomyelitis. J Immunol 2000;164(1):419–26.

[111] Procaccini C, De Rosa V, Pucino V, Formisano L, Matarese G. Animal models of multiple sclerosis. Eur J Pharmacol 2015; pii: S0014-2999(15)00259-9.

[112] O'Connor AB, Schwid SR, Herrmann DN, Markman JD, Dworkin RH. Pain associated with multiple sclerosis: systematic review and proposed classification. Pain 2008;137 (1):96–111.

[113] Crawford DK, Mangiardi M, Song B, Patel R, Du S, Sofroniew MV, et al. Oestrogen receptor beta ligand: a novel treatment to enhance endogenous functional remyelination. Brain 2010;133 (10):2999–3016.

[114] Kumar S, Patel R, Moore S, Crawford DK, Suwanna N, Mangiardi M, et al. Estrogen receptor β ligand therapy activates PI3K/Akt/mTOR signaling in oligodendrocytes and promotes remyelination in a mouse model of multiple sclerosis. Neurobiol Dis 2013;56:131–44.

[115] Moore SM, Khalaj AJ, Kumar S, Winchester Z, Yoon J, Yoo T, et al. Multiple functional therapeutic effects of the estrogen receptor β agonist indazole-Cl in a mouse model of multiple sclerosis. Proc Natl Acad Sci U S A 2014;111(50):18061–6.

[116] Nath N, Khan M, Paintlia MK, Singh I, Hoda MN, Giri S. Metformin attenuated the autoimmune disease of the central nervous system in animal models of multiple sclerosis. J Immunol 2009;182(12):8005–14.

[117] Paintlia AS, Paintlia MK, Mohan S, Singh AK, Singh I. AMP-activated protein kinase signaling protects oligodendrocytes that restore central nervous system functions in an experimental autoimmune encephalomyelitis model. Am J Pathol 2013;183(2):526–41.

mTOR IN THE ENDOCRINE SYSTEM AND DISORDERS OF METABOLISM

21

mTOR in Metabolic and Endocrine Disorders

Marta M. Swierczynska and Michael N. Hall

Biozentrum, University of Basel, Basel, Switzerland

21.1 INTRODUCTION

To survive when nutrients are scarce and to grow when nutrients are abundant, organisms balance energy-producing catabolic and energy-consuming anabolic processes. In mammals, the metabolic response to nutrients, at the cell, organ, or whole organism level, is controlled by mammalian target of rapamycin (mTOR). In nutrient-rich conditions, mTOR activates multiple anabolic processes that ultimately lead to mass accumulation and growth. When nutrients are limited, mTOR is rapidly inhibited to restrict anabolism [1]. In multicellular organisms, metabolism is centrally controlled by the coordinated actions of the central nervous system (CNS) and endocrine glands. mTOR signaling also modulates cellular responses to various hormones secreted by endocrine glands. This is of particular importance in metabolic organs, such as liver, skeletal muscle, and adipose tissue, where hormonal responses affect whole-body energy balance [2].

Most common metabolic and endocrine disorders, among them obesity, insulin resistance, diabetes, and dyslipidemia, result from aberrant energy homeostasis. Not surprisingly, they are often accompanied by deregulated mTOR signaling in metabolic organs [3–5]. Thus, mTOR may be an attractive therapeutic target in the treatment of metabolic and endocrine aberrations [3]. In this chapter, we summarize findings concerning the role of mTOR signaling in energy metabolism, both at the single-cell and whole-body level. Furthermore, we discuss how deregulation of mTOR signaling may contribute to the development of metabolic and endocrine diseases.

21.2 THE mTOR COMPLEXES

mTOR is a serine/threonine kinase that forms two structurally and functionally distinct multiprotein complexes, namely mTOR complex 1 (mTORC1) and mTOR complex 2 (mTORC2). The core components of mTORC1 are mTOR, regulatory-associated protein of mTOR (Raptor), and mammalian lethal with sec-13 protein 8 (mLST8). The core components of mTORC2 include mTOR, rapamycin-insensitive companion of TOR (Rictor), mammalian stress-activated map kinase-interacting protein 1 (mSIN1), and mLST8. Raptor and Rictor are essential for mTORC1 and mTORC2 activity, respectively [6–10]. mTORC1 is activated in response to growth factors and nutrients, and inhibited in response to low cellular energy levels. The best-characterized downstream targets of mTORC1 are p70 ribosomal S6 kinase (S6K), eukaryotic translation initiation factor 4E (eIF4E)-binding protein (4E-BP), and UNC-51-like kinase 1 (ULK1). Thus, their phosphorylation status is often used as a readout for mTORC1 activity [11]. mTORC2 is activated by growth factors alone and via association with ribosomes [12]. Its downstream targets are AGC kinase family members, including Akt, serum/glucocorticoid-regulated kinase (SGK), and protein kinase C (PKC) [13]. The most commonly used readout for mTORC2 activity is Akt-Ser473 phosphorylation [13,14]. As the name implies, mTOR is the target of the immunosuppressive and anticancer drug rapamycin. Rapamycin forms a complex with FK506-binding protein that directly binds and allosterically inhibits mTOR in mTORC1 but not in mTORC2 [8,9,15,16]. However, prolonged rapamycin treatment can indirectly inhibit mTORC2 in certain cell types by preventing mTORC2 assembly [17].

21.3 REGULATION OF mTOR SIGNALING

mTOR integrates multiple signals to coordinate processes involved in cell growth. Regulation of mTORC1 is well understood, whereas regulation of mTORC2 is relatively poorly characterized. It is known that mTORC2 is activated by phosphoinositide 3-kinase (PI3K) in response to growth factors, and requires mTORC2 association with the ribosome [12,13] (Figure 21.1A). However, the detailed molecular mechanism of mTORC2 activation remains to be determined. It is well established that mTORC1 is regulated by growth factors, nutrients, and energy status. This regulation has been the topic of many excellent reviews [1,11,19–22], and will therefore be discussed here only briefly. In this section, we focus on less

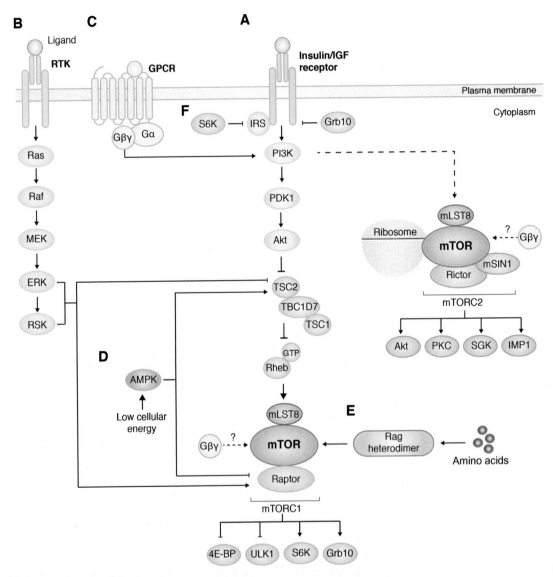

FIGURE 21.1 **Signaling upstream of mTOR: classical inputs.** (A) Binding of insulin and IGFs to their cognate receptors leads to PI3K stimulation and PDK1-dependent Akt activation. Akt phosphorylates TSC2 at multiple sites to inhibit its GAP activity toward Rheb. GTP-bound Rheb binds and activates mammalian mTORC1. Insulin/IGF-dependent PI3K stimulation also promotes association of mTORC2 with the ribosome and mTORC2 activation. (B) Growth factors binding to RTKs stimulate mTORC1 signaling through activation of the MAPK cascade. The effector kinases of this pathway, ERK and RSK, promote mTORC1 activity through inhibition of TSC2 and phosphorylation of Raptor. (C) Activation of GPCRs affects mTORC1 activity by multiple mechanisms. First, GPCR signaling controls the MAPK cascade (not depicted, discussed in detail in Ref. [18]). Second, βγ subunits of G proteins promote mTOR activity by activating PI3K. Third, βγ subunits also directly bind mTOR and promote its activity via an unknown mechanism. (D) In response to low-cellular energy levels (manifested as increased AMP/ATP ratio), AMPK activates TSC2 and inhibits Raptor leading to mTORC1 attenuation. (E) Amino acids stimulate mTORC1 through activation of Rag heterodimers and recruitment of mTORC1 to the lysosomal surface where it interacts with Rheb. (F) Activated mTORC1 phosphorylates S6K and Grb10 to feedback inhibit signaling downstream of insulin receptor.

characterized mTOR regulation, such as activation by lipids and by hormones acting via nuclear and G protein-coupled receptors (GPCRs).

21.3.1 Hormones and Growth Factors Acting Through the TSC Complex

Various hormones and growth factors activate mTORC1 signaling by inhibiting the tuberous sclerosis complex (TSC) that consists of tuberous sclerosis complex 2 (TSC2; also known as tuberin), TSC1 (hamartin), and TBC1 domain family member 7 [23,24]. The TSC complex is a GTPase activating protein (GAP) for the small GTPase Rheb and thus stimulates conversion of active Rheb-GTP into inactive Rheb-GDP [25,26]. Rheb-GTP binds and activates mTORC1 directly [27] (Figure 21.1A).

Insulin and growth factors activate mTORC1 by binding their cognate receptor tyrosine kinases (RTKs) and activating the PI3K-phosphoinositide-dependent kinase 1 (PDK1)—Akt pathway [28]. Akt phosphorylates TSC2 at multiple sites, leading to inhibition of the TSC complex [29—31]. This in turn allows for Rheb activity and mTORC1 activation (Figure 21.1A).

Thyroid hormone receptors (TRs) also transmit signals to mTORC1 via the PI3K—Akt axis [32,33]. This finding is surprising since, unlike membrane RTKs, TRs are nuclear receptors that classically activate or repress transcription of target genes in response to hormone binding. However, TRs have also been proposed to have non-nuclear functions such as at the plasma membrane, cytoplasm, or cellular organelles [34]. In human fibroblasts, treatment with triiodothyronine (T3) increases mTORC1 activity and leads to increased expression of the calcineurin inhibitor ZAKI-4α. This effect is mediated by TRβ1 that binds the p85α regulatory subunit of PI3K to activate PI3K in a T3-dependent manner [32]. T3-induced ZAKI-4α expression is inhibited in the presence of rapamycin and PI3K inhibitor wortmanin, as well as in cells expressing dominant-negative TRβ1 that does not bind T3 [32]. Thus, TRβ1-mediated PI3K and mTORC1 activation is required for T3-induced ZAKI-4α expression [32] (Figure 21.2A). Similarly, in primary rat myocytes, TRα1 directly binds PI3K and stimulates its activity [33]. This in turn leads to Akt and mTORC1 activation and increased protein synthesis [33] (Figure 21.2A). This mechanism may provide an explanation for the hypertrophic effect of thyroid hormones on cardiac muscle. In support of this finding, rapamycin treatment prevents thyroid-hormone-induced cardiac hypertrophy [35]. Mice treated with thyroxine (T4; the prohormone of T3) display increased cardiac Akt and mTORC1 activity and an enlarged heart. Treatment of animals with rapamycin prevents T4-induced hypertrophy [35]. Taken together, these data suggest that mTOR is a mediator of nongenomic TR actions.

Estrogen receptor also binds the p85α subunit of PI3K and thereby activates PI3K [36]. Like TR, estrogen receptor is a nuclear receptor that activates target genes but has also been proposed to regulate various nongenomic processes [36]. In cultured cells stimulated with 17β-estradiol, activated estrogen receptor directly binds and activates PI3K [36]. This in turn activates Akt and nitric oxide synthase (NOS). Treatment of cells with the PI3K inhibitor wortmanin completely abrogates estradiol-induced NOS activity [36]. Whether estrogen-receptor-dependent PI3K activation leads to mTORC1 activation has not been determined, although recent findings support this possibility. Treatment of primary mouse hippocampal neurons with estradiol leads to increased mTORC1 activity, and this effect is completely abolished by the estrogen receptor inhibitor ICI or rapamycin [37]. Treatment of neurons with ICI also reduces neuronal and dendritic growth. This phenotype can be rescued either by overexpressing Rheb or by expressing constitutively active S6K [37] (Figure 21.2B). These observations may explain the neuroprotective action of estradiol during brain development. Whether mTORC1 mediates estrogen receptor signaling in other tissues remains to be determined.

Aldosterone, a steroid hormone acting through nuclear mineralocorticoid receptor, has been reported to activate mTORC1 signaling. Systemic treatment of rats with aldosterone increases mTORC1 activity in kidney and induces renal fibrosis [38]. Concomitant treatment with rapamycin prevents aldosterone-induced fibrosis by decreasing synthesis of extracellular matrix components and proinflammatory chemokines [38]. The mechanism of aldosterone-mediated mTORC1 activation has not been investigated. However, in human neutrophils, aldosterone-dependent expression of vascular endothelial growth factor A (VEGF-A) is partly reduced upon treatment with a PI3K inhibitor [39]. This suggests that mineralocorticoid receptor might activate PI3K in a similar manner as TRs and estrogen receptors.

Growth factors, polypeptide hormones, neurotransmitters, and chemokines can activate mTORC1 indirectly by activating the MAPK pathway (Figure 21.1B). The MAPK pathway and its effectors, including extracellular signal-regulated kinase (ERK) and p90 ribosomal S6 kinase (RSK), phosphorylate TSC2 and thereby inhibit TSC activity. Additionally, ERK and RSK phosphorylate Raptor and promote mTORC1 activity towards 4E-BP [40]. Interestingly, aldosterone-induced VEGF-A expression is also partly reduced upon treatment with MAPK pathway inhibitors [39]. This raises the possibility that aldosterone can activate the mTORC1 pathway also via the MAPK cascade.

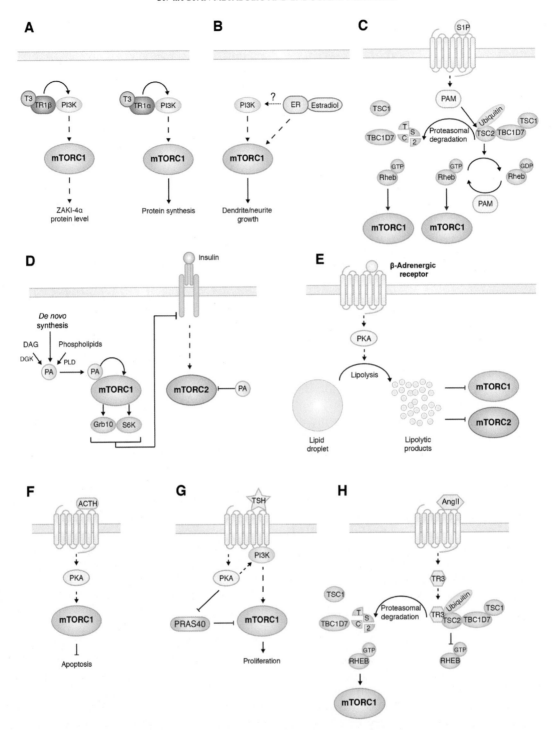

FIGURE 21.2　**Signaling upstream of mTOR: nonclassical inputs.** (A) Thyroid hormone receptors (TRs) stimulate mTORC1 by binding and activating PI3K. (B) Estrogen receptor (ESR) bound to estradiol activates mTORC1 via an unknown mechanism. Although ESR directly binds and stimulates PI3K, it has not been established whether this interaction also results in mTORC1 activation. (C) Sphingosine-1-phosphate (S1P) stimulates mTORC1 activity by activating E3 ubiquitin ligase PAM that targets TSC2 for proteasomal degradation. PAM has also been proposed to be a GEF for Rheb. (D) Phosphatidic acid (PA) is generated via *de novo* synthesis from lysophosphatidic acid, by hydrolysis of phospholipids by phospholipase D (PLD), and by phosphorylation of diacylglycerols (DAGs) by diacylglycerol kinase (DGK). PA stimulates mTORC1 activity by directly binding to mTOR and allosterically activating its kinase activity. Increased PA levels attenuate mTORC2 indirectly via the negative feedback loop and directly by disrupting interaction between mTOR and Rictor. (E) In adipocytes, activation of β-adrenergic receptor induces PKA-dependent lipolysis. Lipolytic products inhibit mTORC1 and mTORC2 signaling by promoting dissociation of mTOR and Raptor or Rictor, respectively. (F) In adrenocortical cells, ACTH receptor signaling leads to PKA-dependent mTORC1 activation. (G) In thyrocytes, TSH stimulation results in PI3K-dependent but Akt-independent activation of mTORC1. Activated TSH receptor also stimulates PKA to promote PI3K activity and block the mTORC1 inhibitor PRAS40. (H) In cardiomyocytes, AngII receptor activates orphan receptor TR3 that directly binds TSC2 and targets it for proteasomal degradation.

21.3.2 Nutrients

Nutrients, in particular amino acids, are a key mTORC1 input as they are required for mTORC1 activation in response to any other signal [19,22]. Amino acids signal to mTORC1 through Ras-related GTP-binding protein homolog (Rag) family GTPases [41,42]. Upon activation, Rag heterodimers recruit mTORC1 to the surface of the lysosome, where it is activated by Rheb (Figure 21.1E).

21.3.3 Cellular Energy

mTORC1 indirectly senses intracellular energy levels. Upon energy depletion (manifested as an increased AMP/ATP ratio), AMP-activated protein kinase (AMPK) phosphorylates TSC2 and Raptor leading to mTORC1 inhibition. In particular, TSC2 phosphorylation leads to TSC activation and inhibition of Rheb [43] whereas Raptor phosphorylation induces 14-3-3 binding and mTORC1 inhibition [44] (Figure 21.1D).

21.3.4 Negative Feedback Loop

Upon activation, mTORC1 phosphorylates S6K and growth factor receptor-bound protein 10 (Grb10) [11]. mTORC1 and S6K phosphorylate and inhibit insulin receptor substrate that is necessary for activation of PI3K and Akt downstream of the insulin receptor [45–50]. mTORC1 phosphorylates and stabilizes Grb10, thereby feedback-inhibiting the PI3K and MAPK pathways [51,52]. Thus, mTORC1-dependent phosphorylation of S6K and Grb10 counteracts insulin receptor signaling and dampens mTORC1 and mTORC2 activity (Figure 21.1F). In addition, S6K phosphorylates Rictor, further attenuating mTORC2 signaling [53,54].

The purpose of the negative feedback loop is possibly to ensure that the response to insulin is transient [4]. Upon chronic mTORC1 activation, the negative feedback loop reduces insulin receptor signaling and mTORC2 activity and leads to insulin resistance [4].

21.3.5 Lipids

Lipids constitute a class of "nonclassical" regulators of mTOR signaling. Phosphatidic acid (PA), a precursor of glycerolipids and signaling molecules, has been reported to directly bind and allosterically activate mTORC1 in cultured cells [55,56]. PA is generated via de novo synthesis from lysophosphatidic acid, by hydrolysis of phospholipids by phospholipase D (PLD), or by phosphorylation of diacylglycerol (DAG) by diacylglycerol kinase (DGK) [57]. All three pathways appear to contribute to PA-dependent up-regulation of mTORC1 activity [58–60]. Up to date, activation of mTORC1 by PA has been studied mostly in the context of cancer development, as PLD is up-regulated in many types of cancer [61]. However, recent reports suggest that the PA–mTORC1 axis might modulate insulin signaling in physiological conditions. In mouse primary hepatocytes, elevated de novo PA synthesis, due to increased expression of enzymes involved in the glycerol-3-phosphate pathway, activates mTORC1. This in turn leads to mTORC2 inhibition and reduced insulin signaling due to the negative feedback loop [62]. Another study has demonstrated that elevated levels of PA affect mTORC2 independently of the negative feedback loop. In cells with increased PA content due to PLD or DGK overexpression, mTORC2 signaling is down-regulated due to disruption of the mTOR–Rictor interaction [63] (Figure 21.2D). These two mechanisms of PA-dependent disruption of mTORC2 and insulin signaling may explain the correlation between triglyceride synthesis and insulin resistance, which is frequently observed in nonadipose tissues [63]. In addition to modulating hepatic insulin signaling, activation of mTORC1 via PA appears to be a regulator of muscle function. In explanted skeletal muscle, mechanical stimuli lead to activation of PLD, an increase in PA levels, and subsequently to mTORC1 activation [64]. However, a recent report suggests that generation of PA via DGK rather than PLD is the main regulator of PA-dependent mTORC1 activity in skeletal muscles [58]. Muscle isolated from mice lacking DGKζ display impaired PA generation and mTORC1 activation in response to mechanical stimulation. In contrast, muscle-specific overexpression of DGKζ leads to muscle hypertrophy, a phenotype that can be suppressed by treating animals with rapamycin [58]. Taken together, these data suggest that PA—via stimulation of mTORC1—may be a regulator of skeletal muscle metabolism. However, DGKζ-deficient animals do not display any overt abnormalities in muscle size or histological appearance, suggesting that the DGKζ-PA–mTORC1 axis is dispensable for maintaining muscle function under basal conditions. Alternatively, other inputs can compensate for DGKζ deficiency. Thus, further studies are required to fully understand the role of PA-mediated mTORC1 regulation in skeletal muscle.

Another lipid reported to modulate mTOR activity is sphingosine-1-phosphate (S1P). In human cancer cell lines and rat primary cells, binding of S1P to its cognate GPCR leads to activation of E3 ubiquitin ligase protein associated with Myc (PAM), which in turn targets TSC2 for proteasomal degradation, activates Rheb, and stimulates mTORC1 activity [65] (Figure 21.2C). The S1P–mTORC1 axis plays a role in T-cell differentiation *in vivo*. Up-regulation of the S1P$_1$ receptor in T-cell precursors results in increased Akt and mTORC1 activities and reduced differentiation and function of regulatory T cells (T$_{reg}$). Inhibition of Akt and mTORC1 restores T$_{reg}$ differentiation [66]. Recently, S1P receptor signaling has been implicated in modulating glucose metabolism in the liver and skeletal muscle. One of the mechanisms of S1P action in the liver is abrogating insulin-induced Akt phosphorylation and reducing glucokinase expression [67]. It remains to be determined whether S1P exerts its effect on glucose metabolism, at least in part, by affecting mTOR signaling.

Mullins et al. have recently reported another mechanism of mTOR regulation by lipids. They have demonstrated that lipids isolated from adipocytes undergoing lipolysis reduce mTORC1 and mTORC2 signaling. Lipolytic products exert their inhibitory effect by disrupting interactions between mTOR and Raptor or Rictor, causing mTORC1 and mTORC2 dissociation [68] (Figure 21.2E). The lipid species responsible for inducing complex dissociation has not been identified. Taken together, lipids appear to regulate mTOR signaling by multiple mechanisms. While the physiological relevance of this regulation requires confirmation, regulation of mTOR by lipids may provide a link between dyslipidemias and disturbed mTOR signaling in vital metabolic organs.

21.3.6 Hormones and Growth Factors Acting Through GPCRs

Stimulation of mTOR signaling by hormones and growth factors has been discussed mainly in the context of signaling via RTKs. However, recent studies suggest that various hormones impinge on mTOR via GPCRs. GPCRs may feed into the mTOR pathway indirectly by activation of the MAPK signaling cascade [18,40]. As discussed above, the MAPK pathway mediates phosphorylation of TSC2 and Raptor, ultimately leading to mTORC1 activation. GPCRs can also activate mTORC1 and mTORC2, via direct binding of Gβγ to the catalytic and regulatory subunits of PI3K [69]. Recently, it has been demonstrated that Gβγ also directly binds mTORC1 and mTORC2 [70]. Accordingly, overexpression of Gβγ increases phosphorylation of mTORC1 and mTORC2 targets [70]. However, it is not clear whether this effect may be due to direct binding to the mTORCs or to PI3K activation (Figure 21.1C).

Many GPCRs act by activating adenylate cyclase, increasing intracellular cAMP levels, and activating protein kinase A (PKA) [71]. Recent reports suggest that PKA influences mTORC1 signaling. In human and mouse adrenocortical cell lines, activation of PKA via stimulation with adrenocorticotropic hormone (ACTH) or with a cAMP-analog leads to activation of mTORC1 signaling [72]. Activation of mTORC1 upon ACTH/cAMP stimulation is inhibited in the presence of the PKA inhibitor H-89 [72] (Figure 21.2F). Transgenic mice expressing a constitutively active version of PKA display increased mTORC1 signaling in adrenocortical cells, thereby conferring resistance to dexamethasone-induced apoptosis. Treatment of animals with rapamycin restores normal levels of apoptosis in the adrenal cortex. Consistently, patients affected by primary pigmented nodular adrenocortical disease caused by activating mutations in PKA display increased mTORC1 signaling and decreased apoptosis in the adrenal cortex [72]. Thus, an ACTH-PKA-mTORC1 axis may play a role in mediating adrenal cortex homeostasis. PKA has been shown to directly affect PI3K activity [69]. Whether this mechanism can explain PKA-dependent mTORC1 activation in adrenocortical cells remains to be determined. Similar to the situation in adrenocortical cells, thyroid-stimulating hormone (TSH)-dependent induction of PKA results in mTORC1 activation in thyrocytes [73,74]. TSH promotes proliferation of thyroid cells by activating cyclin D3-cyclin-dependent kinase 4 (CDK4). Stimulation of thyroid cells with TSH or cAMP appears to increase mTORC1 activity in a PI3K-dependent and Akt-independent manner [73,74]. TSH-induced mTORC1 activation results, at least in part, from PKA-dependent phosphorylation of proline-rich Akt substrate of 40 kDa (PRAS40) that alleviates the inhibitory effect that PRAS40 exerts on mTORC1 [74]. Accordingly, rapamycin treatment decreases TSH-induced CDK4 activity and reduces thyroid cell proliferation [74] (Figure 21.2G). In contrast to its action in adrenocortical and thyroid cells, increased cAMP and PKA signaling have an inhibitory effect on mTOR activity in white adipocytes. In these cells, stimulation of beta-adrenergic receptors and concomitant increase in PKA activity promotes lipolysis and release of lipids from lipid droplets that in turn cause dissociation of both mTORC1 and mTORC2 [68] (Figure 21.2E). This mechanism may explain the opposing effects of insulin and catecholamines on glucose uptake in adipose tissue. Taken together, PKA appears to be an mTOR regulator.

In addition to stimulating the MAPK and PKA pathways, GPCRs can induce mTORC1 activity by another mechanism, as reported for the angiotensin II (AngII) receptor. AngII stimulation increases mTORC1 activity in both cardiomyocytes *in vivo* and heart muscle cell lines *in vitro*, an effect that depends on activation of the orphan receptor TR3. Cells with reduced TR3 expression and TR3 knockout animals fail to up-regulate mTORC1 activity in response to AngII, whereas TR3 overexpression increases mTORC1 signaling. TR3 directly binds TSC2 and targets it for proteasome-dependent degradation, thus alleviating its negative effect on mTORC1 [75] (Figure 21.2H). Taken together, it appears that GPCRs affect mTOR signaling by multiple mechanisms, although physiological relevance remains to be determined in many cases. Many hormones involved in regulation of metabolism, including glucagon, catecholamines, TSH, and ACTH, exert their effects through GPCR stimulation. Thus, deciphering GPCR-mediated modulation of mTOR signaling crosstalk may lead to better understanding of the role of mTOR in metabolic and endocrine disorders.

21.4 CELLULAR PROCESSES REGULATED BY mTOR

21.4.1 Protein Synthesis

Protein synthesis is the most energy-consuming process in the cell and therefore is tightly regulated. One way by which mTORC1 controls protein biosynthesis is via ribosome biogenesis. mTORC1 stimulates expression of ribosomal biogenesis (RiBi) genes that encode factors required for rRNA synthesis, cleavage, post-transcriptional modification, assembly with ribosomal proteins, and transport [76]. mTORC1 does so by phosphorylating and activating S6K, which in turn phosphorylates ribosomal protein S6. This in turn leads to expression of RiBi genes by a poorly understood mechanism. Mice expressing phospho-deficient S6 protein display a 25–35% reduction in expression of RiBi transcripts [76]. Phosphorylation of S6 may not be the only mechanism by which the mTOR−S6K axis promotes ribosome biogenesis, as double knockout mice lacking S6K1 and S6K2 display a 75% reduction in expression of RiBi genes [76]. Indeed, mTORC1 affects ribosome biogenesis also by stimulating transcription and activity of RNA polymerase I and polymerase III, enzymes responsible for transcription of rDNA (reviewed in Ref. [77]). mTORC1 also contributes to ribosome biogenesis and translation by preferentially promoting translation of a subset of mRNAs that contain a 5′Terminal OligoPirymidine track (5′TOP) which encodes components of translation machinery such as ribosomal proteins and elongation factors [78]. The mechanism by which mTORC1 affects 5′TOP-dependent translation is not well understood (reviewed in Ref. [79]). In addition to controlling ribosome biogenesis, mTORC1 regulates protein synthesis by promoting cap-dependent translation. It does so by phosphorylation of 4E-BP. Hypophosphorylated 4E-BP binds and inhibits the eIF4E. Upon phosphorylation, 4E-BP releases eIF4E which then associates with eIF4G and stimulates translation initiation [80].

The role of mTORC2 in the regulation of protein synthesis is not well defined. mTORC2 is activated upon association with the ribosome via an unknown mechanism [12], and cotranslationally phosphorylates substrates. mTORC2-dependent cotranslational phosphorylation of Akt and PKC prevents their ubiquitin-dependent degradation [81]. Phosphorylation of insulin-like growth factor 2 (IGF2) mRNA binding protein 1 (IMP1) by mTORC2 strongly increases IMP1's binding to IGF2 mRNA. This interaction promotes IGF2 translation and ultimately leads to increased IGF2 production [82]. Whether other substrates are cotranslationally phosphorylated by mTORC2 remains to be determined.

21.4.2 Lipid Synthesis

Lipids are essential components of cellular membranes that serve as signaling molecules and can be used to store energy. mTOR signaling promotes lipid biogenesis and accumulation. mTORC1 stimulates *de novo* lipid biosynthesis by activating master regulators of lipogenesis such as sterol regulatory element-binding protein 1 and 2 (SREBP1/2) [83,84]. SREBPs are synthesized as precursors that reside in the endoplasmic reticulum (ER). When sterol levels are high, SREBPs are retained on ER membranes. Upon decrease of sterol levels, the SREBPs are transported to the Golgi where they are proteolytically processed. This results in the release of a soluble, active form that can enter the nucleus and activate transcription [85]. mTORC1 signaling, through an S6K-dependent mechanism, promotes processing and nuclear accumulation of SREBPs, and thereby expression of genes controlling steroid and fatty acid biosynthesis [83]. Another mechanism by which mTORC1 stimulates SREBP activity relies on the phosphorylation and inhibition lipin 1, a PA phosphatase that decreases SREBP activity [86]. mTORC1 directly phosphorylates lipin 1, which prevents its nuclear accumulation. This in turn allows accumulation of SREBP in the nucleus by an unknown mechanism [86]. Mice with a deletion of *Raptor* specifically in liver cells, display reduced hepatic SREBP1c activity and decreased expression of lipogenic genes [87]. Thus, mTORC1 activity is necessary for hepatic lipogenesis. However, recent data suggest that it is not sufficient. *TSC1*-knockout mice are characterized by constitutive mTORC1 activity, yet display impaired SREBP activity and lipogenesis [88,89]. This phenotype most likely results from the attenuation of Akt activity through the negative feedback loop (see above). Interestingly, this feedback loop is partly S6K-independent, as liver-specific *TSC1* and *S6K* double knockout mice still display attenuated Akt signaling and reduced hepatic lipid content [90]. Impaired lipogenesis in *TSC1*-knockout animals can be explained by the fact that mTORC2, whose signaling is inhibited by the negative feedback loop, is also a regulator of lipogenesis. Liver-specific *Rictor* knockout mice that lack functional mTORC2 display decreased lipid synthesis due to reduced SREBP levels [91,92]. In contrast, mice with liver-specific *PTEN* deletion and increased mTORC2 and Akt activities, are characterized by increased expression of SREBP1 and lipid accumulation [93]. Taken together, mTORC1 and mTORC2 synergize to increase SREBP activity and thus to promote lipid synthesis.

Recent findings suggest that in addition to lipid biosynthesis, mTORC1 positively affects lipid storage in lipid droplets. In cultured intestinal epithelial cells, increased mTORC1 signaling due to leptin stimulation leads to lipid droplet accumulation. This effect is dose-dependent and reversed by rapamycin treatment [94]. However, the mechanism of mTORC1-dependent lipid droplet formation, as well as its relevance *in vivo*, requires further studies. Indirectly, mTORC1 may contribute to accumulation of lipid droplets by inhibiting lipolysis, a process that leads to the release of free fatty acid (FFA) and glycerol from neutral lipid stores. In cultured adipocytes, mTORC1 activation decreases expression of the two major lipases adipose triglyceride lipase (ATGL) and hormone-sensitive lipase, and thereby reduces the rate of lipolysis [95]. Inhibition of mTORC1 activity has the opposite effect, that is, elevated ATGL levels and increased lipolysis [95]. This function of mTORC1 appears to be mediated, at least in part, via 4E-BP as 4E-BP1/2 null MEFs have lower ATGL levels and increased fat accumulation [96], and mice lacking 4E-BP1 and 4E-BP2 display reduced lipolysis [97].

The role of mTOR in lipid metabolism has been investigated mostly in the context of lipid biosynthesis in the liver and adipose tissue. However, recent findings raise the interesting possibility that mTORC1 promotes the synthesis of steroid hormones. In cultured adrenocortical cells, inhibition of mTORC1 activity with rapamycin leads to a decrease in aldosterone and cortisol production [98,99]. Whether mTORC1 affects adrenal steroidogenesis directly or indirectly, that is by regulating substrate availability, has not been established. Furthermore, the physiological relevance of these observations remains to be determined.

21.4.3 Nucleotide Synthesis

Nucleotides play an essential role in multiple cellular processes. They are precursors of nucleic acids, serve as energy carriers, and constitute essential components of signal transduction pathways. Recent findings suggest that mTORC1 modulates biosynthesis of these important molecules. mTORC1 stimulates *de novo* pyrimidine synthesis by activating Gln-dependent carbamoyl-phosphate synthase, Asp carbamoyltransferase, dihydroorotase (CAD). CAD is a multifunctional enzyme that controls the first three steps of pyrimidine synthesis. Phosphorylation of CAD by S6K ultimately stimulates DNA and RNA production [100,101]. mTORC1 can stimulate nucleotide synthesis by yet another mechanism. In cultured cells, activation of mTORC1 increases expression of genes encoding components of the pentose phosphate pathway (PPP), at least partly via activation of SREBP [83]. Among the metabolic products of PPP is ribose 5-phosphate that can be directly used for nucleotide synthesis. However, whether mTORC1-dependent up-regulation of PPP results in increased nucleotide synthesis has not been established. Up to now, there are no reports supporting the role of mTORC2 in nucleotide synthesis.

21.4.4 Energy Production

Biosynthesis pathways activated by mTOR are highly energy-demanding. To account for the increased energy expenditure, mTOR stimulates ATP production. The major source of cellular energy is glucose. The first step of glucose utilization, glycolysis, takes place in the cytosol, while the later steps, the citric acid cycle and oxidative phosphorylation, take place in mitochondria. mTOR controls many aspects of glucose utilization. First, mTOR affects glucose uptake. In cultured adipocytes, prolonged exposure to insulin increases expression of glucose transporter 1 (GLUT1) via mTORC1 [102]. Similarly, embryonic fibroblasts and renal cells in which mTORC1 is hyperactivated, due to loss of TSC2, display a rapamycin-sensitive increase in GLUT1 expression and glucose uptake [103]. GLUT1 is the major GLUT in most cell types. It would be of interest to determine whether GLUT1 activity is affected by mTORC1 signaling *in vivo*. mTORC2 activity is required for GLUT4-mediated glucose transport in adipose tissue. Mice lacking *Rictor* in adipocytes are glucose-intolerant due to impaired insulin-stimulated GLUT4 translocation to the plasma membrane [104]. Similarly, knockout of *Rictor* in skeletal muscles results in reduced GLUT4 translocation to the plasma membrane upon insulin stimulation and impaired glucose uptake [105]. Whether mTORC2 activity is required for GLUT-mediated glucose uptake in the liver remains to be determined. Second, mTOR positively regulates glycolysis. Tumor cells in which mTORC1 is activated due to loss of TSC2 display increased glycolytic flux and glucose-dependence [106,107]. The underlying mechanism has been elucidated recently. mTORC1 increases levels of hypoxia inducible factor 1α (HIF1α), which in turn increases transcription of multiple glycolytic genes [83]. In cell-based models, mTOR activation is sufficient to increase HIF1α levels, at least in part through inhibition of 4E-BP and subsequent activation of cap-dependent translation [83]. Similar to mTORC1, mTORC2 is a positive regulator of glycolysis. In the liver, mTORC2-dependent Akt activation promotes expression of glucokinase, the rate-limiting enzyme of glycolysis. The importance of this mechanism is highlighted by the fact that liver-specific *Rictor* knockout mice display reduced glucokinase activity and impaired hepatic glycolysis, a phenotype that is suppressed by expression of constitutively active Akt [91]. Finally, mTOR regulates mitochondrial biogenesis and oxidative functions by multiple mechanisms [5]. mTORC1 is found in a mitochondrial fraction and its activity positively correlates with mitochondrial function [108]. In cultured cells, inhibition of mTORC1 lowers mitochondrial membrane potential, oxygen consumption, and ATP synthesis capacity [108]. mTORC1 is a regulator of oxidative metabolism and mitochondrial function in skeletal muscle *in vivo*. Knockout of *mTOR* or *Raptor* in muscle leads to impaired oxidative metabolism, altered mitochondrial regulation, and ultimately to dystrophy [109,110]. In contrast, hyperactivation of mTORC1 due to *TSC1* deletion increases muscle oxidative activity [111]. mTORC1 exerts its positive effect on mitochondrial function, at least in part, by increasing levels of the key transcriptional regulators of mitochondrial function, PPAR-γ coactivator 1α and promoting its association with the transcription factor Ying-Yang 1 (YY1) [111–113]. This, in turn, increases expression of genes involved in oxidative metabolism. The importance of this mechanism is underscored by the observation that treatment of muscle cells with rapamycin significantly reduces mitochondrial DNA content and oxygen consumption, an effect that is mimicked by YY1 deletion [111–113]. mTORC1 also positively affects the activity of the estrogen receptor-related receptor α (ERRα), an orphan nuclear receptor regulating genes involved in mitochondrial biogenesis and oxidative metabolism [114]. Activated mTORC1 prevents proteasome-dependent degradation of ERRα, allowing for increased expression of genes encoding enzymes involved in the citric acid cycle [114]. Independent of its effect on transcription of genes involved in mitochondrial function, mTORC1 also selectively promotes translation of nuclear-encoded mitochondria-related mRNAs [115]. In contrast to its positive role in skeletal muscle, mTORC1 negatively regulates mitochondrial activity in adipose tissue. Mice with adipocyte-specific deletion of *Raptor* display enhanced mitochondrial respiration in white adipose tissue (WAT) [116]. A possible explanation for this phenotype may be related to the physiological role of adipose tissue, which is to store energy. Thus, when nutrients are abundant, the main function of adipose mTORC1 would be to promote lipid storage in lipid droplets rather than to stimulate energy production [5].

mTORC2 appears to negatively affect mitochondrial respiration. Upon activation, mTORC2 localizes to contact sites between the ER and mitochondria, so-called mitochondria-associated ER membrane (MAM), which is important for mitochondrial biogenesis and metabolism. mTORC2 recruits hexokinase 2 to MAM in an Akt-dependent manner and increases glycolysis. As a result, mitochondrial inner membrane potential and mitochondrial oxidative activity are reduced [117]. Consistent with the above, ablation of mTORC2 in Myf5-positive muscle and brown adipose tissue precursors increases mitochondrial activity [118].

21.4.5 Autophagy

mTOR promotes cellular growth not only by stimulating anabolic processes and energy production but also by inhibiting autophagy, a major degradation process in cells. Under physiological conditions, autophagy recycles damaged macromolecules and organelles. During times of nutrient deprivation, autophagy is induced to provide molecular building blocks and energy. Autophagy entails formation of autophagosomes, double-membrane vesicles that engulf a portion of cytoplasm and/or organelles. The autophagosomes then fuse with lysosomes to deliver their contents for degradation within the lysosomal lumen [21]. mTORC1 regulates autophagy at the transcriptional level by phosphorylating the transcription factor EB (TFEB), which leads to its sequestration in the cytoplasm. Upon mTORC1 inhibition, TFEB translocates into the nucleus where it promotes transcription of lysosomal and autophagy genes [119]. Autophagy is initiated by AMPK-dependent phosphorylation and activation of ULK1, which is essential for autophagosome formation. mTORC1 inhibits this process by directly phosphorylating ULK1, thereby preventing its interaction with AMPK. Additionally, mTORC1 phosphorylates and inhibits a positive regulator of ULK1, autophagy-related protein 13 (ATG13) [120–123]. In growing cells, mTORC1 phosphorylates and blocks the autophagy inhibitor death-associated protein 1 (DAP1). Upon nutrient deprivation, DAP1 is rapidly dephosphorylated and activated, supposedly as a mechanism to prevent excessive autophagy in starved cells [124].

A role for mTORC2 in the regulation of autophagy has not been shown. In yeast, TORC2 (mTORC2 ortholog) is a positive regulator of autophagy. TORC2 signals to general control nonderepressible 2 that in turn stimulates expression of genes involved in amino acid synthesis and autophagy [125].

21.5 mTOR AND METABOLIC AND ENDOCRINE DISORDERS

21.5.1 Obesity

Obesity is a disorder in which excessive fat accumulation adversely affects human health [126]. Obesity is associated with several comorbidities, including type 2 diabetes, cardiovascular diseases, nonalcoholic steatohepatitis, liver failure, and certain forms of cancer. With its increasing prevalence, obesity is currently one of the most important public health problems [126]. mTOR activity is implicated in obesity development at many levels.

Excessive food intake contributes to increased adiposity. Feeding behavior is controlled by the CNS, in particular the hypothalamus [127]. The hypothalamus senses the level of circulating nutrients and hormones to adjust feeding behavior and peripheral metabolism accordingly [128]. The arcuate nucleus (ARC) of hypothalamus is of particular importance in this process. ARC neurons that coexpress neuropeptide Y (NPY) and agouti-related peptide (AgRP) respond to ghrelin and stimulate food intake. In contrast, ARC neurons that express pro-opiomelanocortin (POMC) and cocaine- and amphetamine-regulated transcript (CART) respond to leptin and decrease feeding behavior [127]. Leptin is a hormone secreted by adipose tissue in direct proportion to fat mass and nutritional status. As a signal of nutritional sufficiency, leptin levels rapidly decrease upon dietary restriction and weight loss [129]. Leptin inhibits AgRP/NPY [130] while activating POMC/CART neurons [131]. Intracerebroventricular (ICV) administration of leptin induces mTORC1 activity in POMC/CART neurons and the anorexic effect of leptin is blunted upon ICV rapamycin treatment [132]. The role of mTORC1 in mediating leptin effects in the hypothalamus is underscored by the fact that S6K knockout mice are resistant to appetite-suppressing leptin action [133], while rats expressing constitutively active S6K in the mediobasal hypothalamus display increased leptin sensitivity [134]. In POMC/CART neurons, leptin-induced mTORC1 phosphorylates S6K, which in turn phosphorylates and inhibits AMPK [135]. This mTORC1-mediated AMPK inhibition is necessary for leptin effects on food intake. Expression of constitutively active AMPK in the mouse hypothalamus leads to increased food intake, while expression of dominant-negative AMPK has the opposite effect [136]. Interestingly, in rats maintained on a high-fat diet, ICV injection of leptin does not decrease food intake. This leptin insensitivity correlates with decreased mTORC1 signaling in the hypothalamus [133]. Thus, aberrant mTORC1 signaling may be implicated in hypothalamic leptin resistance in diet-induced obesity. mTORC2 also appears to play an important role in regulating POMC/CART neurons. mTORC2 ablation in POMC/CART neurons due to *Rictor* deletion leads to hyperphagia and obesity [137]. Whether mTORC2 synergizes with mTORC1 to modulate leptin responses remains to be determined.

In contrast to leptin, ghrelin increases food intake. This gastrointestinal peptide hormone is secreted upon starvation and its circulating levels decrease after feeding [138]. Ghrelin stimulates AgRP/NPY while inhibiting POMC/CART neurons [139]. In rats, ICV ghrelin administration increases mTORC1 activity in AgRP/NPY neurons and the positive effect of ghrelin on food intake is abrogated upon rapamycin treatment [140]. Curiously, knock out of mTORC1 specifically in AgRP/NPY neurons does not affect feeding behavior [141].

In addition to leptin and ghrelin, the ARC is also influenced by thyroid hormones. Alteration in thyroid hormone levels in humans is associated with changes in feeding behavior. Hyperthyroid patients display hyperphagia, while hypothyroid patients decrease food intake [142]. Thyroid-hormone-induced hyperphagia is associated with increased mTORC1 activity in AgRP/NPY neurons. Accordingly, ICV injection of rapamycin reverses hyperphagia in a rat model of hyperthyroidism [143]. The above taken together suggests that mTOR in the ARC plays a role in regulating feeding behavior but possibly only under certain conditions. Disturbances in hypothalamic mTOR signaling may contribute to the obesity development either by promoting to the anorexic action of leptin and/or by potentiating the hyperphagic effect of ghrelin and thyroid hormones.

Overfeeding or chronically high availability of nutrients leads to a persistent hyperactivation of mTORC1 in peripheral tissues [144]. This may contribute to obesity by enhancing fat deposition in WAT, as mTORC1 plays a role in adipose homeostasis. Rapamycin inhibits differentiation of mouse [116,145,146] and human [147] adipocytes *in vitro*. It does so by decreasing the expression of peroxisome-proliferator activated receptor-γ (PPAR-γ) and CCAAT/enhancer binding protein-α (C/EBP-α), two key transcription factors in adipogenesis [116,146,148]. Hyperactivation of mTORC1 in adipocytes elicits the opposite phenotype [149]. Mice with adipose tissue-specific deletion of

Raptor are resistant to diet-induced obesity and display reduced white adipose mass due to fewer and smaller adipocytes [116]. Interestingly, these mice have normal PPAR-γ and C/EBP-α levels [116]. This may suggest that mTORC1 induces PPAR-γ and C/EBP-α expression during adipocyte differentiation but promotes mature adipocyte function through a different mechanism. Reduced fat mass and resistance to diet-induced obesity in adipose-specific *Raptor* knockout mice is attributed to enhanced energy expenditure due to elevated mitochondrial respiration [116]. In addition to white adipocytes that contain a single large lipid droplet, mammals also have brown adipocytes with multiple small lipid droplets, increased mitochondrial count, and distinct physiology [150]. While the main function of white adipocytes is to store energy in the form of triglycerides, brown adipocytes generate heat via uncoupling protein 1 (UCP1) in response to cold exposure or excessive feeding. Most fat depots contain both types of adipocytes [150]. Recently, mTORC1 has been implicated in a sort of trans-differentiation of brown and white adipocytes [151]. In mice, hyperactivation of mTORC1 in adipose tissue due to *TSC1* deletion leads to conversion of brown adipocytes into cells with morphology and gene expression pattern reminiscent of white adipocytes [151]. In contrast, deletion of *Raptor* increases UCP1 expression [116]. Thus, excessive mTORC1 activity in adipose tissue may contribute to obesity development by promoting differentiation of energy-storing white adipocytes and by inhibiting differentiation of energy-consuming brown adipocytes. The importance of mTORC1 in adipocyte function in the adult organism is underscored by the observation that systemic rapamycin treatment reduces fat cell number and lipid content in mice and rats [152,153]. This phenotype most likely results from downregulation of genes involved in lipid uptake and synthesis in adipocytes [152]. Accordingly, rapamycin treatment protects mice against diet-induced obesity [154]. When fed a high-fat diet, rapamycin-treated animals do not gain as much weight as vehicle-treated controls, and display fewer and smaller adipocytes [154]. Thus, mTORC1 activity is required for adipose tissue expansion upon feeding of a high-calorie diet.

Data concerning the role of mTORC1 in adipose physiology in humans are limited. Interestingly, kidney transplant patients treated with rapamycin have lower body mass index (BMI) compared to patients maintained on cyclosporine [153]. BMI is calculated using a formula combining weight and height of a patient, and serves as a simple measure of body fat content [126]. Whether the lower BMI results from inhibition of mTORC1 in adipose tissue or from changes in feeding behavior due to mTORC1 inhibition in the hypothalamus has not been determined.

A recent report further points towards the involvement of mTORC1 signaling in obesity development in humans. Participants in the Honolulu Heart Program, a homogeneous population of American men of Japanese ancestry, contain single nucleotide polymorphisms (SNPs) in *RAPTOR* that correlate with overweight/obesity [155]. However, how these SNPs affect *RAPTOR* expression and whether *Raptor* SNPs correlate with obesity in other populations remain to be determined. Taken together, aberrations in mTOR signaling may contribute to obesity by regulating both feeding behavior and lipid accumulation in adipocytes. The involvement of mTOR in pathophysiology of obesity in humans requires further studies.

21.5.2 Insulin Resistance and Diabetes

Impaired physiological response to insulin (insulin resistance) is a common pathogenic consequence of obesity that may lead to development of type 2 diabetes mellitus (T2DM). Insulin resistance comprises both reduced glucose uptake (by the liver, skeletal muscle, and adipose tissue) and impaired insulin-mediated suppression of hepatic glucose output. Not surprisingly, development of insulin resistance is preceded by deregulation of mTOR signaling in metabolic organs. Deregulation of mTOR in only one tissue suffices to precipitate changes in whole-body metabolism and energy balance.

The decrease in circulating glucose after a meal relies mostly on glucose uptake and its storage as glycogen in skeletal muscle [156]. Although this process is controlled mainly by Akt [156], mTORC1 also appears to be involved. Muscle-specific *Raptor* knockout mice display increased Akt signaling, due to lack of the negative feedback loop (see above), and have increased glycogen content, presumably as a result of increased glucose uptake [109]. Surprisingly, these animals are glucose-intolerant [109]. This suggests that intact mTORC1 in skeletal muscle is necessary to maintain whole-body glucose homeostasis. On the other hand, muscle-specific *Rictor* knockout display decreased Akt activity and insulin-stimulated glucose uptake [105]. Surprisingly, their systemic insulin tolerance is unaffected and they present with very mild impairment in glucose tolerance [105]. Taken together, these data suggest that mTORC1 and mTORC2 in skeletal muscle have complementary roles in regulating whole-body glucose metabolism.

Similar to muscle, disruption of mTOR signaling in adipose tissue affects systemic glucose tolerance. Upon diet-induced obesity, adipose-specific *Raptor* knockout animals have improved glucose homeostasis due to increased insulin signaling in adipose tissue and skeletal muscle [116]. The mechanism for improved muscle sensitivity in these animals remains to be determined. In contrast, adipose-specific *Rictor* knockout animals display whole-body insulin resistance and hyperinsulinemia [104,157]. Initially, hyperinsulinemia is sufficient to compensate for systemic insulin resistance and these animals have normal glucose tolerance. However, with time they become glucose-intolerant [104]. Adipose-specific *Rictor* knockout mice have reduced glucose uptake, fail to suppress lipolysis in response to insulin, and thereby display high levels of plasma FFAs and glycerol [104]. This might further exacerbate glucose intolerance as FFA stimulate *de novo* glucose production in the liver [158] and impair muscle insulin signaling [159]. In animal models, diet-induced obesity is accompanied by increased mTORC1 activity in adipose tissue and insulin resistance due to the negative feedback loop [45,50,160]. In contrast, adipocytes isolated from obese humans with insulin resistance display lower levels of *mTOR, Rheb*, and *4E-BP* mRNA in omental fat depots [161]. Although it was not shown that the lower mRNA levels correlate with lower protein levels, such mRNA changes were not observed in omental fat of insulin-sensitive subjects [161]. Adipocytes isolated from subcutaneous fat depot of obese insulin-resistant patients have reduced S6K and mTOR protein levels compared to noninsulin-resistant subjects [162]. Furthermore, adipocytes isolated from insulin-resistant subjects have impaired mTORC1 signaling in response to insulin. This attenuation of mTORC1 signaling is accompanied by increased autophagy and mitochondria impairment [162]. In adipocytes from nondiabetic subjects, the degree of mTORC1 activation in response to insulin correlates with whole-body insulin sensitivity [162]. The apparent discrepancy in mTORC1 activity observed in insulin-resistant mice (high activity) and humans (low activity) remains to be explained.

In liver, insulin signaling activates glucose uptake and, by inhibiting glycogenolysis and gluconeogenesis, inhibits glucose output. Hepatic insulin resistance plays an important role in the hyperglycemia of T2DM patients. As already mentioned, mTORC2 activates Akt and thereby increases glycolysis and hepatic glucose flux [91]. Furthermore, mTORC2-dependent Akt activation leads to inhibition of FoxO and impedes transcription of gluconeogenic genes. Accordingly, liver-specific *Rictor* knockout animals are characterized by constitutive gluconeogenesis, impaired glycolysis, hyperglycemia, and compensatory hyperinsulinemia [91]. These defects are reminiscent of T2DM, suggesting that impaired hepatic mTORC2 signaling may play a role in the pathophysiology of diabetes [2]. Similarly, animals with hepatic mTORC1 hyperactivation due to *TSC1* deletion display reduced hepatic glycolytic flux

and impaired glucose tolerance [89,163]. The *TSC1*-knockout phenotype is likely due to the negative feedback loop and thereby mTORC2 attenuation and decreased Akt signaling.

The pancreas initially compensates for systemic insulin resistance by increasing insulin production. This compensation mechanism involves β-cell hypertrophy and formation of new β cells from progenitors. With time, the pancreas fails to sustain insulin production, leading to hyperglycemia and T2DM [3]. mTORC1 is a crucial regulator of β-cell homeostasis. Up-regulation of mTORC1 signaling in β cells due to *TSC2* deletion or *Rheb* overexpression leads to increased β-cell mass, hyperinsulinemia, and improved glucose tolerance [164−166]. Interestingly, chronic mTORC1 activation in pancreatic islet cells leads to insulin resistance in β cells (again, due to the negative feedback loop), decreased survival, and increased apoptosis. This in turn leads to hypoinsulinemia and hyperglycemia [165]. Obese mice have increased pancreatic mTORC1 signaling [165]. This suggests that failure of the pancreas that ultimately leads to T2DM may be caused by obesity-related chronic hyperactivation of mTORC1 in β cells and impaired Akt signaling due to the negative feedback loop. The importance of the negative feedback loop and concomitant attenuation of mTORC2 signaling in the development of T2DM is underscored by the observation that pancreatic *Rictor* knockout mice display reduced β-cell mass, decreased insulin production, hyperglycemia, and impaired glucose tolerance [167].

Up to 50% of patients treated with rapamycin or its analogs (rapalogs) display hyperglycemia [168]. Similarly, chronic rapamycin treatment impairs glucose homeostasis in mice [169] and rats [152]. The underlying mechanism is not completely understood. Given the role of mTORC1 in β-cell function, one possible explanation for impaired glucose tolerance upon rapamycin treatment is reduced insulin secretion. Indeed, chronic rapamycin treatment reduces β-cell mass in rats [152] and the T2DM model *Psammomys obesum* [170]. Impaired glucose tolerance could also stem from reduced hepatic glucose uptake due to disruption of mTORC2 signaling upon prolonged rapamycin treatment. Accordingly, rapamycin does not enhance hyperglycemia in liver-specific *Rictor* knockout mice [169]. Considering the phenotype of muscle-specific *Raptor* knockout mice [109], it is possible that impaired glucose tolerance in rapamycin-treated patients results from disrupted mTORC1 signaling in skeletal muscle. The relative contributions of pancreatic impairment, reduced hepatic glucose uptake, and disrupted muscle mTORC1 signaling to the development of hyperglycemia upon rapalog treatment has not been determined.

21.5.3 Dyslipidemias and Nonalcoholic Fatty Liver Disease

Disturbances in the production and clearance of plasma lipoproteins are often observed in obese or diabetic patients [171]. Deregulation of mTOR signaling may contribute to the dyslipidemias by promoting lipid synthesis in the liver and by impairing lipid mobilization and uptake in peripheral tissues. In cultured hepatocytes, mTORC1 mediates insulin-induced inhibition of expression of apolipopoprotein B-100 (ApoB-100), a major scaffolding protein for very low- and low-density lipoproteins (VLDL and LDL, respectively) [172]. Thus, hepatic mTORC1 affects lipoprotein assembly. Accordingly, plasma levels of ApoB-100, cholesterol, and triglyceride are elevated in patients treated with rapamycin [173]. mTORC1 has also been implicated in regulating the expression of LDL receptor (LDLR) as well as SR-BI, the major receptor for high-density lipoproteins (HDL). Mice treated with rapamycin display reduced hepatic expression of LDLR and increased expression of proprotein convertase subtilisin/kexin type 9 (PCSK9), a negative regulator of LDLR [174]. In contrast, animals with hepatic hyperactivation of mTORC1 have lower expression of PCSK9 and increased LDLR levels. The effect of rapamycin on LDLR levels is abolished in PCSK9-deficient mice, suggesting that mTORC1 affects LDLR mainly through inhibition of PCSK9 [174]. In human umbilical cord vein endothelial cells, mTORC1 promotes the expression of SR-B1 but, surprisingly, is not required for HDL uptake [175]. Whether mTORC1 regulates SR-BI-mediated HDL uptake *in vivo* has not been examined.

mTORC1 affects peripheral lipoprotein metabolism by regulating lipoprotein lipase (LPL), an enzyme that hydrolyzes lipoprotein triglycerides and promotes the uptake of FFAs in adipose tissue and skeletal muscle [176]. In cultured rat adipocytes, rapamycin strongly reduces insulin-induced LPL activity [176]. The importance of mTOR in regulating lipoprotein metabolism is highlighted by the fact that rapalogs induce hyperlipidemia in 40—75% of patients [168]. Treatment with mTORC1 inhibitors increases both HDL and LDL levels as well as plasma cholesterol and triglyceride concentrations [168,173]. The hyperlipidemia most likely does not result from increased hepatic lipid output but rather from impaired peripheral lipoprotein catabolism (reviewed in Ref. [177]). Taken together, these data suggest that mTORC1 positively regulates uptake of lipoprotein lipids in the liver and peripheral tissues. As mTORC1 activity is up-regulated in many tissues upon obesity development [50,160], this could explain excess fat deposition in peripheral organs as often observed in obese individuals.

Obesity, T2DM, and dyslipidemia are frequently associated with nonalcoholic fatty liver disease (NAFLD) that can progress to cirrhosis and liver failure [178]. Although many contributing factors have been identified, pathogenesis of NAFLD is still not completely understood. The initial step of NAFLD development involves deposition of triglycerides in the liver, a process that is controlled by both mTOR complexes (discussed earlier in this chapter). Accordingly, enhanced hepatic mTORC1 activity in obese rodents is associated with increased lipid storage in the liver [45]. Mice with impaired hepatic mTORC1 signaling are resistant to NAFLD and hypercholesterolemia upon high-fat diet challenge [86,160]. Interestingly, liver-specific *TSC1*-knockout mice are resistant to age- and diet-induced NAFLD due to attenuation of Akt signaling [88,89] and to an Akt-independent mechanism [90]. This suggests that physiological (i.e., chronic but not constitutive) mTORC1 stimulation in obese mice promotes NAFLD development, whereas nonphysiological, constitutive mTORC1 signaling has the opposite effect. In humans, NAFLD is associated with increased hepatic mTORC1 signaling [179]. Microarray and immunohistochemical analyses show that the mTORC1 pathway is hyperactivated in livers of NAFLD/cirrhotic patients [179]. Liver cirrhosis involves the growth of fibrotic tissue that replaces necrotic liver parenchyma. In the rat model of liver fibrosis, treatment with rapalogs decreased fibrosis up to 70% [180]. Thus, it would be interesting to test whether mTOR inhibition improves already pre-existing NAFLD and prevents its progression to cirrhosis.

21.6 CONCLUSIONS AND FUTURE PERSPECTIVES

We have reviewed the importance of mTOR signaling in cellular and organismal metabolism. mTOR impinges on cellular metabolism by regulating macromolecular synthesis, mitochondrial function, and other metabolic processes. mTOR affects whole-body metabolism by regulating the function of metabolic organs in response to nutrients and circulating hormones. The importance of mTOR in organismal homeostasis is highlighted by the fact that aberrant mTOR signaling in a single organ is sufficient to precipitate changes that affect whole-body metabolism. Due to the intricate relationship between mTOR signaling in different organs, systemic mTOR inhibition may have counteracting effects in the treatment of metabolic disorders. Thus, deciphering cellular events up- and downstream of mTOR in individual organs and identifying mechanisms of mTOR deregulation will allow better treatments for metabolic and endocrine diseases.

References

[1] Kim SG, Buel GR, Blenis J. Nutrient regulation of the mTOR complex 1 signaling pathway. Mol Cells 2013;35:463–73.

[2] Cornu M, Albert V, Hall MN. mTOR in aging, metabolism, and cancer. Curr Opin Genet Dev 2013;23:53–62.

[3] Laplante M, Sabatini DM. mTOR signaling in growth control and disease. Cell 2012;149:274–93.

[4] Howell JJ, Manning BD. mTOR couples cellular nutrient sensing to organismal metabolic homeostasis. Trends Endocrinol Metab 2011;22:94–102.

[5] Albert V, Hall MN. mTOR signaling in cellular and organismal energetics. Curr Opin Cell Biol 2015;33:55–66.

[6] Kim DH, Sarbassov DD, Ali SM, King JE, Latek RR, Erdjument-Bromage H, et al. mTOR interacts with raptor to form a nutrient-sensitive complex that signals to the cell growth machinery. Cell 2002;110:163–75.

[7] Hara K, Maruki Y, Long X, Yoshino K, Oshiro N, Hidayat S, et al. Raptor, a binding partner of target of rapamycin (TOR), mediates TOR action. Cell 2002;110:177–89.

[8] Sarbassov DD, Ali SM, Kim DH, Guertin DA, Latek RR, Erdjument-Bromage H, et al. Rictor, a novel binding partner of mTOR, defines a rapamycin-insensitive and raptor-independent pathway that regulates the cytoskeleton. Curr Biol 2004;14:1296–302.

[9] Jacinto E, Loewith R, Schmidt A, Lin S, Ruegg MA, Hall A, et al. Mammalian TOR complex 2 controls the actin cytoskeleton and is rapamycin insensitive. Nat Cell Biol 2004;6:1122–8.

[10] Loewith R, Jacinto E, Wullschleger S, Lorberg A, Crespo JL, Bonenfant D, et al. Two TOR complexes, only one of which is rapamycin sensitive, have distinct roles in cell growth control. Mol Cell 2002;10:457–68.

[11] Shimobayashi M, Hall MN. Making new contacts: the mTOR network in metabolism and signalling crosstalk. Nat Rev Mol Cell Biol 2014;15:155–62.

[12] Zinzalla V, Stracka D, Oppliger W, Hall MN. Activation of mTORC2 by association with the ribosome. Cell 2011;144:757–68.

[13] Cybulski N, Hall MN. TOR complex 2: a signaling pathway of its own. Trends Biochem Sci 2009;34:620–7.

[14] Sarbassov DD, Guertin DA, Ali SM, Sabatini DM. Phosphorylation and regulation of Akt/PKB by the rictor-mTOR complex. Science 2005;307:1098–101.

[15] Sabatini DM, Erdjument-Bromage H, Lui M, Tempst P, Snyder SH. RAFT1: a mammalian protein that binds to FKBP12 in a rapamycin-dependent fashion and is homologous to yeast TORs. Cell 1994;78:35–43.

[16] Brown EJ, Albers MW, Shin TB, Ichikawa K, Keith CT, Lane WS, et al. A mammalian protein targeted by G1-arresting rapamycin-receptor complex. Nature 1994;369:756–8.

[17] Sarbassov DD, Ali SM, Sengupta S, Sheen JH, Hsu PP, Bagley AF, et al. Prolonged rapamycin treatment inhibits mTORC2 assembly and Akt/PKB. Mol Cell 2006;22:159–68.

[18] Goldsmith ZG, Dhanasekaran DN. G protein regulation of MAPK networks. Oncogene 2007;26:3122–42.

[19] Laplante M, Sabatini DM. Regulation of mTORC1 and its impact on gene expression at a glance. J Cell Sci 2013;126:1713–19.

[20] Jewell JL, Russell RC, Guan KL. Amino acid signalling upstream of mTOR. Nat Rev Mol Cell Biol 2013;14:133–9.

[21] Inoki K, Kim J, Guan KL. AMPK and mTOR in cellular energy homeostasis and drug targets. Annu Rev Pharmacol Toxicol 2012;52:381–400.

[22] Hara K, Yonezawa K, Weng QP, Kozlowski MT, Belham C, Avruch J. Amino acid sufficiency and mTOR regulate p70 S6 kinase and eIF-4E BP1 through a common effector mechanism. J Biol Chem 1998;273:14484–94.

[23] Dibble CC, Elis W, Menon S, Qin W, Klekota J, Asara JM, et al. TBC1D7 is a third subunit of the TSC1-TSC2 complex upstream of mTORC1. Mol Cell 2012;47:535–46.

[24] van Slegtenhorst M, Nellist M, Nagelkerken B, Cheadle J, Snell R, van den Ouweland A, et al. Interaction between hamartin and tuberin, the TSC1 and TSC2 gene products. Hum Mol Genet 1998;7:1053–7.

[25] Tee AR, Manning BD, Roux PP, Cantley LC, Blenis J. Tuberous sclerosis complex gene products, Tuberin and Hamartin, control mTOR signaling by acting as a GTPase-activating protein complex toward Rheb. Curr Biol 2003;13:1259–68.

[26] Inoki K, Li Y, Xu T, Guan KL. Rheb GTPase is a direct target of TSC2 GAP activity and regulates mTOR signaling. Genes Dev 2003;17:1829–34.

[27] Long X, Lin Y, Ortiz-Vega S, Yonezawa K, Avruch J. Rheb binds and regulates the mTOR kinase. Curr Biol 2005;15:702–13.

[28] Wullschleger S, Loewith R, Hall MN. TOR signaling in growth and metabolism. Cell 2006;124:471–84.

[29] Inoki K, Li Y, Zhu T, Wu J, Guan KL. TSC2 is phosphorylated and inhibited by Akt and suppresses mTOR signalling. Nat Cell Biol 2002;4:648–57.

[30] Potter CJ, Pedraza LG, Xu T. Akt regulates growth by directly phosphorylating Tsc2. Nat Cell Biol 2002;4:658–65.

[31] Manning BD, Tee AR, Logsdon MN, Blenis J, Cantley LC. Identification of the tuberous sclerosis complex-2 tumor suppressor gene product tuberin as a target of the phosphoinositide 3-kinase/akt pathway. Mol Cell 2002;10:151–62.

[32] Cao X, Kambe F, Moeller LC, Refetoff S, Seo H. Thyroid hormone induces rapid activation of Akt/protein kinase B-mammalian target of rapamycin-p70S6K cascade through phosphatidylinositol 3-kinase in human fibroblasts. Mol Endocrinol 2005;19:102–12.

[33] Kenessey A, Ojamaa K. Thyroid hormone stimulates protein synthesis in the cardiomyocyte by activating the Akt-mTOR and p70S6K pathways. J Biol Chem 2006;281:20666–72.

[34] Bassett JH, Harvey CB, Williams GR. Mechanisms of thyroid hormone receptor-specific nuclear and extra nuclear actions. Mol Cell Endocrinol 2003;213:1–11.

[35] Kuzman JA, O'Connell TD, Gerdes AM. Rapamycin prevents thyroid hormone-induced cardiac hypertrophy. Endocrinology 2007;148:3477–84.

[36] Simoncini T, Hafezi-Moghadam A, Brazil DP, Ley K, Chin WW, Liao JK. Interaction of oestrogen receptor with the regulatory subunit of phosphatidylinositol-3-OH kinase. Nature 2000;407:538–41.

[37] Varea O, Escoll M, Diez H, Garrido JJ, Wandosell F. Oestradiol signalling through the Akt-mTORC1-S6K1. Biochim Biophys Acta 2013;1833:1052–64.

[38] Wang B, Ding W, Zhang M, Li H, Gu Y. Rapamycin attenuates aldosterone-induced tubulointerstitial inflammation and fibrosis. Cell Physiol Biochem 2015;35:116–25.

[39] Walczak C, Gaignier F, Gilet A, Zou F, Thornton SN, Ropars A. Aldosterone increases VEGF-A production in human neutrophils through PI3K, ERK1/2 and p38 pathways. Biochim Biophys Acta 2011;1813:2125–32.

[40] Mendoza MC, Er EE, Blenis J. The Ras-ERK and PI3K-mTOR pathways: cross-talk and compensation. Trends Biochem Sci 2011;36:320–8.

[41] Kim E, Goraksha-Hicks P, Li L, Neufeld TP, Guan KL. Regulation of TORC1 by Rag GTPases in nutrient response. Nat Cell Biol 2008;10:935–45.

[42] Sancak Y, Peterson TR, Shaul YD, Lindquist RA, Thoreen CC, Bar-Peled L, et al. The Rag GTPases bind raptor and mediate amino acid signaling to mTORC1. Science 2008;320: 1496–501.

[43] Inoki K, Zhu T, Guan KL. TSC2 mediates cellular energy response to control cell growth and survival. Cell 2003;115:577—90.

[44] Gwinn DM, Shackelford DB, Egan DF, Mihaylova MM, Mery A, Vasquez DS, et al. AMPK phosphorylation of raptor mediates a metabolic checkpoint. Mol Cell 2008;30:214—26.

[45] Tremblay F, Brule S, Hee Um S, Li Y, Masuda K, Roden M, et al. Identification of IRS-1 Ser-1101 as a target of S6K1 in nutrient- and obesity-induced insulin resistance. Proc Natl Acad Sci USA 2007;104:14056—61.

[46] Shah OJ, Hunter T. Turnover of the active fraction of IRS1 involves raptor-mTOR- and S6K1-dependent serine phosphory-lation in cell culture models of tuberous sclerosis. Mol Cell Biol 2006;26:6425—34.

[47] Tzatsos A, Kandror KV. Nutrients suppress phosphatidylinositol 3-kinase/Akt signaling via raptor-dependent mTOR-mediated insulin receptor substrate 1 phosphorylation. Mol Cell Biol 2006;26:63—76.

[48] Harrington LS, Findlay GM, Gray A, Tolkacheva T, Wigfield S, Rebholz H, et al. The TSC1-2 tumor suppressor controls insulin-PI3K signaling via regulation of IRS proteins. J Cell Biol 2004;166:213—23.

[49] Shah OJ, Wang Z, Hunter T. Inappropriate activation of the TSC/Rheb/mTOR/S6K cassette induces IRS1/2 depletion, insulin resistance, and cell survival deficiencies. Curr Biol 2004;14:1650—6.

[50] Um SH, Frigerio F, Watanabe M, Picard F, Joaquin M, Sticker M, et al. Absence of S6K1 protects against age- and diet-induced obesity while enhancing insulin sensitivity. Nature 2004;431: 200—5.

[51] Hsu PP, Kang SA, Rameseder J, Zhang Y, Ottina KA, Lim D, et al. The mTOR-regulated phosphoproteome reveals a mechanism of mTORC1-mediated inhibition of growth factor signaling. Science 2011;332:1317—22.

[52] Yu Y, Yoon SO, Poulogiannis G, Yang Q, Ma XM, Villen J, et al. Phosphoproteomic analysis identifies Grb10 as an mTORC1 substrate that negatively regulates insulin signaling. Science 2011;332:1322—6.

[53] Dibble CC, Asara JM, Manning BD. Characterization of Rictor phosphorylation sites reveals direct regulation of mTOR com-plex 2 by S6K1. Mol Cell Biol 2009;29:5657—70.

[54] Julien LA, Carriere A, Moreau J, Roux PP. mTORC1-activated S6K1 phosphorylates Rictor on threonine 1135 and regulates mTORC2 signaling. Mol Cell Biol 2010;30:908—21.

[55] Fang Y, Vilella-Bach M, Bachmann R, Flanigan A, Chen J. Phosphatidic acid-mediated mitogenic activation of mTOR sig-naling. Science 2001;294:1942—5.

[56] Yoon MS, Sun Y, Arauz E, Jiang Y, Chen J. Phosphatidic acid activates mammalian target of rapamycin complex 1 (mTORC1) kinase by displacing FK506 binding protein 38 (FKBP38) and exerting an allosteric effect. J Biol Chem 2011;286:29568—74.

[57] Foster DA, Salloum D, Menon D, Frias MA. Phospholipase D and the maintenance of phosphatidic acid levels for regulation of mammalian target of rapamycin (mTOR). J Biol Chem 2014;289:22583—8.

[58] You JS, Lincoln HC, Kim CR, Frey JW, Goodman CA, Zhong XP, et al. The role of diacylglycerol kinase zeta and phospha-tidic acid in the mechanical activation of mammalian target of rapamycin (mTOR) signaling and skeletal muscle hypertrophy. J Biol Chem 2014;289:1551—63.

[59] Fang Y, Park IH, Wu AL, Du G, Huang P, Frohman MA, et al. PLD1 regulates mTOR signaling and mediates Cdc42 activation of S6K1. Curr Biol 2003;13:2037—44.

[60] Blaskovich MA, Yendluri V, Lawrence HR, Lawrence NJ, Sebti SM, Springett GM. Lysophosphatidic acid acyltransferase beta regulates mTOR signaling. PLoS One 2013;8:e78632.

[61] Kang DW, Choi KY, Min do S. Functional regulation of phos-pholipase D expression in cancer and inflammation. J Biol Chem 2014;289:22575—82.

[62] Zhang C, Wendel AA, Keogh MR, Harris TE, Chen J, Coleman RA. Glycerolipid signals alter mTOR complex 2 (mTORC2) to diminish insulin signaling. Proc Natl Acad Sci USA 2012;109: 1667—72.

[63] Zhang C, Hwarng G, Cooper DE, Grevengoed TJ, Eaton JM, Natarajan V, et al. Inhibited insulin signaling in mouse hepato-cytes is associated with increased phosphatidic acid but not diacylglycerol. J Biol Chem 2015;290:3519—28.

[64] Hornberger TA, Chu WK, Mak YW, Hsiung JW, Huang SA, Chien S. The role of phospholipase D and phosphatidic acid in the mechanical activation of mTOR signaling in skeletal muscle. Proc Natl Acad Sci USA 2006;103:4741—6.

[65] Maeurer C, Holland S, Pierre S, Potstada W, Scholich K. Sphingosine-1-phosphate induced mTOR-activation is mediated by the E3-ubiquitin ligase PAM. Cell Signal 2009;21:293—300.

[66] Liu G, Burns S, Huang G, Boyd K, Proia RL, Flavell RA, et al. The receptor S1P1 overrides regulatory T cell-mediated immune suppression through Akt-mTOR. Nat Immunol 2009;10:769—77.

[67] Fayyaz S, Japtok L, Kleuser B. Divergent role of sphingosine 1-phosphate on insulin resistance. Cell Physiol Biochem 2014;34:134—47.

[68] Mullins GR, Wang L, Raje V, Sherwood SG, Grande RC, Boroda S, et al. Catecholamine-induced lipolysis causes mTOR complex dissociation and inhibits glucose uptake in adipocytes. Proc Natl Acad Sci USA 2014;111:17450—5.

[69] Vadas O, Burke JE, Zhang X, Berndt A, Williams RL. Structural basis for activation and inhibition of class I phosphoinositide 3-kinases. Sci Signal 2011;4:re2.

[70] Robles-Molina E, Dionisio-Vicuna M, Guzman-Hernandez ML, Reyes-Cruz G, Vazquez-Prado J. Gbetagamma interacts with mTOR and promotes its activation. Biochem Biophys Res Commun 2014;444:218—23.

[71] Pierce KL, Premont RT, Lefkowitz RJ. Seven-transmembrane receptors. Nat Rev Mol Cell Biol 2002;3:639—50.

[72] de Joussineau C, Sahut-Barnola I, Tissier F, Dumontet T, Drelon C, Batisse-Lignier M, et al. mTOR pathway is activated by PKA in adrenocortical cells and participates *in vivo* to apoptosis resistance in primary pigmented nodular adrenocortical disease (PPNAD). Hum Mol Genet 2014;23:5418—28.

[73] Suh JM, Song JH, Kim DW, Kim H, Chung HK, Hwang JH, et al. Regulation of the phosphatidylinositol 3-kinase, Akt/protein kinase B, FRAP/mammalian target of rapamycin, and ribosomal S6 kinase 1 signaling pathways by thyroid-stimulating hormone (TSH) and stimulating type TSH receptor antibodies in the thyroid gland. J Biol Chem 2003;278:21960—71.

[74] Blancquaert S, Wang L, Paternot S, Coulonval K, Dumont JE, Harris TE, et al. cAMP-dependent activation of mammalian target of rapamycin (mTOR) in thyroid cells. Implication in mitogenesis and activation of CDK4. Mol Endocrinol 2010;24:1453—68.

[75] Wang RH, He JP, Su ML, Luo J, Xu M, Du XD, et al. The orphan receptor TR3 participates in angiotensin II-induced car-diac hypertrophy by controlling mTOR signalling. EMBO Mol Med 2013;5:137—48.

[76] Chauvin C, Koka V, Nouschi A, Mieulet V, Hoareau-Aveilla C, Dreazen A, et al. Ribosomal protein S6 kinase activity controls the ribosome biogenesis transcriptional program. Oncogene 2014;33:474—83.

[77] Iadevaia V, Liu R, Proud CG. mTORC1 signaling controls mul-tiple steps in ribosome biogenesis. Semin Cell Dev Biol 2014;36:113—20.

[78] Thoreen CC, Chantranupong L, Keys HR, Wang T, Gray NS, Sabatini DM. A unifying model for mTORC1-mediated regulation of mRNA translation. Nature 2012;485:109–13.

[79] Meyuhas O, Kahan T. The race to decipher the top secrets of TOP mRNAs. Biochim Biophys Acta 2015;1849(7):801–11.

[80] Gingras AC, Raught B, Gygi SP, Niedzwiecka A, Miron M, Burley SK, et al. Hierarchical phosphorylation of the translation inhibitor 4E-BP1. Genes Dev 2001;15:2852–64.

[81] Oh WJ, Wu CC, Kim SJ, Facchinetti V, Julien LA, Finlan M, et al. mTORC2 can associate with ribosomes to promote cotranslational phosphorylation and stability of nascent Akt polypeptide. EMBO J 2010;29:3939–51.

[82] Dai N, Christiansen J, Nielsen FC, Avruch J. mTOR complex 2 phosphorylates IMP1 cotranslationally to promote IGF2 production and the proliferation of mouse embryonic fibroblasts. Genes Dev 2013;27:301–12.

[83] Duvel K, Yecies JL, Menon S, Raman P, Lipovsky AI, Souza AL, et al. Activation of a metabolic gene regulatory network downstream of mTOR complex 1. Mol Cell 2010;39:171–83.

[84] Porstmann T, Santos CR, Griffiths B, Cully M, Wu M, Leevers S, et al. SREBP activity is regulated by mTORC1 and contributes to Akt-dependent cell growth. Cell Metab 2008;8:224–36.

[85] Ye J, DeBose-Boyd RA. Regulation of cholesterol and fatty acid synthesis. Cold Spring Harb Perspect Biol 2011;3a004754.

[86] Peterson TR, Sengupta SS, Harris TE, Carmack AE, Kang SA, Balderas E, et al. mTOR complex 1 regulates lipin 1 localization to control the SREBP pathway. Cell 2011;146:408–20.

[87] Wan M, Leavens KF, Saleh D, Easton RM, Guertin DA, Peterson TR, et al. Postprandial hepatic lipid metabolism requires signaling through Akt2 independent of the transcription factors FoxA2, FoxO1, and SREBP1c. Cell Metab 2011;14:516–27.

[88] Yecies JL, Zhang HH, Menon S, Liu S, Yecies D, Lipovsky AI, et al. Akt stimulates hepatic SREBP1c and lipogenesis through parallel mTORC1-dependent and independent pathways. Cell Metab 2011;14:21–32.

[89] Kenerson HL, Yeh MM, Yeung RS. Tuberous sclerosis complex-1 deficiency attenuates diet-induced hepatic lipid accumulation. PLoS One 2011;6:e18075.

[90] Kenerson HL, Subramanian S, McIntyre R, Kazami M, Yeung RS. Livers with constitutive mTORC1 activity resist steatosis independent of feedback suppression of Akt. PLoS One 2015;10: e0117000.

[91] Hagiwara A, Cornu M, Cybulski N, Polak P, Betz C, Trapani F, et al. Hepatic mTORC2 activates glycolysis and lipogenesis through Akt, glucokinase, and SREBP1c. Cell Metab 2012;15: 725–38.

[92] Yuan M, Pino E, Wu L, Kacergis M, Soukas AA. Identification of Akt-independent regulation of hepatic lipogenesis by mammalian target of rapamycin (mTOR) complex 2. J Biol Chem 2012;287:29579–88.

[93] Ishii H, Horie Y, Ohshima S, Anezaki Y, Kinoshita N, Dohmen T, et al. Eicosapentaenoic acid ameliorates steatohepatitis and hepatocellular carcinoma in hepatocyte-specific Pten-deficient mice. J Hepatol 2009;50:562–71.

[94] Fazolini NP, Cruz AL, Werneck MB, Viola JP, Maya-Monteiro CM, Bozza PT. Leptin activation of mTOR pathway in intestinal epithelial cell triggers lipid droplet formation, cytokine production and increased cell proliferation. Cell Cycle 2015;14:2667–76.

[95] Chakrabarti P, English T, Shi J, Smas CM, Kandror KV. Mammalian target of rapamycin complex 1 suppresses lipolysis, stimulates lipogenesis, and promotes fat storage. Diabetes 2010;59:775–81.

[96] Singh M, Shin YK, Yang X, Zehr B, Chakrabarti P, Kandror KV. 4E-BPs control fat storage by regulating the expression of Egr1 and ATGL. J Biol Chem 2015;290:17331–8.

[97] Le Bacquer O, Petroulakis E, Paglialunga S, Poulin F, Richard D, Cianflone K, et al. Elevated sensitivity to diet-induced obesity and insulin resistance in mice lacking 4E-BP1 and 4E-BP2. J Clin Invest 2007;117:387–96.

[98] Su H, Gu Y, Li F, Wang Q, Huang B, Jin X, et al. The PI3K/AKT/mTOR signaling pathway is overactivated in primary aldosteronism. PLoS One 2013;8:e62399.

[99] De Martino MC, van Koetsveld PM, Feelders RA, Sprij-Mooij D, Waaijers M, Lamberts SW, et al. The role of mTOR inhibitors in the inhibition of growth and cortisol secretion in human adrenocortical carcinoma cells. Endocr Relat Cancer 2012;19: 351–64.

[100] Robitaille AM, Christen S, Shimobayashi M, Cornu M, Fava LL, Moes S, et al. Quantitative phosphoproteomics reveal mTORC1 activates de novo pyrimidine synthesis. Science 2013;339: 1320–3.

[101] Ben-Sahra I, Howell JJ, Asara JM, Manning BD. Stimulation of de novo pyrimidine synthesis by growth signaling through mTOR and S6K1. Science 2013;339:1323–8.

[102] Taha C, Liu Z, Jin J, Al-Hasani H, Sonenberg N, Klip A. Opposite translational control of GLUT1 and GLUT4 glucose transporter mRNAs in response to insulin. Role of mammalian target of rapamycin, protein kinase b, and phosphatidylinositol 3-kinase in GLUT1 mRNA translation. J Biol Chem 1999;274:33085–91.

[103] Buller CL, Loberg RD, Fan MH, Zhu Q, Park JL, Vesely E, et al. A GSK-3/TSC2/mTOR pathway regulates glucose uptake and GLUT1 glucose transporter expression. Am J Physiol Cell Physiol 2008;295:C836–43.

[104] Kumar A, Lawrence Jr. JC, Jung DY, Ko HJ, Keller SR, Kim JK, et al. Fat cell-specific ablation of rictor in mice impairs insulin-regulated fat cell and whole-body glucose and lipid metabolism. Diabetes 2010;59:1397–406.

[105] Kumar A, Harris TE, Keller SR, Choi KM, Magnuson MA, Lawrence Jr. JC. Muscle-specific deletion of rictor impairs insulin-stimulated glucose transport and enhances Basal glycogen synthase activity. Mol Cell Biol 2008;28:61–70.

[106] Jiang X, Kenerson HL, Yeung RS. Glucose deprivation in tuberous sclerosis complex-related tumors. Cell Biosci 2011;1:34.

[107] Sun Q, Chen X, Ma J, Peng H, Wang F, Zha X, et al. Mammalian target of rapamycin up-regulation of pyruvate kinase isoenzyme type M2 is critical for aerobic glycolysis and tumor growth. Proc Natl Acad Sci USA 2011;108: 4129–34.

[108] Schieke SM, Phillips D, McCoy Jr. JP, Aponte AM, Shen RF, Balaban RS, et al. The mammalian target of rapamycin (mTOR) pathway regulates mitochondrial oxygen consumption and oxidative capacity. J Biol Chem 2006;281: 27643–52.

[109] Bentzinger CF, Romanino K, Cloetta D, Lin S, Mascarenhas JB, Oliveri F, et al. Skeletal muscle-specific ablation of raptor, but not of rictor, causes metabolic changes and results in muscle dystrophy. Cell Metab 2008;8:411–24.

[110] Risson V, Mazelin L, Roceri M, Sanchez H, Moncollin V, Corneloup C, et al. Muscle inactivation of mTOR causes metabolic and dystrophin defects leading to severe myopathy. J Cell Biol 2009;187:859–74.

[111] Bentzinger CF, Lin S, Romanino K, Castets P, Guridi M, Summermatter S, et al. Differential response of skeletal muscles to mTORC1 signaling during atrophy and hypertrophy. Skelet Muscle 2013;3:6.

[112] Cunningham JT, Rodgers JT, Arlow DH, Vazquez F, Mootha VK, Puigserver P. mTOR controls mitochondrial oxidative function through a YY1-PGC-1alpha transcriptional complex. Nature 2007;450:736—40.

[113] Blattler SM, Verdeguer F, Liesa M, Cunningham JT, Vogel RO, Chim H, et al. Defective mitochondrial morphology and bioenergetic function in mice lacking the transcription factor Yin Yang 1 in skeletal muscle. Mol Cell Biol 2012;32:3333—46.

[114] Chaveroux C, Eichner LJ, Dufour CR, Shatnawi A, Khoutorsky A, Bourque G, et al. Molecular and genetic crosstalks between mTOR and ERRalpha are key determinants of rapamycin-induced nonalcoholic fatty liver. Cell Metab 2013;17:586—98.

[115] Morita M, Gravel SP, Chenard V, Sikstrom K, Zheng L, Alain T, et al. mTORC1 controls mitochondrial activity and biogenesis through 4E-BP-dependent translational regulation. Cell Metab 2013;18:698—711.

[116] Polak P, Cybulski N, Feige JN, Auwerx J, Ruegg MA, Hall MN. Adipose-specific knockout of raptor results in lean mice with enhanced mitochondrial respiration. Cell Metab 2008;8: 399—410.

[117] Betz C, Stracka D, Prescianotto-Baschong C, Frieden M, Demaurex N, Hall MN. Feature article: mTOR complex 2-Akt signaling at mitochondria-associated endoplasmic reticulum membranes (MAM) regulates mitochondrial physiology. Proc Natl Acad Sci USA 2013;110:12526—34.

[118] Hung CM, Calejman CM, Sanchez-Gurmaches J, Li H, Clish CB, Hettmer S, et al. Rictor/mTORC2 loss in the Myf5 lineage reprograms brown fat metabolism and protects mice against obesity and metabolic disease. Cell Rep 2014;8:256—71.

[119] Settembre C, Zoncu R, Medina DL, Vetrini F, Erdin S, Erdin S, et al. A lysosome-to-nucleus signalling mechanism senses and regulates the lysosome via mTOR and TFEB. EMBO J 2012;31:1095—108.

[120] Hosokawa N, Hara T, Kaizuka T, Kishi C, Takamura A, Miura Y, et al. Nutrient-dependent mTORC1 association with the ULK1-Atg13-FIP200 complex required for autophagy. Mol Biol Cell 2009;20:1981—91.

[121] Ganley IG, Lam du H, Wang J, Ding X, Chen S, Jiang X. ULK1.ATG13.FIP200 complex mediates mTOR signaling and is essential for autophagy. J Biol Chem 2009;284:12297—305.

[122] Kim J, Kundu M, Viollet B, Guan KL. AMPK and mTOR regulate autophagy through direct phosphorylation of Ulk1. Nat Cell Biol 2011;13:132—41.

[123] Jung CH, Jun CB, Ro SH, Kim YM, Otto NM, Cao J, et al. ULK-Atg13-FIP200 complexes mediate mTOR signaling to the autophagy machinery. Mol Biol Cell 2009;20:1992—2003.

[124] Koren I, Reem E, Kimchi A. DAP1, a novel substrate of mTOR, negatively regulates autophagy. Curr Biol 2010;20:1093—8.

[125] Vlahakis A, Graef M, Nunnari J, Powers T. TOR complex 2-Ypk1 signaling is an essential positive regulator of the general amino acid control response and autophagy. Proc Natl Acad Sci USA 2014;111:10586—91.

[126] Kopelman PG. Obesity as a medical problem. Nature 2000;404:635—43.

[127] Morton GJ, Cummings DE, Baskin DG, Barsh GS, Schwartz MW. Central nervous system control of food intake and body weight. Nature 2006;443:289—95.

[128] Catania C, Binder E, Cota D. mTORC1 signaling in energy balance and metabolic disease. Int J Obes (Lond) 2011;35:751—61.

[129] Kershaw EE, Flier JS. Adipose tissue as an endocrine organ. J Clin Endocrinol Metab 2004;89:2548—56.

[130] Takahashi KA, Cone RD. Fasting induces a large, leptin-dependent increase in the intrinsic action potential frequency of orexigenic arcuate nucleus neuropeptide Y/Agouti-related protein neurons. Endocrinology 2005;146:1043—7.

[131] Hill JW, Williams KW, Ye C, Luo J, Balthasar N, Coppari R, et al. Acute effects of leptin require PI3K signaling in hypothalamic proopiomelanocortin neurons in mice. J Clin Invest 2008;118: 1796—805.

[132] Cota D, Proulx K, Smith KA, Kozma SC, Thomas G, Woods SC, et al. Hypothalamic mTOR signaling regulates food intake. Science 2006;312:927—30.

[133] Cota D, Matter EK, Woods SC, Seeley RJ. The role of hypothalamic mammalian target of rapamycin complex 1 signaling in diet-induced obesity. J Neurosci 2008;28:7202—8.

[134] Blouet C, Ono H, Schwartz GJ. Mediobasal hypothalamic p70 S6 kinase 1 modulates the control of energy homeostasis. Cell Metab 2008;8:459—67.

[135] Dagon Y, Hur E, Zheng B, Wellenstein K, Cantley LC, Kahn BB. p70S6 kinase phosphorylates AMPK on serine 491 to mediate leptin's effect on food intake. Cell Metab 2012;16:104—12.

[136] Minokoshi Y, Alquier T, Furukawa N, Kim YB, Lee A, Xue B, et al. AMP-kinase regulates food intake by responding to hormonal and nutrient signals in the hypothalamus. Nature 2004;428:569—74.

[137] Kocalis HE, Hagan SL, George L, Turney MK, Siuta MA, Laryea GN, et al. Rictor/mTORC2 facilitates central regulation of energy and glucose homeostasis. Mol Metab 2014;3: 394—407.

[138] Kojima M, Kangawa K. Ghrelin: structure and function. Physiol Rev 2005;85:495—522.

[139] Ferrini F, Salio C, Lossi L, Merighi A. Ghrelin in central neurons. Curr Neuropharmacol 2009;7:37—49.

[140] Martins L, Fernandez-Mallo D, Novelle MG, Vazquez MJ, Tena-Sempere M, Nogueiras R, et al. Hypothalamic mTOR signaling mediates the orexigenic action of ghrelin. PLoS One 2012;7:e46923.

[141] Albert V, Cornu M, Hall MN. mTORC1 signaling in Agrp neurons mediates circadian expression of Agrp and NPY but is dispensable for regulation of feeding behavior. Biochem Biophys Res Commun 2015;464:480—6.

[142] Lopez M, Alvarez CV, Nogueiras R, Dieguez C. Energy balance regulation by thyroid hormones at central level. Trends Mol Med 2013;19:418—27.

[143] Varela L, Martinez-Sanchez N, Gallego R, Vazquez MJ, Roa J, Gandara M, et al. Hypothalamic mTOR pathway mediates thyroid hormone-induced hyperphagia in hyperthyroidism. J Pathol 2012;227:209—22.

[144] Yang Z, Ming XF. mTOR signalling: the molecular interface connecting metabolic stress, aging and cardiovascular diseases. Obes Rev 2012;13(Suppl. 2):58—68.

[145] Yeh WC, Bierer BE, McKnight SL. Rapamycin inhibits clonal expansion and adipogenic differentiation of 3T3-L1 cells. Proc Natl Acad Sci USA 1995;92:11086—90.

[146] Gagnon A, Lau S, Sorisky A. Rapamycin-sensitive phase of 3T3-L1 preadipocyte differentiation after clonal expansion. J Cell Physiol 2001;189:14—22.

[147] Bell A, Grunder L, Sorisky A. Rapamycin inhibits human adipocyte differentiation in primary culture. Obes Res 2000;8:249—54.

[148] Kim JE, Chen J. Regulation of peroxisome proliferator-activated receptor-gamma activity by mammalian target of rapamycin and amino acids in adipogenesis. Diabetes 2004;53:2748—56.

[149] Zhang HH, Huang J, Duvel K, Boback B, Wu S, Squillace RM, et al. Insulin stimulates adipogenesis through the Akt-TSC2-mTORC1 pathway. PLoS One 2009;4:e6189.

[150] Cinti S. Transdifferentiation properties of adipocytes in the adipose organ. Am J Physiol Endocrinol Metab 2009;297: E977—86.

[151] Xiang X, Lan H, Tang H, Yuan F, Xu Y, Zhao J, et al. Tuberous sclerosis complex 1-mechanistic target of rapamycin complex 1 signaling determines brown-to-white adipocyte phenotypic switch. Diabetes 2015;64:519—28.

[152] Houde VP, Brule S, Festuccia WT, Blanchard PG, Bellmann K, Deshaies Y, et al. Chronic rapamycin treatment causes glucose intolerance and hyperlipidemia by upregulating hepatic gluconeogenesis and impairing lipid deposition in adipose tissue. Diabetes 2010;59:1338—48.

[153] Rovira J, Marcelo Arellano E, Burke JT, Brault Y, Moya-Rull D, Banon-Maneus E, et al. Effect of mTOR inhibitor on body weight: from an experimental rat model to human transplant patients. Transpl Int 2008;21:992—8.

[154] Chang GR, Chiu YS, Wu YY, Chen WY, Liao JW, Chao TH, et al. Rapamycin protects against high fat diet-induced obesity in C57BL/6J mice. J Pharmacol Sci 2009;109:496—503.

[155] Morris BJ, Carnes BA, Chen R, Donlon TA, He Q, Grove JS, et al. Genetic variation in the raptor gene is associated with overweight but not hypertension in American men of Japanese ancestry. Am J Hypertens 2015;28:508—17.

[156] Shepherd PR, Kahn BB. Glucose transporters and insulin action—implications for insulin resistance and diabetes mellitus. N Engl J Med 1999;341:248—57.

[157] Cybulski N, Polak P, Auwerx J, Ruegg MA, Hall MN. mTOR complex 2 in adipose tissue negatively controls whole-body growth. Proc Natl Acad Sci USA 2009;106:9902—7.

[158] Fanelli C, Calderone S, Epifano L, De Vincenzo A, Modarelli F, Pampanelli S, et al. Demonstration of a critical role for free fatty acids in mediating counterregulatory stimulation of gluconeogenesis and suppression of glucose utilization in humans. J Clin Invest 1993;92:1617—22.

[159] Belfort R, Mandarino L, Kashyap S, Wirfel K, Pratipanawatr T, Berria R, et al. Dose-response effect of elevated plasma free fatty acid on insulin signaling. Diabetes 2005;54:1640—8.

[160] Khamzina L, Veilleux A, Bergeron S, Marette A. Increased activation of the mammalian target of rapamycin pathway in liver and skeletal muscle of obese rats: possible involvement in obesity-linked insulin resistance. Endocrinology 2005;146:1473—81.

[161] MacLaren R, Cui W, Simard S, Cianflone K. Influence of obesity and insulin sensitivity on insulin signaling genes in human omental and subcutaneous adipose tissue. J Lipid Res 2008;49:308—23.

[162] Ost A, Svensson K, Ruishalme I, Brannmark C, Franck N, Krook H, et al. Attenuated mTOR signaling and enhanced autophagy in adipocytes from obese patients with type 2 diabetes. Mol Med 2010;16:235—46.

[163] Cornu M, Oppliger W, Albert V, Robitaille AM, Trapani F, Quagliata L, et al. Hepatic mTORC1 controls locomotor activity, body temperature, and lipid metabolism through FGF21. Proc Natl Acad Sci USA 2014;111:11592—9.

[164] Rachdi L, Balcazar N, Osorio-Duque F, Elghazi L, Weiss A, Gould A, et al. Disruption of Tsc2 in pancreatic beta cells induces beta cell mass expansion and improved glucose tolerance in a TORC1-dependent manner. Proc Natl Acad Sci USA 2008;105:9250—5.

[165] Shigeyama Y, Kobayashi T, Kido Y, Hashimoto N, Asahara S, Matsuda T, et al. Biphasic response of pancreatic beta-cell mass to ablation of tuberous sclerosis complex 2 in mice. Mol Cell Biol 2008;28:2971—9.

[166] Hamada S, Hara K, Hamada T, Yasuda H, Moriyama H, Nakayama R, et al. Upregulation of the mammalian target of rapamycin complex 1 pathway by Ras homolog enriched in brain in pancreatic beta-cells leads to increased beta-cell mass and prevention of hyperglycemia. Diabetes 2009;58:1321—32.

[167] Gu Y, Lindner J, Kumar A, Yuan W, Magnuson MA. Rictor/mTORC2 is essential for maintaining a balance between beta-cell proliferation and cell size. Diabetes 2011;60:827—37.

[168] Pallet N, Legendre C. Adverse events associated with mTOR inhibitors. Expert Opin Drug Saf 2013;12:177—86.

[169] Lamming DW, Ye L, Katajisto P, Goncalves MD, Saitoh M, Stevens DM, et al. Rapamycin-induced insulin resistance is mediated by mTORC2 loss and uncoupled from longevity. Science 2012;335:1638—43.

[170] Fraenkel M, Ketzinel-Gilad M, Ariav Y, Pappo O, Karaca M, Castel J, et al. mTOR inhibition by rapamycin prevents beta-cell adaptation to hyperglycemia and exacerbates the metabolic state in type 2 diabetes. Diabetes 2008;57:945—57.

[171] Gallagher EJ, LeRoith D. Obesity and diabetes: the increased risk of cancer and cancer-related mortality. Physiol Rev 2015;95:727—48.

[172] Sidiropoulos KG, Meshkani R, Avramoglu-Kohen R, Adeli K. Insulin inhibition of apolipoprotein B mRNA translation is mediated via the PI-3 kinase/mTOR signaling cascade but does not involve internal ribosomal entry site (IRES) initiation. Arch Biochem Biophys 2007;465:380—8.

[173] Morrisett JD, Abdel-Fattah G, Hoogeveen R, Mitchell E, Ballantyne CM, Pownall HJ, et al. Effects of sirolimus on plasma lipids, lipoprotein levels, and fatty acid metabolism in renal transplant patients. J Lipid Res 2002;43:1170—80.

[174] Ai D, Chen C, Han S, Ganda A, Murphy AJ, Haeusler R, et al. Regulation of hepatic LDL receptors by mTORC1 and PCSK9 in mice. J Clin Invest 2012;122:1262—70.

[175] Fruhwurth S, Krieger S, Winter K, Rosner M, Mikula M, Weichhart T, et al. Inhibition of mTOR down-regulates scavenger receptor, class B, type I (SR-BI) expression, reduces endothelial cell migration and impairs nitric oxide production. Biochim Biophys Acta 2014;1841:944—53.

[176] Kraemer FB, Takeda D, Natu V, Sztalryd C. Insulin regulates lipoprotein lipase activity in rat adipose cells via wortmannin- and rapamycin-sensitive pathways. Metabolism 1998;47:555—9.

[177] Verges B, Walter T, Cariou B. Endocrine side effects of anti-cancer drugs: effects of anti-cancer targeted therapies on lipid and glucose metabolism. Eur J Endocrinol 2014;170:R43—55.

[178] Angulo P. Nonalcoholic fatty liver disease. N Engl J Med 2002;346:1221—31.

[179] Kubrusly MS, Correa-Giannella ML, Bellodi-Privato M, de Sa SV, de Oliveira CP, Soares IC, et al. A role for mammalian target of rapamycin (mTOR) pathway in non alcoholic steatohepatitis related-cirrhosis. Histol Histopathol 2010;25:1123—31.

[180] Patsenker E, Schneider V, Ledermann M, Saegesser H, Dorn C, Hellerbrand C, et al. Potent antifibrotic activity of mTOR inhibitors sirolimus and everolimus but not of cyclosporine A and tacrolimus in experimental liver fibrosis. J Hepatol 2011;55:388—98.

22

Chronic mTOR Inhibition by Rapamycin and Diabetes: What is the Role of Mitochondria?

Liang-Jun Yan[1] and Zhiyou Cai[2]

[1]Department of Pharmaceutical Sciences, UNT System College of Pharmacy,
University of North Texas Health Science Center, Fort Worth, TX, USA [2]Department of Neurology, Renmin Hospital,
Hubei University of Medicine, Shiyan, Hubei Province, People's Republic of China

ABBREVIATIONS

AMPK	adenine monophosphate-activated kinase
HIFα	hypoxia-inducible factor α
CoQ	Coenzyme Q
CAD	carbamoyl-phosphate synthetase 2, aspartate transcarbamylase, and dihydroorotase
mTOR	mechanistic target of rapamycin
mTORC	mechanistic target of rapamycin complex
mSIN1	mammalian stress-activated protein kinase-interacting protein 1
mLST8	mammalian lethal with sec-13
NAD	nicotinamide adenine dinucleotide
NADH	Reduced nicotinamide adenine dinucleotide
Raptor	regulatory associated protein of mTOR
Rictor	rapamycin-insensitive companion of mTOR
Deptor	DEP-domain-containing mTOR-interacting protein
PRAS40	proline-rich Akt substrate 40 kDa
Tel2	telomere maintenance 2
Ttil1	Tel2 interacting protein 1
SREBPs	sterol regulatory element-binding proteins
PPAR-γ	peroxisome proliferator-activated receptor gamma
eIF4E	eukaryotic translation initiation factor 4E
4EBP	4E-binding protein
S6K	protein S6 kinase
IRS	insulin receptor substrate
PI3K	phosphatidylinositol 3 kinase
PDK	phosphoinositide-dependent protein kinase
TIF-1A	transcription initiating factor 1A
PGC1	PPAR-γ coactivator
TFEB	transcriptional factor EB
Dap1	death-associated protein 1
RHEB	Ras homolog enriched in brain
TSC1/TSC2	tuberous sclerosis 1 and 2

22.1 INTRODUCTION

The mammalian (or mechanistic) target of rapamycin (mTOR) is a protein kinase [1,2]. Its signaling process is essential for growth, proliferation, and development [3,4]. However, mTOR signaling is also linked to the pathogenesis of many diseases [5–8]. Rapamycin, as a specific mTOR inhibitor and an FDA-approved drug for immunosuppression following organ transplantation and for anticancer therapies [8–10], often induces a new onset of diabetes after long-term administration [11–13]. Such a chronic adverse effect severely compromises the therapeutic value of rapamycin, and has elicited great interest in elucidating its diabetogenic mechanisms. In this chapter, following overviews of glucose metabolism in normal and diabetic conditions and of the mTOR signaling pathways, we argue that mitochondria could play a major role in the pathogenesis of rapamycin-induced diabetes in terms of a redox imbalance disorder between NADH (reduced nicotinamide adenine dinucleotide) and NAD+ (nicotinamide adenine dinucleotide), which is essential in metabolism, stress sensing, and cell signaling [14–19]. As NADH fluxes through mitochondrial complex I to regenerate NAD+ [20–23], we focus our discussion on the potential role of mitochondrial complex I in chronic mTOR inhibition and diabetes. It is conceivable that complex I could well be involved in the pathogenesis of diabetes by chronic mTOR inhibition by rapamycin.

22.2 DIABETES AND CHRONIC HYPERGLYCEMIA

Diabetes mellitus is an aberrant glucose metabolic disease; and chronic hyperglycemia is the hallmark of diabetes mellitus [24–29]. There are two main types of diabetes. One is insulin-dependent, which is usually called type 1 diabetes; the other is insulin-independent, which is usually called type 2 diabetes [30–33]. In type 1 diabetes, there is a net decrease in insulin secretion or lack of insulin due to impaired pancreatic β-cell function [34–36]. In type 2 diabetes, however, insulin resistance is believed to be the initiating factor that gradually leads to impairment in β-cell function and eventual β-cell death [37–41]. Therefore, there is also a net decrease in insulin secretion in overt type 2 diabetes. Regardless of the type of diabetes, diabetic complications such as retinopathy, neuropathy, and nephropathy are all thought to be driven by chronic hyperglycemia [20,25,41–43].

22.3 GLUCOSE COMBUSTION UNDER NORMAL PHYSIOLOGICAL CONDITIONS AND IN DIABETES

Glucose is the major fuel for cellular bioenergetics. Under physiological conditions, glucose is mainly burned through the glycolytic pathway and the Krebs cycle [21]. Through these pathways, glucose is first phosphorylated by glucokinase or hexokinase, a committed step that traps glucose inside a cell. This is followed by glycolysis that splits glucose into two molecules of pyruvate (Figure 22.1). Pyruvate is then transported into mitochondria wherein it is decarboxylated to form acetyl-CoA. Acetyl-CoA then enters into the Krebs cycle to be completely combusted to produce CO_2 and H_2O. Electrons derived from glucose combustion are mainly stored in NADH and $FADH_2$, both of which are used by the electron transport chain for the synthesis of adenosine triphosphate (ATP) that is required for all cellular activities [21]. Therefore, regeneration of NAD+, and FAD^{2+} to a lesser extent, that can continue to accept electrons from glucose combustion is crucial for cell survival [44–46]. In addition to being used for ATP production, glucose is also used via the pentose phosphate pathway [47] for the generation of NADPH and 5-ribose that are required for anabolism and synthesis of nucleotides, respectively. Moreover, glucose can also be used for lipogenesis using acetyl-CoA as the building blocks [21]. Therefore, glucose is the fundamental molecule for life.

In diabetes, however, glucose metabolism is highly aberrant. Many pathways that are usually dormant in normal conditions are activated or up-regulated by high

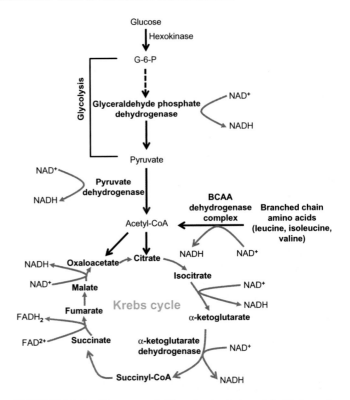

FIGURE 22.1 Glucose metabolism via glycolysis and the Krebs cycle under normal physiological condition. As the main purpose of glucose combustion is to produce NADH for ATP production during oxidative phosphorylation, the steps whereby NADH is made from NAD+ are also shown. NADH, reduced nicotinamide adenine dinucleotide; NAD+, nicotinamide adenine dinucleotide; ATP, adenosine triphosphate.

levels of blood glucose [20]. These pathways deviate from the conventional glycolysis and Krebs cycle pathways and thus are often called branching-off glucose metabolic pathways [20]. As shown in Figure 22.2, these branching-off pathways include the polyol pathway [50,51], the hexosamine pathway [52,53], the advanced glycation pathway [54], the protein kinase C (PKC) activation pathway [24,55], and the α-ketoaldehyde formation pathway [20]. All these pathways are able to elevate cellular reactive oxygen species (ROS) production [56], which in turn can inactivate the glycolytic enzyme glyceraldehyde 3-phosphate dehydrogenase [57,58], resulting in further diversion of glucose to the branching-off pathways and rendering the occurrence of chronic glucotoxicity in diabetes that is bound to cause a myriad of diabetic complications.

22.4 REDOX IMBALANCE IN DIABETES AND ITS COMPLICATIONS

In diabetes, there is an excess supply of NADH due to oversupply of glucose [20,59]. The burden of excess NADH in diabetes mainly comes from the polyol

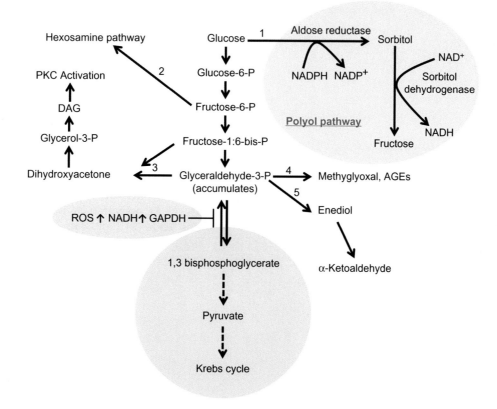

FIGURE 22.2 The branching-off pathways that are activated or up-regulated by chronic hyperglycemia in diabetes. Shown are five pathways that all can lead to ROS production and oxidative stress. These pathways are the major mechanisms of diabetic glucotoxicity. Among these pathways, the polyol pathway is considered as a major pathway that can perturb redox balance due to its production of NADH at the expense of NADPH via two reactions catalyzed by aldose reductase and sorbitol dehydrogenase, respectively [48,49]. Note that GAPDH can be inhibited by excess NADH or inactivated by ROS, which would divert more glucose to these non-conventional pathways and aggravate chronic diabetic glucotoxicity. NADH, reduced nicotinamide adenine dinucleotide; ROS, reactive oxygen species. NADPH, nicotinamide adenine dinucleotide phosphate; GAPDH, glyceraldehyde 3-phosphate dehydrogenase.

pathway [50,60–63] that involves aldose reductase and sorbitol dehydrogenase [51,64]. This pathway not only elevates NADH levels, but also elevates cellular sorbitol content that can cause osmotic stress [65–68]. It has been estimated that more than 30% of the glucose pool in diabetes is utilized by the polyol pathway [69], making it a significant pathway in NADH production in chronic hyperglycemia. On the other hand, as the five branching-off pathways all generate ROS [56] that can damage DNA; poly adenosine diphosphate (ADP) ribose polymerase is activated. This enzyme uses NAD+ as its substrate and is often overactivated in diabetes [70–72], leading to depletion of NAD+ and cell death [73,74]. When the NAD+ level lowers, a condition of pseudohypoxia develops [20,75–79], which is detrimental to cell survival. Therefore, in diabetes, there is a severe imbalance between NADH and NAD+ with NADH in excess [21] (see further below and the rectangular area in Figure 22.9 for further details). This is a condition that is also known as reductive stress [20,77,80,81].

22.5 mTOR SIGNALING

The mechanistic target of rapamycin (mTOR) is an evolutionarily conserved protein serine/threonine kinase [82,83]. It has two functionally distinct complexes, mechanistic target of rapamycin complex (mTORC) 1 and mTORC2 [84]. While mTORC1 has been well studied and is established to be a sensor of nutrient and growth hormones responsible for development and proliferation [84–86], the function of mTORC2 is less clear. It is now generally agreed that mTORC2 is insensitive to acute rapamycin inhibition, but is responsive to chronic rapamycin inhibition [87].

mTOR is the core component of mTORC1 and mTORC2 [88]. In each complex, mTOR serves as a catalytic protein kinase (Figure 22.3). mTORC1, as shown in Figure 22.3, is composed of the following components: mTOR, Raptor (regulatory associated protein of mTOR), PRAS40 (proline-rich Akt substrate 40 kDa), MLsT8 (mammalian lethal with sec-13), Ttil1 (Tel2 interacting protein 1), Tel2 (telomere maintenance 2),

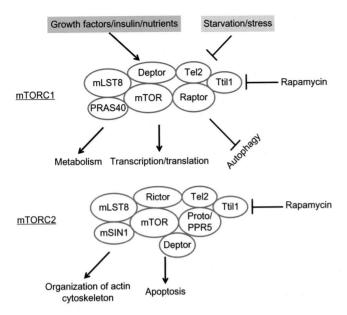

FIGURE 22.3 mTOR is a key component of two complexes mTORC1 and mTORC2. Shown are the essential protein components in each complex. The downstream biological processes are also shown for each complex, as are the upstream signals that can activate or inhibit mTORC1. As mTORC2 signaling is not well understood, no upstream signaling events are shown. mTORC1, mechanistic target of rapamycin complex 1; mTORC2, mechanistic target of rapamycin complex 2.

and Deptor (DEP-domain-containing mTOR-interacting protein). The component Raptor is in charge of mTOR assembly [84]. The function of mTORC1 is believed to regulate transcription, translation, metabolism, and autophagy [89,90]. In contrast to mTORC1, the mTORC2 complex contains Rictor (rapamycin-insensitive companion of mTOR) instead of Raptor. Rictor competes with Raptor for the binding of mTOR kinase. Additionally, mTORC2 also contains mSIN1 (mammalian stress-activated protein kinase-interacting protein 1) and Protor/PPR5 that are not in mTORC1. The function of mTORC2 is believed to regulate apoptosis and the organization of actin-cytoskeleton (Figure 22.3).

mTOR can be activated by nutrients such as glucose and amino acids [91], and can also receive signals from ROS, DNA damage, and lipid peroxidation [87,92]. There are many downstream substrates. As shown in Figure 22.4, these substrates include S6K (protein S6 kinase), PPARγ (peroxisome proliferator-activated receptor gamma), HIF-1α (hypoxia-inducible factor α), 4E-BP, and SREBPs (sterol regulatory element-binding proteins). As has not been elucidated well so far, the number of mTORC2 substrates is much smaller. The known ones include Akt, serum- and glucocorticoid-regulated kinase (SGK), and PKC-α [84,93].

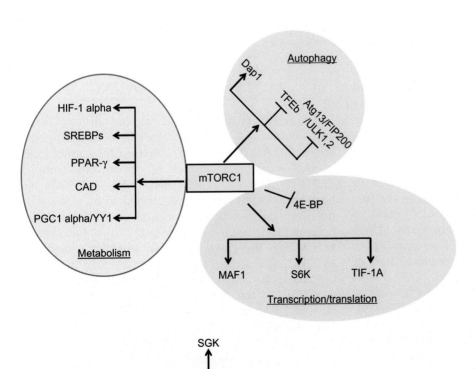

FIGURE 22.4 Functional consequences of activation of mTORC1 and mTORC2. mTORC1 signaling can regulate numerous cellular processes including autophagy, metabolism, protein synthesis, and proliferation. mTORC2 signaling is less well elucidated, but is believed to regulate the activities of Akt, PKC-α, and SGK. mTORC1, mechanistic target of rapamycin complex 1; mTORC2, mechanistic target of rapamycin complex 2; PKC-α, protein kinase C-alpha; SGK, serum- and glucocorticoid-regulated kinase.

22.6 mTOR SIGNALING AS A DOUBLE-EDGED SWORD

During growth and development, mTOR signaling is absolutely required and is beneficial as cells need to sense nutrients for proliferation and growth [94]. However, once the growth and development process is complete, mTOR signaling is usually detrimental [94]. In fact, the mTOR signaling pathways tend to go aberrant after growth and development is complete [95–97]. Therefore, inhibition of mTOR by rapamycin can prolong the lifespan of organisms such as yeast [98], *C. elegans* [99], *Drosophila* [100], and mouse [101,102]. This is also true in the case of immune suppression following organ transplantation [103], whereby the mTOR activity needs to be brought under control [104], which otherwise usually leads to tissue rejection [3]. The problem here is that such an inhibition can often lead to new onset of diabetes in the transplant patients [103]. Therefore, the mTOR signaling pathway is really a double-edged sword and there is increasing interest in studying the mechanisms of diabetes induced by chronic mTOR inhibition by rapamycin. Additionally, mTOR also tends to deregulate itself during aging [105–108] and tumorigenesis [109–113] and in age-related neurodegenerative diseases such as Alzheimer's disease and Parkinson's disease [7,82,83,114–116].

22.7 CHRONIC mTOR INHIBITION BY RAPAMYCIN AND DIABETES

mTOR can regulate adaptation of β cells to hyperglycemia [117] and can also slow progression of diabetic complications upon rapamycin's action [118,119], but as described above, its chronic inhibition can also induce diabetes [120,121]. In animal studies, chronic inhibition of mTOR signaling can lead to diabetic hyperglycemia [122,123]. This apparently contradicts with the findings that rapamycin inhibition of mTOR extends the lifespan in several organisms as described above [98–102]. It has been suggested that the effects of rapamycin are duration-dependent, that is, a short duration of rapamycin ingestion can alleviate age-associated metabolic stress, while a long duration of rapamycin treatment can overactivate mTOR and induce insulin resistance and diabetes [123,124].

How it is that rapamycin that has been shown to extend lifespan can also induce diabetes that seemingly would shorten lifespan? In the mouse, it has been shown that rapamycin induces glucose intolerance by decreasing β-cell mass and insulin levels [123]. In this same study, rapamycin was also found to decrease cellular sensitivity to insulin stimulation in insulin-responsive cells such as hepatocytes [123]. In

adipose tissue of rats, chronic rapamycin ingestion can also induce glucose intolerance and dyslipidemia [125], which is further found to be caused by enhancing gluconeogenesis in the liver and by impairing lipid deposition and storage in adipose tissue [125]. Additionally, in skeletal muscle cells, mTOR inhibition by chronic rapamycin ingestion also induces insulin resistance imposed by impaired Akt signaling and decreased glucose transportation across the plasma membrane [126]. It should be noted that rapamycin-induced metabolic syndrome could be reversed after a few weeks of rapamycin withdrawal [127], and both the induced metabolic syndrome and the reversal from hyperglycemia to euglycemia are independent of gender and genetic background [127].

22.8 mTOR AND INSULIN RESISTANCE

As insulin signaling can go aberrant, so can mTOR signaling [128,129]. In fact, there is a tight link between the two pathways [130,131]. mTOR is a downstream substrate in the insulin signaling pathway, which has been demonstrated by the observation that the lifespan extension in insulin signaling mutant could not be further extended by mTOR inhibition by rapamycin [132]. As shown in Figure 22.5, Akt activated by insulin binding to its receptor can abrogate the constitutive inhibition of Rheb (Ras homolog enriched in brain)-GTP by TSC1/TSC2 (tuberous sclerosis 1 and 2). Without the inhibition by TSC1/TSC2, Rheb-GTP is active and thus activates mTORC1 that then activates S6K or inactivates 4E-BP [133,134], both of which are involved in promoting cell growth, proliferation, and metabolism. It should be noted that phosphorylation of S6K can inhibit PI3K (phosphatidylinositol 3 kinase) by increasing IRS1 (insulin receptor substrate 1) phosphorylation via a negative feedback mechanism [87,135]. This has been suggested to be a major mechanism of insulin resistance involving mTOR signaling [136,137]. Similarly, excess amino acids can also inhibit PI3K and induce insulin resistance by the same mechanism [138]. During energy crises, AMPK (adenine monophosphate-activated kinase) is activated and phosphorylates TSC2, which inhibits mTORC1 signaling, leading to hampering of cell growth and proliferation [139,140].

22.9 ROLE OF MITOCHONDRIA IN CHRONIC mTOR INHIBITION AND DIABETES

While it is known that the mitochondrion is a target of mTOR signaling [141,142] and mitochondrial

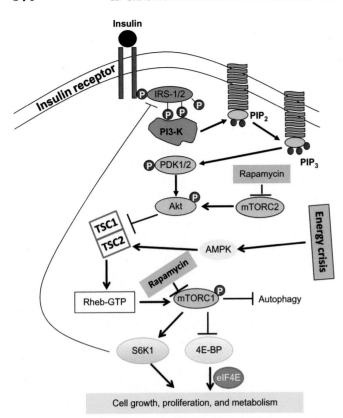

FIGURE 22.5   mTORC signaling and insulin resistance. mTORC is a component in the insulin signaling cascade. TSC1 is a constitutive inhibitor of Rheb, its phosphorylation by Akt abolishes this inactivation and thus activates Rheb, which in turn activates mTORC1. mTORC1 activation then regulates S6K1 and 4E-BP (4E-binding protein) that can promote cell growth, proliferation, and metabolism. As S6K1's phosphorylation can inhibit IRS1 via a negative feedback mechanism, this feedback inhibition has been suggested to be the main mechanism for insulin resistance. Note that mTORC2 can also activate Akt. During energy crisis, activated AMPK can phosphorylate TSC2 and thus inhibits mTORC1 signaling, hampering cell growth and proliferation. mTORC1, mechanistic target of rapamycin complex 1; TSC1 and 2, tuberous sclerosis 1 and 2; IRS1, insulin receptor substrate 1; AMPK, adenine monophosphate-activated kinase.

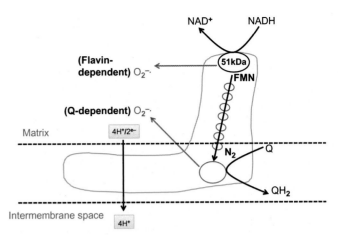

FIGURE 22.6   Mitochondrial complex I (NADH: ubiquinone oxidoreductase) is the major site for NAD + regeneration. This complex has three functions: NADH oxidation, proton pumping, and superoxide production. The rate of superoxide production is proportional to that of NADH oxidation. There are two known sites in complex I that can generate superoxide, either flavin-dependent or quinone-dependent. Electrons from NADH are transported to CoQ via a series of iron—sulfur clusters from different protein subunits within the complex. NADH, reduced nicotinamide adenine dinucleotide; NAD+ , nicotinamide adenine dinucleotide; CoQ, coenzyme Q.

function is impaired in mice treated with rapamycin [143], the detailed mechanisms by which mitochondrial function is regulated by mTOR signaling remain elusive. We believe that redox imbalance [144,145] could be responsible for the pathogenesis of diabetes induced by chronic mTOR inhibition.

In diabetes, the polyol pathway is activated by a high level of glucose [60,62,78,146,147]. This pathway, while attempting to dispose of excess glucose, also consumes NADPH and produces NADH via two-step reactions (Figure 22.2) [20], therefore, elevating the overall cellular level of NADH (Figure 22.2). On the other hand, diabetic hyperglycemia can induce ROS production [56], leading to DNA damage that activates poly ADP ribose polymerase [148,149]. This enzyme

uses NAD + as its substrate and hence can potentially deplete cellular NAD + [148]. The net outcome of the activation of the polyol pathway and poly ADP ribose polymerase is thus a redox imbalance, reflected by elevated levels of the NADH/NAD + ratio [21]. As one of the major functions of mitochondria is to regenerate NAD+ from NADH by complex I (Figure 22.6), the NADH/NAD+ imbalance in rapamycin-induced diabetes could certainly involve complex I dysfunction or dysregulation. This reasoning will need to be tested in future investigations.

22.10 MITOCHONDRIAL COMPLEX I AND ELECTRON TRANSPORT CHAIN

As mitochondrial complex I (also known as NADH/ ubiquinone oxidoreductase) is in charge of elevating cellular NAD+ content, we would like to devote a little more space here to our discussion of this enzyme [150−152]. Complex I is the first component of the mitochondrial electron transport chain and is the initial electron entry point to the electron transport chain [153,154]. It has at least 45 subunits in mammalian mitochondria and transfers electrons from NADH to Coenzyme Q (CoQ) [152,155,156]. Therefore, complex I is the major enzyme for elevating the cellular level of NAD+ [157]. Moreover, complex I is one of the major sites in mitochondria that are able to produce superoxide anion [158], a precursor of all the other ROS including

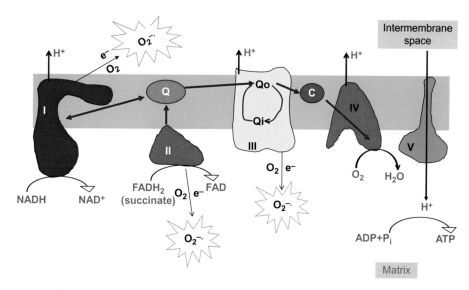

FIGURE 22.7 Components of mitochondrial electron transport chain and oxidative phosphorylation. Shown are complexes I to V. In this electron transport chain, complex I and complex II transport electrons from NADH and FADH₂, respectively, to CoQ that then passes the electrons to cytochrome c and eventually to molecular oxygen. During this process, protons are pumped out into the inner membrane space forming a proton gradient that drives the synthesis of ATP via complex V. Note that complex I is the major site for superoxide production, along with complexes II and III. In complex III, the Q cycle (Q_0 to Q_i) involved in oxidation of ubiquinol (to ubiquinone) and reduction of cytochrome c [162,163] is also shown. NADH, reduced nicotinamide adenine dinucleotide; CoQ, coenzyme Q; ATP, adenosine triphosphate.

hydrogen peroxide, hydroxyl radical, and peroxynitrite (in the presence of nitric oxide) [159–161]. Figure 22.7 shows a cartoon of complex I and the other electron transport chain components as well as complex V. As can be seen in this figure, electrons from both complexes I and II are transported to CoQ, which then passes electrons to cytochrome c via complex III. Cytochrome c then passes electrons to oxygen via complex IV, forming water. In the meantime, protons are pumped out via complexes I, III, and IV into the inner membrane space, forming a proton gradient across the inner membrane that then drives ATP synthesis via complex V. As the electron transport chain is not a perfect system, up to 5% of electrons can leak out of the chain [161], leading to one electron reduction of oxygen to form superoxide anion [161]. It should be noted that besides complex I, complexes II and III are also able to produce superoxide [164,165]. Therefore, the electron transport chain, in particular, complex I, is a major source of cellular oxidative stress [166,167]. In fact, complex I dysfunction and its superoxide production have been thought to be involved in numerous diseases such as diabetes [168,169], cancer [150,170], Alzheimer's disease [171–173], and Parkinson's disease [174–177]. While mitochondria are known to be a key component in β-cell insulin secretion [178–180], the role of complex I in β-cell function and dysfunction has not been elucidated. Nonetheless, without complex I's normal function, insulin secretion from β cells may become aberrant. This is because ATP production driven by glucose combustion is a trigger for β-cell insulin secretion.

22.11 GLUCOSE METABOLISM AND β-CELL INSULIN SECRETION

Unlike cells in many other tissues, β-cell glucose metabolism employs a supply-driven mechanism [181,182], meaning that the higher the blood glucose, the more the insulin secretion. Note that this supply-driven mechanism only operates under normal physiological conditions. The major reason that β cells can handle a high level of blood glucose is because of its glucose transporter and glucokinase that have high Km values for glucose and cannot be readily saturated by glucose [27,182–184]. Therefore, the more glucose goes in, the more glucose gets phosphorylated, which leads to more ATP production via mitochondrial oxidative phosphorylation. When the ATP level is high, so is the ratio between ATP and ADP, which in turn leads to closure of the K_{ATP} channel on the β-cell plasma membrane. This closure opens the calcium channel that is also on the plasma membrane, leading to calcium influx that triggers the release of insulin from β cells [185–188] (Figure 22.8). As one can imagine, without complex I, insulin secretion would not be possible as ATP production by mitochondria absolutely requires complex I.

In diabetes, however, β-cell insulin secretion will not be able to compensate for the ever-increasing levels of blood glucose and β cell will eventually be exhausted and die [189,190], leading to deterioration of the diabetic situation. How this exhaustion process is related to mTOR chronic inhibition and diabetes remains unknown.

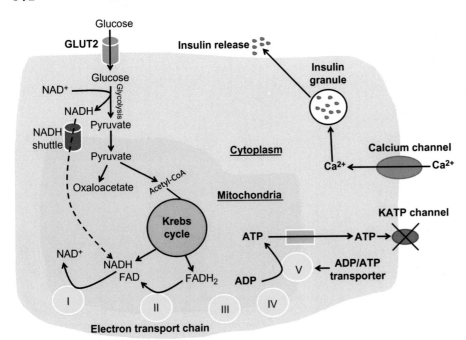

FIGURE 22.8 Glucose metabolism and β-cell insulin secretion. Shown is a well-accepted mechanism by which ATP production by glucose combustion via glycolysis and tricarboxylic acid cycle drives insulin secretion. When blood glucose level is high, so is the level of ATP content. This high level of ATP (or ATP/ADP ratio) leads to K_{ATP} channel closure, which in turn leads to Ca^{2+} channel opening that triggers calcium influx. It is this calcium influx that leads to insulin release out of the β cell. In this pathway, complex I should play an important role. ATP, adenosine triphosphate; ADP, adenosine diphosphate.

22.12 POSSIBLE ROLE OF MITOCHONDRIAL COMPLEX I IN DIABETES INDUCED BY CHRONIC mTOR INHIBITION

A hypothetic model for the role of complex I in mTOR/rapamycin diabetes is depicted in Figure 22.9. While many enzyme systems inside a cell can generate NADH by glucose combustion (as shown in Figure 22.1), there are not many enzyme systems that can make NAD+ from NADH. Therefore, as described above, complex I is the major site for NAD+ recycling from NADH [154,191] (Figure 22.6), in addition to its proton-pumping function and superoxide generation capacity [151,152,158,192,193]. In our hypothetic model, as shown in Figure 22.9, diabetes can be caused by redox imbalance between NADH and NAD+ due to oversupply of NADH. One direct consequence of NADH oversupply is the elevated electron pressure imposed on mitochondrial complex I [21]. On one hand, more NADH oxidation by complex I will yield more ROS as the rate of ROS production by complex I is proportional to that of NADH oxidation [154,194]. On the other hand, activation of poly ADP ribose polymerase can deplete NAD+ [72,195]. Additionally, an elevated ratio of NADH to NAD+ could also mean impaired NADH oxidation [196–198] at complex I. Therefore, if complex I function can be enhanced or manipulated to handle the overflow of NADH, the redox imbalance between NADH and NAD+ could then be restored, which would alleviate or prevent occurrence of rapamycin-induced diabetes. Furthermore, investigating complex I mechanisms of redox imbalance may

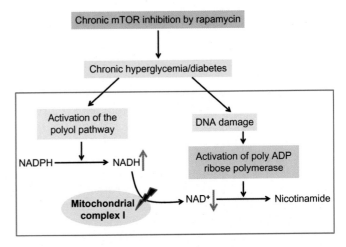

FIGURE 22.9 Proposed redox imbalance mechanisms of chronic mTOR inhibition by rapamycin in pathogenesis of diabetes. Chronic inhibition of mTOR by rapamycin could impair complex I function, leading to NADH accumulation that promotes oxidative stress. Besides the glycolysis and the Krebs cycle, activation of the polyol pathway by chronic hyperglycemia would further increase NADH accumulation. In the meantime, activation of poly ADP ribose polymerase can deplete the NAD+ pool that is already at a low level. Therefore, NADH/NAD ratio is highly perturbed due to an impaired NADH flow through complex I. Note that how complex I function is regulated by mTOR signaling is largely unknown. NADH, reduced nicotinamide adenine dinucleotide; mTOR, mechanistic target of rapamycin; NAD+, nicotinamide adenine dinucleotide; ADP, adenosine diphosphate.

provide insights into novel strategies for preventing the adverse effects of diabetes induced by mTOR chronic inhibition that is absolutely required for therapeutic purposes, such as after organ transplantation. Moreover, a fundamental link between complex I dysfunction and mTOR signaling may also be established using rapamycin-induced diabetes in rodents.

22.13 CAN RAPAMYCIN-INDUCED DIABETES IN RODENTS SERVE AS AN ANIMAL MODEL OF DIABETES?

Good animal models of diabetes are crucial for our understanding of the pathogenic mechanisms of diabetes and its complications [199–203]. As rapamycin is known to be β-cell toxic [204] and diabetogenic [12,121,205] and chronic mTOR inhibition by rapamycin creates a diabetic state reflected by elevated levels of blood glucose and β-cell functional impairment [6], it is conceivable that chronic mTOR inhibition by rapamycin could serve as a good animal model of diabetes. Although it has been reported that diabetes created as such is reversible after rapamycin withdrawal [127], the underlying mechanisms have not been well delineated. Nevertheless, if this model proves to be useful, it would lend a great platform for not only the study of the mTOR signaling pathway in diabetes, but also for screening potential antidiabetic compounds such as polyphenols and phytochemicals derived from plants. Such a rapamycin diabetes model will also complement those existing rodent models including streptozotocin- and alloxan-induced diabetes models [203,206–211].

22.14 SUMMARY AND PERSPECTIVES

Chronic mTOR inhibition by rapamycin can induce development of diabetes. Diabetic hyperglycemia can activate non-conventional glucose utilization pathways such as the polyol pathway, the advanced glycation pathway, and the hexosamine pathway that all eventually lead to elevated levels of ROS that can damage DNA, lipids, and proteins, resulting in progressive worsening of insulin resistance and impairment of β-cell function. The key event in diabetes is glucotoxicity [212] driven by the redox imbalance between NADH and NAD+ [24,25,59] due to glucose oversupply that activates the NADH-generating polyol pathway and the NAD+-depleting enzyme poly ADP ribose polymerase. As mitochondrial complex I is directly responsible for NAD+ regeneration from NADH, its dysregulation by chronic mTOR inhibition may be implicated in rapamycin diabetes. Elucidating the mechanisms of mitochondria and complex I dysfunction in rapamycin-induced diabetes may help better utilization of rapamycin as a therapeutic agent.

References

[1] Tremblay F, Jacques H, Marette A. Modulation of insulin action by dietary proteins and amino acids: role of the mammalian target of rapamycin nutrient sensing pathway. Curr Opin Clin Nutr Metab Care 2005;8(4):457–62.

[2] Tremblay F, Gagnon A, Veilleux A, Sorisky A, Marette A. Activation of the mammalian target of rapamycin pathway acutely inhibits insulin signaling to Akt and glucose transport in 3T3-L1 and human adipocytes. Endocrinology 2005;146 (3):1328–37.

[3] McDaniel ML, Marshall CA, Pappan KL, Kwon G. Metabolic and autocrine regulation of the mammalian target of rapamycin by pancreatic beta-cells. Diabetes 2002;51(10):2877–85.

[4] Wullschleger S, Loewith R, Hall MN. TOR signaling in growth and metabolism. Cell 2006;124(3):471–84.

[5] Ashworth RE, Wu J. Mammalian target of rapamycin inhibition in hepatocellular carcinoma. World J Hepatol 2014;6 (11):776–82.

[6] Leibowitz G, Cerasi E, Ketzinel-Gilad M. The role of mTOR in the adaptation and failure of beta-cells in type 2 diabetes. Diabetes Obes Metab 2008;10(Suppl 4):157–69.

[7] Cai Z, Chen G, He W, Xiao M, Yan LJ. Activation of mTOR: a culprit of Alzheimer's disease? Neuropsychiatr Dis Treat 2015;11:1015–30.

[8] Stallone G, Infante B, Grandaliano G, Gesualdo L. Management of side effects of sirolimus therapy. Transplantation 2009;87 (8 Suppl):S23–6.

[9] Rao RD, Buckner JC, Sarkaria JN. Mammalian target of rapamycin (mTOR) inhibitors as anti-cancer agents. Curr Cancer Drug Targets 2004;4(8):621–35.

[10] Sehgal SN, Molnar-Kimber K, Ocain TD, Weichman BM. Rapamycin: a novel immunosuppressive macrolide. Med Res Rev 1994;14(1):1–22.

[11] Romagnoli J, Citterio F, Nanni G, Favi E, Tondolo V, Spagnoletti G, et al. Incidence of posttransplant diabetes mellitus in kidney transplant recipients immunosuppressed with sirolimus in combination with cyclosporine. Transplant Proc 2006;38(4):1034–6.

[12] Johnston O, Rose CL, Webster AC, Gill JS. Sirolimus is associated with new-onset diabetes in kidney transplant recipients. J Am Soc Nephrol 2008;19(7):1411–18.

[13] Teutonico A, Schena PF, Di Paolo S. Glucose metabolism in renal transplant recipients: effect of calcineurin inhibitor withdrawal and conversion to sirolimus. J Am Soc Nephrol 2005;16 (10):3128–35.

[14] Michan S. Calorie restriction and NAD(+)/sirtuin counteract the hallmarks of aging. Front Biosci (Landmark Ed) 2014;19:1300–19.

[15] Cerutti R, Pirinen E, Lamperti C, Marchet S, Sauve AA, Li W, et al. NAD(+)-dependent activation of Sirt1 corrects the phenotype in a mouse model of mitochondrial disease. Cell Metab 2014;19(6):1042–9.

[16] Caton PW, Richardson SJ, Kieswich J, Bugliani M, Holland ML, Marchetti P, et al. Sirtuin 3 regulates mouse pancreatic beta cell function and is suppressed in pancreatic islets isolated from human type 2 diabetic patients. Diabetologia 2013;56(5):1068–77.

[17] Chiarugi A, Dolle C, Felici R, Ziegler M. The NAD metabolome—a key determinant of cancer cell biology. Nat Rev Cancer 2012;12(11):741–52.

[18] Yang T, Sauve AA. NAD metabolism and sirtuins: metabolic regulation of protein deacetylation in stress and toxicity. AAPS J 2006;8(4):E632–43.

[19] Yang H, Yang T, Baur JA, Perez E, Matsui T, Carmona JJ, et al. Nutrient-sensitive mitochondrial NAD+ levels dictate cell survival. Cell 2007;130(6):1095–107.

[20] Yan LJ. Pathogenesis of chronic hyperglycemia: from reductive stress to oxidative stress. J Diabetes Res 2014;2014:137919.

[21] Luo X, Li R, Yan LJ. Roles of pyruvate, NADH, and mitochondrial complex I in redox balance and imbalance in β cell function and dysfunction. J Diabetes Res 2015.

[22] Felici R, Lapucci A, Cavone L, Pratesi S, Berlinguer-Palmini R, Chiarugi A. Pharmacological NAD-boosting strategies improve mitochondrial homeostasis in human complex I-mutant fibroblasts. Mol Pharmacol 2015;87(6):965–71.

[23] Pittelli M, Felici R, Pitozzi V, Giovannelli L, Bigagli E, Cialdai F, et al. Pharmacological effects of exogenous NAD on mitochondrial bioenergetics, DNA repair, and apoptosis. Mol Pharmacol 2011;80(6):1136–46.

[24] Bensellam M, Laybutt DR, Jonas JC. The molecular mechanisms of pancreatic beta-cell glucotoxicity: recent findings and future research directions. Mol Cell Endocrinol 2012;364(1-2):1–27.

[25] Del Prato S. Role of glucotoxicity and lipotoxicity in the pathophysiology of Type 2 diabetes mellitus and emerging treatment strategies. Diabet Med 2009;26(12):1185–92.

[26] Dedoussis GV, Kaliora AC, Panagiotakos DB. Genes, diet and type 2 diabetes mellitus: a review. Rev Diabet Stud 2007;4(1):13–24.

[27] Somesh BP, Verma MK, Sadasivuni MK, Mammen-Oommen A, Biswas S, Shilpa PC, et al. Chronic glucolipotoxic conditions in pancreatic islets impair insulin secretion due to dysregulated calcium dynamics, glucose responsiveness and mitochondrial activity. BMC Cell Biol 2013;14:31.

[28] Leibowitz G, Kaiser N, Cerasi E. Beta-cell failure in type 2 diabetes. J Diabetes Invest 2011;2(2):82–91.

[29] Leibowitz G, Bachar E, Shaked M, Sinai A, Ketzinel-Gilad M, Cerasi E, et al. Glucose regulation of beta-cell stress in type 2 diabetes. Diabetes Obes Metab 2010;12(Suppl. 2):66–75.

[30] Cai D. Neuroinflammation and neurodegeneration in overnutrition-induced diseases. Trends Endocrinol Metab 2013;24(1):40–7.

[31] Gupta D, Krueger CB, Lastra G. Over-nutrition, obesity and insulin resistance in the development of beta-cell dysfunction. Curr Diabetes Rev 2012;8(2):76–83.

[32] Prentki M, Nolan CJ. Islet beta cell failure in type 2 diabetes. J Clin Invest 2006;116(7):1802–12.

[33] Maiese K. mTOR: driving apoptosis and autophagy for neurocardiac complications of diabetes mellitus. World J Diabetes 2015;6(2):217–24.

[34] Funk SD, Yurdagul Jr. A, Orr AW. Hyperglycemia and endothelial dysfunction in atherosclerosis: lessons from type 1 diabetes. Int J Vasc Med 2012;2012:569654.

[35] Eiselein L, Schwartz HJ, Rutledge JC. The challenge of type 1 diabetes mellitus. ILAR J 2004;45(3):231–6.

[36] Tuch B, Dunlop M, Proietto J. Diabetes research: a guide for postgraduates. The Netherlands: Harwood Academic Publishers; 2000.

[37] Larsen MO. Beta-cell function and mass in type 2 diabetes. Dan Med Bull 2009;56(3):153–64.

[38] Butler AE, Janson J, Bonner-Weir S, Ritzel R, Rizza RA, Butler PC. Beta-cell deficit and increased beta-cell apoptosis in humans with type 2 diabetes. Diabetes 2003;52(1):102–10.

[39] Abdul-Ghani MA, DeFronzo RA. Oxidative stress in type 2 diabetes. In: Miwa S, Beckman KB, Muller FL, editors. Oxidative stress in aging. Totowa, New Jersey, USA: Humana Press; 2008. p. 191–212.

[40] DeFronzo RA. Insulin resistance: a multifaceted syndrome responsible for NIDDM, obesity, hypertension, dyslipidaemia and atherosclerosis. Neth J Med 1997;50(5):191–7.

[41] Kaiser N, Leibowitz G, Nesher R. Glucotoxicity and beta-cell failure in type 2 diabetes mellitus. J Pediatr Endocrinol Metab 2003;16(1):5–22.

[42] Poitout V, Robertson RP. Minireview: secondary beta-cell failure in type 2 diabetes—a convergence of glucotoxicity and lipotoxicity. Endocrinology 2002;143(2):339–42.

[43] Roseman HM. Progression from obesity to type 2 diabetes: lipotoxicity, glucotoxicity, and implications for management. J Manag Care Pharm 2005;11(6 Suppl B):S3–11.

[44] Dukes ID, McIntyre MS, Mertz RJ, Philipson LH, Roe MW, Spencer B, et al. Dependence on NADH produced during glycolysis for beta-cell glucose signaling. J Biol Chem 1994;269(15):10979–12.

[45] Eto K, Suga S, Wakui M, Tsubamoto Y, Terauchi Y, Taka J, et al. NADH shuttle system regulates K(ATP) channel-dependent pathway and steps distal to cytosolic Ca(2+) concentration elevation in glucose-induced insulin secretion. J Biol Chem 1999;274(36):25386–92.

[46] Eto K, Tsubamoto Y, Terauchi Y, Sugiyama T, Kishimoto T, Takahashi N, et al. Role of NADH shuttle system in glucose-induced activation of mitochondrial metabolism and insulin secretion. Science 1999;283(5404):981–5.

[47] Wamelink MM, Struys EA, Jakobs C. The biochemistry, metabolism and inherited defects of the pentose phosphate pathway: a review. J Inherited Metab Dis 2008;31(6):703–17.

[48] Chung SS, Ho EC, Lam KS, Chung SK. Contribution of polyol pathway to diabetes-induced oxidative stress. J Am Soc Nephrol 2003;14(8 Suppl 3):S233–6.

[49] Chung SS, Chung SK. Aldose reductase in diabetic microvascular complications. Curr Drug Targets 2005;6(4):475–86.

[50] Tang WH, Martin KA, Hwa J. Aldose reductase, oxidative stress, and diabetic mellitus. Front Pharmacol 2012;3:87.

[51] Hodgkinson AD, Sondergaard KL, Yang B, Cross DF, Millward BA, Demaine AG. Aldose reductase expression is induced by hyperglycemia in diabetic nephropathy. Kidney Int 2001;60(1):211–18.

[52] Beyer AM, Weihrauch D. Hexosamine pathway activation and O-linked-N-acetylglucosamine: novel mediators of endothelial dysfunction in hyperglycemia and diabetes. Vascul Pharmacol 2012;56(3-4):113–14.

[53] Yanagida K, Maejima Y, Santoso P, Otgon-Uul Z, Yang Y, Sakuma K, et al. Hexosamine pathway but not interstitial changes mediates glucotoxicity in pancreatic beta-cells as assessed by cytosolic Ca2+ response to glucose. Aging 2014;6(3):207–14.

[54] Nowotny K, Jung T, Hohn A, Weber D, Grune T. Advanced glycation end products and oxidative stress in type 2 diabetes mellitus. Biomolecules 2015;5(1):194–222.

[55] Derubertis FR, Craven PA. Activation of protein kinase C in glomerular cells in diabetes. Mechanisms and potential links to the pathogenesis of diabetic glomerulopathy. Diabetes 1994;43(1):1–8.

[56] Robertson RP. Chronic oxidative stress as a central mechanism for glucose toxicity in pancreatic islet beta cells in diabetes. J Biol Chem 2004;279(41):42351–4.

[57] Brownlee M. Biochemistry and molecular cell biology of diabetic complications. Nature 2001;414(6865):813–20.

[58] Giacco F, Brownlee M. Oxidative stress and diabetic complications. Circ Res 2010;107(9):1058–70.

[59] Watson JD. Type 2 diabetes as a redox disease. Lancet 2014;383(9919):841–3.

[60] Lee AY, Chung SS. Contributions of polyol pathway to oxidative stress in diabetic cataract. FASEB J 1999;13(1):23–30.

[61] Lo AC, Cheung AK, Hung VK, Yeung CM, He QY, Chiu JF, et al. Deletion of aldose reductase leads to protection against cerebral ischemic injury. J Cereb Blood Flow Metab 2007;27(8):1496–509.

[62] Ng TF, Lee FK, Song ZT, Calcutt NA, Lee AY, Chung SS, et al. Effects of sorbitol dehydrogenase deficiency on nerve conduction in experimental diabetic mice. Diabetes 1998;47(6):961–6.

[63] Tang WH, Wu S, Wong TM, Chung SK, Chung SS. Polyol pathway mediates iron-induced oxidative injury in ischemic-reperfused rat heart. Free Radic Biol Med 2008;45(5):602–10.

[64] Yabe-Nishimura C. Aldose reductase in glucose toxicity: a potential target for the prevention of diabetic complications. Pharmacol Rev 1998;50(1):21–33.

[65] Bagnasco SM, Murphy HR, Bedford JJ, Burg MB. Osmoregulation by slow changes in aldose reductase and rapid changes in sorbitol flux. Am J Physiol 1988;254(6 Pt 1): C788–92.

[66] Bagnasco SM, Uchida S, Balaban RS, Kador PF, Burg MB. Induction of aldose reductase and sorbitol in renal inner medullary cells by elevated extracellular NaCl. Proc Natl Acad Sci U S A 1987;84(6):1718–20.

[67] Burg MB. Coordinate regulation of organic osmolytes in renal cells. Kidney Int 1996;49(6):1684–5.

[68] Burg MB, Ferraris JD. Intracellular organic osmolytes: function and regulation. J Biol Chem 2008;283(12):7309–13.

[69] Fantus IG. The pathogenesis of the chronic complications of the diabetes mellitus. Endocrinol Rounds 2002;2(4):1–8.

[70] Mouchiroud L, Houtkooper RH, Auwerx J. NAD(+) metabolism: a therapeutic target for age-related metabolic disease. Crit Rev Biochem Mol Biol 2013;48(4):397–408.

[71] Du X, Matsumura T, Edelstein D, Rossetti L, Zsengeller Z, Szabo C, et al. Inhibition of GAPDH activity by poly(ADP-ribose) polymerase activates three major pathways of hyperglycemic damage in endothelial cells. J Clin Invest 2003;112 (7):1049–57.

[72] Dolle C, Rack JG, Ziegler M. NAD and ADP-ribose metabolism in mitochondria. FEBS J 2013;280(15):3530–41.

[73] Pieper AA, Brat DJ, Krug DK, Watkins CC, Gupta A, Blackshaw S, et al. Poly(ADP-ribose) polymerase-deficient mice are protected from streptozotocin-induced diabetes. Proc Natl Acad Sci U S A 1999;96(6):3059–64.

[74] Masutani M, Suzuki H, Kamada N, Watanabe M, Ueda O, Nozaki T, et al. Poly(ADP-ribose) polymerase gene disruption conferred mice resistant to streptozotocin-induced diabetes. Proc Natl Acad Sci U S A 1999;96(5):2301–4.

[75] Yasunari K, Kohno M, Kano H, Yokokawa K, Horio T, Yoshikawa J. Aldose reductase inhibitor prevents hyperproliferation and hypertrophy of cultured rat vascular smooth muscle cells induced by high glucose. Arterioscler Thromb Vasc Biol 1995;15(12):2207–12.

[76] Gomes AP, Price NL, Ling AJ, Moslehi JJ, Montgomery MK, Rajman L, et al. Declining NAD(+) induces a pseudohypoxic state disrupting nuclear-mitochondrial communication during aging. Cell 2013;155(7):1624–38.

[77] Ido Y, Williamson JR. Hyperglycemic cytosolic reductive stress 'pseudohypoxia': implications for diabetic retinopathy. Invest Ophthalmol Vis Sci 1997;38(8):1467–70.

[78] Williamson JR, Chang K, Frangos M, Hasan KS, Ido Y, Kawamura T, et al. Hyperglycemic pseudohypoxia and diabetic complications. Diabetes 1993;42(6):801–13.

[79] Teodoro JS, Rolo AP, Palmeira CM. The NAD ratio redox paradox: why does too much reductive power cause oxidative stress? Toxicol Mech Methods 2013;23(5):297–302.

[80] Williamson JR, Kilo C, Ido Y. The role of cytosolic reductive stress in oxidant formation and diabetic complications. Diabetes Res Clin Pract 1999;45(2-3):81–2.

[81] Lipinski B. Evidence in support of a concept of reductive stress. Br J Nutr 2002;87(1):93–4 discussion 94

[82] Cai Z, Yan LJ. Rapamycin, autophagy, and Alzheimer's disease. J Biochem Pharmacol Res 2013;1(2):84–90.

[83] Cai Z, Zhao B, Li K, Zhang L, Li C, Quazi SH, et al. Mammalian target of rapamycin: a valid therapeutic target through the autophagy pathway for Alzheimer's disease? J Neurosci Res 2012;90(6):1105–18.

[84] Parkhitko AA, Favorova OO, Khabibullin DI, Anisimov VN, Henske EP. Kinase mTOR: regulation and role in maintenance of cellular homeostasis, tumor development, and aging. Biochemistry (Mosc) 2014;79(2):88–101.

[85] Kim SG, Buel GR, Blenis J. Nutrient regulation of the mTOR complex 1 signaling pathway. Mol Cells 2013;35(6):463–73.

[86] Albert V, Hall MN. mTOR signaling in cellular and organismal energetics. Curr Opin Cell Biol 2015;33:55–66.

[87] Perluigi M, Di Domenico F, Butterfield DA. mTOR signaling in aging and neurodegeneration: at the crossroad between metabolism dysfunction and impairment of autophagy. Neurobiol Dis 2015.

[88] Haissaguerre M, Saucisse N, Cota D. Influence of mTOR in energy and metabolic homeostasis. Mol Cell Endocrinol 2014;397(1-2):67–77.

[89] Asnaghi L, Bruno P, Priulla M, Nicolin A. mTOR: a protein kinase switching between life and death. Pharmacol Res 2004;50(6):545–9.

[90] Balasubramanian S, Johnston RK, Moschella PC, Mani SK, Tuxworth Jr WJ, Kuppuswamy D. mTOR in growth and protection of hypertrophying myocardium. Cardiovasc Hematol Agents Med Chem 2009;7(1):52–63.

[91] Howell JJ, Manning BD. mTOR couples cellular nutrient sensing to organismal metabolic homeostasis. Trends Endocrinol Metab 2011;22(3):94–102.

[92] Calamaras TD, Lee C, Lan F, Ido Y, Siwik DA, Colucci WS. The lipid peroxidation product 4-hydroxy-trans-2-nonenal causes protein synthesis in cardiac myocytes via activated mTORC1-p70S6K-RPS6 signaling. Free Radic Biol Med 2015;82:137–46.

[93] Perl A. mTOR activation is a biomarker and a central pathway to autoimmune disorders, cancer, obesity, and aging. Ann N Y Acad Sci 2015;1346(1):33–44.

[94] Blagosklonny MV. TOR-centric view on insulin resistance and diabetic complications: perspective for endocrinologists and gerontologists. Cell Death Dis 2013;4:e964.

[95] Nguyen LH, Brewster AL, Clark ME, Regnier-Golanov A, Sunnen CN, Patil VV, et al. mTOR inhibition suppresses established epilepsy in a mouse model of cortical dysplasia. Epilepsia 2015;56(4):636–46.

[96] Fernandez D, Perl A. mTOR signaling: a central pathway to pathogenesis in systemic lupus erythematosus? Discov Med 2010;9(46):173–8.

[97] Agulnik M. New developments in mammalian target of rapamycin inhibitors for the treatment of sarcoma. Cancer 2012;118 (6):1486–97.

[98] Powers 3rd RW, Kaeberlein M, Caldwell SD, Kennedy BK, Fields S. Extension of chronological life span in yeast by decreased TOR pathway signaling. Genes Dev 2006;20 (2):174–84.

[99] Robida-Stubbs S, Glover-Cutter K, Lamming DW, Mizunuma M, Narasimhan SD, Neumann-Haefelin E, et al. TOR signaling and rapamycin influence longevity by regulating SKN-1/Nrf and DAF-16/FoxO. Cell Metab 2012;15(5):713–24.

[100] Bjedov I, Toivonen JM, Kerr F, Slack C, Jacobson J, Foley A, et al. Mechanisms of life span extension by rapamycin in the fruit fly Drosophila melanogaster. Cell Metab 2010;11(1): 35–46.

[101] Harrison DE, Strong R, Sharp ZD, Nelson JF, Astle CM, Flurkey K, et al. Rapamycin fed late in life extends lifespan in genetically heterogeneous mice. Nature 2009;460(7253):392–5.

[102] Miller RA, Harrison DE, Astle CM, Fernandez E, Flurkey K, Han M, et al. Rapamycin-mediated lifespan increase in mice is

dose and sex dependent and metabolically distinct from dietary restriction. Aging Cell 2014;13(3):468–77.

[103] Vodenik B, Rovira J, Campistol JM. Mammalian target of rapamycin and diabetes: what does the current evidence tell us? Transplant Proc 2009;41(6 Suppl):S31–8.

[104] Kaplan B, Qazi Y, Wellen JR. Strategies for the management of adverse events associated with mTOR inhibitors. Transplant Rev (Orlando) 2014;28(3):126–33.

[105] Lamming DW. Diminished mTOR signaling: a common mode of action for endocrine longevity factors. Springerplus 2014;3:735.

[106] Ehninger D, Neff F, Xie K. Longevity, aging and rapamycin. Cell Mol Life Sci 2014;71(22):4325–46.

[107] Yang F, Chu X, Yin M, Liu X, Yuan H, Niu Y, et al. mTOR and autophagy in normal brain aging and caloric restriction ameliorating age-related cognition deficits. Behav Brain Res 2014;264:82–90.

[108] Gharibi B, Farzadi S, Ghuman M, Hughes FJ. Inhibition of Akt/mTOR attenuates age-related changes in mesenchymal stem cells. Stem Cells 2014;32(8):2256–66.

[109] Carr TD, Feehan RP, Hall MN, Ruegg MA, Shantz LM. Conditional disruption of rictor demonstrates a direct requirement for mTORC2 in skin tumor development and continued growth of established tumors. Carcinogenesis 2015;36 (4):487–97.

[110] Cheng H, Walls M, Baxi SM, Yin MJ. Targeting the mTOR pathway in tumor malignancy. Curr Cancer Drug Targets 2013;13(3):267–77.

[111] Habib SL, Liang S. Hyperactivation of Akt/mTOR and deficiency in tuberin increased the oxidative DNA damage in kidney cancer patients with diabetes. Oncotarget 2014;5 (9):2542–50.

[112] Edlind MP, Hsieh AC. PI3K-AKT-mTOR signaling in prostate cancer progression and androgen deprivation therapy resistance. Asian J Androl 2014;16(3):378–86.

[113] Laplante M, Sabatini DM. mTOR signaling in growth control and disease. Cell 2012;149(2):274–93.

[114] Maiese K, Chong ZZ, Shang YC, Wang S. Targeting disease through novel pathways of apoptosis and autophagy. Expert Opin Ther Targets 2012;16(12):1203–14.

[115] Xu Y, Liu C, Chen S, Ye Y, Guo M, Ren Q, et al. Activation of AMPK and inactivation of Akt result in suppression of mTOR-mediated S6K1 and 4E-BP1 pathways leading to neuronal cell death in in vitro models of Parkinson's disease. Cell Signal 2014;26(8):1680–9.

[116] Sarkar S. Regulation of autophagy by mTOR-dependent and mTOR-independent pathways: autophagy dysfunction in neurodegenerative diseases and therapeutic application of autophagy enhancers. Biochem Soc Trans 2013;41 (5):1103–30.

[117] Xie J, Herbert TP. The role of mammalian target of rapamycin (mTOR) in the regulation of pancreatic beta-cell mass: implications in the development of type-2 diabetes. Cell Mol Life Sci 2012;69(8):1289–304.

[118] Lieberthal W, Levine JS. The role of the mammalian target of rapamycin (mTOR) in renal disease. J Am Soc Nephrol 2009;20 (12):2493–502.

[119] Lloberas N, Cruzado JM, Franquesa M, Herrero-Fresneda I, Torras J, Alperovich G, et al. Mammalian target of rapamycin pathway blockade slows progression of diabetic kidney disease in rats. J Am Soc Nephrol 2006;17(5):1395–404.

[120] Schindler CE, Partap U, Patchen BK, Swoap SJ. Chronic rapamycin treatment causes diabetes in male mice. Am J Physiol Regul Integr Comp Physiol 2014;307(4):R434–43.

[121] Fraenkel M, Ketzinel-Gilad M, Ariav Y, Pappo O, Karaca M, Castel J, et al. mTOR inhibition by rapamycin prevents beta-cell adaptation to hyperglycemia and exacerbates the metabolic state in type 2 diabetes. Diabetes 2008;57(4):945–57.

[122] Blagosklonny MV. Rapamycin-induced glucose intolerance: hunger or starvation diabetes. Cell Cycle 2011;10(24):4217–24.

[123] Yang SB, Lee HY, Young DM, Tien AC, Rowson-Baldwin A, Shu YY, et al. Rapamycin induces glucose intolerance in mice by reducing islet mass, insulin content, and insulin sensitivity. J Mol Med (Berl) 2012;90(5):575–85.

[124] Krebs M, Brunmair B, Brehm A, Artwohl M, Szendroedi J, Nowotny P, et al. The mammalian target of rapamycin pathway regulates nutrient-sensitive glucose uptake in man. Diabetes 2007;56(6):1600–7.

[125] Houde VP, Brule S, Festuccia WT, Blanchard PG, Bellmann K, Deshaies Y, et al. Chronic rapamycin treatment causes glucose intolerance and hyperlipidemia by upregulating hepatic gluconeogenesis and impairing lipid deposition in adipose tissue. Diabetes 2010;59(6):1338–48.

[126] Deblon N, Bourgoin L, Veyrat-Durebex C, Peyrou M, Vinciguerra M, Caillon A, et al. Chronic mTOR inhibition by rapamycin induces muscle insulin resistance despite weight loss in rats. Br J Pharmacol 2012;165(7):2325–40.

[127] Liu Y, Diaz V, Fernandez E, Strong R, Ye L, Baur JA, et al. Rapamycin-induced metabolic defects are reversible in both lean and obese mice. Aging 2014;6(9):742–54.

[128] Rozengurt E. Mechanistic target of rapamycin (mTOR): a point of convergence in the action of insulin/IGF-1 and G protein-coupled receptor agonists in pancreatic cancer cells. Front Physiol 2014;5:357.

[129] Nyman E, Rajan MR, Fagerholm S, Brannmark C, Cedersund G, Stralfors P. A single mechanism can explain network-wide insulin resistance in adipocytes from obese patients with type 2 diabetes. J Biol Chem 2014;289(48):33215–30.

[130] Kleinert M, Sylow L, Fazakerley DJ, Krycer JR, Thomas KC, Oxboll AJ, et al. Acute mTOR inhibition induces insulin resistance and alters substrate utilization in vivo. Molecular metabolism 2014;3(6):630–41.

[131] Cheng Z, Tseng Y, White MF. Insulin signaling meets mitochondria in metabolism. Trends Endocrinol Metab 2010;21 (10):589–98.

[132] Donati A, Recchia G, Cavallini G, Bergamini E. Effect of aging and anti-aging caloric restriction on the endocrine regulation of rat liver autophagy. J Gerontol A Biol Sci Med Sci 2008;63 (6):550–5.

[133] Karlsson E, Perez-Tenorio G, Amin R, Bostner J, Skoog L, Fornander T, et al. The mTOR effectors 4EBP1 and S6K2 are frequently coexpressed, and associated with a poor prognosis and endocrine resistance in breast cancer: a retrospective study including patients from the randomised Stockholm tamoxifen trials. Breast Cancer Res 2013;15(5):R96.

[134] Tucci P. Caloric restriction: is mammalian life extension linked to p53? Aging 2012;4(8):525–34.

[135] Huang K, Fingar DC. Growing knowledge of the mTOR signaling network. Semin Cell Dev Biol 2014;36:79–90.

[136] Ueno M, Carvalheira JB, Tambascia RC, Bezerra RM, Amaral ME, Carneiro EM, et al. Regulation of insulin signalling by hyperinsulinaemia: role of IRS-1/2 serine phosphorylation and the mTOR/p70 S6K pathway. Diabetologia 2005;48(3):506–18.

[137] Jia G, Aroor AR, Martinez-Lemus LA, Sowers JR. Overnutrition, mTOR signaling, and cardiovascular diseases. Am J Physiol Regul Integr Comp Physiol 2014;307(10):R1198–206.

[138] Tremblay F, Marette A. Amino acid and insulin signaling via the mTOR/p70 S6 kinase pathway. A negative feedback

mechanism leading to insulin resistance in skeletal muscle cells. J Biol Chem 2001;276(41):38052−60.

[139] Inoki K, Zhu T, Guan KL. TSC2 mediates cellular energy response to control cell growth and survival. Cell 2003;115 (5):577−90.

[140] Inoki K, Kim J, Guan KL. AMPK and mTOR in cellular energy homeostasis and drug targets. Annu Rev Pharmacol Toxicol 2012;52:381−400.

[141] Andreux PA, Houtkooper RH, Auwerx J. Pharmacological approaches to restore mitochondrial function. Nat Rev Drug Discov 2013;12(6):465−83.

[142] Morita M, Gravel SP, Hulea L, Larsson O, Pollak M, St-Pierre J, et al. mTOR coordinates protein synthesis, mitochondrial activity and proliferation. Cell Cycle 2015;14(4):473−80.

[143] Cunningham JT, Rodgers JT, Arlow DH, Vazquez F, Mootha VK, Puigserver P. mTOR controls mitochondrial oxidative function through a YY1-PGC-1alpha transcriptional complex. Nature 2007;450(7170):736−40.

[144] Churbanova IY, Sevrioukova IF. Redox-dependent changes in molecular properties of mitochondrial apoptosis-inducing factor. J Biol Chem 2008;283(9):5622−31.

[145] Chiu J, Dawes IW. Redox control of cell proliferation. Trends Cell Biol 2012;22(11):592−601.

[146] Iwata K, Nishinaka T, Matsuno K, Kakehi T, Katsuyama M, Ibi M, et al. The activity of aldose reductase is elevated in diabetic mouse heart. J Pharmacol Sci 2007;103(4):408−16.

[147] Yasunari K, Kohno M, Kano H, Minami M, Yoshikawa J. Aldose reductase inhibitor improves insulin-mediated glucose uptake and prevents migration of human coronary artery smooth muscle cells induced by high glucose. Hypertension 2000;35(5):1092−8.

[148] Virag L, Szabo C. The therapeutic potential of poly(ADP-ribose) polymerase inhibitors. Pharmacol Rev 2002;54 (3):375−429.

[149] Oka S, Hsu CP, Sadoshima J. Regulation of cell survival and death by pyridine nucleotides. Circ Res 2012;111(5):611−27.

[150] Sharma LK, Fang H, Liu J, Vartak R, Deng J, Bai Y. Mitochondrial respiratory complex I dysfunction promotes tumorigenesis through ROS alteration and AKT activation. Hum Mol Genet 2011;20(23):4605−16.

[151] Cooper JM, Mann VM, Krige D, Schapira AH. Human mitochondrial complex I dysfunction. Biochim Biophys Acta 1992;1101(2):198−203.

[152] Carroll J, Fearnley IM, Skehel JM, Shannon RJ, Hirst J, Walker JE. Bovine complex I is a complex of 45 different subunits. J Biol Chem 2006;281(43):32724−7.

[153] Andrews B, Carroll J, Ding S, Fearnley IM, Walker JE. Assembly factors for the membrane arm of human complex I. Proc Natl Acad Sci U S A 2013;110(47):18934−9.

[154] Hirst J. Mitochondrial complex I. Annu Rev Biochem 2013;82:551−75.

[155] Hirst J, Carroll J, Fearnley IM, Shannon RJ, Walker JE. The nuclear encoded subunits of complex I from bovine heart mitochondria. Biochim Biophys Acta 2003;1604(3):135−50.

[156] Carroll J, Ding S, Fearnley IM, Walker JE. Post-translational modifications near the quinone binding site of mammalian complex I. J Biol Chem 2013;288(34):24799−808.

[157] Santidrian AF, Matsuno-Yagi A, Ritland M, Seo BB, LeBoeuf SE, Gay LJ, et al. Mitochondrial complex I activity and NAD+/NADH balance regulate breast cancer progression. J Clin Invest 2013;123(3):1068−81.

[158] Pryde KR, Hirst J. Superoxide is produced by the reduced flavin in mitochondrial complex I: a single, unified mechanism that applies during both forward and reverse electron transfer. J Biol Chem 2011;286(20):18056−65.

[159] Ferrer-Sueta G, Radi R. Chemical biology of peroxynitrite: kinetics, diffusion, and radicals. ACS Chem Biol 2009;4 (3):161−77.

[160] Goldstein S, Merenyi G. The chemistry of peroxynitrite: implications for biological activity. Methods Enzymol 2008;436:49−61.

[161] Ames BN, Shigenaga MK. Oxidants are a major contributor to aging. Ann N Y Acad Sci 1992;663:85−96.

[162] Liu SS. Mitochondrial Q cycle-derived superoxide and chemiosmotic bioenergetics. Ann N Y Acad Sci 2010;1201:84−95.

[163] Cramer WA, Hasan SS, Yamashita E. The Q cycle of cytochrome bc complexes: a structure perspective. Biochim Biophys Acta 2011;1807(7):788−802.

[164] Turrens JF, Alexandre A, Lehninger AL. Ubisemiquinone is the electron donor for superoxide formation by complex III of heart mitochondria. Arch Biochem Biophys 1985;237 (2):408−14.

[165] Siebels I, Drose S. Q-site inhibitor induced ROS production of mitochondrial complex II is attenuated by TCA cycle dicarboxylates. Biochim Biophys Acta 2013;1827(10):1156−64.

[166] Yan LJ, Levine RL, Sohal RS. Oxidative damage during aging targets mitochondrial aconitase. Proc Natl Acad Sci USA 1997;94:11168−72.

[167] Venditti P, Di Stefano L, Di Meo S. Mitochondrial metabolism of reactive oxygen species. Mitochondrion 2013;13(2):71−82.

[168] Croston TL, Thapa D, Holden AA, Tveter KJ, Lewis SE, Shepherd DL, et al. Functional deficiencies of subsarcolemmal mitochondria in the type 2 diabetic human heart. Am J Physiol Heart Circ Physiol 2014;307(1):H54−65.

[169] Sharma V, Sharma I, Singh VP, Verma S, Pandita A, Singh V, et al. mtDNA G10398A variation provides risk to type 2 diabetes in population group from the Jammu region of India. Meta Gene 2014;2:269−73.

[170] Brandon M, Baldi P, Wallace DC. Mitochondrial mutations in cancer. Oncogene 2006;25(34):4647−62.

[171] Sullivan PG, Brown MR. Mitochondrial aging and dysfunction in Alzheimer's disease. Prog Neuropsychopharmacol Biol Psychiatry 2005;29(3):407−10.

[172] Kim SH, Vlkolinsky R, Cairns N, Fountoulakis M, Lubec G. The reduction of NADH ubiquinone oxidoreductase 24- and 75-kDa subunits in brains of patients with Down syndrome and Alzheimer's disease. Life Sci 2001;68(24):2741−50.

[173] Schapira AH. Human complex I defects in neurodegenerative diseases. Biochim Biophys Acta 1998;1364(2):261−70.

[174] Zhou Q, Liu C, Liu W, Zhang H, Zhang R, Liu J, et al. Rotenone induction of hydrogen peroxide inhibits mTOR-mediated S6K1 and 4E-BP1/eIF4E pathways, leading to neuronal apoptosis. Toxicol Sci 2015;143(1):81−96.

[175] Marella M, Seo BB, Nakamaru-Ogiso E, Greenamyre JT, Matsuno-Yagi A, Yagi T. Protection by the NDI1 gene against neurodegeneration in a rotenone rat model of Parkinson's disease. PLoS One 2008;3(1):e1433.

[176] Schapira AH, Cooper JM, Dexter D, Clark JB, Jenner P, Marsden CD. Mitochondrial complex I deficiency in Parkinson's disease. J Neurochem 1990;54(3):823−7.

[177] Perier C, Bove J, Vila M, Przedborski S. The rotenone model of Parkinson's disease. Trends Neurosci 2003;26(7):345−6.

[178] Huypens P, Pillai R, Sheinin T, Schaefer S, Huang M, Odegaard ML, et al. The dicarboxylate carrier plays a role in mitochondrial malate transport and in the regulation of glucose-stimulated insulin secretion from rat pancreatic beta cells. Diabetologia 2011;54(1):135−45.

[179] Antinozzi PA, Ishihara H, Newgard CB, Wollheim CB. Mitochondrial metabolism sets the maximal limit of fuel-stimulated insulin secretion in a model pancreatic beta cell: a

survey of four fuel secretagogues. J Biol Chem 2002;277 (14):11746—55.

[180] Maechler P. Mitochondrial function and insulin secretion. Mol Cell Endocrinol 2013;379(1-2):12—18.

[181] Reyes A, Cardenas ML. All hexokinase isoenzymes coexist in rat hepatocytes. Biochem J 1984;221(2):303—9.

[182] Efrat S, Tal M, Lodish HF. The pancreatic beta-cell glucose sensor. Trends Biochem Sci 1994;19(12):535—8.

[183] Matschinsky F, Liang Y, Kesavan P, Wang L, Froguel P, Velho G, et al. Glucokinase as pancreatic beta cell glucose sensor and diabetes gene. J Clin Invest 1993;92(5):2092—8.

[184] Matschinsky FM. Regulation of pancreatic beta-cell glucokinase: from basics to therapeutics. Diabetes 2002;51(Suppl. 3): S394—404.

[185] Maechler P, Wollheim CB. Mitochondrial signals in glucose-stimulated insulin secretion in the beta cell. J Physiol 2000;529 (Pt 1):49—56.

[186] Cline GW. Fuel-stimulated insulin secretion depends upon mitochondria activation and the integration of mitochondrial and cytosolic substrate cycles. Diabetes Metab J 2011;35 (5):458—65.

[187] MacDonald PE, Joseph JW, Rorsman P. Glucose-sensing mechanisms in pancreatic beta-cells. Philos Trans R Soc Lond B Biol Sci 2005;360(1464):2211—25.

[188] Prentki M, Matschinsky FM, Madiraju SR. Metabolic signaling in fuel-induced insulin secretion. Cell Metab 2013;18 (2):162—85.

[189] Satin LS, Butler PC, Ha J, Sherman AS. Pulsatile insulin secretion, impaired glucose tolerance and type 2 diabetes. Mol Aspects Med 2015.

[190] Zou CY, Gong Y, Liang J. Metabolic signaling of insulin secretion by pancreatic beta-cell and its derangement in type 2 diabetes. Eur Rev Med Pharm Sci 2014;18(15):2215—27.

[191] Matsuzaki S, Humphries KM. Selective inhibition of deactivated mitochondrial complex I by biguanides. Biochemistry 2015;54(11):2011—21.

[192] Murphy MP. How mitochondria produce reactive oxygen species. Biochem J 2009;417(1):1—13.

[193] Triepels RH, Van Den Heuvel LP, Trijbels JM, Smeitink JA. Respiratory chain complex I deficiency. Am J Med Genet 2001;106(1):37—45.

[194] Hirst J, King MS, Pryde KR. The production of reactive oxygen species by complex I. Biochem Soc Trans 2008;36(Pt 5):976—80.

[195] Ying W. NAD+/NADH and NADP+/NADPH in cellular functions and cell death: regulation and biological consequences. Antioxid Redox Signal 2008;10(2):179—206.

[196] Xu HN, Tchou J, Chance B, Li LZ. Imaging the redox states of human breast cancer core biopsies. Adv Exp Med Biol 2013;765:343—9.

[197] Hoppel CL, Kerr DS, Dahms B, Roessmann U. Deficiency of the reduced nicotinamide adenine dinucleotide dehydrogenase component of complex I of mitochondrial electron transport. Fatal infantile lactic acidosis and hypermetabolism with skeletal-cardiac myopathy and encephalopathy. J Clin Invest 1987;80(1):71—7.

[198] Buckler KJ. Effects of exogenous hydrogen sulphide on calcium signalling, background (TASK) K channel activity and mitochondrial function in chemoreceptor cells. Pflugers Arch 2012;463(5):743—54.

[199] Wei M, Ong L, Smith MT, Ross FB, Schmid K, Hoey AJ, et al. The streptozotocin-diabetic rat as a model of the chronic complications of human diabetes. Heart Lung Circ 2003;12 (1):44—50.

[200] Skovso S. Modeling type 2 diabetes in rats using high fat diet and streptozotocin. J Diabetes Invest 2014;5(4):349—58.

[201] Sakata N, Yoshimatsu G, Tsuchiya H, Egawa S, Unno M. Animal models of diabetes mellitus for islet transplantation. Exp Diabetes Res 2012;2012:256707.

[202] Ghasemi A, Khalifi S, Jedi S. Streptozotocin-nicotinamide-induced rat model of type 2 diabetes (review). Acta Physiol Hung 2014;101(4):408—20.

[203] Wu J, Yan LJ. Streptozotocin-induced type 1 diabetes in rodents as a model for studying mitochondrial mechanisms of diabetic beta cell glucotoxicity. Diabetes Metab Syndr Obes 2015;8:181—8.

[204] Barlow AD, Nicholson ML, Herbert TP. Evidence for rapamycin toxicity in pancreatic beta-cells and a review of the underlying molecular mechanisms. Diabetes 2013;62 (8):2674—82.

[205] Chang GR, Wu YY, Chiu YS, Chen WY, Liao JW, Hsu HM, et al. Long-term administration of rapamycin reduces adiposity, but impairs glucose tolerance in high-fat diet-fed KK/HlJ mice. Basic Clin Pharmacol Toxicol 2009;105(3):188—98.

[206] Wu MS, Liang JT, Lin YD, Wu ET, Tseng YZ, Chang KC. Aminoguanidine prevents the impairment of cardiac pumping mechanics in rats with streptozotocin and nicotinamide-induced type 2 diabetes. Br J Pharmacol 2008;154(4):758—64.

[207] Yin D, Tao J, Lee DD, Shen J, Hara M, Lopez J, et al. Recovery of islet beta-cell function in streptozotocin- induced diabetic mice: an indirect role for the spleen. Diabetes 2006;55 (12):3256—63.

[208] Junod A, Lambert AE, Stauffacher W, Renold AE. Diabetogenic action of streptozotocin: relationship of dose to metabolic response. J Clin Invest 1969;48(11):98.

[209] Wu J, Luo X, Yan LJ. Two dimensional blue native/SDS-PAGE to identify mitochondrial complex I subunits modified by 4-hydroxynonenal (HNE). Front Physiol 2015;6:98.

[210] Szkudelski T. The mechanism of alloxan and streptozotocin action in B cells of the rat pancreas. Physiol Res 2001;50 (6):537—46.

[211] Lenzen S. The mechanisms of alloxan- and streptozotocin-induced diabetes. Diabetologia 2008;51(2):216—26.

[212] Weir GC, Marselli L, Marchetti P, Katsuta H, Jung MH, Bonner-Weir S. Towards better understanding of the contributions of overwork and glucotoxicity to the beta-cell inadequacy of type 2 diabetes. Diabetes Obes Metab 2009;11(Suppl. 4):82—90.

23

mTOR in Diabetic Nephropathy and Retinopathy

Rosa Fernandes[1,2] *and Flávio Reis*[1,2]

[1]Laboratory of Pharmacology and Experimental Therapeutics, Institute for Biomedical Imaging and Life Sciences (IBILI), Faculty of Medicine, University of Coimbra, Coimbra, Portugal [2]Center for Neuroscience and Cell Biology, Institute for Biomedical Imaging and Life Sciences (CNC.IBILI) Consortium, University of Coimbra, Coimbra, Portugal

23.1 INTRODUCTION

There were 347 million people with diabetes all over the world in 2013, and WHO estimates that diabetes will be the seventh leading cause of death by 2030 [1]. The rising numbers of diabetes prevalence are mainly fueled by the modern lifestyle associated with the economic growth and industrialization, leading to sedentary habits related to mechanized transport and lack of free time, as well as unhealthy diets, with a nutritional transition to processed high-calorie foods. The major consequence of these changes is an excessive accumulation of body fat due to the imbalance between energy intake and expenditure. With the worldwide increase in the prevalence of diabetes in recent years, diabetic micro- and macrovascular complications, including diabetic nephropathy (DN) and diabetic retinopathy (DR), are responsible for worrying rates of morbidity and mortality in diabetic patients, with significant social and economic costs [2].

DN, a common complication in patients with both type 1 and type 2 diabetes mellitus (T1DM and T2DM), is the leading cause of end-stage renal disease (ESRD) in many regions of the world, including in the United States and Western Europe; in fact, approximately one-third of all diabetic individuals are affected by DN [3,4]. Simultaneously, DR remains the leading cause of blindness in the working-age population of most developed countries, affecting nearly 90% of patients with T1DM and more than 60% of people with T2DM 20 years disease onset [5,6].

The precise causes underlying the initiation and progression of both diabetic complications remain to be fully elucidated, but various factors have been postulated, including genetic predisposition, age, obesity, hyperglycemia, hypertension, and dyslipidemia [7,8]. There are no treatment strategies available that specifically target the pathogenesis of both DN and DR, instead they control high blood pressure and glucose levels, in an effort to alleviate the problem. Nonetheless, mounting evidence suggests that various interconnecting biochemical pathways are the key links between hyperglycemia and the progression of these complications, including inflammation, oxidative stress, hypoxia, angiogenesis, and proliferation/fibrosis [7,8]. Therefore, many studies have attempted to elucidate the myriad abnormalities in molecular signaling pathways that may combine to produce these pathologic changes associated with DN and DR, so that new and more effective therapies and preventative strategies might be developed [9–11]. Through these efforts, the general understanding of the pathogenic signaling factors that lead to progressive DN and DR has expanded considerably over the past decade.

Mammalian target of rapamycin (mTOR) lies at the center of an intricate signaling network with serious implications for a large number of diverse and important biological events, including development, metabolism, regenerative processes, and aging, in response to distinct stimuli, including nutrients, mitogens, hormones (namely insulin), and cytokines [12,13]. Moreover, deregulated mTOR activity plays

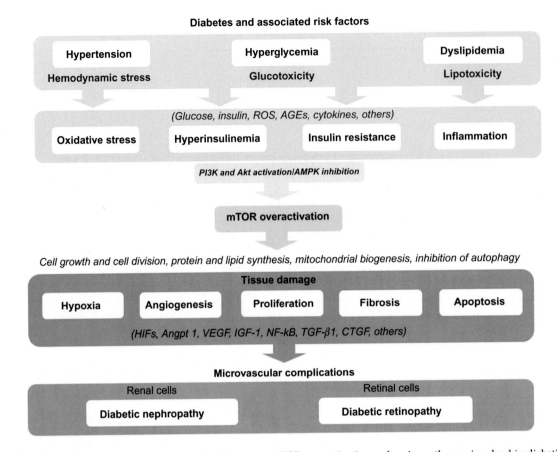

FIGURE 23.1 Schematic diagram of the proposed links between mTOR overactivation and major pathways involved in diabetic nephropathy and retinopathy pathophysiology. AGEs, advanced glycation end products; Akt, protein kinase B; AMPK, AMP-activated protein kinase; Angpt 1, angiopoietin 1; CTGF, connective tissue growth factor; HIFs, hypoxia-inducible factors; IGF-1, insulin-like growth factor-1; mTOR, mammalian target of rapamycin; NF-κB, nuclear factor kappa B; PI3K, phosphoinositide 3-kinase; ROS, reactive oxygen species; TGF-β1, transforming growth factor beta 1; VEGF, vascular endothelial growth factor.

a central role in the progression of several highly prevalent diseases including metabolic syndrome, diabetes, and cancer [14–16]. Therefore, it is not surprising that elucidation of the pathways governing the activity and effects of mTOR is currently a rapidly expanding area of research. Activation of mTOR in diabetes is due, at least in part, to the effects of hyperglycemia, by the combined effects of Akt (or protein kinase B) activation and inhibition of AMP-activated protein kinase (AMPK) [9]. There is now increasing evidence that mTOR signaling pathways are interrelated with several mechanisms involved in DN and DR progression, including oxidative stress, inflammation, hypoxia, proliferation, angiogenesis, and proliferation/fibrosis (Figure 23.1), in such a way that could be viewed as a potential drug target to both DN and DR.

This chapter reviews the mTOR signaling pathway in the kidney and retina, focusing on the role of mTOR overactivation for the progression of DN and DR, as well as the possible therapeutic implications of its modulation.

23.2 OVERVIEW OF mTOR COMPONENTS AND SIGNALING PATHWAYS

mTOR, a serine/threonine protein kinase, member of the phosphatydilinositol 3-kinase (PI3K) cell survival pathway, is one of the key players of cellular metabolism that is coupled with nutrient abundance and energy [12,13]. mTOR regulates cell growth and metabolism in response to nutrients (aminoacids, glucose and fatty acids), mitogens (like insulin-like growth factor-1 [IGF-1] and vascular endothelial growth factor [VEGF]), hormones (namely insulin), and cytokines [17–20]. mTOR interacts with several proteins to form two complexes, known as mTOR complex 1 (mTORC1) and 2 (mTORC2), that differ in their composition, regulation, and functions [21]. mTORC1 associates with Raptor (regulatory-associated protein of mTOR), whereas mTORC2 associates with Rictor (rapamycin-insensitive companion of mTOR), as well as with other proteins. Both multiprotein complexes include mLST8 (mammalian lethal with Sec 13 protein 8) and the negative regulator Deptor (DEP-domain containing mTOR-interacting protein). In addition, mTORC1 associates with the inhibitory protein PRAS40 (proline-rich Akt substrate of 40 kDa), which itself is inhibited by Akt. mTORC2 requires mSIN1 (mammalian stress-activated protein kinase-interacting protein) for its function [21]. Rapamycin binding to FKBP12 (FK506-binding protein of 12 kDa) acutely inhibits the function of mTOR, predominantly of mTORC1.

The mTOR-containing complexes also have different sensitivities to rapamycin as well as upstream and downstream signaling networks. mTORC1 is rapamycin-sensitive and is generally associated with cell growth and division [22], controlling many important cellular processes, including protein and lipid synthesis, mitochondrial biogenesis, and autophagy [23]. Its effect is largely mediated through the downstream effector proteins, the p70S6 kinases (S6K1 and S6K2), the EIF4EBP1 (eukaryotic translation initiation factor 4E-binding protein-1), lipin 1, and the PGC-1α (peroxisome proliferator activated receptor gamma coactivator 1 alpha), leading to the translation of mRNAs and the synthesis of proteins necessary for cell growth, cell cycle proliferation, and metabolism [24,25]. Akt activates mTORC1 indirectly by phosphorylation and inactivation of TSC2 (tuberous sclerosis protein 2), which suppresses the activity of the Rheb GTPase, an activator of mTORC1. The upstream modulators of mTORC2 are as yet almost unknown, but insulin and related pathways have been suggested as the main activators. mTORC2 activation is mediated by the phosphorylation of Akt and other AGC-family kinases such as serum- and glucocorticoid-induced protein kinase (SGK) and protein kinases C (PKCs), playing an important role in cell survival and organization of the cytoskeleton [22,26].

Impaired mTOR activity has been implicated in the pathogenesis of many human diseases, including cancer, obesity, type 2 diabetes, and neurodegeneration [14–16,27], and accumulated evidence suggests the involvement in the pathophysiology of DN and DR, as further discussed in this chapter.

23.3 mTOR SIGNALING IN DN

23.3.1 Overview of DN Pathophysiology

DN is typically defined by macroalbuminuria (>300 mg in a 24-h collection) or microalbuminuria and abnormal renal function, accessed by serum creatinine and clearance, as well as by glomerular filtration rate (GFR). The main clinical features are progressive albuminuria, decline in GFR, hypertension, and the high risk of cardiovascular morbidity and mortality [7].

DN pathogenesis is complex and multifactorial, involving the direct effect of hyperglycemia in glomerular, tubular, vascular, and interstitial renal cells. Besides hyperglycemia, genetic predisposition, age, obesity, hypertension, and dyslipidemia can also contribute to the induction and/or progression of DN. Kidney fibrosis, clinically starting by an increase in GFR and microalbuminuria, is a pivotal pathological process. However, it is now known that oxidative stress, inflammation, and fibrosis are the key links in the development of DN [28–32]. Chronic hypoxia, another important factor in DN development, causes cellular stress, including oxidative, nitrosative, carbonyl, and of the endoplasmic reticulum (ER) [33]. Under conditions of hyperglycemia and hypertension, diabetes induces excessive production of various reactive oxygen species (ROS), which can cause direct and indirect damage to the renal interstitium and, consequently, renal vascular sclerosis, increased vascular permeability, and structural and functional kidney damage. The major sources of ROS are the activation of nicotinamide adenine dinucleotide phosphate (NADP$^+$) and PKC, the increased formation of advanced glycation end products and the polyol pathway [34–37]. ROS stimulates several downstream mediators, particularly extracellular regulated protein kinases, p38 mitogen-activated protein kinases (p38 MAPK), nuclear factor kappa B (NF-κB) and activator protein-1 (AP-1), whose cellular responses contribute to the development of DN [30]. In addition, various fibrogenic cytokines, including the tumor necrosis factor alpha (TNF-α) and the interleukin 1 beta (IL-1β), are released by inflammatory cells (lymphocytes, macrophages, dendritic cells, and mast cells), evoking epithelial-to-mesenchymal transition (EMT), extracellular matrix (ECM) accumulation, and kidney damage [31,38].

Kidney fibrosis, which is characterized by thickening of basement membrane, glomerular hypertrophy, and expansion of mesangial matrix, is promoted by oxidative stress and inflammation [39,40]. Insistent injury, caused by hypoxia, oxidative stress, and/or inflammation, induces the release of profibrotic cytokines by inflammatory cells and the expression of α-smooth muscle actin by activated fibroblasts, thus contributing to excessive production and deposition of ECM in the renal tissue. In addition, fibroblast-derived transforming growth factor beta (TGF-β) activates the Smad pathways, which upregulates the transcription of target genes and the production of excessive ECM, thus accelerating mesangial expansion and kidney fibrosis [32,41]. The vicious cycle of renal oxidative stress–inflammation–fibrosis may contribute to the progression of DN to ESRD.

Despite the knowledge achieved during recent years, much remains to be discovered regarding the precise molecular pathways underlying kidney tissue dysfunction and lesions, which will be crucial to develop strategies able to efficiently combat chronic kidney disease (CKD). Growing evidence exists that the mTOR pathway plays a pivotal role in the mechanisms associated with the progression of CKD evoked by diabetes or by other causes [42,43]. Some of the strongest evidences have been obtained from a number of studies using animal models, in which mTOR inhibition by rapamycin significantly ameliorated renal inflammation and fibrosis, improving renal function. The following sections highlight the participation of mTOR in kidney physiology and pathophysiology, focusing on the role that mTOR overexpression might play in DN.

23.3.2 mTOR Signaling in the Kidney

Despite the longstanding use of mTOR inhibitors as immunosuppressive agents to prevent allograft rejection in transplanted patients, as well as an anticancer agent against renal cell carcinoma, the precise roles played by mTOR in the distinct renal cells remain to be entirely clarified, regardless of the advances of the last years.

It has been clear for many years that mTOR has a fundamental role in the regulation of two fundamental cellular events: cell proliferation and cell growth. mTOR acts as a metabolic sensor, monitoring the cell and its immediate environment for acceptability of the conditions and ingredients required for proliferation and growth [42,44]. While growth factors, nutrients, cellular energy stores, and oxygen availability are factors that stimulate cell growth and proliferation, requiring mTOR activation, conditions such as deficiency of cellular energy stores or DNA damage are not advantageous to proliferation and growth, and are accompanied by suppression of mTOR activity [44]. Accumulated evidence suggests that mTOR, particularly the mTORC1 complex, might play a role in both glomerular and tubulointerstitial renal mechanisms, during glomerular development as well as in adulthood renal functions [42,43].

Podocyte constitutes the most vulnerable part of all kidney compartments. A cascade of injuries occurs after kidney insults (including metabolic, immunological, or toxic), starting with foot process effacement, podocyte loss, and eventually glomerulosclerosis [45], which is accompanied by evolution of proteinuria. Although podocyte regeneration was previously viewed as a rare clinical condition [46], recent findings, mainly obtained from transgenic models, have suggested the existence of mechanisms of podocyte adaptation [47,48]. Podocyte hypertrophy, in particular, has received particular attention, and mTOR has been

indicated as the key regulator of podocyte size control [49–51]. mTORC1 signaling deficiency during glomerular development was associated with smaller podocytes compared with wild-type littermates, causing a condition resembling secondary focal segmental glomerulosclerosis (FSGS) [50]. A transgenic rat model expressing dominant negative EIF4EBP1, which acts downstream of mTORC1 to modulate cap-dependent translation and cell hypertrophy, reinforced the hypothesis and clearly indicated that mTOR signaling sets the threshold for the adaptive growth capacity of podocytes [51]. Thus, mTOR-dependent podocyte size control seems to be able to compensate for podocyte losses after injury [52]. Interestingly, mTOR inhibition by rapamycin causes podocyte damage and proteinuria, as reported in transplanted patients under immunosuppressive therapy with this mTOR inhibitor, as well as in animal models [53–60], thus suggesting that in situations where mTORC1 sustains an adaptive compensatory action in response to any underlying glomerular stress/injury, rapamycin might present a "second hit" leading to proteinuria.

Parietal epithelial cells, the most abundant cell type within glomerular crescents, have been viewed as potential progenitor cells of podocytes [61]. It has been suggested that mTORC1 could regulate cell growth and division underlying the proliferative response that occurs during crescent formation in several forms of glomerulonephritis; in particular, it has been speculated that mTORC1 could mediate effects downstream of the epidermal growth factor receptor, namely hypertrophy and proliferation [62].

In contrast to mTORC1, very little is known regarding the potential function of mTORC2 in glomerular function, which might be explained by the fact that rapamycin has long been regarded as a highly specific mTORC1 inhibitor. However, recent work indicates that prolonged application of rapamycin can also inhibit mTORC2 [22,63,64], thus hypothesizing that some of the functions that have been only attributed to mTORC1 might also have been mediated by mTORC2. Therefore, recent research points to a role for mTORC2 in podocyte survival, as suggested by the abolished mTORC2-Akt2 survival signal in patients receiving rapamycin, thus contributing to the deleterious effects of rapamycin in some clinical settings [65]. In addition, studies using genetic models of podocyte-specific mTORC2 and/or mTORC1 KO have suggested that mTORC2 plays an important role in podocyte stress surveillance and survival [50]. Reinforcing this supposition, it has been demonstrated that mTORC2 and its downstream target Akt2 are essential for survival of remaining podocytes in response to nephron reduction [65]. In fact, the mTORC2-Akt2 activation could also be observed in biopsy tissue from kidney transplant patients, and rapamycin was able to prevent the activation of mTORC2-Akt2 in these patients, accompanied by increased glomerular apoptosis [65]. Thus, in opposition to what was previously expected, mTORC2-Akt2 seems to also play a role in podocyte survival, contributing to rapamycin-induced proteinuria, in concert with mTORC1.

Further research is needed to unravel the exact mechanisms underlying the glomerular (podocyte and/or parietal epithelial-cell-specific) actions of mTORC1 and mTORC2, as well as how they could be therapeutically modulated to improve the management of renal diseases and reduce glomerular side effects and proteinuria reported with the mTOR inhibitors currently available.

Even though mTORC1 is a well-known drug target for renal transplantation and renal cell carcinoma, much remains to be discovered concerning the physiological role of mTORC1 and mTORC2 in kidney tubules. Patients treated with sirolimus presented with hypophosphatemia and hypokalemia [66], but mTORC1 inhibition in vivo does not seem to influence the apical phosphate reabsorption machinery consisting of NaPi-IIa, NaPi-IIc, and Pit-2 in the proximal tubule [67,68]. As phosphaturia has been viewed as a consistent finding, one could hypothesize whether mTORC1 is able to modulate the basolateral efflux pathways in proximal tubular cells or affects a so-far-unidentified hormonal component of phosphate homeostasis. mTORC1 ablation in the proximal tubule might assist in unraveling the correct possibility. An inhibition of mTORC1s anabolic action could be speculated, although undoubted confirmation is still needed. In addition, owing to the lack of specific mTORC2 inhibitors there is no consistent information concerning its in vivo role in renal tubules, and only could be speculated on its putative in vivo effects based on in vitro data. It has been suggested that mTORC2 might be involved in the regulation of Na^+ balance [69,70], which will be clinically relevant for salt-sensitive hypertension or volume overload occurring with congestive heart failure, but this needs to be properly proved in vivo.

23.3.3 Overactivation of mTOR in DN

DN is clinically characterized by gradual worsening of albuminuria, together with GFR decline, progressively leading to ESRD, in a process that involves glomerular podocyte damage and loss, followed sequentially by fibrosis of renal glomerulus and of tubulointerstitial region cells.

All kidney cell types are affected by diabetes, including podocytes and mesangial, endothelial, and tubulointerstitial cells. The relevance of podocyte damage in the formation and progression of DN has been reinforced during the past decade, due to a broad range of molecular studies. Nevertheless, complementary perspectives on the pathogenesis of DN have been reported. In fact, endothelial damage within the glomerulus seems to be already present when podocyte damage starts, in a stage of normoalbuminuria. In addition, in contrast to what was previously thought (a resultant effect of glomerular protein leakage), it seems that tubulointerstitial injury plays a crucial role in DN progression. Actually, the glomerular filtration barrier and tubulointerstitial compartment are now viewed as a combined dynamic entity where a lesion in one cell type is reflected in the other, thus causing dysfunction of the tissue/organ, viewed as an entire unit.

Several studies have focused on the elucidation of the molecular signaling mechanisms underlying the development of DN, in order to identify drug targets and effective therapies. Increased expression of mesangial cell matrix protein or reduced matrix metalloproteinase expression have been implicated in DN. In addition, signaling abnormalities may contribute to podocyte damage and loss, an early event in DN that further causes glomerulosclerosis [71,72]. Indeed, injured podocytes transmit signals to the mesangial cells to react increasing ECM synthesis or decreasing degradation of damaged podocytes. In fact, mesangial expansion is as good a predictor of progressive nephropathy in T2DM patients [73]. In addition, the crucial role of glomerular endothelial cells in the pathogenesis of glomerulosclerosis has been reinforced in recent years, as suggested by the existence of a considerable crosstalk between endothelial and mesangial cells. Lastly, signaling abnormalities in the tubulointerstitial cells most probably contribute to fibrosis and induce EMT [74]. Figure 23.2 overviews the putative pathways linking mTOR overactivation and DN initiation and/or development.

The mTOR signaling pathway in the kidney has been increasingly studied and new exciting findings could have a major therapeutic impact in several conditions, including in DN [75]. In fact, mounting evidence from experimental and clinical studies supports a role for mTOR in the pathogenesis of DN, but the mechanism by which mTOR is activated remains under debate, particularly in T2DM patients.

In a general point of view, under conditions of excessive calorie intake mTOR is activated, before the development of overt hyperglycemia. Chronic ingestion of excess nutrients has been shown to induce a persistent activation of mTOR, followed by development of obesity and insulin resistance [76,77]. Both mTORC1 and mTORC2 promote fat deposition in white adipose tissue as well as in the liver and muscle [78,79]. This is reinforced by rapamycin-induced hyperlipidemia, attributed to accumulation of lipids in

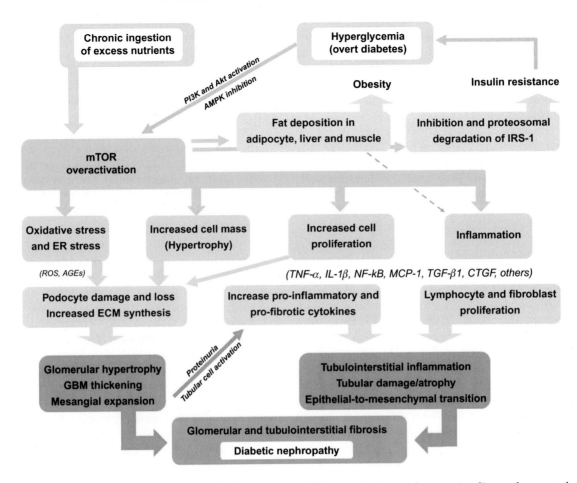

FIGURE 23.2 Schematic diagram of the proposed links between mTOR overactivation and major signaling pathways and molecular mediators governing diabetic nephropathy initiation and progression. AGEs, advanced glycation end products; Akt, protein kinase B; AMPK, AMP-activated protein kinase; CTGF, connective tissue growth factor; ECM, extracellular matrix; GBM, glomerular basement membrane; IRS-1, insulin receptor substrate-1; MCP-1, monocyte chemoattractant protein-1; mTOR, mammalian target of rapamycin; NF-κB, nuclear factor kappa B; PI3K, phosphoinositide 3-kinase; ROS, reactive oxygen species; TGF-β1, transforming growth factor beta 1; TNF-α, tumor necrosis factor alpha; IL-1β, interleukin 1 beta.

the adipose tissue and to increased hepatic synthesis of triglycerides [80,81]. Another consequence of nutrient excess and mTOR overactivity is insulin resistance, which is evoked and/or amplified by a feedback loop on the insulin receptor, leading to inhibition of its downstream effectors and consequent attenuation of insulin-dependent signal transduction; in fact, mTOR inhibits insulin receptor substrate-1 (IRS-1) by phosphorylation of IRS-1 by Raptor, a component of the mTORC1 complex, as well as by proteosomal degradation of IRS-1 [76,82] (Figure 23.2). Once diabetes develops, hyperglycemia further contributes to mTORC1 activation, as viewed by phosphorylation of S6K1/2 and 4E-BP1, which seems to be mediated, at least partially, by inhibition of AMPK as well as by activation of PI3K and Akt, as suggested by distinct *in vitro* and *in vivo* animal studies [83,84].

Regarding specifically kidney tissue, mTOR activation seems to be associated with both the glomerular and tubulointerstitial pathological features of DN. Concerning diabetic glomerulosclerosis, overactivation of mTOR plays a role in the three major glomerular manifestations of DN that culminate in global glomerulosclerosis, namely cellular hypertrophy, glomerular basement membrane (GBM) thickening, and mesangial matrix accumulation (Figure 23.2). Renal enlargement and glomerular hypertrophy (which contribute to podocyte injury that seems to have a primary role in the development of proteinuria and glomerulosclerosis) precede nephron loss and occur in association with an increased GFR. Several studies using animal models of DN have suggested a role for mTORC1, mediated by S6K1, in the pathogenesis of DN, particularly due to renal hypertrophy, which was substantially attenuated by rapamycin [85,86].

It is now known that podocyte hypertrophy is a key and early feature of the glomerular hypertrophy that precedes the irreversible structural changes of glomerulosclerosis in DN [87]. However, while transient mTORC1 activation seems to compensate in certain settings for lost podocytes by increasing the size of the remaining podocytes, persistently activated mTORC1 might cause cell hypertrophy ultimately leading to podocyte loss and glomerulosclerosis. The role of mTOR in podocyte function and development of DN is now better known due to two exciting studies using podocyte-specific genetic deletion of critical components of the mTOR signaling pathway [49,50]. Godel et al. have demonstrated proteinuria developed within a few weeks of birth in mice with a podocyte-specific deletion of both alleles of the gene expressing Raptor, which was followed by progressive glomerulosclerosis [50]. The authors also demonstrated that proteinuria and glomerular disease were less severe in streptozotocin-induced diabetic mice lacking only one podocyte-specific Raptor allele, when compared with the wild-type controls, thus suggesting, on one hand, that complete Raptor absence abolishes podocyte integrity and function and, on the other, that reduction of Raptor expression (i.e., mTORC1 activity) protects mice from DN [50]. A complementary study performed by Inoki et al. has shown that diabetic mice with a podocyte-specific genetic ablation of TSC1, which is a negative regulator of mTORC1, increased mTORC1 activity by podocytes; on the contrary, non-diabetic mice with ablation of podocyte-specific TSC1 presented glomerular disease and many features of DN, including podocyte loss, GBM thickening, mesangial expansion, and proteinuria, together with mislocalization of slit diaphragm proteins and induction of epithelial—mesenchymal transformation [49]. Furthermore, the genetic reduction of podocyte-specific mTORC1 activity by single-allele deletion of Raptor in the spontaneously diabetic mice suppressed the development of DN.

The studies of Godel and Inoki [49,50], using two distinct models of diabetes, showed that podocyte-specific reduction of mTORC1 activity is able to protect mice from DN development, while upregulation of mTORC1 activity in non-diabetic mice caused a glomerular disease closely resembling DN [49,50,88], providing strong evidence that reduction of mTORC1 activity might represent an attractive therapeutic strategy against DN development. However, the finding that elimination of raptor expression in podocytes resulted in significant proteinuria is a weakness of this strategy that deserves further attention. In fact, the development of proteinuria has been described in animal models and in humans under rapamycin treatment [53—60]. In addition, other putative side effects of rapamycin should be considered, including hyperglycemia, insulin resistance, and dyslipidemia, due to changes in glucose and lipid metabolism in pancreas, liver, adipocyte tissue, and muscle, as we and others have previously reported in animal and human studies [89—96].

The precise downstream mechanisms of mTOR signaling in podocytes contributing to podocyte loss are still only very moderately understood. Nevertheless, a direct association between podocyte dysfunction and mTORC1-induced ER stress was reported; in addition, the podocyte loss was abolished in podocyte-specific TSC1 knock-out (KO) mice by the reduction of ER stress [49]. The mTORC1-induced ER stress was also demonstrated in rat models of minimal change disease [97,98]. Additionally, hyperglycemia leads to mTOR-dependent increased NADPH oxidase type 1 and 4 activity in podocytes in vitro [99], suggesting a role of mTOR-dependent cellular stress signaling pathways, namely production of ROS and ER stress. Another important effect of mTORC1 is the inhibition of autophagy, which then leads to degenerative glomerulopathy [100,101], despite evidence that autophagy is not the major downstream pathway of mTOR in mediating podocyte injury and loss.

Kidney tissues from type 2 diabetic db/db mice showed activation of mTORC1 that coincides with renal hypertrophy and matrix expansion, with increased expression of type IV collagen, fibronectin, and laminin [102,103], which was inhibited by rapamycin [104]. The role of mTOR as a mediator of renal changes in DN was also evaluated using streptozotocin-induced diabetic rats that present early activation of mTOR within the kidney; when treated with rapamycin, there was a reduction of mTOR activity and amelioration of the glomerular changes typical of DN, including hypertrophy, basement membrane thickening, and mesangial matrix accumulation, accompanied by a decrease in albuminuria [86,105,106]. Besides DN, rapamycin treatment also ameliorated pathological phenotypes of many different renal diseases in rodent models, including minimal change disease, FSGS, membranous nephropathy, and crescentic glomerulonephritis [97,107—111].

Interstitial fibrosis is a key feature of DN and mTOR is likely to be involved in the tubulointerstitial fibrotic changes associated with DN. In fact, mTOR stimulates the proliferation of fibroblasts and the synthesis of collagen, as well as the expression of profibrotic cytokines, such as TGF-β1 and connective tissue growth factor (CTGF), which are key players in tubulointerstitial damage in DN [86,105,106,112,113]. In addition, mTOR seems to be also involved in the EMT, with fibroblast migration through the tubular basement membrane into the interstitium (Figure 23.2), a process that is inhibited by rapamycin in animal models of DN and of other forms of kidney disease [114—117].

Together with the beneficial effects of rapamycin on the glomerulus, other studies using animal models of DN demonstrated a reduction of interstitial inflammation and fibrosis, independently of changes in blood glucose or blood pressure, as represented by reduced release of proinflammatory (namely monocyte chemotactic protein-1: MCP-1) and of profibrotic cytokines (namely TGF-β1 and CTGF) [75,105,118]. In the streptozotocin-induced model of DN, rapamycin was able to significantly reduce the influx of inflammatory cells, mainly lymphocytes and macrophages [86,105,106]. The anti-inflammatory effect of rapamycin was also viewed by the reduction of release of proinflammatory cytokines and chemokines within the kidney, including MCP-1, RANTES, IL-8, and fractaline [86,105,106].

An increase in protein synthesis occurs not only by stimulation of transcription and translation but also by suppression of inhibitory molecules, including AMPK and Deptor, both inhibitors of mTOR activity, and kinase glycogen synthase 3 beta, that inhibits the activity of eukaryotic initiation factor 2B epsilon, as recently reviewed by Mariappan [119]. However, the signaling mechanisms counteracting prohypertrophic processes remain to be elucidated.

Most studies so far have been based on pharmacological inhibition of mTORC1 by rapamycin, a specific and potent inhibitor of mTORC1. While evidence exists for the activation of mTORC2 in animal models of diabetes, the role of mTORC2 in the pathogenesis of DN remains uncertain. However, recent studies demonstrated that mTORC2 is able to regulate the phosphorylation of the antiapoptotic proteins Akt/PKB and SGK and may promote cell survival, as well as control translation and processing of nascent polypeptides [120–123]. Further research focus on the role of mTORC2 on DN, as well as on the interaction between both complexes, might open new windows and opportunities for therapeutic interventions targeting DN.

23.4 mTOR SIGNALING IN DR

23.4.1 Overview of DR Pathophysiology

The progression rate of DR can be predicted by a fasting plasma glucose level ≥7.0 mmol/L, a 2-hour plasma glucose (2hPG) value ≥11.1 mmol/L in a 75 g oral glucose tolerance test or a glycated hemoglobin (HbA1c) value ≥6.5% [124].

Although a 1% reduction in achieved mean HbA1c is associated with a 37% decrease in the risk of microvascular complications and a 21% decreased risk of death related to diabetes and decreased the duration-specific prevalence of background (non-proliferative) retinopathy, the tight glycemic control and blood pressure control appear not to be enough in attenuating the prevalence of proliferative DR [125]. Both forms, non-proliferative and proliferative DR, result from long-term accumulated damage to the small blood vessels that nourish the retina. Non-proliferative DR is associated at early stages with capillary basement membrane thickening, increased vascular permeability, pericyte and endothelial cell death, capillary occlusion, microaneurysms, leukostasis, hemorrhage, and some extent neuronal cell death [126]. Proliferative DR, the more advanced form of the disease, occurs when new fragile blood vessels begin to grow (neovascularization) in the retina and into the vitreous. These abnormal blood vessels may leak blood into the vitreous in response to a rise in the blood pressure, causing swelling of retinal tissue (diabetic macular edema) and clouding of vision. The new blood vessels can also cause retinal detachment, often associated with severe and irreversible vision loss. Although clinical trials have confirmed a strong relationship between chronic hyperglycemia and the development and progression of DR [127,128], the molecular mechanisms that underlie endothelial dysfunction in diabetes are still unclear.

Several metabolic pathways, namely, the diacylglycerol—PKC, non-enzymatic glycosylation, polyol pathway, hexosamine pathway flux and oxidative stress are activated by hyperglycemia and contribute to both the development and progression of DR [126]. Oxidative stress activation has been associated with increased mitochondrial superoxide production in endothelial cells and inflammatory mediators and dysregulated angiogenesis [129]. It is also linked to apoptosis of pericytes by the induction of the highly reactive oxoaldehyde, methylglyoxal [130]. Furthermore, apotosis of retinal pericytes in diabetics was found to be associated with activation of NF-κB [131].

More recently, DR has been associated with low-grade chronic inflammation and neurodegeneration [132]. In the early stages of the disease, inflammatory mediators promote increased vascular permeability and leukocyte adhesion [132–134]. Proinflammatory cytokines (such as TNF-α and IL-1β) and adhesion molecules were reported to be increased in the retina, vitreous, and serum of diabetic patients and rats [134–138]. As the inflammation persists certain cytokines may accelerate apoptosis of retinal vascular cells by activation of the NF-κB and exacerbate vascular dysfunction [135].

Although the more advanced DR is associated with neovascularization, the disease originates with microvascular occlusion and capillary non-perfusion, which culminate in widespread inner-retinal ischemia. The cellular response to hypoxia involves activation of hypoxia-inducible factor 1 (HIF-1). This transcription factor helps cells to survive and efficiently adapt to low oxygen. However, hyperglycemia disrupts, at least in part, the ability of tissue to respond to hypoxia, by destabilizing HIF-1α [139,140]. HIF-1α induces the expression of several growth factors and genes including VEGF, angiopoietin-2 (Ang-2), VEGF receptor Flt-1 (VEGFR-1), basic fibroblast growth factor, platelet-derived growth factor, nitric oxide synthases (NOS), and IGF-1, which are associated with vessel destabilization and neovascularization [6,140]. Reports have shown that patients with DR present high levels of VEGF in the vitreous and retina [141,142]. This growth factor is considered a key mediator in breakdown of the inner blood—retinal barrier in the early stages of DR and neovascularization in the more advanced proliferative DR. In fact, increased vascular permeability in diabetes involves an increased VEGF level that promotes the phosphorylation of the tight junction protein occludin, leading to its ubiquitination and tight junction trafficking in retinal endothelial cells [6,142,143].

23.4.2 Deregulation of the mTOR Signaling Pathway in DR

Hypoxia plays an important role in the development and progression of diabetic DR. Hypoxic conditions are associated with a rapid accumulation of the transcription factor HIF-1α and induction of hypoxia-inducible genes that are involved in blood—retinal barrier breakdown and angiogenesis, like VEGF. VEGF synthesis is stimulated via the insulin/mTOR [144,145] pathway in retinal pigment epithelial cells [146]. It has also been demonstrated that hyperglycemia induces VEGF protein expression via eIF4E and its binding proteins 4E-BP1 and 4E-BP2 [147]. Diabetes in mice deficient in 4E-BP1 and 4E-BP2 proteins did not induce an increase in VEGF expression. As these proteins are downstream targets of mTORC1, the mTOR pathway seems to be implicated in the retinal changes in DR. Figure 23.3 presents the putative pathways linking mTOR overactivation and DR initiation and/or development.

The binding of VEGF to its receptor tyrosine kinases VEGFR-1, VEGFR-2, and VEGFR-3 activates several signaling cascades, including PKC, MAPK, and Akt [143,148], leading to endothelial proliferation, differentiation, migration, vascular permeability, reorganization of the actin cytoskeleton, and endothelial inflammation (Figure 23.3). Furthermore, the binding of VEGF to VEGFR-2 is also linked to angiogenesis [149]. One of the VEGFR-2 downstream signaling pathways that is activated is PI3K/Akt, which subsequently activates mTOR [150,151]. Silibinin, a mTOR inhibitor, inhibits VEGF secretion in a hypoxiadependent manner through a PI3K/Akt/mTOR pathway, preventing edema and neovascularization in a rat model of age-related macular degeneration [152]. In the same way, mTOR inhibitors could modulate the production and secretion of the retinal vasoactive factor VEGF induced by HIF-1α, controlling abnormal vascular permeability and angiogenesis in DR.

Oxidative stress, a major mechanism in DR pathophysiology, leads to increased mitochondrial superoxide production in endothelial cells and triggers inflammatory mediators and dysregulated angiogenesis [129]. Microglia, the resident immune cells in the retina, serves the first line of defense or disease occurs as they modulate the inflammatory processes. However, microglia act in concert with the activities of neurons, glial, and vascular cells. The activation of microglia in response to neural damage involves proliferative, morphological, immunoreactive, and migratory changes [132]. Studies have suggested that the severity of the microglia response is dependent on the gravity of the neural damage; additionally, the microglial response can be either neural-protective or toxic. Since it is well established that chronic inflammation plays a role in DR and microglia are the resident immune cells in the retina, it is likely that microglia activation may play a role in the neurotoxicity and tissue damage in diabetic retina, by exacerbating the inflammatory state, with the recruitment of leukocytes, causing vascular breakdown and directly inducing glial dysfunction and neuronal cell death through the release of cytotoxic substances [153]. It is well known that hyperglycemia-induced ROS can cause activation of microglia, with translocation of NF-κB to the nucleus, and consequent induction and release of proinflammatory and/or cytotoxic factors, such as TNF-α, IL-1β, IL-6 and IL-8, NO, inducible NOS, cycloxygenase-2, vascular cell adhesion molecule 1, and intercellular adhesion molecule 1. These factors have been associated with increased vascular permeability or leukostasis in DR (Figure 23.3). Yoshida et al. reported activated NF-κB in several retinal cell types, including pericytes, vascular endothelial cells, macrophages, and microglia in a hypoxia-induced mouse model of neovascularization [154]. These authors found that inhibition of NF-κB by the antioxidant resveratrol, a naturally occurring phenol that increases the action of superoxide dismutase, resulted in decreased serum levels of IL-1β, IL-6, and TNF-α. More recently, Zhong et al. [155] have shown that resveratrol inhibited lipopolysaccharideinduced proinflammatory enzymes and proinflammatory cytokines via downregulation phosphorylation of NF-κB, cAMP response element-binding protein, and MAPK family, in a mTOR-dependent manner [156,157]. The immunomodulatory feature of mTOR inhibition could be used to suppress the production of proinflammatory mediators via activation of NF-κB.

Overall, the recognition that mTOR is a key mediator for cytokine-induced inflammatory responses and that it is involved in hypoxia-induced neovascularization suggests that PI3K/Akt/mTOR pathway inhibition would be an attractive treatment option in the management of DR.

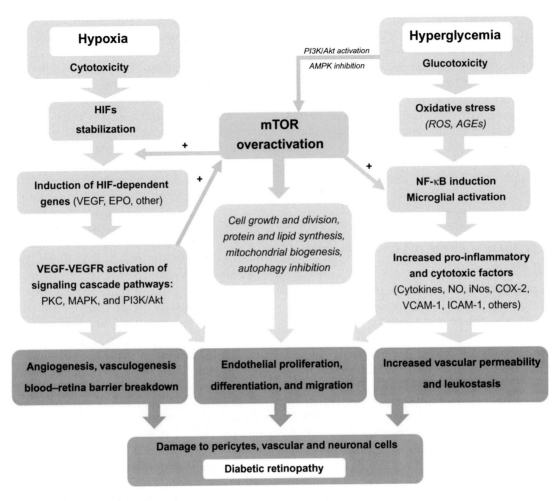

FIGURE 23.3 Schematic diagram of the proposed links between mTOR overactivation and major signaling pathways and molecular mediators governing diabetic retinopathy initiation and progression. AGEs, advanced glycation end products; Akt, protein kinase B; AMPK, AMP-activated protein kinase; COX-2, cycloxygenase-2; EPO, erythropoietin; HIFs, hypoxia-inducible factors; ICAM-1, intercellular adhesion molecule 1; iNOS, inducible nitric oxide synthase; MAPK, mitogen-activated protein kinases; mTOR, mammalian target of rapamycin; NF-κB, nuclear factor kappa B; NO, nitric oxide; PI3K, phosphoinositide 3-kinase; PKC, protein kinase C; ROS, reactive oxygen species; VCAM-1, vascular cell adhesion molecule 1; VEGF, vascular endothelial growth factor; VEGFR, vascular endothelial growth factor receptor.

23.5 CONCLUDING REMARKS AND FUTURE DIRECTIONS

Much progress has been made in recent years to unveil the signaling pathways underlying the development of both DN and DR; however, they are still undoubtedly insufficient to allow considerable progress on prevention and/or treatment of patients. The elucidation of the precise molecular targets and further development of specific drugs to them could significantly improve the treatment based on the current armamentarium of non-specific therapies.

The mTOR signaling pathway is currently a rapidly expanding area of research, which is explained by the exciting discoveries of recent years regarding its involvement in several important biological events, including development, metabolism, regenerative processes, and aging, as well as the possibility of a crucial role played by inappropriate mTOR activity in the progression of several highly prevalent diseases including obesity, diabetes, and cancer. In addition, mTOR overactivation in renal and retinal cells has been viewed as a putative major pathophysiological contribution to initiation and/or progression of DN and DR, both diseases involving excess nutrients and hyperglycemia-driven oxidative stress and inflammation, as well as other mechanisms interlinked with mTOR signaling upstream and downstream regulation, including hypoxia, angiogenesis, proliferation, and fibrosis.

mTOR complexes 1 and 2 (but mainly mTORC1) have been recognized to regulate a variety of renal epithelial processes, including podocyte size (hypertrophy and/or proliferation), EMT, and tubulointerstitial inflammation, which underlies the development of glomerular and tubular damage/fibrosis; in addition, rapamycin has been shown to decrease hyperglycemia-induced increase in mTOR activity, inhibiting or ameliorating renal changes associated with DN, including mesangial expansion, glomerular basement thickening, and release of proinflammatory cytokines or chemokines by monocytes and macrophages; however, rapamycin is associated with some metabolic and renal side effects, namely insulin resistance and proteinuria, which could compromise its widespread use in some conditions. Concurrently, mTOR seems to be also involved in the mechanisms linking retinal cell hypoxia, oxidative stress, inflammation, angiogenesis, and proliferation, and underlying damage to pericytes, vascular, and neuronal cells in DR.

Thus, specific mTOR targeting has emerged as a putative therapeutic strategy to treat both microvascular diabetic complications. However, before effective strategies could be developed and implemented, some important knowledge gaps deserve further elucidation. In particular, further insights are still needed into the upstream regulation of mTOR, the identification of kidney- and retina-specific downstream targets of mTOR, as well as the precise role played by the regulatory proteins that interacts with mTOR in both mTORC1 and mTORC2 complexes, such as Deptor, in order to unveil the impact of the mTORC1–mTORC2 interconnection, thus opening new avenues for management of these diabetic complications.

Acknowledgments

The authors' research is supported by the Portuguese Foundation for Science and Technology (FCT) (Strategic Project Pest-C/SAU/UI3282/2013 and UID/NEU/04539/2013) and COMPETE-FEDER.

References

[1] Danaei G, Finucane MM, Lu Y, Singh GM, Cowan MJ, Paciorek CJ, et al. National, regional, and global trends in fasting plasma glucose and diabetes prevalence since 1980: systematic analysis of health examination surveys and epidemiological studies with 370 country-years and 2.7 million participants. Lancet 2011;378: 31–40.

[2] IDF. Diabetes Atlas. 6th ed. Brussels, Belgium: International Diabetes Federation; 2013.

[3] Forbes JM, Cooper ME. Mechanisms of diabetic complications. Physiol Rev 2013;93:137–88.

[4] Rask-Madsen C, King GL. Vascular complications of diabetes: mechanisms of injury and protective factors. Cell Metab 2013;17:20–33.

[5] Aldebasi YH, Reddy PR, Nair VG, Ahmed MI. Screening for diabetic retinopathy: the optometrist's perspective. Clinical Optometry 2015;7:1–14.

[6] Jacot JL, Sherris D. Potential therapeutic roles for inhibition of the PI3K/Akt/mTOR pathway in the pathophysiology of diabetic retinopathy. J Ophthalmol 2011;2011:589813.

[7] Ahmad J. Management of diabetic nephropathy: recent progress and future perspective. Diabetes Metab Syndr 2015; pii, S1871-S4021.

[8] Safi SZ, Qvist R, Kumar S, Batumalaie K, Ismail IS. Molecular mechanisms of diabetic retinopathy, general preventive strategies, and novel therapeutic targets. Biomed Res Int 2014;2014:801269.

[9] Inoki K, Kim J, Guan KL. AMPK and mTOR in cellular energy homeostasis and drug targets. Annu Rev Pharmacol Toxicol 2012;52:381–400.

[10] Brosius III FC, Alpers CE. New targets for treatment of diabetic nephropathy: what we have learned from animal models. Curr Opin Nephrol Hypertens 2013;22:17–25.

[11] Rangasamy S, McGuire PG, Das A. Diabetic retinopathy and inflammation: novel therapeutic targets. Middle East Afr J Ophthalmol 2012;19:52–9.

[12] Sabatini DM. mTOR and cancer: insights into a complex relationship. Nat Rev Cancer 2006;6:729–34.

[13] Bhaskar PT, Hay N. The two TORCs and Akt. Dev Cell 2007;12:487–502.

[14] Fraenkel M, Ketzinel-Gilad M, Ariav Y, Pappo O, Karaca M, Castel J, et al. mTOR inhibition by rapamycin prevents beta-cell adaptation to hyperglycemia and exacerbates the metabolic state in type 2 diabetes. Diabetes 2008;57:945–57.

[15] Khamzina L, Veilleux A, Bergeron S, Marette A. Increased activation of the mammalian target of rapamycin pathway in liver and skeletal muscle of obese rats: possible involvement in obesity-linked insulin resistance. Endocrinology 2005;146:1473–81.

[16] Foster DA. Phosphatidic acid signaling to mTOR: signals for the survival of human cancer cells. Biochim Biophys Acta 2009;1791:949–55.

[17] Tremblay F, Marette A. Amino acid and insulin signaling via the mTOR/p70 S6 kinase pathway. A negative feedback mechanism leading to insulin resistance in skeletal muscle cells. J Biol Chem 2001;276:38052–60.

[18] Hands SL, Proud CG, Wyttenbach A. mTOR's role in ageing: protein synthesis or autophagy? Aging (Albany NY) 2009;1:586–97.

[19] Cornu M, Albert V, Hall MN. mTOR in aging, metabolism, and cancer. Curr Opin Genet Dev 2013;23:53–62.

[20] Zoncu R, Efeyan A, Sabatini DM. mTOR: from growth signal integration to cancer, diabetes and ageing. Nat Rev Mol Cell Biol 2011;12:21–35.

[21] Wang X, Proud CG. mTORC1 signaling: what we still don't know. J Mol Cell Biol 2011;3:206–20.

[22] Laplante M, Sabatini DM. mTOR signaling in growth control and disease. Cell 2012;149:274–93.

[23] Bar-Peled L, Sabatini DM. Regulation of mTORC1 by amino acids. Trends Cell Biol 2014;24:400–6.

[24] Ichikawa A, Nakahara T, Kurauchi Y, Mori A, Sakamoto K, Ishii K. Rapamycin prevents N-methyl-D-aspartate-induced retinal damage through an ERK-dependent mechanism in rats. J Neurosci Res 2014;92:692–702.

[25] Peterson TR, Sengupta SS, Harris TE, Carmack AE, Kang SA, Balderas E, et al. mTOR complex 1 regulates lipin 1 localization to control the SREBP pathway. Cell 2011;146:408–20.

[26] Parrales A, Lopez E, Lee-Rivera I, Lopez-Colome AM. ERK1/2-dependent activation of mTOR/mTORC1/p70S6K regulates thrombin-induced RPE cell proliferation. Cell Signal 2013;25:829–38.

[27] Sarkar S, Rubinsztein DC. Small molecule enhancers of autophagy for neurodegenerative diseases. Mol Biosyst 2008;4:895–901.

[28] Singh DK, Winocour P, Farrington K. Oxidative stress in early diabetic nephropathy: fueling the fire. Nat Rev Endocrinol 2011;7:176–84.

[29] Navarro-Gonzalez JF, Mora-Fernandez C, Muros de Fuentes M, Garcia-Perez J. Inflammatory molecules and pathways in the pathogenesis of diabetic nephropathy. Nat Rev Nephrol 2011;7:327—40.

[30] Wada J, Makino H. Inflammation and the pathogenesis of diabetic nephropathy. Clin Sci (Lond) 2013;124:139—52.

[31] Liu Y. Cellular and molecular mechanisms of renal fibrosis. Nat Rev Nephrol 2011;7:684—96.

[32] Gomes KB, Rodrigues KF, Fernandes AP. The role of transforming growth factor-beta in diabetic nephropathy. Int J Med Genet 2014;2014, Article ID 180270, 6 pages.

[33] Miyata T, de Strihou C. Diabetic nephropathy: a disorder of oxygen metabolism? Nat Rev Nephrol 2010;6:83—95.

[34] Tojo A, Asaba K, Onozato ML. Suppressing renal NADPH oxidase to treat diabetic nephropathy. Expert Opin Ther Targets 2007;11:1011—18.

[35] Noh H, King GL. The role of protein kinase C activation in diabetic nephropathy. Kidney Int Suppl 2007;S49—53.

[36] Brownlee M. Biochemistry and molecular cell biology of diabetic complications. Nature 2001;414:813—20.

[37] Nishikawa T, Edelstein D, Du XL, Yamagishi S, Matsumura T, Kaneda Y, et al. Normalizing mitochondrial superoxide production blocks three pathways of hyperglycaemic damage. Nature 2000;404:787—90.

[38] Kolset SO, Reinholt FP, Jenssen T. Diabetic nephropathy and extracellular matrix. J Histochem Cytochem 2012;60:976—86.

[39] Mega C, de Lemos ET, Vala H, Fernandes R, Oliveira J, Mascarenhas-Melo F, et al. Diabetic nephropathy amelioration by a low-dose sitagliptin in an animal model of type 2 diabetes (Zucker diabetic fatty rat). Exp Diabetes Res 2011;2011:162092.

[40] Marques C, Mega C, Goncalves A, Rodrigues-Santos P, Teixeira-Lemos E, Teixeira F, et al. Sitagliptin prevents inflammation and apoptotic cell death in the kidney of type 2 diabetic animals. Mediators Inflamm 2014;2014:538737.

[41] Hills CE, Squires PE. The role of TGF-beta and epithelial-to mesenchymal transition in diabetic nephropathy. Cytokine Growth Factor Rev 2011;22:131—9.

[42] Lieberthal W, Levine JS. Mammalian target of rapamycin and the kidney. II. Pathophysiology and therapeutic implications. Am J Physiol Renal Physiol 2012;303:F180—91.

[43] Grahammer F, Wanner N, Huber TB. mTOR controls kidney epithelia in health and disease. Nephrol Dial Transplant 2014;29(Suppl. 1):i9—i18.

[44] Lieberthal W, Levine JS. Mammalian target of rapamycin and the kidney. I. The signaling pathway. Am J Physiol Renal Physiol 2012;303:F1—10.

[45] Tharaux PL, Huber TB. How many ways can a podocyte die? Semin Nephrol 2012;32:394—404.

[46] Grahammer F, Wanner N, Huber TB. Podocyte regeneration: who can become a podocyte? Am J Pathol 2013;183:333—5.

[47] Guo JK, Marlier A, Shi H, Shan A, Ardito TA, Du ZP, et al. Increased tubular proliferation as an adaptive response to glomerular albuminuria. J Am Soc Nephrol 2012;23:429—37.

[48] Wiggins RC. The spectrum of podocytopathies: a unifying view of glomerular diseases. Kidney Int 2007;71:1205—14.

[49] Inoki K, Mori H, Wang J, Suzuki T, Hong S, Yoshida S, et al. mTORC1 activation in podocytes is a critical step in the development of diabetic nephropathy in mice. J Clin Invest 2011;121:2181—96.

[50] Godel M, Hartleben B, Herbach N, Liu S, Zschiedrich S, Lu S, et al. Role of mTOR in podocyte function and diabetic nephropathy in humans and mice. J Clin Invest 2011;121:2197—209.

[51] Fukuda A, Chowdhury MA, Venkatareddy MP, Wang SQ, Nishizono R, Suzuki T, et al. Growth-dependent podocyte failure causes glomerulosclerosis. J Am Soc Nephrol 2012;23:1351—63.

[52] Inoki K, Huber TB. Mammalian target of rapamycin signaling in the podocyte. Curr Opin Nephrol Hypertens 2012;21:251—7.

[53] Amer H, Cosio FG. Significance and management of proteinuria in kidney transplant recipients. J Am Soc Nephrol 2009;20:2490—2.

[54] Torras J, Herrero-Fresneda I, Gulias O, Flaquer M, Vidal A, Cruzado JM, et al. Rapamycin has dual opposing effects on proteinuric experimental nephropathies: is it a matter of podocyte damage? Nephrol Dial Transplant 2009;24:3632—40.

[55] Letavernier E, Bruneval P, Mandet C, Duong Van Huyen JP, Peraldi MN, Helal I, et al. High sirolimus levels may induce focal segmental glomerulosclerosis de novo. Clin J Am Soc Nephrol 2007;2:326—33.

[56] Munivenkatappa R, Haririan A, Papadimitriou JC, Drachenberg CB, Dinits-Pensy M, Klassen DK. Tubular epithelial cell and podocyte apoptosis with de novo sirolimus based immunosuppression in renal allograft recipients with DGF. Histol Histopathol 2010;25:189—96.

[57] Sereno J, Parada B, Rodrigues-Santos P, Lopes PC, Carvalho E, Vala H, et al. Serum and renal tissue markers of nephropathy in rats under immunosuppressive therapy: cyclosporine versus sirolimus. Transplant Proc 2013;45:1149—56.

[58] Sereno J, Nunes S, Rodrigues-Santos P, Vala H, Rocha-Pereira P, Fernandes J, et al. Conversion to sirolimus ameliorates cyclosporine-induced nephropathy in the rat: focus on serum, urine, gene, and protein renal expression biomarkers. Biomed Res Int 2014;2014:576929.

[59] Sereno J, Vala H, Nunes S, Rocha-Pereira P, Carvalho E, Alves R, et al. Cyclosporine A-induced nephrotoxicity is ameliorated by dose reduction and conversion to sirolimus in the rat. J Physiol Pharmacol 2015;66:285—99.

[60] Vogelbacher R, Wittmann S, Braun A, Daniel C, Hugo C. The mTOR inhibitor everolimus induces proteinuria and renal deterioration in the remnant kidney model in the rat. Transplantation 2007;84:1492—9.

[61] Appel D, Kershaw DB, Smeets B, Yuan G, Fuss A, Frye B, et al. Recruitment of podocytes from glomerular parietal epithelial cells. J Am Soc Nephrol 2009;20:333—43.

[62] Chiu T, Santiskulvong C, Rozengurt E. EGF receptor transactivation mediates ANG II-stimulated mitogenesis in intestinal epithelial cells through the PI3-kinase/Akt/mTOR/p70S6K1 signaling pathway. Am J Physiol Gastrointest Liver Physiol 2005;288:G182—94.

[63] Sarbassov DD, Ali SM, Sengupta S, Sheen JH, Hsu PP, Bagley AF, et al. Prolonged rapamycin treatment inhibits mTORC2 assembly and Akt/PKB. Mol Cell 2006;22:159—68.

[64] Lamming DW, Ye L, Katajisto P, Goncalves MD, Saitoh M, Stevens DM, et al. Rapamycin-induced insulin resistance is mediated by mTORC2 loss and uncoupled from longevity. Science 2012;335:1638—43.

[65] Canaud G, Bienaime F, Viau A, Treins C, Baron W, Nguyen C, et al. AKT2 is essential to maintain podocyte viability and function during chronic kidney disease. Nat Med 2013;19:1288—96.

[66] Morales JM, Wramner L, Kreis H, Durand D, Campistol JM, Andres A, et al. Sirolimus does not exhibit nephrotoxicity compared to cyclosporine in renal transplant recipients. Am J Transplant 2002;2:436—42.

[67] Kempe DS, Dermaku-Sopjani M, Frohlich H, Sopjani M, Umbach A, Puchchakayala G, et al. Rapamycin-induced phosphaturia. Nephrol Dial Transplant 2010;25:2938—44.

[68] Haller M, Amatschek S, Wilflingseder J, Kainz A, Bielesz B, Pavik I, et al. Sirolimus induced phosphaturia is not caused by

inhibition of renal apical sodium phosphate cotransporters. PLoS One 2012;7:e39229.

[69] Lu M, Wang J, Ives HE, Pearce D. mSIN1 protein mediates SGK1 protein interaction with mTORC2 protein complex and is required for selective activation of the epithelial sodium channel. J Biol Chem 2011;286:30647–54.

[70] Pearce LR, Sommer EM, Sakamoto K, Wullschleger S, Alessi DR. Protor-1 is required for efficient mTORC2-mediated activation of SGK1 in the kidney. Biochem J 2011;436:169–79.

[71] Pagtalunan ME, Miller PL, Jumping-Eagle S, Nelson RG, Myers BD, Rennke HG, et al. Podocyte loss and progressive glomerular injury in type II diabetes. J Clin Invest 1997;99:342–8.

[72] Wharram BL, Goyal M, Wiggins JE, Sanden SK, Hussain S, Filipiak WE, et al. Podocyte depletion causes glomerulosclerosis: diphtheria toxin-induced podocyte depletion in rats expressing human diphtheria toxin receptor transgene. J Am Soc Nephrol 2005;16:2941–52.

[73] Meyer TW, Bennett PH, Nelson RG. Podocyte number predicts long-term urinary albumin excretion in Pima Indians with type II diabetes and microalbuminuria. Diabetologia 1999;42:1341–4.

[74] Oldfield MD, Bach LA, Forbes JM, Nikolic-Paterson D, McRobert A, Thallas V, et al. Advanced glycation end products cause epithelial-myofibroblast transdifferentiation via the receptor for advanced glycation end products (RAGE). J Clin Invest 2001;108:1853–63.

[75] Mori H, Inoki K, Masutani K, Wakabayashi Y, Komai K, Nakagawa R, et al. The mTOR pathway is highly activated in diabetic nephropathy and rapamycin has a strong therapeutic potential. Biochem Biophys Res Commun 2009;384:471–5.

[76] Tzatsos A, Kandror KV. Nutrients suppress phosphatidylinositol 3-kinase/Akt signaling via raptor-dependent mTOR-mediated insulin receptor substrate 1 phosphorylation. Mol Cell Biol 2006;26:63–76.

[77] Um SH, D'Alessio D, Thomas G. Nutrient overload, insulin resistance, and ribosomal protein S6 kinase 1, S6K1. Cell Metab 2006;3:393–402.

[78] Zhang HH, Huang J, Duvel K, Boback B, Wu S, Squillace RM, et al. Insulin stimulates adipogenesis through the Akt-TSC2-mTORC1 pathway. PLoS One 2009;4:e6189.

[79] Soukas AA, Kane EA, Carr CE, Melo JA, Ruvkun G. Rictor/TORC2 regulates fat metabolism, feeding, growth, and life span in Caenorhabditis elegans. Genes Dev 2009;23:496–511.

[80] Houde VP, Brule S, Festuccia WT, Blanchard PG, Bellmann K, Deshaies Y, et al. Chronic rapamycin treatment causes glucose intolerance and hyperlipidemia by upregulating hepatic gluconeogenesis and impairing lipid deposition in adipose tissue. Diabetes 2010;59:1338–48.

[81] Morrisett JD, Abdel-Fattah G, Hoogeveen R, Mitchell E, Ballantyne CM, Pownall HJ, et al. Effects of sirolimus on plasma lipids, lipoprotein levels, and fatty acid metabolism in renal transplant patients. J Lipid Res 2002;43:1170–80.

[82] Haruta T, Uno T, Kawahara J, Takano A, Egawa K, Sharma PM, et al. A rapamycin-sensitive pathway down-regulates insulin signaling via phosphorylation and proteasomal degradation of insulin receptor substrate-1. Mol Endocrinol 2000;14:783–94.

[83] Feliers D, Duraisamy S, Faulkner JL, Duch J, Lee AV, Abboud HE, et al. Activation of renal signaling pathways in db/db mice with type 2 diabetes. Kidney Int 2001;60:495–504.

[84] Lee MJ, Feliers D, Mariappan MM, Sataranatarajan K, Mahimainathan L, Musi N, et al. A role for AMP-activated protein kinase in diabetes-induced renal hypertrophy. Am J Physiol Renal Physiol 2007;292:F617–27.

[85] Chen JK, Chen J, Thomas G, Kozma SC, Harris RC. S6 kinase 1 knockout inhibits uninephrectomy- or diabetes-induced renal hypertrophy. Am J Physiol Renal Physiol 2009;297:F585–93.

[86] Sakaguchi M, Isono M, Isshiki K, Sugimoto T, Koya D, Kashiwagi A. Inhibition of mTOR signaling with rapamycin attenuates renal hypertrophy in the early diabetic mice. Biochem Biophys Res Commun 2006;340:296–301.

[87] Herbach N, Schairer I, Blutke A, Kautz S, Siebert A, Goke B, et al. Diabetic kidney lesions of GIPRdn transgenic mice: podocyte hypertrophy and thickening of the GBM precede glomerular hypertrophy and glomerulosclerosis. Am J Physiol Renal Physiol 2009;296:F819–29.

[88] Fogo AB. The targeted podocyte. J Clin Invest 2011;121:2142–5.

[89] Reis F, Parada B, Teixeira de Lemos E, Garrido P, Dias A, Piloto N, et al. Hypertension induced by immunosuppressive drugs: a comparative analysis between sirolimus and cyclosporine. Transplant Proc 2009;41:868–73.

[90] Sereno J, Romao AM, Parada B, Lopes P, Carvalho E, Teixeira F, et al. Cardiorenal benefits of early versus late cyclosporine to sirolimus conversion in a rat model. J Pharmacol Pharmacother 2012;3:143–8.

[91] Lopes P, Fuhrmann A, Sereno J, Pereira MJ, Nunes P, Pedro J, et al. Effects of cyclosporine and sirolimus on insulin-stimulated glucose transport and glucose tolerance in a rat model. Transplant Proc 2013;45:1142–8.

[92] Fuhrmann A, Lopes P, Sereno J, Pedro J, Espinoza DO, Pereira MJ, et al. Molecular mechanisms underlying the effects of cyclosporin A and sirolimus on glucose and lipid metabolism in liver, skeletal muscle and adipose tissue in an in vivo rat model. Biochem Pharmacol 2014;88:216–28.

[93] Lopes PC, Fuhrmann A, Carvalho F, Sereno J, Santos MR, Pereira MJ, et al. Cyclosporine A enhances gluconeogenesis while sirolimus impairs insulin signaling in peripheral tissues after 3 weeks of treatment. Biochem Pharmacol 2014;91:61–73.

[94] Lopes PC, Fuhrmann A, Sereno J, Espinoza DO, Pereira MJ, Eriksson JW, et al. Short and long term in vivo effects of cyclosporine A and sirolimus on genes and proteins involved in lipid metabolism in Wistar rats. Metabolism 2014;63:702–15.

[95] Pereira MJ, Palming J, Rizell M, Aureliano M, Carvalho E, Svensson MK, et al. mTOR inhibition with rapamycin causes impaired insulin signalling and glucose uptake in human subcutaneous and omental adipocytes. Mol Cell Endocrinol 2012;355:96–105.

[96] Pereira MJ, Palming J, Rizell M, Aureliano M, Carvalho E, Svensson MK, et al. The immunosuppressive agents rapamycin, cyclosporin A and tacrolimus increase lipolysis, inhibit lipid storage and alter expression of genes involved in lipid metabolism in human adipose tissue. Mol Cell Endocrinol 2013;365:260–9.

[97] Ito N, Nishibori Y, Ito Y, Takagi H, Akimoto Y, Kudo A, et al. mTORC1 activation triggers the unfolded protein response in podocytes and leads to nephrotic syndrome. Lab Invest 2011;91:1584–95.

[98] Luo ZF, Feng B, Mu J, Qi W, Zeng W, Guo YH, et al. Effects of 4-phenylbutyric acid on the process and development of diabetic nephropathy induced in rats by streptozotocin: regulation of endoplasmic reticulum stress-oxidative activation. Toxicol Appl Pharmacol 2010;246:49–57.

[99] Eid AA, Ford BM, Bhandary B, de Cassia Cavaglieri R, Block K, Barnes JL, et al. Mammalian target of rapamycin regulates Nox4-mediated podocyte depletion in diabetic renal injury. Diabetes 2013;62:2935–47.

[100] Hartleben B, Godel M, Meyer-Schwesinger C, Liu S, Ulrich T, Kobler S, et al. Autophagy influences glomerular disease susceptibility and maintains podocyte homeostasis in aging mice. J Clin Invest 2010;120:1084—96.

[101] Cina DP, Onay T, Paltoo A, Li C, Maezawa Y, De Arteaga J, et al. Inhibition of MTOR disrupts autophagic flux in podocytes. J Am Soc Nephrol 2012;23:412—20.

[102] Ha TS, Barnes JL, Stewart JL, Ko CW, Miner JH, Abrahamson DR, et al. Regulation of renal laminin in mice with type II diabetes. J Am Soc Nephrol 1999;10:1931—9.

[103] Ziyadeh FN. Role of transforming growth factor beta in diabetic nephropathy. Exp Nephrol 1994;2:137.

[104] Sataranatarajan K, Mariappan MM, Lee MJ, Feliers D, Choudhury GG, Barnes JL, et al. Regulation of elongation phase of mRNA translation in diabetic nephropathy: amelioration by rapamycin. Am J Pathol 2007;171:1733—42.

[105] Yang Y, Wang J, Qin L, Shou Z, Zhao J, Wang H, et al. Rapamycin prevents early steps of the development of diabetic nephropathy in rats. Am J Nephrol 2007;27:495—502.

[106] Lloberas N, Cruzado JM, Franquesa M, Herrero-Fresneda I, Torras J, Alperovich G, et al. Mammalian target of rapamycin pathway blockade slows progression of diabetic kidney disease in rats. J Am Soc Nephrol 2006;17:1395—404.

[107] Huber TB, Walz G, Kuehn EW. mTOR and rapamycin in the kidney: signaling and therapeutic implications beyond immunosuppression. Kidney Int 2011;79:502—11.

[108] Rangan GK, Coombes JD. Renoprotective effects of sirolimus in non-immune initiated focal segmental glomerulosclerosis. Nephrol Dial Transplant 2007;22:2175—82.

[109] Bonegio RG, Fuhro R, Wang Z, Valeri CR, Andry C, Salant DJ, et al. Rapamycin ameliorates proteinuria-associated tubulointerstitial inflammation and fibrosis in experimental membranous nephropathy. J Am Soc Nephrol 2005;16:2063—72.

[110] Naumovic R, Jovovic D, Basta-Jovanovic G, Miloradovic Z, Mihailovic-Stanojevic N, Aleksic T, et al. Effects of rapamycin on active Heymann nephritis. Am J Nephrol 2007;27:379—89.

[111] Kurayama R, Ito N, Nishibori Y, Fukuhara D, Akimoto Y, Higashihara E, et al. Role of amino acid transporter LAT2 in the activation of mTORC1 pathway and the pathogenesis of crescentic glomerulonephritis. Lab Invest 2011;91:992—1006.

[112] Shegogue D, Trojanowska M. Mammalian target of rapamycin positively regulates collagen type I production via a phosphatidylinositol 3-kinase-independent pathway. J Biol Chem 2004;279:23166—75.

[113] Brenneisen P, Wenk J, Wlaschek M, Krieg T, Scharffetter-Kochanek K. Activation of p70 ribosomal protein S6 kinase is an essential step in the DNA damage-dependent signaling pathway responsible for the ultraviolet B-mediated increase in interstitial collagenase (MMP-1) and stromelysin-1 (MMP-3) protein levels in human dermal fibroblasts. J Biol Chem 2000;275:4336—44.

[114] Liu Y. Epithelial to mesenchymal transition in renal fibrogenesis: pathologic significance, molecular mechanism, and therapeutic intervention. J Am Soc Nephrol 2004;15:1—12.

[115] Lamouille S, Derynck R. Cell size and invasion in TGF-beta-induced epithelial to mesenchymal transition is regulated by activation of the mTOR pathway. J Cell Biol 2007;178:437—51.

[116] Wu MJ, Wen MC, Chiu YT, Chiou YY, Shu KH, Tang MJ. Rapamycin attenuates unilateral ureteral obstruction-induced renal fibrosis. Kidney Int 2006;69:2029—36.

[117] Diekmann F, Rovira J, Carreras J, Arellano EM, Banon-Maneus E, Ramirez-Bajo MJ, et al. Mammalian target of rapamycin inhibition halts the progression of proteinuria in a rat model of reduced renal mass. J Am Soc Nephrol 2007;18:2653—60.

[118] Wittmann S, Daniel C, Stief A, Vogelbacher R, Amann K, Hugo C. Long-term treatment of sirolimus but not cyclosporine ameliorates diabetic nephropathy in the rat. Transplantation 2009;87:1290—9.

[119] Mariappan MM. Signaling mechanisms in the regulation of renal matrix metabolism in diabetes. Exp Diabetes Res 2012;2012:749812.

[120] Alessi DR, Pearce LR, Garcia-Martinez JM. New insights into mTOR signaling: mTORC2 and beyond. Sci Signal 2009;2:pe27.

[121] Sarbassov DD, Guertin DA, Ali SM, Sabatini DM. Phosphorylation and regulation of Akt/PKB by the rictor-mTOR complex. Science 2005;307:1098—101.

[122] Oh WJ, Wu CC, Kim SJ, Facchinetti V, Julien LA, Finlan M, et al. mTORC2 can associate with ribosomes to promote cotranslational phosphorylation and stability of nascent Akt polypeptide. EMBO J 2010;29:3939—51.

[123] Xie X, Guan KL. The ribosome and TORC2: collaborators for cell growth. Cell 2011;144:640—2.

[124] Goldenberg R, Punthakee Z. Definition, classification and diagnosis of diabetes, prediabetes and metabolic syndrome. Can J Diabetes 2013;37(Suppl. 1):S8—11.

[125] Brown JB, Pedula KL, Summers KH. Diabetic retinopathy: contemporary prevalence in a well-controlled population. Diabetes Care 2003;26:2637—42.

[126] Cai J, Boulton M. The pathogenesis of diabetic retinopathy: old concepts and new questions. Eye (Lond) 2002;16:242—60.

[127] Matthews DR, Stratton IM, Aldington SJ, Holman RR, Kohner EM. Risks of progression of retinopathy and vision loss related to tight blood pressure control in type 2 diabetes mellitus: UKPDS 69. Arch Ophthalmol 2004;122:1631—40.

[128] White NH, Cleary PA, Dahms W, Goldstein D, Malone J, Tamborlane WV. Beneficial effects of intensive therapy of diabetes during adolescence: outcomes after the conclusion of the Diabetes Control and Complications Trial (DCCT). J Pediatr 2001;139:804—12.

[129] Giacco F, Brownlee M. Oxidative stress and diabetic complications. Circ Res 2010;107:1058—70.

[130] Kim J, Son JW, Lee JA, Oh YS, Shinn SH. Methylglyoxal induces apoptosis mediated by reactive oxygen species in bovine retinal pericytes. J Korean Med Sci 2004;19:95—100.

[131] Romeo G, Liu WH, Asnaghi V, Kern TS, Lorenzi M. Activation of nuclear factor-kappaB induced by diabetes and high glucose regulates a proapoptotic program in retinal pericytes. Diabetes 2002;51:2241—8.

[132] Grigsby JG, Cardona SM, Pouw CE, Muniz A, Mendiola AS, Tsin AT, et al. The role of microglia in diabetic retinopathy. J Ophthalmol 2014;2014:705783.

[133] Miyamoto K, Khosrof S, Bursell SE, Rohan R, Murata T, Clermont AC, et al. Prevention of leukostasis and vascular leakage in streptozotocin-induced diabetic retinopathy via intercellular adhesion molecule-1 inhibition. Proc Natl Acad Sci U S A 1999;96:10836—41.

[134] Abu el Asrar AM, Maimone D, Morse PH, Gregory S, Reder AT. Cytokines in the vitreous of patients with proliferative diabetic retinopathy. Am J Ophthalmol 1992;114:731—6.

[135] Krady JK, Basu A, Allen CM, Xu Y, LaNoue KF, Gardner TW, et al. Minocycline reduces proinflammatory cytokine expression, microglial activation, and caspase-3 activation in a rodent model of diabetic retinopathy. Diabetes 2005;54:1559—65.

[136] Joussen AM, Poulaki V, Mitsiades N, Kirchhof B, Koizumi K, Dohmen S, et al. Nonsteroidal anti-inflammatory drugs prevent early diabetic retinopathy via TNF-alpha suppression. FASEB J 2002;16:438—40.

[137] Goncalves A, Leal E, Paiva A, Teixeira Lemos E, Teixeira F, Ribeiro CF, et al. Protective effects of the dipeptidyl peptidase IV

inhibitor sitagliptin in the blood-retinal barrier in a type 2 diabetes animal model. Diabetes Obes Metab 2012;14:454—63.

[138] Goncalves A, Marques C, Leal E, Ribeiro CF, Reis F, Ambrosio AF, et al. Dipeptidyl peptidase-IV inhibition prevents blood-retinal barrier breakdown, inflammation and neuronal cell death in the retina of type 1 diabetic rats. Biochim Biophys Acta 2014;1842:1454—63.

[139] Catrina SB, Okamoto K, Pereira T, Brismar K, Poellinger L. Hyperglycemia regulates hypoxia-inducible factor-1alpha protein stability and function. Diabetes 2004;53:3226—32.

[140] Bento CF, Fernandes R, Matafome P, Sena C, Seica R, Pereira P. Methylglyoxal-induced imbalance in the ratio of vascular endothelial growth factor to angiopoietin 2 secreted by retinal pigment epithelial cells leads to endothelial dysfunction. Exp Physiol 2010;95:955—70.

[141] Miller JW, Adamis AP, Shima DT, D'Amore PA, Moulton RS, O'Reilly MS, et al. Vascular endothelial growth factor/vascular permeability factor is temporally and spatially correlated with ocular angiogenesis in a primate model. Am J Pathol 1994;145:574—84.

[142] Harhaj NS, Felinski EA, Wolpert EB, Sundstrom JM, Gardner TW, Antonetti DA. VEGF activation of protein kinase C stimulates occludin phosphorylation and contributes to endothelial permeability. Invest Ophthalmol Vis Sci 2006;47:5106—15.

[143] Murakami T, Frey T, Lin C, Antonetti DA. Protein kinase cbeta phosphorylates occludin regulating tight junction trafficking in vascular endothelial growth factor-induced permeability *in vivo*. Diabetes 2012;61:1573—83.

[144] Forsythe JA, Jiang BH, Iyer NV, Agani F, Leung SW, Koos RD, et al. Activation of vascular endothelial growth factor gene transcription by hypoxia-inducible factor 1. Mol Cell Biol 1996;16:4604—13.

[145] Vinores SA, Xiao WH, Aslam S, Shen J, Oshima Y, Nambu H, et al. Implication of the hypoxia response element of the Vegf promoter in mouse models of retinal and choroidal neovascularization, but not retinal vascular development. J Cell Physiol 2006;206:749—58.

[146] Liu NN, Zhao N, Cai N. Suppression of the proliferation of hypoxia-Induced retinal pigment epithelial cell by rapamycin through the /mTOR/HIF-1alpha/VEGF/ signaling. IUBMB Life 2015;67:446—52.

[147] Schrufer TL, Antonetti DA, Sonenberg N, Kimball SR, Gardner TW, Jefferson LS. Ablation of 4E-BP1/2 prevents hyperglycemia-mediated induction of VEGF expression in the rodent retina and in Muller cells in culture. Diabetes 2010;59:2107—16.

[148] Jin J, Yuan F, Shen MQ, Feng YF, He QL. Vascular endothelial growth factor regulates primate choroid-retinal endothelial cell proliferation and tube formation through PI3K/Akt and MEK/ERK dependent signaling. Mol Cell Biochem 2013;381:267—72.

[149] Claesson-Welsh L, Welsh M. VEGFA and tumour angiogenesis. J Intern Med 2013;273:114—27.

[150] Yuan TL, Choi HS, Matsui A, Benes C, Lifshits E, Luo J, et al. Class 1A PI3K regulates vessel integrity during development and tumorigenesis. Proc Natl Acad Sci USA 2008;105: 9739—44.

[151] Graupera M, Guillermet-Guibert J, Foukas LC, Phng LK, Cain RJ, Salpekar A, et al. Angiogenesis selectively requires the p110alpha isoform of PI3K to control endothelial cell migration. Nature 2008;453:662—6.

[152] Lin CH, Li CH, Liao PL, Tse LS, Huang WK, Cheng HW, et al. Silibinin inhibits VEGF secretion and age-related macular degeneration in a hypoxia-dependent manner through the PI-3 kinase/Akt/mTOR pathway. Br J Pharmacol 2013;168:920—31.

[153] Yang LP, Sun HL, Wu LM, Guo XJ, Dou HL, Tso MO, et al. Baicalein reduces inflammatory process in a rodent model of diabetic retinopathy. Invest Ophthalmol Vis Sci 2009;50:2319—27.

[154] Yoshida A, Yoshida S, Ishibashi T, Kuwano M, Inomata H. Suppression of retinal neovascularization by the NF-kappaB inhibitor pyrrolidine dithiocarbamate in mice. Invest Ophthalmol Vis Sci 1999;40:1624—9.

[155] Zhong LM, Zong Y, Sun L, Guo JZ, Zhang W, He Y, et al. Resveratrol inhibits inflammatory responses via the mammalian target of rapamycin signaling pathway in cultured LPS-stimulated microglial cells. PLoS One 2012;7:e32195.

[156] Soufi FG, Vardyani M, Sheervalilou R, Mohammadi M, Somi MH. Long-term treatment with resveratrol attenuates oxidative stress pro-inflammatory mediators and apoptosis in streptozotocin-nicotinamide-induced diabetic rats. Gen Physiol Biophys 2012;31:431—8.

[157] Palsamy P, Subramanian S. Ameliorative potential of resveratrol on proinflammatory cytokines, hyperglycemia mediated oxidative stress, and pancreatic beta-cell dysfunction in streptozotocin-nicotinamide-induced diabetic rats. J Cell Physiol 2010;224:423—32.

mTOR AND CANCER

24

Lymphangioleiomyomatosis (LAM): Molecular Insights into mTOR Regulation Lead to Targeted Therapies

Wendy K. Steagall, Connie G. Glasgow, Gustavo Pacheco-Rodriguez and Joel Moss

Cardiovascular and Pulmonary Branch, National Heart, Lung, and Blood Institute, National Institutes of Health, Bethesda, MD, USA

24.1 LYMPHANGIOLEIOMYOMATOSIS AND TUBEROUS SCLEROSIS COMPLEX

Lymphangioleiomyomatosis (LAM) is a multisystem disease that occurs almost exclusively in women and is characterized by cystic lung destruction, recurrent pneumothoraces, pleural effusions, ascites, lymphangioleiomyomas, and angiomyolipomas (AMLs) [1]. Pulmonary LAM is caused by the proliferation of neoplastic cells (LAM cells) with characteristics of smooth muscle cells and melanocytes [2,3]. In the lung, these cells are found in the walls of cysts and along blood vessels, lymphatics, and bronchioles, resulting in airway obstruction and lymphatic damage (Figure 24.1). Two types of LAM lung cells have been identified: the proliferative spindle-shaped cells and the more differentiated epithelioid cells [2]. While both types of cells express the smooth muscle antigens α-actin, vimentin, and desmin, only the epithelioid cells react with the monoclonal antibody HMB-45, which recognizes gp100, a premelanosomal protein encoded by the *Pmel17* gene [5].

There are two forms of LAM: sporadic and inherited. The sporadic form occurs in 3.3–7.7 per one million women [6], while the inherited form may occur in women with tuberous sclerosis complex (TSC), an autosomal dominant disorder with variable penetrance that is characterized by hamartomatous growths in the central nervous system, skin, liver, heart, and eyes [7].

Both forms of LAM are characterized by mutations or dysregulation of the tumor suppressor TSC genes, *TSC1* (encoding hamatin), or more frequently in *TSC2* (encoding tuberin) [8–10]. Mutations in these genes lead to activation of the serine/threonine kinase mechanistic/mammalian target of rapamycin (mTOR), resulting in abnormal cell growth and proliferation. In addition to mutations in the *TSC2* gene, LAM cells often also have loss of heterozygosity (LOH) in the *TSC2* gene, consistent with Knudson's two-hit hypothesis of tumor development [8,10–12] (Figure 24.2). Inactivation of the second allele of *TSC2* may also occur by a second somatic mutation in *TSC2* or by gene silencing due to promoter methylation [13]. LAM cells can metastasize and may be isolated from body fluids such as blood, chyle, bronchoalveolar lavage fluid, and urine [14–16]. Recipient LAM cells have been detected in donor lungs after lung transplantation, consistent with the ability of the LAM cell to migrate and metastasize [17,18]. In agreement with this observation, circulating cells have been isolated from transplant patients [19].

24.2 TSC GENES AND PROTEINS

The tumor suppressor genes *TSC1* (9q34) and *TSC2* (16p13) were discovered in 1997 [20] and 1993 [21], respectively, and the *TSC2* gene was linked to LAM in

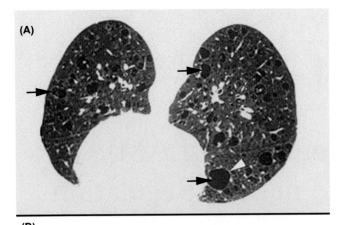

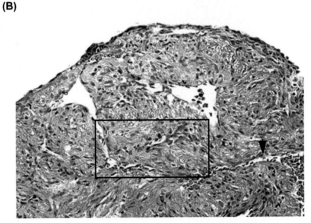

FIGURE 24.1 Pulmonary LAM disease. (A) High-resolution computed tomography scan of the lungs of a patient with LAM. Many cysts (arrows) with visible thin walls (white arrowhead) are distributed throughout the lungs. (B) Section of a lung from a LAM patient stained with hematoxylin and eosin. A proliferative area of LAM cells is shown in the box. There are channels that resemble lymphatic vessels (arrowhead). Previously published in Ref. [4].

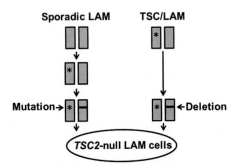

FIGURE 24.2 Development of LAM cells follows Knudson's two-hit hypothesis. In sporadic LAM, a mutation arises in one of the two alleles of TSC2, followed by a deletion (LOH) in the same gene on the other allele. In LAM associated with TSC, a germline mutation exists in one allele, and tumor development follows LOH in the other allele. Loss of function of TSC1 or TSC2 may also occur due to a second somatic mutation or by gene silencing due to promoter methylation.

2000 through the identification of somatic *TSC2* mutations in AMLs from sporadic LAM patients [9]. A third subunit, Tre2-Bub2-Cdc16 (TBC) 1 domain family, member 7 (TBC1D7), of the complex was identified in 2012 [22] on chromosome 6.

Hamartin (TSC1) is a widely expressed 1164-amino-acid, 130-kDa protein (Figure 24.3). Analysis of the hamartin primary sequence (1164 amino acids) with SMART [23,24] indicated the presence of the hamartin domain (1–719 aa) and two coiled-coil regions (724–886 and 925–995). However, experimental data and previous analysis showed that hamartin contained a tuberin-binding domain near the N-terminus, a central coiled-coil region, and a sequence that allowed interactions with the ezrin-radixin-moezin (ERM) family of actin-binding proteins near the C-terminus. The tuberin-binding domain also contained transmembrane and Rho GTPase-activating domains [25–27].

Tuberin (TSC2) is a 1807-amino-acid, 198-kDa protein with a leucine-zipper, two coiled-coil domains, a hamartin-binding domain, two transcription-activating domains (TAD), a calmodulin-binding domain (CaM), and a region responsible for GTPase-activating (GAP) activity near the C-terminus [25,26] (Figure 24.3). Interestingly, *in silico* analysis of the sequence showed a region encompassing amino acids 51–472 (DUF3384 domain), characteristic of tuberin but not studied in detail. This region contained the hamartin-binding domain, a leucine-zipper, and a coiled-coil region. The tuberin domain (550–904) is followed by the GAP region near the C-terminus.

TBC1D7 is a 34-kDa protein, with a poorly conserved TBC domain, lacking two motifs that contain residues essential for the Rab-GAP activity seen in TBC proteins. Nevertheless, TBC1D7 inactivates Rab17 [28]. TBC1D7 binds to the TSC1/2 complex through TSC1, and TSC1 stabilizes both TSC2 and TBC1D7 [29].

The TSC1/TSC2/TBC1D7 complex acts as a hub that coordinates growth factor stimulation, energy levels, and oxygen availability, which then signals to mTOR (Figure 24.4). It negatively regulates Rheb (Ras homolog enriched in brain), a small GTP-binding protein, that, when activated, stimulates the kinase activity of mTOR. The GAP activity of tuberin for Rheb results in conversion of active Rheb-GTP to inactive Rheb-GDP. The intact "Rhebulator" complex of TSC1/TSC2/TBC1D7 is necessary for full GAP activity of TSC2 for Rheb [29].

Tuberin may be phosphorylated by Akt (protein kinase B), RSK1 (p90 ribosomal S6 kinase 1), and AMPK (AMP-activated protein kinase). Growth factor signals through PI3K (phosphoinositide 3-kinase) lead to phosphorylation of Akt, which then negatively regulates TSC1/TSC2/TBC1D7 activity by phosphorylation

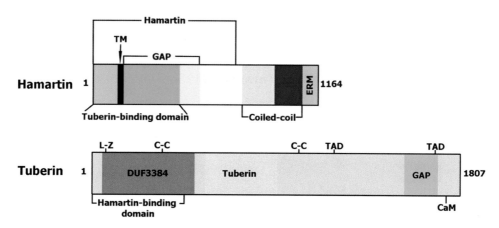

FIGURE 24.3 Schematic representations of the domains of hamartin and tuberin. Hamartin is a 1164-amino-acid protein with a hamartin domain (1–719 aa) and two coiled-coil regions (724–886 and 925–995). Hamartin also contains a tuberin-binding domain near the N-terminus, a central coiled-coil region, and a domain that interacts with ERM proteins near the C-terminus. The tuberin-binding domain also contains transmembrane (TM) and Rho GAP domains. Tuberin is a 1807-amino-acid protein with a leucine-zipper (L-Z), two coiled-coil (C-C) domains, hamartin-binding domain, two TAD, CaM, and a region responsible for GAP activity at the C-terminus. There is a region encompassing amino acids 51–472 (DUF3384 domain), containing the hamartin-binding domain, a leucine-zipper, and a coiled-coil region. The tuberin domain (550–904) is followed by the GAP region near the C-terminus.

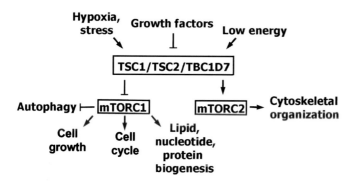

FIGURE 24.4 TSC1/TSC2/TBC1D7 form a hub, which integrates growth factor stimulation, energy levels, and oxygen availability, that then signals mTOR, thereby regulating autophagy, cell growth, cell cycle, cytoskeletal organization, and lipid, nucleotide, and protein biogenesis.

of tuberin at S939, S1086, and T1422 [30]. Activation of RSK1 through the Raf/MAPK/ERK pathway also negatively regulates TSC1/TSC2/TBC1D7 activity by phosphorylation of S1798 of tuberin [31]. Tuberin may also be phosphorylated by MK2 (p38 mitogen-activated protein kinase-activated protein kinase 2) at S1210, which then promotes binding of 14-3-3 proteins to tuberin and inhibition of TSC1/TSC2/TBC1D7 [32]. When conditions are favorable, inhibition of the TSC1/TSC2/TBC1D7 complex promotes cell growth and proliferation.

When substrates are limiting, the TSC1/TSC2/TBC1D7 complex may be activated to prevent cell growth and proliferation. AMPK is activated by high intracellular AMP concentrations that occur under low-energy conditions. AMPK can then phosphorylate tuberin at T1227 and S1345, thereby activating the protein [33].

24.3 mTOR

The TSC1/TSC2/TBC1D7 complex is a negative regulator of mTOR, which is involved in cell growth (increase in cell mass) and proliferation (increase in cell number), including cell cycle progression and general protein biogenesis. mTOR is a serine/threonine kinase, which is a member of the PI3K-related kinase family. Domains in mTOR include the N-terminal and central HEAT repeats that are involved in protein–protein interactions, followed by a FAT (FRAP, ATM, and TRRAP) domain, FRB (FKBP12-rapamcyin binding domain) region, the kinase catalytic domain, and the FATC (FAT C-terminus) region [34,35] (Figure 24.5). The FAT and FATC regions may affect kinase activity [35].

Two forms of mTOR complexes have been identified, which contain distinct proteins with defined functions (Figure 24.5). mTORC1 is composed of raptor (regulatory-associated protein of mTOR), Deptor (DEP domain-containing mTOR-interacting protein), mLST8 (mammalian lethal with SEC13 protein 8), Tti1, Tel2 [36], and PRAS40 (proline-rich Akt substrate, 40 kDa), whereas mTORC2 contains rictor (rapamycin-insensitive companion of mTOR), protor1/2 (protein observed with rictor), mSin1 (mammalian stress-activated protein

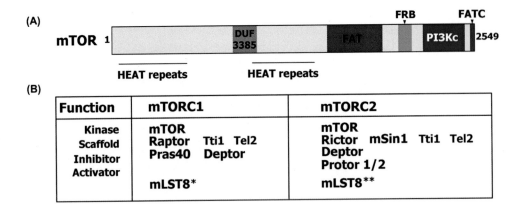

FIGURE 24.5 Schematic representation of the domains of mTOR and the composition of the mTORC1 and mTORC2 complexes. (A) mTOR domains include clusters of HEAT repeats, a FAT domain, FRB region, kinase catalytic domain, and a FATC domain. *In silico* analysis shows a DUF3385 region without defined function but structurally recognized in many proteins. (B) mTORC1 and mTORC2 are composed of multiple proteins with scaffolding, inhibitory, and activator functions. mTORC1 contains raptor, deptor, mLST8, Tti1, Tel2, and PRAS40, whereas mTORC2 contains rictor, protor1/2, mSin1, Tti1, Tel2, mLST8, and deptor. mLST8 (*) may not be critical for mTORC1 function but is critical for mTORC2 activity (**).

kinase-interacting protein 1), Tti1, Tel2, mLST8, and Deptor [35,37,38]. Both mTORC1 and mTORC2 exist as dimers [39,40]. The differences in the functions of mTORC1 and mTORC2 may lie with the scaffolding proteins, raptor and rictor [41].

24.4 mTORC1

mTORC1 coordinates growth factor signals and nutrient availability, for example, insulin, hypoxia, glucose, amino acids, to ensure that precursors are available when protein translation and cell growth are promoted [42,43]. Growth factors and hormones activate mTORC1 through the PI3K/Akt and the Ras/MAPK (mitogen-activated protein kinase) pathways. Stimulation of these pathways results in phosphorylation of tuberin and inhibition of the TSC1/TSC2/TBC1D7 complex. Inhibition of the complex decreases GAP activity toward Rheb and allows the GTP-bound form of Rheb to accumulate and activate mTORC1 (Figure 24.6). Phosphorylation of tuberin by Akt allows for binding of 14-3-3 proteins, which sequester tuberin in the cytosol [44].

As cell growth needs ample nutrients, mTORC1 is also regulated by amino acid availability, such that growth factors cannot simulate mTORC1 without available amino acids [45]. In the presence of amino acids, Rag GTPase (Ras-related small GTP-binding proteins) heterodimers promote the localization of mTORC1 to the lysosome, where it can also interact with Rheb [42] (Figure 24.7). The Ragulator is a complex tethered to the lysosome by the p18 protein and also contains LAMTOR4, LAMTOR5, and MP1-p14. This complex acts as a guanine-exchange factor toward

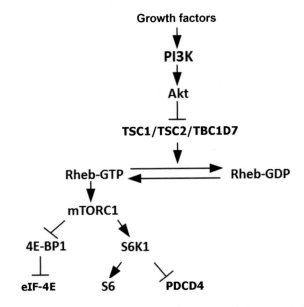

FIGURE 24.6 Inhibition of TSC1/TSC2/TBC1D7 by growth factor stimulation leads to cell growth and proliferation. The presence of growth factors stimulates the PI3K/Akt pathway, leading to phosphorylation of tuberin by Akt and inhibition of the TSC1/TSC2/TBC1D7 complex. This inhibition decreases GAP activity toward Rheb, allowing Rheb-GTP to accumulate and activate mTORC1. Activation of mTORC1 results in phosphorylation of S6K and 4E-BP1, leading to stimulation of protein, lipid, and nucleotide synthesis needed for cell growth and proliferation.

RagA and RagB. The activated form of GTP-RagA/B and GDP-RagC/D can interact with raptor, bringing mTORC1 to the lysosomal surface. Activated mTORC1 regulates protein, lipid, and nucleotide synthesis, which promote cell growth (Figure 24.8). Due to mutations in *TSC1* or *TSC2*, mTOR is constitutively active in LAM cells [46].

24.4.1 mTORC1 and Protein Synthesis

mTORC1 phosphorylates and activates S6K 1/2 (ribosomal S6 kinases 1 and 2), which then phosphorylate and regulate proteins involved in translation, such as ribosomal protein S6, the initiation factor eIF-4B, and PDCD4 (programmed cell death 4), which allow 5′ UTR unwinding and promote the translation initiation step (Figure 24.6). S6K 1/2 also phosphorylate and inhibit eEF2 (eukaryotic elongation factor 2) kinase, releasing the negative regulation on eEF2 and promoting translation elongation. One of the markers of LAM cells is phosphorylated S6 [46−48].

The 4E-BPs regulate the translation of a subset of mRNA by preventing eIF-4F (eukaryotic initiation factor 4F) complex assembly [49] by competing with eIF-4G for binding to eIF-4E. mTORC1 phosphorylates 4E-BP1, resulting in its release from eIF-4E. The cap-binding protein eIF-4E plus the RNA helicase eIF-4A and the scaffolding protein eIF-4G form the eIF-4F complex, which assembles at the 5′ mRNA cap structure, stimulates recruitment of the ribosome to the transcript, and allows translation of mRNAs with a structured 5′ UTR.

PDCD4 is a tumor suppressor protein that binds both eIF-4A and eIF-4G, and suppresses translation by inhibiting the helicase activity of eIF-4A and/or by competing with eIF-4G for binding to eIF-4A, preventing the formation of the eIF-4F complex. Upon stimulation with growth factors, PDCD4 is phosphorylated by S6K1 and then degraded by the ubiquitin ligase SCF$^{\beta TRCP}$ (SKP1-CUL1-F-box protein) [50].

The 5′ TOP (terminal oligopyrimidine) mRNAs are also highly sensitive to mTORC1 activity. These mRNAs encode components of the translation machinery, including ribosomal proteins and elongation factors. The intermediaries between mTORC1 and 5′ TOP mRNA translation were believed to be 4E-BP and S6K 1/2, but are now thought to be LARP1 (La-related protein 1). LARP1 promotes 5′ TOP mRNA translation by associating with the 5′ TOP mRNA and the translation machinery [51,52].

24.4.2 mTORC1 and Lipid Synthesis

mTORC1 positively regulates the activities of sterol regulatory element binding protein 1 (SREBP1) and peroxisome proliferator-activated receptor γ [37,53], transcription factors that control the expression of proteins involved in lipid and cholesterol homeostasis. SREBPs exist as inactive precursors in the endoplasmic reticulum and are processed to active forms by Golgi proteases. The active forms then translocate to the nucleus. mTORC1 acts on SREBP1 through S6K1, which promotes SREBP1 processing through an unknown mechanism [54].

Plasma from LAM patients has a different lipid profile than that seen in healthy women [55], with elevated levels of four lysophosphatidylcholine (LPC) species (C16:0, C18:0, C18:1, C20:4). More LPC was also found in *TSC2*-deficient cells as compared to control cells. Interestingly, the levels of LPC in *TSC2*-deficient cells were not decreased by incubation with rapamycin, an inhibitor of mTORC1, suggesting that these effects on lipids in LAM were mTORC1 independent.

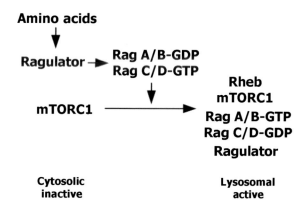

FIGURE 24.7 Growth factor stimulation of mTORC1 cannot occur without available amino acids. In the presence of amino acids, GTP-RagA/B and GDP-RagC/D interact with raptor, bringing mTORC1 to the lysosomal surface where Rheb is located, allowing the interaction of Rheb and mTORC1.

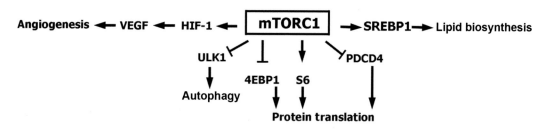

FIGURE 24.8 Downstream pathways of mTORC1. mTORC1 regulates angiogenesis, autophagy, protein translation, and lipid biosynthesis.

24.4.3 mTORC1 and Nucleotide Synthesis

mTORC1 also stimulates nucleotide synthesis to control cell proliferation. S6K phosphorylates CAD (carbamoyl-phosphate synthetase 2, aspartate transcarbamylase, and dihyroorotase), an enzyme complex that catalyzes several steps in *de novo* pyrimidine synthesis [56,57]. This phosphorylation promotes the oligomerization of CAD and thereby stimulates pyrimidine synthesis.

24.4.4 mTORC1 and HIF/VEGF

Hypoxia-inducible factors (HIF) are dimeric transcription factors consisting of a subunit with oxygen-dependent instability (i.e., HIF1α, HIF2α, HIF3α) and a more stable subunit (HIF1β) [58]. Under hypoxic conditions, the highly degradable subunit binds with the more stable subunit to form a stable heterodimer capable of transcriptionally activating genes, including vascular endothelial growth factor (VEGF), which promote cell survival [59]. In cancer cells, HIF activation can lead to tumor growth and metastasis [60–62]. mTORC1 induces HIF1α expression by up-regulating translation [63].

In $Tsc2^{-/-}$ cells from mouse embryonic fibroblasts (MEFs) and AMLs from TSC patients, loss of TSC2 function up-regulated HIF1α in an mTOR-dependent manner with a resulting increase in expression of HIF1α target genes, including VEGF. Treatment of cells with rapamycin only partially down-regulated VEGF, indicating the possible involvement of an mTORC1-independent mechanism [64].

Many LAM patients experience exercise-induced hypoxemia, even with only mild impairment of lung function as measured by the forced expiratory volume in one second (FEV_1) and diffusing capacity of the lung for carbon monoxide (DL_{CO}) [65]. Hypoxemia, through HIF induction, can increase erythropoietin (EPO) expression, promoting the survival and differentiation of erythropoietic cells and the proliferation of nonerythropoietic cells, such as smooth muscle cells. Elevated red cell count was associated with a more rapid decline in lung function in LAM patients, and EPO increased proliferation of $TSC2^{-/-}$ cells, suggesting a role for EPO in disease progression [66].

VEGF contributes to angiogenesis by its antiapoptotic activity and by increasing cell permeability [67]. In cancer, VEGF-C/D expression correlates with vascular invasion, lymph vessel and lymph node involvement, distant metastasis, and poor clinical outcomes [68]. VEGF-A [69], VEGF-C [70], and VEGF-D [71], as well as VEGFR-3 [70], were present in LAM lung lesions. AMLs are highly vascularized tumors, suggesting that the TSC1/2 complex plays a role in the regulation of angiogenesis. Serum levels of VEGF-D were elevated in LAM patients compared to healthy women [71,72], and VEGF-D was present in chylous effusions and LAM lung lesions. Evidence suggests the lymphatic circulation serves as a means to disseminate LAM disease. LAM cells, enveloped by endothelial cells, are found in the walls and lumen of lymphatic vessels, and in the chylous pleural and peritoneal effusions [70,73]. In LAM, VEGF-C/D induced lung LAM cells to proliferate through activation of VEGFR-3 via the PI3K/Akt/mTOR/S6 pathway [74].

24.4.5 mTORC1 and Autophagy

In macrophagy, an autophagosome forms a double membrane around cellular material and travels along microtubules to lysosomes [75]. There, the autophagosome fuses with the lysosome, resulting in degradation of its contents by lysosomal enzymes. The primary initiator of autophagosome formation is the ULK1 complex, consisting of autophagy-related (ATG) proteins [76]. In nontumor cells, autophagy is a mechanism of tumor suppression, clearing unnecessary and/or harmful cellular material. In established tumor cells, autophagy can promote tumor survival by maintaining energy production under metabolic stress [75].

mTOR is a negative regulator of autophagy. When the TSC1/TSC2 complex is functional, mTORC1 phosphorylates ATG13 within the ULK1 complex, which binds the complex to mTORC1. This action inhibits ULK1 phosphorylation of BECLIN-1, the next step in autophagosome formation [77]. Metabolic stress conditions or lack of growth factor stimulation inhibits mTOR and induces autophagy. Usuki et al. [78] noted increased amounts of Bcl-2 and Mcl-1 (apoptosis inhibitors) in LAM cells compared to normal cells in LAM lung tissue.

Tissue homeostasis is, in part, maintained through the regulation of the delicate balance between proapoptotic and antiapoptotic factors [79,80]. The proapoptotic caspase family is initially expressed as inactive zymogens, and through proteolytic cleavage forms mature enzymatic proteins [81]. The apoptotic pathway is activated by the recruitment of caspase-9 to an apoptosome complex. There, activated caspase-9 stimulates the proteolytic activation of other caspases, resulting in cell death [79,82]. Upon proapoptotic cell signaling, survivin, a member of the inhibitors of apoptosis (IAP) family of proteins [80], is translocated from the mitochondria to the cytosol where, by direct binding [82] or in synergistic association with hepatitis B X-interacting protein and/or X-linked IAP [79], caspase-9 is inactivated, inhibiting apoptosis. Survivin expression promotes tumor growth by stimulating angiogenesis through up-regulation of VEGF expression, enhancing cell survival (antiapoptosis), and chemoresistance. Regulation of survivin function occurs via multiple signaling transduction pathways, including PI3K/Akt-mTOR [83,84] and MAP/ERK [85]. Disruption of these pathways can result in the overexpression of survivin and decreased caspase-mediated cell death. Inhibition of apoptosis is critical to tumor growth; however, overexpression of survivin may lead to resistance to drug therapies [84] and is generally labeled as a negative prognostic marker in cancer patients [79]. Cells from an AML of a TSC patient expressed survivin [86].

24.5 mTORC2

Interestingly, while the TSC1/TSC2/TBC1D7 complex inhibits mTORC1 activity, it stimulates mTORC2 activity, through a physical association between TSC2 and rictor [87,88]. The GAP domain of TSC2 is not necessary for the effects of TSC2 on mTORC2 [88]. mTORC2 phosphorylates the hydrophobic motif (HM) and the turn motif (TM) of Akt and protein kinase Cα (PKCα) and the HM of serum and glucocorticoid-induced protein kinase 1. PKCα is involved in the regulation of cell shape and motility and is a mediator of the interaction between mTORC2 and the actin cytoskeleton [87,89,90]. mTORC2 signals through Rho and Rac GTPase to regulate F-actin assembly and disassembly [90].

Insulin stimulates PI3K signaling, resulting in the binding of mTORC2 to the ribosome [91]. The binding and activation of mTORC2 to the 80S ribosome occurs whether or not the ribosome is translating. The phosphorylation of HM of Akt is induced by growth factors, while the phosphorylation of TM is constitutive. Without TM phosphorylation, Akt needs the chaperone Hsp90 to prevent its ubiquitination and degradation. TM phosphorylation of Akt by mTORC2 occurs during translation, while mTORC2 is bound to the large ribosomal subunit [92]. mTORC2 interacts with the tunnel exit protein rpL23a, which would allow mTORC2 to phosphorylate Akt as it emerges from the ribosome.

24.6 EFFECTS OF PHOSPHATIDIC ACID ON mTOR AND RAPAMYCIN

Phospholipase D (PLD) catalyzes the hydrolysis of phosphatidylcholine to phosphatidic acid (PA) and choline [93]. PA binds the FRB domain of mTOR and is necessary for mTOR activity [94]. The formation and/or stability of the mTOR complexes (mTOR with raptor or rictor) are dependent on PA [95,96]. mTORC2 is less sensitive to rapamycin treatment and is only suppressed with long-term treatment. It may be that the differences in sensitivity to rapamycin between mTORC1 and mTORC2 are due to differences in the stability of these complexes [93]. If PA has a lower affinity for mTORC1 than it does for mTORC2, PA and mTORC1 may be more susceptible to dissociation, and respond to lower concentrations of rapamycin.

24.7 mTORC1 AND mTORC2 AND SMALL GTPases

TSC1 binds ERM (actin-binding proteins) [27] and regulates Rho activation through its NH$_2$-terminus Rho-activating domain, which overlaps the TSC2-binding domain. The Rho GTPase family includes RhoA, Rac, and Cdc42 and is responsible for the regulation of actin cytoskeletal remodeling and cell adhesion and migration [97]. TSC1 inhibits Rac1, which leads to increased RhoA activation and the formation of stress fiber and focal adhesions. Without TSC1, Rac1 is not inhibited and RhoA is not activated, resulting in stress fiber dissembly and focal adhesion remodeling [98]. Binding of TSC2 to TSC1 modulates the TSC1-dependent activation of RhoA. LAM cells isolated from the lung showed invasiveness, increased migration rate, and RhoA activation [98,99]. Goncharova et al. [100] demonstrated in Tsc2-null ELT3 cells from the Eker rat and LAMD (LAM-derived) cells from explanted lungs of LAM patients after transplantation that mTORC2 was necessary for proliferation and survival of these cells and involved in the activation of RhoA GTPase.

AMPK inhibits anabolic processes that consume energy and induces catabolic processes that produce ATP. AMPK phosphorylates and activates TSC2, leading to inhibition of mTORC1. Loss of tuberin leads to constitutively active Rheb-GTPase, which can then activate AMPK [101,102]. AMPK phosphorylates p27KipI (cyclin-dependent kinase inhibitor) resulting in sequestration of p27 in the cytoplasm where it then promotes cell migration and survival. Nuclear p27 prevents cell cycle progression by binding and inhibiting the kinase activity of cyclin A-Cdk2 and cyclin E-Cdk2, thereby inhibiting proliferation [101,102].

24.8 RAPAMYCIN TREATMENT OF LAM

Rapamycin (also called sirolimus) is an antifungal macrolide antibiotic that serves as an immunosuppressive agent. It binds to FKBP12 (12 kDa FK506 binding protein) and together they bind mTORC1 [39]. The initial binding of rapamycin—FKBP12 to mTORC1 weakens the mTOR—raptor interaction and may block access to the active site of larger substrates. The mTORC1 complex may then disintegrate over time due to instability caused by the binding of the rapamycin—FKBP12 complex or due to binding of a second rapamycin complex to the mTORC1 dimer [39,40]. Rapamycin may also have more difficulty inhibiting the mTOR-dependent phosphorylation of proteins with preferred target sites [35].

Some phosphorylation sites of 4E-BP and ULK1 show reduced sensitivity to rapamycin. Cells may have a high level of basal phosphorylation of 4E-BP1 on Thr37/46 that is not influenced by growth factor stimulation or rapamycin inhibition of mTORC1. Some cells exhibit further phosphorylation of 4E-BP1 on other threonines in response to growth factors that is sensitive to rapamycin. Thus, phosphorylation of 4E-BP1 can be both constitutive and rapamycin-insensitive and inducible and rapamycin-sensitive [103].

24.8.1 *In Vivo* Studies of Rapamycin—Eker Rat

The Eker rat, an animal model of TSC [104,105], has a *Tsc2* germline mutation. *Tsc2*$^{Ek/+}$ rats developed renal epithelial tumors that exhibited LOH of the *Tsc2* locus, thereby exhibiting a two-hit inactivation of *Tsc2*. When these rats were administered rapamycin over 3 days, tumor apoptosis and a decrease in the proliferative index were seen [105]. About 50% of the *Tsc2*$^{Ek/+}$ rats also developed pituitary adenomas. These tumors exhibited *Tsc2* LOH, increased phosphorylation of 4E-BP1 and S6, and were found more frequently in rats that died prematurely. After 1 week on rapamycin, these rats gained weight and showed an improvement in neurologic status, both of which were reversed when rapamycin was discontinued. Symptoms reversed upon restarting rapamycin. Interestingly, these rats still had pituitary tumors at autopsy.

Renal tumors were also affected by rapamcyin treatment, with a >90% tumor volume regression as measured by ultrasound. While "tumor scars" were seen after rapamcyin treatment, there were also unaffected lesions without evidence of necrosis or apoptosis and increased levels of phosphorylated S6. When rats were treated with rapamycin between the ages of 2—4 months to determine if rapamycin could prevent tumors, the treatment did not alter the frequency of microscopic precursor lesions, but did affect lesions ≥1 mm in size. These data suggested that mTOR may have a role in tumor progression, not tumor initiation. Other *Tsc2* pathways may be involved in tumor initiation [105].

24.8.2 Clinical Trials of Rapamycin and Rapalogs

Rapamycin is an immunosuppressive agent approved by the Food and Drug Administration. Because it can inhibit mTOR, rapamycin has been studied in various clinical trials to examine the effects on disease manifestations of LAM and TSC. One of the first studies [106] examined the effect of rapamycin on the volume of AMLs in patients with LAM, LAM/TSC, or TSC. In this nonrandomized, open-label, phase 1—2 trial, patients received rapamycin for 1 year and then were followed for an additional year without drug. AML volumes decreased nearly 50% while on rapamycin, and the FEV$_1$ increased. Unfortunately, both improvements reversed upon discontinuation of therapy. A second nonrandomized, open-label, phase 2 trial [107] also examined the effect of rapamycin on AML volume and followed patients for 2 years on drug. Again, AMLs shrunk with rapamycin therapy. During the second year of therapy, there was little further shrinkage of AMLs compared to the first year, although the tumor response was maintained by continuous drug use. A randomized, placebo-controlled study [108] designed to examine the effects of rapamycin on lung function in LAM looked at pulmonary function of patients given rapamycin for a year, followed by a year off, compared to patients given placebo. The patients given rapamycin had significantly improved FEV$_1$ compared to patients given placebo, and significant improvement from baseline to 1 year in forced vital capacity (FVC), functional residual capacity, serum VEGF-D, and quality of life and functional performance. There were no differences between the rapamycin group and the placebo group in the change in 6-min walking distance or DL$_{CO}$. When rapamycin was withdrawn for the second year of study, decline in lung function resumed. A randomized, double-blind, placebo-controlled trial [109] using everolimus (a rapamycin derivative) also showed significant reduction in AML volume with drug use. Rapamycin was recently approved by the FDA for use in pulmonary LAM.

24.8.3 Observational Studies of Rapamycin and Rapalogs

A study examining the effect of rapamycin on lung function, chylous effusions, and lymphangioleiomyomas in LAM patients found that, after approximately 2.5 years on rapamycin, both FEV$_1$ and DL$_{CO}$ increased significantly as compared to the rate of decrease before rapamycin therapy [110]. The patients with chylous effusions and lymphangioleiomyomas experienced almost complete resolution of these conditions.

Similarly, rapamycin treatment resolved chylothorax seen in six of seven patients after 5 months of treatment [111], while treatment of five patients with everolimus [112] shrunk or resolved lymphangioleiomyomas in all five cases. A study examining the effect of rapamycin therapy over 5 years confirmed that there was significant reduction in rates of yearly decline in FEV$_1$ [113], suggesting that the effects of rapamycin on lung function were sustained as long as the drug was continued.

Similar to some neoplasms [114], circulating LAM cells have been isolated from blood, urine, chylous effusions, and bronchoalveolar lavage fluid [14–16]. Detection rates of these cells from blood and urine significantly decreased after approximately 2.2 years of rapamycin therapy [16]. Interestingly, while the detection rates drop with therapy, it was still possible to isolate circulating LAM cells even from patients on rapamycin [16].

24.9 OTHER PATHWAYS THAT PROMOTE LAM PATHOLOGY

24.9.1 Feedback Loops

While the mTORC1/S6K pathway promotes cellular growth, it also mediates negative feedback loops that are able to restrain upstream signaling and control growth [103]. When the mTORC1/S6K pathway is inhibited, there may be an overactivation of upstream signaling that opposes the antiproliferative effects of the inhibitor. The first example of a negative feedback loop involves the IRS (insulin receptor substrate) docking proteins, IRS-1 and IRS-2, which have a role in insulin/IGF signaling through activation of PI3K. Activation of mTORC1/S6K leads to phosphorylation of IRS-1 on multiple residues by both mTORC1 and S6K. These phosphorylations inactivate IRS-1 and control further PI3K/Akt activation. If mTORC1 is inhibited by rapamycin, there is less suppression of IRS-1 activity and increased PI3K/Akt activation [103].

Inhibition by rapamycin also results in increased phosphorylation of Akt by PDK1 and mTORC2. S6K phosphorylates SIN1, a component of mTORC2. Phosphorylated SIN1 dissociates from mTORC2, resulting in suppression of mTORC2 kinase activity. Therefore, when S6K is down-regulated by rapamycin, mTORC2 kinase activity may increase [103].

Another negative feedback loop involves ERK activation and depends on the function of the PI3K/RAS/RAF pathway. Subunits of PI3K, p85α and p110δ, are transcriptionally up-regulated by long-term exposure to rapamycin. Similarly, PDGFRβ has increased expression and phosphorylation upon exposure to rapamycin, which may also lead to increased ERK signaling [103].

mTORC1 also phosphorylates GRB10 (growth factor receptor-bound protein 10), which enhances its stability. GRB10 mediates the degradation of the insulin/IGF1 receptors by ubiquitination, resulting in suppression of insulin and IGF1 signaling. Inhibition of mTORC1 by rapamycin would suppress this feedback loop [103].

24.9.2 Phospholipase D and Rapamycin

When PLD activity is high, more PA is formed, resulting in an increased resistance to rapamycin [115,116]. It is possible that rapamycin treatment may select cells with high PLD activity [93], which have increased cell migration and invasion [117]. Thus, rapamycin treatment may have the unintended consequence of selecting cells with increased malignant potential.

24.9.3 Anti-EGFR Antibodies

Epidermal growth factor receptors (EGFRs) are a family of receptor tyrosine kinases capable of activating numerous signaling pathways, including PI3K/AKT and mTOR [118–120]. Lesma et al. [13,121–123] have examined the effects of EGFR on the proliferation of LAM cells, using smooth-muscle-like cells cultured from an AML of a patient with TSC and a chylous effusion from a patient with LAM/TSC. Interestingly, EGF was required for cell growth and proliferation of the $TSC2^{-/-}$ cells from the AML, and this effect was dependent on the lack of tuberin. The addition of anti-EGFR antibodies to the AML cell cultures led to almost complete cell death in less than 2 weeks. The data from these studies looking at blockade of the EGF receptors (exposure to anti-EGFR) and IGF1 receptors (exposure to rapamycin) suggested that proliferation of human $TSC2^{-/-}$ cells is a result of a convergence of the RAS/MAPK pathway and PI3K pathways to activate mTOR. Both receptor inhibitors reduced the phosphorylation of SK6 [121,122]. The proliferation of LAM/TSC cells isolated from chylous effusions was also EGF-dependent. In addition, these cells were capable of anchorage-independent survival, which was negatively affected by exposure to rapamycin and anti-EGFR antibody [13].

Mouse models were created by the endonasal administration of human $TSC2^{-/-}$ AML or chylous effusion smooth muscle cell cultures [13,123]. The dye-labeled abnormal smooth muscle cells infiltrated the lymph nodes and lung. In the chylous effusion model [13], lung nodules were formed. In both mouse models, injection of anti-EGFR antibodies promoted a reversal in lung degeneration and a reduction in phosphorylated S6.

24.9.4 Statins

Statins inhibit geranylgeranylation of Rho GTPases and farnesylation of Ras and Rheb GTPases [124]. 3-Hydroxy-3-methyl-glutaryl-coenzyme A reductase is necessary for the geranylgeranylation of Rho GTPases, a post-translational modification that promotes membrane attachment [125,126]. It has been hypothesized that the combination of rapamycin and statins may be cytocidal for LAM and $TSC2$-null cells. Simvastatin, with or without rapamycin, inhibited cell growth and promoted apoptosis in $TSC2$-null cells [100]. Atorvastatin inhibited the proliferation of $Tsc2^{-/-}$ MEFs and ELT3 smooth muscle cells from the Eker rat and inhibited phosphorylation of mTOR, S6 kinase, and S6 [127]. Another study showed no inhibition of tumor growth by atorvastatin when examining renal and liver tumors of $Tsc2^{+/-}$ mice [128]. In these mice, atorvastatin reduced serum cholesterol, and decreased levels of phosphorylated S6 and GTP-RhoA, but perhaps did not completely eliminate GTP-bound GTPases, resulting in insufficient inhibition of Rheb effects on mTORC1 activation [128]. Atochina-Vasserman et al. [129] examined the effects of atorvastatin versus simvastatin on the growth of $TSC2$-null cells. Rapamycin and atorvastatin had an additive effect on cell inhibition, while rapamycin and simvastatin had a synergistic effect. Simvastatin decreased levels of Akt and S6 phosphorylation, and increased cleaved caspase-3 levels. Atorvastatin had none of these effects, showing that the kind of statin may matter. Atorvastatin alone or in combination with rapamycin did not decrease tumor volume or increase survival in a nude mouse model with subcutaneous $Tsc2^{-/-}$ tumors [130]. Farnesylthiosalicylic acid (FTS; salirasib), a farnesylcysteine mimetic that competes with farnesylated proteins for association with the cell membrane [131], inhibited Rheb activity in ELT3 cells and proliferation and migration of these cells [132]. FTS also inhibited tumor growth in a nude mouse model of subcutaneously injected ELT3 cells [132].

A retrospective study examined the effects of statins on pulmonary function of patients with LAM. LAM patients on statins had a more rapid rate of decline in DL_{CO} than their matched counterparts not taking statins [133]. The study did not subcategorize statins, which may account for some of the findings. Atorvastatin has been found to increase expression of VEGF-D [134], which may be a confounding factor. Another retrospective study on the use of both rapamycin and simvastatin in LAM patients found no differences in pulmonary function compared to rapamycin alone [135].

24.9.5 Autophagy Inhibitors

$TSC2$-null cells have low levels of autophagy under basal and stress conditions [136]; administration of rapamycin, however, induces autophagy. In $Tsc2^{-/-}$ MEFs, rapamycin completely inhibited S6K, but partially inhibited phosphorylation of 4E-BP1 and autophagy [137]. In $Tsc2^{-/-}$ MEFs under bioenergetic stress, constitutive activation of mTOR inhibited autophagy and increased cell death [138]. Rapamycin inhibited activation of mTOR which increased autophagy (promoting tumor growth through the increased cell survival effect of autophagy) and decreased cell death. Therefore, a drug targeting autophagy (chloroquine, plaquenil) combined with rapamycin may affect proliferation through inhibition of S6K and promote cell death through inhibition of autophagy [136,139,140]. Resveratrol, in combination with rapamycin, has also been found to reduce rapamycin-induced autophagy, prevent rapamycin-induced Akt reactivation, and promote apoptosis of $TSC2$-null cells [141,142].

A subset of cells from the AML of a TSC patient expressed survivin [86]. In this case, cells expressed survivin when exposed to conditions that inactivated mTOR or inhibited the PI3K pathway, such as treatment with rapamycin, thereby increasing antiapoptotic activity. These results are in contrast to the effects of rapamycin on survivin expression in chronic lymphocytic leukemia [85] and prostate epithelial cells [143], where translation of survivin is inhibited by rapamycin, leading to increased sensitivity to drug treatment. Survivin is undetectable in noncancerous adult tissue [82] and therefore makes an interesting drug target, with the expectation of minimal adverse events.

24.9.6 Src Inhibitor Saracatinib

A further complication of autophagy inhibition in $TSC2$-null cells is the activation of Src with increased phosphorylation on Tyr416 [144]. Activation of Src led to increased expression and activation of the transcription factor Snail, which inhibited E-cadherin expression, and increased expression of matrix metalloproteinase (MMP) 9, two hallmarks of epithelial—mesenchymal transition. The Src inhibitors dasatinib or saracatinib reduced expression of both Snail and MMP9, while rapamycin had no effect. The inhibitors reduced both migration and invasiveness of $TSC2^{-/-}$ EEF (Eker rat embryonic fibroblasts) cells [144].

Finally, saracatinib reduced the number of $TSC2^{-/-}$ EEF cells found in the lungs of CB17-SCID mice after intravenous injection, thus appearing to reduce the metastatic potential of the TSC2-null cells [144].

24.9.7 VEGFR Inhibitor Sorafenib

Sorafenib is a kinase inhibitor that inhibits VEGFR-1, VEGFR-2, and VEGFR-3 [145]. Sorafenib alone did not decrease tumor volume or increase survival time of nude mice bearing subcutaneous $Tsc2^{-/-}$ tumors [130]. While rapamycin decreased tumor volume and increased survival of these mice, the combination of rapamycin and sorafenib significantly improved the survival of the mice over rapamycin alone, suggesting that the combination of an mTOR inhibitor plus a VEGF signaling inhibitor may be a treatment for LAM.

24.9.8 MMPs Inhibitor Doxycycline

LAM lung biopsy specimens showed destruction of lung parenchyma, including structural damage to elastic fibers and collagen fibrils, and increased expression of MMP2, MMP9, and MMP1. MMP2 was often colocalized with type IV collagen [146,147]. MMP2 expression and activity were increased in TSC2-deficient cells and this increase was not affected by rapamcyin [148]. The antibiotic doxycycline is an MMP inhibitor. A patient with LAM treated with doxycycline showed improved pulmonary function and a decrease in the levels of MMPs found in urine [149]. In a study examining the effect of doxycycline on ELT3 cells and cells derived from AMLs from patients with LAM or TSC [150], cell number was decreased, apoptosis was increased, and changes were observed in cell morphology. In a second study, incubation of TSC2-null MEFs and cell cultures from lungs of patients with LAM [151] with doxycycline resulted in decreased MMP levels and metabolic activity, with no effect on proliferation. Doxycycline alone, or in combination with rapamycin, did not decrease tumor volume or increase survival in a nude mouse model with subcutaneous $Tsc2^{-/-}$ tumors [130]. A 2-year randomized placebo-controlled trial of doxycycline with 15 patients showed no effect on the rate of decline of FEV_1, FVC, T_{LCO}, total lung capacity, shuttle-walk distance, or quality of life scores [152], suggesting that doxycycline may not be effective in the treatment of LAM.

24.9.9 Estradiol

LAM is a disease that predominantly affects women, and disease progression may worsen with pregnancy or with the use of oral contraceptives [153,154]. Symptoms may ease with oophorectomy or postmenopause [155,156]. These data suggested that estradiol may play a role in LAM progression. Estrogen receptor alpha and progesterone receptor were expressed by LAM and AML cells [157]. The growth of a primary cell culture from an AML of a LAM patient was stimulated by both estradiol and tamoxifen, a selective estrogen receptor modulator acting as an estrogen agonist in these cells [158]. Estradiol increased p44/42 MAPK phosphorylation and also increased expression of c-myc, indicating both cytoplasmic signaling and transcriptional effects of estradiol. In another study of the effects of estradiol and tamoxifen on liver hemangioma in $Tsc1^{+/-}$ mice [159], estrogen increased the frequency and severity of these lesions in mice, while tamoxifen reduced both, suggesting that in this tissue tamoxifen was an estrogen antagonist. Estrogen also stimulated ELT3 cell growth, while tamoxifen inhibited it [160]. Estradiol increased the size of the tumors in a nude mouse xenograft model using ELT3 cells, while tamoxifen decreased tumor size [160].

Treatment of either female or male CB17-SCID mice with estradiol after injection with TSC2-null ELT3 cells resulted in an increased number of pulmonary metastases versus placebo-treated mice [161]. Estradiol treatment enhanced the survival of the injected circulating ELT3 cells, and resulted in a twofold increase in the seeding of the lung. Estradiol activated p42/44 MAPK in the ELT3 cells in both metastases and the primary xenograft tumors, and promoted the survival of circulating cells by increasing resistance to anoikis, matrix deprivation-induced apoptosis. Treatment of mice with the MEK1/2 inhibitor CI-1040 and estradiol reduced the number of circulating ELT3 cells and blocked lung metastases. Treatment of mice with RAD001, an mTORC1 inhibitor, prevented both primary tumor development and lung metastasis, with or without estradiol [161].

Li et al. [162] examined the extracellular matrix of the primary tumors in mice treated with estradiol and found a 60% reduction in type IV collagen compared to placebo-treated mice. Estradiol increased the expression and activity of MMP2 in ELT3 cells in vitro and increased the concentration of MMP2 in tumors from mice. MMP2 can degrade type IV collagen. Mice treated with faslodex, an estradiol receptor antagonist, and estradiol had no lung metastases, with no effect noticed on primary tumor growth. Faslodex also restored extracellular matrix organization and reduced MMP2 expression and activity.

Progesterone has been proposed as hormonal therapy to slow lung disease progression in LAM. A retrospective study examining the effect of progesterone on 348 patients with LAM found that the rates of decline in DL$_{CO}$ were significantly higher in patients treated with progesterone than in untreated patients [163], suggesting that progesterone did not slow the decline of lung function in LAM. Interestingly, LAM patients also have a higher prevalence of meningiomas than is seen in the general population [164]. As progesterone may be mitogenic for meningiomas, progesterone may not be beneficial to patients with LAM. The expression of the progesterone receptor in lungs from LAM patients was frequently higher than that of the estrogen receptor [165]. Sun et al. [166] examined the effect of progesterone on *TSC2*-null AML cells and ELT3 cells and found that under oxidative stress, it activated ERK1/2 and Akt and, with estradiol, increased proliferation, reduced the cellular levels of reactive oxygen species, and enhanced cell survival. Interestingly, the combination of progesterone and estradiol was more potent than estradiol alone in promoting lung metastases in the CB17-*SCID* mouse model injected with ELT3 cells [166].

mTORC1 activation and estrogen-ERK2 signaling acting in concert stimulated migration and invasion of *TSC2*-null AML cells [167]. Estradiol stimulated migration and invasion of these cells along with a biphasic activation of ERK2, with the first activation being transcription-independent, while the second was transcription-dependent. This activation was associated with a mesenchymal to epithelial transition from a spindle-like phenotype to a cuboidal-like phenotype. Sustained ERK2 signaling led to expression of the late response gene, *Fra1*, which in turn led to expression of zinc finger E-box-binding homeobox 1/2 (ZEB1/2), resulting in increased cell migration and invasion [168]. mTORC1, which is known to enhance the translation of mRNA with highly structured 5′ UTR, was found to promote translation of *Fra1* [167]. Therefore, the estradiol-ERK and the mTORC1 pathways converged to promote migration of *TSC2*-null cells.

Estradiol promoted survival of *TSC2*-deficient cells under conditions of oxidative stress by promoting the pentose phosphate pathway [169]. While *TSC2*-deficient cells had low Akt activation due to feedback inhibition, estradiol was able to reactivate Akt. This reactivation of Akt activity led to glucose uptake through glucose transporters GLUT1 or GLUT4, which translocated to the plasma membrane upon estradiol stimulation. Estradiol also increased the expression of glucose-6-phosphate dehydrogenase, which is the first rate-limiting enzyme of the pentose phosphate pathway.

Finally, estradiol enhanced prostaglandin synthesis and cyclooxygenase-2 (COX-2) expression in *TSC2*-deficient cells through mTORC2 [170]. Prostaglandins have been linked to cancer development [171]. Aspirin, a COX-2 inhibitor, reduced proliferation of *TSC2*-deficient cells [170]. Celecoxib, another COX-2 inhibitor, suppressed renal cystadenomas of *Tsc2*$^{+/-}$ mice as compared to vehicle control, while aspirin reduced tumor growth in an ELT3 xenograft mouse model [170]. Expression of adipocyte phospholipase A2, which initiated prostaglandin biosynthesis by acting on membrane phospholipids, was up-regulated in pulmonary LAM nodules, but independently of mTOR, Akt, and MEK1/2 signaling pathways [172]. Stimulation of *TSC2*-deficient cells with PGE$_2$ or PGI$_2$, prostaglandin metabolites, increased the growth of these cells [172].

24.9.10 Interferons

The expression of type II interferon (IFN) γ was reduced in *Tsc1*- and *Tsc2*-null MEFs [173], and treatment of these cells with IFNγ inhibited proliferation and increased apoptosis. *Tsc1*- or *Tsc2*-null cells showed an increase in expression and phosphorylation of Stat1 (Ser 727) and in the phosphorylation of Stat3 (Tyr 705). IFNγ treatment reduced phosphorylated Stat3 and increased phosphorylated Stat1 (Tyr 701). Rapamycin treatment induced IFNγ secretion, and the combination of IFNγ and rapamycin had a synergistic effect on apoptosis in *Tsc1*- or *Tsc2*-null cells. When the IFNγ-Jak-Stat1 pathway was examined in LAM lung and AML tissues, IFNγ expression was not seen compared to normal tissues. In LAM, Stat1 and pStat1 Ser 727 were increased, as were Stat3 and pStat3 Tyr705 [174]. Studies examining the combination treatment of CCI-779 (a rapamycin analog) and IFNγ on kidney lesions of *Tsc2*$^{+/-}$ mice showed that treatment must be initiated after the appearance of the lesions in order to be effective [175]. Once lesions appear, 2 months of treatment with CCI-779 alone or with IFNγ reduced the tumor volume and the average number of lesions per kidney. In the 2-month study [175], IFNγ alone was not effective, although it was effective in a study where IFNγ was administered for 10 months [176,177].

While IFNγ treatment inhibited normal cell proliferation, it had little effect on LAM cells or ELT3 cells [178]. These cells showed activation of STAT3 with phosphorylation of Tyr705 and nuclear localization of STAT3; the depletion of STAT3 from these cells inhibited their proliferation and induced apoptosis. Interestingly, IFNγ stimulated STAT3 phosphorylation in LAM and ELT3 cells. Rapamycin reduced IFNγ-stimulated STAT3 phosphorylation but had little effect on basal STAT3 phosphorylation, suggesting that mTOR was involved in IFNγ-stimulated STAT3 phosphorylation. Rapamycin and IFNγ synergistically inhibited cell proliferation. IFNγ also activated STAT1 in the *TSC2*-null cells. It is possible that the effects of IFNγ on cell growth depend on the relative activation of STAT1, which is antiproliferative and proapoptotic, versus STAT3, which induces tumorgenesis. Rapamycin, by reducing the activation of STAT3 by IFNγ, may shift the balance to STAT1-mediated effects.

The effects of type I IFNβ on ELT3 and LAM cells have been examined [179]. IFNβ was expressed in LAM lung and in LAM cell cultures and inhibited proliferation of LAM and ELT3 cells, but not to the same extent as normal human bronchial fibroblasts from the same patient. This decreased sensitivity by the *TSC2*-null cells to IFNβ was not due to changes in expression of IFN receptors or STAT1 phosphorylation and/or nuclear translocation as compared to wild-type cells. Rapamycin plus IFNβ induced greater inhibition of *TSC2*-null cell proliferation, perhaps additively, and through an increase in apoptosis.

24.10 CONCLUSION

LAM is a multisystem disease, with cystic lung destruction, lymphatic involvement, and abdominal tumors. LAM is associated with the proliferation of LAM cells, which have mutations and loss of function of *TSC1*- or *TSC2*-encoded proteins, leading to constitutive activation of mTOR and abnormal cell growth and proliferation. While treatment with rapamycin results, to some extent, in stabilization of lung function, reduction in tumor volume, and resolution of chylous effusions to a certain extent, these effects reverse when treatment is discontinued as rapamycin seems to be cytostatic, rather than cytocidal, for LAM cells. Many different treatments have been proposed, to be used separately or in conjunction with rapamycin, to counteract the signaling pathways stimulated in LAM directly or as a consequence of rapamycin treatment (Table 24.1). Some of these treatments have been tested in clinical trials (Table 24.1) or are currently under study (Table 24.2).

TABLE 24.1 Treatments in LAM and TSC

Treatment	Target	Result in *TSC2*-null cells	Result in animal models	Result in LAM and TSC patients
Rapamycin	mTOR	Inhibited cell growth/proliferation [48,180,181]	Decrease in tumor volume and number [104,105,130,175–177]	Improvements in lung function; decreases in AML volume; has effect as long as drug is continued [106–113]
Anti-EGFR antibodies	EGFR	Inhibited cell growth/proliferation of *TSC2*−/− AML cells [122]	Decrease in lung nodules/degeneration [182]	Not done
Statins	GTPases Rho and Rheb	Simvastatin inhibited cell growth, promoted apoptosis of *TSC2*-null cells [100]; simvastatin, but not atorvastatin, inhibited proliferation of *TSC2*-null and LAMD cells [129]; atorvastatin inhibits proliferation of *Tsc2*−/− MEFs and ELT3 cells [127]	Simvastatin increased survival and decreased tumor volume in nude mice with subcutaneous injection of ELT3 cells [100]; atorvastatin treatment resulted in no reduction in tumor size in *Tsc2*+/− mice [128]; atorvastatin treatment resulted in no decrease in tumor volume or increase in survival of nude mice with *Tsc2*−/− tumors [130]	Retrospective studies showed no improvement in lung function [133,135]; see Table 24.2
Salirasib	Rheb	Inhibited proliferation and migration of ELT3 cells [132]	Inhibited tumor growth in nude mouse model of subcutaneously injected ELT3 cells [132]	Not done
Chloroquine/plaquenil/resveratrol	Autophagy	Resveratrol plus rapamycin promoted apoptosis of *TSC2*-deficient cells [141]	Resveratrol plus rapamycin reduced tumor size and resulted in fewer lung metastases in CB17-*SCID* mice injected with ELT3 cells [141,142]	Not done; see Table 24.2
Dasatinib/saracatinib	Src kinase	Reduced migration and invasiveness of *TSC2*−/− cells [144]	Reduced metastatic potential of *TSC2*-null cells [144]	Not done; see Table 24.2
Doxycycline	MMPs	Decreased MMP levels and cell metabolic activity [151]; decreased cell number and increased apoptosis [150]	No decrease in tumor volume or increase in survival of nude mice with *Tsc2*−/− tumors [130]	Case study showed increase in lung capacity and oxygen saturation [149]; randomized placebo-controlled study showed no effect on FEV₁ decline [152]

(Continued)

TABLE 24.1   (Continued)

Treatment	Target	Result in *TSC2*-null cells	Result in animal models	Result in LAM and TSC patients
Progesterone	Progesterone receptor	Progesterone did not affect proliferation of LAM-derived cells; modestly increased proliferation of ELT3 cells; and in combination with estradiol, significantly increased proliferation of both types of cells [166]	Progesterone plus estradiol promotes lung metastasis of *Tsc2*-deficient cells more potently than estradiol alone [166]	Retrospective study with progesterone showed no improvement of lung function [163]
Faslodex	Estrogen receptor	Not done	No effect on primary *TSC2*-deficient tumors; blocked lung metastases and improved survival in the presence of estradiol [162]	Not done
Tamoxifen	Estrogen receptor agonist or antagonist	Increased proliferation of LAM cells from AML [158]; decreased proliferation of ELT3 cells [160]	Reduced incidence and severity of liver hemangiomas in *Tsc1*$^{+/-}$ mice [159]; increased tumor latency and decreased tumor size in nude mouse xenograft model [160]	Not done
Letrozole	Aromatase	Not done	Not done	See Table 24.2
Celecoxib, aspirin	COX-1, COX-2 COX-2	Aspirin decreased proliferation of *TSC2*-deficient cells [170]	Celecoxib suppressed renal lesions in *Tsc2*$^{+/-}$ mice [170] Aspirin decreased tumor size in mice with ELT3 xenograft tumors [170]	Not done
Interferons	Stat1 and Stat3 expression	IFNβ promotes proapoptotic activity [179]; IFNγ induces apoptosis [173]; IFNγ has little effect on LAMD or ELT3 cell growth [178]	Long-term administration of IFNγ decreased tumor volume [175–177]	Not done
Sorafenib	VEGFR-1, VEGFR-2, VEGFR-3	Not done	No effect on tumor volume or mouse survival alone; combination with rapamycin showed smaller tumor volume, improved survival versus rapamycin alone [130]	Not done

TABLE 24.2   Ongoing Clinical Trials in LAM (from ClinicalTrials.gov)

Trial number	Name	Outcome
NCT02116712	The Tolerability of Saracatinib in Subjects with LAM (SLAM-1)	Safety
NCT01353209	Trial of Aromatase Inhibition in LAM (TRAIL)	Change in FEV$_1$ versus placebo
NCT01687179	Safety Study of Sirolimus and Hydroxychloroquine in Women with LAM (SAIL)	Safety
NCT02061397	Safety of Simvastatin in LAM and TSC (SOS)	Safety

References

[1] Taveira-DaSilva AM, Moss J. Management of lymphangioleiomyomatosis. F1000Prime Rep 2014;6:116.

[2] Ferrans VJ, Yu ZX, Nelson WK, Valencia JC, Tatsuguchi A, Avila NA, et al. Lymphangioleiomyomatosis (LAM): a review of clinical and morphological features. J Nippon Med Sch 2000;67:311–29.

[3] Matsumoto Y, Horiba K, Usuki J, Chu SC, Ferrans VJ, Moss J. Markers of cell proliferation and expression of melanosomal antigen in lymphangioleiomyomatosis. Am J Respir Cell Mol Biol 1999;21:327–36.

[4] Pacheco-Rodriguez G, Kumaki F, Steagall WK, Zhang Y, Ikeda Y, Lin JP, et al. Chemokine-enhanced chemotaxis of lymphangioleiomyomatosis cells with mutations in the tumor suppressor TSC2 gene. J Immunol 2009;182:1270−7.

[5] Valencia JC, Steagall WK, Zhang Y, Fetsch P, Abati A, Tsukada K, et al. Antibody alphaPEP13h reacts with lymphangioleiomyomatosis cells in lung nodules. Chest 2015;147:771−7.

[6] Harknett EC, Chang WY, Byrnes S, Johnson J, Lazor R, Cohen MM, et al. Use of variability in national and regional data to estimate the prevalence of lymphangioleiomyomatosis. Q J Med 2011;104:971−9.

[7] Curatolo P, Bombardieri R. Tuberous sclerosis. Handb Clin Neurol 2008;87:129−51.

[8] Smolarek TA, Wessner LL, McCormack FX, Mylet JC, Menon AG, Henske EP. Evidence that lymphangiomyomatosis is caused by TSC2 mutations: chromosome 16p13 loss of heterozygosity in angiomyolipomas and lymph nodes from women with lymphangiomyomatosis. Am J Hum Genet 1998;62:810−15.

[9] Carsillo T, Astrinidis A, Henske EP. Mutations in the tuberous sclerosis complex gene TSC2 are a cause of sporadic pulmonary lymphangioleiomyomatosis. Proc Natl Acad Sci USA 2000;97:6085−90.

[10] Yu J, Astrinidis A, Henske EP. Chromosome 16 loss of heterozygosity in tuberous sclerosis and sporadic lymphangioleiomyomatosis. Am J Respir Crit Care Med 2001;164:1537−40.

[11] Knudson Jr. AG. Mutation and cancer: statistical study of retinoblastoma. Proc Natl Acad Sci USA 1971;68:820−3.

[12] Berger AH, Knudson AG, Pandolfi PP. A continuum model for tumour suppression. Nature 2011;476:163−9.

[13] Lesma E, Ancona S, Sirchia SM, Orpianesi E, Grande V, Colapietro P, et al. TSC2 epigenetic defect in primary LAM cells. Evidence of an anchorage-independent survival. J Cell Mol Med 2014;18:766−79.

[14] Crooks DM, Pacheco-Rodriguez G, DeCastro RM, McCoy Jr. JP, Wang JA, Kumaki F, et al. Molecular and genetic analysis of disseminated neoplastic cells in lymphangioleiomyomatosis. Proc Natl Acad Sci USA 2004;101:17462−7.

[15] Cai X, Pacheco-Rodriguez G, Fan QY, Haughey M, Samsel L, El-Chemaly S, et al. Phenotypic characterization of disseminated cells with TSC2 loss of heterozygosity in patients with lymphangioleiomyomatosis. Am J Respir Crit Care Med 2010;182:1410−18.

[16] Cai X, Pacheco-Rodriguez G, Haughey M, Samsel L, Xu S, Wu HP, et al. Sirolimus decreases circulating lymphangioleiomyomatosis cells in patients with lymphangioleiomyomatosis. Chest 2014;145:108−12.

[17] Bittmann I, Rolf B, Amann G, Lohrs U. Recurrence of lymphangioleiomyomatosis after single lung transplantation: new insights into pathogenesis. Hum Pathol 2003;34:95−8.

[18] Karbowniczek M, Astrinidis A, Balsara BR, Testa JR, Lium JH, Colby TV, et al. Recurrent lymphangiomyomatosis after transplantation: genetic analyses reveal a metastatic mechanism. Am J Respir Crit Care Med 2003;167:976−82.

[19] Steagall WK, Zhang L, Cai X, Pacheco-Rodriguez G, Moss J. Genetic heterogeneity of circulating cells from patients with lymphangioleiomyomatosis with and without lung transplantation. Am J Respir Crit Care Med 2015;191:854−6.

[20] van Slegtenhorst M, de Hoogt R, Hermans C, Nellist M, Janssen B, Verhoef S, et al. Identification of the tuberous sclerosis gene TSC1 on chromosome 9q34. Science 1997;277:805−8.

[21] Consortium, T.E.C.T.S. Identification and characterization of the tuberous sclerosis gene on chromosome 16. Cell 1993;75:1305−15.

[22] Dibble CC, Elis W, Menon S, Qin W, Klekota J, Asara JM, et al. TBC1D7 is a third subunit of the TSC1-TSC2 complex upstream of mTORC1. Mol Cell 2012;47:535−46.

[23] Schultz J, Milpetz F, Bork P, Ponting CP. SMART, a simple modular architecture research tool: identification of signaling domains. Proc Natl Acad Sci USA 1998;95:5857−64.

[24] Letunic I, Doerks T, Bork P. SMART: recent updates, new developments and status in 2015. Nucleic Acids Res 2014;43:D257−60.

[25] Rosner M, Freilinger A, Hengstschlager M. Proteins interacting with the tuberous sclerosis gene products. Amino Acids 2004;27:119−28.

[26] Krymskaya VP. Tumour suppressors hamartin and tuberin: intracellular signalling. Cell Signal 2003;15:729−39.

[27] Lamb RF, Roy C, Diefenbach TJ, Vinters HV, Johnson MW, Jay DG, et al. The TSC1 tumour suppressor hamartin regulates cell adhesion through ERM proteins and the GTPase Rho. Nat Cell Biol 2000;2:281−7.

[28] Frasa MAM, Koessmeier KT, Ahmadian MR, Braga VMM. Illuminating the functional and structural repertoire of human TBC/RABGAPs. Nat Rev Mol Cell Biol 2012;13:67−73.

[29] Santiago Lima AJ, Hoogeveen-Westerveld M, Nakashima A, Maat-Kievit A, van den Ouweland A, Halley D, et al. Identification of regions critical for the integrity of the TSC1-TSC2-TBC1D7 complex. PLoS One 2014;9:e93940.

[30] Inoki K, Li Y, Zhu T, Wu J, Guan KL. TSC2 is phosphorylated and inhibited by Akt and suppresses mTOR signalling. Nat Cell Biol 2002;4:648−57.

[31] Huang J, Manning BD. The TSC1-TSC2 complex: a molecular switchboard controlling cell growth. Biochem J 2008;412:179−90.

[32] Li Y, Inoki K, Vacratsis P, Guan KL. The p38 and MK2 kinase cascade phosphorylates tuberin, the tuberous sclerosis 2 gene product, and enhances its interaction with 14-3-3. J Biol Chem 2003;278:13663−71.

[33] Inoki K, Zhu T, Guan KL. TSC2 mediates cellular energy response to control cell growth and survival. Cell 2003;115:577−90.

[34] Sauer E, Imseng S, Maier T, Hall MN. Conserved sequence motifs and the structure of the mTOR kinase domain. Biochem Soc Trans 2013;41:889−95.

[35] Yang HJ, Rudge DG, Koos JD, Vaidialingam B, Yang HJ, Pavletich NP. mTOR kinase structure, mechanism and regulation. Nature 2013;497:217−23.

[36] Kaizuka T, Hara T, Oshiro N, Kikkawa U, Yonezawa K, Takehana K, et al. Tti1 and Tel2 are critical factors in mammalian target of rapamycin complex assembly. J Biol Chem 2010;285:20109−16.

[37] Laplante M, Sabatini DM. mTOR signaling at a glance. J Cell Sci 2009;122:3589−94.

[38] Laplante M, Sabatini DM. mTOR signaling in growth control and disease. Cell 2012;149:274−93.

[39] Yip CK, Murata K, Walz T, Sabatini DM, Kang SA. Structure of the human mTOR complex I and its implications for rapamycin inhibition. Mol Cell 2010;38:768−74.

[40] Jain A, Arauz E, Aggarwal V, Ikon N, Chen J, Ha T. Stoichiometry and assembly of mTOR complexes revealed by single-molecule pulldown. Proc Natl Acad Sci USA 2014;111:17833−8.

[41] Takei N, Nawa H. mTOR signaling and its roles in normal and abnormal brain development. Front Mol Neurosci 2014;7:28.

[42] Zheng X, Liang Y, He Q, Yao R, Bao W, Bao L, et al. Current models of mammalian target of rapamycin complex 1 (mTORC1) activation by growth factors and amino acids. Int J Mol Sci 2014;15:20753−69.

[43] Cargnello M, Tcherkezian J, Roux PP. The expanding role of mTOR in cancer cell growth and proliferation. Mutagenesis 2015;30:169−76.

[44] Cai SL, Tee AR, Short JD, Bergeron JM, Kim J, Shen J, et al. Activity of TSC2 is inhibited by AKT-mediated phosphorylation and membrane partitioning. J Cell Biol 2006;173:279−89.

[45] Sancak Y, Peterson TR, Shaul YD, Lindquist RA, Thoreen CC, Bar-Peled L, et al. The rag GTPases bind raptor and mediate amino acid signaling to mTORC1. Science 2008;320:1496−501.

[46] Goncharova EA, Goncharov DA, Spaits M, Noonan DJ, Talovskaya E, Eszterhas A, et al. Abnormal growth of smooth muscle-like cells in lymphangioleiomyomatosis: role for tumor suppressor TSC2. Am J Respir Cell Mol Biol 2006;34:561−72.

[47] Robb VA, Astrinidis A, Henske EP. Frequent [corrected] hyperphosphorylation of ribosomal protein S6 [corrected] in lymphangioleiomyomatosis-associated angiomyolipomas. Mod Pathol 2006;19:839−46.

[48] Goncharova EA, Goncharov DA, Eszterhas A, Hunter DS, Glassberg MK, Yeung RS, et al. Tuberin regulates p70 S6 kinase activation and ribosomal protein S6 phosphorylation. A role for the TSC2 tumor suppressor gene in pulmonary lymphangioleiomyomatosis (LAM). J Biol Chem 2002;277:30958−67.

[49] Dowling RJO, Topisirovic I, Alain T, Bidinosti M, Fonseca BD, Petroulakis E, et al. mTORC1-mediated cell proliferation, but not cell growth, controlled by the 4E-BPs. Science 2010;328:1172−6.

[50] Dorrello NV, Peschiaroli A, Guardavaccaro D, Colburn NH, Sherman NE, Pagano M. S6K1- and bTRCP-mediated degradation of PDCD4 promotes protein translation and cell growth. Science 2006;314:467−71.

[51] Fonseca BD, Zakaria C, Jia JJ, Graber TE, Svitkin Y, Tahmasebi S, et al. La-related protein 1 (LARP1) represses terminal oligopyrimidine (TOP) mRNA translation downstream of mTOR complex 1 (mTORC1). J Biol Chem 2015;290:15996−6020.

[52] Tcherkezian J, Cargnello M, Romeo Y, Huttlin EL, Lavoie G, Gygi SP, et al. Proteomic analysis of cap-dependent translation identifies LARP1 as a key regulator of 5′ TOP mRNA translation. Genes Dev 2014;28:357−71.

[53] Porstmann T, Santos CR, Griffiths B, Cully M, Wu M, Leevers S, et al. SREBP activity is regulated by mTORC1 and contributes to Akt-dependent cell growth. Cell Metab 2008;8:224−36.

[54] Duvel K, Yecies JL, Menon S, Raman P, Lipovsky AI, Souza AL, et al. Activation of a metabolic gene regulatory network downstream of mTOR complex 1. Mol Cell 2010;39:171−83.

[55] Priolo C, Ricoult SJ, Khabibullin D, Filippakis H, Yu J, Manning BD, et al. TSC2 loss increases lysophosphatidylcholine synthesis in lymphangioleiomyomatosis. Am J Respir Cell Mol Biol 2015;53:33−41.

[56] Robitaille AM, Christen S, Shimobayashi M, Cornu M, Fava LL, Moes S, et al. Quantitative phosphoproteomics reveal mTORC1 activates de novo pyrimidine synthesis. Science 2013;339: 1320−3.

[57] Ben-Sahra I, Howell JJ, Asara JM, Manning BD. Stimulation of de novo pyrimidine synthesis by growth signaling through mTOR and S6K1. Science 2013;339:1323−8.

[58] Yang Y, Sun M, Wang L, Jiao B. HIFs, angiogenesis, and cancer. J Cell Biochem 2013;114:967−74.

[59] Harris AL. Hypoxia—a key regulatory factor in tumour growth. Nat Rev Cancer 2002;2:38−47.

[60] Shibaji T, Nagao M, Ikeda N, Kanehiro H, Hisanaga M, Ko S, et al. Prognostic significance of HIF-1 alpha overexpression in human pancreatic cancer. Anticancer Res 2003;23:4721−7.

[61] Sumiyoshi Y, Kakeji Y, Egashira A, Mizokami K, Orita H, Maehara Y. Overexpression of hypoxia-inducible factor 1alpha and p53 is a marker for an unfavorable prognosis in gastric cancer. Clin Cancer Res 2006;12:5112−17.

[62] Schindl M, Schoppmann SF, Samonigg H, Hausmaninger H, Kwasny W, Gnant M, et al. Overexpression of hypoxiainducible factor 1alpha is associated with an unfavorable prognosis in lymph node-positive breast cancer. Clin Cancer Res 2002;8:1831−7.

[63] Karar J, Maity A. PI3K/AKT/mTOR pathway in angiogenesis. Front Mol Neurosci 2011;4:51.

[64] Brugarolas JB, Vazquez F, Reddy A, Sellers WR, Kaelin Jr. WG. TSC2 regulates VEGF through mTOR-dependent and -independent pathways. Cancer Cell 2003;4:147−58.

[65] Taveira-DaSilva AM, Stylianou MP, Hedin CJ, Kristof AS, Avila NA, Rabel A, et al. Maximal oxygen uptake and severity of disease in lymphangioleiomyomatosis. Am J Respir Crit Care Med 2003;168:1427−31.

[66] Ikeda Y, Taveira-DaSilva AM, Pacheco-Rodriguez G, Steagall WK, El-Chemaly S, Gochuico BR, et al. Erythropoietin-driven proliferation of cells with mutations in the tumor suppressor gene TSC2. Am J Physiol Lung Cell Mol Physiol 2011;300: L64−72.

[67] Tran J, Rak J, Sheehan C, Saibil SD, LaCasse E, Korneluk RG, et al. Marked induction of the IAP family antiapoptotic proteins survivin and XIAP by VEGF in vascular endothelial cells. Biochem Biophys Res Commun 1999;264:781−8.

[68] Stacker SA, Baldwin ME, Achen MG. The role of tumor lymphangiogenesis in metastatic spread. Faseb J 2002;16:922−34.

[69] Watz H, Engels K, Loeschke S, Amthor M, Kirsten D, Magnussen H. Lymphangioleiomyomatosis-presence of receptor tyrosine kinases and the angiogenesis factor VEGF-A as potential therapeutic targets. Thorax 2007;62:559.

[70] Kumasaka T, Seyama K, Mitani K, Sato T, Souma S, Kondo T, et al. Lymphangiogenesis in lymphangioleiomyomatosis: its implication in the progression of lymphangioleiomyomatosis. Am J Surg Pathol 2004;28:1007−16.

[71] Seyama K, Kumasaka T, Souma S, Sato T, Kurihara M, Mitani K, et al. Vascular endothelial growth factor-D is increased in serum of patients with lymphangioleiomyomatosis. Lymphat Res Biol 2006;4:143−52.

[72] Glasgow CG, Avila NA, Lin JP, Stylianou MP, Moss J. Serum vascular endothelial growth factor-D levels in patients with lymphangioleiomyomatosis reflect lymphatic involvement. Chest 2009;135:1293−300.

[73] Kumasaka T, Seyama K, Mitani K, Souma S, Kashiwagi S, Hebisawa A, et al. Lymphangiogenesis-mediated shedding of LAM cell clusters as a mechanism for dissemination in lymphangioleiomyomatosis. Am J Surg Pathol 2005;29:1356−66.

[74] Issaka RB, Oommen S, Gupta SK, Liu G, Myers JL, Ryu JH, et al. Vascular endothelial growth factors C and D induces proliferation of lymphangioleiomyomatosis cells through autocrine crosstalk with endothelium. Am J Pathol 2009;175:1410−20.

[75] Patel AS, Morse D, Choi AM. Regulation and functional significance of autophagy in respiratory cell biology and disease. Am J Respir Cell Mol Biol 2013;48:1−9.

[76] Mizushima N. The role of the Atg1/ULK1 complex in autophagy regulation. Curr Opin Cell Biol 2010;22:132−9.

[77] Alers S, Loffler AS, Wesselborg S, Stork B. Role of AMPKmTOR-Ulk1/2 in the regulation of autophagy: cross talk, shortcuts, and feedbacks. Mol Cell Biol 2012;32:2−11.

[78] Usuki J, Horiba K, Chu SC, Moss J, Ferrans VJ. Immunohistochemical analysis of proteins of the Bcl-2 family in pulmonary lymphangioleiomyomatosis: association of Bcl-2 expression with hormone receptor status. Arch Pathol Lab Med 1998;122:895−902.

[79] Mita AC, Mita MM, Nawrocki ST, Giles FJ. Survivin: key regulator of mitosis and apoptosis and novel target for cancer therapeutics. Clin Cancer Res 2008;14:5000−5.

[80] Salvesen GS, Duckett CS. IAP proteins: blocking the road to death's door. Nat Rev Mol Cell Biol 2002;3:401—10.

[81] Wurstle ML, Laussmann MA, Rehm M. The central role of initiator caspase-9 in apoptosis signal transduction and the regulation of its activation and activity on the apoptosome. Exp Cell Res 2012;318:1213—20.

[82] Shin S, Sung BJ, Cho YS, Kim HJ, Ha NC, Hwang JI, et al. An anti-apoptotic protein human survivin is a direct inhibitor of caspase-3 and -7. Biochemistry 2001;40:1117—23.

[83] Vaira V, Lee CW, Goel HL, Bosari S, Languino LR, Altieri DC. Regulation of survivin expression by IGF-1/mTOR signaling. Oncogene 2007;26:2678—84.

[84] Zhao P, Meng Q, Liu LZ, You YP, Liu N, Jiang BH. Regulation of survivin by PI3K/Akt/p70S6K1 pathway. Biochem Biophys Res Commun 2010;395:219—24.

[85] Kanwar JR, Kamalapuram SK, Kanwar RK. Targeting survivin in cancer: the cell-signalling perspective. Drug Discov Today 2011;16:485—94.

[86] Carelli S, Lesma E, Paratore S, Grande V, Zadra G, Bosari S, et al. Survivin expression in tuberous sclerosis complex cells. Mol Med 2007;13:166—77.

[87] Huang J, Wu S, Wu CL, Manning BD. Signaling events downstream of mammalian target of rapamycin complex 2 are attenuated in cells and tumors deficient for the tuberous sclerosis complex tumor suppressors. Cancer Res 2009;69: 6107—14.

[88] Huang J, Dibble CC, Matsuzaki M, Manning BD. The TSC1-TSC2 complex is required for proper activation of mTOR complex 2. Mol Cell Biol 2008;28:4104—15.

[89] Sarbassov DD, Ali SM, Kim DH, Guertin DA, Latek RR, Erdjument-Bromage H, et al. Rictor, a novel binding partner of mTOR, defines a rapamycin-insensitive and raptor-independent pathway that regulates the cytoskeleton. Curr Biol 2004;14:1296—302.

[90] Jacinto E, Loewith R, Schmidt A, Lin S, Ruegg MA, Hall A, et al. Mammalian TOR complex 2 controls the actin cytoskeleton and is rapamycin insensitive. Nat Cell Biol 2004;6:1122—8.

[91] Zinzalla V, Stracka D, Oppliger W, Hall MN. Activation of mTORC2 by association with the ribosome. Cell 2011;144: 757—68.

[92] Oh WJ, Wu CC, Kim SJ, Facchinetti V, Julien LA, Finlan M, et al. mTORC2 can associate with ribosomes to promote cotranslational phosphorylation and stability of nascent Akt polypeptide. EMBO J 2010;29:3939—51.

[93] Foster DA. Phosphatidic acid signaling to mTOR: signals for the survival of human cancer cells. Biochim Biophys Acta 2009;1791:949—55.

[94] Fang Y, Vilella-Bach M, Bachmann R, Flanigan A, Chen J. Phosphatidic acid-mediated mitogenic activation of mTOR signaling. Science 2001;294:1942—5.

[95] Toschi A, Lee E, Xu L, Garcia A, Gadir N, Foster DA. Regulation of mTORC1 and mTORC2 complex assembly by phosphatidic acid: competition with rapamycin. Mol Cell Biol 2009;29:1411—20.

[96] Sarbassov DD, Ali SM, Sengupta S, Sheen JH, Hsu PP, Bagley AF, et al. Prolonged rapamycin treatment inhibits mTORC2 assembly and Akt/PKB. Mol Cell 2006;22:159—68.

[97] Murali A, Rajalingam K. Small Rho GTPases in the control of cell shape and mobility. Cell Mol Life Sci 2014;71:1703—21.

[98] Goncharova EA, Goncharov DA, Lim PN, Noonan D, Krymskaya VP. Modulation of cell migration and invasiveness by tumor suppressor TSC2 in lymphangioleiomyomatosis. Am J Respir Cell Mol Biol 2006;34:473—80.

[99] Astrinidis A, Cash TP, Hunter DS, Walker CL, Chernoff J, Henske EP. Tuberin, the tuberous sclerosis complex 2 tumor suppressor gene product, regulates Rho activation, cell adhesion and migration. Oncogene 2002;21:8470—6.

[100] Goncharova EA, Goncharov DA, Li H, Pimtong W, Lu S, Khavin I, et al. mTORC2 is required for proliferation and survival of TSC2-null cells. Mol Cell Biol 2011;31:2484—98.

[101] Lacher MD, Pincheira R, Zhu Z, Camoretti-Mercado B, Matli M, Warren RS, et al. Rheb activates AMPK and reduces p27Kip1 levels in Tsc2-null cells via mTORC1-independent mechanisms: implications for cell proliferation and tumorigenesis. Oncogene 2010;29:6543—56.

[102] Lacher MD, Pincheira RJ, Castro AF. Consequences of interrupted Rheb-to-AMPK feedback signaling in tuberous sclerosis complex and cancer. Small GTPases 2011;2:211—16.

[103] Rozengurt E, Soares HP, Sinnet-Smith J. Suppression of feedback loops mediated by PI3K/mTOR induces multiple overactivation of compensatory pathways: an unintended consequence leading to drug resistance. Mol Cancer Ther 2014;13:2477—88.

[104] Yeung RS. Lessons from the eker rat model: from cage to bedside. Curr Mol Med 2004;4:799—806.

[105] Kenerson H, Dundon TA, Yeung RS. Effects of rapamycin in the Eker rat model of tuberous sclerosis complex. Pediatr Res 2005;57:67—75.

[106] Bissler JJ, McCormack FX, Young LR, Elwing JM, Chuck G, Leonard JM, et al. Sirolimus for angiomyolipoma in tuberous sclerosis complex or lymphangioleiomyomatosis. N Engl J Med 2008;358:140—51.

[107] Davies DM, de Vries PJ, Johnson SR, McCartney DL, Cox JA, Serra AL, et al. Sirolimus therapy for angiomyolipoma in tuberous sclerosis and sporadic lymphangioleiomyomatosis: a phase 2 trial. Clin Cancer Res 2011;17:4071—81.

[108] McCormack FX, Inoue Y, Moss J, Singer LG, Strange C, Nakata K, et al. Efficacy and safety of sirolimus in lymphangioleiomyomatosis. N Engl J Med 2011;364:1595—606.

[109] Bissler JJ, Kingswood JC, Radzikowska E, Zonnenberg BA, Frost M, Belousova E, et al. Everolimus for angiomyolipoma associated with tuberous sclerosis complex or sporadic lymphangioleiomyomatosis (EXIST-2): a multicentre, randomised, double-blind, placebo-controlled trial. Lancet 2013;381:817—24.

[110] Taveira-DaSilva AM, Hathaway O, Stylianou M, Moss J. Changes in lung function and chylous effusion in patients with lymphangioleiomyomatosis treated with sirolimus. Ann Intern Med 2011;154:797—805.

[111] Ando K, Kurihara M, Kataoka H, Ueyama M, Togo S, Sato T, et al. Efficacy and safety of low-dose sirolimus for treatment of lymphangioleiomyomatosis. Respir Invest 2013;51:175—83.

[112] Mohammadieh AM, Bowler SD, Silverstone EJ, Glanville AR, Yates DH. Everolimus treatment of abdominal lymphangioleiomyoma in five women with sporadic lymphangioleiomyomatosis. Med J Aust 2013;199:121—3.

[113] Yao J, Taveira-DaSilva AM, Jones AM, Julien-Williams P, Stylianou M, Moss J. Sustained effects of sirolimus on lung function and cystic lung lesions in lymphangioleiomyomatosis. Am J Respir Crit Care Med 2014;190:1273—82.

[114] Pantel K, Speicher MR. The biology of circulating tumor cells. Oncogene 2015:1—9.

[115] Chen Y, Zheng Y, Foster DA. Phospholipase D confers rapamycin resistance in human breast cancer cells. Oncogene 2003;22:3937—42.

[116] Hornberger TA, Chu WK, Mak YW, Hsiung JW, Huang SA, Chien S. The role of phospholipase D and phosphatidic acid in the mechanical activation of mTOR signaling in skeletal muscle. Proc Natl Acad Sci USA 2006;103:4741—6.

[117] Zheng Y, Rodrik V, Toschi A, Shi M, Hui L, Shen Y, et al. Phospholipase D couples survival and migration signals in

stress response of human cancer cells. J Biol Chem 2006;281:15862−8.

[118] Moghal N, Sternberg PW. Multiple positive and negative regulators of signaling by the EGF-receptor. Curr Opin Cell Biol 1999;11:190−6.

[119] Wieduwilt MJ, Moasser MM. The epidermal growth factor receptor family: biology driving targeted therapeutics. Cell Mol Life Sci 2008;65:1566−84.

[120] Hynes NE, Lane HA. ERBB receptors and cancer: the complexity of targeted inhibitors. Nat Rev Cancer 2005;5:341−54.

[121] Lesma E, Grande V, Carelli S, Brancaccio D, Canevini MP, Alfano RM, et al. Isolation and growth of smooth muscle-like cells derived from tuberous sclerosis complex-2 human renal angiomyolipoma: epidermal growth factor is the required growth factor. Am J Pathol 2005;167:1093−103.

[122] Lesma E, Grande V, Ancona S, Carelli S, Di Giulio AM, Gorio A. Anti-EGFR antibody efficiently and specifically inhibits human TSC2-/- smooth muscle cell proliferation. Possible treatment options for TSC and LAM. PLoS One 2008;3:e3558.

[123] Lesma E, Eloisa C, Isaia E, Grande V, Ancona S, Orpianesi E, et al. Development of a lymphangioleiomyomatosis model by endonasal administration of human TSC2-/- smooth muscle cells in mice. Am J Pathol 2012;181:947−60.

[124] Sirtori CR. The pharmacology of statins. Pharmacol Res 2014;88:3−11.

[125] Zhou Q, Liao JK. Pleiotropic effects of statins—basic research and clinical perspectives. Circ J 2010;74:818−26.

[126] Riganti C, Aldieri E, Doublier S, Bosia A, Ghigo D. Statins-mediated inhibition of Rho GTPases as a potential tool in anti-tumor therapy. Mini Rev Med Chem 2008;8:609−18.

[127] Finlay GA, Malhowski AJ, Liu Y, Fanburg BL, Kwiatkowski DJ, Toksoz D. Selective inhibition of growth of tuberous sclerosis complex 2 null cells by atorvastatin is associated with impaired Rheb and Rho GTPase function and reduced mTOR/S6 kinase activity. Cancer Res 2007;67:9878−86.

[128] Finlay GA, Malhowski AJ, Polizzi K, Malinowska-Kolodziej I, Kwiatkowski DJ. Renal and liver tumors in Tsc2(+/-) mice, a model of tuberous sclerosis complex, do not respond to treatment with atorvastatin, a 3-hydroxy-3-methylglutaryl coenzyme A reductase inhibitor. Mol Cancer Ther 2009;8:1799−807.

[129] Atochina-Vasserman EN, Goncharov DA, Volgina AV, Milavec M, James ML, Krymskaya VP. Statins in lymphangioleiomyomatosis. Simvastatin and atorvastatin induce differential effects on tuberous sclerosis complex 2-null cell growth and signaling. Am J Respir Cell Mol Biol 2013;49:704−9.

[130] Lee N, Woodrum CL, Nobil AM, Rauktys AE, Messina MP, Dabora SL. Rapamycin weekly maintenance dosing and the potential efficacy of combination sorafenib plus rapamycin but not atorvastatin or doxycycline in tuberous sclerosis preclinical models. BMC Pharmacol 2009;9:8.

[131] Blum R, Cox AD, Kloog Y. Inhibitors of chronically active Ras: potential for treatment of human malignancies. Recent Patents Anticancer Drug Discov 2008;3:31−47.

[132] Makovski V, Haklai R, Kloog Y. Farnesylthiosalicylic acid (salirasib) inhibits Rheb in TSC2-null ELT3 cells: a potential treatment for lymphangioleiomyomatosis. Int J Cancer 2012;130:1420−9.

[133] El-Chemaly S, Taveira-DaSilva A, Stylianou MP, Moss J. Statins in lymphangioleiomyomatosis: a word of caution. Eur Respir J 2009;34:513−14.

[134] Loboda A, Jazwa A, Jozkowicz A, Molema G, Dulak J. Angiogenic transcriptome of human microvascular endothelial cells: effect of hypoxia, modulation by atorvastatin. Vascul Pharmacol 2006;44:206−14.

[135] Taveira-DaSilva AM, Jones AM, Julien-Williams PA, Stylianou M, Moss J. Retrospective review of combined sirolimus and simvastatin therapy in lymphangioleiomyomatosis. Chest 2015;147:180−7.

[136] Yu J, Parkhitko A, Henske EP. Autophagy: an 'Achilles' heel of tumorigenesis in TSC and LAM. Autophagy 2011;7:1400−1.

[137] Thoreen CC, Sabatini DM. Rapamycin inhibits mTORC1, but not completely. Autophagy 2009;5:725−6.

[138] Ng S, Wu Y-T, Chen B, Zhou J, Shen H-M. Impaired autophagy due to constitutive mTOR activation sensitizes TSC2-null cells to cell death under stress. Autophagy 2011;7:1173−86.

[139] Yu J, Parkhitko AA, Henske EP. Mammalian target of rapamycin signaling and autophagy: roles in lymphangioleiomyomatosis therapy. Proc Am Thorac Soc 2010;7:48−53.

[140] Taveira-DaSilva AM, Moss J. Optimizing treatments for lymphangioleiomyomatosis. Expert Rev Respir Med 2012;6:267−76.

[141] Alayev A, Sun Y, Snyder RB, Berger SM, Yu JJ, Holz MK. Resveratrol prevents rapamycin-induced upregulation of autophagy and selectively induces apoptosis in TSC2-deficient cells. Cell Cycle 2014;13:371−82.

[142] Alayev A, Berger SM, Kramer MY, Schwartz NS, Holz MK. The combination of rapamycin and resveratrol blocks autophagy and induces apoptosis in breast cancer cells. J Cell Biochem 2015;116:450−7.

[143] Song K, Shankar E, Yang J, Bane KL, Wahdan-Alaswad R, Danielpour D. Critical role of a survivin/TGF-beta/mTORC1 axis in IGF-I-mediated growth of prostate epithelial cells. PLoS One 2013;8:e61896.

[144] Tyryshkin A, Bhattacharya A, Eissa NT. SRC kinase is a novel therapeutic target in lymphangioleiomyomatosis. Cancer Res 2014;74:1996−2005.

[145] Domblides C, Gross-Goupil M, Quivy A, Ravaud A. Emerging antiangiogenics for renal cancer. Expert Opin Emerg Drugs 2013;18:495−511.

[146] Hayashi T, Fleming MV, Stetler-Stevenson WG, Liotta LA, Moss J, Ferrans VJ, et al. Immunohistochemical study of matrix metalloproteinases (MMPs) and their tissue inhibitors (TIMPs) in pulmonary lymphangioleiomyomatosis (LAM). Hum Pathol 1997;28:1071−8.

[147] Matsui K, Takeda K, Yu ZX, Travis WD, Moss J, Ferrans VJ. Role for activation of matrix metalloproteinases in the pathogenesis of pulmonary lymphangioleiomyomatosis. Arch Pathol Lab Med 2000;124:267−75.

[148] Lee PS, Tsang SW, Moses MA, Trayes-Gibson Z, Hsiao LL, Jensen R, et al. Rapamycin-insensitive up-regulation of MMP2 and other genes in tuberous sclerosis complex 2-deficient lymphangioleiomyomatosis-like cells. Am J Respir Cell Mol Biol 2010;42:227−34.

[149] Moses MA, Harper J, Folkman J. Doxycycline treatment for lymphangioleiomyomatosis with urinary monitoring for MMPs. N Engl J Med 2006;354:2621−2.

[150] Chang WY, Clements D, Johnson SR. Effect of doxycycline on proliferation, MMP production, and adhesion in LAM-related cells. Am J Physiol Lung Cell Mol Physiol 2010;299:L393−400.

[151] Moir LM, Ng HY, Poniris MH, Santa T, Burgess JK, Oliver BG, et al. Doxycycline inhibits matrix metalloproteinase-2 secretion from TSC2-null mouse embryonic fibroblasts and lymphangioleiomyomatosis cells. Br J Pharmacol 2011;164:83−92.

[152] Chang WY, Cane JL, Kumaran M, Lewis S, Tattersfield AE, Johnson SR. A 2-year randomised placebo-controlled trial of doxycycline for lymphangioleiomyomatosis. Eur Respir J 2014;43:1114−23.

[153] Cohen MM, Freyer AM, Johnson SR. Pregnancy experiences among women with lymphangioleiomyomatosis. Respir Med 2009;103:766—72.

[154] Sauter M, Sigl J, Schotten KJ, Gunthner-Biller M, Knabl J, Fischereder M. Association of oestrogen-containing contraceptives with pulmonary lymphangioleiomyomatosis in women with tuberous sclerosis complex—findings from a survey. Eur J Contracept Reprod Health Care 2014;19:39—44.

[155] Banner AS, Carrington CB, Emory WB, Kittle F, Leonard G, Ringus J, et al. Efficacy of oophorectomy in lymphangioleiomyomatosis and benign metastasizing leiomyoma. N Engl J Med 1981;305:204—9.

[156] Yano S, Kobayashi K, Tokuda Y, Touge H, Ikeda T, Ishikawa S, et al. Postmenopausal progression of pulmonary lymphangioleiomyomatosis. Respir Med Extra 2007;3:71—3.

[157] Matsui K, Takeda K, Yu ZX, Valencia J, Travis WD, Moss J, et al. Downregulation of estrogen and progesterone receptors in the abnormal smooth muscle cells in pulmonary lymphangioleiomyomatosis following therapy. An immunohistochemical study. Am J Respir Crit Care Med 2000;161:1002—9.

[158] Yu J, Astrinidis A, Howard S, Henske EP. Estradiol and tamoxifen stimulate LAM-associated angiomyolipoma cell growth and activate both genomic and nongenomic signaling pathways. Am J Physiol Lung Cell Mol Physiol 2004;286: L694—700.

[159] El-Hashemite N, Walker V, Kwiatkowski DJ. Estrogen enhances whereas tamoxifen retards development of Tsc mouse liver hemangioma: a tumor related to renal angiomyolipoma and pulmonary lymphangioleiomyomatosis. Cancer Res 2005;65:2474—81.

[160] Howe SR, Gottardis MM, Everitt JI, Walker C. Estrogen stimulation and tamoxifen inhibition of leiomyoma cell-growth *in-vitro* and *in-vivo*. Endocrinology 1995;136:4996—5003.

[161] Yu JJ, Robb VA, Morrison TA, Ariazi EA, Karbowniczek M, Astrinidis A, et al. Estrogen promotes the survival and pulmonary metastasis of tuberin-null cells. Proc Natl Acad Sci USA 2009;106:2635—40.

[162] Li C, Zhou X, Sun Y, Zhang E, Mancini JD, Parkhitko A, et al. Faslodex inhibits estradiol-induced extracellular matrix dynamics and lung metastasis in a model of lymphangioleiomyomatosis. Am J Respir Cell Mol Biol 2013;49:135—42.

[163] Taveira-DaSilva AM, Stylianou MP, Hedin CJ, Hathaway O, Moss J. Decline in lung function in patients with lymphangioleiomyomatosis treated with or without progesterone. Chest 2004;126:1867—74.

[164] Moss J, DeCastro R, Patronas NJ, Taveira-DaSilva A. Meningiomas in lymphangioleiomyomatosis. JAMA 2001;286: 1879—81.

[165] Gao L, Yue MM, Davis J, Hyjek E, Schuger L. In pulmonary lymphangioleiomyomatosis expression of progesterone receptor is frequently higher than that of estrogen receptor. Virchows Arch 2014;464:495—503.

[166] Sun Y, Zhang E, Lao T, Pereira AM, Li C, Xiong L, et al. Progesterone and estradiol synergistically promote the lung metastasis of tuberin-deficient cells in a preclinical model of lymphangioleiomyomatosis. Horm Cancer 2014;5:284—98.

[167] Gu X, Yu JJ, Ilter D, Blenis N, Henske EP, Blenis J. Integration of mTOR and estrogen-ERK2 signaling in lymphangioleiomyomatosis pathogenesis. Proc Natl Acad Sci USA 2013;110:14960—5.

[168] Shin S, Dimitri CA, Yoon SO, Dowdle W, Blenis J. ERK2 but not ERK1 induces epithelial-to-mesenchymal transformation

via DEF motif-dependent signaling events. Mol Cell 2010;38:114—27.

[169] Sun Y, Gu X, Zhang E, Park MA, Pereira AM, Wang S, et al. Estradiol promotes pentose phosphate pathway addiction and cell survival via reactivation of Akt in mTORC1 hyperactive cells. Cell Death Dis 2014;5:e1231.

[170] Li C, Lee PS, Sun Y, Gu X, Zhang E, Guo Y, et al. Estradiol and mTORC2 cooperate to enhance prostaglandin biosynthesis and tumorigenesis in TSC2-deficient LAM cells. J Exp Med 2014;211:15—28.

[171] Fernandes J, Cobucci RN, Jatoba CA, de Medeiros Fernandes TA, de Azevedo JW, de Araujo JM. The role of the mediators of inflammation in cancer development. Pathol Oncol Res 2015;21:527—34.

[172] Li C, Zhang E, Sun Y, Lee PS, Zhan Y, Guo Y, et al. Rapamycin-insensitive up-regulation of adipocyte phospholipase A2 in tuberous sclerosis and lymphangioleiomyomatosis. PLoS One 2014;9:e104809.

[173] El-Hashemite N, Zhang H, Walker V, Hoffmeister KM, Kwiatkowski DJ. Perturbed IFN-gamma-Jak-signal transducers and activators of transcription signaling in tuberous sclerosis mouse models: synergistic effects of rapamycin-IFN-gamma treatment. Cancer Res 2004;64:3436—43.

[174] El-Hashemite N, Kwiatkowski DJ. Interferon-gamma-Jak-Stat signaling in pulmonary lymphangioleiomyomatosis and renal angiomyolipoma: a potential therapeutic target. Am J Respir Cell Mol Biol 2005;33:227—30.

[175] Messina MP, Rauktys A, Lee L, Dabora SL. Tuberous sclerosis preclinical studies: timing of treatment, combination of a rapamycin analog (CCI-779) and interferon-gamma, and comparison of rapamycin to CCI-779. BMC Pharmacol 2007;7:14.

[176] Lee L, Sudentas P, Donohue B, Asrican K, Worku A, Walker V, et al. Efficacy of a rapamycin analog (CCI-779) and IFN-gamma in tuberous sclerosis mouse models. Genes Chromosomes Cancer 2005;42:213—27.

[177] Lee L, Sudentas P, Dabora SL. Combination of a rapamycin analog (CCI-779) and interferon-gamma is more effective than single agents in treating a mouse model of tuberous sclerosis complex. Genes Chromosomes Cancer 2006;45:933—44.

[178] Goncharova EA, Goncharov DA, Damera G, Tliba O, Amrani Y, Panettieri RA, et al. Signal transducer and activator of transcription 3 is required for abnormal proliferation and survival of TSC2-deficient cells: relevance to pulmonary lymphangioleiomyomatosis. Mol Pharmacol 2009;76:766—77.

[179] Goncharova EA, Goncharov DA, Chisolm A, Spaits MS, Lim PN, Cesarone G, et al. Interferon beta augments tuberous sclerosis complex 2 (TSC2)-dependent inhibition of TSC2-null ELT3 and human lymphangioleiomyomatosis-derived cell proliferation. Mol Pharmacol 2008;73:778—88.

[180] Zhang HB, Cicchetti G, Onda H, Koon HB, Asrican K, Bajraszewski N, et al. Loss of Tsc1/Tsc2 activates mTOR and disrupts PI3K-Akt signaling through downregulation of PDGFR. J Clin Invest 2003;112:1223—33.

[181] Grzegorek I, Zuba-Surma E, Chabowski M, Janczak D, Szuba A, Dziegiel P. Characterization of cells cultured from chylous effusion from a patient with sporadic lymphangioleiomyomatosis. Anticancer Res 2015;35:3341—52.

[182] Lesma E, Chiaramonte E, Ancona S, Orpianesi E, Di Giulio AM, Gorio A. Anti-EGFR antibody reduces lung nodules by inhibition of EGFR-pathway in a model of lymphangioleiomyomatosis. Biomed Res Int 2015;2015:315240.

mTOR Pathway in Renal Cell Carcinoma

Matteo Santoni[1] and Francesco Massari[2]

[1]Clinica di Oncologia Medica, AOU Ospedali Riuniti, Polytechnic University of the Marche Region, Ancona, Italy
[2]Medical Oncology, Azienda Ospedaliera Universitaria Integrata, University of Verona, Verona, Italy

25.1 ROLE OF THE MAMMALIAN TARGET OF RAPAMYCIN PATHWAY IN RENAL CELL CARCINOMA

The mammalian target of rapamycin (mTOR) signaling pathway is implicated in the modulation of cell growth, differentiation, survival, and metabolism, as well as in the regulation of angiogenesis. This pathway is crucial for the development of renal diseases [1]. Indeed, mTOR is involved in the process of recovery from renal ischemia-reperfusion injury [2], in the progression of chronic kidney disease [3], and in the pathogenesis of autosomal dominant polycystic kidney disease [4]. Moreover, it has been shown that mTOR is frequently deregulated in human tumors, including renal cell carcinoma (RCC) (Figure 25.1), thus representing a promising therapeutic target for this tumor [5].

Mutations in the mTOR pathway seem to be correlated with the risk of RCC. Shu et al. investigated the association between single nucleotide polymorphisms (SNPs) of the mTOR pathway and the risk of developing RCC. They evaluated 190 SNPs from 22 genes in the mTOR pathway, finding that six SNPs located in the AKT3 gene were correlated with the risk of RCC [6].

In 2013, the Cancer Genome Atlas Research Network analyzed RCC tumors obtained from 413 nephrectomy specimens by whole-genome and -exome sequencing. They confirmed the presence of mutations in the *von Hippel—Lindau* (*VHL*) tumor suppressor gene, which represents a fundamental step in clear cell RCC (ccRCC) carcinogenesis. Deletions into the *VHL* gene are usually large and located in 3p25 chromosomal region. As a consequence, near tumor suppressor genes, such as BRCA1-associated protein-1, Polybromo-1, and

Set domain-containing 2 can be lost, thus contributing to the development of RCC [7]. Among detected alterations, the authors frequently found mutations in the PI3K/AKT/mTOR pathway [7].

Several techniques have been proposed to detect mTOR pathway activation status in RCC. In 2014, Fiorini and her colleagues reported differences in evaluating mTOR pathway status by immunohistochemistry, Western blot, and immunofluorescence in RCC tissue specimens from 16 patients. Notably, none of the investigated molecules, including mTOR, phospho-mTOR, phospho-p70S6k, phospho-S6Rb, and phospho-4E-binding protein-1 (4E-BP1), was simultaneously assessed by all the techniques (Figure 25.2) [8].

On this scenario, several agents targeting different elements of the PI3K/AKT/mTOR pathway have demonstrated to be effective both *in vitro* and *in vivo* RCC models.

25.2 mTOR INHIBITORS IN RCC

Rapamycin was the first mTOR inhibitor discovered. It is a macrolide antibiotic originally derived from *Streptomyces hygroscopicus*, that acts by binding to the FK506 binding protein 12 (FKBP12) and inhibits the interaction of raptor with mTOR, thus disrupting mTORC1 complex assembly [9]. The subsequent inhibition of p70S6K reduces the translation of several mRNA encoding for ribosomal proteins and elongation factors, thus decreasing protein synthesis, impeding cell-cycle progression through the G1/S transition and resulting in a mid-to-late G1 arrest. Rapalog-mediated mTORC1 inhibition can reduce the phosphorylation of the mTORC2 functional repressor

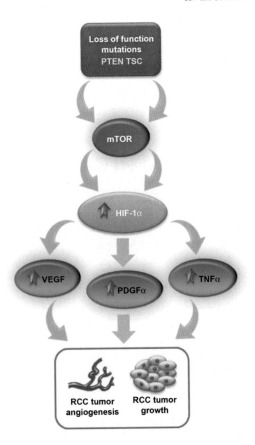

FIGURE 25.1 mTOR pathway in renal cell carcinoma. This figure underlines the crucial role of HIF-1α in regulating the mTOR-induced RCC growth and tumor angiogenesis.

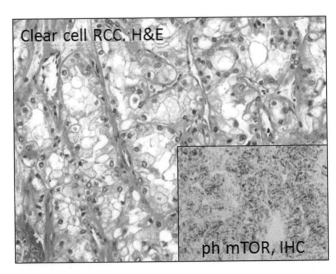

FIGURE 25.2 Clear cell RCC, morphologic view at ×20 HPF (hematoxylin and eosin staining). Details of ph mTOR membranous immunoexpression in neoplastic cells.

(Rictor). As a consequence, mTORC2 kinase activity results increased, thus triggering AKT signal [10,11].

The antitumor effect of rapamycin has been shown in various cancer cell lines [12] and the drug was registered for its dominant immunosuppressive properties, given its ability to inhibit the activation and proliferation of T lymphocytes after stimulation of antigens and cytokines (IL2, IL4, IL15), and the production of antibodies [13].

The rationale for investigating the potential role of rapamycin as an anticancer agent is related to the antiproliferative effects of rapamycin and its derivatives (rapalogs); in particular, three synthetic analogs have been developed, with improved pharmacokinetic (PK) properties and reduced immunosuppressive effects: temsirolimus (CCI-779), everolimus (RAD001), and ridaforolimus (AP23573/MK-8669, formerly deforolimus). While temsirolimus and everolimus are commonly used in the clinical setting, ridaforolimus has no clinical indication yet. Rapalogs share the same mechanism of action as rapamycin, forming a complex with FKBP12 and inhibiting the mTORC1, without impairing mTORC2 function [14].

25.2.1 Temsirolimus

Temsirolimus, a dihydroxymethyl propionic acid ester of rapamycin, is an intravenous mTORC1 inhibitor that works by suppressing the mTOR-mediated phosphorylation of S6K1 and 4E-BP1, decreasing expression of several key proteins involved in the cell-cycle proliferation [15]. It was seen in different preclinical studies that temsirolimus stimulated significant cancer growth inhibitory activity, subsequently confirmed in clinical trials. Temsirolimus was found to be well tolerated in phase I trials; the dose-limiting toxicities were mucositis, depression, thrombocytopenia, and hyperlipidemia [16], and it showed antitumor activity in patients with advanced RCC that had already received prior treatment [17].

Currently, temsirolimus is used for the treatment of advanced RCC with poor prognostic factor; it received FDA approval in 2007 for the treatment of advanced RCC as the first mTORC1 inhibitor [18].

The approval of temsirolimus for the treatment of metastatic RCC arises from the GLOBAL study, an international, multicenter, open-label, three-arm phase III trial that enrolled patients with mRCC. In this study temsirolimus was compared when administered alone or combined with IFN-α, and to a third condition of IFN-α alone [19]. This trial enrolled 626 treatment-naive patients with advanced RCC and a poor prognosis. These patients were defined as poor prognosis as having at least three of the poor prognostic factors, these reflect the five factors identified by Motzer et al. [20], with the addition of multiple sites of metastases [21]: Karnofsky PS inferior to 70%, hemoglobin inferior to lower limit normal, disease-free interval inferior to 1 year from time of initial diagnosis, corrected serum Ca^{2+} superior to 10 mg/dl, lactate dehydrogenase superior to $1.5 \times$ upper limit of normal, number of metastatic sites (1 vs. ≥ 2), prior radiation therapy. The primary endpoint was overall survival (OS) in the intent-to-treat population and secondary endpoints included the progression-free survival (PFS), overall response rate (ORR), and response duration. The temsirolimus-alone arm demonstrated an improvement in median OS compared with IFN-α alone (10.9 vs. 7.3 months; hazard ratio (HR), 0.73; $P = 0.0078$). A significantly longer median PFS time for the temsirolimus group (median, 5.5 vs. 3.1 months; HR, 0.66; unadjusted $P = 0.001$) was also found. The combination of temsirolimus and IFN-α did not prolong OS compared with IFN-α alone (8.4 vs. 7.3 months, respectively). Other interesting data are those relating the secondary endpoints of ORR and response duration. In the study it was demonstrated that the ORR was 8.6% in the temsirolimus versus 4.8% in the IFN-α arm ($P = 0.12$). All of the responses were partial. Median response duration was 11.1 months for temsirolimus, compared to 7.4 months for IFN-α-treated individuals. The observation of improved OS and PFS with temsirolimus led to its approval for use in patients with poor prognosis, advanced RCC [22]. The approved dose is 25 mg IV once weekly.

This agent is the only one to have demonstrated prolonged OS in a phase III trial and the evidence for its efficacy is particularly strong due to the fact that, upon progression, the poor-prognosis patients did not receive subsequent active therapy, therefore the OS endpoint was not altered by crossover of patients to other active treatments.

Another important randomized phase III trial is the INTORSECT study, that compared temsirolimus with sorafenib in patients who had progressed on sunitinib [23]. This trial enrolled patients with both clear cell and non-ccRCC who had progressed following sunitinib. The study failed to demonstrate an improvement in the primary endpoint, PFS, between the arms. The median PFS following independent radiological review was 4.3 months for temsirolimus and 3.9 months for sorafenib (HR 0.87; 95% confidence interval (CI): 0.71—1.07; $P = 0.19$). A very interesting fact is that sorafenib was statistically superior in regards to OS when compared to temsirolimus. The median OS was 16.6 versus 12.3 months for sorafenib and temsirolimus, respectively (HR 1.31; 95% CI: 1.05—1.63; $P = 0.014$).

25.2.2 Everolimus

Everolimus (RAD001), which differs from rapamycin for an O-(2-hydroxyethyl) chain substitution at position C-40, is an orally available rapamycin-related mTORC1 inhibitor, with a better PK profile compared

to rapamycin, characterized by a shorter half-life (28 h instead of 60), a slightly higher bioavailability, and a higher correlation of bioavailability with the administered dosage [24]. In preclinical models the potent antitumor activity of everolimus was associated with a significant suppression of S6K1 and regulation of 4E-BP1 activity [25], carrying to its evaluation in the clinical setting. The 10-mg daily everolimus dosing demonstrated a profound effect on target inhibition, with a tolerable safety profile and predictable PK in patients with advanced solid tumors [26].

Everolimus is indicated for the treatment of patients with advanced RCC that have received prior therapy with sunitinib or sorafenib [27].

The first encouraging data of activity of RAD001 were initially evaluated in a phase II study in 41 patients with clear cell renal cell cancer that it allowed subsequently its evaluation in a large phase III trial [28].

The RECORD-1 trial (Renal Cell cancer treatment with Oral RAD001 given Daily—ClinicalTrials.gov identifier NCT00410124), a randomized, double-blind, placebo-controlled phase III study, compared everolimus alone versus placebo in patients with mRCC. The primary endpoint, median PFS, was significantly longer with everolimus than with placebo (4.9 vs. 1.9 months; $P < 0.0001$). An important proportion of patients in the placebo arm crossed over to everolimus and consequently an insignificant difference in OS between arms was found (median 14.8 vs. 14.4 months; HR 0.87). However, an analysis accounting for crossover demonstrated that survival was 1.9 times longer (95% CI: 0.5—8.5) with everolimus compared to placebo. Among patients treated with everolimus have been detected three partial responses and 171 with stable disease. Nevertheless, most of the included patients who received everolimus were not really second-line. In particular, only 89 patients (21%) received everolimus in the second-line setting. The other 82 patients received everolimus in fifth line, 104 patients in fourth line, and 141 patients in third line. Additionally, second-line everolimus after failure of sunitinib was only administered to 43 patients, with a median PFS of 4.6 versus 1.8 months in 13 patients treated with placebo after failure of sunitinib (HR 0.22, 95% CI 0.09—0.55) [29—31]. Based on these data, everolimus is now considered one option for the second-line treatment of mRCC. The approved dose is 10 mg per os every day.

Another randomized, non-inferiority phase II trial (RECORD-3) compared alternate sequences of drugs in patients with metastatic RCC [32]. The primary objective of the study was to assess PFS non-inferiority of first-line everolimus compared with first-line sunitinib. In this study combined PFS for the two sequences of treatment was also compared. In the trial, 471 patients without prior systemic therapy were enrolled and

TABLE 25.1   mTOR Inhibitors in the Treatment of Advanced RCC

Study	Agents	Line of therapy	mPFS (months)	P value	mOS (months)	P value
GLOBAL	Temsirolimus/IFN temsirolimus alone IFN alone	First	3.7	<0.001	8.4	NS[a]
			3.8		10.9	P < 0.008[a]
			1.9		7.3	
INTORSECT	Temsirolimus sorafenib	Second	4.28	0.193	16.64	P = 0.014
			3.91		12.27	
RECORD-1	Everolimus	Second	4.9	0.001	14.8	P = 0.162
	Placebo		1.9		14.4	

[a]*Comparison to interferon alone.*
mPFS, median progression-free survival; mOS, median overall survival; NS, not significant.

randomized either to a group where sunitinib was received until disease progression, at which time everolimus was initiated, or to a group where everolimus was received initially, followed by sunitinib. Patients sustained a first-line therapy until the evidence of progression and then were crossed over to another treatment. Although the primary objective was to assess the PFS of the first-line treatment, secondary endpoints also evaluated OS and the combined first- and second-line PFS endpoints. Non-inferiority of first-line everolimus was not achieved: as the median PFS was 7.9 months (95% CI: 5.6–8.2) for first-line everolimus and 10.7 months (8.2–11.5) for first-line sunitinib. Attractively, the sequence of sunitinib followed by everolimus demonstrated a trend towards improved combined PFS and median OS compared to the reverse sequence of agents. The combined PFS was 25.8 for the sunitinib followed by everolimus condition versus 21.1 for the everolimus followed by sunitinib condition. Similarly, sunitinib followed by everolimus demonstrated a trend of improved median survival over the reverse sequence, with the median OS being 32.0 months and 22.4 months, respectively.

At the 2015 American Society of Clinical Oncology Annual Meeting the data of a randomized phase II open-label study, ASPEN Trial, were presented that compared everolimus 10 mg daily and sunitinib 50 mg daily (days 1–28 in 6-week cycles) in non-clear-cell renal cell carcinoma (nccRCC). The primary endpoint was PFS and sunitinib significantly prolonged PFS compared to everolimus in previously untreated metastatic nccRCC [33] (Table 25.1).

25.2.3 Combination Therapies with mTOR Inhibitor in mRCC

Enthusiasm to combine targeted agents in mRCC declined as both significant toxicity and minimal benefits have been observed with some combinations [34–38].

An example of a study of combination therapy with mTOR inhibitor in metastatic RCC is the TORAVA trial, a randomized phase II study that compared temsirolimus and bevacizumab to sunitinib or IFN-α and bevacizumab in untreated patients with mRCC [35]. Median PFS was 8.2 months for the bevacizumab and temsirolimus arm, 8.2 months for the sunitinib arm, and 16.8 months for the bevacizumab and IFN-α arm. During the development of the study several toxicities were observed and the trial was stopped because of adverse events (AEs) such as fatigue, skin disorders, proteinuria, and hypertension. Grade 3 or worse AEs, including hemorrhage, venous or arterial thromboembolism, pulmonary interstitial syndrome and gastrointestinal perforation were reported in 77% of the patients in the temsirolimus plus bevacizumab group, compared with 60% in the sunitinib group and 70% in the IFN-α plus bevacizumab group. Serious AEs were reported in 39 of the patients in the temsirolimus plus bevacizumab group, compared with 13 and 18 in the other two groups. Over half the patients stopped the combination treatment for reasons other than progression.

Another study of combination therapy with mTOR inhibitor is the BEST trial, a randomized phase II study, that compared bevacizumab monotherapy 10 mg/kg intravenously every 2 weeks, bevacizumab 10 mg/kg intravenously every 2 weeks plus temsirolimus 25 mg intravenously every week, bevacizumab 5 mg/kg intravenously every 2 weeks plus sorafenib 200 mg orally twice daily (on days 1–5, 8–12, 15–19, and 22–26) and sorafenib 200 mg twice daily plus temsirolimus 25 mg intravenously weekly in untreated patients with mRCC [36]. The median PFS was 7.5 months for bevacizumab alone, 7.6 months for bevacizumab plus temsirolimus, 9.2 months for bevacizumab plus sorafenib, and 7.4 months for sorafenib plus temsirolimus. The AEs were similar among treatment arms, but the activity of sorafenib, temsirolimus, and bevacizumab administered in doublet combinations

TABLE 25.2 mTOR Inhibitors in the Combination Therapy for Advanced Renal Cell Carcinoma

Study	Combination therapy	Control arm	Results
INTORACT (phase IIIb)	Temsirolimus plus bevacizumab	Bevacizumab plus interferon	No differences
			More AE
RECORD-2 (phase II)	Everolimus plus bevacizumab	Bevacizumab plus interferon	No differences
			More AE
BEST (phase II)	Temsirolimus plus bevacizumab	Bevacizumab	No differences
	Bevacizumab plus sorafenib		
	Temsirolimus plus sorafenib		More AE
TORAVA (phase II)	Temsirolimus plus bevacizumab	Bevacizumab plus interferon	No differences
		sunitinib	More AE

Note: AE, adverse events.

TABLE 25.3 Ongoing Trials on mTOR Inhibitors

Inhibitor type	Agents	Phase	Trial Id number	Design
mTORC1 inhibitors	Ridaforolimus	I/IIa	NCT00112372	Patients with advanced solid tumors including mRCC
mTORC1/2 inhibitors	AZD2014	II	NCT01793636	AZD2014 vs. everolimus in mRCC patients who failed prior VEGF-binding or TKI therapy
	AZD8055	I	NCT00973076	Patients with advanced solid tumors including mRCC
	OSI-027	I	NCT00698243	Patients with advanced solid tumors including mRCC
	INK128	I	NCT01058707	Patients with advanced solid tumors including mRCC

VEGF, vascular endothelial growth factor; mRCC, metastatic renal cell carcinoma.

did not significantly improve median PFS in comparison with bevacizumab monotherapy.

Other examples of study of combination therapy with mTOR inhibitor are: the INTORACT trial, that compared temsirolimus plus bevacizumab and bevacizumab plus interferon, and the RECORD-2 Trial that compared everolimus plus bevacizumab and bevacizumab plus interferon; neither of these studies demonstrated the efficacy of the combination therapy [37,38]. Furthermore, at the 2015 American Society of Clinical Oncology Annual Meeting the interesting data from a randomized phase II open-label study that compared lenvatinib 18 mg daily plus everolimus 5 mg daily, lenvatinib 24 mg daily and everolimus 10 mg daily were presented [39]. The primary endpoint was PFS with lenvatinib ± everolimus versus everolimus alone. The median PFS was 14.6 months for lenvatinib plus everolimus, 7.4 months for lenvatinib alone, and 5.5 months for everolimus alone. The AEs were similar among treatment arms and the combination lenvatinib plus everolimus seemed to be highly effective as second-line treatment in mRCC (Table 25.2).

25.2.4 Emerging mTOR/PI3K/AKT Inhibitors in RCC

Several compounds are presently under evaluation in patients with RCC. Some of these agents are designed to selectively target the mTORC1 (ridaforolimus) or to simultaneously inhibit both mTORC1 and mTORC2 (AZD2014, AZD8055, OSI-027, INK128). In addition, agents targeting PI3K (BKM120, XL147, BAY80-6946) or its specific isoforms (BYL719, INK1117), as well as dual PI3K and mTORC1/2 inhibitors (BEZ235, PF-05212384, XL765, GSK2126458, GDC-0980, SF1126) have been developed. Moreover, pan-AKT inhibitors are also under study in this setting (MK-2206, AZD5363, GSK2141795, GDC-0068). The complete list of these compounds and of the characteristics of ongoing clinical trials are reported in Table 25.3—25.5.

Of note, another class of mTOR inhibitor has been developed to directly block the activity of mTOR in a FKBP-independent manner. These agents, named TOR-KIs (TOR-kinase inhibitors), have been shown to inhibit protein synthesis and induce apoptosis more

TABLE 25.4   Ongoing Trials on PI3K Inhibitors

Inhibitor type	Agents	Phase	Trial Id number	Design
Pan-PI3K inhibitors	BKM120	I	NCT01470209	In combination with everolimus in patients with advanced solid tumors including mRCC
		I	NCT01283048	In combination with bevacizumab in patients with mRCC
	XL147	I	NCT00486135	Patients with advanced solid tumors including mRCC
	BAY80-6946	I	NCT01460537	In combination with gemcitabine or cisplatin plus gemcitabine in patients with advanced solid tumors including mRCC
		I	NCT01411410	In combination with paclitaxel in patients with advanced solid tumors including mRCC
Isoform-specific	BYL719	Ib	NCT01928459	Patients with advanced solid tumors including mRCC
PI3K inhibitors	INK1117	I	NCT01449370	Patients with advanced solid tumors including mRCC
Dual PI3K/ mTORC1/2 inhibitors	BEZ235	I/II	NCT01453595	Patients with refractory mRCC
		I	NCT01482156	In combination with everolimus in mRCC patients
	PF-05212384	I	NCT01347866	In combination with MEK inhibitor PD-0325901 in patients with advanced solid tumors including mRCC
	XL765	I	NCT00485719	Patients with advanced solid tumors including mRCC
	GSK2126458	I	NCT01248858	In combination with MEK inhibitor GSK1120212 in patients with advanced solid tumors including mRCC
	GDC-0980	II	NCT01442090	GDC-0980 vs. everolimus in mRCC patients who failed prior VEGF-binding or TKI therapy
	SF1126	I	NCT00907205	Patients with advanced solid tumors including mRCC

Note: VEGF, vascular endothelial growth factor; mRCC, metastatic renal cell carcinoma.

TABLE 25.5   Ongoing Trials on AKT Inhibitors

Inhibitor type	Agents	Phase	Trial Id number	Design
Pan-AKT inhibitors	MK-2206	I	NCT01480154	In combination with hydroxychloroquine in patients with advanced solid tumors including mRCC
		II	NCT01239342	MK-2206 vs. everolimus in patients with refractory mRCC
	AZD5363	I	NCT01895946	Comparison of two formulations in patients with advanced solid tumors including mRCC
	GSK2141795	I	NCT01138085	In combination with MEK inhibitor GSK1120212 in patients with advanced solid tumors including mRCC
	GDC-0068	I	NCT01562275	In combination with MEK inhibitor GSK1120212 in patients with advanced solid tumors including mRCC

Note: mRCC, metastatic renal cell carcinoma.

than rapalogs [40–42]. TOR-KIs include WYE354, WYE132, PP30, PP242, AZD8055, and Torin 1 and should be investigated and compared to rapalogs in prospective clinical trials.

Furthermore, the results of a phase III trial (NCT01668784) comparing everolimus and anti-programmed death-1 agent nivolumab in advanced ccRCC patients pretreated with EGFR-TKIs will be presented at the end of 2015 and will shed light on the future therapeutic scenario for these patients.

25.3 RESISTANCE TO mTOR INHIBITORS

The efficacy of mTOR inhibitors is limited by the development of drug resistance. This is caused by the presence of feedback loops and by the crosstalk with other signaling pathways [43]. Several mechanisms, including PI3K/AKT, hypoxia-inducible factor (HIF), ERK/MAPK, and autophagy contribute to the development of such resistance in RCC patients [44] and are described in the sections below.

25.3.1 PI3K/AKT and PTEN

Rapamycin and its analogs promote a complex sequence of events that lead to the development of drug resistance. Among them, an essential role is played by the feedback activation of AKT [45]. This event enhances cell proliferation and survival and angiogenesis through the phosphorylation of downstream signaling proteins, such as glycogen synthase kinase 3, forkhead box O, B cell lymphoma 2 (Bcl-2) antagonist of cell death (BAD), the E3 ubiquitin-protein ligase MDM2, and p27 [46].

Phosphatase and tension homolog deleted on chromosome ten (PTEN) acts as a negative regulator of the PI3K/AKT axis. In preclinical models, PTEN loss correlates with increased sensitivity to mTOR inhibitors [47]. Interestingly, high levels of the PTEN may coexist with enhanced AKT activation, suggesting that other mechanisms may lead to the attenuation of PTEN activity [48]. This evidence may partially explain the lack of a correlation between the baseline levels of PTEN and the outcome of mRCC patients treated with temsirolimus [49].

25.3.2 Hypoxia-Inducible Factor

HIF majorly contributes to RCC tumorigenesis [50]. It is a transcription factor formed by an α subunit and a β subunit. HIF-α is bound by the VHL tumor suppressor protein [51]. During hypoxia or VHL mutant conditions (which are present in the majority of ccRCC tumors [52,53]), unhydroxylated HIF-α cannot bind to VHL, accumulates into the cell [54] and translocates into the nucleus where it promotes the transcription of target genes [55].

Differences between HIF-1α and HIF-2α in response to mTORC1/2 have been reported. Indeed, while HIF-1α expression is mTORC1/2-dependent, HIF-2α is only dependent on mTORC2 [56]. Based on this finding, the employment of mTORC1 inhibitors is limited by the lack of activity on mTORC2 and consequently on HIIF-2α, which is crucial for RCC development and progression [57,58].

25.3.3 p27

The cyclin-dependent kinase (Cdk) inhibitor p27 is involved in regulating cell-cycle progression. In RCC, p27 is frequently altered. Signaling by PI3K/AKT has been shown to regulate the S-phase kinase associated protein 2 (SKP-2), which forms a complex that mediates the proteosomal degradation of p27 [59]. It has been demonstrated that mTORC2 promotes cell proliferation via the SKP-2-mediated reduction of p27 levels [60]. The use of rapamycin and its analogs can inhibit the down-regulation of p27 [61], suggesting that low p27 levels can be associated with reduced response of RCC cells to these agents.

25.3.4 ERK/MAPK Signaling Pathway

The inhibition of mTORC1 contributes to the PI3K-dependent activation of the ERK/MAPK pathway [45]. It has been reported that the inhibition of PI3K mediated by rapalogs correlates with inhibited ERK activation. Accordingly, use of the MEK1/2 inhibitor U0126 has been shown to enhance the anticancer activity of everolimus [45]. Thus, several MEK inhibitors are under study in mRCC patients, either alone or combined with mTOR/PI3K/AKT inhibitors.

25.3.5 Bcl-2 and Survivin Proteins

The Bcl-2 family is a protein family that includes members with antiapoptotic (i.e., Bcl-2 and Bcl-2 related protein A1) and proapoptotic (i.e., BAK) activity. The imbalance between these proteins modulates the tumor cell susceptibility to apoptosis [62]. In RCC, Bcl-2 overexpression has been shown to inhibit rapalog-induced apoptosis [63].

As for survivin, it is an inhibitor of apoptosis often overexpressed in a vast majority of tumors, including RCC [64]. Different survivin splice variants have been implicated in progression and clinical behavior of RCC [65]. Similarly to Bcl-2, overexpression of survivin is able to inhibit the apoptosis of RCC cells induced by temsirolimus [66].

25.3.6 Reactive Oxygen Species

Reactive oxygen species (ROS) can favor the process of carcinogenesis through inducing genomic instability and can modulate cell proliferation, differentiation, and migration by acting as second messengers in signal transduction [67,68]. In addition, ROS are key factors for the production and release of interleukin 6 (IL-6) and 8 (IL-8) [69], involved in RCC progression and angiogenesis, as well as for the activation of mTOR feedback compensatory loops.

In ccRCC, tumor cells lacking the *VHL* gene produce a higher amount of ROS due to the impairment of the oxidative phosphorylation systems [70]. Increased ROS can directly stimulate several members of the mTOR pathway, such as PI3K and AKT [71], leading to cell proliferation and contributing to acquired resistance to mTOR inhibitors [72]. Moreover, ROS can indirectly contribute to cell survival and drug resistance by preventing the ubiquitination and degradation of HIF-1α [73,74].

25.3.7 Autophagy

Autophagy is a ubiquitous cellular self-degradative process used by cells to eliminate cytoplasmic components via an autophagosomal–lysosomal pathway under particular conditions. Hypoxia, growth factor withdrawal, and other forms of metabolic stress can trigger autophagy in physiological conditions to sustain cellular biosynthesis and energy supply [75]. In RCC, autophagy is also involved in ccRCC resistance to immune response. Indeed, the autophagy sensor inositol 1,4,5-trisphosphate receptor type 1 has been shown to act as a potent regulator of natural killer-mediated lysis by a mechanism involving HIF-2α stabilization [76].

Under glucose starvation, AMPK promotes autophagy through the direct activation of autophagy-initiating kinase Ulk1 (a homolog of yeast ATG1) by phosphorylating Ser 317 and Ser 777. As a consequence, ULK1 can activate FIP200, a fundamental step for the formation of autophagosomes, thus initiating the autophagic process [77–79] (Figure 25.3). On the other hand, mTOR can prevent Ulk1 activation by phosphorylating Ulk1 Ser 757 under normal nutrient conditions [80]. The pharmacological inhibition of mTOR through rapalogs can mimic hypoxia, leading to the dissociation of mTOR from the complex ATG13-ULK1 and ULK2 and promoting the autophagic flux and acquired resistance to this therapeutic approach. Accordingly, the knockdown of Atg7, another key autophagic machinery protein or the use of chloroquine, an autophagy blocker, can enhance the *in vitro* and *in vivo* cytotoxic effects of mTOR inhibitors [81].

Interestingly, sunitinib has been shown to induce cell death by disrupting autophagy, while pazopanib induced complete autophagy in *in vitro* models of bladder cancer [82]. Taken together, these data suggest that the predictive and prognostic role of autophagy biomarkers should be assessed in future clinical studies.

25.3.8 Aurora Kinase and Notch Pathways

Aurora kinases are frequently overexpressed or mutated in RCC [83]. It has been shown that silencing of Aurora kinases leads to inhibited proliferation and metastasis in ccRCC [84].

Panobinostat (LBH589) is a pan-deacetylases (DACs) inhibitor able to cause G2–M arrest and apoptosis of RCC [85]. The suppression of Aurora-A kinase expression using shRNA leads to RCC growth inhibition and enhances the sensitivity of RCC cells towards rapalogs or TOR-KIs activity [86].

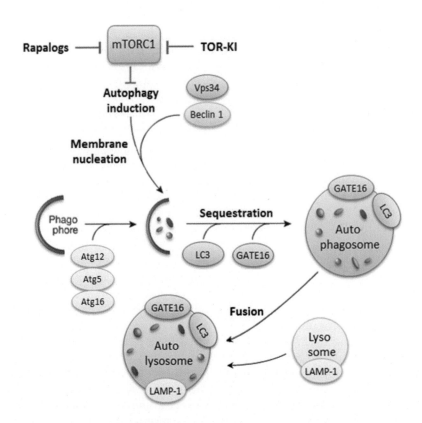

FIGURE 25.3　Induction of autophagy by mTOR inhibitors.

25.4 PREDICTIVE AND PROGNOSTIC FACTORS IN PATIENTS TREATED WITH mTOR INHIBITORS

Since mTOR inhibitors plays a crucial role in the management of ccRCC, there is an increasing need for predictive and prognostic biomarkers of tumor response. Li et al. evaluated by immunochemistry the expression levels of phosphorylated AKT, mTOR, eukaryotic initiation factor 4E (eIF4E) binding protein-1 (4E-BP1) and 40S ribosomal protein S6 (S6RP) in tumor samples from 18 Chinese mRCC patients enrolled in the phase Ib trial of everolimus in VEGFR-TKI-refractory (clinicaltrials.gov, NCT01152801). They showed that patients with positive expression of phospho-mTOR had a higher clinical benefit rate and PFS than patients with negative expression. In the same view, patients with phospho-S6RP had longer PFS than patients with phospho-S6RP negative expression [87]. In addition, phospho-4E-BP1 was significantly associated with the PFS of patients treated with everolimus in the study led by Nishikawa et al. [88].

More recently, Bodnar et al. presented the results of a prospective evaluation of potential biomarkers of the PI3K/AKT/mTOR pathway in patients with mRCC who received everolimus. They found that PI3KCA gene polymorphism, lactate dehydrogenase (LDH), and histologic grade were independent predictive factors associated with patient outcome. In addition, corrected calcium level and the PIK3CA gene variant rs6443624 were independent prognostic factors [89]. These results should be confirmed and validated in larger prospective studies, they open the way to the design of personalized strategies for mRCC patients.

The mTOR pathway is a key regulator of immune response. Indeed, the mTOR pathway can regulate T-lymphocyte functions and the maturation of myeloid DC from monocytes [90]. mTOR inhibitors inhibit T-cell proliferation [91], promote CD4Foxp3 Tregs and selectively regulate CD4 T-effector-cell differentiation [90]. In 2014, we have shown that the neutrophil to lymphocyte ratio (NLR) can be a reliable prognostic factor in mRCC patients treated with everolimus as second- or subsequent-lines after progression during VEGFR-TKI therapy. In this study, patients with NLR ≥ 3 had poor OS and PFS compared to patients with NLR < 3. Based on these findings, the role of neutrophils in primary and acquired resistance to rapalogs should be further investigated [92].

mTOR is also involved in lipid biosynthesis and homeostasis. It has been showed that serum lipids can modulate the response to targeted agents in RCC patients. In this scenario, Santini and his group investigated the correlation between total serum cholesterol (C), triglycerides (T), body mass index, fasting blood glucose level, and blood pressure and the response to everolimus administered as second- or third-line therapy in 177 patients with metastatic RCC. They reported that median time to progression (TTP) was longer in patients with triglyceride increase (10 vs. 6, $P = 0.030$), cholesterol increase (8 vs. 5, $P = 0.042$), or both (10.9 vs. 5.0, $P = 0.003$). Moreover, the concomitant rise of C + T was an independent predictor of TTP at multivariate analysis, but not of OS [93].

25.5 CONCLUSIONS

The introduction of mTOR inhibitors in the therapeutic landscape of RCC has led to relevant advances in the management of these patients. mTOR inhibitors are commonly well tolerated and patients treated with these agents may experience stable disease and even cancer regression. The novel frontier of targeting the mTOR pathway in RCC consists in the identification of the feedback loops and other pathways implicated in the development of resistance to rapalogs. Simultaneous targeting of different components of the PI3K/AKT/mTOR pathway or combining mTOR inhibitors with agents directed towards molecules implicated in acquired drug resistance may provide promising results in terms of clinical benefit. This parallels with the necessity of reliable and validated biomarkers of response to the agents in mRCC, in order to optimize patient selection and outcome and to avoid unnecessary toxicities in this subpopulation.

References

[1] Lieberthal W, Levine JS. The role of the mammalian target of rapamycin (mTOR) in renal disease. J Am Soc Nephrol 2009;20:2493–502.

[2] Lieberthal W, Fuhro R, Andry CC, Rennke H, Abernathy VE, Koh JS, et al. Rapamycin impairs recovery from acute renal failure: role of cell-cycle arrest and apoptosis of tubular cells. Am J Physiol Renal Physiol 2001;281:F693–706.

[3] Yang Y, Wang J, Qin L, Shou Z, Zhao J, Wang H, et al. Rapamycin prevents early steps of the development of diabetic nephropathy in rats. Am J Nephrol 2007;27:495–502.

[4] Tao Y, Kim J, Schrier RW, Edelstein CL. Rapamycin markedly slows disease progression in a rat model of polycystic disease. J Am Soc Nephrol 2005;16:46–51.

[5] Hidalgo M, Rowinsky EK. The rapamycin-sensitive signal transduction pathway as a target for cancer therapy. Oncogene 2000;19:6680–6.

[6] Shu X, Lin J, Wood CG, Tannir NM, Wu X. Energy balance, polymorphisms in the mTOR pathway, and renal cell carcinoma risk. J Natl Cancer Inst 2013;105:424–32.

[7] Cancer Genome Atlas Research Network. Comprehensive molecular characterization of clear cell renal cell carcinoma. Nature 2013;499:43–9.

[8] Fiorini C, Massari F, Pedron S, Sanavio S, Ciccarese C, Porcaro AB, et al. Methods to identify molecular expression of mTOR pathway: a rationale approach to stratify patients affected by clear cell renal cell carcinoma for more likely response to mTOR inhibitors. Am J Cancer Res 2014;9:907–15.

[9] Oshiro N, Yoshino K, Hidayat S, Tokunaga C, Hara K, Eguchi S, et al. Dissociation of raptor from mTOR is a mechanism of rapamycin-induced inhibition of mTOR function. Genes Cells 2004;9:359–66.

[10] Dibble CC, Asara JM, Manning BD. Characterization of Rictor phosphorylation sites reveals direct regulation of mTOR complex 2 by S6K1. Mol Cell Biol 2009;29:5657–70.

[11] Julien LA, Carriere A, Moreau J, Roux PP. mTORC1-activated S6K1 phosphorylates Rictor on threonine 1135 and regulates mTORC2 signaling. Mol Cell Biol 2010;30:908–21.

[12] Seufferlein T, Rozengurt E. Rapamycin inhibits constitutive p70s6k phosphorylation, cell proliferation, and colony formation in small cell lung cancer cells. Cancer Res. 1996;56:3895–7.

[13] Dumont FJ, Su Q. Mechanism of action of the immunosuppressant rapamycin. Life Sci 1996;58:373–95.

[14] Zhou H, Luo Y, Huang S. Updates of mTOR inhibitors. Anticancer Agents Med Chem 2010;10:571–81.

[15] Rini BI. Temsirolimus, an inhibitor of mammalian target of rapamycin. Clin Cancer Res 2008;14:1286–90.

[16] Raymond E, Alexandre J, Faivre S, Vera K, Materman E, Boni J, et al. Safety and pharmacokinetics of escalated doses of weekly intravenous infusion of CCI-779, a novel mTOR inhibitor, in patients with cancer. J Clin Oncol 2004;22:2336–47.

[17] Atkins MB, Hidalgo M, Stadler WM, Logan TF, Dutcher JP, Hudes GR, et al. Randomized phase II study of multiple dose levels of CCI-779, a novel mammalian target of rapamycin kinase inhibitor, in patients with advanced refractory renal cell carcinoma. J Clin Oncol 2004;22:909–18.

[18] Torisel prescribing information. Available at: <http://www.accessdata.fda.gov/>; 2011.

[19] Hudes G, Carducci M, Tomczak P, Dutcher J, Figlin R, Kapoor A, Global ARCC Trial, et al. Temsirolimus, interferon alfa, or both for advanced renal-cell carcinoma. N Engl J Med 2007;356:2271–81.

[20] Motzer RJ, Bacik J, Murphy BA, Russo P, Mazumdar M. Interferon-alfa as a comparative treatment for clinical trials of new therapies against advanced renal cell carcinoma. J Clin Oncol 2002;20:289–96.

[21] Mekhail TM, Abou-Jawde RM, Boumerhi G, Malhi S, Wood L, Elson P, et al. Validation and extension of the Memorial Sloan-Kettering prognostic factors model for survival in patients with previously untreated metastatic renal cell carcinoma. J Clin Oncol 2005;23:832–41.

[22] Kwitkowski VE, Prowell TM, Ibrahim A, Farrell AT, Justice R, Mitchell SS, et al. FDA approval summary: temsirolimus as treatment for advanced renal cell carcinoma. Oncologist 2010;15:428–35.

[23] Hutson TE, Escudier B, Esteban E, Bjarnason GA, Lim HY, Pittman KB, et al. Randomized phase III trial of temsirolimus versus sorafenib as second-line therapy after sunitinib in patients with metastatic renal cell carcinoma. J Clin Oncol 2014;32:760–7.

[24] Kirchner GI, Meier-Wiedenbach I, Manns MP. Clinical pharmacokinetics of everolimus. Clin Pharmacokinet 2004;43:83–95.

[25] Boulay A, Zumstein-Mecker S, Stephan C, Beuvink I, Zilbermann F, Haller R, et al. Antitumor efficacy of intermittent treatment schedules with the rapamycin derivative RAD001 correlates with prolonged inactivation of ribosomal protein S6 kinase 1 in peripheral blood mononuclear cells. Cancer Res 2004;64:252–61.

[26] O'Donnell A, Faivre S, Burris HA, Rea D, Papadimitrakopoulou V, Shand N, et al. Phase I pharmacokinetic and pharmacodynamic study of the oral mammalian target of rapamycin inhibitor everolimus in patients with advanced solid tumors. J Clin Oncol 2008;26:1588–95.

[27] Afinitor prescribing information. Available at: <http://www.accessdata.fda.gov/>; 2012.

[28] Amato RJ, Jac J, Giessinger S, Saxena S, Willis JP. A phase 2 study with a daily regimen of the oral mTOR inhibitor RAD001 (everolimus) in patients with metastatic clear cell renal cell cancer. Cancer 2009;115:2438–46.

[29] Motzer RJ, Escudier B, Oudard S, Hutson TE, Porta C, Bracarda S, RECORD-1 Study Group, et al. Phase 3 trial of everolimus for metastatic renal cell carcinoma: final results and analysis of prognostic factors. Cancer 2010;116:4256–65.

[30] Motzer RJ, Escudier B, Oudard S, Hutson TE, Porta C, Bracarda S, RECORD-1 Study Group, et al. Efficacy of everolimus in advanced renal cell carcinoma: a double-blind, randomised, placebo-controlled Phase III trial. Lancet 2008;372:449–56.

[31] Zustovich F, Lombardi G, Nicoletto O, Pastorelli D. Second-line therapy for refractory renal-cell carcinoma. Crit Rev Oncol Hematol 2012;83:112–22.

[32] Motzer RJ, Barrios CH, Kim TM, Falcon S, Cosgriff T, Harker WG, et al. Phase II randomized trial comparing sequential first-line everolimus and second-line sunitinib versus first-line sunitinib and second-line everolimus in patients with metastatic renal cell carcinoma. J Clin Oncol 2014;32:2765–72.

[33] Armstrong AJ, Broderick S, Eisen T, Stadler WM, Jones RJ, Garcia JA, et al. Final clinical results of a randomized phase II international trial of everolimus vs sunitinib in patients with metastatic non-clear cell renal cell carcinoma (ASPEN). J Clin Oncol 2015;33: abstr 4507

[34] Porta C, Szczylik C, Escudier B. Combination or sequencing strategies to improve the outcome of metastatic renal cell carcinoma patients: a critical review. Crit Rev Oncol Hematol 2012;82:323–37.

[35] Négrier S, Gravis G, Pérol D, Chevreau C, Delva R, Bay JO, et al. Temsirolimus and bevacizumab, or sunitinib, or interferon alfa and bevacizumab for patients with advanced renal cell carcinoma (TORAVA): a randomised phase 2 trial. Lancet Oncol 2011;12:673–80.

[36] Flaherty KT, Manola JB, Pins M, McDermott DF, Atkins MB, Dutcher JJ, et al. BEST: a randomized phase II study of vascular endothelial growth factor, RAF kinase, and mammalian target of rapamycin combination targeted therapy with bevacizumab, sorafenib, and temsirolimus in advanced renal cell carcinoma—a trial of the ECOG-ACRIN Cancer Research Group (E2804). J Clin Oncol 2015;33:2384–91.

[37] Rini BI, Bellmunt J, Clancy J, Wang K, Niethammer AG, Hariharan S, et al. Randomized phase III trial of temsirolimus and bevacizumab versus interferon alfa and bevacizumab in metastatic renal cell carcinoma: INTORACT trial. J Clin Oncol 2014;32:752–9.

[38] Ravaud A, Barrios CH, Alekseev B, Tay MH, Agarwala SS, Yalcin S, et al. RECORD-2: phase II randomized study of everolimus and bevacizumab versus interferon α-2a and bevacizumab as first-line therapy in patients with metastatic renal cell carcinoma. Ann Oncol 2015;26:1378–84.

[39] Motzer R, Hutson T, Glen H, Michaelson D, Molina AM, Eisen T, et al. Randomized phase II, three-arm trial of lenvatinib (LEN), everolimus (EVE), and LEN + EVE in patients (pts) with metastatic renal cell carcinoma (mRCC). J Clin Oncol 2015;33: abs. 4506

[40] Chresta CM, Davies BR, Hickson I, Harding T, Cosulich S, Critchlow SE, et al. AZD8055 is a potent, selective, and orally

bioavailable ATP−competitive mammalian target of rapamycin kinase inhibitor with *in vitro* and *in vivo* antitumor activity. Cancer Res 2010;70:288−98.

[41] Thoreen CC, Kang SA, Chang JW, Liu Q, Zhang J, Gao Y, et al. An ATP−competitive mammalian target of rapamycin inhibitor reveals rapamycin-resistant functions of mTORC1. J Bio Chem 2009;284:8023−32.

[42] Yu K, Toral−Barza L, Shi C, Zhang WG, Lucas J, Shor B, et al. Biochemical, cellular, and *in vivo* activity of novel ATP-competitive and selective inhibitors of the mammalian target of rapamycin. Cancer Res 2009;69:6232−40.

[43] Markman B, Dienstmann R, Tabernero J. Targeting the PI3K/Akt/mTOR pathway−beyond rapalogs. Oncotarget 2010;1:530−43.

[44] Santoni M, Pantano F, Amantini C, Nabissi M, Conti A, Burattini L, et al. Emerging strategies to overcome the resistance to current mTOR inhibitors in renal cell carcinoma. Biochim Biophys Acta 2014;1845:221−31.

[45] Carracedo A, Ma L, Teruya−Feldstein J, Rojo F, Salmena L, Alimonti A, et al. Inhibition of mTORC1 leads to MAPK pathway activation through a PI3K−dependent feedback loop in human cancer. J Clin Invest 2008;118:3065−74.

[46] Manning BD, Cantley LC. AKT/PKB signaling: navigating downstream. Cell 2007;129:1261−74.

[47] Podsypanina K, Lee RT, Politis C, Hennessy I, Crane A, Puc J, et al. An inhibitor of mTOR reduces neoplasia and normalizes p70/S6 kinase activity in Pten$^{+/-}$ mice. Proc Natl Acad Sci USA 2001;98:10320−5.

[48] He L, Fan C, Gillis A, Feng X, Sanatani M, Hotte S, et al. Co-existence of high levels of the PTEN protein with enhanced Akt activation in renal cell carcinoma. Biochim Biophys Acta 2007;1772:1134−42.

[49] Figlin RA, de Souza P, McDermott D, Dutcher JP, Berkenblit A, Thiele A, et al. Analysis of PTEN and HIF-1alpha and correlation with efficacy in patients with advanced renal cell carcinoma treated with temsirolimus versus interferon-alpha. Cancer 2009;115:3651−60.

[50] Semenza GL. Targeting HIF-1 for cancer therapy. Nat Rev Cancer 2003;3:721−32.

[51] Ohh M, Park CW, Ivan M, Hoffman MA, Kim TY, Huang LE, et al. Ubiquitination of hypoxia−inducible factor requires direct binding to the beta-domain of the von Hippel−Lindau protein. Nat Cell Biol 2000;2:423−7.

[52] Banks RE, Tirukonda P, Taylor C, Hornigold N, Astuti D, Cohen D, et al. Genetic and epigenetic analysis of von Hippel−Lindau (VHL) gene alterations and relationship with clinical variables in sporadic renal cancer. Cancer Res 2006;66:2000−11.

[53] Kaelin Jr. WG. The von Hippel−Lindau tumor suppressor protein and clear cell renal carcinoma. Clin Cancer Res 2007;13:680s−4s.

[54] Sang N, Fang J, Srinivas V, Leshchinsky I, Caro J. Carboxyl-terminal transactivation activity of hypoxia−inducible factor 1 alpha is governed by a von Hippel−Lindau protein-independent, hydroxylation-regulated association with p300/CBP. Mol Cell Biol 2002;22:2984−92.

[55] Toschi A, Lee E, Gadir N, Ohh M, Foster DA. Differential dependence of hypoxia-inducible factors 1 alpha and 2 alpha on mTORC1 and mTORC2. J Biol Chem 2008;283:34495−9.

[56] Guertin DA, Sabatini DM. Defining the role of mTOR in cancer. Cancer Cell 2007;12:9−22.

[57] Elorza A, Soro−Arnáiz I, Meléndez-Rodríguez F, Rodríguez-Vaello V, Marsboom G, de Cárcer G, et al. HIF2α acts as an mTORC1 activator through the amino acid carrier SLC7A5. Mol Cell 2012;48:681−91.

[58] Roberts AM, Watson IR, Evans AJ, Foster DA, Irwin MS, Ohh M. Suppression of hypoxia-inducible factor 2alpha restores p53 activity via Hdm2 and reverses chemoresistance of renal carcinoma cells. Cancer Res 2009;69:9056−64.

[59] Tsvetkov LM, Yeh KH, Lee SJ, Sun H, Zhang H. p27(Kip1) ubiquitination and degradation is regulated by the SCF(Skp2) complex through phosphorylated Thr187 in p27. Curr Biol 1999;9:661−4.

[60] Shanmugasundaram K, Block J, Nayak BK, Livi CB, Venkatachalam MA, Sudarshan S. PI3K regulation of the SKP−2/p27 axis through mTORC2. Oncogene 2013;32:2027−36.

[61] Luo Y, Marx SO, Kiyokawa H, Koff A, Massague J, Marks AR. Rapamycin resistance tied to defective regulation of p27Kip1. Mol Cell Biol 1996;16:6744−51.

[62] Majumder PK, Febbo PG, Bikoff R, Berger R, Xue Q, McMahon LM, et al. mTOR inhibition reverses Akt-dependent prostate intraepithelial neoplasia through regulation of apoptotic and HIF-1-dependent pathways. Nat Med 2004;10:594−601.

[63] Hosoi H, Dilling MB, Shikata T, Liu LN, Shu L, Ashmun RA, et al. Rapamycin causes poorly reversible inhibition of mTOR and induces p53-independent apoptosis in human rhabdomyosarcoma cells. Cancer Res. 1999;59:886−94.

[64] Altieri DC. Survivin, versatile modulation of cell division and apoptosis in cancer. Oncogene 2003;22:8581−9.

[65] Mahotka C, Krieg T, Krieg A, Wenzel M, Suschek CV, Heydthausen M, et al. Distinct *in vivo* expression patterns of survivin splice variants in renal cell carcinomas. Int J Cancer 2002;100:30−6.

[66] Mahalingam D, Medina EC, Esquivel II JA, Espitia CM, Smith S, Oberheu K, et al. Vorinostat enhances the activity of temsirolimus in renal cell carcinoma through suppression of survivin levels. Clin Cancer Res 2010;16:141−53.

[67] Chen CL, Chan PC, Wang SH, Pan YR, Chen HC. Elevated expression of protein kinase C delta induces cell scattering upon serum deprivation. J Cell Sci 2010;123:2901−13.

[68] Tochhawng L, Deng S, Pervaiz S, Yap CT. Redox regulation of cancer cell migration and invasion. Mitochondrion 2013;13:246−53.

[69] Fitzgerald JP, Nayak B, Shanmugasundaram K, Friedrichs W, Sudarshan S, Eid AA, et al. Nox4 mediates renal cell carcinoma cell invasion through hypoxia-induced interleukin 6- and 8-production. PLoS One 2012;7:e30712.

[70] Hervouet E, Simonnet H, Godinot C. Mitochondria and reactive oxygen species in renal cancer. Biochimie 2007;89:1080−8.

[71] Son YO, Wang L, Poyil P, Budhraja A, Hitron JA, Zhang Z, et al. Cadmium induces carcinogenesis in BEAS-2B cells through ROS-dependent activation of PI3K/AKT/GSK-3beta/beta-catenin signaling. Toxicol Appl Pharmacol 2012;264:153−60.

[72] Okoh VO, Felty Q, Parkash J, Poppiti R, Roy D. Reactive oxygen species via redox signaling to PI3K/AKT pathway contribute to the malignant growth of 4-hydroxy estradiol-transformed mammary epithelial cells. PLoS One 2013;8:e54206.

[73] Sasabe E, Yang Z, Ohno S, Yamamoto T. Reactive oxygen species produced by the knockdown of manganese-superoxide dismutase up-regulate hypoxia-inducible factor-1alpha expression in oral squamous cell carcinoma cells. Free Radic Bio Med 2010;48:1321−9.

[74] Hervouet E, Cizkova A, Demont J, Vojtiskova A, Pecina P, Franssen−van Hal NL, et al. HIF and reactive oxygen species regulate oxidative phosphorylation in cancer. Carcinogenesis 2008;29:1528−37.

[75] Levine B, Kroemer G. Autophagy in the pathogenesis of disease. Cell 2008;132:27−42.

[76] Messai Y, Noman MZ, Janji B, Hasmim M, Escudier B, Chouaib S. The autophagy sensor ITPR1 protects renal carcinoma cells from NK-mediated killing. Autophagy 2015;: In press

[77] Chang YY, Neufeld TP. An Atg1/Atg13 complex with multiple roles in TOR-mediated autophagy regulation. Mol Biol Cell 2009;20:2004—14.

[78] Hosokawa N, Hara T, Kaizuka T, Kishi C, Takamura A, Miura Y, et al. Nutrient—dependent mTORC1 association with the ULK1-Atg13-FIP200 complex required for autophagy. Mol Biol Cell 2009;20:1981—91.

[79] Jung CH, Jun CB, Ro SH, Kim YM, Otto NM, Cao J, et al. ULK—Atg13—FIP200 complexes mediate mTOR signaling to the autophagy machinery. Mol Biol Cell 2009;20:1992—2003.

[80] Mizushima N. The role of the Atg1/ULK1 complex in autophagy regulation. Curr. Opin Cell Biol 2010;22:132—9.

[81] Bray K, Mathew R, Lau A, Kamphorst JJ, Fan J, Chen J, et al. Autophagy suppresses RIP kinase-dependent necrosis enabling survival to mTOR inhibition. PLoS One 2012;7:e41831.

[82] Santoni M, Amantini C, Morelli MB, Liberati S, Farfariello V, Nabissi M, et al. Pazopanib and sunitinib trigger autophagic and non-autophagic death of bladder tumour cells. Br J Cancer 2013;109:1040—50.

[83] Kurahashi T, Miyake H, Hara I, Fujisawa M. Significance of Aurora-A expression in renal cell carcinoma. Urol Oncol 2007;25:128—33.

[84] Li Y, Zhou W, Wei L, Jin J, Tang K, Li C, et al. The effect of Aurora kinases on cell proliferation, cell cycle regulation and metastasis in renal cell carcinoma. Int. J Oncol 2012;41:2139—49.

[85] Cha TL, Chuang MJ, Wu ST, Sun GH, Chang SY, Yu DS, et al. Dual degradation of aurora A and B kinases by the histone deacetylase inhibitor LBH589 induces G2—M arrest and apoptosis of renal cancer cells. Clin Cancer Res 2009;15:840—50.

[86] Terakawa T, Miyake H, Kumano M, Fujisawa M. Growth inhibition and enhanced chemosensitivity induced by down-regulation of Aurora-A in human renal cell carcinoma Caki-2 cells using short hairpin RNA. Oncol Lett 2011;2:713—17.

[87] Li S, Kong Y, Si L, Chi Z, Cui C, Sheng X, et al. Phosphorylation of mTOR and S6RP predicts the efficacy of everolimus in patients with metastatic renal cell carcinoma. BMC Cancer 2014;14:376.

[88] Nishikawa M, Miyake H, Harada K, Fujisawa M. Expression level of phosphorylated-4E-binding protein 1 in radical nephrectomy specimens as a prognostic predictor in patients with metastatic renal cell carcinoma treated with mammalian target of rapamycin inhibitors. Med Oncol 2014;31:792.

[89] Bodnar L, Stec R, Cierniak S, Synowiec A, Wcisło G, Jesiotr M, et al. Clinical usefulness of PI3K/Akt/mTOR genotyping in companion with other clinical variables in metastatic renal cell carcinoma patients treated with everolimus in the second and subsequent lines. Ann Oncol 2015;26:1385—9.

[90] Powell JD, Pollizzi KN, Heikamp EB, Horton MR. Regulation of immune responses by mTOR. Annu Rev Immunol 2012;30:39—68.

[91] Breslin EM, White PC, Shore AM, Clement M, Brennan P. LY294002 and rapamycin co-operate to inhibit T-cell proliferation. Br J Pharmacol 2005;144:791—800.

[92] Santoni M, De Giorgi U, Iacovelli R, Conti A, Burattini L, Rossi L, et al. Pre-treatment neutrophil-to-lymphocyte ratio may be associated with the outcome in patients treated with everolimus for metastatic renal cell carcinoma. Br J Cancer 2013;109:1755—9.

[93] Pantano F, Santoni M, Procopio G, Rizzo M, Iacovelli R, Porta C, et al. The changes of lipid metabolism in advanced renal cell carcinoma patients treated with everolimus: a new pharmacodynamic marker? PloS One 2015;10:e0120427.

26

Metabolic Shunt Pathways, Carcinoma, and mTOR

Norisuke Shibuya, Ken-ichi Inoue and Keiichi Kubota

Second Department of Surgery, Dokkyo Medical University, Mibu, Japan

26.1 mTORC1 AND CANCER

Mammalian target of rapamycin complex 1 (mTORC1) acts as an intracellular nutrient sensor, and plays an important role in carcinogenesis [1,2]. Point mutations in the mTOR gene *per se* have been observed in intestinal adenocarcinoma or renal cancer [3–5]. Tuberous sclerosis complex 2 (TSC2), a causative gene of tuberous sclerosis, is also a known tumor suppressor that negatively regulates mTORC1 signaling (Figure 26.1) [6,7]. Moreover, TSC1/TSC2 protein complex is a signaling hub which integrates various oncogenic signals such as phosphatidylinositol-4,5-bisphospate 3 kinase (PI3K)/protein kinase B (AKT) (Figure 26.1) [8,9] and Ras/mitogen-activated protein kinase pathways (Figure 26.1) [10–12]. PI3K/AKT is a well-established oncogenic signal which is activated in various mechanisms, including inactivation of upstream tumor suppressor phosphatase and tensin homolog deleted from chromosome 10 (Figure 26.1) [13–15]. Thus, the TSC2–mTORC1 axis is an integrative switch of oncogenic signaling cascades. Indeed, AKT1/2 depend for their functions as driver oncogenes on mTORC1 activation [16]. There are six biological hallmark cancer cells acquire during tumorigenesis [17]: (1) sustaining proliferative signaling, (2) evading growth suppressors, (3) resisting cell death, (4) enabling replicative immortality, (5) inducing angiogenesis, and (6) activating invasion and metastasis [17]. Among them, the best-characterized hallmark to which mTORC1 contributes is inducing angiogenesis [18–20]. mTORC1 activation stabilizes hypoxia inducible factor 1-alpha (HIF1-α) (Figure 26.1) [19,21,22], which is another established oncogene

implicated with angiogenesis or Warburg effects (see below) [23,24]. The important substrate proteins of mTOR kinase include eukaryotic translation initiation factor 4E-binding protein 1 (4E-BP1) and ribosomal protein S6 kinase β 1/2 (S6K1/S6K2), both of which play important roles in mRNA translation [25]. 4E-BP1 is a negative regulator of a rate-limiting enzyme of mRNA translation, eukaryotic translation initiation factor 4E (eIF4E) [26,27]. The phosphorylation of 4E-BP1 dissociates from eIF4E, allowing for the recruitment of other initiation factors, to the 5′ end of mRNA [26,27]. In cancer cells, eIF4E enhances cell proliferation and survival, as well as angiogenesis, by directing selective translation of mRNAs such as cyclin D1, B-cell lymphoma 2, B-cell lymphoma-extra large, survivin and vascular endothelial growth factor (VEGF) [28–33]. S6K1 and 2, which are other substrate kinases, have both redundant and isoform-specific functions [25]. Among them, S6K1 contributes more significantly to the mTORC1 functions in growth regulation. Ribosomal protein S6 (RPS6) is often used as a biomarker of mTORC1 inhibition and might not be a crucial target of S6K1 for translational control [34]. In addition to RPS6, S6K1 regulates other target substrates such as eukaryotic elongation factor-2 kinase [35,36], eukaryotic translation initiation factor 4B (eIF4B) [37,38], S6K1 Aly/REF-like substrate [39,40] and programmed cell death 4 (PDCD4) [41–43]. Among them, PDCD4 is especially important because it was originally identified as a tumor suppressor gene [44]. Although such evidence was mostly obtained from a hypothesis-driven approach (i.e., based on the premise that mTORC1 is an oncogene), unbiased data-driven evidence also supports the hypothesis. Thus,

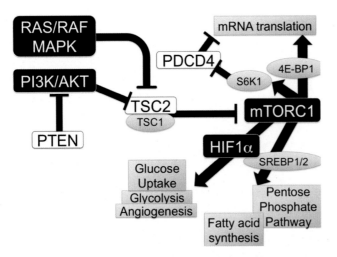

FIGURE 26.1 **mTORC1 plays pivotal roles in the oncogene pathway network.** Black font with white background denotes tumor suppressor. White font with black background denotes oncogene. Note that tumor suppressor and oncogene are nested each other, indicating that the pathway crosstalk is a genetically proven oncogene network. mTORC1 exerts oncogenic functions through selective mRNA translation. In addition, newly identified functions in glycolysis and PPP are discussed in this review.

Ruggero and his collaborators used a ribosomal profiling technique powered by next-generation sequencing [45]. They verified the effects of newly developed catalytic inhibitor of mTOR (INK128) on aggressive prostate cancer cell line PC3 [45]. The RNA profile approach enabled them to discriminate the two different consequences of mTOR inhibition, namely, mRNA translation and the secondary change of gene transcription. Gene ontology analysis of INK128 effects revealed that genes for invasion and metastasis are highly enriched [45]. They subsequently focused on two genes, MeTastasis-Associated Protein 1 [46,47] and Y-box Binding protein 1 [48,49] and concluded that 4E-BP1 is responsible for the regulation. Together, through the two target substrates 4E-BP1 and S6K1, mTOR controls selective translation of mRNA to protein, and its translation control is closely related to cancer progression.

26.2 RAPAMYCIN ANALOGS AND THEIR CLINICAL BENEFITS

The compound rapamycin, from which mTOR derives its name, was approved by the Food and Drug Administration (FDA) of the United States Department of Health and Human Service in 1999, as an immunosuppressant drug after allogenic organ transplantation [50]. A first-in-class rapamycin analog temsirolimus (CCI-779) was approved in 2007, as an anticancer drug in advanced renal cell carcinomas [51]. In comparison

among temsirolimus, interferon-alpha (IFN-α)- and temsirolimus + IFN-α-treated groups (randomly assigned), the temsirolimus-treated group had significantly prolonged overall survival (OS) compared to the IFN-α-treated group but the combination effects were negligible [51]. The effect of another rapamycin analog, everolimus (RAD001), was subsequently investigated in a phase III, randomized, double-blind, placebo-controlled trial (RECORD-1 study). Orally available everolimus significantly prolonged progression-free survival (PFS) from 1.9 to 4 months compared to placebo [52,53]. However, a randomized phase III trial of temsirolimus versus sorafenib (VEGF receptor inhibitor) revealed that sorafenib prolonged OS significantly (temsirolimus and sorafenib: 12.3 and 16.6 months, respectively), indicating that direct angiogenesis inhibition has superior efficacy against aggressive renal cancers [54]. Anticancer effects of other solid tumors were mostly disappointing [55–57] though, rapamycin analogs are still being intensively studied in clinical settings. In a phase III RADIANT-3 trial against Pancreatic NeuroEndocrine Tumors (PNET), everolimus prolonged PFS compared to placebo control, regardless of prior chemotherapy use [58–61]. In breast cancers, after various phase II trials [62–64], a phase III BORERO-3 trial was carried out for everolimus [65]. Thus, HER2 (receptor tyrosine kinase)-positive-advanced breast carcinoma patients were randomly assigned and double-blindly treated. Everolimus was combined with targeted chemotherapy using trastuzumab (anti-HER2 monoclonal antibody) and with conventional antimitotic chemotherapy vinorelbine [65]. Addition of everolimus prolonged PFS from 5.78 to 7.00 months compared to placebo (trastuzumab + vinorelbine) and the effect was statistically significant, though adverse effects also increased from 20% to 42%. Another rapamycin analog, temsirolimus, was also thoroughly investigated in a phase III trial against metastatic breast cancers [66,67], which failed to reproduce the prior promising results of the phase II trial [68]. In hepatocellular carcinomas (HCC), a phase III EVOLVE-1 trial was done for estimation of efficacy of everolimus [69,70]. Sorafenib (low-molecular-weight inhibitor against tyrosine kinase)-resistant HCC patients were randomly assigned but everolimus did not improve OS in malignant cases [69,70]. Again, the results were disappointing in spite of the prior promising data [71–74]. However, some argue that the failure could be due to intrinsic study design, which does not necessarily mean a lack of antitumor function in everolimus [75]. Neither temsirolimus nor everolimus fulfilled criteria to proceed in a phase III trial in lung cancers [76–79]. Similarly, temsirolimus failed to proceed to a phase III trial in ovarian cancers [80]. In advanced gastric cancer, however, a phase III

GRANITE-1 trial was carried out [81,82]. Although everolimus did not significantly prolong OS compared to placebo control, it prolonged PFS from 1.4 to 1.7 months. The anticancer effects of another rapamycin analog, ridaforolimus, were tested in a phase III trial against metastatic sarcomas [83], with a significant but small degree of prolongation of OS (from 85.3 to 90.6 weeks). In other solid tumors, phase II trials of rapamycin analogs have been completed in esophageal cancers [84], colorectal carcinomas (CRCs) [84,85], prostate cancers [86–88], bladder cancers [89], glioblastomas [90,91] and neuroblastomas [92]. Together, rapamycin analogs display antitumor effects against particular cancer types and clinical usage is approved in some of them (advanced renal cell carcinoma, PNET, and metastatic breast cancers).

26.3 WARBURG EFFECT AND THE PENTOSE PHOSPHATE PATHWAY

Differentiated cells metabolize glucose through two steps, namely, glycolysis and oxidative phosphorylation. With sufficient oxygen, pyruvic acid generated during glycolysis enters the mitochondria, where the reaction proceeds through oxidative phosphorylation to carbon dioxide [93]. Under anaerobic conditions, pyruvic acid is discharged out of the cell after being converted to lactic acid, and energy production is dependent on glycolysis. It has long been known that tumor cells selectively utilize glycolysis even under an aerobic microenvironment, in a phenomenon called the Warburg effect or aerobic glycolysis [94,95]. Recent research has shown that the Warburg effect is not a disruption of normal glucose metabolism, but rather a deliberate use of the glycolytic system [96–98]. One of the implications of aerobic glycolysis is to decrease intracellular reactive oxygen species (ROS) [99–101]. Cancer cells are metabolically active and if they utilize oxidative phosphorylation for energy production, its byproduct ROS increases beyond the permissible level. Genes for subunits of succinate dehydrogenase (SDH) complex are frequently mutated in familial paraganglioma or pheochromocytoma [102,103]. Succinate is a substate of SDH and its cytoplasmic accumulation inhibits proryl hydroxylase domain-containing protein 2 and subsequent stabilization of HIF1-α [104]. HIF1-α, in turn, induces mitochondrial protease LON Peptidase 1 (LONP1) and degrades the cytochrome oxidase IV (COX4-1) [105]. HIF1-α replaces COX4 subunit from COX4-1 to COX4-2 to decrease the byproduct ROS [105]. Thus, mitochondria are a major source of intracellular ROS and their activity (oxidative phosphorylation, in particular) is deliberately modulated to regulate the ROS level. Another implication of aerobic

glycolysis, which has recently been attracting more attention, is activation of the pentose phosphate pathway (PPP) [106,107]. During rapid proliferation, cells consume and assimilate large quantities of glucose and glutamine because of its need to actively biosynthesize nucleic acids, amino acids, and lipids that are structural bases of the cells [108,109]. The PPP is located as a shunt pathway of the glycolytic system, and positively and negatively regulated by tumor suppressor TP53 [110–112]. Pyruvate kinase isoenzyme type M2 (PKM2) has been considered to be driver of the Warburg effect [113,114], but recent evidence indicates the causative link between PKM2 oxidation and PPP activation [115,116]. PPP contains two different functional arms, oxidative and nonoxidative arms [106,107]. The critical enzyme of the nonoxidative arm is transketolase (TKT) and transaldolase 1 (TALDO1). Cooperation of TKT and TALDO1 recombines glycolysis metabolites into the final product, ribose 5-phosphate (R5P), contributing to the synthesis of new nucleic acids such as DNA and RNA (Figure 26.2) [98,107]. Interestingly, the nonoxidative arm of PPP is critial for K-Ras^{G12D}-driven tumorigenesis [118]. The oxidative arm of PPP, on the other hand, recycles back NADP$^+$ to NADPH through successive reactions (Figure 26.2). The final product of the oxidative arm, ribulose 5-phosphate, flows into the nonoxidative arm (Figure 26.2) [93,106,107,119]. A significant portion of intracellular NADPH (reduced form) is recycled from NADP$^+$ (oxidized form) by the rate-limiting PPP enzymes, glucose-6-phosphate dehydrogenase (G6PD) and phosphogluconate dehydrogenase (PGD) (Figure 26.2). Similar to ATP, NADPH has reducing

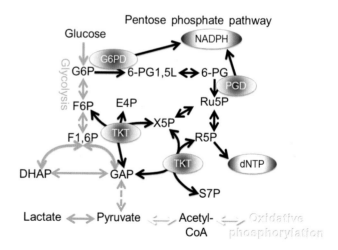

FIGURE 26.2 **Schematic representation of glucose metabolism and PPP.** PPP is a bypass of glycolysis and is biologically significant for two reasons. One is to produce R5P, which is converted to dNTP during DNA replication/RNA transcription. The other is to produce NADPH, which is important for fatty acid synthesis and redox regulation. *Figure is modified from Shimizu et al. [117].*

power, and is essential for fatty acid synthesis [120]. In addition, NADPH plays an important role in regulating the intracellular redox potential, as described in detail below.

26.4 REGULATION OF REDOX POTENTIAL BY TRANSCRIPTION FACTOR NRF2: IMPLICATIONS FOR PPP ACTIVATION

The Keltch-like ECH-associated protein 1 (KEAP1)-nuclear factor (erythroid-derived 2) like 2 (NRF2) system plays a central role in the cellular defense mechanisms against oxidative stress [121–124]. NRF2 is a master transcription factor that activates a battery of genes involved in various cellular defense systems [121,122]. KEAP1 functions as a sensor of oxidative stress, and is an inhibitory factor that in normal conditions degrades NRF2 as a ubiquitin ligase [125,126]. In the presence of oxidative stress, KEAP1 undergoes structural changes by modification of cysteine residues, resulting in the release of NRF2 [127]. However, NRF2 has been reported to be constitutively stabilized in many cancer cells, irrespective of the presence or absence of oxidative stress [128–132]. Thus, the KEAP1–NRF2 axis is recognized as an oncogene pathway. NRF2 target genes are so extensive that one can overview how cells evolved cellular defense systems simply by annotating NRF2 targets. Among those annotations, glutathione and thioredoxin systems are especially important for redox regulation [133–135]. Glutathione and thioredoxin are synthesized as redox regulatory peptides, by coupling with a reaction in which they are converted from the reduced form to the oxidized form, thus neutralizing ROS [133–135]. NRF2 induces transcription of a group of enzymes (glutathione peroxidase GPX2, glutathione S-transferase GST alpha1-3, GST mu1-3, GST pi1) that catalyze the reaction to neutralize ROS, as well as glutathione- and thioredoxin-generating reactions (glutamate-cysteine ligase regulatory subunit GCLM, GCLC, glutathione reductase GSR, cystine/glutamate exchange transporter xCT, thioredoxin TXN1, thioredoxin reductase TXNRD1, peroxiredoxin PRDX1) [124,136–139]. Another important annotation of NRF2 target genes is NADPH production. Motohashi and colleagues identified G6PD, PGD, malic enzyme 1, and isocitrate dehydrogenase 1 (IDH1), which are NADPH-generating enzymes, as new gene targets of NRF2 [140]. The significance of the fact that NRF2, which responds to oxidative stress, induces NADPH-producing enzymes is in the recycling of glutathione and thioredoxin (Figure 26.3). In other words, glutathione and thioredoxin are converted into their oxidized forms after

neutralizing ROS, and, therefore, the reducing power of NADPH needs to be utilized for them to be recycled back to the reduced form (Figure 26.3) [133–135]. In other words, depletion of intracellular NADPH hampers the recycling of glutathione and thioredoxin, causing an increase in intracellular oxidative stress. In fact, patients who have mutations in the G6PD gene are unable to handle oxidative stress in their erythrocytes, and have a shorter erythrocyte half-life [141]. For cancer cells, ROS functions as signals promoting carcinogenesis, such as promoting epithelial–mesenchymal transition (EMT) and cell motility, angiogenesis, and inflammation [142–144]; however, excessive production of ROS is harmful and toxic for cells [133,145]. Moreover, tumor initiating cells (TICs, or also termed cancer stem cells) are thought to keep the intracellular ROS level lower than that of more differentiated counterparts (i.e., differentiated, still cancerous) [146,147]. In this context, it is worth citing the recent discovery that CD44 splicing valiant CD44v9, which is an established marker for TICs, physically binds to xCT [148–150]. xCT is especially important for redox regulation because it is critical for both glutathione and thioredoxin systems [151–153]. TICs and their differentiated offspring develop defense systems against ROS, by mutual feedback regulation among glutathione and thioredoxin systems [151–153]. After surgical resection, residual cancer cells must fight against extracellular

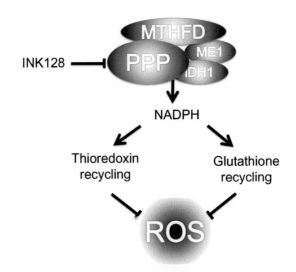

FIGURE 26.3 **Schematic representation of the cancer cell strategy to neutralize ROS.** Glutathione and thioredoxin are peptides for redox regulation. They are oxidized by coupling reaction to neutralize ROS. Both molecules are reduced again (recycling), using the reducing power of NADPH. Gastrointestinal cancer cells augment the transcription of PPP enzymes, prompting us to investigate them as cancer-specific drug targets. mTOR catalytic inhibitor decreased the synthesis of PPP enzymes and decreased the ratio of (reduced glutathione)/(oxidized glutathione) in colorectal cancer cell line COLO201.

ROS in metastatic microenvironments (mostly foreign and harsh) and intracellular ROS generated as a result of chemotherapy [154–158]. Indeed, many tumors display augmented glutathione metabolism and the PPP, suggesting their function as a cellular defense system [107,133]. Transcription factor Snail, a master regulator of EMT, negatively regulates gene for fructose-1,6-biphosphatase and, as a result, glucose flux is diverted to PPP from the flow to mitochondrial respiratory complex [159]. The activation of PPP, in turn, decreases intracellular ROS and augments the phenotype of TICs and EMT. When we examined the surgical specimens of hepatocellular carcinoma, nuclear translocation of NRF2 and increased G6PD synthesis were specifically observed in the cancerous parts, and the two factors showed significant correlation [117]. Compared to another NADPH-producing enzyme (IDH1), G6PD showed a more prominent tumor-specific increase ([117] and manuscript in preparation). This suggests the possibility that G6PD could be a molecular target, as a part of the redox potential regulation system in cancer cells.

26.5 mTOR AND PENTOSE PHOSPHATE PATHWAY

Through a combination of genomic, metabolomics, and bioinformatic approaches, Manning and colleagues discovered that mTORC1 controls glycolysis and the PPP in mouse embryonic fibroblasts (MEFs) [160]. mTORC1 signaling induces glucose uptake and glycolytic enzyme genes by stabilizing HIF1-α (Figure 26.1). Simultaneously, mTORC1 induces the expression of PPP enzymes such as G6PD and PGD, by activating the transcriptional factor Sterol regulatory element-binding protein 1/2 (SREBP1/2) [160,161]. These genes are consistently down-regulated in $Tsc1^{-/-}$ as well as in $Tsc2^{-/-}$ MEFs and inhibited by rapamycin [160]. The significance of the transcriptional changes is confirmed by altered metabolites in $Tsc1^{-/-}$ as well as in $Tsc2^{-/-}$ MEFs. On the other hand, the aforementioned study of ribosome profiling identified another annotation for INK128 (mTOR catalytic inhibitor)-inhibited genes. While the majority of genes inhibited is annotated for protein synthesis, a substantial portion of them is for glucose metabolism [45]. These genes include the PPP enzymes such as TKT and glycolysis enzyme glyceraldehyde 3-phosphate dehydrogenase, signifying that the PPP is regulated at the mRNA translation level as well. What is a biological meaning of mTORC1 in PPP regulation? Some regulatory components of mTORC1 are redox-sensitive [162–164], it is not surprising if mTORC1 actively regulates redox status through PPP regulation. As stated above,

mTORC1 exerts oncogenic function mostly through enhanced angiogenesis but we speculate that the multifunctional signaling hub could be linked with Warburg effects and PPP as well. Below, we introduce our latest findings, focusing on CRCs.

26.6 CRCs: IMPACT AND CLINICAL SIGNIFICANCE OF PPP ACTIVITY AND PHARMACOLOGICAL INHIBITION BY mTOR CATALYTIC INHIBITOR, INK128

We observed the augmented synthesis of PPP enzymes in almost all surgical specimens of CRC examined [165]. When we treated CRC cell lines with INK128, we observed a decreased synthesis of PPP enzymes (G6PD and TKT). This finding is consistent with the aforementioned report showing that INK128 inhibits TKT synthesis in prostate cancer cell lines [45], and supports our working hypothesis. Interestingly, INK128 showed a drop in glucose consumption and lactic acid discharge in all of the CRC cell lines that were used. This signifies that INK128 inhibits aerobic glycolysis, that is, the Warburg effect. The COLO201 and DLD-1 cell lines were used to measure the reduced glutathione (GSH) and oxidized glutathione (GSSG) ratio after pharmacological inhibition of mTOR. In COLO201, INK128 was found to decrease the GSH/GSSG ratio, but in DLD-1, a significant effect could not be confirmed. Finally, we investigated the *in vivo* antitumor effects of INK128 using CRC surgical specimens, which could be successively transplanted subcutaneously in nude mice. TICs or cancer stem cells have characters, when serially transplanted in immune-deficient mice, successfully forming tumors of similar histology [166]. We established such patient-derived tumor xenografts which contain a substantial amount of TICs. TICs are an indicator of malignancy, and we measured the effects of the drug from two indices. One index (colonization) was whether a specimen, when re-engrafted, formed tumor in another mouse, and the other (tumor growth) is the speed at which an engrafted tumor grows in size. When the effects of INK128 were investigated in three malignant cases (xenografts), we observed that colonization was inhibited in two of the three cases, and tumor growth was inhibited in two of the three cases. In one of the two cases where colonization was inhibited, no tumor engraftment was observed for 14 days after successive transplantation. We ran a primary culture of almost all samples of CRCs that were resected at the university hospital from 2013 to 2014, and recorded subcutaneous engraftment in nude mice; however, among these, successive transplantation was possible only in approximately 10% of the cases. Subcutaneous engraftment in nude mice was considered as mimicking circumstances in which distantly metastasized cells are colonized and reappear (recurrence). The fact that a certain effect of INK128 was observed in such an experimental system implies the possibility that mTOR inhibition or PPP inhibition could exhibit a certain effect in eliminating the TICs. Thus, INK128 could be useful for preventing the recurrence of CRCs after surgical resection.

26.7 FUTURE CHALLENGES

In the experimental system using MEFs, SREBP1/2 was identified as a transcription factor that couples the actions of mTOR and the PPP. Being a master regulator of lipid synthesis, it would be reasonable for SREBF1/2 to activate the PPP [160]. This is because NADPH is essential as a reducing power during the fatty acid synthesis [120], and activation of the PPP as an energy source is expected. Therefore, it is also possible that a completely different molecular mechanism is involved in the inhibition of the PPP by INK128 in a different context, such as in CRCs, and the schema of mTOR-SREBF1/2-PPP may not be applicable in all contexts. For example, mTOR could control PPP through redox regulator NRF2 activation [167]. Moreover, there are individual differences in the effects of INK128, as was observed in GSH/GSSG-decreasing effects in COLO201/DLD-1, and the anti-cancer effects on patient-derived xenografts. In some cases, subcutaneous growth in nude mice was effectively suppressed, whereas in other cases, no growth-inhibiting effect was observed. On the other hand, cases have also been observed where colonization was inhibited despite the absence of a growth-inhibiting effect. This suggests the possibility that in future clinical trials, the results in determining the effects could vary depending on what the index is. Therefore, it is extremely important to assess the effect of INK128 on inhibiting postoperative cancer recurrence in addition to reducing tumor bulk.

References

[1] Zoncu R, Efeyan A, Sabatini DM. mTOR: from growth signal integration to cancer, diabetes and ageing. Nat Rev Mol Cell Biol 2011;12:21–35.

[2] Guertin DA, Sabatini DM. Defining the role of mTOR in cancer. Cancer Cell 2007;12:9–22.

[3] Sato T, Nakashima A, Guo L, Coffman K, Tamanoi F. Single amino-acid changes that confer constitutive activation of mTOR are discovered in human cancer. Oncogene 2010;29:2746–52.

[4] Grabiner BC, Nardi V, Birsoy K, Possemato R, Shen K, Sinha S, et al. A diverse array of cancer-associated MTOR mutations are hyperactivating and can predict rapamycin sensitivity. Cancer Discov 2014;4:554–63.

[5] Gerlinger M, Rowan AJ, Horswell S, Larkin J, Endesfelder D, Gronroos E, et al. Intratumor heterogeneity and branched evolution revealed by multiregion sequencing. N Engl J Med 2012;366:883–92.

[6] Crino PB, Nathanson KL, Henske EP. The tuberous sclerosis complex. N Engl J Med 2006;355:1345–56.

[7] Tee AR, Fingar DC, Manning BD, Kwiatkowski DJ, Cantley LC, Blenis J. Tuberous sclerosis complex-1 and -2 gene products function together to inhibit mammalian target of rapamycin (mTOR)-mediated downstream signaling. Proc Natl Acad Sci USA 2002;99:13571–6.

[8] Manning BD, Tee AR, Logsdon MN, Blenis J, Cantley LC. Identification of the tuberous sclerosis complex-2 tumor suppressor gene product tuberin as a target of the phosphoinositide 3-kinase/akt pathway. Mol Cell 2002;10:151–62.

[9] Inoki K, Li Y, Zhu T, Wu J, Guan K. TSC2 is phosphorylated and inhibited by Akt and suppresses mTOR signalling. Nat Cell Biol 2002;4:648–57.

[10] Ballif BA, Roux PP, Gerber SA, MacKeigan JP, Blenis J, Gygi SP. Quantitative phosphorylation profiling of the ERK/p90 ribosomal S6 kinase-signaling cassette and its targets, the tuberous sclerosis tumor suppressors. Proc Natl Acad Sci USA 2005;102:667–72.

[11] Ma L, Chen Z, Erdjument-Bromage H, Tempst P, Pandolfi PP. Phosphorylation and functional inactivation of TSC2 by Erk: implications for tuberous sclerosisand cancer pathogenesis. Cell 2005;121:179–93.

[12] Roux PP, Ballif BA, Anjum R, Gygi SP, Blenis J. Tumor-promoting phorbol esters and activated Ras inactivate the tuberous sclerosis tumor suppressor complex via p90 ribosomal S6 kinase. Proc Natl Acad Sci USA 2004;101:13489–94.

[13] Maehama T, Dixon JE. The tumor suppressor, PTEN/MMAC1, dephosphorylates the lipid second messenger, phosphatidylinositol 3,4,5-trisphosphate. J Biol Chem 1998;273:13375–8.

[14] Cantley LC, Neel BG. New insights into tumor suppression: PTEN suppresses tumor formation by restraining the phosphoinositide 3-kinase/AKT pathway. Proc Natl Acad Sci USA 1999;96:4240–5.

[15] Stambolic V, Suzuki A, De La Pompa J L, Brothers GM, Mirtsos C, Sasaki T, et al. Negative regulation of PKB/Akt-dependent cell survival by the tumor suppressor PTEN. Cell 1998;95:29–39.

[16] Skeen JE, Bhaskar PT, Chen C, Chen WS, Peng X, Nogueira V, et al. Akt deficiency impairs normal cell proliferation and suppresses oncogenesis in a p53-independent and mTORC1-dependent manner. Cancer Cell 2006;10:269–80.

[17] Hanahan D, Weinberg RA. Hallmarks of cancer: the next generation. Cell 2011;144:646–74.

[18] Guba M, Von Breitenbuch P, Steinbauer M, Koehl G, Flegel S, Hornung M, et al. Rapamycin inhibits primary and metastatic tumor growth by antiangiogenesis: involvement of vascular endothelial growth factor. Nat Med 2002;8:128–35.

[19] Thomas GV, Tran C, Mellinghoff IK, Welsbie DS, Chan E, Fueger B, et al. Hypoxia-inducible factor determines sensitivity to inhibitors of mTOR in kidney cancer. Nat Med 2006;12:122–7.

[20] Phung TL, Ziv K, Dabydeen D, Eyiah-Mensah G, Riveros M, Perruzzi C, et al. Pathological angiogenesis is induced by sustained Akt signaling and inhibited by rapamycin. Cancer Cell 2006;10:159–70.

[21] Bernardi R, Guernah I, Jin D, Grisendi S, Alimonti A, Teruya-Feldstein J, et al. PML inhibits HIF-1α translation and neoangiogenesis through repression of mTOR. Nature 2006;442:779–85.

[22] Hudson CC, Liu M, Chiang GG, Otterness DM, Loomis DC, Kaper F, et al. Regulation of hypoxia-inducible factor 1alpha

expression and function by the mammalian target of rapamycin. Mol Cell Biol 2002;22:7004—14.

[23] Semenza GL. Targeting HIF-1 for cancer therapy. Nat Rev Cancer 2003;3:721—32.

[24] Semenza GL. HIF-1: upstream and downstream of cancer metabolism. Curr Opin Genet Dev 2010;20:51—6.

[25] Ma XM, Blenis J. Molecular mechanisms of mTOR-mediated translational control. Nat Rev Mol Cell Biol 2009;10:307—18.

[26] Gingras AC, Kennedy SG, O'Leary MA, Sonenberg N, Hay N. 4E-BP1, a repressor of mRNA translation, is phosphorylated and inactivated by the Akt(PKB) signaling pathway. Genes Dev 1998;12:502—13.

[27] Gingras AC, Gygi SP, Raught B, Polakiewicz RD, Abraham RT, Hoekstra MF, et al. Regulation of 4E-BP1 phosphorylation: a novel two-step mechanism. Genes Dev 1999;13:1422—37.

[28] Dever TE. Gene-specific regulation by general translation factors. Cell 2002;108:545—56.

[29] Mader S, Lee H, Pause A, Sonenberg N. The translation initiation factor eIF-4E binds to a common motif shared by the translation factor eIF-4 gamma and the translational repressors 4E-binding proteins. Mol Cell Biol 1995;15:4990—7.

[30] Yang SX, Hewitt SM, Steinberg SM, Liewehr DJ, Swain SM. Expression levels of eIF4E, VEGF, and cyclin D1, and correlation of eIF4E with VEGF and cyclin D1 in multi-tumor tissue microarray. Oncol Rep 2007;17:281—7.

[31] Li S, Takasu T, Perlman DM, Peterson MS, Burrichter D, Avdulov S, et al. Translation factor eIF4E rescues cells from Myc-dependent apoptosis by inhibiting cytochrome c release. J Biol Chem 2003;278:3015—22.

[32] Graff JR, Konicek BW, Vincent TM, Lynch RL, Monteith D, Weir SN, et al. Therapeutic suppression of translation initiation factor eIF4E expression reduces tumor growth without toxicity. J Clin Invest 2007;117:2638—48.

[33] Graff JR, Konicek BW, Carter JH, Marcusson EG. Targeting the eukaryotic translation initiation factor 4E for cancer therapy. Cancer Res 2008;68:631—4.

[34] Ruvinsky I, Sharon N, Lerer T, Cohen H, Stolovich-Rain M, Nir T, et al. Ribosomal protein S6 phosphorylation is a determinant of cell size and glucose homeostasis. Genes Dev 2005;19:2199—211.

[35] Wang X, Li W, Williams M, Terada N, Alessi DR, Proud CG. Regulation of elongation factor 2 kinase by p90(RSK1) and p70 S6 kinase. EMBO J 2001;20:4370—9.

[36] Browne GJ, Proud CG. A novel mTOR-regulated phosphorylation site in elongation factor 2 kinase modulates the activity of the kinase and its binding to calmodulin. Mol Cell Biol 2004;24:2986—97.

[37] Holz MK, Ballif BA, Gygi SP, Blenis J. mTOR and S6K1 mediate assembly of the translation preinitiation complex through dynamic protein interchange and ordered phosphorylation events. Cell 2005;123:569—80.

[38] Shahbazian D, Roux PP, Mieulet V, Cohen MS, Raught B, Taunton J, et al. The mTOR/PI3K and MAPK pathways converge on eIF4B to control its phosphorylation and activity. EMBO J 2006;25:2781—91.

[39] Richardson CJ, Bröenstrup M, Fingar DC, Jülich K, Ballif BA, Gygi S, et al. SKAR is a specific target of S6 kinase 1 in cell growth control. Curr Biol 2004;14:1540—9.

[40] Ma XM, Yoon S, Richardson CJ, Jülich K, Blenis J. SKAR links pre-mRNA splicing to mTOR/S6K1-mediated enhanced translation efficiency of spliced mRNAs. Cell 2008;133:303—13.

[41] Dorrello NV, Peschiaroli A, Guardavaccaro D, Colburn NH, Sherman NE, Pagano M. S6K1- and betaTRCP-mediated degradation of PDCD4 promotes protein translation and cell growth. Science 2006;314:467—71.

[42] Schmid T, Jansen AP, Baker AR, Hegamyer G, Hagan JP, Colburn NH. Translation inhibitor Pdcd4 is targeted for degradation during tumor promotion. Cancer Res 2008;68:1254—60.

[43] Carayol N, Katsoulidis E, Sassano A, Altman JK, Druker BJ, Platanias LC. Suppression of programmed cell death 4 (PDCD4) protein expression by BCR-ABL-regulated engagement of the mTOR/p70 S6 kinase pathway. J Biol Chem 2008;283:8601—10.

[44] Cmarik JL, Min H, Hegamyer G, Zhan S, Kulesz-Martin M, Yoshinaga H, et al. Differentially expressed protein Pdcd4 inhibits tumor promoter-induced neoplastic transformation. Proc Natl Acad Sci USA 1999;96:14037—42.

[45] Hsieh AC, Liu Y, Edlind MP, Ingolia NT, Janes MR, Sher A, et al. The translational landscape of mTOR signalling steers cancer initiation and metastasis. Nature 2012;485:55—61.

[46] Toh Y, Pencil SD, Nicolson GL. A novel candidate metastasis-associated gene, mta1, differentially expressed in highly metastatic mammary adenocarcinoma cell lines. cDNA cloning, expression, and protein analyses. J Biol Chem 1994;269:22958—63.

[47] Nicolson GL, Nawa A, Toh Y, Taniguchi S, Nishimori K, Moustafa A. Tumor metastasis-associated human MTA1 gene and its MTA1 protein product: role in epithelial cancer cell invasion, proliferation and nuclear regulation. Clin Exp Metastasis 2003;20:19—24.

[48] Bargou RC, Jürchott K, Wagener C, Bergmann S, Metzner S, Bommert K, et al. Nuclear localization and increased levels of transcription factor YB-1 in primary human breast cancers are associated with intrinsic MDR1 gene expression. Nat Med 1997;3:447—50.

[49] Evdokimova V, Tognon C, Ng T, Ruzanov P, Melnyk N, Fink D, et al. Translational activation of snail1 and other developmentally regulated transcription factors by YB-1 promotes an epithelial-mesenchymal transition. Cancer Cell 2009;15:402—15.

[50] Calne R, Lim S, Samaan A, Collier DSJ, Pollard S, White D, et al. Rapamycin for immunosuppression in organ allografting. Lancet 1989;334:227.

[51] Hudes G, Carducci M, Tomczak P, Dutcher J, Figlin R, Kapoor A, et al. Temsirolimus, interferon alfa, or both for advanced renal-cell carcinoma. N Engl J Med 2007;356:2271—81.

[52] Motzer RJ, Escudier B, Oudard S, Hutson TE, Porta C, Bracarda S, et al. Efficacy of everolimus in advanced renal cell carcinoma: a double-blind, randomised, placebo-controlled phase III trial. Lancet 2008;372:449—56.

[53] Motzer RJ, Escudier B, Oudard S, Hutson TE, Porta C, Bracarda S, et al. Phase 3 trial of everolimus for metastatic renal cell carcinoma. Cancer 2010;116:4256—65.

[54] Hutson TE, Escudier B, Esteban E, Bjarnason GA, Lim HY, Pittman KB, et al. Randomized phase III trial of temsirolimus versus sorafenib as second-line therapy after sunitinib in patients with metastatic renal cell carcinoma. J Clin Oncol 2014;32:760—7.

[55] Easton J, Houghton P. mTOR and cancer therapy. Oncogene 2006;25:6436—46.

[56] Faivre S, Kroemer G, Raymond E. Current development of mTOR inhibitors as anticancer agents. Nat Rev Drug Discov 2006;5:671—88.

[57] Granville CA, Memmott RM, Gills JJ, Dennis PA. Handicapping the race to develop inhibitors of the phosphoinositide 3-kinase/Akt/mammalian target of rapamycin pathway. Clin Cancer Res 2006;12:679—89.

[58] Shah M, Lombard-Bohas C, Ito T, Wolin E, Van Cutsem E, Sachs C, et al. Everolimus in patients with advanced pancreatic neuroendocrine tumors (pNET): impact of somatostatin analog use on progression-free survival in the RADIANT-3 trial. J Clin Oncol 2011;29:4010.

[59] Pommier R, Wolin E, Panneerselvam A, Saletan S, Winkler R, Van Cutsem E. Impact of prior chemotherapy on progression-free survival in patients (pts) with advanced pancreatic neuro-endocrine tumors (pNET): results from the RADIANT-3 trial. J Clin Oncol 2011;29:4103.

[60] Ito T, Okusaka T, Ikeda M, Igarashi H, Morizane C, Nakachi K, et al. Everolimus for advanced pancreatic neuroendocrine tumours: a subgroup analysis evaluating Japanese patients in the RADIANT-3 trial. Jpn J Clin Oncol 2012;42:903–11.

[61] Yao J, Shah M, Ito T, Lombard-Bohas C, Wolin E, Van Cutsem E. A randomized, double-blind, placebo-controlled, multicenter phase III trial of everolimus in patients with advanced pancreatic neuroendocrine tumors (PNET)(RADIANT-3). Ann Oncol 2010;21: viii4

[62] Ellard SL, Clemons M, Gelmon KA, Norris B, Kennecke H, Chia S, et al. Randomized phase II study comparing two schedules of everolimus in patients with recurrent/metastatic breast cancer: NCIC Clinical Trials Group IND.163. J Clin Oncol 2009;27:4536–41.

[63] Bachelot T, Bourgier C, Cropet C, Ray-Coquard I, Ferrero JM, Freyer G, et al. Randomized phase II trial of everolimus in combination with tamoxifen in patients with hormone receptor-positive, human epidermal growth factor receptor 2-negative metastatic breast cancer with prior exposure to aromatase inhibitors: a GINECO study. J Clin Oncol 2012;30: 2718–24.

[64] Piccart-Gebhart MJ, Noguchi S, Pritchard KI, Burris HA, Rugo HS, Gnant M, et al. Everolimus for postmenopausal women with advanced breast cancer: updated results of the BOLERO-2 phase III trial. J Clin Oncol 2012;30: Abstract 559

[65] André F, O'Regan R, Ozguroglu M, Toi M, Xu B, Jerusalem G, et al. Everolimus for women with trastuzumab-resistant, HER2-positive, advanced breast cancer (BOLERO-3): a randomised, double-blind, placebo-controlled phase 3 trial. Lancet Oncol 2014;15:580–91.

[66] Wolff AC, Lazar AA, Bondarenko I, Garin AM, Brincat S, Chow L, et al. Randomized phase III placebo-controlled trial of letrozole plus oral temsirolimus as first-line endocrine therapy in postmenopausal women with locally advanced or metastatic breast cancer. J Clin Oncol 2013;31:195–202.

[67] Chow L, Sun Y, Jassem J, Baselga J, Hayes D, Wolff A, et al. Phase 3 study of temsirolimus with letrozole or letrozole alone in postmenopausal women with locally advanced or metastatic. Breast Cancer 2006;100:S286.

[68] Baselga J, Fumoleau P, Gil M, Colomer R, Roche H, Cortes-Funes H, et al. Phase II, 3-arm study of CCI-779 in combination with letrozole in postmenopausal women with locally advanced or metastatic breast cancer: preliminary results. Proc Am Soc Clin Oncol 2004;23:A544.

[69] Zhu AX, Kudo M, Assenat E, Cattan S, Kang Y, Lim HY, et al. EVOLVE-1: phase 3 study of everolimus for advanced HCC that progressed during or after sorafenib. J Clin Oncol 2014;32: Abstract 172.

[70] Zhu AX, Kudo M, Assenat E, Cattan S, Kang Y, Lim HY, et al. Effect of everolimus on survival in advanced hepatocellular carcinoma after failure of sorafenib: the EVOLVE-1 randomized clinical trial. JAMA 2014;312:57–67.

[71] Sahin F, Kannangai R, Adegbola O, Wang J, Su G, Torbenson M. mTOR and P70 S6 kinase expression in primary liver neoplasms. Clin Cancer Res 2004;10:8421–5.

[72] Sieghart W, Fuereder T, Schmid K, Cejka D, Werzowa J, Wrba F, et al. Mammalian target of rapamycin pathway activity in hepatocellular carcinomas of patients undergoing liver transplantation. Transplantation 2007;83:425–32.

[73] Zhu AX, Abrams TA, Miksad R, Blaszkowsky LS, Meyerhardt JA, Zheng H, et al. Phase 1/2 study of everolimus in advanced hepatocellular carcinoma. Cancer 2011;117:5094–102.

[74] Chen L, Shiah H, Chen C, Lin Y, Lin P, Su W, et al. Randomized, phase I, and pharmacokinetic (PK) study of RAD001, an mTOR inhibitor, in patients (pts) with advanced hepatocellular carcinoma (HCC). J Clin Oncol 2009;27: Abstract 4587.

[75] Llovet JM. Liver cancer: time to evolve trial design after everolimus failure. Nat Rev Clin Oncol 2014;11:506–7.

[76] Tarhini A, Kotsakis A, Gooding W, Shuai Y, Petro D, Friedland D, et al. Phase II study of everolimus (RAD001) in previously treated small cell lung cancer. Clin Cancer Res 2010;16:5900–7.

[77] Price KA, Azzoli CG, Krug LM, Pietanza MC, Rizvi NA, Pao W, et al. Phase II trial of gefitinib and everolimus in advanced non-small cell lung cancer. J Thorac Oncol 2010;5:1623–9.

[78] Pandya KJ, Dahlberg S, Hidalgo M, Cohen RB, Lee MW, Schiller JH, et al. A randomized, phase II trial of two dose levels of temsirolimus (CCI-779) in patients with extensive-stage small-cell lung cancer who have responding or stable disease after induction chemotherapy: a trial of the Eastern Cooperative Oncology Group (E1500). J Thorac Oncol 2007;2:1036–41.

[79] Kotsakis A, Tarhini A, Petro D, Flaugh R, Vallabhaneni G, Belani C, et al. Phase II study of RAD001 (everolimus) in previously treated small cell lung cancer (SCLC). J Clin Oncol 2009;27: Abstract 8107.

[80] Behbakht K, Sill MW, Darcy KM, Rubin SC, Mannel RS, Waggoner S, et al. Phase II trial of the mTOR inhibitor, temsirolimus and evaluation of circulating tumor cells and tumor biomarkers in persistent and recurrent epithelial ovarian and primary peritoneal malignancies: a Gynecologic Oncology Group study. Gynecol Oncol 2011;123:19–26.

[81] Van Cutsem E, Yeh K, Bang Y, Shen L, Ajani JA, Bai Y, et al. Phase III trial of everolimus (EVE) in previously treated patients with advanced gastric cancer (AGC): GRANITE-1. J Clin Oncol 2012;30: Abstract LBA3

[82] Ohtsu A, Ajani JA, Bai YX, Bang YJ, Chung HC, Pan HM, et al. Everolimus for previously treated advanced gastric cancer: results of the randomized, double-blind, phase III GRANITE-1 study. J Clin Oncol 2013;31:3935–43.

[83] Demetri GD, Chawla SP, Ray-Coquard I, Le Cesne A, Staddon AP, Milhem MM, et al. Results of an international randomized phase III trial of the mammalian target of rapamycin inhibitor ridaforolimus versus placebo to control metastatic sarcomas in patients after benefit from prior chemotherapy. J Clin Oncol 2013;31:2485–92.

[84] Wainberg ZA, Soares HP, Patel R, DiCarlo B, Park DJ, Liem A, et al. Phase II trial of everolimus in patients with refractory metastatic adenocarcinoma of the esophagus, gastroesophageal junction and stomach: possible role for predictive biomarkers. Cancer Chemother Pharmacol 2015;76:61–7.

[85] Spindler KG, Sorensen MM, Pallisgaard N, Andersen RF, Havelund BM, Ploen J, et al. Phase II trial of temsirolimus alone and in combination with irinotecan for KRAS mutant metastatic colorectal cancer: outcome and results of KRAS mutational analysis in plasma. Acta Oncol 2013;52:963–70.

[86] Templeton AJ, Dutoit V, Cathomas R, Rothermundt C, Bärtschi D, Dröge C, et al. Phase 2 trial of single-agent everolimus in chemotherapy-naive patients with castration-resistant prostate cancer (SAKK 08/08). Eur Urol 2013;64:150–8.

[87] Nakabayashi M, Werner L, Courtney KD, Buckle G, Oh WK, Bubley GJ, et al. Phase II trial of RAD001 and bicalutamide for castration-resistant prostate cancer. BJU Int 2012;110:1729–35.

[88] Seront E, Rottey S, Sautois B, Kerger J, D'Hondt LA, Verschaeve V, et al. Phase II study of everolimus in patients with locally advanced or metastatic transitional cell carcinoma of the urothelial tract: clinical activity, molecular response, and biomarkers. Ann Oncol 2012;23:2663−70.

[89] Milowsky MI, Iyer G, Regazzi AM, Al-Ahmadie H, Gerst SR, Ostrovnaya I, et al. Phase II study of everolimus in metastatic urothelial cancer. BJU Int 2013;112:462−70.

[90] Lee EQ, Kuhn J, Lamborn KR, Abrey L, DeAngelis LM, Lieberman F, et al. Phase I/II study of sorafenib in combination with temsirolimus for recurrent glioblastoma or gliosarcoma: North American Brain Tumor Consortium study 05-02. Neuro Oncol 2012;14:1511−18.

[91] Wen PY, Chang SM, Lamborn KR, Kuhn JG, Norden AD, Cloughesy TF, et al. Phase I/II study of erlotinib and temsirolimus for patients with recurrent malignant gliomas: North American Brain Tumor Consortium trial 04-02. Neuro Oncol 2014;16:567−78.

[92] Geoerger B, Kieran MW, Grupp S, Perek D, Clancy J, Krygowski M, et al. Phase II trial of temsirolimus in children with high-grade glioma, neuroblastoma and rhabdomyosarcoma. Eur J Cancer 2012;48:253−62.

[93] Michal G, Schomburg D. Biochemical pathways: an atlas of biochemistry and molecular biology. Hoboken, NJ: John Wiley & Sons; 2013.

[94] Warburg O. On the origin of cancer cells. Science 1956;123: 309−14.

[95] Gatenby RA, Gillies RJ. Why do cancers have high aerobic glycolysis? Nat Rev Cancer 2004;4:891−9.

[96] Fantin VR, St-Pierre J, Leder P. Attenuation of LDH-A expression uncovers a link between glycolysis, mitochondrial physiology, and tumor maintenance. Cancer Cell 2006;9:425−34.

[97] Levine AJ, Puzio-Kuter AM. The control of the metabolic switch in cancers by oncogenes and tumor suppressor genes. Science 2010;330:1340−4.

[98] Vander Heiden MG, Cantley LC, Thompson CB. Understanding the Warburg effect: the metabolic requirements of cell proliferation. Science 2009;324:1029−33.

[99] Kim JW, Dang CV. Cancer's molecular sweet tooth and the Warburg effect. Cancer Res 2006;66:8927−30.

[100] Hsu PP, Sabatini DM. Cancer cell metabolism: Warburg and beyond. Cell 2008;134:703−7.

[101] Fulda S, Galluzzi L, Kroemer G. Targeting mitochondria for cancer therapy. Nat Rev Drug Discov 2010;9:447−64.

[102] Gottlieb E, Tomlinson IP. Mitochondrial tumour suppressors: a genetic and biochemical update. Nat Rev Cancer 2005;5: 857−66.

[103] Kroemer G, Pouyssegur J. Tumor cell metabolism: cancer's Achilles' heel. Cancer Cell 2008;13:472−82.

[104] Selak MA, Armour SM, MacKenzie ED, Boulahbel H, Watson DG, Mansfield KD, et al. Succinate links TCA cycle dysfunction to oncogenesis by inhibiting HIF-α prolyl hydroxylase. Cancer Cell 2005;7:77−85.

[105] Fukuda R, Zhang H, Kim J, Shimoda L, Dang CV, Semenza GL. HIF-1 regulates cytochrome oxidase subunits to optimize efficiency of respiration in hypoxic cells. Cell 2007;129:111−22.

[106] Riganti C, Gazzano E, Polimeni M, Aldieri E, Ghigo D. The pentose phosphate pathway: an antioxidant defense and a crossroad in tumor cell fate. Free Radic Biol Med 2012;53:421−36.

[107] Patra KC, Hay N. The pentose phosphate pathway and cancer. Trends Biochem Sci 2014;39:347−54.

[108] Dang CV. Glutaminolysis: supplying carbon or nitrogen or both for cancer cells? Cell cycle 2010;9:3884−6.

[109] Chen J, Russo J. Dysregulation of glucose transport, glycolysis, TCA cycle and glutaminolysis by oncogenes and tumor suppressors in cancer cells. Biochim Biophys Acta Rev Cancer 2012;1826:370−84.

[110] Bensaad K, Tsuruta A, Selak MA, Vidal MNC, Nakano K, Bartrons R, et al. TIGAR, a p53-inducible regulator of glycolysis and apoptosis. Cell 2006;126:107−20.

[111] Jiang P, Du W, Wang X, Mancuso A, Gao X, Wu M, et al. p53 regulates biosynthesis through direct inactivation of glucose-6-phosphate dehydrogenase. Nat Cell Biol 2011;13:310−16.

[112] Cosentino C, Grieco D, Costanzo V. ATM activates the pentose phosphate pathway promoting anti-oxidant defence and DNA repair. EMBO J 2011;30:546−55.

[113] Hitosugi T, Kang S, Vander Heiden MG, Chung TW, Elf S, Lythgoe K, et al. Tyrosine phosphorylation inhibits PKM2 to promote the Warburg effect and tumor growth. Sci Signal 2009;2:ra73.

[114] Chaneton B, Gottlieb E. Rocking cell metabolism: revised functions of the key glycolytic regulator PKM2 in cancer. Trends Biochem Sci 2012;37:309−16.

[115] Anastasiou D, Poulogiannis G, Asara JM, Boxer MB, Jiang JK, Shen M, et al. Inhibition of pyruvate kinase M2 by reactive oxygen species contributes to cellular antioxidant responses. Science 2011;334:1278−83.

[116] Luo W, Semenza GL. Emerging roles of PKM2 in cell metabolism and cancer progression. Trends Endocrinol Metab 2012;23:560−6.

[117] Shimizu T, Inoue K, Hachiya H, Shibuya N, Shimoda M, Kubota K. Frequent alteration of the protein synthesis of enzymes for glucose metabolism in hepatocellular carcinomas. J Gastroenterol 2014;49:1324−32.

[118] Ying H, Kimmelman AC, Lyssiotis CA, Hua S, Chu GC, Fletcher-Sananikone E, et al. Oncogenic Kras maintains pancreatic tumors through regulation of anabolic glucose metabolism. Cell 2012;149:656−70.

[119] Fan J, Ye J, Kamphorst JJ, Shlomi T, Thompson CB, Rabinowitz JD. Quantitative flux analysis reveals folate-dependent NADPH production. Nature 2014;510:298−302.

[120] Volpe J, Vagelos P. Saturated fatty acid biosynthesis and its regulation. Annu Rev Biochem 1973;42:21−60.

[121] Jaramillo MC, Zhang DD. The emerging role of the Nrf2-Keap1 signaling pathway in cancer. Genes Dev 2013;27:2179−91.

[122] Kansanen E, Kuosmanen SM, Leinonen H, Levonen A. The Keap1-Nrf2 pathway: mechanisms of activation and dysregulation in cancer. Redox Biol 2013;1:45−9.

[123] Uruno A, Motohashi H. The Keap1−Nrf2 system as an *in vivo* sensor for electrophiles. Nitric Oxide 2011;25:153−60.

[124] Taguchi K, Motohashi H, Yamamoto M. Molecular mechanisms of the Keap1−Nrf2 pathway in stress response and cancer evolution. Genes Cells 2011;16:123−40.

[125] Wakabayashi N, Itoh K, Wakabayashi J, Motohashi H, Noda S, Takahashi S, et al. Keap1-null mutation leads to postnatal lethality due to constitutive Nrf2 activation. Nat Genet 2003;35:238−45.

[126] Kobayashi A, Kang MI, Okawa H, Ohtsuji M, Zenke Y, Chiba T, et al. Oxidative stress sensor Keap1 functions as an adaptor for Cul3-based E3 ligase to regulate proteasomal degradation of Nrf2. Mol Cell Biol 2004;24:7130−9.

[127] Itoh K, Chiba T, Takahashi S, Ishii T, Igarashi K, Katoh Y, et al. An Nrf2/small Maf heterodimer mediates the induction of phase II detoxifying enzyme genes through antioxidant response elements. Biochem Biophys Res Commun 1997;236:313−22.

[128] Singh A, Misra V, Thimmulappa RK, Lee H, Ames S, Hoque MO, et al. Dysfunctional KEAP1-NRF2 interaction in non-small-cell lung cancer. PLoS Med 2006;3:e420.

[129] Padmanabhan B, Tong KI, Ohta T, Nakamura Y, Scharlock M, Ohtsuji M, et al. Structural basis for defects of Keap1 activity provoked by its point mutations in lung cancer. Mol Cell 2006;21:689–700.

[130] DeNicola GM, Karreth FA, Humpton TJ, Gopinathan A, Wei C, Frese K, et al. Oncogene-induced Nrf2 transcription promotes ROS detoxification and tumorigenesis. Nature 2011;475:106–9.

[131] Shibata T, Kokubu A, Gotoh M, Ojima H, Ohta T, Yamamoto M, et al. Genetic alteration of Keap1 confers constitutive Nrf2 activation and resistance to chemotherapy in gallbladder cancer. Gastroenterology 2008;135:1358–1368.e4.

[132] Shibata T, Ohta T, Tong KI, Kokubu A, Odogawa R, Tsuta K, et al. Cancer related mutations in NRF2 impair its recognition by Keap1-Cul3 E3 ligase and promote malignancy. Proc Natl Acad Sci USA 2008;105:13568–73.

[133] Gorrini C, Harris IS, Mak TW. Modulation of oxidative stress as an anticancer strategy. Nat Rev Drug Discov 2013;12:931–47.

[134] Meister A, Anderson ME. Glutathione. Annu Rev Biochem 1983;52:711–60.

[135] Holmgren A. Thioredoxin. Annu Rev Biochem 1985;54: 237–71.

[136] McGrath-Morrow S, Lauer T, Yee M, Neptune E, Podowski M, Thimmulappa RK, et al. Nrf2 increases survival and attenuates alveolar growth inhibition in neonatal mice exposed to hyperoxia. Am J Physiol Lung Cell Mol Physiol 2009;296:L565–73.

[137] Thimmulappa RK, Mai KH, Srisuma S, Kensler TW, Yamamoto M, Biswal S. Identification of Nrf2-regulated genes induced by the chemopreventive agent sulforaphane by oligonucleotide microarray. Cancer Res 2002;62:5196–203.

[138] Meister A. Selective modification of glutathione metabolism. Science 1983;220:472–7.

[139] Sasaki H, Sato H, Kuriyama-Matsumura K, Sato K, Maebara K, Wang H, et al. Electrophile response element-mediated induction of the cystine/glutamate exchange transporter gene expression. J Biol Chem 2002;277:44765–71.

[140] Mitsuishi Y, Taguchi K, Kawatani Y, Shibata T, Nukiwa T, Aburatani H, et al. Nrf2 redirects glucose and glutamine into anabolic pathways in metabolic reprogramming. Cancer Cell 2012;22:66–79.

[141] Alving AS, Carson PE, Flanagan CL, Ickes CE. Enzymatic deficiency in primaquine-sensitive erythrocytes. Science 1956;124: 484–5.

[142] Gupta SC, Hevia D, Patchva S, Park B, Koh W, Aggarwal BB. Upsides and downsides of reactive oxygen species for cancer: the roles of reactive oxygen species in tumorigenesis, prevention, and therapy. Antioxid Redox Signal 2012;16:1295–322.

[143] Nogueira V, Hay N. Molecular pathways: reactive oxygen species homeostasis in cancer cells and implications for cancer therapy. Clin Cancer Res 2013;19:4309–14.

[144] Wu W. The signaling mechanism of ROS in tumor progression. Cancer Metastasis Rev 2006;25:695–705.

[145] Wang J, Yi J. Cancer cell killing via ROS. Cancer Biol Ther 2008;7:1875–84.

[146] Diehn M, Cho RW, Lobo NA, Kalisky T, Dorie MJ, Kulp AN, et al. Association of reactive oxygen species levels and radioresistance in cancer stem cells. Nature 2009;458:780–3.

[147] Clevers H. The cancer stem cell: premises, promises and challenges. Nat Med 2011;:313–19.

[148] Ishimoto T, Nagano O, Yae T, Tamada M, Motohara T, Oshima H, et al. CD44 variant regulates redox status in cancer cells by stabilizing the xCT subunit of system xc− and thereby promotes tumor growth. Cancer Cell 2011;19:387–400.

[149] Liu C, Kelnar K, Liu B, Chen X, Calhoun-Davis T, Li H, et al. The microRNA miR-34a inhibits prostate cancer stem cells and metastasis by directly repressing CD44. Nat Med 2011;17:211–15.

[150] Dalerba P, Dylla SJ, Park IK, Liu R, Wang X, Cho RW, et al. Phenotypic characterization of human colorectal cancer stem cells. Proc Natl Acad Sci USA 2007;104:10158–63.

[151] Wu G, Fang YZ, Yang S, Lupton JR, Turner ND. Glutathione metabolism and its implications for health. J Nutr 2004;134:489–92.

[152] Conrad M, Sato H. The oxidative stress-inducible cystine/glutamate antiporter, system xc−: cystine supplier and beyond. Amino Acids 2012;42:231–46.

[153] Harris IS, Treloar AE, Inoue S, Sasaki M, Gorrini C, Lee KC, et al. Glutathione and thioredoxin antioxidant pathways synergize to drive cancer initiation and progression. Cancer Cell 2015;27:211–22.

[154] Watson J. Oxidants, antioxidants and the current incurability of metastatic cancers. Open Biol 2013;3:120144.

[155] Conklin KA. Chemotherapy-associated oxidative stress: impact on chemotherapeutic effectiveness. Integr Cancer Ther 2004;3:294–300.

[156] Nguyen DX, Bos PD, Massagué J. Metastasis: from dissemination to organ-specific colonization. Nat Rev Cancer 2009;9:274–84.

[157] Psaila B, Lyden D. The metastatic niche: adapting the foreign soil. Nat Rev Cancer 2009;9:285–93.

[158] Scott J, Kuhn P, Anderson AR. Unifying metastasis—integrating intravasation, circulation and end-organ colonization. Nat Rev Cancer 2012;12:445–6.

[159] Schieber MS, Chandel NS. ROS links glucose metabolism to breast cancer stem cell and EMT phenotype. Cancer Cell 2013;23:265–7.

[160] Düvel K, Yecies JL, Menon S, Raman P, Lipovsky AI, Souza AL, et al. Activation of a metabolic gene regulatory network downstream of mTOR complex 1. Mol Cell 2010;39:171–83.

[161] Yecies JL, Manning BD. Transcriptional control of cellular metabolism by mTOR signaling. Cancer Res 2011;71:2815–20.

[162] Sarbassov DD, Sabatini DM. Redox regulation of the nutrient-sensitive raptor-mTOR pathway and complex. J Biol Chem 2005;280:39505–9.

[163] Dames SA, Mulet JM, Rathgeb-Szabo K, Hall MN, Grzesiek S. The solution structure of the FATC domain of the protein kinase target of rapamycin suggests a role for redox-dependent structural and cellular stability. J Biol Chem 2005;280:20558–64.

[164] Yoshida S, Hong S, Suzuki T, Nada S, Mannan AM, Wang J, et al. Redox regulates mammalian target of rapamycin complex 1 (mTORC1) activity by modulating the TSC1/TSC2-Rheb GTPase pathway. J Biol Chem 2011;286:32651–60.

[165] Shibuya N, Inoue K, Tanaka G, Akimoto K, Kubota K. Augmented pentose phosphate pathway plays critical roles in colorectal carcinomas. Oncology 2015;88:309–19.

[166] Nguyen LV, Vanner R, Dirks P, Eaves CJ. Cancer stem cells: an evolving concept. Nat Rev Cancer 2012;12:133–43.

[167] Bray K, Mathew R, Lau A, Kamphorst JJ, Fan J, Chen J, et al. Autophagy suppresses RIP kinase-dependent necrosis enabling survival to mTOR inhibition. PLoS One 2012;7:e41831.

Index